1 MONTH OF
FREE
READING

at

www.ForgottenBooks.com

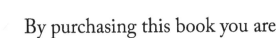

By purchasing this book you are eligible for one month membership to ForgottenBooks.com, giving you unlimited access to our entire collection of over 1,000,000 titles via our web site and mobile apps.

To claim your free month visit:

www.forgottenbooks.com/free465006

ISBN 978-0-266-79439-4
PIBN 10465006

PREAMBLE TO AMENDMENT TO MOTOR VEHICLE SAFETY STANDARD NO. 208

Occupant Crash Protection

(Docket No. 75–33; Notice 2)

This notice amends Standard No. 208, *Occupant Crash Protection*, to permit certain U.S. Postal Service vehicles to meet the requirements of the standard that were in effect until January 1, 1976, instead of the new requirements that became effective on that date.

The NHTSA proposed this modification of Standard No. 208 (49 CFR 571.208) in a notice published December 31, 1975 (40 FR 60075). The occupant protection requirements in the standard until January 1, 1976, specified either a Type 1 or Type 2 seat belt assembly at the driver's position of the light delivery vehicles used by the Postal Service on delivery routes. The Postal Service's safety research organization developed a seat belt design that met the requirements and resulted in improved usage by vehicle operators.

The newer requirements now in effect for the light delivery vehicles in question require the same seat belt assembly installations as in most passenger cars, including a Type 2 seat belt assembly with non-detachable shoulder belt at each front outboard designated seating position. The Service judges that installation of Type 2 seat belts at the driver's position with non-detachable shoulder portion will decrease the percentage of seat belt use by their mail delivery personnel.

The Postal Service indicated its support for the proposal. Ford Motor Company objected to the basis of the vehicle category as a "single user exemption." The agency, while in agreement that categorization based on the status of a single user is not generally utilized, recognizes the distinctive scope and nature of U.S. Postal Service operations. The Service is a part of the Federal government, its delivery activities are unique in scope and variety, and the organization has an active safety research effort that addresses the particular enviromnent of mail delivery by motor vehcile. No other comments were received. The agency concludes that the new requirements for Type 2 seat belt assemblies at the driver's position in this limited category of vehicle are not justified, because their interference with the many entries and exits from the vehcile may discourage usage.

In consideration of the foregoing, S.4.2.2 of Standard No. 208 (49 CFR 571.208) is amended by the addition of the phrase "vehicles designed to be exclusively sold to the U.S. Postal Service," following the phrase "motor homes."

Effective date: March 18, 1976. Because this amendment creates no additional requirements for any person, and in view of the Postal Service's need to contract for vehicles with appropriate seat belt assemblies at the earliest opportunity, an immediate effective date is found to be in the public interest.

(Sec. 103, 119, Pub. L. 89–563, 80 Stat. 718 (15 U.S.C. 1392, 1407); delegation of authority at 49 CFR 1.50.)

Issued on March 10, 1976.

James P. Gregory
Administrator

41 F.R. 11312
March 18, 1976

PREAMBLE TO AMENDMENT TO MOTOR VEHICLE SAFETY STANDARD NO. 208

Occupant Crash Protection

(Docket No. 74-14; Notice 6)

This notice amends Standard No. 208, Occupant Crash Protection, to continue until August 31, 1977, the present three options available for occupant crash protection in passenger cars.

This extension of the present occupant crash protection options of Standard No. 208 (49 CFR 571.208) was proposed July 19, 1976 (41 FR 29715), along with several other subjects that will be the subject of a future notice. Vehicle manufacturers supported the proposal but requested that the options be extended indefinitely instead of being limited to a 1-year extension. Mr. Benjamin Redmond advocated the use of an interlock system to increase usage of active belt systems. Ms. Lucie Kirylak expressed a preference for active occupant crash protection systems. The National Motor Vehicle Safety Advisory Council did not take a position on the proposal.

The Secretary of Transportation has initiated a process for the establishment of future occupant crash protection requirements under Standard No. 208 (41 FR 24070, June 14, 1976). The Secretary's proposal addresses the long term issues involved, and this 1-year extension of requirements is intended to provide the time necessary to reach that decision. Because a 1-year extension is consistent with the process that has been established and because a longer extension was not proposed for comment, the NHTSA declines to extend the existing requirements as recommended by the manufacturers.

Other matters proposed in the notice that underlies this action will be treated at a later date, following the receipt of comments that are due on October 20, 1976.

The NHTSA notes that no effective date was proposed for the other matters addressed by the proposal. Those matters involve modification of the existing passive protection options so that they conform to the proposal of the Department of Transportation, and to reduce somewhat the femur force requirement. Also, further specification of dummy positioning in the vehicle was addressed. The agency proposes an immediate effective date for these changes, because they represent relaxation of the requirements. However, the views of interested persons, particularly Volkswagen (which is certifying compliance under one passive option), are solicited by October 20, 1976.

In consideration of the foregoing, the heading and text of S4.1.2 of Standard No. 208 (49 CFR 571.208) are amended by changing the date "August 31, 1976" to "August 31, 1977" wherever it appears.

Effective date: August 26, 1976.

(Secs. 103, 119, Pub. L. 89-563, 80 Stat. 718 (15 U.S.C. 1392, 1407); delegation of authority at 49 CFR 1.50.)

Issued on August 26, 1976.

John W. Snow
Administrator

41 F.R. 36494
August 30, 1976

PREAMBLE TO AMENDMENT TO MOTOR VEHICLE SAFETY STANDARD NO. 208

Occupant Crash Protection

(Docket No. OST 44; Notice 77–3)

This notice amends Standard No. 208, *Occupant Crash Protection*, to extend indefinitely the current occupant crash protection requirements for passenger cars.

In a notice published June 14, 1976 (41 FR 24070), I proposed five alternative courses of action for future occupant crash protection requirements under Standard No. 208 (49 CFR 571.208). Based on an analysis of comments received, a decision was reached to call upon the automobile manufacturers to join the Federal government in conducting a large-scale demonstration program to exhibit the effectiveness of passive restraint systems. The reasoning that underlines that decision is contained in a December 6, 1976, document ("The Secretary's Decision Concerning Motor Vehicle Occupant Crash Protection") that is hereby incorporated by reference in this notice. The effect of that decision on Standard No. 208 is to require the continuation of the current requirements for passenger cars, as proposed in the first of the five alternative courses of action.

The first alternative was written as a three-year extension (to August 31, 1979), although the preamble discussion made clear that the length of the extension was open to discussion. It is now apparent that a continuation of the existing requirements is best effectuated by a deletion of any termination date. This action accords with the intent of the first alternative to maintain current occupant crash protection requirements for the indefinite future. Because this action represents a continuation of existing manufacturing practices, it is the Department's finding that no new significant economic or environmental impacts result from this amendment.

I have directed the National Highway Traffic Safety Administration (NHTSA) to propose comparable changes in the requirements for multipurpose passenger vehicles and light trucks. The NHTSA has also been directed to take final action on the substantive changes to Standard No. 208 that were proposed in its notice of July 19, 1976 (41 FR 29715).

The Department hereby closes OST Docket No. 44, which is transferred to the NHTSA's docket on occupant crash protection. I want to make it clear, however, that by closing OST Docket No. 44 and amending Standard No. 208 to extend indefinitely the current occupant crash protection requirements for passenger cars, I have not in any way foreclosed a future Secretary or Administrator of NHTSA from instituting at any time a rulemaking to amend Standard No. 208 either to place a terminate date on Standard No. 208 or to mandate passive restraints on some or all passenger cars.

In consideration of the foregoing, the heading and text of S4.1.2 of Standard No. 208 (49 CFR 571.208) are amended in part to read as follows:

S4.1.2 *Passenger cars manufactured on or after September 1, 1973.* Passenger cars manufactured on or after September 1, 1973, shall meet the requirements of S4.1.2.1, S4.1.2.2, or S4.1.2.3. * * *.

Effective date: January 19, 1977.

(Secs. 103, 119, Pub. L. 89–563, 80 Stat. 718 (15 U.S.C. 1392, 1407.)

Issued on January 19, 1977.

William T. Coleman, Jr.
Secretary of Transportation
42 F.R. 5071
January 27, 1977

PREAMBLE TO AMENDMENT TO MOTOR VEHICLE SAFETY STANDARD NO. 208

Occupant Crash Protection

(Docket No. 74–14; Notice 9)

This notice amends Standard No. 208, *Occupant Crash Protection*, to extend indefinitely the current occupant crash protection requirements for light trucks and multipurpose passenger vehicles. The question of future requirements for occupant crash protection is presently being considered by the Secretary of Transportation, and thus the current requirements for light trucks and multipurpose passenger vehicles should be continued for the indefinite future.

Effective date: June 2, 1977.

Addresses: Requests for reconsideration should refer to the docket number and be submitted to: Docket Section, Room 5108, National Highway Traffic Safety Administration, 400 Seventh Street, S.W., Washington, D.C. 20590.

For further information contact:

Guy Hunter
Motor Vehicle Programs
National Highway Traffic Safety
Administration
Washington, D.C. 20590
(202-426-2265)

The requirements of Standard No. 208 (49 CFR 571.208) have been implemented in three stages. The current stage for trucks and multipurpose passenger vehicles (MPV's) with a gross vehicle weight rating (GVWR) of 10,000 pounds or less specifies a choice of three means to provide occupant protections (S4.2.2) and is scheduled to end August 14, 1977. After that date many of these vehicles would be required by S4.2.3 of Standard No. 208 to provide occupant crash protection by means that require no action by vehicle occupants (commonly known as passive protection). In the original promulgation of Standard No. 208 in its present form (36 FR 4600; March 10, 1971) it was established that thic modification of occupant protection should follow a similar modification of protection in passenger cars by two years, to provide manufacturers with time to assimilate and benefit from passenger car experience.

The issue of future occupant protection in passenger cars is being decided at this time, in a notice of proposed rulemaking issued by the Secretary of Transportation (42 FR 15935; March 24, 1977). Thus, light truck and MPV manufacturers have not had the benefit of experience with new systems in passenger cars as originally anticipated. In view of this fact and the fact that they are not prepared to meet requirements other than the existing performance options after August 14, 1977, the agency has decided to continue the existing requirements indefinitely.

This action does not preclude future rulemaking to modify occupant crash protection for the affected vehicles, but notice and opportunity for comment will be provided prior to further action.

Because this action represents a continuation of existing manufacturing practices, it is the agency's finding that no new significant economic or environment impacts result from this amendment.

The lawyer principally responsible for the preparation of this document is Tad Herlihy of the NHTSA Office of Chief Counsel.

The economic and inflationary impacts of this rulemaking have been carefully evaluated in accordance with OMB Circular A-107, and an Inflation Impact Statement is not required.

In view of the fact that future occupant protection requirements are not established and manufacturers are prepared only to meet existing occupant protection requirements after August 1977, the agency finds that notice and public procedure on this amendment to continue existing requirements is unnecessary and contrary to the public interest in knowning next model year's requirements as soon as possible. The agency also finds that this amendment may become effective immediately, because the amendment relieves a restriction.

In consideration of the foregoing, Standard No. 208 (49 CFR 571.208) is amended. . . .

(Sec. 103, 119, Pub. L. 89-563, 80 Stat. 718 (15 U.S.C. 1392, 1407) ; delegation of authority at 49 CFR 1.50.)

Issued on May 27, 1977.

Joan Claybrook
Administrator

42 F.R. 28135
June 2, 1977

PREAMBLE TO AMENDMENT TO MOTOR VEHICLE SAFETY STANDARD NO. 208

Occupant Crash Protection

(Docket No. 74–14; Notice 11; Docket No. 73–8; Notice 7)

This notice amends occupant crash protection Standard No. 208 and its accompanying test dummy specification to further specify test procedures and injury criteria. The changes are minor in most respects and reflect comments by manufacturers of test dummies and vehicles and the NHTSA's own test experience with the standard and the test dummy.

Date: Effective date July 5, 1978.

Addresses: Petitions for reconsideration should refer to the docket number and be submitted to: Docket Section, Room 5108, Nassif Building, 400 Seventh Street, S.W., Washington, D.C. 20590.

For further information contact:

Mr. Guy Hunter
Motor Vehicle Programs
National Highway Traffic Safety
Administration
Washington, D.C. 20590
(202) 426–2265

Supplemental information: Standard No. 208, *Occupant Crash Protection* (49 CFR 571.208), is a Department of Transportation safety standard that requires manufacturers to provide a means of restraint in new motor vehicles to keep occupants from impacting the vehicle interior in the event a crash occurs. The standard has, since January 1968, required the provision of seat belt assemblies at each seating position in passenger cars. In January 1972 the requirements for seat belts were upgraded and options were added to permit the provision of restraint that is "active" (requiring some action be taken by the vehicle occupant, as in the case of seat belts) or "passive" (providing protection without action being taken by the occupant).

In a separate notice issued today (42 FR 34289; FR Reg. 77-19137), the Secretary of Transportation has reached a decision regarding the future occupant crash protection that must be installed in passenger cars. The implementation of that decision will involve the testing of passive restraint systems in accordance with the test procedures of Standard No. 208, and this notice is intended to make final several modifications of that procedure which have been proposed for change by the NHTSA. This notice also responds to two petitions for reconsideration of rulemaking involving the test dummy that is used to evaluate the compliance of passive restraint systems.

DOCKET 74–14; NOTICE 05

Notice 5 was issued July 15, 1976 (41 FR 29715; July 19, 1976) and proposed that Standard No. 208's existing specification for passive protection in frontal, lateral, and rollover modes (S4.1.2.1) be modified to specify passive protection in the frontal mode only, with an option to provide passive protection or belt protection in the lateral and rollover crash modes. Volkswagen had raised the question of the feasibility of small cars meeting the standard's lateral impact requirements: A 20-mph impact by a 4,000-pound, 60-inch-high flat surface. The agency noted the particular vulnerability of small cars to side impact and the need to provide protection for them based on the weight of other vehicles on the highway, but agreed that it would be difficult to provide passive lateral protection in the near future. Design problems also underlay the proposal to provide a belt option in place of the existing passive rollover requirement.

Ford Motor Company argued that a lateral option would be inappropriate in Standard No. 208 as long as the present dummy is used for measurement of passive system performance.

This question of dummy use as a measuring device is treated later in this notice. General Motors Corporation (GM) supported the option without qualification, noting that the installation of a lap belt with a passive system "would provide comparable protection to lap/shoulder belts in side and rollover impacts." Chrysler did not object to the option, but noted that the lap belt option made the title of S4.1.2.1 ("complete passive protection") misleading. Volkswagen noted that its testing of belt systems without the lap belt portion showed little loss in efficacy in rollover crashes. No other comments on this proposal were received. The existing option S4.1.2.1 is therefore adopted as proposed so that manufacturers will be able to immediately undertake experimental work on passive restraints on an optional basis in conformity with the Secretary's decision.

There were no objections to the agency's proposal to permit either a Type 1 or Type 2 seat belt assembly to meet the requirements, and thus it is made final as proposed.

The NHTSA proposed two changes in the injury criteria of S6 that are used as measures of a restraint system's qualification to Standard No. 208. One change proposed an increase in permissible femur force limits from 1,700 pounds to 2,250 pounds. As clarification that tension loads are not included in measurement of these forces, the agency also proposed that the word "compressive" be added to the text of S6.4. Most commenters were cautionary about the changes, pointing out that susceptibility to fracture is time dependent, that acetabular injury could be exacerbated by increased forces, and that angular applications of force were as likely in the real world as axial forces and would more likely fracture the femur.

The agency is aware of and took into account these considerations in proposing the somewhat higher femur force limit. The agency started with the actual field experience of occupants of GM and Volkswagen vehicles that have been shown to produce femur force readings of about 1,700 pounds. Occupants of these vehicles involved in crashes have not shown a significant incidence of femur fracture. The implication from this experience that the 1,700-pound figure can safely be raised somewhat is supported in

work by Patrick on compressive femur forces of relatively long duration. The Patrick data (taken with aged embalmed cadavers) indicate that the average fracture load of the patella-femur-pelvis complex is 1,910 pounds. This average is considered conservative, in that cadaver bone structure is generally weaker than living human tissue. While these data did not address angular force applications, the experience of the GM and Volkswagen vehicle occupants does suggest that angular force application can go higher than 1,700 pounds.

The agency does not agree that the establishment of the somewhat higher outer limit for permissible femur force loads of 2,250 pounds is arbitrary. What is often ignored by the medical community and others in commenting on the injury criteria found in motor vehicle safety standards is that manufacturers must design their restraint systems to provide greater protection than the criteria specified, to be certain that each of their products will pass compliance tests conducted by the NHTSA. It is a fact of industrial production that the actual performance of some units will fall below nominal design standards (for quality control and other reasons). Volkswagen made precisely this point in its comments. Because the National Traffic and Motor Vehicle Safety Act states that each vehicle must comply (15 U.S.C. § 1392(a)(1)(a)), manufacturers routinely design in a "compliance margin" of superior performance. Thus, it is extremely unlikely that a restraint system designed to meet the femur force load criterion of 2,250 pounds will in fact be designed to provide only that level of performance. With these considerations in mind, the agency makes final the changes as proposed.

While not proposed for change, vehicle manufacturers commented on a second injury criterion of the standard: A limitation of the acceleration experienced by the dummy thorax during the barrier crash to 60g, except for intervals whose cumulative duration is not more than 3 milliseconds (ms). Until August 31, 1977, the agency has specified the Society of Automotive Engineers' (SEA) "severity index" as a substitute for the 60g–3ms limit, because of greater familiarity of the industry with that criterion.

General Motors recommended that the severity index be continued as the chest injury criterion until a basis for using chest deflection is developed in place of chest acceleration. GM cited data which indicate that chest injury from certain types of blunt frontal impact is a statistically significant function of chest deflection in humans, while not a function of impact force or spinal acceleration. GM suggested that a shift from the temporary severity index measure to the 60g–3ms measurement would be wasteful, because there is no "strong indication" that the 60g–3ms measurement is more meaningful than the severity index, and some restraint systems might have to be redesigned to comply with the new requirement.

Unlike GM, Chrysler argued against the use of acceleration criteria of either type for the chest, and rather advocated that the standard be delayed until a dummy chest with better deflection characteristics is developed.

The Severity Index Criterion allows higher loadings and therefore increases the possibility of adverse effects on the chest. It only indirectly limits the accelerations and hence the forces which can be applied to the thorax. Acceleration in a specific impact environment is considered to be a better predictor of injury than the Severity Index.

NHTSA only allowed belt systems to meet the Severity Index Criterion of 1,000 instead of the 60g–3ms criterion out of consideration for lead-time problems, not because the Severity Index Criterion was considered superior. It is recognized that restraint systems such as lap-shoulder belts apply more concentrated forces to the thorax than air cushion restraint, and that injury can result at lower forces and acceleration levels. It is noted that the Agency is considering rulemaking to restrict forces that may be applied to the thorax by the shoulder belt of any seat belt assembly (41 FR 54961; December 16, 1976).

With regard to the test procedures and conditions that underlie the requirements of the standard, the agency proposed a temperature range for testing that would be compatible with the temperature sensitivity of the test dummy. The test dummy specification (Part 572, *Anthropomorphic Test Dummy*, 49 CFR Part 572) contains calibration tests that are conducted at any temperature between 66° and 78° F. This is because properties of lubricants and nonmetallic parts used in the dummy will change with large temperature changes and will affect the dummy's objectivity as a test instrument. It was proposed that the Standard No. 208 crash tests be conducted within this temperature range to eliminate the potential for variability.

The only manufacturers that objected to the temperature specification were Porsche, Bayerische Motoren Werke (BMW), and American Motors Corporation (AMC). In each case, the manufacturers noted that dynamic testing is conducted outside and that it is unreasonable to limit testing to the few days in the year when the ambient temperature would fall within the specified 12-degree range.

The commenters may misunderstand their certification responsibilities under the National Traffic and Motor Vehicle Safety Act. Section 108(b)(2) limits a manufacturer's responsibility to the exercise of "due care" to assure compliance. The NHTSA has long interpreted this statutory "due care" to mean that the manufacturer is free to test its products in any fashion it chooses, as long as the testing demonstrates that due care was taken to assure that, if tested by NHTSA as set forth in the standard, the product would comply with the standard's requirements. Thus, a manufacturer could conduct testing on a day with temperatures other than those specified, as long as it could demonstrate through engineering calculations or otherwise, that the difference in test temperatures did not invalidate the test results. Alternatively a manufacturer might choose to perform its preparation of the vehicle in a temporarily erected structure (such as a tent) that maintains a temperature within the specified range, so that only a short esposure during acceleration to the barrier would occur in a higher or lower temperature. To assist any such arrangements, the test temperature condition has been limited to require a stabilized temperature of the test dummy only, just prior to the vehicle's travel toward the barrier.

In response to an earlier suggestion from GM, the agency proposed further specificity in the clothing worn by the dummy during the crash test. The only comment was filed by GM, which argued that any shoe specification other than weight would be unrelated to dummy performance and therefore should not be included in the specification. The agency disagrees, and notes that the size and shape of the heel on the shoe can affect the placement of the dummy limb within the vehicle. For this reason, the clothing specifications are made final as proposed, except that the requirement for a conforming "configuration" has been deleted.

Renault and Peugeot asked for confirmation that pyrotechnic pretensioners for belt retractors are not prohibited by the standard. The standard's requirements do not specify the design by which to provide the specified protection, and the agency is not aware of any aspect of the standard that would prohibit the use of pretensioning devices, as long as the three performance elements are met.

With regard to the test dummy used in the standard, the agency proposed two modifications of Standard No. 208: a more detailed positioning procedure for placement of the dummy in the vehicle prior to the test, and a new requirement that the dummy remain in calibration without adjustment following the barrier crash. Comments were received on both aspects of the proposal.

The dummy positioning was proposed to eliminate variation in the conduct of repeatable tests, particularly among vehicles of different sizes. The most important proposed modification was the use of only two dummies in any test of front seat restraints, whether or not the system is designed for three designated seating positions. The proposal was intended to eliminate the problem associated with placement of three 50th-percentile male dummies side-by-side in a smaller vehicle. In bench seating with three dummies, the system would have to comply with a dummy at the driver's position and at either of the other two designated seating positions.

GM supported this change, but noted that twice as many tests of 3-position bench-seat vehicles would be required as before. The company suggested using a simulated vehicle crash as a means to test the passive restraint at the center seat position. The agency considers this approach unrepresentative of the actual crash pulse

and vehicle kinematic response (e.g., pitching, yawing) that occur during an impact. To the degree that GM can adopt such an approach in the exercise of "due care" to demonstrate that the center seating position actually complies, the statute does not prohibit such a certification approach.

Ford objected that the dummy at the center seat position would be placed about 4 inches to the right of the center of the designated seating position in order to avoid interference with the dummy at the driver's position. While the NHTSA agrees that a small amount of displacement is inevitable in smaller vehicles, it may well occur in the real world also. Further, the physical dimensions of the dummy preclude any other positioning. With a dummy at the driver's position, a dummy at the center position cannot physically be placed in the middle of the seat in all cases. In view of these realities, the agency makes final this aspect of the dummy positioning as proposed.

GM suggested the modification of other standards to adopt "2-dummy" positioning. The compatibility among dynamic tests is regularly reviewed by the NHTSA and will be again following this rulemaking action. For the moment, however, only those actions which were proposed will be acted on.

As a general matter with regard to dummy positioning, General Motors found the new specifications acceptable with a few changes. GM cautioned that the procedure might not be sufficiently reproducible between laboratories, and Chrysler found greater variation in positioning with the new procedures than with Chrysler's own procedures. The agency's use of the procedure in 15 different vehicle models has shown consistently repeatable results, as long as a reasonable amount of care is taken to avoid the effect of random inputs (see "Repeatability of Set Up and Stability of Anthropometric Landmarks and Their Influence on Impact Response of Automotive Crash Test Dummies." Society of Automotive Engineers, Technical Paper No. 770260, 1977). The agency concludes that, with the minor improvements cited below, the positioning procedure should be made final as proposed.

The dummy is placed at a seating position so that its midsagittal plane is vertical and longitudinal. Volkswagen argued against use of the midsagittal plane as a reference for dummy placement, considering it difficult to define as a practical matter during placement. The agency has used plane markers and plane lines to define the midsagittal plane and has experienced no significant difficulty in placement of the dummy with these techniques. For this reason, and because Volkswagen suggested no simpler orientation technique, the agency adopts· use of the midsagittal plane as proposed.

Correct spacing of the dummy's legs at the driver position created the largest source of objection by commenters. Ford expressed concern that an inward-pointing left knee could result in unrealistically high femur loads because of femur-to-steering column impacts. GM asked that an additional 0.6 inch of space be specified between the dummy legs to allow for installation of a device to measure steering column displacement. Volkswagen considered specification of the left knee bolt location to be redundant in light of the positioning specification for the right knee and the overall distance specification between the knees of 14.5 inches.

The commenters may not have understood that the 14.5- and 5.9-inch dimensions are only initial positions, as specified in S8.1.11.1.1. The later specification to raise the femur and tibia centerlines "as close as possible to vertical" without contacting the vehicle shifts the knees from their initial spacing to a point just to the left and right of the steering column.

As for GM's concern about instrumentation, the agency does not intend to modify this positioning procedure to accommodate instrumentation preferences not required for the standard's purposes. GM may, of course, make test modifications so long as it assures, in the exercise of due care, that its vehicles will comply when tested in accordance with the specification by the agency.

In the case of a vehicle which is equipped with a front bench seat, the driver dummy is placed on the bench so that its midsagittal plane intersects the center point of the plane described by the steering wheel rim. BMW pointed out that the center plane of the driver's seating position may not coincide with the steering wheel center and that dummy placement would therefore be unrealistic. Ford believed that the specification of the steering wheel reference point could be more precisely specified.

The agency believes that BMW may be describing offset of the driver's seat from the steering wheel in bucket-seat vehicles. In the case of bench-seat vehicles, there appears to be no reason not to place the dummy directly behind the steering wheel. As for the Ford suggestion, the agency concludes that Ford is describing the same point as the proposal did, assuming, as the agency does, that the axis of the steering column passes through the center point described. The Ford description does have the effect of moving the point a slight distance laterally, because the steering wheel rim upper surface is somewhat higher than the plane of the rim itself. This small distance is not relevant to the positioning being specified and therefore is not adopted.

In the case of center-position dummy placement in a vehicle with a drive line tunnel, Ford requested further specification of left and right foot placement. The agency has added further specification to make explicit what was implicit in the specifications proposed.

Volkswagen suggested that the NHTSA had failed to specify knee spacing for the passenger side dummy placement. In actuality, the specification in S8.1.11.1.2 that the femur and tibia centerlines fall in a vertical longitudinal plane has the effect of dictating the distance between the passenger dummy knees.

The second major source of comments concerned the dummy settling procedure that assures uniformity of placement on the seat cushion and against the seat back. Manufacturers pointed out that lifting the dummy within the vehicle, particularly in small vehicles and those with no rear seat space, cannot be accomplished easily. While the NHTSA recognizes that the procedure is not simple, it is desirable to improve the uniformity of dummy response and it has been accomplished by the NHTSA in several small cars (e.g., Volkswagen Rabbit, Honda Civic, Fiat Spider, DOT HS–801–754). Therefore, the requests of GM and Volkswagen to retain the method that does not involve lifting has been denied. In response to Renault's question, the dummy can be lifted manually by a strap routed beneath the buttocks. Also, Volkswagen's request for more variability in the application of rearward force is denied because, while difficult to achieve, it is desirable to maintain uniformity in dummy placement. In response to the requests of several manufacturers, the location of the 9-square-inch push plate has been raised 1.5 inches, to facilitate its application to all vehicles.

Volkswagen asked with regard to S10.2.2 for a clarification of what constitutes the "lumbar spine" for purposes of dummy flexing. This refers to the point on the dummy rear surface at the level of the top of the dummy's rubber spine element.

BMW asked the agency to reconsider the placement of the driver dummy's thumbs over the steering wheel rim because of the possibility of damage to them. The company asked for an option in placing the hands. The purpose of the specification in dummy positioning, however, is to remove discretion from the test personnel, so that all tests are run in the same fashion. An option under these circumstances is therefore not appropriate.

Ultrasystems, Inc., pointed out two minor errors in S10.3 that are hereby corrected. The upper arm and lower arm centerlines are oriented as nearly as possible in a vertical plane (rather than straight up in the vertical), and the little finger of the passenger is placed "barely in contact" with the seat rather than "tangent" to it.

Two corrections are made to the dummy positioning procedure to correct obvious and unintended conflicts between placement of the dummy thighs on the seat cushion and placement of the right leg and foot on the acceleration pedal.

In addition to the positioning proposed, General Motors suggested that positioning of the dummy's head in the fore-and-aft axis would be beneficial. The agency agrees and has added such a specification at the end of the dummy settling procedure.

In a matter separate from the positioning procedure, General Motors, Ford, and Renault requested deletion of the proposed requirement that the dummy maintain proper calibration following a crash test without adjustment. Such a procedure is routine in test protocols and the agency considered it to be a beneficial addition to the standard to further demonstrate the credibility of the dummy test results. GM, however, has pointed out that the limb joint adjustments for the crash test and for the calibration of the lumbar bending test are different, and that it would be unfair to expect continued calibration without adjustment of these joints. The NHTSA accepts this objection and, until a means for surmounting this difficulty is perfected, the proposed change to S8.1.8 is withdrawn.

In another matter unrelated to dummy positioning, Volkswagen argued that active belt systems should be subject to the same requirements as passive belt systems, to reduce the cost differential between the compliance tests of the two systems. As earlier noted the NHTSA has issued an advance Notice of Proposed Rulemaking (41 FR 54961, December 16, 1976) on this subject and will consider Volkswagen's suggestion in the context of that rulemaking.

Finally, the agency proposed the same belt warning requirements for belts provided with passive restraints as are presently required for active belts. No objections to the requirement were received and the requirement is made final as proposed. The agency also takes the opportunity to delete from the standard the out-of-date belt warning requirements contained in S7.3 of the standard.

RECONSIDERATION OF DOCKET 73–8; NOTICE 04

The NHTSA has received two petitions for reconsideration of recent amendments in its test dummy calibration test procedures and design specifications (Part 572, *Anthropomorphic Test Dummy*, 49 CFR Part 572). Part 572 establishes, by means of approximately 250 drawings and five calibration tests, the exact specifications of the test device referred to earlier in this notice that simulates the occupant of a motor vehicle for crash testing purposes.

Apart from requests for a technical change of the lumbar flexion force specifications, the petitions from General Motors and Ford contained a repetition of objections made earlier in the rulemaking about the adequacy of the dummy as an objective measuring device. Three issues were raised: lateral response characteristics of the dummy, failure of the dummy to meet the five subassembly calibration limits, and the need for a 'whole systems" calibration of the assembled dummy. Following receipt of these comments, the agency published notification in the *Federal Register* that it would entertain any other comments on the issue of objectivity (42 FR 28200; June 2, 1977). General comments were received from Chrysler Corporation and American Motors, repeating their positions from earlier comments that the dummy does not qualify as objective.

The objectivity of the dummy is at issue because it is the measuring device that registers the acceleration and force readings specified by Standard No. 208 during a 30-mph impact of the tested vehicle into a fixed barrier. The resulting readings for each vehicle tested must remain below a certain level to constitute compliance. Certification of compliance by the vehicle manufacturer is accomplished by crash testing representative vehicles with the dummy installed. Verification of compliance by the NHTSA is accomplished by crash testing one or more of the same model vehicle, also with a test dummy installed. It is important that readings taken by different dummies, or by the same dummy repeatedly, accurately reflect the forces and accelerations that are being experienced by the vehicle during the barrier crash. This does not imply that the readings produced in tests of two vehicles of the same design must be identical. In the real world, in fact, literally identical vehicles, crash circumstances, and test dummies are not physically attainable.

It is apparent from this discussion that an accurate reflection of the forces and accelerations experienced in nominally identical vehicles does not depend on the specification of the test dummy alone. For example, identically specified and responsive dummies would not provide identical readings unless reasonable care is exercised in the preparation and placement of the dummy. Such care is analogous to that exercised in positioning a ruler to assure that it is at the exact point where a measurement is to commence. No one would blame a ruler for a bad measurement if it were carelessly placed in the wrong position.

It is equally apparent that the forces and accelerations experienced in nominally identical vehicles will only be identical by the greatest of coincidence. The small differences in body structure, even of mass-produced vehicles, will affect the crash pulse. The particular deployment speed and shape of the cushion portion of an inflatable restraint system will also affect results.

All of these factors would affect the accelerations and forces experienced by a human occupant of a vehicle certified to comply with the occupant restraint standard. Thus, achievement of identical conditions is not only impossible (due to the inherent differences between tested vehicles and underlying conditions) but would be unwise. Literally identical tests would encourage the design of safety devices that would not adequately serve the variety of circumstances encountered in actual crash exposure.

At the same time, the safety standards must be "stated in objective terms" so that the manufacturer knows how its product will be tested and under what circumstances it will have to comply. A complete lack of dummy positioning procedures would allow placement of the dummy in any posture and would make certification of compliance virtually impossible. A balancing is provided in the test procedures between the need for realism and the need for objectivity.

The test dummy also represents a balancing between realism (biofidelity) and objectivity (repeatability). One-piece cast metal dummies could be placed in the seating positions and instrumented to register crash forces. One could argue that these dummies did not act at all like a human and did not measure what would happen to a human, but a lack of repeatability could not be ascribed to them. At the other end of the spectrum, an extremely complex and realistic surrogate could be substituted for the existing Part 572 dummy, which would act realistically but differently each time, as one might expect different humans to do.

The existing Part 572 dummy represents 5 years of effort to provide a measuring instrument that is sufficiently realistic and repeatable to serve the purposes of the crash standard. Like any measuring instrument, it has to be used with care. As in the case of any complex instrumentation, particular care must be exercised in its proper use, and there is little expectation of literally identical readings.

The dummy is articulated, and built of materials that permit it to react dynamically, similarly to a human. It is the dynamic reactions of the dummy that introduce the complexity that makes a check on repeatability desirable and necessary. The agency therefore devised five calibration procedures as standards for the evaluation of the important dynamic dummy response characteristics.

Since the specifications and calibration procedures were established in August 1973, a substantial amount of manufacturing and test experience has been gained in the Part 572 dummy. The quality of the dummy as manufactured by the three available domestic commercial sources has improved to the point where it is the agency's judgment that the device is as repeatable and reproducible as instrumentation of such complexity can be. As noted, GM and Ford disagree and raised three issues with regard to dummy objectivity in their petitions for reconsideration.

Lateral response characteristics. Recent sled tests of the Part 572 dummy in lateral impacts show a high level of repeatability from test to test and reproducibility from one dummy to another ("Evaluation of Part 572 Dummies in Side Impacts"—DOT HS 020 858). Further modification of the lateral and rollover passive restraint requirements into an option that can be met by installation of a lap belt makes the lateral response characteristics of the dummy largely academic. As noted in Notice 4 of Docket 73-8 (42 FR 7148; February 7, 1977), "Any manufacturer that is concerned with the objectivity of the dummy in such [lateral] impacts would provide lap belts at the front seating positions in lieu of conducting the lateral or rollover tests."

While the frontal crash test can be conducted at any angle up to 30 degrees from perpendicular to the barrier face, it is the agency's finding that the lateral forces acting on the test instrument are secondary to forces in the midsagittal plane and do not operate as a constraint on vehicle and restraint design. Compliance tests conducted by NHTSA to date in the 30-degree oblique impact condition have consistently generated similar dummy readings. In addition, they are considerably lower than in perpendicular barrier impact tests, which renders them less critical for compliance certification purposes.

Repeatability of dummy calibration. Ford questioned the dummy's repeatability, based on its analysis of "round-robin" testing conducted in 1973 for Ford at three different test laboratories (Ford Report No. ESRO S–76–3 (1976)) and on analysis of NHTSA calibration testing of seven test dummies in 1974 (DOT–HS–801–861).

In its petition for reconsideration, Ford equated dummy objectivity with repeatability of the calibration test results and concluded "it is impracticable to attempt to meet the Part 572 component calibration requirements with test dummies constructed according to the Part 572 drawing specification."

The Ford analysis of NHTSA's seven dummies showed only 56 of 100 instances in which all of the dummy calibrations satisfied the criteria. The NHTSA's attempts to reproduce the Ford calculations to reach this conclusion were unsuccessful, even after including the HO3 dummy with its obviously defective neck. This neck failed badly 11 times in a row, and yet Ford apparently used these tests in its estimate of 56 percent compliance. This is the equivalent of concluding that the specification for a stop watch is inadequate because of repeated failure in a stop watch with an obviously defective part. In this case, the calibration procedure was doing precisely its job in identifying the defective part by demonstrating that it did not in fact meet the specification.

The significance of the "learning curve" for quality control in dummy manufacture is best understood by comparison of three sets of dummy calibration results in chronological order. Ford in earlier comments relied on its own "round-robin" crash testing, involving nine test dummies. Ford stated that none of the nine dummies could pass all of the component calibration requirements. What the NHTSA learned through follow-up questions to Ford was that three of the nine dummies were not built originally as Part 572 dummies, and that the other six were not fully certified by their manufacturers as qualifying as Part 572 dummies. In addition, Ford instructed its contractors to use the dummies as provided whether or not they met the Part 572 specifications.

In contrast, recent NHTSA testing conducted by Calspan (DOT–HS–6–01514, May and June 1977 progress reports) and the results of tests conducted by GM (USG 1502, Docket 73–8, GR 64) demonstrate good repeatability and reproducibility of dummies. In the Calspan testing a total of 152 calibration tests were completed on four dummies from two manufacturers. The results for all five calibration tests were observed to be within the specified performance criteria of Part 572. The agency concludes that the learning curve in the manufacturing process has reached the point where repeatability and reproducibility of the dummy has been fully demonstrated.

Interestingly, Ford's own analysis of its round-robin testing concludes that variations among the nine dummies were not significant to the test results. At the same time, the overall acceleration and force readings did vary substantially. Ford argued that this showed unacceptable variability of the test as a whole, because they had used "identical" vehicles for crash testing. Ford attributed the variations in results to "chance factors," listing as factors placement of the dummy, postural changes during the ride to the barrier, speed variations, uncertainty as to just what part of the instrument panel or other structure would be impact loaded, instrumentation, and any variations in the dynamics of air bag deployment from one vehicle to another.

The agency does not consider these to be uncontrolled factors since they can be greatly reduced by carefully controlling test procedures. In addition, they are not considered to be unacceptable "chance factors" that should be eliminated from the test. The most important advantage of the barrier impact test is that it simulates with some realism what can be experienced by a human occupant, while at the same time limiting variation to achieve repeatability. As discussed, nominally identical vehicles are not in fact identical, the dynamics of deployment will vary from vehicle to vehicle, and humans will adopt a large number of different seated positions

in the real world. The 30-mph barrier impact requires the manufacturer to take these variables into account by providing adequate protection for more than an overly structured test situation. At the same time, dummy positioning is specified in adequate detail so that the manufacturer knows how the NHTSA will set up a vehicle prior to conducting compliance test checks.

"*Whole systems*" *calibration.* Ford and GM both suggested a "whole systems" calibration of the dummy as a necessary additional check on dummy repeatability. The agency has denied these requests previously, because the demonstrated repeatability and reproducibility of Part 572 dummies based on current specification is adequate. The use of whole systems calibration tests as suggested would be extremely expensive and would unnecessarily complicate compliance testing.

It is instructive that neither General Motors nor Ford has been specific about the calibration tests they have in mind. Because of the variables inherent in a high energy barrier crash test at 30 mph, the agency judges that any calibration readings taken on the dummy would be overwhelmed by the other inputs acting on the dummy in this test environment. The Ford conclusion from its round-robin testing agrees that dummy variability is a relatively insignificant factor in the total variability experienced in this type of test.

GM was most specific about its concern for repeatability testing of the whole dummy in its comments in response to Docket 74–14; Notice 01:

Dummy whole body response requirements are considered necessary to assure that a dummy, assembled from certified components, has acceptable response as a completed structure. Interactions between coupled components and subsystems must not be assumed acceptable simply because the components themselves have been certified. Variations in coupling may lead to significant variation in dummy response.

There is a far simpler, more controlled means to assure oneself of correct coupling of components than by means of a "whole systems" calibration. If, for example, a laboratory wishes to assure itself that the coupling of the dummy

PART 571; S 208—PRE 81

neck structure is properly accomplished, a simple statically applied input may be made to the neck prior to coupling to obtain a sample reading, and then the same simple statically applied input may be repeated after the coupling has been completed. This is a commonly accepted means to assure that "bolting together" the pieces is properly accomplished.

Lumbar spine flexion. The flexibility of the dummy spine is specified by means of a calibration procedure that involves bending the spine through a forward arc, with specified resistance to the bending being registered at specified angles of the bending arc. The dummy's ability to flex is partially controlled by the characteristics of the abdominal insert. In Notice 04, the agency increased the level of resistance that must be registered, in conjunction with a decision not to specify a sealed abdominal sac as had been proposed. Either of these dummy characteristics could affect the lumbar spine flexion performance.

Because of the agency's incomplete explanation for its actions, Ford and General Motors petitioned for reconsideration of the decision to take one action without the other. Both companies suggested that the specification of resistance levels be returned to that which had existed previously. The agency was not clear that it intended to go forward with the stiffer spine flexion performance, quite apart from the decision to not specify an abdomen sealing specification. The purpose for the "stiffer" spine is to attain more consistent torso return angle and to assure better dummy stability during vehicle acceleration to impact speed.

To assure itself of the wisdom of this course of action, the agency has performed dummy calibration tests demonstrating that the amended spine flexion and abdominal force deflection characteristics can be consistently achieved with both vented and unvented abdominal inserts (DOT HS–020875 (1977)).

Based on the considered analysis and review set forth above, the NHTSA denies the petitions of General Motors and Ford Motor Company for further modification of the test dummy specification and calibration procedures for reasons of test dummy objectivity.

In consideration of the foregoing, Standard No. 208 (49 CFR 571.208) is amended as proposed with changes set forth below, and Part 572 (49 CFR Part 572) is amended by the addition of a new sentence at the end of § 572.5, *General Description,* that states: "A specimen of the dummy is available for surface measurements, and access can be arranged through: Office of Crashworthiness, National Highway Traffic Safety Administration, 400 Seventh Street, S.W., Washington, D.C. 20590."

In accordance with Department of Transportation policy encouraging adequate analysis of the consequences of regulatory action (41 FR 16200; April 16, 1976), the Department has evaluated the economic and other consequences of this amendment on the public and private sectors. The modifications of an existing option, the simplification and clarification of test procedures, and the increase in femur force loads are all judged to be actions that simplify testing and make it less expensive. It is anticipated that the "two dummy" positioning procedure may occasion additional testing expense in some larger vehicles, but not the level of expense that would have general economic effects.

The effective date for the changes has been established as one year from the date of publication to permit Volkswagen, the only manufacturer presently certifying compliance of vehicles using these test procedures, sufficient time to evaluate the effect of the changes on the compliance of its products.

The program official and lawyer principally responsible for the development of this amendment are Guy Hunter and Tad Herlihy, respectively.

(Sec. 103, 119, Pub. L. 89–563, 80 Stat. 718 (15 U.S.C. 1392, 1407); delegation of authority at 49 CFR 1.50.)

Issued on June 30, 1977.

Joan Claybrook
Administrator

42 F.R. 34299
July 5, 1977

PREAMBLE TO AMENDMENT TO MOTOR VEHICLE SAFETY STANDARD NO. 208

Occupant Crash Protection

(Docket No. 74-14; Notice 10)

The existing motor vehicle safety standard for occupant crash protection in new passenger cars is amended to require the provision of "passive" restraint protection in passenger cars with wheelbases greater than 114 inches manufactured on and after September 1, 1981, in passenger cars with wheelbases greater than 100 inches on and after September 1, 1982, and in all passenger cars manufactured on or after September 1, 1983. The low usage rate of active seat belt systems negates much of their potential safety benefit. However, lap belts will continue to be required at most front and all rear seating positions in new cars, and the Department will continue to recommend their use to motorists. It is found that upgraded occupant crash protection is a reasonable and necessary exercise of the mandate of the National Traffic and Motor Vehicle Safety Act to provide protection through improved automotive design, construction, and performance.

Dates: Effective date September 1, 1981.

Addresses: Petitions for reconsideration should refer to the docket number and be submitted to: Docket Station, Room 5108—Nassif Building, 400 Seventh Street, S.W., Washington, D.C. 20590.

For further information contact:

Tad Herlihy
Office of Chief Counsel
National Highway Traffic Safety
Administration
Washington, D.C. 20590
(202) 426-9511

Supplementary Information:

Considerations Underlying the Standard

Under the National Traffic and Motor Vehicle Safety Act, as amended (the Act) (15 U.S.C. 1381 et seq.), the Department of Transportation is responsible for issuing motor vehicle safety standards that, among other things, protect the public against unreasonable risk of death or injury to persons in the event accidents occur. The Act directs the Department to consider whether a standard would contribute to carrying out the purposes of the Act and would be reasonable, practicable, and appropriate for a particular type of motor vehicle (15 U.S.C. 1392(f)(3)). The standard must, as formulated, be practicable, meet the need for motor vehicle safety, and be stated in objective terms (15 U.S.C. 1392(a)). The Senate Committee drafting the statute stated that safety would be the overriding consideration in the issuance of standards. S. Rep. No. 1301, 89th Cong., 2d Sess (1966) at 6.

The total number of fatalities annually in motor vehicle accidents is approximately 46,000 (estimate for 1976), of which approximately 25,000 are estimated to be automobile front seat occupants. Two major hazards to which front seat occupants are exposed are ejection from the vehicle, which increases the probability of fatality greatly, and impact with the vehicle interior during the crash. Restraint of occupants to protect against these hazards has long been recognized as a means to substantially reduce the fatalities and serious injuries experienced at the front seating positions.

One of the Department's first actions in implementing the Act was promulgation in 1967 of Standard No. 208, *Occupant Crash Protection* (49 CFR 571.208), to make it possible for vehicle occupants to help protect themselves against the hazards of a crash by engaging seat belts. The standard requires the installation of lap and shoulder seat belt assemblies (Type 2) at front outboard designated seating positions (except in convertibles) and lap belt assemblies (Type 1) at all other designated seating positions. The standard became effective January 1, 1968.

While it is generally agreed that when they are worn, seat belt assemblies are highly effective in preventing occupant impact with the vehicle interior or ejection from the vehicle, only a minority of motorists in the United States use seat belts. For all types of belt systems, National Highway Traffic Safety Administration (NHTSA) studies show that about 20 percent of belt systems are used (DOT HS 6 01340 (in process)). The agency's calculations show that only about 2,600 deaths (and corresponding numbers of injuries) of front seat occupants were averted during 1976 by the restraints required by Standard No. 208 as it is presently written.

Two basic approaches have been developed to increase the savings of life and mitigation of injury afforded by occupant restraint systems. More than 20 nations and two provinces of Canada have enacted mandatory seat belt use laws to increase usage and thereby the effective lifesaving potential of existing seat belt systems. The other approach is to install automatic passive restraints in passenger cars in place of, or in conjunction with, active belt systems. These systems are passive in the sense that no action by the occupant is required to benefit from the restraint. Passive restraint systems automatically provide a high level of occupant crash protection to virtually 100 percent of front seat occupants.

The two forms of passive restraint that have been commercially produced are inflatable occupant restraints (commonly known as air bags) and passive belts. Air bags are fabric cushions that are rapidly filled with gas to cushion the occupant against colliding with the vehicle interior when a crash occurs that is strong enough to register on a sensor device in the vehicle. The deployment is accomplished by the rapid generation or release of a gas to inflate the bag. Passive belt systems are comparable to active belt systems in many respects, but are distinguished by automatic deployment around the occupant as the occupant enters the vehicle and closes the door.

HISTORY OF STANDARD NO. 208

Because of the low usage rates of active belt systems and because alternative technologies were becoming available, the initial seat belt requirements of Standard No. 208 were upgraded in 1970 to require passive restraints by 1974 (35 FR 16927; November 3, 1970). Most passenger car manufacturers petitioned for judicial review of this amendment (*Chrysler v. DOT*, 472 F.2d 659 (6th Cir. 1972)). The Sixth Circuit's review upheld the mandate in most respects but remanded the standard to the agency for further specification of a test dummy that was held to be insufficiently objective for use as a measuring device in compliance tests. The court stated with regard to two of the statutory criteria for issuance of motor vehicle safety standards:

We conclude that the issue of the relative effectiveness of active as opposed to passive restraints is one which has been duly delegated to the Agency, with its expertise, to make; we find that the Agency's decision to require passive restraints is supported by substantial evidence, and we cannot say on the basis of the record before us that this decision does not meet the need for motor vehicle safety. 472 F.2d at 675.

. . . we conclude that Standard 208 is practicable as that term is used in this legislation. 472 F.2d at 674.

As for objective specification of the test dummy device, a detailed set of specifications (49 CFR Part 572) was issued in August 1973 (38 FR 20449; August 1, 1973) and updated with minor changes in February 1977 (42 FR 7148; February 7, 1977). A full discussion of the test dummy specifications is set forth in a rulemaking issued today by the NHTSA concerning technical aspects of Standard No. 208 (42 FR 34299; FR Doc. 77-19138).

In March 1974, the Department made the finding that the test dummy is sufficiently objective to satisfy the *Chrysler* court remand (39 FR 10271; March 19, 1974). In the same notice, mandatory passive restraints were again proposed. Based on the comments received in response to that notice, the passive restraint mandate was once again proposed in a modified form in June 1976 (41 FR 24070; June 14, 1976). In the interim, General Motors Corporation manufactured, certified, and sold approximately 10,000 air-bag-equipped full-size Buicks, Oldsmobiles, and Cadillacs. Volkswagen has manufactured and sold approximately 65,000 passive-belt-equipped Rabbit model passenger cars. Volvo Corporation has also introduced a relatively small number of air-bag-equipped vehicles into service. Ford Motor Company had earlier manufactured 831 air-bag-equipped Mercurys. These vehicles were manufactured under one of two options placed in the standard in 1971 to permit optional production of vehicles with passive restraint systems in place of seat belt assemblies otherwise required. In 1972, the standard was also amended to require an "ignition interlock" system on front seat belts to force their use before the vehicle could be started. This requirement, effective in September 1973, was revoked in October 1974 in response to a Congressional prohibition on its specification (Pub. L. 93-492, § 109 (October 27, 1974)).

The Department's final action on its June 1976 proposal ("The Secretary's Decision Concerning Motor Vehicle Occupant Crash Protection," hereinafter "the December 1976 decision") continued the existing requirements of the standard (42 FR 5071; January 27, 1977) and created a demonstration program to familiarize the public with passive restraints. The Department negotiated contracts with four automobile manufacturers for the production of up to 250,000 passive-equipped vehicles per year for introduction into the passenger car fleet in model years 1980-1. Mercedes-Benz agreed to manufacture 2,250 such passenger cars, and Volkswagen agreed to manufacture 125,000 of its passive-belt-equipped Rabbit models. Ford agreed to participate by "establishing the capability of manufacturing" 140,000 compact model passenger cars, and General Motors agreed to "establish production capacity" to manufacture 300,000 intermediate size passenger cars. The December 1976 decision was based on the finding that, although passive restraints are technologically feasible at reasonable cost and would prevent 9,000 fatalities annually when fully integrated into the fleet, possible adverse reaction by an uninformed public after the standard took effect could inspire their prohibition by Congress with substantial attendant economic waste and incalculable harm to the cause of highway safety. This finding was based in large part on the Department's experience

with the ignition interlock on 1974- and 1975-model passenger cars, which was prohibited by Congress in response to industry and public opposition.

Early in 1977, the Department reconsidered the December 1976 decision because public acceptance or rejection of passive restraints is not one of the statutory criteria which the Department is charged by law to apply in establishing standards. In addition, the demonstration program introduced a minimum 3-year delay in implementation of mandatory passive restraints. The Department questioned the premise that passive restraint systems would foster consumer resistance as had the ignition interlock system. While the ignition interlock system forced action by the motorist as a condition for operating an automobile, passive restraints eliminate the need for any action by the occupant to obtain their crash protection benefits.

A third reason for reassessment of the December 1976 decision was the certainty that an increasing proportion of the passenger car fleet will be small cars, in response to the energy situation and the automotive fuel economy program established by the Energy Policy and Conservation Act. The introduction of these new, smaller vehicles on the highway holds the prospect of an increase in the fatality and injury rate unless countermeasures are undertaken.

Based on this reconsideration, the Department proposed (42 FR 15935; March 24, 1977) that the future crash protection requirements of Standard No. 208 take one of three forms: (1) continuation of the present requirements, (2) mandatory passive restraints at one or more seating positions of passenger cars manufactured on or after September 1, 1980, or (3) continuation of the existing requirements in conjunction with proposed legislation to establish Federal or State mandatory seat belt use laws.

The proposal for an occupant restraint system other than seat belts invoked a provision of the Act (15 U.S.C. § 1400(b)) that requires notification to Congress of the action. The Act also requires that a public hearing be held at which any member of Congress or any other interested person could present oral testimony. The proposal was transmitted to the Congress on March

21, 1977, with an invitation to appear at a public hearing chaired by the Secretary on April 27 and 28, 1977, in Washington, D.C. A transcript of this meeting, along with written comments on the March 1977 proposal, are available in the public docket.

DISCUSSION OF ISSUES

The March 1977 proposal of three possible courses of action for future occupant crash protection is grounded in a large, complex administrative record that has been developed in the 8 years since passive restraints were first contemplated by the Department. Interested persons are invited to review the NHTSA public docket that has been compiled under designations 69–7, 73–8, and 74–14. Consideration of the issues and questions that have arisen during the years of rulemaking can be found in the preambles to the Department's numerous rulemaking notices on passive restraints. Although many of the comments on the March 1977 proposal raised issues that have been discussed in previous notices, the significant issues will be addressed here again, in light of the most recent information available to the Department.

The need for rulemaking action. An important reason to consider anew the occupant crash protection issue is the basic and positive changes that the automobile will undergo in the years ahead. Until recently, the basic characteristics of automobiles sold to the American public have evolved for the most part in response to the forces of the market place. High premium was placed upon styling, roominess, and acceleration performance. In a cheap-energy society, relatively little attention was paid to efficiency of operation. Nor, until relatively recently, was serious consideration given to minimizing the adverse impact of the automobile upon air quality.

Recent circumstances, however, have drastically altered the situation, and have made it abundantly clear that the automobile's characteristics must reflect broadly defined societal goals as well as those advanced by the individual car owner. The President has announced a new national energy policy that recognizes a compelling need for changes in the American lifestyle. Congress has implemented statutory programs to improve

the fuel economy of automobiles, as one result of which this Department has just issued demanding fuel economy standards for 1981 through 1984 passenger cars. Right now, the Congress is deliberating over amendments to the Clean Air Act which will impose relatively stringent emissions requirements effective over the same time frame.

The trend toward smaller cars to improve economy and emissions performance contains a potential for increased hazard to the vehicles' occupants. But technology provides the means to protect against this hazard, and this Department's statutory mandate provides authority to assure its application. The Report of the Federal Interagency Task Force on Motor Vehicle Goals for 1980 and Beyond indicated that simultaneous achievement of ambitious societal goals for the automobile in the areas of fuel economy, emissions, and safety is technologically feasible. Integrated test vehicles developed by this Department confirm that finding and, further, demonstrate that the resulting vehicles need not unduly sacrifice the other functional and esthetic attributes traditionally sought by the American car buyer.

Moreover, the socially responsive automobile of the 1980's need not bring a penalty in economy of ownership. The just-issued passenger car fuel economy standards are calculated to reduce the overall costs of operating an automobile by $1,000 over the vehicle's lifetime. In the case of improved safety performance, the occupant restraint improvements specified in this notice can be expected to pay for themselves in reduced first-person liability insurance premiums during the life of the vehicle.

The issue of occupant crash protection has been outstanding too long, and a decision would have been further delayed while the demonstration programs was conducted. A rigorous review of the findings made by the Department in December 1976 demonstrates that they are in all substantial respects correct as to the technological feasibility, practicability, reasonable cost, and lifesaving potential of passive restraints. The decision set forth in this notice is the logical result of those findings.

In reassessing the December 1976 decision, the Department has considered each available means to increase crash protection in arriving at the most rational approach. As proposed, the possibility of "driver-side only" passive protection was considered, but was rejected because of the unsatisfactory result of having one front-seat passenger offered protection superior to that offered other front-seat passengers in the same vehicle. On balance, there was found to be little cost or lead-time advantage to this approach. The possibility of reinstituting a type of safety belt interlock was rejected because the agency's authority was definitively removed by the Congress less than three years ago and there is no reason to believe that Congress has changed its position on the issue since that time.

Mandatory belt use laws. One of the means proposed in the March notice to achieve a large reduction in highway deaths and injuries is Federal legislation to induce State enactment of mandatory seat belt use laws, either by issuance of a highway safety program standard or by making State passage of such laws a condition for the receipt of Federal highway construction money.

The prospects for passage of mandatory seat belt use laws by more than a few States appear to be poor. None of the commenters suggested that passage of such laws was likely. A public opinion survey sponsored by the Motor Vehicle Manufacturers Association and conducted by Yankelovich, Skelly, and White, Inc. indicated that a 2-to-1 majority nationwide opposes belt use laws. Many such bills have been presented; no State has enacted one up to now. Also, Congress denied funding for a program to encourage State belt use laws in 1974, suggesting that it does not look favorably upon Federal assistance in the enactment of these laws.

More recently, Congress removed the Department's authority to withdraw Federal safety funding in the case of States that do not mandate the use of motorcycle helmets on their highways (Pub. L. 94-280, Sec. 208(a), May 5, 1976). The close parallel between requiring helmet use and requiring seat belt use argues against the likelihood of enactment of belt use laws.

These strong indications that Congress would not enact a belt use program in the foreseeable future demonstrate, in large measure, why the success of other nations in enacting laws is not parallel to the situation in the United States. In the belt use jurisdictions most often compared to the United States (Australia and the Provinces of Canada), the laws were enacted at the State or Province level in the first instance, and not at the Federal level. In the Department's judgment, the most reasonable course of action to obtain effective belt use laws in the United States will be to actively encourage their enactment in one or more States. An attempt to impose belt use laws on citizens by the Federal government would create difficulties in Federal-State relations, and could damage rather than further the interests of highway safety.

Effectiveness of passive restraints. The December 1976 decision concluded that the best estimates of effectiveness in preventing deaths and injuries of the various types of restraint systems under consideration were as set forth in Table I. Using the effectiveness estimates from Table I, the projection of benefits attributable to various restraint systems is summarized in Table II. Several comments concerning the effectiveness of passive restraint systems were submitted in response to the March 1977 proposal.

Insurance company commenters generally supported the Department's estimates. General Motors, however, disputed the validity of the estimates in the December 1976 decision, arguing that the results experienced by the approximately 10,000 GM vehicles sold the public indicated a much lower level of effectiveness. It made comparisons between accidents involving those cars and other accidents with conventional cars, selected to be as similar as possible in type and severity. On the basis of this study, GM stated that the data indicate that the "current air cushion-lap belt system, if available in all cars, would save less than the nearly 3,000 lives that can be saved by only *20* percent active lap/shoulder belt use."

The Department finds the methods used in the General Motors study to be of doubtful value in arriving at an objective assessment of the experience of the air-bag-equipped vehicles. General Motors is a vastly interested party in these proceedings, and the positions that it adopts are necessarily those of an advocate for a particular

PART 571; S 208—PRE 87

result. This is in no sense a disparagement; advocacy of desired outcomes by interested parties is an essential part of the administrative process. But if a study advanced by an interested advocate is to be seriously considered from a "scientific" viewpoint, it must be carefully designed to avoid dilution of its objectivity by the bias of the sponsoring party. The GM study fails that test. Its foundation is a long series of qualitative judgments, which are made by employees of the party itself. An equally serious fault is that the basic body of accident data from which the comparison accidents are selected is not available to the public, so that countering analyses cannot be made by opposing parties, nor can the judgments in the original study be checked. General Motors had previously submitted to an earlier Standard No. 208 docket a study of restraint system effectiveness based on similarly qualitative judgments by its own employees (69–07–GR–256–01). The shoulder belt effectiveness figures arrived at in that study were about one-half of what are now generally recognized to be the actual values. While this later study utilizes a somewhat different methodology, it suffers from the same flaws in its failure to preclude dilution of its objectivity by the bias of its sponsor.

Economics and Science Planning, Inc., submitted three studies that made estimates of air bag effectiveness. In one, the estimate of air bag effectiveness was at least as high as the theoretical projections made in Table II. In another, a very low estimate of air bag effectiveness was made—from 15 to 25 percent.

The Insurance Institute for Highway Safety submitted another estimate of air bag effectiveness based on the experience with the GM cars in highway use. A selection was made of accidents in which the air bag was designed to operate, based on frontal damage, direction of impact, and age of occupant. In these accidents, air bags were determined to have reduced fatalities by 66 percent, as compared to 55 percent for three-point belts. However, the narrow selection of accidents limits the application of the figures derived in the IIHS study.

The Department considers that the most reliable method of evaluating the experience of the air-bag-equipped cars at this time is to compare the number of injuries, at various levels, sustained by their occupants with the number that is experienced in the general population of vehicles of this type. The vehicles in question are not a sampling of the general vehicle population: they are relatively new, and mostly in the largest "luxury" size class. Some adjustment must be made for these factors.

The adjustment for the size of the vehicles has been made by multiplying the overall injury figures by a factor of 0.643, which has been found in one study (Joksch, "Analysis of Future Effects of Fuel Storage and Increased Small Car Usage Upon Traffic Deaths and Injuries," General Accounting Office, 1975) as the ratio of fatalities per year for this size of vehicles to the figure for the general population. The newness of the vehicles has a double-edged aspect: newer vehicles are evidently driven more miles per year than older ones, but they also appear to experience fewer accidents per mile traveled (Dutt and Reinfurt, "Accident Involvement and Crash Injury Rates by Make, Model, and Year of Car," Highway Safety Research Center, 1977). These two factors can be accounted for if it is assumed that they cancel each other, by using vehicle years, rather than vehicle miles, as the basis of comparison. With these adjustments, the expected number of all injuries of AIS–2 (an index of injury severity) and above in severity for conventional vehicles equivalent to the air-bag-equipped fleet during the period considered was 91. The actual number experienced was 38, indicating an effectiveness factor for these injury classes of 0.58.

A possibility of bias in these estimates exists in that injuries that have occurred in the air bag fleet may not have been reported, despite the three-level reporting system (owners, police, and dealers) that has been established. This bias is less likely to be present in frontal accidents, where the air bag is expected to (and generally does) deploy. For frontal accidents only, the number of injuries expected is 60, or 66 percent of the total ("Statistical Analysis of Seat Belt Effectiveness in 1973–1975 Model Cars Involved in Towaway Crashes," Highway Safety Research Center, 1976); only 29 have been experienced, indicating an effectiveness factor of 0.52.

These figures confirm (and in fact exceed) the effectiveness estimates of the December 1976 decision. For injuries of higher severity levels, the numbers experienced are much too small to be statistically significant.

The various assumptions and adjustments that must be made to arrive at a valid "expected" figure, and the possibility that some injuries were unreported, leaves substantial room for uncertainty and argument as to the true observed effectiveness of the restraint systems. Nevertheless, the results of the field experience are encouraging. Even if the observed-effectiveness figures arrived at by these calculations were high by a factor of 2, they would still substantially confirm the estimates of the December 1976 decision. Considering all the arguments on both sides of the issues, the Department concludes that the observed experience of the vehicles on the road equipped with air bags does not cast doubt on the effectiveness estimates in the December 1976 decision.

It has been argued that the Department should not issue a passive restraint standard in the absence of statistically significant real world data which confirm its estimates of effectiveness. Statistical "proof" is certainly desirable in decisionmaking, but it is often not available to resolve public policy decisions. It is also clear from the legislative history of the Act that the Department was not supposed to wait for the widespread introduction of a technology before it could be mandated. The Senate report for example refers to the "failure of safety to sell" in automobiles, and describes how the Department was intended to push the manufacturers into adopting new safety technology that would not be introduced voluntarily (S. Rep. 1301, 89th Cong. 2nd Sess. 4 (1966)). The *Chrysler* case found that "The explicit purpose of the Act is to enable the Federal Government to impel automobile manufacturers to develop and apply new technology to the task of improving the safety design of automobiles as readily as possible." (472 F.2d at 671.)

Cost of passive restraints. Passive belts have been estimated in the past by the Department to add $25 to the price of an automobile, relative to the price of cars with present active belt systems.

The increased operating cost over the life of a vehicle with passive belts is estimated to be $5. These figures are assumed valid for purposes of this review, and were not contsted in the comments received.

This Department, General Motors, Ford, DeLorean, and Minicars all have produced estimates of the passenger car price increase due to the inclusion of air bags. These are sufficiently detailed and current to be compared, and are set forth in Table III. The Department estimate has been raised somewhat above its previous ones because of the $14 increase in the price of the components of an air bag system quoted by a supplier.

The General Motors estimates have been revised from previous estimates in several respects. Research and development, engineering, and tooling expenses are no longer amortized entirely in the first year, but are spread over 3 years (other estimates spread these costs over 5 years). The allowance for removal of active belt hardware has been reduced to conform more closely to the Department's estimates. The newer figures reflect a somewhat more complex system, including new sensors. Of the $81 spread between the Department and the GM estimates, all but $11 can be attributed to differences in the following areas: GM's estimate of dealer profit which is based on sticker prices (rather than actual sale price), GM's shorter amortization period, added complexity of the 1977 system over the 1976 system, and the cost of major modifications of the vehicle which the agency questions. The remaining $11 difference must be considered as disagreement concerning the elements of cost shown in the table.

The Ford estimate is the same as previously submitted. Forty-two dollars of the difference from the Department estimate is a higher profit figure arising from Ford's use of sticker prices rather than actual price of sale, which gives the dealer less mark-up. A substantial amount of difference is for a complex electronic diagnostic module, extra sensors that the Department does not view as necessary, and the use of a knee bolster instead of a cheaper knee air bag. Thirty-nine dollars represents unreconciled differences.

Operating costs consist mainly of the cost of replacing a deployed bag, fuel cost, and mainte-

PART 571; S 208—PRE 89

nance. Ford also includes an amount for periodic inspection. The Department estimate for replacement cost differs from the GM and Ford estimates almost entirely as a result of the lower estimate for the first cost of the system. The fuel costs differ primarily as a result of different weight figures for the passive systems, which may be design choices of the manufacturers. The Department's evaluation of manufacturers' cost objections is being placed in the public docket as required by § 113 of the Act.

If, as projected, passive restraints are effective in saving lives and reducing injuries, as compared to existing belt systems at present use rate, the insurance savings that will result will offset a major portion, and possibly all, of the cost to the consumer of the systems. There may be some doubt on this point that arises from skepticism concerning the behavior of insurers.

The vast majority of auto occupant injuries beyond the minor level result in automobile, health, or life insurance claims. In some States, insurers may lack a degree of flexibility in the adjustment of premiums because of pressures from insurance commissions. However, the evidence indicates that premiums are fundamentally based on claims experience.

In its comments to the docket, Nationwide Mutual Insurance Companies estimated that savings in insurance premiums should average $32.50 per insured car per year, if all cars were equipped with air bags. Of this amount, 75 percent is the result of an assumed savings of 24.6 percent in the bodily injury portion of automobile insurance premiums, 21 percent from a 1.5 percent reduction in health insurance premiums (30 percent of the 5 percent of the premiums that pay for auto-related injuries), and the remainder from savings in life insurance premiums. The American Mutual Insurance Alliance and Allstate referred to existing 30 percent discounts in first-party coverage and concluded that comparable reductions would be expected to follow a mandate of passive restraints.

It has been argued that these savings would be largely offset by the increased cost of collision and property damage insurance due to the increased cost of repairing a car with a deployed air bag. This claim appears to be largely un-

founded. Using figures based on field tests, it is estimated that each year 300,000 automobiles will be in accidents of sufficient severity to deploy the air bag. (Cooke, "Usage of Occupant Crash Protection Systems," NHTSA, July 1976, #74-14-GR-30, App. A.) Accepting vehicle manufacturer estimates, it is further assumed that the cost of replacing an air bag will be 2.5 times the original equipment cost. If a car more than 6 years old is involved in an air-bag-deploying accident, it is assumed scrapped rather than being repaired. Combining these assumptions with the estimated $112 cost of installing a full front air cushion in a new vehicle gives a total annual cost of replacement of $50.4 million, or a per car cost of less than 51 cents per year. Increases in collision premiums should, therefore, not exceed $1 per car per year. It is noted that deployment in non-crash cases would be covered by "comprehensive" insurance policies.

The $32.50 annual insurance savings estimated by Nationwide would be sufficient to pay for the added operating cost (around $4 per year) of an air-bag-equipped car with enough left over to more than pay for the initial cost of the system. Discounting at the average interest rate on new car loans measured in real terms (6 percent), the air bag would almost recover the initial cost in 4 years, with a savings over operating cost of $107.

Economic and Science Planning, Inc. (ESP) has submitted a differing estimate, that insurance savings with full implementation of passive restraints would be only $3.60, rather than $32.50 per year. About one-half of the difference arises from ESP's assumption that seat belt usage would voluntarily rise to the 44 percent level by 1984. This seems highly improbable, based on experience to date.

Moreover, that assumption does not support the deletion of projected insurance savings resulting from passive restraints, but suggests that other courses of action (such as whatever might be done to increase belt usage to 44 percent) might also produce savings. The remaining differences are based on such factors as the portion of injury costs that is paid for by insurance. If the assumptions of ESP are allowed to remain, the savings per year would be about $16, and the present value of auto-lifetime savings would be $120.

Side effects of air bag installation. Some concerns were expressed in the comments about air bags that might be grouped as possible undesirable side effects. One of these was injuries that might be caused by design deployment. There is no question that any restraint system that must decelerate a human body from 30 mph or more to rest within approximately 2 feet can cause injury. Belt systems often cause bruises and abrasions in protecting occupants from more serious injuries. The main question is whether any injuries caused by air bags are generally within acceptable limits, and are significantly less severe than those that would have been suffered had the occupants in question not been restrained by the air bags. The evidence from the vehicles on the road indicates that this is indeed the case. The injuries cited by GM as possibly caused or aggravated by air bag deployment are in the minor to moderate (AIS–1 and –2) category. From this it can be concluded that injuries caused by design deployment, though worthy of careful monitoring with a view to design improvements by manufacturers, do not provide a serious argument against a passive restraint requirement.

A closely related question that has caused concern in the past is whether air bags pose an unreasonable danger to occupants who are not in a normal seating position, such as children standing in front of a dashboard or persons who have been moved forward by panic braking. Much development work has been devoted to this problem in the past, to design systems that minimize the danger to persons who are close to the inflation source. The most important change in this area has probably been the general shift away from inflation systems that depend on stored high-pressure gas, in favor of pyrotechnic gas generators. With these systems the flow of gas can be adjusted to make the rate slower at the beginning of inflation, so that an out-of-position occupant is pushed more gently out of the way before the maximum inflation rate occurs.

With one exception, there have been no cases where out-of-position occupants have been found

to be seriously injured in crashes in which air bags have deployed. Five of the crashes involving GM vehicles have involved children in front seating positions (although not necessarily out of position), and others have involved children unbelted in the rear seat.

The only exception has been the death of an infant that was lying laterally on the front seat unrestrained. Apparently during panic braking that proceeded the crash, the infant was thrown from the seat. While this constitutes an out-of-position situation technically, it is not the type of circumstance in which the air bag contributes to injury of the out-of-position occupant.

Inadvertent actuation of an air bag may be a particular concern to the public, as noted by both General Motors and Ford. The sudden deployment of an air bag in a non-crash situation would generally be a disconcerting experience. The experience with vehicles on the road, and tests that have been performed on 40 subjects who were not aware that there were air bags in their vehicles, indicate the loss of control in such situations should be rare: none has occurred in the incidents up to now. There is little question, however, that inadvertent actuation could cause loss of control by some segments (aged, inexperienced, distracted) of the driving population, and it must be viewed as a small but real cost of air bag protection.

The frequency of inadvertent actuation is therefore of special concern. The Ford fleet of air-bag-equipped cars (about 800 vehicles that have been on the road since late 1972, with around 500 now taken out of service) has experienced no inadvertent actuations at all. The General Motors fleet, about 10,000 sold mostly to private buyers during 1974 and 1975, has experienced three inadvertent actuations on the road. Six others have occurred in the hands of mechanics and body shop personnel, two in externally caused fires or explosions, and one from tampering in a driveway. The Volvo fleet of 75 vehicles has experienced none. It is believed that the causes of the GM inadvertent deployments are understood, and that the means of eliminating or considerably reducing the likelihood of all the known causes of inadvertent deployments have been found. These include shielding of the squibs (the device to ignite the propellant ma-

PART 571; S 208—PRE 91

terial in the bag inflators) against electromagnetic radiation, automatically disarming the system through the ignition system when the car is not in operation, and routing wiring so that it is less accessible to tampering or degradation.

If the figures for the combined fleets are projected onto the U.S. vehicle population, they would amount to around 7,000 on-the-road inadvertent actuations annually, or one for every 15,000 vehicles. The chances of an individual experiencing one. as a vehicle occupant during his or her lifetime would be on the order of 1 in 200. This estimate probably overstates the likelihood of occurrence since the inadvertent actuations in the GM cars to date are believed to be due to design deficiencies that are correctable. Thus, although it will probably continue to be a public concern, the infrequency with which inadvertent actuation occurs leads to the conclusion that it does not constitute a weighty argument against a passive restraint requirement.

Some private individuals expressed, in their comments, concern over possible ear damage, or injuries that might be caused to persons with smoking materials in their mouths, or wearing eyeglasses. Although some early tests with oversized cushions of prototype design produced some temporary hearing losses, later designs have reduced the sound pressures to the point where ear damage is no longer a significant possibility. With respect to eyeglasses and smoking materials, the results from the vehicles on the road have been favorable. Of the occupants that had been involved in air cushion deployments as of a recent date, 71 had been smoking pipes or wearing eyeglasses or other facial accessories. None of these received injuries beyond the minor (AIS-1) level. From this it can be concluded that these circumstances do not create particular hazards to occupants of air-bag-equipped vehicles.

Toyo Kogyo and some private individuals questioned whether air bags might experience reliability problems in high-mileage and older vehicles. The fact that air bags have only one moving part, and most of the critical components rest in sealed containers during their non-deployment life, indicates that they should perform well in this regard. The systems in the vehicles in the field, some of which have been in use for almost 5 years, have demonstrated extremely good durability, with no apparent flaws. Manufacturers use sophisticated techniques such as accelerated test cycles to assure a high level of reliability.

Reliability of restraint systems is, of course, absolutely necessary. Unlike the failure of accident prevention systems such as lights and brakes where failure does not necessarily result in harm to occupants, the failure of a restraint system when needed in a serious crash almost certainly means injury will result. Vehicle and component manufacturers are fully aware of this and take the special precautions to ensure reliability which might not be taken for less critical systems. The Department is equally aware of it and has monitored manufacturer efforts to date to ensure failsafe performance of crash-deployed systems. As an example, copies of reliability information request letters from the Department to manufacturers preparing for the demonstration program or otherwise involved in air bag systems have been made public in the docket.

The projections of reliability to date are, of necessity, based on pilot production volumes, and cannot demonstrate fully that reliability problems associated with mass production will never occur. So that manufacturers can avoid these types of reliability problems, the Department has settled on a phase-in of the requirements which is described later in greater detail.

General Motors and the National Automobile Dealers Association commented that product liability arising from air bag performance would be a major expense. The insurance company commenters, on the other hand, suggested that the presence of air bags in vehicles could reduce auto companies' product liability.

The new risk of liability, attached to a requirement for passive restraints, does not differ from the risk attached to the advent of any device or product whether mandated by the Federal government or installed by a manufacturer by its own choice. Just as liability might arise because of the malfunctioning of a seat belt system or braking system, liability may also arise because of the malfunctioning of a passive restraint system. The mandating of a requirement by the Federal government has, in fact, often served to limit liability, since most jurisdictions accord great weight to evidence showing that a device has met Federal standards.

There is little evidence that the mandating of passive restraints will lead to increases in product liability insurance premiums. Although the advent of new technology has often been accompanied by an increase in products liability insurance, it is unclear how much of the increase is attributable to increased risk and how much to inflation. Officials of the Department of Commerce and at least two major insurance companies doubt that Federal passive restraint requirements will lead to increased risk and insurance premiums. They point out that Federal requirements are imposed to make products safer, and safe products are less likely to cause injury.

It is noteworthy that the Allstate Insurance Company agreed to sell product liability insurance for the GM cars which were to be equipped with passive restraint systems pursuant to the demonstration program, at a rate no greater than the product liability insurance rate for cars not equipped with passive restraint systems.

Small cars. An important consideration in the decision concerning passive restraints is their suitability and availability for small cars, which because of the energy shortage will comprise an increasing segment of the vehicle population in future years. Passive belts have been sold as standard equipment in over 65,000 Volkswagen cars, and must be viewed as a proven means of meeting a passive restraint requirement. Some vehicle body designs may require some modification for their installation, but passive belts could be used as restraints for most bucket-seat arrangements at moderate cost with present technology.

Some manufacturers have expressed doubt that a large proportion of their customers would find passive belts acceptable, because of their relatively obtrusive nature and the resistance shown by the U.S. public to wearing seat belt systems, i.e., belts that occupants must buckle and unbuckle. These manufacturers submitted no supporting market surveys. Further, there is reason to believe that the experience with active belt systems is not an accurate indicator of the experience to be expected with passive belts. The Department anticipates that some manufacturers

will install passive belts in the front seats of small cars having only two front seats. Passive belts would not confront the occupants of those seats with the current inconvenience of having to buckle a belt system to gain its protection or of having to unbuckle that system to get out of their cars. Unlike the interlock active belt systems of several years ago, the passive belt systems will have no effect on the ability of drivers to start their cars.

Nevertheless, the question of the acceptability of passive belts may make the suitability of air bags for small cars an important one. Although the shorter crush distance of small cars may impose more stringent limits on air bag deployment time, the evidence from studies conducted by the Department with air bags in small cars is that there are no insuperable difficulties in meeting the 30-mph crash requirements of Standard 208 in cars as small as 2000 pounds gross vehicle weight rating with existing air bag designs (see, for example, "Small Car Driver Inflatable Restraint System Evaluation Program," Contract DOT–HS–6–01412, Status Report April 15, 1977).

The "packaging" problems of installing air bag systems are greater for small cars than for larger ones. They occupy space in the instrument panel area that might otherwise be utilized by other items such as air conditioning ducts, glove compartment, or controls and displays. Toyo Kogyo (Mazda) and Honda indicated that their instrument panels might have to be displaced 4 inches rearward, that some engine compartment and wheelbase changes might be needed, and that some dash-mounted accessories might have to be deleted or mounted elsewhere. This type of problem is expected to be important to the existing choice between air bag and passive belt systems.

It is not the role of the government to resolve these problems since, in the Department's judgment, they reflect design choices of the manufacturers. No manufacturer has claimed, much less demonstrated, that it would be impracticable to install air bags in small cars without increasing vehicle size. Occupation of instrument panel space is certainly one of the unquantified costs of air bags, however, and the cost is more onerous in a small car than in a large one. At the same time, small car makers may choose to use the less costly passive belt system. The evidence presented to date indicates that small-car manufacturers would be able to meet a passive restraint requirement by making reasonable design compromises without increasing vehicle size.

Lead time and production readiness. There was considerable discussion in the comments to the docket about the ability of the automobile industry to develop the production readiness to provide passive restraint systems for all passenger cars. The installation of passive restraint systems requires the addition of new hardware and modification of vehicle structures in such a way that the system provides performance adequate to meet the standard and a high level of safety and reliability on the road. A new industrial capacity will have to be generated to supply components for air bag systems. Major capital expenditures will have to be made by the vehicle industry to incorporate air bag systems into production models. The Department estimates that the total capital required for tooling and equipment for the production of passive restraint systems in new cars is approximately $500 million.

Establishment of an industry to produce components for air bag systems centers on the production of the inflator component. Five major companies have indicated an interest in producing inflators for air bags. The propellant presently being considered for use is sodium azide. The primary source of sodium azide, Canadian Industries Ltd., has a capacity of around 1 million pounds per year, sufficient for only about 800,000 full front seat air bag systems. Thus, additional capacity of 10 million pounds or more of sodium azide will have to be generated, or alternative propellants would have to be used. The Department's analysis of the capital requirements and lead time to develop sufficient capacity indicates that adequate propellant can be available for annual production levels of several million units in less than three years. The production of inflators (from several sources) can reach several million units within two to three years of the receipt of firm orders, including design specifications, from the automobile manufacturers. A new capacity has already been generated to supply the demonstration program which is being pursued at this time.

The vehicle manufacturers face substantial work to incorporate air bags in their production. In the case of domestic manufacturers alone, the instrument panels of approximately half of the new cars that will be manufactured in the early 1980's will have to be completely redesigned to provide space for the passenger bag and structure to accept the loading on the passenger bag. In some cases, relocation of the instrument cluster is needed to facilitate visibility over the bag module in the steering hub.

The burden placed on the vehicle manufacturers to redesign the instrument panel and related components to accept air bags can be reduced considerably by phasing in the passive restraint requirements over several years. With phased introduction, the redesigning of instrument panels and other components can be done at roughly the same pace that these components would ordinarily be redesigned, although perhaps not within the manufacturer's preferred schedule.

The rulemaking docket contained a number of references to additional reasons for phased introduction of new systems like passive restraints: to establish quality systems in production, to obtain experience with these systems in the hands of a more limited segment of the public, and to obtain feedback on the performance and reliability of the systems. If production levels are relatively small at the beginning of a mandated requirement, any unforeseen issues that arise are made more manageable by the limited number of vehicles affected. A major automotive supplier, Eaton Corporation, stressed this aspect of production feasibility over all others.

Based on its evaluation, the Department has determined that a lead time of four full years should precede the requirement for the production of the first passive-equipped passenger cars. This lead time accords with General Motors' requested lead time to accomplish the change for all model lines. Equally important, the 4-year lead time represents a continuation to its logical conclusion of the early voluntary production of passive restraints represented by the December 1976 decision. The continued opportunity for early, gradual, and voluntary introduction of passive restraints to the public in relatively small numbers offers a great deal of benefit in assuring the orderly implementation of a mandatory passive restraint requirement. Experience with the limited quantities of early passive-restraint-equipped vehicles can confirm in the public's mind the value of these systems prior to mandatory production. Because of the value of such a voluntary phase-in approach to both the manufacturer and the public, the Department anticipates that the manufacturers which were parties to the earlier demonstration program agreements will continue their current preparations for voluntary production of passive restraints. The Department also expects that other manufacturers will undertake to produce limited quantities prior to the effectivity of the mandate. The Department intends to vigorously support the efforts of manufacturers to foster sales on a voluntary basis, both through major public information programs and through efforts to encourage their purchase by Federal, other government agencies, and private-fleet users.

The Department also intends to initiate an intensive monitoring program to oversee the implementation plans of both vehicle manufacturers and their suppliers. The purpose of the monitoring program will be not only to confirm that adequate levels of reliability and quality are being achieved in implementing designs to comply with the standard, but also to provide assurance to the public that the issues that have been raised on passive restraint reliability are being resolved under the auspices of the Secretary of Transportation.

In addition to a long lead time, the Department considers that the mandate should be accomplished in three stages, with new standard-and luxury-size cars (a wheelbase of more than 114 inches) meeting the requirement on and after September 1, 1981, new intermediate- and compact-size cars (a wheelbase of more than 100 inches) also meeting the requirements on and after September 1, 1982, and all new passenger cars meeting the requirement on and after September 1, 1983.

Wheelbase was chosen as a measure to delineate the phasing requirements because it is a well-defined quantity that does not vary significantly within a given car line. With the downsizing of most automobiles made in the United States, wheelbases are being reduced by four to six inches on most standard-intermediate- and com-

pact-size cars. As a result, in the period of phased implementation (the 1982 through 1984 model years) standard-size cars will generally have wheelbases in a range of 115'' to 120'', intermediate-size cars will have wheelbases in a range of 107'' to 113'', and compact-cars will generally have wheelbases in a range of 102'' to 108''. Subcompact-size cars will continue to have wheelbases below 100''.

The determination of which car sizes to include in each year of the phased implementation was made in consideration of the effect on each manufacturer and the difficulty involved in engineering passive restraints into each size class of automobile. Because of the extensive experience with passive restraints in full-size cars, and the space available in the instrument panels of these cars to receive air bag systems, this size car was deemed to be most susceptible to early implementation.

The gradual phase-in schedule is intended to permit manufacturers to absorb the impact of introducing passive restraint systems without undue technological or economic risk at the same time they undertake efforts to meet the challenging requirements imposed by emissions and fuel economy standards for automobiles in the early 1980's.

OTHER CONSIDERATIONS

Section 104(b) of the Act directs that the Secretary consult with the National Motor Vehicle Safety Advisory Council on motor vehicle safety standards. The Council has announced in an April 26, 1977, letter to the Department that "The Council feels that the time has come to move ahead with a fully passive restraint standard." The Council stated that it was recommending passive protection in the lateral and rollover modes as well as the frontal mode proposed by the Department. The Department therefore will take under consideration the Council recommendation, with a view to expanding the passive restraint requirement as new technology is advanced. The Council also recommended that mandatory seat belt use laws should also be promoted until the entire vehicle fleet is equipped with passive restraints. As noted, the Department intends to encourage States to enact such laws in their jurisdictions.

It is noted that the National Transportation Safety Board supported the mandate of passive restraints, with a cautionary note to preserve the present performance specification that permits meeting the requirement by means of passive belts as well as inflatable passive restraints.

The United Auto Workers Union, which represents the vast majority of the workers whose industry is affected by the mandate, has also advocated mandatory passive restraints to the Department.

The Council on Wage and Price Stability (the Council) supported the mandate of passive restraints, based on the assumptions that no serious technical problems exist with either the air bag or the passive belt system concept and that the Department's cost estimates are substantially correct. The Council based its support on the comparative costs of achieving benefits under the three approaches, finding passive restraints to be the most cost effective.

The Council urged that passive belt systems continue to be permitted as meeting the performance requirements of the standard, because they represent the least costly passive restraint system currently commercially available. Standard No. 208 has always been and continues to be a performance standard, and any device that provides the performance specified may be used to comply with the standards. With regard to passive belt systems, it is important that they remain available, particularly in the case of smaller-volume manufacturers who may not care to provide air bag type protection because of its engineering and tooling costs relative to production volume.

In accordance with S 102(2)(C) of the National Environmental Policy Act of 1969 (42 U.S.C. 4332(2)(C)), as implemented by Executive Order 11514 (3 CFR, 1966–1970 Comp., p. 902) and the Council on Environmental Quality's Guidelines of April 23, 1971 (36 FR 7724), the Department has carefully considered all environmental aspects of its three proposed approaches. A Draft Environmental Impact Statement (DEIS) was published March 25, 1977, and comments have been received and analyzed. The Final Environmental Impact Statement (FEIS) is released today. Petitions for reconsideration based on issues and information raised in the FEIS may be filed for the next 30 days (49 CFR Part 553.35).

There was substantial agreement by commenters with the agency's conclusions about impacts on the consumption of additional natural resources, the generation of pollutants in the manufacturing process and in transporting the system throughout the vehicle's life, and on solid waste disposal problems. In response to the comments of General Motors and others on the DEIS, several estimates were revised. In the Department's view, the two most significant consequences of a passive restraint mandate are the use of large amounts of sodium azide as the generator of gas for air bags, and the increased consumption of petroleum fuel by automobiles because of the added weight of air bags.

Sodium azide is a substance that is toxic and that can burn extremely rapidly. The agency is satisfied that the material can be used safety both in an industrial setting and in motor vehicles during its lifetime, due to inaccessibility and strength of the sealed canisters in which it is packed. The problem is to assure a proper means of disposal. Junked vehicles that are shredded have batteries and gas tanks removed routinely, and the air bag could be easily deployed by an electric charge at the same time. A hazard remains, however, for those vehicles that are simply abandoned. However, the agency judges that the chemical's relative inaccessibility will discourage attempts to tamper with it. The proportion of abandoned cars is less than 15 percent of those manufactured. The Department will work with the Environmental Protection Agency to develop appropriate controls for the disposal of air bag systems employing sodium azide.

The additional weight of inflatable passive restraints was judged to increase the annual consumption of fuel by automobiles by 0.71 percent (about 521 million gallons annually). While this increase is not insignificant, the Department believes that it is fully justified by the prospective societal benefits of passive restraints. The Department took full account of the impact of a passive restraint standard in its recent proceeding to set fuel economy standards for 1981–1984 passenger automobiles.

PART 571; S 208—PRE 96

In accordance with Department policy encouraging adequate analysis of the consequences of regulatory action (41 FR 16200, April 16, 1976), the Department has evaluated the economic and other consequences of this amendment on the public and private sectors. The basic evaluation is contained in a document ("Supplemental Inflation Impact Evaluation") that was developed in conjunction with the Department's June 1976 proposal of mandatory passive restraints. That evaluation has been reviewed and a supplement to it represents the Department's position on the effect of this rulemaking on the nation's economy.

The standard, as set forth below, allows manufacturers two options for compliance. First, a manufacturer may provide passive occupant crash protection in frontal modes only. If this option is chosen, the manufacturer must also provide lap belts at all seating positions in the automobile. The lap belts are provided to give crash protection in side and rollover crashes, and have a demonstrated effectiveness in these crash modes.

A second option for manufacturers is to provide full passive protection for front seat ocenpants in three crash modes: frontal, side and rollover. If a manufacturer can achieve this performance, it would not have to provide seat belts in the front seat. Under this option, lap belts would continue to be required for all rear seating positions.

The Department has found that use of any seat belt installed in accordance with the standard is necessary to enhance the safety of vehicle occupants. Thus, the Department continues to advocate the use of all seat belts installed at all seating positions in motor vehicles, regardless of whether the vehicle is also equipped with passive restraints.

In consideration of the foregoing, Standard No. 208 (49 CFR 571.208) is amended. . . .

Effective date finding: Under § 125 of the Act, an amendment of Standard No. 208 that specifies occupant restraint other than belt systems shall not become effective under any circumstances until the expiration of the 60-day review period provided for by Congress under that section "unless the standard specifies a later date." Section 125 also provides that the standard does not become effective at all if a concurrent resolution of disapproval is passed by Congress during the review period. The Department's view of this section is that a "later date" can be established at the time of promulgation of the rule, subject to the possibility of reversal by the concurrent resolution.

The amendment is therefore issued, to become effective beginning September 1, 1981, for those passenger cars first subject to the new requirements. The reasons underlying the effective dates set forth in the standard have been discussed above. The establishment of the effective dates is accomplished at this time to provide the maximum time available for preparations to meet the requirements. The Congressional review period will be completed prior to the commitment of significant new resources by manufacturers to meet the upcoming requirements of the standard.

The program official and lawyer principally responsible for the development of this rulemaking document are Carl Nash and Tad Herlihy, respectively.

(Secs. 103, 119, Pub. L. 89-563, 80 Stat. 718 (15 U.S.C. 1392, 1407))

Issued on June 30, 1977.

Brock Adams
Secretary of Transportation

42 F.R. 34299
July 5, 1977

PREAMBLE TO AMENDMENT TO
MOTOR VEHICLE SAFETY STANDARD NO. 208

Occupant Restraint Systems
(Docket No. 74-14; Notice 12)

With the exception of minor perfecting amendments, this notice denies petitions for reconsideration of the Department's decision to require the provision of automatic occupant crash protection in future passenger cars, commencing in some models on September 1, 1981, and in all models by September 1, 1983. Six petitions for reconsideration and one application for stay of the standard's effective date pending judicial review were filed by parties that disagreed with aspects of the DOT decision to upgrade occupant crash protection as a reasonable and necessary exercise of the mandate of the National Traffic and Motor Vehicle Safety Act (the Act) to provide protection through improved automobile design, construction, and performance. This notice denies the petitions and establishes the automatic crash protection requirements and effective dates of S4.1.2 and S4.1.3 as final for purposes of judicial review under § 105(a)(1) of the Act as to any person who will be adversely affected by them. One petition for reconsideration of a related rulemaking action ("Notice 11") is granted in this notice.

Effective date: December 5, 1977.

For further information contact:

Mr. Ralph Hitchcock, Motor Vehicle Programs, National Highway Traffic Safety Administration, Washington, D.C. 20590 (202-426-2212).

Supplementary information: On June 30, 1977 (42 FR 34289; July 5, 1977) the DOT upgraded the existing occupant restraint requirements of Standard No. 208, *Occupant Crash Protection,* to require the provision of automatic crash protection in passenger cars with wheelbases

greater than 114 inches manufactured on or after September 1, 1981, in passenger cars with wheelbases greater than 100 inches manufactured on or after September 1, 1982, and in all passenger cars manufactured on or after September 1, 1983. In place of the lap/shoulder seat belt combinations provided in the front seats of most of today's passenger cars, the standard mandates a performance standard for crash protection that must be met by means that require no action by the vehicle occupant. The automatic protection must be provided in the frontal mode—specifically, when the vehicle impacts a fixed collision barrier at any speed up to and including 30 mph and at any angle not more than 30 degrees to the left or right of perpendicular, the test dummies installed at the front seating positions must remain in the vehicle and be protected against specified head, chest, and femur injuries by passive means (means that require no action by the vehicle occupants). A manufacturer may meet lateral and rollover crash requirements by the provision of active or passive belt systems.

This amendment to the existing standard invoked a provision of the Act (15 U.S.C. 1400(b)) that provides for a 60-day Congressional review of the action. A resolution of disapproval from both Houses of Congress was specified as necessary to disapprove the action. Hearings were held by both the Senate and the House in September 1977, and votes were conducted in October 1977. The House Committee on Interstate and Foreign Commerce adopted its Subcommittee's adverse report on the disapproval resolution and voted to table it. The Senate also voted to table the disapproval resolution by a vote of 65 to 31. The 60-day review period ended October 14, 1977.

Six petitions for reconsideration of the decision were filed by interested parties, along with an application for stay of the effective date of the decision pending disposition of a petition for judicial review of the standard filed by the Pacific Legal Foundation on September 1, 1977. One petition requested an effective date change in a related rulemaking action.

Disposition of Petitions

Effectiveness. A central factor in the Department's decision to upgrade occupant crash protection requirements was a determination that passive restraint technology could substantially reduce fatalities and injuries in crashes.

Comprehensive analyses of the effectiveness of passive restraints in preventing fatalities and reducing injuries appear in the preamble to the decision, the "Explanation of Rule Making Action" that accompanied the decision, and in underlying research and analyses that were conducted by and for the Department's National Highway Traffic Safety Administration (NHTSA) and placed in the public rule making docket throughout the Standard's eight-year rule making history.

The estimates of restraint system effectiveness are based on extensive field data with active safety belt restraint systems, evaluated in conjunction with thousands of crash and sled tests comparing the performance of various active and passive restraint technologies in occupant protection with each other and with the performance of unrestrained occupants. The analyses show that air bags and passive belt systems are approximately equivalent in overall protective ability to combined lap and shoulder belts when worn. However, usage of passive restraints will be substantially higher than the 20-percent usage rate of active safety belts observed at present.

General Motors (GM) petitioned for suspension of the decision while an organization not involved in the passive restraint issue "audits" the DOT and GM effectiveness estimates. A moderate amount of field experience with the GM 1974-1976 air bag fleet of 10,000 vehicles is now available, and GM sought to obtain an effectiveness estimate from the field data by comparing injuries in the air bag accidents that have oc-

curred with injuries in accidents of comparable severity found in GM insurance company files. Based on this methodology, GM concluded that air bags are little more effective than no restraint at all.

Analysis of GM's "matching case" methodology indicates a failure to correct their statistical conclusions for known differences between the air bag and insurance file fleets. For example, because air bags were only offered in GM's full size and luxury cars, the occupants of the air bag cars were older than the general population of motorists represented in the matching insurance files by an average of about 12 years. Older persons are more susceptible to injury in crashes than the generally younger population of American motorists. This age bias alone could result in an underestimation of air bag effectiveness of about 30 percent.

A further source of error in the GM methodology results from matching the air bag crashes with a range of similar crashes in the insurance files. For example, consider an air bag car crash into a pole resulting in 17 inches of crush to the front of the car. This case was matched against 'similar' crashes into poles or trees of non-air bag cars with between 14 and 20 inches of crush. Since the insurance files contain many more lower speed crashes than higher speed crashes, the comparison group of "similar" crashes will always contain a range of severity that is biased toward less severe crashes. When air bag crashes are matched in this way, a downward bias is introduced that could reduce estimates of air bag effectiveness by 50 to 100 percent.

DOT finds that proper analytical corrections for age distribution and downward severity of the case matching technique yields an air bag effectiveness value of about 40 percent for AIS-3 or greater injuries. The Department's decision in June 1977 (Table I) estimated air bag effectiveness for AIS-3 injuries at 30 percent and for AIS-4 to 6 at 40 percent.

A more direct and definitive comparison can be made of passive and active restraint effectiveness using field data on the accident experience of 80,000 VW Rabbits with passive belt systems that have been sold in the U.S. These data show that the rate of fatalities in Rabbits equipped with passive belts is less than one-third of the rate for Rabbits of the same years of manufacture equipped with active lap/shoulder belt systems.

Economics and Science Planning, Inc. (ESP), asked that the passive restraint decision be modified to require passive belts in all 2-front-seating-position passenger cars on and after September 1, 1981, with passive requirements for other cars to follow only after further evaluation of air bag effectiveness. The seating-position distinction recognizes that passive belts may not be practical yet for 3-passenger bench-seat configurations. ESP's basis for advocating passive belts is the preliminary data on experience with passive-belt-equipped Volkswagen Rabbits.

Standard No. 208 is a performance standard that can be met by several designs, including the air bag and passive belt that have already been shown to be commercially feasible. The same performance would be required of any system chosen by the manufacturer.

ESP's preference for passive belts is grounded in its air bag analysis which, in the Department's opinion, seriously underestimates air bag effectiveness. ESP compared the experience of accident-involved 1973, 1974, and 1975 model cars equipped with seat belts (DOT-HS-5-01255-1) (RSEP study) with accident-involved air bag cars from the 10,000-car GM fleet now in highway service.

In attempting the comparison ESP made two major errors. Because the towaway mileage figures for the air-bag fleet are not known, ESP simply speculated what this critical factor would be, with no credible grounds for the validity of its estimates. The other error was to compare the two data sets, ignoring relevant differences in the ratio of urban to rural exposure, the proportions of vehicles of various sizes in the sets, the crash modes and severity of the crashes, and the age and sex of the vehicle occupants involved. When ESP corrected its analysis, in a later submission to the Department, to eliminate these errors, it obtained results that tend to support the DOT estimates.

The ESP petition for deferral of air-bag-type passive restraints is also grounded in the unfounded assertion that seat belt usage is or can be expected in the future to rise to 44 percent. ESP relies on a finding from the RSEP study that belt use was as high as 44 percent in 1974 and 1975 model cars observed during 1974 and the first part of 1975. But this isolated finding cannot be used out of context as a general predictor of belt usage rates. Most of these vehicles were originally equipped with ignition interlocks and sequential warning systems, many of which had not yet been disabled and thus induced occupants to buckle up. Subsequent observations confirm that belt usage in those model year cars has now dropped to less than 30 percent. In the most recent model year cars (1976 and 1977 models) with only a brief reminder system, usage is only about 20 percent (DOT-HS-6-01340).

ESP suggested that future belt usage could be higher than DOT observations, based on its belief that usage is higher (1) in rural areas where DOT observations were not concentrated, (2) in high-risk situations because drivers perceive a risk and take appropriate action, and (3) in small cars that will become a higher proportion of the fleet in the future. This speculation has no basis in fact. The RSEP study shows belt usage to be higher in urban areas where DOT observations were concentrated, tending if anything to bias the observation in favor of high usage rates. The same study provides evidence that belt usage is no more likely in higher risk situations. Usage was lower for vehicles that sustained higher levels of damage. The higher belt usage in smaller cars is more likely attributable to the general attitudes of existing small car buyers than simply to occupancy of a smaller vehicle.

Chrysler, Ford, and AMC alluded to air bag effectiveness but raised no points that have not already been addressed as a part of the passive restraint decision at the time of its issuance. No basis in these petitions exists upon which to reconsider the decision.

Implementation schedule. The Center for Auto Safety (the Center) and Ralph Nader petitioned for modification of the effective date and phase-in to make the requirements become effective for all cars on September 1, 1980. The Center argued that installation in that time period is technically feasible, that compliance of large cars first, and less crashworthy small cars last, contradicts the

Act's mandate to reduce death and injury, that phase-in of requirements by wheelbase length is not authorized by the Act, and that insufficient notice of the implementation schedule was provided by the Department.

The introduction of passive restraint systems in all new cars will require the design, testing, and manufacture of components for a variety of passive restraint systems, in many variations to accommodate all sizes and models of passenger automobiles sold in the domestic market. Parties to the rulemaking generally agreed what tasks are necessary to redesign new automobiles to accommodate passive belts and air bags. However, some disputed the length of time needed to accomplish these tasks effectively and in an orderly manner for all cars sold in the United States during the time frame from now into the early 1980's.

A comprehensive discussion of the considerations underlying the establishment of the standard's implementation schedule appears in the *Production Readiness and Introduction Schedule* section of the "Explanation of Rule Making Action" underlying the decision.

The Department estimates that the new requirements will apply to approximately 2.8 million five- and six-passenger full size cars in September 1981, an additional 4.9 million intermediate and compact cars in September 1982, and an additional 3.2 million sub-compact and mini-compact cars in September 1983.

Depending on the amount of research and development conducted to date, the product lines, and the resources of the various manufacturers, lead time required by each will vary significantly. Some manufacturers have done preparatory development work toward the installation of passive systems, and some have done very little. Thus, the varying capabilities and state of the development programs of most manufacturers must be considered in establishing technically feasible lead times, and not simply the capability of the most or least advanced.

Facilities for manufacturing air bag inflator components in large numbers do not exist and must be developed. The development of this new industrial capacity cannot be expected to coincide fully with the development and planning activi-

ties of the vehicle manufacturers alone, because component supplier investments will probably not be made without the suppliers having firm orders. This is particularly so where the passive restraint requirements have been issued and remanded several times over the last seven years. Vehicle manufacturers generally do not order components from the suppliers until they have developed, tested, and settled on the configurations necessary to meet the standard in their products. The serial nature of development, design, testing, and tooling processes for mass production strongly affects lead time requirements.

The NHTSA estimates that the lead time for the major and secondary design changes (such as to the instrument panel, steering column, door structure, and "B" pillar) that would be required to place air bags or passive belts in new automobiles can vary from less than 26 months to more than 38 months for a typical large manufacturer.

Another factor affecting lead time is the period of time needed to develop a large scale production capacity for pyrotechnic propellant materials. Based on existing inflator technology and production capacity, the Department estimates that approximately 3 years will be necessary to produce sufficient inflators for the entire annual production of passenger cars without an extraordinary commitment from this industry. The development of large scale inflator manufacturing capacity is likely to occur only after the design and initial testing of air bag systems by the auto manufacturers.

A final and extremely important factor that must be considered in establishing lead time requirements is the necessity to assure that systems furnished to comply with the standard will provide trouble-free, durable, and marketable characteristics in service. Reduction in lead time, or inefficient use of lead time, may increase the probability of defects occurring in service.

From these considerations, it is apparent that installation of either air bags or passive belts would not be practical for all new automobiles within less than 3 years as requested by the Center. To provide reasonable opportunity for development, design, testing, and tooling of passive restraint systems with adequate durability, quality, reliability, and overall performance,

48 months of lead time is justified. This is particularly true for smaller-volume manufacturers who have done little passive restraint development work and are only now studying specific designs for their 1982 and 1983 model year products.

It should be noted that the lead time authorized is required by the facts and circumstances presented in this particular and complex rulemaking and in no way is to be considered as a precedent for the calculation of lead time in any other standard which may later be promulgated by the agency.

The Center also advocated that the changes necessary to install passive restraints should occur at the same time instead of being phased-in over three years. The Center suggested that accommodation of the manufacturers' preferences, specifically their plans to meet future emissions and fuel economy requirements, had dictated the 3-phase implementation. This is not the case. The major vehicle redesign and retooling for materials conservation, fuel economy, and emissions that has been and will occur through the early 1980's must be considered in reaching any determination about the technical and economic feasibility of automotive regulatory actions of DOT. A thorough evaluation of the consequences of this passive restraint decision requires no less.

However, the requirements for improved occupant restraints were not subordinated to the attainment of fuel economy or emissions requirements. The preamble to DOT's fuel economy rulemaking makes clear that downward adjustment in the fuel economy levels was made to accommodate the weight of passive restraints. As earlier explained, a 4-year lead time was judged to be reasonable and appropriate to assure that a satisfactory product could be developed by most manufacturers in the United States market for most of their products.

The decision to require only a portion of production to comply in the first year further recognizes the limit on the available tooling industry capacity to accomplish major changes, and the demands this industry will face within the next several years because of an unprecedented combination of regulatory requirements and commercial pressures. A manufacturer with several

vehicle offerings ordinarily undertakes major product changes in only a portion of its production at one time. Assuming a 4-year cycle within the industry for substantial changes, for example, it is evident that only about one-fourth of the engineering and tooling capacity resources necessary to change the entire production are in place and available for use in any one year. The lead times provided are based on reasonable utilization of available tooling and the objective that reliable and effective passive restraint systems be developed.

The longer lead time allowed for smaller cars is also intended to provide the alternatives to small-car manufacturers for the installation of air bag systems in lieu of the simpler passive belt systems. The development of either type of occupant crash protection for smaller cars presents a greater engineering challenge than for large cars, and some makers of smaller cars have significantly smaller engineering resources than do the makers of the majority of larger cars. The Department intended to provide sufficient lead time so that the most effective designs can be fully considered and tested before production decisions must be reached. The agency considers that its analysis, reported in the "Explanation of Rule Making Action," provides ample justification for a phase-in as the practicable approach to meeting the need for motor vehicle safety in upgrading automobile occupant crash protection.

The Center argued that a phase-in of requirements in stages that distinguish among vehicles on the basis of a design characteristic (wheelbase length) is not authorized by the Act. The Center argued that "type" distinction does not include wheelbase distinctions. The Center asserts that the DOT believes it has only "across-the-board" authority to implement standards, and that Congress acquiesced in this view by not providing DOT additional phase-in authority in the 1974 amendments to the Act.

The Department has repeatedly utilized "type" distinctions based on design in carrying out the Act. The basic vehicle type distinctions used to distinguish the phasing of requirements among passenger cars, multipurpose passenger vehicles, and light trucks are not expressly authorized by the Act. DOT established the distinction to rationally implement the Act. The wheelbase distinction has been used in the bumper safety standard No. 215, *Exterior Protection*, to implement upgraded requirements as expeditiously as possible. This regular practice contradicts the assertion that DOT itself believes it has "across-the-board" authority only. The DOT 1974 request for "percentage of production" phase-in authority in no way applies to the question of phase-in authority based on design distinctions such as wheelbase length, weight, or chassis type, that the Department already had.

Congress has in fact implicitly approved phase-in based on design distinction by its 1974 ratification of Standard No. 301-75, *Fuel System Integrity*, in a form that contains a gross vehicle weight rating (GVWR) phase-in criterion. Such design distinctions have been relied on by DOT and acquiesed in by Congress, the industry, and the public since the Act's inception.

Finally, the agency does not agree that the legislative history cited by the Center supports the proposition that phase-ins are illegal. The quoted statement by Senator Magnuson states that standards will apply to every vehicle, but does not address the question of *when* they would. The refusal by Congress to authorize phase-in by "customary model change" criteria in no way excludes the authority to phase-in by design distinction. The Senate Report language addresses particular vehicle changes that take more than a year to implement, and simply notes that the DOT is authorized to set later dates for those changes. This passage does not address the question of later dates for a particular category of vehicle.

The Center asserted that inadequate notice of the implementation schedule had been provided by the Department, because the September 1981 date was adopted in place of the proposed 1980 date, and because the wheelbase phase-in was adopted in place of the proposed phase-in by occupant position. While conceding that "every precise change ultimately adopted need not be published", the Center believed that inadequate opportunity was made available to the public to address the implementation schedule.

The Department has fully considered the Center's objection in the light of its public notices,

hearings, and the rulemaking record on Standard 208. The question is whether the public has had sufficient notice of the issue (the timing of mandatory passive restraint installation). As a general matter, some changes from the proposal are inherent in the notice and comment process so that the rulemaker can benefit from comments and modify the rulemaking without having to repropose every time new information is learned.

In this case, the notice proposed a timing schedule, and the notice indicated that the implementation was tentative, even suggesting a phase-in at occupant positions as an alternative timing approach. The Draft Environmental Impact Statement described phase-in alternatives, and many parties in their written and oral comments raised the issue of the timing for the mandate. The Center itself commented on timing which demonstrates that they were sufficiently aware of the issue to comment on it.

Implementation of the Standard

An important element in implementing the passive restraint requirements is to ensure that they are introduced in significant numbers prior to the time they are required by mandate. While passive belt systems are already in use in substantial numbers on the Volkswagen Rabbit (about 80,000 cars), relatively few air bag systems are in highway service. The two major reasons to have passive restraints voluntarily produced prior to the mandate are to familiarize the public with passive restraint technology and to work out early problems in production systems that could interfere with orderly implementation of the mandate and jeopardize success of the program.

The Department is taking steps to provide for voluntary early introduction. In addition to Volkswagen, GM and Ford have indicated plans to introduce passive belts as an option as early as the 1979 and 1980 model years, respectively. Ford and GM have also announced the intention of making an air bag option available in one or more models in the 1981 model year, one year before the mandate. The Department commends this initiative and is encouraging these companies to expand this commitment to introduce air bags voluntarily in the 1980 model year and in other than full-size cars. The Department will continue to monitor the performance of voluntarily introduced systems, both air bags and passive belts, as it has to date.

In support of manufacturers' efforts to market air bags earlier than the mandate, the Department has contacted the General Services Administration, State and local government operators of fleet vehicles, the insurance companies, rental fleet owners, taxi operators, and other institutional users of passenger cars to encourage the purchase of air bag cars. This is the most direct inducement to the manufacturers to make air bags available earlier than the initial September 1981 effective date. Complementary activities to assist the early introduction of the systems are: (1) a DOT public education campaign that is already underway throughout the country, (2) monitoring component and vehicle manufacturers' implementation programs to assure proper attention is given to cost, reliability, and effectiveness, and (3) continued research, development, and evaluation of passive restraint systems to insure that the best overall passive restraint technology is available to manufacturers and the public, both now and in the future.

Other Issues

The Pacific Legal Foundation filed a petition for review of the rule in the Court of Appeals for the District of Columbia. It then asked the Department to stay the effective date of the rule for a period of time equal to the length of judicial review.

The Foundation, in its application for a stay, listed in general terms a number of items it said the Department failed to consider or evaluate appropriately. The Department did, however, review and assess all of those items before announcing the rule. It discussed many of them extensively in the preamble to the rule and the accompanying "Explanation of Rule Making Action". Upon receiving the application for a stay the Department reconsidered all of those items and it finds that the Pacific Legal Foundation's list of objections has no merit.

The Foundation argued that the Department should stay the rule pending judicial review because manufacturers will make capital expenditures preparing to comply with the rule in model year 1982 and if the Court then overturns the rule, manufacturers may abandon the passive restraint program and pass on these preparation expenses to new car buyers. The Foundation thus asks the Department to balance a possible loss of a relatively small amount of money against a certain loss of lives and increase in injuries. The Department does not know how much time the Court will need to review the rule, but each year's continuance of the rule will add only a few dollars to the price of a new car while each year's delay of the rule will ultimately cost the public thousands of preventable fatalities and many more thousands of preventable serious injuries. The potential harm the Pacific Legal Foundation seeks to avoid through a stay is trivial compared to the cost of a stay in lives that cannot be restored, injury that cannot be repaired, and suffering that cannot be erased. This rule has already remained unresolved for too long. The Department denies Pacific Legal Foundation's application for a stay.

Some manufacturers repeated many of their earlier objections, all of which were extensively addressed in the preamble that accompanied the decision and the supplementary "Explanation of Rule Making Action". Not only were these issues fully ventilated in the rulemaking action, but they were also extensively treated in the hearings and subsequent reports of the Senate and House Commerce Committees as a part of their review of the standard. The Department does not consider repetitious petitions as a part of the reconsideration process (49 CFR § 553.21) and accordingly denies them.

One new issue raised was Ford's complaint that the NHTSA response on test dummy objectivity had misinterpreted Ford data on testing conducted in 1973. While the Ford dummy test program performed in 1973 may have been an ambitious attempt to investigate all of the variables involved in a vehicle crash test, subsequent development and test programs to reduce sources of test variability have made the Ford test series obsolete. As noted in the preamble to Notice 11, dummy manufacturers have gained experience in the manufacture of dummies, the Part 572 specifications and test procedures have been further defined, and the dummy positioning procedures in Standard No. 208 have been modified for bench-seat cars to eliminate the problem noted in the Ford tests of fitting 3 dummies side-by-side in the test.

Ford did not contest the more recent findings (DOT–HS–6–01514) of hard-seat sled tests of pairs of dummies with belts, air bags, and unrestrained, showing coefficients of variation on the pooled data basis for head accelerations from 1.2 percent to 10.7 percent, for chest acceleration from 1.6 percent to 8.5 percent, and for femur compressive force from 3.51 percent to 24.2 percent. Similar results were obtained in sled test oblique impacts (DOT–HS–802–570). In the face of this unrebutted conclusive evidence of the repeatability of current commercial dummy production, the agency finds the test instrument and associated procedure to be objective.

It has been brought to the attention of the Department that the NHTSA's decision to continue indefinitely the existing requirements for multipurpose passenger vehicles and light trucks was imperfectly stated. A corrective amendment of S4.2.2 is accomplished by this notice.

Volkswagen petitioned to have a longer transition period between the existing requirements for dummy positioning and the upcoming ones published in Notice 11 (42 FR 34299, July 5, 1977), because the company will not be able to evaluate the new requirements by July 5, 1978, yet must continue to certify its passive-belt-equipped Rabbit model. The Automobile Importers Association and General Motors suggested that compliance with either the old or new requirements, at the manufacturer's option, be permitted immediately. The NHTSA considers optional procedures more desirable than specifying the old procedures longer than one year as suggested by Volkswagen. Under optional procedures, Volkswagen can continue its certification of the Rabbit model, effecting a transition at any time, while the manufacturers undertaking new development efforts can immediately utilize the new procedures. To accomplish this, the effective dates of the requirements of Notices 10 and 11 are changed to become effective immediately,

with modifications of the language as necessary to preserve the old procedures as an option until September 1, 1981. These minor adjustments are accomplished in this notice.

Ford noted that the dummy head adjustment procedure of S10.4 was not consistent with dummy construction, which positions the head automatically. The NHTSA had intended that the dummy head and neck system be shimmed to compensate for different seat back angles in vehicles being tested. Because of the relative difficulty in accomplishing this in relation to the amount of specificity gained thereby, the NHTSA hereby deletes S10.4 as requested by Ford.

For the reasons stated above and after full consideration of the petitions by all parties submitted, the Department of Transportation denies petitions for reconsideration of its June 30, 1977, decision to require the installation of automatic crash protection in future passenger cars. The requirements set forth at 42 FR 34289 and 42 FR 34299 (July 5, 1977) are final for purposes of review in accordance with § 105(a) of the Act.

In consideration of the foregoing, Standard No. 28 (49 CFR 571.208) is amended. . . .

Effective date finding: Because the amendments provide an option and do not create additional requirements for any person, it is found that an immediate effective date is in the public interest so that manufacturers may take advantage of the new option as rapidly as possible.

The program official and lawyer principally responsible for the development of this rulemaking document are Ralph Hitchcock and Tad Herlihy, respectively.

(Sec. 103, 119, Pub. L. 89–563, 80 Stat. 718 (15 U.S.C. 1392, 1407)

Issued on December 5, 1977.

Brock Adams
Secretary of Transportation
42 F.R. 61466
December 5, 1977

Occupant Crash Protection in Passenger Cars, Multipurpose Passenger Vehicles, Trucks and Buses

(Docket No. 74–14; Notice 14)

Action: Final rule.

Summary: The purpose of this notice is to amend Safety Standard No. 208, Occupant Crash Protection, to provide for the optional use by motor vehicle manufacturers of alternatives to latches for releasing occupants from passive seatbelt systems in emergencies and to allow means other than pushbuttons to operate the emergency release mechanisms of passive belt systems. The amendment is based on a proposal issued in response to a petition from General Motors Corp. to allow manufacturers greater latitude in designing emergency release mechanisms for passive belt systems. The amendment will allow manufacturers to experiment with various emergency release mechanisms aimed at encouraging passive belt use by motorists, prior to the effective date of passive restraint requirements specified in this standard.

Effective date: November 13, 1978.

Address: Petitions for reconsideration should refer to the docket number and notice number and be submitted to: Docket Section, Room 5108, Nassif Building, 400 Seventh Street SW., Washington, D.C. 20590.

For further information contact:

Guy Hunter, Office of Vehicle Safety Standards, National Highway Traffic Safety Administration, Washington, D.C. 20590, 202–426–2265.

Supplementary information: Safety Standard No. 208, 49 CFR 571.208, currently specifies that a seatbelt assembly installed in a passenger car shall have a latch mechanism that re-leases at a single point by pushbutton action. General Motors petitioned for relief from this requirement for passive belts, following the issuance of the final rule requiring passenger cars to be equipped with passive restraints (air bags, passive belts, or other means of passive, i.e., automatic, protection) (42 FR 34289, July 5, 1977). The petition described a "spool release" design General Motors would like to use on one of its passive belt systems. The system would include a shoulder belt that would not detach at either end. Rather, the design would allow the belt to "play out" or unwind from the retractor in an emergency, allowing sufficient slack for the door to be opened and the occupant to exit from the vehicle. The purpose of such a "spool release" design is to minimize the disconnection of the passive belt system by motorists. Under the current latch mechanism and pushbutton requirements for belts, a passive belt system could be easily disconnected by a buckle release identical to buckles on current active belt systems (i.e., belts that motorists must manually put into place). As long as the belt remains disconnected, the "passivity" of the system would be destroyed for future use.

In response to the GM petition, the NHTSA issued a proposal to amend standard 208 to allow alternative release mechanisms for passive belts (43 FR 21912, May 22, 1978). As noted in that proposal, the NHTSA is very concerned about the usage rate of passive belts by motorists since it appears that there may be many new cars in the 1980's equipped with these systems. If motorists who would prefer air bags in a particular car line can only obtain passive belts from the manufacturer the defeat rate of the belts could be high. The agency is, therefore, interested in fostering any passive belt design that is effective and that minimizes the rate of disconnection. The notice pointed out, however, that there are other factors to be considered in the proposed change.

The original purpose of the latch mechanism and pushbutton requirements of standard 208 was to insure uniformity of buckle design for the purpose of facilitating routine fastening and unfastening of active belts, encouraging belt use by making the belts as convenient as possible and facilitating the exiting of vehicle occupants in emergency situations. Since the proposed amendment would allow various types of release mechanisms, the agency was concerned that the resulting nonuniformity might have adverse consequences in emergency egress situations from passive belts. In order to examine the implications of the General Motors petition thoroughly, the proposal sought public comments on four specific questions concerning the efficacy and advisability of allowing alternative release mechanisms to latches for passive belt systems. The four questions were as follows:

1. "How should the NHTSA or the vehicle manufacturers monitor the efficacy of and public reaction to various systems for discouraging disconnection of passive belts (such as the latch mechanism with a 4–8 second audible/visible warning system that operates if the belt is not connected when the ignition is turned on, a latch mechanism with additional warning or interlock systems voluntarily installed by a vehicle manufacturer, or a lever operated spool release as requested by General Motors)?"

2. "Are there safety or other considerations that would make it inadvisable to allow the spool release at this time as an option to vehicle manufacturers which install passive belts?"

3. "Compared with a passive belt system equipped with the currently required latch mechanism, would a passive belt system equipped with a spool release whose actuation lever is located between the seats have substantial disadvantages for emergency exit or extraction from a vehicle

that would offset any possible increase in usage in the passive belts?"

4. "If the NHTSA decides to permit the use of alternative occupant release mechanisms, should such use be permitted indefinitely or only for a finite period, e.g., several years, to allow field testing of the various systems? If a finite period were to be established, when should it begin and end?"

All 15 comments to the May 22, 1978, notice supported the intent of the proposed change to allow alternative release mechanisms for passive belts. Most commenters agreed that a nonseparable passive belt should discourage disconnection by motorists and that this should be given higher priority consideration than possible adverse effects such a belt might have on emergency occupant egress. Volkswagen did express some concern that the benefits achieved by increased belt usage might be somewhat offset if problems with emergency exiting arise, but agreed that more flexibility in passive belt design should be allowed to encourage belt use.

Volkswagen urged the use of the passive belt system utilized on its Deluxe Rabbit—a pushbutton release latch mechanism guarded by an ignition interlock. The company stated that this type system is simple and works well in emergency situations regardless of the condition of the retractor or the positioning of the webbing (potential problems of a "spool release" type design). Volkswagen pointed out that a system that is too complex will require close monitoring to insure effectiveness.

While the Volkswagen system has shown high use rates in the field, there is a possibility that widespread use of this type system could lead to adverse public reaction because of the interlock feature. As pointed out by the Alliance of American Insurers in its support of the proposed amendment, there could be a second public "backlash" from a return to the use of starter interlocks, even if placed on the vehicle voluntarily by the manufacturer. Alliance stated that the "spool release" system proposed by General Motors should be preferable to the interlock from a public acceptance standpoint.

The Center for Auto Safety and the Prudential Property & Casualty Insurance Co. both commented that "spool release" type mechanisms should be self-restoring to insure that in subsequent uses of the vehicle the passive belt is ready to provide the automatic protection for which it was designed. The self-restoring feature would automatically retract the belt after the manual release has been activated to allow the belt to "play-out." The NHTSA believes that both self-restoring "spool release" designs and manual restoration designs have distinct advantages. The automatic restoration does not require the vehicle user to have any knowledge of the system to reactivate the passive belt. However, a manual restoration design would be less complex and would probably be more reliable. The manual design could be coupled with audible and visible warnings to indicate when the lock-up portion of the retractor is inoperative. The amendment set forth in this notice allows both types of restoration systems for "spool release" passive belt designs.

The majority of commenters argued that the proposed amendment should be effective indefinitely, and not merely during the interim period until the passive restraint requirements become effective. The comments stated that manufacturers should be given the greatest possible design latitude to encourage the early introduction of innovative passive belt systems that are designed to minimize disconnection by motorists. The industry noted that manufacturers will be hesitant to initiate such new programs and passive belt designs if alternative release designs are allowed only for an interim period. Further, the commenters stated that an interim rule would not allow time for an adequate examination of the effectiveness of the various new designs that might be developed. The agency has concluded that these arguments have merit. Accordingly, this amendment is effective indefinitely.

Several comments stated that the new passive belt designs should be standardized, so that the public will understand their use and problems of emergency occupant egress will be minimized. While the agency agrees that uniformity in release design is advantageous, it is not practical to standardize systems that are only in the development stage. Further, if manufacturers are not given latitude in their passive belt designs, the purpose of this amendment would be defeated. It is unclear at this time which passive belt systems will be the most effective in encouraging belt use and at the same time be accepted by the public. The agency will, of course, monitor all new passive belt systems as closely as possible, and efforts to standardize systems could be made in the future.

Ford Motor Co. commented that the revision of standard No. 208 as requested in the General Motors petition would provide greater latitude than presently exists, but that the requested wording is restrictive in that it would inhibit the development of methods of release other than those specifically related to the retractor. Ford requested that the proposed revision include language permitting manufacturers the greatest possible design latitude. The agency emphasized in the previous notice that the proposal was tentative as to the language and substance of an amendment that might be adopted in response to the General Motors petition. Accordingly, this amendment is broader than that proposed in the General Motors petition and does not limit the types of passive belt designs that may be developed.

In order to insure that vehicle occupants are aware if their passive belts are inoperable because a release mechanism has been activated, this amendment specifies that the warning light, "Fasten Belts," remain illuminated until the belt latch mechanism has been fastened or the release mechanism has been deactivated. This warning light of indefinite duration is in addition to the 4- to 8-second audible warning signal currently required by the standard. The agency believes a continuous warning light is essential since this amendment will allow various types of unfamiliar release systems for passive belts.

In summary, the agency has concluded that manufacturers should be given considerable latitude in designing emergency release mechanisms for passive belt systems. This will permit the development of innovative systems aimed at limiting passive belt disconnection by motorists. Otherwise, the use rate of passive belt systems could be as low as the current use rate for active belt systems. This amendment will allow manufacturers to experiment with various passive belt designs before the effective date of the passive restraint requirements and determine which designs are the most effective and at the same time acceptable to the public.

The agency does not believe that the use of alternative release mechanisms will cause serious occupant egress problems if manufacturers take precautions to instruct vehicle owners how the systems work through the owner's manual and through their dealers. While uniformity in release mechanisms is certainly important for purposes of emergency occupant egress, the agency has concluded that this consideration is at least temporarily outweighed by the importance of insuring passive belts are not disconnected. The agency will, however, monitor all new passive belt designs to assure that the release mechanisms are simple to understand and operate. If the

methods of operation of the various release mechanisms are self-evident, the problem of lack of uniformity in design will be less important in terms of emergency occupant egress.

The agency has concluded that this amendment will have no adverse economic or environmental impacts.

The engineer and lawyer primarily responsible for the development of this rule are Guy Hunter and Hugh Oates, respectively.

(Sec. 103, 119, Pub. L. 89–563, 80 Stat. 718 (15 U.S.C. 1392, 1407), delegation of authority at 49 CFR 1.50.)

Issued on November 1, 1978.

Joan Claybrook
Administrator
43 F.R. 52493
November 13, 1978

SUMMARY: This notice responds to petitions for reconsideration of the November 29, 1979, notice (44 F.R. 68470) amending Standard No. 208, *Occupant Crash Protection*. In response to petitions from the Motor Vehicle Manufacturers Association and Chrysler Corporation, the agency is deleting the requirement for emergency-locking or automatic-locking seat belt retractors at the outboard seating positions of the second seat in forward control vehicles. The effect of this deletion is to permit manufacturers to continue to use manual adjusting devices for the seat belts at those seating positions.

EFFECTIVE DATE: March 27, 1980.

FOR FURTHER INFORMATION CONTACT: Mr. William E. Smith, Office of Vehicle Safety Systems, National Highway Traffic Safety Administration, 400 Seventh Street, S.W., Washington, D.C. 20590. (202–426–2242)

SUPPLEMENTARY INFORMATION: On November 29, 1979, NHTSA published a notice amending Standard No. 208, *Occupant Crash Protection* (44 F.R. 68470). The amendment deleted the exemption for forward control vehicles from several of the occupant restraint system requirements of the standard. (A forward control vehicle is one with a short front end. More than half of the engine is located to the rear of the forward point of the windshield base and the steering wheel hub is in the forward quarter of the vehicle.)

Chrysler Corporation and the Motor Vehicle Manufacturers Association (MVMA) filed petitions for reconsideration concerning the amendment. They argued that the November 1978 notice of proposed rulemaking for the amendment only proposed a change in the requirements for the safety belt systems in the front seat of forward control vehicles and did not give adequate notice about a change in the requirements for belts in the second seat of forward control vehicles (43 F.R. 52264). They said that the amendment adopted in the final rule requires forward control vehicles to have lap and shoulder belts in the front outboard designated seating positions and have automatic-locking or emergency-locking retractors at the outboard designated seating positions of the second seat of the vehicle.

The petitioners have correctly described the requirements added by the amendment. The amendment applies the requirements of § 4.2.2 of Standard No. 208 to all forward control vehicles manufactured after September 1, 1981. Section 4.2.2 requires a manufacturer to meet one of the following three occupant crash protection requirements: § 4.1.2.1, complete automatic protection, § 4.1.2.2, head-on automatic protection or § 4.1.2.3, lap and shoulder belt protection system. Manufacturers choosing to comply with § 4.1.2.3 must install seat belt assemblies meeting the adjustment requirements of § 7.1 of the standard. The provisions of § 7.1 require that the seat belt assemblies installed at the outboard seating positions of the front and second seats adjust by means of an emergency-locking or automatic-locking retractor. Seat belt assemblies installed at all other seating positions can adjust either by an emergency-locking or automatic-locking retractor or by a manual adjusting device. Prior to the November 1979 amendment of Standard No. 208, forward control vehicles did not have to meet the requirements of § 4.2.1.3 but instead could meet § 4.2.1.2, which did not require the use of emergency-locking or automatic-locking retractors in the outboard seating positions of those vehicles.

The agency's November 1978 notice of proposed rulemaking was addressed to the specific portion of Standard No. 208 exempting forward control vehicles from the shoulder belt requirements. The final rule eliminating the exemption inadvertently changed the requirements for the second seats of light trucks and vans as well. Therefore, the agency is amending the standard to retain the current seat belt requirement for the second seat in light trucks and vans. The agency notes that one manufacturer (GM) of forward control vehicles voluntarily equips its vehicles with automatic-locking retractors and urges Chrysler to do the same. The agency will consider eliminating the remaining forward control exemptions from Standard No. 208 in future rulemaking.

The principal authors of this notice are Mr. William E. Smith, Office of Vehicle Safety Systems, and Mr. Stephen L. Oesch, Office of Chief Counsel.

Issued on March 18, 1980.

Joan Claybrook,
Administrator,

45 F.R. 20103
March 27, 1980

PREAMBLE TO AN AMENDMENT TO FEDERAL MOTOR VEHICLE SAFETY STANDARD NO. 208

Occupant Crash Protection

(Docket Nos. 1-18 and 74-14; Notices 16 and 18)

ACTION: Final rule (correction).

SUMMARY: The purpose of this notice is to correct an amendment to Safety Standard No. 208, *Occupant Crash Protection*, that was issued September 27, 1979 (44 F.R. 55579). That notice amended the seat belt warning system requirements of the standard to specify the use of the seat belt telltale symbol that is specified in Safety Standard No. 101-80, *Controls and Displays*. In that amendment, certain warning system requirements, which had previously been deleted from Standard No. 208, were incorrectly reinserted in the standard. This notice corrects those errors. Further, this amendment makes clear that the telltale symbol of Standard No. 101-80 will supersede certain existing requirements in Standard No. 208 after Standard No. 101-80 becomes effective September 1, 1980.

DATES: These amendments are effective on July 14, 1980.

FOR FURTHER INFORMATION CONTACT: Mr. Hugh Oates, Office of Chief Counsel, National Highway Traffic Safety Administration, 400 Seventh Street, S.W., Washington, D.C. 20590. (202-426-2992)

SUPPLEMENTARY INFORMATION: The seat belt warning system requirements of Safety Standard No. 208, *Occupant Crash Protection* (49 CFR 571.208), currently specify that under certain conditions, when seat belts are not fastened, the words "Fasten Belts" or "Fasten Seat Belts" shall be displayed on the vehicle dashboard. On June 26, 1978, the NHTSA published Safety Standard No. 101-80 (49 CFR 571.101-80) to establish new uniform requirements for the location, identification, and illumination of controls and displays in

motor vehicles. That standard specifies a telltale symbol that is to be illuminated when a vehicle's front seat belts have not been fastened. The standard is to become effective September 1, 1980.

On September 27, 1979, the agency amended Safety Standard No. 208 to permit the optional use of the seat belt telltale symbol specified in Safety Standard No. 101-80 prior to the effective date of that standard (44 F.R. 55579). However, that amendment failed to clarify that, after the effective date of Standard No. 101-80 (September 1, 1980), the telltale symbol will be required to be used in a vehicle's belt warning system. This notice clarifies that point.

When the seat belt telltale symbol was added to Safety Standard No. 208, the amendment inaccurately stated the pertinent sections of the standard that were to be modified. Further, paragraph S4.5.3.3(b) (1) inadvertently omitted language concerning the audible warning. This notice adds the omitted language for that paragraph and, additionally, deletes the parenthetical "(1)" in the paragraph heading. Since there is no longer a subparagraph "(2)," the heading should be specified as "S4.5.3.3(b)."

The 1979 amendment also incorrectly added two sections to the warning system requirements that had previously been deleted from the standard, S7.3.1 and S7.3a. This mistake occurred because the warning system requirements are incorrectly codified in Title 49 of the Code of Federal Regulations. On July 5, 1977 (42 F.R. 34299), Safety Standard No. 208 was amended to delete section S7.3 and to redesignate section S7.3a as S7.3 (as the sections were numbered at that time). When this amendment was codified in the Code of Federal Regulations, however, only paragraph S7.3 was deleted, not the entire section (S7.3 through S7.3.5.4). Instead, S7.3a was transposed

PART 571; S 208—PRE-113

as S7.3 and S7.3.1 through S7.3.5.4 remained. Unfortunately, these deleted sections were used as a reference when the seat belt telltale symbol amendment was added to Standard No. 208. This notice also corrects that error.

Issued on July 7, 1980.

Michael M. Finkelstein,
Associate Administrator for Rulemaking.

45 F.R. 47151
July 14, 1980

PREAMBLE TO AN AMENDMENT TO FEDERAL MOTOR VEHICLE SAFETY STANDARD NO. 208

Occupant Crash Protection

(Docket No. 74-14; Notice 19)

ACTION: Final rule.

SUMMARY: This notice amends Safety Standard No. 208, *Occupant Crash Protection,* to specify additional performance requirements for both manual and automatic safety belt assemblies installed in motor vehicles with a Gross Vehicle Weight Rating (GVWR) of 10,000 pounds or less. These performance requirements are specified in order to prevent the installation of particularly inconvenient and uncomfortable belt assemblies and to ensure that people are not discouraged from using belts because of their design or performance. This amendment does not include several provisions that were contained in the notice or proposed rulemaking preceding this rule. Based on comments received in response to the proposal, the agency has determined that only certain of the specifications should become mandatory at the present time. Consideration involving cost, lead-time and the encouragement of innovative seat belt designs have led the agency to conclude that the other provisions should be issued only as performance guidelines that manufacturers should follow where possible, or find alternative means to accomplish the same ends. The performance guidelines will be published in a separate *Federal Register* notice.

DATE: Effective date: September 1, 1982.

ADDRESS: Any petitions for reconsideration should refer to the docket number and notice number and be submitted to: National Highway Traffic Safety Administration, 400 Seventh Street, S.W., Washington, D.C. 20590.

FOR FURTHER INFORMATION CONTACT: Mr. Robert Nelson, Office of Vehicle Safety Standards, National Highway Traffic Safety Administration, Washington, D.C. 20590. (202-426-2264)

SUPPLEMENTARY INFORMATION: Safety Standard No. 208, *Occupant Crash Protection* (49 CFR 571.208), currently requires most motor vehicles to be equipped with safety belts at each designated seating position. Beginning in September 1981, and phasing in over the following two years, new passenger cars will have to provide automatic occupant crash protection (i.e., occupant restraint that requires no action by occupants, such as fastening seat belts, to be effective). Many new automobiles will be equipped with automatic belts to comply with the automatic restraint requirements (automatic belts move into place around a vehicle occupant automatically when he or she enters the car and closes the door). The requirements specified in this amendment are designed to remove some of the most egregious disincentives to use of current belt designs to ensure that both the automatic belts and the manual belts installed in future vehicles will be comfortable and convenient to use.

The requirements specified in this notice are applicable to seat belt assemblies installed in all vehicles with a GVWR of 10,000 pounds or less, except for Type 2 manual belts (lap and shoulder combination belts) installed in front seating positions in passenger cars through the 1983 model year. As noted in the proposal preceding this amendment (44 F.R. 77210), Type 2 manual belts will be phased out in passenger cars when the automatic restraint requirements of Standard No. 208 become effective. Accordingly, the agency believes that manufacturers should be allowed to focus their efforts and resources regarding comfort and convenience on manual belts in vehicles other than passenger cars and on developing the Type 1 manual belts (lap belts) which will be installed in rear seats in passenger cars and in some front seats in conjunction with air bags and single diagonal automatic belts.

PART 571; S 208—PRE-115

As stated in the notice of proposed rulemaking the discomfort and inconvenience of current seat belt designs are among the most prominent factors resulting in the current low rate of safety belt use (approximately 11 percent). The proposal cited various studies which conclude that comfort and convenience play a determinative role in whether people continue to use the safety belts installed in their vehicles after they first try them (DOT HS-801-594; DOT HS-803-370). Some of the problems identified in these studies include: many belts are difficult to reach; many belts do not fit properly (e.g., they cross the occupant's neck); the pressure of many shoulder belts is felt to be excessive, particularly by women; many belts are difficult to buckle; and many belts become too tight after they have been worn for several minutes and their users have moved around.

In order to alleviate the most serious of these problems, the notice of proposed rulemaking sought to establish a variety of relatively simple, objective performance requirements that would improve the comfort and convenience of seat belt systems. Specifications involving the following performance areas were therefore proposed: torso belt occupant fit; belt retraction; adjustable buckles for certain belts; belt/seat cushion clearance; torso belt body contact pressure; automatic locking retractors (ALR's) were to be restricted; "comfort clips" were to be precluded; latchplate accessibility; webbing guides; convenience hooks for belt webbing clearance between webbing and the occupant's head; and specifications for motorized belt systems.

There were 38 comments in response to the proposal from vehicle manufacturers, seat belt assembly manufacturers, public interest groups and consumers. All comments were considered and the most significant are discussed in this section. In response to those comments, and for reasons set forth more fully below, the agency has concluded that this amendment will only include specifications relating to: latchplate accessibility; seat belt guides; adjustable buckles for certain belts; shoulder belt pressure; convenience hooks; belt retraction; and comfort devices. The other provisions of the proposal will be issued to the public only as performance guidelines which manufacturers may voluntarily follow if they choose. Those guidelines will be issued in a separate *Federal Register* notice.

Proposed Provisions Not Included in This Amendment

(The following section sets forth the major comments to the proposed provisions that are not being included in this amendment. A general discussion of the agency's response to these comments follows after the summary.)

There were nine comments to the proposed amendment from concerned citizens. Five of these consumers supported the proposed rulemaking and stated that they have experienced extreme comfort and convenience problems with their seat belt systems. Three citizens opposed the proposal on the basis that the rulemaking represents unwarranted government interference. Finally, one commenter objected to the technical nature of the proposal, stating that the specifications were difficult to understand.

Almost all vehicle manufacturers supported the concept of the proposal that seat belt assemblies should be convenient to use and comfortable to wear. However, most manufacturers disagreed with the agency's contention that there is a demonstrable relationship between seat belt comfort and convenience and belt usage rates and that improving comfort and convenience will improve those rates. Additionally, most manufacturers did not agree that the specifications proposed by the agency would lead to belt designs that are appreciably more comfortable and convenient. For example, Ford Motor Company stated that although it does not deny that there may be some correlation between comfort and convenience and wearing rates at the extremes (i.e., for very comfortable belts or belts that are particularly uncomfortable), there is no objective evidence that a measurable relationship exists between comfort and convenience and wearing rates. Ford also stated that certain of the proposed requirements would not accommodate a large number of vehicle occupants (e.g., Ford stated that the fit zone specified in the proposal would only ensure that belts properly fit 60 percent of the population. The proposal stated the agency's belief that the fit zone would ensure over 90 percent of the population had comfortable belts). The Motor Vehicle Manufacturers Association stated that experience has shown that the incorporation of features in belt systems to improve their comfort and convenience has not resulted in increased seat belt use, and that comfort and convenience are highly subjective con-

PART 571; S 208—PRE-116

cepts that are not readily quantifiable. Chrysler Corporation stated that comfort and convenience improvements alone will not result in a substantial increase in belt use. Chrysler stated that the only way to improve seat belt use is to enact mandatory seat belt use laws. Volkswagen of America stated that the proposed modifications would actually eliminate several of the most promising existing automatic seat belt designs because of design restrictions. General Motors Corporation cited a study conducted for it by MOR, Inc., which indicated that removal of all perception of discomfort and inconvenience in belt systems would result in only a 1.7 percent increase in seat belt usage. The NHTSA proposal indicated that usage could be increased about 8 percent, and took exception to the MOR study. General Motors argued that the NHTSA has not adequately demonstrated, however, why the conclusions in the MOR study are invalid. American Motors Corporation stated that manufacturers already incorporate adequate comfort and convenience features in their belt systems and that regulatory action is, therefore, not warranted in this case.

The American Seat Belt Council, Hamill Manufacturing Company and other commenters supported the rationale of the proposal totally. Hamill stated that comfort and convenience is of paramount importance to 75–80 percent of the non-user segment of the driver population, who already perceive that seat belts are effective in mitigating the risk of death and injury in vehicle crashes but are dissuaded from using the belts because of perceived inconvenience and discomfort. Volvo of America Corporation acknowledged that comfort and convenience is one factor that influences usage, but stated that the major reason for the low rates of seat belt use is lack of motivation on the part of the motoring public.

In addition to the general negative comments concerning the relationship between seat belt comfort and convenience and wearing rates, many commenters (vehicle manufacturers) argued that certain of the proposed specifications would adversely affect belt effectiveness in vehicle crashes. For example, several manufacturers argued that the comfort zone for belt webbing specified in the proposal would require belt anchorages in some vehicle models to be in locations that are not the optimum location for belt performance in restraining victims in a crash situation.

Torso Belt Occupant Fit (Manual and Automatic Belts)

To alleviate problems of torso belt fit such as rubbing of the occupant's neck, the proposal specified a zone in which the torso belt would have to lie on a test dummy placed in a vehicle. The zone was established to ensure that belts are installed so that the torso belt crosses the occupant's shoulder and chest approximately midway between the neck and shoulder tip, and crosses the sternum approximately midway between the breasts. The proposed requirements specified geometric criteria to describe the required chest-crossing envelope.

The motor vehicle manufacturers were unanimous in their opposition to the proposed torso belt fit requirement. Their objections were primarily related to: the location of the specified compliance zone on the Part 572 test dummy; the location of the test dummy in the vehicle; the width of the compliance zone on the Part 572 test dummy; and the test procedure to determine compliance.

Manufacturers argued that the test procedure is not objective and repeatable because of the complexities and variability associated with locating the dummy in a specific position in the vehicle. They also argued that the procedure for placing the belt around the test dummy (the "rocking" procedure) is not objectively stated. Most manufacturers argued that the 3-inch width of the fit zone specified in the proposal is too design restrictive. Additionally, Ford argued that its tests show that the 3-inch zone would only assure proper fit on approximately 60 percent of the driving population (the agency stated in the proposal that 90 percent of the population would have the proper fit with the proposed specifications). Ford did not substantiate how it arrived at this conclusion, however. Manufacturers argued that the fit zone should be at least 3.6 inches wide and possibly as much as five inches wide in order to ensure repeatability of the compliance procedure. Manufacturers stated that the location of the compliance zone on the test dummy would not necessarily place the belt in the optimum position for effectiveness in crashes in certain vehicle models. They based this assumption on the fact that in certain current vehicle models both the belt anchorages would have to be moved to place the belt in the specified zone. The manufacturers argued that these new anchorage locations would degrade belt performance in some instances.

Clearance Between Webbing and Seat Cushion (Automatic Belts)

As noted in the notice of proposed rulemaking, the shift from manual to automatic belts may initially lead to confusion on the part of some persons. The lower end of many automatic shoulder belt designs is attached between the two front seating positions. The upper end is attached to the rear upper corner of the front door. If the lap belt or torso belt of an automatic belt system is designed so that it lies on the seat cushion or against the seatback cushion(s) when the belt system is reeled-out in its open-door position, some people are likely to be confused about how to get into the vehicle. Additionally, if the belt is lying on or hanging slightly above the seat cushion, it is likely to pull against clothing in an irritating fashion as the occupant tries to sit down. These factors led the agency to propose minimum specifications for webbing/seat clearance (three inches) so that people would not be encouraged to disconnect automatic belts because of the inconvenience.

Most manufacturers opposed the minimum specification for webbing/seat clearance. The comments stated that there is no safety rationale for the requirement because any misconception concerning the proper way to enter the vehicle would be removed after the occupant became familiar with the vehicle. Peugeot stated that experience has shown that the occupant can easily push the strap aside for a moment in order to enter the vehicle. The company argued that the proposed requirement is tantamount to requiring the installation of an automatic mechanism to move the belt system's top anchor's position. (Note: In response to this specific comment, the agency would not consider a belt system that had to be manually moved out of the way by the occupant to be an "automatic" system that would satisfy the requirements of the standard; see 39 F.R. 14594, April 25, 1974). Several manufacturers stated the minimum specification could degrade belt effectiveness in a crash. These manufacturers argued that the specification would preclude a belt, particularly a lap belt, from fitting securely around the occupant. This could result in the occupant "submarining" under the belt during a crash.

Motorized Track Systems— Webbing/Head Clearance

Some automatic belt designs rely on overhead, motorized track-puller systems instead of the opening of the door to move the webbing automatically out of the occupant's way when getting in and out of the vehicle. These systems pull the webbing toward the dashboard when the vehicle is opened and then pull it toward the rear of the vehicle to deploy around the occupant after the door is closed. If such a system is used, the vehicle design should be such that the belt webbing does not pass too close to the occupant's head during its movement. Webbing that passes too close to or brushes the occupant's face or head could be annoying or disconcerting (perceived as hazardous by the intended user) and cause the occupant to defeat the automatic belt system (by unbuckling or cutting the belt, for example). The proposal specified a webbing/head clearance envelope that was intended to ensure that a moving torso belt would not come within a certain specified distance of an occupant's head and face.

Industry objected to this proposed requirement on the basis that many small vehicle models could not comply with the requirement without substantial changes to the vehicle structure (i.e., because of limited head room in these small cars). Toyota Motor Company stated that an automatic belt design it has already introduced in the market would have to be withdrawn if this proposed requirement were finalized because there is not sufficient room in its vehicle model to obtain the specified clearance. Volkswagen stated that any specification for webbing/head clearance should only specify that the webbing cannot touch the occupant's face while it is articulating, and that a minimum distance specification is too design restrictive. General Motors stated that the spherical zone specified in the proposal falls outside the vehicle on some GM body styles, and would thus preclude motorized belt systems in these vehicles.

Rate of Movement of Motorized Belts

The agency stated its belief in the proposal that motorized belt systems will be unacceptable to the public if the rate of belt movement is too slow, since the occupant would be delayed in exiting the vehicle. Systems that move too rapidly might also be unacceptable since they could be viewed by vehicle occupants as a possible hazard. Each of these problems could lead vehicle occupants to defeat the automatic belt system. Therefore, the proposal specified minimum and maximum times allowed for belts to move forward and backward on motorized track systems (between 1.5 and 1.9 seconds from start to stop).

Manufacturers stated that this proposed specification should be deleted because of the variation in performance of motorized systems due to environmental conditions. The comments pointed out that ambient temperature greatly affects motor speeds and battery conditions and that the movement time, therefore, could not be held stable. Several commenters argued that a single movement time is impractical because of the wide variety of vehicle sizes and the varying distances a belt system would have to move. The commenters stated that if such a requirement is retained it should be stated as a rate rather than total times allowed. In this way, the movement of all systems would be uniform even though it would take longer for the belt webbing to move down the track in a large vehicle than in a small vehicle.

Agency Response to Comments on Unadopted Proposals

The agency does not agree with the general negative response of most vehicle manufacturers regarding the relationship between seat belt comfort and convenience and belt use. Likewise, the agency believes that the specification in the notice of proposed rulemaking would greatly improve the comfort and convenience of seat belt systems, particularly the new automatic belt systems that will be introduced in the future. Although the agency agrees that many factors influence belt use, it continues to believe that belts which are inconvenient to use and uncomfortable to wear will be used less regardless of these other factors. The research studies cited in the notice of proposed rulemaking clearly establish that there is a definite problem with many current seat belt designs, and that seat belt systems can be improved with relatively minor changes. Removing the most egregious problems with seat belt designs will, at a minimum, remove an impediment that currently thwarts other programs designed to increase seat belt use. For example, seat belt education campaigns will have little effect if people attempt to wear the belts but find them inconvenient and uncomfortable.

The agency also does not agree with many of the comments regarding specific provisions included in the proposal. Proper torso belt fit is an extremely important aspect of ensuring that belts are comfortable to wear and do not cross the neck or face. The problems cited by the industry with the proposed specification and test procedure are problems the agency believes can be solved. While it is true that some vehicle models may require significant modifications to comply with the fit zone, the agency believes that this is due primarily to the fact that in the past vehicles have been designed with little attention given to how the belt system will fit when installed in the vehicle. Belt systems are typically added as an afterthought long after the vehicle's structural design has been completed, with no systematic effort to coordinate a particular belt design to a particular structural design.

The industry's comments that webbing/seat clearance for automatic belts will not be a problem after occupants learn how to get into the vehicle only address part of the problem. In the months since issuance of the proposal, the agency has observed many prototype and production automatic belt designs. These observations have demonstrated that webbing/seat clearance is extremely important to ensure that the belt webbing does not scrub across the occupant's clothing when entering the vehicle. Some of the designs that were observed had such minimal clearance that buttons and shirt pocket contents were snagged by the belt system as an occupant entered the vehicle. This is obviously a problem that would encourage disconnection of the belt system. In addition, if the webbing/seat clearance is so minimal that the person has to manually move the belt out of the way to enter the automobile, the system is not really "automatic" and would not satisfy the automatic restraint requirements of the standard. The agency has concluded that these problems outweigh the perception problem discussed in the proposal. Consequently, the agency believes that the 3-inch specification in the proposal is inadequate and a greater clearance is desirable. While it is true that greater clearance may require innovative designs, the agency believes these are problems that can and should be solved.

Although these basic disagreements do exist between the NHTSA and vehicle manufacturers, the agency does believe that many of the specific comments to the proposal have merit. Also, the agency is aware that many of the problems cited by the industry are legitimate concerns. The agency is cognizant of the fact that there are a multitude of vehicle configurations that would have to be dealt with in complying with all of the provisions included in the notice of proposed rulemaking. In certain situations it may be true that strict compliance with the provisions as originally specified might compromise belt effectiveness in crashes to a limited degree, if applied to existing, unchanged structural configurations. Most manufacturers stated that the injury criteria of the standard could be met under the specifications of the proposal, but that in some instances the margin of safety would not be as great. Obviously, the agency does not want belt system performance to be degraded in the attempt to make belts comfortable and convenient enough that they will be used. However, the agency does not believe that such a compromise is necessary if belt system design and vehicle structural design are coordinated at the outset.

The agency has also considered the numerous comments concerning the leadtime that would be necessary to implement the proposed requirements in certain vehicle models, as well as the costs associated with making the changes after design plans have already been completed.

These considerations and the factors mentioned below have led the agency to conclude that requirements for torso belt fit, webbing/seat clearance, webbing/head clearance, and motorized belt track speed should not be included in this final rule. The agency believes that manufacturers should be encouraged to rapidly develop innovative automatic belt designs that will coordinate belt comfort and convenience and belt effectiveness to the greatest extent possible. In some vehicle configurations, particularly in smaller cars, strict compliance with the proposed specifications mentioned earlier may hamper these efforts. While the agency believes that it is possible and desirable to design comfortable and convenient safety belts meeting all of the proposed specifications, it does not wish to retard the introduction of automatic restraints because of minor technical problems in particular vehicle configurations. If all of the proposed requirements were issued in this final rule, additional leadtime would have to be given because of the special problems in a few vehicle models. The agency believes it is preferable to encourage voluntary compliance with some of the proposed provisions so that a majority of vehicles can be introduced at an earlier date with the comfort and convenience features incorporated.

The agency also intends to continue development of the proposed specifications in order to refine comfort zones and test procedures. Although the provisions as proposed would represent an important improvement in seat belt comfort and convenience if incorporated in current vehicle designs, comments from the industry have led the agency to conclude that some modifications and adjustments in the specifications may be desirable. Instead of delaying the introduction of improvements in seat belt design while the agency continues this development work, it has been determined that it is wiser to urge voluntary compliance with the major provisions included in the proposals so that they may be introduced as soon as possible. As automatic belts are introduced in the market, valuable data will be received concerning consumer perception of comfort and convenience. These data will be helpful to both the agency and the industry in further improving the belt systems.

Another factor influencing the decision not to include the proposed specifications in this final rule is the fact that there are automatic belt designs currently in production that do not comply with all the provisions proposed. The agency does not wish to preclude the continual production of these designs because, for example, they are ¼ inch outside the torso belt fit zone. This is particularly true since the automatic belts currently on the road were introduced voluntarily by the manufacturers prior to the effective date of the standard.

As stated earlier, the agency does urge manufacturers to voluntarily incorporate the performance specifications that were proposed but that are not included in this final rule. The agency believes all of the provisions deal with seat belt design features that substantially affect the comfort and convenience of seat belt systems, and therefore help determine whether a particular belt system will be worn. The agency also believes that the provisions adequately specify performance criteria and that manufacturers can design systems that are in conformity with the specifications and that also optimize belt effectiveness in crash situations. Although some variations may be required for specialized vehicle configurations, the great majority of the specifications should prove to be extremely helpful to manufacturers attempting to develop seat belt designs that are comfortable to wear and convenient to use.

In order to aid both seat belt manufacturers and vehicle manufacturers, the NHTSA will publish in a later *Federal Register* notice suggested performance guidelines for torso belt fit, belt/head clearance, belt/seat cushion clearance, and speed of motorized belt track systems. The agency will also include in that notice tabulation of all research reports, studies and other data concerning the improvement of seat belt comfort and convenience that are available at the National Highway Traffic Safety Administration. The agency urges all manufacturers to use the information that is available and to incorporate these performance guidelines so that vehicle occupants will not be discouraged from using seat belts because of their discomfort or inconvenience.

Provisions Included In This Amendment

In addition to the provisions discussed already, the notice of proposed rulemaking included specifications dealing with seat belt guides, torso belt pressure, latch plate accessibility, adjustable buckles for certain belts having emergency-locking retractors, convenience hooks for automatic belts, emergency-locking retractors in lap belts, belt retraction and belt comfort devices. The proposed provisions relating to these topics were intended to alleviate some of the most serious problems with current seat belt designs. Most manufacturers agreed that there are problems in these areas, although there was not total agreement on all of the remedies specified in the proposal. After considering the comments, the agency has concluded that improvements in these areas can and should be made. The changes required by this amendment are not burdensome and can be accomplished rapidly. The major objections of the industry to the proposal related primarily to the proposed provisions that are not being included in this amendment (discussed earlier in this notice).

Seat Belt Guides

Seat belt webbing and buckles in motor vehicles often fall or are pushed down behind the seat. Consequently, occupants are discouraged or actually precluded from using the belts. Therefore, the proposal specified that belt webbing at any designated seating position shall pass through flexible stiffeners or other guides in the seat cushion to ensure that the belts are easily accessible to occupants. The provision also specified that belt buckles and

latchplates are to remain above the rear cushions at all times, even in folding or tumbling seats, and that all buckles are to be "free-standing" to allow one-hand buckling. These provisions were included in response to a petition for rulemaking submitted some time ago by the Center for Auto Safety.

The American Seat Belt Council supported the proposed requirements for both seat belt guides and "free-standing" buckles. Vehicle manufacturers requested that several changes be made in the specification or that it be deleted altogether. Volkswagen stated that it would be difficult to comply with the requirement for seats that both fold and tumble and for seats designed to convert into beds. The agency believes that suitable designs can be developed to ensure that belts remain above seats that both fold and tumble. Two vehicles were furnished by Volkswagen which showed two different rear seat configurations. The agency determined that belts could be developed for either that would comply with the provision. However, one design configuration would require seat-mounted belts, with a considerable increase in cost for the belts and increased weight for the vehicle. Based on its consideration of available designs and their costs, NHTSA has concluded that the cost of requiring seats that both fold and tumble seats to comply with the requirement may not be justified. Therefore, this type of seat is not subject to this amendment.

Several manufacturers stated that the proposed requirement should not apply to fixed seats since the purpose of the requirement can be accomplished without guides or conduits for fixed seats. The agency disagrees. The problem addressed in this proposed requirement has been most prevalent with fixed seats. Latchplates and buckles that get lost behind fixed seat cushions are more difficult to retrieve than buckles behind movable seats. While it is true that fixed seats can be designed so that there is little clearance between seat backs and seat cushions, buckles and latchplates can still be forced down behind the seat when a person sits on the seat.

The proposal specified that the belt latchplate and buckle must remain in fixed positions in relation to the seat cushion and vehicle interior. Several manufacturers pointed out that the belt hardware could not remain in a "fixed" position with adjustable seats. The agency agrees that this aspect of the provision was inaccurately stated.

The intent of the provision was only to require that the belt hardware pass through guides or conduits to maintain the location of the buckle and latchplate on top of the seat cushion. The provision is modified accordingly in this amendment.

Several manufacturers also objected to the specification for the "freestanding" buckles and "one-hand" buckling on the basis that the criteria is design restrictive and not stated in objective terms. The agency continues to believe that these provisions would increase the convenience of buckling a seat belt. Nevertheless, after considering the comments, the agency has decided that the specification would be difficult to enforce and may be too design restrictive in some instances. Additionally, a majority of vehicle manufacturers have already begun using stiffeners and other devices to make buckling of belts more simple. If this trend continues, a provision regarding this aspect of belt performance will not be necessary. Therefore, the agency is not including a requirement for "freestanding" buckles in the amendment at this time. The agency does urge, however, manufacturers to voluntarily design their belt system so that buckles are "freestanding" or of some other design that facilitates easy buckling by consumers.

Torso Belt Body Contact Pressure (Manual and Automatic Belts)

NHTSA research indicates that occupants are likely to complain about belt pressure if the torso belt net contact force is greater than .7 pound. Therefore, the proposal specified that the torso portion of any belt system shall not create a contact pressure exceeding that of a belt with a total net contact force of .7 pound.

Most manufacturers objected to the belt contact force limitation. Many commenters stated that the agency has not adequately demonstrated that .7 pound of belt webbing force is the optimum upper limit in all seating configurations. In lieu of the proposed limitation, various manufacturers suggested force limitations ranging from 1 pound to 11 pounds. Manufacturers also argued that the .7-pound pressure does not allow for engineering tolerances. Ford stated that its tests using the proposed procedure indicate that test variability amounts to ±.3 pound. Other manufacturers stated that the proposed force level is so low that it would be difficult to also meet the proposed requirement that belts retract completely when un-

buckled by the vehicle occupant, i.e., the retractor forces would have to be too low to meet the "self stow" provisions. Chrysler Corporation and General Motors stated that a more precise test procedure for measuring belt contact force is needed. This comment was echoed by several foreign manufacturers.

The agency does not agree with most of these objections. In a detailed study conducted by Man Factors, Inc., webbing retractor forces were varied in an experimental belt system mounted in a production vehicle. A series of male and female test subjects experienced each force level during on-the-road driving tests and reported whether the pressure felt was satisfactory or too great. That study showed that belt pressure greater than 0.7 pound was unacceptable to more than 60 percent of the test subjects. Therefore, manufacturers' comments that belt pressure should be as high as 1 to 11 pounds have little, if any, credence. Regarding other comments, the study that was conducted to determine maximum tolerable belt pressure was not conducted for a myriad of seating configurations since a given belt pressure will likely be either acceptable or unacceptable to an occupant regardless of the seating configuration. In automobiles that presently meet this pressure requirement, retraction has not been found to be a problem. Their belts retract in compliance with the proposed retraction requirements. The agency believes that comments stating that a test procedure should be included in the standard to measure the belt pressure have merit. Therefore, this amendment specifies a .7-pound maximum pressure limitation and includes a procedure for measuring belt pressure.

Latch Plate Accessibility

As noted in the proposal, one of the most inconvenient aspects of using many current seat belt designs is the difficulty that seated occupants have in reaching back to grasp the belt latchplate when the belt is unbuckled and in its retracted position. The greater the difficulty in reaching the latchplate to buckle the belt, the more likely that belt usage will cease or never begin. Poor accessibility of latchplates results from two main factors: Location of the latchplate beyond the convenient reach of some seated vehicle occupants, and inadequate clearance between the seats and side of the vehicle to allow easy grasping of the latchplate.

The proposal specified requirements to define limits on reach distance for latchplates and to prescribe minimum clearances for arm and hand access.

There were several comments from the vehicle manufacturers recommending changes in the proposed specifications. The proposed test procedures for this provision specified that the vehicle seat is to be placed in its forwardmost position when testing for compliance with the reach envelope (the position in which there would presumably be the most problems). Ford Motor Company stated that the requirement should be modified to specify that the seat be located in the mid-track position since a 50th percentile adult would not normally have the seat in the forwardmost position (the proposal specified that a 50th percentile dummy be used to test for compliance with the reach envelope). The NHTSA agrees that some difficulty may be encountered in placing the 50th percentile test dummy in the forwardmost seat adjustment position. If this occurs, there is nothing that would preclude manufacturers from removing the test dummy's legs, since legs are irrelevant to the arm reach envelope. However, the agency believes that the requirement should specify that the seat be in its forwardmost adjustment position since many current latchplates are blocked with the seat in this position although they are not when the seat is in its mid-position. Since a significant number of vehicle occupants will have the seat in the forwardmost position (particularly women), the agency believes that the latchplate should be within easy reach for these occupants or they will be discouraged from wearing the belt system.

One manufacturer stated that it is not clear from the proposal whether the latchplate access specifications would apply to all seats or to just the front outboard seating positions. The requirement applies only to the front outboard seats, and the specification is modified in this amendment to clarify this point. Several commenters stated that the size of the test block used to measure latchplate access should be modified and that the block should be designed to articulate to represent the forearm and wrist of a human being. The agency does not agree with this recommendation. This size of the test block was designed to account for the limitation of the human arm and hand as they would articulate through various openings (in this case, between the seat and vehicle structure). The

dimension was based on a detailed study conducted by Man Factors (See DOT-HS-7-01617, December 1978). The agency also believes that the test apparatus would be unnecessarily complicated if specifications were included for articulation. For these reasons, the test block specification and test procedure is unchanged in this notice, except for minor technical changes in the string dimensions and the deletion of one illustration (Figure 3) that was included in the proposal. These minor technical changes are in response to comments and are included for clarification purposes.

Convenience Hooks for Automatic Belts

Some automatic belt designs might include a manual "convenience hook" located, for example, on the dashboard near the A-pillar, which would enable occupants to manually move the belt webbing totally out of the way as they are about to exit the vehicle. These devices would only be permitted as additional equipment since automatic belts must operate automatically, i.e., manual hooks could not be used as the sole means of moving the belt webbing out of the occupant's way. The proposal specified that if manufacturers install such "convenience hooks," the hook must automatically release the belt webbing so that it will deploy around the occupant prior to the vehicle being driven. The proposal specified that the hook would have to automatically release the webbing when

(a) The vehicle ignition switch is moved to the "on" or "start" position.

(b) The vehicle's drive train is engaged.

Manufacturers did not object to the proposed requirements for "convenience hooks," although there were several comments that the provision needs clarification. Jaguar Rover Triumph, Inc. stated that it is not clear from the proposal whether conditions (a) and (b) mentioned in the preceding paragraphs are sequential or alternatives. This notice modifies the language of the requirement to clarify that the "hook" must release the belt webbing when the ignition switch is in the "on" or "start" position *and* the vehicle's drive train is engaged at the same time (i.e., when both condition (a) and (b) exist at the same time). An optional condition "(c)" is added in response to a comment by American Honda Motor Co. to allow vehicles with manual transmissions to have the "hook" release the webbing when the ignition is on *and* the vehicle's parking brake is released at the same time.

Belt Retraction

Many persons find seat belts inconvenient because the belt webbing will not retract completely to its stowed position when the system is unbuckled, so that the webbing is an obstacle when the occupant is trying to exit the vehicle. Therefore, the proposal included a specification to ensure that belts do retract completely and automatically when they are unbuckled. While there were no serious objections to the proposed requirement, several manufacturers requested changes in the test procedures. For example, it was requested that manufacturers be allowed to remove the arms on the test dummy during the compliance test since the belt webbing can get hung-up on the dummy's arms while retracting. The agency believes that this suggestion has merit since a human occupant can move his arm out of the way when a seat belt is retracting and that flexibility cannot be incorporated in the test dummies currently available. Manufacturers also requested that the test be conducted with the vehicle door open, since some systems are designed to automatically retract when the door latch is released (i.e., the retraction force is stronger in this mode). The agency agrees with this suggestion also, and it is incorporated in this notice.

Automatic Locking Retractors

Seat belts incorporating automatic locking retractors (ALR's) in the lap belt portion of the system have been identified as a major item of complaint by vehicle occupants because of the feature's discomfort and inconvenience. Many vehicle occupants report that belts incorporating the ALR's tighten excessively under normal driving conditions, making it necessary to unbuckle and refasten the lap belt to relieve pressure on the pelvis and abdomen. This discomfort causes many persons to stop using their belts.

Belt systems having ALR's have also been found very inconvenient to use, particularly if the ALR is incorporated as part of the latchplate assembly. During the process of putting the belt on, the occupant must extend the belt in a single continuous movement to a length sufficient to allow buckling. Otherwise, the retractor locks before sufficient webbing has been withdrawn to accomplish buckling, and the belt has to be fully retracted before the occupant can repeat the donning process. Many persons have found this characteristic of

ALR's extremely irritating and consequently have avoided use of the belt. In addition, ALR's inhibit the driver's normal movement to pay tolls, reach the glove compartment, etc. With emergency locking retractors (ELR's) instead of automatic locking retractors, these problems would be alleviated.

Safety Standard No. 208 currently requires lap belts at outboard seating positions to be equipped with either automatic locking retractors or emergency locking retractors, in order to assure that belts are sufficiently tightened to be effective during a crash. However, this effectiveness feature can be achieved by ELR's without the concomitant discomfort and inconvenience associated with ALR's. Therefore, the proposal sought to eliminate ALR's as an alternative in the standard for front outboard designated seating positions.

The proposal also specified that emergency locking retractors for the lap belt portion of the belt system at the front outboard passenger's position shall be equipped with a manual locking device so that child restraint systems can be properly secured. Since emergency locking retractors allow some movement when the belt is fastened, the agency and some child safety experts were concerned that the child restraint system could slide out of position prior to a crash if the retractor cannot be manually locked.

Few manufacturers objected to the requirement that lap belts at front outboard designated seating positions be equipped with emergency locking retractors. However, nearly all manufacturers objected to the requirement that these emergency locking retractors be equipped with a manual locking device for securing child restraint systems. Ford Motor Company stated that the manual lock requirement is design restrictive and will preclude the installation of continuous loop manual belts and certain three-point automatic belts. Also, Ford stated that the proposed requirement is inconsistent with another proposal precluding any device that allows the introduction of slack in a belt system (e.g., comfort devices). Ford argued that the manual lock could be used to introduce excessive slack in the belt when worn by an adult. Toyota Motor Company stated that an emergency locking retractor is definitely superior to an automatic locking retractor from the standpoint of comfort and convenience. Toyota argued, however, that its tests with the GM child seat (braking, fast cornering, driving on rough roads) have demonstrated that the performance of emergency locking retractors in restraining this child seat is satisfactory without a manual locking device.

The Motor Vehicle Manufacturers Association pointed out that the Economic Commission of Europe (which sets international motor vehicle safety standards) does not even permit manual locking devices on emergency locking retractors. Volkswagen of America stated that the proposed requirement would impair the operation of these belts by allowing too much slack in the system, and argued that parents should be encouraged to place their child restraints in rear seating positions that have automatic locking retractors. General Motors argued that the agency's data is totally inconclusive in demonstrating that emergency locking retractors without locking devices cannot adequately secure child restraint systems. General Motors cited its own tests which it states demonstrated child restraints are adequately secured with emergency locking retractors. Finally, several manufacturers stated that the manual locking devices could pose a hazard in emergency situations if the emergency locking retractor is located on the vehicle door. These commenters pointed out that the vehicle door would be impossible to open from the outside if the retractor is locked.

After considering these comments, the agency has decided that while emergency locking retractors should be required for lap belts at front outboard designated seating positions, these retractors should not be required to have manual locking devices. The agency believes that the points raised in the comments represent legitimate concerns. Further, agency tests conducted after the issuance of the proposal indicate that there may not be a substantial problem with Type 2 belts incorporating emergency locking retractors restraining child seats. However, the agency is planning to conduct further research regarding the use of Type 1 belts with ELR's to secure child restraints. Additionally, the agency recently issued a proposal to amend Safety Standard No. 210, *Seat Belt Anchorages*, to require that lap belt anchorages be present at front outboard seating passenger positions that are not equipped with lap belts (e.g., vehicles equipped with a two-point, single diagonal automatic belt). Therefore, if that proposal is adopted, parents wishing to place child seats in front seating positions in the affected vehicles can purchase a lap belt having an automatic locking retractor or a manual webbing adjusting device. In light of these considerations, and the cost of installing manual locking devices on emergency locking retractors, the manual locking device of the proposal is not adopted.

The proposal also included a provision to allow manual adjustment devices on seat belt assemblies in rear seating positions that have emergency locking retractors. Although automatic locking retractors are allowed in rear seating positions, some manufacturers are currently installing emergency locking retractors. These manufacturers have requested that manual webbing adjustment devices be allowed on these belt systems, specifically for facilitating the securement of child restraint systems. Nearly all commenters agreed with this provision and it is included in this amendment.

In summary, although manual locking devices are not being required on emergency locking retractors in front seating positions, these devices or manual webbing adjustment devices are being allowed in rear seating positions. The manual webbing adjustment device would not be permitted in front seating positions, but manufacturers would be permitted to voluntarily install manual locking devices on belts in front seating positions.

Devices That Introduce Slack in Belt Webbing

Some current seat belt designs include devices that are intended to relieve shoulder belt pressure. These "comfort clips," "window-shade" devices, or other tension-relieving devices can reduce the effectiveness of belts in crash situations if the occupant uses the device to put excessive slack in the belt webbing, i.e., so that the belt is not snugly against the occupant. Therefore, the proposal included a provision to prohibit any device, either manual or automatic, that would permit the introduction of slack in the upper torso restraint. The proposal stated that such devices would not be necessary to relieve the discomfort caused by excessive belt pressure since the proposal also included a limitation on belt pressure.

Several manufacturers objected to an outright ban on tension-relieving devices. The American Seat Belt Council stated that an appropriate performance requirement should be developed that will allow a small, controlled amount of slack in belt systems. General Motors stated that its tension-relieving devices allow some slack but that this slack could not be introduced inadvertently. General Motors argued that such devices should be allowed provided the slack is cancelled when the vehicle door is opened, i.e., so that there is no slack at all when an occupant uses the belt on a subsequent occasion. The commenters argued that some persons do not like any belt pressure at all, not even the .7 pounds that would be the maximum allowed under the proposed belt pressure provisions.

The agency believes there is some merit to these arguments particularly in regard to automatic belt systems that are required to comply with the injury criteria of Safety Standard No. 208. Therefore, tension-relieving devices are not prohibited in this amendment in automatic belt systems provided the belt system can comply with the injury criteria of the standard with the belt placed in any position to which it can be adjusted. This means that if six inches of slack can be introduced in the automatic belt system by means of the tension-relieving device, the belt must be able to comply with the injury criteria with the belt webbing in that position. Since manual seat belt systems are not required to comply with the injury criteria of the standard generally, they would also not be required to comply just because they include tension-relieving devices. The agency does urge manufacturers to voluntarily limit the amount of slack that can be introduced in their manual belt systems, however.

Seat Belt Warning System

The proposal included a provision for a new sequential seat belt warning system in all motor vehicles which are not passenger cars and which have a gross vehicle weight rating of 10,000 pounds or less.

Safety Standard No. 208 currently requires a visual and audible warning system to remind vehicle occupants to fasten their manual safety belts. The present standard requires a warning system which activates, for a period of 4 to 8 seconds, a reminder light each time the vehicle ignition is operated, and an audible warning if the driver's lap belt is not in use. Studies of manual seat belt usage in passenger vehicles have shown that a sequential logic system which incorporates a visible reminder light of continuous duration and a 4- to 8-second audible warning could produce usage rates significantly greater than those obtained with the warn-

ing systems currently required. The sequential logic warning system activates unless buckling of a person's belt occurred after the person sat down in his seat. Under the current 208 requirement, the warning system can be permanently defeated if the belt is buckled and pushed behind the seat cushion and left there during subsequent occasions on which the vehicle is used.

Only the American Seat Belt Council supported the requirement for a sequential warning system. The vehicle manufacturers uniformly objected to the requirement, stating that such a system would cost $25 to $35 per vehicle (this is much higher than the agency's estimated cost figure). Also, manufacturers disputed the agency's data and argued that there is no documentation demonstrating that a sequential warning system will substantially increase belt use in vehicles other than passenger cars.

The agency agrees that the data relied upon in the proposal dealt primarily with sequential warning systems in passenger cars (The Phoenix Study, DOT-HS-801-953). There is no conclusive evidence that such a system would also improve seat belt use in light trucks and vans to a comparable degree. Although the agency is convinced that an effective warning system similar to or like that proposed would result in some increased seat belt use in these other vehicles, the agency has concluded that manufacturers should be allowed to voluntarily install such systems under an implementation schedule suited to particular vehicle models in order to minimize costs. Therefore, the proposed requirement is not included in this amendment. Specifications for a sequential warning system will, however, be included in the voluntary performance guidelines that will be issued in the near future, however, for the benefit of manufacturers that are interested in such a system.

The proposal also included a specification for warning systems for automatic seat belts, to ensure that motorized systems are locked into place before the vehicle begins moving. If for some reason the motorized belt has not returned and locked into its protective mode, the occupant would be alerted by the continuous light and by a 4-to 8-second audible warning. Although several manufacturers objected to this requirement, again primarily because of cost, the agency believes such a requirement is essential for motorized automatic belt systems. It is therefore included in this amendment.

The proposal also included an illustration chart specifying the weights and dimensions of various human body sizes (e.g., 5th percentile female). The comments to the proposal indicated that some persons were confused about inclusion of the chart. Some commenters interpreted the figures in the chart to represent a change in the Part 572 dummy dimension. The chart was included in the proposal to be republished in the standard since it had been inadvertently deleted by the *Code of Federal Regulations* some time ago. The chart, however, was not intended to make any changes in the Part 572 test dummy.

In order to give manufacturers sufficient lead time to implement the changes required by this notice, and to minimize the cost of such changes, the effective date of this amendment is September 1, 1982.

Note—The agency has determined that this amendment does not qualify as a significant regulation under Executive Order 12221, "Improving Government Regulations," and the Departmental guidelines implementing that order. Therefore, a regulatory analysis is not required. A regulatory evaluation concerning the amendment has been prepared and placed in the public docket under the docket number and notice number of this *Federal Register* notice.

Issued on December 31, 1980.

Joan Claybrook,
Administrator.

46 F.R. 2064
January 8, 1981

PREAMBLE TO AN AMENDMENT TO
FEDERAL MOTOR VEHICLE SAFETY STANDARD NO. 208

Occupant Crash Protection
(Docket No. 208; Notice 21)

ACTION: Final rule.

SUMMARY: The purpose of this notice is to amend Safety Standard No. 208, Occupant Crash Protection, to delay for one year the effective date of the first phase of the automatic restraint requirements of the standard. Prior to this notice, the automatic restraint requirements were scheduled to become effective for large cars on September 1, 1981 (model year 1982), for mid-size cars on September 1, 1982 (model year 1983), and for small cars on September 1, 1983 (model year 1984). As amended by this notice, the requirement for equipping large cars with automatic restraints will not take effect until September 1, 1982, or model year 1983.

This one-year delay in the automatic restraint requirements is being specified in light of dramatic changes in production plans for the model-year 1982 fleet (fewer large cars and more small cars) and because the economic and other justifications for the existing phase-in schedule have changed drastically since the standard was adopted in 1977.

The one-year delay will also allow the Department sufficient time to re-evaluate the entire automatic restraint standard as required by the Presidential Executive Order 12291 (February 17, 1981). The Department is simultaneously issuing a notice of proposed rulemaking in today's issue of the *Federal Register* discussing further possible changes in the automatic restraint standard.

DATES: The new effective date of the automatic restraint requirements for large cars is September 1, 1982.

ADDRESSES: Any petitions for reconsideration should refer to the docket number and notice number of this notice and be submitted to: Docket Section, Room 5109, Nassif Building, 400 Seventh Street, S.W., Washington, D.C. 20590.

FOR FURTHER INFORMATION CONTACT:

Mr. Michael Finkelstein, Office of Rulemaking, National Highway Traffic Safety Administration, Washington, D.C. 20590 (202-426-1810)

SUPPLEMENTARY INFORMATION: On February 12, 1981, the Department of Transportation issued a notice of proposed rulemaking to delay for one year the first phase of the automatic restraint requirements of Safety Standard No. 208, Occupant Crash Protection, (46 FR 12033). Automatic restraints are systems that require no action by vehicle occupants, such as buckling a seat belt, to be effective. Two existing systems that qualify as automatic restraints are air cushion restraints (air bags) and automatic seat belts (belts which automatically envelop an occupant when entering the vehicle and closing the door).

The automatic restraint requirements were added to Standard No. 208 on July 5, 1977 (42 FR 34289), and require installation in accordance with the following schedule:

• For full-size cars (wheelbase greater than 114 inches) beginning September 1, 1981 (1982 model year);

• For mid-size cars (wheelbase not more than 114 inches but greater than 100 inches) beginning September 1, 1982 (1983 model year);

• For small cars (wheelbase less than 100 inches) beginning September 1, 1983 (1984 model year).

The February notice issued by the Department proposed to alter this phase-in schedule by deferring the first phase (large cars) for one year, from model 1982 to model year 1983. The proposal noted that such a change may be appropriate because of the effects of implementation in model year 1982 on large car manufacturers, because of the added significance which those effects assume due to the change in economic circumstances since the schedule was adopted in 1977, and because of the undermining by subsequent events of the rationale underlying the original phase-in schedule. (See the notice of proposed rulemaking for a full discussion of the facts which led to the proposed alteration of the phase-in schedule.)

Comments Upon Proposal

The responses to the proposal were equally divided between those commenters adamantly opposed to any delay in the automatic restraint requirement and those commenters in favor of both the delay and a total revocation of the requirements. The comments and data supporting these factions were as diametrically opposed as the competing economic interests involved, in this instance the automobile and the insurance industries. Following is a summary of the major comments submitted in response to the proposal. A more detailed summary of representative comments is included as an appendix at the end of this notice.

The automobile insurance industry was unanimously against the proposed delay in the first phase of the automatic restraint requirements, unless the standard is also amended to require an earlier implementation of automatic restraints for small cars (i.e., a delay and reversal of the current schedule). The commenting insurance companies stated that the automatic restraint requirements will save thousands of lives and prevent hundreds of thousands of serious injuries. They argued that the proposed delay of the 1982 requirements would, therefore, result in a significant number of fatalities and injuries that would not otherwise occur. These companies also argued that the monetary savings that would result from the proposed delay are so small that they would not significantly help the ailing automobile industry. The commenters pointed specifically to the fact that most of the capital expenditures have already been made for installing automatic restraints on 1982-model large cars.

In urging a reversal of the implementation schedule, the insurance companies noted the dramatically increasing number of small cars, and pointed to insurance research which shows small cars are inherently more dangerous for occupants than large cars. (NHTSA statistics show that a person is eight times more likely to be killed in a small car than in a full-size car in a crash between the two.) Since small cars will represent a majority of the 1983-model passenger car fleet, the companies argued that more lives could ultimately be saved if automatic restraints are required on small cars in that model year, than under the existing implementation schedule.

Many of these same sentiments were also voiced by consumer groups and health organizations, the majority of which were also opposed to the proposed delay of the MY 1982 requirements. Like the insurance companies, most of these groups asserted that usage rates for automatic belts will be relatively high and that the automatic restraint standard as a whole will save thousands of lives.

Several consumer groups and air bag component suppliers stated that they could support the proposed delay provided there is also a requirement that vehicle manufacturers at least offer air bags as options on some of their model lines. These groups are concerned that further delay of the automatic restraint standard will drive the remaining air bag component suppliers out of the market and that, as a result, the life-saving potential of air bags will be lost.

The insurance industry and a majority of the consumer groups argued that the benefits of the 1982-model year requirements outweigh the costs. A detailed analysis by Professor William Nordhaus of Yale University was submitted on behalf of several insurance companies. This analysis concludes that the economic costs of the proposed delay would be approximately five times greater than the benefits, for a net cost of $200 million. These figures are based on computations regarding the societal costs of deaths and injuries that would result without the MY 1982 automatic restraint requirement.

Several of the commenting insurance companies and consumer groups also argued that as a matter of law and statutory authority the Department cannot rely on the general economic health of the automobile industry to justify a delay in the automatic restraint standard. The National Traffic and Motor Vehicle Safety Act (the Vehicle Safety Act) (15 U.S.C §1381, *et seq.*) provides that motor vehicle safety standards shall be "reasonable" and "practicable." These commenters noted that the legislative history of the Vehicle Safety Act indicates that in promulgating standards, safety shall be the overriding consideration. The commenters contend that the current poor economic condition of the automobile industry does not make the 1982 model-year requirements impracticable.

In addition to comments from the above groups and organizations, the Department also received comments from numerous private citizens, who were equally divided in their support or opposition to the proposed delay.

The proposed delay in the 1982 model-year requirements was unanimously supported by the automobile industry, both foreign and domestic. In addition, most manufacturers urged the Department to reconsider the entire standard, to provide additional leadtime for all phases of the implementation schedule, or to revoke the automatic restraint requirements altogether. Regarding a possible reversal of the current implementation schedule, nearly all of the foreign automobile manufacturers joined Chrysler Corporation and American Motors in stating that it would be impossible to install automatic restraints on 1983-model small passenger cars because of insufficient leadtime.

In support of a complete rescission of the automatic restraint requirements, the vehicle manufacturers made several arguments. The manufacturers believe that automatic seat belts will be so unacceptable to the public that they will create a consumer "backlash" greater than that caused by ignition interlock devices required by NHTSA to be installed on 1974-75 models. These devices made it impossible to start the vehicle unless front seat belts were fastened, and were specifically precluded by the Congress by amendment to the Vehicle Safety Act in 1974.

The manufacturers contend that automatic seat belts will produce such a reaction because of their coercive nature and obtrusiveness. They also contend that automatic belts must be designed so that they are easily detachable (and presumably thereby more acceptable to the public). In such case, they argue that the usage

rate for automatic belts would be no greater than for current manual belts, and that the increased cost of automatic belts would not be justified.

Auto manufacturers also argued that the extremely high price of air bags makes them impractical, and allege that few will be installed on future passenger cars. Consequently, they contend, the only benefits attributable to the automatic restraint standard will be those derived from automatic belts, which for the above reasons will not be effective.

Only two vehicle manufacturers, Ford Motor Company and General Motors, produce any significant number of large cars. Therefore, the existing automatic restraint requirements for 1982 models would only directly affect these two companies.

Ford Motor Company supported the proposed delay and stated that it considers its original 1982-model, three-point automatic belt designs to be "out of date" because of their release concepts (i.e., they include a feature to frustrate release and thus defeat of the system). Ford believes this could lead to significant public dissatisfaction with MY 1982 automatic belts. In response to this concern, Ford had decided to add a conventional release buckle to this three-point belt, so that it can be detached by those motorists who refuse to wear a belt. Ford's submission stated that the company projects that as many as 100,000 purchasers would switch to mid-size cars in the 1982 model year rather than buying large cars with an automatic belt. Ford plans to redesign its automatic belts, but states that such a program has major leadtime implications which would make it impractical to install improved automatic belts in small cars before September 1, 1983.

General Motors Corporation stated that its planned 1982-model automatic belt designs are easily detachable (i.e., there will be a buckle release mechanism without an interlock or other mechanism to discourage defeat of the system). With this type belt, according to GM, the impact on safety will depend upon voluntary use of the automatic belt, so use would not likely be any greater than with current manual belt systems. Therefore, General Motors argues that the proposed delay should have only a minimal adverse safety impact.

General Motors stated that the proposed delay would result in a net increased sales revenue to the company of $760 million, and that the company could realize a savings of approximately $13 million in capital investment for the 1982 model-year program. General Motors explained the $760 million figure with the following rationale:

Automatic belts will be regarded by many as an unnecessary inconvenience, and they will deprive purchasers of six passenger seating capacity. Thus, 1982 full-size cars equipped with such a restraint will be at a competitive disadvantage in that consumers can avoid the penalties of increased cost and reduced accommodation either by purchasing vehicles not subject to passive restraint requirements in that year, or by deferring their purchases. The proposed delay will allow the consumer to purchase a full size car in 1982, without a cost penalty, which fully meets his needs and expectations.

General Motors' concern in this regard derives from the fact that large cars with automatic seat belts will be able to have only two front seating positions, since no company has developed an automatic belt system for the center seat position. With the automatic restraint requirements delayed, General Motors would be able to install bench front seats with three seating positions in its large cars. General Motors estimates that the reduced seating capacity thus caused by automatic belts will result in 120,000 fewer large car sales: 50,000 purchasers will shift from large cars to GM mid-size cars, and 70,000 potential purchasers will defer buying a new large car in the 1982 model year if they cannot obtain a six-passenger large car. General Motors contends that these factors will result in a revenue loss to the company of $760 million if the automatic restraint requirements are not delayed.

Rationale For Agency Decision

The agency has given thorough consideration to all comments submitted in response to the proposed delay of the first phase of the automatic restraint requirements, and carefully analyzed all such information and data in the Record of this proceeding. The wide diversity among factual, analytical and policy-related positions urged by those supporting and those opposing the proposed delay illustrates the degree to which this proceeding involves questions for which there are currently no concrete answers.

For example, the usage rate of automatic belts will be extremely dependent on the exact design of a particular belt system. Consumer expectations (for example, that six-seat cars will be available), consumer acceptance (for example, the purchase of cars with automatic belt systems which cost more than current belt systems) and actual rates of usage are values crucial to the Department's decision-making process. These factors, which are dependent on the desires and reaction of the American public, cannot be quantified or predicted with certainty.

On the basis of the record herein, the Department has concluded that the applicability of FMVSS 208 in MY 1982 to large cars would be impracticable and unreasonable. Requiring such compliance would reduce sales and profits, and increase unemployment, for the manufacturers of such vehicles. The Department believes that it is in the public interest to avoid these unnecessary costs and impacts by providing an additional year of leadtime.

The February 12, 1981 notice detailed many of the specific reasons which led to the proposed delay. As specified in that notice, many of the factual assumptions and premises which led to adoption of the phase-in schedule have been proven wrong by subsequent events. The economic situation of the industry and of consumers and the economy as a whole have drastically changed since the standard was adopted in 1977.

The current phase-in schedule for automatic restraints was intended to permit manufacturers to introduce automatic restraints without undue technological or economic risk. Such risks would otherwise have had to have been assumed contemporaneously with the risks involved in having to meet the requirements imposed by emission and fuel economy standards applicable to automobiles in the early 1980's.

Large cars were chosen for the first phase of the schedule because at that time there was more experience with air bags in such full-size cars. A phased schedule to cover progressively smaller cars, in stages, was adopted to provide manufacturers with a chance to gain similar levels of experience in smaller cars. To ensure that manufacturers would in fact have the maximum flexibility to choose between equipping smaller cars with air bags or automatic belts, those cars were to be phased in last. This justification for a phased implementation schedule is no longer valid. Gasoline shortages, price increases (especially those occurring since the Iranian oil cut-off in 1979), and continuing uncertainty about levels of future petroleum supplies have led to dramatic increases in production plans for small cars. The small car share of new production is growing at a much faster pace than was anticipated by the Department when the automatic restraint requirements were issued.

In 1977, the Department projected that new car production in the model year 1982-1985 period would be approximately 24 percent large cars, 53 percent mid-size cars, and 23 percent small cars. However, NHTSA now estimates that actual production of large cars will be about 11 percent in model year 1982 while mid-size and small cars are expected to increase commensurately in that model year.

Thus, under the state of facts now facing the Department, about 11 percent of the 1982 model-year cars would be required to have automatic restraints under the 208 standard.

This major shift in the absolute and relative numbers of cars which would be subject to the first year of the standard will have important adverse impacts upon the benefits to be achieved by the first year of application of the standard. Consumer acceptance of the automatic restraints now anticipated to be used in the 1982 model-year cars is likely to be substantially less than was assumed in 1977. There will be more than a million fewer vehicles with automatic restraints than was previously expected. With fewer cars equipped with automatic restraints, the vehicles which are so equipped will be far more vulnerable to negative consumer reaction.

The Department has long recognized that any costly, arguably coercive restraint system will cause a certain percentage of the population to react negatively. The factors leading to such negative reaction will be magnified as the percentage of new 1982-model cars equipped with automatic restraints decreases. Adverse consumer preferences leading to deferral of the purchase of large cars, or to shifts to the purchase of mid-size cars, will predictably occur.

Concern about providing additional leadtime to adapt air bags to small cars is also less important

now as a result of changes in facts occurring since 1977. When the standard was issued, the Department assumed that manufacturers would equip a great majority of their vehicles (75%) with air bags in preference to belt systems. However, most manufacturers now indicate that they intend to offer air bags on very few of their large cars, and on almost none of their smaller cars. Almost all 1982 model-year cars are planned to use automatic belts.

The absence of any opportunity to select between automatic restraint systems will materially affect public acceptance of the automatic restraint standard. General Motors has pointed out that two automatic belt designs recently offered as options on its Chevette line produced very low purchaser interest, even though the cost was minimal and the car line was in high demand. GM states that fewer than 13,000 of 415,000 1980-model Chevettes sold were equipped with the automatic belt option, despite the fact that the option was offered at no cost to most purchasers, GM salesmen were to be given an additional commission of $25 for each sale, and over $1 million was spent on advertising and marketing.

Similar low interest has been shown in an automatic belt system offered as an option on General Motor's 1981 Cadillac.

The poor consumer acceptance of these automatic belt options substantiates the Department's assumption that automatic belts installed on only a limited percentage of a particular model-year fleet will have difficult public acceptance problems.

The public acceptance of 1982-model automatic restraints is a valid concern of the Department and is of primary importance in determining the reasonableness and practicability of the standard, and whether there is good cause for the delay. As stated by the Court of Appeals in *Pacific Legal Foundation* v. *Department of Transportation*, 593 F.2d 1338 (D.C. Cir.), *cert. denied*, 444 U.S. 830 (1979):

> We believe that the agency cannot fulfill its statutory responsibility unless it considers popular reaction. Without public cooperation there can be no assurance that a safety system can "meet the need for motor vehicle safety." And it would be difficult to term 'practicable' a system, like the ignition interlock, that so annoyed motorists that they deactivate it.

The Department is unable to conclude from its current data, taking into account the large number of private citizens who took the time and effort to file comments reflecting their opposition to automatic restraints, that the 1982 automatic belt designs planned by the manufacturers will receive "public cooperation."

The proposal stated that the changed economic circumstances may make the current implementation schedule for automatic restraints impracticable. Several commenters argued that the general economic situation of the automobile industry is not a legitimate criterion for determining whether a safety standard is practicable under the National Traffic and Motor Vehicle Safety Act. The legislative history of the Vehicle Safety Act clarifies that economic considerations may be considered in determining the "practicability" of a particular safety standard:

> This would require consideration of all relevant factors, including technological ability to achieve the goal of a particular standard as well as consideration of economic factors. (H.R. Rep. No. 776, 89th Cong., 2d Sess. (1966) at 16.)

One commenter stated that the term "practicable" must be viewed as relating solely to the economic and technological capability of the industry to meet the timetables established by the particular safety standard in question, and not to the general economic health of the industry. The Department disagrees with this reading of the Vehicle Safety Act and its legislative history.

The reasonableness and practicability of the current phase-in schedule cannot be determined in a vacuum. What is reasonable and practicable for a healthy firm or industry may not be for an ailing one. The proposal noted the current financial difficulties of the automobile industry. Vehicle sales remain at depressed levels and unemployment in the domestic industry is extremely high. Approximately 200,000 workers have been indefinitely laid off, and more have been temporarily laid off. These losses come at a time when the domestic manufacturers are spending unprecedented sums to meet the continuing demand for more fuel efficient cars.

The Department concludes further that economic hardship to the affected industry and individual companies must be balanced with all other considerations in determining the "reasonableness" and "practicability" of a particular safety standard. None of the individual factors involved in the deliberations may properly be applied without regard to the other factors. This proposition holds both in promulgating a standard and in retaining a standard when relevant factors have materially changed since the standard was first adopted.

The same commenter also argued that the Department had not shown "good cause" for proposing to delay the effective date of the automatic restraint requirements, in light of the requirements of the Motor Vehicle Safety Act that the leadtime for the effective date of safety standards shall be no longer than one year, unless the Secretary finds, for good cause shown, that an earlier or later effective date is in the public interest (15 U.S.C. 1392).

The leadtimes associated with the existing implementation schedule were much longer than one year. These were upheld by the Court in the Pacific Legal Foundation case, *supra.* In that case, the court relied heavily on the inability of the manufacturers to comply with the requirement in one year's time, and on the need for considering the likelihood that the public will accept the change:

> When dealing with a "technology-forcing" rule like Standard 208, the agency must consider the abilities of producers to comply with the new requirement and of the public to grasp the need for the change.

As was stated earlier, the Department is now concerned that 1982-model large cars might be seriously unacceptable to a large portion of the public.

The Department concludes that "good cause" exists for the proposed delay. The public interest in the economic viability of the industry and, with respect to the proposed delay, the particular circumstances of the manufacturers of the vehicles involved, requires that inequitable burdens and unnecessary costs be avoided where possible in implementing FMVSS 208. Large cars are not expected to be produced beyond MY 1985. Application of the standard to large cars in advance of smaller cars would thus involve such burdens and could involve such costs.

In addition to these considerations, the Department believes that the proposed delay must be viewed as a separate regulatory action insofar as leadtime is concerned. The leadtime specifications for the existing implementation schedule were upheld by the court in *Pacific Legal Foundation*. The proposed delay represents a new consideration of the factors which will determine whether automatic restraints are reasonable and practicable for large cars in the 1982 model year, with primary attention being given to acceptability of these systems by the public.

Opponents of the proposed delay have pointed to the adverse safety impacts that might result, stating specifically that the safety benefits of the 1982 model-year requirements outweigh the costs. The Department's proposal stated that a delay of the first phase requirement could over the ten-year life of the vehicles involved result in a loss of 600 lives, and the accrual of 4,300 more injuries than would have occurred without the delay. After reviewing the information submitted in response to the proposal and analyzing more current data, however, the Department now concludes that its earlier estimate of adverse effects is invalid.

First, the assertion that 600 lives would be lost was based upon earlier estimates of benefits that would arise from 100 percent usage of automatic restraint systems. This calculation in turn had been based primarily on 1977 assumptions that air bags would be the technology of choice. As stated earlier, the Department now knows that very few air bags are planned for the 1982 model-year.

Unlike air bags, estimates of benefits arising from compliance with the automatic restraint standard by means of automatic belts must be based upon projected usage rates. The most optimistic expectations of automatic belt use for the 1982 model-year now appear to be a usage rate of 60 percent. Moreover, given the planned design of the 1982-model automatic belts, NHTSA now believes that a much lower usage rate will in fact occur. Both General Motors and Ford plan automatic belt designs which have a release buckle identical to the buckle on current manual belt systems. Motorists would therefore be able to disconnect the proposed belts with the same ease with which current active belt systems can be released. NHTSA believes it is likely that a large percentage of motorists would adopt this usage pattern, and detach the automatic belts.

Usage could thus in fact turn out to be low, and approach levels similar to that of current manual belt systems (7%).

The final regulatory analysis thus now includes a range of possible usage rates for 1982-model automatic belts, in analyzing possible benefits to be foregone by deferring the MY 82 standard for one year. If usage rates for the automatic belts otherwise required for that model-year were to be 15 percent, more than double the rate of use of current manual belts, retention of the 1982 requirements might save a total of 75 lives over the projected ten-year life of the large cars involved. If usage rates were to occur at the level of 60 percent, this number could possibly increase to as many as 490 lives over the same ten-year period.

NHTSA now believes that the potential usage of 1982-model automatic belt designs would more likely be near the bottom end of this scale. NHTSA data on observed usage rates for the belt systems employed in some models of the Volkswagen Rabbit, for example, are relevant. All such belts are optional, and were chosen by the purchaser either as a separate option or as a part of the "Deluxe" package. Moreover, the VW system employs an interlock mechanism, so that the engine may not be started if the system is not in place. Despite these factors, usage rates have been observed to be only 81%. That is, of the purchasers who specifically selected this optional system, nearly 20% thereafter in practice enter their vehicles, start their engines, and then deliberately disconnect the belt system when driving.

Moreover, actual accident data relating to such vehicles show even lower usage rates, of 55-57%. (See Regulatory Analysis, at V-11, 13 for discussion.)

After analyzing the data submitted in response to the proposal, the Department has determined that the one-year delay will result in a cost savings to consumers of approximately $105 million. Capital investment savings for the industry will be about $30 million. Net income available for reinvestment would be increased to $292 million by the delay. Over 13,000 jobs will be saved in the automobile manufacturer and supplier industry, a savings of $159 million. The basis for these figures is explained in detail in the final regulatory analysis. Given the current economic situation of the American public and the domestic automobile industry, these savings are significant, particularly when viewed in conjunction with the Department's belief that the safety impact of the delay can be minimized.

While some measure of safety benefits will be foregone by this delay, the Department has concluded that such benefits are relatively minor. Moreover, the Department believes that any such loss of safety benefits can be offset with a coordinated effort by all parties involved. The Department believes that an intense seat belt use education campaign, joined by the Department, industry and consumer groups and targeted directly at the 1982 model-year cars, has the potential of affording even greater safety benefits than would otherwise accrue.

Finally, such a targeted campaign to increase the use of existing manual seat belts will provide further data on the viability of such strategies in increasing active seat belt use. Such information would be especially valuable for future rulemaking purposes, since it would in any event be at least ten years before all cars in the passenger fleet would be expected to be equipped with new safety equipment. Such information would enable the Department, State and local governments, and other interested parties to determine how to make the best use of their scarce resources to increase actual usage of the millions of manual seat belts that will remain on the nation's highways for years to come.

Summary of Agency Conclusion

The Department has determined that the existing schedule for the first year of implementation of FMVSS 208 is no longer reasonable or practicable. The assumptions leading to the 1977 rule are no longer valid. There will be few, if any, air bags installed in passenger cars because manufacturers have chosen automatic belts as the preferred means of compliance with the standard. The number of small cars on the road is increasing drastically and these cars are more unsafe than large cars. Yet, under the current implementation schedule, small cars are to be equipped with automatic restraints last.

The delay of the first phase of the automatic restraint requirements will enable the Department to adequately reassess the most viable alternatives for the occupant crash protection standard. The Department is publishing simultaneously with this final rule a Notice of Proposed Rulemaking addressing alternatives to this standard, and attention is specifically directed to that proposal.

The Department is taking these actions because courts have found that the Department has a statutory responsibility to reexamine its safety standards in light of changing circumstances and new data. In those circumstances, the Department is required to make necessary revisions and schedule changes to ensure that the standards are practicable, reasonable and appropriate. As noted above, key assumptions underlying the issuance of the automatic restraint requirements in 1977 have been substantially undermined by subsequent events.

The delay and reevaluation of FMVSS 208 is also consistent with Executive Order 12291, which directs all executive branch agencies to delay final rules to the extent necessary to reevaluate those rules under criteria specified in the Order.

This amendment has been evaluated as a major rule under the guidelines of new Executive Order 12291 and a final regulatory analysis is being placed in the public docket simultaneously with the publication of this notice. The major findings of that analysis have been discussed in the body of this notice.

The effect of the one-year delay has been evaluated in accordance with the National Environmental Policy Act of 1969. It has been determined that this action is not a major Federal action significantly affecting the quality of the human environment. An evaluation of the environmental consequences of the amendment is included in the regulatory analysis. Further information regarding environmental issues concerning automatic restraints, especially air bags, can be found in the environmental impact statements published in conjunction with the 1977 automatic restraint standard.

The regulatory analysis also includes a discussion of the Department's consideration of the possible impact of this amendment on small entities. The analysis shows that the one-year delay will have a minimal effect on the automatic seat belt-related firms, since it is likely that most of the 1982-model large cars will continue to be equipped with conventional manual type seat belts. Generally, however, the same firms produce both automatic and manual belts, and none of these direct suppliers qualify as "small businesses" under the Regulatory Flexibility Act.

The effect of the delay on air bag suppliers is less certain. Neither Ford or General Motors would have installed air bags in 1982 vehicles regardless of the delay. The analysis determined that some suppliers of the air bag components will be adversely affected by the delay to some extent and that a few of these qualify as small businesses. However, it is doubtful that a substantial number of small businesses will be adversely affected by the delay to a significant degree.

The analysis also considered the effect of the delay on the small governmental units and other small fleet purchasers of cars. Since large cars are not generally sought for fleet purposes, the amendment is likely to have only a minimal effect on all types of small fleet purchasers.

In consideration of the foregoing, Safety Standard No. 208, Occupant Crash Protection (49 CFR 571.208) is amended as follows:

Section S4.1.2 is amended to read:

"S4.1.2 *Passenger cars manufactured from September 1, 1973, to August 31, 1983*. Each passenger car manufactured from September 1, 1973, to August 31, 1982, inclusive, shall meet the requirements of S4.1.2.1, S4.1.2.2, or S4.1.2.3. Each passenger car manufactured from September 1, 1982, to August 31, 1983, inclusive, shall meet the requirements of S4.1.2.1, S4.1.2.2, or S4.1.2.3, except that a passenger car with a wheelbase of more than 100 inches shall meet the requirements specified in S4.1.3. A protection system that meets requirements of S4.1.2.1 or S4.1.2.2 may be installed at one or more designated seating positions of a vehicle that otherwise meets the requirements of S4.1.2.3."

(Secs. 103, 119, Pub. L. 89-563, 80 Stat. 718 (15 U.S.C. 1392, 1407).)

Issued on April 6, 1981.

Andrew L. Lewis, Jr.
Secretary of Transportation

46 FR 21172
April 9, 1981

APPENDIX
DETAILED DISCUSSION OF COMMENTS

A. Comments Opposing the Delay

The insurance industry argued that the automatic restraint requirements will save thousands of lives and prevent hundreds of thousands of serious injuries. The League ,Insurance Companies stated that the proposed one-year delay would be "tragic and costly," adding that "there is a legitimate place for regulation when the need is great, the cost-benefit is demonstrably high, and the structure of the market place requires uniformity to be imposed on all manufacturers."

Allstate Insurance Companies argued that the growing proportion of small cars will increase deaths and injuries by 35 percent during the next four years, and that the only way to reverse this trend is by implementation of the automatic restraint standard. Allstate also argued that the proposal's analysis of the economic consequences of the scheduled implementation is based only on conjecture. The company stated that there is no substantial evidence of record that the proposed delay would provide any significant financial assistance to car makers. According to Allstate, however, the proposed delay would result in needless deaths and injuries at huge costs to society at large and to insurances-buying customers. Allstate concluded that it could only support a one-year delay in the automatic restraint requirements if the delay is coupled with a requirement that small cars comply with the standard in model year 1983 (i.e., one year earlier than the existing schedule). This sentiment was also expressed by the Alliance of American Insurers and the League Insurance Companies. Alliance stated that a move to install automatic restraints on small cars first is consistent with insurance research which shows small cars to be inherently more dangerous to occupants than large cars, and that such a change could also afford domestic manufacturers some economic relief.

State Farm Mutual Automobile Insurance Company attacked the proposed delay of the automatic restraint requirements on several grounds. First, State Farm argued that the record in this rulemaking proceeding demonstrates that full implementation of the automatic restraint standard will save thousands of lives and avoid tens of thousands of crippling and maiming injuries. The company pointed to the Department's analysis which found that the proposed delay would cost the nation 600 deaths and approximately 4,300 injuries over the lifetime of the 1982-model large cars, and stated that a delay is not justified under any cost/benefit calculations. State Farm also argued that the proposed delay is inconsistent with the overriding mandate of the National Traffic and Motor Vehicle Safety Act (15 U.S.C. 1381, et seq.) and that "the controlling statutes do not permit the Secretary to defer otherwise supportable life-saving regulations solely on the basis of current economic conditions in the auto industry."

State Farm concludes that the current economic situation of the auto industry does not make the implementation of the current automatic restraint schedule impractical. First, nearly all of the necessary capital commitments for automatic restraint implementation for large cars have already been made. Second, the variable costs associated with installing automatic restraints on 1982-model large cars are insignificant to the industry. State Farm also argued that the balance of costs against benefits does not support the proposed delay; rather, it supports an acceleration of the existing schedule if anything. The company cited a recent study by Professor William Nordhaus (discussed below) which contends that the annual economic costs of the proposed deferral of the model year 1982 requirements relative to the current schedule are five times greater than the economic benefits to the auto industry.

It is State Farm's position that as a matter of law and statutory power, the Department cannot rely on the general economic health of the automobile industry to justify a delay in the implementation of the life-saving automatic restraint standard. The comment cites the Senate report concerning the Vehicle Safety Act which stated that safety is "the overriding consideration" in carrying out the purposes of the Act (S. Rep. No. 1301, 89th Cong., 2d Sess. 6 (1966)). State Farm argues that economic considerations in rulemaking by the Department and NHTSA under the Vehicle Safety Act must relate to the costs and benefits of the standard itself and not to the general health of the auto industry: "... if the Secretary were to implement the proposed delay in this rulemaking on the basis of the general employment, production, and economic status of the auto industry, he would be acting arbitrarily and capriciously and outside the scope of his statutory authority." The legal memorandum submitted in support of State Farm's contentions included the following argument:

If the general economic condition of the auto industry could justify suspending implementation of the automatic restraint standard in the face of such cost and benefit data, the industry's economic condition could also be used to justify suspension or elimination of other safety standards. The industry's current problems could thus be used to effectively nullify the National Traffic and Motor Vehicle Safety Act.

Professor William Nordhaus of Yale University submitted comments concerning the economic ramifications of the proposed delay in the first phase of the automatic restraint requirements. (The submission was sponsored by Allstate, Kemper, Nationwide, and State Farm Insurance Companies.) (For a full discussion of the methodology and bases for these calculations, one should refer to the Nordhaus submission filed at the National Highway Traffic Safety Administration under Docket 74-14, Notice 20. NHTSA's response to this analysis is set forth in detail in the Appendix to the Regulatory Impact Analysis.) The basic conclusions contained in the Nordhaus comment are as follows (verbatim):

1. The current passive restraint requirement (FMVSS 208) has very substantial net benefits

compared to current lap and shoulder belt usage. According to the economic analysis presented here, the current rule has net benefits of approximately $10 billion for model years 1982-85. The substantial economic gain from passive restraints should not be ignored in debates on fine-tuning the phase-in.

2. Using standard analysis, the ranking of options in terms of net benefits is as follows (with the first having the highest net benefits and the last the lowest net benefits):

(1) Simultaneous 1983 implementation (all cars equipped with passive restraints in 1983).

(2) Delay and reversal (small cars in 1983, intermediate cars in 1984, large cars in 1985).

(3) The current rule (large cars in 1982, intermediate cars in 1983, and small cars in 1984).

(4) The proposed delay (large and intermediate cars in 1983, small cars in 1984).

(5) General rollback (large cars in 1983, intermediate cars in 1984, large cars in 1985).

3. A sensitivity analysis shows the ranking of alternatives is unchanged under a wide range of alternative assumptions.

4. Any deferral of requirements to install passive restraints on any size automobile has net costs unless it is "traded in" on an acceleration of requirements on a larger number, or a smaller sized, set of automobiles.

5. In terms of the costs and benefits of different options, there is no justification for either the proposed delay or for a general rollback. In particular, the economic costs of the proposed delay are approximately 5 times greater than the benefits, for a net cost of over $200 million. The net costs of the general rollback are significantly greater, in the order of $4.5 billion.

6. There appears to be strong economic justification for the simultaneous 1983 option if it is technically feasible.

7. The analysis indicates that the delay and reversal option has the highest net benefits of any of the four considered in the proposal and regulatory analysis. The superior net benefit of delay and reversal arises because the reversal of the requirement to small cars first affects a larger number of automobiles more quickly and because the net economic benefits per vehicle are greater for small cars than for large and intermediate cars.

8. The estimated impact of the proposed delay on the automobile industry is minuscule. There will be little or no improvement in the "health" of the domestic automobile industry from the proposed delay. For this reason, nonregulatory considerations discussed in the notice (the effect on imports, the conditions of the automobile industry, or freedom-of-choice arguments) should not, from an economic point of view, enter in this rulemaking.

The proposed delay of the automatic restraint requirements was also opposed by various consumer groups and health-related organizations, including: the Consumer Federation of America, the National Spinal Cord Injury Foundation, the Epilepsy Foundation of America, the Consumers Union, the Automotive Occupant Protection Association, the National Safety Council, the Houston Independent School District, the American College of Surgeons, the Georgia Department of Human Resources, the New York Department of Transportation, and the Center for Auto Safety. The National Safety Council conceded that the economic situation of the auto industry is serious, but stated that any adjustment of the implementation schedule for automatic restraints should also include consideration of an earlier implementation for small cars, since the need for protection is much greater in these vehicles.

The Automotive Occupant Protection Association stated that it could support the proposed delay of the automatic restraint requirements for one year, as well as a reversal of the implementation schedule, provided there is a requirement for the major automobile manufacturer to offer optional air bag systems on at least one model line. The association is concerned that further delay of the automatic restraint standard could drive the remaining air bag supplier manufacturers out of the business, and the life-saving potential of air bags could be lost. The Epilepsy Foundation of America echoed this sentiment and stated that "consumers deserve a guarantee that would assure the air bag option will be available in any model they wish to purchase."

The Consumers Union argued that the auto industry's financial condition should not be used to justify "less safe automobiles." Moreover, according to the Union, the proposed delay is unlikely to significantly alleviate the financial problems facing domestic automobile manufacturers.

The Center for Auto Safety argued that the proposed delay of the first-phase automatic restraint requirements will not help the auto industry solve its current economic problems. In addition, the Center stated that the projected savings of 600 lives and 4,300 injuries associated with the first-phase requirements represents an economic gain of approximately $170 million, and this far outweighs any savings to the industry. In regard to a possible reversal of the existing implementation schedule for automatic restraints, the Center stated that automatic belts can be installed on all small cars with a leadtime as short as one year because automatic belts are so well developed.

Comments were also received from two manufacturers which supply air bag system components, Thiokol and Rocket Research Company. Rocket Research stated that it could support the proposed delay and reversal of the implementation schedule provided any such change also contains a requirement that the major manufacturers "tool for and offer for sale" air bag systems on at least one car line. The company stated that without such a guarantee there is little incentive for air bag suppliers to remain in the business. Rocket Research stated that an indefinite delay of the automatic restraint requirements over the next five years would amount to a business loss of 23 percent. The company also stated that cost savings accruing to General Motors and Ford because of the one year delay (estimated in the proposal to be approximately 37 million dollars) would be reduced if air bag programs are delayed or eventually canceled since both Rocket Research and Hamill Manufacturing Company have substantial claims against the two companies for capital expenditures to build and equip production plants to make air bag modules. (Rocket Research stated that these claims are based on letters of agreement and contingent liability statements.)

Thiokol stated that the model year 1982 automatic restraint requirements for large cars resulted in the first major production program for Thiokol, and that substantial funds have been expended for manpower, tooling and facilities to meet this requirement. According to Thiokol, a one-year delay in the program would add substantial additional expenses and result in a reduction of manpower, facility use and vendor capability. In response to questions contained in the notice of proposed rulemaking, Thiokol stated that another year of delay would discourage rather than encourage further design improvements and research efforts in automatic restraint systems.

B. Comments Favoring the Delay

The Pacific Legal Foundation supported the proposed one-year delay, and stated four primary reasons why such a delay is warranted.

1. The proposed delay would create additional time for the Department of Transportation to implement an adequate evaluation program for air bags.
2. The proposed delay would give the American public an additional year of freedom to choose their means of occupant protection.
3. The proposed delay would allow additional time for the public to familiarize itself with passive restraints [which have been or will be voluntarily installed prior to a mandatory effective date].
4. The proposed delay would reduce the likelihood of costly Congressional action on the passive restraint standard after its implementation.

The proposed delay of the first phase of the automatic restraint requirements was unanimously supported by all commenting automobile manufacturers, both domestic and foreign. Additionally, most manufacturers urged the Department to reconsider the entire standard and to provide additional leadtime for all phases of the implementation schedule or to revoke the automatic restraint requirements altogether. Regarding a possible reversal of the current implementation schedule, nearly all of the foreign automobile manufacturers stated that it would be impossible to install automatic restraints on 1983 model small passenger cars because of insufficient leadtime.

Chrysler Corporation also urged that the automatic restraint requirements be withdrawn entirely. The company argued that automatic belts will be disconnected by many motorists and that purchasers will turn to models that are not equipped with automatic belts. Chrysler predicts that automatic belts would create a consumer "backlash" greater than that resulting from ignition interlocks (devices installed on 1974-75 models which made it impossible to start the vehicle unless the seatbelt was fastened).

In lieu of automatic restraints, Chrysler urged the Department to mount a national educational effort to increase the use of current manual seat belt systems: "Increased usage of these systems is the most cost effective and immediate method of reducing injuries and fatalities in motor vehicle accidents." Regarding a possible reversal of the current implementation schedule, Chrysler stated that it would be impossible at this time to advance automatic belt installation for small cars prior to the 1984 model year.

American Motors Corporation recommended that a delay in effective date of the automatic restraint requirements be adopted for all cars to permit a re-evaluation of all issues. The company particularly does not support a reversal of the implementation schedule so that small cars would be phased in first, since the company will rely on technology developed for or by other automobile manufacturers after it is proven in actual volume production. American Motors also recommended that if a new phase-in schedule is adopted, at least a one-year delay for low-volume manufacturers (e.g., less than 200,000 sales) be included in the change.

Foreign vehicle manufacturers produce few, if any, large passenger cars (i.e., cars with wheelbases over 114 inches), but all the foreign manufacturers supported the proposed delay of the first phase of the automatic restraint requirements. However, these manufacturers were unanimously against any reversal of the existing implementation schedule that would require small passenger cars to be equipped with automatic restraints a year earlier than currently required.

Fiat Motors of North America recommended that the entire automatic restraint schedule be delayed for one year (i.e., each phase delayed one year). The company stated that if its small cars were not required to comply until model year 1985, it would give the company more time to develop appropriate automatic belt designs for its convertibles. Fiat stated that it is currently having difficulty with its convertibles in terms of finding adequate automatic belt attachments and fittings for existing vehicle structures. Fiat stated that it would prefer to see the automatic restraint standard revoked and mandatory seat belt use laws implemented.

Nissan Motor Company stated that it would not be possible to equip its small cars with automatic restraints by September 1, 1982. Nissan's objection does not relate to capital expenditure or retail price increase, but rather, to "the lack of proper leadtime needed to develop acceptable, reliable and high quality vehicles for the consumer." Nissan argued that automatic belts already face a tough challenge in winning consumer acceptance without forcing the imposition of hastily developed designs.

Toyota Motor Company also stated that it could not comply with a change in the effective date for small cars from September 1, 1983, to September 1, 1982. Toyota stated that if such a change is adopted, it would have to drop from production certain of its volume passenger car lines for the 1983 model year, thereby limiting the freedom of choice of the customers who wish to purchase Toyota cars.

Volvo of America Corporation requested that the implementation schedule for automatic restraints be amended to reflect the fact that the current market situation has forced the industry to be flexible with respect to model year introductions. Volvo refers specifically to the desire of some manufacturers to continue model lines past the September 1 effective dates for the three phases of the current implementation schedule, and to discontinue these lines at the beginning of the new calendar year. Volvo argues that tooling for installation of automatic restraints on model lines that will be discontinued six months after the effective date of the standard is cost prohibitive. Consequently, without a change in the implementation schedule, manufacturers would be required to cease production of certain models sooner than they would like.

Volvo recommends that the implementation schedule be amended to provide that the effective dates for the three phases is "September 1 or the date of production start of the new model year if this date falls between September 1 and December 31."

Rolls-Royce Motors produces three models that would have to be equipped with automatic restraints by September 1, 1981, under the existing schedule. Rolls-Royce originally planned to offer air bags in these models but changed plans after General Motors announced in 1979 that it would delay the introduction of air bags. Consequently, Rolls-Royce states that it got a late start with automatic belts and the automatic belt system it has planned for the 1982 models is not developed to a degree of refinement normally associated with Rolls-Royce cars. In support of the proposed one-year delay in the automatic restraint requirements, Rolls-Royce made the following comment:

Refinement, weight and cost will all be subject to continuous development anyway but one year extra leadtime would permit full development of the system before the customer is charged a cost premium for the restraint system.

(NOTE: Allstate Insurance Company requested that a public hearing be held on the one-year delay in the large car requirement. However, due to the limited time available before the previous effective date of this requirement, the agency must deny this request. The issues on which this decision is based are primarily technical and economic, lending themselves well to written presentations. Interested parties have taken full advantage of the opportunity to provide their views in writing in this proceeding. Further, an additional opportunity for comment on issues relating to the automatic restraint standard is provided in the notice of proposed rulemaking issued today.)

PREAMBLE TO AN AMENDMENT TO FEDERAL MOTOR VEHICLE SAFETY STANDARD NO. 208

Occupant Crash Protection

(Docket No. 74-14; Notice 25)

ACTION: Final rule.

SUMMARY: The purpose of this notice is to amend Federal Motor Vehicle Safety Standard No. 208, Occupant Crash Protection, to rescind the requirements for installation of automatic restraints in the front seating positions of passenger cars. Those requirements were scheduled to become effective for large and mid-size cars on September 1, 1982, and for small cars on September 1, 1983.

The automatic restraint requirements are being rescinded because of uncertainty about the public acceptability and probable usage rate of the type of automatic restraint which the car manufacturers planned to make available to most new car buyers. This uncertainty and the relatively substantial cost of automatic restraints preclude the agency from determining that the standard is at this time reasonable and practicable. The reasonableness of the automatic restraint requirements is further called into question by the fact that all new car buyers would be required to pay for automatic belt systems that may induce only a few additional people to take advantage of the benefits of occupant restraints.

The agency is also seriously concerned about the possibility that adverse public reaction to the cost and presence of automatic restraints could have a significant adverse effect on present and future public acceptance of highway safety efforts.

Under the amended standard, car manufacturers will continue to have the current option of providing either automatic or manual occupant restraints.

DATES: The rescission of the automatic restraint requirements of Standard No. 208 is effective December 8, 1981. Any petitions for reconsideration must be received by the agency not later than December 3, 1981.

ADDRESS: Any petitions for reconsideration should refer to the docket number and notice number of this notice and be submitted to: Administrator, National Highway Traffic Safety Administration, 400 Seventh Street, S.W., Washington, D.C. 20590.

FOR FURTHER INFORMATION CONTACT: Mr. Michael Finkelstein, Associate Administrator for Rulemaking, National Highway Traffic Safety Administration, 400 Seventh Street, S.W., Washington, D.C. 20590. (202-426-1810)

SUPPLEMENTARY INFORMATION: On April 9, 1981, the Department of Transportation published a notice of proposed rulemaking (NPRM) setting forth alternative amendments to the automatic restraint requirements of Standard No. 208 (46 F.R. 21205). The purpose of proposing the alternatives was to ensure that Standard No. 208 reflects the changes in circumstances since the automatic restraint requirements were issued (42 F.R. 34289; July 5, 1977) and to ensure that the standard meets the requirements of the National Traffic and Motor Vehicle Safety Act of 1966 and Executive Order 12291, "Federal Regulations" (February 17, 1981).

Background and NPRM

The automatic restraint requirements were adopted in 1977 in response to the high number of passenger car occupants killed annually in crashes and to the persistent low usage rate of manual belts. The manual belt is the type of belt which is found in most cars today and which the occupant must place around himself or herself and buckle in order to gain its protection. Then, as now, there were two types of automatic restraints, i.e., restraints that require no action by vehicle occupants, such as buckling a belt, in order to be effective. One type is the air cushion restraint (air bag) and the other is the automatic belt (a belt which automatically envelopes an occupant when the occupant enters a vehicle and closes the door).

In view of the greater experience with air bags in large cars and to spread out capital investments, the Department established a large-to-small car compliance schedule. Under that schedule, large cars were required to begin compliance on September 1, 1981, mid-size cars on September 1, 1982, and small cars on September 1, 1983.

On April 6, 1981, after providing notice and opportunity for comment, the Department delayed the compliance date for large cars from September 1, 1981, to September 1, 1982. As explained in the April 6, final rule, that delay was adopted

. . . because of the effects of implementation in model year 1982 on large car manufacturers, because of the added significance which those effects assume due to the change in economic circumstances since the schedule was adopted in 1977, and because of the undermining by subsequent events of the rationale underlying the original phase-in schedule.

Simultaneous with publishing the one-year delay in the effective date for large cars, the Department also issued a proposal for making further changes in the automatic restraint requirements. This action was taken in response to a variety of factors that raised questions whether the automatic restraint requirements represented the most reasonable and effective approach to the problem of the low usage of safety belts. Among these factors were the uncertainty about public acceptability of automatic restraints in view of the absence of any significant choice between automatic belts and air bags and the nature of the automatic belt designs planned by the car manufacturers, the consequent uncertainties about the rate of usage of automatic restraints, and the substantial costs of air bags even if produced in large volumes.

The three principal proposals were reversal of phase-in sequence, simultaneous compliance, and rescission. The reversal proposal would have changed the large-to-small car order of compliance to a requirement that small cars commence compliance on September 1, 1982, mid-size cars on September 1, 1983, and large cars on September 1, 1984. The proposal for simultaneous compliance would have required all size classes to begin compliance on the same date, March 1, 1983. The rescission proposal would have retained the manufacturers' current option of equipping their cars with either manual or automatic restraints.

In addition, the Department proposed that, under both the first and second alternatives, the automatic restraint requirements be amended so that such restraints would not be required in the front center seating position.

Following the close of the period for written comments on the April NPRM, NHTSA decided, at its discretion, to hold a public meeting on the alternatives. The purpose of the meeting was to permit interested parties to present their views and arguments orally before the Administrator and ensure that all available data were submitted to the agency. The notice announcing the meeting indicated that participants at the hearing would be permitted to supplement their previous comments. The notice also urged participants to consider the issues raised in former Secretary Coleman's June 14, 1976 proposal regarding occupant restraints and in former Secretary Adams' March 24, 1977 proposal regarding automatic restraints.

Rationale for Agency Decision

The decision to rescind the automatic restraint requirements was difficult for the agency to make. NHTSA has long pursued the goal of achieving substantial increases in the usage of safety belts and other types of occupant restraints. Former Secretary Adams clearly believed that he had ensured the achievement of that goal in July 1977 when he promulgated the automatic restraint requirements. Now that goal appears as elusive as ever. Instead of being equipped with automatic restraints that will protect substantially greater numbers of persons than current manual belts, most new cars would have had a type of automatic belt that might not have been any more acceptable to the public than manual belts. The usage of those automatic belts might, therefore, have been only slightly higher than that of manual belts. While most of the anticipated benefits have virtually disappeared, the costs have not. Vehicle price increases would have amounted to approximately $1 billion per year.

This turn of events may in part reflect the failure of the Department in the years following 1977 to conduct a long term effort to educate the public about the various types of restraints and the need to use them. The need for such an undertaking was seen by former Secretary Coleman in announcing his decision in 1976 to conduct an automatic restraint demonstration project prior to deciding whether to mandate automatic restraints. His instruction that NHTSA undertake significant new steps to promote safety belt usage was never effectively carried out. The result of such an effort could have been that a substantial portion of the public would have been receptive to a variety of automatic restraint designs. As a result of concern over public acceptance, manufacturers have designed their automatic restraints to avoid creating a significant adverse reaction. Unfortunately, the elements of design intended to minimize adverse reaction would also minimize the previously anticipated increases in belt usage and safety benefits of requiring new cars to have automatic restraints instead of manual belts.

The uncertainty regarding the usage of the predominant type of planned automatic restraint has profound implications for the determinations which NHTSA must make regarding a standard under the National Traffic and Motor Vehicle Safety Act. NHTSA has a duty under the Vehicle Safety Act and E.O. 12291 to review the automatic restraint requirements in light of changing events and to ensure that the requirements continue to meet the criteria which each Federal Motor Vehicle Safety Standard must satisfy. If the criteria cannot be satisfied, the agency must make whatever changes in the standard are warranted. The agency must also have the flexibility to modify its standards and programs in its efforts to find effective methods for accomplishing its safety mission.

The agency believes that the post-1977 events have rendered it incapable of finding now, as it was able to do in 1977, that the automatic restraint requirements would meet all of the applicable criteria in the Vehicle Safety Act. Section 103(a) of the Vehicle Safety Act requires that each Federal Motor Vehicle Safety Standard meet the need for safety and be practicable and objective. Each standard must also be reasonable, practicable and appropriate for each type of vehicle or equipment to which it applies (Section 103(f) (3)). To meet the need for safety, a standard must be reasonably likely to reduce deaths and injuries. To be found practicable, the agency must conclude that the public will in fact avail themselves of the safety devices installed pursuant to the standard. (*Pacific Legal Foundation* v. *Department of Transportation*, 593 F. 2d 1338, at 1345-6 (D.C. Cir. 1979)). To be reasonable and practicable, a standard must be economically and technologically feasible, and

the costs of implementation must be reasonable. (S. Rep. No. 1301, 89th Cong., 2d Sess. 6 (1966).)

In reaching the decision announced by this notice, NHTSA has reviewed the enormous record compiled by this agency over the past decade on automatic restraints. Particular attention was paid to the information and issues relating to the notices which the Agency or Department has issued regarding automatic restraints since 1976. All comments submitted in response to the April 1981 proposal by proponents and opponents of the automatic restraint requirements have been thoroughly considered. A summary of the major comments is included as an appendix to this notice. The agency's analysis of those comments may be found in this notice and the final regulatory impact analysis. A copy of the analysis has been placed in the public docket.

Usage of automatic restraints and safety benefits. As in the case of the comments submitted concerning the one-year delay in automatic restraint requirements for large cars, the commenters on the April 1981 proposal expressed sharply divergent views and arguments and reached widely differing conclusions concerning the likely usage rates and benefits of the automatic restraints planned for installation in response to the automatic restraint requirements. The wide distance between the positions of the proponents and opponents of these requirements stems primarily from the lack of any directly relevant data on the most important issue, i.e., the public reaction to and usage rate of detachable automatic belts. These disagreements once again demonstrate the difficulty in reaching reliable conclusions due to the uncertainty created by the lack of adequate data.

In issuing the automatic restraint requirements in 1977, NHTSA assumed that the implementation of those requirements would produce substantial benefits. According to the analysis which NHTSA performed in that year, automatic restraints were expected to prevent 9,000 deaths and 65,000 serious injuries once all cars on the road were equipped with those devices. That prediction was premised on several critical assumptions. Most important among the assumptions were those concerning the safety benefits of automatic restraints—reductions in death and injury—which in turn are a function of the types of automatic restraints to be placed in each year's production of new cars.

The agency assumed that the combination of air bags and lap belts would be approximately 66 percent effective in preventing fatalities and that automatic belts would have a 50 percent level of effectiveness. The agency assumed also that air bags would be placed in more than 60 percent of new cars and that automatic belts would be placed in the remaining approximately 40 percent. The agency's analysis predicted that air bags would provide protection in virtually all crashes of sufficient severity to cause deployment of the air bags. It was further assumed that the automatic belts would be used by 60 to 70 percent of the occupants of those cars.

As to public reaction, the agency anticipated that the public would, as a whole, accept automatic restraints because it could choose between the two types of those restraints. Those not wanting automatic belts would select an air bag. Partly as a function of the expected large volume of air bag installation, the agency projected that the cost of air bags would be only slightly more than $100 (in 1977 dollars) more than manual belts.

As part of its efforts to monitor and facilitate implementation of the automatic restraint requirements, the agency continued its gathering of data about the use and effectiveness of air bags and of automatic belts with use-inducing features, the only type of automatic belt available to the public. With respect to automatic belts, this effort was carried out through a contract with Opinion Research Corporation. Under that contract, observations were made of seat belt usage during the two year period beginning November 1977. These observations provided data on usage of manual and automatic belts in model year 1975-79 VW Rabbits and of manual belts in model year 1978-79 GM Chevettes. As a result of voluntary decisions by VW and GM, a number of the Rabbits and Chevettes were equipped with automatic belts. The observation data showed usage rates of about 36 percent for manual belts and about 81 percent for automatic belts in the Rabbits. The observed rate of manual belt usage in Chevettes was 11 percent. There were insufficient numbers of model year 1978-79 Chevettes equipped with automatic belts to develop reliable usage figures.

Several telephone surveys were also made under contract with Opinion Research. The first survey involved owners of model year 1979 VW Rabbits and GM Chevettes equipped with automatic belts and was conducted during 1979. This survey showed that 89 percent of Rabbit owners and 72 percent of Chevette owners said that they used their automatic belts. A second survey was conducted in late 1979 and early 1980. It covered owners of model year 1980 Rabbits and Chevettes. The usage rates found by the second survey were almost identical to those in the first survey.

Now, however, the validity of the benefit predictions in 1977 and the relevancy of the extensive data gathered by NHTSA on air bags and on automatic belts with use-inducing features have been substantially if not wholly undermined by drastic changes in the types of automatic restraints that would have been installed under the automatic restraint requirements. Instead of installing air bags in approximately 60 percent of new cars, the manufacturers apparently planned to install them in less than 1 percent of new cars. Thus, automatic belts would have been the predominant means of compliance, and installed in approximately 99 percent of new cars. Thus, the assumed life-saving potential of air bags would not have been realized.

Manufacturers have stated that they chose belt systems for compliance because of the competitive disadvantage of offering the relatively expensive, inadequately understood air bag when other manufacturers would have been providing automatic belts. These explanations seem credible.

The other drastic change concerns the type of automatic belt to be installed. Although some aspects of the car manufacturers' automatic belt plans are still tentative, it now appears reasonably certain that if the automatic restraint requirements were implemented, the overwhelming majority of new cars would be equipped with automatic belts that are detachable, unlike the automatic belts in Rabbits and Chevettes. Most planned automatic belts would be like today's manual lap and shoulder belts in that they can be easily detached and left that way permanently.

Again, this design choice would appear to have arisen out of concern that without such features emergency exit could be inhibited, and, in part as a result of a perception of this fact, public refusal to accept new designs would be widespread. The agency shares this concern, and has since 1977 required that all such belts provide for emergency exit. Agency concerns on this point have been validated by recent related attitudinal research, discussed below.

In its final rule delaying the initial effective date of the automatic restraint requirements, the April 1981 proposal and the associated documents analyzing the impacts of those actions, NHTSA expressly confronted the lack of usage data directly relevant to the type of automatic belts now planned to be installed in most new cars. The agency stated that there were several reasons why the available data was of limited utility in attempting to make any reliable predictions about the usage of easily detachable automatic belts. The most important reason, which has already been noted, is that the predominant type of planned automatic belt would not have had features to ensure that these belts are not detached.

Second, all of the available data relate to only two subcompacts, the Rabbit and the Chevette. Due to a combination of owner demographics and a correlation between driver perception of risk and the size of the car being driven, belt usage rates are typically higher in small cars than in larger ones. Therefore, the usage rates for the two subcompacts cannot simply be adopted as the usage rates for automatic belts in all car size classes.

Third, most of the Rabbit and Chevette owners knew that their new car would come with an automatic belt and had it demonstrated for them, even if many state that they did not consciously choose that type of belt. Having voluntarily invested in automatic restraints, they are more likely to use those restraints than someone who is compelled to buy them.

The significance of the fundamental difference between the nondetachable and detachable automatic belt bears further discussion. The Rabbit automatic belts are, as a practical matter, not permanently detachable since they are equipped with an ignition interlock. If the belt is disconnected, the interlock prevents the starting of the car. Each successive use would therefore require reconnection before engine start. The Chevette automatic belts also were initially equipped with an ignition interlock. Beginning in model year 1980, the Chevette belts were made both practically and literally nondetachable. They consist of a continuous, nondetachable shoulder belt. Additional webbing can be played out to produce slack in the belt; however, the belt remains attached at both ends.

By contrast, the automatic belts now planned for most cars do not have any effect on the starting of the cars and are easily detachable. Some belt designs may be detached and permanently stowed as readily as the current manual lap and shoulder belts. Once a detachable automatic belt is detached, it becomes identical to a manual belt. Contrary to assertions of some supporters of the standard, its use thereafter requires the same type of affirmative action that is the stumbling block to obtaining high usage levels of manual belts. If the car owners perceive the belts as simply a different configuration of the current manual belts, this stumbling block is likely to remain. They may treat the belt as a manual one and thus never develop the habit of simply leaving the belt attached so that it can act as an automatic belt.

The agency recognizes the possibility that the exposure of some new car purchasers to attached automatic belts may convert some previously occasional users of manual belts to full time belt users. Present attitudinal survey data clearly establish the existence of a population of such occupants who could be influenced by some external factor to convert to relatively constant users. However, the agency believes that many purchasers of new cars having detachable automatic belts would not experience the potential use-inducing character of attached automatic belts unless they had taken the initiative themselves to attach the belts.

Thus, the change in car manufacturers' plans has left the agency without any factual basis for reliably predicting the likely usage increases due to detachable automatic belts, or for even predicting the likelihood of any increase at all. The only tentative conclusion that can be drawn from available data is that the installation of *nondetachable* automatic belts in other subcompacts could result in usage rates near those found in Rabbits and Chevettes. Even that use of the Rabbit and Chevette data may be questionable, however, given the element of voluntarism in the purchase of automatic belts by many of the Rabbit and Chevette owners. Thus, the data on automatic belt use in Rabbits and Chevettes may do little more than confirm the lesson of the model year 1974-75 cars equipped with manual belts and ignition interlocks, i.e., that the addition to a belt system of a feature that makes the belt nondetachable or necessitates its attachment before a car can be started can substantially increase the rate of belt usage.

In estimating automatic belt usage rates for the purposes of the April final rule and proposal, the agency recognized the substantial uncertainty regarding the effects of easily detachable automatic belts on belt usage. NHTSA attempted to compensate for the lack of directly relevant data by using two different techniques to predict a potential range of usage.

One technique was to assume a consistent multiplier effect, whereby belt usage in cars of all size classes would be assumed to be more than slightly double as it had in Rabbits. A doubling of the current 10–11 percent manual belt usage rate projected over the general car fleet would mean a 22 percent rate could be achieved with the installation of automatic belts. The other technique was to assume that there would be a consistent additive effect, whereby the same absolute percentage point increase in belt usage would occur as there had been in the case with Rabbits. Use of this method would result in a predicted 50 percentage point increase in belt usage, over the entire fleet, from the current 10–11 percent to approximately 60 percent.

The agency used the results of these two techniques in an attempt to construct a range of possible increases in belt usage. Thus, a range of 15 to 60 percent was used in both the final regulatory impact analysis for the April rulemaking to defer the effective date for one year and the preliminary analysis for the current action. The figure of 15 percent was derived by doubling the observed 7 percent usage levels in the large type cars affected by the deferral. A figure of 22 percent would have been more appropriate as the low end of the range for the current action, since it would represent a doubling of the current usage rate of the car fleet as a whole. This latter figure has been used in addressing this question in the current final regulatory analysis.

Although the agency had no definitive way of resolving the uncertainty about the usage of detachable automatic belts, the agency estimated that belt usage with automatic belts would most likely fall near the lower end of either range. This estimate was based on a variety of factors. Most relate to the previously discussed limitations in the relevancy of the observations and surveys of Rabbit and Chevette owners. In addition, those data were on their face inconsistent with data regarding automatic belt usage in crashes involving Rab-

bits. Those crash data indicated a usage rate of 55–57 percent instead of the better than 80 percent rate indicated by the observation study and telephone surveys.

Thus, the agency made the preliminary judgment in its impact analyses that the switch from manual belts to detachable automatic belts could approximately double belt usage. However, the April 1981 final rule noted that the actual belt usage might be lower, even substantially so. With respect to cars with current low usage rates, that notice stated that the usage rate of detachable automatic belts might only approach levels similar to those currently achieved with manual belts.

The commenters on the April 1981 NPRM did not present any new factual data that could have reduced the substantial uncertainty confronting the agency. Instead, the commenters relied on the same data examined by the agency in its impact analyses.

The commenters were sharply divided on the question of usage rates. Proponents of the automatic restraint requirements did not in their analyses address the significance of the use-inducing nature of the nondetachable automatic belts in the Rabbits and Chevettes or the demographic factors relating to those car purchasers. Instead, they asserted that the usage rates achieved in Rabbits and Chevettes would, with slight adjustments, also be achieved in other car size classes. In reaching this conclusion, they asserted that the usage rate increases of automatic belts shown by Rabbit and Chevette owners were the same regardless of whether the automatic belts were purchased knowingly or unknowingly. There was an exception to this pattern of comment among the proponents. One public spokesperson for an interest group acknowledged that automatic belts could be designed in a way that they so closely resembled manual belts that their usage rates would be the same.

Opponents of the automatic restraint requirements, relying on the similarity of detachable automatic belts to manual belts, predicted that the automatic belts would not have any substantial effect on belt usage. The opponents of the requirements also dismissed the experience of the Rabbit and Chevette owners on the grounds that the automatic belts in those cars had been voluntarily purchased and were nondetachable.

While the public comments did not provide the agency with any different or more certain basis for estimating belt usage than it already had, they did induce the agency to reexamine its assumption about the possible automatic belt usage rates. Although it is nearly impossible to sort out with precision the individual contributions made by nondetachability, interlocks, car size, demographics and other factors, NHTSA believes that the usage of automatic belts in Rabbits and Chevettes would have been substantially lower if the automatic belts in those cars were not equipped with a use-inducing device inhibiting detachment.

In the agency's judgment, there is a reasonable basis for believing that most of the increase in automatic belt Rabbits and Chevettes is due to the nondetachability feature, whether an interlock or other design feature, of their belt systems. Necessitating the attachment of belts by the addition of interlocks to 1974–75 cars resulted in an increase in manual belt usage by as much as 40 percent in cars subject to that requirement. A similar effect in the case of the Rabbit would account for four-fifths of the increase observed in the automatic belt vehicles. A significant portion of the remaining increase could in fact be attributable to the fact many owners of automatic belt Rabbits and Chevettes knowingly and voluntarily bought the automatic belts. By the principle of self-selection, these people would be more inclined to use their belts than the purchasers of 1974–75 Rabbits who did not have any choice regarding the purchase of a manual belt equipped with an interlock. This factor would not, of course, be present in the fleet subject to the standard.

The most appropriate way of accounting for the detachability problem and other limitations on the validity of that Rabbit and Chevette data would be to recognize that the levels of usage resulting from both the point estimates are based on uncertain conclusion and adjust each appropriately. The agency's estimate in the final regulatory impact analysis for the April 1981 final rule that usage would likely fall near the lower end of the range had the effect of substantially adjusting downward the usage rate (60 percent) produced by the technique relying on the absolute percentage point increase (50 percentage points) in belt usage in automatic belt Rabbits and Chevettes. A similar adjustment could also be made in the usage rate (15 percent) indicated by the multiplier technique.

Throughout these sequential analyses, the agency has examined the extremely sparse factual data, applied those factors which are known to externally affect usage rates, and defined for analytical purposes the magnitude of potential safety effects. Aside from the initial data points, all such analyses in all cases necessarily involve exercises of discretion and informed judgment. Resultant conclusions are indications of probable usage which always have been and always must be relied upon by the agency in the absence of additional objective data.

The agency believes that the results produced by both techniques must be adjusted to account for the effects of detachability and the other factors affecting usage rates. Therefore, as the April 1981 final rule recognized, the incremental usage attributable to the automatic aspect of the subject belts may be substantially less than 11 percent.

The agency's analysis of the public comments and other available information leads it to conclude that it cannot reliably predict even a 5 percentage point increase as the minimum level of expected usage increase. The adoption of a few percentage points increase as the minimum would, in the agency's judgment, be more consistent with the substantial uncertainty about the usage rate of detachable automatic belts. Based on the data available to it, NHTSA is unable to assess the probability that the actual incremental usage would fall nearer a 0 percentage point increase or nearer some higher value like a 5 or 10 percentage point increase.

Thus, the agency concludes that the data on automatic belt usage in Rabbits and Chevettes does not provide a sufficient basis for reliably extrapolating the likely range of usage of detachable automatic belts by the general motoring public in all car size classes. Those data are not even sufficient for demonstrating the likelihood that those belts would be used in perceptibly greater numbers than the current manual belts. If the percentage increase is zero or extremely small due to the substantial similarity of the design and methods of using detachable automatic belts and manual belts, then the data regarding manual belt usage would be as reliable a guide to the effects of detachable automatic belts on belt usage as data regarding usage of nondetachable automatic belts. Indeed, the manual belt data may even be a more reliable guide since the data are based on usage by the general motoring public in cars from all size and demographic classes.

In view of the uncertainty about the incremental safety benefits of detachable automatic belts, it is difficult for the agency to determine that the automatic restraint requirements in their present form meet the need for safety.

In concluding that for this reason detachable automatic belts may contribute little to achieving higher belt usage rates, the question then arises whether the agency should amend the standard to require that automatic belts have a use-inducing feature like that of the Rabbit and Chevette automatic belts. NHTSA believes that such features would increase belt usage. The agency does not, however, believe that such devices should be mandated, for the reasons discussed in detail below.

Costs of automatic restraints. In view of the possibly minimal safety benefits and substantial costs of implementing the automatic restraint requirements, the agency is unable to conclude that the incremental costs of the requirements are reasonable. The requirements are, in that respect, impracticable. While the car manufacturers have already made some of the capital expenditures necessary to comply with the automatic restraint requirements, they still face substantial, recurring variable costs. The average price increase per car is estimated to be $89. The costs of air bags and some designs of automatic belts would be substantially higher. With a total annual production of more than 10 million cars for sale in this country, there would be a price effect of approximately $1 billion.

While the car manufacturers might be able to pass along some or all of their costs to consumers, the necessary price increases would reduce sales. There might not be any net revenue loss since the extra revenue from the higher prices could offset the revenue loss from the lower volume of sales. However, those sale losses would cause net employment losses. Additional sales losses might occur due to consumer uncertainty about or antipathy toward the detachable automatic belts which do not stow so unobtrusively as current manual lap and shoulder belts.

Consumers would probably not be able to recoup their loss of disposable income due to the higher car prices. There does not appear to be any certainty that owners of cars with detachable automatic belts would receive offsetting discounts in insurance costs. Testimony and written comments submitted to the agency indicate premium reductions generally are available only to owners of cars equipped with air bags, not automatic belts. Some large insurance companies do not now offer discounts to any automatic restraint-equipped cars, even those with air bags. If insurance cost discounts were to be given owners of cars having detachable automatic belts, such discounts would be given only after the automatic belts had produced significant increases in belt usage, and in turn significant decreases in deaths and serious injuries. The apparent improbability of any economic effect approaching the magnitude of the consumer cost means that the discounts would not likely materialize on a general basis.

Insurance company statements at the August 1981 public meeting reaffirmed this belief as they state that they could not now assure reductions in insurance premiums but would have to first collect a considerable amount of claim data.

Finally, the weight added to cars by the installation of automatic belts would cause either increased fuel costs for consumers or further new car price increases to cover the incorporation of offsetting fuel economy improvements.

The agency does not believe that it would be reasonable to require car manufacturers or consumers to bear such substantial costs without more adequate assurance that they will produce benefits. Given the plans of the car manufacturers to rely primarily on detachable automatic belts and the absence of relevant data to resolve the usage question, implementation of the automatic restraint requirements amounts to an expensive federal regulatory risk. The result if the detachable automatic belts fail to achieve significant increases in belt usage could be a substantial waste of resources.

The agency believes that the costs are particularly unreasonable in view of the likelihood that other alternatives available to the agency, the states and the private sector could accomplish the goal of the automatic restraint requirements at greatly reduced cost. Like those requirements, the agency's planned educational campaign is addressed primarily to the substantial portion of the motoring public who are currently occasional users of manual belts.

Effect on public attitude toward safety. Although the issue of public acceptance of automatic restraints has already been discussed as it relates to the usage rate of detachable automatic restraints, there remains the question of the effect of automatic restraints on the public attitude toward safety regulation in general. Whether or not there would be more than minimal safety benefits, implementation of the automatic restraint requirements might cause significant long run harm to the safety program.

No regulatory policy is of lasting value if it ultimately proves unacceptable to the public. Public acceptability is at issue in any vehicle safety rulemaking proceeding in which the required safety equipment would be obtrusive, relatively expensive and beneficial only to the extent that significant portions of the motoring public will cooperate and use it. Automatic belt requirements exhibit all of those characteristics. The agency has given the need for public acceptability of automatic restraints substantial weight since it will clearly determine not only the level of safety benefits but also the general public attitude toward related safety initiatives by the government or the private sector.

As noted above, detachable automatic belts may not be any more acceptable to the public than manual belts at any given point in time. If the detachable automatic belts do not produce more than negligible safety benefits, then regardless of the benefits attributable to the small number of other types of automatic restraints planned to be installed, the public may resent being required to pay substantially more for the automatic systems. Many if not most consumers could well conclude that the automatic belts would in fact provide them with no different freedom of choice about usage or levels of protection than manual belts currently offer. As a result, it is not unreasonable to conclude that the public may regard the automatic restraint requirements as an expensive example of ineffective regulation.

Thus, whether or not the detachable automatic belts might have been successful in achieving higher belt usage rates, mandates requiring such belts could well adversely affect public attitude toward the automatic restraint requirements in particular and safety measures in general. As noted in more detail in the 1976 Decision of Secretary Coleman:

Rejection by the public would lead to administrative or Congressional reversal of a passive restraint requirement that could result in hundreds of millions of dollars of wasted resources, severe damage to the nation's economy, and, equally important, a poisoning of popular sentiment toward efforts to improve occupant restraint systems in the future.

It can only be concluded that the public attitude described by the Secretary at that time is at least as prevalent today. The public might ultimately have sought the legislative rescission of the requirements. Action-forcing safety measures have twice before been overturned by Congress. In the mid-1970's, Congress rescinded the ignition interlock provision and provided that agency could not require the States to adopt and enforce motorcycle helmet use laws. Some people might also have cut the automatic belts out of their cars, thus depriving subsequent owners of the cars of the protection of any occupant restraint system. These are serious concerns for an agency charged by statute with taking steps appropriate for addressing safety problems that arise not only in the short term but also the long term. The agency must be able to react effectively to the expected increases in vehicle deaths and injuries during the 1980's.

Equity. Another relevant factor affecting the reasonableness of the automatic restraint requirements and of their costs is the equity of the distribution of such costs among the affected consumers. Responsible regulatory policy should generally strive to ensure that the beneficiaries of regulation bear the principal costs of that regulation. The higher the costs of a given regulation, the more serious the potential equity problem. The automatic restraint requirements of the standard would have required the current regular user of manual belts not only to pay himself for a system that affords him no additional safety protection, but in part to subsidize the current nonuser of belts who may or may not be induced by the automatic restraints to commence regular restraint usage.

Option of Adopting Use-Compelling Features. As noted above, some commenters have suggested that the only safety belts which are truly "passive" are those with use-compelling features. Such commenters have recommended that the agency amend the standard so as to require such features. For example, an ignition interlock which prohibits the car from starting unless the belt is secured is a use-compelling feature. Another example is a passive belt design which is simply not detachable, because no buckle and latch release mechanism is provided. While NHTSA agrees that such use-compelling features could significantly increase usage of passive belts, NHTSA cannot agree that use-compelling features could be required consistent with the interests of safety. In the case of the ignition interlock, NHTSA clearly has no authority to require such a use-compelling feature. The history of the Congressional action which removed this authority from NHTSA suggests that Congress would look with some disfavor upon any similar attempt to impose a use-compelling feature on a belt system.

But, even if NHTSA were to require that passive belts contain use-compelling features, the agency believes that the requirement could be counterproductive. Recent attitudinal research conducted by NHTSA confirms a widespread, latent and irrational fear in many members of the public that they could be trapped by the seat belt after a crash. Such apprehension may well be contributing factors in decisions by many people not to wear a seat belt at all. This apprehension is clearly a question which can be addressed through education, but pending its substantial reduction, it would be highly inappropriate to impose a technology which by its very nature could heighten or trigger that concern.

In addition, the agency believes there are compelling safety reasons why it should not mandate use-compelling features on passive belts. In the event of accident, occupants wearing belts suffer significantly reduced risk of loss of consciousness, and are commonly able to extricate themselves with relative ease. However, the agency would be unable to find the cause of safety served by imposing any requirement which would further complicate the extrication of any occupant from his or her car, as some use-compelling features would. NHTSA's regulations properly recognize the need for all safety belts to have some kind of release mechanism, either a buckle and latch mechanism or a spool-out release which feeds at least 6 feet of belt long enough to extricate a car occupant.

Alternative methods of increasing restraint usage. Finally, the agency believes that it is possible to induce increased belt usage, and enhance public understanding and awareness of belt mechanisms in general, by means that are at least as effective but much less costly than the installation of millions of detachable automatic belts.

In the decision noted above, Secretary Coleman noted the obligation of the Department of Transportation to undertake efforts to encourage the public to use occupant restraints, active or passive. Toward this point, Secretary Coleman directed the Administrator of NHTSA to undertake significant new steps to promote seat belt usage during the demonstration program. This instruction of the Secretary was not effectively carried out and, unfortunately, we do not enjoy today the benefits of a prolonged Departmental campaign to encourage seat belt usage. Had such a program been successfully carried out, increased seat belt usage could have saved many lives each year, beginning in 1977.

Rather than allowing the Coleman demonstration program and its accompanying education effort to come to fruition, the Department reconsidered Secretary Coleman's 1976 decision during 1977. At the conclusion of the reconsideration period, the Department reversed that decision, and amended the standard to require the provision of automatic restraints in new passenger cars, in accordance with a phased-in schedule.

The benefits of any such belt use enhancement efforts could have already substantially exceeded those projected for the automatic restraint requirements of this standard. Over the next ten years, the requirements of the standard would have addressed primarily those occasional belt users amenable to change who buy new cars during the mid and late 1980's.

Prior to the initiation of rulemaking in February of this year, the Department had resolved to undertake a major educational effort to enhance voluntary belt usage levels. Such efforts will be closely coordinated with new and preexisting major initiatives at the State level and in the private sector, many of which were discussed at the public meeting on the present rulemaking. These efforts will address not only those users/purchasers amenable to change, but also those currently driving and riding in cars, multipurpose passenger vehicles and trucks on the road today. The potential for immediate impact is thus many times greater. Further, with the much greater number of persons directly impacted, educational efforts would need to raise safety belt usage in the vehicles on the road during the 1980's by only a few percentage points to achieve far greater safety benefits than the automatic restraint requirements could have achieved during the same time period.

This is in no sense to argue or suggest that nonregulatory alternatives are or should be considered in all cases appropriate to limit Federal regulation. However, the existence of such efforts, and their relevance to calculations of benefits in the present case, must be and has been considered to the extent discussed herein.

Summary of Agency Conclusion

As originally conceived, the automatic restraint requirement was a far reaching technology forcing regulation that could have resulted in a substantial reduction in injuries and loss of life on our highways.

As it would be implemented in the mid-1980's, however, the requirement has turned into a billion dollar Federal effort whose main technological advance would be to require seat belts that are anchored to the vehicle door rather than the vehicle body, permitting these belts to be used either as conventional active belts or as automatic belts.

To gain this advantage, under the standard as drafted, consumers would see the end of the six passenger car and an average vehicle price increase on the order of $89 per car. The almost certain benefits that had been anticipated as a result of the use of air bag technology have been replaced by the gravely uncertain benefit estimates associated with belt systems that differ little from existing manual belts.

In fact, with the change in manufacturers' plans that in essence replaced air bags with automatic belts, the central issue in this proceeding has become whether automatic belts would induce higher belt usage rates than are occurring with manual belts.

Many of the comments in the course of this rulemaking were directed specifically at the question of belt use. Most addressed themselves to the information in the docket on the usage witnessed in the VW Rabbit and Chevette equipped with automatic belts.

The Agency's own analysis of the available information concludes that it is virtually impossible to develop an accurate and supportable estimate of future belt use increases based upon the Rabbit and Chevette automatic belt observations. The Agency further believes that it is impossible to disaggregate the roles that demographics, use inducing devices, and automatic aspects of the belt played in the observed increases.

Faced with this level of uncertainty, and the wide margins of possible error, the agency is simply unable to comply with its statutory mandate to consider and conclude that the automatic restraint requirements are at this time practicable or reasonable within the meaning of the Vehicle Safety Act. On the other hand, the agency is not able to agree with assertions that there will be absolutely no increase in belt use as a result of automatic belts. Certainly, while a large portion of the population appears to find safety belts uncomfortable or refuses to wear them for other reasons, there is a sizeable segment of the population that finds belts acceptable but still does not use them. It is plausible to assume that some people in this group who would not otherwise use manual belts would not disconnect automatic belts.

It is this same population that will generate all of the benefits that result directly and solely from this regulation. This is a population that can also be reached in other ways. The Agency, State governments and the private sector are in the process of expanding and initiating major national belt use educational programs of unprecedented scale. While undertaken entirely apart from the pending proceeding, the fact remains that this effort will predominantly affect the same population that the automatic belts would be aimed at.

On the one hand, it could be argued that, the success of any belt use program would only be enhanced by the installation of automatic belts. Individuals who can be convinced of the utility of safety belts would presumably have an easier time accepting an automatic belt. On the other hand, there is little evidence that the standard itself will materially increase usage levels above those otherwise achievable.

However, the agency is not merely faced with uncertainty as to the actual benefits that would result from detachable automatic safety belts. When the uncertain nature of the benefits is considered together with the risk of adverse safety consequences that might result from the maintenance of this regulation, the agency must conclude that such retention would not be reasonable, and would not meet the need for motor vehicle safety.

It is useful to summarize precisely what the agency believes these risks might be. The principal risk is that adverse public reaction could undermine the effectiveness of both the standard itself and future or related efforts.

The agency also concludes, however, that retention would present serious risk of jeopardizing other separate efforts to increase manual belt usage by the Federal government, States and the private sector. A public that believes it is the victim of too much government regulation by virtue of the standard might well resist such parallel efforts to enhance voluntary belt usage. Further, to the extent that States begin to consider belt use laws as an option, a Federal regulation addressing the same issue could undermine those attempts as well.

While one cannot be certain of the adverse effects on net belt usage increases, it would be irresponsible to fail to consider them. A decision to retain the regulation under any of the schedules now being considered would not get automatic belts on the road until 1983 and would not apply to the entire fleet of new cars until 1984. By the end of the 1984 model year, under most options, there would have been fewer than 20 million vehicles equipped with automatic belts on the road.

By the same time, however, there will be upward of 150 million vehicles equipped with only manual belts, drivers and occupants of which will have been exposed to interim belt usage encouragement efforts.

Agency analysis indicates that external efforts of whatever kind that increase usage by only 5 percent, will save more than 1,300 lives per year beginning in 1983. Installation of automatic belts could save an equal number of lives in 1983 only with 95 percent belt usage.

Further, even if one is convinced that automatic belts can double belt usage and alternative efforts would only increase usage by 5 percent, it would not be until 1989 that total life savings attributable to automatic belts installed under the automatic restraint requirements would reach the total life savings achieved through such other efforts.

NHTSA fully recognizes that neither outcome is a certainty. Much closer to the truth is that both outcomes are uncertain. However, neither is significantly more likely than the other. That being the case, to impose the $1 billion cost on the public does not appear to be reasonable.

It is particularly unreasonable in light of the fact that the rescission does not foreclose the option to again reopen rulemaking if enhanced usage levels of both manual and automatic belts do not

materialize. Long before there would have been any substantial number of vehicles on the road mandatorily equipped with automatic belts as a result of this standard, NHTSA will conclusively know whether other efforts to increase belt use have succeeded either in achieving acceptable usage levels or in increased public understanding and acceptance of the need for further use-inducing or automatic protection alternatives. If so obviously no further action would be needed. If such is not the case, rulemaking would again be a possibility. Any such rulemaking, following even partially successful efforts to increase belt use, would be much less likely to face public rejection.

It has been said that the Vehicle Safety Act is a "technology-forcing" statute. The agency concurs completely.

However, the issue of automatic restraints now before the agency is not a "technology-forcing" issue. The manual seat belt available in every car sold today offers the same, or more, protection than either the automatic seat belt or the air bag. Instead, the agency today faces a decision to force people to accept protection that they do not choose for themselves. It is difficult to conclude that the Vehicle Safety Act is, or in light of past experience could become, a "people-forcing" statute.

NHTSA cannot find that the automatic restraint requirements meet the need for motor vehicle safety by offering any greater protection than is already available.

After 12 years of rulemaking, NHTSA has not yet succeeded in its original intent, the widespread offering of automatic crash protection that will produce substantial benefits. The agency is still committed to this goal and intends immediately to initiate efforts with automobile manufacturers to ensure that the public will have such types of technology available. If this does not succeed, the agency will consider regulatory action to assure that the last decade's enormous advances in crash protection technology will not be lost.

Impact Analyses

NHTSA has considered the impacts of this final rule and determined that it is a major rulemaking within the meaning of E.O. 12291 and a significant rule within the meaning of the Department of Transportation regulatory policies and procedures. A final regulatory impact analysis is being placed in the public docket simultaneously with the publication of this notice. A copy of the analysis may be obtained by writing to: National Highway Traffic Safety Administration, Docket Section, Room 5109, 400 Seventh Street, S.W., Washington, D.C. 20590.

The agency's determination that the rule is major and significant is based primarily upon the substantial savings in variable manufacturing costs and in consumer costs that result from the rescission of the automatic restraint requirements. These costs would have amounted to approximately $1 billion once all new cars became subject to the requirements. The costs would have recurred annually as long as the requirements remained in effect. There is also a recurring savings in fuel costs of approximately $150 million annually. Implementation of the automatic restraint requirements would have increased the weight of cars and reduced their fuel economy. In addition, the car manufacturers will be able to reallocate $400 million in capital investment that they would have had to allocate for the purpose of completing their efforts to comply with the automatic restraint requirements.

The agency finds it difficult to provide a reliable estimate of any adverse safety effects of rescinding the automatic restraint requirements. There might have been significant safety loss if the installation of detachable automatic belts resulted in a doubling of belt usage and if the question were simply one of the implementation or rescission of the automatic restraint requirements. The April 1981 NPRM provided estimates of the additional deaths that might occur as a result of rescission. However, those estimates included carefully drafted caveats. The notice expressly stated that the impacts of rescission would depend upon the usage rate of automatic belts and of the effectiveness of the agency's educational campaign. The agency has now determined that there is no certainty that the detachable automatic belts would produce more than a several percentage point increase in usage. The small number of cars that would have been equipped with automatic belts having use-inducing features or with air bags would not have added more than several more percentage points to that amount. Further, any potential safety losses associated with the rescission must be balanced against the expected results of the agency's planned educational program about

safety belts. That campaign will be addressed to the type of person who might be induced by the detachable automatic belts to begin regular safety belt usage, i.e., the occasional user of manual belts. Since that campaign will affect occasional users in all vehicles on the road today instead of only those in new cars, the campaign can yield substantially greater benefits than the detachable automatic belts even with a much lower effectiveness level.

The agency has also considered the impact of this action on automatic restraint suppliers, new car dealers and small organizations and governmental units. Since the agency certifies that the rescission would not have a significant effect on a substantial number of small entities, a final regulatory flexibility analysis has not been prepared. However, the impacts of the rescission on the suppliers, dealers and other entities are discussed in the final Regulatory Impact Analysis.

The impact on air bag manufacturers is likely to be minimal. Earlier this year, General Motors, Ford and most other manufacturers cancelled their air bag programs for economic reasons. These manufacturers planned instead to rely almost wholly on detachable automatic belts. Therefore, it is not accurate to say, as some commenters did, that rescission of the automatic restraint requirements will "kill" the air bag. Rescission will not affect the air bag manufacturers to any significant degree. Further, the agency plans to undertake new steps to promote the continued development and production of air bags.

The suppliers of automatic belts are generally the same firms that supply manual belts. Thus, the volume of sales of these firms is not expected to be affected by the rescission. However, there will be some loss of economic activity that would have been associated with developing and producing the more sophisticated automatic belts.

The effects of the rescission on new car dealers would be positive. Due to reduced new car purchase prices and more favorable reaction to manual belts than to automatic belts, sales increases of 395,000 cars were estimated by GM and 235,000 cars by Ford. While these figures appear to be overstated, the agency agrees that rescission will increase new car sales.

Small organizations and governmental units would be benefited by the reduced cost of purchasing and operating new cars. Given the indeterminacy of the usage rate that detachable automatic belts would have achieved, it is not possible to estimate the effects, if any, of the rescission on the safety of persons employed by these groups.

In accordance with the National Environmental Policy Act of 1969, NHTSA has considered the environmental impacts of the rescission and the alternatives proposed in the April 1981 NPRM. The option selected is disclosed by the analysis to result in the largest reductions in the consumption of plastics, steel, glass and fuel/energy. A Final Environmental Impact Statement is being placed in the public docket simultaneously with the publication of this notice.

This amendment is being made effective in less than 180 days because the date on which the car manufacturers would have to make expenditure commitments to meet the automatic restraint requirements for model year 1983 falls within that 180-day period.

In consideration of the foregoing, Federal Motor Vehicle Safety Standard No. 208, *Occupant Crash Protection* (49 CFR 571.208), is amended as set forth below.

§ 571.208 [Amended]

1. S4.1.2 is amended by revising it to read:

S4.1.2 *Passenger cars manufactured on or after September 1, 1973.* Each passenger car manufactured on or after September 1, 1973, shall meet the requirements of S4.1.2.1, S4.1.2.2 or S4.1.2.3. A protection system that meets the requirements of S4.1.2.1 or S4.1.2.2 may be installed at one or more designated seating positions of a vehicle that otherwise meets the requirements of S4.1.2.3.

2. The heading of S4.1.2.1 is amended by revising it to read:

S4.1.2.1 *First option—frontal/angular automatic protection system.*

*   *   *   *   *   *   *   *

3. S4.1.3 is removed.

S4.1.3 [Removed]

(Secs. 103, 119, Pub. L. 89–563, 80 Stat. 718 (15 Stat. 1392, 1407); delegation of authority at 49 CFR 1.50)

Issued on October 23, 1981.

Raymond A. Peck, Jr.,
Administrator.

**46 F.R. 53419
October 29, 1981**

Editorial

Note—This appendix will not appear in the *Code of Federal Regulations*.

Following is a summary of the major comments submitted in response to the April 9, 1981 notice of proposed rulemaking. A more detailed summary of comments has been placed in NHTSA Docket No. 74-14; Notice 22. This summary is organized in broad terms according to the interest groups from which the comments were received.

Insurance Companies

All commenting insurance companies strongly favored retention of the automatic restraint requirements. Many favored maintaining the present implementation schedule (i.e., September 1, 1982, for large and medium-sized cars and September 1, 1983, for small cars), although several companies stated they would support a change to require that small cars are phased in first or a simultaneous implementation date. Several insurance companies stated that air bags offer the best technology for saving lives and reducing injuries. These companies pointed out that repeated surveys have indicated that consumers appear to favor air bags, even if higher costs are likely. Several insurers argued that a retreat from the standard represents a breach of the Secretary's statutory obligation to reduce traffic accidents and deaths and injuries which result from them. One company argued that a delay in the standard (i.e., the delay and reversal alternative) would produce no measurable economic benefit to car makers and might possibly result in an economic loss to them. Nearly all the companies argued that the standard is cost-beneficial and represents the optimum approach to resolving this country's most pressing public health problem. Many companies stated that reduced insurance premiums resulting from the lives saved and injuries prevented by automatic restraints would help offset the cost of those systems to consumers.

A majority of the insurance companies argued that seat belt use campaigns will not be effective in raising the current use rate of manual belts significantly. The companies pointed to the failures of all past campaigns to have any substantial impact on use rates. On the other hand, these companies believe that the use rate of automatic belts will be significant. The companies point to the cur-

rent use data for automatic belts on VW Rabbits and Chevettes as evidence that automatic belt use will be significant. The companies believe that seat belt use campaigns should only be complementary to automatic restraints, not a substitute.

Several insurance companies pointed to the huge economic losses resulting from traffic accidents. One company stated that these losses mount to over $1 billion dollars per year and result in recurring costs because of continuing medical problems such as epilepsy and quadriplegia. One company cited Professor William Nordhaus's analysis of the consequences of rescinding the standard as being equivalent to society's loss if the tuberculosis vaccine had not been developed, or if Congress repealed the Clean Air Act. In his submission on behalf of the insurance companies, Professor Nordhaus stated that fatalities will increase by 6,400 each year and injuries by 120,000 if the standard is rescinded. One company argued that the standard is cost-beneficial if automatic belt use rates increase usage only 5 percent. However, this company stated that use rates as high as 70 percent could be expected, and that the costs of rescinding the standard could reach as much as $2 billion dollars per year. This company also argued that the economic condition of the vehicle industry is no excuse for any delay in the standard and is not a statutorily justified reason for rescinding the standard.

Consumer Groups and Health Organizations

There were many consumer groups and health-related organizations which strongly urged that the automatic restraint requirements be maintained and that there be no further delays in the implementation schedule. Most of these groups argued that the cost of both air bags and automatic belts are greatly exaggerated by vehicle manufacturers. One group stated that the three alternative proposals are "naive and exhibit a callous disregard for human lives that flouts the agency's mandated safety mission." This group argued that a worse alternative is to rescind the standard and rely on education programs to increase the use of manual belts, since seat belt campaigns have failed repeatedly in this country. The group stated that the simultaneous implementation alternative in March 1983 ignores the industry's background of introducing safety changes only at the beginning of a new model year. Regarding a reversed phase-in schedule, the group stated that the requirement that small cars have automatic restraints by September 1, 1982, would not likely provide sufficient lead time for small car manufacturers. Additionally, with approximately 2 to 1 difference in seat belt use in small cars versus larger cars, it is not at all clear that the proposed reversal would make up for the delay in implementation in the larger cars in terms of lives saved. The group argued that the best alternative is to maintain the existing implementation schedule.

Several consumer groups argued that the center seating position should not be eliminated from the requirements for several reasons. First, they argued, this position is likely to be occupied by children. Second, the center seat requirement is one factor that will lead to the installation of air bags in some vehicles since current automatic belt designs cannot be applied to the center seat. Nearly all consumer groups argued that benefits of the automatic restraint standard far outweigh the costs.

One association stated that the air bag supplier industry could be forced out of business if substantial modifications and further delays are made to the standard. This would mean, the association argued, that the life-saving air bag technology could be lost forever. The association would support some modifications to the standard if there were some clear commitment by the Department that some car models would be required to offer the consumer the choice of air bags. The group noted that air bag suppliers have indicated that a sufficient production volume would result in air bag systems priced in the $200 to $300 dollar range.

Various health groups and medical experts argued that the pain and suffering resulting from epilepsy and paraplegia, as well as mental suffering and physical disfigurement, could be greatly reduced by the automatic restraint standard. These persons argued that the standard should be implemented as soon as possible.

One consumer oriented group did not support the automatic restraint standard. That foundation argued that the standard is not justified, particularly if it is complied with by means of air bags. The group stated that air bag effectiveness is overestimated since the agency does not include non-frontal crashes in its statistics. The organiza-

tion argued that in many situations air bags are actually unsafe. This group also argued that the public acceptability of automatic seat belts is uncertain, and that a well-founded finding of additional safety benefits by the Department is required in order to justify retention of the standard.

Vehicle Manufacturers

The vehicle manufacturers, both foreign and domestic, were unanimously opposed to retention of the automatic restraint standard. Most manufacturers stated the predominate means of complying with the standard would be with automatic belts, and that such belts are not likely to increase usage substantially. This is because most automatic belts will be designed to be easily detachable because of emergency egress considerations and to avoid a potential backlash by consumers that would be counterproductive to the cause of motor vehicle safety. The domestic manufacturers argued that the public would not accept coercive automatic belts (i.e., automatic belts with interlocks or some other use-inducing feature). Eliminating any coercive element produces, in effect, a manual belt, which will be used no more than existing manual systems.

The domestic manufacturers also argued that air bags would not be economically practicable and would, therefore, be unacceptable to the public. One manufacturer noted that current belt users will object strenuously to paying additional money for automatic belts that will not offer any more protection than their existing belts.

One manufacturer argued that the injury criteria specified in the standard is not representative of real injuries and should be replaced with only static test requirements for belt systems. The company argued that there are many problems with test repeatability under the 208 requirements.

All manufacturers of small cars stated that it would be impossible for them to comply with the standard by September 1, 1982, i.e., under the reversal proposal. These manufacturers stated that there is insufficient lead time to install automatic restraints in small cars by that date, and several foreign manufacturers stated they would not be able to sell their vehicles in that model year if the schedule is reversed. Most of the manufacturers, both domestic and foreign, stated that it is also too late to install automatic restraints in their small cars even six months earlier than the existing schedule, i.e., under the March 1983 simultaneous implementation proposal. Many manufacturers supported a simultaneous implementation if the standard is not rescinded, but requested that the effective date be September 1, 1983, or later. The manufacturers argued that an effective date for small cars prior to September 1, 1983, would not allow enough time to develop acceptable, reliable and high quality automatic belts.

Nearly all vehicle manufacturers believe that an intensive seat belt education campaign can be just as effective as automatic restraints and without the attendant high costs of automatic restraints. Additionally, most foreign manufacturers recommended that mandatory seat belt use laws be enacted in lieu of automatic restraints.

One foreign manufacturer requested that any effective date for automatic restraints be "September 1 or the date of production start of the new model year if this date falls between September 1 and December 31." The company stated that this would allow manufacturers to continue production for several months of models that would then be phased out of production. However, a domestic vehicle manufacturer argued that this would give foreign manufacturers an unfair competitive advantage, and that current practice of September 1 effective dates should be retained.

Most manufacturers supported the proposal to exclude the center seating position from the automatic restraint requirements, in order to give manufacturers more design flexibility. However, the two domestic manufacturers which would be most affected by such an exception stated that it is too late for them to make use of such an exception for 1983 models. The two companies stated that such an exception would have benefits in the long run, however, and would allow them to continue production of six-seat passenger cars in the mid-1980's.

Suppliers and Trade Groups

Suppliers of air bag system components supported continuation of the automatic restraint requirements. One commenter stated that having to buckle-up is an act which requires a series of psychological and physical reactions which are responsible for the low rate of manual seat belts. Also, this company stated that educational campaigns to increase belt use will not work.

One motor vehicle trade group stated that a study by the Canadian government has established the superiority of manual seat belt systems. This group argued that the automatic restraint requirements cannot be justified because any expected benefits are speculative.

One trade group voiced its concern about sodium azide (an air bag propellant) as it pertains to possible hazards posed to the scrap processing industry.

A group representing seat belt manufacturers stated that the most effective way of guaranteeing belt use is through mandatory belt use laws. That group believes that belt usage can be increased through public education, and that simple, easy to use automatic belts such as are currently on the VW Rabbit will also increase belt usage. This group did not support a simultaneous implementation date for automatic restraints, stating that this could put a severe strain on the supplier industry. The group did support elimination of the automatic restraint requirements for center seating positions.

An automobile association recommended equipping small cars with automatic restraints first. The association stated that a reversed phase-in schedule would protect a significantly large segment of the public at an earlier date, would reduce a foreign competitive advantage (under the existing schedule), and would give needed economic relief to large car manufacturers. This organization also recommended that, as an alternative, automatic restraints be required only at the driver's position. This would achieve three-quarters of the reductions in deaths and serious injuries now projected for full-front seat systems, yet cost only half as much.

Congressional Comments

Mr. Timothy E. Wirth, Chairman of the House Subcommittee on Telecommunications, Consumer Protection and Finance, made the following comments:

—The automatic restraint requirements would produce benefits to society far in excess of costs.

—The Committee findings strongly point to the necessity of requiring the installation of automatic crash protection systems, at a minimum, on a substantial portion of the new car fleet at the earliest possible date. Mr. Wirth suggested that the effective date for small cars be September 1, 1982, and for intermediate and large cars September 1, 1983.

—The economic conditions of the automobile industry should not be relevant to the NHTSA's decision on matters of safety. NHTSA's decision must be guided solely by safety-related concerns.

—The agency should not discount its own findings indicating high use of automatic belts (referring to the existing VW and Chevette automatic belt use data).

In a joint letter to the Secretary, eighteen Congressmen urged that the automatic restraint requirements be maintained. This letter noted that over 50,000 people are killed each year on the highways and stated: "While the tragedy of their deaths cannot be measured in economic terms, the tragedy of their serious injuries cost all of us billions of dollars each year in higher insurance costs, increased welfare payments, unemployment and social security payments and rehabilitation costs paid to support the injured and the families of those who have been killed." The letter stressed the Congressmen's belief that the automatic crash protection standard would produce benefits to society far in excess of its cost.

In a letter addressed to Administrator Peck, fifty-nine Congressmen urged that the automatic restraint standard be rescinded. That letter stated: "The 208 standard persists as one of the more controversial federal regulations to be forced on the automobile industry. . . . The industry continues to spend hundreds of thousands of dollars every day in order to meet this standard, despite considerable evidence that any safety benefits realized by enforcing the standard would be minimal."

Private Citizens

In addition to comments from the above groups and organizations, the agency also received general comments from numerous private citizens. These comments were almost equally divided in their support or opposition to the automatic restraint standard.

Raymond A. Peck, Jr.
Administrator

46 F.R. 53419
October 29, 1981

PREAMBLE TO AN AMENDMENT TO
FEDERAL MOTOR VEHICLE SAFETY STANDARD NO. 208

Occupant Crash Protection
(Docket No. 74-14; Notice 24)

ACTION: Final rule; partial response to petitions for reconsideration.

SUMMARY: The purpose of this notice is to delay for one year the effective date of the comfort and convenience requirements for seat belts in Safety Standard No. 208, *Occupant Crash Protection.* Standard No. 208 was amended January 8, 1981, to promote the installation of more comfortable and convenient belts by specifying additional performance requirements for both manual and automatic seat belts installed in motor vehicles with a Gross Vehicle Weight Rating (GVWR) of 10,000 pounds or less. Petitions for reconsideration of these new performance requirements were received from seven vehicle manufacturers.

The agency has determined that the recent rescission of the automatic restraint requirements of Standard 208 has made it necessary to review the comfort and convenience requirements in their entirety. The changed circumstances have made it difficult to respond to the substantive issues raised in the petitions for reconsideration at this time. Since the requirements are currently scheduled to become effective September 1, 1982, the agency has concluded that it is necessary to extend the effective date until September 1, 1983, to give the agency sufficient time to re-evaluate these requirements.

EFFECTIVE DATE: The new effective date for the existing comfort and convenience requirements is September 1, 1983.

SUPPLEMENTARY INFORMATION: On January 8, 1981, Safety Standard No. 208, *Occupant Crash Protection* (49 CFR 571.208), was amended to specify performance requirements to promote the comfort and convenience of both manual and automatic safety belts installed in vehicles with a GVWR of 10,000 pounds or less (46 F.R. 2064). Type 2 manual belts (lap and shoulder combination belts) installed in front seating positions in passenger cars were excepted from these additional performance requirements since it was assumed such belts would be phased out in passenger cars as the automatic restraint requirements of Standard No. 208 became effective.

Seven petitions for reconsideration of the January 8, 1981 amendment were received from vehicle manufacturers. These petitions requested that the requirements be revoked entirely, or that at least various modifications be made and that the effective date be delayed.

Since the receipt of these petitions for reconsideration, the agency has revoked the automatic restraint requirements of the standard (46 F.R. 53419, October 29, 1981). This rescission alters the circumstances which must be considered in determining appropriate requirements for seat belt comfort and convenience. Therefore, it is difficult for the agency to respond to the substantive issues raised in the petitions for reconsideration at the current time. Many of the issues that were raised are no longer pertinent and many of the rationales discussed by the agency when the requirements were first established must be re-evaluated. Therefore, the agency has determined that the comfort and convenience requirements should be reviewed in their entirety.

In light of these conclusions, the agency has decided that it is necessary to delay the effective date of the current comfort and convenience requirements for at least a year (from September 1, 1982, to September 1, 1983). This will give the agency sufficient time to re-evaluate the requirements and the petitions for reconsideration in light of the changed circumstances. Further, manufacturers should not be required to comply with the requirements by September 1, 1982, since they may be altered substantially.

The agency intends to respond to the substantive issues raised in the petitions for reconsideration at a later date. Moreover, the agency is considering additional changes to the comfort and convenience requirements which would encourage and ensure maximum possible technical improvements and enhancements are included in future seat belt designs.

The NHTSA has considered the economic and other impacts of this one-year delay in effective date and determined that the rule is neither a major rule within the meaning of Executive Order 12291 nor a significant rule within the meaning of the Department of Transportation's regulatory procedures. A regulatory evaluation concerning the one-year delay has been placed in the public docket. This evaluation supplements the regulatory evaluation which was prepared when the regulation was issued in January 1981.

The agency has also analyzed the delay for purposes of the National Environmental Policy Act and has determined that it will not have a significant impact on the quality of the human environment.

No regulatory flexibility analysis has been prepared on this final rule since the proposal underlying this final rule and the January 8, 1981 final rule was issued before the effective date of the Regulatory Flexibility Act.

In consideration of the foregoing, the effective date of the comfort and convience requirements of 49 CFR 571.208 that were issued January 8, 1981 (46 F.R. 2064) is hereby delayed from September 1, 1982 to September 1, 1983.

Issued on February 11, 1982.

_____
Raymond A. Peck, Jr.
Administrator

47 F.R. 7254
February 18, 1982

PREAMBLE TO AN AMENDMENT TO
FEDERAL MOTOR VEHICLE SAFETY STANDARD NO. 208

**Federal Motor Vehicle Safety Standards;
Occupant Crash Protection**

[Docket No. 74-14; Notice 28]

ACTION: Final rule.

SUMMARY: The purpose of this notice is to amend the fuel loading test conditions of Safety Standard No. 208, *Occupant Crash Protection*. The amendment is in response to a petition for rulemaking submitted by Mercedes-Benz of North America. Standard No. 208 currently specifies that vehicles are to be crash tested with their maximum capacity of fuel. Several other NHTSA safety standards only require fuel tanks to be filled from 90 to 95 percent of capacity. This amendment makes the fuel loading conditions of Standard No. 208 consistent with these other standards. This change will enable manufacturers to simultaneously determine compliance with several standards during the same crash tests, thereby reducing compliance test costs. In connection with this change, this notice also adds a definition for "fuel tank capacity" to the agency's general definition list in 49 CFR Part 571.3.

EFFECTIVE DATE: October 28, 1982.

SUPPLEMENTARY INFORMATION: The fuel tank loading condition in Safety Standard No. 208, *Occupant Crash Protection* (49 CFR 571.208) differs from that used in several other NHTSA safety standards. Paragraph S8.1.1(a) of Standard No. 208 currently specifies that a passenger car is to be loaded "to its unloaded vehicle weight plus its rated cargo and luggage capacity weight" prior to conducting a barrier crash test. The term "unloaded vehicle weight" is defined in 49 CFR 571.3 as "the weight of a vehicle with maximum capacity of all fluids necessary for operation of the vehicle...." Therefore, under the current test conditions of the standard, fuel tanks are to be filled to 100 percent capacity. The fuel loading conditions of Safety Standards Nos. 301, *Fuel System Integrity*; 212, *Windshield Mounting*; and 219, *Windshield Zone Intrusion*, specify that fuel tanks are only loaded from 90 to 95 percent of capacity.

On January 28, 1982, the agency proposed to amend the loading conditions of Standard No. 208 to make them consistent with those of Standards Nos. 301, 212 and 219 (47 F.R. 4098). The proposed amendment was issued in response to a petition for rulemaking submitted by Mercedes-Benz of North America, which asked that the fuel loading conditions of Standard No. 208 be amended to be consistent with Safety Standard No. 301. Mercedes pointed out that such an amendment would serve to harmonize the two standards and would eliminate the current need for running separate barrier crash tests for the two standards. The company stated that tests being conducted to evaluate occupant crash protection systems yield data which cannot be used to evaluate the integrity of fuel systems because of the variation in fuel tank loading conditions.

Seven parties commented on the proposed change. All of them were vehicle manufacturers which supported lowering the fuel loading conditions of Standard No. 208. All the manufacturers noted that the proposed change would standardize test conditions for the standards employing dynamic crash testing, and would thereby reduce costs. After reviewing these comments, the agency has determined that the standard should be amended as proposed.

As noted in the proposal, the agency believes that filling fuel tanks from 90 to 95 percent capacity for Standard 208 testing will be sufficiently representative of the maximum fuel loading that will occur on the highway. Vehicles are seldom driven with their fuel tanks filled to 100 percent capacity. Moreover, the difference in overall vehicle weight because of the 5 to 10 percent less fuel with this amendment should have no significant effect on the test results of Standard No. 208. Therefore, the change does not significantly reduce the stringency of the standard and realistically maintains the intended purpose of the loading conditions.

The agency also believes it is important to facilitate simultaneous testing for various safety standards, where possible, in order to minimize testing costs. Since Standard Nos. 301, 212, and 219 only require fuel tanks to be loaded from 90 to 95 percent capacity, the agency has determined that Standard No. 208 should be amended to be consistent. In this case, testing costs can be reduced without jeopardizing safety whatsoever.

In its comment, General Motors Corporation suggested that the amendment also include a definition of "fuel tank capacity," so that there will be no questions concerning the proper procedure for filling fuel tanks prior to testing. General Motors' suggestion was prompted by a discussion in the preamble of the proposal concerning what constitutes the "capacity" of a fuel tank. That discussion was included because the agency had previously received several questions asking whether the vapor volume of a fuel tank is included in determining capacity. The discussion clarified the agency's position that "capacity" does not include vapor volume.

The agency believes that General Motors' suggestion has merit. Therefore, a definition for "fuel tank capacity" is added by this amendment to 49 CFR 571.3, the agency's general definition section. The term is defined as the volume of fuel that can be pumped into a previously unfilled tank through the filler pipe with the vehicle on a level surface, but excluding the vapor volume of the tank and the volume of the tank filler pipe. The definition is being added to 49 CFR 571.3, rather than to Standard No. 208, so that it is clear the same term is applicable to all safety standards which specify fuel loading in terms of tank capacity (i.e., Standards Nos. 301, 212, and 219 as well as Standard No. 208).

The agency has determined that this definition can be added to 49 CFR 571.3 without notice and opportunity to comment since it is merely an interpretive amendment and is therefore within the exceptions to rulemaking procedures specified in the Administrative Procedure Act (5 U.S.C. 553 (b) (3) (A)). In fact, the addition of this definition is merely a codification of previous NHTSA interpretations.

Issued on October 5, 1982.

Raymond A Peck, Jr.
Administrator
47 F.R. 47839
October 28, 1982

PREAMBLE TO AN AMENDMENT TO
FEDERAL MOTOR VEHICLE SAFETY STANDARD NO. 208
Occupant Crash Protection
[Docket No. 74-14; Notice 30]

ACTION: Final rule.

SUMMARY: The purpose of this notice is to delay for two years the effective date of the comfort and convenience requirements for seat belts in Safety Standard No. 208, *Occupant Crash Protection*. These requirements were issued January 8, 1981, to promote the installation of more comfortable and convenient belts by specifying additional performance requirements for both manual and automatic belts installed in motor vehicles with a Gross Vehicle Weight Rating (GVWR) of 10,000 pounds or less. The requirements were originally scheduled to become effective September 1, 1982, but in partial response to petitions for reconsideration, and in light of the agency's rescission of the automatic restraint requirements of Standard No. 208, were delayed for one year to September 1, 1983.

The agency has now concluded that a further delay is necessary because of concerns that have arisen within the agency regarding the efficacy and level of stringency of certain of the requirements, and because of the unsettled state of future plans for seat belt designs. The two-year delay set forth in this notice will give the agency sufficient time to complete its review of performance characteristics of restraint design that would lead to enhanced comfort and convenience for users, and to resolve the many questions that have developed regarding particular provisions.

ADDRESS: Any petitions for reconsideration should refer to the docket number and notice number of this notice and be submitted to: Docket Section, Room 5109, National Highway Traffic Safety Administration, 400 Seventh Street, S.W., Washington, D.C. 20590. Docket hours are from 8 a.m. to 4 p.m., Monday through Friday.

DATES: Any petitions for reconsideration of this rule must be received within 30 days after the date of publication of this notice. The new effective date for the seat belt comfort and convenience requirements is September 1, 1985.

SUPPLEMENTARY INFORMATION: On January 8, 1981, Safety Standard No. 208, *Occupant Crash Protection* (49 CFR 571.208), was amended to specify additional performance requirements to enhance the comfort and convenience of both manual and automatic safety belts installed in vehicles with a GVWR of 10,000 pounds or less (46 FR 2064). Type 2 manual belts (combination lap and shoulder belts) installed in front outboard seating positions in passenger cars were excepted from these additional requirements because it was then assumed that these belts would be phased out of production in passenger cars as the automatic restraint requirements of Standard No. 208 became effective. However, the agency rescinded the automatic restraint requirements on October 29, 1981 (46 FR 53419). This rescission altered basic assumptions that had been made when the comfort and convenience requirements were first issued. Likewise, it altered the belt designs which manufacturers would be installing in future cars.

In partial response to petitions for reconsideration that were received concerning the comfort and convenience requirements, the agency delayed the effective date of the requirements for one year because of the changed circumstances surrounding the rescission of the automatic restraint requirements (47 FR 7254). The agency noted that it was difficult to respond to the substantive issues raised in the petitions for reconsideration, at that time, because many of the issues are no longer pertinent and because many of the rationales discussed by the agency when the requirements were first established must be re-evaluated.

During the agency's review of the comfort and convenience requirements following the one-year delay, questions arose concerning the efficacy and appropriate level of stringency of certain of the requirements. It became evident that the agency needed additional time to re-evaluate the comfort and convenience requirements in their entirety. Thus, on November 15, 1982, the agency proposed an additional two-year delay, to September 1, 1985 (47 FR 51432).

As noted in the proposal, agency experts have identified concerns about various countervailing safety consequences that could develop depending on the final form of the requirements. For example, tension-relieving devices on belt systems can reduce belt pressure and increase comfort, but there is a concern that the increased belt slack due to misuse could reduce belt effectiveness. The proposal pointed out that the agency must have time to complete its evaluation and resolution of these and other similar conflicting considerations.

Eleven comments were received in response to the proposed two-year delay, and only one of these objected to the proposal. The State of Idaho Transportation Department strongly recommended against a further delay on the basis that this would hinder current national and State level education efforts to encourage the voluntary use of seat belts. All of the vehicle manufacturers which commented vigorously supported the proposed delay, as did the American Seat Belt Council. Three manufacturers, however, urged the agency to delay the requirements indefinitely, rather than to September 1, 1985. These manufacturers agreed that the agency needs additional time to re-evaluate the comfort and convenience requirements in their entirety, but they are concerned that the two-year period proposed would then leave no lead time for manufacturers prior to the effective date. One manufacturer stated, "A new effective date should not be specified before the final requirements are established."

The agency understands the manufacturers' concerns regarding lead time. There were many issues raised in the petitions for reconsideration to which the agency has not yet responded (e.g., objectivity of the requirements, test repeat ability, conflicts with the requirements of other safety standards). However, the agency believes that a specific effective date, September 1, 1985, is preferable to an indefinite delay since it gives all parties, including

the agency, a time frame within which to work. The agency will, of course, evaluate whether there is adequate lead time for manufacturers after all the issues have been resolved in this rulemaking, and modify the effective date accordingly if that is necessary.

In spite of the concerns raised by the Idaho Department of Transportation, the agency has concluded that a two-year delay in the effective date of the comfort and convenience requirements is necessary. As noted in the proposal, the issues involved in this proceeding have been clouded in uncertainty since the regulation was first adopted. Safety belt designs are currently in a state of flux. Therefore, it is not certain exactly what type of restraints will be on the road in the foreseeable future. For this reason, the agency has determined that it would be wise to delay the comfort and convenience requirements, to give the agency sufficient time to re-evaluate the requirements in light of evolving belt systems and avoid imposing possibly unnecessary costs. For example, one commenter to the proposal stated that it had been experimenting with a particular seat belt design for nearly two years and is still uncertain whether the design will consistently meet the somewhat conflicting requirements (in Standard No. 208) for full belt retraction, 0.7 pound chest force limitation and the retractive force requirements of Safety Standard No. 209 (49 CFR 571.209). The agency needs additional time to evaluate these and other similar problems.

Finally, as noted in the proposal, the agency believes that it is impossible at the current time to determine how to achieve or induce effective improvements in the comfort and convenience of belt systems until the occupant crash protection standard can be reviewed in its entirety. The two-year delay will allow the agency time to complete its evaluation of all the current provisions in terms of expected applicabilty, effectiveness, overall safety consequences and appropriate level of detail.

The agency does not believe that this delay will retard the introduction of new improved belt systems, in terms of comfort and convenience. One vehicle manufacturer which commented on the proposal specifically stated that it "plans to proceed voluntarily with a variety of improvements in seat belt comfort and convenience for 1984 and future models regardless of the proposed delay in effective date." The agency encourages other manufacturers to also voluntarily introduce improved comfort and convenience features in their belt designs during this interim period in which the agency is resolving the issues associated with the Standard No. 208 requirements.

The agency has examined the impacts of this amendment and determined that it is not major within the meaning of Executive Order 12291 or significant according to the Department of Transportation regulatory policies and procedures. The agency has prepared a final regulatory evaluation concerning the amendment, which has been placed in the Docket. (A free copy may be obtained by contacting the Docket Section.) That evaluation shows that the safety impact of the proposed delay will not be significant. The precise magnitude of the impact cannot be quantified because the agency has not been able to successfully address in quantified terms the larger question of the effects of the comfort and convenience requirements. That adverse impact will be minimized as a result of the improved seat belt designs that are currently being introduced by manufacturers on a voluntary basis, partly in response to the dialogue generated by the proposal and adoption of the comfort and convenience requirements. The agency believes that manufacturers will experiment further during the two-year delay with innovative designs aimed at increasing the comfort and convenience of belt systems. This effort will at least partially offset any negative impacts that the delay might otherwise cause. The proposed delay will provide slight cost savings for both manufacturers and consumers.

NHTSA has also considered the impacts of this amendment under the Regulatory Flexibilty Act. I hereby certify that amending Standard No. 208 to delay the effective date of the comfort and convenience requirements will not have significant economic impact on a substantial number of small entities for the reasons just discussed. The only small entities that would be affected would be small manufacturers or small organizations or governmental units that purchase vehicles. The effect would not be significant since the cost savings made possible by the delay would be slight.

Issued on May 27, 1983

Diane K. Steed,
Acting Administrator.

48 F.R. 24717

June 2, 1983

PREAMBLE TO AN AMENDMENT TO
FEDERAL MOTOR VEHICLE SAFETY STANDARD NO. 208

Occupant Crash Protection;
Automatic Occupant Restraint Requirement

[Docket No. 74-14; Notice 31]

ACTION: Suspension of rule.

SUMMARY: This notice suspends the automatic occupant restraint requirements of Safety Standard No. 208. Occupant Crash Protection. This action permits the agency time for the further review contemplated by the recent Supreme Court decision that found NHTSA's rescission of the requirement to be arbitrary and capricious. This suspension is issued without a prior opportunity for notice and comment; the rule might otherwise be deemed effective on September 1, 1983. However, public comment on the suspension is requested and the suspension will be revised or revoked, if appropriate, in response to the comments received.

DATES: *Suspension*—The mandatory automatic restraint requirement of Standard No. 208 is suspended until September 1, 1984. This suspension is effective on September 1, 1983.

SUPPLEMENTARY INFORMATION: On October 29, 1981 (49 FR 53419), the Department of Transportation's National Highway Traffic Safety Administration (NHTSA) published a notice rescinding the automatic restraint requirements of Safety Standard No. 208, Occupant Crash Protection. (The language of Standard 208 as it was codified prior to the rescission is contained in Appendix A to this notice.) On June 1, 1982, the U.S. Court of Appeals for the D.C. Circuit found the agency's action to be arbitrary and capricious and overturned the agency's action. *(State Farm Mutual Automobile Insurance Co. v. Department of Transportation,* 680 F.2d 206.) On August 4, 1982, the Court of Appeals issued an order stay-ing the effective date of the requirement until September 1, 1983.

In June 1983, the United States Supreme Court rejected the scope of review used by the lower court, but also found the rescission to be arbitrary and capricious. The Supreme Court vacated the judgment of the Court of Appeals and remanded the case to that Court with directions to remand it to NHTSA for further consideration consistent with the Supreme Court's opinion. *(Motor Vehicle Manufacturers Association v. State Farm Mutual Automobile Insurance Co.* (No. 82-354; June 24, 1983)).

Because the Supreme Court vacated the judgment of the Court of Appeals, it could be argued that the rescission of the automatic restraint requirement technically continues in effect pending the further agency review contemplated by the Supreme Court. However, if that were not the case, compliance with the rule could be considered to be required by September 1, 1983. In order to clarify this situation, the Department has determined that it is appropriate to issue this notice suspending the effect date of the requirement.

The Supreme Court stated that the agency has sufficient justification to suspend Standard 208 pending any further consideration in accordance with the Court's decision. The Department believes that further consideration is necessary and, as part of our review efforts, it is our intention to issue a notice of proposed rulemaking (NPRM) by October 15, 1983. We intended to expedite this rulemaking and reach a final decision as quickly as possible and well before the end of the one-year suspension. At that time, we will establish an appropriate effective date either for the rule that was rescinded, if we decide to retain it, or for any other action that we take, including re-rescission of the rule.

We believe that it would be inappropriate to require compliance with the rule during this short review period. Neither consumers nor manufacturers should be required to incur additional expenses to comply with a requirement that is being actively reviewed.

Moreover, there is substantial evidence showing that a September 1, 1983, effective date is not practicable. After the D.C. Circuit entered its of August 4, 1982, reinstating the automatic restraint requirement on September 1, 1983, NHTSA obtained current information from vehicle and automatic restraint equipment manufacturers concerning their ability to comply with a September 1, 1983, effective date. After reviewing and analyzing the letters and affidavits submitted by the manufacturers, NHTSA concluded, in an October 1, 1982, submission to the D.C. Circuit Court, that a September 1, 1983, effective date was not achievable at that time and that a significantly longer time period would be needed before practicable compliance with the automatic restraint requirements could be achieved. Based on that data, the Department has concluded that it would not be practicable for vehicle manufacturers to comply with the September 1, 1983, requirement because there is not sufficient leadtime for them to make all the necessary design, development, testing, and production preparations by that date.

Because it is not practicable for the manufacturers to comply by September 1, 1983, the Department also has determined that notice and public procedure on this notice of suspension are impracticable, unnecessary, and contrary to the public interest. The recency of the Supreme Court decision and the imminence of the deadline for compliance with the rule justify this determination. We wish to stress, however, that we are providing an opportunity for public comment on this suspension immediately subsequent to its issuance. After reviewing the public comment that is recieved, the Department will determine whether this suspension should be revised or revoked and we will issue a document stating our final decision.

This suspension may be made effective immediately upon publication in the **Federal Register** because it relieves a restriction.

This suspension is a major action within the meaning of Executive Order 12291 and a significant action under the Department's Regulatory Policies and Procedures. The benefits and costs of the automatic restraint requirements have been carefully reviewed in the prior final regulatory impact analysis dated October 1981, which has been placed in the docket for the automatic restraint rulemaking. That analysis also provides an assessment of the impact of this suspension. The prior regulatory impact analysis also discusses the impact of the rescission of the automatic restraint requirements on small businesses and governmental entities. Based on that prior analysis, I hereby certify that this suspension will not have a significant economic impact on a substantial number of small entities. The Department has also evaluated this suspension in accordance with the National Environmental Policy Act and has determined that this action is not a major Federal action significantly affecting the quality of the human environment.

Issued in Washington, D.C. on August 30, 1983.

James H. Burnley, IV,
Acting Secretary of Transportation

Appendix A

The text of S4.1.3 of Standard No. 208, *Occupant Crash Protection,* (49 CFR Part 571.208) that was rescinded on October 29, 1981 (46 FR 53419) reads as follows:

S4.1.3 *Passenger cars manufactured on or after September 1, 1983.* Each passenger car manufactured on or after September 1, 1983 shall—

(a) At each front designated seating position meet the frontal crash protection requirements of S5.1 by means that require no action by vehicle occupants;

(b) At each rear designated seating position have a Type 1 or Type 2 seat belt assembly that conforms to Standard No. 209 and S7.1 and S7.2; and

(c) Either—

(1) Meet the lateral crash protection requirement of S5.2 and the roll-over crash protection requirements of S5.3 by means that require no action by vehicle occupants; or

(2) At each front designated seating position have a Type 1 or Type 2 seat belt assembly that conforms to Standard No. 209 and S7 through 7.3, and meet the requirements of S5.1 with front test dummies as required by S5.1, restrained by the Type 1 or Type 2 seat belt assembly (or the pelvic portion of any Type 2 seat belt assembly which has a detachable upper torso belt) in addition to the means that require no action by the vehicle occupant.

48 F.R. 39908
September 1, 1983

PART 571; S208-PRE 170

PREAMBLE TO AN AMENDMENT TO
FEDERAL MOTOR VEHICLE SAFETY STANDARD NO. 208
Occupant Crash Protection
[Docket No. 74-14; Notice No. 36]

ACTION: Final Rule

SUMMARY: This rule requires the installation of automatic restraints in all new cars beginning with model year 1990 (September 1, 1989) unless, prior to that time, State mandatory belt usage laws are enacted that cover at least two-thirds of the U.S. population. The requirement would be phased in by an increasing percentage of production over a 3-year period beginning with model year 1987 (September 1, 1986). To further encourage the installation of advanced technology, the rule would treat cars equipped with such technology other than automatic belts as equivalent to 1.5 vehicles during the phase-in.

DATES: The amendments made by this rule to the text of the *Code of Federal Regulations* are effective August 16, 1984.

The principal compliance dates for the rule, unless two-thirds of the population are covered by mandatory use laws, are:

September 1, 1986 — for phase-in requirement.

September 1, 1989 — for full implementation requirement.

In addition:

February 1, 1985 — for center seating position exemption from automatic restraint provisions.

SUMMARY OF THE FINAL RULE

After a thorough review of the issue of automobile occupant protection, including the long regulatory history of the matter; the comments on the Notice of Proposed Rulemaking (NPRM) and the Supplemental Notice of Proposed Rulemaking (SNPRM); the extensive studies, analyses, and data on the subject; and the court decisions that have resulted from law suits over the different rulemaking actions, the Department of Transportation has reached a final decision that it believes will offer the best method of fulfilling the objectives and purpose of the governing statute, the National Traffic and Motor Vehicle Safety Act. As part of this decision, the Department has reached three basic conclusions:

- Effectively enforced State mandatory seatbelt use laws (MULs) will provide the greatest safety benefits most quickly of any of the alternatives, with almost no additional cost.

- Automatic occupant restraints provide demonstrable safety benefits, and, unless a sufficient number of MULs are enacted, they must be required for the most frequently used seats in passenger automobiles.

- Automatic occupant protection systems that do not totally rely upon belts, such as airbags or passive interiors, offer significant additional potential for preventing fatalities and injuries, at least in part because the American public is likely to find them less intrusive; their development and availability should be encouraged through appropriate incentives.

As a result of these conclusions, the Department has decided to require automatic occupant protection in all passenger automobiles based on a phased-in schedule beginning on September 1, 1986, with full implementation being required by September 1, 1989, unless, before April 1, 1989, two-thirds of the population of the United States are covered by MULs meeting specified conditions. More specifically, the rule would require the following:

- Passenger cars manufactured for sale in the United States after September 1, 1986, will have to have automatic occupant restraints based on the following phase-in schedule:

  - Ten percent of all automobiles manufactured after September 1, 1986.

  - Twenty-five percent of all automobiles manufactured after September 1, 1987.

  - Forty percent of all automobiles manufactured after September 1, 1988.

  - One hundred percent of all automobiles manufactured after September 1, 1989.

- The requirement for automatic occupant restraints will be rescinded if MULs meeting specified conditions are passed by a sufficient number of States before April 1, 1989 to cover two-thirds of the population of the United States.

- During the phase-in period, each passenger automobile that is manufactured with a system that provides automatic protection to the driver without automatic belts will be given an extra credit equal to one-half of an automobile toward meeting the percentage requirement.

- The front center seat of passenger cars will be exempt from the requirement for automatic occupant protection.

- Rear seats are not covered by the requirements for automatic protection.

BACKGROUND

INTRODUCTION

The Supreme Court Decision

On October 23, 1981, the National Highway Traffic Safety Administration (NHTSA) issued an order pursuant to section 103 of the National Traffic and Motor Vehicle Safety Act, 15 U.S.C. 1392, amending Federal Motor Vehicle Safety Standard No. 208, *Occupant Crash Protection* (49 CFR 571.208; "FMVSS 208"), by rescinding the provisions that would have required the front seating positions in all new cars to be equipped with automatic restraints (46 FR 53419; October 29, 1981).

On June 24, 1983, the Supreme Court held that NHTSA's rescission of the new automatic restraint requirements was arbitrary and capricious. *Motor Vehicle Manufacturer's Association* v. *State Farm Mutual Automobile Insurance Co.*, 103 S.Ct. 2856. The agency had rescinded because it was unable to find that more than minimal safety benefits would result from the manufacturers' plans to comply with the requirement through the installation of automatic belts. In particular, the Court found the agency had failed to present an adequate basis and explanation for rescinding the requirement. The Court also stated that the agency must either consider the matter further or adhere to or amend the standard along the lines that its "reasoned analysis" and explanation supports.

By a five to four vote, the Court held that the agency had been too quick in dismissing the benefits of detachable automatic belts. The Court stated that the agency's explanation of its rescission was not sufficient to enable the Court to conclude that the agency's action was the product of reasoned decision making. The Court found that the agency had not taken account of the critical difference between detachable automatic belts and current manual belts. "A detached passive belt does require an affirmative act to reconnect it, but—unlike a manual seatbelt, the passive belt, once reattached, will continue to function automatically unless again disconnected."

The Court unanimously found that, even if the agency was correct that detachable automatic belts would yield few benefits, that fact alone would not justify rescission. Instead, it would justify only a modification of the requirement to prohibit compliance by means of that type of automatic restraint. The Court also unanimously held that having concluded that detachable automatic belts would not result in signficantly increased usage, NHTSA should have considered requiring that automatic belts be continuous (i.e., nondetachable) instead of detachable, or that FMVSS 208 be modified to require the installation of airbags.

The 1983 Suspension

On September 1, 1983, the Department suspended the automatic restraint requirement for 1 year to ensure that sufficient time was available for considering the issues raised by the Supreme Court's decision (48 FR 39908).

The NPRM

On October 14, 1983, the Department issued a notice of proposed rulemaking (NPRM) (48 FR 48622) asking for comment on a range of alternatives, including the following:

- *Retain the automatic occupant protection requirements of FMVSS 208.* Under this alternative, the substantive automatic occupant protection requirement of FMVSS 208 would be retained, but a new compliance date would have to be established. Compliance could be any type of automatic restraint, including detachable belts.

- *Amend the automatic occupant protection requirements of FMVSS 208.* Numerous alternatives were proposed. For example, an amendment could require compliance by airbags only or by airbags or nondetachable automatic belts only. Subalternatives included automatic protection for the full front seat, the outboard seating positions, or the driver only. An additional alternative would have required that cars be manufactured with an airbag retrofit capability.

- *Rescind the automatic occupant protection requirements of FMVSS 208.* The Department could again rescind the requirements if its analysis led it to that conclusion. The Supreme Court decision does not bar rescission after the Department "consider[s] the matter further."

The NPRM also proposed other actions that could be taken in conjunction with, or as a supplement to, the above alternatives. They were as follows:

- *Conduct a demonstration program.* Such a program could be along the voluntary lines suggested by Secretary Coleman in 1976 and would be accompanied by a temporary suspension of FMVSS 208's automatic occupant protection requirements. It would be designed to acquaint the public with the automatic restraint technologies so as to reduce the possibility of adverse public reaction and to obtain additional data to refine effectiveness estimates.

- *Seek mandatory State safety belt usage laws.* The Department could seek Federal legislation that would either establish a seatbelt use requirement or provide incentives for the States to adopt and enforce such laws. If large numbers of persons wore existing manual belts, there would be less need for automatic restraints.

- *Seek legislation mandating consumer option.* Under this alternative, the Department would seek Federal legislation requiring manufacturers to provide consumers the option of purchasing any kind of restraint system: airbag, automatic belt, or manual belt.

Following the issuance of the NPRM, the Department held public meetings in Los Angeles, Kansas City, and Washington, D.C. One hundred fifty-two people testified at these hearings. The public comment period on the NPRM closed on December 9, 1983. The Department received over 6,000 comments on that NPRM by the close of the comment period. Since then, the Department has received an additional 1,800 comments. Some of these comments raised issues or led to the identification of other alternatives on which the Department wanted to receive further public comment.

The SNPRM

As a result of the desire for additional public comment, the Department issued a supplemental notice of proposed rulemaking (SNPRM) on May 10, 1984 (49 FR 20460).

The SNPRM asked for comment on issues involving the following areas: the public acceptance of automatic restraints, the usage rates and the effectiveness of the various restraint systems, the benefits that would be derived from the various alternative means of protecting automobile front seat occupants, including potential insurance premium savings, and the testing procedures that would be required for automatic restraints. The SNPRM also sought comment on four additional proposed alternatives for occupant crash protection:

- *Automatic restraints with waiver for mandatory use law States.* Under this proposal, automatic restraints would be required in all cars manufactured after a set date, but this requirement would be waived for vehicles sold to residents of a State which had passed a mandatory safety belt use law (MUL).

- *Automatic restraints unless three-fourths of States pass mandatory use laws.* Under this proposal, automatic restraints would be required in all cars manufactured after a set date, unless three-fourths of the States had passed mandatory use laws before that date.

- *Mandatory demonstration program.* This alternative involves a mandatory demonstration program, which was suggested by the Ford Motor Company. Each automobile manufacturer would be required to equip an average of 5 percent of its cars with automatic restraints over a 4-year period.

- *Driver's-side airbags in small cars.* Under this alternative, airbags would be required only for small cars and only for the driver's position in those cars.

The comment period on the SNPRM closed on June 13, 1984. The Department received over 130 comments.

The Statute

Pursuant to the National Traffic and Motor Vehicle Safety Act of 1966, as amended, the Department of Transportation is directed to "reduce traffic accidents and deaths and injuries to persons resulting from traffic accidents." The Act authorizes the Secretary of Transportation to issue motor vehicle safety standards that "shall be practicable, shall meet the need for motor vehicle safety, and shall be stated in objective terms." In issuing these standards, the secretary is directed to consider "relevant available motor vehicle safety data," whether the proposed standard "is reasonable, practicable and appropriate for the particular type of motor vehicle...for which it is prescribed," and the "extent to which such standards will contribute to carrying out the purposes" of the Act.

The Safety Problem

Occupants of front seats in passenger cars account for almost half of the deaths that occur annually in motor vehicle accidents (including pedestrian fatalities). In recent years (1981-83), an average of approximately 22,000 persons have been killed annually in the front seats of passenger cars; another 300,000 suffered moderate to severe injuries and more than 2 million had minor injuries. Approximately 55 percent of these fatalities and injuries occur in frontal impacts and another 25 percent occur in side impacts. Table 1 shows the number of fatalities, by seating position, for 1975-1982, while Table 2 shows data for injuries, by severity and seating position, for 1982, the latest year for which such a breakdown is available. Table 3 provides estimates of similar data for 1990 to illustrate the impact of any rulemaking. For the 1990 data, it was assumed (for purposes of this rulemaking analysis only) that manual belt usage rates would remain the same as current rates.

To fully understand the benefits of various occupant restraint systems, it is helpful to recognize the frequency with which various front seating positions are used in cars involved in injury-producing accidents. As Tables 1 and 2 illustrate, three-fourths of all front seat occupant fatalities and serious injuries are experienced by drivers and almost all of the remainder are passengers in

TABLE 1
FRONT SEAT PASSENGER CAR FATALITIES WITH KNOWN SEATING POSITION

	Driver	Front Middle	Front Right	Other Front	Total
1975	16,270	644	5,601	21	22,536
Percent	72.2	2.9	24.8	0.1	100
1976	16,375	602	5,714	24	22,715
Percent	72.1	2.7	25.1	0.1	100
1977	16,967	577	5,992	14	23,550
Percent	72.0	2.5	25.4	0.1	100
1978	18,224	627	6,180	16	25,047
Percent	72.7	2.5	24.7	0.1	100
1979	18,267	513	5,968	6	24,754
Percent	73.8	2.1	24.1	—	100
1980	17,966	526	6,012	9	24,513
Percent	73.3	2.2	24.5	—	100
1981	17,722	460	5,844	6	24,032
Percent	73.8	1.9	24.3	—	100
1982	15,225	373	5,202	16	20,816
Percent	73.1	1.8	25.0	0.1	100

TABLE 2
DISTRIBUTION OF FRONT SEAT PASSENGER CAR OCCUPANT INJURIES BY SEVERITY LEVEL

Injury Severity	Driver	Front Middle	Front Right	Other Front	Total
Minor	1,388,519	29,914	515,786	2,526	1,936,745
Moderate	187,660	6,467	47,417	1,604	243,148
Serious	45,627	289	16,100	0	62,016
Severe	5,592	0	2,411	0	8,003
Critical	3,233	0	728	0	3,961
Percent of Minor Injuries	71.7	1.5	26.6	0.2	100.0
Percent of Moderate to Critical Injuries	76.3	2.1	21.0	0.6	100.0

the right outboard seat. Thus, automatic protection is likely to have three times the level of benefits for drivers as for front seat passengers. Additionally, not only are occupants of the center seat rarely involved in fatal or injury-producing crashes, but their involvement is declining as shown in the tables. This decline is thought to be occurring, at least in part, because of the decline in the number of automobiles manufactured with bench-style front seats.

TABLE 3
PROJECTIONS OF FATALITIES AND INJURIES FOR 1990

	Driver	Front Middle	Front Right	Total
Fatalities	18,050	370	6,140	24,560
Percent	73.5	1.5	25.0	100.0
Moderate to Critical Injuries	290,000	5,000	75,000	370,000
Percent	78.5	1.5	20.0	100.0
Minor Injuries	2,110,000	40,000	800,000	2,950,000
Percent	71.5	1.5	27.0	100.0

Current Occupant Restraint Technology

Manual belts

Manual belts are safety belts that will provide protection in a crash in the occupant places the belt around himself or herself and attaches it. Manual belts can come in two types: lap belts that fit around the pelvic region and combined lap and shoulder belts, which are found in the majority of all new cars sold today. Manual shoulder belts are equipped with inertial reels that allow the belt webbing to play out so that the occupant can reach forward freely in the occupant compartment under normal conditions, but lock the belt in place if a crash occurs. To remind occupants to use their belts, FMVSS 208 requires the installation of a brief (4-8 seconds) audible and visible reminder.

Automatic belts

The automatic belt is similar in many respects to a manual belt but differs in that it is attached at one end between the seats in a two front seat car and at the other end to the interior of the door or, in the case of a belt with a motorized anchorage, to the door frame. The belt moves out of the way when the door is opened and automatically moves into place around the occupant when the door is closed. Thus, the occupant need take no action to gain the protective benefits of the automatic belt.

Automatic belts differ significantly in their design. Some designs consist of a single diagonal shoulder belt (2-point belt) with a knee bolster located under the dashboard to prevent the occupant from sliding forward under the belt. Other designs include both a lap and a shoulder belt (3-point belt).

The designs differ also in the features and devices included to encourage belt use by motorists and at the same time allow for emergency egress if the car door cannot be opened following a crash. Several designs are described below.

One design takes advantage of the opportunity for the manufacturer to include, on a strictly voluntary basis, an ignition interlock. The belt in that design detaches from the door, but must be reattached before the car can be started the next time. This type of automatic belt (2-point belt with knee bolster) has been installed in more than 390,000 Volkswagen (VW) Rabbits over an 8-year period beginning in 1975. It was also installed on a small number 1978-79 General Motors (GM) Chevettes. It is still available as an option on Rabbits.

Another design is similar in that the belt detaches, but there is no ignition interlock. The belt may be detached and left that way without affecting the starting of the car. This was the type of automatic belt that most manufacturers had planned to use in complying with the automatic restraint requirement before the agency issued its rescission order. It was briefly offered by General Motors as a consumer option on a Cadillac model.

A third type of automatic belt is a continuous belt that does not detach at either end. Some continuous belts use a spool release, which plays out additional webbing length. Sufficient slack is created by an emergency release lever so that the motorist can lift the belt out of his or her way and exit in an emergency. Another type of continuous belt with a spool release mechanism is the motorized belt. The belt's outer anchorage is not fixed to the door but runs along a track in the interior side of the door's window frame. When the door is opened, the anchorage moves forward along the track, pulling the belt out of the occupant's way. When the door is closed, the process is reversed so that the belt is placed around the seated occupant. This type of continuous belt, which is a two-point system with a knee bolster *and* which contains a manual lap belt, has been installed in all Toyota Cressidas for the last several model years and enhances occupant ingress and egress.

Another type of continuous belt was installed on a small number of 1980 Chevettes. The belt consisted of a single length of webbing that passed through a ring near the occupant's inboard hip and served both as a lap and shoulder belt. The end of the lap belt that was connected to the lower rear corner of the door could be detached from door.

However, the end could not be pulled through the ring. Thus, the effect of detaching the lap belt was to create an elongated shoulder belt. The extra slack in the belt system enabled occupants to get out of their belt in the event of an emergency.

Airbags

Airbags are fabric cushions that are very rapidly inflated with gas to cushion the occupant and prevent him or her from colliding with the vehicle interior when a crash occurs that is strong enough to trigger a sensor in the vehicle. (Generally, the bag will inflate at a barrier equivalent impact speed of about 12 miles per hour.) After the crash, the bag quickly deflates to permit steering control or emergency egress.

In 1973-76, General Motors produced approximately 11,000 full-sized Chevrolets, Buicks, Oldsmobiles, and Cadillacs equipped with airbags. During the same period, Ford installed airbags in 831 Mercurys. A small number were installed in Volvos also. Today, only a single manufacturer, Mercedes Benz, is offering airbags in the United States. That company began offering airbag-equipped cars in the country beginning with the 1984 model year; it has been selling airbag cars outside the United States since late 1980. Since then, it has sold approximately 22,000 of those cars worldwide, with most sales occurring within the last year or so. GSA has contracted with Ford Motor Company to build 5,000 cars equipped with driver's side airbags. Delivery on these cars is expected to begin in Model Year 1985.

Other Automatic Occupant Protection Technologies

The automatic occupant protection provisions of FMVSS 208 do not specify that particular technologies, such as automatic belts or airbags, be used to comply with the standard. Rather, the standard requires a level of safety performance that can be met by any technology chosen by the manufacturer. Although safety belts and airbags are the most widely discussed technologies, the use of "passive interiors" as a means of compliance is also generating interest.

Under this approach, improvements are made to the vehicle structure, steering column, and interior padding so as to minimize potential occupant injuries. Thus, a "restraint" system of any kind, is unnecessary for occupant protection in frontal crashes. GM has been actively pursuing "passive interiors."

SUMMARY OF THE PUBLIC COMMENTS

INTRODUCTION

In this section of the preamble we have summarized the public comments on the Department's October 19, 1983, NPRM and the May 14, 1984, SNPRM. We have presented the summaries under headings that generally relate to the headings used in the subsequent portions of the preamble. Some of the comments are very generally stated and may relate to more than one issue. Because of the large number of public comments, we have provided a representative sample of the comments made and the commenters who made them. Subsequent portions of the preamble discuss the issues and alternatives and present the Department's position and response to the public comments. The comments are analyzed and responded to in more detail in the Department's Final Regulatory Impact Analysis (FRIA).

OCCUPANT PROTECTION SYSTEMS

Usage

Vehicle manufacturers generally agreed that mandating automatic belts would increase usage initially. However, based on their expectation of installing detachable automatic belts if required to install some type of automatic protection, some car manufacturers generally predicted that use would fall close to the current levels for manual belts once the belts were disconnected for the first time. GM believes this to be true for detachable automatic belts, and for nondetachable automatic belts as well. Honda also believes that, while there would be an initial increase in restraint usage if automatic belts were mandated, long-term usage of automatic belts might not be higher than current usage of manual belts. The key determinants

would be the comfort and convenience of automatic belts. The other manufacturers believed that automatic belts would probably produce some small usage increase. Chrysler stated that usage for automatic belts would be less than 10 percentage points higher than current usage for manual belts. Ford commented that the use of nondetachable automatic belts would initially be higher than the usage level for detachable automatic belts, but that over the long term it would fall to the same level. Ford said further that occasional belt users would use automatic belts more often than they currently use their manual belts, but that the overall level of usage would not significantly rise.

The car manufacturers generally believe that nondetachable automatic belts would not be practicable since consumers would object strongly to them and, therefore, would defeat and possible disable them. The manufacturers concluded that there would be little or no increase in usage over manual belt rates.

The Pacific Legal Foundation (PLF) said that mechanically compelled use by unwilling occupants would be no more likely to succeed than legally compelled use by such persons.

On the other hand, the American Seat Belt Council (ASBC) believes that usage of automatic belts would be 50 percent, which is roughly halfway between the current driver usage of 14 percent for manual belts and 80 percent for automatic belts with ignition interlocks. Professor William Nordhaus of Yale University believes that use of automatic belts would increase by 33 percentage points. John Graham of Harvard University found that expert opinion varies on the extent to which automatic belts would increase usage. His survey of seven experts found that detachable automatic belts would increase usage by 10 percentage points with an 80 percent confidence interval of 5 to 40 percentage points.

The issue of use inducing features or reminder mechanisms was raised by several commenters. ASBC believes that a continuous buzzer could double usage, and that buzzers, chimes and lights would all increase usage over levels that could be observed in vehicles without such features. VW stated that a continuous buzzer might be as effective as an interlock. On the other hand, Ford stated that while a continuous buzzer would induce some nonusers to wear their safety belts, driver irritation and actions to permanently defeat the system could also be anticipated.

Effectiveness

Manual Belts

The vehicle manufacturers generally stated that current manual lap and shoulder belts are more effective (when used) than either automatic belts or airbags. However, the combination of an airbag and manual lap and shoulder belts was acknowledged to be the most effective system of all.

The Automobile Importers of America (AIA) estimated manual belt effectiveness at 50 percent. Honda expressed the view that, based upon results of its 35 mile per hour crash testing, manual belts may be more effective than airbags in terms of chest acceleration and femur load injury criteria.

Most commenters on the SNPRM believed that the agency's range of effectiveness estimates for manual belts is too low. ASBC concluded that the estimate is too low because the agency estimate of lives saved from manual belt usage is approximately half the value previously cited by the agency. Renault argued that manual belt effectiveness data should not be adjusted to account for the presumably more cautious driving behavior of belt users, since belt use may lead some individuals to drive faster in the belief that they are better protected. VW provided a procedure for calculating manual belt effectiveness from NHTSA's Fatal Accident Reporting System (FARS) data, which led to a very high effectiveness estimate. Ford concluded that the agency's analysis would support a higher range of manual belt effectiveness (50-60 percent). Ford also challenged agency conclusions that manual belts are more effective in preventing moderate to serious injuries than fatalities and

that manual belts are not likely to be effective in accidents involving a velocity change of over 35 miles per hour.

Automatic Belts

The manufacturers stated that automatic belts may be less effective than manual belts. Similarly, the National Automobile Dealers Association (NADA) argued that automatic belts may be less effective than current manual belts if the automatic belt is attached to the door. VW and State Farm disagreed, saying that automatic belts are as effective as manual belts.

Volvo argued that nondetachable automatic belts may be less effective than detachable automatic belts due to a "film spool effect." This effect may occur in 1-door models, if the amount of webbing must be increased to allow entrance of passengers into the rear seat area.

The Insurance Institute for Highway Safety (IIHS) criticized the agency's effectiveness estimates for automatic belts, saying there was no support for the agency's conclusion that such belts, compared with manual belts, may increase the probability of occupant ejection. IIHS also suggested that the agency consider data that show that automatic belts may reduce the probability of the occurrence of head injuries. VW also challenged the conclusion that automatic belts could permit higher rates of occupant ejection. Ford argued that the agency should use a range instead of a point estimate for the fatality reduction of automatic belts. Ford also questioned the agency's conclusion that 3-point automatic belts should be as effective as manual belts, due to the lack of data supporting such a conclusion and the fact that manual belts can be more securely adjusted than automatic belts.

Professor William Nordhaus criticized the agency's adjustment of automatic belt effectiveness data to account for the lower accident experience of drivers who had elected to use belts as compared to nonusers of safety belts. The agency had concluded that as increasing numbers of current nonusers of manual belts were brought into the population of automatic belt wearers, the overall effectiveness of automatic belts would be decreased. Professor Nordhaus argued that the agency overestimated the magnitude of this effect. Professor Nordhaus also argued that automatic belts need not be less effective than current manual belts. In making this argument, he relied on agency crash test data and somewhat different data than those found by the agency to be most probative.

Airbags

Many consumer groups and health organizations indicated their belief that the reliability and effectiveness of airbags has been researched and tested to a far greater extent than any other item of vehicle safety equipment, and that the effectiveness of these devices is "unquestionable."

Allstate stated that airbags are more effective than belts in protecting against head and facial injuries. That company stated that while some of the dummies wearing belts "survive" 35 mph crashes under the injury test criteria, they sustained head and facial injuries far in excess of those produced with airbags at comparable speeds. Allstate noted, also, that belts were not dynamically tested as automatic restraints would be. Citing its field experience, Allstate said that airbags are effective not only in reducing deaths and injuries in frontal crashes but also in reducing injuries in side impact crashes. Allstate challenged the accuracy of the agency's NPRM estimate of airbag effectiveness, pointing out that that analysis was based on the use of restraint technology that is more than 10 years old. Allstate noted that GM itself had admitted that that technology was "obsolete." IIHS stated that, based on its analysis, airbags should be at least 34 percent effective in reducing fatalities.

Ford argued that the number of airbag cars that have been produced to date is too small to adequately answer questions about effectiveness.

PLF expressed the view that the agency really had no evidence that airbags are effective. That group argued that the agency erred in saying that the effectiveness of airbags is probably understated in the field data. According to PLF, DOT cannot know about all of the fatalities that have occurred in accidents involving airbag equipped cars. The group stated that the Department's estimate of airbag effectiveness is overstated to the extent that there are such undetected fatalities. Further, the group believes that the claim of the agency in the Preliminary Regulatory Impact Analysis (PRIA) that the large size of the cars equipped with airbags leads to an understating and obscuring of the potential effectiveness of airbags in smaller size cars is no more reasonable a conclusion than one that the large size of these cars masks the deficiencies of airbags by offering greater protection to out-of-position occupants and allowing longer deployment times for airbags. This group also asked DOT to provide an updated analysis of injury data for the fleet of airbag cars.

The National Head Injury Foundation stated that the airbag offers unique protection against head injury which even the automatic belt does not.

PLF and VW suggested that the presence of airbags might induce drivers to take greater risks while driving in reliance on the perceived increased protection. PLF argued that these increased risks could easily offset any gains in protection available as a result of the airbags. Professor Orr of Indiana University raised the same point, arguing that the "risk compensation" theory is sound but that the magnitude of its effect was unknown. IIHS submitted a study showing that the implementation of a safety belt use law in a Canadian province did not result in any increased risk taking by drivers. The study looked at the frequency with which certain risky maneuvers were made before and after the law was implemented and found no significant difference. John Graham stated that, based on several studies he has undertaken, any risk-compensation effect is significantly lower than the magnitude of benefits derived from the safety improvements.

Several vehicle manufacturers expressed their view that an airbag is relatively ineffective by itself, and should be viewed as a supplement to a belt system. The Motor Vehicle Manufacturers Association (MVMA) emphasized its view that airbags are effective in frontal crashes only.

In their SNPRM comments, several commenters addressed the agency's estimated range of effectiveness for airbags. IIHS concluded that the range is conservative but not unreasonable at the middle and high ends. They cautioned, however, that it would be inappropriate to compare the effectiveness of airbags in relation to safety belts by using the low end of the airbag effectiveness range and the middle or high end of the safety belt range. Mercedes Benz commented that its new "supplemental restraint system," which employs an airbag, has worked according to design in all accident situations in which vehicles equipped with the system have been involved.

PLF and VW also said that the Department's effectiveness studies were subjective. PLF argued that DOT was using precisely the same type of analysis that GM had offered and NHTSA had rejected in the 1977 rulemaking on automatic restraints. That group stated that DOT failed to explain this change of view. The PLF also criticized the agency's studies on airbag effectiveness for failing to take into account data for all vehicles using airbags, i.e., the non-GM Air Cushion Restraint System (ACRS) cars. Renault expressed the view that airbag effectiveness could not exceed 20 percent, due to the inability of airbags to provide protection in nonfrontal and ejection accident situations.

Ford argued that notwithstanding the limited amount of actual field data on airbag cars, those data cannot be totally dismissed in arriving at an estimate of airbag effectiveness. Ford also suggests updating field data to include Fatal Accident Reporting System data through 1983, instead of only through 1981 as was done in the PRIA. Ford found two of NHTSA's studies based on the National Crash Severity Study (NCSS) data to provide reasonable estimates of airbag effectiveness but found the third study to be flawed. Ford argued that the latter study was restricted to data from crashes in which airbags would be most likely to be effective. Ford also challenged a fourth agency study, on injury reducing effectiveness, based on field data, since it tended to show airbags to be most effective in accident situations in which the airbag is unlikely to deploy. Ford also stated that there appeared to be no basis for the agency's effectiveness range for airbags used in conjunction with safety belts.

Benefits

Several major insurance companies commissioned Professor William Nordhaus of Yale University to provide an updated economic analysis of alternative approaches to automatic crash protection. In response to the NPRM, Professor Nordhaus concluded that automatic crash protection would have net economic benefits to the nation of between $2.7 and $4.1 billion per year, while rescission would cost the nation $33 billion. Professor Nordhaus stated that every year of delay increases fatalities by approximately 5,000 and increases moderate to critical injuries by at least 70,000. His analysis also concluded that the impact of retaining the rule on profits or jobs in the automobile industry, as well as on the national economy, would be miniscule. He stated that automatic crash protection would be cost beneficial even if automatic belts increased restraint usage by only eight percentage points and even if airbags cost $825.

Many consumer and health organizations expressed concern that the agency had understated the benefits that would be associated with automatic restraints through their prevention of deaths and injuries. IIHS noted that the agency was relying on police reports to calculate the number of injuries from vehicle accidents. The group submitted evidence that only 70 percent of injuries resulting from vehicle accidents and treated in hospital emergency units were reported to the police. The evidence was taken from a study comparing car accident treatments in northeastern Ohio emergency rooms with police reports of accidents. To compensate for this underreporting of vehicle accident related injuries, this group suggested that the agency multiply its projected number of injuries by 1.4 to give a more accurate indication of the number of vehicular nonfatal injuries that could be expected. Such a step would, of course, increase the benefits associated with automatic restraints. Another group was also concerned that the agency had underestimated the minimum level of effectiveness of airbags and submitted an analysis showing that airbags would have a minimum effectiveness of 35 percent, instead of the 20 percent minimum used by the agency in the PRIA.

Several of the health organizations commenting on the proposal emphasized that the agency ought to reconsider the human costs of the head and spinal injuries suffered by persons in car accidents. One group submitted data projecting 66,000 head injuries annually as a result of vehicle accidents, with nine percent of those injured persons either dying in the hospital or discharged to chronic institutional care. Another 8 percent would be discharged but subject to follow-up medical attention. Many of these victims are young people who have to readjust to life with these injuries, which prevent them from performing even simple tasks they once did for themselves. These impacts are not readily quantifiable in dollars, according to these groups, but are just as significant as economic impacts for the people with family members who have suffered serious head and spinal injuries.

VW asked for an explanation of the methodology used in calculating Table 3 of the SNPRM, since the baseline of fatalities if no restrains were used seems to change with each listed effectiveness rate. This comment also noted that if mandatory usage laws are in effect by 1988, and 70 percent buckle up, the airbags' benefits would not equal the benefits of the mandatory use laws until the 21st century.

Professor Nordhaus states that using NHTSA's effectiveness rates for the various types of restraint systems shows both automatic belts and airbags to be highly cost beneficial, and that further delays cost the country at least $24 billion annually. He also stated that the benefits of mandatory belt use laws are so speculative as to necessarily remove those options from any serious consideration.

IIHS stated that DOT's projected airbag usage rate of 98 percent *a fortiori* means that airbags are the most beneficial alternative, because DOT has consistently recognized that the benefits of any of the restraint systems depend almost completely on the usage rates. IIHS repeated its contention that belt nonusers constitute such a disproportionate number of crash-involved occupants that actual reductions in deaths and injuries will be noticeably lower than would be projected for that level of belt use until the usage rate approaches 100 percent.

The insurance companies stated that several companies now have in effect 30 percent premium reductions for first and third party bodily injury liability for cars with automatic restraints. They contended, however, that the benefits associated with this rulemaking are not lower insurance premiums. In their view, the benefits are the prevention or reduction in seriousness of thousands of fatalities and serious injuries annually.

Public Acceptance

State Farm stated that it considered public acceptability of restraint systems to be a very important issue. It argued that a regulatory alternative could not be rejected on the grounds of insufficient public acceptability if the benefits of the alternative would exceed the costs of that alternative. It argued further that the legislative history of the Vehicle Safety Act made it clear that safety was the overriding consideration in implementing the Act. Thus, more weight should be given to the safety benefits of a contemplated safety requirement than to the public acceptability of the devices used to comply with that requirement.

State Farm also said that public reaction has regulatory significance as a legal and practical matter only if it is translated into behavior; that is, if people disable automatic restraints. If not, public acceptability meets the statutory criteria. Public opinion surveys over the last decade, including the 1983 GM and IIHS surveys, show public support

for mandatory automatic restraints, "All studies of usage rates of automatic belts show levels of incremental usage far above break-even levels."

Contradictory evidence was provided on the attitude of the public toward automatic restraints. Consumer Alert provided a public opinion poll showing that fewer than 15 percent of the respondents wanted mandatory automatic restraints. Public Citizen submitted a public opinion poll which it viewed as showing a clear preference for automatic restraints, especially airbags. IIHS cited a recent public opinion poll indicating that 56 percent of the respondents favored requiring automatic restraints on new cars as standard equipment and 37 percent favored requiring that the type of restraint be offered as an option. AAA stated that while consumers may not rush to purchase automatic restraints as options if manual belts were original equipment, they would accept automatic restraints as original equipment, particularly if they could choose between the various types of automatic restraints. Other groups argued that the increased protection against facial, spinal, and head injuries afforded by airbags would result in consumers choosing airbags as the preferred automatic restraint, if they are allowed to make that choice. Most of these groups indicated that airbags are less intrusive than automatic belts, and would therefore be more readily accepted by the public.

The manufacturers said that nondetachable belts would raise consumer acceptance problems since they are more coercive than current belts. This expectation is based in part on the interlock experience of 1974. NADA said that the experience with VW Rabbits, Toyota Cressidas, and GM Chevettes indicates a lack of consumer acceptance of automatic belt systems and that the GM experience with airbag cars shows a similar lack of consumer acceptance.

Mercedes, on the other hand, said that its system had met with "favorable market acceptance" in Europe and projected it would be accepted in the U.S. VW said, contrary to dealer statements, that it did not believe its automatic belts had been defeated in the sense of being destroyed but only that the interlock had been defeated, perhaps by dealers themselves.

MVMA submitted a memorandum of law with which GM and VW agreed. Ford and AMC also agreed, adding comments. MVMA restated the State Farm argument saying that State Farm believes the Act forbids NHTSA from considering adverse public reaction to a mandatory automatic requirement except to the extent that the public will disable the equipment. MVMA believes the State Farm position is not consistent with the legislative history of the Act, judicial precent, or prior positions of DOT. MVMA says that public acceptability is part of the "all relevant factors" consideration under the Act. Two 1974 congressional actions shed light on what is acceptable: the ignition interlock ban and congressional review of a mandatory automatic restraint rule (MVMA cites the Senate debate on the 1974 Federal highway aid bill on the congressional review issue). MVMA claims Secretary Coleman's decision was made with these factors in mind. Matters of future probability, as raised in the Coleman decision, are relevant to an agency decision even though they cannot be precisely measured.

GM agreed, adding that public acceptability is not a narrow issue.

VW also agreed, stating that public acceptability is a two-faceted problem: State Farm's concern over consumers defeating or destroying the restraint systems and public popularity are equally important. Consumer backlash could result from an expensive or coercive system, such as an ignition interlock. VW claims that airbags have been oversold; fatalities would continue and DOT's credibility would be questioned.

Ford agreed, stating that public acceptance involves far broader issues than disabling unwelcome equipment. Ford asks what percentage of front seat occupants would defeat automatic restraints and whether there would be enough benefits to justify the systems. Ford's best projection is that manual and automatic usage will be equivalent over the long run; that is, positive and negative belt use inducement factors for automatic belts will balance out to produce usage rates equivalent to those for active belts. Ford said also that comfort, entry and egress, and the defeatability of automatic belt systems are still unknowns; therefore, a field test is needed.

Chrysler said the State Farm position is too narrow. There must be widespread public perception that benefits are worth the price. It predicted that the automatic restraint requirement would suffer the same fate as the ignition interlock.

Toyota said the State Farm position is inappropriate. The public may press for legislative rescission of an automatic restraint requirement, even

though the public does not or cannot disable the system, citing the ignition interlock experience.

BL Technology Ltd. said that public acceptability and usage should be considered together. It said that the NHTSA definition of public acceptance is correct, i.e., "tolerance and use of restraint system," whether manual or automatic. BL suggests that the U.S. try mandatory seat belt use laws coupled with effective enforcement.

Renault accepts the State Farm interpretation but pointed out that a belt is needed with an airbag. Renault said that public acceptance and use of automatic belts will remain limited.

PLF and Consumer Alert said there is no mandate for an automatic restraint requirement. The issue of public acceptance is not limited to the sole question of deactivating mandatory automatic restraints; it encompasses all factors which may affect DOT's implementation of the Vehicle Safety Act. They said an automatic restraint requirement could cause the public to forestall buying new cars, which would delay the introduction of automatic protection and reduce safety by increasing the age of the total vehicle population. They also said DOT should consider risk compensation by those forced to wear belts or buy bags, citing John Adams' 1982 SAE paper, which PLF claims DOT has ignored. Experience in other countries is also cited to show that restrained occupants are less likely to be involved in fatalities.

IIHS said that earlier evidence submitted by them and others shows that automatic restraints, especially airbags, are acceptable.

Allstate supports State Farm on the acceptance issue. Allstate argues that if public acceptability is a controlling factor, then we cannot continue with the present manual seat belt requirements, due to low usage levels. They said there is no doubt that airbags have the most public acceptance; automatic belts have greater acceptance than manual belts. Therefore, DOT should reinstate the previous automatic restraint standard.

The American Insurance Association supports the State Farm interpretation. It said DOT should require automatic restraints because they only require toleration by the public to be effective. The standard for public acceptance should be public acquiescence, not public preference.

The National Association of Independent Insurors (NAII) said the DOT record shows that mandatory airbags are acceptable.

NADA said State Farm is correct in suggesting

that public acceptance should be given a "narrow, legal interpretation." They argued that there are four indicia for determining public acceptance, each with substantial evidence: (1) The public has expressed opposition to coercive occupant restraint devices, e.g., the ignition interlock. The record shows people will disable automatic belts. (2) The cost indicates that airbags will not be replaced; therefore, they will be disabled after one use. (3) A significant number of consumers are unwilling or unable to purchase new vehicles equipped with automatic restraint devices. (4) Consumers will buy vehicles without automatic restraints, such as vans or pickup trucks, or used cars.

Cost and Leadtime

A number of manufacturers provided cost estimates for automatic restraints. The incremental consumer costs of adding a full airbag system were estimated at $838 by GM, $807 by Ford and $800 by Chrysler. Jaguar provided an estimate of $1800.

Breed Corporation submitted an estimate of $140 for its all-mechanical airbag design, assuming a volume of one million units. According to Breed, this estimate has been independently verified by technical experts familiar with auto industry practices, procedures and pricing mechanisms. The estimate does not include necessary vehicle modifications, such as adding knee bolsters. Romeo Kojyo provided an estimate of $150 for a driver airbag retrofit kit, exclusive of installtion and assuming an annual volume of one million units. Ralph Rockow, president of Dynamic Science, stated that airbags could be produced at an incremental consumer price of $185. The Automotive Occupant Protection Association incorporated the Rockow estimate in its comment and provided a detailed breakdown of costs for a $185 full front passenger system at a production volume of two million units annually.

The incremental consumer costs of adding automatic belts were estimated at $45 by General Motors and Richard Lohr, a cost estimating consultant, $115 by Chrysler, $150 by Jaguar and Honda, and $200 or more by Nissan and Renault. Peugeot provided an estimate of $350 for a motorized automatic belt system.

Numerous manufacturers provided comments on required leadtime. In commenting on an automatic belt requirement, GM stated that while 1¾ years is adequate for models already designed,

three years are necessary for new designs or non-detachable automatic belts. Chrysler, Mazda and Peugeot also stated that 3 years are needed for automatic belts. Renault said that 24 months were needed for belts, while AMC said 30 to 36 months. Nissan provided an estimate of 30 to 42 months and Ford provided a figure of 4 years. VW said it could comply immediately for some models but would need 4 years for all models.

GM's estimate for a airbag requirement was 3 years for large cars and longer for small cars. Chrysler stated that 4 to 5 years would be needed to implement a requirement for full front airbags. AMC stated that 3 to 3½ years would be necessary for such a requirement, while Ford said 4 years. Renault said 3 years were needed while Saab claimed 58 months were necessary.

The National Safety Council said the automatic restraint requirement should be made effective September 1985, or 1 year thereafter at the latest. Mr. Lohr, a cost estimator, provided an estimate for automatic belts of 18 months, while the Automatic Occupant Protection Association (AOPA) stated that 18-30 months leadtime would be sufficient.

Two studies were submitted to the docket that analyzed the overall economic effects of an automatic restraint requirement. One study was by Dr. Barbara Richardson, of the University of Michigan, and was sponsored by MVMA. The other study was by Professor William Nordhaus and was sponsored by several major insurance companies.

Dr. Richardson concluded that a requirement for airbags costing between $300 and $800 per car would have severe detrimental effects on the automotive industry and the economy as a whole. Dr. Richardson stated that a short-run reduction in vehicle sales of 2.7 percent to 9.7 percent would occur, as well as an increase in unemployment of between 62,000 and 197,000 persons. She also concluded that gross national product (GNP), wages, disposable income, and personal consumption would decrease.

Professor Nordhaus concluded that an automatic restraint requirement would have a minimal effect on the automobile industry and the national economy as a whole. According to his analysis, an automatic restraint rule would result in an increase instead of a decrease in jobs in the automobile and supply industries.

NADA said the dealership operating costs and costs of automatic repair and service would increase.

Insurance Premium Changes

Numerous insurance industry commenters stated that implementation of an automatic crash protection requirement would provide significant economic benefits in the form of insurance premium reductions. Some commenters provided specific estimates of savings. Others argued more generally that an automatic restraint requirement would result in cost savings and that those savings would be reflected in insurance premium reductions. According to insurance commenters, a number of insurance companies have for some time been offering premium discounts for medical payment coverage for cars equipped with automatic restraints. Those commenters indicated that some discounts apply to all types of automatic restraints, while others are restricted to airbags.

Nationwide stated that installation of airbags in all automobiles would reduce private first- and third-party liability premiums by 24.6 percent of $31 annually per insured car. Using the Nationwide data, Professor William Nordhaus, in his NPRM comments, estimated that owners of cars equipped with automatic belts would experience consumer insurance cost savings of $24 per year. Professor Nordhaus estimated that, for vehicles equipped with automatic belts, taking into account consumer cost of the automatic belt, fuel cost and insurance cost, the total direct financial impact over the life of the vehicle would be to lower the cost of operating an automobile by about $60. According to Professor Nordhaus, this underestimates true total consumer savings as it omits non-insurance costs, lost wages, medical costs borne by the consumer and pain and suffering. New York State Insurance Superintendent Corcoran stated that, for average New York premiums, an all airbag requirement would result in insurance savings of $66 per year.

State Farm stated that while it does not now offer a discount to policy holders with automatic restraint equipped vehicles, the substantial financial benefits resulting from an automatic restraint requirement would be reflected in its rates, although it could not give an quantified estimate of that reduction. According to State Farm, its consistent policy in making insurance pricing decisions is to base them upon actual observed on-the-road insurance experience. State Farm also stated that, while that practice remains its policy, in other cases it has responded to competitive pressures where discounts have been made available,

and it expects that the same thing would occur in this instance. Several other companies also emphasized that premium reductions would result as fatalities and injuries are reduced by automatic restraints. Emphasizing the relationship between premiums and loss experience, Nationwide noted that since August 1981, it has lowered auto insurance rates in 19 jurisdictions, despite continuing inflation. Insurance Superintendent Corcoran stated that he would mandate reductions in New York to assure that savings to insurers are reflected in premium rate changes to the public and assumes that all other regulators would do the same. Since his comments were submitted, New York has enacted legislation authorizing the superintendent to require such premium reductions.

Not all commenters were certain that insurance costs would be reduced. Dr. Barbara Richardson, of the University of Michigan, stated that estimates of insurance premium changes resulting from airbags range from a large decrease over the lifetime of a vehicle to a net increase in insurance cost. In addition, one insurance company, the Automobile Club of Michigan, expressed concern that the PRIA's estimates of additional insurance costs for airbags, based on replacement frequencies and costs, were substantially understated. The Automobile Club and the General Motors Acceptance Corporation (GMAC) argued that the agency forgot to include increases in insurance premiums to reflect the greater value of cars equipped with airbags.

The commenting insurance companies, including State Farm, also indicated that insurance premium reductions would occur in States that enacted safety belt usage laws, to the extent that real world experience justified such reductions. The American Automobile Association (AAA) of Michigan said it would lower personal injury premiums by 20 percent upon enactment of a seatbelt use law. Commenters indicated that some companies now offer an incentive of increased benefits at no additional cost if manual belts are worn. Commenters pointed out difficulties in implementing a discount program for seatbelt usage, since verification of such usage, both generally and in the case of specific accidents, is not easy to obtain.

In response to the SNPRM, State Farm referred to the discounts offered for 5 mph bumpers as an example of the industry's quick reaction to reduce rates when new safety features are introduced. Citing the D.C. Circuit's decision in *State Farm v. DOT*, State Farm argued that insurance companies' practices have no significance for the decision that DOT has to make. It argued that if this concern were relevant, insurers have already given premium discounts for automatic restraint cars. It further argued that the issue of premium reductions is irrelevant to the conclusion that an automatic restraint rule will be cost beneficial. It said this is so "because a proper cost-benefit analysis weighs the costs and benefits of a standard to society as a whole. That balance cannot be determined from an analysis of the insurance effects of a rule, since there are enormous societal losses that go uncompensated under any insurance coverage." Finally, State Farm argued that DOT has a statutory obligation to require implementation of new technology where necessary to further the Safety Act and that consideration is different from the actuarial considerations that determine whether an insurance company will offer a premium discount.

The American Insurance Association (AIA) said that the industry has previously addressed the issue of insurance reductions. AIA pointed out that many of its members currently offer a 30 percent discount for medical payment and/or no-fault coverage for automatic restraint equipped vehicles. It referred to Nationwide's estimate of a potential annual premium savings per insured car that would equal $31 if all cars had airbags. AIA also noted that Nationwide and United Services Automobile Association (USAA) currently provide incentives for wearing manual belts.

Nationwide criticized the agency for allegedly ignoring Nationwide's previous testimony on insurance premium reductions. Nationwide said that, for the past 10 years, it has provided a 30 percent discount for first-party injury coverages for cars equipped with airbags. It further noted that, in its DOT testimony in 1976, it submitted its estimate of premium savings and its methodology for deriving that estimate. Nationwide updated that estimate to 1982, and said the potential insurance saving per policy holder is $31 annually. That estimate is for a full front-seat airbag system; Nationwide said that it is currently studying what discount it would give to a driver-side only system. It expects to offer a 25 percent discount on first-party medical coverage.

Nationwide also pointed out that, since 1963, it has offered extra medical insurance coverage, at no cost, to policyholders wearing their safety belts; last year it began providing a $10,000 death benefit and doubled medical payments coverage at

no extra cost to policyholders wearing belts.

Allstate said that since 1974 it has had a 30 percent discount on first-party injury coverages for airbag-equipped cars. It said that if airbags were installed in the entire fleet, there would be a 30 percent reduction in all insurance premiums, including medical payments, no-fault personal injury protection, death benefits, uninsured motorist coverage and bodily injury liability protection. Allstate said it could not provide an estimate of the insurance cost savings for automatic belts.

NAII pointed to prior testimony by USAA and Allstate providing details of insurance savings and observed that Nationwide specifically responded to the Secretary's questions at the public hearing concerning savings. NAII provided an attachment summarizing the prior industry testimony on the insurance savings issue.

NAII criticized the SNPRM's suggestion that insurers are not providing incentives for belt use. It cited Nationwide's policy and Leon Robertson's study that found that insurance incentives have not increased belt use. It also cited a 1980 National Academy of Sciences report done for DOT which questioned whether insurance incentives would be effective.

The Kemper Group said it currently offers a discount of up to 30 percent on first-party medical payment and no-fault auto insurance rates for cars with automatic belts or airbags. Kemper said that the cost of replacing an airbag could raise the physical damage insurance cost, but the increase would be minimal compared to the costs of the deaths and injuries that could be avoided with airbags.

Aetna estimated that the reduction in first-party no-fault, medical payments and uninsured motorist coverage premiums would be 25 to 30 percent for airbag equipped cars. As the percentage of automatic restraint equipped cars increases in the fleet, Aetna said there could be a similar reduction in third-party bodily injury premiums.

Conversely, Mercedes said "no company to our knowledge has reduced its rates on Mercedes-Benz Supplementary Restraint System (SRS) equipped vehicles" and Volkswagen stated that, to their knowledge, "no major insurance company offers a discount to owners of automatic restraint equipped vehicles," despite the fact that VW has been approached by insurers ostensibly for that purpose. VW said it has provided information to insurance companies because it desires to see its customers who have purchased automatic belt equipped Rabbits rewarded through lower insurance premiums.

Other Issues

Product Liability

The Automotive Service Council of Michigan raised the issue of the potential liability of independent repair shops that would service automatic restraint equipped vehicles. In addition, individual new car dealers and NADA raised the issue of whether the use of automatic restraints will increase a dealer's product liability costs. William C. Turnbull, President of NADA, testified that:

The reliability of passive restraint systems, particularly airbags, has been a matter of grave concern to dealers and consumers alike. No mass-produced product can ever be "failsafe." Components deteriorate due to passage of time, usage and climate. There are reports of inadvertent airbag deployments in the past. We fear that, with any widespread usage of airbags, incidences of inadvertent deployments and system failure will occur, with perhaps tragic consequences to vehicle occupants. In such cases, dealers may be the innocent victims of product liability lawsuits.

However, Willi Reidelbach of Mercedes-Benz, which is currently marketing an airbag-equipped car in Europe and the U.S., testified that he was not aware of any product liability concerns expressed by Mercedes dealers about the airbag system.

Several insurers provided comments on the potential of automatic restraints to reduce product liability claims and the availability and cost of manufacturer product liability insurance. Mr. Donald Schaffer, Senior Vice President, Secretary, and General Counsel of Allstate, testified that:

Our product liability people believe that the airbag equipped cars, if you insure the total vehicle, will produce better experience than the non-airbag cars because the airbag reliability factors are much higher than anything on the car. They are much higher than the brake failure rates or anything else.

Mr. Schaffer also testified that at the time of Secretary Coleman's proposed demonstration program, Allstate was Ford's product liability insurer and had informed Ford that there would be no increase in its product liability insurance costs if Ford built an airbag fleet. He also testified that Allstate entered into a written agreement with General Motors that "we would write all of their product liability insurance for cars in the Coleman demonstration fleet at the same price they were getting from their regular product liability insurer per unit for non-airbag cars of the same make and model year."

NAII also addressed the product liability concerns raised by manufacturers and dealers. NAII said that:

The potential for product liability suits is always present for any manufacturer or seller of consumer goods. That threat is present at the current time for anyone in the distribution chain. We in the insurance industry expect that savings (not increased costs) would accrue to manufacturers and dealers, as a result of automatic crash protection systems being installed in all cars, as lives are saved and injuries are reduced, thus reducing potential litigation over safety deficiencies.

Another potential source of manufacturer liability was raised by Stephen Teret, representing the National Association for Public Health Policy. Teret argued that:

If a reasonable means of protection is being denied to the motoring public, that denial should lead to liability, even if the liability can be imposed on each and every car manufacturer. People whose crash injury would have been averted had the car been equipped with an airbag can sue the manufacturer to recover the dollar value of that injury.

Sodium Azide

The Institute of Scrap Iron and Steel (ISIS) and the Automotive Dismantlers and Recyclers Association (ADRA) said that they were concerned about potential health hazards posed to their employees by sodium azide contained in airbag systems. Both ISIS and ADRA noted that sodium azide is toxic and a mutagent and that there is a general correlation between mutagenicity and carcinogenicity. In addition, they raised the issue of possible air canister explosions during the recycling and scrapping process.

To reduce potential hazards they recommended a number of actions:

1) Place a warning on the vehicles with airbags so their employees can easily identify them.

2) Design airbag systems so that they can be deployed by remote control or so that they can be easily removed from a vehicle.

3) Provide financial incentives, such as a bounty or fee, for removing the airbag canister.

Breed System

The Breed Corporation estimates the cost to the consumer of a Breed airbag system for the driver and one passenger to be $140 installed, based on an initial production rate of 1 million units annually. Breed states that its cost estimates have been independently verified by technical experts familiar with auto industry practices, procedures and pricing mechanisms. Breed says that the system still requires a "good" year of research before it can be put into production.

Ford and GM expressed doubts about the readiness and performance of the Breed System.

Breed urged DOT to require car makers to design airbag cavities in steering wheels and dashboards to facilitate the retrofitting of cars with airbags.

Automatic Belt Detachability

Virtually all commenters who addressed the issue of detachability expressed concerns that nondetachable belts should not be required. The vehicle manufacturers generally agreed that the public, especially the hard core belt nonusers, would react adversely to nondetachable automatic belts. They also doubted that the difference in the long run usage rates for detachable belts and for nondetachable belts would be significant.

GM suggested that its experience with the 1980 Chevette shows that the public will not accept nondetachable belts. According to GM, general annoyance and fear of entrapment will lead many hard core nonusers to defeat that type of belt. As to detachable automatic belts, GM says that the inertia effect cited in the *State Farm* decision can be expected to operate only until the belts are first detached. While there would be an initial increase in usage, in the long run neither detachable nor nondetachable automatic belts would yield any increase in usage. Ford agreed that fear of entrapment would produce some adverse reaction to nondetachable automatic belts. Ford stated that detachable automatic belts would produce some undefinable amount of usage increase. While nondetachable belts would produce higher increases in the short run, in the long run the usage rate for nondetachable belts would fall to the level of the usage of detachable belts. Honda commented that nondetachable belts would not be accepted by the public because of entry and exit problems, entrapment fears and poor appearance. Nissan anticipated no difference in the long-run usage rates of detachable and nondetachable belts. VW said that the high usage rate of their automatic belt is due largely to the interlock. Without the interlock, VW said, the usage rate would be between that for manual belts and the current VW Rabbit automatic belt system. VW suggested also that it was important in designing an automatic belt to locate the release mechanism near the window so that persons assisting an injured occupant could release the belt. ASBC predicted that 10 to 20 percent of car occupants are hard-core nonusers who will cut out nondetachable belts. The Council said that, in the long run, usage of detachable belts would fall between current manual belt usage rates and the rates for automatic belts in cars on the road today, i.e., usage would be about 50 percent. IIHS submitted a survey indicating that 68 percent would never detach a detachable belt, 21 percent would occasionally and 8 percent would do so permanently. John Graham stated that his survey of experts indicated that detachable automatic belts would increase usage by 10 percentage points and that 55 percent of motorists would dismantle nondetachable belts.

Alternatives

Retain

Most of the manufacturers indicated that they would comply by installing detachable automatic belts, since those belts would facilitate emergency escape from a vehicle after a crash and would face the least consumer resistance due to their lower price (compared to airbags) and the fact that they can be detached by occupants who do not choose to use safety belts for whatever reason.

Several insurance companies argued that the agency is required by law, based on the record, to implement some form of an automatic restraint requirement. According to State Farm, the effect of the Supreme Court's decision in *State Farm* is to require the Department to go forward with an automatic restraint requirement unless it has a rational basis for concluding that effective automatic restraint technology is not within reach of the car manufacturers. That company argued that the record amply demonstrates the existence of such technology.

Allstate argued that the record demonstrates that cost beneficial technology exists which, when included in all new cars, could save up to 10,000 lives each year and prevent more than 100,000 serious injuries annually. Allstate also argued that under the decisions of the United States Court of Appeals and the United States Supreme Court in the *State Farm* case, the Department lacks authority to look beyond that fact. That company stated that in its view, all proposed options that do not include the implementation of some form of automatic restraint requirement must, under the law, be rejected.

Similarly, NAII urged that the case for automatic protection has been fully documented. According to NAII, further delays for studies, demonstrations and so on are totally unwarranted and would only result in many more needless deaths and injuries. Such delays would also be inconsistent with the mandate of the Supreme Court.

Almost all commenting insurance companies favored implementation of the automatic restraint requirement as soon as possible. These commenters generally argued that the requirement is cost beneficial and would save many thousands of lives and prevent tens of thousands of injuries annually. Several insurance companies stated that airbags offer the greatest possible safety benefits. However, the insurance companies generally urged that such issues as requiring compliance by means of airbags only or barring compliance with detachable automatic belts should be considered only after a general automatic restraint requirement has been implemented. Allstate stated that the airbag-only requirement is preferable, but said that simple retention of the automatic restraint requirement is acceptable.

IIHS supported retention, noting, as did various commenters associated with medical and health organizations, that public health measures depending for their success upon repeated cooperation of the intended individual beneficiaries, as would mandatory belt-use laws, have historically had limited effectiveness.

Insurance Superintendent Corcoran of New York State maintained that it has been clearly established that, for whatever reasons, people do not generally use their manual belts, and efforts to modify this behavior have been unsuccessful for the past 15 years. He believed that it is incumbent on DOT to mandate automatic restraints as the only means for increasing usage.

The manufacturers said that if automatic belts are less effective than manual belts, then persons who regularly use manual belts would end up paying more in the future for an inferior restraint system, raising fairness questions. Most of the companies indicated that, if the automatic restraint requirement were retained, they would use detachable automatic belts to comply, since those systems facilitate emergency escape from a vehicle after a crash and would face the least consumer resistance due to their lower price (compared to airbags) and the fact that they can be detached by occupants who do not choose to use safety belts for whatever reason. However, if such belts were left detached by most occupants, little safety benefit would be gained through their installation.

PLF and Consumer Alert and vehicle manufacturers argued that DOT should concentrate on educating the public about the value of manual belts in providing protection in the event of a crash. Once the public is convinced of the need to buckle up, fatalities and injuries will decline without having to mandate expensive new equipment in cars.

GM argued that implementation of the automatic restraint requirement would divert engineering resources away from the development of more publicly acceptable alternatives, such as the "built-in" safety of energy absorbing interiors. Increasing safety through the redesign of vehicle interiors instead of the installation of add-on devices like occupant restraints would benefit unbelted as well as belted occupants at a cost far below that of airbags.

Amend

Airbag Only

Several health organizations argued that the agency should mandate airbags because that type of automatic restraint is the least intrusive for the occupant and because young drivers were the least likely to buckle manual belts and the most likely to try to defeat automatic belts. The Center for Auto Safety (CFAS) argued that small car occupants need the protection of airbags. The organization suggested that belts properly fit less than 50 percent of the population.

Many consumer groups and health organizations supported agency action that would mandate the installation of airbags in at least some new cars. To avoid the Congressional intervention that they thought might follow adoption of a requirement for nondetachable automatic belts, some consumer groups and health organizations urged adoption of either a requirement for airbags only or a requirement for airbags or nondetachable automatic belts.

The manufacturers objected to an airbag-only requirement for several reasons. First, it was stated that an airbag is effective only in single impact, frontal crashes, and does not protect against occupant ejection from vehicles. The manufacturers view airbags as supplemental protection devices, to be used in conjunction with safety belts. The manufacturers also expressed concern as to the real world reliability of airbags, the difficulties in applying airbag technology to small cars, the effects of airbag inflation on out-of-position occupants (particularly small children), the potential adverse environmental impacts of using sodium azide as a propellant to inflate the airbag, and product liability impacts. The economic effects of an airbag only requirement were a major concern of the manufacturers. The additional cost of that restraint system was projected to raise vehicle prices significantly, adversely affecting industry sales and thereby employment and profitability.

Some commenters, including MVMA, argued that adopting an automatic restraint requirement that specified the installation of a specific type of restraint, i.e., airbags, would violate the requirement of the Safety Act that safety standards be stated in terms of performance instead of design.

Congressman Dingell questioned the legal authority for an airbag-only requirement in light of *Chadha*, which declared the legislative veto to be unconstitutional. The Congressman suggested that if the legislative veto provision were invalid, then because of the absence of any severability provision and because of the importance attached by Congress to the veto provision, the exception to the prohibition in the Vehicle Safety Act against non-belt standards must fall with the veto provision.

One public interest group (PLF) and one economist, Professor LLoyd Orr, argued that airbags would encourage motorists to drive less safely since they would be given more safety than they desire and would compensate accordingly. Their argument is based on the "risk compensation hypo-thesis," which states, for example, that given better brakes, a driver is likely to follow more closely, negating some of the benefits associated with the safer braking system. The IIHS and John Graham, another economist, presented data which contradicted the above hypothesis. Those data concern the behavior of drivers in Newfoundland which indicate that safety belt users were not any more likely than nonusers to make risky driving maneuvers. John Graham referred to papers he had authored, criticizing the concept of "risk compensation hypothesis."

Airbags and Nondetachable Automatic Seatbelts

Some consumer groups and health organizations argued that permitting readily detachable automatic belts would only encourage those consumers not already in the habit of wearing belts to detach the belts and would result in a minimal increase in protection for car occupants. These groups urged therefore that the agency mandate that automatic belts not be easily detachable.

Some consumer groups and health organizations argued that automatic belts should be detachable to allow ready escape in emergency situations and to permit those confirmed nonusers of seatbelts (estimated by these groups at 10 to 20 percent of the population) to deactivate the belts for themselves by something other than permanent means, such as cutting the belts. These groups argued that nondetachable automatic belts would lead to Congressional action overturning the entire automatic restraint standard just as Congress had overturned the ignition interlock requirement in 1974. The car manufacturers opposed this option because it would limit their flexibility by requiring the installation of the most expensive and/or controversial types of automatic restraints. Manufacturers also argued that, given a choice, they would not produce nondetachable automatic belts because of anticipated adverse consumer reaction and difficulty in emergency egress with such systems.

Passive Interiors

GM stated that, since the original issuance of FMVSS 208, there have been significant advances in the state of the art of occupant protection. These advances have been made available in large part because of the increased use of advanced computer technology in the design and development of new vehicles. GM has implemented a Vehicle Safety Improvement Program which is aimed at increasing the "built-in" safety of its vehicles for restrained and unrestrained occupants.

GM said that the purpose of the "built-in" safety strategy is to maximize the reduction in total harm resulting from vehicle crashes. It argued that "no promising technology should be excluded simply because it either cannot meet arbitrary laboratory requirement or can only meet them on selected types of vehicles. Nor should new and promising technologies be discouraged because they are not envisioned in a regulatory scheme." GM urged that implementation of FMVSS 208 would "impede, or at lesat greatly dilute the effects that are needed to increase the state-of-the-art of other promising occupant protection technology."

In its comments on the SNPRM, GM suggested that DOT consider a more flexible approach to reducing deaths and injuries. They propose a three-step approach consisting of:

1) Retain the current requirements of FMVSS 208, but give manufacturers the option of meeting it with manual belts;

2) If a manufacturer chooses to comply with Standard 208 using manual belts, test the vehicle as follows:

   (a) fastened manual belts must satisfy the same dynamic criteria as airbags or automatic belts, and

   (b) the vehicle would be subjected to a 25 mph barrier crash with unfastened manual belts. The same injury criteria would be used to evaluate acceptable performance in this test as is used in the 30 mph test above; and

3) Approve various changes in the Standard 208 test procedures, most notably using the Hybrid III dummy, instead of the Hybrid II.

GM stated that this option would offer protection to all unbelted front seat occupants, not just the 5 percent of current non-users who would use automatic belts. GM estimated that this step would yield a 12 percent reduction in fatalities and serious injuries, which is equivalent to attaining 36 percent manual belt usage.

Small Cars

Several car manufacturers expressed concern about the difficulty of applying airbag technology to small cars. The shorter "crush space" between the fronts of small cars and the passenger compartments of those cars means that small cars decelerate faster in a frontal crash, leaving less time for an airbag system to sense the crash and inflate the airbag. The limited time means that the airbags must inflate more rapidly than in a large car, raising concerns as to airbag induced injuries, particularly to out-of-position occupants. GM expressed the view that the faster airbag inflation rate needed for small cars, in conjunction with the thicker airbag needed to decelerate the faster moving occupant of a small car, could cause fatal lesions in out-of-position occupants.

Honda expressed the view that airbags provide inferior protection as compared to manual belts in small cars at crash speeds above 30 miles per hour. Attempts to improve airbag performance in small cars through the use of a knee bolster were not particularly successful, since the resulting limited available space in such cars made entry inconvenient and the weight of the knee bar adversely affected fuel economy.

IIHS noted that two studies compared the effectiveness of airbags and manual lap/shoulder belts in small cars. One study, using Ford Pintos, showed that airbags performed slightly better than belts. The other study, using Renault R-12's, showed that the two types of restraints performed approximately the same, according to IIHS.

GM agreed that small cars needed the highest priority, but argued that the rapid inflation rate required to meet a 30 mph test poses an unacceptable risk to out-of-position occupants.

State Farm said that the analysis by Professor William Nordhaus of Yale University showed that it is significantly more cost beneficial to require installation of automatic restraints in both outboard seating positions and to require automatic protection for all size cars.

NADA restated its general opposition to any mandated automatic restraint and said that it was specifically opposed to a driver airbag-only option for small cars. NADA said that such a standard would be a design standard in violation of the Vehicle Safety Act and current airbag technology is not adequate for small cars.

Ford estimated that the cost of a driver-side airbag system would be about $600, which represents a large cost increase for vehicles at the lower end of the price range. Ford also questions the effectiveness of airbags in any size vehicle, the public acceptability of airbags, and the authority of the agency to issue an airbag-only standard.

VW also opposed driver-side airbags for small cars, saying that the technology is not proven for those vehicles and the Department should set performance and not design standards.

AMC supported the concept of requiring driver-side-only automatic restraints. AMC, however, said that airbags should not only be required on small cars since it "was not aware of any technical information that suggests that restraint requirements are fundamentally variable as a function of car size."

Nissan argued that requiring airbags for small cars is unfair to purchasers of those cars "because people buy small cars for economic reasons and the small car buyer should not be singled out to pay for expensive devices." Nissan also argued that if drivers assume that the airbag provides sufficient protection, then they might stop wearing their manual belts which are needed for protection in rollover and other accidents.

Toyota restated its general opposition to mandated automatic restraints and its specific opposition to a design (airbag) standard rather than a performance standard. It further argued that airbag technology has not been developed for small cars.

Allstate said that automatic protection should not be limited to small cars, but should be available on all cars.

The American Safety Belt Council (ASBC) said that a lap belt should also be required for a driver-only airbag. It recommended that for the right front passenger position, an automatic belt should be required.

Honda said that more development time is needed and that the added cost of airbags will substantially increase the cost of small cars.

Renault said airbag technology for small cars has not advanced far enough. It recommended waiting for the results of the Breed research program.

Jack Martens recommended that all cars with a wheelbase of less than 101 inches be equipped with airbags and with either manual or automatic belts for all front seat positions. Cars greater than 101 inches would be equipped with either nondetachable automatic lap and shoulder belts or airbags.

Public Citizen argued that if drivers of small cars can readily be protected then it is even more unreasonable not to protect the passenger in small cars and drivers and passengers in all cars.

IIHS supported mandating driver-side airbags in all cars, if it would lead to full front airbags.

Center Seating Position

Ford suggested that six-seat cars would probably no longer be produced if the center front seating position were required to be equipped with an automatic restraint. There is no known practical design for an automatic belt system that could be used for a three-position front seat. Hence, the only known automatic restraint system that could be used for the center position would be an airbag. Citing its concern about the hazards it believes would be posed by airbags to an out-of-position occupant, Ford indicated that it would probably choose to eliminate the front center seating position. The American Automobile Association (AAA), Chrysler, AMC and Consumers Union agreed that the center position should be excluded, noting that the agency's 1982 data show that 98.1 percent of front seat fatalities occur to persons sitting either in the driver's seat or in the passenger's seat next to the right door.

One commenter strongly urged that the front center seating position not be excluded from the automatic protection requirements since young children are the most frequent occupants of this position and thus would be the ones who would suffer the most from the absence of automatic protection.

Rescind

Those commenters who favored rescission opposed adoption of the other alternatives and vice-versa. Since this section of the preamble discusses each alternative separately, the views of commenters who favored one alternative are not necessarily included as negative comments to the other alternatives.

Generally, rescission was favored by all automobile manufacturers and by all new car dealers. Insurance companies and health associations all favored some form of retention and thus opposed the rescission alternative.

Most of the individual commenters opposed automatic restraints, especially airbags, on the basis of excessive government interference, high cost, and fear about the failure of airbags to operate properly. A very substantial number of these commenters were GM stockholders or employees.

Automobile manufacturers favored the standard's rescission on several grounds; that it was not as effective or cost-effective as mandatory belt use laws, that it unnecessarily would add to vehicle costs without commensurate benefits and that the technologies available for compliance would be rejected by the public as being too costly or intrusive.

For instance, Ford said that it could not support mandatory passive restraints by either amending or reinstating FMVSS 208 because of serious questions on restraint effectiveness and consumer acceptance.

GM said that detachable automatic belts are unlikely to increase belt usage and nondetachable belts would be rejected by the public. Because of technical concerns regarding airbags, particularly for out-of-position occupants in small cars, and because reinstatement would divert engineering resources from the development of passive interiors, GM believes the automatic occupant protection requirements should be rescinded.

The Automobile Importers of America (AIA) favored the adoption of mandatory use laws and said that questions of consumer acceptance, particularly regarding airbag technology and consumers' fear of entrapment, still need to addressed.

BMW said that the passive restraint issue should be "decided in the free market" and not by regulation.

One airbag supplier, Breed, recommended that the agency retain the current manufacturer option of installing either manual or automatic restraints. The commenter believed that this approach would impose minimal costs on the car manufacturers. After this supplier's airbag has been proven in more field tests, it believed that many car manufacturers would elect to provide airbags as readily available options.

The automobile dealers urged rescission because they thought that car purchasers are unlikely to accept automatic restraints. NADA cited the VW and Toyota experience with automatic belts and GM's experience with automatic belts and airbags as support for this contention. NADA also said automatic restraints would have an adverse impact on sales.

Most insurance companies and most consumer, medical and safety organizations opposed rescission or suspension, whether taken as a single action or in conjunction with a demonstration program or seeking legislation to mandate a consumer option, but organizations such as the Pacific Legal Foundation favored rescission. The PLF argued that the data did not support the Department's analysis of the effectiveness of automatic restraints.

State Farm said that a decision to rescind would be arbitrary and capricious. They referenced Professor Nordhaus' study as showing that rescission would impose enormous net costs on society. Nordhaus said that, for every year during which no automatic protection is required, it will cost society $2 to 2.5 billion. The American Association for Automotive Medicine said that "from a public health perspective, maximum protection requiring no action by the occupant is obviously preferable and desirable."

Congressman John Dingell argued that as long as the Department applied a reasoned analysis, rescission is possible and the best course to follow. Congressman Timothy Wirth contended that the statute requires that DOT move forward as promptly and expeditiously as possible to the implementation of meaningful automatic crash protection.

Joan Claybrook, of Public Citizen, said that there is more information on the benefits of automatic restraints than on any standard ever issued by NHTSA. Consumers Union "strongly" urged DOT "to promulgate promptly" FMVSS 208.

Demonstration Program

Ford argued that the effectiveness of automatic restraints could be determined only after a large-scale demonstration program is conducted. It proposed a program for the installtion of automatic restraints in 5 percent of the new car fleet over a 4-year period. The comments of several other manufacturers suggested that they would not oppose a demonstration program.

Ford said that the SNPRM misstated its proposed demonstration program requirement as at least 5 percent of each manufacturer's annual production for four years. Ford corrects this to mean an average of 5 percent of annual production manufactured for sale in the U.S. over a period of 4 years. Ford continues to believe that its proposal is the most effective means to resolve the stalemate on how best to improve occupant protection.

In response to the SNPRM, AMC said that a demonstration/test program similar to Ford's proposal is absolutely necessary prior to any effective date for requirement of automatic restraints. In the interim, the automatic restraint requirements should be suspended and a rule drafted so that rescission would occur if the findings of the test program were negative. AMC supports a demonstration program, but it does not feel that a mandatory program should necessarily be imposed on all low-volume car manufacturers. In some cases, the minimum added information to be gained would be more than overshadowed by excessive resultant cost. A five percent program for a 2- to 4-year test period would be acceptable, utilizing various automatic restraint systems for the driver only. AMC could launch such a program between early 1987 and fall 1987.

VW endorses a demonstration program and proposes an alternative plan, which would give credit to manufacturers that have already produced large numbers of automatic restraint cars. VW also said that any demonstration program should permit automatic belts to continue to be permitted. VW said that DOT should take into account the fact that costs will be higher for smaller manufacturers and that DOT has proposed no mechanism to "guarantee" that the public will buy automatic restraints.

Chrysler prefers mandatory seat belt use laws. If there is a demonstration program, companies would need adequate time to evaluate test results regarding airbag performance and public acceptability. Chrysler will cooperate in such a program, with up to 5 percent of its production for FY 1987 and 1988, provided that it applies to all domestic and foreign manufacturers. Chrysler believes there should be an automatic restraint for the driver only and that the program should only require a manufacturer's "best effort" to sell 5 percent of its total production, all on one car line, with appropriate pricing to validate public acceptance.

Volvo said the idea has some merit, but any airbag system should be for the driver only. The five percent figure should apply to total vehicle sales, not to a percentage of each car line.

Renault said that the program would produce concrete evidence in an uncertain area and that it should apply to foreign manufacturers selling more than one million vehicles per year in the U.S.

Honda said the program should be voluntary and include ways to encourage use of manual belts. Honda believes there are R&D problems that must be solved prior to an automatic restraint mandate. Honda opposes the requirement of two kinds of tooling on production lines and views the 5 percent requirement as unreasonable, regardless of demand.

Lotus said that since it imports only 300 cars into the U.S., at 5 percent, there would be 15 Lotus autos involved. It suggests an exemption for manufacturers selling less than 10,000 cars per year in the U.S. It points out this this is the small manufac-

turer definition used by EPA, and that DOT has overlooked the impact of this proposal on small entities, including manufacturers and dealers.

BMW would not be adverse to the program, if the manufacturer has a choice of driver-only systems, a choice of restraint type and vehicle models, and the initiation of the program was not earlier than September 1986.

Mazda suggested that DOT limit the program to high-volume production vehicles and to models produced in volumes exceeding 200,000 units per year. This will permit recovery of investment and development costs.

Peugot said that the demonstration program is the best approach. Peugot believes that conclusions can be drawn 4 years after implementation and that the program must take into account both manual and automatic restraints. The only disadvantages of the demonstration program are economic, but this can be alleviated by letting the manufacturer choose 5 percent of each model, or 5 percent of one model.

The American Seat Belt Council said that the program should be used only for airbags to determine market suitability. Any automatic belt system should be permitted to be detachable.

The Pacific Legal Foundation (PLF) said that if DOT is to proceed with the automatic occupant protection issue, it should use the demonstration program to acquire a data base.

General Motors (GM) said that a mandatory automatic restraint demonstration program does not answer the basic question of whether the public will accept or use automatic belts or accept the higher cost of airbags.

AMC said in response to the NPRM that it was inappropriate to require a small company like AMC to participate in a demonstration program.

Toyota was generally opposed to a demonstration program. However, if one were undertaken, the DOT program should: (1) contain performance, not design, requirements; (2) permit the manufacturer to select the car lines to be affected; and (3) have the same requirements for all manufacturers, small and large.

Nissan said that the problem with the program is that sales projections of any percentage are impossible to forecast. Only customer preference can dictate the numbers sold. But if the program is mandated, then: (1) Nissan would need 30 months leadtime; (2) it should permit either automatic or 3-point belts; (3) let the manufacturers decide the type of restraint on any mode; and (4) it agrees with Ford on amending the test injury criteria.

NADA said that automatic restraints have not been proven to be more effective than manual belts and that a demonstration program was a counterproductive idea due to delays in implementation (21 to 42 months) and assessments (6 to 8 years), which would divert manufacturer resources. It would also have an adverse effect on franchised dealers, who would have to attempt to sell the automatic restraint equipped cars.

IIHS opposed the program because it does not meet the statutory responsibility of DOT. There would be no economies of scale; therefore, higher costs could result. However, if it were done very quickly, the program could be a useful supplement to this rulemaking. IIHS reiterated its belief that a mandatory automatic restraint standard was needed as soon as possible.

Allstate said that a demonstration program could delay the safety needs of the public for 7 years, 4 for the demonstration, and 3 for lead-time to equip the rest of the fleet.

State Farm said such an alternative was unlawful, irrational, arbitrary, and capricious. Adoption of the Ford proposal would impose a costly, harmful and unjustified delay.

The National Association of Independent Insurers (NAII) opposed the program as a form of delay.

The Center for Auto Safety (CFAS) said the demonstration is outside the limit of DOT's statutory authority, as illustrated by former Secretaries Volpe's and Brinegar's requests to the Congress for explicit authority for a standard's phase-in based on percentage of production. The CFAS said that NHTSA has recognized that percentage phase-in is of questionable legality, citing the DOT brief in PLF v. Adams, 593 F.2d 1338 (D.C. Cir. 1979).

Public Citizen said that a demonstration was not authorized by the Act.

The Breed Corporation said that a mandatory demonstration program, since it would result in a safety standard which did not apply to all motor vehicles of a particular type, would be unlawful.

Mandatory Belt Use Laws

General

Almost all car manufacturers supported belt use laws in lieu of some form of automatic restraint requirement. They stated that these laws would be the most effective and least costly approach. The automobile dealers also supported these laws. Most individuals who opposed automatic restraints and supported an alternative named belt used laws as that alternative.

The American Seat Belt Council said that belt use laws would be the most effective approach, but expressed the belief that some sort of financial incentive would be necessary to get individual States to consider passage of such laws. Congressman Dingell supported belt use laws and noted his bill to encourage state enactment of them.

Many vehicle manufacturers and other commenters noted that belt usage laws would begin producing benefits over the entire fleet of cars on the road as soon as the laws became effective. By contrast, they noted, the benefits associated with automatic protection would accrue only as new vehicles equipped with automatic protection were added to the fleet of vehicles in use. It would take at least 10 years for car equipped with that type of protection to fully replace nonautomatic cars. Because of this factor, many commenters suggested that the agency mandate automatic restraints, to provide that protection to occupants of new cars, and seek belt usage laws, to provide increased protection to occupants of older cars.

The Motor Vehicle Manufacturers Association (MVMA) and several individual manufacturers stated that the minimum criteria specified in the SNPRM for belt usage laws deny State legislatures the flexibility to design belt use laws consistent with the demographics, motor vehicle statutes, and law enforcement practices of the individual States. These commenters suggested that rather than DOT specifying the means which must be used to achieve the goal of increased belt usage, it should simply specify the desired end (in terms of the percentage of front seat occupants wearing their belts) and allow the State legislatures to select the most effective means to that end for their particular State.

Several insurance companies opposed safety belt use laws as a substitute for the automatic restraint requirement because all front seat occupants of a car equipped with automatic restraints would be protected while a belt use law would protect only those front seat occupants who complied with it. The insurance companies, Congressman Wirth, and Public Citizen argued also that safety belt use laws were not an alternative that would satisfy the Safety Act or the State Farm decision. However, the insurance industry generally favored these laws as a supplement to an automatic restraint requirement.

Although virtually all medical and health organizations opposed substituting safety belt use laws for the automatic restraint requirement, they noted that recent experience in Canada and Great Britain has shown that introduction of these laws produced sizable reductions in injuries and deaths.

Both the Insurance Institute for Highway Safety (IIHS) and the Pacific Legal Foundation (PLF) submitted studies indicating that while belt use laws do increase usage, the resulting reductions in deaths and injuries are proportionately smaller than increases in usage. These studies led both groups to conclude tentatively that the population with the greatest likelihood of being in vehicle accidents is also the least likely to comply with belt use laws. A similar point was made by New York Insurance Superintendent Corcoran. Hence, both groups urged DOT not to overstate the benefits that would result from belt use laws. Ralph Nader opposed safety belt use laws as an alternative because of his belief that such laws would not be adopted by the States and would not be complied with by those who most need to buckle up.

As to the question of the likelihood of enactment of state safety belt use laws, IIHS said the closest analogy was not the child restraint use laws or the recent wave of more stringent drunk driving laws, but the motorcycle helmet use laws that have been repealed or weakened in a signficant number of States.

Several commenters including the National Association of Governors' Highway Safety Representatives (NAGHSR) stated that the DOT approach was fundamentally wrong in that it sets automatic restraints and belt usage laws as an either/or proposition. These commenters argued that both of these requirements are needed to ensure maximum use of restraints by front seat passengers. Further, these commenters asked why the Federal government was intruding on the States' prerogative to shape the usage laws by specifying minimum criteria.

The Governor of Wyoming stated that there was little or no chance of ever passing a belt usage law in that State, and recited a list of enforcement problems which would be posed for that State if it were to pass a belt usage law.

The insurance companies generally argued that DOT's options of pursuing belt usage laws were illegal as an abdication of DOT's statutory responsibilities. The proposals in the SNPRM, it was argued, would result in a lack of uniformity nationwide. As a practical matter, these commenters believed that either of the options which would eliminate the requirement for automatic restraints if States passed belt usage laws would encourage manufacturers to develop the cheapest automatic restraints which would satisfy the standard, since it was possible that the manufacturers would never be required to put these restraints in their vehicles and they would thus wish to minimize any investment losses. It was also stated that these systems would be the least effective automatic restraints. The insurance companies noted the serious enforcement problems which belt usage laws would impose on the States. IIHS stated that there is no evidence anywhere in this record to support the claims that belt usage laws would be obeyed without vigorous enforcement, and such enforcement would be a headache for the States. Their researchers found that in New York, where an administrative regulation requires holders of learner's permits to wear their belts while driving, 39 percent, 32 percent, and 6 percent of drivers with learner's permits actually wore their belts at three different locations. Further, IIHS noted that, as of the time of their docket submission, no State had yet passed a belt usage law and such laws were being considered in only 11 States.

Volvo responded to the claim that belt usage laws would not protect those who are most likely to be in accidents, and that therefore belt use laws will not achieve the reductions in deaths and injuries which would accompany a particular level of belt use. Volvo argued that these drivers would also be the most likely to defeat any automatic belts, and so would not be protected by those restraints, and the most likely to be in rollover crashes, in which they would not be protected by airbags.

SNPRM Alternative: No Automatic Restraints Required in a State That Passes a MUL

The manufacturers generally opposed this alternative on the grounds that it would create major distribution problems, it would create serious enforcement problems for the States (for instance, will residents of a State be permitted to cross the border to purchase a car equipped with the restraint system they want?), and it would force the manufacturers to produce two different types of otherwise identical vehicles.

The State of Washington asked why DOT would

waive an automatic restraint requirement, and stated that it believed the existence of automatic restraints would be as much of an incentive to pass a mandatory belt use law as would a waiver. Similarly, NAGHSR stated that the waiver would be an administrative nightmare for the States, and that this waiver would make it difficult for a consumer to purchase a car with automatic restraints if the State has a mandatory use law.

NADA stated that this alternative would create uncertainty and a patchwork pattern of automatic restraint requirements, which would cripple product planning, pricing, advertising, and distribution.

A Michigan legislator and the Michigan secretary of state supported this proposal, saying the most effective protection available to front seat occupants is the manual belt already in the vehicle.

SNPRM Alternative: Automatic Restraints Required Unless 75 Percent of States Pass Mandatory Belt Use Laws by a Certain Date

The manufacturers strongly objected to this alternative, since they would be forced to immediately begin investing time and money on a device which might never be needed. They said that this alternative would raise car prices even if the automatic restraints were never required. The manufacturers also stated that the progress reports were an unnecessary burden since a manufacturer that was not prepared to install automatic restraints when those were required would be completely forced out of the market until such time as it could install automatic restraints. That is incentive enough to ensure that the manufacturers will be ready to install those restraints.

Ford would change this alternative to suspend FMVSS 208 while a good faith effort is made to pass mandatory use laws, and, if this is unsuccessful, specify an effective date for FMVSS 208. Volkswagen (VW) suggests setting an effective date on a sliding scale after seeing if enough States pass mandatory use laws. For instance, if 10 percent of the States have not passed mandatory use laws in two years, Standard 208 would become effective three years after that date, if 25 percent had not passed mandatory use laws in 4 years, Standard 208 would become effective 3 years after that date, and so forth. American Motors Corporation (AMC) would amend the alternative to specify no automatic restraints when 75 percent of the driving public is subject to mandatory use laws or when 75 percent are using the manual belts in

their vehicles.

The National Automobile Dealers Association (NADA) stated that there is no basis for imposing automatic restraints, whether or not 75 percent of the States pass a mandatory belt use law.

The insurance companies wondered how DOT had decided that residents of 25 percent of the States could be left without enhanced occupant protection in their cars when the record was so clear on the need for enhanced protection. The National Association of Governor's Highway Safety Representatives (NAGHSR) stated that Federal intrusion was not needed to get States to pass mandatory use laws.

Two Michigan officials stated that the 75 percent figure should be lowered, since it was doubtful that it could be achieved, and argued that greater flexibility should be allowed to the States.

Test Procedures

Repeatability

Most automobile manufacturers raised several issues concerning the automatic occupant protection provisions of FMVSS 208. Statements were made that the test procedures, in general, fail to meet the "objective" criterion of the statute. Suggestions were also offered to change the procedures, the anthropomorphic test dummy, and the standard's injury prevention criteria.

Manufacturers stated that the test procedures do not produce repeatable results. Relying on data from the agency's New Car Assessment Program (NCAP) repeatability tests, the manufacturers argued that there is substantial, uncontrollable variability in the test results. As a result, they argue that the standard is not practicable.

NHTSA's New Car Assessment Program, which is an experimental program designed to develop consumer ratings of vehicle crashworthiness, is similar in test procedure to FMVSS 208 in that is uses instrumented Part 572 test dummies to ascertain potential injuries to human occupants in a frontal barrier crash. The program differs from FMVSS 208 in that its purpose is to rate cars. Therefore, there is no minimum level of performance specified as in FMVSS 208, and the tests are conducted at 35 mph instead of the safety standard's specification of 30 mph.

In 1983, NHTSA conducted tests to determine the repeatability of test results from the NCAP. Twelve Chevrolet Citations were tested in three different laboratories (four in each laboratory) to help determine the magnitude of variability surrounding a single test result. GM supplemented the agency's program by crashing an additional four Citations at their own facilities.

In commenting on the October 1983 NPRM, AMC referenced the NCAP repeatability tests and stated that based on the high degree of variability in injury criteria test results, the FMVSS 208 test procedures were "unacceptable" and lacked the necessary objectivity required by a safety standard. To compensate for this large variability, AMC suggested the agency use a "design-to-conform" approach as a means of compliance.

Chrysler also stated its concern over test repeatability and variability, as evidenced in the NCAP program, and argued that testing airbags under the current test procedure could lead to even greater variability. Chrysler suggested testing airbags with a belt, exempting the front center seat from any passive requirements, eliminating the 30-degree oblique test and waiving all injury criteria.

Volkswagen referenced the NCAP repeatability program and concluded from its results that the current test procedures were "not appropriate," particularly for safety belts. VW argued that the test procedures, and the dummy, were developed for testing compliance with airbags. It suggests that the procedures be revised to only use dynamic testing if a vehicle is equipped with airbags.

GM also spoke of excessive variability and stated that the test procedures must be improved. GM urged NHTSA to approve its petition to use the Hybrid III dummy as an alternative test device and to develop different compliance tests for different technological safety improvements.

Ford claimed that the test procedures are neither objective nor practicable and, based on the NCAP tests, manufacturers would have to "overdesign" their vehicles to ensure that all vehicles were in compliance. Ford stated that the procedures do not comply with the Court's ruling in the *Chrysler* case that test procedures must be capable of producing identical results when test conditions are exactly duplicated. Ford argued that repeatable results are impossible to achieve with the current FMVSS 208 test procedures. The company supplied results of early 1970's sled tests to show that variability was inherent in the test procedures and test dummy and was not solely related to vehicle-to-vehicle differences. Ford suggested that test

variability could be compensated for by using a design to conform approach, eliminating the 30-degree oblique test, not dynamically testing automatic belts, changing the FMVSS 210 anchorage location requirements, and testing airbags with a belt.

MVMA emphasized their concern that the NPRM failed to address the issue of test repeatability. Its concern was based on the NCAP test results. MVMA urged the agency to publish a supplemental notice to address the issue.

Several commenters to the NPRM suggested that there was no reason to be concerned over test procedures or repeatability. Byron Bloch, an automotive safety consultant, pointed out that cars are designed using crash tests and sophisticated dummies and he supplied the text of a GM advertisement to that effect.

The Insurance Institute for Highway Safety reviewed the results of the NCAP repeatability test program and concluded that these tests "produced repeatable results when the correct procedures were adhered to. . ."

Allstate Insurance Company claimed that the current test procedures assure individual purchasers of automatic restraints of protection and that the agency should also test manual belts dynamically.

Because of the above, the issue of repeatability, as well as other test procedure concerns, was raised in the SNPRM. In the SNPRM, the Department stated that it believed that the Part 572 test dummy was not a major source of the variability found in the NCAP repeatability tests, that the proposed adoption of two of the NCAP procedures into FMVSS 208 would further reduce variability, and that additional changes in the test procedures to reduce variability were not necessary. Any remaining variability was assumed to be due llargely to vehicle-to-vehicle differences, which are outside the control of the Department.

In commenting on the SNPRM, auto manufacturers took exception to the Department's conclusions.

Ford reiterated its prior arguments about repeatability and criticized the agency for not clearly setting out what are the proposed NCAP changes to the 208 standard. It characterized what it understood to be the revisions to the NCAP test procedures as minor, subjective, and unverified. Ford said that the agency was still conducting its repeatability research study and questioned how the agency could conclude that the test dummy is not a major source of variability.

Ford further argued that the agency had not shown that the "test device and test procedure are separable in their influence on test results from the performance of the vehicle, so that any variability in test results 'must be' attributable to vehicle-to-vehicle differences in manufacture or performance."

Ford also argued that overdesign should be used only to compensate for manufacturing variances, which can be estimated and controlled for by the manufacturer and that overdesign should not be required of manufacturers because of deficiencies in test procedures.

Ford concluded that the test procedures were "flawed," that variability was inherent in barrier crashes and was likely "irreducible," and that the current procedures, with their large associated test result variability, placed a manufacturer in "unacceptable jeopardy" in terms of assuring compliance with the standard.

The company also claimed that "comparable variability," to that observed in the NCAP Citation tests would be expected for other models. It based its conclusion on the coefficient of variation (COV) of 33 Mercury airbag sled tests, scaled to 35 mph, and seven Volvo barrier crash tests.

GM said that the driver HIC results of the NCAP repeatability tests, which incorporated the test procedure changes proposed in the SNPRM, already demonstrate that the range of variability is too large. GM argued that the amount of variability is not due to vehicle differences. It referred to a series of controlled sled tests it conducted, in which the coefficient of variation of the HIC data was as high as 11 percent for the driver and 8 percent for the passenger. For the NCAP series, the COV was 21 percent for the driver and 11 percent for the passenger. GM said that a comparison of the two data sets shows that the major portion of the variability is test-related, not vehicle-related.

GM argued that because of the variability, the amount of overdesign needed to provide a reasonable certainty of compliance would be impracticable. It said that the design level of HIC protection could not be justified in terms of a "minimum" safety requirement. GM said that it does "not believe that a practicable dynamic test requirement can be devised to provide manufacturers with the assurance of 'certainty' specified by the *Paccar* court. The only solution may be the one suggested by that court: "...it must propose some

alternative method for those manufacturers which, if followed, it will recognize as fulfilling the due care requirement.'"

Mazda commented that the NCAP repeatability study dealt with a compact size vehicle, which has more available crush space than a subcompact. It recommended that a similar repeatability study is necessary for subcompact vehicles. Mazda agreed with NHTSA that adoption of the NCAP test procedures would eliminate some of the existing variability, although further refinements are possible.

American Motors said that adopting the NCAP modified test procedures cannot be expected to reduce test variability since the modifications are minor. AMC said that there are other test variables, such as safety belt tension and actual dummy position just prior to impact, that have a similar effect on dummy positioning, but those variables are not controlled for in the test procedure.

AMC also claimed that because of the lack of repeatability in the FMVSS 208 test procedures, the standard does not meet the requested statutory criteria. AMC believes the above because the unreliability of test results demonstrated in the NCAP program are "indicative" that a similar level of variability will exist in FMVSS 208.

Peugot stated that it "can but reluctantly accept as valid a test procedure" with a COV of 21 percent. It suggested that the level of performance (e.g., HIC criterion of 1000) be raised by the amount of variation.

Chrysler, based on the NCAP data, concluded that the test procedures are not capable of producing identical results when a given vehicle is repeatedly tested. They believe the current procedures only measure a manufacturer's ability to conduct the test and do not measure the adequacy of the restraint system. Chrysler said that because differences in dummy foot placement and ambient temperature make a difference in test results, the test is not practicable. Chrysler also argued that the agency must develop a test which takes into account the inherent crash variability of the vehicle itself.

Volvo said that the modified NCAP procedures only address a portion of the variability and that it has not been demonstrated that the new positioning requirements will in fact result in a repeatable positioning of the test dummy. It noted that the procedures do not ensure that the same webbing location is used in each test. Volvo also said that because of the effect of temperature on dummy performance, either the permitted range for crash testing must be narrowed or new materials be used in dummy construction. Volvo also said the NCAP repeatability program shows that there is a certain amount of unreliability in the signals obtained from the accelerometers and that different laboratories have used different methods to process crash data.

Volvo also supplied the results of 10 sled tests in which there was a stable crash pulse and no contact between the dummy's head and vehicle interior, thus eliminating most vehicle-to-vehicle parameters. The mean HIC was 466.5 with a COV of 12.5 percent.

Nissan said that under the current test procedures, it is difficult to maintain the same relative positioning of the test dummy for several tests. It recommended that the agency maintain the same initial relative measurements between the dummy and steering wheel and instrument panel for each test of a particular model. It also said that the positioning of the seatbelt should correlate to design measurements submitted to the agency by manufacturers. It urged changing the seat position requirement (it is currently set at the mid-position) since passengers in small cars tend to move the seat rearward. Nissan recommended that the measurement between the hip point and ankle should be constant for the positioning of the seat.

Toyota said there are still unresolved problems concerning the variability in electronic crash data collection systems. It also recommended that the test procedure specify the "timing of dummy installation prior to crash. . . . Such timing will affect test results depending upon the extent of the breaking-in (sic) between the dummy's hip and the seat materials."

Mercedes said that the Part 572 dummy is not sufficiently repeatable for compliance test purposes, that the Hybrid III dummy provides no improvement in this regard and that adoption of the NCAP test procedures is a step in the right direction.

Volkswagen also contended that the variances resulting from the NCAP repeatability tests were too large for compliance test purposes of a safety standard. VW argued that overdesign to comply with FMVSS 208 has nothing to do with improved safety but only costs the company time, effort, and money in overcoming the inherent variability in the test itself.

Renault said that the current COV of 21 percent (which permits a variation of 63 percent) is too large; it said the COV should not exceed 10 percent. It said that as long as the COV remains at 21 percent, the HIC limit should be raised by 63 percent.

MVMA again reiterated its concern over test variance and said that FMVSS 208 is not objective.

IIHS said that overdesign is standard industry practice and current test data show that compliance is "easily achieveable."

Allstate again contrasted the lack of any dynamic testing of seatbelts with the detailed test procedures for testing of automatic restraints. It cited the *Public Citizen* v. *Steed* decision on tire treadwear grading (UTQGS) for the proposition that "no test procedures. . .are going to approach perfection." Allstate said that it seemed "strange" for the Department to be concerned over "minute details" of test procedures and to refuse to implement FMVSS 208 because of minor test details would be absurd. Allstate said that the test procedures were developed over many years and have proven highly acceptable.

State Farm concurred with the SNPRM analysis of crash test variability and cited the UTQGS decision as undercutting the manufacturers' arguments.

State Farm concluded that FMVSS 208 is both practicable and objective, that the test procedures have been subject to court challenge and have been improved, and that the results of the NCAP repeatability program were conducted at 35 mph, not 30 mph as in FMVSS 208, where the vehicle must absorb 36 percent more energy. They said testing at 30 mph should result in less variance as well as lower readings.

British Leyland suggested "that at this point in the rulemaking process, the subject of test procedures is not supremely important for discussion. . ."

Design to Conform

Because manufacturers believe that the variability in test results, particularly HIC, is so large that extensive overdesign would be required to ensure that all vehicles would comply with the standard, the concept of "design to conform" was suggested as a more appropriate measure of compliance.

Both Ford and American Motors suggested this concept in response to the NPRM. Ford said that to overcome the unacceptable jeopardy of being in noncompliance, as a result of the test procedure's lack of objectivity, compliance should be based on the design-to-conform concept, similar to that used in FMVSS 108. AMC favored the design-to-conform approach for the same reason as Ford, and also said that excessive variability was the same reason design-to-conform was adopted in standard 108.

In the SNPRM, the Department sought public comment on whether an approach which required a manufacturer to show that a vehicle was "designed to conform" to FMVSS 208, instead of requiring actual conformity with the standard's requirements, could be reconciled with the Sixth Circuit Court of Appeals decision in *Chrysler Corp.* v. DOT, 472 F.2d 659 (6th Cir. 1972), wherein the Court stated that compliance should be "obtained from measuring instruments as opposed to the subjective opinions of human beings," 472 F.2d at 676, and that "compliance be made by specified measuring instruments; there is no room for an agency investigation in this procedure." 472 F.2d at 678. Since the design-to-conform approach would require the manufacturer to justify to NHTSA that it had taken reasonable steps in the vehicle's design and testing to certify that it had been designed to conform to the standard's requirements, it appeared that adoption of this proposal would introduce unacceptable levels of subjectivity, contrary to the *Chrysler* court's direction, into what was heretofore an objective compliance procedure. Comments were also sought on the potential effects on vehicle design and construction under a design to conform approach.

Responses to the SNPRM by manufacturers showed agreement with the concept of design to conform as applied to FMVSS 208. Ford argued that if Standard 121, regarding air-braked heavy trucks (subsequently overturned by the courts) had had a design to conform provision, "it might well have been judged to be practicable, for manufacturers would have had the assurance that bona fide results of their own compliance tests would have to be taken into account in determining whether their products were in fact noncompliant." It said that dictum in *Wagner Electric* supports the lawfulness of a design to conform alternative to a strict compliance scheme.

Ford said that adopting a design to conform approach would not "materially" affect a vehicle's design and that its main effect would be to permit a manufacturer to not be judged in noncompliance based on failure to meet the specified injury criteria in a single test, if the manufacturer had *bona fide* test results to verify that the designed level of performance had been achieved.

GM also supported the design to conform con-

cept. GM argued that such a concept does not contravene the *Paccar* decision. It said design to conform is "compatible with the court's finding that all relevant factors must be considered in establishing a standard and would not require manufacturers to overcompensate for test variability to assure compliance."

GM added that a design to conform requirement would not materially change a manufacturer's approach to assuring conformity with FMVSS 208. GM believes that a manufacturer would still be required to demonstrate that the performance of its design would meet the requirement. GM also said that the philosophy of adopting design to conform in FMVSS 108 was based on the recognition of test variabilities and thus applies equally well to this standard.

VW said that it was uncertain about the effect of adopting design to conform language in the standard. VW contrasted what it called the accurate and precise test of Standard 108 with the variable test procedure of Standard 208. VW also believes that the Department essentially operates under such a concept.

Mercedes, Renault, and MVMA supported adoption of a design-to-conform standard.

Peugeot termed the concept "interesting" and said that NHTSA's concern was understandable. Peugeot suggested that an in-depth study of the "reasonable steps" a manufacturer should take might be necessary.

Jack Martens, an automotive safety consultant, opposed a switch to the design-to-conform standard arguing that there will no longer be any means to ensure that the vehicle as purchased meets the performance requirement.

Thirty Degree (30°) Oblique Test

In commenting on the NPRM, both Chrysler and Ford suggested deleting the oblique test requirement in the standard. Ford argued that the test is redundant, since dummy readings are lower than in perpendicular barrier crashes, that it not only adds to development costs and time but also increases test result variability, and that it is a hindrance to airbag development. Chrysler's recommendation for deletion also was in the context of airbag development.

Although not directly addressing the test requirement, Renault said that airbags are not as effective as manual belts in oblique crashes and that their effectiveness limit corresponds to the 30° barrier impact conditions. Beyond 30°, Renault believes, airbag effectiveness is slight or nonexistent.

Puegeot claimed that airbags are less effective than manual belts at oblique crashes of 25 to 30 degrees, while Allstate said that the field experience with airbags indicates that they will be effective in crashes at frontal angles of 30° or greater.

The Department, in the May 10, 1984, SNPRM, voiced its own concerns over the necessity of the 30° oblique test to assure proper passive restraint performance. NHTSA test data indicate that the instrumented dummy readings in such tests are consistently lower than in direct frontal barrier crashes due to a less severe crash pulse. Although the original rationale for the requirement appeared to be ensure that car occupants were protected in oblique crashes, the data available to NHTSA indicated that the 30° test was unnecessary to achieve that goal. That is, the protection was provided regardless of whether or not the test was conducted. The elimination of the oblique test was proposed in the SNPRM and specific data were sought to support commenters' positions on the issue.

Most of the auto manufacturers and several other commenters offered remarks on the proposal. However, the manufacturers' opinions were split into three categories—in favor, against, or retain the oblique test but eliminate the direct frontal barrier crash requirement.

Ford restated its belief that the oblique test is redundant and merely adds to the cost of testing, adversely affects leadtime and adds more unpredictability to the testing.

Ford referenced material it had submitted to NHTSA previously which contained data on 30° angular vs. frontal tests. These data related to Ford's 33-car barrier crash tests of 1972 Mercury airbag vehicles. Ford's February 1976 report on the subject, "Airbag Crash Test Repeatability" (ESRO Report No: S-76-3), stated that the results of the angular crashes were lower in magnitude and had less variability than the frontal crashes. In 12 frontal tests, average driver and passenger HIC values were 479 and 462, respectively. In angular tests, the respective means for HIC were 185 and 330, well below the values in the frontal crashes.

Favoring the deletion of the oblique test, due to its stated redundancy and its adding to costs, leadtime, and variability, were BMW, Volvo, Nissan, Mercedes, Honda, and Mazda. Mazda supplied data which showed a driver HIC of 779 and a passenger value of 758 in a frontal crash test using an experi-

mental two-point passive belt while the corresponding values in the angular test were 488 and 302. Mercedes also stated that the oblique test is an obstacle to producing airbags.

Peugeot and Renault supported retention of the oblique test, arguing that it is more representative of the majority of actual crashes, and deletion of the perpendicular test. They stated this would be harmonized with a European regulation (WP 29/R237/REV 1).

Two manufacturers opposed the elimination of the test outright, while a third expressed concern over deleting the oblique test for airbag-equipped cars.

GM opposed deletion of the oblique test. It said that while "most angular tests would result in lower injury numbers than obtained from a perpendicular barrier test, angular tests are more representative of the variety of frontal crashes that actually occur in the field."

GM further stated that it was their experience that the oblique test is "important in the evaluation of airbag performance."

Saab also opposed its deletion, terming the proposal "a way to cover up for a weakness in the airbag system." Saab stated that a test requirement must cover a large part of real world accidents.

VW supported, with reservation, the proposal to delete the 30 degree oblique test. VW recommended dropping the perpendicular test since the forthcoming Economic Commission for Europe (ECE) regulation on crash protection will only have an oblique test. VW said that an oblique test should be retained for vehicles which do not include upper torso belts, that is, airbag equipped cars.

The CFAS opposed deletion of the oblique test since it could compromise occupant protection.

IIHS supported the deletion of the oblique test if its elimination will promote the use of airbags.

The Breed Corporation favored the deletion of the oblique test, citing confidential data it had seen from manufacturers.

Adequacy of the Part 572 Dummy

In its December 1983 response to the NPRM, GM said that better diagnostic tools are needed to assure improved occupant safety, including better dummies. GM argued these tools should lead to improved test result repeatability. According to GM, the Part 572 dummy "is deficient as a tool on which to base assessments of the potential of all occupant protection technologies." GM believes their development of the Hybrid III dummy provides for such assessments and, as part of their response, petitioned NHTSA to permit the use of the Hybrid III dummy as an alternative test device (i.e., as a substitute for the Part 572 dummy) in measuring compliance with FMVSS 208.

Although not responding directly to the relative adequacy of the GM Hybrid III dummy, the Department concluded, in the SNPRM, "that the test dummy [i.e., the Part 572 dummy] is a repeatable test device and is not a major source of the variability found in NHTSA's 35 mph repeatability test series." It was further stated that NHTSA would address the merits of GM's petition to permit the use of the Hybrid III as an alternative test device in a separate rulemaking action at a later date.

Several manufacturers took exception to the Department's conclusion that the Part 572 dummy was a repeatable test instrument and met the appropriate statutory criteria. Peugeot said that the current dummy is one cause of test result variability and thus it does not meet the statutory criteria. But, since manufacturers need some reference test instrument, Peugeot said that even though its use is questionable, "it must be maintained."

American Motors described the dummy as "a state-of-the-art compromise — it lacks in reasonable measurement fidelity."

Volvo said that "the present Part 572 test dummy has serious limitations with respect to its use for determining compliance with FMVSS 208." Volvo believes design and material improvements are necessary to make the dummy more durable, repeatable, and trouble-free.

Toyota said that there was "uncertainty of the influence of [the] Part 572 dummy tolerances on crash test results" while Ford said that although the calibration of the dummy is repeatable, its performance in barrier crashes may not be. Ford questioned the Department's conclusion that the dummy is not a major source of variability.

GM again reiterated the potential benefits of the Hybrid III dummy and called for quick action on its petition, saying that a delay could hamper installation of new technology in its vehicles.

This view was supported by Nissan which said it believes the Hybrid III demonstrates greater repeatability than does the Part 572 dummy. Nissan believes the Hybrid III has a more controlled twisting motion and offers a greater degree of control and stability.

Mercedes disagreed with the conclusion that the Part 572 dummy satisfies all legal criteria because

it is "not sufficiently repeatable for compliance test purposes." Mercedes also stated that "the Hybrid III provides no improvement in this regard."

Conversely, Renault said that it agreed with NHTSA that "the present Part 572 dummy is not the major cause of the dispersion of results."

Adoption of NCAP Test Procedures

As a result of its repeatability test program, NHTSA amended the test procedures (IP 212-02) for the New Car Assessment Program to reduce any variability associated with the test procedures themselves. Since the NCAP procedures are more specific than the current FMVSS 208 requirements (in terms of dummy foot placement, placement in the seat, etc.) and since the test procedure is an integral part of complying with the standard, it was proposed in the SNPRM that the NCAP test procedures, aside from those aspects solely related to the consumer rating program such as the need for high-speed cameras, testing at 35 mph, etc., be adopted in FMVSS 208. It was argued that the increased specificity of these procedures would further reduce any variability associated with the test procedures themselves.

Most manufacturers favored, or at least took no exception to, the adoption of the NCAP procedures, although many felt it would do little to reduce variability. AMC said that the changes associated with adopting the NCAP procedures were "very minor" and could not be expected to significantly reduce variability. AMC contended that other sources of test procedure variability, such as safety belt tension and actual dummy position just prior to impact, are still not accounted for in the NCAP procedures.

Volvo said that the procedures were "a step in the right direction" but doubted whether variability would be reduced significantly by their adoption. Volvo said that other sources of variability, such as belt geometry and identical dummy positioning, still exist.

Nissan did not comment on the adoption of the procedures themselves, but also stated that dummy positioning may not be properly specified. To aid in this regard, Nissan recommended that dummy placement be further specified by dimensions of dummy-to-car part distances.

Toyota deemed the adoption incomplete and said that the timing of dummy installation prior to impact and the extent of the breaking-in between the dummy's hip and the seat materials was also important.

Mercedes, as did Volvo, said that the NCAP procedures were "moving in the right direction."

Conversely, VW said it "has no confidence that the changes proposed will cause a significant reduction in... variability" and that the Department has not provided any data to show that variability will be reduced. The lack of data to support the contention of reduced variability was also cited by MVMA and Ford.

While Honda said that the NCAP test procedures were "inadequate" to reduce variability, Renault stated it had "no objection" to their incorporation in FMVSS 208. Mazda agreed that there would be some reduction in variability with their adoption. Renault also asked whether all these types of problems are solved by their adoption.

MVMA, Ford, and GM also claim that the latest revisions to the NCAP test procedures, dummy foot placement and seat placement, were already incorporated when the repeatability tests were conducted by NHTSA; thus, no reduction in variability from the values shown in those tests could be expected from their adoption. Ford also contended that adequate public notice was not provided on this issue since the precise NCAP procedures to be incorporated in FMVSS 208 were never specified.

Head Injury Criteria (HIC) Measurements

The SNPRM sought public comment on whether HIC should be measured in the absence of the dummy's head contacting the vehicle interior. It was pointed out in the notice that the historic derivation of HIC was based on the head striking something. It was also noted in the SNPRM that NHTSA had permitted, for belt systems, the compliance with the HIC criterion only when head contact was made and only for the duration of head contact. The Department pointed out that because of some conflicting data and because it believed that a noncontact HIC criterion could act as a surrogate for neck injury, it was not proposing to change the standard.

Peugeot, AMC, Volvo, Mercedes, VW, Renault, MVMA, Ford, GM, and Mazda favored eliminating measurement of HIC in the absence of head contact. Only Allstate opposed this, claiming that it prevents cervical and spinal injuries. BMW, VW, and Mercedes also favored raising the HIC criterion, even if there is dummy head contact, to a

level of 1500, as proposed in a petition to NHTSA by the Committee on Common Market Automobile Constructors (CCMC).

Peugeot said that they believe HIC is not a good criterion to protect against neck injury and that further research needs to be done on the subject. This view was supported by Volvo, Renault, and Ford. Peugeot, Honda, and GM also said that there is no basis to use a different—for example, 1500—value for HIC in the absence of head contact. They believe HIC should not be measured at all in such circumstances.

Volvo said that the origin of HIC was based on forehead impacts and only for accelerations in the anterior-posterior components. Volvo said it was little wonder, as HIC is now used in FMVSS 208 for noncontact accelerations, including those in lateral directions, that HIC readings have little real-world relevance. AMC and Chrysler also claimed little relevance between HIC and the potential for real-world injury. Conversely, IIHS submitted data, based on calculation of HIC and associated real-world injuries to baseball players who were struck in the head, that there is a real-world relevance of HIC and that serious injuries, even death, occur at HIC values of 1,000. The CFAS also said that higher HICs would compromise occupant protection.

Ford, although agreeing that noncontact head accelerations can produce injury, claimed that there was no correlation between the likelihood of such brain injuries and HIC values, nor was there any relation between neck injuries and HIC.

In commenting on HIC in general, Peugeot and Renault asked that HIC values based on dummy head-to-knee contacts also be eliminated from measurement because the dummy's knee is much harder than the human knee, leading to higher values of HIC than would be expected in actual crashes.

Testing of Safety Belts

Commenting on the NPRM, Chrysler, VW, and Ford said that there was no need to dynamically test automatic safety belts, and that the static test requirements of FMVSS 209 and FMVSS 210, as currently related to manual belts, be applied instead. It was argued that current manual belts, which are not tested dynamically, have been proven effective as evidenced by worldwide data. Thus, the companies argue, there is no reason to test automatic belts any differently than manual belts. Dynamic testing of belts only adds to development time and costs without resulting in a higher level of safety. Recognizing the problem of assuring prevention of submarining for two-point automatic belts, VW suggested that a compliance test be added for knee bolsters. Ford also suggested that the anchorage location requirements of FMVSS 210 be waived for automatic belts.

Allstate said that the fact that manual belts are not dynamically tested results in the consumer having no assurance that the restraint system in a particular vehicle will perform as it is supposed to and, thus, is the "safety scandal of the century."

No new comments were offered on this subject in responding to the SNPRM except from Jack Martens, who said that replacing the dynamic test requirement of FMVSS 208 for automatic belts with the static tests of standards 209 and 210 could result in lower quality levels for restraints. Instead, he agreed with Allstate that manual belts be dynamically tested for compliance.

Impact Test Speed

In responding to the SNPRM, GM proposed an additional set of test criteria for NHTSA to consider. GM said that if some form of passive requirements should be retained, then in addition to the current test procedures in FMVSS 208 for automatic restraints, an additional alternative of complying with manual belts, at two test speeds, should be provided. GM's proposal would permit compliance with manual belts if all FMVSS 208 criteria were met at 30 mph, with the manual belts buckled around the test dummies, *and* all criteria were also met at 25 mph, with the dummies unrestrained (i.e., belts unbuckled). GM believes this proposal would allow both consumers and manu-

facturers to choose between active and passive restraints while improving overall motor vehicle safety. GM also asked that the Hybrid III, or equivalent dummy in terms of biofidelity, be permitted as the test instrument.

GM claims safety benefits for their proposal equivalent to 36 percent belt usage. Their estimate is based on the reduction of total harm (which is a surrogate for the weighting of various severities of injuries by their dollar consequences) of 12 percent, which is derived by calculating the percent reduction of harm which occurs at 25 mph assuming that all current injuries were reduced in severity by one AIS level. Since GM believes that no more than a 5 percent increase in belt usage would occur with passive belts, and since the 85 percent of individuals who currently do not use their safety belts would benefit by their proposal, total safety benefits oculd be nearly 17 times higher. GM further states that although they only calculated benefits for reductions in harm due to frontal crashes, benefits could also be extended to other crash modes.

GM envisions that its proposal would result in greater manufacturer flexibility in offering improved occupant safety than does the current FMVSS 208 criteria and would subsequently result in the development of a variety of occupant safety technologies, such as "safer" steering columns, interior padding, door latches to prevent ejection, windshield glazing, etc. GM stated in its NPRM response that reimposition of FMVSS 208 without changes so as to permit such "built-in" safety to be developed could result in the reduction of the firm's efforts in this area due to diversion of engineering resources.

ANALYSIS OF THE DATA

USAGE OF OCCUPANT PROTECTION SYSTEMS

General

Restraint systems will only have safety value if they are used by occupants or are in a state of readiness such that they provide protection from harm when required to do so. The following paragraphs describe these characteristics of the various restraint systems.

Manual Belts

Various changes have been required over the last 15 years to seatbelt designs to improve manual belt usage (replacing separate lap and shoulder belts and buckles with an integrated lap and shoulder belt having a single buckle and adding an inertial reel to give occupants freedom of movement) and to remind occupants to use their belts (adding brief audible and visible reminders). Nevertheless, the rate of manual belt usage has not changed substantially over the 15-year history of FMVSS 208 (except during the brief period around 1973 when interlocks and continuous buzzers were used).

Based on recent NHTSA data, the overall safety belt usage rate for front seat occupants is 12.5 percent. This information also showed that usage varies significantly by seating position—14 percent for drivers, 8.4 percent for passengers in the right front seat, and 5 percent for passengers in the center seat.

Departmental studies have noted other interesting statistics about usage of manual belts:

* People involved in more severe accidents use their restraint systems less often than the general driving public. (One theory is that belt wearers are more cautious and less prone to severe accidents.)

* Import car occupants have substantially higher seatbelt usage than domestic car occupants. (For example: usage in domestic subcompacts was 12.3 percent, while in import subcompacts usage was 22.1 percent in 1981-82.)

* Seatbelt usage increases as car size decreases. (In 1981-82, usage was 16.8 percent in subcompacts, 10.5 percent in compacts, 7.4 percent in intermediates and 5.4 percent in full-size cars.)

* Usage is higher in newer cars than in older cars. (In 1981-82, the usage in MY 81-82 cars was 16.0 percent; the usage in MY 79-80 cars was 13.6 percent.)

Automatic Belts

Usage rates for automatic belts vary substantially depending on the particular type of belt design and on the method of measuring usage. (Around 500,000 American fleet automobiles have been equipped with automatic belts; they include some 1975-1984 VW Rabbits and 1978-1980 GM Chevettes, and the 1981-1984 Toyota Cressidas.) Studies of usage rates of existing automatic restraints are not necessarily applicable to systems that would be used to comply with an automatic restraint requirement. For example, nearly 80 percent of the existing systems (in VW Rabbits) are voluntarily equipped with starter interlocks (which DOT is prohibited by law from requiring), some owners purchased the systems voluntarily, disconnection and storage of the belts on some systems was very easy, some were installed only on rental vehicles (drivers may be atypical and, also, may not try to take long-term action to defeat the system), and some involved the more expensive motorized (with easier ingress and egress) systems. Based on the record of this and previous rulemakings, manufacturers are unlikely to equip automatic restraint vehicles with either interlocks or motorized systems. The most likely system, given that manufacturers have freedom of choice, may be the detachable automatic belt. Since this is the system for which little field experience exists, application of the current usage data to a future fleet of all automatic belt equipped vehicles may not be appropriate.

Current usage estimates for the VW Rabbit range from about 50 percent based on accident data to 80 percent based on traffic observations to 90 percent from telephone surveys. Chevette usage, based on an extremely small number of observations, is about 70 percent (a similar value is derived from telephone surveys), while Cressida belt usage appears to exceed 90 percent (observations and telephone surveys.)

The Department's estimate of future usage is based on an analysis of existing systems and surveys of usage and attitudes. Essentially, the Department tried to determine whether certain features of automatic belts might overcome some of the reasons people do not use manual belts, while recognizing the wide range of belt systems likely to be produced under a mandate. Our current estimate for automatic belt use covers a broad range: 20 to 70 percent. We expect usage rates for automatic belts to be higher than current manual belt usage because of the automatic nature of the belt, which would overcome some of the stated reasons for not buckling up: laziness, forgetfulness, and not wanting to be bothered. Although precise estimates are impossible, it seems reasonable that some increment of increased usage should be imputed to nondetachable belts, since some effort would be required to deactivate them.

There is no way to know precisely where within the range the automatic seatbelt usage rate would actually fall. The actual rate will depend on many considerations, such as comfort and convenience (including ease of entry and exit) and appearance. Education programs and proven on-the-road effectiveness could also affect usage.

Airbags

Impact protection benefits for airbags do not depend on usage since the occupant does not have to do anything. (However, as discussed elsewhere in this preamble, for greater protection, a lap belt should also be used.) As to whether airbags will deploy when they should, the Department believes that airbag technology is reliable and that airbags would function properly (they will not activate inadvertently and they will activate when they should) in virtually all instances. The automobile manufacturers agree. Two manufacturers stated their goal for reliability of airbags to be at least 99.99 percent.

Although usage is not a factor with airbags, "readiness" is. In the Department's Final Regulatory Impact Analysis (FRIA), based on an analysis of the number of automobiles involved in accidents, the Department determined that, if all automobiles were equipped with airbags and none of the airbags were repaired after an accident, 1.2 percent of the fleet would be without airbags at all times. This figure would be slightly higher if there were inadvertent deployments and they were not repaired. The Department has no reliable methodology for determining what percent of these airbags would, if fact, not be repaired. Because it would be very difficult to dismantle or remove an airbag—much more difficult than a belt system—and because it is not obtrusive, the Department estimates that only a small percent of car owners—perhaps 1 percent—would defeat the airbag. If, as a result of these two problems, 2 percent of all automobiles were without airbags at any one time, airbags would still be ready to deploy in 98 percent of the fleet. Thus, for analysis purposes, the Department estimates that airbag readiness would be 98 percent.

As explained in the next section, a lap belt or a lap/shoulder belt should be worn with an airbag to obtain maximum protection in side and roll-over accidents, as well as in frontal crashes. Because of this, questions arise over the usage rate of the belt system supplied with an airbag. (The Department does not know whether manufacturers would supply lap/shoulder belts or just lap belts.) One argument is that belt use would decline because people would believe that airbags give ample protection. On the other side, it is contended that usage will increase if just lap belts are provided because the shoulder belt portion makes the belt uncomfortable to some people and lap belt usage in the past was near 20 percent. Education may help overcome the "decrease" argument, but habit (people are unlikely to change their habits) may also overcome the "increase" argument. As a result, in its benefit calculations, the Deparment has assumed that current belt usage will continue with respect to the belts accompanying airbags (12.5 percent).

*Other Automatic Occupant Protection
Technologies*

As with airbags, passive interiors do not have a "usage" rate applicable to them. However, unlike airbags, there are no deployment, replacement, or inactivation problems associated with them. Thus, the readiness factor of other known technologies is assumed to be 100 percent. As with airbags, lap belts or lap/shoulder belts might be required for protection in other crash modes (i.e., side, rear, rollover).

Effectiveness of Occupant Protection Systems

General

The safety benefits to be derived from any occupant restraint system are a function of both the usage (or readiness) of the system and its effectiveness, when used, to reduce injuries or deaths. Effectiveness of an occupant restraint system is expressed as a percentage reduction in injuries or deaths when compared to the situation when an occupant is unrestrained. If, in 100 crashes, a system would prevent the death of 60 percent of the occupants who would have been killed if they were unrestrained, then it would be rated as 60 percent of the occupants who would have been killed if they were unrestrained, then it would be rated as 60 percent effective in reducing fatalities. It is important to note two points in this regard: (1) some crashes are so severe that no occupant protection system could prevent death or injury; (2) when a device prevents a fatality or serious injury that otherwise would have occurred, the individual may suffer a less serious injury instead. (As a result, a device that is more effective at reducing serious injuries, may appear less effective, statistically, at reducing minor injuries.)

The Department's estimates for the effectiveness of the various occupant restraint systems are presented in Table 4.

Finally, it should be noted that, in general, the Department has less confidence in the effectiveness estimates for minor injuries than for more severe injuries due to reporting problems; many people do not report minor injuries or do not know they are injured until the next day and thus the injuries may not appear on police reports (the main source of injury data). While the relative effectiveness of the various systems should be unaffected, there is some doubt about whether the overall level of effectiveness for minor injuries is accurate.

TABLE 4
SUMMARY OF EFFECTIVENESS ESTIMATES
(All Accident Directions)

Injury	Manual Lap Belts	Manual Lap and Shoulder Belts	Auto-matic Belts	Air-bags Alone	Airbags and Lap Belts	Airbags and Lap/Shoulder Belts
Fatal	30-40	40-50	35-50	20-40	40-50	45-55
Moderate to critical	25-35	45-55	40-55	25-45	45-55	50-60
Minor	10	10	10	10	10	10

Manual Belts

The effectiveness of manual belts is based on a comprehensive analysis of accident data, involving thousands of accidents. The estimates take into account various factors, such as the fact that occupants who wear their belts are generally involved in less severe accidents then unrestrained occupants. If factors such as this were not "controlled," the raw data would over estimate effectiveness. Although "controlling" the data helps, it cannot pinpoint an exact effectiveness estimate. For that reason, ranges were used. Nevertheless, the Department has the greatest confidence in the estimates of manual belt effectiveness.

Automatic Belts

To determine the effectiveness of automatic belts, the Department reviewed a number of different data sources: analyses of accidents involving existing automatic belt systems, crash tests, and a study by the Canadian Government, referred to below. Since most of the available accident data involve a 2-point automatic belt with a knee bolster, the Department's conclusions on the effectiveness of all types of automatic belts lack a statistically reliable base. In addition, in our analysis of accident data involving VW Rabbits with automatic belts, the Department was unable to determine with certainty the usage rates of the automatic belts. Because of the lack of firm usage data, effectiveness could not be estimated with as much confidence as was done for manual belts.

Furthermore, recent research by the Canadian Government has indicated that the absence of a lap belt may result in the 2-point automatic belt being less effective in preventing ejection. In addition, the door mounted, 2-point belt may have little capability of preventing ejection of an occupant in the event of an accidental door opening during a collision. However, even a 3-point automatic belt will not prevent all fatalities involving ejection, since some fatalities occur as a result of impacting interior components before ejection, while others occur as a result of occupant contact with objects outside the vehicle after partial ejection. Moreover, the door mounted belt in the 2-point system may actually prevent door openings in many instances, since the "loading" of the belt (which is attached to the door) can tend to keep the door closed during a crash.

Three-point automatic belts should be as effective as manual belts, and the Department's estimates for effectiveness of automatic belts reflect this. Automatic belt effectiveness estimates have been adjusted downward by 5 percent at the lower end of the range because there is some evidence that 2-point belts may be less effecitve than 3-point belts.

Airbags

Because of limited field experience with airbags, estimating the effectiveness of these devices is very difficult. There are so few cars equipped with airbags and so few cases or serious or fatal injuries that the field experience has no statistical meaning. Based on field experience through December 31, 1983, (excluding prototype and test fleet vehicles) and a front seat fatality count of 10, the computed airbag and manual belt effectiveness (as used in the equivalent cars) for fatalities is now the same. This means that airbags would not save any more lives than the belt systems as used in those cars. But because the data base is so small, we cannot place any confidence in this effectiveness figure. Based on a normal "confidence interval" (statistical certainty) of 90 percent, all that can be stated based on the field data is that airbags could range from being 46 percent more effective than the manual belts as used in the same cars to 70 percent less effective. Small changes in the number of fatalities would have drastic changes in these effectiveness estimates. Also, the comparisons are to manual belt usage in equivalent 1972-1976 cars. Belt usage is these cars was high compared to usage in later models, because they had, first, continuous light and buzzer reminders and, then, interlock systems. The airbag and equivalent manual belt cars also were very large and had low fatality rates. Finally, the accidents—small in

number—were frequently atypical and involved a greater than normal number of circumstances where a restraint system could not provide protection (such as a drowning). All of these factors indicate that the "true" effectiveness could be significantly higher than in this small fleet.

Current estimates of airbag effectiveness are based principally upon four new analyses which have recently been conducted by NHTSA. The three studies concerned with fatality effectiveness all use the National Crash Severity Study (NCSS), a major accident data collection program designed to result in a nationally representative sample. Effectiveness was estimated by partitioning the NCSS accidents into various subgroups by distinguishing characteristics and then making judgments about whether an airbag could prevent the fatalities that occurred in that subgroup. A fourth study estimated moderate to critical injury effectiveness by comparing injury rates sustained in the airbag fleet cars to a comparable non airbag group in the NCSS file.

We have relied on these new studies primarily because they are based on a relatively large, representative set of unrestrained fatal accident cases. These data, as well as the now available 8-year census of fatal accidents, were unavailable to NHTSA when the automatic occupant protection requirements were first promulgated in 1977. Thus, effectiveness estimates which are not derived from field experience now have a large file of accident data upon which to be based. Further, NHTSA assembled a task force comprised of experts in the field of restraint design, crash testing and accident data analyses to ensure that the resulting estimates represented a consensus of varying judgments and expertise.

However, it must be noted that even these new analyses have a significant degree of uncertainty associated with them. For the most part, they rely on judgments about airbag performance based on limited field experience and controlled crash testing. This technique has obvious limitations, because death and injury in highway accidents are very unpredictable.

There is little disagreement that airbags will function very well in noncatastrophic, frontal or near frontal collisions up to speeds approaching 45 mph and will offer little or no protection in rear end collisions. The real issue concerns airbag effectiveness in side or angle impacts, rollover, and catastrophic frontal crashes. Because the Department is undecided on airbag effectiveness in the latter three situations, a wide range of estimated effectiveness for airbags has been provided. The lower portion of the range (20 to 25 percent) is generally consistent with the assumption that airbags will have fairly low effectiveness in side and roll-over crashes. With progressively more optimistic assumptions regarding their performance in these types of crashes, the overall effectiveness estimate approaches the higher end of the range (40 percent). The 20 to 40 percent range fully encompasses the above dichotomy of assumptions. The zero percent field experience figure is discounted because of its statistical unreliability, crash test data showing superior performance of airbags at higher speeds than for manual belts, and statements to the docket.

Other Occupant Protection Technologies
Effectiveness estimates for other technologies are currently unavailable.

Conclusions
Some conclusions can be drawn from the general effectiveness data that have been developed. First,

the most effective system is an airbag plus a lap and shoulder belt. To obtain maximum protection in not only frontal, but also side and roll over accidents, occupants of cars with airbags and lap belts must use a lap belt to supplement the airbag. An airbag plus a lap belt provides an equivalent level of effectiveness to a manual lap and shoulder belt system. Finally, an airbag alone is less effective than a manual lap and shoulder belt or automatic belt, when those systems are used.

Benefits of Occupant Restraint Systems

Safety Benefits
With its estimates for usage and effectiveness, the Department can determine benefits by multiplying the product of those two estimates by the fatality or injury figure. The final result is the number of fatalities or injuries prevented. Table 5 shows the incremental benefits; i.e., the benefits over and above those accruing from current levels of restraint usage. The numbers provided in Table 5 are annual benefits assuming full implementation. They are based on *all* cars on the road having the restraint system noted (which would

TABLE 5
ANNUAL INCREMENTAL REDUCTION IN FATALITIES AND INJURIES

	Fatalities			Moderate-Critical Injuries		
	Low	Mid-Point	High	Low	Mid-Point	High
Airbags only	3,780	6,190	8,630	73,660	110,360	147,560
Airbags with Lap Belts (12.5% usage)	4,410	6,670	8,960	83,480	117,780	152,550
Airbag with Lap/Shoulder Belts (12.5% usage)	4,570	6,830	9,110	85,930	120,250	115,030
Automatic Belts						
Usage						
20%	520	750	980	8,740	12,180	15,650
70%	5,030	6,270	7,510	86,860	105,590	124,570
Mandatory Belt Use Laws (Manual Belts)						
Usage						
40%	2,830	3,220	3,590	47,740	53,440	59,220
70%	5,920	6,720	7,510	110,430	112,410	124,570

not be the case until at least 10 years after full implementation). Mixes of restraint systems, for example, half of the cars with airbags and half with automatic belts, would lead to results between the values shown for those systems. The numbers also reflect the mid points, as well as the extremes, of the effectiveness ranges provided in Table 4. For these calculations, belt usage with airbags was assumed to be at current levels of restraint usage. The Department has also provided data on the benefits of airbags even if belts were not used. A range of benefits is provided for automatic belts and mandatory belt use laws, because of uncertainty over usage rates.

Another aspect of the analysis of benefits is the difference in short-term benefits of the different alternatives. Roughly one-tenth of the American fleet of automobiles is replaced every year. Although some automobiles are kept beyond 10 years, the Department generally assumes that, ten years after a rule requiring a safety device on new automobiles has been implemented, the device would be in place in virtually the entire American fleet. In this regard, mandatory seatbelt use laws that are enforced can have a distinct advantage in that they can be applied to all automobiles in the existing fleet immediately rather than only new cars. Since the precise date at which different States would pass and implement a mandatory belt use law can not be judged, it is difficult to predict with certainty when benefits would accrue and what the level of those benefits would be.

However, comparisons can be made based upon reasonable assumptions. For example, if all states pass a mandatory belt use law and usage throughout the nation increased to 70 percent or more within three years, the short-term benefits (over the next 10 years) would be 2.5 times higher for such laws than those associated with airbags or with automatic belts at the 70 percent usage level. As the amount of time necessary to pass the laws increases, or the number of States passing such legislation decreases, or if usage does not increase to 70 percent, the shortrun (and longrun) benefits of mandatory belt usage would decrease compared to the benefits of airbags (and possible automatic belts if they are used at high levels). Nevertheless, the benefits of mandatory belt use compared to the introduction of automatic restraints are substantial.

Table 6 compares benefits for the first 10 and 15 years after the introduction of automatic restraints into the fleet with those associated with mandatory belt use laws. Three use-law scenarios are examined. If all States quickly pass a mandatory belt use law and usage increased to 70 percent or more, short term benefits (over the next 10 years) would be about 2.5 times higher than benefits with airbags or automatic belts with 70 percent usage. Thus, unless all cars had airbags, or automatic belt usage approached 70 percent, the longrun (15 years) benefits of automatic restraints would be unlikely to approach those associated with rapid passage of State belt use laws. The short-run safety benefits of such laws are always likely to be higher.

Conversely, if a large number of States do not pass a law, or it takes a long time to get the State laws passed, or usage does not increase to 70 percent, then the shortrun and longrun benefits of mandatory belt usage and automatic restraints may be equal.

Insurance Savings

The potential reduction in fatalities and injuries that would result from mandating automatic restraints could produce a corresponding decrease in funeral, medical, and rehabilitation expenses. A reduction in these expenses could, in turn, result in reductions in premiums for any insurance that covers them. (Automobile insurance premiums could also increase to cover added expenses due to accidents or thefts involving airbag equipped automobiles. This is discussed later in the preamble.) The Department cannot be certain that consumers would receive any premium reductions or, if they would, what their magnitude might be. Most insurance industry representatives are reluctant to provide quantitative estimates of potential savings to consumers. However, at least one company provided an independent estimate and one State official assured the Department that he will mandate such reductions in his State.

The Department, based on the potential safety benefits discussed previously and an estimate of the portion of premiums associated with front seat occupant fatalities, estimates that the discounted value of automobile insurance savings (assuming a 10 percent discount rate and a 10-year vehicle life) could be, based on the midpoints of the effectiveness ranges, $95 for cars equipped with airbags. Spread over the entire vehicle fleet (including uninsured vehicles), the discounted value is $89. For belt systems the savings would depend upon usage rates but could be as high as $85 per insured car

TABLE 6
TIME PHASE ANALYSIS OF FATALITY BENEFITS

Year	Air Bag With Automatic Belt:		Mandatory Belt Use Law: 40-70% Usage		
	12.5% Usage of Lap Belt	20-70% Usage	Scenario 1[1]	Scenario 2[2]	Scenario 3[3]
1	400	50-380	3,220-6,720	2,160-4,500	680-1,650
2	1,000	110-940	3,220-6,720	2,160-4,500	730-2,100
3	1,590	180-1,500	3,220-6,720	2,160-4,500	790-2,540
4	2,180	250-2,050	3,220-6,720	2,160-4,500	840-2,980
5	2,730	310-2,570	3,220-6,720	2,160-4,500	890-3,400
6	3,230	360-3,030	3,220-6,720	2,160-4,500	930-3,770
7	3,690	410-3,470	3,220-6,720	2,160-4,500	970-4,120
8	4,130	460-3,880	3,220-6,720	2,160-4,500	1,010-4,450
9	4,560	510-4,280	3,220-6,720	2,160-4,500	1,250-4,770
10	4,960	560-4,660	3,220-6,720	2,160-4,500	1,090-5,070
TOTAL (1-10)	28,470	3,200-26,760	32,330-67,200	21,600-45,000	8,980-34,850
11	5,340	600-5,010	3,220-6,720	2,160-4,500	1,120-5,350
12	5,660	640-5,320	3,220-6,720	2,160-4,500	1,160-5,600
13	5,900	660-5,550	3,220-6,720	2,160-4,500	1,170-5,780
14	6,090	680-5,720	3,220-6,720	2,160-4,500	1,190-5,920
15	6,240	700-5,860	3,220-6,720	2,160-4,500	1,200-6,030
TOTAL (1-15)	57,700	6,480-54,220	48,300-100,800	32,400-67,500	14,820-63,530

[1]Scenario 1 — It is assumed that all States have mandatory belt use laws which are in effect at the time that an automatic occupant protection standard becomes effective for new cars.

[2]Scenario 2 — It is assumed that 57 percent of the population is subject to mandatory belt use laws which are in effect at the time that an automatic occupant protection standard becomes effective for new cars.

[3]Scenario 3 — It is assumed that 20 percent of the population is subject to mandatory belt use laws which are in effect at the time that an automatic occupant protection standard becomes effective for new cars. The remaining 80 percent of the population would have cars equipped with automatic belts, with usage in the 20-70 percent range.

and $79 when spread over all cars, if usage rose to 70 percent; at 2 percent usage, the figures would be $10 and $9, respectively.

The Department's analysis also showed that between $49 million and $1,100 million could be saved annually in health, life, and worker's compensation insurance and governmental payments for social services such as Medicare, Medicaid, disability insurance, etc. The discounted value of these insurance and governmental payment savings expressed on a per vehicle basis would be in the range of $2 to $61.

Table 7 summarizes the insurance savings that couls result from a requirement 'for automatic occupant restraints. These potential insurance savings do not account for some offsetting insurance premium increases for airbag equipped cars, which are discussed later.

Public Acceptance of Occupant Protection Systems

The public acceptance of safety devices likely to be installed in compliance with Federal motor vehicle safety standards is one of the factors which must be considered by the Department in establishing those standards. In *Pacific Legal Foundation* v. *DOT*, the court found that in order for a safety standard to be practicable and meet the need for safety, the safety devices to be installed pursuant to the standard must be acceptable to the

TABLE 7
SUMMARY OF POTENTIAL SAVINGS
ON INSURANCE PREMIUMS FROM AUTOMATIC RESTRAINT REQUIREMENTS

Savings ($)	Per Vehicle Annual Savings ($)	Per Vehicle Lifetime Savings ($)	Total Annual Savings (M) 1990 Fleet
Air Bags			
Automobile Insurance	9-17	62-115	1108-2046
Health Insurance	4-8	29-54	521-962
Life Insurance	0-1	3-7	62-136
Total	13-26	94-176	1691-3144
Automatic Belts (For 20 Percent Assumed Usage)			
Automobile Insurance	1-2	5-14	89-243
Health Insurance	0-1	2-7	42-114
Life Insurance	0	0-1	7-14
Total	1-3	7-22	138-371
Automatic Belts (For 70 Percent Assumed Usage)			
Automobile Insurance	10-14	65-94	1146-1676
Health Insurance	5-7	31-44	539-788
Life Insurance	1	4-6	71-105
Total	16-22	100-144	1756-2570

public. The Department has attempted to determine the likely public attitudes toward manual and automatic restraints and mandatory safety belt usage laws based on public opinion surveys. In analyzing these surveys, the Department recognizes that the usefulness of the surveys as predictors of future public attitudes is limited by several factors. One is the public's lack of experience with automatic restraints on which to base its opinions. In view of the increase in favorable attitudes toward automatic belts by owners of automatic belt cars between the time of initial ownership and a later time, the Department believes that gradual exposure of the public to automatic restraints will increase the acceptability of those restraint systems above the levels indicated in the surveys. Equally important, most of the surveys are more than several years old. Since public opinion appears subject to change in relatively short periods of time in this area, as is evidenced by the

fairly rapid enactment of child restraint usage laws in most States, there is additional reason to believe that these surveys may not accurately reflect future public attitudes and perhaps not even current public opinion.

Awareness/Knowledge of Automatic Restraints

The extent of the survey respondent's knowledge about automatic restraints is important in assessing the validity of the surveys as predictors of public reaction to automatic restraints. The less knowledgeable the respondents are, the less weight can be given to the survey results. Several surveys made in the late 1970's and early 1980's show that considerably higher percentages of the people surveyed were aware of airbags than automatic belts. The figures for airbags were 62 to 93 percent of the respondents, while those for automatic belts were much smaller.

Government's Role in Making Automatic Restraints Available

There were a variety of deficiencies in the surveys which included questions about public attitudes toward a government requirement for airbags or automatic restraints. For example, most surveys did not attempt to ascertain the degree of the respondents' knowledge of airbags and did not inform respondents about the cost of automatic restraints. Eight of the 12 surveys which attempted to ascertain public attitudes found that respondents favored a Federal requirement. Based on its analysis of those surveys, the Department concluded that while many people do not favor such a requirement on all new cars, there is also a substantial number who state their willingness to purchase cars with automatic restraints. Thus, initial public reaction will be divided. Public education and the performance of automatic restraints will be the key factors in determining the long run public acceptance of automatic restraints.

How Much Would the Public Pay for Airbags?

The surveys on the willingness of the public to purchase airbags indicate that only a small percentage appears willing to pay more than $400 or would expect to pay less than $100 for any airbag system. The majority of respondents cluster around the $200 to $300 levels, covering a range of approximately $150 to $350. Toward the upper end of this cost range, the driving public is roughly evenly divided in its willingness to buy airbags. This suggests that a substantial potential market for airbags exists and that a significant portion of the public would opt for them if they were priced within the $150 to $350 range and available in sufficient quantities.

Attitudes Toward Manual Belts, Automatic Belts, and Airbags

The surveys generally indicate that the public views automatic belts as superior to manual belts in comfort and convenience and that these characteristics would apparently override some of the reasons respondents give for not using manual belts. Those reasons include not wanting to be bothered with belt usage and being lazy and forgetful. At the same time, some of the reasons for not using manual belts appear equally applicable to automatic belts, e.g., fear of entrapment, doubting the value of safety belts, and not wanting to be restrained.

Airbags were rated highest on comfort, conve-nience and appearance and were perceived to be safer than other restraint systems by infrequent belt users. Primary concerns expressed about airbags relate to reliability, whether they will work when needed or deploy accidentally, and cost.

Public Attitudes Toward a Mandatory Safety Belt Usage Law

Surveys made in the 1970's indicate that the public is divided on the issue of mandatory belt usage laws when the concept of sanctions is not mentioned; two 1983 surveys found the public to favor mandatory use laws. When the possibility of sanctions was mentioned as part of several surveys taken in the 1970's, there was increased opposition to mandatory use laws. Since the newest of these surveys involving sanctions is 6 years old, the Department does not have a current reading of nationwide public opinion.

Public Opinion Surveys—Docket Submissions

Two public opinion surveys on occupant restraint issues were submitted to the docket, one by GM and the other by IIHS. Since both surveys included questions whose wording appears to have affected the answers, the Department does not believe that the answers to those questions can be regarded as accurately reflecting current public attitudes. For example, some questions failed to mention either the benefits or the costs of automatic restraints. In addition, there are reasons for questioning the representatives of the sample of respondents.

As to whether there should be airbags in new cars, the GM survey found that 51 percent of the respondents favored installtion if the price were $100. That number dropped to 35 percent if the price were $320 and to 19 percent if the price were $500. The GM survey also asked whether the respondents would favor installation of automatic belts at an additional cost of $100. Thirty-eight percent answered affirmatively.

IIHS' survey asked whether airbags and automatic belts should be standard or optional equipment. Fifty-six percent favored installation as standard equipment and 40 percent as optional equipment. When the 44 percent who did not believe that automatic restraints should be standard equipment were asked if manufacturers should be required to offer those restraints as options, 84 percent answered affirmatively.

Of the two surveys, only the IIHS survey directly queried the respondents about their preference for automatic restraints at various price levels. At a cost of $100 over the cost of manual belts, 30 percent favored automatic belts' over manual belts and at a cost of $150, 25 percent did so. Similarly, at a cost of $100 for airbags 55 percent favored airbags over manual belts. The percentage fell to 47 percent at $200 and 42 percent at $350.

Both surveys asked about preferences for airbag requirements versus a safety belt usage law. The GM survey found that 28 percent would most like to see a combination of a belt usage law and a 65 mph speed limit on the Interstate System, 24 percent preferred airbags in all cars, and 16 percent favored a belt usage law by itself. To measure dislikes, the GM survey asked which requirement the respondents would least like to see enforced. Airbags were picked by 44 percent, a belt usage law by 14 percent, and the combination of a belt usage law and a 65 mph speed limit by 11 percent. The IIHS survey showed a preference of 2 to 1 in favor of an airbag requirement over a belt usage law. The results of both surveys in these areas were at least in part due to the particular information provided the respondents and to the wording of the questions.

The Department does not believe that it is necessary to resolve the dispute between the commenters over the precise role of public acceptability in establishing safety requirements. The nature and significance of public acceptability issues varies greatly depending on the particular factual circumstances of individual rulemakings. Since *Pacific Legal Foundation v. DOT*, it has been beyond dispute that public acceptability must be considered in rulemaking under the Act. The Department agrees that public acceptability involves more than considering consumer preferences. As Allstate noted, if preferences alone were a controlling factor, then that would call into question the current provisions under which manual belts are installed in new cars. However, the Department also agrees that behavior other than disabling occupant restraint systems may be relevant in considering public acceptability. The Department believes that its consideration of public acceptability would satisfy whatever definition might be applied in assessing its actions.

Based on the likelihood that the car manufacturers will install detachable automatic belts or airbags instead of nondetachable automatic belts, the Department does not believe that there will be a significant reduction in benefits due to persons disabling automatic restraints. Neither the detachable automatic belt nor the airbag have the intrusive or coercive qualities that the combination of manual belts and ignition interlocks had in 1974. However, the Department recognizes the need for the public to become accustomed to the technology and the need for protection, and believes that an across-the-board mandate too quickly could engender adverse public reaction. The Department's decision to gradually phase in the requirements of this rule will help build public acceptance of this rule. Additionally, although the added costs of automatic restraints will theoretically have some effect on new car sales, those effects, as discussed in the FRIA, would not be substantial.

Costs and Lead Time
for Occupant Protection Systems

Equipment

General

Table 8 provides the Department's estimates for the incremental increase in equipment and fuel costs and required lead time for automatic belts and airbags. The increment is the cost over that of the current manual lap/shoulder belts. The Department estimates that installation of airbags in compact and larger cars would require 3 to 4 years lead time and automatic belts in all cars would require 2 to 3 years; installtion could begin sooner for a small fraction of annual production, and is likely to take even longer for airbags in small cars. Greater detail on the estimates is provided in the Department's Final Regulatory Impact Analysis.

The costs of manual and automatic belts and airbags are based on tear-down studies and comments to the docket. The cost for belts are believed to be typical of high volume production costs; the estimates for airbags are based on production of 1 million units, which is believed to be representative fo full production system costs if airbags were widely used.

Table 9 presents industry estimates on costs and lead time. It shows investment costs separately because of its effect on cash flow. Investment costs are not, however, additive to equipment; they are already included in equipment costs.

Manual Lap and Shoulder Belts

Based on Departmental analyses, the increase in a new car's price attributable to the addition of a

TABLE 8
PER VEHICLE COST IMPACTS

	Incremental Cost	Lifetime Energy Costs	Total Cost Increase	Required Leadtime
Manual Belt System	Base			
Automatic Belt System (2 pt or 3 pt non-power high volume)	$40	$11	$51	24-36 Mo.
Air Bag—Driver Only (High volume)	$220	$12	$232	36 Mo.*
Air Bag—Full Front (High volume)	$320	$44	$364	36-48 Mo.*

*For compact-sized and larger cars

manual lap and shoulder belt to the front outboard seating positions and a manual lap belt to the front center position is approximately $64, based on a production volume of one million units per year. The added weight for the manual belt would increase fuel usage at a cost of $22 over the life of the car.

Industry estimates for the cost of existing manual seatbelts ranged from $50 (Honda and Peugeot for two seating positions) to $90 (Nissan for two positions). GM and Chrysler said seatbelts for three positions cost $65 (GM said $59 for two positions).

Automatic Belts

For the various designs of automatic belts having a fixed anchorage on the door, the increase in a new car purchase price over that for a car with manual seatbelts has been estimated at $40. Added fuel costs over the life of the car would be $11. Some manufacturers may offer motorized belt systems, such as Toyota currently offers in its Cressida. Incremental cost increases for such systems are estimated by manufacturers to be as high as $300 to $400, but the NHTSA teardown study of the Cressida system shows incremental consumer cost increases of only $115 for such systems. Although motorized systems may lead to higher usage levels because of their convenience, they were not required under FMVSS 208 prior to its rescission in 1981, and are not required by this amendment to the rule.

Of the major automakers, only GM provided a detailed cost estimate in its comments to the rulemaking docket. GM's estimate was for a high volume, four-door sedan with two front seats and 3-point detachable automatic belts with single door-mounted retractors. No provision was necessary for knee bolsters. Their estimate, as well as that of an experienced cost estimator (Lohr) was $45, similar to our estimate of $40. The NHTSA tear-down studies of the Rabbit and Chevette systems, including modifications to fit other cars, yielded costs of $11 to $34. Other manufacturers' estimates are higher than NHTSA's because of "extras" (i.e., equipment not required under FMVSS 208; providing manual lap belts with 2-point automatic belts, knee bolsters with 3-point belts or extra retractors to "hide" detached automatic belts) and different assumptions about markups (profit and overhead) over actual variable costs.

The NHTSA teardown studies were adjusted to account for a mix of 2- and 3-point belts as well as for provision of items *not* required by the standard, but which could increase usage or safety. Two items that fit in the latter categories are motorized systems and the provision of manual lap belts with 2-point automatic belts. These additions increase the tear-down study estimates to $40.

The NHTSA estimate of incremental weight associated with automatic belts is 5 pounds. This compares with GM's estimate of no increase in weight with such systems, VW's estimate of

TABLE 9
INDUSTRY STATEMENTS* INCREMENTAL ON COSTS OF OCCUPANT RESTRAINT SYSTEMS AND LEAD TIME
($ 1983)

	Equipment Cost of Consumer per Vehicle ($)			Investment Cost** ($ Millions)		Fuel Cost (lbs)		Lead Time (mos.)	
	Automatic Belts	Airbags		Automatic Belts	Airbags	Automatic Belts	Airbags	Automatic Belts	Airbags
		Driver	Full Front						
GM	$45	510[2]	838[2]	125	573[8]	0	56	36	36
Ford	—	—	807[3]	—	—	25	40	36-48	48
Chrysler	115	500[1]	800[1]	37	89[8]	—	—	36	36-60
AMC	—	—	—	—	—	—	—	30-36	36-42
Mercedes	—	880[5]	.	—	—	—	—	—	—
Renault	200	—	1,000[7]	1.5	—	—	—	24	36
Jaguar	150	900	1,800[7]	—	31	—	35	—	—
VW	—	—	—	—	—	7	—	48	—
Saab	—	—	—	—	—	—	—	30	58
Nissan	130-150	—	—	—	—	—	—	30-42	—
Honda	150-170	—	—	5	—	—	—	36-48	—
Mazda	—	—	—	—	—	—	—	36	36
Peugeot	380	—	—	—	—	—	—	36	36
American Seat Belt Council	—	—	—	—	—	—	—	24	—
AOPA	—	—	185[4]	—	—	—	—	18-30	18-30
Breed	—	45[4]	141[4]	—	—	—	—	18	—
Lohr	45	—	—	little	—	—	—	—	—
Romeo Kojyo	—	150[6]	—	—	—	—	—	—	—

* A "—" indicates no data was submitted or the commenter claimed it was confidential.
** Already included in equipment costs. Also shown separately because of effect on cash flow.
[1] At 1 million units
[2] At 3 million units
[3] At 200,000 units
[4] At 2 million units
[5] Includes pretensioned passenger belt plus driver lap/shoulder belt
[6] Retrofit; does not include installation
[7] Estimate
[8] For driver only airbags, GM said that investment costs would be $428 million and Chrysler said $12 million.

7 pounds and Ford's 25 pound estimate. Assuming an equal increment of secondary weight, NHTSA estimates that the total 10 pound weight increase would result in $11 extra in fuel consumption over the vehicle's lifetime.

Airbags
The Department estimates that the vehicle price increase resulting from the installtion of airbags in all three front seating positions of cars would be $320 over the cost of a car with manual lap and shoulder seatbelts, based on a production volume of one million units. The replacement cost for a deployed airbag is estimated to be $800. There would also be a fuel penalty of $44 over the life of the car, above that for a car with manual lap and shoulder belts. The cost for a driver-only airbag and lap belt is estimated to be $220, plus a $12 fuel penalty.

The price of airbags is sensitive to volume

changes. At annual volumes of less than 300,000 units, full front airbags may cost anywhere from $400 to $1,500 per car. For volumes of 10,000 units per year or less, the latter figure is most representative. A successful, all mechanical airbag system (such as the Breed system) may reduce the unit price of a full front airbag system to about $250 at an annual volume of one million units.

NHTSA's airbag tear-down study involved a 1979 Ford and a 1981 Mercedes Benz driver and passenger airbag system. The systems were disassembled into their component parts and, using automotive engineering cost estimating techniques, a NHTSA contractor estimated a variable or "piece" cost of each component exclusive of any fixed overhead expenses incurred in the production of airbag systems. These estimates are similar to those supplied by the actual airbag manufacturers through their association. The estimates that were developed include our best estimate of the cost of required vehicle modifications. The estimates also include certain component modifications suggested by the contractors for high volume production. Estimates were developed for annual production volumes of 300,000, 1,000,000 and 2½ million for both systems. In arriving at a unit retail price, unit variable costs were marked up by a factor of 1.33 to arrive at "wholesale" or "dealer" cost and a dealer discount of 12 percent was assumed.

The difference between the Department's estimates and industry's estimates is basically due to differences in design and pricing assumptions. For example, one major cost difference involves the price of the diagnostic module and associated electronics. In its comments to the docket, Ford indicated that it believes that military specification grade electronics are necessary in view of product liability considerations; we have assumed that automotive grade electronics will suffice, although we recognize that, initially, manufacturers may resort to military specification grade electronics until the reliability of automotive grade electronics is proven sufficiently. Significant differences also exist in the number of required crash sensors, module costs (NHTSA used supplier quoted costs) and vehicle markups. The Department also found the estimates provided by the major U.S. manufacturers for driver-only airbag costs difficult to justify at their stated volumes. For example, even recognizing that there are vast differences in basic design between Mercedes and GM vehicles, Mercedes appears to be charging its

customers a price 25 percent higher than GM's estimate for a driver-only system even though the Mercedes system is optional and sold at an annual volume which is 42 times lower than that estimated by GM.

Other Occupant Protection Technologies

Costs for other technologies are currently unavailable.

Investment

Investment costs, which are defined as outlays for property, plant, machinery, equipment, and special tools to be used in the production of automatic occupant restraint systems, are estimated to be $1.3 billion if airbags were required in all new cars and $500 million if automatic belts were required. These estimates are for the multiyear period prior to full implementation of an automatic restraint requirement. Industry's estimate for these expenses are contained in Table 9.

The implementation of automatic occupant restraint requirements should not substantially alter the magnitude of planned capital spending over the next several years, since domestic manufacturers alone are investing nearly $10 billion annually.

Insurance

If airbags were required in all automobiles, collision and property damage liability insurance policies would have to absorb additional costs for replacing deployed airbags, for the value airbags add to vehicles that are "totaled", and for the added cost that would result when some damaged vehicles are considered "totaled" instead of repairable because of the added cost of replacing the airbag. The Department estimates that the maximum expected loss, because of a requirement for airbags in the entire automobile fleet, that would be borne by collision insurance policies would be approximately $177 million per year. For property damage liability policies, the cost would be $118.2 million.

Comprehensive insurance policies will also have to absorb additional costs for the value that airbags add to vehicles that are stolen or damaged by such things as fire and flood. The cost to insurance companies for these vehicles would be increased by the average depreciated value of airbags in the vehicles. The Department estimates that the maximum loss that would be covered by this insurance would be approximately $55 million per year.

These additional losses from airbags may cause annual premium increases, per insured vehicle, of about $2.60 per vehicle per year or $16.60 over a vehicle's lifetime. Table 10 shows these costs.

TABLE 10
SUMMARY OF POTENTIAL AUTOMOBILE PHYSICAL DAMAGE PREMIUM COSTS RESULTING FROM AIRBAGS
($ 1982)

	Per Insured Vehicle Annual Cost	Per Insured Vehicle Lifetime Cost	Per Vehicle Annual Cost	Per Vehicle Lifetime Cost	Total Annual Costs, Millions
Collision	1.90	13.45	1.31	8.85	177.2
Property Damage Liability	.94	6.35	.88	5.95	118.1
Comprehensive	.54	3.65	.41	2.77	55.4
TOTAL[1]			2.60	17.57	350.7

[1] No total is provided for per insurance vehicle figures because each type of insurance covers a different number of vehicles. The addition of these numbers would therefore be meaningless.

Economic Impact

In response to the comments about the potential economic impact of any rulemaking, the Department's analysis indicates that, with a labor force of over 115 million projected for the mid-1980's, it would be difficult to conclude that a restraint system costing the consumer no more than $500 would result in any measurable impact on national employment. Any perceptible effect on GNP is unlikely. Finally, as to the consumer price index, the Bureau of Labor Statistics generally considers higher consumer costs due to safety equipment as quality improvements, not inflationary increases, having no effect on the consumer price index. The projection of effects on the GNP and the price index have one thing in common: the relative changes are small. Long-term effects on auto sales are expected to be minor and auto industry revenue and employment would be expected to *increase*. In any event, any significant changes would result only from an all airbag requirement, not from the installation of automatic belts.

There are also positive economic benefits associated with automatic occupant protection. Based on the previously mentioned estimates of lives saved and injuries avoided (see Table 5), and the economic losses associated with those casualties as contained in a recent NHTSA study, "The Economic Cost to Society of Motor Vehicle Accidents" (January 1983), as much as $2.4 billion in protection. Although we do not wish to—and cannot—place a value on human life or injury, there are some costs associated with those deaths and injuries that can be measured, and only these are included in the study. Because they do not include such things as pain and suffering or loss of consortium, they will obviously understate total benefits of the life savings and injury reducing potential of occupancy restraint systems.

ANALYSIS OF THE ALTERNATIVES

General

Introduction

Numerous alternatives have been considered as part of the response to the Supreme Court decision on automatic occupant restraints. Before analyzing each of the specific alternatives, this portion of the paper first looks at some of the general pros and cons of each automatic protection system. It also discusses the pros and cons of other general features of many of the alternatives: a demonstration program, mandatory State seatbelt use laws, legislation to require that the consumer be given the option of buying an automatic restraint system, airbag retrofit capability, passive interiors, and the center seat issue.

Airbags

Airbags offer a distinct advantage over other occupant restraints in that they ensure a usage rate of nearly 100 percent for both drivers and passengers. Used alone, they do offer protection, but, to equal the effectiveness of a manual lap and shoulder belt, airbags must be used with lap belts. Lap belts in airbag equipped cars would probably be used only at a level near the current level of seatbelt use, 12.5 percent. Because manual belt use is so low, however, airbags would provide much greater safety benefits.

Airbags with lap belts also provide protection at higher speeds than safety belts do, and they will provide better protection against several kinds of extremely debilitating injuries (e.g., brain and facial injuries) than safety belts. They also generally spread the impact of a crash better than seatbelts, which are more likely to cause internal injuries or broken bones in the areas of the body where they restrain occupants in severe crashes. However, the airbag does not provide protection at less than 10 to 12 miles per hour, nor does it provide protection in rollover or read-end crashes. Its level of effectiveness in side crashes is uncertain, hence the large range of effectiveness estimates for airbags. To attain protection in these nonfrontal crashes, a lap belt, or lap/should belt must be worn.

Full front airbags also can provide protection for the center seating position. No other automatic restraint system can do this, because, as with manual seatbelt systems, a shoulder belt cannot practicably be offered for the center seat.

The use of airbags would overcome possible public objections to the obtrusiveness of continuous automatic belts, lessen concerns about entrapment and avoid problems of shoulder belt comfort and convenience. Although there are significant public concerns about the alleged hazards associated with airbags, the Department believes that many of these (e.g., inadvertent activation, sodium azide, and lack of assurance that they work when needed) are unfounded.

The public might also be very concerned about the cost associated with airbags—especially current belt users who may argue that they would be getting very little additional protection at much greater cost. The cost of airbags is one of their biggest disadvantages.

One problem with respect to costs is the wide disparity between the Department's cost estimates and industry's. Although the Department can explain its estimate and the reasons for the differences, it cannot control the price at which the system is offered to the consumer. Thus, although the Department believes full front airbags need cost no more than $320, they could, especially in the near term, cost much more, since airbag costs are very sensitive to production volume. Any alternative that does not result in the use of a large number (for example, 300,000) of airbags may result in their per unit costs being very high.

Repair shop owners have raised concerns about their potential liability if an automobile's airbag fails to work after repair work was done on the car. The Department believes this concern is over stated; the introduction of an airbag into an automobile is no different from the introduction of other safety features that may not work after repair work is done on an automobile. Moreover, the insurance companies have indicated in their testimony and docket comments that there would be very little if any increase in premiums to provide insurance protection against such risks. Indeed, some insurance companies testified that product liability claims should decrease with automatic restraints. The expected reduction in deaths and injuries should result in fewer claims, for example, alleging that the brakes or steering were defective. Although some consumers might view airbags as a panacea and bring suit if subsequently injured, such "nuisance" suits are unlikely to be successful and, thus, should be short-lived.

Concerns were also raised about the dangers of sodium azide, the gas generant in most airbag systems. The sodium azide pellets are hermetically sealed and the potential of exposing motorists to a harmful dose is remote. Additional concerns involved the dangers posed by persons tampering with unfired sodium azide canisters and by the scrapping of cars with unfired canisters. While the Department believes that disposal problems can be resolved, further action on this issue is required, and the Department will work with automobile manufacturers and scrappers to ensure the safe retirement of airbag equipped vehicles. Although it is possible that individuals may tamper with or try to steal an unfired sodium azide canister, the Department believes that this is highly unlikely. The amount of sodium azide contained in the canister is small and it is more readily available through other sources. Other items in the automobile—anti-freeze, gasoline, battery acid, or flares—are either more poisonous or explosive.

Dealers are also concerned that car sales will decline with an all airbag fleet. They fear that potential buyers may stay out of the market, hoping to buy in later model years when an all airbag

decision would have been overturned by subsequent agency or congressional action. However, as discussed in the FRIA, the price increases associated with an all airbag new car fleet, would, at most, result in one to three postponed sales per dealership. In the long term, lost sales would not, on average, be expected to exceed one per dealer. Since airbags are not being required by this amendment to FMVSS 208, a consumer need not purchase an airbag-equipped vehicle unless he or she so desires. Thus, there should not be any reduction in sales resulting from the fact that airbags are one of several systems made available to consumers.

Another concern involves the technical problem of out-of-position occupants in small cars. The out-of-position occupant problem primarily affects children less than 3 years old. (The size of the child and the speed with which the bag must open in small cars are the primary reasons for the problem.) Overall, the safety benefits are greater for an out-of-position occupant with an airbag than without one. Moreover, technical modifications (e.g., sensors that could detect an out-of-position occupant and adjust the opening of the airbag to account for the occupant's position) and child restraint laws should lessen the problem. Nevertheless, the Department can not state for certain that airbags will never cause injury or death to a child. This situation is similar to current safety belts where the benefits are well-known, but they do on occasion cause injuries that otherwise would not have occurred. Again, the Department is not mandating the use of airbags.

In addition, manufacturers have commented that space limitations in small cars would inhibit the installation of current airbag systems and adversely affect their effectiveness. While this problem can be resolved, more time would be needed. At least 4 years lead time would be needed if airbags were required in small cars.

Still another issue is raised by some manufacturers who contend that tests required under the rule are not sufficiently repeatable to enable manufacturers to assure themselves of compliance. They argue that they get too wide a variation in results when they test the same automobile under the test procedure. To protect against some cars not passing the test, they say they will have to design the restraint systems to a more stringent standard then should be necessary. Although difficulties in the testing procedures are still of concern to the manufacturers, we believe that the testing device and testing procedure have matured greatly in the last decade. Furthermore, based on the result of NHTSA's NCAP tests, most cars (albeit with manual belts) already meet the injury prevention criteria of FMVSS 208, at 35 mph—a 36 percent more severe crash than required by the standard (with is a 30 mph test). Compliance by airbags is even less of a problem since the injury levels of the test dummy tend to be well below the maxima of the standard (much lower than for belt systems), providing a large margin of safety. In summary, we do not think that test repeatability is such a severe problem as to preclude an airbag or other occupant restraint standard, although the Department will subsequently address possible improvements in this area.

Some people are concerned that the failure to issue a rule that will result in at least some airbags being placed in automobiles might mean the end of the development of airbag technology. In this regard, it must be remembered that some improvements—such as those made by the Breed Corporation—have come about without regulation. Moreover, four manufacturers are currently planning to offer driver-only airbags in their automobiles, even though not required. It is, therefore, possible that others may follow suit to meet the competition. Most important, the Department believes that this rule will result in the use of airbags in a far larger number of automobiles than is the case today.

It should be noted that improvements are possible in the airbag system that might overcome some of the remaining problems. For example, the airbag system being developed by Breed might make airbags available at less cost than current airbag systems.

Some may argue that consumer fears and dislike of airbags may come close to generating a level of public disapproval equivalent to the seatbelt interlock system. On the other hand, the unobtrusiveness of the system may result in the airbag generating the least disapproval.

Nondetachable Automatic Seatbelts

The usage rate for nondetachable automatic belts should be higher than that for manual belts, but some people will certainly find them uncomfortable, combersome, and obtrusive. Others will fear entrapment. Although they are much less costly than airbags and not much more expensive than manual seatbelts, these concerns with nondetachable belts might hamper automobile sales.

Finally, it is possible that, in an emergency, people may find nondetachable belts harder to get out of than detachable belts. Although data do not exist on this issue, the Department has long expressed concern about the possibility that an unfamiliar egress mechanism could impede emergency exit. In the early 1970's, DOT issued a rule requiring all automatic belts to be detachable to permit emergency exit. Even in a later amendment in 1978 allowing the "spool-release" feature on continuous belts, NHTSA continued to express some concerns about ease of exit in case of emergency. The Department believes, however, that current designs of continuous belts will not create a safety problem.

Perhaps the most serious concern with respect to nondetachable belts is that the public's dislike of them may lead to defeat of the system (e.g., by cutting the belt). A number of surveys have found that 10 to 20 percent of the public might do so. This would result in not only the original owner but subsequent owners and passengers being deprived of any occupant restraint system. Since the average car has two to three owners during the useful life, belt availability could decrease to nearly 50 percent for a 10-year-old car.

Nondetachable belts are probably the most coercive type of automatic restraint. Combining this with the fears of entrapment and the concerns over obtrusiveness could cause enough public clamor to result in the same type of problem that arose out of the interlock requirement in the mid 1970's when Congress forbade the Department from requiring that device. (In the NHTSA authorization bill of 1980, which barely failed enactment, there was a provision to ban nondetachable seatbelts.)

Nondetachable belts would also force manufacturers to eliminate the center front seat (by the use of bucket seats and consoles). There is no commercially developed technology to provide an automatic belt for the center seat; even if it were exempted from the requirement for an automatic restraint, occupants would have a difficult time getting by the nondetachable belts to reach the center seat.

Another problem with nondetachable belts is that they make it difficult to install a child restraint seat properly.

Detachable Automatic Seatbelts

Detachable belts should alleviate some consumer concern about automatic belts and government involvement in the consumer's decision about belt usage. Although it is easy not to use the automatic feature (by detaching the belt and leaving it stowed), the availability of the automatic feature would make it easier to overcome some of the problems of manual seatbelt usage.

Detachable belts would also be only slightly more expensive than manual belts, but the additional expenditure would be made for what are likely to be relatively small safety benefits, if usage does not increase substantially over than for manual belts. In this regard, however, it must be remembered that NHTSA rescinded the automatic restraint requirement in 1981 because it found that detachable automatic belts would be installed in most cars and thought that those belts might not increase belt usage enough to justify them. The Supreme Court, in reviewing this action, then found that the evidence in the record indicated a possible doubling of usage with automatic belts. The Court also said that the inertia factor provided grounds for believing that seatbelt use by the 20 to 50 percent who wear their belts occasionally would increase substantially. The manufacturers also now agree that detachable belts will increase usage, at least initially.

Demonstration Program

Although we may gain more data on usage and effectiveness, the main purpose of a demonstration program would be to obtain detailed data on the issue of public acceptability of automatic occupant restraints. To the extent consumer purchases under a demonstration program would be voluntary, data that were gathered on usage or effectiveness would be too small to determine the reaction of the general population under an automatic occupant restraint mandate. To obtain statistically reliable data within a reasonably short period of time, a large number of automobiles would have to be included in the program. If such a program were to be conducted, the Department believes that it should include provision for producing at least 500,000 cars per year over a 4 year period with airbags, detachable and nondetachable automatic belts. The three types of automatic restraints would be divided evenly among the cars produced. This should provide statistically reliable results in 4 to 5 years from the date the first car is sold. (If the program is limited to airbags, 250,000 cars should be manufactured per year over a four year period. This would provide results in about 4 to 5

years.) The program could be conducted in essentially the same fashion as envisioned by Secretary Coleman when he announced his plans in 1976 to conduct a demonstration program. At that time, the Department negotiated contracts with four car manufacturers for the production of up to 250,000 automatic restraint equipped cars per year for model years 1980 and 1981.

During our recent public hearings, Ford indicated support for a mandatory demonstration program. Other manufacturers are receptive to a voluntary program, but only as an alternative to an automatic restraint requirement, and only under several conditions regarding the manufacturer's freedom to choose the type of restraint and model, test procedure changes, etc. Several manufacturers would not voluntarily participate in any demonstration program.

Three methods could be considered for conducting a demonstration program: (1) a voluntary contract program such as that suggested by Secretary Coleman; (2) use existing National Traffic and Motor Vehicle Safety Act authority to mandate such a program; and (3) seek Federal legislation. A mandated demonstration program would be difficult to justify under the Safety Act. Ford now believes that such authority exists, but the Department thinks that new legislation would be necessary. It is unclear whether Congress would provide the necessary legislation or any funding that might be required. Moreover, the time necessary to obtain any legislation would have to be added to the time necessary to conduct an effective program. There also may be serious objection to a demonstration program after so many years of attempted rulemaking, and especially so many years after Secretary Coleman's efforts.

Mandatory State Safety Belt Usage Laws

A number of analyses of seatbelt use in countries that have mandatory use laws show that such laws do increase usage. Survey results, based on responses from officials in foreign countries, show that when seatbelt usage was required and the requirement was properly enforced, usage increased dramatically and remained high. Tables 11 and 12 clearly illustrate these dramatic increases. Table 11 provides data available to the Department on 17 nations that have passed MULs; the table shows the difference in usage rates before and after the enactment of MULs. In addition, a number of Canadian provinces have enacted MULs. Those

TABLE 11
CHANGES IN SEAT BELT USAGE RATES UNDER MANDATORY USE LAWS

Country	Effective Date of Law	Belt Usage	
		Before	After
Australia	1-72	30%	73-87%
New Zealand	6-72	40%	89%
France	7-73	20-25%	95% highways
			75% country roads
			50% night in cities
			35% day and night in built up areas
Puerto Rico	1-74	5%	14%
Sweden	1-75	22%	75%
Belgium	6-75	17%	87%
Netherlands	6-75	11% urban	58% urban
		24% rural	75% rural
Finland	7-75	30% highways on weekdays	68% highways on weekdays
		9% urban traffic	53% urban traffic
Israel	7-75	6%	70%
Norway	9-75	13% urban	77% urban
		35% rural	88% highway
Denmark	1-76	25%	70%
Switzerland	1-76		
	(repealed 10-77)	19% city streets	75% city streets
	Reenacted 11-80	35% highways	81% highways
		42% expressways	88% expressways
West Germany	1-76	55% autobahns	77% autobahns
		32% country roads	64% country roads
		20% city streets	47% city streets
		33% weighted average	58% weighted average
Austria	7-76	10% urban	20% urban
		25% rural	30% rural
South Africa	12-77	10%	62%
Ireland	2-79	20%	45%
Great Britain	1-83	40%	95%

provinces and the data on their experience are contained in Table 12. (More detail on the information in these tables can be found in the FRIA.)

The data in these two tables clearly illustrate the significant effect MULs have on seatbelt usage. As Table 11 shows, usage rates ranged from 5 to 40 percent before MULs went into effect, and from 14 to 95 percent after enactment. Usage typically at least doubled and in some cases increased three times or more. The average usage for the 17 countries in the table was 23 percent before mandatory belt usage and 66 percent after, an increase of 43 percentage points.

The Peat, Marwick, Mitchell and Company (PMM) study from which most of the data included in Table 11 were obtained concluded that the main factors that influence the frequency with which individuals wear their seatbelts under MULs are: (1) the level of enforcement applied by the police; (2) the natural propensity of indivuduals to be law abiding; and/or (3) the individual's personal perspective regarding their own safety.

Given the geographical proximity of Canada to the U.S. and the many similarities between our societies, the Canadian experience with MULs is especially valuable. MULs are in effect in seven prov-

TABLE 12
CHANGES IN DRIVER SEAT BELT USAGE IN CANADA UNDER MANDATORY USE LAWS

Province	Effective Date of Law	Use Before	Use in 1983
Ontario	1-76	23%	60%
Quebec	8-76	18%	61%
Saskatchewan	1-77	32%	54%
British Columbia	10-77	37%	67%
Newfoundland	7-82	9%	76%
New Brunswick	6-83	4%	68%
Manitoba	1-84	12%	12%

Averages weighted by Traffic Counts at Data Collection Sites:

Provinces with Mandatory Use Laws			61%
Provinces with No Mandatory Use Laws			15%
Unweighted Average Usage Before			
Laws Passed (Excl. Manitoba)		21%	

inces in Canada, but, since Manitoba's did not go into effect until January 1984, data are not yet available from the province. Usage rates before MULs went into effect for the six other provinces averaged 21 percent. Usage rates for those six averaged 61 percent in 1983. This is an increase of 40 percent under MULs. The PMM and other studies of MULs, which are more fully discussed in the FRIA, have concluded that success is dependent on how well the public is prepared for these laws, the severity of sanctions, and on the diligence of enforcement. For this reason, the Department has established critieria in the amended rule to ensure an appropriate level of educational, sanction, and enforcement efforts.

The 1982 background paper on "Mandatory Passive Restraint Systems in Automobiles," prepared by the Congressional Office of Technology Assessment, stated that "Mandatory belt use laws are potentially the most effective approach to ensuring passenger restraint. Experience in other industrialized countries suggests that a mandatory law might result in usage rates exceeding those achievable with passive belts because so many passive belts would be detached. Nevertheless, in today's political climate in the United States, mandatory seatbelt-use laws seem unrealistic." The Department agrees with the potential for belt use laws, but feels that the political climate and public attitudes have changed significantly since then, making the possibility of enactment of such laws considerably higher.

Currently, one State legislature, New York's, has passed a mandatory use law which provides for a $50 fine, allows waivers for medical reasons only, and requires the Governor to conduct a public education program in conjunction with the law. Eleven other States are reported as actively considering seatbelt usage laws.

A number of statewide and nationwide surveys have been taken in the United States to determine the public acceptability of mandatory State belt use laws. Surveys taken in 1979 or earlier generally indicate that the public is strongly divided on mandatory seatbelt use laws. However, public attitudes about automobile safety have changed markedly over the past few years, in part because of the grass roots uprising in opposition to drunk driving. The public now strongly supports laws and innovative enforcement action to reduce the needless deaths and injuries caused by drinking and driving. This movement has spilled over into other highway safety areas such as safety belt and child safety seat usage. Evidence of this attitudinal change can be seen in the fact that 46 States and the District of Columbia have enacted child safety seat laws since the beginning of 1981 (bringing the total to 48), the New York State Legislature's recent enactment of the adult MUL law, and the significant progress made toward the enactment of

MULs in other States—notably Illinois, Minnesota, and Michigan. Recent surveys taken by several States found 66 percent in favor of mandatory belt usage laws in Michigan, 69 percent in Delaware, 52 percent in New York, and 56 percent in Ohio.

Many of the commenters who support such legislation stress the need to have public education programs before the actual enactment of the laws and Federal incentive grants as an effective impetus to stimulate the States. Indeed, the success of the mandatory law in Great Britain can be attributed to an intensive public information and education program conducted during the 2 preceding years before enactment of the law.

Legislation to Require Consumer Option

The option would ensure that consumers were given the widest possible choice of both whether to purchase an automatic occupant restraint and, depending on the requirement, what type of automatic restraint. Unlike the current market situation, those who wish to purchase an automatic occupant restraint system could do so. This would probably not be as effective in generating safety benefits as a requirement for automatic restraints in all cars. Those drivers who are involved in more serious accidents are probably the ones least likely to purchase such systems. Depending on how "controlling" the legislation that was adopted was, numerous other problems could develop. For example, dealers might not stock vehicles with automatic restraints, requiring consumers to wait a long time so as to "force" many people to purchase manual safety belts. In addition, the small number of automatic restraints produced under this alternative would likely mean high prices per unit due to a lack of economy of scale. There also would be significant costs imposed on manufacturers because of extra design and tooling costs, if it were necessary to provide more than one type of automatic restraint for each model. As a result, the overall costs for manufacturers and consumers might far outweigh the benefits, and if an insufficient volume of different types of restraints were produced, there might not even be enough data to permit further evaluation of the different types of systems.

Airbag Retrofit Capability

Requiring an airbag retrofit capability would make it easier for owners of automobiles to have airbags installed in their cars in the "aftermarket." It would also allow purchase of an airbag by a sec-ond or third owner, if the original owner failed to purchase one. This would be especially valuable if systems like Breed's airbag eventually proved successful. However, it could be argued that only the more safety conscious consumers are likely to purchase such airbags; the high risk drivers might not take advantage of the option. In addition, all automobiles would become more expensive, even if the airbags were installed in relatively few cars, and the cost of airbags could be very high if they are purchased in low volumes that do not permit economies of scale. Moreover, this alternative would not ensure that airbags would be available to consumers who wish to have them installed.

Passive Interiors

GM has been doing research to develop "passive interiors"—to build in safety by improving such things as the steering columns and padding. It believes this would be better than automatic occupant restraints and contends that it cannot afford to do both. Although an attractive alternative, this approach is still being developed, and even GM is not willing to say that it will meet FMVSS 208 in the immediate future. Moreover, FMVSS 208 does not require airbags or automatic belts; GM's passive interior concept is an acceptable compliance method, which should be encouraged. It holds the promise of being a low cost, nonobtrusive method of complying with the standard.

GM also asked that the Department consider dropping the barrier standard from 30 mph to 25 mph for passive interiors. The Department has virtually no data on what dimunution in safety would occur if the lower standard were to be used. Thus, it has no basis for making such a change.

Nevertheless, the Department encourages further research in this area. From the limited test data available, it is generally evident that it is within the state-of-the-art to pass FMVSS 208 criteria at 25 mph (using unrestrained Hybrid III dummies). General Motors, in their docket submission, indicated that the Oldsmobile Omega and the Pontiac Fiero have passed the injury prevention criteria of FMVSS 208 at 30 mph. Nissan engineers indicated in 1974 that the 260Z would come close to meeting the FMVSS 208 criteria at 25 mph. In a NHTSA test of a Ford Crown Victoria, the driver dummy's performance met the FMVSS 208 injury criteria in a 30 mph barrier test. However, even though these vehicles met the FMVSS 208 criteria, none of the manufacturers have expressed confidence in their ability to so certify to the government. Nonetheless, the Department remains optimistic about further development of this technology.

Center Seating Position

Intertwined with most of the alternatives is the issue of what to do about the center seating position. Automatic seatbelts (and even 3-point manual belts) cannot be used for the center seat. As a result, the only automatic protection available for front center seat occupants is an airbag or passive interiors. If automatic seatbelts were used to comply with a requirement for automatic occupant restraints, the center seat would have to be eliminated as an occupant position, unless it were exempted from coverage. Moreover, even if it were exempted from coverage, if nondetachable belts were required, occupants would have a difficult time getting to the center seat. Finally, even if airbags were used to meet a requirement for automatic restraints, at least one commenter (Ford) indicated that the center seat position might be eliminated due to the problem of out-of-position occupants.

If the center seat were exempted from coverage and detachable belts (or airbags) were used to provide automatic protection for the outboard seats, the center seat could still be used because the automatic belts are detachable. If they are detached to let a passenger sit on the center seat, the question then arises as to how often they would be reattached. In this regard, a recent study by Market Opinion Research is noteworthy. It indicated that the interaction between the driver and the passengers was a significant factor affecting belt usage; i.e., if the driver wore a belt, this made it more likely that a passenger would. Since passengers normally enter the front seat from the passenger side of the automobile, the driver's automatic belt would not have to be disconnected for them to enter. Therefore, if the driver does not disconnect his belt, the fact that the passenger side automatic belt is disconnected to permit entrance to the center seat may not have a serious adverse effect; since the driver is wearing his belt, it may encourage reconnection of the right front belt and/or the use of the center seat lap belt. Conceivably, center seat lap belt usage could increase compared to the expected usage in cars with only manual belts.

If the center seat were not exempted, the loss of the center seat would affect both manufacturers and consumers. In arguing for exempting the cen-

ter seat, Consumer's Union and the AAA pointed out that consumers would lose vehicle utility due to a reduction in the maximum seating capacity. Manufacturers could be affected if customers opt to purchase smaller cars if they lose the center seat in larger cars. This could cause a loss of profits, since larger cars yield more profit per unit than smaller ones.

The indirect safety effects are quite complex. Moving a child, for example, from the center seat to a back seat has the advantage of significantly improving the child's safety, but the disadvantage of possibly leading to a driver who may frequently turn around to check a child in the back seat. There are also fuel economy and safety implications, if two cars are necessary when one would have otherwise been sufficient for a particular trip. The issue is made even more complex by the fact that some center seat passengers may move to the right seat and others may move to the back seat, if the right seat is already occupied. The front right seat is statistically the least safe position in the automobile, but sitting in the back is slightly safer for adults than sitting in the front.

On the other side, only one-third of the cars sold in 1982 were six seat cars, and that number has been declining as cars are being downsized. (Recent trends, however, indicate some increasing consumer preference for larger cars). An estimated 1.5 percent of front seat fatalities and injuries involve the front center seat occupant. Automatic restraints for the front center seating position would not yield as many benefits as when FMVSS 208 was originally imposed in 1977 and would not provide the same benefits per dollar spent as providing protection for the two outboard seats.

Although the center seat is rarely used, about one-third of its present occupants are children. For that reason, many are concerned about the equity of not providing automatic protection to this position. However, with child restraint laws becoming effective in 48 States and the District of Columbia, this argument loses a great deal of its merit.

Rationale for Adoption of the Rule

The Requirement for Automatic Occupant Restraints

The final rule requires, in accordance with the phase-in schedule, that automatic occupant protection be provided in passenger cars. The requirement can be complied with through any of the occupant protection technologies discussed earlier in the preamble, if those systems meet the testing requirements of the rule; i.e., manufacturers may comply with the rule by using automatic detachable or nondetachable belts, airbags, passive interiors, or other systems that will provide the necessary level of protection.

The requirement also only applies to the outboard seating positions of passenger cars. The center seat in those cars that have one is exempt from the requirement for automatic occupant protection. In addition, the requirement does not apply to other than passenger cars; for example, trucks, tractors, or multipurpose vehicles such as jeeps are not covered by the rule.

The National Traffic and Motor Vehicle Safety Act of 1966, as amended, directs the Department of Transportation to reduce fatalities and injuries resulting from traffic accidents. In its decision in the *State Farm* case, the Supreme Court held that, in carrying out its responsibilities under the Safety Act, the Department "must either consider the matter further or adhere to or amend Standard 208 along the lines which its analysis supports" 103 S. Ct. at 2862. In a number of instances throughout its opinion, the Court indicated where it found NHTSA's 1981 rescission to be inadequately supported or explained. The Department has now completed its further review of this matter, giving special consideration to the Supreme Court's decision.

Based on this review, the Department has determined that the data presented in this preamble and more fully analyzed in the Department's Final Regulatory Impact Analysis support the following conclusions:

- After assessing the data now available to it, the Department has revised its 1981 analysis concerning the likelihood of increased usage if automatic detachable belts are installed to meet FMVSS 208; it cannot project either widespread usage, or a widespread refusal to use such systems by automobile occupants.

- While it is clear that airbags will perform as expected in virtually all cases, it is also clear that the effectiveness of the airbag system is substantially diminished if the occupant does not use a belt. Consumer acceptability is difficult to predict, with the major variables being cost, fear, and the unobtrusiveness of airbags.

- Nondetachable automatic belts may result in

sharply increased usage, but there may also be substantial consumer resistance to them.

- The installation of automatic occupant protection in passenger cars may significantly reduce both fatalities and injuries.

- The costs of the existing automatic restraint systems are reasonable, and the potential benefits in lives saved, injuries reduced in severity and costs avoided are substantial.

- Technologically, the systems are feasible and practicable.

Even if we assume the lower level of the range for the effectiveness of automatic belts (35 percent) and very little increase in usage (an increase on only 7½ percent over the current 12½ percent usage rate for manual belts places us at the 20 percent level used in Table 5), there still would be significant incremental annual reductions in deaths and injuries as a result of an automatic occupant restraint rule complied with entirely by the installation of belts; 520 fatalities and 8,740 moderate to critical injuries would be prevented. Using the higher effectiveness figure (50 percent) and still only 20 percent usage, we would come close to doubling the benefits; 980 fatalities and 15,650 moderate to critical injuries would be prevented annually. If usage increases to 70 percent, 5,030 to 7,510 deaths and 86,860 to 124,570 injuries would be prevented annually.

With respect to airbags, even assuming low effectiveness and no use of lap belts, the record supports the conclusion they would provide significant incremental reductions in deaths and injuries. Airbags without a lap belt could save 3,780 to 8,630 lives and prevent 73,660 to 147,560 injuries each year. With lap belts used at the current manual belt usage rate (12.5 percent), the evidence in the record indicates that airbags could save 4,410 to 8,960 lives and prevent 83,480 to 152,550 injuries.

The potential reduction in fatalities and injuries that would result from automatic restraints could produce a corresponding decrease in funeral, medical, and rehabilitation expenses. A reduction in these expenses could, in turn, result in reductions in premiums for any insurance that covers them and a reduction in the burden on taxpayers of various medical, rehabilitation, and welfare costs.

As discussed earlier, collision and property damage liability and comprehensive insurance policies will have to absorb some additional costs to the extent that airbags are used.

In attempting to provide any relationship between costs and benefits of occupant protection systems, three important points must be kept in mind: (1) The statute directs us to "reduce... deaths and injuries," and, in doing so, to consider whether the standard we issue "is reasonable, practicable and appropriate." The Supreme Court noted in the *State Farm* case that it is "correct to look at the costs as well as the benefits of Standard 208," 103 S. Ct. at 2873, but we should also "bear in mind that Congress intended safety to be the preeminent factor under the Motor Vehicle Safety Act." *Id.* (The Senate Report said safety was "the paramount purpose." The House Report called it "the overriding consideration.") (2) The net result of any calculations will only provide information on *measurable* benefits. They would not represent the full benefits of reducing fatalities and injuries because the Department cannot measure the intangible value of a human life or a reduced injury. It cannot adequately measure, for example, the value of pain and suffering or loss of consortium. (3) The data developed on usage and effectiveness are not always precise and in many instances involve broad ranges. As a result, they can have an effect on figures derived from them and the various relationships that ensue.

With this in mind and recognizing that insurance premium reductions alone only identify a portion of the economic benefits resulting from an automatic occupant protection rule, it is interesting to note some breakeven points for the cost related to automatic belts using low and high effectiveness estimates. The breakeven point occurs when lifetime cost (retail price increases and additional fuel cost) equals lifetime insurance premium reductions. At the high effectiveness level, the breakeven point occurs at the 32 percent usage level. At the low effectiveness level, the breakeven point occurs at the 44 percent usage level. Thus, by increasing current usage by approximately 20 to 30 percent, automatic belts will pay for themselves simply based on estimated insurance premium reductions. Inclusion of noninsurance benefits would lower these breakeven points, perhaps significantly.

Although airbag systems do not attain similar breakeven points based just on insurance premium reductions, it is interesting to note that a significant portion of airbag costs would be paid for just

by insurance premium reductions. The estimated lifetime cost of a full front airbag system is $364, including increased fuel cost; the lifetime insurance premium reductions are estimated to range from $76 to $158 assuming 12.5 percent usage of the lap belt.

By issuing a performance standard rather than mandating the specific use of one device such as airbags or prohibiting the use of specific devices such as nondetachable belts, the Department believes that it will provide sufficient latitude for industry to develop the most effective systems. The ability to offer alternative devices should enable the manufacturers to overcome any concerns about public acceptability by permitting some public choice. If there is concern, for example, about the comfort or convenience of automatic belts, the manufacturers have the option of providing airbags or passive interiors. For those who remain concerned about the cost of airbags, automatic belts provide an alternative. This approach also has the advantage of not discouraging the development of other technologies. For example, the development of passive interiors can be continued and offered as an alternative to those who have objections to automatic belts or airbags.

Because one manufacturer has already begun to offer airbags and three others have indicated plans to do so, the Department expects that airbags will be offered on some cars in response to this requirement. Moreover, the continued development of lower cost airbag systems, such as the system being developed by Breed, may result in their use in even larger numbers of automobiles. By encouraging the use of such alternatives to automatic belts through this rulemaking, the Department expects that more effective and less expensive technologies will be developed. In fact, the Department believes it is in the public interest to encourage the development of technologies other than automatic belts to reduce the chance that the purchaser of an automobile will have no other option. *See* 103 S. Ct. at 2864. Thus, the rule is designed to encourage nonbelt technologies during the phase-in period. The Department's expectation is that manufacturers who take advantage of this "weighting" will continue to offer such nonbelt systems should the standard be fully reinstated. It also expects that improvements in automatic belt systems will be developed as more manufacturers gain actual experience with them.

Center Seat

The Department has also decided to exempt the center seat of cars from the requirement for automatic occupant protection. This has been done for a number of reasons described in more detail earlier in this preamble. First, limitations in current automatic belt technology would probably result in the elimination of the center seat for most cars if it were required to be protected. Balancing the loss of vehicle utility, and the numerous effects that this could have, with the limited number of occupants of the center seat and, thus, the limited benefits to be gained from protecting it, warrant exempting its coverage. It should be noted that different protection by seating position already exists as rear seat requirements differ from front seat requirements; the center front seat itself is already exempt from the requirement to provide shoulder belts. Thus, there is ample precedent for this action.

Mandatory Use Law Alternative

The rule requires the rescission of the automatic occupant protection requirement if two-thirds of the population of the United States are residents of States that have passed MULs meeting the requirements set forth in the regulation. The requirement would be rescinded as soon as a determination could be made that two-thirds of the population are covered by such statutes. However, if two-thirds of the population are not covered by MULs that take effect by September 1, 1989, the manufacturers will be required to install automatic protection systems in all automobiles manufactured after September 1, 1989. As discussed in an earlier section, use of the three-point seatbelt (which our analysis indicates is exceeded in its effectiveness range only by an airbag with a three-point belt) is the quickest, least expensive way by far to significantly reduce fatalities and injuries. "We start with the accepted ground that if used, seatbelts unquestionably would save many thousands of lives and would prevent tens of thousands of crippling injuries." 103 S. Ct. at 2871. As set out in detail earlier in the preamble, coverage of a large percentage of the American people by seatbelt laws that are enforced would largely negate the incremental increase in safety to be expected from an automatic protection requirement.

The rule also contains minimum criteria for each State's MUL to be included in the ·determination by the secretary that imposition of an automatic

protection standard is no longer required. Those minimum criteria are as follows:

• A requirement that each outboard front seat occupant of a passenger car, which was required by Federal regulation, when manufactured, to be equipped with front seat occupant restraints, have those devices properly fastened about their bodies at all times while the vehicle is in forward motion.

• A prohibition of waivers from the mandatory use of seatbelts, except for medical reasons;

• An enforcement program that complies with the following minimum requirements:

• An enforcement program that complies with the following minimum requirements:

  • *Penalties.* A penalty of $25 (which may include court costs) or more for each violation of the MUL, with a separate penalty being imposed for each person violating the law.

  • *Civil litigation penalties.* The violation of the MUL by any person when involved in an accident may be used in mitigating any damages sought by that person in any subsequent litigation to recover damages for injuries resulting from the accident. This requirement is satisfied if there is a rule of law in the State permitting such mitigation.

  • *The establishment of prevention and education programs to encourage compliance with the MUL.*

  • *The establishment of a MUL evaluation program by the State.* Each State that enacts a MUL will be required to include information on its experiences with those laws in the annual evaluation report on its Highway Safety Plan (HSP) that it submits to NHTSA and FHWA under 23 U.S.C. 402.

• An effective date of not later than September 1, 1989.

The data in Table 5 indicate the important safety benefits that can be derived from an effective MUL. The relative benefits of a MUL compared to an automatic occupant restraint rule are dependent on two unknowns: the percentage of cars equipped with each restraint and the usage or readiness rates for them. For example, if most cars were equipped with automatic belts and seatbelt usage increased 15 to 20 percent, some people would consider the automatic occupant restraint rule quite successful. A MUL would more than match the safety benefits of this rule, however, even if it was only half as successful as the data indicate foreign MULs have been. Unlike an automatic occupant restraint, MULs achieve these safety benefits without adding any cost to the car.

Moreover, a MUL can save more lives immediately. It covers all cars as soon as it is passed and put into effect. An automatic occupant restraint rule requires lead time before the manufacturers can begin installing the devices, and then it would take 10 years before most of the American fleet was replaced with cars with the automatic restraints.

At the same time, the Department recognizes that MULs must be enacted before they can have any effect. Although a number of States are considering MULs, only one State legislature has passed one that is applicable to the general population. Many commenters have argued that the possibility that MULs may be passed is an insufficient basis for the Department of Transportation to decide not to issue an automatic occupant restraint rule; such inaction would violate the Department's obligations under the National Traffic and Motor Vehicle Safety Act.

This rule allows the Department to meet the concerns over the obstacles to enactment of MULs and still be able to take advantage of their benefits if they are enacted. To the extent that automatic protection systems encounter substantial consumer resistance, it encourages State legislatures to seriously consider what some may view as a more attractive alternative. Regardless of the ultimate course the country takes, the end result will be a significant improvement in automobile safety, which is the Department's goal.

This approach avoids many of the problems associated with the other MUL proposal set forth in the SNPRM. That alternative would have resulted in waivers being granted on an individual, State-by-State basis, for those States that passed MULs. The chosen approach eliminates the need to "regulate" the sale of manual belt automobiles to prevent them from being purchased by people in States without MULs. In addition, under the rule, consumers should not have to delay purchases of cars if they want to avoid automatic protection systems. Before September 1, 1989, they will have a choice, since not all cars will be manufactured with automatic protection systems. After

that, either MULs will be in effect or automatic protection will be required in all cars. Under the other SNPRM MUL alternative, some consumers might have delayed the decision to buy a car while waiting for their State to pass an MUL.

Under this aspect of the regulation, the Department will review each State MUL as it is passed to determine whether it meets the minimum criteria established by the regulation. If, at any time before April 1, 1989, the Secretary determines that the total population covered by MULs that meet the minimum criteria of the regulation reaches or exceeds two thirds of the population of the United States, the Secretary will declare the rule rescinded. If, on April 1, 1989, the Department's information indicates that two-thirds of the population are not covered by MULs, the Department will publish a notice asking for public comment on these data. If no new data are presented to the Department establishing that, prior to April 1, 1989, two-thirds of the population were covered by MULs, the automatic occupant protection requirement will remain in effect.

Some have argued that as soon as the rule is rescinded, one or more States may rescind their MULs. The Department must presume the good faith of State legislators. It also believes that the advantages of MULs will be so clear that it would be extremely difficult and unlikely that any State would rescind its statute. The Department's position on this matter is fortified by the success of MULs in foreign jurisdictions and the fact that only one of those jurisdictions has ever withdrawn a MUL, and that nation subsequently reinstated the law. Furthermore, it would be completely impractical to tie reinstatement/rescission in short cycles to the action of one or two State legislatures. The Department will, of course, continue to monitor the general issue of the protection of automobile occupants and, in accordance with its statutory responsibilities, take whatever action is deemed necessary in the future to ensure that the objective of the National Traffic and Motor Vehicle Safety Act are met.

If the automatic occupant protection requirements are rescinded because of the passage of MULs, up to one-third of the population may have no automatic occupant protection systems in their automobiles and their States may not pass MULs. However, as discussed at length above, there are disadvantages to each of the automatic restraint systems. No approach will completely eliminate deaths and injuries. The National Traffic and Motor Vehicle Safety Act's very purpose is "reduc[ing] traffic accidents and deaths and injuries to persons resulting from traffic accidents." 15 U.S.C. §1381. Coverage of two-thirds or more of the American people by MULs will be a major achievement and is clearly consistent with the Act, and it will result in a more substantial reduction in deaths and injuries more quickly and at a lower cost than any other practical alternative. In the interim, this rule will have required the automobile manufacturers to make automatic protection systems available on an unprecedented scale.

A number of points must be kept in mind while considering the relative merits of an automatic restraint as compared to MULs: (1) MULs immediately cover the entire fleet of automobiles within the State. We do not have to wait 10 or more years for a system to become installed in the entire fleet. (2) The Department expects that, under a simple automatic occupant restraint requirement, the primary method of compliance would have been through the use of automatic belts. Although automatic seatbelts would likely result in some increased usage, MULs, based on foreign experience, should result in higher usage rates. (3) Although automatic belts are relatively inexpensive in terms of the significant safety benefits they achieve, MULs have no cost increment over the existing system. (4) If only two-thirds of the population are covered by MULs and the MULs result in what the Department estimates to be the lowest possible usage rate based on our analysis of foreign experience — 40 percent of the occupants — they will still result in a reduction in fatalities of from 1,900 to 2,400 and a reduction in moderate to critical injuries of 32,000 to 40,000 on an annual basis. This compares to automatic restraints, which, if installed in all automobiles, would result in a reduction in fatalities of between 520 and 980 and a reduction in moderate to critical injuries of between 8,740 and 15,650 at 20 percent usage, after they are installed in all automobiles. Moreover, during the first 10 years, MULs would save a total of from 19,000 to 24,000 lives and prevent from 320,000 to 400,000 moderate to critical injuries. During those same 10 years, while they were being installed in the American fleet, automatic belts at 20 percent usage, for example, would save a total of between 2,900 and 5,400 lives and prevent between 48,000 and 86,000 moderate to critical injuries. Thus, the overall safety benefits of the rule should exceed the benefits of a simple automatic protection requirement, *even if one-third of the population are not covered.* (5) We also expect that residents of MUL States will develop the habit of wearing seatbelts and will wear them even in non-MUL States. Residents of non-MUL States will be required to wear them while traveling in MUL States. This should increase the protection level somewhat.

In addition to the tremendous safety benefits of MULs, we also have the advantage of providing some local option in the decision-making. If enough States prefer MULs to automatic occupant protection, they can pass such laws and the requirement will be rescinded. We believe that offering this "option" should lessen any public resistance to an automatic occupant protection requirement. Having some ability to choose one alternative over the other should make both alternatives more acceptable. As noted earlier, public acceptance is an appropriate and important concern of the Department in its rulemaking under the National Traffic and Motor Vehicle Safety Act. Some commenters argued that automatic restraints should be used in conjunction with and not as an alternative to MULs. This argument ignores both the public acceptability concerns set forth above and the incentive for passage of such laws — to the extent there is significant consumer resistance to automatic protection devices — created by the Department's approach.

A number of commenters disagreed with the SNPRM proposal to establish criteria for the MULs. They argued that the criteria should be left to State governments and that establishment of criteria by the Department of Transportation might discourage a number of States from enacting MULs. Although the Department understands this concern, it believes that, under the National Traffic and Motor Vehicle Safety Act, in order for it to accept MULs as an alternative to requiring automatic crash protection, MULs must provide a level of safety equivalent to that which would be expected to be provided under existing technology by the automatic systems. The Department, therefore, believes it is imperative that it establish minimum criteria that will ensure that the MULs will achieve a usage level high enough to provide at least an equivalent level of safety. Otherwise, for example, a State could pass an MUL that permitted so many waivers or exceptions as to be meaningless.

The Department would like to note that, rather than requiring a State to amend an existing MUL, the Department will consider granting a waiver from the minimum requirements for an MUL for any State that, before August 1, 1984, has passed an MUL that substantially complies with these requirements.

In the SNPRM, the Department asked whether a rule such as the one the Department has adopted should be based on the number of States passing MULs or the population that is covered by the MULs.

The Department has decided to base the final rule on the percentage of the population rather than the number of States for the following reasons. If three-quarters of the States passed MULs, it might result in as little as 41 to 42 percent of the population being covered. The Department believes that the percentage of the people who are covered is the important aspect of any MUL alternative. As the Department has already clearly explained, the valuable safety benefits of MULs warrant encouraging their enactment.

It is the position of the Department that it has both the legal authority and the justification to require automatic occupant protection in all passenger automobiles. It is also the Department's position that it has the legal authority and the justification for rescinding the automatic occupant restraint requirement if two-thirds of the population are covered by MULs before September 1, 1989. It believes that either alternative would provide tremendous safety benefits; both meet all the standards of the Act and both carry out the objective and purpose of the statute.

The Phase-In

The rule requires the manufacturers to follow a phase-in schedule for compliance with the automatic occupant protection requirements. A minimum of 10 percent of all cars manufactured after September 1, 1986, must have automatic occupant crash protection. After September 1, 1987, the percentage is raised to 25 percent; after September 1, 1988, it is raised to 40 percent; and after September 1, 1989, all new cars must have automatic occupant crash protection.

To enable the manufacturers to determine at the beginning of the model year how many automobiles must be manufactured with automatic crash protection, the percentage of automobiles to be covered will be based upon each manufacturer's average number of automobiles produced in the United States during the prior three model years. If, for example, the manufacturer sold 3 million cars in model year 1984, 3.2 million in model year 1985, and 3.7 million in model year 1986, its 3-year average would be 3.3 million automobiles; for model year 1987 (beginning September 1, 1986) it would have to equip 10 percent of 3.3 million — 330,000 automobiles — with automatic occupant crash protection systems.

The Department decided to phase in the requirement for automatic occupant crash protection for a number of reasons.

First, by phasing-in, some automatic protection systems will be available earlier than if implementation were delayed until the systems could be installed in all automobiles. The earliest the Department could have required automatic protection in 100 percent of the fleet would have been September 1, 1987. Manufacturers' comments to the docket on lead time for automatic belts ranged from immediately, for some cars such as the VW Rabbit, on which automatic belts are now offered as an option, to 3 to 4 years for all cars. Estimates for airbags ranged from 2 years for driver side airbags on some models on which these devices were already planned to be offered as options (some Mercedes, BMW, and Volvo car lines) to 5 years for airbags for some companies (e.g., Chrysler and Saab). Differences in lead times among manufacturers are due to such factors as the number of model lines a company has, previous research and development efforts and supplier considerations. The 36 months lead time needed for automatic belts, *inter alia*, is required to develop spool-out features and other components on some nondetachable belts in order to maximize consumer acceptability in terms of entry/egress. Detachable belts could require vehicle modifications to strengthen belt attachment points on the door or integrate door and roof strength to accommodate the belt anchorage. While some driver airbags could be introduced with 24 months lead time, available evidence suggests that many vehicle models will require major modifications to the steering wheel and column and extensive instrument panel modifications or redesign, including glove box relocation, for passenger airbags. Testing of occupant kinematics on the passenger side is also required. Because of the number of models involved, differing car sizes and available industry resources, it is the Department's judgment that at least a 48-month leadtime would have been required for full front airbags.

If the Department had required full compliance by September 1, 1987, it is very likely all of the manufacturers would have had to comply through the use of automatic belts. Thus, by phasing-in the requirement, the Department makes it easier for manufacturers to use other, perhaps better, systems such as airbags and passive interiors.

Phasing-in also permits consumers and the Department to develop more information about the benefits of these systems, thus enhancing the opportunity to overcome any public resistance to automatic protection. Over the first 3 years, consumers will have a choice as to whether they purchase an automobile with automatic protection. Since they will not be forced to accept them, the Department expects that they will be more likely to be openminded about their benefits.

Another advantage of phasing-in the requirement for automatic protection is that is possible that by the time two-thirds of the population are covered by MULs, the manufacturers will have made progress in designing and producing these systems at a lower cost and a significant number of consumers will continue to demand them from the manufacturers as either standard or optional equipment.

The specific percentages used for the phase-in were chosen because they balance technological feasibility with the need to encourage technological innovation. These percentages should also provide the gradual phase-in that the Department believes will help build up public acceptance.

To ensure compliance with the phase-in requirement, it will be necessary for each manufacturer to submit a report to the Department of Transportation within 60 days of the end of each model year certifying that it has met the applicable percentage requirement. The report would have to separately identify, by Vehicle Identification Number (VIN) number, those cars that the manufacturer has equipped with automatic seatbelts and those cars that it has equipped with automatic airbags or some form of occupant protection technology. The Department will issue an NPRM on this matter in the very near future. In the event that a manufacturer fails to comply with the percentage requirement under the phase-in schedule, the Department has appropriate enforcement authority, e.g., civil penalties.

Thus, the use of a phase-in appropriately takes into account the abilities of the different manufacturers to comply with the requirement, encourages the use of different, and perhaps better, means of compliance, and provides the public with an opportunity to better understand the value of automatic protection. The phase-in will permit the manufacturers to ensure that whatever system they use is effective, trouble-free, and reliable. By starting off with a relatively small percentage and building up to full compliance, the phase-in will provide the manufacturers with a better opportunity to manage unforeseen development and production problems and, as a result, also make it less likely that consumers will develop adverse impressions based upon earlier experience.

Some commenters suggested that the manufacturers would use the cheapest system to comply with an automatic restraint requirement under our SNPRM MUL alternatives. They said the short time allowed for passage of MULs would force the manufacturers to choose the least expensive alternative so that they would lose little in investments if sufficient numbers of MULs passed. The Department does not agree with this contention. It believes that competition, potential liability for any deficient systems and pride in one's product would prevent this. The phase-in schedule should provide adequate time to design and produce high quality systems.

The Credit for Nonbelt Restraints

The rule also permits manufacturers to receive extra credit during the phase-in period if they use something other than an automatic belt to provide the automatic protection to the driver. For each car in which they do so, they will receive credit for an extra one-half automobile towards their percentage requirement. It will be the manufacturer's option whether to use the same nonbelt technology to provide the automatic protection to the passenger; however, such protection must be automatic—the manufacturer may not use a manual belt for the right front seat. As a result of this option, manufacturers will be able to get extra credit for the use of airbags, passive interiors, or other systems that meet the test requirements of the rule.

There are a number of reasons for the Department's decision to permit this option. First, it believes that the primary system that would be used under this "extra credit" alternative would be the airbag. As the data in Table 5 clearly illustrate, airbags should provide very significant safety benefits. Even though fewer cars would be equipped with automatic protection if extra credit is given for airbag automobiles, airbags — when used with belts — are very effective. In addition, the Department believes that there is a definite advantage in the initial stages of compliance with this rule to encourage the use of various automatic protection technologies. This should promote the development of what may be better alternatives to automatic belts than would otherwise be developed. If enough alternative devices are installed in automobiles during the phase-in period, it will also enable the Department to develop a sufficient data base to compare the various alternatives to determine whether any future modifications to the rule to make it more effective are necessary or appropriate.

Both the Act and the Supreme Court's decision last year provide the Department with the necessary flexibility to establish safety standards that are tailored to engineering realities. Recognizing some of the technological problems, for example, that have been discussed earlier with respect to airbags and small cars and coupling this with a desire to comply with the statutory safety objectives with the best possible systems, the Department believes it appropriate to establish a regulatory scheme that provides enough flexibility for the best possible systems to be developed.

Rationale for Not Adopting Other Alternatives

Retain

We have determined, for reasons more fully explained in the prior section — "Rationale for Adoption of the Rule," not to simply retain the existing requirements for automatic occupant crash protection. Simply retaining the existing rule would result in the use of detachable automatic seatbelts in nearly all (i.e., 98 or 99 percent) cars. The amended rule the Department has adopted will encourage more effective solutions to the nation's safety problems, and it should result in the prevention of even more deaths and injuries.

Amend

Airbags Only

Despite the potentially large safety benefits that would result from the use of airbags, there are a number of reasons why the Department has determined that airbags should not be required in all cars.

Costs. As we have discussed in more detail elsewhere in this preamble, the Department has estimated that airbags will cost $320 more per car than manual belts. They will also increase fuel costs by $39 over the life of the car. In addition, the replacement cost for a deployed airbag is estimated to be $800. Because of the high cost of airbags, physical damage and comprehensive insurance premiums will also increase, adding over $18 to the lifetime cost of the vehicle. On the other hand, automatic belts would only add $40 for the equipment, $11 in increased fuel costs, and would not adversely affect physical damage and comprehensive insurance premiums. Thus, although airbags may provide greater safety benefits, when used with belts, and potentially larger injury premium reductions than automatic belts, they are unlikely to be as cost effective.

Moreover, there is still a great discrepancy between the Department's airbag cost estimates and those of industry, while the Department's estimates for the cost of automatic belts are much closer to those of industry. If, despite the Department's ability to fully justify our cost estimates, airbags are priced much higher than it has estimated, it will further compound this problem.

Finally, the high cost of replacing an airbag may lead to its not being replaced after deployment. The result would be no protection for the front seat occupants of such an automobile.

Technical Problems. Several technical problems concerning airbags have been mentioned by manufacturers, consumers, and the vehicle scrapping industry. One technical concern involves the alleged dangers of sodium azide. Some commenters claim that sodium azide, the solid propellant which is ignited and converts to nitrogen gas to inflate the air cushion, is hazardous. It is claimed that it is an explosive, is mutagenic, toxic, and an environmental hazard. As explained in the FRIA, sodium azide is not an explosive. Rather it ignites, under controlled conditions, to form harmless nitrogen gas. Furthermore, studies have continually shown that it is not mutagenic or carcinogenic in mammals, due to its inactivation by the liver. Sodium azide can be toxic, but its transport in hermetically sealed containers does not pose a hazard to manufacturers, dealers, repairmen, or consumers. The scrapping of vehicles with undeployed airbag canisters does have to be done under controlled conditions so as to avoid adverse environmental effects and, although the risk is small, the Department will continue to work with manufacturers and the vehicle scrapping industry in this area.

Another concern involves the technical problem of out-of-position occupants in small cars. Manufacturers claim that little development work has been done with airbags for small (e.g., subcompact or smaller) cars and that a particular problem in these vehicles is how to protect small children, who are not properly restrained, from the more rapidly deploying air cushion in such vehicles. The Department believes that this problem can be mitigated and that technical solutions are available, as described in the FRIA. However, the lack of experience in this area, as well as the lack of experience for some companies in any form of airbag development, make the Department reluctant to mandate across-the-board airbags.

Some people have argued that the failure to issue a rule that will require at least some airbags might mean the end of the development of airbag technology. In this regard, it must be remembered that some improvements — such as those made by Breed Corporation — have come about without regulation. Moreover, three manufacturers — Mercedes, Volvo, and BMW — are currently planning to offer driver only airbags in their automobiles even though not required, and Ford will produce driver airbags for 5,000 U.S. General Services Administration cars next year. It is, therefore, possible and likely that others may follow suit to meet the competition. The extra credit provided during the phase-in should encourage manufacturers to equip at least some of their cars with airbags.

Public Acceptability. Airbags engendered the largest quantity of, and most vociferously worded, comments to the docket. Some people have serious fears or concerns about airbags. If airbags were required in all cars, these fears, albeit unfounded, could lead to a backlash affecting the acceptability of airbags. This could lead to their being disarmed, or, perhaps, to a repeat of the interlock reaction. Some people are, for example, fearful of the dangers of the sodium azide used to deploy the airbag. People are also concerned that the airbag will inadvertently deploy and cause an accident or that it will not work at the time of an accident. Some people are also concerned because they feel less secure in an automobile unless they have a 3-point belt wrapped around them (and if the Department requires a 2-point belt with an airbag, the costs will be even higher) and are thus unsure that they will be protected at the time of an accident.

Although the Department believes that these concerns can be adequately addressed, these consumer perceptions must be recognized as real concerns. It may be easier to overcome these concerns if airbags are not the only way of complying with an automatic occupant protection requirement. Under the rule being issued, if people have concerns about airbags, they can purchase automobiles that use automatic belts. The real world experience that will come with the production of airbag equipped cars during the phase-in period should help to mitigate these fears.

Effectiveness. Airbags are not designed to provide protection at barrier equivalent impact speeds less than approximately 12 mph. In addition, in order to provide protection comparable to that of a 3-point belt, they must be used in conjunction with at least a lap belt. Despite this, the overall benefits provided by an airbag, because of its extremely high "usage" rate, may be much better than those provided by automatic belts. Widespread use of both systems is the only way to develop definitive data.

Performance Standards. Several commenters questioned the Department's authority to issue an airbag only standard, claiming it would be a "design" standard. Even if the Department could legally issue a performance standard that could only be met by an airbag under present technology, it believes that by taking away the manufacturers' discretion to comply with an automatic occupant restraint requirement through the use of a variety of technologies, it creates a number of problems. First, by restricting the manufacturers, the Department runs the risk of killing or seriously retarding development of more effective, efficient occupant protection systems. With real world experience, the Department may find, for example, that automatic belts would be used by much higher percentages of occupants than currently anticipated. The manufacturers also would not be able to develop better automatic belt systems that may be more acceptable and, therefore, used by larger numbers of people. This may result in automatic belts that save as many lives but at a much lower cost than airbags. Similarly, the development of passive interiors, being pursued by GM, would be stymied under such an option. The Department believes an airbag only decision would unnecessarily stifle innovation in occupant protection systems.

In addition, if airbags were not mandated in every car, people may be more willing to give them a chance to prove themselves than they would be if they were forced to buy them. If consumers are concerned about automatic belts, it may cause manufacturers to make greater efforts to lower the costs of airbags to make them more acceptable as an alternative.

Airbags and/or Nondetachable Seatbelts

The rationale provided in the preceding sections for adopting the new rule and for not retaining the old rule or amending it to require airbags in all cars essentially provides the basis for the Department's decision not to amend the old rule to require either airbags or nondetachable belts or just nondetachable belts; (i.e., would not permit the use of detachable belts to comply with the automatic protection requirements). It is also concerned that nondetachable belts may be too inconvenient and restrictive, resulting in serious adverse public reaction if required in all cars. (See the discussion on nondetachable belts in the first part of the "Analysis of the Alternatives.")

Limited Seating Positions

Several of the alternatives would have required all or some particular type of automatic protection for specified seating positions. For example, airbags would have been required for only the driver position under one alternative. As explained under the section on "Rationale for Adoption of the Rule," the Department has determined that the data on center seats warrants exempting that position from automatic protection requirements. It also has decided that, during the phase-in period, it is appropriate to give "extra credit" for providing automatic protection to the driver through non-belt technology, such as airbags and passive interiors, to provide an incentive for developing and producing these other, possibly better, systems. The Department has determined that existing data, discussed earlier in the preamble and in the FRIA, does not warrant exempting the front right seat or providing any other special protection to the driver.

Small Cars

The SNPRM raised for comments the alternative of providing airbag protection for the drivers of small cars and questioned the safety justification for this. We have not received data that indi-

cate that small cars are always less safe than large cars. For that reason, we have no justification for requiring any special protection for small cars.

Rescind

After a full review of the rulemaking docket and performing the Analysis contained in our FRIA, we have concluded that the Supreme Court decision in the *State Farm* case precludes us from rescinding the automatic occupant protection requirements at this time based on the present record in this rulemaking.

The Supreme Court noted that "an agency changing its course by rescinding a rule is obligated to supply a reasoned analysis for the change *beyond that which may be required when an agency does not act in the first instance.*" 103 S. Ct. at 2866 (emphasis supplied).

To avoid having its actions labeled "arbitrary and capricious," the Supreme Court said that "the agency must examine the relevant data and articulate a satisfactory explanation for its action including a 'rational connection between the facts found and the choice made.'" 103 S. Cr. at 2866-67.

The Supreme Court also held that, if automatic belts are not justifiable, the agency should have considered requiring airbags in all automobiles. The Court found that:

> Given the effectiveness ascribed to airbag technology by the agency, the mandate of the Safety Act to achieve traffic safety would suggest that the logical response to the faults of detachable seatbelts would be to require the installation of airbags. 103 S. Ct. at 2869.

It added that:

> Given the judgment made in 1977 that airbags are an effective and cost-beneficial life-saving technology, the mandatory passive restraint rule may not be abandoned without any consideration whatsoever of an airbags-only-requirement. 103 S. Ct. at 2871.

The primary issue concerning automatic belts is the anticipated usage of the detachable belts. Although the Department cannot establish with certainty the level of usage it can expect with automatic belts, the information gathered during the comment periods on the current rulemaking NPRM and SNPRM does assist DOT in answering the Supreme Court's finding that:

[T]here is no direct evidence in support of the agency's finding that detachable automatic belts cannot be predicted to yield a substantial increase in usage. The empirical evidence on the record, consisting of surveys of drivers of automobiles equipped with passive belts, reveals more than a doubling of the usage rate experienced with manual belts. 103 S. Ct. at 2872.

Although some would argue that the belts will merely be detached after most drivers or passengers first enter the car and never used more than current manual belts are used, no evidence has been found to support this. In responding to NHTSA's 1981 rescission argument that "it cannot reliably predict even a 5 percentage point increase as the minimum level of increased usage," the Supreme Court said:

But this and other statements that passive belts will not yield substantial increases in seatbelt usage apparently take no account of the critical difference between detachable automatic belts and current manual belts. A detached passive belt does require an affirmative act to reconnect it, but—unlike a manual seatbelt—the passive belt, once attached, will continue to function automatically unless again disconnected. Thus, inertia—a factor which the agency's own studies have found significant in explaining the current low usage rates for seatbelts—works in favor of, not against, use of the protective device. Since 20 to 50% of motorists currently wear seatbelts on some occasions, there would seem to be grounds to believe that seatbelts used by occasional users will be substantially increased by detachable passive belts. Whether this is in fact the case is a matter for the agency to decide, but it must bring its expertise to bear on the question. 103 S. Ct. at 2872.

Although the Department believes that the existing automatic belt usage data is not generally applicable to the entire vehicle population, there is an *absence of data* that indicate that there will be no increase in usage associated with detachable automatic belts. The record of this rulemaking only has assertions that this will be so, but it lacks support for those assertions.

The Supreme Court has made it clear that it believes the better arguments support increased usage. Not only does the Department have no new evidence to counter this, but, for the first time, the manufacturers have acknowledged that, at least initially, automatic detachable belts will result in an increase in usage. The Department also now believes that some level of increase will occur based on the reasons people give for not using manual belts (e.g., "forget" or are "lazy"). Thus, it has no evidence that the belts will not be used, but merely questions about how large an increase will occur. The Supreme Court said:

[An agency may not] merely recite the terms "substantial uncertainty" as justification for its actions. The agency must explain the evidence which is available, and must offer a "rational connection between the facts found and the choice made."...Generally, one aspect of that explanation would be a justification for rescinding the regulation before engaging in a search for further evidence. 103 S. Ct. 2871.

It could also be argued that the public will not accept automatic belts because of such problems as their obtrusiveness and inconvenience. Although an argument about public acceptability can be made, strong data on which to base it do not exist. As is discussed in more detail elsewhere in this preamble, the public opinion surveys that have been taken are flawed to the extent that they will not withstand close scrutiny and support a rescission decision that has already been struck down once by the Supreme Court.

The Supreme Court also found that, if detachable belts were unacceptable to the agency, than it "failed to articulate the basis for not requiring nondetachable belts under Standard 208." 103 S. Ct. at 2873. The Court added that, "while the agency is entitled to change its view on the acceptability of continuous passive belts, it is obligated to explain its reasons for doing so." 103 S. Ct. at 2873. Finally, the Court said that:

The agency also failed to offer any explanation why a continuous passive belt would engender the same adverse public reaction as the ignition interlock, and, as the Court of Appeals concluded "every indication on the record points the other way."...We see no basis for equating the two devices: the continuous belt, unlike the ignition interlock, does not interfere with the operation of the vehicle. 103 S. Ct. at 2873-74.

Again, "substantial uncertainty," 103 S. Ct. at 2871, will not suffice and there is no substantive evidence in the rulemaking record to refute the point made by the Court.

The Department has no new evidence that nondetachable belts are not an acceptable means for reducing deaths and injuries. Although there are some comments in the current docket that some people will dislike tham and may even cut them or otherwise destroy them, it is primarily speculation; there is no clear data. Moreover, even if 20 or 30 or even 40 or 50 percent of the people find some method for defeating the belt, the evidence in the record indicates that it will still result in a significant reduction in deaths and injuries for the remainder who do not.

Some people expressed concern about emergency egress from nondetachable belts. The Supreme Court had the following to say on this:

...NHTSA did not suggest that the emergency release mechanisms used in nondetachable belts are any less effective for emergency egress than the buckle release system used in detachable belts. In 1978, when General Motors obtained the agency's approval to install a continuous passive belt, it assured the agency that nondetachable belts with spool releases were as safe as detachable belts with buckle releases. 103 S. Ct at 2873.

Manufacturers commented that it would likely be more difficult to extricate oneself from a nondetachable as compared to detachable automatic belt. However, they did not claim that it represented an "unsafe" condition, and again, there is no new evidence to buttress their concerns.

Finally, there are a number of attractive arguments that are based in part on the following theme: the presence of the government in the middle of the debate over passive restraints has distorted the activities of both automobile manufacturers and insurance companies; if the marketplace had been allowed to work, insurance incentives would have led to the voluntary adoption of one or more systems by the manufacturers. Whether these arguments are correct or not, they cannot be considered in a vacuum. In fact, the context provided by the Supreme Court is quite harsh:

For nearly a decade, the automobile industry waged the regulatory equivalent of war against the airbag and lost—the inflatable restraint

was proven sufficiently effective. Now the automobile industry has decided to employ a seatbelt system which will not meet the safety objectives of Standard 208. This hardly constitutes cause to revoke the standard itself. Indeed the Motor Vehicle Safety Act was necessary because the industry was not sufficiently responsive to safety concerns. The Act intended that safety standards not depend on current technology and could be "technology-forcing" in the sense of inducing the development of superior safety design. 103 S. Ct. at 2870. (Footnotes omitted).

The history of this rulemaking, the *State Farm* decision, and the rulemaking record have put us in a position where rescission of the automatic occupant restraint requirements—unless there is a very substantial increase in use of seatbelts in the future—cannot be justified. On the other hand, as discussed in detail elsewhere in the preamble, such a substantial increase as a result of the widespread enactment of MULs would provide increased safety benefits much more quickly and at a much lower cost, thus making rescission clearly justifiable. As the Supreme Court said, "We start with the accepted ground that if used, seatbelts unquestionably would save thousands of lives and would prevent tens of thousands of crippling injuries." 103 S. Ct. at 2871. It also noted that the Department originally began the passive restraint rulemaking exercise because "[i]t soon became apparent that the level of seatbelt usage was too low to reduce traffic injuries to an acceptable level" 103 S. Ct. at 2862. The data set out elsewhere in this preamble and in the Final Regulatory Impact Analysis demonstrate the dramatic reductions in deaths and injuries that widespread usage of the manual belt systems would achieve. Thus, the Department has concluded that if two-thirds or more of the American people are covered by such laws, the need for an automatic occupant restraint requirement would be obviated.

Demonstration Program

Because of the length of time a demonstration program would take, the Department believes that it would be necessary to justify rescission of the old rule under this alternative. It also believes that the phase-in portion of the amended rule will achieve the public education/acceptance aspects of any demonstration program.

Other Mandatory Use Law Alternatives

The Department's rationale for not adopting the other MUL alternatives is explained more fully in the preceding sections. These other alternatives are generally deficient in one of two respects: they either make it necessary for the Department to justify rescission under current circumstances or the requirements they impose are much too burdensome.

Under the alternative raised in the NPRM, the Department would have sought the enactment of MULs. The Department could not be certain that a sufficient number of MULs would pass or that, if passed, they would contain the necessary provisions concerning penalties, enforcement, sanctions, education, and waivers. As a result, the Department could not determine whether the necessary level of benefits would be achieved.

Under the other SNPRM alternative, the Department would have waived the requirement for automatic restraints in individual States that enacted MULs. This alternative would have re-

quired the "regulation" of the sale of the manual belt cars to ensure that they were not covered by people not covered by MULs. It also would have had adverse market impacts if consumers delayed their purchases of cars, in anticipation of their States passing MULs, in order to avoid purchasing automatic restraints.

Legislation to Require Consumer Option

As with some of the previous alternatives, this approach would require the Department to justify rescission of the old rule. In addition, it would place a tremendous economic burden on the manufacturers to have to be able to provide a variety of systems on each model. It would, in turn, raise the cost of all automobiles for the consumer.

Airbag Retrofit Capability

This, too, would require justification for rescission. It would also result in increasing the cost of all cars even if no one ever retrofitted a car.

TESTING PROCEDURES

Repeatability

The single most significant repeatability issue related to test procedures, as reflected in comments to the docket, was that of the repeatability of the barrier crash test results. Nearly all manufacturers claim that because test result differences are encountered in repeated tests of the same car, and since these differences are large, they can not be certain that *all* their vehicles will be in compliance even when their development and compliance tests show that the vehicles are. These large differences, or test variability, place a manufacturer in jeopardy, it is claimed, because NHTSA, while checking for compliance, may find a single vehicle with test results exceeding the maximum values in the standard, even though the manufacturer's results are to the contrary. Thus, they stated, they might have to recall vehicles and make vehicle modifications (which they claim they would not know how to make) even though the vehicles actually comply with the standard. The auto companies say that the test result variances are essentially due to deficiencies in the test procedures themselves as well as the prescribed Part 572 test dummy.

Because of these alleged deficiencies, the argument goes, the standard is neither "objective" nor "practicable" as required by statute. Manufacturers cite court decisions in *Chrysler Corp.* v. *DOT* 472 F.2d 659 (6th Cir. 1972) and *Paccar, Inc.* v. *NHTSA*, 573 F.2d 632 (9th Cir. 1978), to argue their point. In *Chrysler*, the court said that for a standard to be "objective"

tests to determine compliance must be capable of producing identical results where test conditions are exactly duplicated, that they be

decisively demonstrable by performing a rational test procedure, and that compliance is based upon the readings obtained from measuring instruments as opposed to the subjective opinion of human beings. 472 F.2d at 676.

Because manufacturers claim that the only way they can assure compliance is to "overdesign" their vehicles (e.g., because of alleged variances in results, to comply with a HIC requirement of 1000 manufacturers would design their vehicles to only have an HIC of 500), resulting in excessive costs without safety benefit, the *Paccar* case has relevance. In overturning a truck braking standard, the Court said that although the standard's test procedures were "objective," they were not "practicable" because variations in test surface skid numbers required manufacturers

not simply to comply with the stated standard, but to over-compensate by testing their vehicles on road surfaces substantially slicker than official regulations require. 573 F.2d at 644.

The Department continues to believe, however, that FMVSS 208 is both objective and practicable. Manufacturers have not supplied for the record data to support their claims of excessive test variability nor have they demonstrated that the bulk of any variability is due to test procedures and instruments as compared to vehicle-to-vehicle differences.

The primary, and for most manufacturers the sole, basis for claims of variability was the Repeatability Test Program conducted by NHTSA under its New Car Assessment Program. NHTSA tested 12 Chevrolet Citations in an attempt to ascertain the reliability of publishing barrier crash test results based on a single test. The results of the testing program for HIC (only HIC was mentioned by manufacturers as a variability "problem") were:

	Mean	Standard Deviation	Coefficient of Variation
Driver	655	137	21%
Passenger	694	77	11%

The manufacturers focused on the COV of the driver HIC values—21 percent—and claimed that this is too large. They claim that with this large a COV, they would have to design their vehicles to achieve an HIC no higher than 560 to assure than 95 percent of their cars, when tested, would have HIC values below 1000.

This argument is faulty for several reasons. First, the NCAP results were based on the testing of a single car—the Citation—at a *higher* test speed (35 mph) than required in FMVSS 208 (30 mph). Passing the FMVSS 208 criteria at 35 mph requires a vehicle to absorb 36 percent more energy—since the energy dissipated in a crash is proportional to the square of the speed—then in the required 30 mph crash. The Department would expect that test result differences would be lower at 30 mph since at 35 mph the design limit of certain structural members has been exceeded. Assuming that the COV at 35 mph would be identical or lower than that at 30 mph is without foundation and is counterintuitive to sound engineering theory.

Second, the NCAP data can only be used to derive a COV, at 35 mph, for the Citation. Extending the Citation results to other vehicles is again without basis. For example, Volvo tested four MY 1983 760 GLE vehicles according to the NCAP procedures (although an additional 3 760 GLEs were tested by a laboratory, MIRA, for Volvo, the NCAP procedures may not have been fully followed by that organization and thus can not be combined with Volvo's own data). The results of the four Volvo tests are:

	Mean	Standard Deviation	COV
Driver	898	71	8%
Passenger	731	27	4%

Here, we see coefficients of variation about 60 percent lower than that shown for the Citation.

Although not as many tests were run as for the Citation, the Volvo 760 GLE results cast doubt as to whether the Citation results can be applied to all vehicles. The Department also points out that even the Citation results for the *passenger*, which tended to be ignored in the docket comments (manufacturers instead tended to focus on the higher COV for the driver) exhibit half the COV cited by the auto companies.

Ford commented that the Volvo data, "though nominally somewhat lower, was not significantly different than that found in the Citation..." Ford, however, used all seven Volvo tests. Since these tests were not all conducted similarly, they are from two different statistical "universes" and cannot be combined for statistical purposes. Nor does Ford disagree that the Volvo results are lower than for the Citation. And, Ford only compared the standard deviation of the Citation and 760 GLE results. Since the *mean* was higher for the 760 GLE than the Citation, and since the COV is equal to the standard deviation divided by the mean, had Ford compared COVs it might have found that these differences *were* statistically significant. Thus, Ford's use of the Volvo is inaccurate in that it: (1) combines two unlike data sets—the MIRA and Volvo 760 GLE tests; (2) fails to examine coefficients of variation, a better descriptor of variance than the standard deviation; and (3) only examines the larger differences associated with driver HIC, and ignores the lower, passenger variances.

Ford also supplied, in response to the SNPRM, data which the company claims shows that their 33 Mercury tests, with airbags, conducted in 1974 also exhibited the same variances. Ford took the results of these tests on MY 1972 Mercurys, which were conducted at 30 mph, and "scaled" them to 35 mph. They claim that after "scaling," the Mercurys exhibited the same standard deviation as the Citation.

The Department has examined the actual 30 mph test results of these Mercurys, contained in Ford's February 1976 report, "Airbag Crash Test Repeatability," ESRO Report No. S-76-3, and finds that the results are not just for frontal barrier tests but also 30 degree angle tests. At least nine of the 24 frontal tests were at the oblique angle. Although FMVSS 208 requires angle tests, the comparison of angle plus frontal results to only frontal results is somewhat inappropriate.

Furthermore, Ford again compares only the standard deviations of driver HICs. After "scaling," Ford shows the driver HIC standard deviation to be 137. However, the standard deviation based on Tables 4-1 of the Ford report show driver HIC standard deviations, without "scaling," in frontal crashes to be only 80, and the COV in frontal crashes, given the mean of 479, is 16.7 percent. As Ford somehow *converted* these values, or some other value representing both frontal and oblique crashes, from 30 mph to 35 mph, Ford is implicitly agreeing with NHTSA that one can not compare statistical results from crash tests conducted at different speeds.

These Departmental positions—that the Citation tests may not be applicable to all cars and that 35 mph test results may not be applicable to results at 30 mph—were raised in the SNPRM wherein the Department stated "We are also interested in comments on the relevance of the Citation variability tests (conducted at 35 mph) to the FMVSS 208 compliance tests (specified to be conducted at 30 mph) and the applicability of the new Citation results to other vehicles." Other than the above cited Ford data, responses were submitted by only GM, which provided data based on 30 mph sled tests which showed COVs of 11 and 8 percent for the driver and passenger, respectively, and Volvo, which also provided sled test data showing a mean of 467 and a COV of 12.5 percent. Further, only Ford claimed that "comparable variability" to that resulting from the Citation tests "would be expected for other vehicle models." Other manufacturers failed to address the issue.

Based on the above, the Department concludes that the Citation test results cannot, without the analysis of data for other vehicles, be applied to other cars models at lower speeds.

The second reason the Department does not accept manufacturer claims of excessive variability is also related to test speeds. Variability by itself is not a crucial factor for a manufacturer to be concerned about. Rather, it is the combination of variability *and* the mean (or average) value which can be cause for concern. For example, assume that a manufacturer is 95 percent confident that all its HIC test results will be within ±150 points of the mean. If the mean value is 900, then the manufacturer may not be certain that all its vehicles will comply with a criterion whose maximum value is 1000. However, if the mean is 500, then the ±150 variation is of little consequence in ascertaining

assurance of compliance.

It is clearly intuitive, due to the 36 percent less energy involved in a 30 mph crash compared to a 35 mph crash, that average test results will be lower at the 30 mph barrier crash speed than at the 35 mph speed used in the NCAP program. No commenter to the docket argued to the contrary. Therefore, the issue of variability can not be examined in isolation but *must* be analyzed in the context of the mean value.

Reexamining the Ford Mercury data, conducted with airbags at 30 mph, the mean HIC value, taken from page 4-20 of the Ford report, is 319.9. With such a low mean, the derived variance is irrelevant for compliance purposes. The Department wishes to point out that: (1) based on its NCAP testing, even with manual belt systems and when tested at 35 mph, 80 percent of the dummy drivers and about 60 percent of the passenger dummies meet the FMVSS 208 injury prevention criteria with mean HICs of 899 and 845, respectively. These percentages would of course increase and the means decrease at 30 mph. And (2) all *airbag* tests shown mean HICs in the 400 to 500 range, a range wherein variability again becomes meaningless for assuring compliance. For instance, tests with airbags for MY 1972 Pintos showed maximum HICs in the 500 to 600 range with the median value less than 400; the maximum and mean for MY 1972 Mercurys were less than 700 and less than 400, respectively; and for MY 1974-76 GM airbag cars the values were under 600 and about 450, respectively.

Thus, mean HICs for automatic belt systems in 30 mph barrier crashes would be lower than the 899 and 845 values observed from the 35 mph NCAP program and for airbag equipped cars would likely be in the 400 to 500 range, making variability a moot issue.

A third reason that the Department believes that variability is not so significant as issue as to preclude the standard's reinstatement is that manufacturers have not demonstrated that the test procedures and test dummy are the major causes of variability. GM and Volvo provided sled test data which showed COVs of about 10 percent. Since sled test provides a steady crash pulse, it was argued that most of, if not all, the variability seen was due to dummy and test procedure variances. Without arguing the point, the Department notes that these manufacturers failed to address the question of whether this 10 percent level of variability, when combined with an expected mean, is unacceptable. For instance, if it is assumed that the mean 30 mph passive belt HIC is 800 — which is not unreasonable given current means of between 845 and 899 at 35 mph — a COV of 10 percent translates into a standard deviation of 80. Since 95.45 percent of all test results fall within the mean ± 2 standard deviations, a manufacturer can be sure than more than 95 percent of its cars will have HICs below 960 (800 + 2[80]) and the manufacturer could be about 98 percent certain that all tested cars will have values below 1,000. A lower mean would increase the above-mentioned percentages.

In the SNPRM, the Department requested comments on what level of variability was deemed "reasonable," given that some variability will always exist. Only Renault provided a quantitative answer, saying the "the variation coefficient must not exceed a maximum of 10 percent." Although Renault provided no further justification for its recommendation, the Department notes it is nearly identical to the variation contributed by the test procedures and dummy, according to Volvo and GM.

Manufacturers generally asserted that the observed variability was not caused by vehicle-to-vehicle differences but by the test procedures and use of Part 572 dummy. In the SNPRM, the Department said that it did not believe that the dummy contributed significantly to test variability. The Department, after reviewing the docket, still retains this conclusion. The 1976 Ford repeatability test report concluded that "that portion of the variability in the test results which can be attributed to differences between the nine part 572 dummies...is small for the HIC measurements and virtually nil for the chest g and femur load measurements." Ford engineers also said in an SAE paper (SAE paper 750935) the "differences in test readings from one test dummy to another were rather small, especially when compared to other factors...In fact, the variance in test readings associated with differences among dummies was essentially zero for chest g and for femur loads." Renault, in response to the SNPRM, said that "the present Part 572 dummy is not the major cause of the dispersion of results."

In its NCAP repeatability program, NHTSA found that differences in dummy calibration results have "no correlation...to dummy response results in the vehicle crash event." (SAE paper 840201, February 1984). NHTSA further noted that the Citation's "structural response...dis-played significant variability" from vehicle-to-vehicle. These differences included variations in engine cradle buckling, floor pan and toe board buckling, and irregular motion of the steering column. NHTSA concluded that "previous safety research has demonstrated that these structural behavior characteristics do have influence on dummy HIC values, possible of major proportions." Because of the large variations among vehicles and the lack of correlation of dummy calibration to HIC results, NHTSA believes that a large part of the test variability is due to vehicle variability.

In summary, the Department finds that FMVSS 208 meets all statutory criteria for objectivity and practicability, that manufacturers have not demonstrated that there would be either excessive variability in total or due to the test procedures alone, and that compliance with FMVSS 208, particularly with airbags, does not represent an insurmountable burden to manufacturers.

Compliance Procedure

Having concluded that any test variability is not sufficient to delay the standard's reinstatement, the Department is still concerned that manufacturers believe themselves to be in unacceptable compliance jeopardy. To reduce this jeopardy, manufacturers suggested that a "design to conform" policy be adopted. They claimed this was neither inconsistent with court decisions regarding the required objectivity of standards nor would it materially affect vehicle design, since they would still have to demonstrate, through crash tests, that their design could achieve the required levels of compliance. Furthermore, it was argued by VW that NHTSA presently operates under this concept.

We agree with VW that, in the event of a non-conforming test result, NHTSA will seek to obtain manufacturer compliance, test data and/or conduct a second compliance test itself, prior to asserting that a particular model is in noncompliance. The Department is unaware of any instance in which NHTSA has sought remedy under the statute for noncompliance with a safety standard based on only a single test result. Thus, for example, if NHTSA found a car with an HIC value of 1050 and, after reviewing manufacturer test data and/or conducting another test, both of which demonstrated compliance, it would likely determine that the manufacturer had exercised "due care" and would not seek remedy under the statute.

However, the *Chrysler* Court disapproved of any agency offering to investigate whether differences in test results (between manufacturer tests and agency compliance tests) were sufficient to determine a noncompliance. The court stated that manufacturers needed objective assurances and there was no room for agency investigations. Thus, the Department recognizes that automobile companies need some guarantee that should one car out of a million, for example, be found to fail the compliance test, that all one million will not have to be recalled.

The guarantee sought by the industry, "design to conform," though, is not acceptable. As pointed out in the SNPRM, the Department believes such an approach introduces unacceptable subjectivity into the determination of compliance with the standard, in contravention to the decisions of the courts to minimize nonobjective determinations of noncompliance. Instead, since NHTSA already exercises discretion in compliance cases, we will seek, through a subsequent Notice to be issued shortly, to provide such assurances without compromising either safety or the necessary statutory objectivity. Essentially, we will propose to amend FMVSS 208 by recognizing that a vehicle shall not be deemed in noncompliance if a manufacturer has exercised "due care" in designing and producing such vehicle. Rather than increase the subjectivity of the compliance process by introducing a "design to conform" concept, NHTSA will explicitly recognize in FMVSS 208 the statutory direction expressed in section 107(b)(2) of the National Traffic and Motor Vehicle Safety Act (15 USC 1397), that the penalties associated with producing a noncomplying vehicle "shall not apply to any person who establishes that he did not have reason to know *in the exercise of due care* that such vehicle . . . is not in conformity with applicable Federal motor vehicle safety standards . . ." (emphasis added).

Test Dummies

As stated earlier, the Department continues to believe that the Part 572 test dummy fully meets all statutory criteria and is not a major source of test result variability. Most manufacturers, however, disagreed. Volvo contended that the dummy has "serious limitations" and must be more durable, repeatable, and trouble-free. Toyota said it could not be sure of the influence of the dummy on test results. Mercedes also said that the Part 572 dummy

is not sufficiently repeatable while Ford said that the dummy's calibration is repeatable but its crash test performance may not be. American Motors said that the Part 572 dummy is a state-of-the-art compromise and lacks in measurement fidelity.

While not claiming that that Part 572 dummy is not repeatable or fails to meet statutory criteria, GM urged NHTSA to approve the use of the Hybrid III dummy as an alternative test device. GM said that the Hybrid III "offers significant improvements over the part 572 dummy relative to biofidelity of frontal head, chest and knee responses, fore-aft neck bending, ankle and knee articulation and automotive seated posture." Nissan agreed that the Hybrid III is a superior dummy which demonstrates greater repeatability. Conversely, Mercedes said that the Hybrid III is not any more repeatable than the Part 572 dummy.

As part of its petition to use the Hybrid III, GM submitted a paper by Mertz ("Anthropomorphic Models," GM USG 2284, Part III, Attachment I, Enclosure 3) which stated that the Part 572 dummy (actually, the Hybrid II dummy, also developed by GM) has "good repeatability, durability, and serviceability." "The Part 572 dummy represents the state-of-the-art of dummy technology in the early 1970's."

Based on the conclusions of the Ford Mercury testing and the agency's NCAP testing, NHTSA has concluded that the dummy does not contribute significantly to test variability. Renault agreed with this conclusion. Industry characterizations of the dummy, as shown above, vary considerably, from the Part 572 being a major cause of variability to it not being a major cause, to the Hybrid III being an improvement, to it not being an improvement. Only a few manufacturers provided data to support their contentions but these data, supplied by Ford, GM, and Volvo, based on sled tests, could neither separate the contribution of variability associated with the dummy alone nor demonstrate why any dummy-induced test result variances were so high as to be unacceptable. Since the Department recognized, in the SNPRM, that some variability will always be present in specifically sought comment on the levels of variance which were deemed "unacceptable." Only Renault replied to this direct question and it did not supply a rationale for its conclusion. In the absence of data to the contrary, the Department continues to believe that the current Part 572 test dummy is adequate to use as a compliance test device in standard 208.

Nevertheless, it is recognized that the Part 572 dummy is more than 10 years old and, we agree with AMC and GM in this regard, is a state-of-the-art compromise. Recognizing that dummy development, especially improved biofidelity—that is, the dummy's replication of actual human motion and potential for injury—is crucial for continued improvements in vehicle safety, NHTSA has been utilizing the Hybrid III dummy in its research and development work, as have GM and other manufacturers. NHTSA recognizes that the Hybrid III dummy does have additional measurement capability over the Hybrid II (Part 572) and, assuming injury criteria can be agreed upon and its repeatability, durability, etc. verified, it could be viewed as an improvement over the Hybrid II. Because of these views, and the data presented in the GM position, NHTSA will address these issues in a separate rulemaking. Because we have concluded that the current Part 572 dummy is fully adequate to use in testing to the injury criteria specified in FMVSS 208, action on the Hybrid III dummy is irrelevant for the purposes of this rulemaking. Should NHTSA decide to permit the use of the Hybrid III as an alternative test device, as GM has petitioned, it would not pose any additional burden on manufacturers since they could still use the current Part 572 dummy for compliance purposes. If NHTSA decides to substitute the Hybrid III for the Hybrid II as the compliance test device specified in Part 572, a gradual phase-in period would be provided so as not to interfere with manufacturer leadtime and the timely implementation of the automatic occupant protection provisions of FMVSS 208.

Injury Criteria

Several manufacturers recommended that the injury criteria associated with potential head injury be adjusted in two ways: (1) to eliminate the measurement of HIC in the absence of head contact, and (2) to increase the HIC in case of a head strike to 1500 from its current level of 1000.

It is recognized by NHTSA that the Head Injury Criterion (HIC) was primarily developed from tests of forehead impacts, resulting in acceleration of the brain in the anterior-posterior (i.e., forward and backward) directions. This was pointed out in the SNPRM, wherein the Department also briefly discussed accident and test data, including information from NHTSA itself, which suggested a very low probability of brain injury in the absence of head contact. However, it was suggested that measuring HIC in noncontact situations could serve as a surrogate for potential neck or other injuries.

Volvo supplemented the above arguments by stating that the use of HIC for other than what was the basis of its development—forehead impacts in the anterior-posterior directions—results in less dummy biofidelity. Volvo suggested that this expanded use of HIC, beyond what it was intended to measure, is inappropriate. They stated that if neck injuries are of concern, then other criteria, related solely to the neck, be used. This position on neck injuries was supported by Peugeot, Renault, Ford, and GM. Mercedes and MVMA also opposed measuring HIC in noncontacts but did not mention its use as a surrogate in potentially preventing neck injuries. Allstate opposed its elimination in such crash situations, claiming it protects occupants from cervical and spinal injuries.

The primary derivation of HIC from head impact tests is not in question. HIC was developed from the Wayne State Tolerance Curve (WSTC) which was itself based on the hypothesis that the dominant head injury mechanism was linear acceleration.

The Department agrees with the commenters, based on its own review of the origins of HIC, that its predictive capability of neck injuries is weak. The Department further agrees that the prevention of neck injuries, through assuring that excessive head motion is prevented, is important for automobile safety since neck injuries account for 78.2 percent of all crash-related noncontact-harm in passenger cars (see SAE Paper 820242, "A Search for Priorities in Crash Protection," Milliaris, et al, February 1982). The Department also notes that the Hybrid III dummy is capable of neck injury measurements, by monitoring the dummy's neck's axial loading, shear load, and bending movement (see GM's petition, USG 2284 Part III, Attachment I, Enclosure 2). Although the Hybrid III's neck biofidelity may be deficient in that its lateral bending response may not be humanlike and its neck too stiff in axial compression, its measurement of fore/aft bending provides superior biofidelity to the Part 572 dummy, which is incapable of direct injury measurements (see ibid, Enclosure 3).

The Department thus believes that prevention of neck injury would be better served by direct dummy measurement, measurement which can be made with the Hybrid III. This position was also expressed by the U.S. delegation to ISO/TC 22/SC

12/WG 6 which stated that "the head injury criterion should not be applied in the event of no head impact...other injury criteria, perhaps based on neck loads..." should be used instead. As part of the subsequent rulemaking mentioned previously, the adoption of neck injury criteria will be proposed. In addition, the issue of noncontact HICs will be further addressed in the context of the current Part 572 dummy. Data relating to the biofidelity of the dummy, in this regard, will be specifically sought.

This issue is not viewed as one which affects the decision regarding FMVSS 208 contained in this notice. Any action by NHTSA in this area should only result in reducing the required test burden, thus additional leadtime should not be required. Action regarding the dummy is viewed by the Department as seeking to continually improve the biofidelity of its anthropomorphic test devices, and is thus separate from, although related to, the 208 decision.

Although several manufacturers requested that the HIC criterion, even when there is a head strike, be raised to 1500, the Department will not take any action on that issue at this time. The 1500 HIC level is the subject of a petition for rulemaking by the CCMC. NHTSA will respond directly to this petition at the same time that it prepares the aforementioned rulemaking action.

Oblique Test Requirement

The SNPRM contained a proposal to eliminate the requirement to test compliance at angles up to 30° from the longitudinal direction. The basis for this proposal was data from Ford's Airbag Crash Test Repeatability report, which consistently showed lower dummy injury readings in angular crashes, especially for HIC and chest g's, and NHTSA test data which agreed with that from Ford. Chrysler, BMW, Volvo, Nissan, Mercedes, Honda, and Mazda agreed with the proposal, claiming that no insight in restraint performance was provided by the test, it was not essential for verifying compliance since test results were lower than in the direct frontal tests, and thus it only contributed to lead time and testing costs. Mazda was the only company to provide data to support its conclusion. Mazda provided the results of a single test which showed lower readings in angular than the frontal crash.

GM and Saab opposed the deletion of the oblique test. GM, in further discussions with NHTSA,

based its objection on the belief that the oblique test is more representative of real world crashes than the frontal test. GM also said that regardless of the agency's decision it would continue to conduct oblique tests; thus, although it believed such tests to be more representative it has no objection to their being deleted from the standard. Saab, in subsequent discussion with NHTSA, did not elaborate on their assertion that deletion of the test would be a "cover-up" for airbag deficiencies nor did VW, when contacted by NHTSA, explain why they believed the test necessary for airbags but not automatic belts.

The Department continues to believe, as expressed in the SNPRM, that the oblique test requirement may not meet the need for motor vehicle safety and thus may unnecessarily add to compliance costs. However, prior to taking final action the Department wishes to have additional test data and/or supporting and dissenting arguments. This information will be sought as part of the notice described earlier, as will comments from the public on the issue of international harmonization of test requirements, as sought by Peugeot and Renault.

Other Test Procedure Issues

The Department still believes that adoption of the NCAP test procedures will reduce test result variability. The added specificity of these procedures, as compared to the current FMVSS 208 compliance criteria, can have no other effect than to reduce variability compared to inconsistent dummy placement, albeit by some unknown amount.

However, we also agree with manufacturer comments concerning the inadequacy of notice as to the specific parts of the NCAP procedure to be adopted. In addition, several commenters suggested other test procedure changes to even further reduce variability. The soon to be issued NPRM will repropose the specific NCAP procedures to be adopted, plus propose additional changes as suggested in comments to Notice 35 of Docket 74-14.

Ford, Chrysler, and VW suggested that if automatic belts are the means of compliance, then the static test requirements of FMVSS 209 and 210, instead of the dynamic test requirements of FMVSS 208, be used to check compliance. The Department disagrees. The concept behind FMVSS 208 is that it is an overall vehicle standard, not just a restraint standard. To simply test the restraint system, statically, would not assure the occupant that injury protection, equivalent to that of other types of restraints which would continue to have to be dynamically tested, was being provided. In this regard, the Department agrees with Allstate that dynamic testing (as is also done for child restraint systems as required by FMVSS 213) is superior to static testing and the requests cited above are responded to in the negative.

The Department also rejects GM's proposal to amend FMVSS 208 by permitting compliance with manual belts if the vehicle complies with the injury criteria at 30 mph with the dummies belted and at 25 mph with the dummies unbelted. The Department does not believe, based on data in its possession on crash tests at 25 mph with unrestrained dummies, that equivalent safety benefits are possible with this proposal. GM's estimate of benefits is not complete in that it is based on vehicles in NHTSA's NCSS file, vehicles which, on average, are of early 1970's vintage. A more complete analysis would be based on the ability of *current* production vehicles to supply such protection. Data available to NHTSA indicate that some current vehicles are capable of supplying automatic occupant protection at speeds up to 25 mph. Without data to the contrary, there is no assurance of the magnitude of safety improvement associated with the GM proposal. Since it has not been demonstrated as an equal alternative, it will not be further considered in this rulemaking, although the Department applauds GM for its work in the area of passive interiors and encourages both it and other companies to continue to provide protection for otherwise unprotected occupants. The Department also notes that nothing in FMVSS 208 precludes compliance through the use of "passive interiors" as being developed by GM. But such compliance must be demonstrated at 30 mph, not 25 mph as GM has suggested.

Finally, Ford requested that convertibles by exempted from the automatic occupant protection requirements. Ford argues that automatic belts are not feasible in convertibles and that the only means of compliance would be airbags, thus resulting in a "design" standard for these vehicles. Since the statute requires that safety standards be "appropriate for the class of vehicles to which they apply," and since convertibles are already exempt from the requirement that all front outboard seating positions have lap and shoulder belts, Ford argues that exemption for convertibles is appropriate. Although we disagree with Ford that providing automatic belts in convertibles is not feasible, it may be not acceptable or appropriate to do so. NHTSA will seek additional guidance from the public on this issue in subsequent rulemaking.

REGULATORY IMPACTS

The Department has considered the impacts of this final rule and determined that it is a major rulemaking within the meaning of E.O. 12291 and a significant rule within the meaning of the Department of Transportation Regulatory Policies and Procedures. A Final Regulatory Impact Analysis is being placed in the public docket simultaneously with the publication of this notice. A copy of the Analysis may be obtained by writing to: National Highway Traffic Safety Administration, Docket Section, Room 5109, 400 Seventh Street, S.W., Washington, D.C. 20590.

The Department's determination that the rule is major and significant is based on the substantial benefits and costs resulting from the requirement for the installation of automatic protection systems. The Department's determinations regarding these matters are discussed elsewhere in this preamble. As noted above, the number of lives saved and injuries prevented will depend on the type of automatic restraints installed in new cars and on the usage and effectiveness of those restraints. Estimates range from 520 to 9,110 lives saved, 8,740 to 155,030 moderate to critical 2 to 5 injuries prevented and 22,760 to 255,770 minor injuries prevented. The total incremental cost increase for a new car would be $51 for automatic belt cars (incremental cost of $40 and lifetime energy costs of $11), $232 for a high volume of cars with driver position airbags (incremental cost of $220 and energy costs of $12), and $364 for a high volume of cars with airbags for all front seat occupants (incremental cost of $320 and energy costs of $44). Assuming 10 million cars sold annually, total economic costs, exclusive of insurance or other savings, would be between $510 million and $3,640 million.

The Department has also assessed the impacts of this final rule on car manufacturers, automatic restraint suppliers, new car dealers, and small organizations and governmental units. Based on that assessment, I certify that this action will not have a significant economic effect on a substantial number of small entities. Accordingly, the Department has not prepared a final regulatory flexibility analysis. However, the impacts of the final rule on suppliers, dealers and other entities are discussed in the FRIA.

The impact on airbag manufacturers is not likely to be significant, but will be positive. The final rule does not require any car manufacturer to install airbags in any new cars. To the extent that car manufacturers respond to the incentive provided by this final rule to install airbags, airbags sales will increase. The Department is not able to assess precisely the extent to which car manufacturers will so respond.

Similarly, the suppliers of automatic belts are not likely to be significantly affected. These are generally the same firms that currently supply manual belts. Therefore, their volume of sales is not expected to increase significantly as a result of this final rule. There may be some economic benefits associated with developing and producing the more sophisticated types of automatic belts.

Since the Department anticipates that most car manufacturers will comply with the final rule by installing detachable automatic belts, the cost impacts on new cars will not be large enough to have a significant effect on new car sales. Similarly, the Department does not expect that the design or operation of the automatic restraints will affect new car sales. The Department expects that the detachable automatic belts will be sufficiently acceptable to the public so that their presence in new cars will not be a factor in the purchasing of new cars.

For the reasons discussed in the preceding paragraph, the Department does not expect that small organizations or governmental units would be significantly affected. The price increases associated with the installation of detachable automatic belts should not affect the purchasing of new cars by these entities. A somewhat greater effect would occur to the extent that any of these entities decide to purchase airbag cars.

In accordance with the National Environmental Policy Act of 1969, the Department has considered the environmental impacts of this final rule. A Final Environmental Impact Statement (FEIS) is being placed in the public docket simultaneously with the publication of this notice. The FEIS focuses on the environmental impacts associated with the alternative having the largest potential impacts. The alternative incorporated in this final rule will have substantially smaller impacts. The Department has concluded that there is no significant effect on the environment. Since most automatic restraints will be automatic belts, the amount of safety belt webbing manufactured should not change significantly.

The Department finds good cause for making this final rule effective more than 1 year from the date of issuance, since the possibility exists that a substantial number of cars would comply with other than belt systems. As discussed earlier in this preamble and in the FRIA, the provision of automatic restraints requires significant vehicle modification. Airbag installation requires steering column changes and instrument panel redesign. The lead time to accomplish these alternatives, based on the time necessary to design and test the structural changes and to order tooling, especially for small cars, is several years. Similarly, a multi-year leadtime is necessary to provide automatic belts due to structural changes in seat and door strength and floor pan reinforcements. Passive interiors can require even longer leadtimes if structural modifications to a vehicle's front end, to better absorb the energy of a 30 mph crash, are necessary. The leadtime provided will provide car manufacturers with an effective choice about the type of automatic restraint they install in their cars. Providing less leadtime would limit their choices and tend to necessitate their selecting detachable automatic belts, the means of compliance with the least certainty as to level of benefits, in place of more advanced technology such as airbags or passive interiors.

THE RULE

PART 571—FEDERAL MOTOR VEHICLE SAFETY STANDARDS

In consideration of the foregoing, Federal Motor Vehicle Safety Standard No. 208, *Occupant Crash Protection*, (49 CFR 571.208), is amended as set forth below.

§571.208 (Amended)

1. S4.1.2 through S4.1.2.2 of Standard No. 208 are revised to read as follows:

S4.1.2 *Passenger cars manufactured on or after September 1, 1973, and before September 1, 1986.* Each passenger car manufactured on or after September 1, 1973, and before September 1, 1986, shall meet the requirements of S4.1.2.1, S4.1.2.2, or S4.1.2.3. A protection system that meets the requirements of S4.1.2.1 or S4.1.2.2 may be installed at one or more designated seating positions of a vehicle that otherwise meets the requirements of S4.1.2.3.

S4.1.2.1 *First option—frontal/angular automatic protection system.* The vehicle shall:

(a) At each front outboard designated seating position meet the frontal crash protection requirements of S5.1 by means that require no action by vehicle occupants;

(b) At the front center designated seating position and at each rear designated seating position have a Type 1 or Type 2 seat belt assembly that conforms to Standard No. 209 and to S7.1 and S7.2; and

(c) Either: (1) Meet the lateral crash protection requirements of S5.2 and the rollover crash protection requirements of S5.3 by means that require no action by vehicle occupants; or

(2) At each front outboard designated seating position have a Type 1 or Type 2 seat belt assembly that conforms to Standard No. 209 and to S7.1 through S7.3, and that meets the requirements of S5.1 with front test dummies as required by S5.1, restrained by the Type 1 or Type 2 seabelt assembly (or the pelvic portion of any Type 2 seat belt assembly which has a detachable upper torso belt) in addition to the means that require no action by the vehicle occupant.

S4.1.2.2 *Second option—head-on automatic protection system.* The vehicle shall:

(a) At each designated seating position have a Type 1 seatbelt assembly or Type 2 seatbelt assembly with a detachable upper torso portion that conforms to S7.1 and S7.2 of this standard.

(b) At each front outboard designated seating position, meet the frontal crash protection requirements of S5.1, in a perpendicular impact, by means that require no action by vehicle occupants;

(c) At each front outboard designated seating position, meet the frontal crash protection requirements of S5.1, in a perpendicular impact, with a test device restrained by a Type 1 seatbelt assembly; and

(d) At each front outboard designated seating position, have a seatbelt warning system that conforms to S7.3.

2. S4.1.3 of Standard No. 208 is revised to read as follows:

S4.1.3 *Passenger cars manufactured on or after September 1, 1986, and before September 1, 1989.*

S4.1.3.1 *Passenger cars manufactured on or after September 1, 1986, and before September 1, 1987.*

S4.1.3.1.1 Subject to S4.1.3.1.2 and S4.1.3.4, each passenger car manufactured on or after September 1, 1986, and before September 1, 1987, shall comply with the requirements of S4.1.2.1, S4.1.2.2 or S4.1.2.3.

S4.1.3.1.2 Subject to S4.1.5, an amount of the cars specified in S4.1.3.1.1 equal to not less than 10 percent of the average annual production of passenger cars manufactured on or after September 1, 1983, and before September 1, 1986, by each manufacturer, shall comply with the requirements of S4.1.2.1.

S4.1.3.2 *Passenger cars manufactured on or after September 1, 1987, and before September 1, 1988.*

S4.1.3.2.1 Subject to S4.1.3.2.2 and S4.1.3.4, each passenger car manufactured on or after September 1, 1987, and before September 1, 1988, shall comply with the requirements of S4.1.2.1, S4.1.2.2 or S4.1.2.3.

S4.1.3.2.2 Subject to S4.1.5, an amount of the cars specified in S4.1.3.2.1 equal to not less than 25 percent of the average annual production of passenger cars manufactured on or after September 1, 1984, and before September 1, 1987, by each manufacturer, shall comply with the requirements of S4.1.2.1.

S4.1.3.3 *Passenger cars manufactured on or after September 1, 1988, and before September 1, 1989.*

S4.1.3.3.1 Subject to S4.1.3.3.2 and S4.1.3.4, each passenger car manufactured on or after September 1, 1988, and before September 1, 1989, shall comply with the requirements of S4.1.2.1, S4.1.2.2 or S4.1.2.3.

S4.1.3.3.2 Subject to S4.1.5, an amount of the cars specified in S4.1.3.3.1 equal to not less than 40 percent of the average annual production of passenger cars manufactured on or after September 1, 1985, and before September 1, 1988, by each manufacturer, shall comply with the requirements of S4.1.2.1.

S4.1.3.4 For the purposes of calculating the number of cars manufactured under S4.1.3.1.2, S4.1.3.2.2, or S4.1.3.3.2 to comply with S4.1.2.1, each car whose driver's seating position will comply with these requirements by means other than any type of seatbelt is counted as 1.5 vehicles.

3. Standard No. 208 is amended by adding the following new sections:

S4.1.4 *Passenger cars manufactured on or after September 1, 1989.* Except as provided in S4.1.5, each passenger car manufactured on or after September 1, 1989, shall comply with the requirements of S4.1.2.1.

S4.1.5 *Mandatory seatbelt use laws.*

S4.1.5.1 If the Secretary of Transportation determines, by not later than April 1, 1989, that State mandatory safety belt usage laws have been enacted that meet the criteria specified in S4.1.5.2 and that are applicable to not less than two-thirds of the total population of the 50 States and the District of Columbia (based on the most recent Estimates of the Resident Population of States, by Age, Current Population Reports, Series P-25, Bureau of the Census), each passenger car manufactured under S4.1.3 or S4.1.4 on or after the date of that determination shall comply with the requirements of S4.1.2.1, S4.1.2.2 or S4.1.2.3.

S4.1.5.2 The minimum criteria for State mandatory safety belt usage laws are:

(a) Require that each front seat occupant of a passenger car equipped with safety belts under Standard No. 208 has a safety belt properly fastened about his or her body at all times when the vehicle is in forward motion.

(b) If waivers from the safety belt usage requirement are to be provided, permit them for medical reasons only.

(c) Provide for the following enforcement measures:

(1) A penalty of not less than $25 (which may include court costs) for each occupant of a car who violates the belt usage requirement.

(2) A provision specifying that the violation of the belt usage requirement may be used to mitigate damages with respect to any person who is involved in a passenger car accident while violating the belt usage requirement and who seeks in any subsequent litigation to recover damages for injuries resulting from the accident. This requirement is satisfied if there is a rule of law in the State permitting such mitigation.

(3) A program to encourage compliance with the belt usage requirement.

(d) An effective date of not later than September 1, 1989.

Sec.103, 119, Pul. L. 89-563, 80 Stat. 718 (15 U.S.C. 1392, 1407)

Issued on July 11, 1984

Elizabeth H. Do
Secretary of Tr
49 F.R. 28962
July 17, 1984

:d by adding the

:factured on or
: as provided in
:ufactured on or
:mply with the re-

e loan.

ransportation de-
:ril 1, 1989, that
e laws have been
ecified in S4.1.5.2
s than two-thirds
:ates and the Dis-
:nost recent Esti-
on of States, by
:rts, Series P-25,
:enger car manu-
: or after the date
:ply with the re-
r S4.1.2.3.

ia for State man-
e:
:at occupant of a
:fety belts under
belt properly fas-
:ll times when the

:ty belt usage re-
:permit them for

:ng enforcement

:n $25 (which may
:pant of a car who
ent.
that the violation
:y be used to miti-
r person who is in-
:nt while violating
who seeks in any
damages for injur-
This requirement
:w in the State per-

:e compliance with

ater than Septem-

PREAMBLE TO AN AMENDMENT TO
FEDERAL MOTOR VEHICLE SAFETY STANDARD NO. 208

Occupant Crash Protection; Improvement of Seat Belt Assemblies
[Docket No. 74-14; Notice 40]

ACTION: Final rule.

SUMMARY: This notice adopts a one-year delay, from September 1, 1985, to September 1, 1986, in the effective date for the safety belt comfort and convenience requirements issued by NHTSA in January 1981. The agency proposed a one-year delay in a notice issued in April of this year. The April notice also proposed several minor modifications to the comfort and convenience requirements, which will be addressed in a subsequent notice.

This notice also denies the petitions submitted by American Motors Corporation and the Motor Vehicle Manufacturers Association for an indefinite delay in the proposed effective date of these amendments. The denial is based on the agency's belief that the substantive issues in the proposal will be quickly resolved in a separate final rule and that delaying the effective date for one year will give the motor vehicle industry sufficient time to meet the modified comfort and convenience requirements.

SUPPLEMENTARY INFORMATION: On January 8, 1981, NHTSA amended Standard No. 208, *Occupant Crash Protection,* to specify additional performance requirements to promote the comfort and convenience of both manual and automatic safety belt systems installed in motor vehicles with a GVWR of 10,000 pounds or less (46 FR 2064). The requirements have not yet become effective. In partial response to seven petitions for reconsideration, the agency extended the effective date of the comfort and convenience requirements for one year, from September 1, 1982, to September 1, 1983 (47 FR 7254). Subsequently, the agency adopted (48 FR 24717) a further extension of the effective date for the requirements to September 1, 1985.

On April 12, 1985, the agency proposed to change the effective date of the comfort and convenience requirements to September 1, 1986, to coincide with the effective date of the Department's July 11, 1984, rule requiring the installation of automatic restraints. This notice also proposed modifications to certain aspects of the comfort and convenience performance requirements in order to clarify the agency's intent and to address the concerns raised in the petitions for reconsideration (50 FR 14580).

After the April 12, 1985, notice of proposed rulemaking was issued, American Motors Corporation and the Motor Vehicle Manufacturers Association petitioned NHTSA to postpone the effective date immediately and indefinitely, until all issues concerning the comfort and convenience requirements are resolved. They stated their belief that a final rule on the former effective date is unlikely to be issued before production of 1986 model year vehicles begins in July 1985; that manufacturers will be uncertain of the standard's applicable requirements; and that it is unreasonable to have this critical timing issue tied to the rulemaking process. Chrysler Corporation, General Motors Corporation and Volkswagen of America, Inc., supported this request in submission to the docket. General Motors stated that deferral is essential to provide time to resolve the many interrelated issues of Notices 37, 38, and 39, as well as to provide time to meet the final requirements flowing from these rulemaking actions. The agency disagrees. Although each of these proposals concerns Standard No. 208, the agency maintains that the issues are separable, as are the notices proposing them.

The agency realizes that September 1, 1985, is an inappropriate effective date for the comfort and convenience requirements because there is insufficient lead time before the beginning of the new model year to comply with the requirements either in the currently adopted version or in the version proposed in April 1985. The agency believes that an effective date of September 1, 1986, provides sufficient time for industry to meet either version of the comfort and convenience requirements. This conclusion is based on NHTSA's own analysis and on the absence of indication in the comments of the other domestic and foreign motor vehicle manufacturers, seat belt manufacturers, and a technical representative that a September 1, 1986, effective date would pose any problems in complying with the proposed requirements. Since its range of choices regarding the substantive differences in the two versions is not large, the agency does not foresee that there will be any changes to the comfort and convenience requirements which would necessitate additional lead time beyond September 1, 1986. Therefore, the agency is adopting that date as the new effective date for all requirements except the one discussed immediately below. However, if the final rule on the substantive issues does include changes for which the industry might need additional lead time, the agency will consider these circumstances and, if necessary, take appropriate steps to adjust the effective date.

In a separate final rule to be issued in the very near future, the agency will respond to the substantive issues raised in the notice of proposed rulemaking and the comments thereon.

In consideration of the foregoing, 49 CFR 571.208 is amended as follows:

1. The authority citation for Part 571 continues to read as follows:

Authority: 15 U.S.C. 1391, 1401, 1403, 1407; delegation of authority at 49 CFR 1.50.

2. S7.1.1.3 is revised to read as follows:

A lap belt installed at any front outboard designated seating position in a vehicle manufactured on or after September 1, 1986, shall meet the requirements of this section by means of an emergency-locking retractor that conforms to Standard No. 209 (571.209) of this chapter.

3. S7.4 is revised to read as follows:

S7.4 *Seat belt comfort and convenience.*

(a) Automatic seat belts installed in any vehicle with a GVWR of 10,000 pounds or less manufactured on or after September 1, 1986, shall meet the requirements of S7.4.1, S7.4.2, and S7.4.3.

(b) Manual seat belts, other than manual Type 2 belts in front seating positions in passengers cars, installed in any vehicle with a GVWR of 10,000 pounds or less manufactured on or after September 1, 1986, shall meet the requirements of S7.4.3, S7.4.4, S7.4.5, and S7.4.6.

Issued on August 19, 1985

Diane K. Steed
Administrator

50 F.R. 34152
August 23, 1985

PREAMBLE TO AN AMENDMENT TO
FEDERAL MOTOR VEHICLE SAFETY STANDARD NO. 208

Occupant Crash Protection
[Docket No. 74-14; Notice 41]

ACTION: Response to Petitions for Reconsideration.

SUMMARY: On July 11, 1984, the Secretary of Transportation issued a final rule requiring automatic occupant protection in all passenger cars based on a phased-in schedule beginning on September 1, 1986, with full implementation being required by September 1, 1989, unless, before April 1, 1989, states covering two-thirds of the population of the United States have enacted mandatory safety belt use laws meeting specified criteria, with such laws becoming effective by September 1, 1989. Subsequently, sixteen interested parties filed petitions for reconsideration of the final rule. This notice responds to the issues raised in those petitions.

EFFECTIVE DATE: October 14, 1985

SUPPLEMENTARY INFORMATION:

Background

On July 11, 1984 (49 FR 28962), the Secretary of Transportation issued a final rule requiring automatic occupant protection in all passenger cars based on a phased-in schedule beginning on September 1, 1986, with full implementation being required by September 1, 1989, unless, before April 1, 1989, states covering two-thirds of the population of the United States have enacted mandatory safety belt use laws (MULs) meeting specified criteria, with such laws becoming effective by September 1, 1989.

More specifically, the rule requires:

• Front outboard seating positions in passenger cars manufactured on or after September 1, 1986, for sale in the United States, will have to be equipped with automatic restraints based on the following schedule:

• Ten percent of all cars manufactured on or after September 1, 1986.

• Twenty-five percent of all cars manufactured on or after September 1, 1987.

• Forty percent of all cars manufactured on or after September 1, 1988.

• One hundred percent of all cars manufactured on or after September 1, 1989.

• During the phase-in period, each car that is manufactured with a system that provides automatic protection to the driver without automatic belts will be given an extra credit equal to one-half car toward meeting the percentage requirement.

• The requirement for automatic restraints will be rescinded if MULs meeting specified conditions are passed by a sufficient number of states before April 1, 1989, to cover two-thirds of the population of the United States.

Sixteen interested parties subsequently petitioned for reconsideration of the standard. The issues raised by the petitioners and the agency's response are discussed below.

Rescind the Standard

One petitioner asked the agency to reconsider the decision not to rescind the automatic restraint requirements of Standard No. 208. He argued that the Secretary's decision was apparently based on a belief that rescission was not a possible result under the Supreme Court decision in *Motor Vehicle Manufacturer's Association v. State Farm Mutual Automobile Insurance Co.* (State Farm). The petitioner further argued that the record in the Standard No. 208 proceeding in fact supports rescission. In particular, the petitioner argued that the rulemaking record shows that air bag technology is not an effective automatic restraint alternative. Quoting from portions of the July 1984 final decision, the petitioner specifically argued that air bag systems require the use of a lap belt and do not provide protection at less than 10-12 mph, the disposal problem related to the gas generation agent in air bag systems needs more action, air bag systems may cause injury to out-of-position occupants, the cost of air bag systems is a major disadvantage, and the use of air bag systems in small cars requires more lead time. The petitioner concluded that few manufacturers will use air bag systems, thus leaving automatic belts as the only automatic restraint alternative. As to automatic belts, the petitioner argued that the record does not show that detachable automatic belts would increase usage. The petitioner specifically argued that there has been no showing that the combination of motorist inertia and automatic belts will increase belt usage.

NHTSA's position is that the *State Farm* decision allows the agency to make a reasoned choice between rescinding or retaining the standard. However, the agency stated in the July 1984 final rule, and still believes, that the rulemaking record does not justify rescission—unless there is a very substantial increase in the use of manual safety belts in the future. The data set forth in the July 1984 final rule demonstrate the dramatic reductions in deaths and injuries that widespread usage of the safety belt systems would achieve. Thus, if twothirds or more of the American people are covered by mandatory use laws, that would increase useage of safety belts, the need for an automatic occupant restraint requirement would be obviated and the rule would be rescinded.

The agency believes that the rulemaking record, taken as a whole, shows that air bag systems are an effective automatic restraint technology. The discussion in the final rule concerning the need to use a safety belt with an air bag system and the ability of such systems to provide protection at low speeds concerned the relative advantages and disadvantages of different restraint technologies. As noted in that discussion, air bag systems have an advantage over other occupant restraints in that they ensure a usage rate of nearly 100 percent for both drivers and passengers. Even without use of a lap belt, an air bag system will offer protection; however, to equal the effectiveness of a manual lap-shoulder belt, air bag systems must be used with a lap belt.

Likewise, while air bag systems do not inflate in low speed crashes, other standards, such as those on energy-absorbing steering columns and instrument panel padding, ensure that occupants will still be provided with protection in low speed collisions. In addition, research data indicate that air bag systems will provide protection at higher speeds than safety belts.

As to potential problems with the disposal of the gas generator, the July 1984 final rule pointed out that as long as appropriate procedures are followed by vehicle recyclers and scrappers, disposal should not pose a problem. Subsequent to issuance of the rule, the agency has had discussions with recyclers and scrappers concerning the joint NHTSA-General Services Administration air bag fleet demonstration program to discuss safe and reasonable disposal procedures. We believe that this effort will lead to further improvements in the safe disposal of the chemical agents in air bag systems.

The July final rule acknowledged concerns about the effects of air bag systems on out-of-position occupants; however, it also explained that technical solutions are available to address the out-of-position occupant problem. The final rule also acknowledged the higher costs of air bag systems in comparison to automatic belts; the high cost of replacing an air bag system, which could lead to its not being replaced after deployment; public uncertainty and concern associated wiht air bag systems; and the longer lead time needed for air bag systems, particularly in small cars. It was a balancing of those factors, plus the factors discussed above, that led to the decision that air bag systems should not be mandated for all cars. However, as discussed in the final rule, the agency believes that air bag systems are an effective restraint technology which, along with other types of automatic restraint technology, will provide demonstrable safety benefits. The provision in the final rule providing manufacturers that use non-belt automatic restraints with extra credit in complying with the phase-in requirements was intended to encourage alternative technologies, including enhanced availability of air bag systems.

As to detachable automatic belts, as discussed in the July 1984 final rule, the agency cannot project either widespread usage for detachable automatic belts or a widespread refusal to use such systems. As discussed by the Supreme Court in the *State Farm* decision, it is reasonable to expect that inertia will work to increase usage, since once an automatic belt is connected, it continues to function automatically until it is disconnected. However, using even the lowest level of the range for the effectiveness of automatic belts and a very little increase in usage (only a 7 1/2 percentage point increase), automatic belts will result in a significant incremental annual reduction in deaths and injuries.

For the above reasons, the agency concluded in July 1984 that automatic restraint systems are reasonable in cost, feasible, and practicable, and the potential benefits in lives saved and injuries reduced in severity are substantial. At that time, the agency stated that rescission, in the absence of a substantial increase in manual belt usage, has not been justified. Since the petitioner has not provided any new data to support rescission, the petition is denied.

Require Automatic Restraints

Several petitioners urged the agency to reconsider the decision to rescind the automatic restraint requirements if two-thirds of the population of the United States is covered by State MULs. They urged the agency to retain the automatic restraint requirement, regardless of what action the States take in adopting MULs.

The petitioners have offered no new evidence to justify modifying the July 11 final rule. As explained in that rule, the Secretary determined that if enough people are covered by State mandatory belt use laws, usage rates will be sufficiently high so that the additional requirement for automatic restraints should not be required. The evidence from Canada and other countries with MULs supports the conclusion that State belt use laws will bring higher usage rates and immediate and inexpensive benefits. The petitioners' requests to mandate automatic restraints even if two-thirds of the population is covered by MULs is therefore denied.

Phase-In Requirements

A number of petitioners asked for several modifications of the phase-in requirements of the standard. Each of the modifications sought by the petitioners is addressed in the following discussions.

Change September 1st Effective Date

One modification was to change the September 1 effective date used for each part of the phase-in. The petitioners argued that they would be precluded from applying any portion of their vehicles produced prior to that date to meet the required percentage of automatic restraint equipped cars. The agency has already, in effect, proposed to grant a portion of the petitioners' request in an-

other notice (Docket 74-14; Notice 38; 50 FR 14602) issued on Standard No. 208. The agency proposal would not change the September 1 effective date, but it does propose that manufacturers be allowed to count any automatic restraint vehicle produced during the one year preceding the first year of the phase-in. In addition, the agency proposes, in Notice 38, to permit manufacturers which exceed the minimum percentage phase-in requirements in the first or second years to count those extra vehicles toward meeting the requirement in the second or third year.

Several petitioners sought a change in the provision of the final rule specifying that the computation of the minimum vehicle production to be equipped with automatic restraints must be based on the average of the production for the three preceding model years. The petitioners argued that if car sales were to drop drastically during the phase-in period, then the number of vehicles that they would have to equip with automatic restraints based on their prior three year sales volume would be a significantly greater percentage of their actual production than intended by the final rule. The agency has already responded to this request in Notice 38 by proposing to adopt an alternative that would permit a manufacturer to equip the required percentage of its vehicles with automatic restraints based on its actual production of passenger cars during each affected year.

Manufactured for Sale in the U.S.

Several petitioners asked the agency to amend the rule to clarify that the rule applies only to cars manufactured for sale in the United States. As discussed in the preamble to the final rule, the determination of the base years' production figures and the calculation of the number of vehicles that must comply with the percentage phase-in requirements of the standard is to be based on vehicles manufactured for sale in the United States. Since all of the agency's safety standards apply only to vehicles manufactured for sale in the United States, the agency does not believe that an amendment to the rule is necessary. Nevertheless, today's preamble should serve as the clarification requested; that the rule applies only to vehicles manufactured for sale in the United States.

Carry-Forward/Carry-Back

A number of petitioners urged the agency to provide manufacturers more flexibility in meeting the phase-in requirements. They proposed that manufacturers be able to carry-forward credits for the number of automatic-restraint equipped vehicles they produce in excess of the required percentage. One petitioner also asked that manufacturers be permitted to carry-back credits earned in the second and third year to the first year.

The agency agrees that it would be appropriate to permit manufacturers that exceed the minimum percentage phase-in requirements in earlier years to count those extra vehicles toward meeting the minimum percentage requirements of later years and has proposed such a carry-forward credit in Notice 38. Such a credit would encourage early introduction of larger numbers of automatic restraints and provide increased safety to the public and flexibility for manufacturers. The agency has decided to deny requests for any carry-back credits because their use would delay the safety benefits of the rule and undermine the purpose of the phase-in, which is to introduce automatic restraints on a prompt and orderly basis.

Definition of Manufacturer

Several petitioners asked the agency to further define the term "manufacturer." The agency has responded to this request in Notice 38 by proposing to permit manufacturers to determine, by contract, which of them will count passenger cars as its own for the purposes of meeting the percentage goals set forth in the phase-in. Notice 38 proposes two rules of attribution in the absence of such a contract. First, a passenger car which is imported for purposes of resale would be attributed to the importer. Second, a passenger car manufactured in the United States by more than one manufacturer, one of which also markets the vehicle, would be attributed to the manufacturer which markets the vehicle. Readers are referred to Notice 38 for a more detailed discussion of the proposed attribution rules.

Credits for Non-Belt Technology

The July 11 final rule provided that manufacturers that used non-belt technology, such as air bags or passive interiors, to meet the automatic restraint requirement for the driver's seating position and any type of automatic restraint at the passenger's seating position during the phase-in period, would receive additional credit. For each car in which they use a non-belt system, they will receive credit for an extra one-half car toward meeting their percentage requirement. One petitioner said that the text of the rule does not achieve the agency's intention, as stated in the July 1984 final rule, to encourage the use of automatic restraints other

than automatic belts, since the rule precludes giving the additional credit for a system that requires the use of a safety belt, whether automatic or manual, to enable the non-belt technology to provide full protection. That petitioner pointed out that all current air bag systems must also use safety belts for full protection; belts are permitted by the standard to be used as an alternative to the use of automatic restraints to meet the lateral and rollover tests. It was not the agency's intention to deny the extra credits to air bag or other systems that also use such safety belt systems to ensure protection in other than frontal crashes. Therefore, the agency is amending the rule to ensure that those systems are eligible for the additional credit.

The agency was also petitioned for another modification to the credit provision. It was asked that manufacturers be allowed, during the phase-in, to receive a one vehicle credit for vehicles which are equipped with non-belt technology at the driver's position and manual safety belts at the front outboard position. The petitioner argued that this would encourage manufacturers to produce driver-side air bag systems or other non-belt system technology sooner than if they had to complete development of passenger-side automatic restraint systems as well, significantly advancing the Secretary's goal in this regard.

The agency has decided to modify the credit provision as requested by the petitioners. The purpose of the phase-in period is to provide a rapid introduction of the lifesaving benefits of automatic restraints and to facilitate the earliest possible introduction of such restraints to permit the public to become familiar with their operation and benefits. The purpose of the credit provision is to encourage the production of a wide variety of such restraints especially in the early years. The agency believes that permitting manufacturers to receive a 1.0 car credit for driver-only non-belt systems with manual belts on the passenger side will encourage the introduction of non-belt technologies into passenger cars, earlier than would otherwise occur.

The agency is aware that one company is currently offering driver-side air bags to the public. Other manufacturers have indicated that they may offer driver-side air bags to the public within the next few years. The agency is aware neither of any manufacturers that currently plan to offer a passenger-side air bag system nor of any firm plans for other types of non-belt automatic protection on the passenger side of vehicles. The longer lead time estimated in the Final Rule to be required for non-belt automatic protection on the passenger side, coupled with the advanced stage of design of vehicles that will be available at the early stage of the phase-in period, mitigates against such full-front non-belt protection being available. Increasing public awareness of the benefits of a variety of automatic protection techniques is one of the primary objectives of the phase-in and credit provisions. Achieving this objective will depend, therefore, on the availability of an adequate number of cars equipped with non-belt protection of the driver's side. We now believe that there are a number of factors that might discourage manufacturers from making such equipment available in significant numbers.

Under the current rule, cars equipped with non-belt driver's-side automatic protection would qualify for credit only if passive protection were made available on the passenger side. As noted above, such protection is most likely to be provided by automatic belts. Some models in which driver's-side air bags are being considered by manufacturers, however, are at an advanced stage of design. It is unlikely the redesign required to equip these cars with automatic belts will be undertaken. Even if these cars could be modified to incorporate automatic belts, manufacturers would be faced with a complex, and expensive, marketing task. Not only would they have to convince customers of the safety and utility of automatic belts, but they must also perform this task for the more expensive air bag. Unwillingness on the part of manufacturers to assume this added task may create a serious disincentive to the prompt offering of air bag technology.

Alternatively, these manufacturers considering driver-side air bags might also elect to meet phase-in requirements by producing a sufficient number of automatic belt equipped cars. Under these circumstances, it is likely that the marketing efforts of the manufacturers during the phase-in will concentrate on marketing the automatic belts, possibly to the detriment of the public's acceptance of the driver-side air bags. As the agency learned in recent research studying the marketing efforts used by General Motors to sell its air bag equipped cars in the mid-1970's effective, affirmative marketing of an air bag system is essential to overcome consumer concerns about such things as the fear of inadvertent deployment, price and post-crash replacement cost. ("A Retrospective Analysis of the General Motors Air Cushion Restraint System Marketing Effort, 1974 to 1976") If cars equipped with driver-only air bags do not count toward compliance with the phase-in, the manufacturers will have less incentive to market the air bags aggressively, and these circumstances may even lead to decisions to drop the early offering of air bags. The agency's goal of encouraging significant public exposure to alternative protection technologies may not be realized. Therefore, the agency has determined that permitting manufacturers to receive a 1.0 car credit during the phase-in by installing driver-only non-belt automatic protection systems in their vehicles will encourage earlier introduction of alternative automatic protection technologies, wider public availability of such technologies, and more effective marketing of such technologies than would be achieved by the original decision. The final rule is amended to permit such vehicles to be counted toward the phase-in requirements.

The agency has fully considered the safety implication of this amendment. An important safety consideration is the number of occupants at the risk of injury at each seating position, not just the number of seating positions that are covered by the automatic restraint requirement. Accident data, presented in the agency's Final Regulatory Impact Analysis, show that there are approximately 2 1/2 to 3 times as many driver injuries and fatalities as there are to front right seat passengers. Therefore, the agency believes that it is reasonable to encourage manufacturers to provide automatic restraint protection as soon as possible to the driver—the person who is most at risk.

Convertibles

Several petitioners asked that convertibles be exempted from the automatic restraint requirements. They argued, for example, that the installation of automatic lap and shoulder belts is not feasible in convertibles, thus air bag systems must be used in those cars. The result, according to the petitioners, is a design standard for convertibles. They also stated that an exemption would be appropriate since convertibles are already exempt from the requirement in Standard No. 208 that all front outboard seating positions have lap and shoulder belts. The agency has already responded to these petitions in Notice 38 by proposing that manufacturers have the option of installing manual lap belts instead of automatic restraints in convertibles. Readers are referred to Notice 38 for a more detailed discussion of the petitions and the reasons for the agency's proposed alternative requirements for convertibles.

Oblique Crash Test

A number of petitioners requested the agency to delete the oblique barrier crash test of Standard No. 208. They argued, among other things, that the test is unnecessary since it generates a lower crash pulse than the frontal crash test. As discussed in detail in Notice 38, the agency is also concerned that the oblique test may not be necessary and has therefore requested commenters to provide additional data on the safety and cost effects of deleting the tests. Readers are referred to Notice 38 for a more detailed discussion of the issues involved in the proposed deletion of the oblique test.

Lead Time

One petitioner requested a change in the two year lead time for the automatic restraint standard. Citing the table on lead time requirements included with the July 11 final rule, the petitioner argued that only one manufacturer, Renault, has said that it can comply in 24 months. The table showed that most companies have said they needed at least 30 to 48 months. The petitioner asked for the lead time to be extended.

The table cited by the petitioner reflects the lead time required by a manufacturer to equip its entire fleet with automatic restraints. The agency agrees that a longer lead time would be necessary if the automatic restraint requirement were simultaneously applied to the entire vehicle fleet. The final rule, however, phases-in the automatic restraint requirement so that only a portion of a manufacturer's fleet must be equipped initially. Based on a study of current automatic restraint equipped vehicles and manufacturers' comments, the agency has determined that automatic belt systems can be added on to existing vehicle designs with approximately 24 months of lead time. The manufacturers generally agree with that estimate. For example, GM said that lead time for models for which detachable belts had previously been designed would be 21 months and Ford said that a driver-side air bag system could be in production for some of its cars within the allotted lead time. The Agency therefore does not believe that additional lead time is necessary for the percentage requirements during the phase-in period and the petition is denied.

AIA raised a separate lead time issue. It said that the July 1984 final rule identified a number of issues, primarily related to test procedures, that would be the subject of further rulemaking. AIA

argued that the implementation schedule for automatic restraints should not begin until those issues are resolved. Any changes due to the unresolved issues are not expected to increase lead time and, indeed, should relieve some burdens associated with preparing for compliance. At this time, the agency believes that the resolution of the remaining issues, which does not involve the imposition of more stringent performance requirements, should be accomplished shortly and therefore is denying AIA's petition.

Repeatability

One petitioner raised arguments about the repeatability of the test procedures used in Standard No. 208 compliance testing. The petitioner's fundamental argument is that the agency's Repeatability Test Program found what the petitioner says is an unacceptable level of variability in the test results and thus, the petitioner argues, the agency has failed to demonstrate that the test procedures can be reproduced, car-to-car and test site-to-test site. The petitioner noted that for a manufacturer to certify its vehicles, it must meet maximum limits for each of eight separate requirements: HIC for driver and passenger dummy heads, "g" loads for driver and passenger chests; and femur loads for each dummy's right and left leg. Because of the test variability, the petitioner said that it cannot confidently predict that its vehicles will comply with the standard. It urged the agency to develop an alternative method of determining compliance with the standard.

The petitioner did not, however, provide any new data which demonstrate that the crash test procedures and the test dummy pose significant repeatability problems. More importantly, the petitioner did not provide new data indicating that the test procedure and the dummy are incapable of measuring compliance with Standard No. 208.

The agency believes that the test procedure, test dummy, and test instrumentation are repeatable within the statutory requirements of objectivity and practicability. The agency does recognize that because of the complexity of the requirements of Standard No. 208, manufacturers are concerned about certifying compliance with each of the requirements of the standard. To address this concern, the agency has proposed in Notice 38 that the rule be amended to state that a vehicle shall not be deemed in noncompliance if its manufacturer establishes that it did not have reason to know in the exercise of due care that the vehicle is not inconformity with the standard.

Comfort and Convenience

Several petitioners asked the agency to answer promptly the pending petitions for reconsideration of the comfort and convenience requirements of Standard No. 208. The agency has already issued a separate notice (Docket 74-14, Notice 37; 50 FR 14580) proposing changes to the comfort and convenience requirements in response to the petitions for reconsideration. Readers are referred to that notice for a detailed discussion of the proposed revisions.

Judicial Review

One petitioner asked the agency to clarify the extent to which a challenge to the legality of the final rule must be made now, rather than when the Secretary makes a determination that two-thirds of the U.S. population is covered by a mandatory belt use law. The reviewability of the final rule and any subsequent agency action is a matter for the courts, not the agency, to decide.

Mandatory Seat Belt Use Law Criteria

A number of petitioners sought reconsideration of the minimum criteria for mandatory safety belt use laws. The agency is still considering the issues raised in those petitions and will respond to them at a later date.

Corrections

MVMA pointed out two minor errors in the text of the final rule. First, in section 4.1.2 of the rule, the word "before" should be used instead of the word "after." Likewise in section 4.1.2.2(b), the word "outboard" is misspelled. Both of those errors are corrected by this notice.

In addition, the agency wants to clarify a conflict between the preamble to the MUL provisions of the final rule and the text of the final rule's provisions on MULs. The preamble to the rule stated that one of the minimum criteria for a MUL was that each front outboard occupant of a passenger car be required to wear a safety belt. The text of the final rule provides that each front seat occupant, which would include the outboard and the center seating positions, would have to be covered

by a MUL. The text of the final rule, requiring a MUL to cover all the front seating positions, is the correct version.

Cost and Benefits

NHTSA has examined the impact of this rulemaking action and determined that it is not major within the meaning of Executive Order 12291 or significant within the meaning of the Department of Transportation's regulatory policies and procedures. A Preliminary Regulatory Evaluation has been prepared on the changes proposed in Notice 38 and discussed in this notice. A copy of that evaluation is available for public inspection and copying in the agency's docket section. The agency has determined that the economic and other effects of the rulemaking action in this notice are so minimal that a full regulatory evaluation is not required. The changes adopted in this action concern minor adjustments to the phase-in requirements, which will give manufacturers more flexibility without imposing any economic costs.

Regulatory Flexibility Act

NHTSA has also considered the effects of this rulemaking action under the Regulatory Flexibility Act. I hereby certify that it will not have a significant economic impact on a substantial number of small entities. Accordingly, the agency has not prepared a regulatory flexibility analysis.

Few if any motor vehicle manufacturers would qualify as small entities. The suppliers of webbing and other manual or automatic restraint components will not likely be significantly affected, since this notice is not making a change in the performance requirements of the standard. Small organizations and governmental units will not be significantly affected since there are no price increases associated with this action.

In consideration of the foregoing, Part 571.208, Occupant Crash Protection, of Title 49 of the Code of Federal Regulations is amended as follows:

1. Section 4.1.3.4 is revised to read as follows:

S4.1.3.4 For the purposes of calculating the numbers of cars manufactured under S4.1.3.1.2, S4.1.3.2.2, or S4.1.3.3.2 to comply with S4.1.2.1:

(a) Each car whose driver's seating position will comply with the requirements of S4.1.2.1(a) by means not including any type of seat belt and whose front right seating position will comply with the requirements of S4.1.2.1(a) is counted as 1.5 vehicles.

(b) Each car whose driver's seating position will comply with the requirements of S4.1.2.1(a) by means not including any type of seat belt and whose front right seating position is equipped with a Type 2 seat belt is counted as a vehicle conforming to S4.1.2.1.

2. The first sentence of section 4.1.2 is revised to read as follows:
Each passenger car manufactured on or after September 1, 1973, and before September 1, 1986, shall meet the requirements of S4.1.2.1, S4.1.2.2 or S4.1.2.3.

3. Section 4.1.2.2(b) is revised to change the word "outbord" to the word "outboard."

Issued on August 27, 1985

Diane K. Steed
Administrator

50 FR 35233
August 30, 1985

PREAMBLE TO AN AMENDMENT TO
FEDERAL MOTOR VEHICLE SAFETY STANDARD NO. 208

Improvement of Seat Belt Assemblies
[Docket No. 74-14; Notice 42]

ACTION: Final rule

SUMMARY: On April 12, 1985, NHTSA issued a notice proposing modifications to certain aspects of the comfort and convenience performance requirements in Standard No. 208, *Occupant Crash Protection.* The agency's purpose was to clarify the intent of the requirements and to address the concerns raised in petitions for reconsideration received from seven vehicle manufacturers regarding the final rule on comfort and convenience issued on January 8, 1981. This notice sets comfort and convenience performance requirements for both manual and automatic safety belt assemblies installed in motor vehicles with a Gross Vehicle Weight Rating of 10,000 pounds or less. The April 12, 1985, notice also proposed to change the effective date of the comfort and convenience requirements. A final rule setting the effective date as September 1, 1986, was issued on August 23, 1985.

EFFECTIVE DATE: September 1, 1986.

SUPPLEMENTARY INFORMATION: On January 8, 1981 (46 FR 2064), NHTSA amended Safety Standard No. 208, *Occupant Crash Protection* (49 CFR 571.208), to specify additional performance requirements to promote the comfort and convenience of both manual and automatic safety belt systems installed in motor vehicles with a GVWR of 10,000 pounds or less. The final rule included specifications relating to the following aspects of safety belt performance and design: latchplate accessibility; safety belt guides; adjustable buckles for certain belts; shoulder belt pressure; convenience hooks; belt retraction; and comfort devices. Type 2 manual belts (lap and shoulder combina-

tion belts) installed in front seating positions in passenger cars were excepted from these additional performance requirements, since it was assumed such belts would be phased out in passenger cars as the automatic restraint requirements of Standard No. 208 became effective.

Seven petitions for reconsideration of the January 8, 1981, amendment were received from vehicle manufacturers. On February 18, 1982 (47 FR 7254), the agency issued a partial response to the petitions for reconsideration by extending the effective date of the comfort and convenience requirements for one year, from September 1, 1982, to September 1, 1983. Subsequently, the agency proposed (47 FR 51432) and then adopted (48 FR 24717) a further extension of the effective date for the requirements until September 1, 1985.

The April 12, 1985 (50 FR 14580), notice proposed to delay the effective date until September 1, 1986, in order to give the industry sufficient leadtime to meet the proposed changes in the rule. A final rule delaying the effective date to September 1, 1986, was issued on August 23, 1985 (50 FR 34152).

As discussed in the April 12, 1985, notice, the agency continues to believe that certain of the performance requirements included in the final rule will tend to enhance safety belt use by providing occupants with safety belts which are more comfortable to wear and more convenient to use. The requirements in this final rule are important to support the agency's program to increase safety belt use in the United States.

This rule makes minor changes to the modifications proposed in April 1985 in response to concerns raised by the commenters. A discussion of these changes is set forth below. (For a complete understanding of the performance requirements discussed in this notice, including the relationship of the requirements to safety belt comfort and convenience, interested persons should refer to both the December 31, 1979 (44 FR 77210), notice of proposed rulemaking and the January 8, 1981 (46 FR 2064), final rule).

Application to Manual Lap/Shoulder Belts in Passenger Cars

The January 1981 final rule exempted manual Type 2 safety belts installed in the front seats of passenger cars from the comfort and convenience requirements. This was done to allow manufacturers to devote their resources to automatic restraints in these vehicles since Type 2 manual belts in the front seats would have been phased out when the automatic restraint requirements became effective. However, the subsequent July 1984 (49 FR 28962) final rule mandating automatic restraints specifies that if States representing two-thirds or more of the nation's population enact qualifying mandatory safety belt usage laws before April 1, 1989, the requirement for automatic protection will no longer apply. The April 1985 notice proposed that, in the event that this occurs, the comfort and convenience requirements would be extended to Type 2 manual belts installed in the front seats of passenger cars, effective September 1, 1989.

Two domestic manufacturers objected to the extension of the comfort and convenience requirements to manual Type 2 safety belts in front outboard seating positions of passenger cars until a decision has been made in 1989 regarding the future of automatic restraints. They stated that there is no justification for setting such a requirement now, which could cause manufacturers to incur design and tooling costs, because manual belts could be phased out in 1989 if an insufficient number of States pass qualifying mandatory safety belt use laws.

The September 1, 1989, effective date provides a leadtime of four years to comply with the comfort and convenience requirements for Type 2 front seat manual belts in passenger cars. The agency is therefore adopting the proposed September 1, 1989, effective date for Type 2 front seat manual belts in passenger cars if the automatic restraint requirement is rescinded.

The agency recognizes that the possibility exists that the industry will have to discontinue manual belts after 1989 if the automatic restraint requirement for all cars becomes effective. However, the agency believes that comfort and convenience technology developed for automatic belts and for Type 2 manual belts in light trucks and multipurpose passenger vehicles (MPV's) should be transferable to passenger cars with a minimum of design and tooling cost with a four-year leadtime. The agency notes that a large number of passenger cars will have been manufactured with manual belts between 1986 and 1989, and the agency believes it is desirable, from a safety standpoint, to have the front outboard seating positions of these cars incorporate comfort and convenience features which will contribute to increased belt usage. The agency therefore encourages manufacturers to begin voluntarily incorporating comfort and convenience features in their Type 2 front seat manual belts. Since the technology is available, the cost to incorporate these features should be minimal, especially if they are made part of the design process for newly introduced vehicles.

Emergency Locking Retractors (ELR) and Child Restraints

Paragraph S7.1.1.3 of Standard No. 208 was amended in the January 1981 final rule to specify that certain lap belts installed at front outboard seating positions are required to have an emergency-locking retractor rather than an automatic-locking retractor (which was previously allowed as an option). Some manufacturers also incorporate emergency-locking retractors in rear seats as well. Automatic-locking retractors are inconvenient to use since they must be extended in a single continuous movement to a length sufficient to allow buckling or they will lock. They also tend to tighten excessively under normal driving conditions, sometimes making it necessary to unbuckle and refasten the lap belt to relieve pressure on the pelvis and abdomen. Neither of these problems exists with the emergency-locking retractor, which allows occupant movement without tightening and which locks only upon rapid occupant movement, vehicle deceleration or impact.

The April 12, 1985, notice proposed a revised version of this requirement. The proposed revision reflected the agency's tentative judgment that use of child restraints in the front outboard passenger position with a lap belt equipped with an emergency-locking retractor could result in the child restraint moving forward during normal, low-speed driving and braking, or pre-crash vehicle maneuvering or braking. (At higher speeds or upon impact, the locking mechanism in existing belt designs would work to restrain the child seat appropriately.) Therefore, the agency proposed that Type 1 safety belts or the lap belt portion of Type 2 belts with emergency-locking retractors, used in any designated seating position other than the driver's position, be equipped with a locking means to prevent forward motion of child restraint devices.

A majority of vehicle manufacturers objected to this proposal. The main arguments they raised were: (1) the locking means could degrade the performance of the belt system for adult passengers; (2) the proposed language would exclude alternative designs, such as owner-installed "locking clips," which could serve the same purpose; (3) the requirement would not be cost effective, because not all vehicle owners need a locking means to secure a child restraint system in the front seat; and (4) the proposed effective date for the requirement, September 1, 1986, does not provide sufficient design and development time for compliance. They also argued that, if this requirement is maintained, it should be delayed until the agency decides whether it will require dynamic testing of manual safety belts.

Two manufacturers of child restraint devices and a child passenger safety association supported the proposed amendment. They stated that the approach cited in the proposal would solve potential problems relating to child seats and ELR's, and would eliminate the need to devise what they termed makeshift solutions.

Child restraint manufacturers stated that some restraint devices, when positioned by safety belt systems which are adjusted by ELR's, become unstable when occupied by very active children. Agency testing of child restraint devices under conditions of low-speed braking and vehicle maneuvers indicates that, although improvements in belt systems could improve the stability of these devices, there are no data to show that low-speed movement of child safety seats is affecting the safety performance of child restraint devices in motor vehicle accidents (Docket 80-18-GR-004).

Because the agency's research did not show that low-speed movement of the seats is actually reducing the effectiveness of child restraints in accidents, and because after-market locking devices are available which achieve the same goal, it has decided not to adopt a manual locking requirement for ELR's at this time. The agency will continue to monitor the potential problems associated with the restraint of child restraint devices by ELR safety belt systems and consider whether to address these problems in future rulemaking actions.

Additional ELR Issues

Regarding S7.1.1.3, one manufacturer asked NHTSA to clarify whether an ELR located at the point of shoulder belt retraction on a Type 2 belt system, which combines the lap and torso belt in a continuous running loop, complies with the requirement. NHTSA confirms that a Type 2 continuous belt system, which incorporates an ELR to control slack in the lap and torso belt portions, would comply with the requirement.

Another manufacturer asked for clarification on the use of lap belts in passenger cars equipped with air bags versus those equipped with single automatic diagonal belts. The requirement of S7.1.1.3 only applies to lap belts installed in a vehicle to comply with Standard No. 208. Thus, a lap belt installed in conjunction with an air bag, in order to meet the lateral and rollover requirements of S4.1.1.2(c)(2), would be required to have an emergency locking retractor. However, a Type 1 lap belt voluntarily installed by a manufacturer in conjunction with a single diagonal automatic belt would not have to comply with the provisions of S7.1.1.3, since the single diagonal automatic belt would fully meet the belt requirements of the standard by itself.

Open-Body Vehicles, MPV's, and ELR's

One manufacturer stated that open-body vehicles should be exempted from the ELR requirement of S7.1.1.3, because these vehicles are designed to perform numerous off-road, heavy-duty tasks, and both the lap and upper torso portions of the belt system are subjected to design criteria far different from typical passenger car belt systems. In particular, occupants may want the belts tightly fastened around them when the vehicle is used on rough terrain. The agency agrees that open-body vehicles do perform numerous off-road, heavy-duty tasks, but they are also commonly used in normal highway driving to perform the same functions as passenger cars, where tight belts may discourage belt use. Furthermore, belt systems are available for open-body vehicles as well as passenger cars, which can function as ELR's for the lap belt or lap belt portion of a combined lap and shoulder belt, and still be capable of being manually or automatically locked by occupants when they want the belt to be tightly fastened around them. These systems can also provide tension relieving and ELR functions for the torso portion of a Type 2 belt system.

Incorporating a single retractor, which can function as either an ALR or an ELR, into a lap belt or the lap belt portion of a Type 2 belt for off-road use, would accommodate the desire of occupants to be tightly restrained when needed and would also provide a more comfortable belt when this is sufficient for normal operation of the vehicle. Such an ALR/ELR feature may be desirable in some vehicles and is currently available in some imported and sports cars. The agency estimates the cost to range from $1.00 to $5.00 per seating position. Alternatively, a locking D-ring in the lap belt, which enables users to snugly fasten the lap belt, could be provided for virtually no increase in cost to the consumer. For these reasons, the agency is not exempting open-body vehicles from the requirement of S7.1.1.3.

Another manufacturer requested an exemption from the requirements of S7.1.1.3 for all multipurpose passenger vehicles, stating with no supporting rationale that the ELR requirement is design restrictive. The agency does not believe that the ELR requirement is design restrictive for the reasons discussed above. In addition, multipurpose passenger vehicles provide the same functions as passenger cars. While some types may also be designed for heavy-duty, off-road use, the same rationale set out in the discussion of open-body vehicles applies to other multipurpose passenger vehicles. The agency concludes that multipurpose passenger vehicles should continue to be subject to the requirement of S7.1.1.3.

Corrections

Two technical corrections are made in this final rule relating to paragraph S7.1.1.3. As proposed in the April 12, 1985, notice paragraph S7.1.1.3(b) exempts manual Type 2 safety belts installed in the front outboard seating position of passenger cars. That exemption was inadvertently omitted from paragraph S7.4(b), which specifies requirements for passenger cars after September 1, 1986. Clarifying language is added to paragraph S7.4(b) in this final rule.

The second technical change clarifies the agency's intent to require passenger cars, manufactured on or after September 1, 1989, to have ELR's

for the lap belts or the lap portion of lap/shoulder belts used in the front outboard seating positions, if the automatic restraint requirement is rescinded. Paragraph S7.1.1.3(b) is revised to include the September 1, 1989, effective date for manual Type 2 belts in the front outboard seats of passenger cars.

Convenience Hooks for Automatic Belts

Some automatic belt design plans include a manual "convenience hook" which enables occupants manually to stow the belt webbing totally out of the way as they are about to exit the vehicle. Paragraph S7.4.1 was included in the January 1981 final rule to ensure that such convenience hooks would not affect compliance with the automatic restraint requirements. Automatic belts installed for compliance with the injury criteria of FMVSS 208 must operate without requiring any manual procedures by the vehicle occupant. Thus, manual hooks could not be a necessary component to move or hold the belt webbing out of the occupant's way since this would defeat the automatic aspect of performance. Paragraph S7.4.1 currently provides that any such hook must automatically release the belt webbing prior to the car being driven.

In response to comments in one petition for reconsideration of the 1981 final rule, the April 1985 proposal contained revised language to make it clear that convenience hooks are intended to release the webbing only when the automatic belt is otherwise operational. One commenter objected to the revision, stating that it would not promote the use of detachable automatic belts which have been disconnected. These objections appear to be based on a misunderstanding of the function of the convenience hook. The convenience hook concept was developed to allow it to be used in conjunction with automatic belt systems which would be in the automatic operational mode. In this way, the convenience hook could promote the use of detachable or nondetachable automatic belts, because the hook would facilitate entering or exiting the vehicle by the front seat occupants, who would then be less prone to detach or mutilate the belt system.

The commenter apparently believed that the "stowage hook," which is used to stow the latchplate of a disconnected, detachable belt, should also be covered by the requirement of S7.4.1. The stowage hook is not a convenience hook; nor is it subject to the provisions of S7.4.1. The commenter's suggestion that the "stowage hook" also release a disconnected detachable belt automatically could, in theory, increase usage, but it might also encourage owners to damage the belt physically or remove it, thus making it unavailable to subsequent owners and vehicle users. In the case of a disconnected automatic belt, the warning system would indicate to the vehicle occupants that the belt is disconnected and remind them to reconnect the belt. For these reasons, the agency denies the suggestion for automatic release of stowage hooks.

Webbing Tension-Relieving Devices

Some safety belt designs include devices intended to relieve shoulder belt pressure. These "windowshade" mechanisms or other tension-relieving devices increase the comfort of the belt, but may reduce the effectiveness of belts in a crash situation if they are misused so as to introduce excessive slack in the belt webbing. The January 1981 final rule specified that any such tension-relieving devices may be used on automatic belt systems only if the system would comply with the injury criteria of the standard with the device adjusted to *any possible position*. (The notice of proposed rulemaking preceding that final rule would have banned tension-relieving devices outright.) The 1981 final rule was adopted in recognition of the fact that tension-relieving devices can improve belt fit and increase belt comfort in certain circumstances, and was intended to allow manufacturers somewhat wider latitude in designing automatic belts, but, as discussed below, would probably have had the effect of banning these devices.

Several manufacturers objected to the wording of the January 1981 final rule on the basis that the belt system would have to meet the injury criteria even when the device had been misused to produce excessive slack in order, essentially, to defeat the system, even if such a usage was not intended by the manufacturer.

In the April 1985 proposal, the agency proposed rewording this provision to require manufacturers to include instructions in their vehicle owner's manual concerning the proper use of any tension-relieving devices incorporated in their automatic belt systems. These instructions must state the maximum amount of slack that can safely be introduced and include a warning to vehicle occupants that if excessive slack is introduced into the system, the protection offered by the belt system would be substantially reduced or even eliminated. The agency will test for compliance with the injury criteria by adjusting the belt within the slack levels recommended by the manufacturer. With one exception, those manufacturers who commented on this proposal supported the revision to allow tension-relieving devices.

However, one domestic manufacturer and a consumer group objected to the provision related to dynamic testing with the tension-relieving device adjusted to the manufacturer's recommended slack position. The manufacturer objected to a dynamic test that would require any slack at all to be introduced into the belt system, on the grounds that uncontrolled variability would be introduced into the dynamic test procedure, which would then lack objectivity. The manufacturer asserted that it might have to eliminate all tension-relieving devices for its safety belts.

The agency's proposed test procedure was intended to accommodate the view that tension-relieving devices increase the comfort of belts while, at the same time, limiting the potential reduction in effectiveness for safety belt systems in which excessive slack is introduced. The agency does not agree that this test procedure would eliminate tension-relieving devices from the marketplace. As mentioned earlier, other manufacturers supported the proposal and did not indicate they would have to remove tension-relieving devices from their belt systems. This commenter did not show that injury levels cannot be controlled within the specified injury criteria by testing at the recommended slack adjustment, as determined by the manufacturer. The recommended slack could be between zero and any level selected by the manufacturer as appropriate to relieve belt pressure without being unsafe. As a practical matter, most tension relievers automatically introduce some slack into the belt for all occupants. Testing without such slack would be unrealistic.

The same commenter objected to the requirement that belt slack be cancelled each time the vehicle door is opened *and* the buckle is released, because this requirement would encourage occupants to disconnect automatic belts. In addition, this commenter stated that the requirement is inconsistent with non-detachable, automatic belts and requested that the belt slack be required to be cancelled each time the door is opened whether or not the buckle is released. The agency believes this request has merit and has revised the requirement to reflect this change.

The consumer group objected to the proposal for automatic belt systems using tension-relieving devices to meet the injury criteria with only the specified amount of slack recommended in the owner's manual. They stated that most owners would not read the instructions in the owner's manual regarding the proper use of the tension-relieving device. It said an occupant could have a false sense of adequate restraint when wearing an automatic belt system adjusted beyond the recommended limit.

The agency's views on allowing the use of tension relievers in automatic safety belts were detailed in the April 1985 notice. The agency specifically noted the effectiveness of a safety belt system could be compromised if excessive slack were introduced into the belt. However, the agency recognizes that a belt system must be used to be effective at all. Allowing manufacturers to install tension-relieving devices makes it possible for an occupant to introduce a small amount of slack to relieve shoulder belt pressure or to get the belt away from the neck. As a result, safety belt use is promoted. This factor could outweigh any loss in effectiveness due to the introduction of a recommended amount of slack in normal use. This is particularly likely in light of the requirement that the belt system, so adjusted, must meet the injury criteria of Standard No. 208 under 30 mph test conditions. Further, the inadvertent introduction of slack into a belt system, which is beyond that for normal use, is unlikely in most current systems. In addition, even if too much slack is introduced, the occupant should notice that excessive slack is present and a correction is needed, regardless of whether he or she has read the vehicle owner's manual.

Torso Belt Body Contact Force

NHTSA research indicates that a substantial number, approximately 60 percent, of occupants are likely to complain about belt pressure if the torso belt net contact force on an occupant is greater than 0.7 pound (DOT HS-805 597). Therefore, the January 8, 1981, final rule specified that the torso portion of any manual or automatic belt system shall not create a contact pressure exceeding that of a belt with a total net contact force of 0.7 pound. Most of the petitions for reconsideration objected to this requirement, but gave no new reasons which would cause the agency to reverse its prior decision on this issue.

The April 1985 proposal contained a revised S7.4.2 which retained the 0.7-pound contact force requirement and proposed applying the requirement to tension relievers. Several commenters objected to the requirement that automatic belt systems with tension-relieving devices must meet the 0.7-pound contact force limit when the tension reliever is deactivated. Both domestic and foreign manufacturers questioned whether imposing this contact force requirement on belt systems with tension relievers would advance safety, because the belt contact force requirement could result in insufficient force to retract webbing reliably in some systems.

The agency has decided to exempt safety belt systems incorporating tension-relieving devices, such as window-shade devices, which can completely relieve belt tension, from the 0.7-pound torso belt contact force requirement. The agency is still concerned that some occupants may introduce belt slack, who otherwise would not, in a belt system incorporating a tension-reliever, if the belt force exceeds 0.7 pound. However, the agency does not want compliance with the body contact force requirement to limit manufacturers' design flexibility in meeting the retraction and other requirements in the rule.

The 0.7-pound contact force limit is retained for belt systems without tension-relieving devices, which have either a constant or variable force. The tension in these belt systems cannot be completely removed, as it can in a belt system incorporating a window-shade or other type of tension reliever. Therefore, the agency believes it is important to limit belt contact force in those systems to promote belt usage.

One manufacturer requested that the 0.7-pound contact force level be increased to ensure belt retraction. Another manufacturer stated that occupants of open-body vehicles may prefer to have the secure feeling of the upper torso belt webbing tight against their chests, i.e., a force greater than 0.7 pound. The company asked that open-body vehicles be excluded from the 0.7-pound limit. As previously noted, manufacturers may use an ALR/ELR belt system or other means to allow occupants to have belts with a tight fit. In addition, the agency believes that such an exclusion, or an increase in the 0.7-pound contact force level, is unnecessary with the modification of S7.4.3 to allow tension-relieving devices in lieu of meeting the 0.7-pound force requirement. Both manufacturers will have the option of meeting this requirement by installing a tension-relieving device in a belt system with a contact force of more than 0.7 pound.

One commenter stated that the standard should be revised to specify requirements for manual belts with tension-relieving devices. The agency did propose requirements for these manual belts in Notice 38, in conjunction with the dynamic tests for manual belts. If the agency does adopt a dynamic test requirement for manual belts, the provision on tension-relievers for manual belts would be expected to be identical to those for automatic belts.

Belt Contact Test Procedures

The April 1985 NPRM proposed that the test dummy be unclothed during the belt contact force test to avoid drag produced by clothing. The agency was concerned that such drag could cause unwanted deviations in the measurement of belt contact force, as specified in S10.6. Three commenters supported the change to remove the dummy clothing for the test. However, two other commenters stated that test variability would be greater with the test dummy's clothes removed based on the variability of skin friction due to changes in test temperature and humidity. They also said that a clothed dummy would more closely represent real world conditions. After consideration of the comments, NHTSA agrees that the clothed test dummy would more closely represent real world conditions. The agency has therefore revised the rule to require testing on a clothed test dummy, using the clothing specified in Part 572.

Two commenters asked that the test for belt contact force set maximum limits for belt release speed. The agency believes that adding a belt release speed requirement would add an unnecessary complication to the test without providing any significant improvement in controlling repeatability.

Several commenters correctly pointed out that the proposed text for S7.4.3 should reference the test procedure of S10.6 instead of S10.8. This notice adopts that correction.

Latchplate Accessibility

One of the most inconvenient aspects of using many current manual safety belt designs is the difficulty that a seated occupant has in reaching back to grasp the belt latchplate when the belt is unbuckled and in its retracted position. The greater the difficulty in reaching the latchplate to buckle the belt, the more likely the occupant will be discouraged from using the belt.

Paragraph S7.4.4 of the January 1981 final rule specified requirements to define limits on the distances an occupant has to reach for latchplates and to prescribe minimum clearances for arm and hand access. The latter requirement was specified in terms of a test block which must be able to move to the latchplate unhindered. The April 12, 1985, proposal contained a revision in the dimensions of the test block, reducing it from 3×4×12 inches to 2½×4×8 inches.

Two manufacturers requested a test procedure revision which would provide for seat cushion deflection in determining access to a latchplate with the test block shown in Figure 4. One suggested that force applied to the test block, not to exceed a certain limit, should be used to allow for seat cushion deflection. The other stated that the requirement should be deleted until such time as a test device that simulates the human hand can be developed to address seat cushion deflection.

The agency believes that reducing the size of the test block is simpler than developing an objective method for measuring and limiting seat cushion deflection. The agency also believes that the test block with its new dimensions, which are based on hand length and thickness dimensions referenced by the Society of Automotive Engineers, is sufficiently representative of the human hand. Therefore, the agency is adopting the new test block in the test procedure.

One manufacturer stated that Figure 3 in Standard No. 208, which gives the location of the reach strings for the latchplate accessibility test, does not state whether the view of the dummy is intended to depict the dummy being tested on the left or right side of the vehicle. Therefore, the implication is that the outboard reach string is always located on the right side of the dummy, according to the manufacturer. The view in Figure 3 is meant to depict the dummy being tested on the right side of the vehicle. The agency would use the string placements in Figure 3 to perform an accessibility test for the right front outboard passenger seating position, because the outboard reach string is located on the right side of the test dummy. This string would be reversed for the driver position, because the outboard side would be located on the left side of the dummy with the dummy facing forward. The string in Figure 3 is labeled "outboard" and the agency believes this explanation is sufficient without changing Figure 3.

Several manufacturers stated that a latchplate accessibility test using the test block representing a human hand to check the clearance between the arm rest and seat cushion should not be necessary, if the belt system is designed so that the latchplate is retained in an accessible area. For example, one manufacturer said that it uses a sliding plastic bar on its belt webbing which positions the latchplate in an accessible area near the upper torso anchorage point. The manufacturer said that the plastic bar prevents the latchplate from sliding down the webbing to a position under the arm rest or between the seat and side of the vehicle. The manufacturer said that it could also use a fixed plastic button to retain the latchplate near the upper torso anchorage. The agency agrees that if a latchplate is permanently retained in an accessible area, reachable by the test block, there is no need to conduct a clearance test between the arm rest and seat cushion.

The purpose of the latchplate accessibility requirement is to address designs in which the latchplate can freely move on the belt webbing. In those cases, the latchplate may initially be located in an accessible area, but the design of the belt may permit the latchplate to slide along the webbing into the area between the seat cushion and the door interior, or below the door arm rest when the belt is retracted. If this situation is likely to occur in normal use with any regularity, such a belt system would be required to comply with the test for accessibility at the point where the latchplate normally slides along the webbing into the area between the seat cushion and the door interior, or below the door arm rest. The agency believes that the addition of language stating that access to the latchplate should be tested with the latchplate in its "normally stowed position" to the requirement should clarify this requirement. If the belt system incorporates a design which ensures that the latchplate cannot move near an arm rest or move down between the seat and the vehicle's side structure, the system will have no problem passing the hand access test.

Several commenters apparently believed that S7.4.4 requires the latchplate to be mounted on the outboard side of a vehicle seat. They said that the requirement was design-restrictive for a Type 1 safety belt assembly because such an assembly could otherwise be designed so that the latchplate is located at either the inboard or outboard position. The requirement was developed to test for access of the latchplate or buckle on belt assemblies which are located outboard of the designated seating position for which the latchplate is installed. This is because access to a latchplate located in that position can be hindered by the vehicle's side structure. The requirement was not intended to specify that the latchplate or buckle be located outboard of a designated seating position. The language of the rule is therefore revised to indicate that the test applies only to latchplates or buckles located outboard of the designated seating position.

One manufacturer recommended that the compliance test for accessibility be made similar to the requirement for safety belt anchorages in Standard No. 210, *Seat Belt Anchorages.* Compliance arcs would be generated from a point on the SAE two-dimensional manikin, whose H-point is positioned at the full-forward position of the design H-point, or on a full-scale design drawing. This commenter stated that such a procedure would eliminate test variability, reduce the compliance test burden, and allow manufacturers to determine compliance while the vehicle is in the advance design stage.

Manufacturers are free to determine compliance with a requirement by any method they choose, while exercising due care. There is no reason to believe that the procedure suggested by the commenter is not compatible with the procedure defined in Standard No. 208. Therefore, it is unnecessary to revise the current test procedure for latchplate accessibility.

Another manufacturer requested that the language of S7.4.4 be amended to specify that the access requirement be met with the seat within the adjustment range of a person whose dimensions range from those of a 50th percentile six-year-old child to those of a 95th percentile adult male. The rationale for the request is that, when securing a child restraint in some of their vehicles, the latchplate is located at a very low height near the floor, after locking. In this situation, the ability of small cars to comply with the latchplate access requirement is severely compromised. To achieve compliance, the seat back would have to be deeply cut away at the outboard side.

The latchplate access requirement is meant to address access problems when the latchplate is in its normally stowed position. It was not meant to address potential access problems with child restraints that might occur in specific vehicles. Therefore, the agency does not believe an amendment is necessary.

Belt Retraction

The April 12, 1985, notice proposed to revise S7.4.5 to allow for the stowage of arm rests on vehicle seats, such as captain's chairs, which must have the outboard arm rests stowed before the occupant can exit the vehicle. One commenter asked the agency to permit all arm rests, which protrude into the door opening in a manner which encumbers egress, to be placed in their stowed position for the retraction test. The agency believes this comment has merit and has revised S7.4.5 to permit the stowage of outboard arm rests if they protrude into the door opening in a manner which encumbers egress. The agency notes that folding arm rests are usually designed that way for the purpose of facilitating egress or ingress by moving them out of the way to a stowed position.

The April notice also proposed to allow tension-relieving devices on the safety belts of open-body vehicles without doors to be manually deactivated for the retraction test. One commenter objected to allowing these tension-relieving devices to be manually, rather than automatically, cancelled. The commenter said that there are belt systems currently available which will automatically cancel a tension-relieving device when the latchplate is released from the buckle.

At the time the agency proposed the requirement for open-body vehicles, it was not aware that there were belt systems which would automatically deactivate tension-relieving devices solely through the action of unbuckling the belt. Therefore, the agency only proposed that belt systems in open-body vehicles be tested with their tension-relieving devices manually deactivated. The agency will consider the commenter's suggestion as one for future rulemaking. The agency notes that manufacturers can voluntarily adopt the use of other automatic means for deactivating the tension-relieving device in open-body vehicles.

The April notice also proposed that the latchplate must retract to its "completely stowed position." Two commenters objected to this proposal saying that determining whether the belt is "completely" stowed is difficult. They believe that, if the stowed position prevents the safety belt from extending out of the vehicle's adjacent open door, the requirement for belt retraction should be satisfied. The agency believes that this comment is reasonable and consistent with the intent of this section to prevent belts from getting dirty as a result of being caught in the door and from hindering ingress or egress of occupants. The language in the rule is revised accordingly.

Seat Belt Guides

The April notice proposed clarifications in the language of S7.4.6.1(a) to increase the accessibility of belt buckles and latchplates and belt webbing to the vehicle occupant, while giving manufacturers flexibility to use stiffeners, guide openings, cables, or conduits of any type. The notice also proposed modifying S7.4.6.1(b) to exempt seats which are movable to serve a dual function.

Two commenters stated that the language in S7.4.6.1(b) did not adequately address seats which are removable or seats which are movable to serve a secondary function. NHTSA believes these comments are valid, because a seat belt latchplate, a buckle, or a portion of the webbing cannot be maintained on top of a seat which has been removed or moved to serve a secondary function. Therefore, the requirement does not apply to seats which are removable or movable so that the space formerly occupied by the seat can be used for a secondary function, such as cargo space. However, the term, secondary function, does not include the movement of a seat to provide a comfortable driving and riding position for different size occupants.

Two manufacturers requested that the words "seat cushion" in S7.4.6.1(b) be amended by adding the words "and/or seat backs." The agency specifically excluded ' seat backs'' from the exemption because there is no evidence that seats with folding seat backs cannot comply with the requirements. Adding movable seat backs to the language in S7.4.6.1(b) could exempt front seats in passenger cars and the second seat in some vehicles, such as station wagons. The agency believes that there is no reason for exempting these seats.

One manufacturer stated that the center safety belt in the rear seat of a motor vehicle should be exempted from the requirement in S7.4.6.1(a) concerning seat guides. This commenter stated that there is little chance of this belt ever becoming "lost" behind the seat due to the abundance of webbing material available for the center rear safety belt; therefore, a webbing guide seems unnecessary. The agency disagrees. The agency believes that the requirements are necessary since they address specific problems associated with belts which are not adjusted by retractors, such as the rear center seat belts. (Center seats are not required to have safety belt retractors, which automatically stow the webbing after the belt is taken off. Instead, they usually have more of the webbing lying on the seat cushion and have a manually adjustable buckle which slides along the webbing so that an occupant can tighten the belt around himself or herself.) Having more of the belt lying on the seat can make the belt more accessible; it can also cause the user to stuff the belt behind the seat cushion to get the webbing out of the way when the center seating position is not being used. In addition, one company, such as the commenter, may provide ample webbing which will lie on the seat cushion, while another company may not. The agency is therefore not exempting center seats.

One manufacturer stated that a 3-point belt assembly, with the lap webbing portion designed to pass between the seat cushion and seat back, will not necessarily have the latchplate positioned on the top of the seat, when the webbing is retracted. It urged that the requirement be revised to read, "maintain the accessibility of the safety belt latchplate or buckle," and to strike the words "or a portion of the safety belt webbing on top of the seat cushion." The agency agrees that the latchplate and buckle do not necessarily have to be located on the seat cushion to be accessible. NHTSA does believe that as long as the webbing is accessible on top of or above the seat, an occupant should be able to retrieve the latchplate and buckle. Therefore, the rule is revised to require that only one of the three belt parts (the seat belt latchplate, the buckle, or seat belt webbing) be maintained on top of or above the seat cushion under normal conditions. Although the other two parts will not be required to be on the seat cushion, the agency has revised the rule to require that they remain accessible under normal conditions.

Another manufacturer stated that the provision that a buckle be accessible in S7.4.6.2 with an adjustable arm rest in any position of adjustment lacked objectivity and should be deleted. The agency does not agree and continues to believe that a simple visual inspection should be sufficient to determine whether or not the buckle is accessible when the arm rest is in the down position.

Warning System Requirements

The purpose of the proposed revision to these requirements in the April notice was to allow for a warning light which activates for at least 60 seconds if condition (A)—the vehicle's ignition switch is moved to the "on" or "start" position, exists simultaneously with condition (B)—the driver's automatic belt is not in use or, if the belt is non-detachable, the emergency release mechanism is in the released position. Specifying a minimum activation time was intended to allow the manufacturer the option of providing for additional warning time. The proposal would also require that condition (C)—the belt webbing of a motorized automatic belt is not in its locked, protective mode at the anchorage point—be indicated only by a continuous or flashing warning light in lieu of a buzzer each time the ignition switch is turned to the "on" position. The light would remain lit as long as condition (C) existed.

Two manufacturers raised concerns about determining when condition (B) exists—the driver belt is not in use or the emergency release mechanism is released—in a motorized belt system. They, in effect, made the point that with certain motorized designs, the April proposal would have required the audible warning required for condition (B) to sound while the belt webbing is moving along its track to its fully locked position. For example, one manufacturer stated that in some motorized belt systems the emergency release belt latch mechanism sensing is done by a proximity switch in the (B) pillar which senses the presence of a magnet in the part attached to the webbing. In this case, the system will sense that the latch is unfastened until the motorized belt is in its fully locked position and, thus, under the proposal, would activate the audible warning during the period that the belt is in motion. This commenter requested that to prevent an audible warning from being given when the mechanism is being operated normally, the manufactuer should be given the option of starting the audible warning period from the time that the belt reaches the fully locked position.

The agency believes that it is important that an audible warning sound when the driver's belt is not in use or the belt's emergency release mechanism is actuated. However, to prevent the sounding of the audible warning when a motorized belt is moving into place, the agency is revising the warning system requirement. The revision provides that, in the case of a motorized belt, the existence of condition (B) is determined once the belt is in its fully locked position. Once a motorized belt has reached its fully locked position, an audible warning must sound if condition (B) exists. The agency wishes to emphasize that all motorized belts, regardless of their design, should have an audible warning that sounds if the driver's belt is not in use or the belt's emergency release mechanism is actuated.

One of the same commenters also said it is planning to use detachable automatic belts in some of its new belt system designs. Its concern is that condition (B), which is determined by the belt latch mechanism not being fastened, would require them to locate the electrical sensor in the emergency release buckle. In a motorized system, the wire harness for the electric sensor would have to be moved along a track, because the "emergency release buckle" slides along the track with the buckle end. The location of the electrical sensor in the buckle makes the wire harness less reliable, because of the constant movement, according to the commenter. After the close of the comment period on the April notice, NHTSA received a petition for rulemaking to amend the requirements of paragraph S4.5.3.3(b) of Standard No. 208 from Chrysler Corporation which raised the same issues. Chrysler petitioned for an alternative means to determine when the belt latch mechanism is not fastened. It asked that the warning requirement be modified to permit actuation of the warning when less than 20 inches of webbing has been withdrawn from the driver's seat belt retractor.

The agency believes the problems identified by the commenter and the Chrysler petition are valid. NHTSA did not intend to imply in the April 1985 notice that the method for determining that the belt latch is not fastened must be by a sensor located in the belt buckle. The agency believes that manufacturers should have maximum design flexibility to develop systems to determine if the latch is not fastened. The condition could be determined by any means, such as a predetermined amount of belt webbing spool-out, or the location of a sensor in the overhead, motorized track area or in the working mechanism of the buckle/latchplate, which would show that the automatic belt is not fastened. The agency does note that if a manufacturer decides to use belt webbing spool-out that it determine the least amount of webbing necessary to go around a person in the driver's position with the seat in its rearmost position. If less than this minimum amount of webbing spools out of the retractor in an attempt to defeat the system, the warning should be activated.

Two manufacturers requested that NHTSA confirm that the same light signal may be activated under both conditions (B) and (C), since the required audible signal suffices to differentiate between the two conditions. The agency agrees that this comment has merit and confirms that the same light signal may be activated under both conditions (B) and (C).

Use of Additional Warnings

One manufacturer sought permission to use additional warnings to supplement those required by the standard. This manufacturer stated that its warning system provided for an audible warning system in addition to the warning light to indicate that condition (A) + (C) exists. Further, the passenger seating position is also equipped with a warning system, which is not required by the standard. The agency notes, again, that a manufacturer is free to provide features in addition to those required by the standard, as long as the standard's requirements are met. No change in the standard is necessary to permit the commenter to install additional features in its warning systems.

Another company stated that, for non-detachable automatic belts, the proposed 60-second visual warning and the 4- to 8-second audible warning may not be sufficient to indicate that the emergency spool release is in the released position. This company believes that the visual warning should remain on for as long as the emergency release mechanism remains in the release or "emergency" position. The agency notes that the requirement specifies a minimum 60-second visual warning and does not limit it to 60 seconds for condition (B). The agency specified a minimum period of time, which is believed sufficient to warn occupants of this condition. Manufacturers have the choice of extending the time for a warning light to more than 60 seconds to indicate that the emergency release mechanism is in the release or emergency condition. Therefore, no change in the language of the standard is required.

Walk-in Van Vehicles

The agency tentatively proposed to exclude walk-in step vans from the safety belt comfort and convenience requirements in the April 12, 1985, notice. By the term, "walk-in vans," NHTSA is referring to city delivery type vehicles used, for example, to deliver parcels or dry cleaning where the drivers can walk directly into the vans without stooping. A consumer group objected to the proposed exemption for walk-in step vans on the basis that NHTSA should promote belt use in these vehicles by making them easier to use. The agency is not persuaded that the increase in belt usage which might result from the redesign of walk-in vans to meet the comfort and convenience requirements would justify the cost of such a modification. Moreover, these vehicles do not normally have a secondary use, for example, as a family vehicle, as do other utility vehicles which are required to meet the comfort and convenience requirements for safety belts. Due to the problems with cost and vehicle redesign, the agency does not believe that it is appropriate to apply the comfort and convenience requirements to these vehicles.

Weights and Dimensions

In the April 12, 1985, notice, the agency proposed a chart of weights and dimensions which included small dimension changes and tolerances for the 50th percentile adult male. One manufacturer commented that the agency has supplied no rationale for these changes and that such dimensional revisions to the Part 572 dummy should be the subject of a separate rulemaking under Part 572. This commenter also objected to inclusion of a seated hip circumference in the chart. The agency notes that the chart of weights and dimensions of vehicle occupants was included in Standard No. 208 as a guide for manufactuers. The seated hip circumference was included in this chart because it is referred to in Standard No. 208. There is no requirement in Part 572 for a seated hip circumference; therefore, this dimension is not a requirement for the Part 572 test dummy. The agency proposed the minor changes to the chart to ensure that the dimensions set forth in the chart agreed with the dimensions specified on drawing SA 150 M002 of the test dummy, which is incorporated by reference in Part 572.5. The agency is therefore adopting the proposed changes.

Another company said that the dimensions of a six-year-old child are contained in the table defining the vehicle occupants. Although it highly recommends safety belt use for a child of this age, this commenter stressed that optimum protection for a person of these dimensions can only be obtained by using an additional special booster cushion equipped with a safety belt guide system. These types of cushions are readily available in the United States. The commenter therefore requested that the standard be amended to permit the commenter to recommend the use of such a cushion in

order to ensure correct positioning of the belt around a six-year-old child. The agency agrees that, in some instances, booster seats do facilitate the use of adult restraints by this size occupant. However, the agency also believes that the average six-year-old child should be suitably accommodated by the adult belt system in such a way that the child is adequately protected from injury and fatality. Therefore, the agency declines to make this change to the standard.

Automatic Safety Belt Interpretation

In 1974 (39 FR 14594), the agency issued an interpretation that it would not consider a belt system which had to be manually moved out of the way by the occupant to be an "automatic" system that would satisfy the requirements of Standard No. 208. In the April 12, 1985, notice, the agency stated its belief that such an interpretation may be overly stringent and requested public comment.

Four commenters argued that the past interpretation was overly stringent, because it would have allowed no manual movement of the belt to accommodate ingress into the vehicle. As a minimum, these commenters stated, such an interpretation should acknowledge that a safety belt design should be considered "passive" or "automatic" if an occupant would normally push the webbing aside upon entering the vehicle. In addition, an automatic belt requiring a slight adjustment for comfort should be considered an automatic restraint system. The commenters urged that any belt design, which would perform its protective restraining function after a normal process of ingress, without separate deliberate action by the vehicle occupant to deploy the restraint system, should be allowed. Finally, the commenters said that to provide an automatic lap and shoulder belt design which would comply with the original interpretation could increase the tendency for the occupant to submarine under the belt. The reason is that the lap belt portion, which would enable an occupant to enter or exit the vehicle without manually moving the belt, could be raised too high. To solve this problem, a very expensive motorized system would be required to move the belts out of the occupant's ingress/egress area.

The agency believes these comments have merit and has revised its interpretation. The concept of an occupant protection system which requires "no action by vehicle occupants," as that term is used in Standard No. 208, is intended to designate a system which will perform its protective restraining function after a normal process of ingress or egress without separate deliberate actions by the vehicle occupant *to deploy* the restraint system. Thus, the agency considers an occupant protection system to be automatic if an occupant has to take no action to deploy the system but would normally slightly push the safety belt webbing aside when entering or exiting the vehicle or would normally make a slight adjustment in the webbing for comfort. The agency believes that the marketplace will help curb use of automatic belt systems which are complicated, or require excessive adjustments before ingress or egress, since prospective purchasers would reject vehicles with such systems. The agency believes that adoption of the comfort and convenience requirements will help ensure that manufacturers provide automatic belt systems which will promote belt usage.

In consideration of the foregoing, 49 CFR 571.208 is amended as follows:

1. The authority citation for Part 571 continues to read as follows:

Authority: 15 U.S.C. 1392, 1401, 1403, 1407; delegation of authority at 49 CFR 1.50.

2. S7.1.1.3 is revised to read:

S7.1.1.3(a) Except as provided in S7.1.1.3(b), a Type 1 lap belt or the lap belt portion of any Type 2 belt installed at any front outboard designated seating position for compliance with this standard in a vehicle (other than walk-in van-type vehicles) manufactured on or after September 1, 1986, shall meet the requirements of S7.1 by means of an emergency-locking retractor that conforms to Standard No. 209 (§ 571.209).

(b) The requirements of S7.1.1.3(a) do not apply to the lap belt portion of any Type 2 belt installed in a passenger car manufactured before September 1, 1989, or to walk-in van-type vehicles.

3. S7.4 is revised to read:

S7.4 *Seat belt comfort and convenience.* (a) Automatic seat belts installed in any vehicle, other than walk-in van-type vehicles, with a GVWR of 10,000 pounds or less, manufactured on or after September 1, 1986, shall meet the requirements of S7.4.1, S7.4.2, and S7.4.3.

(b) Except as provided in S7.4(c), manual seat belts, other than manual Type 2 belt systems installed in the front outboard seating position in passenger cars, installed for compliance with this standard in any vehicle which has a GVWR of 10,000 pounds or less, and is manufactured on or after September 1, 1986, shall meet the requirements of S7.4.3, S7.4.4, S7.4.5, and S7.4.6. Manual Type 2 seat belts in the front outboard seating positions of passenger cars manufactured on or after September 1, 1989, shall meet the requirements of S7.1.1.3(a), S7.4.3, S7.4.4, S7.4.5, and S7.4.6, if the automatic restraint requirements are rescinded pursuant to S4.1.5.

(c) The requirements of S7.4(b) do not apply to manual belts installed in walk-in van-type vehicles.

4. S7.4.1 is revised to read:

S7.4.1 *Convenience hooks.* Any manual convenience hook or other device that is provided to stow seat belt webbing to facilitate entering or exiting the vehicle shall automatically release the webbing when the automatic belt system is otherwise operational and shall remain in the released mode for as long as (a) exists simultaneously with (b), or, at the maufacturer's option, for as long as (a) exists simultaneously with (c)—

(a) The vehicle ignition switch is moved to the "on" or "start" position;

(b) The vehicle's drive train is engaged;

(c) The vehicle's parking brake is in the released mode (nonengaged).

5. S7.4.2 is revised to read:

S7.4.2 *Webbing tension-relieving device.* Each automatic seat belt assembly that includes either manual or automatic devices that permit the introduction of slack in the webbing of the shoulder belt (e.g., "comfort clips" or "window-shade" devices) shall comply with the occupant crash protection requirements of S5 of this standard with the belt webbing adjusted to introduce the maximum amount of slack that is recommended by the vehicle manufacturer in the vehicle owner's manual to be introduced into the shoulder belt under normal use conditions. The vehicle owner's manual shall explain how the device works and shall specify the maximum amount of slack (in inches) which is recommended by the vehicle manufacturer in the owner's manual to be introduced into the shoulder belt under normal use conditions. These instructions shall also warn that introducing slack beyond the specified amount could significantly reduce the effectiveness of the belt in a crash. Any belt slack that can be introduced into the belt system by means of any tension-relieving device or design shall be cancelled each time the safety belt is unbuckled or the adjacent vehicle door is opened except for belt systems in open-body vehicles with no doors.

6. S7.4.3 is revised to read as follows:

S7.4.3 *Belt contact force.* Except for seat belt assemblies which incorporate a webbing tension-relieving device that complies with S7.4.2, the upper torso webbing of any seat belt assembly, when tested in accordance with S10.6, shall not exert more than 0.7 pound of contact force when measured normal to and one inch from the chest of an anthropomorphic test dummy, positioned in accordance with S10 in the seating position for which that assembly is provided, at the point where the centerline of the torso belt crosses the midsagittal line on the dummy's chest.

7. The first sentence of S7.4.4 is revised to read as follows:

S7.4.4 *Latchplate access.* Any seat belt assembly latchplate which is located outboard of a front outboard seating position in accordance with S4.1.2, shall also be located within the outboard reach envelope of either the outboard arm or the inboard arm described in S10.5 and Figure 3 of this standard, when the latchplate is in its normal stowed position. There shall be sufficient clearance between the vehicle seat and the side of the vehicle interior to allow the test block defined in Figure 4 unhindered transit to the latchplate or buckle.

8. S7.4.5 is revised to read as follows:

S7.4.5 *Retraction.* When tested under the conditions of S8.1.2 and S8.1.3, with the anthropomorphic test dummies whose arms have been removed and which are positioned in accordance with S10 and restrained by the belt systems for those positions, the torso and lap belt webbing of any of those seat belt systems shall automatically retract when the adjacent vehicle door is in the open position, or when the seat belt latchplate is released, to a stowed position. That position shall prevent any part of the webbing or hardware from being pinched when the adjacent vehicle door is closed. A belt system with a tension-relieving device in an open-bodied vehicle with no doors shall fully retract when the tension-relief device is manually deactivated. For the purpose of the retraction requirement, outboard armrests may be placed in their stowed positions if they are on vehicle seats which

must have the armrests in the stowed position to allow an occupant to exit the vehicle.

9. S7.4.6.1 is revised to read as follows:

S7.4.6.1(a) Any manual seat belt assembly whose webbing is designed to pass through the seat cushion or between the seat cushion and seat back shall be designed to maintain one of the following three seat belt parts (the seat belt latchplate, the buckle, or the seat belt webbing) on top of or above the seat cushion under normal conditions (i.e., conditions other than when belt hardware is intentionally pushed behind the seat by a vehicle occupant). In addition, the remaining two seat belt parts must be acessible under normal conditions.

(b) The requirements of S7.4.6.1(a) do not apply to: (1) seats whose seat cushions are movable so that the seat back serves a function other than seating, (2) seats which are removable, or (3) seats which are movable so that the space formerly occupied by the seat can be used for a secondary function.

10. S4.5.3.3(b) is revised to read as follows:

S4.5.3.3(b) In place of a warning system that conforms to S7.3 of this standard, be equipped with the following warning system: At the left front designated seating position (driver's position), a warning system that activates a continuous or intermittent audible signal for a period of not less than 4 seconds and not more than 8 seconds and that activates a continuous or flashing warning light visible to the driver for not less than 60 seconds (beginning when the vehicle ignition switch is moved to the "on" or the "start" position) when condition (A) exists simultaneously with condition (B), and that activates a continuous or flashing warning light, visible to the driver, displaying the identifying symbol for the seat belt telltale shown in Table 2 of Standard No. 101 (49 CFR 571.101), or, at the option of the manufacturer if permitted by Standard No. 101, displaying the words "Fasten Seat Belts" or "Fasten Belts," for as long as condition (A) exists simultaneously with condition (C).

(A) The vehicle's ignition switch is moved to the "on" position or to the "start" position.

(B) The driver's automatic belt is not in use, as determined by the belt latch mechanism not being fastened or, if the automatic belt is non-detachable, by the emergency release mechanism being in the released position. In the case of motorized automatic belts, the determination of use shall be made once the belt webbing is in its locked protective mode at the anchorage point.

(C) The belt webbing of a motorized automatic belt system is not in its locked, protective mode at the anchorage point.

11. The first sentence of S10.5 is amended to delete "S7.4.7" and to insert in its place "S7.4.4."

12. S10.6 is amended to read as follows:

S10.6 To determine compliance with S7.4.3 of this standard, position the anthropomorphic test dummy in the vehicle in accordance with S8.1.11, and under the conditions of S8.1.2, S8.1.3, and S8.1.9. Close the vehicle's adjacent door, pull 12 inches of belt webbing from the retractor and then release it, allowing the belt webbing to return to the dummy's chest. Pull the belt webbing three inches from the dummy's chest and release until the webbing is within one inch of the dummy's chest and measure belt pressure.

13. Figure 4 of this standard is modified as follows:

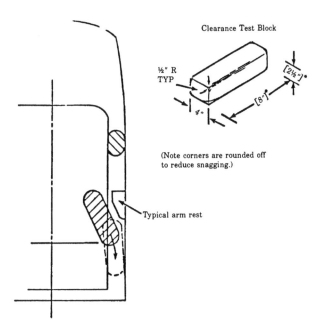

Clearance Test Block

½" R TYP

(Note corners are rounded off to reduce snagging.)

Typical arm rest

Figure 4—USE OF CLEARANCE TEST BLOCK TO DETERMINE HAND/ARM ACCESS

14. The weights and dimensions of the vehicle occupants referred to in this standard and specified in S7.1.13 are modified to read as follows:

an occupant has to take no action to deploy the system but would normally slightly push the seat belt webbing aside when entering or exiting the

	50th-percentile 6-year-old child	5th-percentile adult female	50th-percentile adult male	95th-percentile adult male
Weight	47.3 pounds	102 pounds	164 pounds $\pm\,3$	215 pounds
Erect sitting height	25.4 inches	30.9 inches	35.7 inches ±10	38 inches
Hip breadth (sitting)	8.4 inches	12.8 inches	14.7 inches $\pm\,7$	16.5 inches
Hip circumference (sitting)	23.9 inches	36.4 inches	42 inches	42.5 inches
Waist circumference (sitting)	20.8 inches	23.6 inches	32 inches $+60$	42.5 inches
Chest depth		7.5 inches	9.3 inches ±20	10.5 inches
Chest circumference:				
(nipple)		30.5 inches		
(upper)		29.8 inches	37.4 inches ±6	44.5 inches
(lower)		26.6 inches		

15. The Note following paragraph S11.8 is revised to read as follows:

Note: The concept of an occupant protection system which requires "no action by vehicle occupants," as that term is used in Standard No. 208, is intended to designate a system which will perform its protective restraining function after a normal process of ingress or egress without separate deliberate actions by the vehicle occupant to deploy the restraint system. Thus, the agency considers an occupant protection system to be automatic if

vehicle or would normally make a slight adjustment in the webbing for comfort.

Issued on November 1, 1985

Diane K. Steed
Administrator

50 FR 46056
November 5, 1985

PREAMBLE TO AN AMENDMENT TO
FEDERAL MOTOR VEHICLE SAFETY STANDARD NO. 208
OCCUPANT CRASH PROTECTION
[Docket No. 74-14; Notice 43]

ACTION: Final rule.

SUMMARY: On April 12, 1985, NHTSA issued a notice proposing a number of amendments to Standard No. 208, *Occupant Crash Protection*. Based on its analysis of the comments received in response to that notice, the agency has decided to take the following actions: retain the oblique crash test for automatic restraint equipped cars, adopt some New Car Assessment Program test procedures for use in the standard's crash tests, provide in the standard for a due care defense with respect to the automatic restraint requirement, and require the dynamic testing of manual lap/shoulder belts in passenger cars. This notice also creates a new Part 585 that sets reporting requirements regarding compliance with the automatic restraint phase-in requirements of the standard.

EFFECTIVE DATE: The amendments made by this notice will take effect on May 5, 1986, except the requirement for dynamic testing of manual safety belts in passenger cars will go into effect on September 1, 1989, if the automatic restraint requirement is rescinded.

SUPPLEMENTARY INFORMATION:

Background

On July 11, 1984 (49 FR 28962), the Secretary of Transportation issued a final rule requiring automatic occupant protection in all passenger cars. The rule is based on a phased-in schedule beginning on September 1, 1986, with full implementation being required by September 1, 1989. However, if before

April 1, 1989, two-thirds of the population of the United States are covered by effective state mandatory safety belt use laws (MULs) meeting specified criteria, the automatic restraint requirement will be rescinded.

More specifically, the rule requires:

• Front outboard seating positions in passenger cars manufactured on or after September 1, 1986, for sale in the United States, will have to be equipped with automatic restraints based on the following schedule:

• Ten percent of all cars manufactured on or after September 1, 1986.

• Twenty-five percent of all cars manufactured on or after September 1, 1987.

• Forty percent of all cars manufactured on or after September 1, 1988.

• One hundred percent of all cars manufactured on or after September 1, 1989.

• During the phase-in period, each car that is manufactured with a system that provides automatic protection to the driver without the use of safety belts and automatic protection of any sort to the passenger will be given an extra credit equal to one-half car toward meeting the percentage requirement. In addition, each car which provides non-belt automatic protection solely to the driver will be given a one vehicle credit.

• The requirement for automatic restraints will be rescinded if MULs meeting specified conditions are passed by a sufficent number of states before April 1, 1989, to cover two-thirds of the population of the United States. The MULs must go into effect no later than September 1, 1989.

In the July 1984 notice, the Secretary identified various issues requiring additional rulemaking. On April 12, 1985, the agency issued two notices setting

PART 571; S208–PRE 285

forth proposals on all of those issues. One notice (50 FR 14589), which is the basis for the final rule being issued today, proposed: reporting requirements for the phase-in, deletion of the oblique test, alternative calculations of the head injury criterion (HIC), allowing the installation of manual belts in convertibles, use of the New Car Assessment Program (NCAP) test procedures, and adoption of a due care defense. The notice also proposed the dynamic testing of manual lap/shoulder belts for passenger cars, light trucks and light vans. The second notice (50 FR 14602) set forth the agency's proposals on the use of the Hybrid III test dummy and additional injury criteria. NHTSA has not yet completed its analysis of the comments and issues raised by the Hybrid III proposal or the proposal regarding convertibles and dynamic testing of safety belts in light trucks and light vans. The agency will publish a separate *Federal Register* notice announcing its decision with regard to these issues when it has completed its analysis.

Oblique Crash Tests

Standard No. 208 currently requires cars with automatic restraints to pass the injury protection criteria in 30 mph head-on and oblique impacts into a barrier. The April 1985 notice contained an extensive discussion of the value of the oblique test and requested commenters to provide additional data regarding the safety and other effects of deleting the requirements.

The responses to the April notice reflected the same difference of opinion found in the prior responses on this issue. Those favoring elimination of the test argue that the test is unnecessary since oblique crash tests generally show lower injury levels. They also said the additional test adds to the cost of complying with the standard—although manufacturers differed as to the extent of costs. Four manufacturers suggested that any cost reduction resulting from elimination of the test would be minimal, in part because they will continue to use the oblique tests in their restraint system developmental programs, regardless of what action the agency takes. Another manufacturer, however, said that while it would continue to use oblique testing during its vehicle development programs, the elimination of the oblique test in Standard No. 208 would result in cost and manpower savings. These savings would result because the parts used in vehicles for certification testing must be more representative of actual production parts than the parts used in vehicles crashed during development tests.

Those favoring retention of the test again emphasized that the test is more representative of real-world crashes. In addition, they said that occupants in systems without upper torso belts, such as some air bag or passive interior systems, could experience contact with the A-pillar and other vehicle structures in the oblique test that they would not experience in a head-on test. Although, again, there were conflicting opinions on this issue—one manufacturer said that oblique tests would not affect air bag design, while other manufacturers argued that the oblique test is necessary to ensure the proper design of air bag systems. The same manufacturer that said air bag design would not be affected by the oblique test, emphasized that vehicles with 2-point automatic belts or passive interiors, "may show performance characteristics in oblique tests that do not show up on perpendicular tests." Similarly, one manufacturer said that oblique tests will not result in test dummy contact with the A-pillar or front door—while another manufacturer argued that in the oblique test contact could occur with the A-pillar in vehicles using non-belt technologies.

After examining the issues raised by the commenters, the agency has decided to retain the oblique tests. There are a number of factors underlying the agency's decision. First, although oblique tests generally produce lower injury levels, they do not consistently produce those results. For example, the agency has conducted both oblique and frontal crash tests on 14 different cars as part of its research activities and NCAP testing. The driver and passenger HIC's and chest acceleration results for those tests show that the results in the oblique tests are lower in 31 of the 38 cases for which data were available. However, looking at the results in terms of vehicles, 6 of the 14 cars had higher results, exclusive of femur results, in either passenger or driver HIC's or chest accelerations in the oblique tests. The femur results in approximately one-third of the measurements were also higher in the oblique tests. Accident data also indicate that oblique impacts pose a problem. The 1982 FARS and NASS accident records show that 14 percent of the fatalities and 22 percent of the AIS 2-5 injuries occur in 30 degree impacts.

The agency is also concerned that elimination of the oblique test could lead to potential design problems in some automatic restraint systems. For example, air bags that meet only a perpendicular impact test could be made much smaller. In such a case, in an oblique car crash, the occupant would roll off the smaller bag and strike the A-pillar or instrument panel. Similarly, the upper torso belt of an automatic belt system could slip off an occupant's shoulder in an oblique crash. In belt system with a tension-relieving device, the system will be tested with the maximum amount of slack recommended by the vehicle manufacturer, potentially increasing the possibility of the upper torso belt slipping off the occupant's shoulder. In the case of passive interiors, an occupant may be able to contact hard vehicle structures, such as the A-pillar, in oblique crashes that would not be contacted in a perpendicular test. If the A-pillar and other hard structures are not designed to provide protection in oblique crashes then there would be no assurance, as there presently is, that occupants would be adequately protected. Thus, the oblique test is needed to protect unrestrained occupants in passive interiors, and to ensure that air bags and automatic or manual safety belts are designed to accommodate some degree of oblique impact.

The agency recognizes that retention of the oblique test will result in additional testing costs for manufacturers. The agency believes, however, that there are a number of factors which should minimize those costs. First, even manufacturers opposing retention of the oblique test indicated that they will continue to perform oblique crash tests to meet their own internal requirements as well as to meet the oblique test requirements of the Standard No. 301, *Fuel System Integrity*. Since the oblique tests of Standard No. 208 and Standard No. 301 can be run simultaneously, the costs resulting from retention of the oblique crash test requirements of Standard No. 208 should not be significant.

Dynamic Testing of Manual Belts

The April notice proposed that manual lap/shoulder belts installed at the outboard seating positions of the front seat of four different vehicle types comply with the dynamic testing requirements of Standard No. 208. Those requirements provide for using test dummies in vehicle crashes for measuring the level of protection offered by the restraint system. The four vehicle types subject to this proposal are passenger cars, light trucks, small van-like buses, and light multipurpose passenger vehicles (MPV's). (The agency considers light trucks, small van-like buses, and light MPV's to be vehicles with a Gross Vehicle Weight Rating (GVWR) of 10,000 pounds or less and an unloaded vehicle weight of 5,500 pounds or less. The 5,500 pound unloaded vehicle weight limit is also used in Standard No. 212, *Windshield Retention*, and Standard No. 219, *Windshield Zone Intrusion*. The limit was adopted in those standards on April 3, 1980

(45 FR 22044) to reduce compliance problems for final-stage manufacturers. Readers are referred to the April 1980 notice for a complete discussion of the 5,500 pound limit.)

Currently, manual belts are not subject to dynamic test requirements. Instead they must be tested in accordance with Standard No. 209, *Seat Belt Assemblies*, for strength and other qualities in laboratory bench tests. Once a safety belt is certified as complying with the requirements of Standard No. 209, it currently may be installed in a vehicle without any further testing or certification as to its performance in that vehicle. The safety belt anchorages in the vehicle are tested for strength in accordance with Standard No. 210, *Seat Belt Assembly Anchorages*.

The April 1985 notice also addressed the issue of tension-relieving devices on manual belts. Tension-relieving devices are used to introduce slack in the shoulder portion of a lap-shoulder belt to reduce the pressure of the belt on an occupant or to effect a more comfortable "fit" of the belt to an occupant. The notice proposed that manufacturers are required to specify in their vehicle owner's manuals the maximum amount of slack they recommend introducing into the belt under normal use condition. Further, the owner's manual would be required to warn that introducing slack beyond the maximum amount specified by the manufacturer could significantly reduce the effectiveness of the belt in a crash. During the agency's dynamic testing of manual belts, the tension-relieving devices would be adjusted so as to introduce the maximum amount of slack specified in the owner's manual.

The agency proposed that the dynamic test requirement for passenger cars take effect on September 1, 1989, and only if the Secretary determines that two-thirds of the population is covered by effective safety belt use laws, thereby rescinding the automatic restraint requirement. Should such a determination be made, it is important that users of manual belts be assured that their vehicles offer the same level of occupant protection as if automatic restraints were in their vehicles. Absent a rescission of the automatic restraint requirement, application of the dynamic testing requirements to manual safety belts in passenger cars would be unnecessary since those belts would not be required in the outboard seating positions of the front seat. In the case of light trucks, light MPV's and small van-like buses, the agency proposed that the dynamic test requirement take effect on September 1, 1989. The proposed effective date for light trucks, light MPV's and van-like buses was not conditional, because those vehicles are not covered by the automatic restraint requirement and will likely continue to have manual safety belts.

Adoption of the requirement

As discussed in detail below, the agency has decided to adopt a dynamic test requirement for safety belts used in passenger cars. The agency is still analyzing the issues raised in the comments about dynamic testing for safety belt systems in other vehicles and will announce its decision about safety belt systems in light trucks, MPV's and buses at a later date.

Most of the commenters favored adopting a dynamic test requirement for manual belts at least with respect to passenger cars, although many of those commenters raised questions about the lead-time needed to comply with the requirement. Those opposing the requirement argued that the field experience has shown that current manual belts provide substantial protection and thus a dynamic test requirement is not necessary. In addition, they argued that dynamic testing would substantially increase a manufacturer's testing costs, and its testing workload. One commenter said that because of the unique nature of the testing, it could not necessarily be combined with other compliance testing done by a manufacturer. The same commenter argued that vehicle downsizing, cited by the agency as one reason for dynamically testing belts, does not create safety problems since the interior space of passenger cars has remained essentially the same as it was prior to downsizing. The commenter also argued there is no field evidence that the use of tension-relieving devices in safety belts, the other reason cited by the agency in support of the need to test dynamically manual safety belts, is compromising the performance of safety belts.

The agency strongly believes that current manual belts provide very substantial protection in a crash. The Secretary's 1984 automatic protection decision concluded that current manual safety belts are at least as effective, and in some cases, more effective than current automatic belt designs. That conclusion was based on current manual safety belts, which are not certified to dynamic tests. However, as discussed in the April 1985 notice, the agency is concerned that as an increasing number of vehicles are reduced in size for fuel economy purposes and as more tension-relieving devices are used on manual belts, the potential for occupant injury increases. The agency agrees that downsizing efforts by manufacturers have attempted to preserve the interior space of passenger cars, while reducing their exterior dimensions. Preserving the interior dimensions of the passenger compartment means that occupants will not be placed closer to instrument panels and other vehicle structures which they could strike in a crash. However, the reduction in exterior dimensions can result in a lessening of the protective crush distance available in a car. Thus the agency believes it is important to ensure that safety belts in downsized vehicles will perform adequately. In the case of tension-relieving devices, agency tests of lap/shoulder belt restrained test dummies have shown that as more slack is introduced into a shoulder belt, the injuries measured on the test dummies increased. Thus, as discussed in detail later in this notice, the agency believes it is important to ensure that safety belts with tension-relievers provide adequate protection when they are used in the manner recommended by vehicle manufacturers. This is of particular concern to the agency since the vast majority of new cars (nearly all domestically-produced cars) now are equipped with such devices. For those reasons, the agency is adopting the dynamic test requirement.

The adoption of this requirement will ensure that each and every passenger car, as compared to the vehicle population in general, offers a consistent, minumum level of protection to front seat occupants. By requiring dynamic testing, the standard will assure that the vehicle's structure, safety belts, steering column, etc., perform as a unit to protect occupants, as it is only in such a test that the synergistic and combination effects of these vehicle component can be measured. As discussed in detail in the Final Regulatory Evaluation (FRE), vehicle ' safety improvements will result from dynamic testing; and, as discussed later in this notice, such improvements can often be made quickly and at low cost.

The agency recognizes that manufacturers may have to conduct more testing than they currently do. However, the dynamic testing of manual belts in passenger cars, as with testing of automatic restraints, can be combined with other compliance tests to reduce the overall number of tests. The agency notes that in its NCAP tests, it has been able to combine the dynamic testing of belts with measuring the vehicle's compliance with other standards. The agency has followed the same practice in its compliance tests. For example, the agency has done compliance testing for Standard Nos. 208, 212, 219, and 301 in one test. The agency would, of course, recognize a manufacturer's use of combined tests as a valid testing procedure to certify compliance with these standards.

Effective Date

Two commenters argued that the requirement should become effective as soon as practical. As discussed in the April 1985 notice, the agency proposed an effective date of September 1, 1989, because it did not want to divert industry resources away from designing automatic restraints for passenger cars. The agency continues to believe it would be inappropriate to divert those resources for the purposes of requiring improvements on manual belt systems that might not be permitted in passenger cars.

Other commenters asked for a delay in the effective date – one asked for a delay until September 1, 1991, while another asked that the effective date be set 2-3 years after the determination of whether a sufficient number of States have passed effective mandatory safety belt use laws. NHTSA does not agree there is a need to delay the effective date beyond September 1, 1989 for passenger cars. Commenters argued that the time span between any decision on rescission of the automatic restraint requirements (as late as April 1, 1989) and the effective date of the dynamic testing of manual belts (September 1, 1985) is too short to certify manual belts.

The agency believes there is sufficient leadtime for passenger cars. Most of the vehicle components in passenger cars necessary for injury reduction management are the same for automatic restraint vehicles and dynamically tested manual belt vehicles. Additionally, as indicated and discussed in the April notice, approximately 40 percent of the passenger cars tested in the agency's 35 mph (NCAP) program meet the injury criteria specified in Standard No. 208, even though a 35 mph crash involves 36 percent more energy than the 30 mph crash test required by Standard No. 208. In addition, the FRE shows that with relatively minor vehicle and/or restraint system changes some safety belt systems can be dramatically improved. This is further evidence that development of dynamically tested manual belts for passenger cars in 30 mph tests should not be a major engineering program. Thus, a delay in the effective date for passenger cars is not needed.

Webbing tension-relieving devices

With one exception, those manufacturers who commented on the proposal concerning tension-relieving devices supported testing safety belts adjusted so that they have the amount of slack recommended by the manufacturer in the vehicle owner's manual. However, one manufacturer and two other commenters objected to the provision related to dynamic

testing with the tension-relieving device adjusted to the manufacturer's maximum recommended slack position. The manufacturer objected to a dynamic test that would require any slack at all to be introduced into the belt system, on the grounds that uncontrolled variability would be introduced into the dynamic test procedure, which would then lack objectivity. The manufacturer asserted that it might have to eliminate all tension-relieving devices for its safety belts.

The agency's proposed test procedure was intended to accommodate tension-relieving devices since they can increase the comfort of belts. At the same time, the proposal would limit the potential reduction in effectiveness for safety belt systems with excessive slack. The agency does not agree that this test procedure need result in the elimination of tension-relieving devices from the marketplace. As mentioned earlier, other manufacturers supported the proposal and did not indicate they would have to remove tension-relieving devices from their belt systems. The commenter opposing the requirement did not show that injury levels cannot be controlled within the specified injury criteria by testing with the recommended amount of slack, as determined by the manufacturer. The recommended slack could be very small or at any level selected by the manufacturer as appropriate to relieve belt pressure and still ensure that the injury reduction criteria of Standard No. 208 would be met. As a practical matter, most tension-relievers automatically introduce some slack into the belt for all occupants. Testing without such slack would be unrealistic.

The two other commenters objected to the proposal that manual belt systems using tension-relieving devices meet the injury criteria with only the specified amount of slack recommended in the owner's manual. They stated that most owners would not read the instructions in the owner's manual regarding the proper use of the tension-relieving device. They said an occupant could have a false sense of adequate restraint when wearing a belt system adjusted beyond the recommended limit.

The agency's views on allowing the use of tension relievers in safety belts were detailed in the April 1985 notice. The agency specifically noted the effectiveness of a safety belt system could be compromised if excessive slack were introduced into the belt. However, the agency recognizes that a belt system must be used to be effective at all. Allowing manufacturers to install tension-relieving devices makes it possible for an occupant to introduce a small amount of slack to relieve shoulder belt pressure or to divert the belt away from the neck. As a result, safety belt use is promoted. This factor should outweigh any loss in effectiveness due to the introduction of a recommended amount of slack in normal use. This is particularly likely in light of the requirement that the belt system, so adjusted, must meet the injury criteria of Standard No. 208 under 30 mph test conditions. Further, the inadvertent introduction of slack into a belt system, which is beyond that for normal use, is unlikely in most current systems. In addition, even if too much slack is introduced, the occupant should notice that excessive slack is present and a correction is needed, regardless of whether he or she has read the vehicle's owner's manual.

Exemption from Standard Nos. 203 and 204

One commenter suggested that vehicles equipped with dynamically tested manual belts be exempt from Standard Nos. 203, *Impact Protection for the Driver from the Steering Control Systems*, and 204, *Steering Column Rearward Displacement*. The agency does not believe such an exemption would be appropriate because both those standards have been shown to provide substantial protection to belted drivers.

Latching procedure in Standard No. 208

One commenter asked that Standard No. 208 be modified to include a test procedure for latching and adjusting a manual safety belt prior to the belt being dynamically tested. NHTSA agrees that Standard No. 208 should include such a procedure. The final rule incorporates the instructions contained in the NCAP test procedures for adjusting manual belts, as modified to reflect the introduction of the amount of slack recommended by the vehicle manufacturer.

Revisions to Standard No. 209

The notice proposed to exempt dynamically tested belts from the static laboratory strength tests for safety belt assemblies set forth in S4.4 of Standard No. 209. One commenter asked that such belts be exempted from the remaining requirements of Standard No. 209 as well.

NHTSA agrees that an additional exemption from some performance requirements of Standard No. 209 is appropriate. Currently, the webbing of automatic belts is exempt from the elongation and other belt webbing and attachment hardware requirements of Standard No. 209, since those belts have to meet the injury protection criteria of Standard No. 208 during a crash. For dynamically-tested manual belts,

NHTSA believes that an exemption from the webbing width, strength and elongation requirements (sections 4.2(a)-(c)) is also appropriate, since these belts will also have to meet the injury protection requirements of Standard No. 208. The agency has made the necessary changes in the rule to adopt that exemption.

The agency does not believe that manual belts should be exempt from the other requirements in Standard No. 209. For example, the requirements on buckle release force should continue to apply, since manual safety belts, unlike automatic belts, must be buckled every time they are used. As with retractors in automatic belts, retractors in dynamically tested manual belts will still have to meet Standard No. 209's performance requirements.

Revisions to Standard No. 210

The notice proposed that dynamically tested manual belts would not have to meet the location requirements set forth in Standard No. 210, *Seat Belt Assembly Anchorages*. One commenter suggested that dynamically tested belts be completely exempt from Standard No. 210; it also recommended that Standard No. 210 be harmonized with Economic Commission for Europe (ECE) Regulation No. 14. Two other commenters suggested using the "out-of-vehicle" dynamic test procedure for manual belts contained in ECE Regulation No. 16, instead of the proposed barrier crash test in Standard No. 208.

The agency does not believe that the "out-of-vehicle" laboratory bench test of ECE Regulation No. 16 should be allowed as a substitute for a dynamic vehicle crash test. The protection provided by safety belts depends on the performance of the safety belts themselves, in conjunction with the structural characteristics and interior design of the vehicle. The best way to measure the performance of the safety belt/vehicle combination is through a vehicle crash test.

The agency has already announced its intention to propose revisions to Standard No. 210 to harmonize it with ECE Regulation No. 14; therefore the commenters' suggestions concerning harmonization and exclusion of dynamically tested safety belts from the other requirements of Standard No. 210 will be considered during that rulemaking. At the present time, the agency is adopting only the proposed exclusion of anchorages for dynamically tested safety belts from the location requirements, which was not opposed by any commenter.

Belt Labelling

One commenter objected to the proposal that dynamically tested belts have a label indicating that they may be installed only at the front outboard seating positions of certain vehicles. The commenter said that it is unlikely that anyone would attempt to install a Type 2 lap shoulder belt in any vehicle other than the model for which it was designed. The agency does not agree. NHTSA believes that care must be taken to distinguish dynamically tested belt systems from other systems, since misapplication of a belt in a vehicle designed for use with a specific dynamically tested belt could pose a risk of injury. If there is a label on the belt itself, a person making the installation will be aware that the belt should be installed only in certain vehicles.

Use of the Head Injury Criterion

The April 1985 notice set forth two proposed alternative methods of using the head injury criterion (HIC) in situations when there is no contact between the test dummy's head and the vehicle's interior during a crash. The first proposed alternative was to retain the current HIC calculation for contact situations. However, in non-contact situations, the agency proposed that a HIC would not be calculated, but instead new neck injury criteria would be calculated. The agency explained that a crucial element necessary for deciding whether to use the HIC calculation or the neck criteria was an objective technique for determining the occurrence and duration of head contact in the crash test. As discussed in detail in the April 1985 notice, there are several methods available for establishing the duration of head contact, but there are questions about their levels of consistency and accuracy.

The second alternative proposed by the agency would have calculated a HIC in both contact and non-contact situations, but it would limit the calculation to a time interval of 36 milliseconds. Along with the requirement that a HIC not exceed 1000, this would limit average head acceleration to 60g's or less.

Almost all of the commenters opposed the use of the first proposed alternative. The commenters uniformly noted that there is no current technique that can accurately identify whether head contact has or has not occurred during a crash test in all situations. However, one commenter urged the agency to adopt the proposed neck criteria, regardless of whether the HIC calculation is modified. There was a sharp division among the commenters on the second proposed alternative. Manufacturers commenting on

the issue uniformly supported the use of the second alternative; although many manufacturers argued that the HIC calculation should be limited to a time interval of approximately 15 to 17 milliseconds (ms), which would limit average head accelerations to 80-85 g's. Another manufacturer, who supported the second alternative, urged the agency to measure HIC only during the time interval that the acceleration level in the head exceeds 60 g's. It said that this method would more effectively differentiate results received in contacts with hard surfaces and results obtained from systems, such as airbags, which provide good distribution of the loads experienced during a crash. Other commenters argued that the current HIC calculation should be retained; they said that the proposed alternatives would lower HIC calculations without ensuring that motorists were still receiving adequate head protection.

NHTSA is in the process of reexamining the potential effects of the two alternatives proposed by the agency and of the two additional alternatives suggested by the commenters. Once that review has been completed, the agency will issue a separate notice announcing its decision.

NCAP Test Procedures

The April 1985 notice proposed adopting the test procedures on test dummy positioning and vehicle loading used in the agency's NCAP testing. The commenters generally supported the adoption of the test procedures, although several commenters suggested changes in some of the proposals. In addition, several commenters argued that the new procedures may improve test consistency, but the changes do not affect what they claim is variability in crash test results. As discussed in the April 1985 notice, the agency believes that the test used in Standard No. 208 does produce repeatable results. The proposed changes in the test procedures were meant to correct isolated problems that occurred in some NCAP tests. The following discussion addresses the issues raised by the commenters about the specific test procedure changes.

Vehicle test attitude

The NPRM proposed that when a vehicle is tested, its attitude should be between its "as delivered" condition and its "loaded" condition. (The "as delivered" condition is based on the vehicle attitude measured when it is received at the test site, with 100 percent of all its fluid capacities and with all its tires inflated to the manufacturer's specifications. For passenger cars, the "loaded" condition is based on the vehicle's attitude with a test dummy in each front outboard designated seating position, plus carrying the cargo load specified by the manufacturer).

One commenter said that the weight distribution, and therefore the attitude, of the vehicle is governed more by the Gross Axle Weight Rating (defined in 49 CFR Part 571.3) than the loading conditions identified by the agency. The commenter recommended that the proposal not be adopted. Another commenter said that the agency should adopt more specific procedures for the positioning of the dummy and the cargo weight. For example, that commenter recommended that the "cargo weight shall be placed in such manner that its center of gravity will be coincident with the longitudinal center of the trunk, measured on the vehicle's longitudinal centerline." The commenter said that unless a more specific procedure is adopted, a vehicle's attitude in the fully loaded condition would not be constant.

The agency believes that a vehicle attitude specification should be adopted. The purpose of the requirement is to ensure that a vehicle's attitude during a crash test is not significantly different than the fully loaded attitude of the vehicle as designed by the manufacturer. Random placement of any necessary ballast could have an effect on the test attitude of the vehicle. If these variables are not controlled, then the vehicle's test attitude could be affected and potential test variability increased.

NHTSA does not agree that the use of the Gross Axle Weight Rating (GAWR) is sufficient to determine the attitude of a vehicle. The use of GAWR only defines the maximum load-carrying capacity of each axle rather than in effect specifying a minimum and maximum loading as proposed by the agency. In addition, use of the GAWR may, under certain conditions, make it necessary to place additional cargo in the passenger compartment in order to achieve the GAWR loading. This condition is not desirable for crash testing, since the passenger compartment should be used for dummy placement and instrumentation and not ballast cargo. Thus the commenter's recommendation is not accepted.

The other commenter's recommendations regarding more specific test dummy placement procedures for the outboard seating positions were already accommodated in the NPRM by the proposed new S10.1.1, *Driver position placement*, and S10.1.2, *Passenger position placement*. Since those proposals adequately describe dummy placement in these positions, they are adopted.

NHTSA has evaluated the commenter's other sug-

gestion for placing cargo weight with its center of gravity coincident with the longitudinal center of the trunk. The agency does not believe that it is necessary to determine the center of gravity of the cargo mass, which would add unnecessary complexity to the test procedure, but does agree that the cargo load should be placed so that it is over the longitudinal center of the trunk. The test procedures have been amended accordingly.

Open window

One commenter raised a question about the requirement in S8.1.5 of Standard No. 208 that the vehicle's windows are to be closed during the crash test. It said adjustment of the dummy arm and the automatic safety belt can be performed only after an automatic belt is fully in place, which occurs only after the door is closed. Therefore, the window needs to be open to allow proper arm and belt placement after the door is closed.

NHTSA agrees that the need to adjust the slack in automatic and dynamically-tested manual belts prior to the crash test may require that the window remain open. The agency has modified the test procedure to allow manufacturers the option of having the window open during the crash test.

Seat back position

One commenter recommended that proposed S8.1.3, *Adjustable seat back placement*, be modified. The notice proposed that adjustable seat backs should be set in their design riding position as measured by such things as specific latch or seat track detent positions. The commenter suggested two options. The first option would be to allow vehicle manufacturers to specify any means they want to determine the seat back angle and the resulting dummy torso angle. As its second option, the commenter recommended that if the agency decides to adopt the proposal, it should determine the "torso angle with a H-point machine according to SAE J826." The commenter said that depending on how the torso angle is established, different dummy torso angles could result in substantial adjustment deviations that can affect seat back placement.

The purpose of the requirement is to position the seat at the design riding position used by the manufacturer. The agency agrees with the commenter that manufacturers should have the flexibility to use any method they want to specify the seat back angle. Thus, the agency has made the necessary changes to the test procedure.

Dummy placement

One commenter made several general comments about dummy placement. It agreed that positioning is very important and can have an influence on the outcome of crash tests. It argued that both the old and the proposed procedures are complicated and impractical to use. The commenter claims this situation will become more complicated if the Hybrid III is permitted, since the positioning must be carried out within a narrow temperature range (3°F) for the test dummy to remain in calibration.

The commenter also believes that the positioning of the dummy should relate to vehicle type. It said that the posture and seating position of a vehicle occupant will not be the same in a van as in a sports car. For example, it said it has tried the proposed positioning procedures and found that they can result in an "unnatural" position for the dummy in a sports vehicle. The commenter argued that this "unnatural" position would then lead to a knee bolster design which would perform well in a crash test, but would likely not provide the same protection to a real occupant because of difference in positioning. The commenter recommended that the old positioning procedure be retained and the new procedure be provided as an option for those manufacturers whose vehicles cannot be adequately tested otherwise.

Because consistency in positioning the dummy is required prior to test, NHTSA believes that a single set of procedures should apply. As discussed in the April 1985 notice, the agency proposed the new procedures because of positioning problems identified in the NCAP testing. Allowing the use of the old positioning procedures could lead to sources of variability, thus negating a major objective of the procedures. The commenter's suggestion is therefore not adopted. The agency also notes that during its NCAP testing, which has involved tests of a wide variety of cars (including sports cars), trucks and MPV's, NHTSA has not experienced the "unnatural" seating position problem cited by the commenter.

Knee pivot bolt head clearance

Two commenters said that the proposal did not specify the correct distance between the dummy's knees, as measured by the clearance between the knee pivot bolt heads. The commenters are correct that the distance should be 11¾ inches rather than the proposed value of 14½ inches. The agency corrected the number in the final rule.

Foot rest

One commenter believes that a driver of cars equipped with foot rests typically will place his or her left foot on the foot rest during most driving and therefore this position should be used to simulate normal usage. The commenter said that using the foot rest will minimize variations in the positioning of the left leg, thus improving the repeatability of the test. In a discussion with the commenter, the agency has learned that the type of foot rest the commenter is referring to is a pedal-like structure where the driver can place his or her foot.

For vehicles without foot rests, the commenter recommended the agency use the same provisions for positioning the left leg of the driver as are used for the right leg of the passenger. It noted that positioning the driver's left leg, as with the passenger's right leg, can be hampered by wheelwell housing that projects into the passenger compartment and thus similar procedures for each of those legs should be used.

NHTSA agrees that in vehicles with foot rests, the test dummy's left food should be positioned on the foot rest as long as placing the foot there will not elevate the test dummy's left leg. As discussed below, the agency is concerned that foot rests, such as pads on the wheelwell, that elevate the test dummy's leg can contribute to test variability. The agency also agrees that the positioning procedures for the driver's left leg and the passenger's right leg should be similar in situations where the wheelwell housing projects into the passenger compartment and has made the necessary changes to the test procedure.

Wheelwell

One commenter believes that the wheelwell should be used to rest the dummy's foot. It said that positioning the test dummy's foot there is particularly appropriate if the wheelwell has a design feature, such as a rubber pad, installed by the manufacturer for this purpose.

NHTSA disagrees that the dummy's foot should be rested on the wheelwell housing. The agency is concerned that elevating the test dummy's leg could lead to test variability by, among other things, making the test dummy unstable during a crash test. Although the wheelwell problem is similar to the foot rest problem, placement of the test dummy's foot on a separate, pedal-like foot rest can be accomplished while retaining the heel of the test dummy in a stable position on the floor. That is not the case with pads located on the wheelwell.

Another commenter also said that the proposed procedure for positioning the test dummy's legs in vehicles where the wheelwell projected into the passenger compartment was unclear as to how the centerlines of the upper and lower legs should be adjusted so that both remain in a vertical longitudinal plane. In particular, it was concerned that in a vehicle with a large wheelhousing, it may not be possible to keep the left foot of the driver test dummy in the vertical longitudinal plane after the right foot has been positioned. It believes that the procedure should specify which foot position should be given priority; it recommended that the position of the right leg be required to remain in the plane, while bringing the left leg as close to the vertical longitudinal plane as possible. The agency agrees that maintaining the inboard leg of the test dummy in the vertical plane is more easily accomplished since it will not be blocked by the wheelwell. The agency has modified the test procedure to specify that when it is not possible to maintain both legs in the vertical longitudinal plane, that the inboard leg must be kept as close as possible to the vertical longitudinal plane and the outboard leg should be placed as close as possible to the vertical plane.

Lower leg angle

One commenter argued that proposed sections on lower leg positioning (S10.1.2.1 (b) and S10.1.2.2 (b)) will not result in a constant positioning of the test dummy's heels on the floor pan, thus causing differences in the lower leg angles. It stated that the lower leg angles will affect the femur load generated at the moment the foot hits the toe pan during a collision. The commenter therefore proposed that the test procedure be revised to include placing a 20 pound load on the test dummy's knee during the foot positioning procedure. The commenter did not, however, explain the basis for choosing a force of 20 pounds.

NHTSA believes that use of the additional weight loading and settling procedure proposed by the commenter will add an unnecessary level of complexity to the test procedure without adding any corresponding benefit. The positioning of the test dummy's heel has not been a problem in the agency's NCAP tests. Accordingly, the agency is not adopting the commenter's recommendation.

Shoulder adjustment

One commenter asked the agency to specify that the shoulders of the test dummy be placed at their lowest adjustment position. While the shoulders are slightly adjustable, the agency believes that specifying an adjustment position is unnecessary. The agency's test experience has shown that the up and down movement of the shoulders is physically limited by the test dummy's rubber "skin" around the openings where the arms are connected to the test dummy's upper torso.

Dummy lifting procedure

One commenter was concerned about the dummy lifting proposed in (Section S10.4.1, Dummy Vertical Upward Displacement). It said that if the dummy lifting method is not standardized, test results could be affected by allowing variability in the position of the dummy's H point (the H point essentially represents the hip joint) through use of different lifting methods. It recommended use of a different chest lifting method to avoid variability in the subsequent positioning of the test dummy H-point.

The agency is not aware of any test data indicating that the use of different lifting methods is a significant source of variability. As long as a manufacturer follows the procedures set forth in S10.4.1 in positioning the test dummy, it can use any lifting procedure it wants.

Dummy settling load

One commenter was concerned about the proposed requirements for dummy settling (S10.4.2, *Lower torso force application*, and S10.4.5, *Upper torso force application*). The commenter believes that the proposals are inadequate because they do not prescribe the area over which to apply the load used to settle the test dummy in the seat. The commenter said that if the proposed 50 pound settling force is applied to an extremely small contact area, then the dummy may be deformed. It recommended that the load be applied to a specified area of 9 square inches on the dummy. In addition, it recommended that the agency specify the duration of the 50 lb. force application during the adjustment of the upper torso; it suggested a period of load application ranging from 5 to 10 seconds.

NHTSA and others have successfully used the proposed settling test procedures in their own tests without having any variability problems. Unless abnormally small contact areas are employed, or extremely short durations are used, standard laboratory practices should not result in any such problems. The agency believes that further specifying the area and timing of the force application is not necessary.

Dummy head adjustment

One commenter pointed out that it is impossible to adjust the head according to S10.6, Head Adjustment, because the Part 572 test dummy does not have a head adjustment mechanism. The agency agrees and has deleted the provision.

Additional dummy settling and shoulder belt positioning procedures

One commenter suggested a substantial revised dummy settling procedure and new procedures for positioning of the shoulder belt. NHTSA believes that its proposed procedures sufficiently address the settling and belt position issues. In addition, the commenter did not provide any data to show that variability would be further reduced by its suggested procedures. A substantial amount of testing would be needed to verify if the commenter's suggested test procedures do, in fact, provide any further decrease in variability than that obtained by the agency's test procedures. For those reasons, the agency is not adopting the commenter's suggestions for new procedures.

Due Care

In the April 1985 notice, the agency proposed amending the standard to state that the due care provision of section 108(b)(2) of the National Traffic and Motor Vehicle Safety Act (15 U.S.C. 1397(b)(2)) applies to compliance with the standard. Thus, a vehicle would not be deemed in noncompliance if its manufacturer establishes that it did not have reason to know in the exercise of due care that such vehicle is not in conformity with the standard.

Commenters raised a number of questions about the proposal, with some saying that the agency needed to clarify what constitutes "due care," others recommending that the agency reconsider the use of "design to conform" language instead of due care and another opposing the use of any due care provision.

A number of commenters, while supporting the use of a due care provision, said that the proposal provides no assurance that a manufacturer's good faith effort will be considered due care. They said that the agency should identify the level of testing and analysis necessary to constitute due care. Another commenter emphasized that in defining due care, the agency must ensure that a manufacturer uses recognized statistical procedures in determining that its products comply with the requirements of the standard.

Another group of commenters requested the agency to reconsider its decision not to use "design to conform" language in the standard; they said that the agency's concerns about the subjectivity of a "design to conform" language are not greater and could well be less than that resulting from use of due care language.

One commenter opposed the use of any due care language in the standard. It argued that the National Traffic and Motor Vehicle Safety Act requires the agency to set objective performance requirements in its standards. When a manufacturer determines that it has not met those performance requirements, then the manufacturer is under an obligation to notify owners and remedy the noncomplying vehicles. It argued that the proposed due care provision, in effect, provides manufacturers with an exemption from the Vehicle Safety Act recall provisions.

As discussed in the July 1984 final rule and the April 1985 notice, the agency believes that the test procedure of Standard No. 208 produces repeatable results in vehicle crash tests. The agency does, however, recognize that the Standard No. 208 test is more complicated than NHTSA's other crash test standards since a number of different injury measurements must be made on the two test dummies used in the testing. Because of this complexity, the agency believes that manufacturers need assurance from the agency that, if they have made a good faith effort in designing their vehicles and have instituted adequate quality control measures, they will not face the recall of their vehicles because of an isolated apparent failure to meet one of the injury criteria. The adoption of a due care provision provides that assurance. For the reasons discussed in the July 1984 final rules, the agency still believes use of a due care provision is a better approach to this issue than use of a design to conform provision.

As the agency has emphasized in its prior interpretation letters, a determination of what constitutes due care can only be made on a case-by-case basis. Whether a manufacturer's action will constitute due care will depend, in part, upon the availability of test equipment, the limitations of available technology, and above all, the diligence evidenced by the manufacturer.

Adoption of a due care defense is in line with the agency's long-standing and well-known enforcement policy on test differences. Under this long standing practice if the agency's testing shows noncompliance and a manufacturer's tests, valid on their face, show complying results, the agency will conduct an inquiry into the reason for the differing results. If the agency concludes that the difference in results can be explained to the agency's satisfaction, that the agency's results do not indicate an unreasonable risk to safety, and that the manufacturer's tests were reasonably conducted and were in conformity with standard, then the agency does not use its own tests as a basis for a finding of noncompliance. Although this interpretation has long been a matter of public record, Congress, in subsequent amendments of the Vehicle Safety Act, has not acted to alter that interpretation. The Supreme Court has said that under those circumstances, it can be presumed that the agency's interpretation has correctly followed the intent of the statute. (*See United States* v. *Rutherford,* 442 U.S. 544, 544 n. 10 (1979))

Phase-In

Attribution rules

With respect to cars manufacturered by two or more companies, and cars manufactured by one company and imported by another, the April 1985 notice proposed to clarify who would be considered the manufacturer for purposes of calculating the average annual production of passenger cars for each manufacturer and the amount of passenger cars manufacturered by each manufacturer that must comply with the automatic restraint phase-in requirements. In order to provide maximum flexibility to manufacturers, while assuring that the percentage phase-in goals are met, the notice proposed to permit manufacturers to determine, by contract, which of them will count, as its own, passenger cars manufactured by two or more companies or cars manufactured by one company and imported by another.

The notice also proposed two rules of attribution in the absence of such a contract. First, a passenger car which is imported for purposes of resale would be attributed to the importer. The agency intended that this proposed attribution rule would apply to both direct importers as well as importers authorized by the vehicle's original manufacturer. (In this context, direct importation refers to the importation of cars which are originally manufactured for sale outside the U.S. and which are then imported without the manufacturer's authorization into the U.S. by an importer for purposes of resale. The Vehicle Safety Act requires that such vehicles be brought into conformity with Federal motor vehicle safety standards.) Under the second proposed attribution rule, a passenger car manufactured in the United States by more than one manufacturer, one of which also

markets the vehicle, would be attributed to the manufacturer which markets the vehicle.

These two proposed rules would generally attribute a vehicle to the manufacturer which is most responsible for the existence of the vehicle in the United States, i.e., by importing the vehicle or by manufacturing the vehicle for its own account as part of a joint venture, and marketing the vehicle. (Importers generally market the vehicles they import.) All commenters on these proposals supported giving manufacturers the flexibility to determine contractually which manufacturer would count the passenger car as its own. The commenters also supported the proposed attribution rules. Therefore, the agency is adopting the provisions as proposed.

Credit for early phase-in

The April 1985 notice proposed that manufacturers that exceeded the minimum percentage phase-in requirements in the first or second years could count those extra vehicles toward meeting the requirements in the second or third years. In addition, manufacturers could also count any automatic restraint vehicles produced during the one year preceding the first year of the phase-in. Since all the commenters addressing these proposals supported them, the agency is adopting them as proposed. The agency believes that providing credit for early introduction will encourage introduction of larger numbers of automatic restraints and provide increased flexibility for manufacturers. In addition, it will assure an orderly build-up of production capability for automatic restraint equipped cars as contemplated by the July 1984 final rule.

One commenter asked the agency to establish a new credit for vehicles equipped with non-belt automatic restraints at the driver's position and a dynamically-tested manual belt at the passenger position. The commenter requested that such a vehicle receive a 1.0 credit. The commenter also asked the agency to allow vehicles equipped with driver-only automatic restraint systems to be manufactured after September 1, 1989, the effective date for automatic restraints for the driver and front right passenger seating positions in all passenger cars. In its August 30, 1985 notice (50 FR 35233) responding to petitions for reconsideration of the July 1984 final rule on Standard No. 208, the agency has already adopted a part of the commenter's suggestion by establishing a 1.0 vehicle credit for vehicles equipped with a non-belt automatic restraint at the driver's position and a manual lap/shoulder belt at the passenger's position. For reasons detailed in the July 1984 final rule, the agency believes that the automatic restraint requirement should apply to both front outboard seating positions beginning on September 1, 1989, and is therefore not adopting the commenter's second suggestion.

Phase-In Reporting Requirements

The April 1985 notice proposed to establish a new Part 585, *Automatic Restraint Phase-in Reporting Requirements*. The agency proposed requiring manufacturers to submit three reports to NHTSA, one for each of the three automatic restraint phase-in periods. Each report, covering production during a 12-month period beginning September 1 and ending August 31, would be required to be submitted within 60 days after the end of such period. Information required by each report would include a statement regarding the extent to which the manufacturer had complied with the applicable percentage phase-in requirement of Standard No. 208 for the period covered by the report; the number of passenger cars manufactured for sale in the United States for each of the three previous 12-month production periods; the actual number of passenger cars manufactured during the reporting production (or during a previous production period and counted toward compliance in the reporting production period) period with automatic safety belts, air bags and other specified forms of automatic restraint technology, respectively; and brief information about any express written contracts which concern passenger cars produced by more than one manufacturer and affect the report.

One commenter questioned the need for a reporting requirement, saying that the requirement was unnecessary since manufacturers must self-certify that their vehicles meet Standard No. 208. The agency believes that a reporting requirement is needed for the limited period of the phase-in of automatic restraints so that the agency can carry out its statutory duty to monitor compliance with the Federal motor vehicle safety standards. During the phase-in, only a certain percentage of vehicles are required to have automatic restraints. It would be virtually impossible for the agency to determine if the applicable percentage of passenger cars has been equipped with automatic restraints unless manufacturers provide certain production information to the agency. NHTSA is therefore adopting the reporting requirement.

The same commenter said that requiring the report to be due 60 days after the end of the production year can be a problem for importers. The commenter said that production records may accompany the vehicle, which may not actually reach the United States until 30 or 45 days after the production year ends. The commenter asked the agency to provide an appeal process to seek an extension of the period to file the report. The agency believes that the example presented by the commenter represents a worst case situation and complying with the 60 day requirement should not be a problem for manufacturers, including importers. However, to eliminate any problems in worst case situations, the agency is amending the regulation to provide that manufacturers seeking an extension of the deadline to file a report must file a request for an extension at least 15 days before the report is due.

Calculation of average annual production

The agency also proposed an alternative to the requirement that the number of cars that must be equipped with automatic restraints must be based on a percentage of each manufacturer's average annual production for the past three model years. The proposed alternative would permit manufacturers to equip the required percentage of its actual production of passenger cars with automatic restraints during each affected year. Since all commenters addressing this proposal supported it, the agency is adopting it as an alternative means of compliance, at the manufacturer's option. In the case of a new manufacturer, the manufacturer would have to calculate the amount of passenger cars required to have automatic restraints based on its production of passenger cars during each of the affected years. Since the agency has decided to adopt the alternative basis for determining the production quota, it has made the necessary conforming changes in the reporting requirements adopted in this notice.

One commenter also requested the agency to clarify whether a manufacturer does have to include its production volume of convertibles when it is calculating the percentage of vehicles that must meet the phase-in requirement. The automatic restraint requirement applies to all passenger cars. Thus, a manufacturer's production figures for passenger car convertibles must be counted when the manufacturer is calculating its phase-in requirements.

Retention of VINs

In order to keep administrative burdens to a minimum, the agency proposed that the required report need not use the VIN to identify the particular type of automatic restraint installed in each passenger car produced during the phase-in period. Since that information could be necessary for purposes of enforcement, however, the agency proposed to require that manufacturers maintain records until December 31, 1991, of the VIN and type of automatic restraint for each passenger car which is produced during the phase-in period and is reported as having automatic restraints. Although direct import cars are not required to have a US-format VIN number, those cars would still have a European-format VIN number and thus direct importers would be required to retain that VIN information. (The agency is considering a petition from Volkswagen requesting that direct import cars be required to have US-format VINs.)

The reason for retaining the information until 1991 is to ensure that such information would then be available until the completion of any agency enforcement action begun after the final phase-in report is filed in 1990. The agency believes this requirement meets the needs of the agency, with minimal impacts on manufacturers, and therefore is adopting it as proposed. One commenter asked whether a manufacturer is required to keep the VIN information as a separate file or whether keeping the information as a part of its general business records is sufficient. As long as the VIN information is retrievable, it may be stored in any manner that is convenient for a manufacturer.

In consideration of the foregoing, 49 CFR Part 571.208 is amended as follows:

The authority citation for Part 571 would continue to read as follows:

Authority: 15 U.S.C. 1392, 1401, 1403, 1407; delegation of authority at 49 CFR 1.50.

1. Section S4.1.3.1.2 is revised to read as follows:

S4.1.3.1.2 Subject to S4.1.3.4 and S4.1.5, the amount of passenger cars, specified in S4.1.3.1.1 complying with the requirements of S4.1.2.1 shall be not less than 10 percent of:

(a) the average annual production of passenger cars manufactured on or after September 1, 1983, and before September 1, 1986, by each manufacturer, or

(b) the manufacturer's annual production of passenger cars during the period specified in S4.1.3.1.1.

2. Section 4.1.3.2.2 is revised to read as follows:

S4.1.3.2.2 Subject to S4.1.3.4 and S4.1.5, the amount of passenger cars specified in S4.1.3.2.1 complying with the requirements of S4.1.2.1 shall be not less than 25 percent of:

(a) the average annual production of passenger cars manufactured on or after September 1, 1984,

and before September 1, 1987, by each manufacturer, or

(b) the manufacturer's annual production of passenger cars during the period specified in S4.1.3.2.1.

3. Section 4.1.3.3.2 is revised to read as follows:

S4.1.3.3.2 Subject to S4.1.3.4 and S4.1.5, the amount of passenger cars specified in S4.1.3.3.1 complying with the requirements of S4.1.2.1 shall not be less than 40 percent of:

(a) the average annual production of passenger cars manufactured on or after September 1, 1985, and before September 1, 1988, by each manufacturer or

(b) the manufacturer's annual production of passenger cars during the period specified in S4.1.3.3.1.

4. Section S4.1.3.4 is revised to read as follows:

S4.1.3.4 *Calculation of complying passenger cars.*

(a) For the purposes of calculating the numbers of cars manufactured under S4.1.3.1.2, S4.1.3.2.2, or S4.1.3.3.2 to comply with S4.1.2.1:

(1) each car whose driver's seating position complies with the requirements of S4.1.2.1(a) by means not including any type of seat belt and whose front right seating position will comply with the requirements of S4.1.2.1(a) by any means is counted as 1.5 vehicles, and

(2) each car whose driver's seating position complies with the requirements of S4.1.2.1(a) by means not including any type of seat belt and whose right front seat seating position is equipped with a manual Type 2 seat belt is counted as one vehicle.

(b) For the purposes of complying with S4.1.3.1.2, a passenger car may be counted if it:

(1) is manufactured on or after September 1, 1985, but before September 1, 1986, and

(2) complies with S4.1.2.1.

(c) For the purposes of complying with S4.1.3.2.2, a passenger car may be counted if it:

(1) is manufactured on or after September 1, 1985, but before September 1, 1987, and

(2) complies with S4.1.2.1, and

(3) is not counted toward compliance with S4.1.3.1.2

(d) For the purposes of complying with S4.1.3.3.2, a passenger car may be counted if it:

(1) is manufactured on or after September 1, 1985, but before September 1, 1988,

(2) complies with S4.1.2.1, and

(3) is not counted toward compliance with S4.1.3.1.2 or S4.1.3.2.2.

5. A new section S4.1.3.5 is added to read as follows:

S4.1.3.5 *Passenger cars produced by more than one manufacturer.*

S4.1.3.5.1 For the purposes of calculating average annual production of passenger cars for each manufacturer and the amount of passenger cars manufactured by each manufacturer under S4.1.3.1.2, S4.1.3.2.2 or S4.1.3.3.2, a passenger car produced by more than one manufacturer shall be attributed to a single manufacturer as follows, subject to S4.1.3.5.2:

(a) A passenger car which is imported shall be attributed to the importer.

(b) A passenger car manufactured in the United States by more than one manufacturer, one of which also markets the vehicle, shall be attributed to the manufacturer which markets the vehicle.

S4.1.3.5.2 A passenger car produced by more than one manufacturer shall be attributed to any one of the vehicle's manufacturers specified by an express written contract, reported to the National Highway Traffic Safety Administration under 49 CFR Part 585, between the manufacturer so specified and the manufacturer to which the vehicle would otherwise be attributed under S4.1.3.5.1.

6. A new section S4.6 is added to read as follows:

S4.6 *Dynamic testing of manual belt systems.*

S4.6.1 If the automatic restraint requirement of S4.1.4 is rescinded pursuant to S4.1.5, then each passenger car that is manufactured after September 1, 1989, and is equipped with a Type 2 manual seat belt assembly at each front outboard designated seating position pursuant to S4.1.2.3 shall meet the frontal crash protection requirements of S5.1 at those designated seating positions with a test dummy restrained by a Type 2 seat belt assembly that has been adjusted in accordance with S7.4.2.

S4.6.2 A Type 2 seat belt assembly subject to the requirements of S4.6.1 of this standard does not have to meet the requirements of S4.2(a)-(c) and S4.4 of Standard No. 209 (49 CFR 571.209) of this Part.

7. S7.4.2 is revised to read as follows:

S7.4.2 *Webbing tension relieving device.* Each vehicle with an automatic seat belt assembly or with a Type 2 manual seat belt assembly that must meet S4.6 installed in a front outboard designated seating position that has either manual or automatic devices permitting the introduction of slack in the webbing of the shoulder belt (e.g., "comfort clips" or "window-shade" devices) shall:

(a) comply with the requirements of S5.1 with the shoulder belt webbing adjusted to introduce the maximum amount of slack recommended by the manufacturer pursuant to S7.4.2.(b);

(b) have a section in the vehicle owner's manual that explains how the tension-relieving device works and specifies the maximum amount of slack (in inches) recommended by the vehicle manufacturer to be introduced into the shoulder belt under normal use conditions. The explanation shall also warn that introducing slack beyond the amount specified by the manufacturer can significantly reduce the effectiveness of the shoulder belt in a crash; and

(c) have an automatic means to cancel any shoulder belt slack introduced into the belt system by a tension-relieving device each time the safety belt is unbuckled or the adjacent vehicle door is opened, except that open-body vehicles with no doors can have a manual means to cancel any shoulder belt slack introduced into the belt system by a tension-relieving device.

8. Section 8.1.1(c) is revised to read as follows:

S8.1.1(c) *Fuel system capacity.* With the test vehicle on a level surface, pump the fuel from the vehicle's fuel tank and then operate the engine until it stops. Then, add Stoddard solvent to the test vehicle's fuel tank in an amount which is equal to not less than 92 and not more than 94 percent of the fuel tank's usable capacity stated by the vehicle's manufacturer. In addition, add the amount of Stoddard solvent needed to fill the entire fuel system from the fuel tank through the engine's induction system.

9. A new section 8.1.1(d) is added to read as follows:

S8.1.1(d) *Vehicle test attitude.* Determine the distance between a level surface and a standard reference point on the test vehicle's body, directly above each wheel opening, when the vehicle is in its "as delivered" condition. The "as delivered" condition is the vehicle as received at the test site, with 100 percent of all fluid capacities and all tires inflated to the manufacturer's specifications as listed on the vehicle's tire placard. Determine the distance between the same level surface and the same standard reference points in the vehicle's "fully loaded condition." The "fully loaded condition" is the test vehicle loaded in accordance with S8.1.1(a) or (b), as applicable. The load placed in the cargo area shall be centered over the longitudinal centerline of the vehicle. The pretest vehicle attitude shall be equal to either the as delivered or fully loaded attitude or between the as delivered attitude and the fully loaded attitude.

10. S7.4.3 is revised by removing the reference to "S10.6" and replacing it with a reference to "S10.7."

11. S7.4.4 is revised by removing the reference to "S10.5" and replacing it with a reference to "S10.6."

12. S7.4.5 is revised by removing the reference to "S8.1.11" and replacing it with a reference to "S10."

13. Section 8.1.3 is revised to read as follows:

S8.1.3 *Adjustable seat back placement.* Place adjustable seat backs in the manufacturer's nominal design riding position in the manner specified by the manufacturer. Place each adjustable head restraint in its highest adjustment position.

14. Sections 8.1.11 through 8.1.11.2.3 are removed.

15. Sections 8.1.12 and 8.1.13 are redesignated 8.1.11 and 8.1.12, respectively.

16. Section 10 is revised to read as follows:

S10 *Test dummy positioning procedures.* Position a test dummy, conforming to Subpart B of Part 572 (49 CFR Part 572), in each front outboard seating position of a vehicle as specified in S10.1 through S10.9. Each test dummy is:

(a) not restrained during an impact by any means that require occupant action if the vehicle is equipped with automatic restraints.

(b) restrained by manual Type 2 safety belts, adjusted in accordance with S10.9, if the vehicle is equipped with manual safety belts in the front outboard seating positions.

S10.1 *Vehicle equipped with front bucket seats.* Place the test dummy's torso against the seat back and its upper legs against the seat cushion to the extent permitted by placement of the test dummy's feet in accordance with the appropriate paragraph of S10. Center the test dummy on the seat cushion of the bucket seat and set its midsagittal plane so that it is vertical and parallel to the centerline of the vehicle.

S10.1.1 *Driver position placement.*

(a) Initially set the knees of the test dummy 11¾ inches apart, measured between the outer surfaces of the knee pivot bolt heads, with the left outer surface 5.9 inches from the midsagittal plane of the test dummy.

(b) Rest the right foot of the test dummy on the undepressed accelerator pedal with the rearmost point of the heel on the floor pan in the plane of the pedal. If the foot cannot be placed on the accelerator pedal, set it perpendicular to the lower leg and place it as far forward as possible in the direction of the geometric center of the pedal with the rearmost point of the heel resting on the floor pan. Except as prevented by contact with a vehicle surface, place the right leg so that the upper and lower leg centerlines fall, as close as possible, in a vertical longitudinal plane without inducing torso movement.

(c) Place the left foot on the toeboard with the rearmost point of the heel resting on the floor pan as close as possible to the point of intersection of the planes described by the toeboard and the floor pan. If the foot cannot be positioned on the toeboard, set it perpendicular to the lower leg and place it as far forward as possible with the heel resting on the floor pan. Except as prevented by contact with a vehicle surface, place the left leg so that the upper and lower leg centerlines fall, as close as possible, in a vertical plane. For vehicles with a foot rest that does not elevate the left foot above the level of the right foot, place the left foot on the foot rest so that the upper and lower leg centerlines fall in a vertical plane.

S10.1.2 *Passenger position placement.*

S10.1.2.1 *Vehicles with a flat floor pan/toeboard.*

(a) Initially set the knees 11¾ inches apart, measured between the outer surfaces of the knee pivot bolt heads.

(b) Place the right and left feet on the vehicle's toeboard with the heels resting on the floor pan as close as possible to the intersection point with the toeboard. If the feet cannot be placed flat on the toeboard, set them perpendicular to the lower leg centerlines and place them as far forward as possible with the heels resting on the floor pan.

(c) Place the right and left legs so that the upper and lower leg centerlines fall in vertical longitudinal planes.

S10.1.2.2 *Vehicles with wheelhouse projections in passenger compartment.*

(a) Initially set the knees 11¾ inches apart, measured between outer surfaces of the knee pivot bolt heads.

(b) Place the right and left feet in the well of the floor pan/toeboard and not on the wheelhouse projection. If the feet cannot be placed flat on the toeboard, set them perpendicular to the lower leg centerlines and as far forward as possible with the heels resting on the floor pan.

(c) If it is not possible to maintain vertical and longitudinal planes through the upper and lower leg centerlines for each leg, then place the left leg so that its upper and lower centerlines fall, as closely as possible, in a vertical longitudinal plane and place the right leg so that its upper and lower leg centerlines fall, as closely as possible, in a vertical plane.

S10.2 *Vehicle equipped with bench seating.* Place a test dummy with its torso against the seat back and its upper legs against the seat cushion, to the extent permitted by placement of the test dummy's feet in accordance with the appropriate paragraph of S10.1.

S10.2.1 *Driver position placement.* Place the test dummy at the left front outboard designated seating position so that its midsagittal plane is vertical and parallel to the centerline of the vehicle and so that the midsagittal plane of the test dummy passes through the center of the steering wheel rim. Place the legs, knees, and feet of the test dummy as specified in S10.1.1.

S10.2.2 *Passenger position placement.* Place the test dummy at the right front outboard designated seating position as specified in S10.1.2, except that the midsagittal plane of the test dummy shall be vertical and longitudinal, and the same distance from the vehicle's longitudinal centerline as the midsagittal plane of the test dummy at the driver's position.

S10.3 *Initial test dummy placement.* With the test dummy at its designated seating position as specified by the appropriate requirements of S10.1 or S10.2, place the upper arms against the seat back and tangent to the side of the upper torso. Place the lower arms and palms against the outside of the upper legs.

S10.4 *Test dummy settling.*

S10.4.1 *Test dummy vertical upward displacement.* Slowly lift the test dummy parallel to the seat back plane until the test dummy's buttocks no longer contact the seat cushion or until there is test dummy head contact with the vehicle's headlining.

S10.4.2 *Lower torso force application.* Using a test dummy positioning fixture, apply a rearward force of 50 pounds through the center of the rigid surface against the test dummy's lower torso in a horizontal direction. The line of force application shall be 6½ inches above the bottom surface of the test dummy's buttocks. The 50 pound force shall be maintained with the rigid fixture applying reaction forces to either the floor pan/toeboard, the 'A' post, or the vehicle's seat frame.

S10.4.3 *Test dummy vertical downward displacement.* While maintaining the contact of the horizontal rearward force positioning fixture with the test dummy's lower torso, remove as much of the 50 pound force as necessary to allow the test dummy to return downward to the seat cushion by its own weight.

S10.4.4 *Test dummy upper torso rocking.* Without totally removing the horizontal rearward force being applied to the test dummy's lower torso, apply a horizontal forward force to the test dummy's shoulders sufficient to flex the upper torso forward until its back no longer contacts the seat back. Rock the test dummy from side to side 3 or 4 times so that the test dummy's spine is at any angle from the vertical in the 14 to 16 degree range at the extremes of each rocking movement.

S10.4.5 *Upper torso force application.* With the test dummy's midsagittal plane vertical, push the upper torso against the seat back with a force of 50 pounds applied in a horizontal rearward direction along a line that is coincident with the test dummy's midsagittal plane and 18 inches above the bottom surface of the test dummy's buttocks.

S10.5 *Placement of test dummy arms and hands.* With the test dummy positioned as specified by S10.3 and without inducing torso movement, place the arms, elbows, and hands of the test dummy, as appropriate for each designated seating position in accordance with S10.3.1 or S10.3.2. Following placement of the arms, elbows and hands, remove the force applied against the lower half of the torso.

S10.5.1 *Driver's position.* Move the upper and the lower arms of the test dummy at the driver's position to their fully outstretched position in the lowest possible orientation. Push each arm rearward, permitting bending at the elbow, until the palm of each hand contacts the outer part of the rim of the steering wheel at its horizontal centerline. Place the test dummy's thumbs over the steering wheel rim and position the upper and lower arm centerlines as close as possible in a vertical plane without inducing torso movement.

S10.5.2 *Passenger position.* Move the upper and the lower arms of the test dummy at the passenger position to fully outstretched position in the lowest possible orientation. Push each arm rearward, permitting bending at the elbow, until the upper arm contacts the seat back and is tangent to the upper part of the side of the torso, the palm contacts the outside of the thigh, and the little finger is barely in contact with the seat cushion.

S10.6 *Test dummy positioning for latchplate access.* The reach envelopes specified in S7.4.4 are obtained by positioning a test dummy in the driver's seat or passenger's seat in its forwardmost adjustment position. Attach the lines for the inboard and outboard arms to the test dummy as described in Figure 3 of this standard. Extend each line backward and outboard to generate the compliance arcs of the outboard reach envelope of the test dummy's arms.

S10.7 *Test dummy positioning for belt contact force.* To determine compliance with S7.4.3 of this standard, position the test dummy in the vehicle in accordance with the appropriate requirements specified in S10.1 or S10.2 and under the conditions of S8.1.2 and S8.1.3. Pull the belt webbing three inches from the test dummy's chest and release until the webbing is within 1 inch of the test dummy's chest and measure the belt contact force.

S10.9 *Manual belt adjustment for dynamic testing.* With the test dummy at its designated seating position as specified by the appropriate requirements of S8.1.2, S8.1.3 and S10.1 through S10.5, place the Type 2 manual belt around the test dummy and fasten the latch. Remove all slack from the lap belt. Pull the upper torso webbing out of the retractor and allow it to retract; repeat this operation four times. Apply a 2 to 4 pound tension load to the lap belt. If the belt system is equipped with a tension-relieving device introduce the maximum amount of slack into the upper torso belt that is recommended by the manufacturer for normal use in the owner's manual for the vehicle. If the belt system is not equipped with a tension relieving device, allow the excess webbing in the shoulder belt to be retracted by the retractive force of the retractor.

17. S11 is removed.

18. S4.1.3.1.1, S4.1.3.2.1, S4.1.3.3.1, S4.1.4 and S4.6.1 are revised by adding a new second sentence to S4.1.3.1.1, S4.1.3.2.1, S4.1.3.3.1 and S4.1.4 and a new second sentence to S4.6.1 to read as follows:

A vehicle shall not be deemed to be in noncompliance with this standard if its manufacturer establishes that it did not have reason to know in the exercise of due care that such vehicle is not in conformity with the requirement of this standard.

19. S8.1.5 is amended to read as follows:

Movable vehicle windows and vents are, at the manufacturer's option, placed in the fully closed position.

20. S7.4 is amended to read as follows:

S7.4. *Seat belt comfort and convenience.*

(a) *Automatic seat belts.* Automatic seat belts installed in any vehicle, other than walk-in van-type vehicles, which has a gross vehicle weight rating of 10,000 pounds or less, and which is manufactured on or after September 1, 1986, shall meet the requirements of S7.4.1, S7.4.2, and S7.4.3.

(b) *Manual seat belts.*

(1) *Vehicles manufactured after September 1, 1986.* Manual seat belts installed in any vehicle, other than manual Type 2 belt systems installed in the front outboard seating positions in passenger cars or manual belts in walk-in van-type vehicles, which have a gross vehicle weight rating of 10,000 pounds or less, shall meet the requirements of S7.4.3, S7.4.4, S7.4.5, and S7.4.6.

(2) *Vehicles manufactured after September 1, 1989.*

(i) If the automatic restraint requirement of S4.1.4 is rescinded pursuant to S4.1.5, then manual seat belts installed in a passenger car shall meet the requirements of S7.1.1.3(a), S7.4.2, S7.4.3, S7.4.4, S7.4.5, and S7.4.6.

(ii) Manual seat belts installed in a bus, multipurpose passenger vehicle and truck with a gross vehicle weight rating of 10,000 pounds or less, except for walk-in van-type vehicles, shall meet the requirements of S7.4.3, S7.4.4, S7.4.5, and S7.4.6.

571.209 *Standard No. 209, Seat belt assemblies.*

1. A new S4.6 is added, to read as follows:

S4.6 *Manual belts subject to crash protection requirements of Standard No. 208.*

(a) A seat belt assembly subject to the requirements of S4.6.1 of Standard No. 208 (49 CFR Part 571.208) does not have to meet the requirements of S4.2 (a)-(c) and S4.4 of this standard.

(b) A seat belt assembly that does not comply with the requirements of S4.4 of this standard shall be permanently and legibly marked or labeled with the following language:

This seat belt assembly may only be installed at a front outboard designated seating position of a vehicle with a gross vehicle weight rating of 10,000 pounds or less.

571.210 *Standard No. 210, Seat Belt Assembly Anchorages.*

1. The second sentence of S4.3 is revised to read as follows:

Anchorages for automatic and for dynamically tested seat belt assemblies that meet the frontal crash protection requirement of S5.1 of Standard No. 208 (49 CFR Part 571.208) are exempt from the location requirements of this section.

PART 585 – AUTOMATIC RESTRAINT PHASE-IN REPORTING REQUIREMENTS

1. Chapter V, Title 49, Transportation, the Code of Federal Regulations, is amended to add the following new Part:

PART 585 – AUTOMATIC RESTRAINT PHASE-IN REPORTING REQUIREMENTS

Secs.
585.1 Scope.
585.2 Purpose.
585.3 Applicability.
585.4 Definitions.
585.5 Reporting requirements.
585.6 Records.
585.7 Petition to extend period to file report.

Authority: 15 U.S.C. 1392, 1407; delegation of authority at 49 CFR 1.50.

585.1 *Scope.*

This section establishes requirements for passenger car manufacturers to submit a report, and maintain records related to the report, concerning the number of passenger cars equipped with automatic restraints in compliance with the requirements of S4.1.3 of Standard No. 208, *Occupant Crash Protection* (49 CFR Part 571.208).

585.2 *Purpose.*

The purpose of the reporting requirements is to aid the National Highway Traffic Safety Administration in determining whether a passenger car manufacturer has complied with the requirements of Standard No. 208 of this Chapter (49 CFR 571.208) for the installation of automatic restraints in a percentage of each manufacturer's annual passenger car production.

585.3 *Applicability.*

This part applies to manufacturers of passenger cars.

585.4 *Definitions.*

All terms defined in section 102 of the National Traffic and Motor Vehicle Safety Act (15 U.S.C. 1391) are used in their statutory meaning.

"Passenger car" is used as defined in 49 CFR Part 571.3.

"Production year" means the 12-month period between September 1 of one year and August 31 of the following year, inclusive.

585.5 *Reporting requirements.*

(a) *General reporting requirements.*

Within 60 days after the end of each of the production years ending August 31, 1987, August 31, 1988, and August 31, 1989, each manufacturer shall submit a report to the National Highway Traffic Safety Administration concerning its compliance with the requirements of Standard No. 208 for installation of automatic restraints in its passenger cars produced in that year. Each report shall –

(1) Identify the manufacturer;

(2) State the full name, title and address of the official responsible for preparing the report;

(3) Identify the production year being reported on;

(4) Contain a statement regarding the extent to which the manufacturer has complied with the requirements of S4.1.3 of Standard No. 208;

(5) Provide the information specified in 585.5(b);

(6) Be written in the English language; and

(7) Be submitted to: Administrator, National Highway Traffic Safety Administration, 400 Seventh Street, S.W., Washington, D.C. 20590.

(b) *Report content.*

(1) *Basis for phase-in production goals.*

Each manufacturer shall provide the number of passenger cars manufactured for sale in the United States for each of the three previous production years, or, at the manufacturer's option, for the current production year. A new manufacturer that is, for the first time, manufacturing passenger cars for sale in the United States must report the number of passenger cars manufactured during the current production year.

(2) *Production.*

Each manufacturer shall report for the production year being reported on, and each preceding production year, to the extent that cars produced during the preceding years are treated under Standard No. 208 as having been produced during the production year being reported on, the following information:

(i) the number of passenger cars equipped with automatic seat belts and the seating positions at which they are installed,

(ii) the number of passenger cars equipped with air bags and the seating positions at which they are installed, and

(iii) the number of passenger cars equipped with other forms of automatic restraint technology, which shall be described, and the seating positions at which they are installed.

(3) *Passenger cars produced by more than one manufacturer.*

Each manufacturer whose reporting of information is affected by one or more of the express written contracts permitted by section S4.1.3.5.2 of Standard No. 208 shall:

(i) Report the existence of each contract, including the names of all parties to the contract, and explain how the contract affects the report being submitted,

(ii) Report the actual number of passenger cars covered by each contract.

585.6 *Records.*

Each manufacturer shall maintain records of the Vehicle Identification Number and type of automatic restraint for each passenger car for which information is reported under 585.5(b)(2), until December 31, 1991.

585.7 *Petition to extend period to file report.*

A petition for extension of the time to submit a report must be received not later than 15 days before expiration of the time stated in 585.5(a). The petition must be submitted to: Administrator, National Highway Traffic Safety Administration, 400 Seventh Street, SW, Washington, DC 20590. The filing of a petition does not automatically extend the time for filing a report. A petition will be granted only if the petitioner shows good cause for the extension and if the extension is consistent with the public interest.

Issued on March 18, 1986

Diane K. Steed
Administrator

51 F.R. 9801
March 21, 1986

PREAMBLE TO AN AMENDMENT TO
FEDERAL MOTOR VEHICLE SAFETY STANDARD NO. 208
Occupant Protection-Improvement of Seat Belt Assemblies
(Docket 74-14; Notice 44)

ACTION: Final rule; response to petitions for reconsideration.

SUMMARY: In November 1985, NHTSA published a final rule setting comfort and convenience performance requirements for both manual and automatic safety belt assemblies installed in motor vehicles with a gross vehicle weight rating of 10,000 pounds or less. This notice responds to two petitions for reconsideration and corrects certain technical and typographical errors in that final rule.

EFFECTIVE DATE: The amendments made by this notice to the text of Standard No. 208 will take effect on June 17, 1986. Manufacturers do not have to comply with the comfort and convenience requirements of S7.4 until September 1, 1986.

SUPPLEMENTARY INFORMATION: The agency published a final rule on November 6, 1985 (50 FR 46056), which modified the comfort and convenience performance requirements in Standard No. 208, *Occupant Crash Protection*. Petitions for reconsideration of that final rule were received from Ford Motor Company (Ford) and General Motors Corporation (GM).

Webbing Tension-Relieving Devices

Both Ford and GM requested modification of the requirement in S7.4.2 of the final rule that any belt slack that can be introduced into an automatic safety belt system by means of any tension-relieving device or design "shall be cancelled each time the safety belt is unbuckled or the adjacent vehicle door is opened except for belt systems in open-body vehicles with no doors." Both petitioners said that the language in the rule could be interpreted as requiring belt slack to be cancelled each time a safety belt is unbuckled, whether or not the adjacent door is open. The petitioners also stated that the language in the amendment did not reflect the agency's intent as expressed in the preamble to the final rule. They urged the agency to amend the requirement so that belt slack in an automatic belt system must

be cancelled only when the adjacent vehicle door is opened.

The agency's intent, as expressed in the preamble (50 FR at 46059), was that belt slack in automatic belt systems must be cancelled each time that the adjacent vehicle door is opened, whether or not the belt is buckled. Anticipating the adoption of a dynamic test requirement for manual belts, the language of the final rule was also intended to give manufacturers increased design flexibility by providing them the option of linking cancellation of tension-relievers in dynamically-tested manual belt systems to, at their choice, either opening of the door or releasing of the belt. Therefore NHTSA is amending the requirement to clarify that for automatic belts, cancellation of the tension-reliever is linked to opening the adjacent vehicle door and for dynamically-tested manual safety belts, a manufacturer has the option of using either opening the door or releasing the belt as the event leading to cancellation of the tension-reliever.

Torso Belt Body Contact Force

In the final rule, the agency exempted certain automatic and manual safety belt systems incorporating tension-relieving devices, such as window-shade devices, from the 0.7-pound torso belt contact force requirement. The reason for this exemption was the agency's concern that compliance with the body contact force requirement could limit manufacturers' design flexibility in meeting the retraction and other requirements in the rule. In their comments on the notice of proposed rulemaking, both foreign and domestic manufacturers had questioned whether imposing a contact force requirement on belt systems with tension-relievers would advance safety. They said that the necessity for complying with the belt contact force requirement could result in the production of some belt systems in which there was insufficient force to retract webbing reliably.

Ford and GM objected to the language in the final rule, because the exemption was limited to safety belt systems "which incorporate a webbing tension-

PART 208—PRE 305

relieving device that complies with S7.4.2." Section 7.4.2 requires automatic belt systems with webbing tension-relieving devices to meet the injury criteria of the standard when the belt is adjusted to have the maximum amount of slack recommended by the vehicle manufacturer. The petitioners stated that they do not believe the reference to S7.4.2 was intended to discourage the use of tension-relief devices on manual seat belt systems or to imply that manual seat belt systems incorporating tension-relief devices should not be eligible for the exemption now accorded automatic seat belt systems.

In the preamble to the final rule (50 FR at 46060), the agency noted that the tension-relieving requirements for manual safety belts were proposed in Notice 38 of this docket, in conjunction with the dynamic tests for manual safety belts. The agency also said that if a dynamic test requirement for manual belts was adopted, the provisions on tension-relievers for manual belts would be expected to be identical to those for automatic belts. On March 21, 1986 (51 FR 9800), the agency published a final rule setting dynamic test requirements for manual safety belts in passenger cars. The March 1986 rule adopted the same requirements for tension-relieving devices in dynamically-tested manual safety belts that were adopted in the November 1985 final rule for automatic belts. (In the March 1986 rule, the agency deferred action on whether to adopt the proposed dynamic testing for manual safety belts in light trucks and vans. If such a requirement is adopted, NHTSA will apply the same requirement on tension-relievers to those manual belts that are applied to other dynamically-tested manual safety belts.)

In the November 1985 final rule, the agency did not intend to preclude the use of tension-relieving devices on non-dynamically-tested manual safety belts or to imply that manual belt systems incorporating tension-relieving devices should not be eligible for the exemption from the belt contact force requirement now accorded automatic safety belt systems and dynamically-tested manual safety belts. The agency has revised the language of S7.4.3 to exempt all belts, whether manual or automatic, incorporating tension-relievers from the belt contact force requirement. The agency encourages manufacturers to provide information in their owner's manual on properly adjusting non-dynamically-tested manual safety belts with tension-relievers.

Belt Retraction

In the final rule, the retraction requirement for manual safety belts stated that torso and lap belt webbing must automatically retract to a stowed posi-

tion, when the adjacent vehicle door is in the open position, or when the seat belt latchplate is released. Both Ford and GM interpreted this requirement to mean that retraction must occur when the latchplate is released whether or not the adjacent door is opened. They requested that the wording be revised to require retraction only when both conditions exist, i.e., release of the latchplate and opening of the adjacent door. They stated that the belt cannot retract until it is unbuckled and that they see no safety need to require retraction before the adjacent door is opened.

As stated in the April 1985 notice of proposed rulemaking, many persons find seat belts inconvenient because the belt webbing will not retract completely to its stowed position when the system is unbuckled, thus creating an obstacle when the occupant is trying to exit the vehicle or soiling the belt if it is caught in the door. The intent of the retraction requirement in the final rule was to provide manufacturers increased flexibility by giving them the option of triggering tension-relief cancellation and belt retraction by either release of the latchplate or opening of the adjacent vehicle door. As noted by the American Safety Belt Council in its comments on the April 1985 NPRM, new safety belt designs are available which will cancel a tension-relieving device and retract the belt when the latchplate is released from the buckle, regardless of whether the door is open or not. The agency did not intend that each condition trigger the retraction mechanism, but instead intended to allow manufacturers the option of using either condition to initiate belt retraction. For these reasons, the agency is amending the requirement to make it clear that manufacturers have the option of determining whether door opening or latchplate release is the mechanism that triggers retraction of a manual safety belt.

The rule will continue to provide that in an open-body vehicle with no doors, a manufacturer has the option to provide either automatic or manual deactivation of a tension-relieving device. Thus, in the retraction test in those vehicles, the agency will deactivate the tension-relieving devices in the manner provided by the manufacturer.

Armrests

The petitioners also requested a further clarification of the language of the final rule on belt retraction. That requirement permits an outboard armrest of a seat to be placed in its stowed position for the purpose of the retraction test, if the armrest must be stowed to allow the seat occupant to exit the vehicle. The agency stated in the preamble to the final rule that it intended to allow the stowage of folding armrests during the retraction test if "they protrude into the door opening in a manner which encumbers egress." (50 FR at 46061).

Ford noted that the common dictionary meaning of "encumber" is "impede," or "hinder," so that egress would be made difficult although not necessarily impossible. Ford said that the language of the final rule limited the stowage of armrests to situations in which armrests, unless stowed, make egress impossible.

To eliminate the possibility of having to make subjective judgments as to whether an armrest "hinders" occupant egress, the agency is modifying the retraction requirement to provide that any folding armrest must be stowed prior to initiation of the retraction test.

Technical Corrections

Ford pointed out a typographical error in amendment 14 of the final rule, which referred to S7.1.13, instead of referring to S7.1.3. The agency has made the necessary correction. Ford also stated that decimal points should be added, where appropriate, to the specified dimensional tolerances in the table of weights and dimensions of vehicle occupants. These corrections would conform the dimensions set forth in the chart, which is in amendment 14 in the final rule, to the corresponding dimensions specified on drawing SA 150 M002 or the test dummy. The agency agrees and has made the necessary corrections.

The comfort and convenience requirements in S7.4 of Standard No. 208 apply to automatic and manual safety belt assemblies installed in any vehicle with a GVWR of 10,000 pounds or less. The title of S7 in this standard, *Seat belt assembly requirements—passenger cars*, is no longer accurate, because the paragraphs of S7., by their terms, apply to passenger cars and several other types of vehicles. Therefore, the title is corrected in this notice to read S7. *Seat belt assembly requirements.* The agency is also amending the retraction requirements of S7.4.5 to make clear that, as proposed in the April 1985 NPRM, the retraction test only applies to the front outboard designated seating positions.

The remaining amendments are made to remove an extra "and" in paragraph S7.4.6.1(a), and to correct a typographical error in S4.5.3.3(b) (change "set" to "seat").

Effective Date

This notice makes minor clarifications and typographical and technical corrections to the text of Standard No. 208. NHTSA has determined that it is in the public interest to have these amendments to the language of the standard go into effect on publication of this notice in the *Federal Register*, since these amendments will provide manufacturers with more flexibility in developing designs to comply with the safety belt comfort and convenience requirements, which will go into effect on September 1, 1986.

In consideration of the foregoing, 49 CFR 571.208 is amended as follows:

1. The title of S7. is revised to read:

S7 **Seat belt assembly requirements.**

2. S7.4.2 is revised to read:

S7.4.2 Webbing tension-relieving device. Each vehicle with an automatic seat belt assembly, or with a Type 2 manual seat belt assembly that must meet S4.6, installed in a front outboard designated seating position that has either manual or automatic tension-relieving devices permitting the introduction of slack in the webbing of the shoulder belt (e.g., "comfort clips" or "window-shade" devices) shall:

(a) Comply with the requirements of S5.1 with the shoulder belt webbing adjusted to introduce the maximum amount of slack recommended by the vehicle manufacturer pursuant to S7.4.2(b).

(b) Have a section in the vehicle owner's manual that explains how the tension-relieving device works and specifies the maximum amount of slack (in inches) recommended by the vehicle manufacturer to be introduced into the shoulder belt under normal use conditions. The explanation shall also warn that introducing slack beyond the amount specified by the manufacturer could significantly reduce the effectiveness of the shoulder belt in a crash; and

(c) Have, except for open-body vehicles with no doors, an automatic means to cancel any shoulder belt slack introduced into the belt system by a tension-relieving device. In the case of an automatic safety belt system, cancellation of the tension-relieving device shall occur each time the adjacent vehicle door is opened. In the case of a manual seat belt required to meet S4.6, cancellation of the tension-relieving device shall occur, at the manufacturer's option, either each time the adjacent door is opened or each time the latchplate is released from the buckle. In the case of open-body vehicles with no doors, cancellation of the tension-relieving device may be done by a manual means.

3. S7.4.3 is revised to read as follows:

S7.4.3 Belt contact force. Except for manual or automatic seat belt assemblies which incorporate a webbing tension-relieving device, the upper torso webbing of any seat belt assembly, when tested in accordance with S10.6, shall not exert more than 0.7 pounds of contact force when measured normal to and one inch from the chest of an anthropomorphic test dummy positioned in accordance with S10 in the seating position for which that assembly is provided, at the point where the centerline of the torso belt crosses the midsagittal line on the dummy's chest.

4. S7.4.5 is revised to read as follows:

S7.4.5 Retraction. When tested under the conditions of S8.1.2 and S8.1.3, with anthropomorphic test dummies whose arms have been removed and which are positioned in accordance with S10 in the front outboard designated seating positions and restrained by the belt systems for those positions, the torso and lap belt webbing of any of those seat belt systems shall automatically retract to a stowed position either when the adjacent vehicle door is in the open position and the seat belt latchplate is

released, or, at the option of the manufacturer, when the latchplate is released. That stowed position shall prevent any part of the webbing or hardware from being pinched when the adjacent vehicle door is closed. A belt system with a tension-relieving device in an open-bodied vehicle with no doors shall fully retract when the tension-relieving device is deactivated. For the purpose of the retraction requirement, outboard armrests, which are capable of being stowed, on vehicle seats shall be placed in their stowed positions.

5. S7.4.6.1(a) is amended by removing the second occurrence of the word "and" in the first sentence.

6. S4.5.3.3(b) is amended by correcting the word "set" to read "seat" and the word "show" to read "shown."

7. Condition (B) of S4.5.3.3(b) is amended by removing the second occurrence of the word "the" and by correcting the word "relases" to read "release."

8. The weights and dimensions of the vehicle occupants referred to in this standard and specified in S7.1.3 are revised to read as follows:

	50th-percentile 6-year-old child	5th-percentile adult female	50th-percentile adult male	95th-percentile adult male
Weight	47.3 pounds	102 pounds	164 pounds ± 3	215 pounds
Erect sitting height	25.4 inches	30.9 inches	35.7 inches $\pm .1$	38 inches
Hip breadth (sitting)	8.4 inches	12.8 inches	14.7 inches $\pm .7$	16.5 inches
Hip circumference (sitting)	23.9 inches	36.4 inches	42 inches	47.2 inches
Waist circumference (sitting)	20.8 inches	23.6 inches	32 inches $\pm .6$	42.5 inches
Chest depth		7.5 inches	9.3 inches $\pm .2$	10.5 inches
Chest circumference:				
(nipple)		30.5 inches		
(upper)		29.8 inches	37.4 inches $\pm .6$	44.5 inches
(lower)		26.6 inches		

Issued on: June 11, 1986.

Diane K. Steed
Administrator
51 F. R. 21912
June 17, 1986

PREAMBLE TO AN AMENDMENT TO
FEDERAL MOTOR VEHICLE SAFETY STANDARD NO. 208

Anthropomorphic Test Dummies
(Docket No. 74-14; Notice 45)

ACTION: Final Rule.

SUMMARY: This notice adopts the Hybrid III test dummy as an alternative to the Part 572 test dummy in testing done in accordance with Standard No. 208, Occupant Crash Protection. The notice sets forth the specifications, instrumentation, calibration test procedures, and calibration performance criteria for the Hybrid III test dummy. The notice also amends Standard No. 208 so that effective October 23, 1986, manufacturers have the option of using either the existing Part 572 test dummy or the Hybrid III test dummy until August 31, 1991. As of September 1, 1991, the Hybrid III will replace the Part 572 test dummy and be used as the exclusive means of determining a vehicle's conformance with the performance requirements of Standard No. 208.

The notice also establishes a new performance criterion for the chest of the Hybrid III test dummy which will limit chest deflection. The new chest deflection limit applies only to the Hybrid III since only that test dummy has the capability to measure chest deflection.

These amendments enhance vehicle safety by permitting the use of a more advanced test dummy which is more human-like in response than the current test dummy. In addition, the Hybrid III test dummy is capable of making many additional sophisticated measurements of the potential for human injury in a frontal crash.

DATES: The notice adds a new Subpart E to Part 572 effective October 23, 1986.

This notice also amends Standard No. 208 so that effective October 23, 1986, manufacturers have the option of using either the existing Part 572 test dummy or the Hybrid III test dummy until August 31, 1991. As of September 1, 1991, the Hybrid III will replace the Part 572 test dummy and be used as the exclusive means of determining a vehicle's conformance with the performance requirements of Standard No. 208. The incorporation by reference of certain publications listed in the regulation is approved by the Director of the Federal Register as of October 23, 1986.

SUPPLEMENTARY INFORMATION: In December 1983, General Motors (GM) petitioned the agency to amend Part 572, *Anthropomorphic Test Dummies*, to adopt specifications for the Hybrid III test dummy. GM also petitioned for an amendment of Standard No. 208, *Occupant Crash Protection*, to allow the use of the Hybrid III as an alternative test device for compliance testing. The agency granted GM's petition on July 20, 1984. The agency subsequently received a petition from the Center for Auto Safety to propose making Standard No. 208's existing injury criteria more stringent for the Hybrid III and to establish new injury criteria so as to take advantage of the Hybrid III's superior measurement capability. The agency granted the Center's petition on September 17, 1984. On April 12, 1985 (50 FR 14602), NHTSA proposed amendments to Part 572 and Standard No. 208 that were responsive to the petitioners and which, in the agency's judgment, would enhance motor vehicle safety. Twenty-eight individuals and companies submitted comments on the proposed requirements. This notice presents the agency's analysis of the issues raised by the commenters. The agency has decided to adopt the use of the Hybrid III test dummy and some of the proposed injury criteria. The agency has also decided to issue another notice on the remaining injury criteria to gain additional information about the potential effects of adopting those criteria.

This notice first discusses the technical specifications for the Hybrid III, its calibration requirements, its equivalence with the existing Part 572 test dummy, and the applicable injury criteria. Finally, it discusses the test procedure used to position the dummy for Standard No. 208 compliance testing and the economic and other effects of this rule.

Test Dummy Drawings and Specifications

Test dummies are used as human surrogates for evaluation of the severity of injuries in vehicle crashes. To serve as an adequate surrogate, a test dummy must be capable of simulating human impact responses. To serve as an objective test device, the test dummy must be adequately defined through technical drawings and performance specifications to ensure uniformity in construction, impact response, and measurement of injury in identical crash conditions.

Virtually all of the commenters, with the exception of GM, said that they have not had sufficient experience with the Hybrid III to offer comments on the validity of the technical specifications for the test dummy. Since the issuance of the notice, GM has provided additional technical drawings and a Society of Automotive Engineers-developed user's manual to further define the Hybrid III. These new drawings do not alter the basic nature of the test dummy, but instead provide additional information which will enable users to make sure that they have a correctly designed and correctly assembled test dummy. The user's manual provides information on the inspection, assembly, disassembly, and use of the test dummy. Having the user's manual available will assist builders and users of the Hybrid III in producing and using the test dummy. GM also provided information to correct the misnumbering of several technical drawings referenced in the notice.

In addition, the agency has reviewed the proposed drawings and specifications. While NHTSA believes the proposed drawings are adequate for producing the test dummy, the agency has identified and obtained additional information which should make production and use of the test dummy even more accurate. For example, the agency has obtained information on the range of motions for each moving body part of the test dummy. Finally, to promote the ease of assembly, NHTSA has made arrangements with GM to ensure that the molds and patterns for the test dummy are available to all interested parties. Access to the molds will assist other potential builders and users of the Hybrid III since it is difficult to specify all of the details of the various body contours solely by technical drawings.

The agency has adopted the new drawings and user manual in this rule and has made the necessary corrections to the old drawings. The agency believes that the available drawings and technical specifications are more than sufficient for producing, assembling, and using the Hybrid III test dummy.

Commercial Availability of the Hybrid III

A number of commenters raised questions about the commercial availability of the Hybrid III test dummy, noting problems they have experienced in obtaining calibrated test dummies and the instrumentation for the neck and lower leg of the Hybrid III. For example, Chrysler said that it had acquired two Hybrid III test dummies, but has been unable to obtain the lower leg and neck instrumentation for five months. Likewise, Ford said that it has been unable to obtain the knee displacement and chest deflection measurement devices for the Hybrid III. It also said that of the test dummies it had received, none had sufficient spine stiffness to meet the Hybrid III specifications. Ford claimed to have problems in retaining a stable dummy posture which would make it difficult to carry out some of the specified calibration tests. Subsequent investigation showed that the instability was caused by out-of-specification rubber hardness of the lumbar spine, and was eliminated when spines of correct hardness were used. In addition, Ford said that the necks and ribs of the test dummy would not pass the proposed calibration procedures. Finally, Ford said that the equipment needed for calibrating the dummy is not commercially available.

Although the commenters indicated they had experienced difficulty in obtaining the instrumentation for the Hybrid III's neck and lower legs, they did not indicate that there is any problem in obtaining the instrumentation needed to measure the three injury criteria presently required by Standard No. 208, the head injury criterion, chest acceleration, and femur loading and which are being adopted by this rule for the Hybrid III. For example, Volkswagen said it had obtained Hybrid III test dummies with sufficient instrumentation to measure the same injury criteria as with the Part 572. VW did say it had ordered the additional test devices and instrumentation for the Hybrid III but was told the instrumentation would not be available for six months.

The agency notes that there are now two commercial suppliers of the Hybrid III test dummy, Alderson Research Labs (ARL) and Humanoid Systems. Humanoid has built nearly 100 test dummies and ALR has produced five prototype test dummies as of the end of December 1985. Both manufacturers have indicated that they are now capable of producing sufficient Hybrid IIIs to meet the demand for those dummies. For example, Humanoid Systems said that while the rate of production is dependent on the number of orders, generally three test dummies per week are produced. Thus, in the case of the basic test dummy, there appears to be sufficient commercial capacity to provide sufficient test dummies for all vehicle manufacturers.

As to test dummy instrumentation, the agency is aware that there have been delays in obtaining the new neck, thorax, and lower leg instrumentation for the Hybrid III. However, as Humanoid commented, while there have been delays, the supplies of the needed parts are expected to increase. Even if the supply of the lower leg instrumentation is slow to develop, this will not pose a problem, since the agency is not adopting, at this time, the proposed lower leg injury criteria. In the case of the neck instrumentation, the supply problem should be minimized because each test facility will only need one neck transducer to calibrate all of its test dummies. The neck instrumentation will not be needed for a manufacturer's crash testing since at this time, the agency is not adopting any neck injury criteria. In the case of the instrumentation for measuring thoracic deflection, the supplier has indicated that it can deliver the necessary devices within 3 months of the time an order is placed. As to Ford's comment about calibration test equipment, the agency notes that current equipment used for calibrating the existing Part 572 test dummy can be used, with minor modification, to calibrate the Hybrid III test dummy.

Calibration Requirements

In addition to having complete technical drawings and specifications, a test dummy must have adequate calibration test procedures. The calibration tests involve a series of static and dynamic tests of the test dummy components to determine whether the responses of the test dummy fall within specified performance requirements for each test. The testing involves instrumenting the head, thorax and femurs to measure the test dummy's responses. In addition, there are tests of the neck, whose structural properties may have considerable influence on the kinematics and impact responses of the instrumented head. Those procedures help ensure that the test dummy has been properly assembled and that, as assembled, it will provide repeatable and reproducible results in crash testing. (Repeatability refers to the ability of the same test dummy to produce the same results when subjected to several identical tests. Reproducibility refers to the ability of one test dummy to provide the same results as another test dummy built to the same specifications.)

Lumbar Spine Calibration Test

The technical specifications for the Hybrid III set out performance requirements for the hardness of the rubber used in the lumbar spine to ensure that the spine will have appropriate rigidity. NHTSA's test data show that there is a direct relationship between rubber hardness and stiffness of the spine and

that the technical specification on hardness is sufficient to ensure appropriate spine stiffness. Accordingly, the agency believes that a separate calibration test for the lumbar spine is not necessary. Humanoid supported the validity of relying on the spine hardness specification to assure adequate stability of the dummy's posture, even though it found little effect on the dummy's impact response. Humanoid's support for this approach was based on tests of Hybrid III dummies which were equipped with a variety of lumbar spines having different rubber hardnesses.

Subsequent to issuance of the notice, the agency has continued its testing of the Hybrid III test dummy. Through that testing, the agency found that commercially available necks either cannot meet or cannot consistently meet all of the calibration tests originally proposed for the neck. To further evaluate this problem, NHTSA and GM conducted a series of round robin tests in which a set of test dummies were put through the calibration tests at both GM's and NHTSA's test laboratories.

The test results, which were placed in the docket after the tests were completed, showed that none of the necks could pass all of the originally specified calibration tests.

In examining the test data, the agency determined that while some of the responses of the necks fell slightly outside of the performance corridors proposed in the calibration tests, the responses of the necks showed a relatively good match to existing biomechanical data on human neck responses. Thus, while the necks did not meet all of the calibration tests, they did respond as human necks are expected to respond.

In discussions with GM, the agency learned that the calibration performance requirements were originally established in 1977 based on the responses of three prototype Hybrid III necks. GM first examined the existing biomechanical data and established several performance criteria that reflected human neck responses. GM then built necks which would meet the biomechanically based performance criteria. GM established the calibration tests that it believed were necessary to ensure that the necks of the prototype test dummies would produce the required biomechanical responses. Although extensive performance specifications may have been needed for the development of specially built prototype necks, not all of the specifications appear to be essential once the final design was established for the mass-produced commercial version. Based on the ability of the commercially available test dummies to meet the biomechanical response criteria, NHTSA believes that the GM-derived calibration requirements should be adjusted to reflect the response characteristics of commercially available test dummies and simplified as much as possible to reduce the complexity of the testing.

Based on the results of the NHTSA-GM calibration test series, the agency is making the following changes to the neck calibration tests. In the flexion (forward bending) calibration test, the agency is:

1. increasing the time allowed for the neck to return to its preimpact position after the pendulum impact test from a range of 109–119 milliseconds to a range of 113–128 milliseconds.

2. changing the limits for maximum head rotation from a range of 67°–79° to a range of 64°–78°.

3. expanding the time limits during which maximum moment must occur from a range of 46–56 milliseconds to 47–58 milliseconds.

4. modifying the limits for maximum moment from a range of 72–90 ft-lbs to a range of 65–80 ft-lbs.

5. increasing the time for the maximum moment to decay from a range of 95–105 milliseconds to a range of 97–107 milliseconds.

In the extension (backward bending) calibration test, the agency is:

1. expanding the time allowed for the neck to return to its preimpact position after the pendulum impact test from a range of 157–167 milliseconds to a range of 147–174 milliseconds.

2. changing the limits for maximum head rotation from a range of 94°–106° to a range of 81°–106°.

3. expanding the time limit during which the minimum moment must occur from a range of 69–77 milliseconds to 65–79 milliseconds.

4. modifying the limits for minimum moment from a range of –52 to –63 ft-lbs to a range of –39 to –59 ft-lbs.

5. increasing the time for the minimum moment to decay from the range of 120–144 milliseconds, contained in GM's technical specifications for the Hybrid III, to a range of 120–148 milliseconds.

In reviewing the NHTSA-GM test data, the agency also identified several ways of simplifying the neck's performance requirements. In each case, the following calibration specifications appear to be redundant and their deletion should not affect the performance of the neck. The agency has thus deleted the requirement for minimum moment in flexion and the time requirement for that moment. For extension, the agency has eliminated the limit on the maximum moment permitted and the time requirement for that moment. The agency has

deleted those requirements since the specification on maximum rotation of the neck in flexion and minimum rotation of the neck in extension appear to adequately measure the same properties of the neck. Similarly, the agency has simplified the test by eliminating the pendulum braking requirement for the neck test, since GM's testing shows that the requirement is not necessary to ensure test consistency. Finally, the agency is clarifying the test procedure by deleting the specification in the GM technical drawings for the Hybrid III calling for two pre-calibration impact tests of the neck. GM has informed the agency that the two pre-calibration tests are not necessary.

Based on the NHTSA-GM calibration test data, the agency is making two additional changes to the neck calibration tests. Both NHTSA and GM routinely control the calibration pendulum impact speed to within plus or minus one percent. Currently available dummy necks are able to meet the calibration response requirements consistently when the pendulum impact speed is controlled to that level Thus, NHTSA believes that the proposed range of allowable velocities (± 8.5 percent) for the pendulum impact is excessive. Reducing the allowable range is clearly feasible and will help maintain a high level of consistency in dummy neck responses. The agency has therefore narrowed the range of permissible impact velocities to the neck to ± 2 percent. This range is readily obtainable with commercially available test equipment. In reviewing the neck calibration test data, GM and NHTSA noted a slight sensitivity in the neck response to temperature variation. In its docket submission of January 27, 1986, GM recommended controlling the temperature during the neck calibration test to 71° ± 1°. NHTSA agrees that controlling the temperature for the neck calibration test will reduce variability, but the agency believes that a slightly wider temperature range of 69° to 72°, which is the same range used in the chest calibration test, is sufficient.

Neck Durability

Nissan commented that, in sled tests of the two test dummies, the neck bracket of one of the Hybrid III test dummies experienced damage after 10 tests, while the Part 572 test dummy had no damage. The agency believes that Nissan's experience may be the result of an early neck design which has been subsequently modified by GM. (See GM letter of September 16, 1985, Docket 74-14, Notice 39, Entry 28.) The agency has conducted numerous 30 mile per hour vehicle impact tests using the Hybrid III test dummy and has not had any neck bracket failures.

Thorax Calibration Test

As a part of the NHTSA-GM calibration test series, both organizations also performed the proposed calibration test for the thorax on the same test dummies. That testing showed relatively small differences in the test results measured between the two test facilities The test results from both test facilities show that the chest responses of the Hybrid III test dummies were generally within the established biomechanical performance corridors for the chest. In addition, the data showed that the Hybrid III chest responses fit those corridors substantially better than the chest responses of the existing Part 572 test dummy. The data also showed that the chest responses in the high speed (22 ft/sec) pendulum impact test more closely fit the corridors than did the chest responses in the low speed (14 ft/sec) test. In addition, the data showed that if a test dummy performed satisfactorily in the low speed pendulum impact test, it also performed satisfactorily in the more severe high speed test.

Based on those results, GM recommended in a letter of January 27, 1986, (Docket No. 74-14, Notice 39, Entry 41) that only the low speed pendulum impact be used in calibration testing of the Hybrid III chest. GM noted that deleting the more severe pendulum impact test "can lead to increasing the useful life of the chest structure."

Based on the test data, the agency agrees with the GM recommendation that only one pendulum impact test is necessary. NHTSA recognizes that using only the low speed pendulum impact will increase the useful life of the chest. However, the agency has decided to retain the high speed rather than the low speed test. While NHTSA recognizes that the high speed test is more severe, the agency believes the high speed test is more appropriate for a number of reasons. First, the data showed that the high speed chest impact responses compared more closely with the biomechanical corridors than the low speed responses. Thus, use of the high speed test will make it easier to identify chests that do not have the correct biofidelity. In addition, since the higher speed test is more severe it will subject the ribcage to higher stresses, which will help identify chest structural degradation. Finally, the high speed impact test is more representative of the range of impacts a test dummy can receive in a vehicle crash test.

Although the NHTSA-GM test data showed that the production version of the Hybrid III chest had sufficient biofidelity, the data indicated that proposed calibration performance requirements should be lightly changed to account for the wider range in calibration test responses measured in commercially available test dummies. Accordingly, the agency is adjusting the chest deflection requirement to increase the allowable range of deflections from 2.51-2.75 inches to 2.5-2.85 inches. In addition, the agency is adjusting the resistive force requirement from a range of 1186-1298 pounds to a range of 1080-1245 pounds. Also, the hysteresis requirement is being adjusted from a 75-80 percent range to a 69-85 percent range. Finally, the agency is clarifying the chest calibration test procedure by deleting the specification in GM's technical drawing for the Hybrid III that calls for two pre-calibration impact tests of the chest. GM has informed the agency that these tests are not necessary. These slight changes will not affect the performance of the Hybrid III chest, since the NHTSA-GM test data showed that commercially available test dummies meeting these calibration specifications had good biofidelity.

Chest Durability

Testing done by the agency's Vehicle Research and Test Center has indicated that the durability of the Hybrid III's ribs in calibration testing is less than that of the Part 572 test dummy. ("State-of-the-Art Dummy Selection, Volume I" DOT Publication No. HS 806 722) The durability of the Hybrid III was also raised by several commenters. For example, Toyota raised questions about the durability of the Hybrid III's ribs and suggested the agency act to improve their durability.

The chest of the Hybrid III is designed to be more flexible, and thus more human-like, than the chest of the Part 572 test dummy. One of the calibration tests used for the chest involves a 15 mph impact into the chest by a 51.5 pound pendulum; an impact condition which is substantially more severe than a safety belt or airbag restrained occupant would experience in most crashes. The chest of the Hybrid III apparently degrades after such multiple impacts at a faster rate than the chest of the Part 572 test dummy. As the chest gradually deteriorates, the amount of acceleration and deflection measured in the chest are also affected. Eventually the chest will fall out of specification and will require either repair or replacement.

In its supplemental comments to the April 1985 notice, GM provided additional information about the durability of the Hybrid III ribs. GM said that it uses the Hybrid III in unbelted testing, which is the most severe test for the dummy. GM said that the Hybrid III can be used for about 17 crash tests before the ribs must be replaced. GM explained that it does not have comparable data for the Part 572 test dummy since it does not use that test dummy in unbelted tests. GM said, however, that it believes that the durability of the Part 572 test dummy ribs in vehicle crash testing would be comparable to that of the Hybrid III.

Having reviewed all the available information, the agency concludes that both the Hybrid III and existing Part 572 test dummy ribs will degrade under severe impact conditions. Although the Hybrid III's more flexible ribs may need replacement more frequently, particularly after being used in unrestrained testing, the Hybrid III's ribs appear to have reasonable durability. According to GM's data, which is in line with NHTSA's crash test experience, the Hybrid III's ribs can withstand approximately 17 severe impacts, such as found in unrestrained testing, before they must be replaced. Ford, in a presentation at the MVMA Hybrid III workshop held on February 5, 1986, noted that one of its belt-restrained Hybrid III test dummies was subjected to 35 vehicle and sled crashes without any failures. The potential lower durability of the ribs in unrestrained testing should be of little consequence if the Hybrid III test dummy is used in air bag or belt testing.

Chest Temperature Sensitivity

The April 1985 notice said NHTSA tests have indicated that the measurements of chest deflection and chest acceleration by the Hybrid III are temperature sensitive. For this reason, GM's specifications for the Hybrid III recognize this problem and call for using the test dummy in a narrower temperature range (69° to 72° F) to ensure the consistency of the measurements. GM has also suggested the use of an adjustment factor for calculating chest deflection when the Hybrid III is used in a test environment that is outside of the temperature range specified for the chest. While this approach may be reasonable to account for the adjustment of the deflection measurement, there is no known method to adjust the acceleration measurement for variations in temperature. For this reason, the agency is not adopting GM's proposed adjustment factor, but is instead retaining the proposed 69° to 72° F temperature range.

A number of commenters addressed the feasibility and practicability of maintaining that temperature range. BMW said that although it has an enclosed crash test facility, it had reservations about its ability to control the test temperature within the proposed range. Daihatsu said that it was not sure it could assure the test dummy's temperature will remain within the proposed range. Honda said that while it had no data on the temperature sensitivity of the Hybrid III, it questioned whether the proposed temperature range was practical. Mercedes-Benz said it is not practicable to maintain the proposed temperature range because the flood lights necessary for high speed filming of crash tests can cause the test dummy to heat up. Nissan said it was not easy to maintain the current 12 degree range specified for the existing Part 572 test dummy and thus it would be hard to maintain the three degree range proposed for the Hybrid III. Ford also said that maintaining the three degree range could be impracticable in its current test facilities.

Other manufacturers tentatively indicated that the proposed temperature range may not be a problem. VW said the temperature range should not be an insurmountable problem, but more experience with the Hybrid III is necessary before any definite conclusions can be reached. Volvo said it could maintain the temperature range in its indoor test facilities, but it questioned whether outdoor test facilities could meet the proposed specification. Humanoid indicated in its comments, that it has developed an air conditioning system individualized for each test dummy which will maintain a stable temperature in the test dummy up to the time of the crash test.

The agency believes that there are a number of effective ways to address the temperature sensitivity of the Hybrid III chest. The test procedure calls for placing the test dummy in an area, such as a closed room, whose temperature is maintained within the required range for at least four hours before either the calibration tests or the use of the test dummy in a crash test. The purpose of the requirement is to ensure that the primary components of the test dummy have reached the correct temperature before the test dummy is used in a test. As discussed below, analytical techniques can be used to determine the temperature within the test dummy, to calculate how quickly the test dummy must be used in a crash test before its temperature will fall outside the required temperature range.

Testing done by the agency with the current Part 572 test dummy, whose construction and materials are similar to the Hybrid III, has determined how long it takes for various test dummy components to reach the required temperature range once the test dummy is placed in a room within that range. ("Thermal Responses of the Part 572 Dummy to Step Changes in Ambient Temperature" DOT Publication No. HS-801 960, June 1976) The testing was done by placing thermocouples, devices to measure temperature, at seven locations within the dummy and conducting a series of heating and cooling experiments. The tests showed that the thermal time constants (the thermal time constant is the time necessary for the temperature differential between initial and final temperatures to decrease from its original value to 37% of the original differential) varied from 1.2 hours for the forehead to 6.2 hours for the lumbar spine. Using this information it is possible to estimate the time it takes a test dummy originally within the required temperature range to fall out of the allowable range once it has been exposed to another temperature. The rib's thermal time constant is 2.9 hours. This means, for example, that if a test dummy's temperature has been stabilized at 70.5° F and then transferred to a test environment at 65° F, it would take approximately 0.8 hours for the rib temperature to drop to 69° F, the bottom end of the temperature range specified in Part 572.

Thus, the NHTSA test results cited above show that the chest can be kept within the range proposed by the agency if the test dummy is placed in a temperature-controlled environment for a sufficient time to stabilize the chest temperature. Once the chest of the test dummy is at the desired temperature, the test data indicate that it can tolerate some temperature variation at either an indoor or outdoor crash test site and still be within the required temperature range as long as the crash test is performed within a reasonable amount of time and the temperature at the crash site, or within the vehicle, or within the test dummy is controlled close to the 69 to 72 degrees F range. Obviously, testing conducted at extremely high or low temperatures can move the test dummy's temperature out of the required range relatively quickly, if no means are used to maintain the temperature of the test dummy within the required range. However, auxiliary temperature control devices can be used in the vehicle or the test environment to maintain a stabilized temperature prior to the crash test. Therefore, the agency has decided to retain the proposed 69 to 72 degrees F temperature range.

Chest Response to Changes in Velocity

The April notice raised the issue of the sensitivity of the Hybrid III's chest to changes in impact velocities. The notice pointed out that one GM study on energy-absorbing steering columns ("Factors Influencing Laboratory Evaluation of Energy-Absorbing Steering Systems," Docket No. 74–14, Notice 32, Entry 1666B) indicated that the Hybrid III's chest may be insensitive to changes in impact

velocities and asked commenters to provide further information on this issue.

Both GM and Ford provided comments on the Hybrid III's chest response. GM said that since the Hybrid III chest is designed to have a more human-like thoracic deflection than the Part 572 test dummy, the Hybrid III's response could be different. GM referenced a study ("System Versus Laboratory Impact Tests for Estimating Injury Hazard" SAE paper 680053) which involved cadaver impacts into energy-absorbing steering columns. The study concluded that the force on the test subject by the steering assembly was relatively constant despite changes in test speeds. GM said that this study indicated that "rather than the Hybrid III chest being insensitive to changes in velocity in steering system tests, it is the Part 572 which is too sensitive to changes in impact velocity to provide meaningful information for evaluating steering systems."

GM also presented new data on chest impact tests conducted on the Hybrid III and Part 572 test dummies. The tests involved chest impacts by three pendulum impact devices with different masses and three impact speeds. GM said that the test results show that "the Hybrid III chest deflection is sensitive to both changes in impact velocity and impactor mass." Ford also noted that the Hybrid III appears sensitive in the range of speed and deflections that are relevant to Standard No. 208 testing with belt-restrained dummies.

Ford noted that the GM testing referenced in the April notice was conducted at higher impact speeds than used in the calibration testing of the Hybrid III. Ford said it agreed with GM that the indicated insensitivity of chest acceleration to speed and load is a reflection of the constant-force nature of the steering column's energy absorption features. After reviewing the information provided by Ford and GM, NHTSA agrees that in an impact with a typical steering column, once the energy-absorbing mechanism begins to function, the test dummy's chest will receive primarily constant force. The lower stiffness of the Hybrid III chests would make it respond in a more human-like manner to these forces than the existing Part 572 test dummy.

Chest Accelerometer Placement

Volvo pointed out that the chest accelerometer of the Hybrid III is located approximately at the center of gravity of the chest, while the accelerometer is higher and closer to the back in the Part 572 test dummy. Volvo said that since the biomechanical tolerance limits for the chest were established using a location similar to that in the Part 572, it questioned whether the acceleration limits should apply to the Hybrid III. Volvo recommended changing the location of the accelerometer in the Hybrid III or using different chest acceleration criteria for the Hybrid III.

The agency recognizes that Hybrid III accelerometer placement should more correctly reflect the overall response of the chest because it is placed at the center of gravity of the chest. However, the dimensional differences between the accelerometer placements in the two test dummies are so small that in restrained crash tests the differences in acceleration response, if any, should be minimal.

Repeatability and Reproducibility

As discussed previously, test dummy repeatability refers to the ability of one test dummy to measure consistently the same responses when subjected to the same test. Reproducibility refers to the ability of two or more test dummies built to the same specifications to measure consistently the same responses when they are subjected to the same test.

Ford said that it is particularly concerned about the repeatability of the chest acceleration and deflection measurements of the Hybrid III and about the reproducibility of the Hybrid III in testing by different laboratories. Ford said that once a test dummy positioning procedure has been established, the agency should conduct a series of 16 car crash tests to verify the repeatability and reproducibility of the Hybrid III.

In its comments, GM provided data showing that the repeatability of the Hybrid III is the same as the existing Part 572 test dummy. Volvo, the only other commenter that addressed repeatability, also said that its preliminary tests show that the Hybrid III has a repeatability comparable to the Part 572. The agency's Vehicle Research and Test Center has also evaluated the repeatability of the Hybrid III and the Part 572 in a series of sled tests. The data from those tests show that the repeatability of the two test dummies is comparable. ("State-of-the-Art Dummy Selection, Volume I" DOT Publication No. HS 806 722.)

GM also provided data showing that the reproducibility of the Hybrid III is significantly better than the Part 572. In its supplemental comments filed on September 16, 1985, GM also said that Ford's proposed 16 car test program was not needed. GM said that "in such test the effects of vehicle build variability and test procedure variability would totally mask any effect of Hybrid III repeatability and reproducibility."

The agency agrees with GM that additional testing is unnecessary. The information Provided by GM and Volvo shows that the repeatability of the Hybrid III is at least as good as the repeatability of the existing Part 572 test dummy. Likewise, the GM data show that the reproducibility of the Hybrid III is better than that of the existing Part 572 test dummy. Likewise, the recent NHTSA-GM calibration test series provides further confirmation that tests by different laboratories show the repeatability and reproducibility of the Hybrid III.

Equivalence of Hybrid III and Part 572

As noted in the April 1985 notice, the Hybrid III and the Part 572 test dummies do not generate identical impact responses. Based on the available data, the agency concluded that when both test dummies are tested in lap/shoulder belts or with air cushions, the differences between the two test dummies are minimal. The agency also said that it knew of no method for directly relating the response of the Hybrid III to the Part 572 test dummy.

The purpose of comparing the response of the two test dummies is to ensure that the Hybrid III will meet the need for safety by adequately identifying vehicle designs which could cause or increase occupant injury. The agency wants to ensure that permitting a choice of test dummy will not lead to a degradation in safety performance.

As mentioned previously, one major improvement in the Hybrid III is that it is more human-like in its responses than the current Part 572 test dummy. The primary changes to the Hybrid III that make it more human-like are to the neck, chest and knee. Comparisons of the responses of the Part 572 and Hybrid III test dummies show that responses of the Hybrid III are closer than the Part 572 to the best available data on human responses. (See Chapter II of the Final Regulatory Evaluation on the Hybrid III.)

In addition to being more human-like, the Hybrid III has increased measurement capabilities for the neck (tension, compression, and shear forces and bending moments), chest (deflection), knee (knee shear), and lower leg (knee and tibia forces and moments). The availability of the extra injury measuring capability of the Hybrid III gives vehicle manufacturers the potential for gathering far more information about the performance of their vehicle designs than they can obtain with the Part 572.

To evaluate differences in the injury measurements made by the Hybrid III and the existing Part 572 test dummy, the agency has reviewed all of the available data comparing the two test dummies. The data come from a variety of sled barrier crash tests conducted by GM, Mercedes-Benz, NHTSA, Nissan, and Volvo. The data include tests where the dummies were unrestrained and tests where the dummies were restrained by manual lap/shoulder belts, automatic belts, and air bags. For example, subsequent to issuance of the April 1985 notice, NHTSA did additional vehicle testing to compare the Part 572 and Hybrid III test dummies. The agency conducted a series of crash tests using five different types of vehicles to measure differences in the responses of the test dummies. Some of the tests were frontal 30 mile per hour barrier impacts, such as are used in Standard No. 208 compliance testing, while others were car-to-car tests. All of the tests were done with unrestrained test dummies to measure their impact responses under severe conditions. The agency's analysis of the data for all of the testing done by NHTSA and others is fully described in the Final Regulatory Evaluation for this rulemaking. This notice will briefly review that analysis.

One of the reasons for conducting the analysis was to address the concern raised by the Center for Auto Safety (CAS) in its original petition and the Insurance Institute for Highway Safety (IIHS) in its comments that the Hybrid III produces lower HIC responses than the existing Part 572 test dummy. As discussed in detail below, the test data do not show a trend for one type of test dummy to consistently measure higher or lower HIC's than the other. Based on these test data, the agency concludes that the concern expressed by CAS and IIHS that the use of the Hybrid III test dummy will give a manufacturer an advantage in meeting the HIC performance requirement of Standard No. 208 is not valid.

In the case of chest acceleration measurements, the data again do not show consistently higher or lower measurements for either test dummy, except in the case of unrestrained tests. In unrestrained tests, the data show that the Hybrid III generally measures lower chest g's than the existing Part 572 test dummy. This difference in chest g's measurement is one reason why the agency is adopting the additional chest deflection measurement for the Hybrid III, as discussed further below.

HIC Measurements

The April 1985 notice specifically invited comments on the equivalence of the Head Injury Criterion (HIC) measurements of the two test dummies. Limited laboratory testing done in a University of California at San Diego study conducted by Dr. Dennis Schneider and others had indicated that the Hybrid III test dummy generates lower acceleration responses than either the Part 572 test dummy or cadaver heads in impacts with padded surfaces. The notice explained that the reasons for those differences had not yet been resolved.

In its comments, GM explained that it had conducted a series of studies to address the Schneider results. GM said that those studies showed that the Schneider test results are "complicated by the changing characteristics of the padding material used on his impact surface. As a result, his tests do not substantiate impactor response difference between the Hybrid III head, the Part 572 head and cadaver heads. After examining our reports, Dr. Schneider agreed with the finding that padding degradation resulting from multiple impact exposures rendered an input-response comparison invalid between the cadaver and the dummies." (The GM and Schneider letters are filed in Docket 74-14, General Reference, Entry 556.)

The agency's Vehicle Research and Test Center has also conducted head drop tests of the current Part 572 and Hybrid III heads. The tests were conducted by dropping the heads onto a two inch thick steel plate, a surface which is considerably more rigid than any surface that the test dummy's head would hit in a vehicle crash test. One purpose of the tests was to assess the performance of the heads in an impact which can produce skull fractures in cadavers. The tests found that the response of the Hybrid III head was more human-like at the fracture and subfracture acceleration levels than the Part 572 head. The testing did show that in these severe impacts into thick steel plates, the HIC scores for the Hybrid III were lower than for the Part 572. However, as discussed below, when the Hybrid III is tested in vehicle crash and sled tests, which are representative of occupant impacts into actual vehicle structures, the HIC scores for the Hybrid III are not consistently lower than those of the Part 572 test dummy.

The agency examined crash and sled tests, done by GM, Mercedes-Benz, NHTSA and Volvo, in which both a Hybrid III and the existing Part 572 test dummy were restrained by manual lap/shoulder belts. (The complete results from those and all the other tests reviewed by the agency are discussed in Chapter III of the Final Regulatory Evaluation on the Hybrid III.) The HIC responses in those tests show that the Hybrid III generally had higher HIC responses than the Part 572 test dummy. Although the data show that the Hybrid III's HIC responses are generally higher, in some cases 50 percent higher than the Part 572, there are some tests in which the Hybrid III's responses were 50 percent lower than the responses of the Part 572.

For two-point automatic belts, the agency has limited barrier crash test data and the direct comparability of the data is questionable. The tests using the existing Part 572 test dummy were done in 1976 on 1976 VW Rabbits for compliance purposes. The Hybrid III tests were done in 1985 by the agency's Vehicle Research and Test Center as part of the SRL-98 test series on a 1982 and a 1984 VW Rabbit. Differences in the seats, safety belts, and a number of other vehicle parameters between these model years and between the test set-ups could affect the results. In the two-point automatic belt tests, the data show that the Hybrid III measured somewhat higher head accelerations than the existing Part 572 test dummy. In two-point automatic belts, the differences appear to be minimal for the driver and substantially larger for the passenger. In air bag sled tests, the Hybrid III's HIC responses were generally lower; in almost all the air bag tests, the HIC responses of both the Hybrid III and the Part 572 test dummies were substantially below the HIC limit of 1,000 set in Standard No. 208. Because of the severe nature of the unrestrained sled and barrier tests, in which the uncontrolled movement of the test dummy can result in impacts with different vehicle structures, there was no consistent trend for either test dummy to measure higher or lower HIC responses than the other.

Chest Measurements

For manual lap/shoulder belts, NHTSA compared the results from GM, Mercedes-Benz, NHTSA, and Volvo sled tests, and GM frontal barrier tests. The NHTSA sled test results at 30 and the Volvo sled test results at 31 mph are very consistent, with the mean Hybrid III chest acceleration response being only 2–3 g's higher than the response of the existing Part 572 test dummy. In the 35 mph Volvo sled tests, the Hybrid III chest acceleration response was up to 44 percent higher than the existing Part 572 response. The GM 30 mph sled and barrier test data were fairly evenly divided. In general, the Hybrid III chest acceleration response is slightly higher than that of the existing Part 572 test dummy. The agency concludes from these data that at Standard No. 208's compliance test speed (30 mph) with manual lap/shoulder belts there are no large differences in chest acceleration responses between the two dummies. In some vehicles, the Hybrid III may produce slightly higher responses and in other vehicles it may produce slightly lower responses.

As discussed earlier, the agency has limited test data on automatic belt tests and their comparability is questionable. The Hybrid III chest acceleration responses are up to 1.5 times higher than those for the existing Part 572 test dummy. Only very limited sled test data are available on air bags alone, air bag plus lap belt, and air bag plus lap/shoulder belt. In all cases, the Hybrid III chest acceleration responses were lower than those for the existing Part 572 test dummy.

For unrestrained occupants, the Hybrid III produces predominantly lower chest acceleration responses than the existing Part 572 test dummy in sled and barrier tests, and in some cases the difference is significant. In some tests, the Hybrid III chest acceleration response can be 40 to 45 percent lower than the Part 572 response, although in other tests the acceleration measured by the Hybrid III can exceed that measured by the Part 572 test dummy by 10 to 15 percent.

In summary, the test data indicate the chest acceleration responses between the Hybrid III and the existing Part 572 test dummy are about the same for restrained occupants, but differ for some cases of unrestrained occupants. This is to be expected since a restraint system would tend to make the two dummies react similarly even though they have different seating postures. The different seating postures, however, would allow unrestrained dummies to impact different vehicle surfaces which would in most instances produce different responses. Since the Hybrid III dummy is more human-like, it should experience loading conditions that are more human-like than would the existing Part 572 test dummy. One reason that the agency is adding a chest deflection criterion for the Hybrid III is that the unrestrained dummy's chest may experience more severe impacts with vehicle structures than would be experienced in an automatic belt or air bag collision. Chest deflection provides an additional measurement of potential injury that may not be detected by the chest acceleration measurement.

Femur Measurements

The test data on the femur responses of the two types of test dummies also do not show a trend for one test dummy to measure consistently higher or lower responses than the other. In lap/shoulder belt tests, GM's sled and barrier tests from 1977 show a trend toward lower measurements for the Hybrid III, but GM's more recent tests in 1982–83 show the reverse situation. These tests, however, are of little significance unless there is femur loading due to knee contact. These seldom occur to lap/shoulder belt restrained test dummies. Also, in none of the tests described above do the measurements approach Standard No. 208's limit of 2250 pounds for femur loads. The air bag test data are limited; however, they show little difference between the femur responses of the two test dummies. As would be expected, the unrestrained tests showed no systematic differences, because of the variability in the impact locations of an unrestrained test dummy.

Injury Criteria

Many manufacturers raised objections to the additional injury criteria proposed in the April 1985 notice. AMC, Ford, and MVMA argued that adopting the numerous injury criteria proposed in the April 1985 notice would compound a manufacturer's compliance test problems. For example, Ford said it "would be impracticable to require vehicles to meet such a multitude of criteria in a test with such a high level of demonstrated variability. Notice 39 appears to propose 21 added pass-fail measurements per dummy, for a total of 25 pass-fail measurements per dummy, or 50 pass-fail measurements per test. Assuming these measurements were all independent of one another, and a car design had a 95% chance of obtaining a passing score on each measurement, the chance of obtaining a passing score on all measurements in any single test for a single dummy would be less than 28% and for both dummies would be less than 8%." Ford, Nissan, VW and Volvo also said that with the need for additional measurements, there will be an increase in the number of tests with incomplete data. BMW, while supporting the use of the Hybrid III as a potential improvement to safety, said that the number of measurements needed for the additional injury criteria is beyond the capability of its present data processing equipment.

VW said there is a need to do additional vehicle testing before adopting any new criteria. It said that if current production vehicles already meet the additional criteria then the criteria only increase testing variability without increasing safety. If current vehicles cannot comply, then additional information is needed about the countermeasures needed to meet the criteria. Honda said there are insufficient data to determine the relationship between actual injury levels and the proposed injury criterion.

As discussed in detail below, the agency has decided to adopt only one additional injury criterion, chest deflection, at this time. The agency plans to issue another notice on the remaining criteria proposed in the April 1985 notice to gather additional information on the issues raised by the commenters.

Alternative HIC Calculations

The April 1985 notice set forth two proposed alternative methods of using the head injury criterion

(HIC) in situations when there is no contact between the test dummy's head and the vehicle's interior during a crash. The first proposed alternative was to retain the current HIC formula, but limit its calculation to periods of head contact only. However, in non-contact situations, the agency proposed that an HIC would not be calculated, but instead new neck injury criteria would be calculated. The agency explained that a crucial element necessary for deciding whether to use the HIC calculation or the neck criteria was an objective technique for determining the occurrence and duration of head contact in the crash test. As discussed in detail in the April 1985 notice, there are several methods available for establishing the duration of head contact, but there are questions about their levels of consistency and accuracy.

The second alternative proposed by the agency would have calculated an HIC in both contact and non-contact situations, but it would limit the calculation to a time interval of 36 milliseconds. Along with the requirement that an HIC not exceed 1,000, this would limit average head acceleration to 60 g's or less for any durations exceeding 36 milliseconds.

Almost all of the commenters opposed the use of the first proposed alternative. The commenters uniformly noted that there is no current technique that can accurately identify whether head contact has or has not occured during a crash test in all situations. However, the Center for Auto Safety urged the agency to adopt the proposed neck criteria, regardless of whether the HIC calculation is modified.

There was a sharp division among the commenters regarding the use of the second alternative; although many manufacturers argued that the HIC calculation should be limited to a time interval of approximately 15 to 17 milliseconds (ms), which would limit average long duration (i.e., greater than 15–17 milliseconds) head accelerations to 80–85 g's. Mercedes-Benz, which supported the second alternative, urged the agency to measure HIC only during the time interval that the acceleration level in the head exceeds 60 g's. It said that this method would more effectively differentiate results received in contacts with hard surfaces and results obtained from systems, such as airbags, which provide good distribution of the loads experienced during a crash. The Center for Auto Safety, the Insurance Institute for Highway Safety and State Farm argued that the current HIC calculation should be retained; they said that the proposed alternative would lower HIC calculations without ensuring that motorists were still receiving adequate head protection.

NHTSA is in the process of reexamining the potential effects of the two alternatives proposed by the agency and of the two additional alternatives suggested by the commenters. Once that review has been completed, the agency will issue a separate notice announcing its decision.

Thorax

At present, Standard No. 208 uses an acceleration-based criterion to measure potential injuries to the chest. The agency believes that the use of a chest deflection criterion is an important supplement to the existing chest injury criterion. Excessive chest deflection can produce rib fractures, which can impair breathing and inflict damage to the internal organs in the chest. The proposed deflection limit would only apply to the Hybrid III test dummy, since unlike the existing Part 572 test dummy, it has a chest which is designed to deflect like a human chest and has the capability to measure deflection of the sternum relative to the spine, as well as acceleration, during an impact.

The agency proposed a three-inch chest deflection limit for systems, such as air bags, which symmetrically load the chest during a crash and a two-inch limit for all other systems. The reason for the different proposed limits is that a restraint system that symmetrically and uniformly applies loads to the chest increases the ability to withstand chest deflection as measured by the deflection sensor, which is centrally located in the dummy.

The commenters generally supported adoption of a chest deflection injury criterion. For example, Ford said it supported the use of a chest deflection criterion since it may provide a better means of assessing the risk of rib fractures. Likewise, the Insurance Institute for Highway Safety said the chest deflection criteria."will aid in evaluating injury potential especially in situations where there is chest contact with the steering wheel or other interior components." IIHS also supported adoption of a three-inch deflection limit for inflatable systems and a two-inch limit for all other systems. However, most of the other commenters addressing the proposed chest deflection criteria questioned the use of different criteria for different restraint systems.

GM supported limiting chest deflections to three-inches in all systems. GM said that it uses a two-inch limit as a guideline for its safety belt system testing, but it had no data to indicate that the two-inch limit is appropriate as a compliance limit.

Renault/Peugeot also questioned the three-inch deflection limit for systems that load the dummy symmetrically and two inches for systems that do not. It said that the difference between those systems should be addressed by relocation of the deflection sensors. It also asked the agency to define what constitutes a symmetrical system. VW also questioned the appropriateness of setting separate limits for chest compression for different types of restraint systems. It recommended adoption of a three-inch limit for all types of restraint systems.

Volvo also raised questions about the appropriateness of the proposed deflection criteria. Volvo said that the GM-developed criteria proposed in the April 1985 notice were based on a comparison of accident data gathered by Volvo and evaluated by GM in sled test simulations using the Hybrid III test dummy. Volvo said that the report did not analyze "whether the chest injuries were related to the chest acceleration or the chest deflection, or a combination of both."

The agency recognizes that there are several different types of potential chest injury mechanisms and that it may not be possible to precisely isolate and measure what is the relevant contribution of each type of mechanism to the final resulting injury. However, there is a substantial amount of data indicating that chest deflection is an important contributing factor to chest injury. In addition, the data clearly demonstrate that deflection of greater than three inches can lead to serious injury. For example, research done by Neathery and others has examined the effects of frontal impacts to cadaver chests with an impactor that represents the approximate dimensions of a steering wheel hub. Neathery correlated the measured injuries with the amount of chest deflection and recommended that for a 50th percentile male, chest deflection not exceed three inches. (Neathery, R. F., "Analysis of Chest Impact Response Data and Scaled Performance Recommendations," SAE Paper No. 741188)

Work by Walfisch and others looked at crash tests of lap/shoulder belt restrained cadavers. They found that substantial injury began to occur when the sternum deflection exceeded 30 percent of the available chest depth ("Tolerance Limits and Mechanical Characteristic of the Human Thorax in Frontal and Side Impact and Transposition of these Characteristics into Protective Criteria," 1982 IRCOBI Conference Proceedings). With the chest of the average man being approximately 9.3 inches deep, the 30 percent limit would translate into a deflection limit of approximately 2.8 inches. Since the chest of the Hybrid III test dummy deflects somewhat less than a human chest under similar loading conditions, the chest deflection limit for systems which do not symmetrically and uniformly load the chest, such as lap/shoulder belts, must be set at a level below 2.8 inches to assure an adequate level of protection.

To determine the appropriate level for non-symmetrical systems, the agency first reviewed a number of test series in which cadaver injury levels were measured under different impact conditions. (All of the test results are fully discussed in Chapter III of the Final Regulatory Evaluation on the Hybrid III.) The impact conditions included 30 mph sled tests done for the agency by Wayne State University in which a pre-inflated, non-vented air bag system symmetrically and uniformly spread the impact load on the chest of the test subject. NHTSA also reviewed 30 mph sled tests done for the agency by the University of Heidelberg which used a lap/shoulder belt system, which does not symmetrically and uniformly spread chest loads. In addition, the agency reviewed 10 and 15 mph pendulum impact tests done for GM to evaluate the effects of concentrated loadings, such as might occur in passive interior impacts. The agency then compared the chest deflection results for Hybrid III test dummies subjected to the same impact conditions. By comparing the cadaver and Hybrid III responses under identical impact conditions, the agency was able to relate the deflection measurements made by the Hybrid III to a level of injury received by a cadaver.

The test results show that when using a relatively stiff air bag, which was pre-inflated and non-vented, the average injury level measured on the cadavers corresponded to an Abbreviated Injury Scale (AIS) of 1.5. (The AIS scale is used by researchers to classify injuries an AIS of one is a minor injury, while an AIS of three represents a serious injury.) In tests with the Hybrid III under the same impact conditions, the measured deflection was 2.7 inches. These results demonstrate that a system that symmetrically and uniformly distributes impact loads over the chest can produce approximately three-inches of deflection and still adequately protect an occupant from serious injury.

The testing in which the impact loads were not uniformly or symmetrically spread on the chest or were highly concentrated over a relatively small area indicated that chest deflection measured on the Hybrid III must be limited to 2-inches to assure those systems provide a level of protection comparable to that provided by systems that symmetrically spread the load. In the lap/shoulder belt tests, the average AIS was 2.6. The measured deflection for the Hybrid III chest in the same type of impact test was 1.6 inches. Likewise, the results from the

pendulum impact tests showed that as the chest deflection measured on the Hybrid III increased, the severity of the injuries increased. In the 10 mph pendulum impacts, the average AIS was 1.3 and the average deflection was 1.3 inches. In the 15 mph pendulum impacts the average AIS rose to 2.8. Under the same impact conditions, the chest deflection measured on the Hybrid III was 2.63 inches.

Based on these test results NHTSA has decided to retain the two-inch limit on chest deflection for systems that do not symmetrically and uniformly distribute impact loads over a wide area of the chest. Such systems include automatic safety belts, passive interiors and air bag systems which use a lap and shoulder belt. For systems, such as air bag only systems or air bag combined with a lap belt, which symmetrically and uniformly distribute chest forces over a large area of the chest, the agency is adopting the proposed three-inch deflection limit. This should assure that both symmetrical and non-symmetrical systems provide the same level of protection in an equivalent frontal crash.

In addition to the biomechanical basis for the chest deflection limits adopted in this notice, there is another reason for adopting a two-inch deflection limit for systems that can provide concentrated loadings over a limited area of the test dummy. The Hybrid III measures chest deflection by a deflection sensor located near the third rib of the test dummy. Tests conducted on the Hybrid III by NHTSA's Vehicle Research and Test Center have shown that the deflection sensor underestimates chest displacement when a load is applied to a small area away from the deflection sensor. (The test report is filed in Docket No. 74-14, General Reference, Entry 606.)

In a crash, when an occupant is not restrained by a system which provides centralized, uniform loading to a large area, such as an air bag system, the thorax deflection sensor can underestimate the actual chest compression. Thus, in a belt-restrained test dummy, the deflection sensor may read two-inches of deflection, but the actual deflection caused by the off-center loading of a belt near the bottom of the ribcage may be greater than two inches of deflection. Likewise, test dummies in passive interior cars may receive substantial off-center and concentrated loadings. For example, the agency has conducted sled tests simulating 30 mile per hour frontal barrier impacts in which unrestrained test dummies struck the steering column, as they would do in a passive interior equipped car. Measurements of the pre- and post-impact dimensions of the steering wheel rim showed that there was substantial non-symmetrical steering wheel deformation, even though these were frontal impacts. (See, e.g.,

"Frontal Occupant Sled Simulation Correlation, 1983 Chevrolet Celebrity Sled Buck," Publication No. DOT HS 806 728, February 1985.) The expected off-center chest loadings in belt and passive interior systems provide a further basis for applying a two-inch deflection limit for those systems to assure they provide protection comparable to that provided by symmetrical systems.

Use of Acceleration Limits for Air Bag Systems

Two commenters raised questions about the use of an acceleration-based criterion for vehicles which use a combined air bag and lap/shoulder belt system. Mercedes-Benz said that acceleration-based criteria are not appropriate for systems that reduce the deflection of the ribs but increase chest acceleration values. Ford also questioned the use of acceleration-based criteria. Ford said that its tests and testing done by Mercedes-Benz have shown that using an air bag in combination with a lap/shoulder belt can result in increased chest acceleration readings. Ford said it knew of no data to indicate that combined air bag-lap/shoulder belt system loads are more injurious than shoulder belt loads alone. Ford recommended that manufacturers have the option of using either the chest acceleration or chest deflection criterion until use of the Hybrid III is mandatory.

As discussed previously, acceleration and deflection represent two separate types of injury mechanisms. Therefore, the agency believes that it is important to test for both criteria. Although the tests by Mercedes-Benz and Ford show higher chest accelerations, the tests also show that it is possible to develop air bag and lap/shoulder belt systems and meet both criteria. Therefore, the agency is retaining the use of the acceleration-based criterion.

Use of Additional Sensors

Mercedes-Benz said the deflection measuring instrumentation of the Hybrid III cannot adequately measure the interaction between the chest and a variety of vehicle components. Mercedes-Benz said that it is necessary to use either additional deflection sensors or strain gauges. Renault/Peugeot recommended that the agency account for the difference between symmetrical systems and asymmetrical systems by relocating the deflection sensor.

The agency recognizes that the use of additional sensors could be beneficial in the Hybrid III to measure chest deflection. However, such technology would require considerable further development before it could be used for compliance purposes. NHTSA believes that, given the current level of technology, use of a single sensor is sufficient for the assessment of deflection-caused injuries in frontal impacts.

Femurs

The April 1985 notice proposed to apply the femur injury reduction criterion used with the Part 572 test dummy to the Hybrid III. That criterion limits the femur loads to 2250 pounds to reduce the possibility of femur fractures. No commenter objected to the proposed femur limit and it is accordingly adopted.

Ford and Toyota questioned the need to conduct three pendulum impacts for the knee. They said that using one pendulum impact with the largest mass impactor (11 pounds) was sufficient. GM has informed the agency that the lower mass pendulum impactors were used primarily for the development of an appropriate knee design. Now that the knee design is settled and controlled by the technical drawings, the tests with the low mass impactors are not needed. Accordingly, the agency is adopting the suggestion from Ford and Toyota to reduce the number of knee calibration tests and will require only the use of the 11-pound pendulum impactor.

Hybrid III Positioning Procedure

The April notice proposed new positioning procedures for the Hybrid III, primarily because the curved lumbar spine of that test dummy requires a different positioning technique than those for the Part 572. Based on its testing experience, NHTSA proposed adopting a slightly different version of the positioning procedure used by GM. The difference was the proposed use of the Hybrid III, rather than the SAE J826 H-point machine, with slightly modified leg segments, to determine the H-point of the seat.

GM urged the agency to adopt its dummy positioning procedure. GM said that users can more consistently position the test dummy's H-point using the SAE H-point machine rather than using the Hybrid III. Ford, while explaining that it had insufficient experience with the Hybrid III to develop data on positioning procedures, also urged the agency to adopt GM's positioning procedure. Ford said that since GM has developed its repeatability data on the Hybrid III using its positioning procedure, the agency should use it as well. Ford also said that the use of GM's method to position the test dummy relative to the H-point should reduce variability.

Based on a new series of dummy positioning tests done by the agency's Vehicle Research and Test Center (VRTC), NHTSA agrees that use of the SAE H-point machine is the most consistent method to position the dummy's H-point on the vehicle seat.

Accordingly, the agency is adopting the use of the H-point machine.

In the new test series, VRTC also evaluated a revised method for positioning the Hybrid III test dummy. The testing was done after the results of a joint NHTSA-SAE test series conducted in November 1985 showed that the positioning procedure used for the current Part 572 test dummy and the one proposed in the April 1985 notice for the Hybrid III does not satisfactorily work in all cars. (See Docket 74-14, Notice 39, Entry 39.) The positioning problems are principally due to the curved lumbar spine of the Hybrid III test dummy. In its tests, VRTC positioned the Hybrid III by using the SAE H-point machine and a specification detailing the final position of the Hybrid III body segments prior to the crash test. The test results showed that the H-point of the test dummy could be consistently positioned but that the vertical location of the Hybrid III H-point is ¼ inch below the SAE H-point machine on average. Based on these results, the agency is adopting the new positioning specification for the Hybrid III which requires the H-point of the dummy to be within a specified zone centered ¼ inch below the H-point location of the SAE H-point machine.

GM also urged the agency to make another slight change in the test procedures. GM said that when it settles the test dummy in the seat it uses a thin sheet of plastic behind the dummy to reduce the friction between the fabric of the seat back and the dummy. The plastic is removed after the dummy has been positioned. GM said this technique allows the dummy to be more repeatably positioned. The agency agrees that use of the plastic sheet can reduce friction between the test dummy and the seat. However, the use of the plastic can also create problems, such as dislocating the test dummy during removal of the plastic. Since the agency has successfully conducted its positioning tests without using a sheet of plastic, the agency does not believe there is a need to require its use.

Ford noted that the test procedure calls for testing vertically adjustable seats in their lowest position. It said such a requirement was reasonable for vertically adjustable seats that could not be adjusted higher than seats that are not vertically adjustable. However, Ford said that new power seats can be adjusted to positions above and below the manually adjustable seat position. It said that testing power seats at a different position would increase testing variability. Ford recommended adjusting vertically adjustable seats so that the dummy's hip point is as close as possible to the manufacturer's design

H-point with the seat at the design mid-point of its travel.

The agency recognizes that the seat adjustment issue raised by Ford may lead to test variability. However, the agency does not have any data on the effect of Ford's suggested solution on the design of other manufacturer's power seats. The agency will solicit comments on Ford's proposal in the NPRM addressing additional Hybrid III injury criteria.

Volvo said that the lumbar supports of its seats influence the positioning of the Hybrid III. It requested that the test procedure specify that adjustable lumbar supports should be positioned in their rearmost position. Ford made a similar request. GM, however, indicated that it has not had any problems positioning the Hybrid III in seats with lumbar supports. To reduce positioning problems resulting from the lumbar supports in some vehicles, the agency is adopting Ford's and Volvo's suggestion.

Test Data Analysis

The Chairman of the Society of Automotive Engineers Safety Test Instrumentation Committee noted that the agency proposed to reference an earlier version of the SAE Recommended Practice on Instrumentation (SAE J211a, 1971). He suggested that the agency reference the most recent version (SAE J211, 1980), saying that better data correlation between different testing organizations would result. The agency agrees with SAE and is adopting the SAE J211, 1980 version of the instrumentation Recommended Practice.

Ford and GM recommended that the figures 25 and 26, which proposed a standardized coordinate system for major body segments of the test dummy, be revised to reflect the latest industry practice on coordinate signs. Since those revisions will help ensure uniformity in data analysis by different test facilities, the agency is making the changes for the test measurements adopted in this rulemaking.

Both GM and Ford also recommended changes in the filter used to process electronically measured crash data. GM suggested that a class 180 filter be used for the neck force transducer rather than the proposed class 60 filter. Ford recommended the use of a class 1,000 filter, which is the filter used for the head accelerometer.

NHTSA has conducted all of the testing used to develop the calibration test requirement for the neck using a class 60 filter. The agency does not have any data showing the effects of using either the class 180 filter proposed by GM or the class 1,000 filter proposed by Ford. Therefore the agency has adopted the use of a class 60 filter for the neck transducer during the calibration test. The agency also used a class 60 filter for the accelerometer mounted on the neck pendulum and is therefore adopting the use of that filter to ensure uniformity in measuring pendulum acceleration.

Optional and Mandatory Use of Hybrid III

AMC, Chrysler, Ford, Jaguar and Subaru all urged the agency to defer a decision on permitting the optional use of the Hybrid III test dummy until manufacturers have had more experience with using that test dummy. AMC said it has essentially no experience with the Hybrid III and urged the agency to postpone a decision on allowing the optional use of that test dummy. AMC said this would give small manufacturers time to gain experience with the Hybrid III.

Chrysler also said that it has no experience with the Hybrid III test dummy and would need to conduct two years of testing to be able to develop sufficient information to address the issues raised in the notice. Chrysler said that it was currently developing its 1991 and 1992 models and has no data from Hybrid III test dummies on which to base its design decisions. It said that allowing the optional use of the Hybrid III before that time would give a competitive advantage to manufacturers with more experience with the test device and suggested indefinitely postponing the mandatory effective date.

Ford said that the effective date proposed for optional use of the Hybrid III should be deferred to allow time to resolve the problems Ford raised in its comments and to allow manufacturers time to acquire Hybrid III test dummies. It suggested deferring the proposed optional use until at least September 1, 1989. Ford also recommended that the mandatory use be deferred. Jaguar also said it has not had experience with the Hybrid III and asked that manufacturers have until September 1, 1987, to accumulate information on the performance of the test dummy. Subaru said that it has exclusively used the Part 572 test dummy and does not have any experience with the Hybrid III. It asked the agency to provide time for all manufacturers to gain experience with the Hybrid III, which in its case would be two years, before allowing the Hybrid III as an alternative.

A number of manufacturers, such as GM, Honda, Mercedes-Benz, Volkswagen, and Volvo, that supported optional use of the Hybrid III, urged the agency not to mandate its use at this time. GM asked the agency to permit the immediate optional use of the Hybrid III, but urged NHTSA to provide more

time for all interested parties to become familiar with the test dummy before mandating its use. Honda said that while it supported optional use, it was just beginning to assess the performance of the Hybrid III and needed more time before the use of the Hybrid III is mandated. Mercedes-Benz also supported the use of the Hybrid III as an alternative test device because of its capacity to measure more types of injuries and because of its improved biofidelity for the neck and thorax. However, Mercedes recommended against mandatory use until issues concerning the Hybrid III's use in side impact, the biofidelity of its leg, durability and chest deflection measurements are resolved. Nissan opposed the mandatory us of the Hybrid III saying there is a need to further investigate the differences between the Hybrid III and the Part 572. Toyota said that it was premature to set a mandatory effective date until the test procedure and injury criteria questions are resolved. Volkswagen supported the adoption of the Hybrid III as an alternative test device, but it opposed mandating its use. Volvo supported the optional use of the Hybrid III. It noted that since NHTSA is developing an advanced test dummy, there might not be a need to require the use of the Hybrid III in the interim.

The agency recognizes that manufacturers are concerned about obtaining the Hybrid III test dummy and gaining experience with its use prior to the proposed September 1, 1991, date for mandatory use of that test dummy. However, information provided by the manufacturers of the Hybrid III shows that it will take no longer than approximately one year to supply all manufacturers with sufficient quantities of Hybrid III's. This means that manufacturers will have, at a minimum, more than four years to gain experience in using the Hybrid III. In addition, to assist manufacturers in becoming familiar with the Hybrid III, NHTSA has been placing in the rulemaking docket complete information on the agency's research programs using the Hybrid III test dummy in crash and calibration tests. Since manufacturers will have sufficient time to obtain and gain experience with the Hybrid III by September 1, 1991, the agency has decided to mandate use of the Hybrid III as of that date.

As discussed earlier in this notice, the evidence shows that the Hybrid III is more human-like in its responses to impacts than the existing Part 572 test dummy. In addition, the Hybrid III has the capability to measure far more potential injuries than the current test dummy. The agency is taking advantage of that capability by adopting a limitation on chest deflection which will enable NHTSA to measure a significant source of injury that cannot be measured on the current test dummy. The combination of the better biofidelity and increased injury-measuring capability available with the Hybrid III will enhance vehicle safety.

Adoption of the Hybrid III will not give a competitive advantage to GM, as claimed by some of the commenters, such as Chrysler and Ford. As the developer of the Hybrid III, GM obviously has had more experience with that test dummy than other manufacturers. However, as discussed above, the agency has provided sufficient leadtime to allow all manufacturers to develop sufficient experience with the Hybrid III test dummy. In addition, as discussed in the equivalency section of this notice, there are no data to suggest that it will be easier for GM or other manufacturers to meet the performance requirements of Standard No. 208 with the Hybrid III. Thus GM and other manufacturers using Hybrid III during the phase-in period will not have a competitive advantage over manufacturers using the existing Part 572 test dummy.

Finally, in its comments GM suggested that the agency consider providing manufacturers with an incentive to use the Hybrid III test dummy. GM said that the agency should consider providing manufacturers with extra vehicle credits during the automatic restraint phase-in period for using the Hybrid III. The agency does not believe it is necessary to provide any additional incentive to use the Hybrid III. The mandatory effective date for use of the Hybrid III provides sufficient incentive, since manufacturers will want to begin using the Hybrid III as soon as possible to gain experience with the test dummy before that date.

Optional use of the Hybrid III may begin October 23, 1986. The agency is setting an effective date of less than 180 days to facilitate the efforts of those manufacturers wishing to use the Hybrid III in certifying compliance with the automatic restraint requirements.

Use of Non-instrumented Test Dummies

Ford raised a question about whether the Hybrid III may or must be used for the non-crash performance requirements of Standard No. 208, such as the comfort and convenience requirements of S7.4.3, 7.4.4, and 7.4.5 of the standard. Ford said that manufacturers should be given the option of using either the Part 572 or Hybrid III test dummy to meet the comfort and convenience requirements. The agency agrees that until September 1, 1991, manufacturers should have the option of using either the Part 572 or Hybrid III test dummy. However, since it is important the crash performance requirements and comfort and convenience

requirements be linked together through the use of a single test dummy to measure a vehicle's ability to meet both sets of requirements. Therefore, beginning on September 1, 1991, use of the Hybrid III will be mandatory in determining a vehicle's compliance with any of the requirements of Standard No. 208.

In addition, Ford asked the agency to clarify whether manufacturers can continue to use Part 572 test dummies in the crash tests for Standard Nos. 212, 219, and 301, which only use non-instrumented test dummies to simulate the weight of an occupant. Ford said that the small weight difference and the small difference in seated posture between the two test dummies should have no effect on the results of the testing for Standard Nos. 212, 219, and 301. The agency agrees that use of either test dummy should not affect the test results for those standards. Thus, even after the September 1, 1991, effective date for use of the Hybrid III in the crash and non-crash testing required by Standard No. 208, manufacturers can continue to use, at their option, either the Part 572 or the Hybrid III test dummy in tests conducted in accordance with Standard Nos. 212, 219, and 301.

Economic and Other Impacts

NHTSA has examined the impact of this rulemaking action and determined that it is not major within the meaning of Executive Order 12291 or significant within the meaning of the Department of Transportation's regulatory policies and procedures. The agency has also determined that the economic and other impacts of this rulemaking action are not significant. A final regulatory evaluation describing those effects has been placed in the docket.

In preparing the regulatory evaluation, the agency has considered the comments from several manufacturers that the agency had underestimated the costs associated with using the Hybrid III. Ford said that the cost estimates contained in the April 1985 notice did not take into account the need to conduct sled tests during development work. Ford said that for 1985, it estimated it will conduct 500 sled tests requiring 1000 test dummy applications. Ford also said that NHTSA's estimate of the test dummy inventory needed by a manufacturer is low. It said that it currently has an inventory of 31 Part 572 test dummies and would expect to need a similar inventory of Hybrid III's. In addition, Ford said that NHTSA's incremental cost estimate of $3,000 per test dummy was low. It said that the cost for monitoring the extra data generated by the Hybrid III is $2,700. Ford said that it also would have to incur costs due to upgrading its data acquisition and data processing equipment.

GM said that NHTSA's estimate of a 30-test useful life for the test dummy substantially underestimates its actual useful life, assuming the test dummy is repaired periodically. It said that some of its dummies have been used in more than 150 tests. GM also said that the agency's assumption that a large manufacturer conducts testing requiring approximately 600 dummy applications each year underestimates the actual number of tests conducted. In 1984, GM said it conducted sled and barrier tests requiring 1179 dummy applications. GM said that the two underestimates, in effect, cancel each other out, since the dummies are usable for at least five times as many tests, but they are used four times as often.

Mitsubishi said that its incremental cost per vehicle is $7 rather than 40 cent as estimated by the agency. Mitsubishi explained the reason for this difference is that the price of an imported Hybrid III is approximately two times the agency estimate and its annual production is about one-tenth of the amount used in the agency estimate. Volvo also said the agency had underestimated the incremental cost per vehicle. Volvo said it conducts approximately 500–600 test dummy applications per year in sled and crash testing, making the incremental cost in the range of $15–18 per vehicle based on its export volume to the United States.

NHTSA has re-examined the costs associated with the Hybrid III test dummy. The basic Hybrid III dummy with the instrumentation required by this final rule costs $35,000 or approximately $16,000 more than the existing 572 test dummy. Assuming a useful life for the test dummy of 150 tests, the total estimated incremental capital cost is approximately $107 per dummy test.

To determine the incremental capital cost per test, the agency had to estimate the useful life of the Hybrid III. Based on NHTSA's test experience, the durability of the existing Part 572 test dummy and the Hybrid III test dummy is essentially identical with the exception of the Hybrid III ribs. Because the Hybrid III dummy chest was developed to simulate human chest deflection, the ribs had to be designed with much more precision to reflect human impact response. This redesign uses less metal and consequently they are more susceptible to damage during testing than the Part 572 dummy.

As discussed previously, GM estimates that the Hybrid III ribs can be used in severe unrestrained testing approximately 17 times before the ribs or the

damping material needs replacement. In addition, GM's experience shows that the Hybrid III can withstand as many as 150 test applications as long as occasional repairs are made. Ford reported at the previously cited MVMA meeting that one of its belt-restrained Hybrid III test dummies underwent 35 crash tests without any degradation. Clearly, the estimated useful life of the test dummy is highly dependent on the type of testing, restrained or unrestrained, it is used for. Based on its own test experience and the experience of Ford and GM cited above, the agency has decided to use 30 applications as a conservative estimate of the useful life of the ribs. Assuming a life of 30 tests before a set of ribs must be replaced at a cost of approximately $2,000, the incremental per test cost is approximately $70.

The calibration tests for the Hybrid III test dummy have been simplified from the original specification proposed in the April 1985 notice. The Transportation Research Center of Ohio, which does calibration testing of the Hybrid III for the agency, vehicle manufacturers and others estimates the cost of the revised calibration tests is $1528. This is $167 less than the calibration cost for the existing Part 572 test dummy.

Numerous unknown variables will contribute to the manufacturers' operating expense, such as the cost of new or modified test facilities or equipment to maintain the more stringent temperature range of 69° F to 72° F for test dummies, and capital expenditures for lab calibration equipment, signal conditioning equipment, data processing techniques and capabilities, and additional personnel. Obviously, any incremental cost for a particular manufacturer to certify compliance with the automatic restraint requirements of Standard No. 208 will also depend on the extent and nature of its current test facilities and the size of its developmental and new vehicle test programs.

In addition to the costs discussed above, Peugeot raised the issue of a manufacturer's costs increasing because the proposed number of injury measurements made on the Hybrid III will increase the number of tests that must be repeated because of lost data. Since the agency is only adding one additional measurement, chest deflection, for the Hybrid III the number of tests that will have to be repeated due to lost data should not be substantially greater for the Hybrid III than for the Part 572.

NHTSA has determined that it is in the public interest to make the optional use of the Hybrid III test dummy effective in 90 days. This will allow manufacturers time to order the new test dummy to use in their new vehicle development work. Mandatory use of the Hybrid III does not begin until September 1, 1991.

In consideration of the foregoing, Part 572, *Anthropomorphic Test Dummies,* and Part 571.208, *Occupant Crash Protection,* of Title 49 of the Code of Federal Regulations is amended as follows:

Part 572–[AMENDED]

1. The authority citation for Part 572 is amended to read as follows:
Authority: 15 U.S.C. 1392, 1401, 1403, and 1407; delegation of authority at 49 CFR 1.50.

2. A new Subpart E is added to Part 572 to read as follows:
Subpart E—Hybrid III Test Dummy
   § 572.30 *Incorporated materials*
   § 572.31 *General description*
   § 572.32 *Head*
   § 572.33 *Neck*
   § 572.34 *Thorax*
   § 572.35 *Limbs*
   § 572.36 *Test conditions and instrumentation*

§ 572.30 *Incorporated Materials*

(a) The drawings and specifications referred to in this regulation that are not set forth in full are hereby incorporated in this part by reference. The Director of the Federal Register has approved the materials incorporated by reference. For materials subject to change, only the specific version approved by the Director of the Federal Register and specified in the regulation are incorporated. A notice of any change will be published in the *Federal Register.* As a convenience to the reader, the materials incorporated by reference are listed in the Finding Aid Table found at the end of this volume of the Code of Federal Regulations.

(b) The materials incorporated by reference are available for examination in the general reference section of Docket 74-14, Docket Section, National Highway Traffic Safety Administration, Room 5109, 400 Seventh Street, S.W., Washington, DC 20590. Copies may be obtained from Rowley-Scher Reprographics, Inc., 1216 K Street, N.W., Washington, DC 20005 ((202) 628-6667). The drawings and specifications are also on file in the reference library of the Office of the Federal Register, National Archives and Records Administration, Washington, D.C.

§ 572.31 *General description*

(a) The Hybrid III 50th percentile size dummy consists of components and assemblies specified in the Anthropomorphic Test Dummy drawing and specifications package which consists of the following six items:

(1) The Anthropomorphic Test Dummy Parts List, dated July 15, 1986, and containing 13 pages, and a Parts List Index, dated April 26, 1986, containing 6 pages,

(2) A listing of Optional Hybrid III Dummy Transducers, dated April 22, 1986, containing 4 pages,

(3) A General Motors Drawing Package identified by GM drawing No. 78051-218, revision P and subordinate drawings,

(4) Disassembly, Inspection, Assembly and Limbs Adjustment Procedures for the Hybrid III dummy, dated July 15, 1986,

(5) Sign Convention for the signal outputs of Hybrid II dummy transducers, dated July 15, 1986,

(6) Exterior Dimensions of the Hybrid III dummy, dated July 15, 1986.

(b) The dummy is made up of the following component assemblies:

Drawing Number	Revision
78051-61 Head Assembly—Complete—	(T)
78051-90 Neck Assembly—Complete—	(A)
78051-89 Upper Torso Assembly—Complete—	(I)
78051-70 Lower Torso Assembly—Without Pelvic Instrumentation Assembly, Drawing No. 78051-59	(C)
86-5001-001 Leg Assembly—Complete (LH)—	
86-5001-002 Leg Assembly—Complete (RH)—	
78051-123 Arm Assembly—Complete (LH)—	(D)
78051-124 Arm Assembly—Complete (RH)—	(D)

(c) Any specifications and requirements set forth in this part supercede those contained in General Motors Drawing No. 78051-218, revision P.

(d) Adjacent segments are joined in a manner such that throughout the range of motion and also under crash-impact conditions, there is no contact between metallic elements except for contacts that exist under static conditions.

(e) The weights, inertial properties and centers of gravity location of component assemblies shall conform to those listed in drawing 78051-338, revision S.

(f) The structural properties of the dummy are such that the dummy conforms to this part in every respect both before and after being used in vehicle test specified in Standard No. 208 of this Chapter (§ 571.208).

§ 572.32 *Head*

(a) The head consists of the assembly shown in the drawing 78051-61, revision T, and shall conform to each of the drawings subtended therein.

(b) When the head (drawing 78051-61, revision T) with neck transducer structural replacement (drawing 78051-383, revision F) is dropped from a height of 14.8 inches in accordance with paragraph (c) of this section, the peak resultant accelerations at the location of the accelerometers mounted in the head in accordance with 572.36(c) shall not be less than 225g, and not more than 275g. The acceleration/time curve for the test shall be unimodal to the extent that oscillations occurring after the main acceleration pulse are less than ten percent (zero to peak) of the main pulse. The lateral acceleration vector shall not exceed 15g (zero to peak).

(c) *Test Procedure.* (1) Soak the head assembly in a test environment at any temperature between 66° F to 78° F and at a relative humidity from 10% to 70% for a period of at least four hours prior to its application in a test.

(2) Clean the head's skin surface and the surface of the impact plate with 1,1,1 Trichlorethane or equivalent.

(3) Suspend the head, as shown in Figure 19, so that the lowest point on the forehead is 0.5 inches below the lowest point on the dummy's nose when the midsagittal plane is vertical.

(4) Drop the head from the specified height by means that ensure instant release onto a rigidly supported flat horizontal steel plate, which is 2 inches thick and 2 feet square. The plate shall have a clean, dry surface and any microfinish of not less than 8 microinches (rms) and not more than 80 microinches (rms).

(5) Allow at least 2 hours between successive tests on the same head.

§ 572.33 *Neck*

(a) The neck consists of the assembly shown in drawing 78051-90, revision A and conforms to each of the drawings subtended therein. .

(b) When the neck and head assembly (consisting of the parts 78051-61, revision T; -84; -90, revision A; -96; -98; -303, revision E; -305; -306; -307, revision X, which has a neck transducer (drawing 83-5001-008) installed in conformance with 572.36(d), is tested in accordance with paragraph (c) of this section, it shall have the following characteristics:

(1) *Flexion* (i) Plane D, referenced in Figure 20, shall rotate, between 64 degrees and 78 degrees, which shall occur between 57 milliseconds (ms) and

64 ms from time zero. In first rebound, the rotation of plane D shall cross 0 degree between 113 ms and 128 ms.

(ii) The moment measured by the neck transducer (drawing 83-5001-008) about the occipital condyles, referenced in Figure 20, shall be calculated by the following formula: Moment (lbs-ft) = M_y + 0.02875 $_x$ F_x where M_y is the moment measured in lbs-ft by the moment sensor of the neck transducer and F_x is the force measure measured in lbs by the x axis force sensor of the neck transducer. The moment shall have a maximum value between 65 lbs-ft and 80 lbs-ft occurring between 47 ms and 58 ms, and the positive moment shall decay for the first time to 0 lb-ft between 97 ms and 107 ms.

(2) *Extension* (i) Plane D, referenced in Figure 21, shall rotate between 81 degrees and 106 degrees, which shall occur between 72 and 82 ms from time zero. In first rebound, the rotation of plane D shall cross 0 degree between 147 and 174 ms.

(ii) The moment measured by the neck transducer (drawing 83-5001-008) about the occipital condyles, referenced in Figure 21, shall be calculated by the following formula: Moment (lbs-ft) = M_y + 0.02875 $_x$ F_x where M_y is the moment measured in lbs-ft by the moment sensor of the neck transducer and F_x is the force measure measured in lbs by the x axis force sensor of the neck transducer. The moment shall have a minimum value between – 39 lbs-ft and – 59 lbs-ft, which shall occur between 65 ms and 79 ms., and the negative moment shall decay for the first time to 0 lb-ft between 120 ms and 148 ms.

(3) Time zero is defined as the time of contact between the pendulum striker plate and the aluminum honeycomb material.

(c) *Test Procedure.* (1) Soak the test material in a test environment at any temperature between 69 degrees F to 72 degrees F and at a relative humidity from 10% to 70% for a period of at least four hours prior to its application in a test.

(2) Torque the jamnut (78051-64) on the neck cable (78051-301, revision E) to 1.0 lbs-ft ± .2 lbs-ft.

(3) Mount the head-neck assembly, defined in paragraph (b) of this section, on a rigid pendulum as shown in Figure 22 so that the head's midsagittal plane is vertical and coincides with the plane of motion of the pendulum's longitudinal axis.

(4) Release the pendulum and allow it to fall freely from a height such that the tangential velocity at the pendulum accelerometer centerline at the instance of contact with the honeycomb is 23.0 ft/sec ± 0.4 ft/sec. for flexion testing and 19.9 ft/sec ± 0.4 ft/sec. for extension testing. The pendulum deceleration vs. time pulse for flexion testing shall

conform to the characteristics shown in Table A and the decaying deceleration-time curve shall first cross 5g between 34 ms and 42 ms. The pendulum deceleration vs. time pulse for extension testing shall conform to the characteristics shown in Table B and the decaying deceleration-time curve shall cross 5g between 38 ms and 46 ms.

Table A
Flexion Pendulum Deceleration vs. Time Pulse

Time (ms)	Flexion deceleration level (g)
10..............................	22.50-27.50
20..............................	17.60-22.60
30..............................	12.50-18.50
Any other time above 30 ms.......	29 maximum

Table B
Extension Pendulum Deceleration vs. Time Pulse

Time (ms)	Extension deceleration level (g)
10..............................	17.20-21.20
20..............................	14.00-19.00
30..............................	11.00-16.00
Any other time above 30 ms.......	22 maximum

(5) Allow the neck to flex without impact of the head or neck with any object during the test.

§ 572.34 *Thorax*

(a) The thorax consists of the upper torso assembly in drawing 78051-89, revision I and shall conform to each of the drawings subtended therein.

(b) When impacted by a test probe conforming to S572.36(a) at 22 fps ± .40 fps in accordance with paragraph (c) of this section, the thorax of a complete dummy assembly (78051-218, revision P) with left and right shoes (78051-294 and -295) removed, shall resist with the force measured by the test probe from time zero of 1162.5 pounds ± 82.5 pounds and shall have a sternum displacement measured relative to spine of 2.68 inches ± .18 inches. The internal hysteresis in each impact shall be more than 69% but less than 85%. The force measured is the product of pendulum mass and deceleration. Time zero is defined as the time of first contact between the upper thorax and pendulum face.

(c) *Test procedure.* (1) Soak the test dummy in an environment with a relative humidity from 10% to 70% until the temperature of the ribs of the test dummy have stabilized at a temperature between 69° F and 72° F.

(2) Seat the dummy without back and arm supports on a surface as shown in Figure 23.

(3) Place the longitudinal centerline of the test probe so that it is .5 ± .04 in. below the horizontal centerline of the No. 3 Rib (reference drawing number 79051-64, revision A-M) as shown in Figure 23.

(4) Align the test probe specified in S572.36(a) so that at impact its longitudinal centerline coincides within .5 degree of a horizontal line in the dummy's midsagittal plane.

(5) Impact the thorax with the test probe so that the longitudinal centerline of the test probe falls within 2 degrees of a horizontal line in the dummy midsagittal plane at the moment of impact.

(6) Guide the probe during impact so that it moves with no significant lateral, vertical, or rotational movement.

(7) Measure the horizontal deflection of the sternum relative to the thoracic spine along the line established by the longitudinal centerline of the probe at the moment of impact, using a potentiometer (ref. drawing 78051-317, revision A) mounted inside the sternum as shown in drawing 78051-89, revision I.

(8) Measure hysteresis by determining the ratio of the area between the loading and unloading portions of the force deflection curve to the area under the loading portion of the curve.

§572.35 *Limbs*

(a) The limbs consist of the following assemblies: leg assemblies 86-5001-001 and -002 and arm assemblies 78051-123, revision D, and -124, revision D, and shall conform to the drawings subtended therein.

(b) When each knee of the leg assemblies is impacted by the pendulum defined in S572.36(b) in accordance with paragraph (c) of this section at 6.9 ft/sec ± .10 ft/sec., the peak knee impact force, which is a product of pendulum mass and acceleration, shall have a minimum value of not less than 996 pounds and a maximum value of not greater than 1566 pounds.

(c) *Test Procedure.* (1) The test material consists of leg assemblies (86-5001-001) left and (-002) right with upper leg assemblies (78051-46) left and

(78051-47) right removed. The load cell simulator (78051-319, revision A) is used to secure the knee cap assemblies (79051-16, revision B) as shown in Figure 24.

(2) Soak the test material in a test environment at any temperature between 66° F to 78° F and at a relative humidity from 10% to 70% for a period of at least four hours prior to its application in a test.

(3) Mount the test material with the leg assembly secured through the load cell simulator to a rigid surface as shown in Figure 24. No contact is permitted between the foot and any other exterior surfaces.

(4) Place the longitudinal centerline of the test probe so that at contact with the knee it is colinear within 2 degrees with the longitudinal centerline of the femur load cell simulator.

(5) Guide the pendulum so that there is no significant lateral, vertical or rotational movement at time zero.

(6) Impact the knee with the test probe so that the longitudinal centerline of the test probe at the instant of impact falls within .5 degrees of a horizontal line parallel to the femur load cell simulator at time zero.

(7) Time zero is defined as the time of contact between the test probe and the knee.

§ 572.36 *Test conditions and instrumentation*

(a) The test probe used for thoracic impact tests is a 6 inch diameter cylinder that weighs 51.5 pounds including instrumentation. Its impacting end has a flat right angle face that is rigid and has an edge radius of 0.5 inches. The test probe has an accelerometer mounted on the end opposite from impact with its sensitive axis colinear to the longitudinal centerline of the cylinder.

(b) The test probe used for the knee impact tests is a 3 inch diameter cylinder that weighs 11 pounds including instrumentation. Its impacting end has a flat right angle face that is rigid and has an edge radius of 0.2 inches. The test probe has an accelerometer mounted on the end opposite from impact with its sensitive axis colinear to the longitudinal centerline of the cylinder.

(c) Head accelerometers shall have dimensions, response characteristics and sensitive mass locations specified in drawing 78051-136, revision A or its equivalent and be mounted in the head as shown in drawing 78051-61, revision T, and in the assembly shown in drawing 78051-218, revision D.

(d) The neck transducer shall have the dimensions, response characteristics, and sensitive axis

locations specified in drawing 83-5001-008 or its equivalent and be mounted for testing as shown in drawing 79051-63, revision W, and in the assembly shown in drawing 78051-218, revision P.

(e) The chest accelerometers shall have the dimensions, response characteristics, and sensitive mass locations specified in drawing 78051-136, revision A or its equivalent and be mounted as shown with adaptor assembly 78051-116, revision D, for assembly into 78051-218, revision L.

(f) The chest deflection transducer shall have the dimensions and response characteristics specified in drawing 78051-342, revision A or equivalent, and be mounted in the chest deflection transducer assembly 87051-317, revision A, for assembly into 78051-218, revision L.

(g) The thorax and knee impactor accelerometers shall have the dimensions and characteristics of Endevco Model 7231c or equivalent. Each accelerometer shall be mounted with its sensitive axis colinear with the pendulum's longitudinal centerline.

(h) The femur load cell shall have the dimensions, response characteristics, and sensitive axis locations specified in drawing 78051-265 or its equivalent and be mounted in assemblies 78051-46 and -47 for assembly into 78051-218, revision L.

(i) The outputs of acceleration and force-sensing devices installed in the dummy and in the test apparatus specified by this part are recorded in individual data channels that conform to the requirements of SAE Recommended Practice J211, JUNE 1980, "Instrumentation for Impact Tests," with channel classes as follows:

(1) Head acceleration—Class 1000
(2) Neck force—Class 60
(3) Neck pendulum acceleration—Class 60
(4) Thorax and thorax pendulum acceleration—Class 180
(5) Thorax deflection—Class 180
(6) Knee pendulum acceleration—Class 600
(7) Femur force—Class 600

(j) Coordinate signs for instrumentation polarity conform to the sign convention shown in the document incorporated by § 572.31(a)(5).

(k) The mountings for sensing devices shall have no resonance frequency within range of 3 times the frequency range of the applicable channel class.

(l) Limb joints are set at 1g, barely restraining the weight of the limb when it is extended horizontally. The force required to move a limb segment shall not exceed 2g throughout the range of limb motion.

(m) Performance tests of the same component, segment, assembly, or fully assembled dummy are separated in time by a period of not less than 30 minutes unless otherwise noted.

(n) Surfaces of dummy components are not painted except as specified in this part or in drawings subtended by this part. PART 571 [Amended]

2. The authority citation for Part 571 continues to read as follows:

Authority: 15 U.S.C. 1392, 1401, 1403, 1407; delegation of authority at 49 CFR 1.50.

3. Section S5 of Standard No. 208 (49 CFR 571.208) is amended by revising S5.1 to read as follows:

§ 571.208 [Amended]

S5. *Occupant crash protection requirements.*

S5.1 Vehicles subject to S5.1 and manufactured before September 1, 1991, shall comply with either, at the manufacturer's option, 5.1(a) or (b). Vehicles subject to S5.1 and manufactured on or after September 1, 1991, shall comply with 5.1(b).

(a) Impact a vehicle traveling longitudinally forward at any speed, up to and including 30 mph, into a fixed collision barrier that is perpendicular to the line of travel of the vehicle, or at any angle up to 30 degrees in either direction from the perpendicular to the line of travel of the vehicle under the applicable conditions of S8. The test dummy specified in S8.1.8.1 placed at each front outboard designated seating position shall meet the injury criteria of S6.1.1, 6.1.2, 6.1.3, and 6.1.4.

(b) Impact a vehicle traveling longitudinally forward at any speed, up to and including 30 mph, into a fixed collision barrier that is perpendicular to the line of travel of the vehicle, or at any angle up to 30 degrees in either direction from the perpendicular to the line of travel of the vehicle, under the applicable conditions of S8. The test dummy specified in S8.1.8.2 placed at each front outboard designated seating position shall meet the injury criteria of S6.2.1, 6.2.2, 6.2.3, 6.2.4, and 6.2.5.

3. Section S5.2 of Standard No. 208 is revised to read as follows:

S5.2 Lateral moving barrier crash.

S5.2.1 Vehicles subject to S5.2 and manufactured before September 1, 1991, shall comply with either, at the manufacturer's option, 5.2.1(a) or (b). Vehicles subject to S5.2 and manufactured on or after September 1, 1991, shall comply with 5.2.1(b).

(a) Impact a vehicle laterally on either side by a barrier moving at 20 mph under the applicable conditions of S8. The test dummy specified in S8.1.8.1 placed at the front outboard designated seating position adjacent to the impacted side shall meet the injury criteria of S6.1.2 and S6.1.3:

(b) When the vehicle is impacted laterally under the applicable conditions of S8, on either side by a barrier moving at 20 mph, with a test device specified in S8.1.8.2, which is seated at the front outboard designated seating position adjacent to the impacted side, it shall meet the injury criteria of S6.2.2, and S6.2.3.

4. Section S5.3 of Standard No. 208 is revised to read as follows:

S5.3 *Rollover* Subject a vehicle to a rollover test under the applicable condition of S8 in either lateral direction at 30 mph with either, at the manufacturer's option, a test dummy specified in S8.1.8.1 or S8.1.8.2, placed in the front outboard designated seating position on the vehicle's lower side as mounted on the test platform. The test dummy shall meet the injury criteria of either S6.1.1 or S6.2.1.

5. Section S6 of Standard No. 208 is revised to read as follows:

S6. *Injury Criteria*

S6.1 Injury criteria for the Part 572, Subpart B, 50th percentile Male Dummy.

S6.1.1 All portions of the test dummy shall be contained within the outer surfaces of the vehicle passenger compartment throughout the test.

S6.1.2 The resultant acceleration at the center of gravity of the head shall be such that the expression:

$$\left[\frac{1}{t_2 - t_1} \int_{t_1}^{t_2} a\,dt \right]^{2.5} t_2 - t_1$$

shall not exceed 1,000, where a is the resultant acceleration expressed as a multiple of g (the acceleration of gravity), and t_1 and t_2 are any two points during the crash.

S6.1.3 The resultant acceleration at the center of gravity of the upper thorax shall not exceed 60 g's, except for intervals whose cumulative duration is not more than 3 milliseconds.

S6.1.4 The compressive force transmitted axially through each upper leg shall not exceed 2250 pounds.

S6.2 *Injury criteria for the Part 572, Subpart E, Hybrid III Dummy*

S6.2.1 All portions of the test dummy shall be contained within the outer surfaces of the vehicle passenger compartment throughout the test.

S6.2.2 The resultant acceleration at the center of gravity of the head shall be such that the expression:

$$\left[\frac{1}{t_2 - t_1} \int_{t_1}^{t_2} a\,dt\right]^{2.5} t_2 - t_1$$

shall not exceed 1,000, where a is the resultant acceleration expressed as a multiple of g (the acceleration of gravity), and t_1 and t_2 are any two point during the crash.

S6.2.3 The resultant acceleration calculated from the thoracic instrumentation shown in drawing 78051-218, revision L, incorporated by reference in Part 572, Subpart E of this Chapter, shall not exceed 60g's, except for intervals whose cumulative duration is not more than 3 milliseconds.

S6.2.4 Compression deflection of the sternum relative to spine, as determined by instrumentation shown in drawing 78051-317, revision A, incorporated by reference in Part 572, Subpart E of this Chapter, shall not exceed 2 inches for loadings applied through any impact surfaces except for those systems which are gas inflated and provide distributed loading to the torso during a crash. For gas-inflated systems which provide distributive loading to the torso, the thoracic deflection shall not exceed 3 inches.

S6.2.5 The force transmitted axially through each upper leg shall not exceed 2250 pounds.

6. Section S8.1.8 of Standard No. 208 is revised to read as follows:

S8.1.8 *Anthropomorphic test dummies*

S8.1 8.1 The anthropomorphic test dummies used for evaluation of occupant protection systems manufactured pursuant to applicable portions of paragraphs S4.1.2, 4.1.3, and S4.1.4 shall conform to the requirements of Subpart B of Part 572 of this Chapter.

S8.1.8.2 Anthropomorphic test devices used for the evaluation of occupant protection systems manufactured pursuant to applicable portions of paragraphs S4.1.2, S4.1.3, and S4.1.4 shall conform to the requirements of Subpart E of Part 572 of this Chapter.

7. Section S8.1.9 of Standard No. 208 is revised to read as follows:

S8.1.9.1 Each Part 572, Subpart B, test dummy specified in S8.1.8.1 is clothed in formfitting cotton stretch garments with short sleeves and midcalf length pants. Each foot of the test dummy is equipped with a size 11EE shoe which meets the config-

uration size, sole, and heel thickness specifications of MIL-S-131192 and weighs 1.25 ± 0.2 pounds.

S8.1.9.2 Each Part 572, Subpart E, test dummy specified in S8.1.8.2 is clothed in formfitting cotton stretch garments with short sleeves and midcalf length pants specified in drawings 78051-292 and -293 incorporated by reference in Part 572, Subpart E, of this Chapter, respectively or their equivalents. A size 11EE shoe specified in drawings 78051-294 (left) and 78051-295 (right) or their equivalents is placed on each foot of the test dummy.

8. Section S8.1.13 of Standard No. 208 is revised to read as follows:

S8.1.13 *Temperature of the test dummy*

S8.1.13.1 The stabilized temperature of the test dummy specified by S8.1.8.1 is at any level between 66 degrees F and 78 degrees F.

S8.1.13.2 The stabilized temperature of the test dummy specified by S8.1.8.2 is at any level between 69 degrees F and 72 degrees F.

9. A new fourth sentence is added to section S8.1.3 to read as follows:

Adjustable lumbar supports are positioned so that the lumbar support is in its lowest adjustment position.

10. A new section S11 is added to read as follows:

S11. *Positioning Procedure for the Part 572 Subpart E Test Dummy*

Position a test dummy, conforming to Subpart E of Part 572 of this Chapter, in each front outboard seating position of a vehicle as specified in S11.1 through S11.6. Each test dummy is restrained in accordance with the applicable requirements of S4.1.2.1, 4.1.2.2 or S4.6.

S11.1 *Head.* The transverse instrumentation platform of the head shall be horizontal within ½ degree.

S11.2 *Arms*

S11.2.1 The driver's upper arms shall be adjacent to the torso with the centerlines as close to a vertical plane as possible.

S11.2.2 The passenger's upper arms shall be in contact with the seat back and the sides of torso.

S11.3 *Hands*

S11.3.1 The palms of the driver test dummy shall be in contact with the outer part of the steering wheel rim at the rim's horizontal centerline. The thumbs shall be over the steering wheel rim and attached with adhesive tape to provide a breakaway force of between 2 to 5 pounds.

S11.3.2 The palms of the passenger test dummy shall be in contact with outside of thigh. The little finger shall be in contact with the seat cushion.

S11.4 *Torso*

S11.4.1 In vehicles equipped with bench seats, the upper torso of the driver and passenger test dummies shall rest against the seat back. The midsagittal plane of the driver dummy shall be vertical and parallel to the vehicle's longitudinal centerline, and pass through the center of the steering wheel rim. The midsagittal plane of the passenger dummy shall be vertical and parallel to the vehicle's longitudinal centerline and the same distance from the vehicle's longitudinal centerline as the midsagittal plane of the driver dummy.

S11.4.2 In vehicles equipped with bucket seats, the upper torso of the driver and passenger test dummies shall rest against the seat back. The midsagittal plane of the driver and the passenger dummy shall be vertical and shall coincide with the longitudinal centerline of the bucket seat.

S11.4.3 *Lower torso*

S11.4.3.1 *H-point*. The H-point of the driver and passenger test dummies shall coincide within ½ inch in the vertical dimension and ½ inch in the horizontal dimension of a point ¼ inch below the position of the H-point determined by using the equipment and procedures specified in SAE J826 (Apr 80) except that the length of the lower leg and thigh segments of the H-point machine shall be adjusted to 16.3 and 15.8 inches, respectively, instead of the 50th percentile values specified in Table 1 of SAE J826.

S11.4.3.2 *Pelvic angle*. As determined using the pelvic angle gage (GM drawing 78051-532 incorporated by reference in Part 572, Subpart E, of this chapter) which is inserted into the H-point gaging hole of the dummy, the angle measured from the horizontal on the 3 inch flat surface of the gage shall be 22½ degrees plus or minus 2½ degrees.

S11.5 *Legs*. The upper legs of the driver and passenger test dummies shall rest against the seat cushion to the extent permitted by placement of the feet. The initial distance between the outboard knee clevis flange surfaces shall be 10.6 inches. To the extent practicable, the left leg of the driver dummy and both legs of the passenger dummy shall be in vertical longitudinal planes. Final adjustment to accommodate placement of feet in accordance with S11.6 for various passenger compartment configurations is permitted.

S11.6 *Feet*

S11.6.1 The right foot of the driver test dummy shall rest on the undepressed accelerator with the rearmost point of the heel on the floor surface in the plane of the pedal. If the foot cannot be placed on the accelerator pedal, it shall be positioned perpendicular to the tibia and placed as far forward as possible in the direction of the centerline of the pedal with the rearmost point of the heel resting on the floor surface. The heel of the left foot shall be placed as far forward as possible and shall rest on the floor surface. The left foot shall be positioned as flat as possible on the floor surface. The longitudinal centerline of the left foot shall be placed as parallel as possible to the longitudinal centerline of the vehicle.

S11.6.2 The heels of both feet of the passenger test dummy shall be placed as far forward as possible and shall rest on the floor surface. Both feet shall be positioned as flat as possible on the floor surface. The longitudinal centerline of the feet shall be placed as parallel as possible to the longitudinal centerline of the vehicle.

S11.7 *Test dummy positioning for latchplate access*. The reach envelopes specified in S7.4.4 are obtained by positioning a test dummy in the driver's seat or passenger's seat in its forwardmost adjustment position. Attach the lines for the inboard and outboard arms to the test dummy as described in Figure 3 of this standard. Extend each line backward and outboard to generate the compliance arcs of the outboard reach envelope of the test dummy's arms.

S11.8 *Test dummy positioning for belt contact force*. To determine compliance with S7.4.3 of this standard, position the test dummy in the vehicle in accordance with the requirements specified in S11.1 through S11.6 and under the conditions of S8.1.2 and S8.1.3. Pull the belt webbing three inches from the test dummy's chest and release until the webbing is within 1 inch of the test dummy's chest and measure the belt contact force.

S11.9 *Manual belt adjustment for dynamic testing*. With the test dummy at its designated seating position as specified by the appropriate requirements of S8.1.2, S8.1.3 and S11.1 through S11.6, place the Type 2 manual belt around the test dummy and fasten the latch. Remove all slack from the lap belt. Pull the upper torso webbing out of the retractor and allow it to retract; repeat this operation four times. Apply a 2 to 4 pound tension load to the lap belt. If the belt system is equipped with a tension-relieving device introduce the maximum amount of slack into the upper torso belt that is recommended by the manufacturer for normal use in the owner's manual for the vehicle. If the belt system is not equipped with a tension-relieving device, allow the excess webbing in the shoulder belt to be retracted by the retractive force of the retractor.

Issued on July 21, 1986

Diane K. Steed
Administrator

51 F.R. 26688
July 25,1986

PREAMBLE TO AN AMENDMENT TO FEDERAL MOTOR VEHICLE SAFETY STANDARD NO. 208

Occupant Crash Protection and Seat Belt Assemblies
(Docket No. 74-14; Notice 46)

ACTION: Final Rule; Response to petitions for reconsideration.

SUMMARY: This notice responds to eight petitions for reconsideration of several of the amendments to Standard No. 208, *Occupant Crash Protection*, that appeared in the *Federal Register* of Friday, March 21, 1986. In response to the petitions, the agency is modifying the test dummy positioning procedures. However, so as not to affect compliance testing done using the old procedures, the agency is permitting manufacturers to use either the old or new procedures for a one-year period. Beginning September 1, 1987, the new procedures would be mandatory. This notice denies a request to extend the September 1, 1989, effective date for dynamic testing of manual lap/shoulder belts in the front seat of passenger cars. (The dynamic test requirement would go into effect on that date only if the automatic restraint requirement is rescinded.) A response to four petitions asking the agency to reinstate certain of the test requirements of Standard No. 209, *Seat Belt Assemblies*, for dynamically-tested manual lap/shoulder belts, and to revise the current exemption for automatic belts, will be addressed separately at a later date.

DATES: The amendments made by this notice are effective on September 5, 1986.

SUPPLEMENTARY INFORMATION: On March 21, 1986 (51 FR 9800), NHTSA published a final rule amending Standard No. 208, *Occupant Crash Protection*. Subsequent to publication of the amendments, eight interested parties timely filed petitions asking the agency to reconsider some of the amendments adopted in that final rule. This notice responds to those petitions.

Test Procedures

The March notice adopted several changes to the test dummy positioning procedures of the standard. Ford Motor Company (Ford) said that the revised test procedures were not objective because of what it termed ambiguities, inconsistencies, and subjective elements in the test procedure provisions. Each of Ford's specific objections are discussed below, in the order that Ford raised them.

Positioning of Manual Belts for Dynamic Testing

Ford noted that the standard provides that in the dynamic test for manual belts, the lap/shoulder belt is to be placed around the test dummy after the dummy's arms and hands have been positioned. Ford said it is impracticable to position properly a lap/shoulder belt on a driver test dummy whose hands are on the steering wheel or on a passenger test dummy whose palms are in contact with its thighs. Ford noted that the agency's New Car Assessment Program (NCAP) test procedures provide for positioning the arms and hands after the safety belt has been positioned.

Ford is correct that the NCAP test procedure provides that the safety belts are to be placed on the test dummy before the arms and hands are placed in their final positions. To eliminate possible safety belt positioning problems, NHTSA is amending the Standard No. 208 positioning requirements to adopt the NCAP requirement.

Positioning of Automatic Belt for Dynamic Testing

Ford also noted that the safety belt positioning procedure applies only to manual belts and asked the agency to specify at what stage during the positioning of the test dummy automatic belts are to be deployed. Ford also asked what adjustment procedures the agency would use with automatic belts.

In NCAP testing, NHTSA has finally positioned both automatic and manual safety belts after the test dummy has been settled in its specified position and before the hands are placed in their final position. The agency has used this procedure because it is simpler than having to position the hands first and then move them in order to place the safety belt on the test dummy. NHTSA is therefore modifying the title of the safety belt positioning procedure to indicate that it applies to the positioning of both manual and automatic safety belts.

In the agency's NCAP testing, the only adjustment NHTSA has made to an automatic belt once it has been deployed on the test dummy is to ensure that the belt is lying flat on the test dummy's shoulder when the belt is in its final position. The agency is adopting the same procedure for the Standard No. 208 compliance test. In addition, as discussed immediately below, the agency will also adjust an automatic belt with a tension-relieving device that can be used to introduce slack in the belt system in accordance with the manufacturer's instructions provided in the vehicle owner's manual. For automatic belts that do not have devices that can be used to introduce slack in the belt system, it should not be necessary to make any further adjustments, other than ensuring the belt is flat on the test dummy's shoulder.

Adjusting Belt Slack

Ford noted that S7.4.2 of the standard requires automatic belts and dynamically-tested manual lap/shoulder belts to be tested with the maximum amount of slack recommended by the manufacturer. It said that the standard does not, however, prescribe a procedure for adjusting the slack of automatic belts with tension-relieving devices.

The purpose of S7.4.2 of the standard is to ensure that automatic and dynamically-tested manual belt systems will perform adequately when they are adjusted to the maximum amount of belt slack recommended by the vehicle manufacturer. S7.4.2(b) of the standard specifically requires manufacturers that use tension-relieving devices to provide information in their owner's manual describing how the tension-reliever works. In addition, the owner's manual must inform vehicle owners of the maximum amount of safety belt slack recommended by the vehicle manufacturer. In conducting its crash tests, the agency will adjust any safety belt tension-relieving devices in accordance with the instructions provided by the vehicle manufacturer in the owner's manual.

Belt Tension Loading

Ford noted that the safety belt positioning procedure specifies applying a 2-to-4 pound tension load to the lap belt of a lap/shoulder belt, but does not specify how the load is to be applied or how the tension is to be measured. Ford asked the agency to clarify the procedure, particularly with regard to whether the load is to be applied to the lap portion of the belt or whether an increasing load is to be placed on the shoulder portion of the belt until the required amount of tension has been reached in the lap portion of the belt.

NHTSA does not believe that the area of application of the belt tension load should have a significant effect on the subsequent performance of the belt in a dynamic test. However, to promote uniformity in application of the load, the agency is amending the standard to provide that the load will be applied to the shoulder portion of the belt adjacent to the latchplate of the belt. If the safety belt system is equipped with two retractors (one for the lap belt and one for the upper torso belt), then the tension load will be applied at the point the lap belt enters the retractor, since the separate lap belt retractor effectively controls the tension in the lap portion of a lap/shoulder belt. The amount of tension will also be measured at the location where the load is applied. Finally, the agency is amending the standard to provide that after the tension load has been applied, the shoulder belt will be positioned flat on the test dummy's shoulder. This will ensure that if the belt is twisted during the application of the tension load, it will be correctly positioned prior to the crash test.

Test Dummy Settling and Leg Positioning

Ford said that it was particularly concerned about the repeatability of the leg placement obtained using the new test procedures. Ford said that the positioning procedures provide for the placement of the test dummy's legs before the test dummy is settled. Ford said that the settling procedure usually results in movement of the test dummy's legs, but the new procedure does not call for readjustment of the leg positions after the test dummy has been settled. Ford requested that the procedure be changed by providing that after test dummy settling and placement of the arms and hands, the test dummy's feet and knees should be repositioned, if necessary. As an alternative approach, Ford suggested that the procedure provide that the test dummy settling be performed prior to adjustment of the legs.

NHTSA agrees that the procedures should be changed to minimize the possibility of inadvertent leg movement during the settling procedure. The agency is therefore adopting Ford's suggestion that the test dummy's feet and legs should be repositioned, if necessary, after the test dummy has been settled and its hands and arms have been positioned.

Initial Knee Spacing for the Driver

Ford and Nissan Motor Company, Ltd., (Nissan) expressed concern that NHTSA had misinterpreted comments made by General Motors Corporation (GM) and Honda Motor Company, Ltd., (Honda) concerning one of the proposed changes to the test dummy positioning procedures in the April 1985 NPRM. In that notice, NHTSA proposed a test dummy initial knee spacing of 14.5 inches for both the driver and passenger test dummies. In their comments on the April 1985 notice, GM and Honda requested that the proposed initial spacing of the passenger test dummy knees be changed from 14.5 inches to 11.75 inches, which would mean that the passenger test dummy legs would be parallel. In the March 1986 final rule, NHTSA adopted the 11.75 inch initial knee spacing change for both the driver and the passenger test dummy.

In their petitions for reconsideration, Ford and Nissan said that they support the change sought by GM and Honda for the initial placement of the passenger's knees. Thus, they requested the agency to apply the 11.75 inch requirement only to the spacing of the passenger's knees and retain the former 14.5 inch requirement for the driver's knees. Ford noted that an 11.75 inch initial knee spacing for the driver is not compatible with the requirement to position the driver's right foot on the accelerator pedal and keep the leg in a vertical plane.

NHTSA misinterpreted GM and Honda's suggested change and therefore believed that the commenters were seeking a change to the initial knee spacing requirement for both the driver and the passenger. NHTSA agrees that a change should not have been made to the initial knee spacing for the driver's knees, since the smaller initial knee spacing requirement is not compatible with the positioning requirement for the driver's right foot. The agency is therefore reinstating the 14.5 inch initial spacing requirement for the driver.

NHTSA emphasizes that, as it stated in the notice adopting the test dummy positioning procedures on July 5, 1977 (42 FR 34301), the knee spacing requirements apply only to the *initial* placement of the knees. The final spacing of the knees depends on the specific configuration of the vehicle's occupant compartment and may vary due to the positioning of the test dummy's feet to accommodate such differing design features as protruding wheelwells, foot rests, and ventilating system ducts. Thus, the agency recognizes that the initial spacing may have to be modified to ensure that the legs and feet are correctly positioned.

Driver Right Leg Positioning

Ford objected to the requirement in S10.1.1(b) that the driver's right leg be placed so that the upper and lower leg centerlines fall, as close as possible, in a vertical longitudinal plane. Ford said the requirement that the legs be in a vertical longitudinal plane is not compatible with the requirement that the driver's foot be placed on the accelerator pedal. Ford said that "in many passenger cars the accelerator pedal is further inboard than the pivot point of the driver's right femur and therefore not in the same longitudinal plane as the dummy's upper leg." Ford further said that requiring the leg to remain in a vertical plane is incompatible with the knee spacing requirement. Ford suggested that a leg position specification is unnecessary since specifying the positions of the foot and knee would adequately define the position of the right leg.

NHTSA recognizes that the initial knee spacing requirement and the requirements on foot placement help to maintain the right leg in a consistent position. However, because of the numerous variations in passenger car interior designs, it may not be possible to maintain the initial knee position and thus a further control is needed to maintain proper placement of the right leg. NHTSA recognizes it may be particularly difficult to place the right leg so that it is in a longitudinal plane, since as Ford pointed out, the right leg may have to be moved to place the foot on the accelerator. On reconsideration, the agency believes that simply requiring the leg to remain in a vertical plane after the right foot has been positioned (instead of a vertical longitudinal plane) should be sufficient to ensure consistent placement of the right leg.

Foot Placement on the Accelerator Pedal

Ford noted that S10.1.1(b) provides that if the driver's right foot can not be placed on the accelerator pedal, it is to be placed as far forward as possible in the direction of the "geometric center" of the pedal. Ford said that a formula is needed to guide technicians in determining the geometric center of an asymmetrically shaped accelerator pedal.

The agency agrees with Ford's underlying point that it is unnecessary to place the foot in the "geometric center" of the accelerator pedal to ensure proper foot placement. The intent of the requirement, which is to provide for consistent placement by different testing organizations, can be achieved by simplifying the requirement by providing that the centerline of the foot is to be placed, as close as possible, in the same plane as the centerline of the pedal.

Driver Left Foot Placement

Ford said it was concerned about the requirements of S10.1.1 for the placement of the driver's left foot in vehicles which have wheelwells that project into the passenger compartment. Ford agreed that in the case of the passenger test dummy, it "may be desirable to avoid placing the passenger dummy's right foot on the wheelwell because such placement can result in head contact with the dummy's knee, but head-to-knee contact is virtually impossible on the driver's side of the vehicle because the steering wheel would block any potential contact. In addition, placement of the driver's left foot is complicated by the presence of brake and clutch pedals, and therefore placement of the driver's left leg to avoid the brake and clutch pedals may have to take precedence over avoiding the wheelhouse projections."

Ford also said that it is not clear from the text of the standard whether the driver's left foot is to be placed inboard of a wheelwell projection. In addition, Ford said that S10.1.1(c) does not clearly specify where the driver's left leg should be positioned in such cases. Ford said "it is unclear whether the foot should be placed perpendicular to the tibia with the heel resting on the floor pan and the sole resting on one end of the brake pedal, or whether the foot may be pivoted around the axis of the tibia to eliminate contact with a brake pedal. It is also unclear whether the entire foot (and leg) may be moved laterally to miss the brake and clutch pedals."

NHTSA agrees with Ford that avoiding the positioning of the passenger's right foot on the wheelwell is more of a concern, since if there is floor buckling, the passenger's right knee can be pushed upward and strike the head. Although the agency has not seen as much floor buckling on the driver's side of the car in its NCAP tests, such buckling can happen. Although the positioning procedures for the driver's left foot and leg and the passenger's right foot and leg are the same as far as the final positioning of those parts is concerned, Ford is correct that the standard does not specifically state that the driver's left foot should not be

placed on the wheelwell. To correct this, the agency has amended the standard to specifically provide that the driver's left foot is not to be placed on the wheelwell.

NHTSA has not experienced in its NCAP testing the difficulty mentioned by Ford in placing the driver test dummy's left foot in the vicinity of the clutch or brake pedals. However, to provide for a consistent positioning if there is pedal interference, the agency is making a minor amendment to the foot positioning procedure. The amendment provides that if there is pedal interference, the driver's left foot should be rotated about the tibia to avoid contact with the pedal. This simple action should avoid most problems. If that is not sufficient, the procedure provides that the left leg should be rotated about the hip in the outboard direction.

Driver Left Leg Placement

Ford noted that the agency did not adopt the requirement proposed in the April 1985 notice that the driver test dummy's left leg be placed in a vertical and longitudinal plane. Instead, in the March 1986 final rule, the agency provided that the driver's left leg need only be placed in a vertical plane. Ford said that if the leg is placed in a vertical plane with the knee 5.9 inches from the mid-sagittal plane, as called for in the initial knee spacing requirement for the driver, the leg will still be in a vertical longitudinal plane. Ford said it was unclear whether the agency intended the leg to remain in a vertical longitudinal plane or whether the 5.9 inch dimension is no longer appropriate.

The requirements are not inconsistent. As emphasized earlier in this notice, the requirement for the knee spacing is an *initial* setting. The agency recognizes that this initial placement will result in the driver's left leg being in a vertical longitudinal plane. However, to accommodate differences in vehicle designs, that spacing can be modified to achieve the other leg and foot placement requirements. The agency is retaining the requirement adopted in the March 1986 final rule that when the driver's left leg is in its final position it must be in a vertical plane.

Foot Rests

Ford said that its new Taurus/Sable models have a driver's foot rest, which is a flat area located low on the wheelwell projection. Ford said that placing the driver test dummy's left foot on the foot rest would mean that the dummy's left heel would be no higher than its right heel. Thus, Ford said that its foot rest is apparently different from the Honda foot rest discussed by NHTSA in the March 1986 notice. Ford asked the agency to clarify whether S10.1.1 of the standard would result in the driver test dummy's foot being placed on the Ford-type foot rest or whether the knee spacing and leg positioning requirements specified elsewhere in S10.1.1 would be controlling.

The foot rest positioning requirement adopted in the March 1986 final rule states that if the foot rest "does not elevate the left foot above the level of the right foot," then the left foot should be placed on the foot rest. If as it appears, the Ford foot rest does not elevate the left foot above the right foot, then the left foot should be placed on the foot rest.

Restraint Use During Testing

Ford said that the provisions of S10 regarding the restraint of the test dummy are inconsistent with the provisions of S4.1.2.1 for the testing of vehicles equipped both with automatic restraints and with manual Type 2 safety belts. The agency has modified S10 to make it consistent with S4.1.2.1. In brief, the new language provides that if a seating position in a vehicle is equipped with an automatic restraint to meet the frontal crash requirement and a manual safety belt to meet the lateral and rollover protection requirements, then the vehicle is subjected to two tests. First, the vehicle must pass a test in which the test dummy is restrained solely by the automatic restraint. In addition, the vehicle must pass a second test in which the test dummy is restrained by the automatic restraint and the manual safety belt as well. To reduce unnecessary testing costs for vehicles equipped with driver-only, non-belt automatic restraint systems, the agency is providing manufacturers with the option of using a passenger test dummy during the Standard No. 208 compliance test.

Placement of the Test Dummy on the Seat

Ford said that the wording of S10.1 is unclear regarding the placement of a test dummy in a seat whose centerline is not positioned in the vertical longitudinal plane of the vehicle. Ford said that in its Econoline van-type vehicles, the centerline of the front passenger's seat is 'oriented a few degrees outboard to comfortably accommodate occupants by avoiding the intrusion of the engine cover on foot placement space. It is unclear whether, in compliance testing, the dummy would be placed in the vertical longitudinal plane passing through the center of the seat cushion, as implied by the wording of S10.1 This would place the

dummy's torso out of alignment with the seat back, and such a position may be unstable. Alternatively, it is unclear whether the dummy would be placed in the vertical longitudinal plane passing through the seating reference point. Or would the dummy's torso be centered in the seat and only the legs placed in vertical longitudinal planes."

The positioning procedures have two purposes; to ensure consistency in dummy placement and to have the test dummy reasonably simulate the posture of a human in the seat. As Ford noted, the seats in its Econoline vehicles are oriented only a few degrees outboard of the vehicle's centerline. Thus, regardless of how the test dummy is positioned, the few degrees difference in orientation should not make a significant difference. It appears unlikely that many persons would even notice a few degrees difference in the seat orientation and it thus would be natural for a person to sit so they are centered in the seat. The agency is modifying the positioning requirements to provide that the test dummy is centered with the centerline of the seat cushion.

Subjective Phrases

Ford said that many of the test dummy positioning requirements contained subjective phrases, such as "to the extent permitted," and "except as prevented." Ford said that these phrases make the procedures ambiguous and can lead to varying interpretations by different testers.

As discussed previously, manufacturers use a wide variety of interior design configurations and the agency has established a positioning procedure that attempts to accommodate those differing configurations. The purpose of such phrases as "to the extent permitted" is to permit reasonable, minor adjustments in the positioning requirements so that a test dummy can be positioned in a vehicle with design features which may make it impossible to position the test dummy in absolute conformance to the test procedure. By allowing for minor, necessary adjustments, the test procedure can be used in all vehicles, regardless of their differing design features.

Test Dummy Upper Torso Rocking

Ford said that the provisions of S10.44 are unclear as to how much force is to be applied to the test dummy's lower torso while the test dummy is being positioned in a seat. Ford asked whether the initial force application of 50 pounds is to be reduced only long enough to allow the test dummy to slide down the seat back into contact with the seat cushion and whether that force is to be maintained until the test dummy's arms and hands are positioned. Ford recommended that the agency specify one specific force and provide that this force should be maintained during the upper torso force application.

The purpose of permitting testers to reduce the horizontal force on the test dummy during the settling procedure is to accommodate seats with differing frictional properties. In a vehicle with "slick" material, the test dummy may easily slide down the seat back without reducing the horizontal force much, if at all. If the seat has high friction material, the horizontal force must be reduced considerably to allow the test dummy to slide down the seat back. NHTSA, however, agrees with Ford that providing for use of a specific force should eliminate another possible source of test variability. NHTSA is thus modifying the settling procedure to provide that a force of 10 to 15 pounds of horizontal rearward force will be applied to the test dummy during the final upper torso positioning procedures (S10.4.4 and S10.4.5).

Test Dummy Position Fixture

Ford also asked the agency to specify the test dummy positioning fixture that will be used in accordance with the requirements of S10.4.2 to position the test dummy. Although the NCAP test procedures specify the use of a specific test positioning fixture, the agency does not believe it is necessary to specify such a device here. NHTSA believes that manufacturers should be permitted the option of devising their own positioning fixtures. This results in a more performance-oriented standard. Thus, the agency is not adopting Ford's recommendation for a specific test procedure but is making a minor change to S10.4.2 to delete any reference to a "dummy positioning fixture."

Arm and Hand Placement

Ford noted that S10.5 calls for placement of the test dummy's arms and hands prior to settling and asked that the requirements be changed to provide for arm and hand placement after settling. Ford also noted that the reference in S10.5 to the arm and hand placement requirements is incorrect.

NHTSA agrees with Ford that the procedure should be changed to provide for arm and hand placement after the test dummy has been settled. The agency has made the necessary change and has also corrected the references in the positioning procedure.

Vehicle Test Attitude

Ford said that the requirements of S8.1.1(d) require the cargo load to be centered over the longitudinal centerline of the vehicle. Ford said that the "longitudinal centerline of the vehicle marks the lateral center of the vehicle, and centering of the cargo on the longitudinal centerline of the vehicle only determines its lateral (side-to-side) position, but not its fore-and-aft position." Ford asked the agency to specify that the cargo be centered over the longitudinal centerline of the vehicle and at the longitudinal center of the cargo area.

Ford also asked the agency to clarify how to determine the longitudinal center of the cargo area in a station wagon or hatchback with a second seat that can be folded down to form a cargo area or in a multipurpose passenger vehicle with readily removable rear seats.

NHTSA agrees with Ford that cargo should be centered on the vertical longitudinal centerline of the vehicle and in the center of the cargo area. In the case of vehicles with a folddown seat or with a readily removable seat, the agency will consider the cargo area as the area that is available with a folddown seat in its upright position and a readily removable seat anchored at its position. The agency will then determine the center of that position and place the cargo there.

Effective Date for New Test Procedures

Ford and the Automobile Importers of America (AIA), asked the agency to reconsider its decision to implement the test dummy positioning procedure changes prior to September 1, 1986. AIA said that while some manufacturers wanted the new procedures to go into effect as soon as possible, the 45-day effective date placed an unreasonable burden on other manufacturers that are currently producing automatic restraints. AIA said that the short effective date did not provide enough time for a manufacturer to determine whether the test procedure changes affect the compliance of its current vehicles. AIA asked the agency to allow the optional use of the test procedures now and set a later mandatory effective date.

By adopting a 45-day effective date, the agency did not intend to jeopardize the compliance testing that has already been done by manufacturers. NHTSA is adopting AIA's suggestion to allow the use, at the manufacturer's option, of either the old or new test procedure during the first year of the phase-in. Beginning September 1, 1987, the use of the new test procedure will become mandatory.

Revisions to Standard No. 210

Ford asked the agency to clarify the revision made to the safety belt anchorage location requirements of S4.3 of Standard No. 210, *Seat Belt Assembly Anchorages*. The March 1986 notice exempted anchorages for automatic belts and dynamically-tested manual belts from the anchorage location requirements of Standard No. 210. Ford asked whether a manufacturer must provide two sets of anchorages in vehicles with dynamically-tested manual lap/shoulder belts that have the anchorages located outside the zone specified in S4.3—one set of anchorages for Type 2 manual belt systems located within the anchorage zone set out in S4.3 of the standard, and the other set of anchorages for the dynamically-tested Type 2 manual belt systems.

NHTSA has recently responded to a petition from GM raising the same issue. In a letter of April 14, 1986, the agency explained that anchorages for Type 2 manual belt systems must be included for vehicles that have automatic or dynamically-tested manual belts located outside of the zone. (The agency's letter is available in the Standard No. 210 interpretation file in the NHTSA docket section.) The agency did, however, grant GM's petition to amend the requirement, saying that GM had raised a number of reasons why the requirements of Standard No. 210 should be changed. NHTSA will shortly issue a notice of proposed rulemaking on this subject.

Labeling of Dynamically-Tested Safety Belts

Ford objected to the adoption, in Standard No. 209, *Seat Belt Assemblies,* of a requirement that dynamically-tested belts have a label identifying the vehicles in which they can be used. Ford said that the required label does not specifically identify the safety belt as a dynamically-tested belt and the label does not suggest that the belt may be safely used only in specific vehicles at specific seats. Ford asked the agency to rescind the labeling requirement.

Ford suggested that the intent of S4.6(b) could be accomplished by requiring the safety belt installation instruction required by S4.1(k) of the standard to specify both the vehicles for which the belt system is to be used and the specific type of seating position for which it is intended.

NHTSA still believes that it is important that a dynamically-tested safety belt be labeled to ensure that it is installed only in the type of vehicle for which it is intended. NHTSA agrees with Ford that

providing the information in the installation instructions would address most of the problem of possible misuse. However, there still may be instances where the instruction would be lost. In addition, the installation instruction requirements apply only to aftermarket belts. There can be situations where a safety belt may be taken from one vehicle and transferred to another. Given these considerations and the importance of alerting motorists that a safety belt may have been designed for use in one particular make and model vehicle, the agency has decided to retain the labeling requirement.

In response to Ford's comment, NHTSA believes that the statement appearing on the label should be changed to require a manufacturer to specify the specific vehicles for which the safety belt is intended and the specific seating position (e.g., "right front") in which it can be used.

Exemption of Dynamically-Tested Safety Belts

The March 1986 rule adopted a requirement that the manual lap/shoulder belts in the front seats of passenger cars must meet a dynamic crash test. The requirement would go into effect for those manual belts on September 1, 1989, if the automatic restraint requirements of the standard are rescinded. Three petitioners, the American Seat Belt Council (ASBC), the Narrow Fabrics Institute (NFI), and Phoenix Trimming Company, asked the agency to reconsider its decision to exempt dynamically-tested manual safety belts from the webbing width and breaking strength requirements of Standard No. 209, *Seat Belt Assemblies.* On August 4, 1986, ASBC petitioned the agency to rescind the current Standard No. 209 exemption for automatic safety belts. The three petitions for reconsideration on dynamically-tested manual safety belts and the new petition for rulemaking on automatic safety belts raise similar issues, which the agency is currently reviewing. The agency will respond to those petitions at a later date.

Effective Date for Dynamic Testing of Manual Lap/Shoulder Belts

Nissan asked the agency for a two-year postponement, from September 1, 1989, to September 1, 1991, of the effective date of the dynamic test requirement for front seat manual lap/shoulder belts in passenger cars. The dynamic test requirement for passenger car manual belts will go into effect only if the automatic restraint requirement for passenger cars is rescinded. Nissan said that if a decision to rescind the automatic restraint requirements is not made until the end of March 1989, it will have only six months in which to develop a manual belt which can meet the dynamic test requirement. Nissan also said that having to develop a dynamically-tested manual safety belt prior to March 1989 places an unreasonable burden on manufacturers since they would have to be simultaneously developing both automatic restraints and dynamically-tested manual belts.

The agency has previously denied, in the March 21, 1986, final rule, a similar request from American Motors Corporation (AMC) for such an extension. In denying AMC's request, the agency noted that most of the vehicle components in passenger cars necessary for injury reduction are the same for automatic restraint vehicles and dynamically-tested manual belt vehicles. In addition, the agency noted that the New Car Assessment Program results show that approximately 40 percent of current model passenger cars can meet the injury criteria of Standard No. 208 in 35-mph crash tests, which involve 36 percent more crash energy than the 30-mph crash test used in Standard No. 208. Nissan has not provided any new data that would justify changing the agency's prior decision and therefore, Nissan's request for an extension of the effective date is denied.

Due Care Defense

The Center for Auto Safety (CFAS) and Ford petitioned the agency to reconsider its decision to adopt a due care defense in Standard No. 208. CFAS said that adoption of the defense contravenes the noncompliance notification and remedy requirements of the National Traffic and Motor Vehicle Safety Act. In addition, CFAS said that the due care defense is not a standard for motor vehicle performance as required by the Vehicle Safety Act and is too broad to accomplish its intended purpose. Ford said that adoption of the due care defense does not sufficiently address its concerns about the objectivity and practicability of the standard's requirements. It urged the agency to adopt a design to conform to the requirement in the standard.

The agency is still reviewing the issues raised by CFAS and Ford about the due care defense. Because the automatic restraint phase-in requirement is imminent, NHTSA has decided to retain the due care provision for the first year of the phase-in, pending the agency's final decision on

this issue. The agency will expedite its review of these issues.

To clarify its interpretation of the due care defense, the agency does want to address one issue raised by the CFAS. In its comments, CFAS offered an example of what it believed was a problem with the due care defense. The CFAS said:

Consider, for example, a scenario in which the agency's compliance test reveals a very high HIC score. The manufacturer's tests show complying results. It turns out that the manufacturer received from a supplier a shipment of poor quality restraint system components that resulted in the poor figure in the agency's test and would cause similarly poor results for most vehicles containing the components from that shipment. The poor quality components were not caught in the manufacturer's quality control program. Perhaps this failure to catch the poor quality component is because their problems only show up in dynamic crash testing. (The due care defense surely will not require manufacturers to crash test a vehicle containing components from each shipment.) Or perhaps the manufacturer's quality control by chance checked only some of the few units in the shipment that were of good quality. Under the due care exemption these vehicles could not be recalled for noncompliance despite clear evidence of a specific problem that will cause high HIC levels.

As stated in the preamble to the March 21, 1986, final rule, the due care defense is meant to address an instance where there is an isolated apparent failure and the manufacturer can demonstrate that it made a good faith effort in designing its vehicles and instituted adequate quality control measures. NHTSA considers the example used by CFAS as an instance in which the agency would *not* accept a due care defense and the vehicles would be subject to the noncompliance notification and remedy provisions of the Vehicle Safety Act. Clearly, the CFAS's example shows there is a significant flaw in the manufacturer's quality control process which affects a widespread number of vehicles. Manufacturers are capable of instituting quality control measures that will adequately test the performance of individual components without having to subject a vehicle containing that component to a crash test, Likewise, quality control measures are available so that manufacturers can statistically check a sufficient number of components to ensure that nearly all of the components of a particular shipment are of the required quality. For these reasons, the agency would *not* accept a due care defense in the example posed by CFAS.

Belt Contact Force Test Procedure

The March 21, 1986, notice renumbered the test dummy positioning procedure for the belt contact force test of the safety belt comfort and convenience requirements. In making that amendment, the following sentence was inadvertently left out: "Close the vehicle's adjacent door, pull 12 inches of belt webbing from the retractor and then release it, allowing the belt webbing to return to the dummy's chest."

Nissan has recently written the agency containing the deletion of the sentence. Nissan said that if the deletion was inadvertent and the requirement was reinstated, then the agency should slightly modify the requirement. Nissan said that in systems where it is not possible to pull out 12 inches of belt webbing, the requirement should provide for pulling out the maximum available length of the belt webbing.

Nissan pointed out that, as stated by the agency in the April 12, 1985, notice proposing amendments to the comfort and convenience requirement, one purpose of pulling out the webbing is to reduce belt drag in the the belt guide components prior to measuring the belt contact force. It further said that maintaining the 12-inch requirement would necessitate a complete redesign of some of the belt systems for its vehicles.

NHTSA agrees that the purpose of the belt webbing pull requirement can be adequately met by pulling out the maximum allowable amount of the belt, when the belt has less than 12 inches of available additional webbing. Pulling the belt in this way will ensure that the belt retractor is working and webbing drag is reduced. Thus, the agency is changing the requirement to provide that prior to measuring the belt contact force the agency will pull out 12 inches of webbing or the maximum amount of webbing available when the maximum amount is less than 12 inches.

The agency recognizes that manufacturers may have relied, in good faith, on the version of the belt contact force test procedure and based their certification of compliance on tests conducted according to that procedure. So as not to invalidate those compliance tests, the agency is amending the standard to allow the manufacturers to conduct the belt contact force test either with or without first pulling the webbing. Beginning September 1, 1987, the old test procedure will become mandatory.

Typographical Errors

The amendments made on March 21, 1986, contained a typographical error which is being corrected in this notice. In S4.1.3.2.2(b), the word "car" is corrected to read "cars."

Costs and Benefits

NHTSA has examined the impact of this rulemaking action and determined that it is not major within the meaning of Executive Order 12291 or significant within the meaning of the Department of Transportation's regulatory policies and procedures. The agency has also determined that the economic and other impacts of this rulemaking action are so minimal that a full regulatory evaluation is not required.

The amendments adopted by this notice make some minor clarifying changes to the test dummy positioning procedures. In addition, the agency is providing increased flexibility to manufacturers by allowing them to use one of two sets of test procedures for a one-year period. Use of either set of test procedures should have only minimal impact on a manufacturer's testing costs.

Regulatory Flexibility Act

NHTSA has also considered the impacts of this rulemaking action under the Regulatory Flexibility Act. I hereby certify that it would not have a significant economic impact on a substantial number of small entities. Accordingly, the agency has not prepared a full regulatory flexibility analysis.

Few, if any, passenger car manufacturers would qualify as small entities and the test procedure changes made by this notice are minimal. Small organizations and governmental units should not be significantly affected since the costs, if any, associated with the test procedure changes should be minimal.

Environmental Effects

NHTSA has analyzed this rulemaking action for the purposes of the National Environmental Policy Act. The agency has determined that implementation of this action will not have any significant impact on the quality of the human environment.

Paperwork Reduction

The information collection requirements of this notice are being submitted to the Office of Management and Budget pursuant to the requirements of the Paperwork Reduction Act (44 U.S.C. 3501 *et seq.*).

Effective Date

NHTSA has determined that it is in the public interest to amend, upon publication of this final rule, the requirement of Standard No. 208 since the test dummy positioning options adopted by this notice affect manufacturer's plans for the 1987 model year.

In consideration of the foregoing, Part 571.208 of Title 49 of the Code of Federal Regulations is amended as follows:

1. In S4.1.3.2.2(b), the word "car" is amended to read "cars."

2. S10 through S10.9 is revised to read as follows:

S10 *Test dummy positioning procedures.* For vehicles manufactured before September 1, 1987, position a test dummy, conforming to Subpart B of Part 572 (49 CFR Part 572), in each front outboard seating position of a vehicle as specified in S10 through S10.9 or, at the manufacturer's option, as specified in S12 through S12.2.3.2. For vehicles manufactured on or after September 1, 1987, position a test dummy, conforming to Subpart B of Part 572 (49 CFR Part 572), in each front outboard seating position of a vehicle as set forth below in S10 through S10.9. Regardless of which positioning procedure is used, each test dummy is restrained during the crash tests of S5 as follows:

(a) In a vehicle equipped with automatic restraints at each front outboard designated seating position that is certified by its manufacturer as meeting the requirements of S4.1.2.1(a) and (c)(1), each test dummy is not restrained during the frontal test of S5.1, the lateral test of S5.2 and the rollover test of S5.3 by any means that require occupant action.

(b)(i) In a vehicle equipped with an automatic restraint at each front outboard seating position that is certified by its manufacturer as meeting the requirements of S4.1.2.1(a) and (c)(2), each test dummy is not restrained during one frontal test of S5.1 by any means that require occupant action. If the vehicle has a manual seat belt provided by the manufacturer to comply with the requirements of S4.1.2.1(c), then a second frontal test is conducted in accordance with S5.1 and each test dummy is restrained both by the automatic restraint system and the manual seat belt, adjusted in accordance with S10.9.

(ii) In a vehicle equipped with an automatic restraint only at the driver's designated seating position, pursuant to S4.1.3.4(a)(2), that is certified by its manufacturer as meeting the requirements of S4.1.2.1(a) and (c)(2), the driver test dummy is not restrained during one frontal test of S5.1 by any means that require occupant action. If the vehicle also has a manual seat belt provided by the manufacturer to comply with the requirements of S4.1.2.1(c), then a second frontal test is conducted in accordance with S5.1 and the driver test dummy is restrained both by the automatic restraint system and the manual seat belt, adjusted in accordance with S10.9. At the option of the manufacturer, a passenger test dummy can be placed in the right front outboard designated seating position during the testing required by this section. If a passenger test dummy is present, it shall be restrained by a manual seat belt, adjusted in accordance with S10.9.

(c) In a vehicle equipped with a manual safety belt at the front outboard designated seating position that is certified by its manufacturer to meet the requirements of S4.6, each test dummy is restrained by the manual safety belts, adjusted in accordance with S10.9, installed at each front outboard seating position.

S10.1 *Vehicle equipped with front bucket seats.* Place the test dummy's torso against the seat back and its upper legs against the seat cushion to the extent permitted by placement of the test dummy's feet in accordance with the appropriate paragraph of S10. Center the test dummy on the seat cushion of the bucket seat and set its midsagittal plane so that it is vertical and parallel to the centerline of the seat cushion.

S10.1.1 *Driver position placement.*

(a) Initially set the knees of the test dummy 14.5 inches apart, measured between the outer surfaces of the knee pivot bolt heads, with the left outer surface 5.9 inches from the midsagittal plane of the test dummy.

(b) Rest the right foot of the test dummy on the undepressed accelerator pedal with the rearmost point of the heel on the floor pan in the plane of the pedal. If the foot cannot be placed on the accelerator pedal, set it initially perpendicular to the lower leg and place it as far forward as possible in the direction of the pedal centerline with the rearmost point of the heel resting on the floor pan. Except as prevented by contact with a vehicle surface, place the right leg so that the upper and lower leg centerlines fall, as closely as possible, in a vertical plane without inducing torso movement.

(c) Place the left foot on the toeboard with the rearmost point of the heel resting on the floor pan as close as possible to the point of intersection of the planes described by the toeboard and the floor pan and not on the wheelwell projection. If the foot cannot be positioned on the toeboard, set it initially perpendicular to the lower leg and place it as far forward as possible with the heel resting on the floor pan. If necessary to avoid contact with the vehicle's brake or clutch pedal, rotate the test dummy's left foot about the lower leg. If there is still pedal interference, rotate the left leg outboard about the hip the minimum distance necessary to avoid the pedal interference. Except as prevented by contact with a vehicle surface, place the left leg so that the upper and lower leg centerlines fall, as closely as possible, in a vertical plane. For vehicles with a foot rest that does not elevate the left foot above the level of the right foot, place the left foot on the foot rest so that the upper and lower leg centerlines fall in a vertical plane.

S10.1.2 *Passenger position placement.*

S10.1.2.1 *Vehicles with a flat floor pan/toeboard.*

(a) Initially set the knees 11.75 inches apart, measured between the outer surfaces of the knee pivot bolt heads.

(b) Place the right and left feet on the vehicle's toeboard with the heels resting on the floor pan as close as possible to the intersection point of the toeboard. If the feet cannot be placed flat on the toeboard, set them perpendicular to the lower leg centerlines and place them as far forward as possible with the heels resting on the floor pan.

(c) Place the right and left legs so that the upper and lower leg centerlines fall in vertical longitudinal planes.

S10.1.2.2 *Vehicles with wheelhouse projections in passenger compartment.*

(a) Initially set the knees 11.75 inches apart, measured between the outer surfaces of the knee pivot bolt heads.

(b) Place the right and left feet in the well of the floor pan/toeboard and not on the wheelhouse projection. If the feet cannot be placed flat on the toeboard, initially set them perpendicular to the lower leg centerlines and then place them as far forward as possible with the heels resting on the floor pan.

(c) If it is not possible to maintain vertical and longitudinal planes through the upper and lower leg centerlines for each leg, then place the left leg so that its upper and lower centerlines fall, as closely as possible, in a vertical longitudinal plane and place the right leg so that its upper and lower

leg centerlines fall, as closely as possible, in a vertical plane.

S10.2 *Vehicle equipped with bench seating.* Place the test dummy's torso against the seat back and its upper legs against the seat cushion, to the extent permitted by placement of the test dummy's feet in accordance with the appropriate paragraph of S10.1.

S10.2.1 *Driver position placement.* Place the test dummy at the left front outboard designated seating position so that its midsagittal plane is vertical and parallel to the centerline of the vehicle and so that the midsagittal plane of the test dummy passes through the center of the steering wheel rim. Place the legs, knees, and feet of the test dummy as specified in S10.1.1.

S10.2.2 *Passenger position placement.* Place the test dummy at the right front outboard designated seating position so that the midsagittal plane of the test dummy is vertical and longitudinal, and the same distance from the vehicle's longitudinal centerline as the midsagittal plane of the test dummy at the driver's position. Place the legs, knees, and feet of the test dummy as specified in S10.1.2.

S10.3 *Initial test dummy hand and arm placement.* With the test dummy at its designated seating position as specified by the appropriate requirements of S10.1 or S10.2, place the upper arms against the seat back and tangent to the side of the upper torso. Place the lower arms and palms against the outside of the upper leg.

S10.4 *Test dummy settling.*

S10.4.1 *Test dummy vertical upward displacement.* Slowly lift the test dummy parallel to the seat back plane until the test dummy's buttocks no longer contact the seat cushion or until there is test dummy head contact with the vehicle's headlining.

S10.4.2 *Lower torso force application.* Apply a rearward force of 50 pounds against the center of the test dummy's lower torso in a horizontal direction. The line of force application shall be 6.5 inches above the bottom surface of the test dummy's buttocks.

S10.4.3 *Test dummy vertical downward displacement.* Remove as much of the 50-pound force as necessary to allow the test dummy to return downward to the seat cushion by its own weight.

S10.4.4 *Test dummy upper torso rocking.* Apply a 10-to-15-pound horizontal rearward force to the test dummy's lower torso. Then apply a horizontal forward force to the test dummy's shoulders sufficient to flex the upper torso forward until its back no longer contacts the seat back. Rock the test dummy from side to side 3 or 4 times so that the test dummy's spine is at any angle from the vertical in the 14-to-16-degree range at the extremes of each rocking movement.

S10.4.5 *Test dummy upper torso force application.* While maintaining the 10-to-15-pound horizontal rearward force applied in S10.4.4 and with the test dummy's midsagittal plane vertical, push the upper torso back against the seat back with a force of 50 pounds applied in a horizontal rearward direction along a line that is coincident with the test dummy's midsagittal plane and 18 inches above the bottom surface of the test dummy's buttocks.

S10.5 *Belt adjustment for dynamic testing.* With the test dummy at its designated seating position as specified by the appropriate requirements of S8.1.2, S8.1.3, and S10.1 through S10.4, place and adjust the safety belt as specified below.

S10.5.1 *Manual safety belts.* Place the Type 1 or Type 2 manual belt around the test dummy and fasten the latch. Pull the Type 1 belt webbing out of the retractor and allow it to retract; repeat this operation four times. Remove all slack from the lap belt portion of a Type 2 belt. Pull the upper torso webbing out of the retractor and allow it to retract; repeat this operation four times so that the excess webbing in the shoulder belt is removed by the retractive force of the retractor. Apply a 2-to-4-pound tension load to the lap belt of a single retractor system by pulling the upper torso belt adjacent to the latchplate. In the case of a dual retractor system, apply a 2-to-4-pound tension load by pulling the lap belt adjacent to its retractor. Measure the tension load as close as possible to the same location where the force was applied. After the tension load has been applied, ensure that the upper torso belt lies flat on the test dummy's shoulder.

S10.5.2 *Automatic safety belts.* Ensure that the upper torso belt lies flat on the test dummy's shoulder after the automatic belt has been placed on the test dummy.

S10.5.3 *Belts with tension-relieving devices.* If the automatic or dynamically-tested manual safety belt system is equipped with a tension-relieving device, introduce the maximum amount of slack into the upper torso belt that is recommended by the manufacturer for normal use in the owner's manual for the vehicle.

S10.6 *Placement of test dummy arms and hands.* With the test dummy positioned as specified by S10.4 and without inducing torso movement, place the arms, elbows, and hands of the test dummy,

as appropriate for each designated seating position in accordance with S10.6.1 or S10.6.2. Following placement of the arms, elbows, and hands, remove the force applied against the lower half of the torso.

S10.6.1 *Driver's position.* Move the upper and the lower arms of the test dummy at the driver's position to their fully outstretched position in the lowest possible orientation. Push each arm rearward permitting bending at the elbow, until the palm of each hand contacts the outer part of the rim of the steering wheel at its horizontal centerline. Place the test dummy's thumbs over the steering wheel rim and position the upper and lower arm centerlines as closely as possible in a vertical plane without inducing torso movement.

S10.6.2 *Passenger position.* Move the upper and the lower arms of the test dummy at the passenger position to the fully outstretched position in the lowest possible orientation. Push each arm rearward, permitting bending at the elbow, until the upper arm contacts the seat back and is tangent to the upper part of the side of the torso, the palm contacts the outside of the thigh, and the little finger is barely in contact with the seat cushion.

S10.7 *Repositioning of feet and legs.* After the test dummy has been settled in accordance with S10.4, the safety belt system has been positioned, if necessary, in accordance with S10.5, and the arms and hands of the test dummy have been positioned in accordance with S10.6, reposition the feet and legs of the test dummy, if necessary, so that the feet and legs meet the applicable requirements of S10.1 or S10.2.

S10.8 *Test dummy positioning for latchplate access.* The reach envelopes specified in S7.4.4 are obtained by positioning a test dummy in the driver's seat or passenger's seat in its forwardmost adjustment position. Attach the lines for the inboard and outboard arms to the test dummy as described in Figure 3 of this standard. Extend each line backward and outboard to generate the compliance arcs of the outboard reach envelope of the test dummy's arms.

S10.9 *Test dummy positioning for belt contact force.*

S10.9.1 *Vehicles manufactured before September 1 1987.* To determine compliance with S7.4.3 of this standard, a manufacturer may use, at its option, either the test procedure of S10.9.1 or the test procedure of S10.9.2. Position the test dummy in the vehicle in accordance with the appropriate requirements specified in S10.1 or S10.2 and under the conditions of S8.1.2 and S8.1.3. Fasten the latch and pull the belt webbing three inches from the test dummy's chest and release until the webbing is within one inch of the test dummy's chest and measure the belt contact force.

S10.9.2 *Vehicles manufactured on or after September 1, 1987.* To determine compliance with S7.4.3 of this standard, position the test dummy in the vehicle in accordance with the appropriate requirements specified in S10.1 or S10.2 and under the conditions of S8.1.2 and S8.1.3. Close the vehicle's adjacent door, pull either 12 inches of belt webbing or the maximum available amount of belt webbing, whichever is less, from the retractor and then release it, allowing the belt webbing to return to the dummy's chest. Fasten the latch and pull the belt webbing three inches from the test dummy's chest and release until the webbing is within one inch of the test dummy's chest and measure the belt contact force.

3. A new section S12 is added to read as follows:

S12. *Optional position procedures for the Part 572, Subpart B test dummy.* The following test dummy positioning procedures for the Part 572, Subpart B test dummy may be used, at the option of a manufacturer, until September 1, 1987.

S12.1 *Dummy placement in vehicle.* Anthropomorphic test dummies are placed in the vehicle in accordance with S12.1.1 and S12.1.2.

S12.1.1 *Vehicle equipped with front bucket seats.* In the case of a vehicle equipped with front bucket seats, dummies are placed at the front outboard designated seating positions with the test device torso against the seat back, and the thighs against the seat cushion to the extent permitted by placement of the dummy's feet in accordance with the appropriate paragraph of S12.1. The dummy is centered on the seat cushion of the bucket seat and its midsagittal plane is vertical and longitudinal.

S12.1.1.1 *Driver position placement.* At the driver's position, the knees of the dummy are initially set 14.5 inches apart, measured between the outer surfaces of the knee pivot bolt heads, with the left outer surface 5.9 inches from the midsagittal plane of the dummy. The right foot of the dummy rests on the undepressed accelerator pedal with the rearmost point of the heel on the floor pan in the plane of the pedal. If the foot cannot be placed on the accelerator pedal, it is set perpendicular to the tibia and placed as far forward as possible in the direction of the geometric center of the pedal with the rearmost point of the heel resting on the floor pan. The plane defined by the femur and tibia centerlines of the right leg is as close as possible to vertical without inducing torso movement and except as prevented by contact with a vehicle surface. The left foot is placed on the

toeboard with the rearmost point of the heel resting on the floor pan as close as possible to the point of intersection of the planes described by the toeboard and the floor pan. If the foot cannot be positioned on the toeboard, it is set perpendicular to the tibia and placed as far forward as possible with the heel resting on the floor pan. The femur and tibia centerlines of the left leg are positioned in a vertical plane except as prevented by contact with a vehicle surface.

S12.1.1.2 *Passenger position placement.* At the right front designated seating position, the femur, tibia, and foot centerlines of each of the dummy's legs are positioned in a vertical plane. The feet of the dummy are placed on the toeboard with the rearmost point of the heel resting on the floor pan as close as possible to the point of intersection of the planes described by the toeboard and the floorpan. If the feet cannot be positioned flat on the toeboard, they are set perpendicular to the tibia and are placed as far forward as possible with the heels resting on the floor pan.

S12.1.2 *Vehicle equipped with bench seating.* In the case of a vehicle which is equipped with a front bench seat, a dummy is placed at each of the front outboard designated seating positions with the dummy torso against the seat back and the thighs against the seat cushion to the extent permitted by placement of the dummy's feet in accordance with the appropriate paragraph of S12.1.1.

S12.1.2.1 *Driver position placement.* The dummy is placed at the left front outboard designated seating position so that its midsagittal plane is vertical and longitudinal, and passes through the center point of the plane described by the steering wheel rim. The legs, knees, and feet of the dummy are placed as specified in S12.1.1.1.

S12.1.2.2 *Passenger position placement.* The dummy is placed at the right front outboard designated seating position as specified in S.12.1.1.2, except that the midsagittal plane of the dummy is vertical, longitudinal, and the same distance from the longitudinal centerline as the midsagittal plane of the dummy at the driver's position.

S12.2 *Dummy positioning procedures.* The dummy is positioned on a seat as specified in S12.2.1 through S12.2.3.2 to achieve the conditions of S12.1.

S12.2.1 *Initial dummy placement.* With the dummy at its designated seating position as described in S12.1 place the upper arms against the seat back and tangent to the side of the upper torso and the lower arms and palms against the outside of the thighs.

S12.2.2 *Dummy settling.* With the dummy positioned as specified in S10.1, slowly lift the dummy in the direction parallel to the plane of the seat back until its buttocks no longer contact the seat cushion or until its head contacts the vehicle roof. Using a flat, square, rigid surface with an area of 9 square inches and oriented so that its edges fall in longitudinal or horizontal planes, apply a force of 50 pounds through the center of the rigid surface against the dummy's torso in the horizontal rearward direction along a line that is coincident with the midsagittal plane of the dummy and 5.5 inches above the bottom surface of its buttocks. Slowly remove the lifting force.

S12.2.2.1 While maintaining the contact of the force application plate with the torso, remove as much force as is necessary from the dummy's torso to allow the dummy to return to the seat cushion by its own weight.

S12.2.2.2 Without removing the force applied to the lower torso, apply additional force in the horizontal, forward direction, longitudinally against the upper shoulders of the dummy sufficient to flex the torso forward until the dummy's back above the lumbar spine no longer contacts the seatback. Rock the dummy from side to side three or four times, so that the dummy's spine is at an angle from the vertical of not less than 14 degrees and not more than 16 degrees at the extreme of each movement. With the midsagittal plane vertical, push the upper half of the torso back against the seat back with a force of 50 pounds applied in the horizontal rearward direction along a line that is coincident with the midsagittal plane of the dummy and 18 inches above the bottom surface of its buttocks. Slowly remove the horizontal force.

S12.2.3 *Placement of dummy arms and hands.* With the dummy positioned as specified in S12.2.2 and without inducing torso movement, place the arms, elbows, and hands of the dummy, as appropriate for each designated seating position in accordance with S12.2.3.1 or S12.2.3.2. Following placement of the limbs, remove the force applied against the lower half of the torso.

S12.2.3.1 *Driver's position.* Move the upper and the lower arms of the dummy at the driver's position to the fully outstretched position in the lowest possible orientation. Push each arm rearward, permitting bending at the elbow, until the palm of each hand contacts the outer part of the rim of the steering wheel at its horizontal centerline. Place the dummy's thumbs over the steering wheel rim, positioning the upper and lower arm centerlines as close as possible in a vertical plane without including torso movement.

S12.2.3.2 *Passenger position.* Move the upper and the lower arms of the dummy at the passenger position to the fully outstretched position in the lowest possible orientation. Push each arm rearward, permitting bending at the elbow, until the upper arm contacts the seat back and is tangent to the upper part of the side of the torso, the palm contacts the outside of the thigh, and the little finger is barely in contact with the seat cushion.

§571.209 Standard No. 209, *Seat Belt Assemblies.*

1. S4.6(b) of §571.209 is revised to read as follows:

(b) A seat belt assembly that meets the requirements of 4.6.1 of Standard No. 208 of this part (§571.208) shall be permanently and legibly marked or labeled with the following statement:

"This dynamically-tested seat belt assembly is for use only in (insert specific seating position(s), e.g., 'front right') in (insert specific vehicle make(s), and model(s))."

Issued on August 29, 1986

Diane K. Steed
Administrator

51 F.R. 29552
August 19, 1986

PREAMBLE TO AN AMENDMENT TO FEDERAL MOTOR VEHICLE SAFETY STANDARD NO. 208

Occupant Crash Protection and Automatic Restraint Phase-in Reporting

(Docket No. 74-14; Notice 47)

ACTION: Final Rule.

SUMMARY: On April 12, 1985, NHTSA published a notice proposing amendments to Standard No. 208, *Occupant Crash Protection.* On March 21, 1986, NHTSA published a final rule that addressed a number of the proposed requirements. This notice announces the agency's decisions on several of the remaining proposals. NHTSA has decided to adopt an exemption from the automatic restraint requirement for convertibles. The exemption would only apply during the phase-in period. In a subsequent rulemaking the agency will determine whether to apply the automatic restraint requirement to convertibles manufactured after September 1, 1989, or whether to apply a dynamic test requirement to the manual safety belts used in those vehicles. The agency is modifying the head injury criterion used in Standard No. 208 compliance testing by adopting a maximum time interval of 36 milliseconds for calculating the HIC values.

EFFECTIVE DATE: The amendments made by this notice will be effective on November 17, 1986.

SUPPLEMENTARY INFORMATION:

On April 12, 1985 (50 FR 14589), NHTSA published a notice, which is the basis for the final rule being issued today, proposing the following amendments to Standard No. 208, *Occupant Crash Protection:* reporting requirements for the phase-in of automatic restraints, deletion of the oblique crash test, use of the New Car Assessment Program (NCAP) test procedures, adoption of a due care defense, alternative calculations of the head injury criterion (HIC), and alternative occupant crash protection requirements for convertibles. The notice also proposed the dynamic testing of manual lap/shoulder belts for passenger cars, light trucks and light van-type vehicles.

On March 21, 1986 (51 FR 9800), NHTSA published a final rule amending Standard No. 208 that retained the oblique crash test for automatic restraint equipped cars, adopted some NCAP test procedures for use in the standard's crash tests, provided for a due care defense with respect to the automatic restraint require-

ment, and required the dynamic testing of manual lap/shoulder belts in passenger cars if the automatic restraint requirement is rescinded. The March 1986 notice also created a new Part 585 setting reporting requirements regarding compliance with the automatic restraint phase-in requirements of the standard. This notice announces the agency's decision on several of the other actions proposed in the April 1985 notice. NHTSA will soon publish a separate notice announcing its decision on dynamic testing of safety belts in light trucks, buses, and multipurpose passenger vehicles.

Convertibles

The April 1985 notice proposed alternative occupant crash protection requirements for convertibles, beginning with model year 1990. The agency proposed that manufacturers have the option of installing manual lap belts, subject to the belt strength requirements of Standard No. 209, *Seat Belt Assemblies*, and the anchorage strength requirements of Standard No. 210, *Seat Belt Assembly Anchorages*, instead of installing automatic restraints subject to the occupant crash protection criteria of Standard No. 208.

As a part of the notice, NHTSA requested data on several specific questions to assist the agency in making a decision. Those questions covered such issues as current and future production figures for convertibles and the cost and practicability of installing various types of automatic restraints. The answers provided by the commenters show that:

- Through 1989, convertibles will average slightly over one percent of annual passenger car production.

- Manufacturers uniformly said that automatic safety belts are not a practical alternative for convertibles. For example, General Motors estimated an automatic lap/shoulder belt would cost $600 for convertibles, with much of that cost needed for structural modifications to the car. It also said that while automatic lap belts may be technically possible, their actual performance could be below that of manual belts because of additional belt "slack" that would be inherent in such designs.

- Manufacturers' estimates of the costs of air bag systems, exclusively for use in convertibles, ranged from $1,200 to $3,500.

- Most manufacturers supported exemption of convertibles from the automatic restraint requirement, saying that the increased costs of automatic restraints would diminish convertible sales. Ford, Toyota, and Volkswagen said that if convertibles had to meet the automatic restraint requirement, they would probably have to discontinue their convertible lines.

- All manufacturers that provided information on the type of safety belt they are installing in their convertibles stated that they use lap/shoulder safety belts, even though the standard currently gives them the option of using only a lap belt. Volkswagen suggested requiring all convertibles to have lap/shoulder belts.

- The Center for Auto Safety (CFAS), Insurance Institute for Highway Safety (IIHS), and State Farm, all of which supported the use of automatic restraints in convertibles, argued that convertibles are "luxury" cars and thus any cost increase associated with automatic restraints would not affect the sales of convertibles. In support of its argument or requiring automatic restraints in convertibles, CFAS also noted that the agency's NCAP data show that, with two exceptions, crash test results in the convertible version of a vehicle were considerably worse than in the "parent" vehicle.

- The National Transportation Safety Board (NTSB) argued that the current provision in the standard allowing manufacturers the option of installing only lap belts in convertibles is inadequate and may not provide sufficient protection in a crash.

After reviewing the comments, NHTSA continues to believe that applying the automatic restraint requirement to convertibles is not reasonable, practicable or appropriate for that vehicle type, at least during the phase-in. The information provided by the commenters shows that use of automatic belts is not reasonable for some models because they would have to make substantial structural redesigns to incorporate a "pylon" or other structure for attaching the upper torso portion of the automatic belt. If manufacturers use air bag systems, then the cost of the system could be substantial enough to severely curtail sales of those models. However, as new types of air bag and other automatic restraint systems are developed, the cost could be reduced. The agency has therefore decided to limit the exemption for convertibles to the phase-in period. NHTSA will re-examine, at a later date, the issue of whether to apply an automatic restraint requirement to convertibles manufactured after September 1, 1989, or to require dynamic testing of the manual safety belts installed in those vehicles.

NHTSA believes that its decision is consistent with its duty, under section 103(f)(3) of the National Traffic and Motor Vehicle Safety Act (15 U.S.C. 1392(f)(3)), to "consider whether any such proposed standard is reasonable, practicable and appropriate for the particular type of motor vehicle . . . for which it is prescribed." The legislative history of the Vehicle Safety Act makes clear that Congress recognized that it might not be appropriate to set the same standards for some vehicle types, such as convertibles, as other vehicle types. In discussing the purpose of section 103(f)(3), the Senate Report stated that:

[T]he committee intends that the Secretary will consider the desirability of affording consumers continued wide range of choices in the selection of motor vehicles. Thus it is not intended that standards will be set which will eliminate or necessarily be the same for small cars *or such widely accepted models as convertibles* and sports cars, so long as all motor vehicles meet basic minimum standards. [Emphasis added.]

NHTSA's decision with regard to convertibles is also consistent with the guidance provided by the U.S. Court of Appeals for the Sixth Circuit in its decision in *Chrysler v. Department of Transportation,* 472 F.2d 659 (1972). In that decision, the court reviewed the legislative history of section 103(f)(3), discussed above, and concluded that the agency did not give sufficient attention to the issue of whether convertibles should be subject to the same occupant crash protection requirements as hard top vehicles. While the court's decision to send the rule back to the agency for further consideration was based primarily on the perceived inadequacy of the test dummy used in compliance tests, the decision was also based on the need for the agency to consider adequately the potential effects of the occupant crash protection rule on convertibles.

The substantial cost impact of requiring convertibles to have automatic restraints, would be true even if convertibles were considered "luxury" cars, since the cost would have to be spread over a very low production volume. For example, although the agency believes that the cost for low volume installation of air bag systems—10,000 to 100,000 cars or less annually—would be smaller than the estimates submitted by some manufacturers, the cost, which ranges from $600 to $1,500 per vehicle, would still be substantial. Although convertible models are priced higher than their sedan counterparts, they are not all "high priced" or "luxury" cars. For example, convertible versions of the Renault Alliance, Chevrolet Cavalier, Chrysler LeBaron, Dodge 600, Ford Mustang LX, and Pontiac Sunbird all sell from $11,000—$13,000.

It is possible that development of new technology may lead to new air bag systems with lower costs. The agency is currently conducting research with the Breed Corporation on an air bag system with an all-mechanical sensor, which has the potential of being produced at a lower cost than current systems with electronic sensors. The preliminary data from the sled and crash tests of the Breed system are promising. However, the system still must be field-tested before the agency will be able to evaluate its effectiveness. Thus, it is still too early to predict whether this research system or other systems can be successfully developed into an effective and low-cost air bag system that can be used in convertibles and other passenger cars.

In the case of "built-in" safety (i.e., use of padding and structural changes to provide protection to unrestrained occupants), the agency notes that only General Motors has done some preliminary work, and GM has not yet indicated that it could certify convertibles or any vehicles to the injury protection criteria of Standard No. 208. Thus, the practicability of this approach across the fleet of convertibles (i.e., for all manufacturers for each of their convertibles) is uncertain at this time. The agency will continue to monitor the development of new automatic belt, air bag and built-in safety systems and review the practicability and appropriateness of those systems for convertibles.

Definition of convertible

Toyota asked the agency to clarify what vehicles are considered to be convertibles; in particular, it asked whether a passenger car with a T-bar roof or a Targa top would be considered a convertible. In several letters of interpretation, the agency has said that a convertible is a vehicle whose A-pillar or windshield peripheral support is not joined at the top with the B-pillar or other rear roof support rearward of the B-pillar by a fixed rigid structural member. Thus, a vehicle with a Targa top would be considered a convertible since it does not have any fixed structural member connecting the tops of the A and B-pillars. However, a vehicle with a T-bar roof would not be considered a convertible since there is a fixed structural member in the vehicle's roof which connects the A and B-pillars.

Changes in reporting requirements

Part 585, Automatic Restraint Phase-in Reporting Requirements, requires manufacturers to provide NHTSA with a yearly report on their compliance with the automatic restraint phase-in requirements of Standard No. 208. Part 585 currently requires manufacturers to provide data on their entire production of passenger cars, including convertibles. Since NHTSA has decided to exempt convertibles from the requirement for mandatory installation of automatic

restraints during the phase-in period, the agency is making a change to Part 585. The agency is amending the reporting requirement so that a manufacturer does not have to count convertibles as a part of its passenger car production volume when it is calculating its phase-in requirement. However, since a manufacturer may decide to install voluntarily automatic restraints in its convertibles, the changes made to the phase-in requirements of Standard No. 208 and the reporting requirements of Part 585 will allow a manufacturer the option to include automatic-restraint equipped convertibles in its passenger car production volume when it is determining its compliance with the automatic restraint phase-in requirement.

Modification of the head injury criterion

In response to a petition from the Committee on Common Market Automobile Constructors and comments from other vehicle manufacturers, the April 1985 notice set forth two proposed alternative methods of using the head injury criterion (HIC) in situations when there is no contact between the test dummy's head and the vehicle's interior during a crash. The agency said that, after considering the comments, it would decide whether to retain the current HIC requirement or to adopt one of the proposed alternatives. As discussed in detail below, the agency has decided to adopt the proposed alternative which will calculate a HIC in both contact and non-contact situations, but limit the calculation to a maximum time interval of 36 milliseconds.

I. First Proposed HIC Alternative.

A. Use HIC only when there is head contact

The first proposed alternative was to retain the current HIC calculation for contact situations, but limited to the actual times that contact occurs. However, in non-contact situations, the agency proposed that a HIC would not be calculated, but instead new neck injury criteria would be calculated. The agency proposed that neck criteria would be calculated differently depending upon whether the existing Part 572 test dummy or the Hybrid III test dummy was used in the crash test. The reason for the proposed difference was that the Hybrid III test dummy has instrumentation in its neck to measure directly shear and tension forces in the neck and the existing Part 572 test dummy does not. The agency proposed to use the Hybrid III's neck instrumentation and set limits on the shear and tension forces in the neck. Since neck forces cannot be measured directly by the existing Part 572 test dummy, the agency proposed to use a surrogate measure for neck forces through the use of head acceleration-based criteria, a calculation that is valid only when the head does not contact any object during a crash test.

The agency explained that a crucial element necessary for deciding whether to use the HIC calculation or the neck criteria was an objective technique for determining the occurrence and duration of head contact in the crash test. As discussed in detail in the April 1985 notice, there are several methods available for establishing the occurrence and/or duration of head contact, but there are questions about their levels of consistency and accuracy.

Almost all of the commenters opposed the use of the first proposed alternative. The commenters uniformly noted that there is no current technique that can accurately and reliably identify whether and exactly when head contact has or has not occurred during a crash test in all situations. The agency agrees that, in the absence of such a method to determine the occurrence and duration of head contact, the first alternative is not appropriate.

B. Apply neck criteria if there is no head contact

As discussed above, the agency proposed a new neck criteria to be used in non-contact situations, however, because of the problems involved in trying to identify when head contact occurs, the agency is not adopting the non-contact proposal. CFAS urged the agency to apply the neck injury criteria in both contact and non-contact situations. It also argued that because the neck has more soft tissue than the head, a lower acceleration threshold should be used. As noted above, with the Part 572 test dummy, the proposed neck injury criteria (based on head acceleration measurements) are valid only when the head does not contact another object, so they should not be used in situations when there is an impact to the head. Similarly, the impossibility of determining, in all situations, when head contact begins and ends precludes the agency from adopting the proposed non-contact neck injury criteria for the Part 572 test dummy. The agency has already indicated that it will consider the issue of neck injury criteria for the Hybrid III test dummy in the separate rulemaking on that test dummy.

II. Second HIC Alternative.

At present, a HIC is calculated for the entire crash duration. The second alternative proposed by the agency would calculate a HIC in both contact and non-contact situations, but it would limit the time duration during which a HIC is calculated. NHTSA proposed a limit on the maximum time duration of the HIC calculation because the current calculation can produce high HIC values for a crash which has a relatively low acceleration level, but a long time duration, and which in all likelihood will not result in brain injuries.

The agency proposed to limit the HIC calculation to a maximum of 36 milliseconds because it determined

that the 36 millisecond limit together with a HIC of 1,000 limit will assure that the acceleration level of the head will not exceed 60 g's for any period greater than 36 milliseconds. The 60 g's acceleration limit was set as a reasonable head injury threshold by the originators of the Wayne State Tolerance Curve, which was used in the development of the HIC calculation. (Readers are referred to the April 12, 1985 notice of proposed rulemaking for information on the development of HIC.)

There was a marked division among the commenters on the second alternative. Manufacturers and their trade associations commenting on the issue uniformly supported the use of the second alternative, although nine of those commenters (AMC, Chrysler, Ford, GM, Motor Vehicle Manufacturers Association (MVMA), Peugeot, Renault, Volvo, and Volkswagen) argued that the HIC calculation should be limited to a time interval of approximately 15 to 17 milliseconds, which would limit average long time duration head accelerations to 80-85 g's. Other commenters (CFAS, IIHS, and State Farm) argued that the current HIC calculations should be retained, they said that the proposed alternative would lower HIC levels without ensuring that motorists were still receiving adequate head protection.

Those favoring the second alternative raised a number of arguments in support of its use. They said that using a time limit for the HIC calculation is appropriate because head contacts with hard surfaces generally have high accelerations, but a short time duration (10 to 15 milliseconds). In the case of head contacts with softer surfaces, such as an airbag system, they said that the time duration of the contact is longer, but the acceleration is much lower, and thus the potential for injury is reduced. Ford pointed to airbag system testing in which human volunteers "experienced average accelerations between 59 to 63 g's for HIC calculation durations of 24 to 30 ms, without any head or neck injury."

Those favoring use of a shorter time duration than 36 ms offered additional arguments. They said that the proposed 36 ms requirement is too stringent because it would not allow the average head acceleration levels during a crash to exceed 60 g's. For example, GM said that the the Wayne State cadaver test data show that the head can withstand acceleration levels of up to 80 g's without injury. GM also said that Wayne State and other test data show that brain injuries and skull fractures in cadavers occur at HIC durations of 15 ms or less and thus there is no basis for considering any time interval longer than 15 ms. Likewise, Volvo said that it does not believe that 60 g is a critical acceleration level. Volvo noted that Standard No. 201; *Occupant Protection in Interior Impact*, permits an acceleration level of up to 80 g's in 15 mile per hour impacts of the instrument panel with a headform.

Mercedes-Benz, which supported the second alternative, urged the agency to measure HIC only during the time interval when the acceleration level in the head exceeds 60 g's. It said that this method would more effectively differentiate results received in contacts with hard surfaces and results obtained from systems, such as airbags, which provide good distribution of the loads experienced during a crash.

Those opposing the proposed second alternative argued that a 36 millisecond time limit is too short and could result in lower HIC scores being calculated than are calculated by the current HIC formula. For example, IIHS noted that a 60 g impact with a time duration of 50 milliseconds would produce a HIC of greater than 1,000 using the calculation methods currently found in the standard. IIHS also said that since some brain injuries can occur at a HIC level of less than 1,000, the agency should not take any action that would, in effect, allow HIC levels of above 1,000. It also urged the agency not to adopt the 36 millisecond limit since there is evidence showing that even mild brain injuries can produce long-term disability and it is not known whether such injuries can be caused without head contact.

A. Rejection of 17 millisecond HIC limit

To evaluate the effect of the 17 millisecond limit suggested by many of the commenters, NHTSA re-examined the biomechanical studies cited by the commenters and looked at the effect of how the recommended time limits would affect the HIC values measured in a 30 mile per hour barrier crash test, which is the compliance test used in Standard No. 208 for different types of restraint systems and also with respect to the New Car Assessment Program (NCAP). After completing this review, NHTSA has concluded, as discussed below, that the use of a 17 millisecond limit is not appropriate in vehicle crash tests.

The agency reviewed the Wayne State laboratory test results cited by several of the commenters in support of adopting a 17 millisecond limit for the HIC calculation. In those tests, cadaver heads were dropped on various hard and padded surfaces. The results from those tests show that those impacts generally produce a single peak acceleration, which ranges from 4 to 13 ms in duration. While NHTSA agrees that a 17 millisecond limit would be appropriate for short duration, single impacts into a hard surface, head acceleration responses in crash tests are considerably different from laboratory drop tests. In a vehicle crash, the duration of head impacts is often considerably longer, the head impact can involve considerably higher forces, and

the head can experience multiple impacts. Given these differences, NHTSA does not believe that a 17 millisecond limit, based on single, short duration laboratory tests, should be adopted.

NHTSA agrees with Ford that the test results from the human volunteer airbag test are important and demonstrate that the probability of injury in longer duration impacts (greater than 15 milliseconds) with moderate accelerations is low. However, NHTSA believes that the air bag tests are limited in their application. Those well-controlled tests using young, healthy males do not necessarily represent the results that would be found using other segments of the population. Likewise, the recommendation by the Wayne State researchers regarding a head acceleration limit of 60 to 80 g's is deduced mostly from tests with healthy 19 to 48 year old male volunteers. As to Volvo's comments about the use of an 80 g criteria in Standard No. 201, the agency notes that the standard places a specific limit on the 80 g criteria by prohibiting the accelerations from exceeding 80 g's for more than 3 continuous milliseconds.

NHTSA believes that it should take a cautious approach in modifying the head injury tolerance level set by the HIC requirement. Any modifications should ensure that a wide range of the population is provided protection. Therefore, the agency believes that it should use a HIC calculation which will not exceed 60 g's during relatively long duration impacts, which is the lower end of the recommended range proposed by the Wayne State researchers for use with HIC.

A review of the effect of a 17 millisecond limit on 291 test results from the 35 mph NCAP test program and the test results from 30 mph barrier impact tests also support the agency's decision not to adopt that suggestion. This analysis yielded the following results:

1. Using the current HIC calculation, this agency noted that the average HIC for the 291 NCAP tests was 1,107 and the percentage of HIC's that exceeded 1,000 was 46 percent. Using a 17 millisecond limit, the average HIC in the 291 NCAP tests dropped to 931 and the percentage of HIC's that exceeded 1,000 fell to 35 percent.

2. The current HIC failure rate of approximately 16 percent for 30 mph belted occupants could be cut to approximately 8 percent,

3. For unrestrained occupants, the average HIC value would drop by 21 percent and their Standard No. 208 compliance failure rate would be reduced by 42 percent,

4. Airbag average HIC values would be reduced by 28 percent, however, this would not affect the Standard No. 208 failure rates, since air bags that function properly produce HIC values well below the 1,000 level.

B. Rejection of Mercedes-Benz HIC limitation

To evaluate the effect of the Mercedes-Benz suggestion to limit the calculation of HIC to instances when the acceleration exceeded 60 g's, the agency recalculated the HIC values for 30 mph 3-point belts (driver and passenger sides), 30 mph unrestrained (driver and passenger), air bag (only), and 35 mph NCAP barrier and barrier equivalent crash tests using the Mercedes-Benz method.

Compared to the 36 ms proposed by NHTSA and 15-17 ms. approach advocated by some commenters, the Mercedes-Benz method would bring about the most significant numerical reduction in HIC scores. At 30 mph, all lap/shoulder belt passenger HIC scores would be reduced to zero (a 100 percent reduction). Using the current HIC calculation, the average HIC for the 291 NCAP tests was 1107 and the percentage of HIC's that exceeded 1,000 was 46 percent. Using a 60 g limit, the average HIC would drop to 808 and the percentage of HIC's that exceeded 1,000 would fall to 32 percent. 30 mph air bag HIC values would be cut by 47.9 percent and unrestrained occupants would experience a 31 to 36 percent reduction of the average HIC score. The data also indicates that failure rates on airbags would not be affected, while approximately 14 percent of the unrestrained passengers would be shifted from failing to passing the HIC 1,000 limit.

The use of a minimum head acceleration threshold or cut-off to define the maximum HIC time duration, as proposed by Mercedes-Benz, provides a means of differentiating between critical and non-critical acceleration peaks, if and when they exceed 60 g's. However, there are a number of problems which led the NHTSA to reject the proposed Mercedes-Benz method. The Mercedes-Benz method only takes into account head accelerations that are greater than 60 g's. Thus, the average head acceleration permitted by the Mercedes-Benz method must be, at a minimum, 60 g's and most likely the average head acceleration permitted by the Mercedes-Benz method would substantially exceed that limit. In contrast, the 36 millisecond alternative adopted by the agency will ensure that the average head acceleration does not exceed the 60 g acceleration limit. In addition, it is unclear from Mercedes-Benz's comments how their method would accommodate multiple non-continuous acceleration peaks in excess of 60 g's. Discriminating between injurious and non-injurious peaks is critical to picking the time duration. If all peaks are to be included, it is unclear from Mercedes' proposal how the time interval would be measured. Given all of these concerns, NHTSA believes that the Mercedes-Benz proposal should not be adopted.

C. Adoption of 36 millisecond HIC limit

As discussed earlier in this notice, the agency proposed a time limit for the HIC calculation because the current method can produce an artificially high HIC for a crash which has a relatively low acceleration level, but a long time duration. To evaluate the effects of the proposal, NHTSA took the NCAP results and recalculated the HIC using the proposed 36 millisecond limit. That analysis shows that the 36 millisecond limit would have only a minor effect on HIC scores recorded in the NCAP tests. As discussed above, using the current HIC calculation, the average HIC for the 291 tests was 1107 and the percentage of HIC's that exceeded 1,000 was 46 percent. Using a 36 millisecond limit, the average HIC dropped slightly to 1061, and the percentage of HIC values that exceeded 1,000 dropped to 41 percent. Thus, the results show that in the NCAP tests, which are conducted at 35 mph, the average HIC value would be only four percent lower when calculated with the 36 millisecond limit. In addition, the results showed that of the 291 NCAP tests, only 38 tests had both a HIC value which exceeded 1,000 and a HIC duration exceeding 36 milliseconds. Of this group of 38 tests, there are only 15 instances in which the 36 millisecond limit results in a new HIC value less than 1,000. Since the NCAP tests at 35 mph involve 36 percent greater energy than the 30 mph tests used in Standard No. 208 compliance testing, the number of HIC values possibly changing from above 1,000 to below 1,000 because of the 36 millisecond limit should be even less in the Standard No. 208 compliance tests.

The agency further examined these 15 instances of HIC's greater than 1,000 being recalculated to be less than 1,000. In 12 of these 15 cases, the original HIC (i.e., without a time limitation) was between 1,000 and 1,074. Again at 30 mph, with 36 percent, less energy involved, it is doubtful if any of these vehicles would have had occupant HIC's greater than 1,000. Thus, in only three cases (one percent of the total involved) would a "fail" have potentially become a "pass," using the 208 criteria. If this same value is associated with 30 mph barrier tests, the risks to safety associated with having a HIC calculation which is founded on a sounder basis than the current calculation are not significant.

To further evaluate the effects of a 36 millisecond limit, the agency specifically examined the potential impact of the new HIC calculation on whether a vehicle will pass or fail the HIC of 1,000 limit set in Standard No. 208. NHTSA recalculated the HICs recorded in a wide variety of 30 mph crash tests, which is the compliance test speed used in Standard No. 208. The tests included vehicles using the following different types of restraint systems: manual lap/shoulder belts, automatic belts, air bags only, and air bag with lap and lap/

PART 571; S 208—PRE 354

shoulder belts. In addition, the agency recalculated the HIC values recorded in 30 mph tests with unrestrained occupants, which would simulate the types of HIC values that could be recorded in vehicles with built-in safety features. (The results of those tests are discussed in Chapter III of the Final Regulatory Evaluation on HIC). The agency's analysis shows that in all the 30 mph tests, the 36 millisecond limit does not change a "failing" HIC into a "passing" HIC. Thus, a vehicle which currently does not comply with the HIC requirement of Standard No. 208 using the prior HIC calculation method also will not comply using the 36 millisecond limit.

Cost and Benefits

NHTSA has examined the impact of this rulemaking action and determined that it is not major within the meaning of Executive Order 12291 or significant within the meaning of the Department of Transportation's regulatory policies and procedures. The agency has prepared a regulatory evaluation that examines the economic and other impacts of this rulemaking action.

The changes in the HIC calculation should not have a significant impact. As discussed in detail above, the agency's analysis of crash test data shows that the 36 millisecond limit does not have any significant effect on changing the HIC values currently recorded in 30 mile per hour compliance crash tests. The extent of the effect of this change on mild brain injuries is unknown. As IIHS noted, there is insufficient data on how such injuries are caused. Thus, the agency cannot assess the role of the current or changed HIC calculation in preventing or reducing such injuries. However, since the agency's crash test analysis shows that a vehicle that currently exceeds a HIC of 1000 in Standard No. 208's 30 mile per hour compliance test will still exceed 1000 using the new 36 millisecond limit, the agency believes that the effect of the 36 millisecond limit on mild brain injuries should be no different than the effect of the current calculation. In addition, NHTSA does not believe that manufacturers will change their vehicle designs because of the slight change in the HIC calculation. Thus, the 36 millisecond limit should not adversely affect safety or a manufacturer's compliance costs.

Likewise, the decision to exempt convertibles during the phase-in period should not have a significant effect. Because convertibles represent a small portion of most manufacturers' production, they do not need to install automatic restraints in their convertibles in order to meet the production requirements during the phase-in. The problems associated with installing automatic restraints in convertibles also make it unlikely that manufacturers would equip their convertibles with such restraints during the phase-in. Thus, the exemp-

tion adopted in this notice should have little effect on the type of restraint system that will be used in convertibles during the phase-in.

Effective Date

NHTSA has determined that it is in the public interest to make the amendments, adopted in today's notice, effective immediately. The change in the HIC calculation can affect manufacturer's plans for the model year beginning September 1, 1986.

49 CFR PART 585—Reporting and recordkeeping requirements.

In consideration of the foregoing, Part 571.208 of Title 49 of the Code of Federal Regulations is amended as follows:

1. The authority citation for Part 571 continues to read as follows:

Authority: 15 U.S.C. 1392, 1401, 1403, 1407., delegation of authority at 49 CFR 1.50.

2. A new S4.1.3.1.3 is added to Part 571.208 to read as follows:

S4.1.3.1.3 A manufacturer may exclude convertibles which do not comply with the requirements of S4.1.2.1, when it is calculating its average annual production under S4.1.3.1.2(a) or its annual production under S4.1.3.1.2(b).

3. A new S4.1.3.2.3 is added to Part 571.208 to read as follows:

S4.1.3.2.3 A manufacturer may exclude convertibles which do not comply with the requirements of S4.1.2.1, when it is calculating its average annual production under S4.1.3.2.2(a) or its annual production under S4.1.3.2.2(b).

4. A new S4.1.3.3.3 is added to Part 571.208 to read as follows:

S4.1.3.3.3 A manufacturer may exclude convertibles which do not comply with the requirements of S4.1.2.1, when it is calculating its average annual production under S4.1.3.3.2(a) or its annual production under S4.1.3.3.2(b).

6. S6.2 of Part 571.208 is revised to read as follows:

S6.2 The resultant acceleration at the center of gravity of the head shall be such that the expression:

$$\left[\frac{1}{t_2 - t_1} \int_{t_1}^{t_2} a\,dt \right]^{2.5} t_2 - t_1$$

shall not exceed 1,000 where a is the resultant acceleration expressed as a multiple of g (the acceleration of gravity), and t_1 and t_2 are any two points in time during the crash of the vehicle which are separated by not more than a 36 millisecond time interval.

Part 585, *Automatic Restraint Phase-In Reporting*

1. Part 585.4 is revised to read as follows:

§ 585.4 **Definitions.**

(a) All terms defined in section 102 of the National Traffic and Motor Vehicle Safety Act (15 U.S.C. 1391) are used in their statutory meaning.

(b) "Passenger car" means a motor vehicle with motive power, except a multipurpose passenger vehicle, motorcycle, or trailer, designed for carrying 10 persons or less.

(c) "Production year" means the 12-month period between September 1 of one year and August 31 of the following year, inclusive.

2. Part 585(b)(1) is revised to read a follows:

(b) *Report content*—(1) *Basis for phase-in production goals.* Each manufacturer shall provide the number of passenger cars manufactured for sale in the United States for each of the 3 previous production years, or, at the manufacturer's option, for the current production year. A new manufacturer that is, for the first time, manufacturing passenger cars for sale in the United States must report the number of passenger cars manufactured during the current production year. For the purpose of the reporting requirements of this Part, a manufacturer may exclude its production of convertibles, which do not comply with the requirments of S4.1.2.1 of Part 571.208 of this Chapter, from the report of its production volume of passenger cars manufactured for sale in the United States.

Issued on October 10, 1986.

Diane K. Steed
Administrator

51 F.R. 37028
October 10, 1986

MOTOR VEHICLE SAFETY STANDARD NO. 208

Occupant Crash Protection in Passenger Cars, Multipurpose Passenger Vehicles, Trucks and Buses

(Docket No. 69-7; Notice No. 9)

S1. Scope. This standard specifies performance requirements for the protection of vehicle occupants in crashes.

S2. Purpose. The purpose of this standard is to reduce the number of deaths of vehicle occupants and the severity of injuries, by specifying vehicle crashworthiness requirements in terms of forces and accelerations measured on anthropomorphic dummies in test crashes, and by specifying equipment requirements for active and passive restraint systems.

S3. Application. This standard applies to passenger cars, multipurpose passenger vehicles, trucks, and buses. In addition, S9, *Pressure vessels and explosive devices,* applies to vessels designed to contain a pressurized fluid or gas, and to explosive devices, for use in the above types of motor vehicles as part of a system designed to provide protection to occupants in the event of a crash.

S4. General requirements.

S4.1 Passenger cars.

S4.1.1 Passenger cars manufactured from January 1, 1972, to August 31, 1973. Each passenger car manufactured from January 1, 1972, to August 31, 1973, inclusive, shall meet the requirements of S4.1.1.1, S4.1.1.2, or S4.1.1.3. A protection system that meets the requirements of S4.1.1.1 or S4.1.1.2 may be installed at one or more designated seating positions of a vehicle that otherwise meets the requirements of S4.1.1.3.

S4.1.1.1 First option—complete passive protection system. The vehicle shall meet the crash protection requirements of S5 by means that require no action by vehicle occupants.

S4.1.1.2 Second option—lap belt protection system with belt warning. The vehicle shall—

(a) At each designated seating position have a Type 1 seat belt assembly or a Type 2 seat belt assembly with a detachable upper torso portion that conforms to S7.1 and S7.2 of this standard.

(b) At each front outboard designated seating position have a seat belt warning system that conforms to S7.3; and

(c) Meet the frontal crash protection requirements of S5.1, in a perpendicular impact, with respect to anthropomorphic test devices in each front outboard designated seating position restrained only by Type 1 seat belt assemblies.

S4.1.1.3 Third option—lap and shoulder belt protection system with belt warning.

S4.1.1.3.1 Except for convertibles and open-body vehicles, the vehicle shall—

(a) At each front outboard designated seating position have a Type 2 seat belt assembly that conforms to Standard No. 209 and S7.1 and S7.2 of this standard, with either an integral or detachable upper torso portion, and a seat belt warning system that conforms to S7.3;

(b) At each designated seating position other than the front outboard positions, have a Type 1 or Type 2 seat belt assembly that conforms to Standard No. 209 and to S7.1 and S7.2 of this standard; and

(c) When it perpendicularly impacts a fixed collision barrier, while moving longitudinally forward at any speed up to and including 30 m.p.h., under the test conditions of S8.1 with anthropomorphic test devices at each front outboard position restrained by Type 2 seat belt assemblies, experience no complete separation of any load-bearing element of a seat belt assembly or anchorage.

S4.1.1.3.2 Convertibles and open-body type vehicles shall at each designated seating position have a Type 1 or Type 2 seat belt assembly that conforms to Standard No. 209 and to S7.1 and S7.2 of this standard, and at each front outboard designated seating position have a seat belt warning system that conforms to S7.3.

S4.1.2 Passenger cars manufactured on or after September 1, 1973, and before September 1, 1986. Each passenger car manufactured on or after September 1, 1973, and before September 1, 1986, shall meet the requirements of S4.1.2.1, S4.1.2.2, or S4.1.2.3.

S4.1.2.1 First option—frontal/angular automatic protection system. The vehicle shall—

(a) At each front outboard designated seating position meet the frontal crash protection requirements of S5.1 by means that require no action by vehicle occupants;

(b) At each front center designated seating position have a Type 1 or Type 2 seat belt assembly that conforms to Standard No. 209 and to S7.1 and S7.2; and

(c) Either—

(1) Meet the lateral crash protection requirements of S5.2 and the rollover crash protection requirements of S5.3 by means that require no action by vehicle occupants; or

(2) At each front outboard designated seating position have a Type 1 or Type 2 seat belt assembly that conforms to Standard No. 209 and to S7.1 through S7.3, and that meets the requirements of S5.1 with front test dummies as required by S5.1, restrained by the Type 1 or Type 2 seat belt assembly (or the pelvic portion of any Type 2 seat belt assembly which has a detachable upper torso belt) in addition to the means that require no action by the vehicle occupant.

S4.1.2.2 Second option—head-on automatic protection system. The vehicle shall—

(a) At each designated seating position have a Type 1 seat belt assembly or a Type 2 seat belt assembly with a detachable upper torso portion that conforms to S7.1 and S7.2 of this standard.

(b) At each front outboard designated seating position, meet the frontal crash protection requirements of S5.1, in a perpendicular impact, by means that require no action by vehicle occupants;

(c) At each front outboard designated seating position, meet the frontal crash protection requirements of S5.1, in a perpendicular impact, with a test device restrained by a Type 1 seat belt assembly; and

(d) At each front outboard designated seating position, have a seat belt warning system that conforms to S7.3.

S4.1.2.3 Third option—lap and shoulder belt protection system with belt warning.

S4.1.2.3.1 Except for convertibles and open-body vehicles, the vehicle shall—

(a) At each front outboard designated seating position have a seat belt assembly that conforms to S7.1 and S7.2 of this standard, and a seat belt warning system that conforms to S7.3. The belt assembly shall be either a Type 2 seat belt assembly with a nondetachable shoulder belt that conforms to Standard No. 209 (S571.209), or a Type 1 seat belt assembly such that with a test device restrained by the assembly the vehicle meets the frontal crash protection requirements of S5.1 in a perpendicular impact.

(b) At any center front designated seating position, have a Type 1 or Type 2 seat belt assembly that conforms to Standard No. 209 (S571.209) and to S7.1 and S7.2 of this standard, and a seat belt warning system that conforms to S7.3; and

(c) At each other designated seating position, have a Type 1 or Type 2 seat belt assembly that conforms to Standard No. 209 (S571.209) and S7.1 and S7.2 of this standard.

S4.1.2.3.2 Convertibles and open-body type vehicles shall at each designated seating position have a Type 1 or Type 2 seat belt assembly that conforms to Standard No. 209 (S571.209) and to S7.1 and S7.2 of this standard, and at each front designated seating position have a seat belt warning system that conforms to S7.3.

S4.1.3 Passenger cars manufactured on or after September 1, 1986, and before September 1, 1989.

S4.1.3.1 Passenger cars manufactured on or after September 1, 1986, and before September 1, 1987.

S4.1.3.1.1 Subject to S4.1.3.1.2 and S4.1.3.4, each passenger car manufactured on or after September 1, 1986, and before September 1, 1987, shall comply with the requirements of S4.1.2.1, S4.1.2.2 or S4.1.2.3.

[A vehicle shall not be deemed to be in compliance with this standard if its manufacturer establishes that it did not have a reason to know in the exercise of due care that such vehicle is not in conformity with the requirement of this standard. (51 F.R. 9801—March 21, 1986. Effective: May 5, 1986)]

S4.1.3.1.2 Subject to S4.1.3.4 and S4.1.5, the amount of passenger cars, specified in S4.1.3.1.1 complying with the requirements of S4.1.2.1, shall not be less than 10 percent of:

(a) the average annual production of passenger cars manufactured on or after September 1, 1983, and before September 1, 1986, by each manufacturer, or

(b) the manufacturer's annual production of passenger cars during the specified in S4.1.3.1.1.

[**S4.1.3.1.3** A manufacturer may exclude convertibles which do not comply with the requirements of S4.1.2.1, when it is calculating its average annual production under S4.1.3.1.2(a) or its annual production under S4.1.3.1.2(b). (51 F.R. 37028—October 17, 1986. Effective: November 17, 1986.)]

S4.1.3.2 Passenger cars manufactured on or after September 1, 1987, and before September 1, 1988.

S4.1.3.2.1 Subject to S4.1.3.2.2 and S4.1.3.4, each passenger car manufactured on or after September 1, 1987, and before September 1, 1988, shall comply with the requirements of S4.1.2.1, S4.1.2.2 or S4.1.2.3.

A vehicle shall not be deemed to be in compliance with this standard if its manufacturer establishes that it did not have a reason to know in the exercise of due care that such vehicle is not in conformity with the requirement of this standard.

S4.1.3.2.2 Subject to S4.1.5, the amount of passenger care specified in S4.1.3.2.1 complying with the requirements of S4.1.2.1 shall be not less than 25 percent of:

(a) the average annual production of passenger cars manufactured on or after September 1, 1984, and before September 1, 1987, by each manufacturer, or

(b) the manufacturer's annual production of passenger cars during the period specified in S4.1.3.2.1.

[**S4.1.3.2.3** A manufacturer may exclude convertibles which do not comply with the requirements of S4.1.2.1, when it is calculating its average annual production under S4.1.3.1.2.2(a) or its annual production under S4.1.3.1.2.2(b). (51 F.R. 37028—October 17, 1986. Effective: November 17, 1986.)]

S4.1.3.3 Passenger cars manufactured on or after September 1, 1988, and before September 1, 1989.

S4.1.3.3.1 Subject to S4.1.3.3.2 and S4.1.3.4, each passenger car manufactured on or after September 1, 1988, and before September 1, 1989, shall comply with the requirements of S4.1.2.1, S4.1.2.2 or S4.1.2.3.

A vehicle shall not be deemed to be in compliance with this standard if its manufacturer establishes

that it did not have a reason to know in the exercise of due care that such vehicle is not in conformity with the requirement of this standard.

S4.1.3.3.2 Subject to S4.1.3.4 and S4.1.5, the amount of passenger cars specified in S4.1.3.3.1 complying with the requirement of S4.1.2.1 shall be not less than 40 percent of:

(a) the average annual production of passenger cars manufactured on or after September 1, 1985, and before September 1, 1988, by each manufacturer or

(b) the manufacturer's annual production of passenger cars during the period specified in S4.1.3.3.1.

[**S4.1.3.3.3** A manufacturer may exclude convertibles which do not comply with the requirements of S4.1.2.1, when it is calculating its average annual production under S4.1.3.3.2(a) or its annual production under S4.1.3.3.2(b). (51 F.R. 37028—October 17, 1986. Effective: November 17, 1986.)]

S4.1.3.4 Calculation of complying passenger cars.

For the purposes of calculating the numbers of cars manufactured under S4.1.3.1.2, S4.1.3.2.2, or S4.1.3.3.2 to comply with S4.1.2.1:

(1) each car whose driver's seating position complies with the requirements of S4.1.2.1(a) by means not including any type of seat belt and whose front right seating position will comply with the requirements of S4.1.2.1(a) by any means is counted as 1.5 vehicles, and

(2) each car whose driver's seating position complies with the requirements of S4.1.2.1(a) by means not including any type of seat belt and whose right front seat seating position is equipped with a manual Type 2 seat belt is counted as one vehicle.

(b) For the purposes of complying with S4.1.3.1.2, a passenger car may be counted if it:

(1) is manufactured on or after September 1, 1985, but before September 1, 1986, and

(2) complies with S4.1.2.1

(c) For the purposes of complying with S4.1.3.2.2, a passenger car may be counted if it:

(1) is manufactured on or after September 1, 1985, but before September 1, 1987,

(2) complies with S4.1.2.1, and

(3) is not counted toward compliance with S4.1.3.1.2

(d) For the purposes of complying with S4.1.3.3.2, a passenger car may be counted if it:

(1) is manufactured on or after September 1, 1985, but before September 1, 1988,

(2) complies with S4.1.2.1, and

(3) is not counted toward compliance with S4.1.3.1.2 or S4.1.3.2.2.

[S4.1.3.5 Passenger cars produced by more than one manufacturer.

S4.1.3.5.1 For the purposes of calculating average annual production of passenger cars for each manufacturer and the amount of passenger cars manufactured by each manufacturer under S4.1.3.1.2, S4.1.3.2.2 or S4.1.3.3.2, a passenger car produced by more than one manufacuter shall be attributed to a single manufacturer as follows, subject to S4.1.3.5.2:

(a) A passenger car which is imported shall be attributed to the importer.

(b) A passenger car manufactured in the United States by more than one manufacturer, one of which also markets the vehicle, shall be attributed to the manufacturer which markets the vehicle.

S4.1.3.5.2 A passenger car produced by more than one manufacturer shall be attributed to any one of the vehicle's manufacturers specified by an express written contract, reported to the National Highway Traffic Safety Administration under 49 CFR Part 585, between the manufacturer so specified and the manufacturer to which the vehicle would otherwise be attributed under S4.1.3.5.1. (51 F.R. 9801—March 21, 1986. Effective: May 5, 1986)]

S4.1.4 Passenger cars manufactured on or after September 1, 1989. Except as provided in S4.1.5, each passenger car manufactured on or after September 1, 1989, shall comply with the requirements of S4.1.2.1.

[A vehicle shall not be deemed to be in compliance with this standard if its manufacturer establishes that it did not have a reason to know in the exercise of due care that such vehicle is not in conformity with the requirement of this standard. (51 F.R. 9801—March 21, 1986. Effective: May 5, 1986)]

S4.1.5 Mandatory seatbelt use laws.

S4.1.5.1 If the Secretary of Transportation determines, by not later than April 1, 1989, that state mandatory safety belt usage laws have been enacted that meet the criteria specified in S4.1.5.2 and that are applicable to not less than two-thirds of the total population of the 50 states and the District of Columbia (based on the most recent Estimates of the Resident Population of States, by

Age, Current Population Reports, Series P-25, Bureau of the Census), each passenger car manufactured under S4.1.3 or S4.1.4 on or after the date of that determination shall comply with the requirements of S4.1.2.1, S4.1.2.2, or S4.1.2.3.

S4.1.5.2 The minimum criteria for state mandatory safety belt usage laws are:

(a) Require that each front seat occupant of a passenger car equipped with safety belts under Standard No. 208 has a safety belt properly fastened about his or her body at all times when the vehicle is in forward motion.

(b) If waivers from the safety belt usage requirement are to be provided, permit them for medical reasons only.

(c) Provide for the following enforcement measures:

(1) A penalty of not less than $25.00 (which may include court costs) for each occupant of a car who violates the belt usage requirement.

(2) A provision specifying that the violation of the belt usage requirement may be used to mitigate damages with respect to any person who is involved in a passenger car accident while violating the belt usage requirement and who seeks in any subsequent litigation to recover damages for injuries resulting from the accident. This requirement is satisfied if there is a rule of law in the State permitting such mitigation.

(3) A program to encourage compliance with the belt usage requirement.

(d) An effective date of not later than September 1, 1989.

S4.2 Trucks and multipurpose passenger vehicles with GVWR of 10,000 pounds or less.

S4.2.1 Trucks and multipurpose passenger vehicles, with GVWR of 10,000 pounds or less, manufactured from January 1, 1972, to December 31, 1975. Each truck and multipurpose passenger vehicle with a gross vehicle weight rating of 10,000 pounds or less, manufactured from January 1, 1972, to December 31, 1975, inclusive, shall meet the requirements of S4.2.1.1 or S4.2.1.2, or at the option of the manufacturer, the requirements of S4.2.2. A protection system that meets the requirement of S4.2.1.1 may be installed at one or more designated seating positions of a vehicle that otherwise meets the requirements of S4.2.1.2.

S4.2.1.1 First option—complete passive protection system. The vehicle shall meet the crash protection requirements of S5 by means that require no action by vehicle occupants.

S4.2.1.2 Second option—belt system. The vehicle shall have seat belt assemblies that conform to Standard 209 installed as follows:

(a) A Type 1 or Type 2 seat belt assembly shall be installed for each designated seating position in convertibles, open-body type vehicles, and walk-in van-type trucks.

(b) In all vehicles except those for which requirements are specified in S4.2.1.2(a), a Type 2 seat belt assembly shall be installed for each outboard designated seating position that includes the windshield header within the head impact area, and a Type 1 or Type 2 seat belt assembly shall be installed for each other designated seating position.

S4.2.2 Trucks and multipurpose passenger vehicles, with GVWR of 10,000 pounds or less, manufactured on or after January 1, 1976. Each truck and multipurpose passenger vehicle, with a gross vehicle weight rating of 10,000 pounds or less, manufactured on or after January 1, 1976, shall meet the requirements of S4.1.2.1, or at the option of the manufacturer, S4.1.2.2 or S4.1.2.3 (as specified for passenger cars), except that forward control vehicles manufactured prior to September 1, 1981, convertibles, open-body type vehicles, walk-in van-type trucks, motor homes, vehicles designed to be exclusively sold to the U.S. Postal Service, and vehicles carrying chassis-mount campers may instead meet the requirements of S4.2.1.2.

S4.2.3 (Reserved)

S4.3 Trucks and multipurpose passenger vehicles, with GVWR of more than 10,000 pounds. Each truck and multipurpose passenger vehicle, with a gross vehicle weight rating of more than 10,000 pounds, manufactured on or after January 1, 1972, shall meet the requirements of S4.3.1 or S4.3.2. A protection system that meets the requirements of S4.3.1 may be installed at one or more designated seating positions of a vehicle that otherwise meets the requirements of S4.3.2.

S4.3.1 First option—complete passive protection system. The vehicle shall meet the crash protection requirements of S5 by means that require no action by vehicle occupants.

S4.3.2 Second option—belt system. The vehicle shall, at each designated seating position, have either a Type 1 or Type 2 seat belt assembly that conforms to Standard No. 209.

S4.4 Buses. Each bus manufactured on or after January 1, 1972, shall meet the requirements of S4.4.1 or S4.4.2.

S4.4.1 First option—complete passive protection system—driver only. The vehicle shall meet the crash protection requirements of S5, with respect to an anthropomorphic test device in the driver's designated seating position, by means that require no action by vehicle occupants.

S4.4.2 Second option—belt system—driver only. The vehicle shall, at the driver's designated seating position, have either a Type 1 or a Type 2 seat belt assembly that conforms to Standard No. 209.

S4.5 Other general requirements.

S4.5.1 Labeling and driver's manual information. Each vehicle shall have a label setting forth the manufacturer's recommended schedule for the maintenance or replacement, necessary to retain the performance required by this standard, of any crash-deployed occupant protection system. The schedule shall be specified by month and year, or in terms of vehicle mileage, or by intervals measured from the date appearing on the vehicle certification label provided pursuant to 49 CFR Part 567. The label shall be permanently affixed to the vehicle within the passenger compartment and lettered in English in block capitals and numerals not less than three thirty-seconds of an inch high. Instructions concerning maintenance or replacement of a system and a description of the functional operation of the system shall be provided with each vehicle, with an appropriate reference on the label. If a vehicle owner's manual is provided, this information shall be included in the manual.

S4.5.2 Readiness indicator. An occupant protection system that deploys in the event of a crash shall have a monitoring system with a readiness indicator. The indicator shall monitor its own readiness and shall be clearly visible from the driver's designated seating position. A list of the elements of the system being monitored by the indicator shall be included with the information furnished in accordance with S4.5.1 but need not be included on the label.

S4.5.3 Automatic belts. Except as provided in S4.5.3.1, a seat belt assembly that requires no action by vehicle occupants (hereinafter referred to as a "passive belt") may be used to meet the crash protection requirements of any option under S4 and in place of any seat belt assembly otherwise required by that option.

S4.5.3.1 An automatic belt that provides only pelvic restraint may not be used pursuant to S4.5.3 to meet the requirements of an option that requires a Type 2 seat belt assembly.

S4.5.3.2 An automatic belt, furnished pursuant to S4.5.3, that provides both pelvic and upper torso restraint may have either a detachable or nondetachable upper torso portion, notwithstanding provisions of the option under which it is furnished.

S4.5.3.3 A passive belt furnished pursuant to S4.5.3 shall—

(a) Conform to S7.1 and have a single emergency release mechanism whose components are readily accessible to a seated occupant.

(b) In place of a warning system that conforms to S7.3 of this standard, be equipped with the following warning system: At the left front designated seating position (driver's position), a warning system that activates a continuous or intermittent audible signal for a period of not less than 4 seconds and not more than 8 seconds and that activates a continuous or flashing warning light visible to the driver for not less than 60 seconds (beginning when the vehicle ignition switch is moved to the "on" or the "start" position) when condition (A) exists simultaneously with condition (B), and that activates a continuous or flashing warning light, visible to the driver, displaying the identifying symbol for the seat belt telltale shown in Table 2 of Standard No. 101 (49 CFR 571.101), or, at the option of the manufacturer if permitted by Standard No. 101, displaying the words "Fasten Seat Belts" or "Fasten Belts" for as long as condition (A) exists simultaneously with condition (C).

(A) The vehicle's ignition switch is moved to the "on" position or to the "start" position.

(B) The driver's automatic belt is not in use, as determined by the belt latch mechanism not being fastened or, if the automatic belt is non-detachable, by the emergency release mechanism being in the released position. In the case of motorized automatic belts, the determination of use shall be made once the belt webbing is in its locked protective mode at the anchorage point.

(C) The belt webbing of a motorized automatic belt system is not in its locked, protective mode at the anchorage point.

S4.5.3.4 An automatic belt furnished pursuant to S4.5.3 that is not required to meet the perpendicular frontal crash protection requirements of S5.1 shall conform to the webbing, attachment hardware, and assembly performance requirements of Standard No. 209.

[S4.6 Dynamic testing of manual belt systems.

S4.6.1 If the automatic restraint requirement of S4.1.4 is rescinded pursuant to S4.1.5, then each passenger car that is manufactured after September 1, 1989, and is equipped with a Type 2 manual seat belt assembly at each front outboard designated seating position pursuant to S4.1.2.3 shall meet the frontal crash protection requirements of S5.1 at those designated seating positions with a test dummy restrained by a Type 2 seat belt assembly that has been adjusted in accordance with S7.4.2.

A vehicle shall not be deemed to be in compliance with this standard if its manufacturer establishes that it did not have a reason to know in the exercise of due care that such vehicle is not in conformity with the requirement of this standard.

S4.6.2 A Type 2 seat belt assembly subject to the requirements of S4.6.1 of this standard does not have to meet the requirements of S4.2(a)—(c) and S4.4 of Standard No. 209 (49 CFR 571.209) of this Part.

S5. Occupant crash protection requirements.

S5.1 Frontal barrier crash. [Vehicle subject to S5.1 and manufactured before September 1, 1991, shall comply with either, at the manufacturer's option, 5.1(a) or (b). Vehicles subject to S5.1 and manufactured on or after September 1, shall comply with 5.1(b).

(a) Impact a vehicle traveling longitudinally forward at any speed, up to and including 30 mph, into a fixed collision barrier that is perpendicular to the line of travel of the vehicle, or at any angle up to 30 degrees in either direction from the perpendicular to the line of travel of the vehicle under the applicable conditions of S8. The test dummy specified in S8.1.8.1 placed at each front outboard designated seating position shall meet the injury criteria of S6.1.1, 6.1.2, 6.1.3, and 6.1.4.

(b) Impact a vehicle traveling longitudinally forward at any speed, up to and including 30 mph, into a fixed collision barrier that is perpendicular to the line of travel of the vehicle, or at any angle up to 30 degrees in either direction from the perpendicular to the line of travel of the vehicle, under the applicable conditions of S8. The test dummy specified in S8.1.8.2 placed at each front outboard designated seating position shall meet the injury criteria of S6.2.1, 6.2.2, 6.2.3, 6.2.4, and 6.2.5. (51 F.R. 26688—July 25, 1986. Effective: October 23, 1986)]

S5.2 Lateral moving barrier crash.

S5.2.1 Vehicles subject to S5.2 and manufactured before September 1, 1991, shall comply with either, at the manufacturer's option, 5.2.1(a) or (b). Vehicles subject to S5.2 and manufactured on or after September 1, 1991, shall comply with 5.2.1(b).

(a) Impact a vehicle laterally on either side by a barrier moving at 20 mph under the applicable conditions of S8. The test dummy specified in S8.1.8.1 placed at the front outboard designated seating position shall meet the injury criteria of S6.1.2 and S6.1.3.

(b) When the vehicle is impacted laterally under the applicable conditions of S8, on either side by a barrier moving at 20 mph, with a test device specified in S8.1.8.2, which is seated at the front outboard designated seating position adjacent to the impacted side, it shall meet the injury criteria of S6.2.2, and S6.2.3.

S5.3 Rollover. Subject a vehicle to a rollover test under the applicable condition of S8 in either lateral direction at 30 mph with either, at the manufacture's option, a test dummy specified in S8.1.8.1 or S8.1.8.2, placed in the front outboard designated seating position on the vehicle's lower side mounted on the test platform The test dummy shall meet the injury criteria of either S6.1.1 or S6.2.1.

S6 Injury criteria.

S6.1 Injury criteria for the Part 572, Subpart B, 50th percentile Male Dummy.

S6.1.1 All portions of the test dummy shall be contained within the outer surfaces of the vehicle passenger compartment throughout the test.

S6.1.2 [The resultant acceleration at the center of gravity of the head shall be such that the expression:

$$\left[\frac{1}{t_2 - t_1} \int_{t_1}^{t_2} a dt \right]^{2.5} t_2 - t_1$$

shall not exceed 1,000 where a is the resultant acceleration expressed as a multiple of g (the acceleration of gravity), and t_1 and t_2 are any two points in time during the crash of the vehicle which are separated by not more than a 36 millisecond time interval. (51 F.R. 37028—October 17, 1986. Effective: November 17, 1986.)]

S6.1.3 The resultant acceleration at the center of gravity of the upper thorax shall not exceed 60 g's, except for intervals whose cumulative duration is not more than 3 milliseconds.

S6.1.4 The compressive force transmitted axially through each upper leg shall not exceed 2250 pounds.

S6.2 Injury Criteria for the Part 572, Subpart E, hybrid III Dummy.

S6.2.1 All portions of the test dummy shall be contained within the outer surfaces of the vehcile passenger compartment throughout the test.

6.2.2 The resultant acceleration at the center of gravity of the head shall be such that the expression:

$$\left[\frac{1}{t_2 - t_1} \int_{t_1}^{t_2} a dt \right]^{2.5} t_2 - t_1$$

shall not exceed 1,000, where a is the resultant acceleration expressed as a multiple of g (the acceleration of gravity), and t_1 and t_2 are any two point during the crash of the vehicle which are separated by not more than a 36 millisecond time interval. (51 F.R. 37028—October 17, 1986. Effective: November 17, 1986.)]

6.2.3 The resultant acceleration calculated from the thoracic instrumentation shown in drawing 78051-218, revision L incorporated by reference in Part 572, Subpart E of this Chapter shall not exceed 60g's, except for intervals whose cumulative duration is not more than 3 milliseconds.

S6.2.4 Compression deflection of the sternum relative to spine, as determined by instrumentation shown in drawing 78051-317, revision A incorporated by reference in Part 572, Subpart E of this Chapter, shall not exceed 2 inches for loadings applied through any impact surfaces except for those systems which are gas inflated and provide distributed loading to the torso during a crash. For gas inflated systems which provide distributed loading to the torso, the thoracic deflection shall not exceed 3 inches.

S6.2.5 The force transmitted axially through each upper leg shall not exceed 2250 pounds.

S6.3 The resultant acceleration at the center of gravity of the upper thorax shall not exceed 60g, except for intervals whose cumulative duration is not more than 3 milliseconds. However, in the case

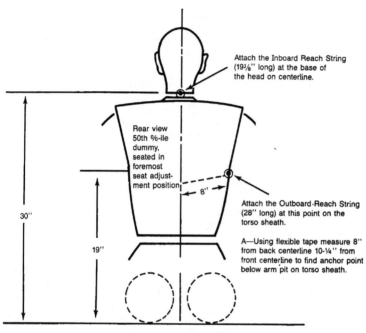

Attach the Inboard Reach String (19⅛" long) at the base of the head on centerline.

Rear view 50th %-ile dummy, seated in foremost seat adjustment position

Attach the Outboard Reach String (28" long) at this point on the torso sheath.

A—Using flexible tape measure 8" from back centerline 10-¼" from front centerline to find anchor point below arm pit on torso sheath.

Seat Plane is 90° to the Torso Line

Figure 3. Location of Anchoring Points for Latchplate Reach Limiting Chains or Strings to Test for Latchplate Accessibility

PART 571; S 208–8

Clearance Test Block

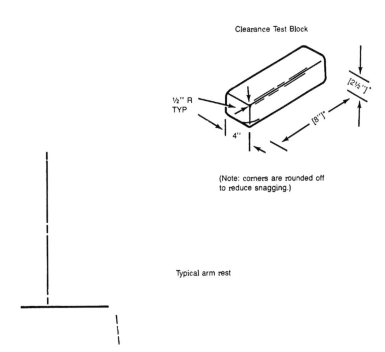

½" R
TYP

4"

[8"]*

[2½"]*

(Note: corners are rounded off
to reduce snagging.)

Typical arm rest

Figure 4. Use of Clearance Test Block to Determine Hand/Arm Access

* (50 F.R. 46056—November 6, 1985. Effective: September 1, 1986)

of a passenger car manufactured before August 31, 1976, or a truck or multipurpose passenger vehicle with a GVWR of 10,000 pounds or less manufactured before August 15, 1977, the resultant acceleration at the center of gravity of the upper thorax shall be such that the severity index calculated by the method described in SAE Information Report J885a, October 1966, shall not exceed 1,000.

S6.4 The compressive force transmitted axially through each upper leg shall not exceed 2,250 pounds.

S7. Seat belt assembly requirements.

S7.1 Adjustment.

S7.1.1 Except as specified in S7.1.1.1 and S7.1.1.2, the lap belt of any seat belt assembly furnished in accordance with S4.1.1 and S4.1.2 shall adjust by means of an emergency-locking or automatic locking retractor that conforms to Standard No. 209 to fit persons whose dimensions range from those of a 50th-percentile 6-year-old child to those of a 95th-percentile adult male and the upper torso restraint shall adjust by means of an emergency-locking retractor or a manual adjusting device that conforms to Standard No. 209 to fit persons whose dimensions range from those of a 5th-percentile adult female to those of a 95th-percentile adult male, with the seat in any position and the seat back in the manufacturer's nominal design riding position. However, an upper torso restraint furnished in accordance with S4.1.2.3.1(a) shall adjust by means of an emergency-locking retractor that conforms to Standard No. 209.

S7.1.1.1 A seat belt assembly installed at the driver's seating position shall adjust to fit persons whose dimensions range from those of a 5th-percentile adult female to those of a 95th-percentile adult male.

S7.1.1.2. (a) A seat belt assembly installed in a motor vehicle other than a forward control vehicle at any designated seating position other than the outboard positions of the front and second seats shall adjust either by a retractor as specified in S7.1.1 or by a manual adjusting device that conforms to S571.209.

(b) A seat belt assembly installed in a forward control vehicle at any designated seating position other than the front outboard seating positions shall adjust either by a retractor as specified in S7.1.1 or by a manual adjusting device that conforms to S571.209.

S7.1.1.3 (a) Except as provided in S7.1.1.3(b), a Type 1 lap belt or the lap belt portion of any Type 2 belt installed at any front outboard designated seating position for compliance with this standard in a vehicle (other than walk-in van-type vehicles) manufactured on or after September 1, 1986, shall meet the requirements of S7.1 by means of an emergency-locking retractor that conforms to Standard No. 209. (S571.209)

(b) The requirements of S7.1.1.3(a) do not apply to the lap belt portion of any Type 2 belt installed in a passenger car manufactured before September 1, 1989, or to walk-in van-type vehicles.

S7.1.1.4 Notwithstanding the other provisions of S7.1-S7.1.3, emergency-locking retractors on belt assemblies located in positions other than front outboard designated seating positions may be equipped with a manual webbing adjustment device capable of causing the retractor that adjusts the lap belt to lock when the belt is buckled.

S7.1.2 The intersection of the upper torso belt with the lap belt in any Type 2 seat belt assembly furnished in accordance with S4.1.1 or S4.1.2, with the upper torso manual adjusting device, if provided, adjusted in accordance with the manufacturer's instructions, shall be at least 6 inches from the front vertical centerline of a 50th-percentile adult male occupant, measured along the centerline of the lap belt, with the seat in its rearmost and lowest adjustable position and with the seat back in the manufacturer's nominal design riding position.

S7.1.3 The weights and dimensions of the vehicle occupants specified in this standard are as follows:

	50th-percentile 6-year-old child	5th-percentile adult female	50th-percentile adult male	95th-percentile adult male
Weight	47.3 pounds	102 pounds	164 pounds ± 3	215 pounds
Erect sitting height	25.4 inches	30.9 inches	35.7 inches $\pm .1$	38 inches
Hip breadth (sitting)	8.4 inches	12.8 inches	14.7 inches $\pm .7$	16.5 inches
Hip circumference (sitting)	23.9 inches	36.4 inches	42 inches	47.2 inches
Waist circumference (sitting)	20.8 inches	23.6 inches	32 inches $\pm .6$	42.5 inches
Chest depth		7.5 inches	9.3 inches $\pm .2$	10.5 inches
Chest circumference:				
(nipple)		30.5 inches		
(upper)		29.8 inches	37.4 inches $\pm .6$	44.5 inches
(lower)		26.6 inches		

(51 F.R. 21912—June 17, 1986. Effective: June 17, 1986)

S7.2 Latch mechanism. A seat belt assembly installed in a passenger car, except an automatic belt assembly, shall have a latch mechanism:

(a) Whose components are accessible to a seated occupant in both the stowed and operational positions;

(b) That releases both the upper torso restraint and the lap belt simultaneously, if the assembly has a lap belt and an upper torso restraint that require unlatching for release of the occupant; and

(c) That releases at a single point by a push-button action.

S7.3 A seat belt assembly provided at the driver's seating position shall be equipped with a warning system that activates, for a period of not less than 4 seconds and not more than 8 seconds (beginning when the vehicle ignition switch is moved to the "on" or the "start" position), a continuous or flashing warning light, visible to the driver, displaying the identifying symbol for the seat belt telltale shown in Table 2 of Federal Motor Vehicle Safety Standard No. 101-80 or, at the option of the manufacturer for vehicles manufactured before September 1, 1980, displaying the words "Fasten Seat Belts" or "Fasten Belts" when condition (a) exists, and a continuous or intermittent audible signal when condition (a) exists simultaneously with condition (b).

(a) The vehicle's ignition switch is moved to the "on" position or to the "start" position.

(b) The driver's lap belt is not in use, as determined at the option of the manufacturers, either by the belt latch mechanism not being fastened, or by the belt not being extended at least 4 inches from its stowed position.

S7.3.1	Deleted
S7.3.2	Deleted
S7.3.3	Deleted
S7.3.4	Deleted
S7.3.5	Deleted
S7.3.5.1	Deleted
S7.3.5.2	Deleted
S7.3.5.3	Deleted
S7.3.5.4	Deleted
S7.3a	Deleted

S7.4 Seat belt comfort and convenience.

(a) *Automatic seat belts.* Automatic seat belts installed in any vehicle, other than walk-in van-type vehicles, which has a gross vehicle weight rating of 10,000 pounds or less, and which is manufactured on or after September 1, 1986, shall meet the requirements of S7.4.1, S7.4.2, and S7.4.3.

(b) *Manual seat belts.*

(1) *Vehicles manufactured after September 1, 1986.* Manual seat belts installed in any vehicle, other than manual Type 2 belt systems installed in the front outboard seating positions in passenger cars or manual belts in walk-in van-type vehicles, which have a gross vehicle weight rating of 10,000 pounds or less, shall meet the requirements of S7.4.3, S7.4.4, S7.4.5, and S7.4.6.

(2) *Vehicles manufactured after September 1, 1989.*

(i) If the automatic restraint requirement of S4.1.4 is rescinded pursuant to S4.1.5, than manual seat belts installed in a passenger car shall meet the requirements of S7.4.2, S7.4.3, S7.4.4, S7.4.5, and S7.4.6.

(ii) Manual seat belts installed in a bus, multipurpose passenger vehicle and truck with a gross vehicle weight rating of 10,000 pounds or less, except for walk-in van-type vehicles, shall meet the requirements of S7.4.3, S7.4.4, S7.4.5, and S7.4.6.

S7.4.1 Convenience hooks. Any manual convenience hook or other device that is provided to stow seat belt webbing to facilitate entering or exiting the vehicle shall automatically release the webbing when the automatic belt system is otherwise operational and shall remain in the released mode for as long as (a) exists simultaneously with (b), or, at the manufacturer's option, for as long as (a) exists simultaneously with (c)—

(a) The vehicle ignition switch is moved to the "on" or "start" position;

(b) The vehicle's drive train is engaged;

(c) The vehicle's parking brake is in the released mode (nonengaged).

S7.4.2 Webbing tension-relieving device. [Each vehicle with an automatic seat belt assembly, or with a Type 2 manual seat belt assembly that must

PART 571; S 208-11

meet S4.6, installed in a front outboard designated seating position that has either manual or automatic devices permitting the introduction of slack in the webbing of the shoulder belt (e.g., "comfort clips" or "window-shade" devices) shall:

(a) comply with the requirements of S5.1 with the shoulder belt webbing adjusted to introduce the maximum amount of slack recommended by the manufacturer pursuant to S7.4.2(b);

(b) have a section in the vehicle owner's manual that explains how the tension-relieving device works and specifies the maximum amount of slack (in inches) recommended by the vehicle manufacturer to be introduced into the shoulder belt under normal use conditions. The explanation shall also warn that introducing slack beyond the amount specified by the manufacturer could significantly reduce the effectiveness of the shoulder belt in a crash; and

(c) have, except for open-body vehicles with no doors, an automatic means to cancel any shoulder belt slack introduced into the belt system by a tension-relieving device. In the case of an automatic safety belt system, cancellation of the tension relieving device shall occur each time the adjacent vehicle door is opened. In the case of a manual seat belt required to meet S4.6, cancellation of the tension-relieving device shall occur, at the manufacturer's option, either each time the latchplate is released from the buckle. In the case of open-body vehicles with no doors, cancellation of the tension-relieving device may be done by a manual means. (51 F.R. 21912—June 17, 1986. Effective: June 17, 1986)]

S7.4.3 Belt contact force. [Except for manual or automatic seat belt assemblies which incorporate a webbing tension-relieving device, the upper torso webbing of any seat belt assembly, when tested in accordance with S10.6, shall not exert more than 0.7 pounds of contact force when measured normal to and one inch from the chest of an anthropomorphic test dummy, positioned in accordance with S10 in the seating position for which that assembly is provided, at the point where the centerline of the torso belt crosses the midsagittal line on the dummy's chest. (51 F.R. 21912—June 17, 1986. Effective: June 17, 1986)]

S7.4.4 Latchplate access. Any seat belt assembly latchplate which is located outboard of a front outboard seating position in accordance with S4.1.2, shall also be located within the outboard reach envelope of either the outboard arm or the inboard arm described in [S10.6] and Figure 3 of this standard, when the latchplate is in its normal stowed position. There shall be sufficient clearance between the vehicle seat and the side of the vehicle interior to allow the test block defined in Figure 4 unhindered transit to the latchplate or buckle.

S7.4.5 Retraction. [When tested under the conditions of S8.1.2 and S8.1.3, with anthropomorphic test dummies whose arms have been removed and which are positioned in accordance with S10 in the front outboard designated seating positions and restrained by the belt systems for those positions, the torso and lap belt webbing of any of those seat belt systems shall automatically retract to a stowed position either when the adjacent vehicle door is in the open position and the seat belt latchplate is released, or, at the option of the manufacturer, when the latchplate is released. That stowed position shall prevent any part of the webbing or hardware from being pinched when the adjacent vehicle door is closed. A belt system with a tension-relieving device in an open-bodied vehicle with no doors shall fully retract when the tension-relief device is deactivated. For the purpose of the retraction requirement, outboard armrests, which are capable of being stowed, on vehicle seats shall be placed in their stowed positions. (51 F.R. 21912—June 17, 1986. Effective: June 17, 1986.)]

S7.4.6 Seat belt guides and hardware.

S7.4.6.1 (a) Any manual seat belt assembly whose webbing is designed to pass through the seat cushion or between the seat cushion and seat back shall be designed to maintain one of the following three seat belt parts (the seat belt latchplate, the buckle, or the seat belt webbing) on top of or above the seat cushion under normal conditions (i.e., conditions other than when belt hardware is intentionally pushed behind the seat by a vehicle occupant). In addition, the remaining two seat belt parts must be accessible under normal conditions.

(b) The requirements of S7.4.6.1(a) do not apply to: (1) seats whose seat cushions are movable so that the seat back serves a function other than seating, (2) seats which are removable, or (3) seats which are movable so that the space formerly occupied by the seat can be used for a secondary function.

S7.4.6.2 The buckle and latchplate of a manual seat belt assembly subject to S7.4.6.1 shall not pass through the guides or conduits provided for in S7.4.6.1 and fall behind the seat when the events listed below occur in the order specified: (a) the belt is completely retracted or, if the belt is nonretractable, the belt is unlatched; (b) the seat is moved to any position to which it is designed to be adjusted; and (c) the seat back, if foldable, is folded forward as far as possible and then moved backward into position. The inboard receptacle end of a seat belt assembly installed at a front outboard designated seating position shall be accessible with the center arm rest in any position to which it can be adjusted (without having to move the armrest).

S8. Test conditions.

S8.1 General conditions. The following conditions apply to the frontal, lateral, and rollover tests.

S8.1.1 "Except as provided in paragraph (c) of this section, the vehicle, including test devices and instrumentation, is loaded as follows:"

(a) *Passenger cars.* A passenger car is loaded to its unloaded vehicle weight plus its rated cargo and luggage capacity weight, secured in the luggage area, plus the weight of the necessary anthropomorphic test devices.

(b) *Multipurpose passenger vehicles, trucks, and buses.* A multipurpose passenger vehicle, truck, or bus is loaded to its unloaded vehicle weight plus 300 pounds or its rated cargo and luggage capacity weight, whichever is less, secured in the load carrying area and distributed as nearly as possible in proportion to its gross axle weight ratings, plus the weight of the necessary anthropomorphic test devices.

S8.1.1(c) Fuel system capacity. With the test vehicle on a level surface, pump the fuel from the vehicle's fuel tank and then operate the engine until it stops. Then, add Stoddard solvent to the test vehicle's fuel tank in an amount which is equal to not less than 92 and not more than 94 percent of the fuel tank's usable capacity stated by the vehicle's manufacturer. In addition, add the amount of Stoddard solvent needed to fill the entire fuel system from the fuel tank through the engine's induction system.

[S8.1.1(d) Vehicle test attitude. Determine the distance between a level surface and a standard reference point on the test vehicle's body, directly above each wheel opening, when the vehicle is in its "as delivered" condition. The "as delivered" condition is the vehicle as received at the test site, with 100 percent of all fluid capacities and all tires inflated to the manufacturer's specifications as listed on the vehicle's tire placard. Determine the distance between the same level surface and the same standard reference points in the vehicle's "fully loaded condition." The "fully loaded condi-

tion" is the test vehicle loaded in accordance with S8.1.1.(a) or (b), as applicable. The load placed in the cargo area shall be centered over the longitudinal centerline of the vehicle. The pretest vehicle attitude shall be equal to either the "as delivered" or "fully loaded" attitude or between the "as delivered" attitude and the "fully loaded" attitude.

S8.1.2 Adjustable seats are in the adjustment position midway between the forwardmost and rearmost positions, and if separately adjustable in a vertical direction, are at the lowest position. If an adjustment position does not exist midway between the forwardmost and rearmost positions, the closest adjustment position to the rear of the midpoint is used.

S8.1.3 Adjustable seat backs replacement. Place adjustable seat backs in the manufacturer's nominal design riding position in the manner specified by the manufacturer. Place each adjustable head restraint in its highest adjustment position.

S8.1.4 Adjustable steering controls are adjusted so that the steering wheel hub is at the geometric center of the locus it describes when it is moved through its full range of driving positions.

S8.1.5 Movable vehicle windows and vents are at the manufacturer's option, placed in the fully closed position.

S8.1.6 Convertibles and open-body type vehicles have the top, if any, in place in the closed passenger compartment configuration.

S8.1.7 Doors are fully closed and latched but not locked.

[S8.1.8 Anthropomorphic test dummies

S8.1.8.1 The anthropomorphic test dummies used for evaluation of occupant protection systems manufactured pursuant to applicable portions of paragraphs S4.1.2, 4.1.3, and S4.1.4 shall conform to the requirements of Subpart B of Part 572 of this Chapter.

S8.1.8.2 Anthropomorphic test devices used for the evaluation of occupant protection systems manufactured pursuant to applicable protions of paragraphs S4.1.2, S4.1.3, and S4.1.4 shall conform to the requirements of Subpart E of Part 572 of this Chapter. (51 F.R. 26688—July 25, 1986. Effective: October 23, 1986)]

S8.1.9.1 [Each Part 572, Subpart B test dummy specified in S8.1.8.1 is clothed in formfitting cotton

PART 571; S 208-13

stretch garments with short sleeves and midcalf length pants. Each foot of the test dummy is equipped with a size 11EE shoe which meets the configuration size, sole, and heel thickness specifications of MIL–S–131192 and weighs 1.25 ± 0.2 pounds.

S8.1.9.2 Each Part 572, Subpart E test dummy specified in S8.1.8.2 is clothed in formfitting cotton stretch garments with short sleeves and midcalf length pants specified in drawings 78051–292 and –293 incorporated by reference in Part 572, Subpart E of this Chapter, respectively or their equivalents. A size 11EE shoe specified in drawings 78051–294 (left) and 78051–295 (right) or their equivalents is placed on each foot of the test dummy. (51 F.R. 26688—July 25, 1986. Effective: October 23, 1986)]

S8.1.10 Limb joints are set at *1g*, barely restraining the weight of the limb when extended horizontally. Leg joints are adjusted with the torso in the supine position.

S8.1.11 Instrumentation does not affect the motion of dummies during impact or rollover.

S8.1.12 The stabilized temperature of the test instrument specified by S8.1.8 is at any level between 66°F. and 78°F.

S8.1.13 Temperature of the test dummy

S8.1.13.1 [The stabilized temperature of the test dummy specified by S8.1.8.1 is at level between 66 degrees F and 75 degrees F.

S8.1.13.2 The stabilized temperature of the test dummy specified by S8.1.8.2 is at any level between 69 degrees F and 72 degrees F. (51 F.R. 26688—July 25, 1986. Effective: October 23,1986]

S8.2 Lateral moving barrier crash test conditions. The following conditions apply to the lateral moving barrier crash test:

S8.2.1 The moving barrier, including the impact surface, supporting structure, and carriage, weighs 4,000 pounds.

S8.2.2 The impact surface of the barrier is a vertical, rigid, flat rectangle, 78 inches wide and 60 inches high, perpendicular to its direction of movement, with its lower edge horizontal and 5 inches above the ground surface.

S8.2.3 During the entire impact sequence the barrier undergoes no significant amount of dynamic or static deformation, and absorbs no significant portion of the energy resulting from the impact, except for energy that results in translational rebound movement of the barrier.

S8.2.4 During the entire impact sequence the barrier is guided so that it travels in a straight line, with no significant lateral, vertical or rotational movement.

S8.2.5 The concrete surface upon which the vehicle is tested is level, rigid and of uniform construction, with a skid number of 75 when measured in accordance with American Society for Testing and Materials Method E–274–65T at 40 mph, omitting water delivery as specified in paragraph 7.1 of that method.

S8.2.6 The tested vehicle's brakes are disengaged and the transmission is in neutral.

S8.2.7 The barrier and the test vehicle are positioned so that at impact—

(a) The vehicle is at rest in its normal attitude;

(b) The barrier is traveling in a direction perpendicular to the longitudinal axis of the vehicle at 20 mph; and

(c) A vertical plane through the geometric center of the barrier impact surface and perpendicular to the surface passes through the driver's seating reference point in the tested vehicle.

S8.3 Rollover test condition. The following conditions apply to the rollover test:

S8.3.1 The tested vehicle's brakes are disengaged and the transmission is in neutral.

S8.3.2 The concrete surface on which the test is conducted is level, rigid, of uniform construction, and of a sufficient size that the vehicle remains on it throughout the entire rollover cycle. It has a skid number of 75 when measured in accordance with American Society of Testing and Materials Method E–274–65T at 40 mph omitting water delivery as specified in paragraph 7.1 of that method.

S8.3.3 The vehicle is placed on a device, similar to that illustrated in Figure 1, having a platform in the form of a flat, rigid plane at an angle of 23° from the horizontal. At the lower edge of the platform is an unyielding flange, perpendicular to the platform with a height of 4 inches and a length sufficient to hold in place the tires that rest against it. The intersection of the inner face of the flange with the upper face of the platform is 9 inches above the rollover surface. No other restraints are used to hold the vehicle in position during the deceleration of the platform and the departure of the vehicle.

PART 571; S 208–14

S8.3.4 With the vehicle on the test platform, the test devices remain as nearly as possible in the posture specified in S8.1.

S8.3.5 Before the deceleration pulse, the platform is moving horizontally, and perpendicularly to the longitudinal axis of the vehicle, at a constant speed of 30 mph for a sufficient period of time for the vehicle to become motionless relative to the platform.

S8.3.6 The platform is decelerated from 30 to 0 mph in a distance of not more than 3 feet, without change of direction and without transverse or rotational movement during the deceleration of the platform and the departure of the vehicle. The deceleration rate is at least 20g for a minimum of 0.04 seconds.

S9. Pressure vessels and explosive devices.

S9.1 Pressure vessels. A pressure vessel that is continuously pressurized shall conform to the requirements of 49 CFR S178.65-2, -6(b), -7, -9(a) and (b), and -10. It shall not leak or evidence visible distortion when tested in accordance with S178.65-11(a) and shall not fail in any of the ways enumerated in S178.65-11(b) when hydrostatically tested to destruction. It shall not crack when flattened in accordance with S178.65-12(a) to the limit specified in S178.65-12(a) (4).

S9.2 Explosive devices. An explosive device shall not exhibit any of the characteristics prohibited by 49 CFR S173.51. All explosive material shall be enclosed in a structure that is capable of containing the explosive energy without sudden release of pressure except through overpressure relief devices or parts designed to release the pressure during actuation.

S10. Test dummy positioning procedures. [For vehicles manufactured before September 1, 1987, position a test dummy, conforming to Subpart B of Part 572 (49 CFR Part 572), in each front outboard seating position of a vehicle as specified in S10 through S10.9 or, at the manufacturer's option, as specified in S12 through S12.3.3.2. For vehicles manufactured on or after September 1, 1987, position a test dummy, conforming to Subpart B of Part 572 (49 CFR Part 572), in each front outboard seating position of a vehicle as set forth below in S10 through S10.9. Regardless of which positioning procedure in used, each test dummy is restrained during the crash tests of S5 as follows:

(a) In a vehicle equipped with automatic restraints at each front outboard designated seating position that is certified by its manufacturer as meeting the requirements of S4.1.2.1(a) and (c) (1), each test dummy is not restrained during the frontal test of S5.1, the lateral test of S5.2 and the rollover test of S5.3 by any means that requires occupant action.

(b) In a vehicle equipped with an automatic restraint at each front outboard seating position that is certified by its manufacturer as meeting the requirements of S4.1.2.1(a) and (c)(2), each test dummy is not restrained during one frontal test of S5.1 by any means that require occupant action. If the vehicle has a manual seat belt provided by the manufacturer to comply with the requirements of S4.1.2.1(c), then a second frontal test is conducted in accordance with S5.1 and each test dummy is restrained both by the authomatic restraint system and the manual seat belt, adjusted in accordance with S10.9.

(ii) In a vehicle equipped with an automatic restraint only at the driver's designated seating position, pursuant to S4.1.3.4(a)(2), that is certified by its manufacturer as meeting the requirements of S4.1.2.1(a) and (c)(2), the driver test dummy is not restrained during one frontal test of S5.1 by any means that require occupant action. If the vehicle also has a manual seat belt provided by the manufacturer to comply with the requirements of S4.1.2.1(c), then a second frontal test is conducted in accordance with S5.1 and the driver test dummy is restrained both by the automatic restraint system and the manual seat belt, adjusted in accordance with S10.9. At the option of the manufacturer, a passenger test dummy can be placed in the right front designated seating postion during the testing required by this section. If a passenger test dummy is present, it shall be restrained by a manual seat belt, adjusted in accordance with S10.9

(c) In a vehicle equipped with a manual safety belt at the front outboard designated seating positions that is certified by its manufacturer to meet the requirements of S4.6, each test dummy is restrained by the manual safety belts, adjusted in accordance with S10.9, installed at each front outboard seating positions. (51 F.R. 31795—September 5, 1986. Effective: September 5, 1986)]

S10.1 Vehicle equipped with front bucket seats. Place the test dummy's torso against the seat back and its upper legs against the seat cushion to the extent permitted by placement of the test dummy's feet in accordance with the appropriate paragraph of S10. Center the test dummy on the seat cushion of the bucket seat and set its midsagittal plane so that it is vertical and parallel to the centerline of the vehicle.

S10.1.1 Driver position placement.

(a) Initially set the knees of the test dummy 11¾ inches apart, measured between the outer surfaces of the knee pivot bolt heads, with the left outer surface 5.9 inches from the midsagittal plane of the test dummy.

(b) Rest the right foot of the test dummy on the undepressed accelerator pedal with the rearmost point of the heel on the floor pan in the plane of the pedal. If the foot cannot be placed on the accelerator pedal, set it perpendicular to the lower leg and place it as far forward as possible in the direction of the geometric center of the pedal with the rearmost point of the heel resting on the floor pan. Except as prevented by contact with a vehicle surface, place the right leg so that the upper and lower leg centerlines fall, as close as possible, in a vertical longitudinal plane without inducing torso movement.

(c) [Place the left foot on the toeboard with the rearmost point of the heel resting on the floor pan as close as possible to the point of intersection of the planes described by the toeboard and the floor pan and not on the wheelwell projection. If the foot cannot be positioned on the toeboard, set it initially perpendicular to the lower leg and place it as far forward as possible with the heel resting on the floor pan. If necessary to avoid contact with the vehicle's brake or clutch pedal, rotate the test dummy's left foot about the lower leg. If there is still pedal interference, rotate the left leg outboard about the hip the minimum distance necessary to avoid the pedal interference. Except as prevented by contact with a vehicle surface, place the left leg so that the upper and lower leg centerlines fall, as close as possible, in a vertical plane. For vehicles with a foot rest that does not elevate the left foot above the level of the right foot, place the left foot on the foot rest so that the upper and lower leg centerlines fall in a vertical plane. (51 F.R. 31765—September 5, 1986. Effective: September 5, 1986)]

S10.1.2 Passenger position placement.

S10.1.2.1 Vehicle with a flat floor pan/toeboard.

(a) Initially set the knees 11¾ inches apart, measured between the outer surfaces of the knee pivot bolt heads.

(b) Place the right and left feet on the vehicle's toeboard with the heels resting on the floor pan as close as possible to the intersection point with the toeboard. If the feet cannot be placed flat on the toeboard, set them perpendicular to the lower leg centerlines and place them as far forward as possible with the heels resting on the floor pan.

(c) Place the right and left legs so that the upper and lower leg centerlines fall in vertical longitudinal planes.

S10.1.2.2 Vehicles with wheelhouse projections in passenger compartment.

(a) Initially set the knees 11¾ inches apart, measured between outer surfaces of the knee pivot bolt heads.

(b) Place the right and left feet in the well of the floor pan/toeboard and not on the wheelhouse projection. If the feet cannot be placed flat on the toeboard, set them perpendicular to the lower leg centerlines and as far forward as possible with the heels resting on the floor pan.

(c) If it is not possible to maintain vertical and longitudinal planes through the upper and lower leg centerlines for each leg, then place the left leg so that its upper and lower centerlines fall, as closely as possible, in a vertical longitudinal plane and place the right leg so that its upper and lower leg centerlines fall, as closely as possible, in a vertical plane.

S10.2 Vehicle equipped with bench seating. Place a test dummy with its torso against the seat back and its upper legs against the seat cushion, to the extent permitted by placement of the test dummy's feet in accordance with the appropriate paragraph of S10.1.

S10.2.1 Driver position placement. Place the test dummy at the left front outboard designated seating position so that its midsagittal plane is vertical and parallel to the centerline of the vehicle and so that the midsagittal plane of the test dummy passes through the center of the steering wheel rim. Place the legs, knees, and feet of the test dummy as specified in S10.1.1.

S10.2.2 Passenger position placement. [Place the test dummy at the right front outboard designated seating position so that the midsagittal plane of the test dummy shall is vertical and longitudinal, and the same distance from the vehicle's longitudinal centerline as the midsagittal plane of the test dummy at the driver's position. Place the legs, knees, and feet of the test dummy as specified in S10.1.2. (51 F.R. 31765¢September 5, 1986. Effective: September 5, 1986)]

S10.3 Initial test dummy placement. With the test dummy at its designated seating position as specified by the appropriate requirements of S10.1 or S10.2, place the upper arms against the seat back and tangent to the side of the upper torso. Place the lower arms and palms against the outside of the upper legs.

S10.4 Test dummy settling.

S10.4.1 Test dummy vertical upward displacement. Slowly lift the test dummy parallel to the seat back plane until the test dummy's buttocks no longer contact the seat cushion or until there is test dummy head contact with the vehicle's headlining.

S10.4.2 Lower torso force application. [Apply a rearward force of 50 pounds against the center of the test dummy's lower torso in a horizontal direction. The line of force application shall be 6½ inches above the bottom surface of the test dummy's buttocks.

S10.4.3 Test dummy vertical downward displacement. Remove as much of the 50 pound force as necessary to allow the test dummy to return downward to the seat cushion by its own weight.

S10.4.4 Test dummy torso rocking. Apply a 10 to 15 pound horizontal rearward force to the test dummy's lower torso. Then apply a horizontal forward force to the test dummy's shoulders sufficient to flex the upper torso forward until its back no longer contacts the seat back. Rock the test dummy from side to side 3 or 4 times so that the test dummy's spine is at any angle from the vertical in the 14 to 16 degree range at the extremes of each rocking movement.

S10.4.5 Test dummy upper torso force application. While maintaining the 10 to 15 pound horizontal rearward force applied in S10.4.4 and with the test dummy's midsagittal plane vertical, push the upper torso back against the seat back with a force of 50 pounds applied in a horizontal rearward direction along a line that is coincident with the test dummy's midsagittal plane and 18 inches above the bottom surface of the test dummy's buttocks.

S10.5 Belt adjustment for dynamic testing. With the test dummy at its designated seating position as specified by the appropriate requirements of S8.1.2, S8.1.3 and S10.1 through S10.4, place and adjust the safety belt as specified below.

S10.5.1 Manual safety belts. Place the Type 1 or Type 2 manual belt around the test dummy and fasten the latch. Pull the Type 1 belt webbing out of the retractor and allow it to retract; repeat this operation four times. Remove all slack from the lap belt portion of a Type 2 belt. Pull the upper torso webbing out of the retractor and allow it to retract; repeat this operation four times so that the excess webbing in the shoulder belt is removed by the retractive force of the retractor. Apply a 2 to 4 pound tension load to the lap belt of a single retractor system by pulling the upper torso belt adjacent to the latchplate. In the case of a dual retractor system, apply a 2 to 4 pound tension load by pulling the lap belt adjacent to its retractor. Measure the tension load as clase as possible to the same location where the force was applied. After the tension load has been applied, ensure that the upper torso belt lies flat on the test dummy's shoulder.

S10.6 Placement of test dummy arms and hands. With the test dummy positioned as specified by S10.4 and without inducing torso movement, place the arms, elbows, and hands of the test dummy, as appropriate for each designated seating position in accordance with S10.6.1 or S10.6.2. Following placement of the arms, elbows and hands, remove the force applied against the lower half of the torso.

S10.6.1 Driver's position. Move the upper and the lower arms of the test dummy at the driver's position to their fully outstretched position in the lowest possible orientation. Push each arm rearward permitting bending a the elbow, until the palm of each hand contacts the outer part of the rim of the steering wheel at its horizontal centerline. Place the test dummy's thumbs over the steering wheel rim and position the upper and lower arm centerlines as close as possible in a vertical plane without inducing torso movement.

S10.6.2 Passenger position. Move the upper and lower arms of the test dummy at the passenger position to the fully outstretched position in the lowest possible orientation. Push each arm rearward, permitting bending at the elbow, until the upper arm contacts the seat back and is tangent to the upper part of the side of the torso, the palm contacts the outside of the thigh, and the little finger is barely in contact witht the seat cushion.

S10.7 Repositioning of feet and legs. After the test dummy has been settled in accordance with S10.4, the safety belt system has been positioned, if necessary, in accordance with S10.5, and the arms and hands of the test dummy have been positioned in accordance with S10.6, reposition the feet and legs of the test dummy, if necessary, so that the feet and legs meet the applicable requirements of S10.1 or S10.2

S10.8 Test dummy positioning for latchplate access. The reach envelopes specified in S7.4.4. are obtained by positioning a test dummy in the driver's seat or passenger's seat in its forwardmost adjustment position. Attach the lines for the inboard and outboard arms to the test dummy as described in Figure 3 of this standard. Extend each line backward and outboard to generate the compliance arcs of the outboard reach envelope of the test dummy's arms.

Test dummy positioning for belt contact force.

S10.9.1 Vehicle manufactured before September 1, 1987. To determine compliance with S7.4.3 of this standard, a manufacturer may use, at its option, either the test procedure of S10.9.1 or the test procedure of S10.9.2. Position the test dummy in the vehicle in accordance with the appropriate requirements specified in S10.1 or S10.2 and under the conditions of S8.1.2 and S8.1.3. Fasten the latch and pull the belt webbing three inches from the test dummy's chest and release until the webbing is within one inch of the test dummy's chest and measure the belt contact force.

S10.9.2 Vehicle manufactured on or after September 1, 1987. To determine compliance with S7.4.3. of this standard, position the test dummy in the vehicle in accordance with the appropriate requirements specified in S10.1 or S10.2 and under the conditions of S8.1.2 and S8.1.3. Close the vehicle's adjacent door, pull either 12 inches of belt webbing or the maximum available amount of belt webbing, whichever is less, from the retractor and then release it, allowing the belt webbing to return to the dummy's chest. Fasten the latch and pull the belt webbing three inches from the test dummy's chest and release until the webbing is within one inch of the test dummy's chest and measure the belt contact force. **(51 F.R. 31765—September 5, 1986. Effective: September 5, 1986.)]**

S11 Positioning procedure for the Part 572 Subpart E Test Dummy.

Position a test dummy, conforming to Subpart E of Part 572 of this Chapter, in each front outboard seating position of a vehicle as specified in S11.1 through S11.6. Each test dummy is restrained in accordance with the applicable requirements of S4.1.2.1, 4.1.2.2 or S4.6.

S11.1 Head. The transverse instrumentation platform of the head shall be horizontal within ½ degree.

S11.2 Arms.

S11.2.1 The driver's upper arms shall be adjacent to the torso with the centerlines as close to a vertical plane as possible.

S11.2.2 The passenger's upper arms shall be in contact with the seat back and the sides of torso.

S11.3 Hands.

S11.3.1 The palms of the driver test dummy shall be in contact with the outer part of the steering wheel rim at the rim's horizontal centerline. The thumbs shall be over the steering wheel rim and attached with adhesive tape to provide a breakaway force of between 2 to 5 pounds.

S11.3.2 The palms of the passenger test dummy shall be in contact with outside of thigh. The little finger shall be in contact with the seat cushion.

S11.4 Torso.

S11.4.1 In vehicles equipped with bench seats, the upper torso of the driver and passenger test dummies shall rest against the seat back. The midsagittal plane of the driver dummy shall be vertical and parallel to the vehicle's longitudinal centerline, and pass through the center of the steering wheel rim. The midsagittal plane of the passenger dummy shall be vertical and parallel to the vehicle's longitudinal centerline and the same distance from the vehicle's longitudinal centerline as the midsagittal plane of the driver dummy.

S11.4.2 In vehicles equipped with bucket seats, the upper torso of the driver and passenger test dummies shall rest against the seat back. The midsagittal plane of the drivers and the passenger dummy shall be vertical and shall concide with the longitudinal centerline of the bucket seat.

S11.4.3 Lower Torso.

S11.4.3.1 H-point. The H-point of the driver and passenger test dummies shall coincide within ½ inch in the vertical dimension and ½ inch in the horizontal dimension of a point ¼ inch below the position of the H-point determined by using the equipment and procedures specified in SAE J826 (Apr 80) except that the length of the lower leg and thigh segments of the H-point machine shall be adjusted to 16.3 and 15.8 inches, respectively, instead of the 50th percentile values specified in Table 1 of SAE J826.

S11.4.3.2 Pelvic angle. As determined using the pelvic angle gage (GM drawing 78051–532 incorporated by reference in Part 572, Subpart E of this chapter) which is inserted into the H-point gaging hole of the dummy, the angle measured from the horizontal on the 3 inch flat surface of the gage shall be 22½ degrees plus or minus 2½ degrees.

S11.5 Legs. The upper legs of the driver and passenger test dummies shall rest against the seat cushion to the extent pemitted by placement of the feet. The initial distance between the outboard knee clevis flange surfaces shall be 10.6 inches. To the extent particable, the left leg of the driver dummy and both legs of the passenger dummy shall be in vertical longitudinal planes. Final adjustment to accommodate placement of feet in accordance with S11.6 for various passenger compartment configurations is permitted.

S11.6 Feet.

S11.6.1 The right foot of the driver test dummy shall rest on the undepressed accelerator with the rearmost point of the heel on the floor surface in the plane of the pedal. If the foot cannot be placed on the accelerator pedal. If the foot cannot be placed on the accelerator pedal, it shall be positioned perpendicular to the tibia and placed as far forward as possible in the direction of the centerline of the pedal with the rearmost point of the heel resting on the floor surface. The heel of the left foot shall be placed as far forward as possible and shall rest on the floor surface. The left foot shall be positioned as flat as possible on the floor surface. The longitudinal centerline of the left foot shall be placed as parallel as possible to the longitudinal centerline of the vehicle.

S11.6.2 The heels of both feet of the passenger test dummy shall be placed as far forward as possible and shall rest on the floor surface. Both feet shall be positioned as flat as possible on the floor

surface. The longitudinal centerline of the feet shall be placed as parallel as possible to the longitudinal centerline of the vehicle.

S11.7 Test dummy positioning for latchplate access. The reach envelopes specified in S7.4.4 are obtained by positioning a test dummy in the driver's seat for passenger's seat in its forwardmost adjustment position. Attach the lines for the inboard and outboard arms to the test dummy as described in Figure 3 of this standard. Extend each line backward and outboard to generate the compliance arcs of the outboard reach envelope of the test dummy's arms.

S11.8 Test dummy positioning for belt contact force. To determine compliance with S7.4.3 of this standard, position the test dummy in the vehicle in accordance with the requirements specified in S11.1 through S11.6 and under the conditions of S8.1.2 and S8.1.3. Pull the belt webbing three inches from the test dummy's chest and release until the webbing is within 1 inch of the test dummy's chest and measure the belt contact force.

S11.9 Manual belt adjustment for dynamic testing. With the test dummy at its designated seating position as specified by the appropriate requirements of S8.1.2, S8.1.3 and S11.1 through S11.6, place the Type 2 manual belt around the test dummy and fasten the latch. Remove all slack from the lap belt. Pull the upper torso webbing out of the retractor and allow it to retract; repeat this operation four times. Apply a 2 to 4 pound tension load to the lap belt. If the belt system is equipped with a tension-relieving device introduce the maximum amount of slack into the upper torso belt that is recommended by the manufacturer for normal use in the owner's manual for the vehicle. If the belt system is not equipped with a tension-relieving device, allow the excess webbing in the shoulder belt to be retracted by the retractive force of the retractor.

[S12 Optional position procedures for the Part 572, Subpart B test dummy. The following test dummy positioning procedures for the Part 572, Subpart B test dummy may be used, at the option of the manufacturer, until September 1, 1987.

S12.1 Dummy placement in vehicle. Anthropomorphic test dummies are placed in the vehicle in accordance with S12.1.1 and S12.1.2.

S12.1.1 Vehicle equipped with front bucket seats. In the case of a vehicle equipped with front bucket seats, dummies are placed at the front outboard designated seating positions with the test device torso against the seat back, and the thighs against the seat cushion to the extent permitted by placement of the dummy's feet in accordance with the appropriate paragraph of S12.1. The dummy is centered on the seat cushion of the bucket seat and its midsagittal plane is vertical and longitudinal.

S12.1.1.1 Driver position placement. At the driver's position, the knees of the dummy are initially set 14.5 inches apart, measured between the outer surfaces of the knee pivot bolt heads, with the left outer surface 5.9 inches from the midsagittal plane of the dummy. The right foot of the dummy rests on the underpressed accelerator pedal with the rearmost point of the heel on the floorpan in the plane of the pedal. If the foot cannot be placed on the accelerator pedal, it is set perpendicular to the tibia and placed as far forward as possible in the direction of the geometric center of the pedal with the rearmost point of the heel resting on the floorpan. The plane defined by the femur and tibia cneterlines of the right leg is as close as possible to vertical without inducing torso movement and except as prevented by contact with a vehicle surface. The left foot is placed on the floorpan as close as possible to the point of intersection fo the planes described by the toeboard and the floorpan. If the foot cannot be positioned on the toeboard, it is set perpendicular to the tibia and placed as far forward as possible with the heel resting on the floorpan. The femur and tibia centerlines of the left leg are positioned in a vertical plane except as prevented by contact with a vehicle surface.

S12.1.1.2 Passenger position placement. At the right front designated seating position, the femur, tibia, and foot centerlines of each of the dummy's legs are positioned in a vertical longitudinal plane. The feet of the dummy are placed on the toeboard with the rearmost point of the heel resting on the floorpan as close as possible to the point of intersection of the planes described by the toeboard and the floorpan. If the feet cannot be positioned flat on the toeboard they are set perpendicular to the tibia and are placed as far forward as possible with the heels resting on the floorpan.

S12.1.2 Vehicle equipped with bench seating. In the case of a vehicle which is equipped with a front bench seat, a dummy is placed at each of the front outboard designated seating positions with the dummy torso against the seat back and the thighs against the seat cushion to the extent permitted by placement of the dummy's feet in accordance with the appropriate paragraph of S12.1.1.

S12.1.2.1 Driver position placement. The dummy is placed at the left front outboard designated seating position so that its midsagittal plane is vertical and longitudinal, and passes through the center point of the plane described by the steering wheel rim. The legs, knees, and feet of the dummy are placed as specified in S12.1.1.1.

S12.1.2.2 Passenger position placement. The dummy is placed at the right front outboard designated seating position as specified in S8.12.1.1.2, except that the midasgittal plane of the dummy is vertical, longitudinal, and the same distance from the longitudinal centerline as the midsagittal plane of the dummy at the driver's position.

S12.2 Dummy positioning procedures. The dummy is positioned on a seat as specified in S12.2.1. throught S12.2.3.2. to achieve the conditions of S12.1.

S12.2.1 Initial dummy placement. With the dummy at its designated seating position as described in S12.1 place the upper arms against the seat back and tangent to the side of the upper torso and the lower arms and palms against the outside of the thighs.

S12.2.2 Dummy settling. With the dummy positioned as specified in S10.1, slowly lift the dummy in the direction parallel to the plane of the seat back until its buttocks no longer contact the seat cushion or until its head contacts the vehicle roof. Using a flat, square, rigid surface with an area of 9 square inches and oriented so that its edges fall in longitudinal or horizontal planes, apply a force of 50 pounds through the center of the rigid surface against the dummy's torso in the horizontal rearward direction along a line that is coincident with the midsagittal plane of the dummy and 5.5 inches above the bottom surface of its buttocks. Slowly remove the lifting force.

S12.2.2.1 While maintaining the contact of the force application plate with the torso, remove as much force as is necessary from the dummy's torso to allow the dummy to return to the seat cushion by its own weight.

S12.2.2.2 Without removing the force applied to the lower torso, apply additional force in the horizontal, forward direction, longitudinally against the upper shoulders of the dummy sufficient to flex the torso forward until the dummy's back above the lumbar spine no longer contacts the seatback. Rock the dummy from side to side three or four times, so that the dummy's spine is at an angle from the vertical of not less than 14 degrees and not more than 16 degrees at the extreme of each movement. With the midsagittal plane vertical, push the upper half of the torso back against the seat back with a force of 50 pounds applied in the horizontal rearward direction along a line that is coincident with the midsagittal plane of the dummy and 18 inches above the bottom surface of its buttocks. Slowly remove the horizontal force.

S12.2.3 Placement of dummy arms and hands. With the dummy positioned as specified in S12.2.2 and without inducing torso movement, place the arms, elbows, and hands of the dummy, as appropriate for each designated seating position in accordance with S12.2.3.1 or S12.2.3.2. Following placement of the limbs, remove the force applied against the lower half of the torso.

S12.2.3.1 Driver's position. Move the upper and the lower arms of the dummy at the driver's position to the fully outstretched position in the lowest possible orientation. Push each arm rearward, permitting bending at the elbow, until the palm of each hand contacts the outer part of the rim of the steering wheel at its horizontal centerline. Place the dummy's thumbs over the steering wheel rim, positioning the upper and lower arm centerlines as close as possible in a vertical plane without including torso movement.

S12.2.3.2 Passenger position. Move the upper and the lower arms of the dummy at the passenger position to the fully outstretched position in the lowest possible orientation. Push each arm rearward, permitting bending at the elbow, until the upper arm contacts the seat back and is tangent to the upper part of the side of the torso, the palm contacts the outside of the thigh, and the little finger is barely in contact with the seat cushion. (51 F.R. 31765—September 5, 1986. Effective: September 5, 1986)]

Interpretation

The concept of an occupant protection system which requires "no action by vehicle occupants," as that term is used in Standard No. 208, is intended to designate a system which will perform its protective restraining function after a normal process of ingress or egress without separate deliberate actions by the vehicle occupant to deploy the restraint system. Thus, the agency considers an occupant protection system to be automatic if an occupant has to take no action to deploy the system but would normally slightly push the seat belt webbing aside when entering or exiting the vehicle or would normally make a slight adjustment in the webbing for comfort.

36 F.R. 4600
March 10, 1971

PREAMBLE TO AMENDMENT TO MOTOR VEHICLE SAFETY STANDARD NO. 209

Seat Belt Assemblies—Passenger Cars, Multipurpose Passenger Vehicles, Trucks, and Buses

Motor Vehicle Safety Standard No. 209 (32 F.R. 2415, as amended 32 F.R. 3310), specifies requirements for seat belt assemblies for use in passenger cars, multipurpose passenger vehicles, trucks and buses, incorporating by reference the requirements of Department of Commerce, National Bureau of Standards, *Standards for Seat Belts for Use in Motor Vehicles* (15 C.F.R. Part 9; 31 F.R. 11528).

The Administrator of the Federal Highway Administration has determined in the interests of clarity and ease of reference that the requirements specified by 15 C.F.R. Part 9 should be incorporated into Standard No. 209 where it is presently incorporated only by reference. Therefore Standard No. 209 is hereby amended by deleting present paragraph S3 and adding new paragraphs S3, S4, and S5, so as to incorporate the requirements of 15 C.F.R. Part 9. Accordingly 15 C.F.R. Part 9 is hereby deleted.

Since this amendment imposes no additional burden on any person and involves no substantive change in the requirements of Standard No. 209, notice and public procedure hereon are unnecessary and good cause is shown that an effective date earlier than 180 days after issuance is in the public interest and the amendment may be made effective less than 30 days after publication in the *Federal Register*. The requirement of former Paragraph S3 of Standard No. 209 that seat belt assemblies shall use the attachment hardware specified in 15 C.F.R. § 9.3(f) "or approved equivalent hardware" has been incorporated into new Paragraph S4.1(f) of Standard No. 209.

This amendment is made under the authority of sections 103, 117(c) and 119 of the National Traffic and Motor Vehicle Safety Act of 1966 (15 U.S.C. secs. 1392, 1405(c), and 1407) and the delegation of authority contained in the Regulations of the Office of the Secretary (49 C.F.R. § 1(c)), and is effective upon publication in the *Federal Register*.

Issued in Washington, D.C., on December 24, 1968.

Lowell K. Bridwell,
Federal Highway Administrator

34 F.R. 115
January 4, 1969

PREAMBLE TO AMENDMENT TO MOTOR VEHICLE SAFETY STANDARD NO. 209

Seat Belt Assemblies in Passenger Cars, Multipurpose Passenger Vehicles, Trucks and Buses

(Docket No. 69–23; Notice No. 2)

This notice amends Federal Motor Vehicle Safety Standard No. 209 in § 571.21 of Title 49 of the Code of Federal Regulations, to upgrade the requirements for seatbelt assemblies for use in passenger cars, multipurpose passenger vehicles, trucks, and buses. As amended, the standard is both an equipment and a vehicle standard. The equipment aspect applies to a seatbelt assembly manufactured on or after the effective date. The vehicle aspect applies to an assembly installed in a vehicle manufactured on or after the effective date, regardless of when the assembly was manufactured.

During the period since the original issuance of Standard No. 209, laboratory tests and experience with actual seatbelt usage have disclosed areas where improvements in performance requirements are necessary. Consequently, a notice of proposed amendments to the standard was published on March 17, 1970 (35 F.R. 4641) to upgrade the performance requirements for seatbelt assemblies. Interested persons were given an opportunity to comment on the contents of the proposed rule. These comments, and other available data, have been carefully considered in the development of these amendments.

Paragraph S4.1(f) of the standard is amended to make it clear that a manufacturer may use bolts other than the specified bolts if the substituted bolts are equivalent.

The standard formerly required a Type 1 or Type 2 seatbelt assembly to be adjustable to fit an occupant with the weight and dimensions of a 95th-percentile adult male. To insure that belt assemblies can be adjusted to fit the range of occupants who may use them, paragraph S4.1(g) is amended to require each Type 1 or Type 2 seatbelt assembly to be adjustable to fit occupants whose weight and dimensions range from those of a 5th-percentile adult female to those of a 95th-percentile adult male. A belt assembly installed for an adjustable seat must conform to the requirements regardless of seat position. Several comments noted that no dimensions were specified in the notice for the various occupants which a belt assembly must fit. To remedy the problem, the standard provides a table of weights and dimensions for 5th-percentile adult females and 95th-percentile adult males.

In the notice, it was proposed to reduce the force required to release seat belt buckles from 30 to 22.5 pounds and to require that the release force for pushbutton-type buckles be applied no closer than 0.125 inch from the edge of the pushbutton access opening. In light of comments received, and other available information, the value of 30 pounds has been retained. The procedure for testing the buckle release force of a pushbutton-type buckle has been amended as proposed, however, to insure that the release force will not be applied so close to the edge of the access opening that the button might tilt in a manner unrepresentative of actual use conditions and thereby exaggerate the release force.

The buckle crush release requirements are amended to extend the standard's crush release requirements to all Type 1 and Type 2 seatbelt buckles, and to require application of the test load to areas of a buckle other than directly over the center of the release mechanism. Experience has indicated that non-pushbutton buckle release mechanisms are also subject to impairment when compressed, and occupants using such buckles are therefore provided equivalent protection by the extension of the buckle crush release requirements. In laboratory tests on pushbutton-type buckles, buckle release or malfunction occurred when a compressive force as low as 275 pounds was applied to a surface area other than the area directly over the pushbutton. The amended test will tend to eliminate buckle designs that are prone to accidental damage, or that release during the initial phase of the accident.

The notice proposed a new buckle latch test procedure in which a specified tensile load was to be applied at 30° to the buckle. In the light of comments received and other information that has become available indicating that the requirement was not justified, the procedure has not been adopted.

In response to comments that the acceleration levels proposed in the notice were too high, the acceleration level above which an emergency-locking retractor must lock has been reduced from 2g, as proposed, to 0.7g, and the acceleration level below which the retractor must not lock has been reduced from 1g to 0.3g. For reasons of occupant convenience, the notice proposed that the required upper limit on acceleration had to be met only when the webbing was extended to the length necessary to fit a 5th-percentile adult female. Upon review it has been determined that the proposed free travel distance could make a belt unsafe for use by a child, and,

further, that an adequate measure of convenience is provided by the requirement that a belt not lock at accelerations of less than 0.3g. Accordingly, the standard does not limit the belt withdrawal range within which the acceleration levels must be met. For similar reasons, the retraction force requirements are required to be met regardless of the amount of belt withdrawal.

As stated in the notice, the hex-bar abrasion test does not adequately simulate the type of webbing abrasion caused by some buckles. The standard as amended retains the hex-bar test, but supplements it with an additional abrasion requirement, under which webbing is required to retain at least 75 percent of its breaking strength after being repeatedly passed through the assembly buckle or manual adjustment device.

Effective date: September 1, 1971.

In consideration of the foregoing, Motor Vehicle Safety Standard No. 209 in § 571.21 of Title 49, Code of Federal Regulations, is amended. . . .

Issued on March 3, 1971.

Douglas W. Toms,
Acting Administrator.

36 F.R. 4607
March 10, 1971

PREAMBLE TO AMENDMENT TO MOTOR VEHICLE SAFETY STANDARD NO. 209

Seat Belt Assemblies for Passenger Cars, Multipurpose Passenger Vehicles, Trucks and Buses

The purpose of this notice is to amend Motor Vehicle Safety Standard No. 209, in § 571.21 of Title 49, Code of Federal Regulations, to clarify the method in which the buckle release force of a Type 3 seat belt assembly is measured.

The standard provides (S4.3(d)(1), S5.2(d) (1)) that the force required to release a Type 3 assembly buckle is measured following the assembly test of S5.3, with a force of 45±5 pounds applied to a torso block restrained by the Type 3 assembly. The test procedure was intended to represent the situation in which the vehicle is inverted and the child is held by the harness. The force applied along the line of the belt is of primary significance, but it appears that the release force of some buckles is significantly increased by the pressure of the torso block on the back of the buckle. This pressure is not regarded as representative of actual conditions, in that the hard surface of the torso block offers much more resistance than would a child's body. To eliminate the effects of such pressure by the torso block, section S5.3(c)(1) of the standard is amended to read as set forth below.

Since this amendment is interpretative and clarifying in intent and imposes no additional burden on any person, notice and public procedure thereon are unnecessary.

Effective date: April 1, 1971.

The major usage of Type 3 seat belt assembly buckles will be on child seating systems that comply with Standard No. 213, effective April 1, 1971. So that the amendment to Standard No. 209 will have maximum effect, good cause is found for establishing an effective date sooner than 180 days after issuance. Since the amendment is interpretative in nature and relieves a restriction, there is also good cause for establishing an effective date sooner than 30 days after issuance.

In consideration of the foregoing, Motor Vehicle Safety Standard No. 209, in § 571.21 of Title 49, Code of Federal Regulations, is amended. . . .

Issued on March 23, 1971.

Douglas W. Toms,
Acting Administrator.

36 F.R. 5973
March 27, 1971

PREAMBLE TO AMENDMENT TO MOTOR VEHICLE SAFETY STANDARD NO. 209

Seat Belt Assemblies in Passenger Cars, Multipurpose Passenger Vehicles, Trucks and Buses

(Docket No. 69–23; Notice No. 3)

Reconsideration and Amendment

The purpose of this notice is to respond to petitions filed pursuant to § 553.35 of Title 49, Code of Federal Regulations, requesting reconsideration of various amendments to Motor Vehicle Safety Standard No. 209, Seat Belt Assemblies, that were published March 10, 1971 (36 F.R. 4607). The petitions are granted in part and denied in part. Requests not expressly discussed in this notice should be considered denied.

1. One of the results of the March 10 amendments was that as of September 1, 1971, the standard would have become a vehicle standard as well as an equipment standard, i.e., vehicles manufactured after the effective date would have had to have equipment conforming to the new requirements. The amendments relating to emergency-locking retractors are such, however, that with normal production tolerances it would be difficult to manufacture retractors that conform to the currently applicable requirements so that they would also conform to the post-September 1 requirements, and vice-versa. This creates an awkward situation, in which retractors supplied to vehicle manufacturers for use on September 1 would have to be made on September 1 and not before.

The vehicle aspect of the standard is therefore being deleted, and the date on which the amended requirements become mandatory is postponed to January 1, 1972, to coincide with the effective date of the new Standard No. 208. To allow for efficient changeover, manufacturers are permitted to manufacture belts to either the current or the amended requirements between September 1, 1971, and January 1, 1972.

2. With respect to the technical amendments to the attachment hardware requirements in S4.1(f), American Safety Equipment Corporation requested that the reference to Standard No. 210 be omitted, so that anchorage nuts, plates, and washers would not have to be supplied if the vehicle has an anchorage that does not require them. The request has been found reasonable, and the standard is amended accordingly.

3. The National Highway Traffic Safety Administration has also evaluated requests by the American Safety Equipment Corporation concerning the range of occupants that a belt must adjust to fit, the test buckle release force test procedure, and the buckle crush resistance test procedure. The amended adjustment requirements (S4.1(g)(1) and (2)) specify more exactly the range of occupants that was intended by the original standard. The importance of having installed belts of proper length for the normal range of occupants outweighs, in the agency's judgment, the effort involved in ascertaining vehicle dimensions. The adjustment requirements are therefore not changed. With respect to the buckle test procedures, the petitioner's requests relating to the clarity of the buckle release procedure and to the need for an explanatory diagram to accompany the crush test are also denied. Although the buckle release test no longer refers to a method for testing lever action buckles, the method was little more than a suggestion and may in some cases have conflicted with the intent of the procedure that the force shall be applied so as to produce maximum releasing effect. The diagram requested to show the buckle crush procedure is not regarded as essential to understanding the procedure and has not been adopted.

4. Although no petition was received directly relating to the subject, the Swedish Trade Commission, on behalf of the Swedish manufactur-

ers, has expressed uncertainty as to how the crush test is to be applied to seat belt assemblies that have a buckle mounted on a rigid or semi-rigid bracket between the front seats. As described by the Commission, one design would tend to bend downwards under the pressure of the test device long before the required force of 400 pounds could be reached. In this case, the buckle will have to be supported from beneath, just as the conventional lap belt has to have some rigid backing in order to reach the 400-pound level. It is anticipated that if additional questions are raised concerning the method of force application to specific buckles, such questions can be answered through administrative interpretation.

5. Several petitions questioned the need to test a vehicle-sensitive emergency-locking retractor by accelerating it "in three directions normal to each other with its central axis oriented horizontally". The pendulum device used in most vehicle-sensitive retractors can sense lateral accelerations and sense the tilt of the vehicle, but it cannot readily sense upward or downward accelerations of the type required by the three-direction test when the retractor is oriented horizontally. It was suggested by Volvo that a retractor that locks when tilted to 35° in any direction should be exempt from the acceleration requirement. Volkswagen recommended accelerating the retractor in the horizontal plane in two directions normal to each other. On reconsideration, the National Highway Traffic Safety Administration has concluded that it is appropriate to relieve such a retractor from the vertical acceleration requirement when it is oriented horizontally and to establish an alternative to the requirement that it lock when accelerated in directions out of the horizontal plane, but that accelerations within the horizontal plane should continue to be required.

Accordingly, S5.2(j) is amended to require a vehicle-sensitive retractor to be accelerated in the horizontal plane in two directions normal to each other. During these accelerations, the retractor will be oriented at the angle in which it is installed in the vehicle. In addition, the retractor must either lock when accelerated in orientations out of the horizontal as prescribed in the March 10 rule or lock by gravity when

tilted in any direction to any angle greater than 45°.

6. One petitioner questioned the correctness of requiring webbing-sensitive retractors to be accelerated in the direction of webbing retraction, rather than in the direction of webbing withdrawal. The usage is necessary because under the test procedures of S5.2(j) it is the *retractor*, and not the webbing, that is accelerated. The acceleration must be in the direction that will reel the webbing out of the retractor—*i.e.*, the direction in which the webbing moves when retracting.

7. An additional question on retractor acceleration levels concerns the distance which a belt must be withdrawn in determining compliance with the requirement that the retractor shall not lock at 0.3g or less (S4.3(j)(ii)). The Hamill Manufacturing Company has requested an amendment to S4.3(j)(ii) to provide that the retractor shall not lock before the webbing extends a short distance at an acceleration of 0.3g. The National Highway Traffic Safety Administration recognizes that many retractors may be velocity-sensitive to some degree as well as acceleration-sensitive. Although a retractor that locks at too low a velocity would be an inconvenience, the NHTSA recognizes that an occupant does not ordinarily accelerate the belt after an initial pull and that the usual velocity involved in withdrawing the belt is low. On reconsideration, the NHTSA has therefore decided to amend S4.3(j)(ii) to provide that the retractor shall not lock before the webbing extends 2 inches at 0.3g.

8. Several petitioners pointed out that the requirements for retractor force specified in S4.3(j)(iii) and (iv) were not appropriate for systems in which a single length of webbing is used to provide both lap and shoulder restraint. In a typical installation of this sort, the webbing passes from a floor-mounted retractor up to a fitting on the B-pillar, then down across the shoulder to a slip joint on the buckle connector, and from there back across the lap to an outboard floor attachment. Although such a system may provide satisfactory restraint, it cannot simultaneously exceed a retractive force of 1.5 pounds on the lap belt and have a retractive

force on the shoulder belt of between 0.45 and 1.1 pounds, and it would therefore fail to conform to the standard as published March 10.

Upon reconsideration, the National Highway Traffic Safety Administration has decided to amend S4.3(j) by establishing retraction forces for 3-point systems that employ a single length of webbing. A new subsection (v) is added that requires such a system to have a retraction force falling within the range 0.45 pounds-1.50 pounds, and (iii) and (iv) are amended so that they do not apply to retractors in such systems. This range was suggested by Volkswagen, Volvo, and Klippan, and is considered to be a reasonable compromise between the need to provide complete retraction of the belt when not in use and the need to limit the force so that it will not be uncomfortable to occupants.

Effective date: January 1, 1972, except that seat belt assemblies manufactured on or after September 1, 1971 and before January 1, 1972, may conform either to the current requirements of Standard No. 209 in 49 CFR 571.21 or to the requirements of Standard No. 209 as amended by this notice and the notice of March 10, 1971 (36 F.R. 4607).

Issued on August 26, 1971.

Charles H. Hartman
Acting Administrator

36 F.R. 17430
August 31, 1971

PREAMBLE TO AMENDMENT TO MOTOR VEHICLE SAFETY STANDARD NO. 209

Seat Belt Assemblies

(Docket No. 73-16; Notice 2)

The purpose of this notice is to amend certain requirements of Motor Vehicle Safety Standard No. 209 (49 CFR 571.209), *Seat belt assemblies*, relating to the width of belt webbing and to the performance of seat belt retractors. The amendments were proposed in a notice published June 20, 1973 (38 FR 16084).

In the June 20 notice, the agency proposed to allow the width of those portions of a combination lap and shoulder belt that do not touch the occupant to be less than the 1.8 inches formerly required by the standard. The Chrysler Corporation, in its comment, suggested that narrower webbing should also be permitted for the type of lap belt that is used by itself. The agency agrees that a lap belt in combination with a shoulder belt (known as Type 2 assembly) is indistinguishable from an independent lap belt (Type 1 assembly), as far as the width of its webbing is concerned, and is therefore amending the standard to permit narrower webbing for non-contact portions of Type 1 belts as well as Type 2 belts.

Chrysler also requested narrower webbing for non-contact portions of children's harnesses (Type 3 assemblies). In view of the close-fitting design of Type 3 assemblies, the agency has not found a benefit to be gained from the use of narrower webbing in the few areas of non-contact. The Type 3 requirements are not being amended at this time. The American Safety Equipment Corporation requested that the contactability of the webbing with occupants be determined with a range of occupants. The agency remains persuaded that the use of a 95th percentile adult male occupant will be sufficient to insure that the narrower webbing will not touch any occupant who uses the seat. The

agency therefore declines to adopt American Safety's suggestion.

The proposed amendment of the emergency-locking retractor requirements of S4.3 drew several comments, not all of them relating to the parts of S4.3 that were proposed to be changed. Mercedes Benz requested revision of the requirement of S4.3(j)(2) that the retractor must not lock before the webbing extends 2 inches under an acceleration of 0.3g or less. The 0.3g requirement had been carried over without change from the previous version of S4.3 and was thought to be a reasonable means of preventing retractors from being inconveniently sensitive. The NHTSA does not find sufficient cause at this time to alter its conclusion concerning the most appropriate minimum level and is therefore retaining the minimum level of 0.3g.

A second issue raised by Mercedes Benz concerns the treatment under section S4.3(j) of a retractor having both vehicle sensitive and webbing sensitive features. It has been the NHTSA's position that with respect to the maximum permissible locking level, a dual-action retractor would conform if it met either of the applicable requirements. Thus, a dual-action retractor whose webbing-sensitive mechanism locks within 1 inch at an acceleration of 0.7g will conform, even though its vehicle-sensitive mechanism is not capable of locking at its required level. With respect to the minimum locking level, however, different considerations apply. The agency's intent in providing a minimum level below which the retractor must not lock is to enhance the convenience of the system. The webbing-sensitive mechanism that locks below 0.3g would be no less inconvenient if coupled with a vehicle sensitive mechanism than it would be if used by itself. The agency has therefore concluded that a dual-action retractor may conform to the maximum locking acceleration level of 0.7g (S4.3(j)(1)) with either mechanism, but that it must conform to both minimum locking level requirements (S4.3(j)(2) and (3)).

The tilt angle of 17° proposed as the minimum locking level for vehicle sensitive retractors was stated by several comments to be too high. Although there was general agreement as to the advisability of using a tilt test rather than an acceleration test, lower tilt angles were suggested, ranging downward to 11°. After considering the comments, the NHTSA has concluded that a moderate downward revision to 15° will prevent retractor lockup in normal road operation and has adopted that angle in S4.3(j)(3). The suggestion by Ford and American Motors that the "retractor drum's central axis" may be difficult to determine in complicated mechanisms has been found to have merit and the requirement as adopted refers to the orientation at which the retractor is installed in the vehicle.

The proposed revisions to the minimum retraction force requirements for retractors attached to upper torso restraints encountered several objections, the principal one being that no one was certain about the meaning of the proposed requirement that the retractor should "retract the webbing fully." The quoted language had been proposed in response to a petition by General Motors requesting amendment of the requirement that the retractor exert a retractive force of not less than 0.45 pound. The GM petition had requested a force of 0.2 pound, but the agency's initial intent, as reflected in the notice, was to grant a potentially greater relief by deleting reference to a specific minimum force. It appears from the confusion in the comments that a contrary result might be produced in some cases, and the agency has fherefore concluded that a simple reduction in the force level to the level requested by GM is the least complicated and most readily enforceable means of lowering the minimum force level. The suggestion by Ford, that the ability to retract is implicit in the definition of retractor and that no minimum force level is required, has some merit, but the agency prefers to retain a measurable minimum level.

There were several questions of interpretation concerning the point at which the retraction force is to be measured. The test procedures of S5.2 provide that the webbing is to be fully extended, passing over any hardware or other material specified for use with the webbing, and that it is then to be retracted and the retraction force measured as the lowest force within plus or minus 2 inches of 75 percent extension. The procedure is intended to measure the ability of the retractor to retract the webbing as installed in the vehicle, and the point of measurement most consistent with this intent is the most distant point of the webbing from the retractor. The NHTSA intends to conduct its measurements in this fashion.

The proposed amendment to S5.2 that would amend the test procedures to reflect the limitation of the 0.3g acceleration level to webbing-sensitive retractors was not objected to and is adopted as proposed.

In consideration of the foregoing, S4.2(a), S4.3(j), and S5.2(j) of Motor Vehicle Safety Standard No. 209, 49 CFR § 571.209, are amended. . . .

Effective date: August 28, 1973. The NHTSA finds it desirable to allow manufacturers to produce seat belt assemblies under the requirements as hereby amended (which generally are relaxed relative to previous requirements) prior to the effective date of the next phase of Standard No. 208 (49 CFR 571.208). It is therefore found for good cause shown that an immediate effective date is in the public interest.

(Sec. 103, 119, Pub. L. 89–563, 80 Stat. 718, 15 U.S.C. 1392, 1407; delegation of authority at 49 CFR 1.51.)

Issued on August 23, 1973.

James B. Gregory
Administrator

**38 F.R. 22958
August 28, 1973**

PREAMBLE TO AMENDMENT TO MOTOR VEHICLE SAFETY STANDARD NO. 209

Seat Belt Assemblies

(Docket No. 73-16; Notice 4)

This notice amends Standard No. 209, *Seat belt assemblies*, 49 CFR 571.209, to reduce the minimum retraction force required of emergency-locking retractors attached to lap belts from 1.5 pounds to 0.6 pounds. This amendment to S4.3 (j)(4) responds to a rulemaking petition submitted by Toyo Kogyo.

A notice of proposed rulemaking published October 2, 1973 (38 F.R. 27303), proposed the modification because the 1.5-pound force could prove excessive for occupant comfort, and experience with the 0.6-pound level in automatic-locking retractors has been satisfactory. Their performance at 0.6 pounds does not support an assertion in one comment to the docket that degradation of the retractor elements over time would result in almost total loss of retractive force. All other comments to the docket were favorable.

In consideration of the foregoing, S4.3(j)(4) of Motor Vehicle Safety Standard No. 209, *Seat belt assemblies*, 49 CFR 571.209, is amended....

Effective date: January 24, 1974. Because the amendment relaxes a requirement and creates no additional burden, it is found for good cause shown that an effective date earlier than one hundred eighty days after issuance is in the public interest.

(Secs. 103, 119, Pub. L. 89-563, 80 Stat. 718, 15 U.S.C. 1392, 1407; delegation of authority at 49 CFR 1.51.)

Issued on January 18, 1974.

James B. Gregory
Administrator

39 F.R. 2771
January 24, 1974

PREAMBLE TO AN AMENDMENT TO FEDERAL MOTOR VEHICLE SAFETY STANDARD NO. 209

Seat Belt Assemblies

(Docket No. 74-9; Notice 7)

ACTION: Final rule; response to petitions for reconsideration.

SUMMARY: This notice responds to five petitions for reconsideration and petitions for rulemaking concerning Standard No. 213, *Child Restraint Systems.* In response to the petitions, the agency is changing the labeling requirements to permit the use of alternative language, modifying the minimum radius of curvature requirement for restraint system surfaces and extending the effective date of the standard from June 1, 1980, to January 1, 1981. In addition, several typographical errors are corrected in Standard No. 209, *Seat Belt Assemblies.*

DATES: The amendments are effective on May 1, 1980. The effective date of the standard is changed from June 1, 1980, to January 1, 1981.

FOR FURTHER INFORMATION CONTACT: Mr. Vladislav Radovich, Office of Vehicle Standards, National Highway Traffic Safety Administration, 400 Seventh Street, S.W., Washington, D.C. 20590 (202-426-2264).

SUPPLEMENTARY INFORMATION: On December 13, 1979 (44 F.R. 72131) NHTSA published in the *Federal Register* a final rule establishing Standard No. 213, *Child Restraint Systems,* and making certain amendments to Standard No. 209, *Seat Belt Assemblies and Anchorages.* Subsequently, petitions for reconsideration were timely filed with the agency by Cosco, General Motors, Juvenile Products Manufacturers Association (JPMA), and Strolee. Subsequent to the time for filing petitions for reconsideration, Strolee also filed a petition for rulemaking to amend the standard. After evaluat-

ing the petitions, the agency has decided to modify, as fully explained below, some of the requirements of Standard No. 213. All other requests for modifications are denied. The agency is also correcting several minor typographical errors in the text of Standard No. 209.

Labeling

Standard No. 213 requires manufacturers to place a permanently mounted label on the restraint to encourage its proper use. General Motors (GM) petitioned for reconsideration of three of the labeling requirements.

Section S5.5.2(f) of the standard requires each child restraint to be labeled with the size and weight ranges of children capable of using the restraint. In its petition, GM said that the requirement could "unnecessarily preclude some children from using the restraint or suggest use by children too large for the restraint." GM also commented that some infant restraints are intended to be used from birth and thus the lower size and weight limitation serves no purpose.

In addition, GM said that stating the upper size limit for infant restraints in terms of seated height rather than in standing height is a more appropriate way to set size limitations for infants. For example, GM said that an infant with a short torso and long legs might be precluded from using the restraint if the limitation is stated in terms of standing height, while an infant with short legs and a torso too long for the restraint would be inappropriately included among ones who could supposedly use the restraint. GM requested that infant restraints be allowed to be labeled with an optional statement limiting use by upper weight and seated height.

PART 571; S 209—PRE 15

NHTSA agrees that specifying a lower weight and size limit is unnecessary for an infant carrier designed to be used from birth and has amended the standard accordingly. The agency has decided not to adopt GM's proposal to state the upper size limit in seating rather than standing height. The purpose of the label is to provide important instructions and warnings in as simple and understandable terms as possible. Standing height, rather than seating height, is a measurement parents are familiar with and which is commonly measured during pediatric examinations. As GM pointed out, it is possible to establish a limit based on standing height which would exclude any infant whose seating height is too high to properly use the restraint. Therefore, the agency will continue to require the upper size limit to be stated in terms of standing height.

GM also requested that manufacturers be allowed to establish a lower usage limit for restraints used for older children based on the child's ability to sit upright rather than on his or her size and weight. GM said the lower limit "is not as dependent upon the child's size as it is on the child's ability to hold its head up (sit upright) by itself. This important capability is achieved at a wide range of child sizes." NHTSA agrees that the type of label GM proposes can clearly inform parents on which children can safely use a restraint and therefore will permit use of such a label.

Section S5.5.2(g) of the standard requires the use of the word "Warning" preceding the statement that failure to follow the manufacturer's instructions can lead to injury to a child. GM requested that the word "Caution" be permitted as an alternative to "Warning." GM said that since 1975 it has used caution in its labels and owners' and service manuals as a lead or signal word where the message conveys instructions to prevent possible personal injury. GM said that the words caution and warning are generally accepted as synonymous.

The agency believes that the word "Warning," when used in its ordinary dictionary sense, is a stronger term that conveys a greater sense of danger than the word "Caution" and thus will emphasize the importance of following the specified instructions. Therefore, the agency will continue to require the use of the word "Warning."

Section S5.5.2(k) of the standard requires restraints to be labeled that they are to be used in a rear-facing position when used with an infant. GM said that while the requirement is appropriate for so-called convertible child restraints (restraints that can be used by infants in a rear-facing position and by children in a forward-facing position), it is potentially misleading when used with a restraint designed exclusively for infants. GM said the current label might imply that the restraints can be used in forward-facing positions with children. GM recommended that restraints designed only for infants be permitted to have the statement, "Place this infant restraint in a rear-facing position when using it in the vehicle." The agency's purpose for establishing the labeling requirement was to preclude the apparent widespread misuse of restraints designed for infants in a forward-facing rather than rear-facing position. Since GM's recommended label will accomplish that goal, the agency is amending the standard to permit its use.

Radius of Curvature

Section S5.2.2.1(c) of the standard requires surfaces designed to restrain the forward movement of a child's torso to be flat or convex with a radius of curvature of the underlying structure of not less than 3 inches. Ford Motor Co. objected to the 3-inch limitation on radius of curvature arguing that measuring the radius of curvature of the underlying structure would eliminate designs that have not produced serious injuries in actual crashes. Ford said the shield of its Tot-Guard has a radius of curvature from 2.2 to 2.3 inches and it had no evidence of serious injury being caused by the shield when the restraint has been properly used.

The purpose of the radius of curvature requirement was to prohibit the use of surfaces that might concentrate impact forces on vulnerable portions of a child's body. It was not the agency's intent to prohibit existing designs, such as the Tot-Guard, which have not produced injuries in actual crashes. Since a 2-inch radius of curvature should therefore not produce injury the agency has decided to change the radius of curvature requirement from 3 to 2 inches.

Although the standard sets a minimum radius of curvature for surfaces designed to restrain the forward movement of a child, it does not set a minimum surface area for that surface. Prototypes of new restraints shown to the agency by some manufacturers indicate that they are voluntarily incorporating sufficient surface areas in their designs. The agency encourages all manufacturers to use surface areas at least equivalent to those of the designs used by today's better restraints.

Occupant Excursion

Section S5.1.3.1 of the standard sets a limit on the amount of knee excursion experienced by the test dummy during the simulated crash tests. It specifies that "at the time of maximum knee forward excursion the forward rotation of the dummy's torso from the dummy's initial seating configuration shall be at least 15° measured in the saggital plane along the line connecting the shoulder and hip pivot points."

Ford Motor Co. objected to the requirements that the dummy's torso rotate at least 15 degrees. Ford said that it is impossible to measure the 15 degree angle on restraints such as the Tot-Guard since the test dummy "folds around the shield in such a manner that there is no 'line' from the shoulder to the hip point." In addition, restraints, such as the Tot-Guard, that enclose the lower torso of the child can conceal the test dummy hip pivot point.

The agency established Ted the knee excursion and torso rotation requirements to prevent manufacturers from controlling the amount of test dummy head excursion by allowing the test dummy to submarine excessively during a crash (i.e., allowing the test dummy to slide too far downward underneath the lap belt and forward, legs first). A review of the agency's testing of child restraints shows that current designs that comply with the knee excursion limit do not allow submarining. Since the knee excursion limit apparently will provide sufficient protection to prevent submarining, the agency has decided to drop the torso rotation requirement. If future testing discloses any problems with submarining, the agency will act to establish a new torso rotation requirement as an additional safeguard.

Head Impact Protection

Section S5.2.3 requires that each child restraint designed for use by children under 20 pounds have energy-absorbing material covering "each system surface which is contactable by the dummy head." Strolee petitioned the agency to amend this requirement because it would prohibit the use of unpadded grommets in the child restraint. Strolee explained that some "manufacturers use grom-

mets to support the fabric portions of a car seat where the shoulder belt and lap belt penetrate the upholstery. These grommets retain the fabric in place and give needed support where the strap comes through to the front of the unit." Because of the use of the grommets in positioning the energy-absorbing padding and belts, the agency does not want to prohibit their use. However, to ensure that use of the grommets will not compromise the head impact protection for the child, the agency will only allow grommets or other structures that comply with the protrusion limitations specified in section S5.2.4. That section prohibits protrusions that are more than ⅝ of an inch high and have a radius of less than ¼ inch. Because this amendment makes a minor change in the standard to relieve a restriction, prior notice and a comment period are deemed unnecessary.

Belt Requirements

Strolee petitioned the agency to amend the requirement that all of the belts used in the child restraint system must be 1½ inches in width. Strolee said that straps used in some restraints to position the upper torso restraints have " 'snaps' so that the parent may release this positioning belt conveniently." Strolee argued that such straps should be exempt from the belt width requirement since "the snap would release far before any loads could be experienced."

The agency still believes that any belt that comes into contact with the child should be of a minimum width so as not to concentrate forces on a limited area of the child. This requirement would reduce the possibility of injury in instances where the snap on a positioning strap failed to open. Strolee's petition is therefore denied.

Strolee has also raised a question about the interpretation of section S5.4.3.3 on belt systems. Strolee asked whether the section requires a manufacturer to provide both upper torso belts, a lap belt and a crotch strap or whether a manufacturer can use a "hybrid" system which uses upper torso belts, a shield, in place of a lap belt, and a crotch strap. The agency's intent was to allow the use of hybrid systems. The agency established the minimum radius of curvature requirements of section S5.2.2.1(c) to ensure that any shield used in place of a lap or other belt would not concentrate forces on a limited area of the child's body. NHTSA has amended section S5.4.3.3 to clarify the agency's intent. Because this is an interpretative amendment, which imposes no new restrictions, prior notice and a comment period are deemed unnecessary.

Height Requirements

Strolee asked the agency to reconsider the requirements for seat back surface heights set in section S5.2.1.1. Strolee argued that the higher seat back required by the standard would restrict the driver's rear vision when the child restraint is placed in the rear seat.

The final rule established a new seat back height requirement for restraints recommended for use by children that weigh more than 40 pounds. To provide sufficient protection for those children's heads, the agency required the seat back height to be 22 inches. The agency explained that the 22-inch requirement was based on anthropometric data showing that the seating height of children weighing 40 or more pounds can exceed 23 inches. The agency still believes that 22-inch requirement is necessary for the protection of the largest child for which the restraint is recommended. NHTSA notes that child restraints can be designed to accommodate the higher seat backs without allowing the overall height of the child restraint to unduly hinder the driver's vision.

Padding

In its petition, JPMA claimed that the standard "calls for the application of outdated specifications" for determining the performance of child restraint padding in a 25-percent compression-deflection test. A review of the most recent edition of the American Society for Testing and Materials (ASTM) handbook shows that the compression-deflection test in two of the three ASTM standards referenced by the agency has not changed. The third standard (ASTM D1565) referenced by the agency has been replaced. However, the replacement standard does not contain a 25 percent compression-deflection test. Therefore, the agency will continue to use the three ASTM standards referenced in the December 1979 final rule.

Effective Date

Cosco, Strolee, and the Juvenile Products Manufacturers Association (JPMA) petitioned the agency for an extension of the June 1, 1980, effective date. They requested that the effective date be changed to at least January 1, 1981, and Strolee requested a delay until March 1, 1981. They argued that the June 1, 1980, effective date does not allow manufacturers sufficient time to develop, test and tool new child restraints.

Testing done for the agency has shown that many of the better child restraint systems currently on the market can meet the injury criteria and occupant excursion limitation set by the standard. Some of those seats would need changes in their labeling, removal of arm rests and new belt buckles and padding to meet the standard. Such relatively minor changes can be made in the time available before the June 1, 1980, effective date.

Several manufacturers have informed the agency that they are designing new restraints to meet the standard. Based on prototypes of those restraints shown to the agency, NHTSA believes that these new restraints may be more convenient to use, less susceptible to misuse and provide a higher overall level of protection than current restraints. Based on leadtime information provided by individual manufacturers and the JPMA, the agency concludes that extending the standard from June 1, 1980, to January 1, 1981, will provide sufficient leadtime. Providing a year's leadtime is in agreement with the leadtime estimates provided by the manufacturers as to the time necessary for design and testing, tooling and buckle redesign.

Compatibility With Vehicle Belts

On December 12, 1979, NHTSA held a public meeting on child transportation safety. At that meeting, several participants commented about the difficulty, and in some cases the impossibility, of securing some child restraint systems with a vehicle lap belt because the belt will not go around the restraint. Testing done by the agency during the development of the recently proposed comfort and convenience rulemaking also confirms that problem. The agency reminds child restraint manufacturers that Standard No. 213, *Child Restraint Systems*, requires all child restraints to be capable of being restrained by a vehicle lap belt.

Corrections

In the final rule published on Standard No..209, *Seat Belt Assemblies*, there were a number of typographical errors, such as listing the lower chest circumference of the 5 percentile female as 36.6 inches rather than the correct figure of 26.6 inches. Those errors have been corrected.

In addition, the final rules for Standards No. 209 and No. 213 inadvertently did not include a requirement on belt resistance to buckle abrasion. The notice of proposed rulemaking for both standards included the belt buckle abrasion requirements, which were not opposed by any of the commenters. The standards have therefore been amended to include that requirement.

The principal authors of this notice are Vladislav Radovich, Office of Vehicle Safety Standards, and Stephen Oesch, Office of Chief Counsel.

Issued on April 23, 1980.

Joan Claybrook
Administrator
45 F.R. 29045
May 1, 1980

PREAMBLE TO AN AMENDMENT TO FEDERAL MOTOR VEHICLE SAFETY STANDARD NO. 209

Seat Belt Assemblies

(Docket No. 80-12; Notice 2)

ACTION: Final rule.

SUMMARY: This notice amends Safety Standard No. 209, *Seat Belt Assemblies*, to exempt seat belts installed in conjunction with automatic restraint systems from the belt elongation requirements of the standard. This amendment is based on a petition for rulemaking submitted by Mercedes-Benz of North America and follows the publication of a proposal. The amendment permits manufacturers to install belt systems incorporating load-limiting devices which are intended to make further reductions in head and upper torso injuries during an accident. Some load-limiting belt systems utilize webbing that elongates more than is currently allowed by Standard No. 209. This amendment would permit this and other type systems to exceed the maximum elongation allowed by the standard.

DATES: This amendment is effective January 12, 1981.

ADDRESSES: Any petition for reconsideration should refer to the docket number and notice number and be submitted to: National Highway Traffic Safety Administration, 400 Seventh Street, S.W., Washington, D.C. 20590.

FOR FURTHER INFORMATION CONTACT: Mr. William Smith, Office of Vehicle Safety Standards, National Highway Traffic Safety Administration, Washington, D.C. 20590 (202-426-2264).

SUPPLEMENTARY INFORMATION: Safety Standard No. 209, *Seat Belt Assemblies* (49 CFR 571.209), specifies performance requirements for seat belts to be used in motor vehicles. One of these performance requirements specifies the maximum amount that the webbing of a belt assembly is permitted to extend or elongate when subjected to certain specified forces (paragraph S4.2(c)). Mercedes-Benz of North America petitioned NHTSA to exempt seat belt assemblies installed in passenger cars in conjunction with air cushion restraint systems from the webbing elongation requirements of the standard. The agency granted that petition and issued a notice of proposed rulemaking to amend the standard on August 4, 1980 (45 F.R. 51626).

Mercedes is considering the use of a belt system that incorporates a load-limiting device. A load-limiter is a seat belt assembly component or feature that controls tension on the seat belt and modulates or limits the force loads that are imparted to a restrained vehicle occupant by the belt assembly during a crash. Load-limiting devices are intended to reduce head and upper torso injuries through increased energy management. A load-limiter can be a separate component of the seat belt system, such as a torsion bar that allows the retractor to reel out additional webbing when a certain designed force level is reached. The load-limiter can also be a feature of the webbing itself, such as webbing that will elongate to certain designed lengths when subjected to particular force levels. Mercedes is interested in using the latter type load-limiting system. However, the webbing in the Mercedes belt system would elongate beyond the limits that are currently specified in Standard No. 209. Mercedes' petition stated that this type belt system should be allowed in vehicles equipped with air cushion restraints since the two systems used in conjunction with one another can be designed to achieve the maximum reduction in head injuries and upper-torso injuries.

Although safety belts protect occupants from life-threatening impacts with the vehicle interior, the forces necessarily generated by the belts upon occupants during a crash can result in upper torso injury. As noted in the notice of proposed rulemaking, data available to the agency indicate that load-limiting belts can reduce these injuries, as well as working in combination with an automatic restraint system to provide protection for impacts with the vehicle interior. The proposal specified that both Type 1 (lap belts) and Type 2 (combination lap and shoulder belts) manual belts having load-limiting devices and used in conjunction with automatic restraints would be exempted from the elongation requirements. Additionally, the proposal specified that such belts would have to be labeled to clarify that they are intended for use only in vehicles equipped with automatic restraint systems.

The proposal limited the use of load-limiting belts to vehicles equipped with automatic restraints since there are currently no dynamic performance requirements or injury criteria for manual belt systems used alone. There are no requirements to ensure that a load-limiting belt system would protect vehicle occupants from impacting the steering wheel, instrument panel and windshield, which would be very likely if the belts elongated beyond the limits specified in Standard No. 209. Therefore, the elongation requirements are necessary to ensure that manual belts used as the sole restraint system will adequately restrain vehicle occupants.

Nine comments were submitted in response to the August 4 proposal, all supporting the exemption for load-limiting belts. Vehicle manufacturers stated that the proposed exemption from the elongation requirements would allow design flexibility and lead to improved occupant restraint systems.

American Motors Corporation (AMC) stated that the exemption for load-limiting belts should only apply to Type 2 manual belts. The company argued that the only available data relates to the ability of Type 2 load-limiting belts to reduce certain head and upper-torso injuries. AMC stated that torso injury is not a function of lap belt loads and that no similar correlation has been made between lap belt loads and pelvic fractures. Therefore, the company believes that the exemption from the elongation re-

quirements for Type 1 belts should be postponed until specific injury patterns can be correlated with lap belt loads.

The agency proposed allowing the exemption for both Type 1 and Type 2 belts in order to give manufacturers broader design latitude to use load-limiting features on all belt systems used in conjunction with automatic restraints. AMC is correct in its statement that more data are available regarding the correlation between Type 2 belts and upper-torso injury than is available regarding load-limiting features on Type 1 belts. However, comments received from Rolls-Royce Motors stated that the company has tested manual Type 1 belts incorporating load-limiting features and found that better results are obtained under the injury criteria of Safety Standard No. 208 (49 CFR 571.208) than with Type 1 belts which must comply with the elongation requirements. In light of this information, and the fact that load-limiting Type 1 belts would only be allowed in conjunction with automatic restraint systems complying with the injury criteria of Standard No. 208, the agency has decided to include Type 1 belts in the exemption. This will allow manufacturers to develop innovative designs to maximize the protection provided by its automatic restraint systems. If future data indicate a problem with Type 1 belts that incorporate load-limiting features, the exemption from the elongation requirements can be reconsidered by the agency.

The August 4, 1980, notice proposed to add a new definition to Standard No. 209 to define "load-limiter," and limited the exemption from the elongation requirements to belts incorporating load-limiters and installed in conjunction with automatic restraints. Volvo of America Corporation commented that the definition of "load-limiter" is very broad and could be interpreted to include all existing belt webbing. Volvo stated that the exemption should, therefore, apply to any Type 1 or 2 belt installed in conjunction with an automatic restraint, and not be limited to load-limiting belts.

While the agency understands Volvo's point that the proposed language may be extremely detailed, we believe the language is necessary to clarify the exemption and to avoid confusion for belt manufacturers. Safety Standard No. 209 is an equipment standard rather than a vehicle standard, and each seat belt assembly must be certified by the belt manufacturer. The proposed language was intended to create a clear distinction between belts complying with elongation requirements of Safety Standard No. 209 and those that incorporate load-limiting features that preclude compliance with the elongation requirements. The proposed language explained which belt systems must be labeled as being for use only in vehicles equipped with automatic restraints. The agency believes this language, including the definition of "load-limiter," is necessary at the current time to clarify the requirements for those persons or manufacturers who may not be totally familiar with the requirements of Safety Standard No. 209. Otherwise, it would not be clear from the standard why certain belts are exempted from the elongation requirements of the standard.

In another comment related to this same subject, General Motors Corporation pointed out that the proposed labeling requirement for load-limiting belts could apply to all Type 1 and 2 belts incorporating load-limiting features even if all current 209 requirements are met. General Motors stated that load-limiting belt systems that can, nevertheless, comply with the elongation requirements of the standard should not be limited in their application to vehicles equipped with automatic restraint systems. The agency agrees with this argument, and the language is changed in this amendment accordingly.

General Motors also questioned the need to require any label at all on load-limiting belts. The proposal specified that such belts would have to be permanently marked or labeled to indicate the assembly may only be installed in vehicles in conjunction with an automatic restraint system. General Motors argued that a label is not necessary to control the installation of load-limiting belts in the proper vehicles. Seat belt manufacturers must currently provide appropriate installation instructions for its equipment. General Motors contends that this requirement, coupled with the fact that replacement belts are generally ordered and installed by a repair facility, will ensure that load-limiting belts are only installed in vehicles equipped with automatic restraints. The agency does not agree with this position. As stated earlier, the agency believes that care must be taken to distinguish load-limiting belt systems from other systems. If there is a label on the belt

itself, a person making the installation will be aware that the belt should only be installed in conjunction with automatic restraints. This should be made obvious to the person making the installation without reference to the installation instructions. Further, none of the other commenters objected to the proposed labeling requirement. American Motors Corporation specifically stated that a label is necessary.

General Motors is correct in its statement that this warning will also be provided in the installation instructions provided by the belt manufacturer. Paragraphs S4.1(1) of Safety Standard No. 209 provides, in part, that the installation instruction sheet provided by the belt manufacturer shall state whether the assembly is for universal installation or for installation only in specifically stated motor vehicles. Therefore, belt manufacturers will be required to specify in the installation instructions that load-limiting belts are only to be installed in combination with automatic restraint systems. The agency believes that at the current time these duplicative warnings, in the instruction sheet and on a belt label, are a necessary precaution to ensure that load-limiting belts are only installed in the proper vehicles. After a majority of vehicles on the road are equipped with automatic restraints, such labeling may no longer be necessary.

Volvo of America Corporation commented that some upper limit on belt elongation may be required for Type 1 manual belts incorporating load-limiting features, although no such limit was specified in the proposal. Volvo pointed out that Type 1 belts installed in conjunction with air cushion restraints will also provide roll-over protection for vehicle occupants. The company is concerned that if no upper limit on elongation is specified, such belts may not provide the intended protection in roll-over accidents.

While the agency agrees that this is a legitimate concern, it does not believe it is necessary to specify such an upper limit at the current time. It is not likely that manufacturers will design load-limiting belt systems that will elongate appreciably beyond the limits specified in Standard No. 209. Presumably, load-limiting belts will be designed to provide actual restraint in conjunction with the automatic restraint system, if the vehicle is to comply with the injury criteria of Safety Standard No.

208. If a load-limiting belt design elongates to the extent that it would provide no protection in roll-over accidents, it would also not provide any protection in frontal crashes. Therefore, it is not likely that manufacturers would permit such extensive elongation in their systems. Moreover, the forces generated in frontal crashes are more severe than those that occur in roll-over accidents, so the elongation that would occur even with load-limiting systems would not be as great in roll-over accidents as in frontal accidents. The agency believes that manufacturers should be given broad latitude in the development of load-limiting belt systems to be used in vehicles equipped with automatic restraints. In light of these considerations, no upper limit on belt elongation is specified in this amendment. Manufacturers should be cognizant of the point made by Volvo, however, during the development of their systems.

The comments of Renault USA included general questions regarding automatic seat belts and the relationship between Safety Standard No. 208 and Safety Standard No. 209. Some confusion apparently exists regarding paragraph S4.5.3.4 of Safety Standard No. 208 and agency interpretations regarding that paragraph. The agency has stated in the past that only automatic belts that are installed to meet the frontal crash protection requirements of S5.1 of Standard No. 208 are exempted from the requirements of Standard No. 209. Yet, the agency has also stated that those portions of Standard No. 209 relating to retractors are applicable to all automatic belts. Renault finds these statements inconsistent.

Paragraph S4.5.3.4 of Standard No. 208 is a general provision which exempts certain automatic belts, those meeting the injury criteria of the standard, from the requirements of Standard No. 209. However, paragraph S4.5.3.3(a) of Standard No. 208 specifically provides that automatic belts shall conform to S7.1 of Standard No. 208, and that paragraph relates to the performance requirements for belt retractors specified in Standard No. 209. It is for this reason that the agency has stated that all automatic belts must comply with the retractor requirements, notwithstanding the general exemption specified in S4.5.3.4.

Renault contends that paragraph S4.5.3.4 is also inconsistent by its own terms since, Renault states, an automatic belt system must always comply with the injury criteria of S5.1 of Standard No. 208. This incorrect Paragraph S4.5.3 of Safety Standard No. 208 specifies that an automatic belt

may be used to meet the crash protection requirements of any option under S4 and in place of any seat belt assembly otherwise required by that option. Therefore, prior to the effective date of the automatic restraint requirements of the standard, automatic belts could be used to satisfy the third option of section S4—the seat belt option. Automatic belts installed under the third option would not be required to comply with the injury criteria of S5.1, since the injury criteria is only specified as a requirement under option 1 and option 2. Manufacturers are permitted, however, to install automatic belts in satisfaction of either option 1 or option 2 and to certify to the injury criteria, if they desire. In summary, automatic belts installed in passenger cars in compliance with the injury criteria of Safety Standard No. 208 are only required to comply with the provisions of Safety Standard No. 209 relating to retractors. They are not required to comply with any other provision in Standard No. 209. Automatic belts installed in passenger cars that are not certified as being in compliance with the injury criteria of Standard No. 208, i.e., those installed under the third option of the standard, are required to comply with all provisions of Standard No. 209. Manual seat belts having load-limiters, installed in vehicles in conjunction with automatic restraints meeting the injury criteria of Standard No. 208, are required to comply with all provisions of Standard No. 209 except the elongation requirements (by this amendment).

The agency has determined that this amendment is not a significant regulation under Executive Order 12221, "Improving Government Regulations," and the Departmental guidelines implementing that Order. Therefore, a regulatory analysis is not required. The exemption specified in this amendment provides manufacturers with broader design alternatives and should have little if any economic or environmental impact. Consequently, the agency has also determined that a regulatory evaluation is not required.

The engineer and lawyer primarily responsible for the development of this rule are William Smith and Hugh Oates, respectively.

Issued on January 5, 1981.

Joan Claybrook
Administrator
46 F.R. 2618
January 12, 1981

Federal Motor Vehicle Safety Standards;
Seat Belt Assemblies

[Docket No. 82-15; Notice 2]

ACTION: Final rule.

SUMMARY: The purpose of this notice is to amend Safety Standard No. 209, *Seat Belt Assemblies,* which incorporates by reference a number of recommended practices and test procedures developed by voluntary standards organizations. This amendment updates those references by incorporating the most recent version of the recommended practices and procedures. This amendment is intended to keep the standard in pace with the technological changes and improvements in the industry.

DATE: This amendment is effective July 30, 1983.

SUPPLEMENTARY INFORMATION: Federal Motor Vehicle Safety Standard No. 209, *Seat Belt Assemblies* (49 CFR 571.209), specifies performance requirements for seat belts used in passenger cars, trucks, buses and multipurpose passenger vehicles (both as original and after-market equipment). Several of the performance requirements of the standard incorporate recommended practices developed by voluntary standards organizations and associations. In addition, the standard specifies that certain, long-established industry test procedures be used in determining whether the seat belts meet those performance requirements. Because of the lengthy and technical nature of the recommended practices and test procedures, the standard incorporates those specifications by reference rather than setting out full texts in Standard No. 209.

Since Standard No. 209 was first issued, along with the incorporated material, some of the referenced practices and procedures have been modified in some respects by the standards organizations, because of technological changes and advancements. In light of these modifications, the agency conducted a review of all the materials incorporated by reference within Standard No. 209 to determine which materials needed to be changed so that their most recent version is incorporated in the standard. That review led to the issuance of a proposal to amend the standard to update all materials incorporated by reference (47 FR 31712, July 22, 1982). Interested persons should consult that notice of proposed rulemaking which sets out in detail the specific sections of the standard that include incorporated material, along with the proposed updated version of that material. As noted in the proposal, the incorporated material was developed by such voluntary standards associations as the American Association of Textile Chemists and Colorists (AATCC), the American Society for Testing and Materials (ASTM) and the Society of Automotive Engineers (SAE).

Nine comments were submitted to the agency in response to the notice of proposed rulemaking, all of which supported the proposed update of materials incorporated by reference in the standard. There were only a few recommended changes in the proposed revisions.

In addition to incorporating the new ASTM corrosion resistance test procedure (paragraph S5.2(a) of the standard), the agency proposed a minor change in the procedure. The ASTM procedure specifies that the seat belt hardware is to be "suitably cleaned" prior to testing. To clarify the extent of cleaning necessary, the agency proposed to specify that any temporary coating placed on the seat belt hardware shall be removed prior to testing. The purpose of the proposed change was to prevent the use of a coating material on the hardware during the corrosion resistance test that would aid the hardware in meeting the requirement, but which would not be found on the hardware when it is in actual vehicle use. Coatings which are applied permanently to the hardware would not have to be removed. The language proposed was as follows:

"Any surface coating or material not intended for permanent retention on the metal parts during service life shall be removed prior to preparation of the test specimen for testing."

Both Ford Motor Company and the Motor Vehicle Manufacturers Association requested changes in this language. Ford argued that the phrases "intended for permanent retention" and "during service life" are unduly restrictive because some anticorrosion coatings are applied to component parts to inhibit their corrosion during shipment to assembly plants and are intended to remain on those parts after assembly of the vehicle and its delivery to the first retail purchaser. Ford noted that such oil coatings may, however, disappear (e.g., dry up) during the service life of the vehicle. (MVMA's concern appeared to be identical to Ford's.)

The agency proposed to clarify the cleaning instructions in the corrosion test procedure because a testing laboratory brought a potential problem to the agency's attention. The laboratory reported that certain seat belt components had been delivered to it for corrosion testing which had been coated with wax. Obviously, such a coating would preclude a true testing of the components' corrosion resistance and the coating would not likely be present throughout the service life of the vehicle (and might in fact be removed during vehicle assembly). While the agency understands the point raised by Ford and MVMA (that oil coatings are intended to remain on the components upon delivery), as Ford pointed out, these coatings will likely dry up during the service life of the vehicle. Therefore, it is the agency's opinion that wax, oil or other coatings that are not permanent should be removed prior to testing since they can skew the test results and misrepresent the corrosion resistance of component parts during actual vehicle use. Consequently, the proposed language is being maintained in this amendment. It should be noted, however, that this test requirement is in no way intended to preclude manufacturers from plac-

ing any coatings, either temporary or permanent, on their seat belt assembly components.

Section S5.1(e) of Standard No. 209 specifies the test procedures for measuring the resistance to light of seat belt assemblies. In May 1980, the agency proposed to alter the test apparatus used for these requirements in light of new dacron materials being used in belt assemblies (45 FR 29102). As a part of that action, the agency proposed to update the one ASTM recommended practice (E42-64) already incorporated in the standard and to add a reference to another ASTM practice (G24-66). The proposal preceding this amendment noted that the agency is awaiting the completion of additional testing before taking final action on the May 1980 proposal and that, if an amendment were adopted, the agency would incorporate the most recent version of both the ASTM recommended practices.

Volkswagen of America pointed out that ASTM G24-66 is not the most recent version of that standard and cited instead G24-73. The Motor Vehicle Manufacturers Association stated that its member companies had not yet had a chance to evaluate the new ASTM procedures and indicated that they could involve significant changes. Both commenters requested that a new proposal be issued before a final amendment involving the resistance to light requirements is issued. The agency realizes that the new ASTM procedures may involve substantial changes in the test procedures and does intend to issue an additional proposal prior to updating that aspect of the Standard No. 209 test procedures (pending completion of additional testing, as noted in the notice of proposed rulemaking).

Two commenters, American Motors Corporation and Ms. Patricia Hill, pointed out a discrepancy between the Occupant Weight and Dimension Charts referenced in S4.1(g)(3) of Standard No. 209 and in S7.1.3 of Standard No. 208, *Occupant Crash Protection* (49 CFR 571.208). The hip breadth (sitting) for the 95th percentile adult male is listed as 16.4 inches in the former and as 16.5 inches in the latter. To remove this discrepancy, this notice amends the chart in Standard No. 209 to agree with the chart in Standard No. 208 (i.e., to read 16.5 inches). (Originally, the chart in Standard No. 208 also listed the hip breadth as 16.4 inches. This was amended January 8, 1981, to be consistent with the dimensions of the Part 572 test dummy (46 FR 2064)).

The American Seat Belt Council noted that a

more recent version of AATCC Test Method 30 (30-81), Resistance to Microorganisms, has been issued than was noted in the proposal (which referenced 30-79). The agency has reviewed this latest version and determined that the only difference between 30-79 and 30-81 is the optional addition of glucose to the test culture used in Test III. The agency agrees with this option and therefore is incorporating AATCC Method 30-81 in this amendment.

The notice of proposed rulemaking preceding this amendment also solicited comments, information and data from the public concerning any current requirements of Standard No. 209 which possibly impose a regulatory burden and have a negligible or inconsequential impact on safety. The agency solicited this information as part of its regulatory review of all existing regulations. All comments to the proposal included suggested changes or revisions to reduce burdens, clarify requirements or to harmonize Standard No. 209 with European standards. These comments are currently being reviewed by the agency under its Regulatory Reform program and may lead to additional rulemaking to reduce or eliminate regulatory burdens imposed by Standard No. 209. (Persons interested in the recommended changes should consult comments to the proposal: Docket 82-15; Notice 1.)

In addition to the amendments discussed earlier, this notice also amends 49 CFR Part 571.5, Matter Incorporated by Reference, to list the address of the American Association of Textile Chemists and Colorists (AATCC). This amendment will assist interested parties in obtaining copies of the AATCC test procedures which are incorporated by reference in Standard No. 209.

The amendments included in this notice are to become effective 30 days after the date of this publication. The Administrator has determined that there is good cause for an effective date sooner than 180 days because this amendment only updates material incorporated by reference and makes no real substantive changes in the standard. Consequently, the burdens on manufacturers will in no way be increased.

Executive Order 12291

The agency has evaluated the economic and other impacts of this final rule and determined that they are neither major as defined by Executive Order 12291 nor significant as defined by the Department of Transportation's regulatory policies and procedures. The final rule only updates references to

recommended practices and test methods already incorporated by reference in Standard No. 209. Because the economic and other effects of this proposal are so minimal, a full regulatory evaluation has not been prepared.

Regulatory Flexibility Act

In accordance with the Regulatory Flexibility Act, the agency has evaluated the effects of this action on small entities. Based on that evaluation, I certify that the final rule will not have a significant economic impact on a substantial number of small entities. Accordingly, no regulatory flexibility analysis has been prepared.

Only a few of the vehicle and parts manufacturers required to comply with Standard No. 209 are small businesses as defined by the Regulatory Flexibility Act. Small organizations and governmental jurisdictions which purchase fleets of motor vehicles would not be significantly affected by the amendments. The final rule merely updates references to test methods and recommended practices incorporated by reference in Standard No. 209. These updates should not impose any costs or other burdens.

PART 571—FEDERAL MOTOR VEHICLE SAFETY STANDARDS

In consideration of the foregoing, the following amendments are made to Title 49, Chapter V, § 571.209, *Seat Belt Assemblies*, and § 571.5, Matter incorporated by reference:

§ 571.209 [Amended]

1. The first sentence of S4.1(f) is revised to read as follows:

*　　*　　*　　*　　*

S4.1 * * *

(f) *Attachment hardware.* A seat belt assembly shall include all hardware necessary for installation in a motor vehicle in accordance with Society of Automotive Engineers Recommended Practice J800c, "Motor Vehicle Seat Belt Installation," November 1973. * * *

*　　*　　*　　*　　*

2. The chart included in S4.1(g)(3) is amended so that the dimension for hip breadth (sitting) for the 95th percentile adult male reads as follows:

S4.1(g) * * *

(3) * * *

Hip breadth (sitting)....12.8 in. ....16.5 in.

3. The last sentence of S4.1(k) is revised to read as follows:

*　　*　　*　　*　　*

S4.1 * * *

(k) *Installation instructions.* * * * The installation instructions shall state whether the assembly is for universal installation or for installation only in specifically stated motor vehicles, and shall include at least those items specified in SAE Recommended Practice J800c, "Motor Vehicle Seat Belt Installations," November 1973.

* * * * *

4. The second sentence of S4.3(a)(1) is revised to read as follows:

* * * * *

S4.3 * * *

(a) *Corrosion resistance.* (1) * * * Alternatively, such hardware at or near the floor shall be protected against corrosion by at least an electrodeposited coating of nickel, or copper and nickel with at least a service condition number of SC2, and other attachment hardware shall be protected by an electrodeposited coating of nickel, or copper and nickel with a service condition number of SC1, in accordance with American Society for Testing and Materials B456-79, "Standard Specification for Electrodeposited Coatings of Copper Plus Nickel Plus Chromium and Nickel Plus Chromium," but such hardware shall not be racked for electroplating in locations subjected to maximum stress.

* * * * *

5. The first sentence of S5.1(b) is revised to read as follows:

* * * * *

S5.1 * * *

(b) *Breaking strength.* Webbing from three seat belt assemblies shall be conditioned in accordance with paragraph (a) of this section and tested for breaking strength in a testing machine of capacity verified to have an error of not more than one percent in the range of the breaking strength of the webbing in accordance with American Society for Testing and Materials E4-79, "Standard Methods of Load Verification of Testing Machines."

* * * * *

6. The first sentence of S5.1(f) is revised to read as follows:

* * * * *

S5.1 * * *

(f) *Resistance to microorganisms.* Webbing at least 20 inches or 50 centimeters in length from three seat belt assemblies shall first be preconditioned in accordance with Appendix A(1) and (2) of American Association of Textile Chemists and Col-

orists Test Method 30-81, 'Fungicides Evaluation on Textiles; Mildew and Rot Resistance of Textiles," and then subjected to Test I, "Soil Burial Test" of that test method.

* * * * *

7. Paragraph (g) of S5.1 is revised to read as follows:

* * * * *

S5.1 * * *

(g) *Colorfastness to crocking.* Webbing from three seat belt assemblies shall be tested by the procedure specified in American Association of Textile Chemists and Colorists Standard Test Method 8-181, "Colorfastness to Crocking: AATCC Crockmeter Method."

* * * * *

8. Paragraph (h) of S5.1 is revised to read as follows:

* * * * *

S5.1 * * *

(h) *Colorfastness to staining.* Webbing from three seat belt assemblies shall be tested by the procedure specified in American Association of Textile Chemists and Colorists (AATCC) Standard Test Method 107-1981, "Colorfastness to Water," except that the testing shall use (1) distilled water, (2) the AATCC perspiration tester, (3) a drying time of four hours, specified in section 7.4 of the AATCC procedure, and (4) section 9 of the AATCC test procedures to determine the colorfastness to staining on the AATCC Chromatic Transference Scale.

* * * * *

9. The first sentence of S5.2(a) is revised and a new sentence is added after the first sentence so that the two sentences read as follows:

S5.2 Hardware.—

(a) *Corrosion Resistance.* Three seat belt assemblies shall be tested in accordance with American Society for Testing and Materials B117-73, "Standard Method of Salt Spray (Fog) Testing." Any surface coating or material not intended for permanent retention on the metal parts during service life shall be removed prior to preparation of the test specimens for testing. * * *

* * * * *

10. The first sentence of S5.2(b) is revised to read as follows:

S5.2 *Hardware.*

* * * * *

(b) *Temperature resistance.* Three seat belt assemblies having plastic or nonmetallic hardware

PART 571; S209-PRE 28

or having retractors shall be subjected to the conditions prescribed in Procedure D of American Society for Testing and Materials D756-78, "Standard Practice for Determination of Weight and Shape Changes of Plastics under Accelerated Service Conditions." * * *

* * * * *

11. The eighth sentence of S5.2(k) is revised to read as follows:

* * * * *

S5.2 * * *

(k) * * * Then, the retractor and webbing shall be subjected to dust in a chamber similar to one illustrated in Figure 8 containing about 2 pounds or 0.9 kilogram of coarse grade dust conforming to the specification given in Society of Automotive Engineering Recommended Practice J726, "Air Cleaner Test Code" Sept. 1979. * * *

In § 571.5, paragraph (b)(5) is redesignated (b)(6) and a new paragraph (b)(5) is added to read as follows:

§ 571.5 *Matter incorporated by reference.*

* * * * *

(b) * * *

(5) *Test methods of the American Association of Textile Chemists and Colorists.* They are published by the American Association of Textile Chemists and Colorists. Information and copies can be obtained by writing to: American Association of Textile Chemists and Colorists, Post Office Box 886, Durham, NC.

(6) * * *

Issued on June 22, 1983

Diane K. Steed,
Acting Administrator.

48 F.R. 30138
June 30, 1983

PREAMBLE TO AN AMENDMENT TO
FEDERAL MOTOR VEHICLE SAFETY STANDARD NO. 209

Seat Belt Assemblies
[Docket No. 80-06; Notice 3]

ACTION: Final rule.

SUMMARY: This notice amends Safety Standard No. 209, *Seat Belt Assemblies*, to alter the test procedure specified under the "resistance to light" requirements of the standard. This amendment is intended to establish an equivalent strength test for both nylon and polyester webbing materials used in seat belt assemblies. This amendment changes the test apparatus for polyester fibers by replacing the currently specified "Corex D" filter with a chemically strengthened or tempered soda-lime glass filter. The "Corex D" filter would still be utilized in testing nylon webbing, since it offers the best correlation with actual outdoor results when dealing with nylon webbing material.

EFFECTIVE DATE: September 18, 1985.

SUPPLEMENTARY INFORMATION: Under Safety Standard No. 209, *Seat Belt Assemblies* (49 CFR 571.209), seat belts must pass a "resistance to light" test (paragraph S4.2(e)). This test measures the strength and durability of the seat belt webbing material after exposure to sunlight. The "resistance to light" test represents an accelerated determination of outdoor exposure or aging. A rapid form of testing is needed so that webbing may be certified in accordance with Standard No. 209 and automotive companies' specifications prior to shipment.

On May 1, 1980, a Notice of Proposed Rulemaking (45 FR 29102) was issued, proposing an amendment to the procedure to be used in "resistance to light" tests. The original standard called for a "Corex D" filter in testing webbing material. The "Corex D" filter was an adequate test appa-

ratus prior to the introduction of polyester webbing material for seat belts. Research had shown that although the specified test apparatus of a carbon arc light source combined with a "Corex D" filter, in general, was an effective method of simulating the effects of sunlight, it did result in the emission of certain radiations that were unrepresentative of the actual effects of natural sunlight. These peculiar radiations, which destroyed polyester but not nylon fibers, made the "Corex D" test procedure inappropriate for measuring the "resistance to light" requirements of seat belts containing polyester webbing material.

The proposed procedure replaced the required "Corex D" filter with a plain soda-lime glass filter in an attempt to create a similar, adequate testing for both nylon and polyester webbing material used in seat belt assemblies. Responses to that notice indicated that the proposed plain soda-lime glass filters were cracking either during the test cycle, due to the intense heat emitted during the 100 hours of test time, or after the test period, during the cool down of the equipment.

The Narrow Fabrics Institute, Inc. requested a delay in the rulemaking process in order to locate a less heat sensitive substitute. On September 16, 1980, the agency informed the Narrow Fabrics Institute, Inc. that the rulemaking process would be delayed until the development of a filter more resistant to thermal shock.

Upon completion of a 2-year search and a 1-year period of evaluation, the Narrow Fabrics Institute submitted a revised test apparatus. The improved filter was a chemically strengthened or tempered soda-lime glass. Testing done by the agency under Contract No. DTNH-22-83-P-02016 confirmed that the new filter maintained the same light transmittance characteristics of the untreated soda-lime glass filter originally proposed, but was free of the previous thermal shock problems. The treated soda-lime glass filter produces an excellent correlation with actual outdoor results, for the proper accelerated degradation of polyester webbing, without the prior breakage difficulties.

A careful evaluation of data compiled over the past few years demonstrates that as to nylon webbing material, the "Corex D" filter still affords the best correlation with actual outdoor results. In light of these various findings, the agency proposed on November 28, 1983 (48 FR 53583) to amend the test procedure to reflect these results.

Four of the five commenters to the docket supported the proposed amendment to Standard No. 209. The other commenter, Renault, made two objections. First, it argued that the carbon arc light used in Standard No. 209 is unrepresentative of real use conditions. It urges the use of an xenon lamp. As stated previously, the use of the carbon arc light with the appropriate filters produces excellent correlation with actual outdoors test of the resistance to light capability of seat belts. The agency, therefore, does not believe it is necessary to propose an amendment to allow the use of an xenon lamp.

Renault also said that Standard No. 209 should not use different test procedures for different materials. It recommended that the agency not require the use of different filters, but instead specify the transmission band and spectral distribution of the radiation used in the test. Finally, Renault said that if the agency decides to require a filter, it should provide a more specific definition of the filter to be used in the testing. In particular, Renault asked that the agency specify the wave length of the light being used.

The agency disagrees with Renault concerning the use of different filters in the resistance to light test. The carbon arc test equipment used in the resistance to light test is a well-established test procedure that has been long used by the motor vehicle and seat belt industries. Tests conducted by the Narrow Fabrics Institute show that the carbon arc test equipment, when used with the appropriate filters, produces results comparable to actual outdoor resistance to light tests. Although the agency has decided to retain the use of the filters, it agrees with Renault that the specific characteristics of the new soda-lime filter need to be

more precisely defined. The agency has obtained information on the transmittance of chemically strengthened soda-lime glass from the principal manufacturer of that device. Based on that information, the agency is amending the standard to specify the transmittance of the soda-lime glass to be used in the resistance to light test of polyester belts.

Update References

In the November 1983 notice, the agency proposed to update one of the American Society for Testing and Materials recommended practices incorporated by reference in the standard. The proposal to incorporate ASTM G23-81 was not opposed by the commenters and is therefore adopted.

PART 571—[AMENDED]

In consideration of the foregoing, paragraph S5.1(e) of Safety Standard No. 209, *Seat Belt Assemblies* (49 CFR 571.209), is amended by revising paragraph (e) to read as follows:

§571.209 Standard No. 209; seat belt assemblies.

S5.1 * * *

(e) *Resistance to Light.* Webbing at least 20 inches or 50 centimeters in length from three seat belt assemblies shall be suspended vertically on the inside of the specimen rack in a Type E carbon-arc light-exposure apparatus described in Standard Practice for Operating Light-Exposure Apparatus (Carbon-Arc Type) With and Without Water for Exposure of Nonmetallic Materials, ASTM Designation: G23-81, published by the American Society for Testing and Materials, except that the filter used for 100 percent polyester yarns shall be chemically strengthened soda-lime glass with a transmittance of less than 5 percent for wave lengths equal to or less than 305 nanometers and 90 percent or greater transmittance for wave lengths of 375 to 800 nanameters. The apparatus shall be operated without water spray at an air temperature of 60 ± 2 degrees Celsius or 140 ± 3.6 degrees Fahrenheit measured at a point 1.0 ± 0.2 inch or 25 ± 5 millimeters outside the specimen rack and midway in height. The temperature sensing element shall be shielded from radiation. The specimens shall be exposed to light from the carbon-arc for 100 hours and then conditioned as prescribed in paragraph (a) of this section. The colorfastness of the exposed and conditioned specimens shall be determined on the Geometric Gray

Scale issued by the American Association of Textile Chemists and Colorists. The breaking strength of the specimens shall be determined by the procedure prescribed in paragraph (b) of this section. The median values for the breaking strengths determined on exposed and unexposed specimens shall be used to calculate the percentage of breaking strength retained.

Issued on August 31, 1984.

Diane K. Steed
Administrator

49 FR 36507
September 18, 1984

MOTOR VEHICLE SAFETY STANDARD NO. 209

Seat Belt Assemblies

(Docket No. 69-23)

S1. Purpose and Scope.

This standard specifies requirements for seat belt assemblies.

S2. Application.

This standard applies to seat belt assemblies for use in passenger cars, multipurpose passenger vehicles, trucks, and buses.

S3. Definitions.

"Seat belt assembly" means any strap, webbing, or similar device designed to secure a person in a motor vehicle in order to mitigate the results of any accident, including all necessary buckles and other fasteners, and all hardware designed for installing such seat belt assembly in a motor vehicle.

"Pelvic restraint" means a seat belt assembly or portion thereof intended to restrain movement of the pelvis.

"Upper torso restraint" means a portion of a seat belt assembly intended to restrain movement of the chest and shoulder regions.

"Hardware" means any metal or rigid plastic part of a seat belt assembly.

"Buckle" means a quick release connector which fastens a person in a seat belt assembly.

"Attachment hardware" means any or all hardware designed for securing the webbing of a seat belt assembly to a motor vehicle.

"Adjustment hardware" means any or all hardware designed for adjusting the size of a seat belt assembly to fit the user, including such hardware that may be integral with a buckle, attachment hardware, or retractor.

"Retractor" means a device for storing part or all of the webbing in a seat belt assembly.

"Nonlocking retractor" means a retractor from which the webbing is extended to essentially its full length by a small external force, which provides no adjustment for assembly length, and which may or may not be capable of sustaining restraint forces at maximum webbing extension.

"Automatic-locking retractor" means a retractor incorporating adjustment hardware by means of a positive self-locking mechanism which is capable when locked of withstanding restraint forces.

"Emergency-locking retractor" means a retractor incorporating adjustment hardware by means of a locking mechanism that is activated by vehicle acceleration, webbing movement relative to the vehicle, or other automatic action during an emergency and is capable when locked of withstanding restraint forces.

"Seat back retainer" means the portion of some seat belt assemblies designed to restrict forward movement of a seat back.

"Webbing" means a narrow fabric woven with continuous filling yarns and finished selvages.

"Strap" means a narrow non-woven material used in a seat belt assembly in place of webbing.

"Type 1 seat belt assembly" is a lap belt for pelvic restraint.

"Type 2 seat belt assembly" is a combination of pelvic and upper-torso restraints.

"Type 2a shoulder belt" is an upper-torso restraint for use only in conjunction with a lap belt as a Type 2 seat belt assembly.

"Load-limiter" means a seat belt assembly component or feature that controls tension on the seat belt to modulate the forces that are imparted to occupants restrained by the belt assembly during a crash.

S4. Requirements.

S4.1 (a) *Single occupancy.* A seat belt assembly shall be designed for use by one, and only one, person at any one time.

(b) *Pelvic restraint.* A seat belt assembly shall provide pelvic restraint whether or not upper torso restraint is provided, and the pelvic restraint shall be designed to remain on the pelvis under all conditions, including collision or roll-over of the motor vehicle. Pelvic restraint of a Type 2 seat belt assembly that can be used without upper torso restraint shall comply with requirements for Type 1 seat belt assembly in S4.1 to S4.4.

(c) *Upper torso restraint.* A Type 2 seat belt assembly shall provide upper-torso restraint without shifting the pelvic restraint into the abdominal region. An upper-torso restraint shall be designed to minimize vertical forces on the shoulders and spine. Hardware for upper-torso restraint shall be so designed and located in the seat belt assembly that the possibility of injury to the occupant is minimized.

A Type 2a shoulder belt shall comply with applicable requirements for a Type 2 seat belt assembly in S4.1 to S4.4, inclusive.

(d) *Hardware.* All hardware parts which contact under normal usage a person, clothing, or webbing shall be free from burrs and sharp edges.

(e) *Release.* A Type 1 or Type 2 seat belt assembly shall be provided with a buckle or buckles readily accessible to the occupant to permit his easy and rapid removal from the assembly. Buckle release mechanism shall be designed to minimize the possibility of accidental release. A buckle with release mechanism in the latched position shall have only one opening in which the tongue can be inserted on the end of the buckle designed to receive and latch the tongue.

(f) *Attachment hardware.* [A seat belt assembly shall include all hardware necessary for installation in a motor vehicle in accordance with Society of Automotive Engineers Recommended Practice J800c, "Motor Vehicle Seat Belt Installation," Novemmber 1973. (48 F.R. 30138—June 30, 1983. Effective: July 30, 1983)] However, seat belt assemblies designed for installation in motor vehicles equipped with seat belt assembly anchorages that do not require anchorage nuts, plates, or washers, need not have such hardware, but shall have 7/16-20 UNF-2A or 1/2-13 UNC-2A attachment bolts or equivalent hardware. The hardware shall be designed to prevent attachment bolts and other parts from becoming disengaged from the vehicle while in service. Reinforcing plates or washers furnished for universal floor installations shall be of steel, free from burrs and sharp edges on the peripheral edges adjacent to the vehicle, at least 0.06 inch in thickness and at

least 4 square inches in projected area. The distance between any edge of the plate and the edge of the bolt hole shall be at least 0.6 inch. Any corner shall be rounded to a radius of not less than 0.25 inch or cut so that no corner angle is less than 135° and no side is less than 0.25 inch in length.

(g) *Adjustment.*

(1) A Type 1 or Type 2 seat belt assembly shall be capable of adjustment to fit occupants whose dimensions and weight range from those of a 5th-percentile adult female to those of a 95th-percentile adult male. The seat belt assembly shall have either an automatic-locking retractor, an emergency-locking retractor, or an adjusting device that is within the reach of the occupant.

(2) A Type 1 or Type 2 seat belt assembly for use in a vehicle having seats that are adjustable shall conform to the requirements of S4.1(g) (1) regardless of seat position. However, if a seat has a back that is separately adjustable, the requirements of S4.1(g) (1) need be met only with the seat back in the manufacturer's nominal design riding position.

(3) The adult occupants referred to in S4.1(g) (1) shall have the following measurements:

	5th-percentile adult female	95th-percentile adult male
Weight	102 pounds	215 pounds.
Erect sitting height	30.9 inches	38 inches.
Hip breadth (sitting)	12.8 inches	16.5 inches.
Hip circumference (sitting)	36.4 inches	47.2 inches.
Waist circumference (sitting)	23.6 inches	42.5 inches.
Chest depth	7.5 inches	10.5 inches.
Chest circumference: (nipple)	30.5 inches	
(upper)	29.8 inches	} 44.5 inches.
(lower)	26.6 inches	

(h) *Webbing.* The ends of webbing in a seat belt assembly shall be protected or treated to prevent raveling. The end of webbing in a seat belt assembly having a metal-to-metal buckle that is used by the occupant to adjust the size of the assembly shall not pull out of the adjustment hardware at maximum size adjustment. Provision shall be made for essentially unimpeded movement of webbing routed between a seat back and seat cushion and attached to a retractor located behind the seat.

(i) *Strap.* A strap used in a seat belt assembly to sustain restraint forces shall comply with the requirements for webbing in S4.2, and if the strap is made from a rigid material, it shall comply with applicable requirements in S4.2, S4.3 and S4.4.

(j) *Marking.* Each seat belt assembly shall be permanently and legibly marked or labeled with year of manufacture, model, and name or trademark of manufacturer or distributor, or of importer if manufactured outside the United States. A model shall consist of a single combination of webbing having a specific type of fiber weave and construction, and hardware having a specific design. Webbings of various colors may be included under the same model, but webbing of each color shall comply with the requirements for webbing in S4.2.

(k) *Installation instructions.* A seat belt assembly or retractor shall be accompanied by an instruction sheet providing sufficient information for installing the assembly in a motor vehicle except for a seat belt assembly installed in a motor vehicle by an automobile manufacturer. [The installation instructions shall state whether the assembly is for universal installation or for installation only in specifically stated motor vehicles, and shall include at least those items specified in SAE Recommended Practice J800c, "Motor Vehicle Seat Belt Installations," November 1973. (48 F.R. 30138—June 30, 1983. Effective: July 30, 1983)]

(l) *Usage and maintenance instructions.* A seat belt assembly or retractor shall be accompanied by written instructions for the proper use of the assembly, stressing particularly the importance of wearing the assembly snugly and properly located on the body, and on the maintenance of the assembly and periodic inspection of all components. The instructions shall show the proper manner of threading webbing in the hardware of seat belt assemblies in which the webbing is not permanently fastened. Instructions for a non-locking retractor shall include a caution that the webbing must be fully extended from the retractor during use of the seat belt assembly unless the retractor is attached to the free end of webbing which is not subjected to any tension during restraint of an occupant by the assembly. Instructions for Type 2a shoulder belt shall include a warning that the shoulder belt is not to be used without a lap belt.

(m) *Workmanship.* Seat belt assemblies shall have good workmanship in accordance with good commercial practice.

S4.2 Requirements for webbing.

(a) *Width.* The width of the webbing in a seat belt assembly shall be not less than 1.8 inches, except for portions that do not touch a 95th-percentile adult male with the seat in any adjustment position and the seat back in the manufacturer's nominal design riding position when measured under the conditions prescribed in S5.1(a).

(b) *Breaking strength.* The webbing in a seat belt assembly shall have not less than the following breaking strength when tested by the procedures specified in S5.1(b): Type 1 seat belt assembly—6,000 pounds or 2,720 kilograms; Type 2 seat belt assembly—5,000 pounds or 2,270 kilograms for webbing in pelvic restraint and 4,000 pounds or 1,810 kilograms for webbing in upper-torso restraint.

(c) *Elongation.* Except as provided in S4.5, the webbing in a seat belt assembly shall not be extended to more than the following elongations when subjected to the specified forces in accordance with the procedure specified in S5.1(c): Type 1 seat belt assembly— 20 percent at 2,500 pounds or 1,130 kilograms; Type 2 seat belt assembly—30 percent at 2,500 pounds or 1,130 kilograms for webbing in pelvic restraint and 40 percent at 2,500 pounds or 1,130 kilograms for webbing in upper-torso restraint.

(d) *Resistance to abrasion.* The webbing of a seat belt assembly, after being subjected to abrasion as specified in S5.1(d) or S5.3(c), shall have a breaking strength of not less than 75 percent of the breaking strength listed in S4.2(b) for that type of belt assembly.

(e) *Resistance to light.* The webbing in a seat belt assembly after exposure to the light of a carbon arc and tested by the procedure specified in S5.1(e) shall have a breaking strength not less than 60 percent of the strength before exposure to the carbon arc and shall have a color retention not less than No. 2 on the Geometric Gray Scale published by the American Association of Textile Chemists and Colorists, Post Office Box 886, Durham, N.C.

(f) *Resistance to micro-organisms.* The webbing in a seat belt assembly after being subjected to micro-organisms and tested by the procedures specified in S5.1(f) shall have a breaking strength not less than 85 percent of the strength before subjection to micro-organisms.

(g) *Colorfastness to crocking.* The webbing in a seat belt assembly shall not transfer color to a crock cloth either wet or dry to a greater degree than class 3 on the AATCC Chart for Measuring Transference of Color published by the American Association of Textile Chemists and Colorists, when tested by the procedure specified in S5.1(g).

(h) *Colorfastness to staining.* The webbing in a seat belt assembly shall not stain to a greater degree than class 3 on the AATCC Chart for Measuring Transference of Color published by the American Association of Textile Chemists and Colorists, when tested by the procedure specified in S5.1(h).

S4.3 Requirements for hardware.

(a) *Corrosion resistance.*

(1) Attachment hardware of a seat belt assembly after being subjected to the conditions specified in S5.2(a) shall be free of ferrous corrosion on significant surfaces except for permissible ferrous corrosion at peripheral edges or edges of holes on underfloor reinforcing plates and washers. [Alternatively, such hardware at or near the floor shall be protected against corrosion by at least an electrodeposited coating of nickel, or copper and nickel with at least a service condition number of SC2, and other attachment hardware shall be protected by an electrodeposited coating of nickel, or copper and nickel with a service condition number of SC1, in accordance with American Society for Testing and Materials B456–79, "Standard Specification for Electrodeposited Coatings of Copper Plus Nickel Plus Chromium and Nickel Plus Chromium," but such hardware shall not be racked for electroplating in locations subjected to maximum stress. (48 F.R. 30138—June 30, 1983. Effective: July 30, 1983)]

(2) Surfaces of buckles, retractors and metallic parts, other than attachment hardware, of a seat belt assembly after subjection to the conditions specified in S5.2(a) shall be free of ferrous or nonferrous corrosion which may be transferred, either directly or by means of the webbing, to the occupant or his clothing when the assembly is worn. After test, buckles shall conform to applicable requirements in paragraphs (d) to (g) of this section.

(b) *Temperature resistance.* Plastic or other nonmetallic hardware parts of a seat belt assembly when subjected to the conditions specified in S5.2(b) shall not warp or otherwise deteriorate to cause the assembly to operate improperly or fail to comply with applicable requirements in this section and S4.4.

(c) *Attachment hardware.*

(1) Eye bolts, shoulder bolts, or other bolts used to secure the pelvic restraint of a seat belt assembly to a motor vehicle shall withstand a force of 9,000 pounds or 4,080 kilograms when tested by the procedure specified in S5.2(c)(1), except that attachment bolts of a seat belt assembly designed for installation in specific models of motor vehicles in which the ends of two or more seat belt assemblies can not be attached to the vehicle by a single bolt shall have a breaking strength of not less than 5,000 pounds or 2,270 kilograms.

(2) Other attachment hardware designed to receive the ends of two seat belt assemblies shall withstand a tensile force of at least 6,000 pounds or 2,720 kilograms without fracture of any section when tested by the procedure specified in S5.2(c)(2).

(3) A seat belt assembly having single attachment hooks of the quick-disconnect type for connecting webbing to an eye bolt shall be provided with a retaining latch or keeper which shall not move more than 0.08 inch or 2 millimeters in either the vertical or horizontal direction when tested by the procedure specified in S5.2(c)(3).

(d) *Buckle release.*

(1) The buckle of a Type 1 or Type 2 seat belt assembly shall release when a force of not more than 30 pounds or 14 kilograms is applied.

(2) A buckle designed for pushbutton application of buckle release force shall have a minimum area of 0.7 square inch or 4.5 square centimeters with a minimum linear dimension of 0.4 inch or 10 millimeters for applying the release force, or a buckle designed for lever application of a buckle release force shall permit the insertion of a cylinder 0.4 inch or 10 millimeters in diameter and 1.5 inches or 38 millimeters in length to at least the midpoint of the cylinder along the cylinder's entire length in the actuation portion of the buckle release. A buckle having other design for release shall have adequate access for two or more fingers to actuate release.

(3) The buckle of a Type 1 or Type 2 seat belt assembly shall not release under a compressive force of 400 pounds applied as prescribed in paragraph S5.2(d)(3). The buckle shall be operable and shall meet the applicable requirements of paragraph S4.4 after the compressive force has been removed.

(e) *Adjustment force.* The force required to decrease the size of a seat belt assembly shall not exceed 11 pounds or 5 kilograms when measured by the procedure specified in S5.2(e).

(f) *Tilt-lock adjustment.* The buckle of a seat belt assembly having tilt-lock adjustment shall lock the webbing when tested by the procedure specified in S5.2(f) at an angle of not less than 30 degrees between the base of the buckle and the anchor webbing.

(g) *Buckle latch.* The buckle latch of a seat belt assembly when tested by the procedure specified in S5.2(g) shall not fail, nor gall or wear to an extent that normal latching and unlatching is impaired, and a metal-to-metal buckle shall separate when in any position of partial engagement by a force of not more than 5 pounds or 2.3 kilograms.

(h) *Nonlocking retractor.* The webbing of a seat belt assembly shall extend from a nonlocking retractor within 0.25 inch or 6 millimeters of maximum length when a tension is applied as prescribed in S5.2(h). A nonlocking retractor on upper-torso restraint shall be attached to the nonadjustable end of the assembly, the reel of the retractor shall be easily visible to an occupant while wearing the assembly, and the maximum retraction force shall not exceed 1.1 pounds or 0.5 kilogram in any strap or webbing that contacts the shoulder when measured by the procedure specified in S5.2(h), unless the retractor is attached to the free end of webbing which is not subjected to any tension during restraint of an occupant by the assembly.

(i) *Automatic-locking retractor.* The webbing of a seat belt assembly equipped with an automatic-locking retractor, when tested by the procedure specified in S5.2(i), shall not move more than 1 inch or 25 millimeters between locking positions of the retractor, and shall be retracted with a force under zero acceleration of not less than 0.6 pound or 0.27 kilogram when attached to pelvic restraint, and not less than 0.45 pound or 0.2 kilogram nor more than 1.1 pounds or 0.5 kilogram in any strap or webbing that contacts the shoulder of an occupant when the retractor is attached to upper-torso restraint. An automatic-locking retractor attached to upper-torso restraint shall not increase the restraint on the occupant of the seat belt assembly during use in a vehicle traveling over rough roads as prescribed in S5.2(i).

(j) *Emergency-locking retractor.* An emergency-locking retractor of a Type 1 or Type 2 seat belt assembly, when tested in accordance with the procedures specified in paragraph S5.2(j)—

(1) Shall lock before the webbing extends 1 inch when the retractor is subjected to an acceleration of 0.7g;

(2) Shall not lock, if the retractor is sensitive to webbing withdrawal, before the webbing extends 2 inches when the retractor is subjected to an acceleration of 0.3g or less;

(3) Shall not lock, if the retractor is sensitive to vehicle acceleration, when the retractor is rotated in any direction to any angle of 15° or less from its orientation in the vehicle;

(4) Shall exert a retroactive force of at least 0.6 pound under zero acceleration when attached only to the pelvic restraint;

(5) Shall exert a retractive force of not less than 0.2 pound and not more than 1.1 pounds under zero acceleration when attached only to an upper-torso restraint;

(6) Shall exert a retractive force of not less than 0.2 pound and not more than 1.5 pounds under zero acceleration when attached to a strap or webbing that restrains both the upper torso and the pelvis.

(k) *Performance of retractor.* A retractor used on a seat belt assembly after subjection to the tests specified in S5.2(k) shall comply with applicable requirements in paragraphs (h) to (j) of this section and S4.4, except that the retraction force shall be not less than 50 percent of its original retraction force.

S4.4 Requirements for assembly performance.

(a) *Type 1 seat belt assembly.* Except as provided in S4.5, the complete seat belt assembly including webbing, straps, buckles, adjustment and attachment hardware, and retractors shall comply with the following requirements when tested by the procedures specified in S5.3(a):

(1) The assembly loop shall withstand a force of not less than 5,000 pounds or 2,270 kilograms; that is, each structural component of the assembly shall withstand a force of not less than 2,500 pounds or 1,130 kilograms.

(2) The assembly loop shall extend not more than 7 inches or 18 centimeters when subjected to a force of 5,000 pounds or 2,270 kilograms; that is, the length of the assembly between anchorages shall not increase more than 14 inches or 36 centimeters.

(3) Any webbing cut by the hardware during test shall have a breaking strength at the cut of not less than 4,200 pounds or 1,910 kilograms.

(4) Complete fracture through any solid section of metal attachment hardware shall not occur during test.

(b) *Type 2 seat belt assembly.* Except as provided in S4.5, the components of a Type 2 seat belt assembly including webbing, straps, buckles, adjustment and attachment hardware, and retractors shall comply with the following requirements when tested by the procedure specified in S5.3(b):

(1) The structural components in the pelvic restraint shall withstand a force of not less than 2,500 pounds or 1,139 kilograms.

(2) The structural components in the upper torso restraint shall withstand a force of not less than 1,500 pounds or 680 kilograms.

(3) The structural components in the assembly that are common to pelvic and upper torso restraints shall withstand a force of not less than 3,000 pounds or 1,360 kilograms.

(4) The length of the pelvic restraint between anchorages shall not increase more than 20 inches or 50 centimeters when subjected to a force of 2,500 pounds or 1,130 kilograms.

(5) The length of the upper torso restraint between anchorages shall not increase more than 20 inches or 50 centimeters when subjected to a force of 1,500 pounds or 680 kilograms.

(6) Any webbing cut by the hardware during test shall have a breaking strength of not less than 3,500 pounds or 1,590 kilograms at a cut in webbing of the pelvic restraint, or not less than 2,800 pounds or 1,270 kilograms at a cut in webbing of the upper-torso restraint.

(7) Complete fracture through any solid section of metal attachment hardware shall not occur during test.

S4.5 Load-limiter.

(a) A Type 1 or Type 2 seat belt assembly that includes a load-limiter is not required to comply with the elongation requirements of S4.2(c), S4.4(a) (2), S4.4(b) (4) or S4.4(b) (5).

(b) A Type 1 or Type 2 seat belt assembly that includes a load-limiter and that does not comply with the elongation requirements of this standard may be installed in motor vehicles only in conjunction with an automatic restraint system as part of a total occupant restraint system.

(c) In addition to the marking requirements specified in S4.1(k), a Type 1 or Type 2 seat belt assembly that includes a load-limiter and that does not comply with the elongation requirements of this standard shall be permanently and legibly marked or labeled with the following words:

"This seat belt assembly may only be installed in vehicles in combination with an automatic restraint system such as an air cushion or an automatic belt."

S4.6 Manual belts subject to crash protection requirements of Standard No. 208.

(a) A seat belt assembly subject to the requirements of S4.6.1 of Standard No. 208 (49 CFR Part 571.208) does not have to meet the requirements of S4.2 (a)-(c) and S4.4 of this standard.

[(b) A seat belt assembly that meets the requirements of 4.6.1 of Standard No. 208 (§ 571.208) shall be permanently and legibly marked or labeled with the following statement:

"This dynamically-tested seat belt assembly is for use only in (insert specific seating position(s), e.g., "front right") in (insert specific vehicle make(s), and model(s))." (51 F.R. 31765—September 5, 1986. Effective: September 5, 1986)]

This seat belt assembly may only be installed at a front outboard designated seating position of a vehicle with a gross vehicle weight rating of 10,000 pounds or less.

S5. Demonstration Procedures.

S5.1 Webbing.

(a) *Width.* The width of webbing from three seat belt assemblies shall be measured after conditioning for at least 24 hours in an atmosphere having relative humidity between 48 and 67 percent and a temperature of $23° \pm 2°$ or $73.4° \pm 3.6°$. The tension during measurement of width shall be not more than 5 pounds or 2 kilograms on webbing from a Type 1 or Type 3 seat belt assembly, and $2,200 \pm 100$ pounds or $1,000 \pm 50$ kilograms on webbing from a Type 2 seat belt assembly. The width of webbing from a Type 2 seat belt assembly may be measured during the breaking strength test described in paragraph (b) of this section.

(b) *Breaking strength.* Webbing from three seat belt assemblies shall be conditioned in accordance with paragraph (a) of this section and tested for breaking strength in a testing machine of capacity verified to have an error of not more than one percent in the range of the breaking strength of the webbing in accordance with American Society for Testing and Materials E4–79, "Standard Methods of Load Verification of Testing Machines."

(c) *Elongation.* Elongation shall be measured during the breaking strength test described in paragraph (b) of this section by the following procedure: A preload between 44 and 55 pounds or 20 and 25 kilograms shall be placed on the webbing mounted in the grips of the testing machine and the needle points of an extensometer, in which the points remain parallel during test, are inserted in the center of the specimen. Initially the points shall be set at a known distance apart between 4 and 8 inches or 10 and 20 centimeters. When the force on the webbing reaches the value specified in S4.2(c), the increase in separation of the points of the extensometer shall be measured and the percent elongation shall be calculated to the nearest 0.5 percent. Each value shall be not more than the appropriate elongation requirement in S4.2(c).

(d) *Resistance to abrasion.* The webbing from three seat belt assemblies shall be tested for resistance to abrasion by rubbing over the hexagon bar prescribed in Figure 2 in the following manner:

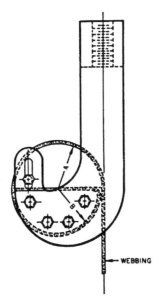

WEBBING

A 1 TO 2 INCHES OR 2.5 TO 5 CENTIMETERS
B A MINUS 0.06 INCH 0.15 CENTIMETER

FIGURE 1

The machine shall be equipped with split drum grips illustrated in Figure 1, having a diameter between 2 and 4 inches or 5 and 10 centimeters. The rate of grip separation shall be between 2 and 4 inches per minute or 5 and 10 centimeters per minute. The distance between the centers of the grips at the start of the test shall be between 4 and 10 inches or 10 and 25 centimeters. After placing the specimen in the grips, the webbing shall be stretched continuously at a uniform rate to failure. Each value shall be not less than the applicable breaking strength requirement in S4.2(b), but the median value shall be used for determining the retention of breaking strength in paragraphs (d), (e), and (f) of this section.

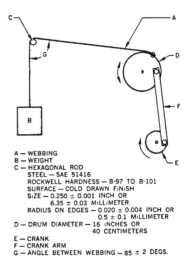

A — WEBBING
B — WEIGHT
C — HEXAGONAL ROD
    STEEL — SAE 51416
    ROCKWELL HARDNESS — B-97 TO B-101
    SURFACE — COLD DRAWN FINISH
    SIZE — 0.250 ± 0.001 INCH OR
         6.35 ± 0.03 MILLIMETER
    RADIUS ON EDGES — 0.020 ± 0.004 INCH OR
         0.5 ± 0.1 MILLIMETER
D — DRUM DIAMETER — 16 INCHES OR
         40 CENTIMETERS
E — CRANK
F — CRANK ARM
G — ANGLE BETWEEN WEBBING — 85 ± 2 DEGS.

FIGURE 2

The webbing shall be mounted in the apparatus shown schematically in Figure 2. One end of the webbing (A) shall be attached to a weight (B) which has a mass of 5.2 ± 0.1 pounds or 2.35 ± 0.05 kilograms, except that a mass of 3.3 ± 0.1 pounds or 1.50 ± 0.05 kilograms shall be used for webbing in pelvic and upper-torso restraints of a belt assembly used in a child restraint system. The webbing shall be passed over the two new abrading edges of the hexagon bar (C) and the other end attached to an oscillating drum (D) which has a stroke of 13 inches or 33 centimeters. Suitable guides shall be used to prevent movement of the webbing along the axis of hexagonal bar C. Drum D shall be oscillated for 5,000 strokes or 2,500 cycles at a rate of 60 ± 2 strokes per minute or 30 ± 1 cycles per minute. The abraded webbing shall be conditioned as prescribed in paragraph (a) of this section and tested for breaking strength by the procedure described in paragraph (b) of this section. The median values for the breaking strengths determined on abraded and unabraded specimens shall be used to calculate the percentage of braking strength retained.

(e) *Resistance to light.* [Webbing at least 20 inches or 50 centimeters in length from three seat belt assemblies shall be suspended vertically on the inside of the specimen rack in a Type E carbon-arc light-exposure apparatus described in Standard Practice for Operating Light-Exposure Apparatus (Carbon-Arc Type) With and Without Water for Exposure of Nonmetallic Materials, ASTM Designation: G23–81, published by the American Society for Testing and Materials, except that the filter used for 100 percent polyester yarns shall be chemically strengthened soda-lime glass with a transmittance of less than 5 percent for wave lengths equal to or less than 305 nanometers and 90 percent or greater transmittance for wave lengths of 375 to 800 nanometers. The apparatus shall be operated without water spray at an air temperature of $60° \pm 2$ degrees Celsius or $140° \pm 3.6$ degrees Fahrenheit measured at a point 1.0 ± 0.2 inch or 25 ± 5 millimeters outside the specimen rack and midway in height. The temperature sensing element shall be shielded from radiation. The specimens shall be exposed to light from the carbon arc for 100 hours and then conditioned as prescribed in paragraph (a) of this section. The colorfastness of the exposed and conditioned specimens shall be determined on the Geometric Gray Scale issued by the American Association of Textile Chemists and Colorists. The breaking strength of the specimens shall be determined by the procedure prescribed in paragraph (b) of this section. The median values for the breaking strengths determined on exposed and unexposed specimens shall be used to calculate the percentage of breaking strength retained. (49 F.R. 36507—September 18, 1984. Effective: September 18, 1985)]

(f) *Resistance to micro-organisms.* Webbing at least 20 inches or 50 centimeters in length from three seat belt assemblies shall first be preconditioned in accordance with Appendix A(1) and (2) of American Association of Textile Chemists and Colorists Test Method 30–81, "Fungicides Evaluation on Textiles; Mildew and Rot Resistance of Textiles," and then subjected to Test I, "Soil Burial Test" of that test method. After soil-burial for a period of 2 weeks, the specimen shall be washed in water, dried and conditioned as prescribed in paragraph (a) of this section. The breaking strengths of the specimens shall be determined by the procedure prescribed in paragraph (b) of thissection. The median values for the breaking strengths determined on exposed and unexposed specimens shall be used to calculate the percentage of breaking strength retained.

NOTE.—This test shall not be required on webbing made from material which is inherently resistant to micro-organisms.

(g) *Colorfastness to crocking.* Webbing from three seat belt assemblies shall be tested by the procedure specified in American Association of Textile Chemists and Colorists Standard Test Method 8—181, "Colorfastness to Crocking: AATCC Crockmeter Method."

(h) *Colorfastness to staining.* Webbing from three seat belt assemblies shall be tested by the procedure specified in American Association of Textile Chemists and Colorists (AATCC) Standard Test Method 107–1981, "Colorfastness to Water," except that the testing shall use (1) distilled water, (2) the AATCC perspiration tester, (3) a drying time of four hours, specified in section 7.4 of the AATCC procedure, and (4) section 9 of the AATCC test procedures to determine the colorfastness to staining on the AATCC Chromatic Transference Scale.

S5.2 Hardware.

(a) *Corrosion resistance.* Three seat belt assemblies shall be tested in accordance with American Society for Testing and Materials

B117-73, "Standard Method of Salt Spray (Fog) Testing." Any surface coating or material not intended for permanent retention on the metal parts during service life shall be removed prior to preparation of the test specimens for testing. The period of test shall be 50 hours for all attachment hardware at or near the floor, consisting of two periods of 24 hours exposure to salt spray followed by 1 hour drying and 25 hours for all other hardware, consisting of one period of 24 hours exposure to salt spray followed by 1 hour drying. In the salt spray test chamber, the parts from the three assemblies shall be oriented differently, selecting those orientations most likely to develop corrosion on the larger areas. At the end of test, the seat belt assembly shall be washed thoroughly with water to remove the salt. After drying for at least 24 hours under standard laboratory conditions specified in S5.1(a) attachment hardware shall be examined for ferrous corrosion on significant surfaces, that is, all surfaces that can be contacted by a sphere 0.75 inch or 2 centimeters in diameter, and other hardware shall be examined for ferrous and nonferrous corrosion which may be transferred, either directly or by means of the webbing, to a person or his clothing during use of a seat belt assembly incorporating the hardware.

NOTE.—When attachment and other hardware are permanently fastened, by sewing or other means, to the same piece of webbing, separate assemblies shall be used to test the two types of hardware. The test for corrosion resistance shall not be required for attachment hardware made from corrosion-resistant steel containing at least 11.5 percent chromium or for attachment hardware protected with an electro-deposited coating of nickel, or copper and nickel, as prescribed in S4.3(a). The assembly that has been used to test the corrosion resistance of the buckle shall be used to measure adjustment force, tilt-lock adjustment, and buckle latch in paragraphs (e), (f) and (g), respectively, of this section, assembly performance in S5.3 and buckle release force in paragraph (d) of this section.

(b) *Temperature resistance.* Three seat belt assemblies having plastic or nonmetallic hardware or having retractors shall be subjected to the conditions prescribed in Procedure D of American Society for Testing and Materials D756-78, "Standard Practice for Determination of Weight and Shape Changes of Plastics under Accelerated Service Conditions." The dimension and weight measurement shall be omitted. Buckles shall be unlatched and retractors shall be fully retracted during conditioning. The hardware parts after conditioning shall be used for all applicable tests in S4.3 and S4.4.

(c) *Attachment hardware.*

(1) Attachment bolts used to secure the pelvic restraint of a seat belt assembly to a motor vehicle shall be tested in a manner similar to that shown in Figure 3. The load shall be applied at an angle of 45 degrees to the axis of the bolt through attachment hardware from the seat belt assembly, or through a special fixture which simulates the loading applied by the attachment hardware. The attachment hardware or simulated fixture shall be fastened by the bolt to the anchor-

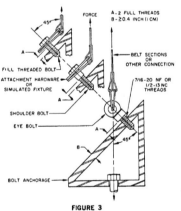

FIGURE 3

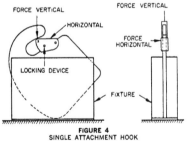

FIGURE 4
SINGLE ATTACHMENT HOOK

(Rev. 9/18/84)

age shown in Figure 3, which has a standard 7/16-20 UNF-2B or 1/2-13 UNC-2B threaded hole in a hardened steel plate at least 0.4 inch or 1 centimeter in thickness. The bolt shall be installed with two full threads exposed from the fully seated position. The appropriate force required by S4.3(c) shall be applied. A bolt from each of three seat belt assemblies shall be tested.

(2) Attachment hardware, other than bolts, designed to receive the ends of two seat belt assemblies shall be subjected to a tensile force of 6,000 pounds or 2,720 kilograms in a manner simulating use. The hardware shall be examined for fracture after the force is released. Attachment hardware from three seat belt assemblies shall be tested.

(3) Single attachment hook for connecting webbing to any eye bolt shall be tested in the following manner: The hook shall be held rigidly so that the retainer latch or keeper, with cotter pin. or other locking device in place, is in a horizontal position as shown in Figure 4. A force of 150 ± 2 pounds or 68 ± 1 kilograms shall be applied vertically as near as possible to the free end of the retainer latch, and the movement of the latch by this force at the point of application shall be measured. The vertical force shall be released, and a force of 150 ± 2 pounds or 68 ± 1 kilograms shall be applied horizontally as near as possible to the free end of the retainer latch. The movement of the latch by this force at the point of load application shall be measured. Alternatively, the hook may be held in other positions, provided the forces are applied and the movements of the latch are measured at the points indicated in Figure 4. A single attachment hook from each of three seat belt assemblies shall be tested.

(d) *Buckle release.*

(1) Three seat belt assemblies shall be tested to determine compliance with the maximum buckle release force requirements, following the assembly test in S5.3. After subjection to the force applicable for the assembly being tested, the force shall be reduced and maintained at 150 pounds on the assembly loop of a Type 1 seat belt assembly, 75 pounds on the components of a Type 2 seat belt assembly, or 45 pounds on a Type 3 seat belt assembly. The buckle release force shall be measured by applying a force on the buckle in a manner and direction typical of those which would be employed by a seat belt occupant. For pushbutton-release buckles, the

force shall be applied at least 0.125 inch from the edge of the push-button access opening of the buckle in a direction that produces maximum releasing effect. For lever-release buckles, the force shall be applied on the centerline of the buckle lever or finger tab in a direction that produces maximum releasing effect.

(2) The area for application of release force on pushbutton actuated buckle shall be measured to the nearest 0.05 square inch or 0.3 square centimeter. The cylinder specified in S4.3(d) shall be inserted in the actuation portion of a lever ·release buckle for determination of compliance with the requirement. A buckle with other release actuation shall be examined for access of release by fingers.

(3) The buckle of a Type 1 or Type 2 seat belt assembly shall be subjected to a compressive force of 400 pounds applied anywhere on a test line that is coincident with the centerline of the belt extended through the buckle or on any line that extends over the center of the release mechanism and intersects the extended centerline of the belt at an angle of 60°. The load shall be applied by using a curved cylindrical bar having a cross section diameter of 0.75 inch and a radius of curvature of 6 inches, placed with its longitudinal centerline along the test line and its center directly above the point on the buckle to which the load will be applied. The buckle shall be latched, and a tensile force of 75 pounds shall be applied to the connected webbing during the application of the compressive force. Buckles from three seat belt assemblies shall be tested to determine compliance with paragraph S4.3(d) (3).

(e) *Adjustment force.* Three seat belt assemblies shall be tested for adjustment force on the webbing at the buckle, or other manual adjusting device normally used to adjust the size of the assembly. With no load on the anchor end, the webbing shall be drawn through the adjusting device at a rate of 20 ± 2 inches per minute or 50 ± 5 centimeters per minute and the maximum force shall be measured to the nearest 0.25 pound or 0.1 kilogram after the first 1.0 inch or 25 millimeters of webbing movement. The webbing shall be precycled 10 times prior to measurement.

(f) *Tilt-lock adjustment.* This test shall be made on buckles or other manual adjusting devices having tilt-lock adjustment normally used to adjust the size of the assembly. Three buckles or devices shall be tested. The base of the adjustment mechanism and the anchor end of the webbing shall be oriented in planes normal to each other. The webbing shall be drawn through the adjustment mechanism in a direction to increase belt length at a rate of 20 ± 2 inches per minute or 50 ± 5 centimeters per minute while the plane of the base is slowly rotated in a direction to lock the webbing. Rotation shall be stopped when the webbing locks, but the pull on the webbing shall be continued until there is a resistance of at least 20 pounds or 9 kilograms. The locking angle between the anchor end of the webbing and the base of the adjustment mechanism shall be measured to the nearest degree. The webbing shall be precycled 10 times prior to measurement.

(g) *Buckle latch.* The buckles from three seat belt assemblies shall be opened fully and closed at least 10 times. [Then the buckles shall be clamped or firmly held against a flat surface so as to permit normal movement of buckle parts, but with the metal mating plate (metal-to-metal buckles) or webbing end (metal-to-webbing buckles) withdrawn from the buckle. (45 F.R. 29045—May 1, 1980. Effective: 5/1/80)] The release mechanism shall be moved 200 times through the maximum possible travel against its stop with a force of 30 ± 3 pounds or 14 ± 1 kilograms at a rate not to exceed 30 cycles per minute. The buckle shall be examined to determine compliance with the performance requirements of S4.3(g). A metal-to-metal buckle shall be examined to determine whether partial engagement is possible by means of any technique representative of actual use. If partial engagement is possible, the maximum force of separation when in such partial engagement shall be determined.

(h) *Nonlocking retractor.* After the retractor is cycled 10 times by full extension and retraction of the webbing, the retractor and webbing shall be suspended vertically and a force of 4 pounds or 1.8 kilograms shall be applied to extend the webbing from the retractor. The force shall be reduced to 3 pounds or 1.4 kilograms when attached to a pelvic restraint, or to 1.1 pounds or 0.5 kilogram per strap or webbing that contacts the shoulder of an occupant when retractor is attached to an upper-torso restraint. The residual extension of the webbing shall be measured by manual rotation of the retractor drum or by disengaging the retraction mechanism. Measurements shall be made on three retractors. The location of the retractor attached to upper-torso restraint shall be examined for visibility of reel during use of seat belt assembly in a vehicle.

NOTE.—This test shall not be required on a nonlocking retractor attached to the free-end of webbing which is not subjected to any tension during restraint of an occupant by the assembly.

(i) *Automatic-locking retractor.* Three retractors shall be tested in a manner to permit the retraction force to be determined exclusive of the gravitational forces on hardware or webbing being retracted. The webbing shall be fully extended from the retractor. While the webbing is being retracted, the average force of retraction within plus or minus 2 inches or 5 centimeters of 75 percent extension (25-percent retraction) shall be determined and the webbing movement between adjacent locking segments shall be measured in the same region of extension. A seat belt assembly with automatic locking retractor in upper torso restraint shall be tested in a vehicle in a manner prescribed by the installation and usage instructions. The retraction force on the occupant of the seat belt assembly shall be determined before and after traveling for 10 minutes at a speed of 15 miles per hour or 24 kilometers per hour or more over a rough road (e.g., Belgian block road) where the occupant is subjected to displacement with respect to the vehicle in both horizontal and vertical directions. Measurements shall be made with the vehicle stopped and the occupant in the normal seated position.

(j) *Emergency-locking retractor.* A retractor shall be tested in a manner that permits the retraction force to be determined exclusive of the gravitational forces on hardware or webbing being retracted. The webbing shall be fully extended from the retractor, passing over or through any hardware or other material specified in the installation instructions. While the webbing is being retracted, the lowest force of retraction within plus or minus 2 inches of 75 percent extension shall be determined. A retractor that is sensitive to webbing withdrawal shall be subjected to an acceleration of 0.3g within a period of 50 milliseconds while the webbing is at 75-percent extension, to determine compliance with S4.3(j) (2). The retractor shall be subjected to an acceleration of 0.7g within a period of 50 milliseconds, while the webbing is at 75-percent extension, and the webbing movement before locking shall be measured under the following conditions: For a retractor sensitive to webbing withdrawal, the retractor shall be accelerated in the direction of webbing retraction while the retractor drum's central axis is oriented horizontally and at angles of 45°, 90°, 135°, and 180° to the horizontal plane. For a retractor sensitive to vehicle acceleration, the retractor shall be—

(1) accelerated in the horizontal plane in two directions normal to each other, while the retractor drum's central axis is oriented at the angle at which it is installed in the vehicle; and,

(2) accelerated in three directions normal to each other while the retractor drum's central axis is oriented at angles of 45°, 90°, 135° and 180° from the angle at which it is installed in the vehicle, unless the retractor locks by gravitational force when tilted in any direction to any angle greater than 45° from the angle at which it is installed in the vehicle.

(k) *Performance of retractor.* After completion of the corrosion-resistance test described in paragraph (a) of this section, the webbing shall be fully extended and allowed to dry for at least 24 hours under standard laboratory conditions specified in S5.1(a). ▌Then, the retractor and webbing shall be subjected to dust in a chamber similar to one illustrated in Figure 8 containing about 2 pounds or 0.9 kilogram of coarse grade dust conforming to the specification given in Society of Automotive Engineering Recommended Practice J726, "Air Cleaner Test Code" Sept. 1979. (48 F.R. 30138—June 30, 1983. Effective: July 30, 1983)▌ The webbing shall be withdrawn manually and allowed to retract for 25 cycles. The retractor shall be mounted in an apparatus capable of extending the webbing fully, applying a force of 20 pounds or 9 kilograms at full extension, and allowing the webbing to retract freely and completely. The webbing shall be withdrawn from the retractor and allowed to retract repeatedly in this apparatus until 2,500 cycles are completed. The retractor and webbing shall then be subjected to the temperature resistance test prescribed in paragraph (b) of this section. The retractor shall be subjected to 2,500 additional cycles of webbing withdrawal and retraction. Then, the retractor and webbing shall be subjected to dust in a chamber similar to one illustrated in Figure 6 containing about 2 pounds or 0.9 kilogram of coarse grade dust conforming to the specification given in SAE Recommended Practice, Air Cleaner Test Code—SAE J726a, published by the Society of Automotive Engineers. The dust shall be agitated every 20 minutes for 5 seconds by compressed air, free of oil and moisture, at a gauge pressure of 80±8 pounds per square inch or 5.6±0.6 kilograms per square centimeter entering through an orifice 0.060±0.004 inch or 1.5±0.1 millimeters in diameter. The web-

bing shall be extended to the top of the chamber and kept exended at all times except that the webbing shall be subjected to 10 cycles of complete retraction and extension within 1 to 2 minutes after each agitation of the dust. At the end of 5 hours, the assembly shall be removed from the chamber. The webbing shall be fully withdrawn from the retractor manually and allowed to retract completely for 25 cycles. An automatic-locking retractor or a nonlocking retractor attached to pelvic restraint shall be subjected to 5,000 additional cycles of webbing withdrawal and retraction. An emergency-locking retractor or a nonlocking retractor attached to upper-torso restraint shall be subjected to 45,000 additional cycles of webbing withdrawal and retraction between 50 and 100 percent extension. The locking mechanism of an emergency-lock-

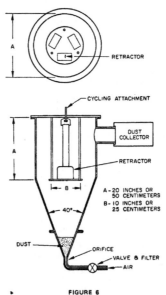

A - 20 INCHES OR 50 CENTIMETERS
B - 10 INCHES OR 25 CENTIMETERS

FIGURE 6

ing retractor shall be actuated at least 10,000 times within 50 to 100 percent extension of webbing during the 50,000 cycles. At the end of test, compliance of the retractors with applicable requirements in S4.3(h), (i), and (j) shall be determined. Three retractors shall be tested for performance.

S5.3 Assembly Performance.

(a) *Type 1 seat belt assembly*. Three complete seat belt assemblies, including webbing, straps, buckles, adjustment and attachment hardware, and retractors, arranged in the form of a loop as shown in Figure 5, shall be tested in the following manner:

(1) The testing machine shall conform to the requirements specified in S5.1(b). A double-roller block shall be attached to one head of the testing machine. This block shall consist of 2 rollers 4 inches or 10 centimeters in diameter and sufficiently long so that no part of the seat belt assembly touches parts of the block other than the rollers during test. The rollers shall be mounted on anti-friction bearings and spaced 12 inches or 30 centimeters between centers, and shall have sufficient capacity so that there is no brinelling, bending or other distortion of parts which may affect the results. An anchorage bar shall be fastened to the other head of the testing machine.

(2) The attachment hardware furnished with the seat belt assembly shall be attached to the anchorage bar. The anchor points shall be spaced so that the webbing is parallel in the two sides of the loop. The attaching bolts shall be parallel to, or at an angle of 45 or 90 degrees to the webbing, whichever results in an angle nearest to 90 degrees between webbing and attachment hardware except that eye bolts shall be vertical, and attaching bolts or nonthreaded anchorages of a seat belt assembly designed for use in specific models of motor vehicles shall be installed to produce the maximum angle in use indicated by the installation instructions, utilizing special fixtures if necessary to simulate installation in the motor vehicle. Rigid adapters between anchorage bar and attachment hardware shall be used if necessary to locate and orient the adjustment hardware. The adapters shall have a flat support face perpendicular to the threaded hole for the attaching bolt and adequate in area to provide full sup-

port for the base of the attachment hardware connected to the webbing. If necessary, a washer shall be used under a swivel plate or other attachment hardware to prevent the webbing from being damaged as the attaching bolt is tightened.

(3) The length of the assembly loop from attaching bolt to attaching bolt shall be adjusted to about 51 inches or 130 centimeters, or as near thereto as possible. A force of 55 pounds or 25 kilograms shall be applied to the loop to remove any slack in webbing at hardware. The force shall be removed and the heads of the testing machine shall be adjusted for an assembly loop between 48 and 50 inches or 122 and 127 centimeters in length. The length of the assembly loop shall then be adjusted by applying a force between 20 and 22 pounds or 9 and 10 kilograms to the free end of the webbing at the buckle, or by the retraction force of an automatic-locking or emergency-locking retractor. A seat belt assem-

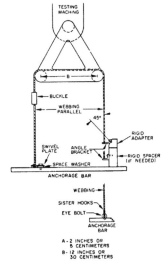

FIGURE 5

bly that cannot be adjusted to this length shall be adjusted as closely as possible. An automatic-locking or emergency-locking retractor when included in a seat belt assembly shall be locked at the start of the test with a tension on the webbing slightly in excess of the retractive force in order to keep the retractor locked. The buckle shall be in a location so that it does not touch the rollers during test, but to facilitate making the buckle release test in S5.2(d) the buckle should be between the rollers or near a roller in one leg.

(4) The heads of the testing machine shall be separated at a rate between 2 and 4 inches per minute or 5 and 10 centimeters per minute until a force of 5,000 ± 50 pounds or 2,270 ± 20 kilograms is applied to the assembly loop. The extension of the loop shall be determined from measurements of head separation before and after the force is applied. The force shall be decreased to 150 ± 10 pounds or 68 ± 4 kilograms and the buckle release force measured as prescribed in S5.2(d).

(5) After the buckle is released, the webbing shall be examined for cutting by the hardware. If the yarns are partially or completely severed in a line for a distance of 10 percent or more of the webbing width, the cut webbing shall be tested for breaking strength as specified in S5.1(b) locating the cut in the free length between grips. If there is insufficient webbing on either side of the cut to make such a test for breaking strength, another seat belt assembly shall be used with the webbing repositioned in the hardware. A tensile force of 2,500 ± 25 pounds or 1,135 ± 10 kilograms shall be applied to the components or a force of 5,000 ± 50 pounds or 2,270 ± 20 kilograms shall be applied to an assembly loop. After the force is removed, the breaking strength of the cut webbing shall be determined as prescribed above.

(6) If a Type 1 seat belt assembly includes an automatic-locking retractor or an emergency-locking retractor, the webbing and retractor shall be subjected to a tensile force of 2,500 ± 25 pounds or 1,135 ± 10 kilograms with the webbing fully extended from the retractor.

(7) If a seat belt assembly has a buckle in which the tongue is capable of inverted insertion, one of the three assemblies shall be tested with the tongue inverted.

(b) *Type 2 seat belt assembly.* Components of three seat belt assemblies shall be tested in the following manner:

(1) The pelvic restraint between anchorages shall be adjusted to a length between 48 and 50 inches or 122 and 127 centimeters, or as near this length as possible if the design of the pelvic restraint does not permit its adjustment to this length. An automatic-locking or emergency-locking retractor when included in a seat belt assembly shall be locked at the start of the test with a tension on the webbing slightly in excess of the retractive force in order to keep the retractor locked. The attachment hardware shall be oriented to the webbing as specified in paragraph (a) (2) of this section and illustrated in Figure 5. A tensile force of 2,500 ± 25 pounds or 1,135 ± 10 kilograms shall be applied on the components in any convenient manner and the extension between anchorages under this force shall be measured. The force shall be reduced to 75 ± 5 pounds

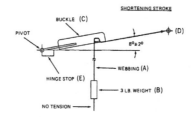

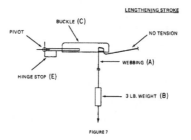

FIGURE 7

or 34 ± 2 kilograms and the buckle release force measured as prescribed in S5.2(d).

(2) The components of the upper-torso restraint shall be subjected to a tensile force of $1,500 \pm 15$ pounds or 680 ± 5 kilograms following the procedure prescribed above for testing pelvic restraint and the extension between anchorages under this force shall be measured. If the testing apparatus permits, the pelvic and upper-torso restraints may be tested simultaneously. The force shall be reduced to 75 ± 5 pounds or 34 ± 2 kilograms and the buckle release force measured as prescribed in S5.2(d).

(3) Any component of the seat belt assembly common to both pelvic and upper-torso restraint shall be subjected to a tensile force of $3,000 \pm 30$ pounds or $1,360 \pm 15$ kilograms.

(4) After the buckle is released in tests of pelvic and upper-torso restraints, the webbing shall be examined for cutting by the hardware. If the yarns are partially or completely severed in a line for a distance of 10 percent or more of the webbing width, the cut webbing shall be tested for breaking strength as specified in S5.1(b) locating the cut in the free length between grips. If there is insufficient webbing on either side of the cut to make such a test for breaking strength, another seat belt assembly shall be used with the webbing repositioned in the hardware. The force applied shall be $2,500 \pm 25$ pounds or $1,135 \pm 10$ kilograms for components of pelvic restraint, and $1,500 \pm 15$ pounds or 680 ± 5 kilograms for components of upper-torso restraint. After the force is removed, the breaking strength of the cut webbing shall be determined as prescribed above.

(5) If a Type 2 seat belt assembly includes an automatic-locking retractor or an emergency-locking retractor, the webbing and retractor shall be subjected to a tensile force of $2,500 \pm 25$ pounds or $1,135 \pm 10$ kilograms with the webbing fully extended from the retractor, or to a tensile force of $1,500 \pm 15$ pounds or 680 ± 5 kilograms with the webbing fully extended from the retractor if the design of the assembly permits only upper-torso restraint forces on the retractor.

(6) If a seat belt assembly has a buckle in which the tongue is capable of inverted insertion, one of the three assemblies shall be tested with the tongue inverted.

(c) *Resistance to buckle abrasion.* Seatbelt assemblies shall be tested for resistance to abrasion by each buckle or manual adjusting device normally used to adjust the size of the assembly. The webbing of the assembly to be used in this test shall be exposed for 4 hours to an atmosphere having relative humidity of 65 percent and temperature of 70° F. The webbing shall be pulled back and forth through the buckle or manual adjusting device as shown schematically in Figure 7. The anchor end of the webbing (A) shall be attached to a weight (B) of 3 pounds. The webbing shall pass through the buckle (C), and the other end (D) shall be attached to a reciprocating device so that the webbing forms an angle of 8° with the hinge stop (E). The reciprocating device shall be operated for 2,500 cycles at a rate of 18 cycles per minute with a stroke length of 8 inches. The abraded webbing shall be tested for breaking strength by the procedure described in paragraph S5.1(b).

44 F.R. 72131
December 13, 1979

PREAMBLE TO AMENDMENT TO MOTOR VEHICLE SAFETY STANDARD NO. 210

Seat Belt Assembly Anchorages—Passenger Cars, Multipurpose Passenger Vehicles, Trucks and Buses

(Docket No. 2–14; Notice No. 4)

An amendment to Motor Vehicle Safety Standard No. 210, Seat Belt Assembly Anchorages, was published on October 1, 1970 (35 F.R. 15293). Thereafter, pursuant to § 553.35 of the procedural rules (49 CFR 553.35, 35 F.R. 5119), petitions for reconsideration were filed by Rolls Royce, Ltd., International Harvester Co., Chrysler Corp., Ford Motor Co., General Motors Corp., the Automobile Manufacturers Association, Toyota Motor Co., Ltd., American Motors, Jeep Corp., Chrysler United Kingdom, Ltd., and Checker Motors Corp.

In response to information contained in the petitions, and other considerations, certain requirements of the standard are hereby amended and the effective date of the standard with respect to passenger cars is postponed until January 1, 1972. The petitions for relief from certain other requirements of the standard are denied.

1. The effective date of the amended standard with respect to passenger cars was to have been January 1, 1971. Each petitioner claimed to be unable to produce vehicles conforming to the amended standard by that date. Those who provided lead time information indicated that several months would be needed, with estimates ranging from March 31, 1971, for Rolls Royce, to January 1, 1972, for a number of manufacturers. A January 1972 effective date would have the advantage of coinciding with the effective date proposed for the closely related interim standard on occupant crash protection (Docket 69–7, Notice 6, 35 F.R. 14941). Since the amendments with respect to passenger cars are intended primarily to enhance the enforceability of the standard rather than to provide new levels of safety, it has been determined that good cause has been shown for establishing an effective date for passenger cars of January 1, 1972.

With a single exception, the requests for postponement of the effective date of the standard with respect to multipurpose passenger vehicles, trucks, and buses, are denied. One of the primary reasons for amending the standard was to extend the protection afforded by seat belts to occupants of these types of vehicles. A postponement of effective date would leave these vehicles completely without anchorage requirements for an additional 6 months. Although manufacturers who have been installing anchorages may find it necessary to reexamine the strength and location of their anchorages, this is not considered a sufficient ground for postponing the effective date.

International Harvester requested a postponement until January 1, 1972, in the date on which upper torso restraint anchorages will be required on seats other than front seats in multipurpose passenger vehicles. On consideration of the lead time difficulties that have been demonstrated by this manufacturer, the Director regards the request as reasonable and has decided to grant the requested postponement.

2. A number of petitions requested reconsideration of the sections dealing with anchorage location. Section S4.3.1.4 of the standard states that "Anchorages for an individual seat belt assembly shall be located at least 13.75 inches apart laterally for outboard seats and at least 6.75 inches apart laterally for other seats."

General Motors stated that several of its vehicles have anchorages for the center seating position that are 6.50 inches apart, that some of the anchorages for outboard seats are less than 13.75 inches apart, and that there is no basis either for setting a minimum spacing, or for setting different minimum spacings for different seating positions. Similar comments were made by AMA, Chrysler, Ford and American Motors.

As originally issued, Standard No. 210 had required anchorages to be "as near as practicable, 15 inches apart laterally." To make the standard more precise and more easily enforceable, the notice of September 20, 1969 (34 F.R. 14658), proposed to delete the qualifying language and to require that anchorages be 15 inches apart laterally. The comments indicated that anchorages for center seating positions, particularly the front positions, would require complete relocation. The available data on the effects of anchorage spacing were not regarded as conclusive enough to justify imposing this burden on the manufacturers, and the spacing for anchorages for inboard locations was accordingly reduced to 6.75 inches in the amended standard. Without clearer biomechanical data, the intent was to adopt the prevailing industry minimum as the standard. The same rationale applied to outboard seating position, where the 15-inch spacing was reduced to 13.75 inches.

It now appears that both spacing employed in the amended standard failed to reflect prevailing locations. The Director is accordingly amending section S4.3.1.4 to establish a minimum spacing of 6.50 inches.

A further problem with the spacing requirement arises from the use of "anchorage" as the reference point for measurement. As long as the standard used the qualifying language "as near as practicable," there was no difficulty. Removal of that phrase by the notice of September 20, 1969, created a problem of interpretation that escaped comment until after issuance of the amended standard. Several petitioners commented that they do not know what point to use for measurement. The director concedes the deficiency, and accordingly amends section S4.3.1.4 to specify that the spacing is "measured between the vertical centerlines of the bolt holes."

In conjunction with its request for a reduction of the spacing requirement, General Motors stated that where structural members between the

anchorage and the seating position have the effect of spreading the seat belt loop apart, the spacing should be measured between the widest contact points on the structure. Since the strength of these structural members is not regulated, there is no assurance that their performance in a crash will be equal to that of properly spaced anchorages. The request offers no improvement in occupant crash protection, and may, in fact, diminish such protection. The request is therefore denied.

3. The amended standard's other location requirements concern the placement of anchorages to achieve desirable seat belt angles. Sections S4.3.1.1 and S4.3.1.3 each use the "nearest belt contact point on the anchorage" as the lower point defining the line whose angle is to be measured. Several petitions expressed uncertainty as to the point described, and on reconsideration the Director agrees that clarification is needed.

In the notice of proposed rule making that preceded the amended standard (34 F.R. 14658, Sept. 20, 1969) the line had been run to the "anchorage". This usage lacked precision, as stated by several comments. In an attempt to define a line that would closely approximate the actual belt angle, the language in question was adopted. The problem lies in the use of the word "anchorage", since in most installations the belt does not actually contact the anchorage. The point intended was, in fact, the nearest contact point of the belt webbing with the hardware that attaches it to the anchorage. In the typical installation, this point would be on an angle plate bolted to the anchorage. Sections S4.3.1.1 and S4.3.1.3 are accordingly amended to use the phrase "the nearest contact point of the belt with the hardware attaching it to the anchorage."

4. The test procedures of S5.1 and S5.2 were the subject of several requests for reconsideration. Most petitioners stated that the test was not representative of crash conditions, and several suggested that it should be displaced by a dynamic test. Times suggested for such a dynamic test ranged from 0.1 second to 1.0 second, and were said to be the tests used by the petitioners, or by one or another of the international standards organizations. The requirement for a 10-second hold period at maximum

load attracted the most strongly adverse comment.

From its inception, Standard No. 210 has contemplated a static test. The notice of proposed rule making of September 20, 1969, proposed a test that was clearly static, in that it involved a slow rate of load application (2 to 4 inches per minute). In response to comments that the rate was too slow, and to avoid problems of interpretation as to where the rate of pull was to be measured, the procedures were amended to specify the rate of load application in time rather than distance, with the full load reached in a period of from 0.1 to 30 seconds. It should be noted that the vehicle must be capable of meeting the requirements when tested at any rate within this range. To insure that the basic strength of the structure would be measured whatever the shape of the load application curve, a hold period of 10 seconds was specified. The procedures of the amended standard do no more than give more specific form to the test contemplated in the original standard.

The postponement of the effective date of the amended standard will provide additional time for passenger car manufacturers to assure themselves of compliance with the standard. After consideration of the issues raised in the petitions for reconsideration, the Director has concluded that the tests prescribed by the standard are reasonable, practicable, and appropriate for the affected motor vehicles. The petitions for reconsideration of sections S5.1 and S5.2 are therefore denied.

5. Two petitioners, Rolls Royce and General Motors, stated that it was not practicable to use the "seat back" in determining the angle of the torso line in S4.3.2, in that the seat back angle may vary according to which of its surfaces is measured. Although there may be instances where the angle of the seat back is difficult to determine, questions arising from such instances can be resolved, if necessary, by administrative interpretation, and it has been decided to retain the reference to "seat back" in section S4.3.2.

6. Several petitioners stated that the substitution of the word "device" for "provision" in the definition of seat belt anchorage appeared to change the meaning of that term. No substantive change was intended, and since the rewording has caused some misunderstanding, the Director has decided to return to the original wording.

7. General Motors also petitioned to reinstate the provision in section S4.3.2 that would allow the upper torso restraint angle to be measured from the shoulder to the anchorage "or to a structure between the shoulder point and the anchorage". The phrase rendered uncertain the effective angle of the belt under stress. The quoted language was deleted in the notice of September 20, 1969, and no sufficient reason has been given for reinstating it. The request is therefore denied.

8. Toyota Motor Co. requested that sections S5.1 and S5.2 be amended to allow use of body blocks equivalent to those specified. Although the standard provides that an anchorage must meet the strength requirements when tested with the specified blocks, manufacturers may use whatever methods they wish to ascertain that their products meet these requirements when so tested, as long as their methods constitute due care. If the Toyota procedures are, in fact, equivalent, there is no need to amend the standard·to accommodate them. The request is therefore denied.

In consideration of the foregoing, Motor Vehicle Safety Standard No. 210, in § 571.21 of Title 49, Code of Federal Regulations is amended. . . .

Effective date. For the reasons given above, it has been determined that the effective date of the amended standard shall be January 1, 1972, for passenger cars. The effective date for multipurpose passenger vehicles, trucks, and buses shall be July 1, 1971, except that the effective date for installation of anchorages for upper torso restraints for seating positions other than front outboard designated seating positions shall be January 1, 1972.

Issued on November 20, 1970.

Charles H. Hartman,
Acting Director.

35 F.R. 18116
Nov. 26, 1970

PREAMBLE TO AMENDMENT TO MOTOR VEHICLE SAFETY STANDARD NO. 210

Seat Belt Assembly Anchorages and Seat Belt Installations;
Reconsideration and Amendment

(Docket No. 2–14; Notice No. 4)

The purpose of this notice is to amend Motor Vehicle Safety Standards No. 208 and 210, with respect to the installation of shoulder belts in multipurpose passenger vehicles exceeding 10,000 pounds GVWR and the provision of anchorages for shoulder belts in vehicles other than passenger cars.

The seat belt installation standard was amended on September 30, 1970, to require installation of seat belts in multipurpose passenger vehicles, trucks, and buses manufactured after July 1, 1971 (35 F.R. 15222). Exemptions from the requirement for shoulder belt installation were provided for certain types and weights of vehicles.

During the course of the subsequent rulemaking activity which led to the issuance of the occupant crash protection standard, it was determined that the larger weight classes of trucks and multipurpose passenger vehicles should not be required to install shoulder belts (35 F.R. 14941, 35 F.R. 16937, 36 F.R. 4600). The standard therefore required lap belts, but not shoulder belts, for vehicles over 10,000 pounds GVWR, effective January 1, 1972. The September 30 amendment, which is to become effective six months earlier than the occupant crash protection rule, had provided a similar exemption for large trucks but not for multipurpose passenger vehicles, with the result that shoulder belts would have been required for many large multipurpose passenger vehicles during the period July 1, 1971–January 1, 1972, but not afterward. To correct this inconsistency, the seat belt installation standard is amended, effective July 1, 1971, to exempt multipurpose passenger vehicles of more than

10,000 pounds GVWR from the shoulder belt requirement.

In accordance with the foregoing, section S3.1 of Standard No. 208, as published September 30, 1970 (35 F.R. 15222) is amended effective July 1, 1971 . . . .

Standard No. 210, *Seat Belt Assembly Anchorages*, presently requires vehicles other than passenger cars to have shoulder belt anchorages installed at front outboard seating positions by July 1, 1971, and at rear outboard seating positions by January 1, 1972 (35 F.R. 15293, 35 F.R. 18116, 36 F.R. 4291). The Recreational Vehicle Institute has petitioned for an amendment of the standard, to delete the requirement for shoulder belt anchorages at positions where shoulder belt installation is not required by Standard No. 208.

It has been found that this petition has merit. The probability of shoulder belt installation by the owners of these vehicles is very small, and the difficulty of anchorage installation, particularly in multipurpose passenger vehicles, is often greater than in passenger cars. The amendment is therefore considered to be in the public interest.

The request by RVI for a postponement of the July 1, 1971, effective date for installation of shoulder belt anchorages has not been found justified, and the petition is in that respect denied.

In accordance with the foregoing, section S4.1.1 of the present Motor Vehicle Safety Standard No. 210 (effective July 1, 1971), and the amended Standard No. 210 as published November 26, 1970 (35 F.R. 18116, effective January 1, 1972), in 49 CFR 571.21, are both amended . . . .

Effective: July 1, 1971
      January 1, 1972

The effective dates of the amendments made by this notice are as indicated above. Because the amendments relieve restrictions and impose no additional burden on any person, notice and request for comments on such notice are found to be unnecessary, and it is found, for good cause shown, that an effective date earlier than 180 days after issuance is in the public interest.

36 F.R. 9869
May 29, 1971

PREAMBLE TO MOTOR VEHICLE SAFETY STANDARD NO. 210

Seat Belt Anchorages

(Docket No. 72–23; Notice 3)

This notice amends Safety Standard No. 210, *Seat Belt Assembly Anchorages*, to eliminate the "buckle cutout" as an optional configuration of the body block test device used for testing the strength of lap-shoulder belt anchorages, and to clarify the illustration (Figure 2) of body blocks used for testing lap belt anchorages. The optional configuration is being deleted because it unnecessarily complicates the test of the anchorages and is no longer being used by manufacturers.

Effective Date: May 18, 1978.

For Further Information Contact:

William E. Smith, Division of Crashworthiness, National Highway Traffic Safety Administration, 400 Seventh Street, S.W., Washington, D.C. 20590 (202–426–2242).

Supplementary Information: Standard No. 210 (49 CFR 571.210) requires seat belt anchorages in motor vehicles to comply with specified strength requirements. The procedure for strength testing is set forth in paragraph S5 of the standard. The tests involve the attachment of a seat belt to the anchorage, followed by the application of force to the seat belt which is thereby transferred to the anchorage itself. Force is applied to Type 1 and Type 2 seat belt assemblies through body blocks that simulate the human vehicles torso. The body blocks are illustrated in Figures 2 and 3 of the standard. This notice modifies Figures 2 and 3 in accordance with the notice of proposed rulemaking issued December 16, 1976 (41 F.R. 54050).

Figure 2 describes the body block used for lap belt anchorage testing, and there has been some confusion concerning certain minor specifications in the Figure. This amendment modifies the drawing in Figure 2 to clarify the description of the body block. The change does not affect the substantive requirements of the standard in any way.

Figure 3 describes the body block used for combination shoulder and lap belt anchorage testing. An optional "buckle cutout" is shown on the surface of the body block in Figure 3, permitting a manufacturer to make an indentation in the face of the body block to accommodate buckle hardware. NHTSA compliance test experience with the cutout demonstrates that the edge of the cutout causes additional stress on the belt webbing and interferes with its movement, thereby interfering with the test of the underlying anchorage. Comments to the proposal favored deletion of the "buckle cutout" option since it is disadvantageous to manufacturers and is no longer being utilized. This amendment, therefore, deletes the optional cutout from Figure 3.

General Motors' comment recommended additional modifications of the drawing in Figure 2. The agency has determined, however, that the suggestion to add shading to define the area of the body block to be covered by foam padding does not significantly alter the clarity of the drawing. General Motors also recommended a substitute test device for the lap-shoulder belt body block. This recommendation will possibly be considered in future rulemaking.

The engineer and lawyer primarily responsible for the development of this notice are William Smith and Hugh Oates, respectively.

Since this amendment does not make any substantive change in the requirements of the standard, it is found that an immediate effective date is in the public interest.

Effective: May 18, 1978

In consideration of the foregoing, Standard No. 210, 49 CFR 571.210, is amended . . . .

(Sec. 103, 119, Pub. L. 89-563, 80 Stat. 718 (15 U.S.C. 1392, 1407); delegation of authority at 49 CFR 1.50).

Issued on May 15, 1978.

Joan Claybrook
Administrator
43 F.R. 21892
May 23, 1978

Seat Belt Assembly Anchorages

(Docket No. 72–23; Notice 5)

Action: Final rule.

Summary: This notice amends Safety Standard No. 210, *Seat Belt Assembly Anchorages*, to eliminate the anchorage location requirements for passive seat belt assemblies that meet the frontal crash protection requirements of Safety Standard No. 208. The purpose of the amendment is to give manufacturers wider latitude in passive belt design in order to facilitate the early introduction of passive restraints in existing passenger car designs. The amendment will allow manufacturers to experiment with various passive belt designs to help determine the optimum relationship between anchorage location and passive belt effectiveness in a variety of crash modes and their comfort and convenience. Anchorage location would still be indirectly controlled by the necessity for passive belts to comply with the Standard No. 208 requirements.

Effective date: November 16, 1978.

Addresses: Petitions for reconsideration should refer to the docket number and notice number and be submitted to: Docket Section, Room 5108—Nassif Building, 400 Seventh Street, S.W., Washington, D.C. 20590.

For further information contact:

William Smith, Office of Vehicle Safety Standards, National Highway Traffic Safety Administration, Washington, D.C. 20590 (202) 426–2242.

Supplementary information: Safety Standard No. 210, *Seat Belt Assembly Anchorages* (49 CFR 571.210), specifies zones and acceptable ranges within which seat belt anchorages must be located to ensure that the anchorages are in the proper location for effective occupant restraint and specifies strength requirements to reduce the likelihood of their failure in a crash. In response to a petition from General Motors Corporation, the NHTSA issued a proposal to delete these anchorage location requirements for passive belt systems that meet the dynamic frontal crash protection requirements of Safety Standard No. 208 (43 FR 22419, May 25, 1978).

The proposal noted that General Motors would like to use a passive belt design whose anchorages, in some vehicles, would lie outside the parameters specified in the standard. GM stated that the anchorage locations of this design are intended to ensure the comfort and convenience of the passive belt so that it will not be disconnected by vehicle users who find current active belts lacking in these qualities. General Motors wanted to introduce this passive belt design prior to the effective date of the passive restraint requirements issued July 5, 1977 (42 FR 34289). As stated in the preamble of the proposal, the agency has determined manufacturers should be given wide latitude in passive belt design in order to facilitate the early introduction of passive systems, since they should save many lives and prevent hundreds to thousands of injuries. Although the current anchorage location requirements were developed primarily for active belt systems, passive belt systems such as the one used on the Volkswagen Rabbit have successfully complied with the anchorage location requirements and met the frontal injury criteria of Standard No. 208 as well. Nonetheless, manufacturers have said they can develop more effective and comfortable passive systems to comply with Standard 208. The agency thinks they should be given the opportunity. Nevertheless, it is the agency's view that research should be conducted to determine the optimum anchorage locations for the various passive belt designs in terms of both passive belt effectiveness and of comfort and convenience for vehicle occupants. Accordingly, the earlier notice proposed the deletion of the anchorage requirements for passive belts until appropriate requirements for these systems can be developed and incorporated in the standard.

Comments in support of the proposed change were received from Chrysler, British Leyland, American Motors, Ford, Volkswagen, General Motors, and the Association Peugeot-Renault. These commenters argued that manufacturers should not be restricted in passive belt design, so that manufacturers can determine which designs are the most effective and at the same time acceptable to the public. The Center for Auto Safety argued against the proposal, however, stating that elimination of the anchorage location requirements may degrade available occupant protection.

The Center for Auto Safety agreed that manufacturers should be allowed flexibility in passive belt design to facilitate the early introduction of passive restraints. However, it argued that elimination of the forward boundary for upper torso belt anchorages may "(1) seriously degrade occupant protection available by allowing the anchorages to be installed in areas likely to be struck by the occupant in a side impact and (2) may result in systems that do not sufficiently restrain the occupant from submarining or moving laterally under the belt." The Center's first concern is that side-impact head injuries will increase if passive belt retractors, buckles, and other hardware are permitted in areas likely to be struck by the occupant's head in a side collision. The comment noted that vehicles equipped with passive belts are not required to meet the lateral impact requirements of Standard No. 208 and that manufacturers would, therefore, have no incentive to design anchorages and other hardware to avoid injuries in non-frontal collisions.

The Center's second concern is that elimination of the anchorage location requirements will allow passive belt designs that lead to more lateral occupant movement and "submarining" in side crashes, thereby increasing side impact injuries. The Center also argued that it should be the responsibility of General Motors to demonstrate the safety consequences of moving passive belt anchorages outside the current range require-

ments, before the agency eliminates the requirements for passive belts. Finally, the Center is concerned that once the exemption is allowed, it might be years before new location requirements for passive belts are specified.

Regarding the Center's first concern, the present requirements do not prohibit the placement of hardware in areas where they could be struck by an occupant's head in a side collision. While manufacturers may not be constrained by present standards from placing hardware where it poses a danger to occupants in side impacts, all manufacturers are on notice that the agency is preparing to propose a side impact standard as delineated in the agency's rulemaking plan. Thus, in anticipation of the upgraded side impact requirements, manufacturers should design their passive belt systems in such a way that they will not compromise side impact protection.

The Center's concern about the potential for increased lateral movement and submarining in side crashes was not supported by any data. The NHTSA is also concerned about side impact injuries. However, the existing location requirements for belt anchorages were not specifically designed to address the problem of lateral occupant motion in non-frontal collisions where the occupant is restrained by a single, diagonal passive upper torso restraint used with a knee bolster.

The notice of proposed rulemaking explicitly stated that the NHTSA intends to issue separate anchorage location requirements for passive belts following research to determine the optimum locations for passive belt effectiveness, comfort and convenience, and that the proposed exemption from the current requirements is only an interim measure. The NHTSA intends to conduct studies to look at the change in injury data resulting from displacement of the upper anchorage point of a single diagonal belt for various sizes of occupants. The research program includes testing that will investigate the "submarining" problem and, during frontal oblique impact simulations, the likelihood of excessive lateral movement. The agency will consider simulated side impact testing during this research program to evaluate potential degradation of occupant protection in this crash mode. The agency will also consider anchorage location during the upgrading of side impact protection requirements. As stated in the recent "Five Year Rulemaking Plan," the improvement of occupant protection in side impacts is one of the NHTSA's highest priorities.

The Center's suggestion that GM demonstrate the safety consequences of passive belt anchorages should be addressed by the NHTSA's intention to look with great care at manufacturers' compliance testing of all passive belt designs to assure that these new systems will, in fact, provide at least the level of overall protection now afforded by conventional restraint systems.

Finally, regarding the Center's concern that new location requirements for passive belt anchorages will not be specified for many years, the notice of proposed rulemaking and this notice make it clear that the exemption is only an interim measure to allow improvements in passive belt designs. It is consistent, however, with the attempt to make FMVSS 208 a performance standard to the greatest extent possible. Nevertheless, should any manufacturer produce passive belt hardware or systems that cause or exacerbate injuries that would not occur with active systems currently in production, the NHTSA's safety defect authority would permit the agency to investigate such systems for possible recall and correction. Manufacturers are hereby put on notice of that fact.

In summary, the NHTSA has concluded that manufacturers should be given wide latitude in passive belt design in order to aid the early introduction of passive restraints and to aid the development of optimum designs in terms of both effectiveness and comfort and convenience. The agency agrees that anchorage location requirements are important for passive belts, but believes that more effective requirements can be developed following further research specifically involving passive belts. To ensure that safe and effective systems are being developed, the agency will be testing many of the new passive systems that will come on the market prior to the 1982 model year. In addition, the agency intends to ask manufacturers to supply data concerning the performance of passive systems in both compliance crash testing and in sled and crash testing in other modes.

The NHTSA has determined that this amendment will have no economic or environmental consequences.

The engineer and lawyer primarily responsible for the development of this notice are William Smith and Hugh Oates, respectively.

In consideration of the foregoing, Federal Motor Vehicle Safety Standard No. 210, *Seat Belt Assembly Anchorages* (49 CFR 571.210), is amended . . . .

AUTHORITY: (Sec. 103, 119, Pub. L. 89–563, 80 Stat. 718 (15 U.S.C. 1392, 1407) ; delegation of authority at 49 CFR 1.50.)

Issued on November 3, 1978.

Joan Claybrook
Administrator

43 F.R. 53440
November 16, 1978

PREAMBLE TO AN AMENDMENT TO
FEDERAL MOTOR VEHICLE SAFETY STANDARD NO. 210

Anchorages for Child Restraint Systems
[Docket No. 80-18; Notice 4]

ACTION: Final Rule.

SUMMARY: To permit the securing of child safety seats, this notice amends Standard No. 210, *Seat Belt Assembly Anchorages*, to require all vehicles with automatic restraint systems at the right front passenger seating position to be equipped with anchorages for a lap belt at that position if the automatic restraint cannot be used to secure a child safety seat. Some automatic belts cannot be used to secure child safety seats since they include only a single, diagonal shoulder belt. The new requirement will enable parents to install a lap belt if they wish to secure a child safety seat in the front right outboard seating position. The amendment also requires vehicle manufacturers to include information in their owner's manuals on child safety and the location of shoulder belt anchorages in the rear seats. The owner's manual must also provide instructions explaining how a lap belt can be installed for use with child safety seats in the front right passenger seating position in vehicles with automatic restraints that cannot be used for securing child restraints.

EFFECTIVE DATE: The effective date for all of the amendments, except for the amendments adding S6 and S7 to the standard, is September 1, 1987. The amendments adding S6 and S7 contain information collection requirements which must be approved by the Office of Management and Budget (OMB). After OMB approval, the agency will publish a notice announcing the effective date of S6 and S7 of the standard.

SUPPLEMENTARY INFORMATION: On December 11, 1980 (45 FR 81625), NHTSA issued a notice of proposed rulemaking to amend Standard No. 210, *Seat Belt Assembly Anchorages*, to require anchorages in certain vehicles for child safety seat tether straps. In addition, the notice proposed requiring vehicles equipped with automatic restraint systems at the right front designated seating position, which cannot be used for the securing of child safety seats, to have separate anchorages at that position for the installation of Type 1 lap belts.

On July 5, 1985 (50 FR 27632) the agency published a notice terminating the portion of the proposed rule concerning anchorages for child safety seat tether straps. As explained in that notice the agency has decided that the appropriate way to reduce problems created by tether misuse is to propose an amendment (50 FR 27633) to Standard No. 213, *Child Restraint Systems* to require all child safety seats to pass a 30 mile per hour simulated crash test without a tether attached. This will ensure that all child safety seats provide an adequate level of safety even if they designed to be used with a tether strap. This notice announces the agency decision on the remaining portion of the proposed rule relating to front passenger seat safety belt anchorages.

Lap Belt Anchorages for Front Seats

A large percentage of the commenters supported the proposed requirement on the basis that some provision is necessary for securing child restraint systems used in front right seating positions, especially in vehicles with single, diagonal automatic belt designs. Several commenters noted that, in particular, infant safety seats are often used in that seat so that the infant is within view and reach of an adult. However, several commenters stated that the proposal did not go far enough. Some commenters recommended that in addition to requiring holes for anchorages, the agency should require anchorage hardware to be installed by vehicle manufacturers so that lap belts could be readily installed by consumers. Other commenters recommended that lap belts be required for these positions in addition to the anchorages.

A few commenters argued that the proposed anchorages should not be required at all because the rear seat is the safest location for the transportation of children and the proposal would encourage parents to place their children in the less safe front seat. Several commenters also requested that the anchorage strength for the lap belt anchorages be set at 3,000 pounds rather than the proposed 5,000 pounds, on the basis that the lap belts would only be used to restrain children, not adults.

The agency agrees that the installation of lap belts in front seating positions not currently having them (vehicles equipped with single, diagonal automatic belts or with nondetachable automatic belts that cannot be used for attachment of child safety seats) would be the optimum situation insofar as securing child safety seats is concerned. Short of this, requiring complete attachment hardware would make the installation of lap belts somewhat easier than if manufacturers only provide anchorage holes. However, both of these approaches involves costs that the agency believes are not justified because of the limited number of vehicle owners who would actually have need of this equipment.

The cost of requiring the actual anchorage hardware in addition to providing threaded anchorage holes would be approximately $.30 for each vehicle, and the cost of requiring the lap belts to be installed would be approximately $14.00 per vehicle. If lap belts or anchorage hardware were required, many owners would be paying for equipment they do not need. The agency does not believe these costs are justified since the presence of the threaded hole will allow those vehicle owners who actually have need of lap belts to easily install them. The agency has therefore decided to require only threaded anchorage holes to be present. With the threaded holes present, the attachment hardware and lap belt can be installed in a short time.

Type of Threaded Holes

Several commenters objected to the proposed requirement that the anchorage holes be threaded to accept one specific type of bolt for attaching a lap belt. They said that Standard No. 209, *Seat Belt Assemblies,* permits the use of several types of bolts and argued that specifying the use of only one type of bolt would be restrictive. The agency agrees that manufacturers should have the same design flexibility as provided by Standard No. 209. Therefore, the final rule provides that manufacturers can thread the anchorage holes to accept any one of the bolts permitted by Standard No. 209.

Anchorage Strength

With regard to anchorage strength, the agency believes that the lap belt anchorages required by this amendment should comply with the 5,000 pound requirement currently specified in Standard No. 210 for Type 1 lap belts, rather than the 3,000 pound requirement recommended by some commenters. It is true that certain "special" lap belts designed only for use by children might not need to meet a 5,000 pound strength requirement. However, since only anchorage holes are required, some persons may install typical lap belts which will be at times, likely used by adults. Adults might also use the "special" lap belt designed only for use by children, thinking that it is intended for use by anyone. For these reasons, the agency believes it is important for the anchorage strength to be sufficient to withstand the 5,000 pound force that could be generated by an adult in a crash. The agency is therefore adopting a 5,000 pound strength requirement.

Information in the Owner's Manual

The notice of proposed rulemaking proposed that the owner's manual in each vehicle provide specific information about protecting children in motor vehicles. It proposed that each owner's manual explain how to use a vehicle lap belt to secure a child safety seat, alert parents that children are safer in the rear seats, particularly in the center rear seat, and have a specific warning about the need to use infant and child safety seats. All 50 States and the District of Columbia now require children to be fastened into child safety seats. The notice also propose that the owner's manual provide information about the proper installation of a lap belt in the front right passenger seating position of a vehicle with an automatic restraint that cannot be used to secure a child safety seat. In addition, the notice proposed that the owner's manual identify the location of the shoulder belt anchorages that are currently required by the standard for outboard rear seating positions.

Several commenters said that recommendations concerning the proper use of lap belts for attachment of child safety seats should be given by the child safety seat manufacturer rather than the vehicle manufacturer. They said that the child safety seat manufacturer is more knowledgeable about the proper use of its product. The agency agrees and notes that all child safety seat manufacturers currently provide such information. Accordingly, vehicle manufacturers will only be required to have a section in the owner's manual referring to the importance of properly using the vehicle belts with child safety seats and will not have to provide specific information about the use of belts with each type of child safety seat.

Other commenters expressed concern about the proposed requirement that vehicle manufacturers state that the center rear seat is the safest position to secure a child safety seat. The commenters noted that many vehicles currently do not have a center rear seat. Other commuters objected to including the information in owner's manuals of vehicles that do not have a rear seat. The agency agrees with these objections and has therefore modified the requirement so that vehicles with no rear seats do not have to include the statement and in vehicles with no center rear seat, a manufacturer only has to state that the rear seat is the safest position. Several commenters argued that the agency should not require manufacturers to provide information in the owner's manual since the agency's noncompliance notification and remedy regulations would then apply. They recommended that the manufacturers voluntarily provide the information.

The agency recognizes that the proposed warning requirement, which would have required manufacturers to use specific wording on child safety in the owner's manual, could lead to situations where manufacturers would have to file petitions for inconsequentiality for minor variations in the wording. At the same time, the agency believes it is important that vehicle owners receive general information on child safety and specific information on installing lap belts at the right front seat. Thus manufacturers will still have to provide information about protecting children. However, the agency has decided against requiring a warning with prescribed wording about child safety in all owner's manuals, so as to give manufacturers the maximum flexibility to incorporate that information effectively.

Finally, the agency is adopting, as proposed, the requirement that the owner's manual provide information about the location of the shoulder belt anchorages for the rear seat. Several commenters said that few people are aware that the anchorages are currently present and therefore do not know that shoulder belts can be installed in rear seats. No commenter objected to this proposal.

Effective Date

The safety belt anchorage requirements included in this amendment become effective September 1, 1987. In response to the notice of proposed rulemaking, various vehicle manufacturers indicated leadtime needs of one year, 18 months, two years and three years. Those estimates, however, reflected the time necessary for designing, tooling and installing tether anchorages rather than for the simpler task of providing additional lap belt anchorages. Standard No. 210 currently requires anchorages for a Type 2 lap-shoulder safety belt (an inboard and an outboard floor anchorage for the lap portion of belt and an outboard anchorage for the upper torso belt) at each front outboard seating position, even if the vehicle is equipped with a single, diagonal automatic belt. However, the inboard anchorage of some diagonal belts is not suitable for attachment of a lap belt since the anchorage is designed only to accommodate an automatic belt. The amendment adopted today would require, for some vehicles, the addition of one more anchorage (an additional inboard anchorage) than currently required. For any vehicles which have a three point nondetachable automatic belt that cannot be used, two additional anchorages may be required. After a careful consideration of all comments and an evaluation of the necessary design changes and tooling requirements, the agency has concluded that a leadtime of one year should be sufficient. However, if the rule were to go into effect in mid-model year, the tooling and other costs associated with the rule will substantially increase. Therefore, the agency has decided that there is good cause for making the rule effective on September 1, 1987. A leadtime of longer than a year is in the public interest since it will serve to reduce the cost of the rule to manufacturers and consumers.

Issued on October 4, 1985.

Diane K. Steed
Administrator

50 FR 41356
October 10, 1985

PREAMBLE TO AN AMENDMENT TO FEDERAL MOTOR VEHICLE
SAFETY STANDARD NO. 210
Seat Belt Assembly Anchorages
(Docket No. 80-18; Notice 5)

ACTION: Final Rule; Response to petitions for reconsideration.

SUMMARY: This notice responds to two petitions for reconsideration of the amendments to Standard No. 210, *Seat Belt Assembly Anchorages,* published on October 10, 1985. Those amendments required manufacturers to provide anchorages for a lap safety belt in automatic-restraint equipped vehicles in which the automatic restraint system cannot be used to restrain a child safety seat. In addition, the amendments required manufacturers to provide certain safety information in their vehicle owner's manual describing how to install the lap belt. Also, the owner's manual was to state that children are safer when properly restrained in the rear seating positions than in front seating positions and that, in a vehicle with a rear seating position, the center rear seating position is the safest. Two manufacturers, American Motors Corporation (AMC) and Toyota Motor Corporation (Toyota), filed timely petitions seeking reconsideration of those amendments. In response to AMC's petition, the agency has amended the lap belt anchorage requirement to make it clear that if a manufacturer voluntarily provides a manual lap or lap/shoulder belt at the front right passenger's seat, it does not have to provide an additional set of anchorages. AMC's remaining requests to permit the use of self-tapping safety belt anchorage bolts and to extend the September 1, 1987, effective date are denied. Toyota's request to delete the requirement that manufacturers state that the center rear seat is the safest seating position is granted.

EFFECTIVE DATE: The amendments made by this notice are effective on August 19, 1986. Manufacturers do not have to comply with the requirements of S4.1.3, S6, and S7 until September 1, 1987.

SUPPLEMENTARY INFORMATION: On October 10, 1985 (43 FR 53364), NHTSA published a final rule amending Standard No. 210, *Seat Belt Assembly Anchorages.* The amendments require

manufacturers to provide anchorages for a lap belt at the front right seat in vehicles manufactured after September 1, 1987, if the vehicle is equipped with an automatic restraint system that cannot be used to restrain a child safety seat. In addition, the amendments require manufacturers to provide safety information in their vehicle owner's manuals on the proper installation of lap belts in vehicles equipped with the supplemental lap belt anchorages. Also, the owner's manual was to state that children are safer when properly restrained in the rear seating positions than in front seating positions and that, in a vehicle with a rear seating position, the center rear seating position is the safest. Two vehicle manufacturers, AMC and Toyota, filed timely petitions seeking reconsideration of those amendments. In the following discussion, NHTSA addresses the issues raised by the petitioners.

Anchorage Requirements

AMC said that the language of the lap belt anchorage requirement of S4.1.3 of the standard could be "construed to mean that the supplemental anchorages might be required, *even if* a lap belt is present." The NHTSA explained in the preamble to the October 1985 final rule that the purpose of the anchorage requirement is to enable vehicle owners to quickly and easily install a lap belt to secure a child safety seat in the front right passenger's seat. The agency agrees with AMC that clearly if a manufacturer has already provided a lap belt at that position, there is no need for the supplemental anchorages. NHTSA has amended the language of the standard to clarify the requirement by providing that a manufacturer can, at its option, provide either the supplemental anchorages or a manual lap or lap/shoulder belt.

Modification of Automatic Belt Systems

AMC also asked the agency to allow manufacturers to provide methods, other than lap belt anchorages, to enable vehicle owners to secure child safety seats. AMC said that one "possible approach would be the adaptation of the automatic restraint system to secure a child restraint. For example, for a two-point automatic belt with a door-mounted emergency release, the manufacturer could include instructions to the owner on the installation of a buckle on the lower outboard anchorage. The automatic belt could then be released from the door, and buckled at the floor to form a lap belt." AMC said that it was "not necessarily recommending the use of these systems, because the questions of cost, adult misuse, etc., all must be addressed."

As NHTSA explained in the preamble to the October 1985 final rule, the purpose of the amendment is to address the problems associated with securing a child safety seat in some types of automatic restraint systems. For example, some automatic safety belts cannot be used to secure child safety seats either because they have only a single diagonal shoulder belt or because they are nondetachable and thus cannot be threaded through the structure of the child safety seat to hold the safety seat in place. By requiring manufacturers to provide threaded anchorage holes in those vehicles, the agency believed that vehicle owners who wanted to install a lap belt at the front right seat could easily and quickly do so by taking the simple step of threading a bolt into the anchorage.

NHTSA agrees with AMC that it would not be necessary to require the additional lap belt anchorages, if the vehicle owner can adjust the automatic belt system so that it can effectively restrain a child safety seat. NHTSA believes that the ease and simplicity of the adjustment is crucial. The agency does not want vehicle owners to have to follow complicated instructions or have to obtain special tools or have to purchase and install special attachment (other than the belt itself) hardware before they can use the automatic belt system to restrain a child safety seat. The more difficult and complicated the procedure is, the greater the possibility that a vehicle owner may improperly adjust the automatic belt system. In contrast, if a vehicle manufacturer has installed the additional hardware necessary to allow the use of the automatic belt to restrain a child safety seat and all a vehicle owner has to do is simply operate the emergency release for the automatic belt and then reconnect it to the attachment hardware provided by the manufacturer, NHTSA believes that vehicle owners can quickly, easily, and safely use the automatic belt to restrain a child safety seat. Thus, the agency is amending the language of S4.1.3 to

provide that a manufacturer does not have to install threaded anchorage holes if it has installed all the necessary hardware needed to adjust the automatic safety belt to secure a child safety seat.

With this amendment, manufacturers now have three options for securing child safety seats in automatic restraint equipped vehicles. First, they can provide an automatic restraint that can be used, with no modifications, to secure a child safety seat. Alternatively, they can provide an automatic restraint that can be modified or adjusted by the vehicle owner to secure a child safety seat, as long as the manufacturer has installed all the hardware necessary to secure the child safety seat. Finally, a vehicle manufacturer has the alternative of, at its option, installing a manual lap or lap/shoulder belt with its automatic restraint system or providing threaded holes so that the vehicle owner can install a manual lap belt. The agency believes that these three alternatives give a substantial amount of flexibility to vehicle manufacturers to determine which approach they want to use and assures that vehicle owners can quickly, easily, and safely use child safety seats in the front right seats of automatic restraint equipped vehicles.

Threaded Holes

The final rule required manufacturers to provide threaded holes that would accept a bolt complying with Standard No. 209, *Seat Belt Assemblies.* AMC explained that it does not use a threaded nut in its safety belt assembly, but instead uses a self-tapping bolt. It said use of the self-tapping bolt eliminates the possibility of cross-threading or misalignment caused by paint on the thread of the nut. AMC asked that the requirement be changed from providing threaded holes to providing holes that will accept any type of safety belt hardware.

NHTSA specified the installation of a threaded hole so that a vehicle owner could quickly, easily, and safely install a lap belt without using special tools or purchasing special attachment hardware. The agency expected that with the threaded holes, a vehicle owner could, if need be, find the appropriate bolt at a hardware store and install the bolt with a simple wrench or pliers. The agency is concerned that a self-tapping bolt of sufficient size and strength to withstand the forces imposed by a safety belt is not commonly available. In addition, it may be more difficult for a vehicle owner to properly align a self-tapping bolt and exert sufficient force to drive the bolt through the steel floor without a special tool. Therefore, NHTSA has decided to deny AMC's request, and instead retain the requirement that manufacturers provide threaded holes.

Leadtime

Saying that its petition sought several changes which will impact the design of its vehicles, AMC requested the agency to provide additional leadtime to implement any changes adopted by the agency. The agency does not believe that any additional leadtime is necessary. As adopted, the rule provided nearly two years of leadtime. AMC has provided no new information to show that it cannot meet the requirements of the rule within that period of time. Therefore, NHTSA has decided to deny AMC's request for additional leadtime.

Owner's Manual Information

The October 1985 final rule requires manufacturers to provide certain information in their owner's manuals about securing child safety seats in their vehicles. Among the requirements is one that, in vehicles with a center rear seat, manufacturers must state in the owner's manual that the center rear seat is the safest. Toyota asked the agency to reconsider that requirement.

Toyota agrees that children are safest when properly restrained in the rear seat, but it said it does not have data to show the center rear seat is always the safest. In addition, Toyota said that in a vehicle with front bucket seats, "depending how a child is restrained in the center rear seating position, he or she could hit against the console box and or the transmission shift lever, which are more solid than the front seatbacks." Finally, Toyota said that the required statement might mislead persons into thinking that the center rear seat is the safest, regardless of how an occupant is restrained.

NHTSA decided to require a statement about the safety of the center seat in the owner's manual based on crash tests and accident data which show that the center rear seat is safer, particularly in side impacts, than other seats. For example, side impact crash tests conducted for the agency have shown that, as would be expected, test dummies closer to the struck side of the vehicle experience larger acceleration than dummies seated away from that side. In addition to experiencing larger accelerations, the test dummies located closer to the side door contacted the interior of the vehicle as it crushed inward during the impact. (See, for example, "Countermeasures for Side Impact," DOT Contract HS 9-02177.)

Likewise, accident data have generally shown that the center rear seat is the safest. For example, data on injuries to unrestrained occupants show that occupants of center seating positions have fewer serious injuries and fatalities than unrestrained occupants in outboard rear seats. (See, "Usage and Effectiveness of Seat and Shoulder Belts in Rural Pennsylvania," DOT Publication HS 801-398). Data on restrained occupants in the rear seats are more limited. The Canadian Ministry of Transport analyzed data on the fatality and injury rates in Ontario and Alberta for the years 1978–1980. The Alberta data show, for example, that restrained children (birth–14 years old) riding in the center rear seat had the lowest rate of major and fatal injuries. Likewise, the Ontario data showed that restrained children (birth–14 years old) riding in the center rear seat had the lowest rate of major and fatal injuries. Likewise, the Ontario data showed that restrained children (birth–14 years old) riding in the center rear seat had the lowest fatality rate. NHTSA acknowledges that because of the small amount of information available on injuries and fatalities to restrained children in the rear seat, the results should not be regarded as conclusive.

NHTSA does not have sufficiently detailed files on real-world crashes to be able to address Toyota's statement that for vehicles with bucket seats it is possible that, depending on how a child is restrained, he or she could strike the console box or other vehicle features that are harder than the seatback. The agency also has not done any crash testing of bucket seat vehicles with child test dummies restrained in the rear seat. The agency agrees, however, that depending on how a child is restrained and the severity of the crash, it is possible for a restrained child in the center rear seat of a bucket seat vehicle to strike a portion of the vehicle's interior in front of the child. Therefore, the agency has decided to grant Toyota's petition and has deleted the requirement in S6(b) that manufacturers state that the center rear seat is the safest seating position. NHTSA anticipates that if a manufacturer has a particular concern about a design feature in its bucket seat equipped vehicles that could be struck by a properly restrained child, the manufacturer would take steps to minimize the risk posed by the design feature.

Navistar International Corporation (Navistar) has recently written the agency concerning the applicability of the owner's manual requirements to vehicles with a gross vehicle weight rating (GVWR) of more than 10,000 pounds. Navistar said that such heavy vehicles are generally property-carrying and service vehicles used for commercial purposes and would seldom, if ever, be carrying children. Navistar also noted that the drivers of those heavy vehicles may never see the owner's manual, since they may not be the owners of the vehicles.

The agency believes that Navistar has raised several good reasons why the owner's manual requirements should be limited to vehicles with a GVWR of 10,000 pounds or less, the class of vehicle which would normally be transporting children in child safety seats. Thus, the agency is amending the standard to limit the owner's manual requirements to vehicles with a GVWR of 10,000 pounds or less.

The agency is also making another minor clarifying change to the owner's manual information requirements.

S6(c) of the standard requires vehicle manufacturers to provide information about the location of the anchorages for shoulder belts in the rear outboard seats in their vehicles under the following conditions. Manufacturers are required to provide the owner's manual information if Standard No. 210 requires them to install shoulder belt anchorages at those positions and they have not installed lap/shoulder belts at those positions as items of original equipment. Since S4.1.1 of Standard No. 210 only requires the installation of shoulder belt anchorages in the rear outboard seats of passenger cars, the agency is amending S6(c) to make clear that this portion of the owner's manual requirements only apply to passenger cars.

For the reasons set out in the preamble, section 571.210 of Title 49 of the Code of Federal Regulations is amended as follows:

1. The authority citation for Part 571 would continue to read as follows:

Authority: 15 U.S.C. 1392, 1401, 1403, 1407; delegation of authority at 49 CFR 1.50.

2. S4.1.3 is amended by revising the first sentence to read as follows:

S4.1.3 Notwithstanding the requirement of paragraph S4.1.1, each vehicle manufactured on or after September 1, 1987, that is equipped with an automatic restraint at the front right outboard designated seating position that cannot be used for securing a child restraint system or cannot be adjusted by the vehicle owner to secure a child restraint system solely through the use of attachment hardware installed as an item of original

equipment by the vehicle manufacturer shall have, at the manufacturer's option, either anchorages for a Type 1 seat belt assembly at that position or a Type 1 or Type 2 seat belt assembly at that position.

3. The first sentence of S6 is revised to read as follows:

S6 *Owner's Manual Information.* The owner's manual in each vehicle with a GVWR of 10,000 pounds or less manufactured after September 1, 1987, shall include:

4. S6(b) is revised to read as follows:

(b) In a vehicle with rear designated seating positions, a statement alerting vehicle owners that, according to accident statistics, children are safer when properly restrained in the rear seating positions than in the front seating positions.

5. S6(c) is revised to read as follows:

(c) In each passenger car, a diagram or diagrams showing the location of the shoulder belt anchorages required by this standard for the rear outboard designated seating positions, if shoulder belts are not installed as items of original equipment by the vehicle manufacturer at those positions.

6. S7 is revised to read as follows:

S7 *Installation Instructions.* The owner's manual in each vehicle manufactured on or after September 1, 1987, with an automatic restraint at the front right outboard designated seating position that cannot be used to secure a child restraint system when the automatic restraint is adjusted to meet the performance requirements of S5.1 of Standard No. 208 shall have:

(a) A statement that the automatic restraint at the front right outboard designated seating position cannot be used to secure a child restraint and, as appropriate, one of the following three statements:

(i) A statement that the automatic restraint at the front right outboard designated seating position can be adjusted to secure a child restraint system using attachment hardware installed as original equipment by the vehicle manufacturer;

(ii) A statement that anchorages for installation of a lap belt to secure a child restraint system have been provided at the front right outboard designated seating position; or

(iii) A statement that a lap or manual lap or lap/shoulder belt has been installed by the vehicle manufacturer at the front right outboard designated seating position to secure a child restraint.

(b) In each vehicle in which a lap or lap/shoulder belt is not installed at the front right outboard designated seating position as an item of original equipment, but the automatic restraint at that position can be adjusted by the vehicle owner to secure a child restraint system using an item or items of original equipment installed in the vehicle by the vehicle manufacturer, the owner's manual shall also have:

(i) A diagram or diagrams showing the location of the attachment hardware provided by the vehicle manufacturer.

(ii) A step-by-step procedure with a diagram or diagrams showing how to modify the automatic restraint system to secure a child restraint system. The instructions shall explain the proper routing of the attachment hardware.

(c) In each vehicle in which the automatic restraint at the front right outboard designated seating position cannot be modified to secure a child restraint system using attachment hardware installed as an original equipment by the vehicle manufacturer and a manual lap or lap/shoulder belt is not installed as an item of original equipment by the vehicle manufacturer, the owner's manual shall also have:

(i) A diagram or diagrams showing the locations of the lap belt anchorages for the front right outboard designated seating position.

(ii) A step-by-step procedure and a diagram or diagrams for installing the proper lap belt anchorage hardware and a Type 1 lap belt at the front right outboard designated seating position. The instructions shall explain the proper routing of the seat belt assembly and the attachment of the seat belt assembly to the lap belt anchorages.

Issued on August 12, 1986

_____
Diane K. Steed
Administrator

51 F.R. 29552
August 19, 1986

MOTOR VEHICLE SAFETY STANDARD NO. 210
Seat Belt Assembly Anchorages—Passenger Cars, Multipurpose Passenger Vehicles, Trucks, and Buses
(Docket No. 2-14; Notice No. 4)

S1. Purpose and scope. This standard establishes requirements for seat belt assembly anchorages to insure their proper location for effective occupant restraint and to reduce the likelihood of their failure.

S2. Application. This standard applies to passenger cars, multipurpose passenger vehicles, trucks, and buses.

S3. Definition. "Seat belt anchorage" means the provision for transferring seat belt assembly loads to the vehicle structure.

S4. Requirements.

S4.1 Type.

S4.1.1 Seat belt anchorages for a Type 2 seat belt assembly shall be installed for each foward-facing outboard designated seating position in passenger cars, other than convertibles, and for each designated seating position for which a Type 2 seat belt assembly is required by Standard No. 208 in vehicles other than passenger cars.

S4.1.2 Seat belt anchorages for a Type 1 or a Type 2 seat belt assembly shall be installed for each designated seating position, except a passenger seat in a bus or a designated seating position for which seat belt anchorages for a Type 2 seat belt assembly are required by S4.1.1.

S4.1.3 [Notwithstanding the requirement of paragraph S4.1.1, each vehicle manufactured on or after September 1, 1987, that is equipped with an automatic restraint at the front right outboard designated seating position that cannot be used for securing a child restraint system or cannot be adjusted by the vehicle owner to secure a child restraint system solely through the use of attachment hardware installed as an item of original equipment by the vehicle manufacturer shall have, at the manufacturer's option, either anchorages for a Type 1 seat belt assembly at that position or a Type 1 or Type 2 seat belt assembly at that position. (51 F.R. 29552—August 19, 1986. Effective: August 19, 1986)]

S4.2 Strength.

S4.2.1 Except for side-facing seats, the anchorage for a Type 1 seat belt assembly or the pelvic portion of a Type 2 seat belt assembly shall withstand a 5,000-pound force when tested in accordance with S5.1.

S4.2.2 The anchorage for a Type 2 seat belt assembly shall withstand 3,000-pound forces when tested in accordance with S5.2.

S4.2.3 Permanent deformation or rupture of a seat belt anchorage or its surrounding area is not considered to be a failure, if the required force is sustained for the specified time.

S4.2.4 Except for common seat belt anchorages for forward-facing and rearward-facing seats, floor-mounted seat belt anchorages for adjacent designated seating positions shall be tested by simultaneously loading the seat belt assemblies attached to those anchorages.

S4.3 Location. As used in this section, "forward" means in the direction in which the seat faces, and other directional references are to be interpreted accordingly. Anchorages for automatic and for dynamically tested seat belt assemblies that meet the frontal crash protection requirements of S5.1 of Standard No. 208 (49 CFR Part 571.208) are exempt from the location requirements of this section.

S4.3.1 Seat belt anchorages for Type 1 seat belt assemblies and the pelvic portion of Type 2 seat belt assemblies.

S4.3.1.1 In an installation in which the seat belt does not bear upon the seat frame, a line from the seating reference point to the nearest contact point of the belt with the hardware attaching it to the anchorage for an nonadjustable seat or from a point 2.50 inches forward of and 0.375 inch above the seating reference point to the nearest contact point of the belt with the hardware attaching it to the anchorage for an adjustable seat in its rearmost position, shall extend forward from the anchorage at an angle with the horizontal of not less than 20° and not more than 75°.

S4.3.1.2 In an installation in which the belt bears upon the seat frame, the seat belt anchorage, if not on the seat structure, shall be aft of the rearmost belt contact point on the seat frame with the seat in the rearmost position. The line from the seating reference point to the nearest belt contact point on the seat frame shall extend forward from that contact point at an angle with the horizontal of not less than 20° and not more than 75°.

S4.3.1.3 In an installation in which the seat belt anchorage is on the seat structure, the line from the seating reference point to the nearest contact point of the belt with the hardware attaching it to the anchorage shall extend forward from that contact point at an angle with the horizontal of not less than 20° and not more than 75°.

S4.3.1.4 Anchorages for an individual seat belt assembly shall be located at least 6.50 inches apart laterally, measured between the vertical centerlines of the bolt holes.

S4.3.2 Seat belt anchorages for the upper torso portion of Type 2 seat belt assemblies. With the seat in its full rearward and downward position and the seat back in its most upright position, the seat belt anchorage for the upper end of the upper torso restraint shall be located within the acceptable range shown in Figure 1, with reference to a two-dimensional manikin described in SAE Stand-

ard J826 (November 1962) whose "H" point is at the seating reference point and whose torso line is at the same angle from the vertical as the seat back.

S5. Test procedures. Each vehicle shall meet the requirements of S4.2 when tested according to the following procedures. Where a range of values is specified, the vehicle shall be able to meet the requirements at all points within the range.

S5.1 Seats with Type 1 or Type 2 seat belt anchorages. With the seat in its rearmost position, apply a force of 5,000 pounds in the direction in which the seat faces to a pelvic body block as described in Figure 2, restrained by a Type 1 or the pelvic portion of a Type 2 seat belt assembly, as applicable, in a plane parallel to the longitudinal centerline of the vehicle, with an initial force application angle of not less than 5° nor more than 15° above the horizontal. Apply the force at the onset rate of not more than 50,000 pounds per second. Attain the 5,000-pound force in not more than 30 seconds and maintain it for 10 seconds.

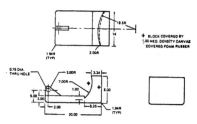

FIGURE 2 - BODY BLOCK FOR LAP BELT ANCHORAGE

S5.2 Seats with Type 2 seat belt anchorages. With the seat in its rearmost position, apply forces of 3,000 pounds in the direction in which the seat faces simultaneously to pelvic and upper torso body blocks as described in Figures 2 and 3, restrained by a Type 2 seat belt assembly, in a plane parallel to the longitudinal centerline of the vehicle, with an initial force application angle of not less than 5° nor more than 15° above the horizontal. Apply the forces at the onset rate of not

FIGURE 1 - LOCATION OF ANCHORAGE FOR UPPER TORSO RESTRAINT

PART 571; S 210-2

more than 30,000 pounds per second. Attain the 3,000-pound forces in not more than 30 seconds and maintain them for 10 seconds.

S6. Owner's Manual Information. [The owner's manual in each vehicle with GVWR of 10,000 pounds or less, manufactured after September 1, 1987, shall include:]

(a) A section explaining that all child restraint systems are designed to be secured in vehicle seats by lap belts or the lap belt portion of a lap-shoulder belt. The section shall also explain that children could be endangered in a crash if their child restraints are not properly secured in the vehicle.

[(b) In a vehicle with rear designated seating positions, a statement alerting vehicle owners that, according to accident statistics, children are safer when properly restrained in the rear seating positions than in the front seating positions.

(c) In each passenger car, a diagram or diagrams showing the location of the shoulder belt anchorages required by this standard for the rear outboard designated seating positions, if shoulder belts are not installed as items of original equipment by the vehicle manufacturer at those positions. (51 F.R. 29552—August 19, 1986. Effective: August 19, 1986)]

S7. Installation Instructions. [The owner's manual in each vehicle manufactured on or after September 1, 1987, with an automatic restraint at the front right outboard designated seating position that cannot be used to secure a child restraint system when the automatic restraint is adjusted to meet the performance requirements of S5.1 of Standard No. 208 shall have:

(a) A statement that the automatic restraint at the front right outboard designated seating position cannot be used to secure a child restraint and, as appropriate, one of the following three statements:

(i) A statement that the automatic restraint at the front right outboard designated seating position can be adjusted to secure a child restraint system using attachment hardware installed as original equipment by the vehicle manufacturer;

(ii) A statement that anchorages for installation of a lap belt to secure a child restraint system have been provided at the front right outboard designated seating position; or

(iii) A statement that a lap or manual lap or lap/shoulder belt has been installed by the vehicle manufacturer at the front right outboard designated seating position to secure a child restraint.

(b) In each vehicle in which a lap or lap/shoulder belt is not installed at the front right outboard designated seating position as an item of original equipment, but the automatic restraint at that position can be adjusted by the vehicle owner to secure a child restraint system using an item or items of original equipment installed in the vehicle by the vehicle manufacturer, the owner's manual shall also have:

(i) A diagram or diagrams showing the location of the attachment hardware provided by the vehicle manufacturer.

(ii) A step-by-step procedure with a diagram or diagrams showing how to modify the automatic restraint system to secure a child restraint system. The instructions shall explain the proper routing of the attachment hardware.

(c) In each vehicle in which the automatic restraint at the front right outboard designated seating position cannot be modified to secure a child restraint system using attachment hardware installed as an original equipment by the vehicle manufacturer and a manual lap or lap/shoulder belt is not installed as an item of original equipment by the vehicle manufacturer, the owner's manual shall also have:

(i) A diagram or diagrams showing the locations of the lap belt anchorages for the front right outboard designated seating position.

(ii) A step-by-step procedure and a diagram or diagrams for installing the proper lap belt anchorage hardware and a Type 1 lap belt at the front right outboard designated seating position. The instructions shall explain the proper routing of the seat belt assembly and the seat belt attachment of the assembly to the lap belt anchorages. (51 F.R. 29552—August 19, 1986. Effective: August 19, 1986)]

Issued on August 12, 1986

Diane K. Steed
Administrator

51 F.R. 29552
August 19, 1986

MOTOR VEHICLE SAFETY STANDARD NO. 211

Wheel Nuts, Wheel Discs, and Hub Caps—Passenger Cars and Multipurpose Passenger Vehicles

S1. Purpose and scope. This standard precludes the use of wheel nuts, wheel discs, and hub caps that constitute a hazard to pedestrians and cyclists.

S2. Application. This standard applies to passenger cars, multipurpose passenger vehicles, and passenger cars and multipurpose passenger vehicle equipment.

S3. Requirements. Wheel nuts, hub caps, and wheel discs for use on passenger cars and multipurpose passenger vehicles shall not incorporate winged projections.

INTERPRETATION

A clarification of the term "wheel nut" as used in the requirements section S3 of Standard No. 211 has been requested. This section states that "wheel nuts, hub caps, and wheel discs for use on passenger cars and multipurpose passenger vehicles shall not incorporate winged projections." A "wheel nut" is an exposed nut that is mounted at the center or hub of a wheel, and not the ordinary small hexagonal nut, one of several which secures a wheel to an axle, and which is normally covered by a hub cap or wheel disc.

Issued on July 22, 1969.

F. C. Turner
Federal Highway Administrator

32 F.R. 2416
February 3, 1967

PREAMBLE TO MOTOR VEHICLE SAFETY STANDARD NO. 212

Windshield Mounting—Passenger Cars

A proposal to amend Part 371 of the Federal Motor Vehicle Safety Standards by adding a Standard No. 212, Windshield Mounting—Passenger Cars, was published as an advance notice of proposed rule making an October 14, 1967 (32 F.R. 14281) and a notice of proposed rule making on December 28, 1967 (32 F.R. 20866).

Interested persons have been given the opportunity to participate in the making of this amendment, and careful consideration has been given to all relevant matter presented.

This new standard requires that, when tested as prescribed, each passenger car windshield mounting must retain either: (1) not less than 75% of the windshield periphery; or (2) not less than 50% of that portion of the windshield periphery on each side of the vehicle longitudinal centerline, if an unrestrained 95th percentile adult male manikin is seated in each outboard front seating position.

Several comments objected to the proposed standard and in some cases urged that more research should be done before any type of windshield mounting is required. The standard, is however, part of an integrated program aimed at accomplishing the widely accepted safety goal of keeping occupants within the confines of the passenger compartment during a crash. One major step in this program is the utilization of the laminated glazing material prescribed in Federal motor vehicle safety standard No. 205, which has resulted in a marked reduction in serious head injury to occupants known to have struck the windshield. The windshield mounting retention requirement prescribed in this standard takes advantage of this improved glazing material and will further minimize the likelihood of occupants being thrown from the vehicle during a crash.

Several comments requested reduction of the 75% retention requirement to 50%. The Administrator concludes that, as an alternative, 50% retention is acceptable if: (1) an unrestrained 95% percentile adult male manikin is seated in each outboard front seating position when the test procedure is performed, and (2) at least 50% of that portion of the windshield periphery on each side of the vehicle longitudinal centerline is retained.

Several comments requested that the phrase "or approved equivalent" be added to the "Demonstration procedures" provision. § 371.11 of the Federal motor vehicle safety standards provides that "an approved equivalent may be substituted for any required destructive demonstration procedure." Consequently, inclusion of the phrase requested is not necessary.

In consideration of the foregoing, § 371.21, of Part 371 of the Federal motor vehicle safety standards is amended by adding Standard No. 212, "Windshield Mounting—Passenger Cars," as set forth below, effective January 1, 1970.

This rule-making action is taken under the authority of sections 103 and 119 of the National Traffic and Motor Vehicle Safety Act of 1966 (P.L. 89-563, 15 U.S.C. §§ 1392 and 1407) and the delegation of authority contained in Part 1 of the Regulations of the Office of the Secretary of Transportation (49 CFR Part 1).

Issued in Washington, D.C. on August 13, 1968.

John R. Jamieson, Deputy
Federal Highway Administrator

33 F.R. 11652
August 16, 1968

PREAMBLE TO AMENDMENT TO MOTOR VEHICLE SAFETY STANDARD NO. 212

Windshield Mounting

(Docket No. 69–29; Notice 5)

This notice amends Motor Vehicle Safety Standard No. 212, 49 CFR 571.212, *Windshield Mounting*, to extend its applicability to multipurpose passenger vehicles, trucks, and buses having a gross vehicle weight rating (GVWR) of 10,000 pounds or less, except for forward control vehicles and open-body type vehicles with folding or removable windshields, and to coordinate its test procedures with those of Standard No. 208, 49 CFR 571.208, *Occupant Crash Protection*.

An advance notice of proposal rulemaking was published September 16, 1969 (34 FR 14438), followed by notices of proposed rulemaking published on August 23, 1972 (37 FR 16979) and January 18, 1974 (39 FR 2274). This notice is based on the latter notice of proposed rulemaking, and responds to the comments submitted thereto.

The final rule retains the proposed rule's extension to multipurpose passenger vehicles, trucks, and buses having a gross vehicle weight rating (GVWR) of 10,000 pounds or less. However, forward control vehicles and open-body vehicles with fold-down windshields are excluded from the application of the standard because of the impracticability of complying with the requirements.

Many manufacturers objected to the requirement in the proposal that the dummies used in the test vehicle not be restrained by active restraint systems. Upon impact in a crash test, unrestrained dummies tend to fly about the passenger compartment, damaging the dummies.

In 1972 the NHTSA proposed the amendment of Standard No. 212 (37 FR 16979) to specify a 75 percent retention requirement using restrained dummies. The purpose of the proposal was to eliminate optional retention requirements and to permit dynamic testing consistent with other safety standards. In 1974 another approach was taken with the NHTSA proposing (39 FR 2274) a 50 percent retention requirement using unrestrained dummies, in anticipation of the passive restraint requirements that were to be included in Standard No. 208. Having the benefit of a large number of comments on both proposals the NHTSA has determined that both are suitable, the 1972 approach for vehicles equipped with active restraints, where dummy damage would be great if the dummy were unrestrained, and the 1974 approach for vehicles equipped with passive restraints, since the dummy would not contact the windshield.

The frontal barrier crash test conditions specified in the final rule are substantially similar to those of Standard No. 208, *Occupant Crash Protection*, Standard No. 219, *Windshield Zone Intrusion*, and Standard No. 301, *Fuel System Integrity*. This will allow compliance testing for these standards in one crash test under certain circumstances. In this way, much of the expense associated with crash testing can be reduced.

Most of the manufacturers who commented on the proposal objected to the requirement that the vehicle be tested at a temperature range of 15° F to 110° F. Some manufacturers objected that the higher temperatures would damage sensitive instrumentation. Others argued that the range should be coordinated with that of Standard No. 301 (49 CFR 571.301) or with ISO regulations. Some asserted that they would have to build expensive test facilities in order to conduct tests at the temperature extremes. The NHTSA has determined that testing over the specified range is necessary, in light of the fact that windshield moldings have significantly different retention capabilities at different temperatures. The NHTSA recognizes that certain additional expenses may be entailed in testing over the specified temperature range. However, the safety need to ensure adequate windshield retention justifies the additional expense.

In consideration of the foregoing, Standard No. 212, 49 CFR 571.212, is amended to read as set forth below.

Effective date: September 1, 1977.

(Sec. 103, 119, Pub. L. 89563, 80 Stat. 718 (15 U.S.C. 1392, 1407); delegation of authority at 49 CFR 1.50)

Issued on: August 23, 1976.

John W. Snow
Administrator

41 F.R. 36493
August 30, 1976

PREAMBLE TO AMENDMENT TO MOTOR VEHICLE SAFETY STANDARD NO. 212

Windshield Mounting

(Docket No. 69–29; Notice 6)

This notice responds to nine petitions for reconsideration of a recent amendment (41 FR 36493, August 30, 1976) of Safety Standard No. 212, *Windshield Mounting*, by extending the effective date of the amendment from September 1, 1977, to September 1, 1978, and by excluding "walk-in van-type" vehicles from the standard's applicability. Other aspects of the petitions for reconsideration are denied.

Dates: The amendment of August 30, 1976, will be effective September 1, 1978. The change in the effective date and the amendment to exclude "walk-in van-type" vehicles from the standard's applicability should be changed in the text of the Code of Federal Regulations, effective August 4, 1977.

For Further Information Contact:

Robert Nelson
National Highway Traffic Safety Administration
Washington, D.C. 20590
(202 426-2802)

Supplementary Information: Safety Standard No. 212, *Windshield Mounting* (49 CFR Part 571.212), was amended August 30, 1976, to modify the performance requirements and test procedures of the standard and to extend the standard's applicability to multipurpose passenger vehicles, trucks, and buses having a gross vehicle weight rating of 10,000 pounds or less. Petitions for reconsideration were received from International Harvester (IH), Jeep Corporation, American Motors Corporation (AMC), Volvo of America Corporation, Toyo Kogyo Co., General Motors Corporation (GM), Rolls Royce Motors, Nissan Motor Co. Ltd., and Leyland Cars.

Requests from some of these petitioners that the new provisions of Standard No. 212 (49 CFR 571.212) be withdrawn entirely are hereby denied, but several modifications are undertaken by the National Highway Traffic Safety Administration (NHTSA), based on a review of the information and arguments submitted.

Nearly all of the petitioners requested that the effective date of the new provisions be changed from September 1, 1977, to September 1, 1978. Petitioners argued that a lead time of one year will be insufficient to accomplish design changes and retooling necessary to adapt passenger-car windshield technology to other vehicle types. Petitioners also pointed out that the specification of a temperature range in the test conditions will require manufacturers to undertake more extensive certification testing than in the past.

The NHTSA has determined that the requests for additional lead time are justified in light of the information submitted regarding design changes that some manufacturers will undertake. The petitions are, therefore, granted in part and the effective date of the new provisions is postponed to September 1, 1978.

In conformity with the agency's 1972 and 1974 proposals (37 FR 16979, August 23, 1972) (39 FR 2274, January 18, 1974), an optional means of meeting the retention requirement (that exists in the present provisions) was eliminated by the August 30, 1976, amendments. This was done to reduce the amount of necessary compliance testing and to encourage "simultaneous" certification testing of separate standards where practicable. As proposed in 1972, the "75-percent alternative" (retention of 75 percent of the windshield periphery—dummies properly restrained) was made mandatory for al vehicles not equipped with passive restraints. In this way, windshield retention tests could be performed at the same time as tests already required for fuel system integrity (49 CFR 571.301-75) that specify restrained dummies.

While some additional weight is added to the vehicle by the required dummies, it is the minimum necessary to permit "simultaneous" testing, and the dummies are restrained so that there is only incidental, if any, contact with the windshield. Thus, the "75-percent alternative" specified in the amendments is, basically, a continuation of the existing requirement that manufacturers have been meeting for years.

The 1974 proposal to adopt the "50-percent option" (retention of 50 percent of the windshield periphery on each side of the windshield—dummies unrestrained) was vigorously objected to by manufacturers because of the damage that could occur to dummies during impact with the windshield. Also, the fuel system integrity standard was made final in a form that required restraining the dummies by safety belts if provided. It was apparent that the "50-percent option" should only become mandatory as proposed for vehicles equipped with passive restraint systems that could protect the dummy against impact damage. In the case of air cushion restraint systems, of course, some contact with the windshield by the cushion or incidental contact by the dummy is expected during the crash test. For this reason, the somewhat less stringent "50-percent option" was made final for vehicles equipped with passive restraints.

AMC argued that this distinction between vehicles is unjustified. The only reason put forward by AMC was that "dummy impact is not a critical factor in determining windshield retention." This reason does not, however, support the AMC request for a reduction in retention performance from the 75-percent level presently being met. Rather, it argues for an increase in the 50-percent level established for those vehicles in which the NHTSA estimated that dummy and restraint contact could affect results. If AMC believes that the distinction is not justified, the agency will review further evidence to increase the 50-percent requirement (for passive-equipped vehicles) to the 75-percent level presently being met in most of today's passenger cars.

Several commenters objected that the final rule differed in some respects from the 1972 and 1974 proposals to amend Standard No. 212, taken separately. AMC, Volvo, and Jeep petitioned to revoke the separate retention requirements for vehicles with different restraint systems, on the grounds that such a distinction had never been proposed. Jeep Corporation also objected to extension of the standard's applicability to MPV's, trucks, and buses because of variations in language from the proposals.

As earlier noted, the requirement for 75-percent retention conforms to the 1972 proposal. The only variation from the 1972 proposal was to implement the performance levels proposed in 1974 for the vehicles that might be equipped with passive restraints. It is the agency's view that "a description of the subjects and issues involved" in the rulemaking action was published in the Federal Register as required by the Administrative Procedure Act (the Act) (5 U.S.C. § 553(b)(2)), permitting opportunity for comment by interested persons. A reading of the cases on this provision of the Act supports the agency's view.

Volvo's petition objected to the fact that the amendments specify the use of restrained dummies in the test procedures. Volvo stated that unrestrained dummies should be used because in actual crash conditions it is the head of an unrestrained occupant that is most likely to impact and substantially load the windshield, since the head of a restrained occupant would not normally contact the windshield.

While Volvo's statement is true, it must be understood that test procedures specified in the standards cannot simulate every element of actual crash conditions. Rather, the procedures are based on a variety of considerations, including test expense and degree of complexity. There were many comments to the prior notices proposing the amendments in question that urged the use of restrained dummies, due to the possibility of damage to the expensive dummies during the barrier crash tests. These comments were taken into consideration prior to issuance of the final rule. Also, the NHTSA concluded that the vehicle deceleration forces are the primary forces affecting windshield retention and not the impact of occupants with the windshield. The restrained dummies are required, primarily, for purposes of permitting simultaneous testing. The NHTSA concludes that the retention requirements and test procedures specified in the amendments will ensure that vehicles are equipped with windshields that provide the needed protection for occupant safety.

Volvo's petition also argued that Standard No. 212 "must include a measurement procedure that weights the various segments of the windshield periphery in a technically accurate maner." Volso points to tests it has conducted which indicate that "when the unrestrained occupant's head impacts and substantially loads the windshield, the loading will most likely occur in the windshield's upper regions and *not* uniformly throughout the windshield."

While it is recognized that the degree of dislodging of the windshield from its mounting may vary at different locations around the periphery of the windshield, sufficient information is not available on which to base varying retention requirements (for different areas of the windshield). Further, the specification of retention requirements in the terms suggested by Volvo was not proposed by the agency in 1972 or 1974. This aspect of Volvo's petition is therefore denied.

Several petitioners objected to the specification of a temperature range in the test conditions and asked that this provision be withdrawn. Rolls Royce Motors argued that the amendment will require additional tests to determine the most critical temperature for windshield retention and stated that this would greatly increase the burden on low-volume manufacturers. General Motors and Jeep Corporation stated that the expansion of the test requirements over a wide temperature range adds to the stringency of the standard without any evidence of a safety need. American Motors petitioned to remove the 15°F to 110°F temperature range from the barrier test conditions on the basis that "it was not specified as a barrier test condition in the proposal for rulemaking," and on the basis that there are laboratory tests that can serve the same purpose.

The NHTSA denies all petitions to withdraw the temperature range from the standard. As stated in the preamble to the final rule, testing over the specified range is necessary in light of the fact that windshield moldings have significantly different retention capabilities at different temperatures. This fact was graphically confirmed by NHTSA compliance testing in which windshields retained at low temperatures were dislodged at higher temperatures (in identical vehicles). Concerning the objection of American Motors, the temperature range was proposed in paragraph S4 of the 1974 proposal to amend Standard No. 212 (39 FR 2274).

General Motors recommended that the temperature range be revised to specify 66°F to 78°F limits, to coordinate the Standard 212 test with the calibration conditions for the Part 572 dummy. General Motors argued that this would reduce the number of barrier crash tests that would be required.

The NHTSA rejects this recommendation. The Part 572 dummies are conditioned in the 66°F–78°F temperature range for calibration purposes in those standards in which the dynamic dummy response is part of the requirements of the standard. Since the response of the dummy is not directly involved in the performance requirements of Standard No. 212, the temperature of the dummies is not significant. Therefore, it is not necessary to restrict the temperature range of Standard No. 212 to correspond to the calibration temperature range of the Part 572 dummies. For purposes of simultaneous testing, manufacturers could devise a means to control the immediate environment of the test dummy within the 66°F–78°F calibration temperature range, independent of the temperature range specified in Standard No. 212.

General Motors also argued that there could be considerable variation in vehicles condition and test results, depending on when and where the vehicle is tested, since there could be an air temperature of 110°F while windshield components are at a much higher temperature due to "sun load." General Motors, therefore, requested that the temperature requirement be clarified to specify that the temperature of the entire vehicle be stabilized between 15°F and 110°F prior to the test.

The NHTSA does not intend that vehicles be tested with the windshield components at temperatures higher than 110°F. For purposes of clarification, paragraph S6.5 of the new provisions is revised to specify that the windshield mounting material, and all vehicle components in direct contact with the mounting material are to be at any temperature between 15°F and 110°F. Presumably this could be accomplished by localized heating or cooling of the vehicle components or by any other method chosen, in the exercise of due care, by a manufacturer.

The August 1976 amendments to Standard No. 212 modified the application section to include multipurpose passenger vehicles, trucks and buses having a gross weight rating of 10,000 pounds or less. "Open-body type" vehicles and "forward control" vehicles were excluded because of the impracticability of applying the barrier crash test to these vehicles. General Motors has pointed out that the NHTSA failed to exclude "walk-in van-type" vehicles, which have essentially the same configuration and amount of front-end crush space as forward control vehicles.

The NHTSA recently addressed this same issue in connection with Standard No. 219, *Windshield Zone Intrusion*, and, in the absence of any objections, amend that standard to exclude walk-in van-type vehicles (41 FR 54945, December 16, 1976). On reconsideration of the extended applicability of Standard No. 212 to these vehicles, the agency concludes that the same rationale applies. Accordingly, applicability of Standard No. 212 to walk-in van-type vehicles is withdrawn.

In consideration of the foregoing, the effective date of the amendment to Standard No. 212 (49 CFR 571.212) published August 30, 1976 (41 FR 36493) is changed from September 1, 1977, to September 1, 1978, and paragraphs S3 and S6.5 of that text are modified. . . .

(Sec. 103, 119, Pub. L. 89-563, 80 Stat. 718 (15 U.S.C. 1392, 1407); delegation of authority at 49 CFR 1.50.)

Issued on June 29, 1977.

Joan Claybrook
Administrator

42 F.R. 34288
July 5, 1977

PREAMBLE TO AN AMENDMENT TO MOTOR VEHICLE SAFETY STANDARD NO. 212

Windshield Mounting; Windshield Zone Intrusion

(Docket No. 79-14; Notice 02)

ACTION: Final Rule.

SUMMARY: This notice amends two safety standards, Standard No. 212, *Windshield Mounting,* and Standard No. 219, *Windshield Zone Intrusion,* to limit the maximum unloaded vehicle weight at which vehicles must be tested for compliance with these standards. This action is being taken in response to petitions from the Truck Body and Equipment Association and the National Truck Equipment Association asking the agency to amend the standards to provide relief from some of the test requirements for final-stage manufacturers. Many of these small manufacturers do not have the sophisticated test devices of major vehicle manufacturers. The agency concludes that the weights at which vehicles are tested can be lessened while providing an adequate level of safety for vehicles such as light trucks and while ensuring that compliance with these standards does not increase their aggressivity with respect to smaller vehicles.

EFFECTIVE DATE: Since this amendment relieves a restriction by easing the existing test procedure and will not impose any additional burdens upon any manufacturer, it is effective (upon publication).

FOR FURTHER INFORMATION CONTACT:

Mr. William Smith, Crashworthiness Division, National Highway Traffic Safety Administration, 400 Seventh Street, S.W., Washington, D.C. 20590 (202-426-2242)

SUPPLEMENTARY INFORMATION:

On August 2, 1979, the National Highway Traffic Safety Administration published a notice of proposed rulemaking (44 FR 45426) relating to two safety standards: Standard Nos. 212, *Windshield Mounting,* and 219 *Windshield Zone Intrusion.* That notice proposed two options for amending the test procedures of the standards that were designed to ease the compliance burdens of small final-stage manufacturers.

The agency issued the proposal after learning that final-stage manufacturers were frequently unable to certify certain vehicles in compliance with these two safety standards. The problem arises because of weight and center of gravity restrictions imposed upon the final-stage manufacturer by the incomplete vehicle manufacturer. (The final-stage manufacturer typically purchases an incomplete vehicle from an incomplete vehicle manufacturer, usually Ford, General Motors or Chrysler.) The incomplete vehicle usually includes the windshield and mounting but does not include any body or work-performing equipment. Since the incomplete vehicle manufacturer installs the windshield, it represents to the final-stage manufacturer that the windshield will comply with the two subject safety standards. In making this representation, however, the incomplete vehicle manufacturer states that the representation is contingent on the final-stage manufacturer's adherence to certain restrictions. Any final-stage manufacturer that does not adhere to the restrictions imposed by the incomplete vehicle manufacturer must recertify the vehicle based upon its own information, analysis, or tests. The major restrictions imposed by the incomplete vehicle manufacturers on the final-stage manufacturer involve weight and center of gravity limitation. In many instances, these limitations have made it impossible for final-stage manufacturers either to rely on the incomplete vehicle manufacturer's certification or to complete vehicles on the same chassis that they were accustomed to using (prior to the extension of the two safety standards to these vehicle types). As a result, the final-stage manufacturer is faced either with buying the same chassis as before and recertifying them or with buying more expensive chassis with higher GVWR's and less stringent weight and center of gravity limitations.

The agency has tried several different ways to alleviate this problem for the final-stage manufacturer. The NHTSA has met with representatives of the major incomplete vehicle manufacturers to encourage them to respond voluntarily by strengthening their windshield structures and reducing the restrictions that they currently impose upon final-stage manufacturers. The agency also discussed the possibility of its mandating these actions by upgrading Standards Nos. 212 and 219. Ford and General Motors indicated that the making of any major changes in these standards could lead to their deciding to discontinue offering chassis for use in the manufacturing of multi-stage vehicles. They said that such chassis were a very small percentage of their light truck sales and that, therefore, they would not consider it worth the cost to them to make any extensive modifications in their vehicles. NHTSA also asked the incomplete vehicle manufacturers to be sure that they have properly certified their existing vehicles and that they are not imposing unnecessarily restrictive limitations upon final-stage manufacturers. To this agency's knowledge, these vehicle manufacturers have neither undertaken any strengthening of their vehicles' windshield structures nor lessened any of their restrictions.

At the same time that the agency was made aware of the final-stage manufacturers' problems of certifying to these standards, the agency was becoming concerned about the possibility that compliance of some light trucks and vans with these standards might have made the vehicles more aggressive with respect to smaller passenger cars that they might impact. According to agency information, if these standards require a substantial strengthening of vehicle frames, the aggressivity of the vehicles is increased. Therefore, as a result of the agency's concern about aggressivity and its desire to address the certification problems of final-stage manufacturers in a manner that would not lead to a cessation of a chassis sales to those manufacturers, the agency issued the August 1979 proposal. The agency hoped that the proposal would allow and encourage incomplete vehicle manufacturers to reduce their weight and center of gravity restrictions, thereby easing or eliminating the compliance test burdens of final-stage manufacturers. The agency believed that this could occur using either option, because either would result in vehicles being tested at lower weights. Currently vehicles are tested under both standards at their unloaded vehicle weights plus 300 pounds.

The first option would have required some vehicles whose unloaded vehicle weights exceeded 4,000 pounds to be tested by being impacted with a 4,000 pound moving barrier. The second option proposed by the agency would have required vehicles to be tested at their unloaded vehicle weight up to a maximum unloaded vehicle weight of 5,500 pounds. This option was suggested to the agency by several manufacturers and manufacturer representatives.

Comments on Notice

In response to the agency's notice, nine manufacturers and manufacturer representatives submitted comments. All of the commenters supported some action in response to the problems of final-stage manufacturers. Most of the commenters also suggested that the agency's second alternative solution was more likely to achieve reductions in the restrictions being imposed by incomplete vehicle manufacturers. The first option would have created a new, unproven test procedure, and manufacturers would have been cautious in easing center of gravity or weight restrictions based upon this test procedure. Accordingly, most commenters were not sure that the first option would achieve the desired results. The consensus was, therefore, that the second option should be adopted.

Some manufacturers recommended that both options be permitted allowing the manufacturer to decide how to test its vehicles. The agency does not agree with this recommendation. Not only would it be more difficult and expensive to enforce a standard that has alternative test procedures, but most manufacturers prefer the 5,500 pound weight limit option. The NHTSA concludes that as a result of the comments supporting the 5,500 pound maximum test weight, that this is an acceptable procedure for testing compliance with these two standards. Therefore, the standards are amended to incorporate this procedure.

The major incomplete vehicle manufacturers commenting on the notice suggested that testing vehicles at a maximum weight of 5,500 pounds might provide some immediate relief. None of the major incomplete vehicle manufacturers provided any information concerning how substantial that relief might be. Ford indicated that any relief might be limited.

The agency believes that the incomplete vehicle manufacturers must accept the responsibility for establishing reasonable restrictions upon their incomplete vehicles. The NHTSA has not been provided with sufficient evidence substantiating the statements of the incomplete vehicle manufacturers that their existing restrictions are reasonable. In fact, some evidence indicates that unnecessarily stringent restrictions are being imposed because incomplete vehicle manufacturers do not want to conduct the necessary testing to establish the appropriate weight and center of gravity restrictions. Since this amendment should reduce the severity of the test procedures, the agency concludes that incomplete vehicle manufacturers should immediately review their certification test procedures and reduce the restrictions being passed on to final-stage manufacturers.

Due to changes in the light truck market, there is reason to believe that the incomplete vehicle manufacturers will be more cooperative than when the agency spoke to them before beginning this rulemaking. At that time, light truck sales were still running well. Now that these sales are down, these manufacturers may be more solicitous of the needs of the final-stage manufacturers. If relief is not provided by the incomplete vehicle manufacturers, then the agency will consider taking additional steps, including the upgrading of Standards Nos. 212 and 219 as they apply to all light trucks.

General Motors (GM) questioned one of the agency's rationales for issuing the notice of proposed rulemaking. GM stated that the agency concludes that this action will provide a more appropriate level of safety for the affected vehicles while the initial extension of these standards to the affected vehicles provides, in GM's view, only a slight increase in the level of safety of the vehicles. GM indicates that since the application of these standards to the affected vehicles provides only slight benefits and since this amendment will

reduce those benefits, the standards should not apply to light trucks and vans. The agency disagrees with this suggestion.

The agency is currently reviewing the applicability of many of its safety standards to determine whether they ought to be extended to light trucks and other vehicles. Accident data clearly indicate the benefits that have resulted from the implementation of safety standards to cars. The fatality rate for passenger cars has decreased substantially since the implementation of a broad range of safety standards to those vehicles. On the other hand, light trucks and vans have not had a corresponding reduction in fatality rates over the years. The agency attributes much of this to the fact that many safety standards have not been applied to those vehicles. Since those vehicles are becoming increasingly popular as passenger vehicles, the agency concludes that safety standards must apply to them.

In response to GM's comment that this reduction in the test requirements for Standard Nos. 212 and 219 will remove all benefits derived by having the standards apply to those vehicles, the agency concludes that GM has misinterpreted the effects of this amendment. This amendment will reduce somewhat the compliance test requirements for those light trucks and vans with unloaded vehicle weights in excess of 5,500 pounds. It will not affect light trucks with unloaded vehicle weights below 5,500 pounds. According to agency information, approximately 25 percent of the light trucks have unloaded vehicle weights in excess of 5,500 while the remainder fall below that weight. As a result of weight reduction to improve fuel economy, it is likely that even more light trucks will fall below the 5,500 pound maximum test weight in the future. Therefore, this amendment will have no impact upon most light trucks and vans. In light of the small proportion of light trucks and vans affected by this amendment and considering the potential benefits of applying these standards to all light trucks and vans, the agency declines to adopt GM's suggestion that the standards be made inapplicable to these vehicles.

With respect to GM's question about the appropriate level of safety for light trucks, the agency's statement in the notice of proposed rulemaking was intended to show that the safety of light trucks and vans cannot be viewed without considering the relative safety of lighter vehicles that they may impact. Accordingly, the level of safety that the agency seeks to achieve by this and other safety standards is determined by balancing the interests of the occupants of passenger cars and heavier vehicles.

GM also questioned the agency's statement that vehicle aggressivity may be increased by imposing too severe requirements on these vehicles. GM suggested that no evidence exists that vehicle aggressivity is increased as a result of complying with these standards.

The agency stated in the proposal that it was concerned that compliance with the standards as they now exist might have increased the aggressivity of the vehicles, thereby harming the occupants of passenger cars that are impacted by these larger, more rigid vehicles. The agency is now beginning to examine the full range of vehicle aggressivity problems. The docket for this notice contains a paper recently presented by a member of our staff to the Society of Automotive Engineers on this subject. The agency tentatively concludes, based upon the initial results of our research and analysis, that vehicle aggressivity could be a safety problem and that the agency considers that possibility in issuing its safety standards. The NHTSA notes that Volkswagen applauds the agency's recognition of the vehicle aggressivity factor in safety.

As to GM's argument that compliance with the standards may not have increased vehicle aggressivity, our information on this point came from the manufacturers. The manufacturers indicated that compliance with Standards 212 and 219 requires strengthening the vehicle frame. This makes a vehicle more rigid. Our analysis indicates that making a vehicle more rigid may also make it more aggressive. Therefore, the agency concludes partially on the basis of the manufacturer's information, that compliance with the safety standards as they are written may have increased the aggressivity of the vehicles.

Ford Motor Company suggested that, rather than change these two particular standards, the agency should amend the certification regulation (Part 568) to state that any vehicle that is barrier tested would be required only to comply to an unloaded vehicle weight of 5,500 pounds or less. Ford suggested that this would standardize all of the tests and provide uniformity.

The agency is unable to accept Ford's recommendation for several reasons. First, the certification regulation is an inappropriate place to put a test requirement applicable to several standards. The tests' requirements of the standards should be found in each standard. Second, the Ford recommendation would result in a reduction of the level of safety currently imposed by Standard No. 301, *Fuel System Integrity*.

As we stated earlier and in several other notices, the agency is legislatively forbidden to modify Standard No. 301 in a way that would reduce the level of safety now required by that standard. Even without this legislative mandate, the agency would not be likely to relieve the burdens imposed by Standard No. 301. That standard is extremely important for the prevention of fires during crashes. Compliance of a vehicle with this standard not only protects the occupants of the vehicle that is in compliance but also protects the occupants of vehicles that it impacts. The agency concludes that the standard now provides a satisfactory level of safety in vehicles, and NHTSA would not be likely to amend it to reduce these safety benefits even if such an amendment were possible.

With respect to fuel system integrity, several manufacturers suggested that the agency had underestimated the impact of that standard upon weight and center of gravity restrictions. These commenters indicated that compliance with that standard requires more than merely adding shielding to the fuel systems of the vehicles. The agency is aware that compliance with that standard in certain instances has imposed restrictions upon manufacturers. Nonetheless, the agency continues to believe that as a result of this amendment, the chassis manufacturers will be able to reduce their weight and center of gravity restrictions while still maintaining the compliance of their vehicles with Standard No. 301.

Chrysler commented that the agency should consider including the new test procedure in Standard No. 204 and all other standards that require barrier testing. The agency has issued a notice on Standard No. 204 (44 FR 68470) stating that it was considering a similar test provision for that standard. The agency also is aware that any barrier test requirement imposed upon vehicles subject to substantial modifications by final-stage manufacturers will create problems for the final-stage manufacturers. Accordingly, the agency will consider the special problems of these manufacturers prior to the the issuance of standards that might affect them and will attempt to make the test requirements of the various standards consistent wherever possible.

The agency has reviewed this amendment in accordance with Executive Order 12044 and concludes that it will have no significant economic or other impact. Since the regulation relieves some testing requirements, it may slightly reduce costs associated with some vehicles. Accordingly, the agency concludes that this is not a significant amendment and a regulatory analysis is not required.

In accordance with the foregoing, Volume 49 of the Code of Federal Regulations Part 571 is amended by adding the following sentence to the end of paragraph S6.1(b) of Standard No. 212 (49 CFR 571.212) and paragraph S7.7(b) of Standard No. 219 (49 CFR 571.219).

Vehicles are tested to a maximum unloaded vehicle weight of 5,500 pounds.

The authors of this notice are William Smith of the Crashworthiness Division and Roger Tilton of the Office of Chief Counsel.

Issued on March 28, 1980.

Joan Claybrook
Administrator

45 F.R. 22044
April 3, 1980

MOTOR VEHICLE SAFETY STANDARD NO. 212

Windshield Mounting

S1. Scope. This standard establishes windshield retention requirements for motor vehicles during crashes.

S2. Purpose. The purpose of this standard is to reduce crash injuries and fatalities by providing for retention of the vehicle windshield during a crash, thereby utilizing fully the penetration-resistance and injury-avoidance properties of the windshield glazing material and preventing the ejection of occupants from the vehicle.

S3. Application. This standard applies to passenger cars and to multipurpose passenger vehicles, trucks, and buses having a gross vehicle weight rating of 10,000 pounds or less. However, it does not apply to forward control vehicles, walk-in van-type vehicles, or to open-body-type vehicles with fold-down or removable windshields.

S4. Definition. "Passive restraint system" means a system meeting the occupant crash protection requirements of S5 of Standard No. 208 by means that require no action by vehicle occupants.

S5. Requirements. When the vehicle traveling longitudinally forward at any speed up to and including 30 mph impacts a fixed collision barrier that is perpendicular to the line of travel of the vehicle, under the conditions of S6, the windshield mounting of the vehicle shall retain not less than the minimum portion of the windshield periphery specified in S5.1 and S5.2.

S5.1 Vehicles equipped with passive restraints.

Vehicles equipped with passive restraint systems shall retain not less than 50 percent of the portion of the windshield periphery on each side of the vehicle longitudinal centerline.

S5.2 Vehicles not equipped with passive restraints. Vehicles not equipped with passive restraint systems shall retain not less than 75 percent of the windshield periphery.

S6. Test conditions. The requirements of S5 shall be met under the following conditions:

S6.1 The vehicle, including test devices and instrumentation, is loaded as follows:

(a) Except as specified in S6.2, a passenger car is loaded to its unloaded vehicle weight plus its cargo and luggage capacity weight, secured in the luggage area, plus a 50th-percentile test dummy as specified in Part 572 of this chapter at each front outboard designated seating position and at any other position whose protection system is required to be tested by a dummy under the provisions of Standard No. 208. Each dummy is restrained only by means that are installed for protection at its seating position.

(b) Except as specified in S6.2, a multipurpose passenger vehicle, truck, or bus is loaded to its unloaded vehicle weight plus 300 pounds or its rated cargo and luggage capacity, whichever is less, secured to the vehicle, plus a 50th-percentile test dummy as specified in Part 572 of this chapter at each front outboard designated seating position and at any other position whose protection system is required to be tested by a dummy under the provisions of Standard No. 208. Each dummy is restrained only by means that are installed for protection at its seating position. The load is distributed so that the weight on each axle as measured at the tire-ground interface is in proportion to its GAWR. If the weight on any axle when the vehicle is loaded to its unloaded vehicle weight plus dummy weight exceeds the axle's proportional share of the test weight, the remaining weight is placed so that the weight on that axle remains the same. For the purposes of this section, unloaded vehicle weight does not include the weight of workperforming accessories. Vehicles are tested to a maximum unloaded vehicle weight of 5,500 pounds.

S6.2 The fuel tank is filled to any level from 90 to 95 percent of capacity.

S6.3 The parking brake is disengaged and the transmission is in neutral.

S6.4 Tires are inflated to the vehicle manufacturer's specifications.

S6.5 The windshield mounting material and all vehicle components in direct contact with the mounting material are at any temperature between 15°F and 110°F.

41 F.R. 36493
August 30, 1976

PREAMBLE TO MOTOR VEHICLE SAFETY STANDARD NO. 213

Child Restraint Systems, Seat Belt Assemblies, and Anchorages

(Docket No. 74-9; Notice 6)

ACTION: Final rule.

SUMMARY: This rule establishes a new Standard No. 213, *Child Restraint Systems,* which applies to all types of child restraints used in motor vehicles. It also upgrades existing child restraint performance requirements by setting new performance criteria and by replacing the current static tests with dynamic sled tests that simulate vehicle crashes and use anthropomorphic child test dummies. The new standard would reduce the number of children under 5 years of age killed or injured in motor vehicle accidents.

DATES: On June 1, 1980, compliance with the requirements of this standard will become mandatory. The current Standard No. 213 is amended to permit, at the manufacturer's option, compliance during the interim period either with the requirements of existing Standard No. 213, Child Seating Systems, or the new Standard No. 213, Child Restraint Systems.

ADDRESSES: Petitions for reconsideration should refer to the docket number and be submitted to: Docket Section, Room 5108, National Highway Traffic Safety Administration, 400 Seventh Street, S.W., Washington, D.C. 20590

FOR FURTHER INFORMATION CONTACT:

Mr. Vladislav Radovich, Office of Vehicle Safety Standards, National Highway Traffic Safety Administration, 400 Seventh Street, S.W., Washington, D.C.20590 (202-426-2264)

SUPPLEMENTARY INFORMATION:

This notice establishes a new Standard No. 213, *Child Restraint Systems.* A notice of proposed rulemaking was published on May 18, 1978 (43 FR 21470) proposing to upgrade and extend the applicability of the existing Standard No. 213, *Child Seating Systems.* The existing standard does not regulate car beds and infant carriers and uses static testing to assess the effectiveness of child restraint systems. The new standard covers all types of child restraint systems and evaluates their performance in dynamic sled tests with anthropomorphic test dummies. On May 18, 1978, NHTSA also published a companion notice of proposed rulemaking proposing to amend Part 572, *Anthropomorphic Test Dummies,* by specifying requirements for two anthropomorphic test dummies representing 3 year and 6 month old children (43 FR 21490) for use in compliance testing under proposed Standard No. 213. The comment closing date for both notices was December 1, 1978.

At the request of the Juvenile Product Manufacturers Association, NHTSA extended the comment closing date until January 5, 1979, for the portions of both proposals dealing with testing with the child test dummies. This extension was granted because manufacturers were reportedly having problems obtaining the proposed test dummies to conduct their own evaluations.

Consumers, public health organizations, child restraint manufacturers and others submitted comments on the proposed standard. The final rule is based on a thorough evaluation of all data obtained in NHTSA testing, data submitted in the comments, and data obtained from other pertinent documents and test reports. Significant comments submitted to the docket are addressed below. The agency will soon issue a final rule on the anthropomorphic test dummy proposal.

Summary of the Final Rule Provisions

The significant portions of the new standard are as follows:

1. The performance of the child restraint system is evaluated in dynamic tests under conditions

S6.2 The fuel tank is filled to any level from 90 to 95 percent of capacity.

S6.3 The parking brake is disengaged and the transmission is in neutral.

S6.4 Tires are inflated to the vehicle manufacturer's specifications.

S6.5 The windshield mounting material and all vehicle components in direct contact with the mounting material are at any temperature between 15°F and 110°F.

41 F.R. 36493
August 30, 1976

PREAMBLE TO MOTOR VEHICLE SAFETY STANDARD NO. 213

Child Restraint Systems, Seat Belt Assemblies, and Anchorages

(Docket No. 74-9; Notice 6)

ACTION: Final rule.

SUMMARY: This rule establishes a new Standard No. 213, *Child Restraint Systems,* which applies to all types of child restraints used in motor vehicles. It also upgrades existing child restraint performance requirements by setting new performance criteria and by replacing the current static tests with dynamic sled tests that simulate vehicle crashes and use anthropomorphic child test dummies. The new standard would reduce the number of children under 5 years of age killed or injured in motor vehicle accidents.

DATES: On June 1, 1980, compliance with the requirements of this standard will become mandatory. The current Standard No. 213 is amended to permit, at the manufacturer's option, compliance during the interim period either with the requirements of existing Standard No. 213, Child Seating Systems, or the new Standard No. 213, Child Restraint Systems.

ADDRESSES: Petitions for reconsideration should refer to the docket number and be submitted to: Docket Section, Room 5108, National Highway Traffic Safety Administration, 400 Seventh Street, S.W., Washington, D.C. 20590

FOR FURTHER INFORMATION CONTACT:

Mr. Vladislav Radovich, Office of Vehicle Safety Standards, National Highway Traffic Safety Administration, 400 Seventh Street, S.W., Washington, D.C. 20590 (202-426-2264)

SUPPLEMENTARY INFORMATION:

This notice establishes a new Standard No. 213, *Child Restraint Systems.* A notice of proposed rulemaking was published on May 18, 1978 (43 FR 21470) proposing to upgrade and extend the applicability of the existing Standard No. 213, *Child Seating Systems.* The existing standard does not regulate car beds and infant carriers and uses static testing to assess the effectiveness of child restraint systems. The new standard covers all types of child restraint systems and evaluates their performance in dynamic sled tests with anthropomorphic test dummies. On May 18, 1978, NHTSA also published a companion notice of proposed rulemaking proposing to amend Part 572, *Anthropomorphic Test Dummies,* by specifying requirements for two anthropomorphic test dummies representing 3 year and 6 month old children (43 FR 21490) for use in compliance testing under proposed Standard No. 213. The comment closing date for both notices was December 1, 1978.

At the request of the Juvenile Product Manufacturers Association, NHTSA extended the comment closing date until January 5, 1979, for the portions of both proposals dealing with testing with the child test dummies. This extension was granted because manufacturers were reportedly having problems obtaining the proposed test dummies to conduct their own evaluations.

Consumers, public health organizations, child restraint manufacturers and others submitted comments on the proposed standard. The final rule is based on a thorough evaluation of all data obtained in NHTSA testing, data submitted in the comments, and data obtained from other pertinent documents and test reports. Significant comments submitted to the docket are addressed below. The agency will soon issue a final rule on the anthropomorphic test dummy proposal.

Summary of the Final Rule Provisions

The significant portions of the new standard are as follows:

1. The performance of the child restraint system is evaluated in dynamic tests under conditions

simulating a frontal crash of an average automobile at 30 mph. The restraint system is anchored with a lap belt and, if provided with the restraint, a supplementary anchorage belt (tether strap). An additional frontal impact test at 20 mph is conducted for restraints equipped with tether straps or arm rests. In that additional test, child restraints with tether straps will be tested with the tether straps detached and child restraints with arm rests will be tested with the arm rest in place but with the child restraint system belts unbuckled. The additional 20 mph tests are intended to ensure a minimum level of safety performance when the restraints are improperly used.

2. To protect the child, limitations are set on the amount of force exerted on the head and chest of the child test dummy during the dynamic testing of restraints specified for children over 20 pounds. Limitations are also set on the amount of frontal head and knee excursions experienced by the test dummy in forward-facing child restraints and harnesses. To prevent a child from being ejected from a rearward-facing restraint, limitations are set on the amount the seat can tip forward and on the amount of excursion experienced by the test dummy during the simulated crash.

3. During the dynamic testing, no load-bearing or other structrual part of any child restraint system shall separate so as to create jagged edges that could injure a child. If the restraint has adjustable positions, it must remain in its pre-test adjusted position during the testing so that the restraint does not shift positions in a crash and possibly injure a child's limbs caught between the shifting parts or allow a child to submarine during the crash (i.e., allow the child's body to slide too far forward and downward, legs first).

4. To prevent injuries to children during crashes from contact with the surface of the restraint, requirements for the size and shape are specified for those surfaces. In addition, protective padding requirements are set for restraints used by children weighing 20 pounds or less.

5. Requirements in Standard No. 209, *Seat Belt Assemblies* (49 CFR 571.209), are applied to the belt restraints used in child restraint systems.

6. The amount of force necessary to open belt buckles and release a child from a restraint system is specified so that children cannot unbuckle themselves, but adults can easily open the buckle.

7. To promote the easy and correct use of all child restraint systems, they are required to attach to the vehicle by means of vehicle seat belts.

8. Warnings for proper use of the restraints must be permanently posted on the restraint so that the warnings are visible when the restraint is installed. Other information, such as the height and weight limits for children using the child restraint, must also be permanently displayed on the restraint but it does not have to be visible when the restraint is installed. The restraint must also have a location for storing an accompanying information booklet or sheet on how to correctly install and use the restraint.

9. A standard seat assembly is used in the dynamic testing to represent the typical vehicle bench seat and thereby aboid the cost of testing child restraints on numerous vehicle seats.

Applicability of Standard No. 213

The provisions of new Standard No. 213 apply to all types of child restraints used in motor vehicles for protection of children weighing up to 50 pounds, such as child seats, infant carriers, child harnesses and car beds. Beginning on June 1, 1980, compliance with the requirements of this standard will become mandatory. The current Standard No. 213 is amended to permit, at the manufacturer's option, compliance during the interim period either with the requirements of existing Standard No. 213, Child Seating Systems, or of the new Standard No. 213, Child Restraint Systems.

Dynamic Testing

The requirements to be met in the dynamic testing of child restraints include: maintaining the structural integrity of the system, retaining the head and knees of the dummy within specified excursion limits (i.e., limits on how far those portions of the body may move forward) and limiting the forces exerted on the dummy by the restraint system. These requirements will reduce the likelihood that the child using a child restraint system will be injured by the collapse or disintegration of the system, or by contact with interior of the vehicle, or by imposition of intolerable forces by the restraint system. As explained below, omission of any of these three requirements would render imcomplete the criteria for the quantitative assessment of the safety of a child restraint system and could very well lead to the design and use of unsafe restraints.

It was suggested in comments by the child restraint manufacturers and their trade association, the Juvenile Products Manufacturers Association (JPMA), that available restraints are performing satisfactorily. According to them, the new standard imposes expensive testing requirements with instrumented dummies which will increase the price of child restraints and discourage the purchasing of child restraints because of their increased costs. Many manufacturers suggested that the agency limit the standard to tests for occupant excursion and restraint system structural integrity in dynamic tests and not require the use of instrumented test dummies to measure crash forces imposed upon a child.

NHTSA recognizes that some child restraints perform relatively well, but the agency's testing has shown that others perform unsatisfactorily. Measuring only the structural integrity of the system and the amount of occupant excursion allowed during the testing does not provide a measurement of the severity of forces imposed on a child during a crash and thus does not provide an accurate assessment of the actual safety of the system. For example, a manufacturer could design a restraint with a surface mounted in front of the child that would allow a small amount of occupant excursion. However, that surface could impose potentially injurious forces on a child. NHTSA believes that the force measurement performance requirements are a crucial and necessary test to adequately judge a restraint system's effectiveness in preventing or reducing injuries. The use of instrumented test dummies and force measurement requirements are crucial elements of Standard No. 208, *Occupant Crash Protection*, which establish performance requirements for automatic restraint systems. NHTSA believes that systems designed specifically for children should have to provide the same high degree of occupant protection.

Several manufacturers (GM, Ford, Questor, and others) and JPMA objected to the proposed head and chest acceleration limits that must not be exceeded in the dynamic testing. They argued that the acceleration limits are based on biomechanical data for adults and there are no data showing their applicability to children. Because of the lack of biomechanical data on children's tolerance to impact forces, NHTSA has conducted tests of child

restraints with live primates to serve as surrogates for three-year-old children. Primates are similar in certain respects to children and, have been used by GM, Ford and others as surrogates in child restraint testing to assess protential injuries to children in crashes. In simulated 30 mph crashes conducted for NHTSA, similar to the test prescribed in the proposed standard, the primates either were not injured or sustained only minor injuries. NHTSA has also conducted child restraint tests using instrumented test dummies representing three-year-old children instead of primates. In the tests, the forces measured on the test dummies, which had not been injurious to the primates, did not exceed the head and chest acceleration criteria proposed in the standard. NHTSA is thus confident that the child restraints which do not exceed these performance criteria in the prescribed tests should prevent or reduce injuries to children in crashes.

Use of instrumented test dummies should not unduly raise the price of child restraints. Since many child restraint systems are already close to compliance, the cost per restraint of any needed design and testing costs should be minimal.

The May 1978 notice would have required restraint systems with adjustable positions to meet the performance requirements of the standards in any of its adjusted positions recommended for use in a motor vehicle. The restraint would have had to remain in its adjusted position during testing. International Manufacturing Co. requested the agency to test adjustable restraints in only their extreme up and down positions. If a manufacturer chooses to offer a seat with a number of adjustable positions which it recommends for use in a motor vehicle, it is important that the seat meet the performance requirements of the standard at any of those positions. Therefore, International's request is denied. NHTSA urges manufacturers not to include any adjustment positions for their restraints which are not to be used in a motor vehicle.

Strollee, Questor and Volvo asked NHTSA to allow adjustable position restraints to change positions during the testing, arguing that controlled change of position can be an effective energy-absorbing method. Allowing changes from one adjustment position to another during a crash can cause injuries to children's hands or fingers caught between the structural elements of the restraint as it changes position. Other effective energy-absorbing methods are available which will not pose a risk of injury to children. Thus, NHTSA is not adopting this suggestion.

Child restraint manufacturers and other interested parties, such as Action for Child Transportation Safety (ACTS), American Academy of Pediatrics, Physicians for Automotive Safety, and Michigan's Office of Highway Safety, urged NHTSA to lengthen the 30 inch head and knee excursion requirements for forward-facing restraints. They argued that some child restraint systems which have been effective in real world crashes will exceed the proposed head excursion limit. NHTSA has reviewed its child restraint tests and determined that during the last few inches of excursion the remaining velocity of the head in impacts with padded surfaces is relatively low. Because slightly increasing the head excursion should not increase the forces imposed upon the child's head, the head excursion limit is changed from 30 to 32 inches.

The May 1978 notice proposed limiting the amount of knee excursion in forward-facing child restraints to 30 inches. The purpose of the knee excursion limit is to prevent manufacturers from controlling the amount of head excursion by designing their restraints so that their occupants submarine excessively during a crash (i.e., so that their bodies slide too far downward and forward, legs first). Many child restraint manufacturers and JPMA asked the agency to lengthen the knee excursion limits. They argued that many restraints, particularly reclining child restraints where the occupant's knees will be further forward than a non-reclining child restraint, cannot pass the knee excursion limit, but do not allow the occupants to submarine. They claimed that the reclining feature is a comfort and convenience device which promotes seat usage since it allows a child to sleep in the restraint. They recommended that the agency establish a separate requirement which would prevent the occupant's torso from straightening out and submarining under the belts. NHTSA has tested several child restraints in the reclining position and determined that the knee excursion can be lengthened to 36 inches without allowing submarining if the dummy's torso has rotated at least 15 degrees forward from its initial starting position when the knees have reached their maximum excursion. Thus, the new standard

incorporates a 36 inch knee excursion limit and requires the test dummy's torso to have rotated at least 15 degrees forward when the knees have reached their maximum excursion.

For rear-facing child restraints (i.e., infant carriers) the May 1978 notice proposed retaining the dummy's head within the confines of the seat and preventing the back support surface of the restraint from tipping forward far enough to allow the angle between it and the vertical to exceed 60 degrees. If the support surface were allowed to tip more, the infant in the restraint could slide head first out of the shoulder straps. GM and Heinrich Von Wimmersperg pointed out that there is a conflict between the description of the confines of rear-facing restraints contained in the text of the standard and the manner in which the confines are defined in one of the figures incorporated in the standard. The text has been modified to correctly identify the confines of the restraint systems. GM also commented that the text of the standard defined the head confinement requirements in reference to the head target points of the infant dummy, although the infant dummy, unlike the 3 year child test dummy, does not have target points. The revised specifications for the infant test dummy do include head target points and therefore the confinement requirement is retained as originally proposed.

Several child restraint manufacturers objected to limiting the forward tipping of rear-facing restraints to 60 degrees. They argued that rear-facing child restraints can tip as much as 70 degrees forward and still retain the child within the restraint. They also argued that a rear-facing restraint will hit the instrument panel in the front seat, or the back of the front seat if the restraint is used in the rear seat, before the restraint tips 60 degrees. NHTSA is retaining a limit on forward tipping since a child restraint can be used in a vehicle with the vehicle's front seat moved to its extreme forward or rearward position. If the child restraint is used in the front seat and the vehicle seat is in the extreme rearward position, the child restraint can tip forward without striking the instrument panel. Likewise, a child restraint used in the rear seat, where the vehicle's front seat is in its extreme forward position, can tip forward without striking the back of the front seat. However, tests done by NHTSA have shown that a restraint can tip forward as much as 70 degrees while still retaining the child within the confines of the restraint. Therefore, the limitation on forward tipping is being changed to 70 rather than 60 degrees.

One child restraint manufacturer, the American Association for Automotive Medicine and Heinrich Von Wimmersperg commented that manufacturers of rear-facing restraints may attempt to comply with the limitation on forward rotation by designing the normal resting angle of the seat in a very vertical alignment or by adding attachments to prop the seat into a vertical position. Either of those approaches can create an uncomfortable seating position for the child. They recommended that the agency establish a minimum resting angle for rear-facing restraints. The agency is not adopting this suggestion at this time. By increasing the amount of forward rotation allowed, the agency should have removed the temptation for manufacturers to design restraint resting angles which would make it easier to comply with the requirement, but would create uncomfortable seating positions for the child.

The May 1978 notice proposed an additional dynamic test at 20 mph for child restraint systems equipped with tether straps with those straps left unattached. A number of commenters (such as Insurance Institute for Highway Safety, ACTS, University of Tennessee, Questor, Bobby Mac, and Michigan's Office of Highway Safety) commented that many people fail to connect the tether. They recommended that this type of restraint be tested at 30 mph with unattached tethers.

The agency is aware of the benefits and disadvantages of child restraints equipped with tethers, which presently account for over 70 percent of the child restraint sales. The agency's testing has shown that in 30 mph frontal tests child restraints with the tethers attached have less occupant excursion and lower head and chest accelerations than shield-type restraints that do not use tethers. Tethered restraints also allow far less occupant excursion in lateral crashes than shield-type restraints. The available accident data on child restraints, which includes consumer letters and accident investigation reports, is limited since the usage of child restraints is low. It does show, however, that tethered restraints, both properly tethered and untethered, have prevented injuries to children in crashes where other vehicle occupants were severely injured.

Because of the performance of properly tethered child restraints under testing and accident conditions, the agency does not want to eliminate those restraints from the market. At the same time, the agency wants to reduce or eliminate the possibility of people not using the tethers that accompany those restraints. Therefore, the agency is requiring all seats equipped with a tether to have a visible label warning people to correctly fasten the tether. In addition, the agency is considering issuing a proposal to require vehicle manufacturers to provide attachments for tether anchorages in all their vehicles. Having such attachments will enable parents to easily and properly attach tethers. The agency is also striving to promote the increased and proper use of child restraints through educational programs. As a part of this effort, NHTSA has conducted a series of regional seminars aimed at helping grass roots organizations educate parents about the importance of child restraints. A NHTSA-sponsored national conference on child restraint safety is scheduled for December 10-12 in Washington, D.C. to further these educational programs.

To ensure that restraints equipped with tethers provide at least a minimum level of protection if they are misused, the agency will require an additional dynamic test at 20 mph for those restraints. When tested with tethers unattached, the restraints must pass all the dynamic test performance requirements of the standard.

Energy Absorption and Distribution

Several manufacturers (Questor, Strollee, Cosco) and JPMA objected to the proposed height requirements for head restraints used to control the rearward movement of a child's head in a crash. The proposal would have slightly increased the requirements currently set in Standard No. 213. They argued that there was no basis for the change, which would require them to redesign their child restraints. The new requirements are based on anthropometric data on children gathered since the standard was originally adopted. NHTSA proposed the new head restraint height requirements in its earlier March 1974 notice of proposed rulemaking on child restraints and many manufacturers have already redesigned their seats to comply with the requirements. Since the new heights more accurately reflect the seating heights of children than the old requirements, the agency is adopting them as proposed. The notice proposed that the top of the head restraint be 22 inches above the seating surface for restraints used by children weighing more than 40 pounds. Questor requested the upper weight be changed to 43 pounds. Since 40 pounds represents the weight of a 50th percentile 5 year old and 23 inches represents its seating height, the requirement is not changed.

Several manufacturers (Cosco, Strollee, Questor) and JPMA raised objections to the proposed requirement that head restraints of child restraint systems have a width of not less than 8 inches. They pointed out that the minimum head restraint width requirement is intended to prevent a child's head from going beyond the width of a head restraint in a lateral or rear impact. They argued that restraints with side supports or "wings" should not have to meet the 8 inch width requirement since the side supports will prevent an occupant's head from moving laterally outside the restraint system. NHTSA agrees that the side supports should help laterally retain the child's head within the restraint during a side or rear impact and therefore is exempting those restraints from the 8 inch minimim width requirement. However, to ensure that child restraints with side supports have sufficient width to accommodate the heads of the largest child using the restraint, the agency has set a 6 inch minimum width for those restraints. In addition, to ensure that side supports are large enough to retain an occupant's head within the restraint, the agency has set a minimum depth requirement of four inches for those supports. Anthropomorphic data show that the head of a 50th percentile 5 year old child measures 7 inches front to rear and is 6 inches in breadth. Therefore, a four inch support should contact a sufficient area of the child's head to restrain it.

Manufacturers also questioned if the 8 inch width requirements is to be measured in restraints with side support from the surface of the padded side support or from the surface of the underlying structure before the padding is added. The wording of the standard is changed to make clear that the distance is measured from the surface of the padding, since the padded surface must be wide enough to accommodate the child's head.

The notice proposed that the minimum head restraint height requirement would not apply to restraints that use the vehicle's seat back to restrain the head, if the target point on the side of the head of the test dummy representing a 3 year old child is raised above the top of the seat back. Ford said that because of permitted differences in the dimensions of different test dummies and test seats, its child restraint will not consistently meet the requirements. Ford asked that the height requirement be changed or the manufacturers be permitted to restrict their restraints to seats with head restraints or to rear seats which have a flat surface immediately behind the seat. The standard allows a manufacturer to specify in its instruction manual accompanying the restraints which seating locations cannot be used with the child restraint. Therefore, no change is necessary, since Ford is allowed to restrict use of its restraint.

Several manufacturers (Cosco, Strollee, Questor) objected to the proposed force distribution requirement set for the sides of child restraint systems. The specifications do not require manufacturers to incorporate side supports in their restraints, they only regulate the surfaces that the manufacturer decides to provide so that they distribute crash forces over the child's torso. The commenters requested that the agency define the term "torso" and explain the reason for setting different side support requirements for systems used by infants weighing less than 20 pounds than for systems used by children weighing 20 pounds or more. In restraints for infants less than 20 pounds, the minimum side surface area requirements are based on anthropometric data for a 6-month-old 50th percentile infant to ensure maximum lateral contact in a side impact. Since the skeletal structure of an infant is just beginning to develop, it is important to distribute impact forces over as large a surface area of the child as possible, rather than concentrating the potentially injurious forces over a small area. For restraints used by children weighing more than 20 pounds and, therefore, having a more developed skeletal structure the minimum surface area requirement is based on anthropometric data for a 50th percentile 3-year-old child to provide restraint for the shoulder and hip areas of the child.

To enable manufacturers to determine their compliance with the torso support requirement, the standard follows the dictionary definition of "torso" and defines the term as referring to the portion of the body of a seated anthropomorphic test dummy, excluding the thighs, that lies between the top of the seating surface and the top of the shoulders of the test dummy.

Several manufacturers (Cosco, Strollee, Questor) and JPMA questioned the basis for prohibiting surfaces with a radius of curvature of less than 3 inches. They and Hamill also asked if the measurement of the curvature is to be made before or after application of foam padding on the underlying surface. The radius of curvature limitation will prevent sharp surfaces that might concentrate potentially injurious forces on the child. It is based on the performance of systems with such a radius of curvature that have not produced injuries in real world crashes. The standard is changed to require the measurement of the radius of curvature to be made on the underlying structure of the restraint, before application of foam padding. Since foam compresses when impacted in a crash, it is important that the structure under the foam be sufficiently curved so it does not concentrate the crash forces on a limited area of the child's body.

For child restraints used by children weighing less than 20 pounds, the notice proposed that surfaces which can be contacted by the test dummy's head during dynamic testing must be padded with a material that meets certain thickness and static compression requirements. A number of manufacturers (Strollee, Cosco, GM and Questor) and JPMA questioned the specifications set for the padding, arguing that there is no need to change from the current materials and the specification of a minimum thickness is design restrictive. Other commenters (Bobby-Mac, Hamill and American Association for Automotive Medicine) requested that the agency establish a test to measure the energy-absorbing capabilities of the underlying structure of the restraint, as well as of the padding.

NHTSA eventually wants to establish dynamic test requirements using instrumented test dummies for restraints used by children weighing 20 pounds or less. Such testing would measure the total energy absorption capability of the padding and underlying structure. At present, there are no instrumented infant test dummies, so the agency is instead specifying long-established static tests of the padding material.

In response to manufacturer comments, the NHTSA has reevaluated the materials currently used in child restraints and determined that those and other widely available materials can apparently provide sufficient energy absorption if used with a specified thickness. The agency has changed the proposed compression-deflection requirements to allow the use of a wider range of materials which should enable manufacturers to provide protective padding for children without having to increase the price of the restraint.

The proposed ban on components, such as arm rests, directly in front of a child which do not restrain the child was objected to by JPMA, and some manufacturers (Strollee, Century Products, International Manufacturing). They argued that arm restraints should not be banned since they promote usage of a child restraint by giving the child an area to rest against or place a book or other plaything. Other manufacturers (Hamill, Bobby-Mac), Michigan's Office of Highway Safety, and the American Academy of Pediatrics supported the ban arguing that arm rests promote misuse by creating the impression that a child can be adequately restrained by merely placing the arm rest in front of the child. The agency is concerned that parents' mistaken beliefs about the protective capability of arm rests may mislead them into not using the harness systems in the restraints.

Therefore, such arm rests or other components only may be installed if they provide adequate protection to a child when the restraint is misused in a foreseeable way because of the presence of the arm rest (i.e., the child is not buckled into the harness that comes with the child restraint system). To measure the performance of child restraints with arm rests and other devices that flip down in front of the child, those restraints will be tested at 20 mph with the component placed in front of the child, but without the child strapped into the restraint system. The restraint must pass the occupant excursion and other dynamic performance requirements in that condition.

Child Restraint Belt Systems

The May 1978 notice proposed three alternatives for the buckle release force required for the harnesses that restrain a child within the restraint. Many manufacturers favored the alternative based on the current Standard No. 213 which establishes a maximum force of 20 pounds, but does not establish a minimum force. In order to promote international harmonization, Volvo endorsed another alternative proposed by the Economic Commission of Europe which would set a minimum force of 2.25 pounds and a maximum of 13.45 pounds. However, Volvo proposed deviating from the ECE proposal and allowing a maximum release force of 20 pounds. Michigan's Office of Highway Safety and the American Seat Belt Council (ASBC) supported the other alternative which, based on a study by the National Swedish Road and Traffic Institute, would have set a 12 pound minimum force and a 20 pound maximum force. ASBC stated that this alternative should prevent a small child from opening the buckle, but not be too strong to prevent a small adult female from opening the buckle. Other commenters, such as ACTS and Borgess Hospital, recommended that the force be set at a level which children could not manage. Borgess noted that their experience with 400 rental child restraints shows that keeping children from unbuckling their restraints is a common problem. Physicians for Automotive Safety recommended that all buckle types be standardized and the release force be set at a level which can be quickly opened in an emergency.

Based on its review of the comments, NHTSA has decided to require buckles with a minimum release force of 12 pounds and a maximum release force of 20 pounds. The effectiveness of a restraint depends on the child being properly buckled at the time of impact. If a child is capable of releasing the buckle, it can inadvertently or purposely defeat the protection of the harness system. Setting a minimum force of 12 pounds should prevent small children from opening the buckle. Setting a maximum of 20 pounds as the release force will enable parents to easily open the buckle. NHTSA encourages manufacturers of child restraints to use push button buckles, similar to those used in automobile belts, so that people unfamiliar with child restraints can readily unbuckle them in emergencies. The agency will consider further rulemaking to standardize the buckle if manufacturers do not voluntarily adopt this approach.

Likewise, NHTSA has already advised child restraint manufacturers that physicians have informed the agency that some children are burned during the summer by over-heated metal buckles or other metal child restraint hardware. NHTSA will monitor manufacturer efforts to eliminate this problem and determine if additional rulemaking is necessary.

The proposal that the belt systems in child restraints meet many of the belt and buckle requirements of Standard No. 209, *Seat Belt Assemblies,* such as those relating to abrasion, resistance to light, resistance to microorganisms, color fastness and corrosion and temperature resistance was not opposed by any of the commenters and is therefore adopted. The buckle release test in Standard No. 209 for child restraint buckles is deleted, since Standard 213 now sets new performance requirements for buckles. Ford noted that the proposal inadvertently dropped a portion of Standard No. 209's abrasion requirements, which have been reincorporated in the final rule.

To prevent the belts from concentrating crash forces over a narrow area of a child's body, the proposal sets a minimum belt width of 1½ inch for any belt that contacts the test dummy during the testings. Hamill requested that pieces of webbing used to position the principal belts that maintain crash loads be exempt from the minimum width requirements. The agency believes that as long as the test dummy, and thus a child, can contact the belts during a crash the belts should be wide enough to spread the crash forces and therefore Hamill's request is denied.

Methods of Installation

Many commenters, including ACTS, American Academy for Pediatrics, Insurance Institute for Highway Safety, and American Seat Belt Council, said that child restraint systems cannot be used with some automatic belt systems, since they do not have a lap belt to secure the child restraint to the seat. They asked the agency to require all automatic belt systems to include lap belts.

The agency considers the compatibility of child restraints with automatic belt systems to be an important issue. One of the purposes of the agency's December 12, 1979, public meeting on child safety and motor vehicles is to obtain the public's views and information on that and other child passenger safety issues to assist the agency in determining whether to commence rulemaking. One rulemaking option currently being considered by the agency is to require vehicle manufacturers to provide anchorages for lap belts in automatic restraint equipped vehicles so that parents wishing to install lap belts can easily do so.

A number of manufacturers are voluntarily taking steps to make automatic belt systems compatible with child restraint systems. For example, GM provides an additional manual belt with its optional automatic lap-shoulder belt system for the front passenger's seat in the 1980 model Chevrolet Chevette to enable parents to secure child restraint systems.

Many of the commenters also asked the agency to require vehicle manufacturers to install anchorages or provide predrilled holes to attach tether anchorages in all their vehicles. They argued such anchorages or holes will make it easy for parents to attach tether straps correctly. As mentioned earlier in this notice, the agency is considering issuing a proposal to require manufacturers to provide attachments for tether anchorages in all their vehicles.

The May 1978 notice proposed that all child restraints be capable of being secured to the vehicle seat by a lap belt. Volvo and Mercedes once again asked the agency to allow the use of "vehicle specific" child restraints (systems uniquely designed for installation in a particular make and model which do not utilize vehicle seat belts for anchorages). As explained in the May 1978 notice, such systems can easily be misused by being placed in vehicles for which they were not specifically designed. Standardizing all restraints by requiring them to be capable of being attached by a lap belt is an important way to prevent misuse.

However, since vehicle specific child restraints can provide adequate levels of protection when installed correctly, NHTSA is not prohibiting the manufacture of such devices. The new standard requires them to meet the performance requirements of the standard when secured by a vehicle lap belt. As long as child restraints can pass the performance requirements of the standard secured only by a lap belt, a manufacturer is free to specify other "vehicle specific" installation conditions.

Labeling

The requirement for having a visible label permanently mounted to the restraint to encourage proper use of child restraints was supported by many of the commenters, including the Center for Auto Safety, ACTS, Insurance Institute for Highway Safety, and Michigan's Office of Highway Safety. Several manufacturers (Century, Cosco, Questor) objected to having a visible label on child restraints, claiming that there is not enough space on some restraints to place all the required information. Other commenters supported the visible labeling requirement but suggested that the visible label only have a single warning telling people to follow the manufacturer's instructions (American Association for Automotive Medicine, Strollee, Hamill). Others suggested placing warnings about the correct use of the restraint on a visible label and placing such information as the height and weight limits for children using the restraint and the manufacturer's certification that it meets all Federal Motor Vehicle Safety Standards on a nonvisible label (GM, PAS).

After reviewing the comments, NHTSA concludes that it is important to have certain warnings in a visible position to serve as a constant reminder on how to correctly use the restraint. Because of the limited space on some restraints, the agency has shortened the labeling requirements to require only those instructions most directly concerned with the safe use of the seat be visible. Thus, depending on its design, the restraint must warn parents to secure the restraint with the vehicle lap belt, snugly adjust all belts provided with the restraint, correctly attach the top tether strap and only use a restraint adjustment position which are intended for use in a motor vehicle.

In response to the agency's request for other instructions that a manufacturer should give parents, several commenters (ACTS, Michigan's Office of Highway Safety, Borgess Hospital) said that a warning on the label is necessary to prevent misuse of infant carriers. They said many people mistakenly place infant carriers in a forward-facing, rather than a rear-facing position. A forward-facing position defeats the purpose of those restraints which are designed to spread the forces of the crash over the infant's back. Because of the importance of preventing this type of misuse, the agency will require the visible label to also remind parents not to use rear-facing infant restraints in any other position.

Information about the height and weight limits of the children for which the restraint is designed, the manufacturer and model of the child restraint, and the month, year and place of manufacture and the certification that the restraint complies with all applicable Federal Motor Vehicle Safety Standards would also have to be provided, but that information does not have to be on a label that is visible when the seat is installed.

Many commenters (GM, Insurance Institute for Highway Safety, Multnomah County Department of Human Services, Physicians for Automotive Safety, Center for Auto Safety, and American Academy of Pediatrics) supported the proposed requirement that manufacturers inform consumers about the primary consequences of not following the manufacturer's warning about the correct use of the restraint. Therefore, the visible label must state the primary consequence of misusing the restraint. The same information would also have to be included in the instruction manual accompanying the restraint.

Ford objected to the requirement that the label have a diagram showing the child restraint installed in a vehicle as specified in the manufacturer's instructions. It said that because of the complexity of the instructions required for proper installation of a restraint with different types of belt systems, it is not practical to place all of the information on a single label. Hamill suggested that because of those same considerations, the agency should only require the diagram to show the proper installation of the restraint at one seating position. Other commenters, such as the American Academy for Pediatrics, supported the use of diagrams on the restraint noting that diagrams can more easily convey information than written instructions.

To promote the correct use of child restraints, NHTSA believes that it is important to have a diagram on the restraint to remind users of the proper method of installation. However, so that the label does not become too unwieldy, the agency will only require manufacturers to provide a diagram showing the restraint correctly installed in the right front seating position with a continuous loop lap/shoulder belt and in the center rear seating position installed with a lap belt. For restraints equipped with top tethers, the diagram must show the tethers correctly attached in both seating positions. It is important to show the correct use of a child restraint with a continuous loop lap/shoulder belt (a type of belt system used on many current cars) since such belts must have a locking clip installed on the belt to safely secure the child restraint.

GM objected to the requirement that the label be in block type, which it said makes the label difficult to read. GM requested that manufacturers be allowed to use 10 point type with either capitals or upper and lower case lettering. GM said that using such type will result in an easier to read label which, in turn, should promote more complete reading of the label by the consumer. Since the type sought by GM should promote the reading of the label, the agency is changing the requirement to allow the use of such type as an option.

Several organizations (ACTS, Center for Auto Safety and Insurance Institute for Highway Safety) asked the agency to establish performance test to accompany the requirement that the label be permanently affixed to the restraint. They pointed out that some current paper labels peel off after the restraint has been used awhile. NHTSA has not conducted the necessary testing to establish such a requirement. NHTSA urges manufacturers, whenever possible, to mold the label into the surface of the restraint rather than use a paper label.

Consumers Union and the Center for Auto Safety suggested that all restraints be graded based on their performance in frontal and lateral crash tests and the grades be posted on all the packaging, labels, and instruction manuals accompanying the child restraint. The grades would indicate the seating position within the vehicle with which the restraint can be safely used. Neither Consumers Union nor the Center suggested any performance requirements for establishing the different grades. Since the proposed grading system is outside of the scope of the proposed rule and the agency has not done the necessary testing to determine the specific tests and performance requirements necessary to establish such grading system, NHTSA will evaluate the suggestion for use in future rulemaking.

Installation Instructions

The May 1978 notice proposed that each restraint be accompanied by instructions for correctly installing the restraint in any passenger seat in motor vehicles. Many commenters (Center for Auto Safety, Borgess and Rainbow Hospitals, University of Tennessee And ACTS) suggested that the requirement for the instructions to accompany the restraint should be more explicit to require the restraint to have a storage location, such as a slot in the restraint or a plastic pouch affixed to the restraint, for permanently storing the instructions. They point out that storing the instructions with the restraint means they will be available for ready reference and will be passed on to subsequent owners of the restraint. NHTSA believes such a requirement would best carry out its intent to require the instructions to be easily available to all users and therefore the suggestion is adopted.

Several manufacturers (Strollee, Cosco) and JPMA objected to the agency's proposed requirement that the instructions state that the center rear seating position is the safest seating position in a vehicle. While not questioning the validity of the accident data showing the center rear seat to be the safest seating position in most vehicles, they argued that the agency should consider the psychological impact of not having the child near the adult. Accident data have consistently shown that the occupants in the rear seat are safer than occupants in the front seat. The same data show that the center rear seating position is the safest seating position in the rear seat. To enable parents to make an informed judgment about how best to protect their children, NHTSA believes that it is important to clearly inform them about the safest seating positions in the vehicle, and is therefore retaining the requirement.

In response to the agency's request for additional suggestions to be included in the instruction manual accompanying the restraint, ACTS suggested that car bed manufacturers informed consumers that the child should be placed with its head near the center of the vehicle. Because orienting a child's head in that way will ensure that it is the maximum distance away from the sides of the vehicle in a side impact, the agency has adopted ACTS suggestion. Tennessee's Office of Urban and Federal Affairs suggested that users should be told to secure child restraints with a vehicle belt when the child restraint is in the vehicle but not in use. Since an unsecured child restraint can become a flying missile in a crash and injure other vehicle occupants, the agency has adopted Tennessee's suggestion.

Test Conditions

The standard specifies requirements for a test assembly representing a vehicle bench seat to be used in the dynamic testing. Bobby-Mac commented that the test seat has a more level seating surface and less support at the forward edge of the seat than the seats in many current cars. These differences mean that a child restraint may experience more excursion on the test seat than on more angled and firmer car seats, Bobby-Mac said. NHTSA agrees that in comparison to some vehicles seats, the test seat may present more demanding test conditions. However, the test seat is representative of many seats used in vehicles currently on the road. Meeting the performance requirement of the standard on the test seat will ensure that child restraints perform adequately on the variety of different seats found in cars on the road.

Several manufacturers (Cosco and Strollee) and JPMA raised questions about the requirement proposed for the crash pulse (i.e., the amount of test sled deceleration required to simulate the crash forces experienced by a car) for the 20 and 30 mph tests. The agency had proposed a range of sled test pulses to allow manufacturers the option of using pneumatic or impact sled testing machines. Since a variety of different sled test pulses would be permitted under the proposal, manufacturers asked the agency to explain what would happen if they and the agency tested a child restraint system using different sled test pulses and produced inconsistent results (i.e., a failure using one pulse and a pass at the other, when both pulses were within the permissible range). JPMA suggested that the agency should consider a restraint as in compliance if the restraint meets all the applicable performance requirements in a test in which the sled test pulse lies entirely within the proposed range.

To provide manufacturers with the certainty they desire, the agency has redefined the sled test pulse requirement to establish a single 20 mph (Figure 3) and a single 30 mph (Figure 2) sled test pulse. Thus, in conducting its compliance testing, NHTSA may not exceed the sled test pulse set for the 20 and 30 mph tests. The sled test pulses chosen by NHTSA are the least severe pulses that meet the acceleration thresholds proposed in the notice of proposed rulemaking. Manufacturers are free to use other sled pulses, as long as the acceleration/time curve of the sled test pulse used is equal to or greater than the acceleration/time curve of the sled test pulse set in the standard.

In response to comments by Ford and others that the durability of the foam used in the standard seat assembly may influence the test results, the agency has changed the standard to specify that the foam in the test seat be changed after each test.

GM pointed out that the instructions for positioning the test dummy within the restraint did not specify when in the positioning sequences any of the restraint's belts should be placed on the test dummy. An appropriate change has been made to specify when the belts should be attached. Ford said that the dummy positioning requirements result in an "unnatural" positioning of the dummy within its Tot-Guard restraint so that the dummy's arms rest on the side of the restraint rather than with its arms on the padded portion of the shield. NHTSA notes that a child in a real-world accident will not necessarily have its arms resting on the shield. Allowing the test dummy's arm to be positioned on the shield may inhibit the dummy's forward movement and make it easier to comply with the limits on test dummy excursion and acceleration set in the standard. Thus, Ford's requested change in the positioning requirements is rejected.

Flammability

The notice proposed requiring child restraints to meet the burn resistance requirements of Standard No. 302, *Flammability of Interior Materials*. The requirement was supported by GM, the American Academy of Pediatrics and the American Seat Belt Council. No commenters opposed the requirement. In supporting the requirement, GM said that the flammability characteristics of child restraints, "which are in close proximity to an occupant," should be "compatible with the flammability characteristics of other parts of the vehicle occupant compartment interior," which already must meet the performance requirements of Standard No. 302. The agency agrees with GM about the desirability of providing all vehicle occupants with the protection of Standard No. 302 and is thus requiring all child restraints to meet the performance requirements of that standard.

Inertial Reels

Several commenters raised questions about the effectiveness of vehicle seat belts equipped with inertial reels in securing child restraints. The American Academy of Pediatrics requested the agency to restrict the use of inertial reels to the driver's seating position. Physicians for Automotive Safety and ACTS pointed out that continuous loop lap/shoulder belts with inertial reels must be used with locking clips to secure a child restraint. They

said that the difficulty of installing such clips deters their use.

Agency research has found that use of inertial reels increases the comfort and convenience of seat belts and thus promotes their use by older children and adults. Thus, the agency will continue to require the use of inertial reels in vehicle belt systems. However, to ensure that inertial reels are compatible with child restraints, the agency will soon begin rulemaking on the comfort and convenience of vehicle belt systems to require that the belts used in the front right outboard seating position have a manual locking device. This requirement will mean that continuous loop and other types of inertial reel belt systems can be easily and effectively used with child restraints. Such manual locking devices will also be permitted with belts used in the rear seats. As previously outlined in this notice, the agency has established several labeling and installation instruction requirements which deal specifically with the correct use of locking clips on continuous loop belts with inertial reels. Those requirements should reduce or eliminate problems associated with using child restraint in current vehicles equipped with inertial reels.

Costs and Benefits

The agency has considered the economic and other impacts of this final rule and determined that this rule is not significant within the meaning of Executive Order 12044 and the Department of Transportation's policies and procedures implementing that order. The agency's assessment of the benefits and economic consequences of this final rule are contained in a regulatory evaluation which has been placed in the docket. Copies of that regulatory evaluation can be obtained by writing NHTSA's docket section, at the address given in the beginning of this notice.

In the 0 to 5 age group, more than 800 children are killed and more than 100,000 children are injured annually as occupants of motor vehicles. Because of the large difference in effectiveness between restraints that can pass the dynamic test of the new standard and those which have passed only a static test, NHTSA projects that there should be 43 fewer deaths and 6,528 fewer injuries per year. Because many restraints have already been upgraded in response to the agency's prior rulemaking proposal, some of the death and injury prevention benefits of the standard have already been realized.

The projected benefits of this standard are limited by the existing low rate of child restraint use. However, the labeling and instruction requirements of this standard should increase the proper usage of child restraints.

Because of NHTSA's 1974 proposal to upgrade child restraints, many manufacturers have currently designed their restraints to meet dynamic test requirements. Therefore, those restraints are only projected to increase in price by approximately $1.00 in order to meet the other requirements of this standard. Restraints that do not currently pass dynamic tests would have a price increase of $16.00 to meet the new requirements. The average sales weighted price increase is $4.25.

Numerous commenters (including National Safety Council, American Academy of Pediatricians, Tennessee Office of Child Development and North Dakota's Department of Public Health)

urged the agency to make the standard effective before the proposed May 1, 1980, effective date. GM and the American Safety Belt Council requested that the effective date be delayed beyond the proposed May 1, 1980. Many manufacturers have already upgraded their restraints to the performance requirements set in this rule. The agency believes that providing six months lead-time, until June 1, 1980, will provide sufficient time for the remaining manufacturers to upgrade their restraints.

The principal authors of this notice are Vladislav Radovich, Office of Vehicle Safety Standards, and Stephen Oesch, Office of Chief Counsel.

Issued on December 5, 1979.

Joan Claybrook
Administrator,

44 F.R. 72131
December 13, 1979

PREAMBLE TO AN AMENDMENT TO MOTOR VEHICLE SAFETY STANDARD NO. 213

Child Restraint Systems; Seat Belt Assemblies

(Docket No. 74-9; Notice 7)

ACTION: Response to petitions for reconsideration.

SUMMARY: This notice responds to five petitions for reconsideration and petitions for rulemaking concerning Standard No. 213, *Child Restraint Systems.* In response to the petitions, the agency is changing the labeling requirements to permit the use of alternative language, modifying the minimum radius of curvature requirement for restraint system surfaces and extending the effective date of the standard from June 1, 1980, to January 1, 1981. In addition, several typographic errors are corrected in Standard No. 209, *Seat Belt Assemblies.*

EFFECTIVE DATE: The amendments are effective on May 1, 1980. The effective date of the standard is changed from June 1, 1980, to January 1, 1981.

FOR FURTHER INFORMATION CONTACT:

Mr. Vladislav Radovich,
Office of Vehicle Standards,
National Highway Traffic Safety Administration
Washington, D.C. 20590 (202-426-2264)

SUPPLEMENTARY INFORMATION: On December 13, 1979, NHTSA published in the FEDERAL REGISTER a final rule establishing Standard No. 213, *Child Restraint Systems,* and making certain amendments to Standard No. 209, *Seat Belt Assemblies and Anchorages.* Subsequently, petitions for reconsideration were timely filed with the agency by Cosco, General Motors, Juvenile Products Manufacturers Association, and Strolee. Subsequent to the time for filing petitions for reconsideration, Strolee also filed a petition for

rulemaking to amend the standard. After evaluating the petitions, the agency has decided to modify, as fully explained below, some of the requirements of Standard No. 213. All other requests for modification are denied. The agency is also correcting several minor typographical errors in the text of Standard No. 209.

LABELING

Standard No. 213 requires manufacturers to place a permanently mounted label on the restraint to encourage its proper use. General Motors (GM) petitioned for reconsideration of three of the labeling requirements.

Section S5.5.2 (f) of the standard requires each child restraint to be labeled with the size and weight ranges of children capable of using the restraint. In its petition, GM said that the requirement could "unnecessarily preclude some children from using the restraint or suggest use by children too large for the restraint." GM also commented that some infant restraints are intended to be used from birth and thus the lower size and weight limitation serves no purpose.

In addition, GM said that stating the upper size limit for infant restraints in terms of seated height rather than in standing height is a more appropriate way to set size limitations for infants. For example, GM said that an infant with a short torso and long legs might be precluded from using the restraint if the limitation is stated in terms of standing height, while an infant with short legs and a torso too long for the restraint would be inappropriately included among ones who could supposedly use the restraint. GM requested that infant restraints be allowed to be labeled with an optional statement limiting use by upper weight and seated height.

NHTSA agrees that specifying a lower weight and size limit is unnecessary for an infant carrier designed to be used from birth and has amended the standard accordingly. The agency has decided not to adopt GM's proposal to state the upper size limit in seating rather than standing height. The purpose of the label is to provide important instructions and warnings in as simple and understandable terms as possible. Standing height, rather than seating height, is a measurement parents are familiar with and which is commonly measured during pediatric examinations. As GM pointed out, it is possible to establish a limit based on standing height which would exclude any infant whose seating height is too high to properly use the restraint. Therefore, the agency will continue to require the upper size limit to be stated in terms of standing height.

GM also requested that manufacturers be allowed to establish a lower usage limit for restraints used for older children based on the child's ability to sit upright rather than on his or her size and weight. GM said the lower limit "is not as dependent upon the child's size as it is on the child's ability to hold its head up (sit upright) by itself. This important capability is achieved at a wide range of child sizes." NHTSA agrees that the type of label GM proposes can clearly inform parents on which children can safely use a restraint and therefore will permit use of such a label.

Section S5.5.2(g) of the standard requires the use of the word "Warning" preceding the statement that failure to follow the manufacturer's instructions can lead to injury to a child. GM requested that the word "Caution" be permitted as an alternative to "Warning." GM said that since 1975 it has used caution in its labels and owners' and service manuals as a lead or signal word where the message conveys instructions to prevent possible personal injury. GM said that the words caution and warning are generally accepted as synonymous.

The agency believes that the word "Warning," when used in its ordinary dictionary sense, is a stronger term that conveys a greater sense of danger than the word "Caution" and thus will emphasize the importance of following the specified instructions. Therefore, the agency will continue to require the use of the word "Warning."

Section S5.5.2(k) of the standard requires restraints to be labeled that they are to be used in a rear-facing position when used with an infant. GM said that while the requirement is appropriate for so-called convertible child restraints (restraints that can be used by infants in a rear-facing position and by children in a forward-facing position), it is potentially misleading when used with a restraint designed exclusively for infants. GM said the current label might imply that the restraint can be use in forward-facing positions with children. GM recommended that restraints designed only for infants be permitted to have the statement, "Place this infant restraint in a rear-facing position when using it in the vehicle." The agency's purpose for establishing the labeling requirement was to preclude the apparent widespread misuse of restraints designed for infants in a forward-facing rather than rear-facing position. Since GM's recommended label will accomplish that goal, the agency is amending the standard to permit its use.

RADIUS OF CURVATURE

Section S5.2.2.1(c) of the standard requires surfaces designed to restrain the forward movement of a child's torso to be flat or convex with a radius of curvature of the underlying structure of not less than 3 inches. Ford Motor Co. objected to the three inch limitation on radius of curvature arguing that measuring the radius of curvature of the underlying structure would eliminate designs that have not produced serious injuries in actual crashes. Ford said the shield of its Tot-Guard has a radius of curvature from 2.2 to 2.3 inches and it had no evidence of serious injury being caused by the shield when the restraint has been properly used.

The purpose of the radius of curvature requirement was to prohibit the use of surfaces that might concentrate impact forces on vulnerable portions of a child's body. It was not the agency's intent to prohibit existing designs, such as the Tot-Guard, which have not produced injuries in actual crashes. Since a 2 inch radius of curvature should therefore not produce injury, the agency has decided to change the radius of curvature requirement from 3 to 2 inches.

Although the standard sets a minimum radius of curvature for surfaces designed to restrain the forward movement of a child, it does not set a minimum surface area for that surface. Prototypes of new restraints shown to the agency by some manufacturers indicate that they are voluntarily incorporating sufficient surface areas in their designs. The agency encourages all manufacturers to use surface areas at least equivalent to those of the designs used by today's better restraints.

OCCUPANT EXCURSION

Section S5.1.3.1 of the standard sets a limit on the amount of knee excursion experienced by the test dummy during the simulated crash tests. It specifies that "at the time of maximum knee forward excursion the forward rotation of the dummy's torso from the dummy's initial seating configuration shall be at least 15° measured in the sagittal plane along the line connecting the shoulder and hip pivot points."

Ford Motor Co. objected to the requirements that the dummy's torso rotate at least 15 degrees. Ford said that it is impossible to measure the 15 degree angle on restraints such as the Tot-Guard since the test dummy "folds around the shield in such a manner that there is no 'line' from the shoulder to the hip point." In addition, restraints, such as the Tot-Guard, that enclose the lower torso of the child can conceal the test dummy hip pivot point.

The agency established the knee excursion and torso rotation requirements to prevent manufacturers from controlling the amount of test dummy head excursion by allowing the test dummy to submarine excessively during a crash (i.e., allowing the test dummy to slide too far downward underneath the lap belt and forward, legs first). A review of the agency's testing of child restraints shows that current designs that comply with the knee excursion limit do not allow submarining. Since the knee excursion limit apparently will provide sufficient protection to prevent submarining, the agency has decided to drop the torso rotation requirement. If future testing discloses any problems with submarining, the agency will act to establish a new torso rotation requirement as an additional safeguard.

HEAD IMPACT PROTECTION

Section 5.2.3 requires that each child restraint designed for use by children under 20 pounds have energy-absorbing material covering "each system surface which is contactable by the dummy head." Strolee petitioned the agency to amend this requirement because it would prohibit the use of unpadded grommets in the child restraint. Strolee explained that some "manufacturers use grommets to support the fabric portions of a car seat where the shoulder belt and lap belt penetrate the upholstery. These grommets retain the fabric in place and give needed support where the strap

comes through to the front of the unit." Because of the use of the grommets in positioning the energy-absorbing padding and belts, the agency does not want to prohibit their use. However, to ensure that use of the grommets will not compromise the head impact protection for the child, the agency will only allow grommets or other structures that comply with the protrusion limitations specified in section S5.2.4. That section prohibits protrusions that are more than ⅜ of an inch high and have a radius of less than ¼ inch. Because this amendment makes a minor change in the standard to relieve a restriction, prior notice and a comment period are deemed unnecessary.

BELT REQUIREMENTS

Strolee petitioned the agency to amend the requirement that all of the belts used in the child restraint system must be 1½ inches in width. Strolee said that straps used in some restraints to position the upper torso restraints have " 'snaps' so that the parent may release this positioning belt conveniently." Strolee argued that such straps should be exempt from the belt width requirement since "the snap would release far before any loads could be experienced."

The agency still believes that any belt that comes into contact with the child should be of a minimum width so as not to concentrate forces on a limited area of the child. This requirement would reduce the possibility of injury in instances where the snap on a positioning strap failed to open. Strolee's petition is therefore denied.

Strolee has also raised a question about the interpretation of section S5.4.3.3 on belt systems. Strolee asked whether the section requires a manufacturer to provide both upper torso belts, a lap belt and a crotch strap or whether a manufacturer can use a "hybrid" system which uses upper torso belts, a shield, in place of a lap belt, and a crotch strap. The agency's intent was to allow the use of hybrid systems. The agency established the minimum radius of curvature requirements of section S5.2.2.1(c) to ensure that any shield used in place of a lap or other belt would not concentrate forces on a limited area of the child's body. NHTSA has amended section S5.4.3.3. to clarify the agency's intent. Because this is an interpretative amendment, which imposes no new restrictions, prior notice and a comment period are deemed unnecessary.

HEIGHT REQUIREMENTS

Strolee asked the agency to reconsider the requirements for seat back surface heights set in section S.5.2.1.1. Strolee argued that the higher seat back required by the standard would restrict the driver's rear vision when the child restraint is placed in the rear seat.

The final rule established a new seat back height requirement for restraints recommended for use by children that weigh more than 40 pounds. To provide sufficient protection for those children's heads, the agency required the seat back height to be 22 inches. The agency explained that the 22 inch requirement was based on anthropometric data showing that the seating height of children weighing 40 or more pounds can exceed 23 inches. The agency still believes that 22 inch requirement is necessary for the protection of the largest child for which the restraint is recommended. NHTSA notes that child restraints can be designed to accommodate the higher seat backs without allowing the overall height of the child restraint to unduly hinder the driver's vision.

PADDING

In its petition, JPMA claimed that the standard "calls for the application of outdated specifications" for determining the performance of child restraint padding in a 25 percent compression-deflection test. A review of the most recent edition of the American Society for Testing and Materials (ASTM) handbook shows that the compression-deflection test in two of the three ASTM standards (ASTM D1565) referenced by the agency has been replaced. However, the replacement standard does not contain a 25 percent compression-deflection test. Therefore, the agency will continue to use the three ASTM standards referenced in the December 1979 final rule.

EFFECTIVE DATE

Cosco, Strolee and the Juvenile Products Manufacturers Association (JPMA) petitioned the agency for an extension of the June 1, 1980, effective date. They requested that the effective date be changed to at least January 1, 1981, and Strolee requested a delay until March 1, 1981. They argued that the June 1, 1980, effective date does not allow manufacturers sufficient time to develop, test and tool new child restraints.

Testing done for the agency has shown that many of the better child restraint systems currently on the market can meet the injury criteria and occupant excursion limitation set by the standard. Some of those seats would need changes in their labeling, removal of arm rests and new belt buckles and padding to meet the standard. Such relatively minor changes can be made in the time available before the June 1, 1980, effective date.

Several manufacturers have informed the agency that they are designing new restraints to meet the standard. Based on prototypes of those restraints shown to the agency, NHTSA believes that these new restraints may be more convenient to use, less susceptible to misuse and provide a higher overall level of protection than current restraints. Based on leadtime information provided by individual manufacturers and the JPMA, the agency concludes that extending the standard from June 1, 1980, to January 1, 1981, will provide sufficient leadtime. Providing a year's leadtime is in agreement with the leadtime estimates provided by the manufacturers as to the time necessary for design and testing, tooling and buckle redesign.

COMPATIBILITY WITH VEHICLE BELTS

On December 12, 1979, NHTSA held a public meeting on child transportation safety. At that meeting, several participants commented about the difficulty, and in some cases the impossibility, of securing some child restraint systems with a vehicle lap belt because the belt will not go around the restraint. Testing done by the agency during the development of the recently proposed comfort and convenience rulemaking also confirms that problem. The agency reminds .child restraint manufacturers that Standard No. 213, *Child Restraint Systems*, requires all child restraints to be capable of being restrained by a vehicle lap belt.

Joan Claybrook
Administrator

45 F.R. 29045
May 1, 1980

PREAMBLE TO AN AMENDMENT TO
FEDERAL MOTOR VEHICLE SAFETY STANDARD NO. 213

Child Restraint Systems
(Docket No. 74-09; Notice 8)

ACTION: Correction.

SUMMARY: On May 1, 1980, the agency published a notice in the *Federal Register* responding to petitions for reconsideration concerning Standard No. 213, Child Restraint Systems. In response to a petition from Ford Motor Co., the agency stated in the preamble of the notice that it was eliminating the torso rotation requirement of the standard. However, the notice inadvertently did not amend the standard to delete that requirement. This notice makes the necessary amendment.

DATES: The amendment is effective upon publication in the *Federal Register*, October 6, 1980.

FOR FURTHER INFORMATION CONTACT:
  Stephen Oesch, Office of Chief Counsel, National Highway Traffic Safety Administration, 400 Seventh Street, S.W., Washington, D.C. (202-426-2992)

SUPPLEMENTARY INFORMATION: On May 1, 1980, the agency published a notice responding to several petitions for reconsideration concerning Standard No. 213, Child Restraint Systems (45 FR 29045).

Among the petitions was one from Ford Motor Co. objecting to the requirement that the test dummy's torso rotate at least 15 degrees during the simulated crash test of the child restraint. Ford argued that it is impossible to measure the 15 degree angle on restraints such as its Tot-Guard which enclose the lower torso of the child and thus conceal one of the pivot points used in measuring the dummy's rotation.

In response to the Ford petition, the agency decided to drop the torso rotation requirement. In the May 1 notice, the agency explained that the purpose of the requirement was to prevent manufacturers from controlling the amount of head excursion by allowing the test dummy to submarine excessively during a crash (i.e., allowing the test dummy to slide too far downward underneath the lap belt and forward, legs first). After further reviewing its child restraint test results, the agency concluded that restraints meeting the knee excursion limit of the standard will provide sufficient protection to prevent such submarining.

Section 5.1.3.1 is revised to read as follows:

S5.1.3.1 Child restraint systems other than rear-facing ones and car beds. In the case of each child restraint system other than a rear-facing child restraint system or a car bed, the test dummy's torso shall be retained within the system and no portion of the test dummy's head shall pass through the vertical transverse plane that is 32 inches forward of point z on the standard seat assembly, measured along the center SORL (as illustrated in Figure 1B), and neither knee pivot point shall pass through the vertical transverse plane that is 36 inches forward of point z on the standard seat assembly, measured along the center SORL.

Issued on September 26, 1980.

Michael M. Finkelstein
Associate Administrator
for Rulemaking

45 FR 67095
October 9, 1980

PREAMBLE TO AN AMENDMENT TO
FEDERAL MOTOR VEHICLE SAFETy STANDARD NO. 213

Child Restraint Systems
(Docket No. 74-09; Notice 9)

ACTION: Final rule.

SUMMARY: This notice amends Standard No. 213, Child Restraint Systems, to allow the use of thinner padding materials in some child restraints. The agency proposed the amendment in response to a petition for rulemaking filed by General Motors Corporation.

DATES: The amendment is effective on December 15, 1980.

ADDRESSES: Petitions for reconsideration should refer to the docket number and be submitted to: Docket Section, Room 5108, National Highway Traffic Safety Administration, 400 Seventh Street, S.W., Washington, D.C. 20590. (Docket hours: 8:00 a.m. to 4:00 p.m.)

FOR FURTHER INFORMATION CONTACT:

Mr. Vladislav Radovich, Office of Vehicle Safety Standards, National Highway Traffic Safety Administration, 400 Seventh Street, S.W., Washington, D.C. 20590 (202-426-2264)

SUPPLEMENTARY INFORMATION: On December 13, 1979, NHTSA issued Standard No. 213, Child Restraint Systems (44 FR 72131). The standard established new performance requirements for child restraints, including requirements for the padding used in child restraint systems recommended for use by children under 20 pounds (i.e., infant carriers).

The padding requirements provide that surfaces of the infant carrier that can be contacted by the test dummy's head during dynamic testing must be padded with a material that meets certain thickness and static compression-deflection requirements. The standard requires that the pad-

ding must have a 25 percent compression-deflection resistance of not less than 0.5 and not more than 10 pounds per square inch (psi). Material with a resistance of between 3 and 10 psi must have a thickness of ½ inch. If the material has a resistance of less than 3 psi, it must have a thickness of at least ¾ inch.

In response to a petition for rulemaking filed by General Motors Corporation (GM), the agency proposed on October 17, 1980 (45 FR 68694) to modify the padding requirements to allow the use of thinner padding. GM's petition said that the compression-deflection resistance of padding is sensitive to the rate at which deflection occurs during the test procedure. As the deflection rate increases during testing, so does the measured resistance of the material. GM said that the padding used in the head impact area of its child seat has a maximum compression-deflection resistance of 3 psi. However, several different deflection rates are permitted by the American Society for Testing and Materials test procedures incorporated into Standard No. 213. GM reported that the measured 25 percent compression-deflection value of the padding it uses can be as low as 1.8 psi.

To accommodate variations attributable to the use of the different deflection rates permitted in the testing, the agency proposed to allow the use of padding with a compression-deflection resistance of 1.8 psi or more to have a minimum thickness of ½ inch.

The notice denied GM's petition to permit the use of padding with a compression-deflection resistance of 0.2 psi and a thickness of ⅝ or ¾ inch.

GM, the only party that commented on the proposal, supported the proposed revision.

GM requested the agency to reconsider its decision to prohibit the use of padding with a compression-deflection resistance of 0.2 psi. GM argued that the field performance of its child

PART 571; S213-PRE 21

restraints shows that current padding material is effective in reducing deaths and injuries.

As explained in the October notice, the agency agrees that child restraints, such as GM's infant carrier, which have an energy-absorbing shell can provide effective protection with padding having a compression-deflection resistance of 0.2 psi. Many infant carriers, however, use rigid plastic shells rather than energy absorbing shells. Manufacturers of the rigid plastic shells currently use padding with a compression-deflection resistance of 0.5 psi. The agency does not want to degrade that level of performance and therefore GM's request is again denied.

COSTS

The agency has assessed the economic and other impacts of the proposed change to the padding requirements and determined that they are not significant within the meaning of Executive Order 12221 and the Department of Transportation's policies and procedures for implementing that order. Based on that assessment, the agency concludes further that the economic and other consequences of this proposal are so minimal that additional regulatory evaluation is not warranted. When Standard No. 213 was published in the *Federal Register* on December 12, 1979, the agency placed in the docket for that rulemaking a regulatory evaluation assessing the effect of the padding requirements set by the standard. The effect of that rule adopted today is to permit the use of some padding materials in a thickness of ½ inch rather than ¾ inches. Such a change will slightly reduce manufacturer padding costs.

The agency finds, for good cause shown, that an immediate effective date for this amendment is in the public interest since it relieves a restriction in the standard that goes into effect on January 1, 1981.

The principal authors of this notice are Vladislav Radovich, Office of Vehicle Safety Standards, and Stephen Oesch, Office of Chief Counsel.

For the reasons set out in the preamble, Part 571 of Chapter V of Title 49, Code of Federal Regulations, is amended as set forth below.

§571.213 [Amended]

1. 49 CFR Part 571 is amended by revising paragraph §S5.2.3.2(b) of §571.213 to read as follows:

*   *   *   *   *

(b) A thickness of not less than ½ inch for materials having a 25 percent compression-deflection resistance of not less than 1.8 and not more than 10 pounds per square inch when tested in accordance with S6.3. Materials having 25 percent compression-deflection resistance of less than 1.8 pounds per square inch shall have a thickness of not less than ¾ inch.

Issued on December 8, 1980.

Joan Claybrook
Administrator

45 FR 82264
December 15, 1980

PREAMBLE TO AN AMENDMENT TO
FEDERAL MOTOR VEHICLE SAFETY STANDARD NO. 213

Child Restraint Systems
(Docket No. 74-09; Notice 11)

ACTION: Technical amendment.

SUMMARY: When the final rule establishing Standard No. 213, *Child Restraint Systems*, was issued, it included a section setting requirements for a diagram to show the proper installation of a child restraint within a vehicle. Although the preamble discussed the installation diagram requirement, the standard inadvertently did not require the diagram to be placed on the restraint. This notice makes the necessary technical amendment to correct the standard.

EFFECTIVE DATE: August 26, 1982.

SUPPLEMENTARY INFORMATION: In May 1978, the agency proposed a substantially upgraded Standard No. 213, *Child Restraint Systems* (43 F.R. 21470). In section 5.5.2(a)-(k) of the standard, the agency proposed requirements for certain warning and installation labaels for child restraints. In particular, section 5.5.2(k) proposed specific requirements for a diagram showing the proper installation of a child restraint in a vehicle. Section 5.5.1 of the standard proposed that all of the labels specified in 5.5.2(a)-(k) would have to be placed permanently on the child restraint.

When the agency issued its final rule, it expanded the labeling requirements for child restraints (44 F.R. 72131). The preamble for the final rule discussed the specifics of the expansion and the reasons for adopting the labeling requirements. Because of the expansion, the installation diagram requirement of section 5.5.2(k) of the proposal was redesignated as section 5.5.2(l) in the final rule. Inadvertently, section 5.5.1 of the standard was not modified to reflect the expansion of the labeling requirements

and thus it continued to specify that only the information found in section 5.5.2(a)-(k) be placed on the child restraint.

Most manufacturers recognized the intent of the agency and have placed the correct installation diagram on their restraints. A number of manufacturers apparently have not included such diagrams on their child restraints.

This notice makes the necessary technical amendment to correct the standard to require the installation diagram to be placed on a child restraint. The effective date of this correction is 45 days after the publication of this notice in the *Federal Register*. This will allow time for the few manufacturers that have not included installation diagrams to prepare the needed diagrams for their child restraints.

The agency has determined that there is good cause for not providing additional notice and opportunity to comment on this technical amendment. The public has previously had notice and opportunity to comment on the installation diagram requirement. This technical amendment merely corrects an error arising from the redesignation of the installation diagram requirement during the rulemaking process.

Issued on July 2, 1982.

Courtney M. Price
Associate Administrator
for Rulemaking

47 F.R. 30077
July 12, 1982

ACTION: Final rule.

SUMMARY: This final rule amends Federal Motor Vehicle Safety Standard No. 213, *Child Restraint Systems,* so that child restraint systems can be certified for use in motor vehicles, or for use in both motor vehicles and aircraft. The requirements for certifying child restraints for use in aircraft were formerly specified in the Federal Aviation Administration's (FAA) Technical Standard Order (TSO) C100, which required that in order for child restraint systems to be certified for use in aircraft, they must first be certified for use in motor vehicles and then pass three additional performance tests. Simultaneously with the effective date of this rule, FAA will rescind the requirements of TSO C100 and take action to permit child restraints certified under the requirements of this rule to be used in aircraft.

The notice of proposed rulemaking which preceded this final rule proposed to add the three performance requirements of the TSO and one additional performance requirement for restraints with tether straps to Standard No. 213. This rule adopts one of the three performance requirements of the TSO, the inversion test, and requires that child restraint manufacturers wishing to certify their products for use in both motor vehicles and aircraft certify that the product complies with the requirements of that test. The other performance requirements proposed in the notice are not incorporated in this rule because a joint testing program conducted by .FAA and NHTSA last year showed these requirements to be redundant. Child restraints which passed the existing higher performance requirements in Standard No. 213 easily passed the requirements of the TSO, which indicates that those TSO requirements are unnecessary to establish that child restraints are effective in the differing environment of the aircraft interior. Accordingly, compliance with those requirements is no longer required to certify child restraints for use in aircraft.

Child restraints which are certified for use in both motor vehicles and aircraft will be required to be labeled in red with the phrase "THIS RESTRAINT IS CERTIFIED FOR USE IN MOTOR VEHICLES AND AIRCRAFT". Child restraints certified only for use in motor vehicles will not be required to change the information currently required by Standard 213 on their labels.

By combining and simplifying the requirements for certifying child restraints for use in motor vehicles and aircraft, FAA and NHTSA hope to encourage more child restraint manufacturers to certify their products for use in both modes of transportation. The ultimate goal of seeking more models of child restraints to be certified for use in both motor vehicles and aircraft is to encourage families traveling by air to use child restraints for their children before, during, and after the air travel portion of their trips.

EFFECTIVE DATE: This rule becomes effective March 30, 1985.

SUPPLEMENTARY INFORMATION: This rule amends Standard No. 213, Child Restraint Systems (49 CFR §571.213), so that child restraint systems can be certified for use in both motor vehicles and aircraft, or simply for use in motor vehicles. These amendments are intended to encourage families traveling by air to use child restraints to protect their children before, during, and after the air travel portion of their trips.

Background

Need for Increased Use of Child Restraints. Parents cannot adequately protect their very young children against the risk of death and injury while riding in motor vehicles or aircraft either by holding them in their lap or by fastening a lap belt around them. The forces generated during sudden stops even at speeds as low as 10-15 miles per hour (mph) make it physically impossible for a parent to hold and protect a child in his or her arms. Using a lap belt is better, but it is still inadequate for this purpose (particularly for children under the age of 1 year) because of the physical dimensions, bone structure, and weight distribution of young children.

The most effective protection that can be afforded these young children are special supplementary seating devices, which are attached to and secured by the lap belt in the vehicle or aircraft. These devices, generically referred to as child restraints, are specifically designed to take into account the physiological differences between young children and older children and adults, and to offer the appropriate protection for these young children exposed to the large energy levels inherent in vehicle crashes.

Efforts to Promote Increased Use of Child Restraints. The NHTSA has been working hard to promote the use of child restraints by more parents. The agency has been advising the various States on the drafting of mandatory child restraint use laws. Such laws have now been enacted in 49 States and the District of Columbia. These laws have significantly increased the sales and use of child restraints, and increased the public awareness of the safety consequences of allowing children to travel unrestrained in motor vehicles.

In addition, the NHTSA has been working to educate the public on the benefits of child restraints. Working with medical professionals, childbrith educational programs and others, the agency has provided information to pediatricians and prospective parents on ways to protect their children in motor vehicles. Further, the agency has developed manuals on how to develop a child restraint loaner program that can assist parents unable to afford their own child restraints.

All of these factors have succeeded in greatly increasing the use of child restraints for children riding in motor vehicles. Currently, restraint uage for infants less than 1 year old is about 68 percent; and for children ages 1 to 4 the rate is 44 percent; based on the agency's continuing survey of restraint usage in 19 cities.

Impediments to Increased Use of Child Restraints.

This heightened use and awareness, combined with the limited number of child restraint models which can be used in both motor vehicles and aircraft, caused confusion and frustration for families traveling by air and car. Both NHTSA and FAA have standards for child restraints. Until recently, of the 42 models of child restraints certified under NHTSA's Standard No. 213 for use in motor vehicles, only five models were also approved under the FAA's standard for use in aircraft. If a family tried to take one of the remaining 37 models of child restraints, they were usually required to check the restraint along with the rest of their luggage. This discouraged families from traveling with the unapproved child restraints, and resulted in the child not having the benefit of the safety seat not only during the takeoff and landing of the aircraft, but also when the family was driving in a motor vehicle on the ground portions of the trip.

From a safety viewpoint, data on injuries and fatalities show that travel by air is much safer than by motor vehicle. For children up to 4 years of age, approximately one fatality and 10 injuries occur yearly during commercial air travel vs. over 600 fatalities and 70,000 injuries to motor vehicle occupants. Consequently, the main benefits from the use of child restraints will be derived from the motor vehicle portion of the trip.

The NHTSA Child Restraint Standard

As an initial step toward ensuring that child restraint systems would offer adequate portection to their occupants, NHTSA issued Standard No. 213 in 1970. That standard, which was issued under the authority granted in the National Traffic and Motor Vehicle Safety Act of 1966, as amended (hereinafter "the Safety Act"; 15 U.S.C. 1381 *et seq.*), became effective in 1971. As then drafted, it specified various static tests to ensure the safe performance of child restraints. However, subsequent data showed that child restraints which passed these static tests might not prove effective at protecting a child in certain vehicle crash situations.

Under the current standard, which became effective January 1, 1981, the performance of child restraint systems is evaluated in dynamic tests under conditions simulating a frontal crash of an average car at 30 mph. The restraint is anchored by a lap belt and, if provided with the restraint, by a supplemental anchorage belt (known as a tether strap). An additional frontal impact test at 20 mph is conducted for restraints equipped with either tether straps or internal harnes and a restraint surface. In that additional test, child restraints with tether straps are tested with the straps detached and child restraints with a restraint surface (e.g., a padded shield) are tested with the restraint surface in place but with the child restraint system's internal harness unbuckled. The additional 20 mph tests are intended to ensure a minimum level of safety performance when the restraints are improperly used. Thus, child restraints with tethers or with a restraint surface are tested at both 20 and 30 mph, while those without tethers or such a surface are tested at 30 mph only. Both the 20 mph and the 30 mph tests are conducted with the child restraint fastened to a seat representing the typical motor vehicle bench seat.

To protect the child, limits are set on the amount of force exerted on the head and chest of a child test dummy during the dynamic testing of restraints specified for children over 20 pounds. Limits are also set on the amount of frontal head and knee excursions experienced by the test dummy in forward-facing child restraints. To prevent a small child from being ejected from a rearward-facing restraint, limits are set on the amount that the seat can tip forward and on the amount of excursion experienced by the test dummy during the simulated crash.

Compliance of child restraints with Standard No. 213 is assured by the requirement in the Safety Act that manufacturers certify compliance for each child restraint. The agency may review the basis for that certification and conduct testing to assure compliance. The Safety Act provides for the assessment of civil penalties for failures to comply with applicable safety standards, and for certifications which the manufacturer in the exercise of due care has reason to know are false or misleading in a material respect.

The FAA Child Restraint Standard

In May 1982, the FAA issued its own child restraint standard, Technical Standard Order (TSO) C100. One of the key factors underlying the development of TSO C100 was child restraint testing conducted by the Civil Aeromedical Institute in 1974. The results of that testing appeared in FAA test report "Child Restraint Systems for Civil Aircraft" (FAA-AM-78-12, March 1978). Another factor was the FAA's determination that differences in the environments of aircraft and motor vehicles necessitated its establishing performance requirements to address the special safety risks posed to young children traveling in aircraft. One of these differences is the tendency of the seat back of aircraft seats to fold forward with the application of a very low force. The FAA determined that there was a need to control the interaction between the young child, especially those facing rearward in a child restraint, and the seat back to ensure that the seat back does not apply unacceptable levels of force onto the child. The FAA also determined that there was a need to address the danger that in-flight turbulence (especially in the upward direction) might throw a child out of his or her child restraint.

Accordingly, the FAA drafted TSO C100 so that it requires each child restraint to meet the requirements of NHTSA's Standard No. 213 and four additional requirements. First, while attached to an aircraft passenger seat with a free-folding seat back by an aircraft safety belt, and occupied by a test dummy, each child restraint must provide protection in an impact producing a 20 mph velocity change. There is no double testing of child restraints with tethers as under Standard No. 213. Such restraints are tested only once in an impact and with their tethers unattached. Second, each child restraint must retain its occupant during an inversion test. Third, each child restraint must withstand the static forces specified in Federal Aviation Regulations §25.561 (14 CFR §25.561), with each of the forces acting separately. Fourth, TSO C100 specifies requirements for marking child restraints with assembly and usage instructions, providing a copy of such instruction to child restraint users and submitting a copy of these instructions and various technical information and test results to the FAA. In addition, the TSO procedures require the establishment and maintenance of a manufacturer quality control system. The quality control system is intended to assure that seats are manufactured in such a way as to meet the standard's performance requirements.

For a child restraint to be approved for use in aircraft, the manufacturer must submit specified information to the FAA along with a certifying statement that the restraint meets the requirements of TSO C100. After the FAA approval is issued, if airlines permit, the restraint can be used for infants or young children during all phases of flight, including takeoff and landing. Once the FAA approved a particular model of child restraint, that agency followed a policy of accepting child restraints of that model that were manufactured prior to the date of approval for use in aircraft during all phases of flight, provided that those earlier child restraints were substantially identical to the approved one and were properly identified as to make and model by a Standard No. 213 certification label.

The result of these differing requirements was that only a few of the child restraints certified for use in motor vehicles were also certified for use in aircraft. In 1983, the National Transportation Safety Board (NTSB) considered the safety problems facing young children traveling in motor vehicles and aircraft and urged that a variety of actions be taken to promote the use of child restraints. It urged that all States adopt laws requiring that infants and young children be placed in child restraints when riding in motor vehicles. It also recommended that the DOT simplify its standards specifying performance requirements for child restraints by combining all technical requirements into a single standard (NTSB Safety Recommendations A-83-1, issued February 24, 1983).

After considering the benefits which would result from the increased use of child restraints, the FAA and the NHTSA jointly concluded that the process of certifying child restraints for use in both motor vehicles and aircraft could and should be simplified and expedited. By combining the separate NHTSA and FAA standards into a single standard under the jurisdiction of a single agency, child restraint manufacturers could avoid the difficulties of dealing with different standards, methods of certification, and testing procedures promulgated by the two agencies. Accordingly, a notice of proposed rulemaking (NPRM) was published at 48 FR 36849, August, 1983.

Details of the NPRM

The NPRM proposed that the NHTSA would be the sole agency responsible for enforcing the new

Standard No. 213, which would be applicable to child restraint systems designed for use in both motor vehicles and aircraft. In essence, the NPRM proposed that the requirements in both agencies standards for child restraints be unchanged and simply combined into an expanded Standard No. 213, with one further performance test added for child restraints to be certified for use in aircraft. This would avoid the problems inherent in dealing with the differing certification procedures of the two agencies and consolidate all of the requirements into one standard.

Under the proposal, manufacturers which elected to certify their child restraints for use on aircraft would have to certify that these restraints could pass those four additional tests. Those manufacturers which did not elect to certify their restraints for use on aircraft would not have to make that certification. The existing requirements in Standard No. 213 applicable to child restraints certified for use in motor vehicles were not proposed to be changed in any way by the NPRM. What was proposed was simply an option for manufacturers to subject their restraints to some additional testing if they wanted to certify those restraints for use on aircraft.

Three of the four additional performance tests proposed to be added to Standard No. 213 for child restraints certified for use on aircraft were drawn almost verbatim from the FAA's child restraint standard. These additional tests were proposed to be required to ensure that child restraints certified for use in aircraft would offer adequate protection to young children in the unique interior environment of aircraft.

The first additional test proposed in the NPRM was a dynamic impact test at 20 mph for all restraints not equipped with a tether strap. The child restraint would be attached to a representative aircraft seat only by the aircraft seat belt attached to the aircraft seat. The child restraint would not be permitted to fail or deform in a manner that could seriously injure or prevent subsequent extrication of the occupant. This test was taken almost verbatim from paragraph (a)(2)(i) of TSO C100.

The second additional test proposed in the NPRM would apply only to child restraints equipped with a tether strap. These restraints would be tested under the same procedures as untethered restraints, except that the impact would be at 30 mph with the tether strap unattached. The same criteria for determining satisfactory performance specified above for untethered restraints would again be used. This requirement was not drawn from TSO C100. However, NHTSA decided to include the requirement because the FAA believed that, since aircraft seats have no place to which the tether strap could be anchored, it was necessary to subject such restraints to a more stringent performance test to ensure that these restraints would offer adequate aircraft safety.

The third test proposed in the NPRM was an inversion test. Its purpose is to ensure that the child restraint could protect the child from air turbulence. The test, drawn directly from the language of paragraph (a)(2)(ii) of TSO C100, would have required the combination of a child restraint, test dummy, and aircraft passenger seat to be rotated to an inverted position and held there without any failure or deformation of the child restraint that would seriously injure or prevent the subsequent removal of the occupant.

The fourth additional test proposed in the NPRM would have required each child restraint to withstand the ultimate inertia forces specified in 14 CFR §25.561, with each of those forces acting separately. This requirement was specified in paragraph (a)(2)(iii) of TSO C100. Engineering analysis would have been acceptable in lieu of actual testing to establish compliance with this proposed requirement.

The procedures to be followed in conducting these tests or analyses were drawn from paragraph (a)(2)(iv) of TSO C100. They provided for the testing or analysis of child restraints to determine their adequacy for protecting the weight and stature of child for which the restraint is designed. The test dummies to be used were those specified in section S7 of Standard No. 213. Other procedural provisions related to the placing of the test dummy in the restraint, the attaching of the restraint to the aircraft seat, and the design of the aircraft seat.

As noted above, the NPRM gave child restraint manufacturers an option either to certify their restraints for use in both motor vehicles and aircraft or to certify the restraints only for use in motor vehicles. Those electing the latter option would have been required by the NPRM to include the statement, "THIS RESTRAINT IS NOT CERTIFIED FOR USE IN AIRCRAFT", on the certification label and operating instructions for the child restraint. This labeling requirement was proposed to ensure that parents seeking to buy restraints for use in both modes of transportation and airline flight attendants would easily ascertain whether a particular child restraint was not certified for use in aircraft.

The NPRM also announced that FAA and NHTSA would jointly test many models of child restraints for compliance with the TSO C100 requirements. The test results generated by this program were made available to the manufacturers of the tested restraints to assist them to certify their child restraints for use in both modes of transportation.

FAA-NHTSA Testing of Child Restraints

The testing program evaluated all 42 models of child restraints currently manufactured and certified as meeting the requirements of Standard No. 213 to determine whether they complied also with the existing requirements of TSO C100. (See DOT HS-806-413) There was some preliminary difficulty in determining how to establish whether a child restraint system had "failed or deformed in a manner that could seriously injure or prevent subsequent extrication of a child occupant," the criterion for determining compliance with the tests in TSO C100. The two agencies agreed to use the performance requirements specified in section S5 of Standard No. 213, but to exclude the head and chest acceleration requirements set forth in section S5.1.2.

All 42 models of child restraints, including the 11 which have tether straps, were subjected to the 20 mph dynamic test while attached to a representative aircraft seat, and all passed by a considerable margin. Similarly, the three tethered child seats and eight tethered booster seats were subjected to a 30 mph impact with the tether unattached, and all again passed by a considerable margin. The performance of the three tethered child seats was not appreciably different than was registered by them in the 20 mph impact test, and the head and knee excursions measured in this test were well under those recorded for the restraints in the Standard No. 213 tests. All 42 models were subjected to the TSO C100 inversion test, and all 42 were deemed to have passed those requirements. Additionally, all 42 models were subjected to the static loading tests at the levels specified in TSO C100, and all 42 passed the test.

All 42 models were also tested to the requirements of "old" Standard No. 213, which required the restraint to withstand inertia loads approximately 3 times greater than those specified in TSO C100. Standard No. 213 was upgraded from these old requirements primarily because of the structural failures which occurred in 30 mph dynamic tests of restraints which met the static load requirements under the old version of the standard. NHTSA believed that any of the restraints which could satisfy the dynamic testing requirements of the new Standard No. 213 would also satisfy the static loading requirements of the old standard. Since the loads required under the old standard were approximately 3 times the level required by the TSO, any devices which could satisfy the old standard would *ipso facto* satisfy the TSO requirements.

In this testing to the levels prescribed under the old standard, 40 of 42 models of child restraints passed. The two restraints which failed the tests did so in only one direction, and at load levels 2½ times those required in the TSO.

The joint testing program made it possible for the manufacturers of every model of child restraint currently produced to seek prompt FAA approval for the restraints under TSO C100. This has expedited the process for certifying current models of child restraints for both aircraft and motor vehicle use. At present 36 models have received TSO approval.

However, the Department of Transportation still believes that it is necessary to proceed with a final rule in this area. As a practical matter, new child restraints will be introduced into the market, and those models would face the same obstacles which were confronted by current models before the completion of the joint testing program. It is poor regulatory policy to subject manufacturers to needless and repetitious testing of the identical product to satisfy slightly differing requirements of two different agencies. These considerations impel FAA and NHTSA to proceed to a final rule at this time, so that the situation which existed prior to the joint testing program does not recur at some future date.

Comments

Most of the more than 20 commenters on the NPRM endorsed the concept of combining the FAA and NHTSA standards into one standard. Some of the commenters expressed qualified support for the concept, but reserved final judgment until the results of the joint testing program were made available to the public.

Only one commenter opposed the basic concept of combining the two standards, and that opposition was based on the belief that NHTSA was neither competent nor properly equipped to regulate items related to avaiation and the aircraft industry. First, NHTSA believes it should be emphasized that this rule was developed with the cooperation and support of the FAA, which certainly has the necessary expertise regarding the aviation industry. Further, child restraints are not items which are uniquely related to aviation and the aircraft industry; most of the lifesaving benefits of child restraints accrue while the young child is riding in a motor vehicle. Finally, both NHTSA and FAA gained new knowledge about the interplay of the aircraft seat, child restraint, and child during a sudden deceleration during the recently completed joint testing program. For these reasons, the agencies believe it is appropriate to go forward with this rulemaking.

Several comments raised issues outside the scope of this rulemaking. These included permissible seat positions for approved child restraints in aircraft, retroactive certification for aircraft use of models recently approved for such use, the extent to which individual airlines must examine the restraint's certification to determine its validity, differences in the various airlines' policies permitting the use of child restraints, and so forth. This rulemaking is addressing only the steps child restraint manufacturers must take to certify their products for use in motor vehicles and aircraft. The procedures regulating the actual use of the restraints in aircraft are not being addressed herein; such procedures will be decided solely by the FAA. These and other questions on the procedures should be addressed to that agency.

The commenters made several objections to each of the four proposed additional requirements, to which compliance would have to be certified if a manufacturer wanted to certify its child restraint for use in aircraft. Regarding the first proposed additional test that child restraints without tether straps be tested in an aircraft seat at a 20 mph impact, these commenters argued that all child restraints certified as complying with Standard No. 213 are already subjected to a 30 mph impact in the more severe environment of a car seat. Accordingly, this argument continued, the proposal to require a lower speed test in a less severe environment would simply add to the testing burden for child restraint manufacturers, without ensuring any higher degree of safety.

One of the child restraint manufacturers correctly noted in its comments that the reason for proposing the 20 mph test in the aircraft seat was the concern that the more flexible back of such a seat could snap forward on impact and hit the child restraint and/or child with additional crash forces and that those additional forces would not be considered in the 30 mph test with the restraint attached to a car seat. This commenter suggested that their own testing and some NHTSA tests in 1982 showed that the back of the aircraft seat does not exert significant forces relative to the crash forces. The commenter concluded that NHTSA should delete this proposed requirement unless the joint testing program showed some evidence that significant forces were actually exerted.

The joint testing program showed that the forces to which the test dummy and restraint are subjected in the 20 mph dynamic test in the aircraft seat were 1/3 to 1/2 less than those to which they were subjected in the 30 mph dynamic test in the car seat. This finding was hardly significant or surprising, given the lower speed at impact.

A far more significant finding was made regarding the amount of the loading imposed by the flexible aircraft seat back on the restrained dummy. For this testing, the aircraft seat back was instrumented with a triaxial accelerometer so that quantitative assessments of the produced forces could be made. Inspection of the acceleration-time histories and the loads measured on the aircraft seat belts revealed that in every test the maximum forces generated by the child restraints (as measured by the test dummy and including the peak head and chest accelerations and the peak belt loads) occurred some 25-40 milliseconds before the occurrence of the peak acceleration of the seat back. Also, the magnitude of the head and chest accelerations imparted to the child seat occupant by the restraining action of child seats were much higher than those imparted later on by the action of the aircraft seat back. These facts indicate that the loads imparted when the seat back struck the child restraint and its occupant are relatively insignificant when compared with the loads imparted by the crash. Confirmation of this was found in the fact that the seat back acceleration had no significant influence on the head and chest accelerations measured in the test dummies. However, the loads measured on the aircraft seat belt were increased during the seat back acceleration. This finding suggests that the load exerted by the

acceleration of the seat back is transferred directly through the structure of the child restraint to the seat belt. This fact would again confirm the view that the seat back acceleration poses no threat to the occupant of a child restraint.

Based on these results, which occurred in each test, NHTSA believes that it has been established that seat back acceleration poses an inconsequential threat to occupants of child restraints, and that any restraint which protects its occupant against the crash forces will adequately protect its ocenpant against the forces generated by the seat back acceleration. Given these conclusions, it is unnecessary to test child restraints for their ability to protect a child against the threat of the folding aircraft seat back. Accordingly, the agency has deleted the requirement that child restraints be certified for use in aircraft capable of protecting a restrained child in a 20 mph impact when attached to an aircraft seat.

Many of the commenters objected to the requirement that tethered restraints be subjected to a 30 mph crash in an aircraft seat with the tether unattached. The rationale for these objections was perhaps best summed up in the NTSB comment. The NTSB stated that it could understand subjecting restraints with tethers to the same test as restraints without tethers, and not permitting the restraints with tethers to have their tether strap attached during the test. Such a proposal would ensure that these restraints could pass the same requirements as other child restraints, and that they could do so under the conditions present in aircraft; i.e., with their tether straps unattached. However, the NTSB continued, it was not justifiable to require these restraints to undergo a more severe test than other restraints. One child restraint manufacturer commented that this 30 mph test requirement would not ensure any higher level of safety on aircraft since the aircraft seats themselves would not withstand a 30 mph impact. This commenter went on to say that in an actual crash at 30 mph, there is as much potential of injury to the child from the failure of the aircraft seat itself as from the failure of the child restraint.

As indicated above in the section summarizing the joint testing program, the tests conducted on child restraints with tethers showed that all of those restraints easily passed this 30 mph crash test requirement, that the results were not much higher than were those measured in the 20 mph tests, and that the results showed an appreciably lower force level for the restraints in this test than were obtained in the Standard No. 213 misuse test. Given the conclusion that the seat back acceleration does not transmit any significant forces to the occupant of the child restraint and the fact that this test imposes lower crash forces than the Standard No. 213 tests, it seems unnecessary to require the child restraint manufacturers to certify compliance with this test. The points made in the comments on this proposal also are convincing, so it has been determined not to incorporate this test in the final rule.

The third proposed additional test was an inversion test whose purpose is to ensure that the child restraints certified for use in aircraft could adequately protect the child against the dangers posed by sudden air turbulence. The commenters who addressed this issue seemed to generally agree that this was a hazard which child restraints for use in aircraft should protect against and that restraints which passed the requirements of Standard No. 213 would not necessarily pass this test. NHTSA also believes that the inversion test was not shown to be redundant of existing test procedures, and has determined that this test should be incorporated in this final rule. The requirements for this inversion test are adopted verbatim from those proposed in the NPRM. Several commenters questioned some of the inversion test procedures and offered suggested alternatives. The agency agrees that some refinements could be made. However, it is necessary first to issue a new NPRM. The NPRM, which proposes to amend the requirements for the inversion test adopted in this rule, discusses these comments further.

The fourth additional test proposed in the NPRM was a static load test. Several commenters questioned the need for the relatively low inertial loads of that test to be applied to the restraints, considering the much greater loads to which the child restraint is subjected in the testing for Standard No. 213. This fact, together with the joint testing results which showed that all currently produced child restraints can withstand loads at least 2½ times greater than those specified in this proposed test, leads NHTSA to conclude that this test is redundant and does not ensure any higher level of safety. Accordingly, it is not adopted in this final rule.

Several commenters addressed the criteria used to determine if a child retraint has passed the two simulated crash tests and the inversion test applicable to restraints for aircraft use. These criteria were that the child restraint system "may not fail nor deform in a manner that could seriously injure or prevent subsequent extrication of a child occupant." Some of the child retraint manufacturers asked precisely how one determines if a restraint has failed or deformed in such a manner. Another commenter opined that those criteria "are so vague and subjective as to be of no substantive value whatsoever."

NHTSA agrees with these commenters' judgment that the criteria for determining compliance could be made more objective. However, the Administrative Procedure Act requires that interested persons be given notice of proposed rulemaking and an opportunity to comment thereon prior to an agency's adopting changed requirements as a final rule (5 U.S.C. 553). This provision of the law prevents the agency from adopting these more objective criteria in this final rule, because the interested persons would not have had an opportunity to comment on those criteria. Accordingly, NHTSA is today publishing a notice of proposed rulemaking to incorporate more objective criteria for the inversion test. This notice has a 45-day comment period, to provide any interested persons with the chance to comment on the changes while allowing the agency to move promptly to incorporate more objective criteria.

Most of the commenters addressed the issues raised by the language proposed to be labeled on child restraints which were certified only for use in motor vehicles. The NPRM proposed that such child restraints have the statement "THIS RESTRAINT IS NOT CERTIFIED FOR USE IN AN AIRCRAFT." A number of commenters opposed this "negative" labeling because it could give consumers the impression that such a restraint was not as safe for motor vehicle use as a restraint which was certified for use in both aircraft and motor vehicles. In fact, both restraints would have been certified as passing the same dynamic tests for use in motor vehicles. Other problems alleged to exist with this labeling scheme were that consumers would not be sure whether a child restraint not bearing such a label could be used safely in aircraft, and that this "negative" labeling could result in older, unlabeled and uncertified seats being used on aircraft. Further, the proposed labeling could make it difficult for flight attendants to determine which restraints were actually approved for use in aircraft, causing delays and frustration for parents wishing to use child restraints on flights. These commenters all requested that the "negative" labeling proposed in the NPRM be replaced with a simple positive statement in the final rule.

NHTSA agrees with these comments. The informational purposes of the labeling requirement would be better served by simple positive declarations. The labeling requirement adopted in the final rule specifies that child restraints certified for use only in motor vehicles recite the same certification that is currently required, with no additional statements, and those restraints certified for use in both motor vehicles and aircraft simply add a statement of that dual certification.

Finally, a child restraint manufacturer asked that the final rule clarify the standard aircraft seat assembly to be used for testing the child restraint. The NPRM stated in section S7.3(b) that a "representative aircraft passenger seat" be used. The term "representative aircraft passenger seat" was defined S5 in the NPRM as either a production seat approved by the FAA or a simulated seat conforming to Drawing Package SAS-100-2000. NHTSA believes this definition is clear, and will result in consistent test results. No further changes to this definition have been made in this final rule.

OMB Clearance

The labeling requirements for child restraints are considered to be information collection requirements, as that term is defined by the Office of Management and Budget (OMB) in 5 CFR Part 1320. OMB has approved the labeling requirements for child restraints certified for use in motor vehicles (OMB No. 2127-0511), but has not approved the labeling requirements for child restraints certified for use in motor vehicles and aircraft. Accordingly, those labeling requirements have been submitted to the OMB for its approval, pursuant to the requirements of the Paperwork Reduction Act of 1980 (44 U.S.C. 3501 *et seq.*). A notice will be published in the *Federal Register* when OMB approves this information collection.

Impacts

NHTSA has analyzed the impacts of this rule and determined that the rule is not "major" within the meaning of Executive Order 12291, but is

"significant" within the meaning of the Department of Transportation regulatory policies and procedures. The rule simplifies and combines the requirements of two existing government regulatios into one regulation. It would not impose any new burdens upon any manufacturer. If a child restraint manufacturer wishes to continue certifying one of its child restraint models for use in motor vehicles only, the requirements for doing so are unchanged and the testing costs would remain at about $3,500. If a child restraint manufacturer wishes to certify a model for use in motor vehicles and aircraft, its testing costs under Standard No. 213 would increase by about $1,500 to a total of about $5,000. However, the total testing costs for certifying a model to this combined Standard No. 213 will be less than the total testing costs for certifying compliance with Standard No. 213 and TSO C100 (estimated at about $8,000). Further, this cost reduction and the need to certify to only one agency's regulation, instead of two agencies' regulations, should provide a slightly reduced cost of compliance for those child restraint manufacturers that choose to certify their products for use in motor vehicles and aircraft. Although these impacts are minimal, a regulatory evaluation has been prepared.

In consideration of the foregoing, the following amendments are made to section 571.213, *Child Restraint Systems*, of Title 49 of the Code of *Federal Regulations.*

1. Section S1 is amended to read as follows:

S1. *Scope.* This standard specifies requirements for child restraint systems used in motor vehicles and aircraft.

2. Section S2 is amended to read as follows:

S2. *Purpose.* The purpose of this standard is to reduce the number of children killed or injured in motor vehicle crashes and in aircraft.

3. Section S3 is amended to read as follows:

S3. *Application.* This standard applies to child restraint systems for use in motor vehicles and aircraft.

4. The definition of "Child restraint system" in section S4 is amended to read as follows:

"Child restraint system" means any device except Type I or Type II seat belts, designed for use in a motor vehicle or aircraft to restrain, seat, or position children who weigh 50 pounds or less.

5. Section S4 is amended by adding the following new definitions in alphabetical order:

"Representative aircraft passenger seat" means either a Federal Aviation Administration approved production aircraft passenger seat or a simulated aircraft passenger seat conforming to Drawing Package SAS-100-2000.

6. Section S5 is amended to read as follows:

S5. *Requirements for child restraint systems certified for use in motor vehicles.* Each child restraint certified for use in motor vehicles shall meet the requirements in this section when, as specified, tested in accordance with S6.1.

7. Section S5.5.2 is revised by the addition of a new paragraph (m) which reads as follows:

(m) Child restraints that are certified as complying with the provisions of section S8 shall be labeled with the statement "THIS RESTRAINT IS CERTIFIED FOR USE IN MOTOR VEHICLES AND AIRCRAFT". This statement shall be in red lettering, and shall be placed after the certification statement required by paragraph (e) of this section.

8. Section S7.3 is revised to read as follows:

S7.3 *Standard seat assemblies.* The standard seat assemblies used in testing under this standard are:

(a) For testing for motor vehicle use, a simulated vehicle use, a simulated vehicle bench seat, with three seating positions, which is described in Drawing Package SAS-100-1000 (consisting of drawings and a bill of materials); and seat.

9. A new section S8 is added to the standard to read as follows:

S8. *Requirements, test conditions, and procedures for child restraint systems manufactured for use in an aircraft.* Each child restraint system manufactured for use in both motor vehicles and aircraft must comply with all of the applicable test requirements specified in section S5 and, when tested in accordance with the conditions and procedures of S8.2, the additional requirements specified in section S8.1.

S8.1 Child containment for conditions of in-flight turbulence must be determined by inversion tests. The combination of a representative aircraft passenger seat, child restraint system, and appropriate test dummy must be rotated from the normal unright position to an inverted position. The combination must remain inverted for at least 3 seconds with neither failure nor deformation that could seriously injure or prevent subsequent extrication of a child occupant. Child containment must be demonstrated for rotation in the forward direction and a sideward direction.

S8.2 Each configuration and mode of installation must be tested for protection of a child of a weight and stature for which the child restraint system is designed. The child occupant must be simulated with an appropriate test dummy as specified in paragraph S7. Placement of each restraint system in a representative aircraft passenger seat and placement of the test dummy must be in accordance with the manufacturer's instructions. Each child restraint system must be attached to the seat by means of an aircraft safety belt without supplementary anchorage belts or tether straps; FAA Technical Standard Order approved safety belt extensions may be used. The representative aircraft passenger seat used in each test must have a seat back that is completely free to fold over.

Issued on August 24, 1984

Diane K. Steed
Administrator

49 FR 34357
August 30, 1984

PREAMBLE TO AN AMENDMENT TO
FEDERAL MOTOR VEHICLE SAFETY STANDARD NO. 213

Child Restraint Systems for Use in
Motor Vehicles and Aircraft

[Docket No. 74-09; Notice 16]

ACTION: Final rule.

SUMMARY: This rule amends the inversion test added to Standard No. 213, *Child Restraint Systems*, to allow those manufacturers which choose to do so to certify their restraints for use in both motor vehicles and aircraft. These amendments specify more objective criteria for the testing procedures and determining compliance with the inversion tests. This rule adopts what was proposed, except that the rate of acceleration and deceleration at the start and finish of the test is now specified. The rule also specifically allows manufacturers the option of using any of the specified aircraft seats and safety belts. In addition, several typographical errors have been corrected.

EFFECTIVE DATE: April 17, 1985.

SUPPLEMENTARY INFORMATION: During the latter half of 1982, the Department of Transportation had two standards for child restraints. Child restraints for use in motor vehicles had to be certified as complying with the requirements of this agency's Standard No. 213 (49 CFR §571.213). That standard specifies performance and labeling requirements applicable to child restraints. Child restraints for use in aircraft had to be certified as complying with the requirements of the Federal Aviation Administration's (FAA) Technical Standard Order C100. That standard required child restraints to satisfy differing performance and labeling requirements if they were to be used in aircraft.

The result of these differing requirements was that only a few of the child restraints certified for use in motor vehicles were also certified for use in aircraft. In early 1983, the National Transporta-

tion Safety Board considered the safety problems posed for young children traveling in motor vehicles and aircraft and urged that a variety of actions be taken to promote increased use of child restraints. One of those recommendations was that the Department of Transportation simplify its two different standards setting forth requirements for child restraints, by combining the standards into a single standard.

After considering the benefits which would result from the increased use of child restraints, the FAA and NHTSA jointly concluded that the process of certifying child restraints for use in both motor vehicles and aircraft could and should be simplified and expedited. By combining the separate NHTSA and FAA standards into a single standard under the jurisdiction of a single agency, child restraint manufacturers could avoid the difficulties of dealing with different standards, methods of certification, and test procedures promulgated by the two different agencies. Accordingly, a notice of proposed rulemaking (NPRM) was published at 48 FR 36849, August 15, 1983.

This notice proposed that NHTSA would be the sole agency responsible for administering the new Standard No. 213, which would be applicable to both child restraints designed for use in motor vehicles and child restraints designed for use in aircraft. In essence, the notice proposed that the requirements in both agencies' standards be adopted *in toto* and simply combined in an expanded version of Standard No. 213. This would eliminate the problems inherent in dealing with the differing certification and testing procedures of the two agencies and consolidate all the requirements into one standard.

After publication of the NPRM, NHTSA and FAA undertook a joint testing program of all 42 models of child restraints being manufactured at that time and certified as complying with the requirements of Standard No. 213. The purpose of the joint testing program was to determine whether these child restraints could also be certified as complying with the FAA standard for child restraints for use in aircraft. The joint testing program showed that some of the FAA requirements proposed to be added to Standard No. 213 were simply less severe tests of performance capabilities which had already been measured in testing ,to satisfy the NHTSA requirements. Hence, those requirements were deemed redundant and not necessary to ensure adequate protection of restraint occupants in aircraft.

NHTSA published a final rule amending Standard No. 213 at 49 FR 34357, August 30, 1984. That rule added one additional test to Standard No. 213 which had to be satisfied by those child-restraint manufacturers which chose to certify their products for use in both motor vehicles and aircraft. The additional test was an inversion test, whose purpose is to ensure that child restraints certified for use in aircraft adequately protect occupants against the dangers posed by sudden air turbulence. The procedures to be followed were adopted exactly as proposed in the NPRM, which was in turn drawn verbatim from the FAA standard.

A number of the comments received in response to the NPRM agreed with the proposal to include an inversion test in Standard No. 213, but questioned the "vagueness and subjectivity" associated with the inversion test as proposed. After reviewing both the proposed criteria and the comments received on that proposal, NHTSA concluded that the test procedure should be clarified. However, the rulemaking procedures of the Administrative Procedure Act (5 U.S.C. 551 et seq.) precluded the agency from adopting the modifications to the test procedure in the final rule. This was because 5 U.S.C. 553 requires that interested persons receive notice of proposed rulemaking, and that such notice shall include either the terms or substance of the proposed rule or a description of the subjects and issues involved. The NPRM did not give the public notice that NHTSA was even considering different criteria from those which were proposed, so the final rule could not adopt such criteria.

To correct this perceived shortcoming of the final rule, NHTSA published another NPRM on the same day as the final rule, at 49 FR 34374, August 30, 1984. That notice proposed to establish the procedures and criteria used by NHTSA and the FAA in the joint testing program as the procedures and criteria to be followed in the inversion test just added to Standard No. 213. Only one commenter responded to this NPRM.

This notice proposed that to prepare for the inversion test, the subject child restraint should be attached to a representative aircraft passenger seat using only an FAA-approved aircraft safety belt and FAA-approved aircraft safety-belt extensions, if needed. A representative aircraft passenger seat was defined as either an FAA-approved production aircraft passenger seat or a simulated aircraft passenger seat conforming to Figure 6.

The commenter stated that this procedure failed to specify objective criteria, as required by section 102(2) of the National Traffic and Motor Vehicle Safety Act (15 U.S.C. 1391(2)), because it was not clear that every FAA-approved production passenger seat is the equivalent of the simulated passenger seat shown in Figure 6. In the same vein, the commenter argued that it was not clear that all FAA-approved safety belts and safety belt extensions were equivalent for the purposes of the inversion test. If they are not equivalent, the commenter argued, the outcome of the inversion test would depend on the particular seat and/or safety belt chosen for the tests. When the outcome of the test is influenced by something other than the properties of what is being tested, the test is not objective. To remedy this, the commenter urged that the inversion test be amended to either specify the exact seat and safety-belt combinations which would be used for testing or specify that the seat and safety belts may be chosen at the manufacturer's option from among any of the specified seats and safety belts.

The inversion test in Standard No. 213 is a qualitative test, the results of which are mainly dependent upon the geometry of the aircraft seat and safety-belt combination. The test results will not be significantly affected by the seat's structural and padding characteristics or by the seat-belt properties. Nevertheless, the commenter is correct in asserting that the properties of the particular aircraft seat and safety belt used in a test *might* make the difference between the restraint passing and failing the test in a very marginal case. The agency wishes to emphasize that this is a possibility, but it has not been demonstrated. In the joint testing program in which all currently produced models of child restraints were tested, all restraints passed the inversion test, using the criteria adopted in this rule.

To address this possibility, the rule adopts the commenter's suggestion that the proposed language be amended to specify that childs restraint manufacturers may at their option select any of the specified passenger seats and aircraft safety belts for use in the inversion test. A complete listing of all FAA-approved aircraft passenger seats and safety belts can be found in the FAA's Advisory Circular AC 20-36, which is updated annually. By adopting this approach, NHTSA is assuming that the simulated passenger seat shown in Figure 6 and each of the FAA-approved passenger seats are equivalent for the purposes of the inversion test, and that the slight differences between those seats will not make a difference in whether a restraint passes or fails the inversion test. A similar assumption is made with respect to each of the FAA-approved safety belts. The agency has adopted a similar approach in some other standards. *See, e.g.,* S3 of Standard No. 214, *Side door strength* (49 CFR §571.214). Should the agency assumption of equivalence be shown to be incorrect, NHTSA would amend the standard to specify those seats and safety belts which must be used for the inversion test. However, there is no reason to be that restrictive at this time.

Once the child restraint and test dummy have been secured in place in the representative aircraft passenger seat, the notice proposed that the seat be rotated around a horizontal axis at a rate of 35 to 45 degrees per second to an angle of 180 degrees, and the rotation would be stopped when it reached an angle of 180 degrees. The commenter stated that this language was indefinite because it did not specify the starting acceleration and stopping deceleration for the rotation. The commenter stated that the test would be more severe if the rotation were begun with a sudden jerk and halted by banging the combination against a stop positioned at 180 degrees than if it were started and stopped more gradually. However, the proposed language does not indicate which of these procedures is to be used for the testing.

NHTSA agrees with the commenter on this point, and the language of this final rule specifies that the inversion test should be conducted to allow not less than 1/2 second and not more than 1 second for the seat to achieve the required rate of rotation and to be stopped from that rate of rotation. These rates of acceleration and deceleration were the ones used in the NHTSA-FAA joint testing program.

The commenter also stated that there were some minor typographical errors in section S8.2.3, S8.2.4, and S8.2.5, and that the explanatory language beneath Figure 6 needed to be slightly clarified. NHTSA has made each of these requested changes in this final rule.

As discussed above, NHTSA has decided to clarify the test procedures and criteria for determining compliance with the inversion test specified in Standard No. 213. These requirements of this inversion test are optional, and need only be followed by those manufacturers which choose to certify their child restraints for use in aircraft as well as in motor vehicles. Manufacturers which choose to certify their products only for use in motor vehicles will not be adversely affected by an early effective date for these amendments. The amendments made by this notice do not change the fundamental performance requirement that those manufacturers which choose to also certify their products for use in aircraft will have to meet; the amendment benefits the manufacturers by clarifying the test procedure. Accordingly, I find good cause for making the amendments in this rule effective upon publication in the *Federal Register.*

The NHTSA has analyzed this rule and determined that it is neither "major" within the meaning of Executive Order 12291 nor "significant" within the meaning of the Department of Transportation regulatory policies and procedures. No additional requirements are imposed for restraints to be certified for use in aircraft, and no additional requirements are imposed for those restraints to be certified only for use in motor vehicles. These amendments simply clarify the testing procedures to be followed for child restraint systems which the manufacturer chooses to certify for use in aircraft. Since the impacts of this rule are minimal, full regulatory evaluation has not been prepared.

In consideration of the foregoing, 49 CFR Part 571.213 is amended to read as follows:

1. Paragraph S4 is amended by revising the definition of "representative aircraft passenger seat" to read as follows:

"Representative aircraft passenger seat" means

either a Federal Aviation Administration-approved production aircraft passenger seat or a simulated aircraft passenger seat conforming to Figure 6.

2. Paragraph S8 is revised to read as follows:

S8. *Requirements, test conditions, and procedures for child-restraint systems manufactured for use in aircraft.*

Each child-restraint system manufactured for use in both motor vehicles and aircraft must comply with all of the applicable requirements specified in section S5 and with the additional requirements specified in S8.1 and S8.2.

S8.1. *Installation instructions.* Each child-restraint system manufactured for use in aircraft shall be accompanied by printed instructions in the English language that provide a step-by-step procedure, including diagrams, for installing the system in aircraft passenger seats, securing the system to the seat, positioning a child in the system when it is installed in aircraft, and adjusting the system to fit the child. In the case of each child restraint which is not intended for use in aircraft at certain adjustment positions, the following statement, with the manufacturer's restrictions inserted, shall be included in the instructions.

DO NOT USE THE – – – ADJUSTMENT POSITION(S) OF THIS CHILD RESTRAINT IN AIRCRAFT.

S8.2. *Inversion test.* When tested in accordance with S8.2.1 through S8.2.5 and adjusted in any position which the manufacturer has not, in accordance with S8.1, specifically warned against using in aircraft, each child-restraint system manufactured for use in aircraft shall meet the requirements of S8.2.1 through S8.2.6. The manufacturer may, at its option, use any seat which is a representative aircraft passenger seat within the meaning of S4.

S8.2.1. A representative aircraft passenger seat shall be positioned and adjusted so that its horizontal and vertical orientation and its seat-back angle are the same as shown in Figure 6.

S8.2.2. The child-restraint system shall be attached to the representative aircraft passenger seat using, at the manufacturer's option, any Federal Aviation Administration-approved aircraft safety belt, according to the restraint manufacturer's instructions for attaching the restraint to an aircraft seat. No supplementary anchorage belts or tether straps may be attached;

however, Federal Aviation Administration-approved safety-belt extensions may be used.

S8.2.3. In accordance with S6.1.2.3.1 through S6.1.2.3.3, place in the child restraint any dummy specified in S7 for testing systems for use by children of the heights and weights for which the system is recommended in accordance with S5.5 and S8.1.

S8.2.4. If provided, shoulder and pelvic belts that directly restrain the dummy shall be adjusted in accordance with S6.1.2.4.

S8.2.5. The combination of representative aircraft passenger seat, child restraint, and test dummy shall be rotated forward around a horizontal axis which is contained in the median transverse vertical plane of the seating-surface portion of the aircraft seat and is located 1 inch below the bottom of the seat frame, at a speed of 35 to 45 degrees per second, to an angle of 180 degrees. The rotation shall be stopped when it reaches that angle and the seat shall be held in this position for 3 seconds. The child restraint shall not fall out of the aircraft safety belt nor shall the test dummy fall out of the child restraint at any time during the rotation or the 3-second period. The specified rate of rotation shall be attained in not less than ½ second and not more than 1 second, and the rotating combination shall be brought to a stop in not less than ½ second and not more than 1 second.

S8.2.6. Repeat the procedures set forth in S8.2.1 through S8.2.4. The combination of the representative aircraft passenger seat, child restraint, and test dummy shall be rotated sideways around a horizontal axis which is contained in the median longitudinal vertical plane of the seating-surface portion of the aircraft seat and is located 1 inch below the bottom of the seat frame, at a speed of 35 to 45 degrees per second, to an angle of 180 degrees. The rotation shall be stopped when it reaches that angle and the seat shall be held in this position for 3 seconds. The child restraint shall not fall out of the aircraft safety belt, nor shall the test dummy fall out of the child restraint at any time during the rotation or the 3 second period. The specified rate of rotation shall be attained in not less than ½ second and not more than 1 second, and the rotating combination shall be brought to a stop in not less than ½ second and not more than 1 second.

3. A new Figure 6 would be added at the end of § 571.213, appearing as follows:

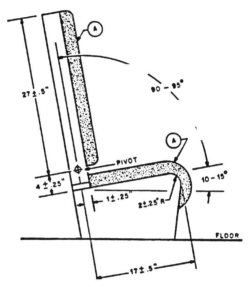

"A" represents a 2- to 3-inch thick polyurethane foam pad, 1.5 to 2.0 pounds per cubic foot density, over 0.020-inch-thick aluminum pan, and covered by 12- to 14-ounce marine canvas. The sheet-aluminum pan is 20 inches wide and supported on each side by a rigid structure. The seat back is a rectangular frame covered with the aluminum sheet and weighing between 14 and 15 pounds, with a center of mass 13 to 16 inches above the seat pivot axis. The mass moment of inertia of the seat back about the seat pivot axis is between 195 and 220 ounce-inch-second[2]. The seat back is free to fold forward about the pivot, but a stop prevents rearward motion. The passenter safety belt anchor points are spaced 21 to 22 inches apart and are located in line with the seat pivot axis.

FIGURE 6: SIMULATED AIRCRAFT PASSENGER SEAT

Issued on April 10, 1985.

Diane K. Steed
Administrator

50 FR 15154
April 17, 1985

FEDERAL MOTOR VEHICLE SAFETY STANDARD NO. 213

Child Restraint Systems
[Docket No. 74-09; Notice 18]

ACTION: Final rule.

SUMMARY: This rule amends Standard No. 213, *Child Restraint Systems,* with respect to the requirements applicable to buckles used in child restraints. The requirement regarding the force necessary to operate the buckle release mechanism in the pre-impact test is changed from the previous minimum level of 12 pounds to a range between 9 and 14 pounds. The maximum release force for the buckle release in the post-impact test is reduced from the previous level of 20 pounds to 16 pounds. Additionally, this rule adds buckle size and buckle latching requirements to the standard. The effect of this rule is to ensure that child restraint buckles are easier for adults to operate, while still ensuring that small children will not be able to open the buckles by themselves.

EFFECTIVE DATE: February 18, 1986.

SUPPLEMENTARY INFORMATION: As an initial step toward ensuring that child restraint systems would offer adequate protection for their occupants, NHTSA issued Standard No. 213 in 1970. That version of the Standard required, among other things, that the buckle release mechanism operate when a force of not more than 20 pounds was applied.

NHTSA issued a new Standard No. 213, *Child Restraint Systems* (49 CFR §571.213) at 44 FR 72131, December 13, 1979. This new Standard substantially upgraded the performance requirements for child restraint systems. It also specified that the buckles must not release when a force of less than 12 pounds was applied to the buckle before conducting the dynamic systems test

required by section S6.1 of Standard No. 213 and must release when a force of not more than 20 pounds was applied after conducting that dynamic systems test. The test for measuring the amount of force needed to release the buckle was to be conducted in accordance with the procedures set forth in section S6.2 of the standard. The purpose of the buckle force requirements is to prevent young children from unbuckling the restraint belt(s), while allowing adults to do so easily.

After the adoption of the standard, the agency received information indicating that the minimum force level needed to release the buckles was too high to permit many adults to easily release the buckles. Some of the buckles tested in the field required more than 20 pounds of pressure to release, according to a report done for the agency by K. Weber and N. P. Allen (Docket No. 74-09-GR-120). This same report concluded that even a force of 20 pounds is difficult for most women to generate with one hand. The agency has also been provided with consumer letters received by one child restraint manufacturer commenting on the difficulty of operating the child restraint harness buckles. The agency itself has received numerous telephone calls from consumers complaining about the size of the release buttons on child restraint belts and the high force levels required to operate them.

The agency's safety concerns over child restraint buckle force release and size stem from the need for convenient buckling and unbuckling of a child and, in emergencies, to quickly remove the child from the restraint. This latter situation can occur in instances of post-crash fires, immersions, etc. A restraint that is difficult to disengage, due to the need for excessive buckle pressure or difficulty in operating the release mechanism because of a very small release button, can unnecessarily endanger the child in the restraint and the adult attempting to release the child.

This amendment is also intended to reduce the everyday misuse rate of child restraint harness and shields. Several studies conducted by Goodell-Grives, Inc., under contract to NHTSA indicate that the harness and shield misuse rate for infant and toddler restraints is between 25 and 40 percent. According to this study and others, misused child restraints may not only fail to protect the child in a crash situation, but may increase injury severity. The December 1984 study asked parents why they were apparently misusing the harness and shields. The misuse did not result from the lack of knowledge about the proper use of the harness and shields, because 95 percent of those parents knew the child restraint was being used incorrectly. Although the buckles were not cited directly, the inconvenience of the harness and shield operation was the most frequent reason given for misuse. This amendment will improve the operational convenience of the harness and shield buckles and thus should increase the correct usage rate of child restraint systems.

Accordingly, NHTSA published a notice of proposed rulemaking (NPRM) at 48 FR 20259, May 5, 1983, which proposed several changes to the buckle release force measurement test procedures. Those changes were intended to facilitate the use of buckles which would require approximately 10 1/2 pounds of force to release. The buckle force release test procedure specified that the buckle was to be tested both before and after the impact testing of the child restraint. In both the pre- and post-impact tests, tension was applied to the buckle prior to measuring the buckle release force. The purpose of applying tension was to simulate the force that would be applied to the buckle by a child hanging upside down in the child restraint.

The first proposed change was to eliminate the tension applied to the buckle in the pre-impact test. While it was considered appropriate for the post-impact test to simulate tension which would be present on the buckle in the event of a rollover crash, it was tentatively concluded that there were no forces whose presence ought to be simulated in the pre-impact test. Therefore, the notice proposed to measure the buckle release force in the pre-impact test with no load applied to the belt buckle, except the load exerted by properly adjusting the belt system around a child.

The second proposed change was to reduce the minimum buckle force permitted in the pre-impact test by three pounds, from 12 pounds to 9 pounds. According to the evidence available to the agency, a minimum buckle force level of 9 pounds is sufficient to prevent children up to the age of approximately 4 from opening the buckle by themselves. Further, the notice proposed to set a force of 12 pounds as the maximum force permitted in the pre-impact test. The NPRM specifically sought comments on whether this 3-pound range was sufficient to account for the amount of buckle force variation which inevitably arises from mass production manufacturing techniques.

The third change was proposed for the post-impact testing of the buckles. The tension previously specified in the standard would still be applied to the buckles before the release force was measured. However, the maximum force needed to release the buckles was proposed to be reduced from 20 pounds to 16 pounds. A higher force level is specified in the post-impact test as compared to the pre-impact test to allow for damage which could occur to the buckles during an actual crash and to allow for the additional belt loading which is possible from a child suspended upside down in the restraint system. The proposed lowering of the maximum force level was intended to permit a large portion of adults to more easily and quickly release the buckle in normal use (thus encouraging routine correct use of the restraints which would provide enhanced child safety) and in emergency post-crash situations.

The NPRM also proposed a change to Standard No. 213 in response to complaints about instances where a child restraint buckle was seemingly securely fastened by a parent, but subsequently popped open. This problem is commonly referred to as false latching. To address this problem, the NPRM proposed to require that child restraint buckles meet the latching requirements in section S4.3(g) of Standard No. 209, Seat Belt Assemblies. These requirements ensure that the design and construction of the buckle release mechanism are sufficiently durable to permit repeated latching and unlatching of the buckle and that the buckle releases when it is falsely latched and a minimum force (in this case, 5 pounds) is applied to it.

The final change proposed in the NPRM related to the size of the buckle release area. The agency believed that some of the problems experienced by parents in fastening and unfastening the child restraint buckles might be attributable to the size of the buckle release mechanism. For instance, the smaller the area of a push button release mechanism, the more difficult it would be to use more than one finger, and hence apply a greater force, to open the buckle. The release mechanisms on some buckles were too small to allow sufficient engagement area for easy release of the buckle, particularly for persons with large hands. Most child restraint buckles use push buttons to release the buckle, so the NPRM proposed that push buttons have a minimum area of 0.6 square inch. The minimum surface area requirements applicable to motor vehicle seat belts were specified for other types of release mechanisms used on child restraint buckles.

The NPRM also requested comments on regulatory and non-regulatory ways in which the issues of belt length and shell width could be addressed. This request was based on the Weber and Allen report referenced above which raised questions about the length of the harness webbing used in child restraints and the seating width of the shells. The researchers noted that use of winter clothing significantly increases the amount of harness webbing needed to accommodate a fully clothed child. They reported that a snowsuit can add six inches to the length necessary for a harness lap belt to accommodate a child. Further, the researchers said that nearly all child restraints are too narrow for the size children they claim to accommodate.

The agency received 16 comments on the NPRM, and the commenters included private citizens, safety advocacy groups, child restraint manufacturers, and the National Transportation Safety Board. All these comments were considered in developing this final rule, and the most relevant ones are specifically addressed in the following discussion.

Pre-Impact Test Buckle Release Force Limit. In the NPRM, the agency specifically sought comments on the feasibility of manufacturing buckles within the 3-pound range. Many of the commenters objected to the proposed 9- to 12-pound release force limits, primarily because the 3-pound range was said to be too narrow based on current manufacturing techniques, to ensure that all buckles would comply with the proposed requirement. Some of these commenters asserted that the proposed 3-pound range would cause the buckle manufacturers to increase buckle prices in order to recoup the costs of the changes in manufacturing techniques and quality control which would have to be implemented to satisfy the proposed require-

ment. One child restraint manufacturer offered a statistical analysis of buckle release force tests in an effort to demonstrate the difficulty of maintaining a 3-pound range with current buckle manufacturing techniques. The manufacturer indicated that buckle release forces can vary up to 3- times the standard deviation for a given sample. The standard deviation for current production buckles is sufficiently large that, given a mean of 10.5 pounds and a range of 3 pounds, some buckles would have release forces outside the range. A different manufacturer submitted data from tests of current buckle designs showing that the release force can vary by as much as 6 pounds for current buckles. Finally, several commenters objected to the proposed 9-pound minimum release force on the grounds that buckles manufactured in compliance with the Canadian child restraint standard, which specifies an 8-pound minimum release and 16-pound maximum release force, would not satisfy the proposed U.S. standard. These commenters further stated that NHTSA should use this opportunity to harmonize this requirement with the Canadian standard.

In response to these comments, NHTSA has reconsidered its proposed 9- to 12-pound range for the buckle release force permitted in the pre-impact testing. The agency has concluded that a 3-pound range in release force would not be feasible with current manufacturing techniques, and the benefits of narrowing the feasible range to 3 pounds do not warrant requiring a change in current manufacturing techniques.

The only research study of which the agency is aware, examining the most appropriate release force range for child restraint buckles, is entitled "Child Restraint Systems," published in 1976 by Peter Arnberg of the National Swedish Road and Traffic Institute. This study, which is available in the General Reference section of Docket No. 74-09, presented the results of testing 80 children aged 2 1/2 to 4 1/2 years and 200 women. This study concluded that child restraint buckles should have a release force of 40 to 60 Newtons (approximately 9 to 13 1/2 pounds).

After analyzing the comments, NHTSA has determined that a 5-pound range in buckle release force is needed to allow for current buckle manufacturing techniques. Based on this determination and the recommendations of the Arnberg study, this rule requires child restraint buckles to have a release force of between 9 and 14 pounds before the buckles are subjected to dynamic testing.

The agency notes that this rule is not precisely harmonized with the Canadian standard for child restraint buckle release forces, which specifies a minimum release force of 8 pounds before dynamic testing and a maximum release force of 16 pounds after dynamic testing. NHTSA has adopted a 9 pound minimum release force because of its concern that 3 1/2- to-4-year-old children could open their child restraint buckles if the release force were 8 pounds, as shown in the Arnberg study. Further, the 14-pound maximum release force before dynamic testing was added in this rule because buckles with a release force of more than 14 pounds are difficult for many women to open in everyday use, as demonstrated in the Arnberg study. The result of these differing requirements in the United States and Canada is that buckles which comply with the Canadian buckles force requirements will not automatically comply with Standard No. 213. However, buckles which comply with Standard No. 213 will also comply with the buckle force requirements of the Canadian standard.

Pre-Impact Buckle Test Procedure. The NPRM proposed a new procedure for this test. The same procedures have been used for measuring the buckle release force in both the pre-impact and the post-impact testing. Briefly stated, the child restraint is installed on a standard seat assembly, the dummy is positioned in the child restraint, a sling is attached to each wrist and ankle of the dummy, and the sling is pulled by a designated force. As noted above, the presence of the dummy and the force applied to the sling simulate a rollover crash situation.

The NPRM proposed, and this final rule adopts, a new test procedure for the pre-impact testing, because there is no need to simulate a rollover crash situation before impact. The NPRM proposed placing the buckle on a hard, flat surface and loading each end of the buckle with a force of 2 pounds before measuring the force required to release the buckle. None of the commenters objected to this basic change in the test procedure, and it is adopted for the reasons stated in the NPRM.

Several commenters did object to the release force application device, which was proposed as a rigid, right-circular cone with an enclosed angle of 90 degrees or less. This device would be used to transfer the release force to the push button release. Some commenters argued that this device would not adequately represent real-world push button actuation. Specifically, they were con-cerned that the pointed device applies the release force over an area considerably smaller than that of a finger or thumb. Other commenters argued in favor of a different release force application device, contending that this device would permanently deface some of the tested buckles.

NHTSA has decided to adopted the proposed conical test device. Its small contact area allows accurate positioning on the release button, which will yield consistently repeatable test results. The buckle release force test procedures proposed in the NPRM, as modified for this final rule, were conducted by the Calspan Corporation in July 1984 during the annual FMVSS No. 213 compliance test procedures. On the basis of these tests, the agency concluded that the amended test procedures simulate real-world actuation of push button release mechanisms because the release force is applied in a manner similar to hand operation and tests with several alternative devices indicated that conical devices produce release force values consistent with those generated by different probes. Manufacturers choosing to test a large number of buckles to be used on their child restraints can place a protective surface between the button and the test device to prevent defacing of the buckles. Those manufacturers who want to use an alternative test device are free to do so, provided that they can correlate the results obtained with that alternative device with results obtained with the specified test device, which will be used by the agency in compliance tests.

The NPRM proposed that the force applied by the test device be "at the center line of the push button 0.125 inches from a movable edge and in the direction that produces maximum releasing effect." Many commenters argued that this procedure needed to be refined to take account of the different release mechanisms. One commenter stated that there are two different types of push button release mechanisms, hinged and floating. A hinged button has one fixed edge and release forces applied near the fixed edge may not activate the release mechanism. Instead, the hinged button is designed to release when force is applied near the center of the button or toward the edge opposite the fixed edge. On the other hand, the floating button has no fixed edges and is designed to release when force is applied near the center of the button. This commenter noted that, while the force application proposed in the NPRM may be suitable for hinged buttons, it would be inappropriate for floating buttons.

PART 571; S213–PRE 44

The agency agrees with the commenters that some further refinements should be made to the test procedures to account for the different types of push buttons. Accordingly, this rule specifies that, for hinged buttons, the force shall be applied according to the procedures proposed in the NPRM. For floating buttons, the force shall be applied at the geometric center of the button. These differing force application points will take into account the differing designs of push buttons, without favoring one or the other design.

Several commenters stated that the NPRM failed to specify any test procedures for buckles designed for the insertion of two or more buckle latch plates, even though a number of buckles on current models of child restraints are designed to secure more than one belt. Further, these commenters noted that, while the NPRM did specify a 2-pound pre-load force should be applied to buckles before conducting the pre-impact buckle release test, it failed to specify the direction in which the force should be applied. To remedy these perceived shortcomings, some of the commenters recommended that the final rule specify that the 2-pound pre-load force be applied along the direction of the latch plate insertion for single latch plate buckles and that the 2-pound force be divided by the number of latch plates and the resultant force applied to each latch plate in the direction of latch plate insertion for multiple latch plate buckles. This final rule adopts this recommendation. The NPRM's intent was that the force be applied along the direction of latch plate insertion, and it is appropriate to make this intent explicit in this final rule. Further, the one pound pre-load force for multiple latch plate buckles is sufficient force to simulate the tension which would be present in properly adjusted belts, yet small enough so as not to simulate other forces which would not be present in normal everyday use.

Along these lines, one commenter suggested that the pre-load force be increased from two to five pounds. This commenter stated that the proposed pre-load force of 2-pounds might not be sufficient to release the buckles, while the 5-pound load would assure that the buckles always release. Further, the commenter noted that Standard No. 209 allows a false latching load of 5-pounds maximum, and that this change would make the two Standards consistent.

NHTSA is not persuaded by these comments, and has not incorporated the suggested change in this final rule. For the pre-impact buckle release force test procedure, the 2-pound pre-load is designed to simulate the separation tension in the harness restraint system during normal use and approximate the buckle loading on a restraint system adjusted for the compliance impact test.

Section S5.2(g) of Standard No. 209, on the other hand, is not intended to approximate forces present during normal buckle operation. That section requires that the buckle latching mechanism be tested for durability and then the latch plate or hasp inserted in any position of "partial" engagement (false latching). When the buckle and latch plate are in this position of "induced" partial engagement, a force of 5 or less shall separate the latch plate from the buckle. The separation of the latch plate is affected without operating the release mechanism. Since this procedure is not intended to simulate normal buckle operation but to test the susceptibility of the buckle to false latching, it would not be appropriate to incorporate its loading into Standard No. 213.

Post-Impact Buckle Test Procedure. As noted above, the NPRM proposed to reduce the maximum force needed to release the buckle after it had been subjected to the impact test from the 20-pound level currently specified to 16 pounds. A higher release force is specified for the post-impact test to account for damage which might occur to the buckle during the impact test and to counter the forces which could be exerted on the buckle by a child hanging upside down in rollover crash conditions. The reason for proposing the lower force was that it was sufficient to account for damage which might occur to the buckle, and such force can be generated by almost all women using only one hand, according to the Arnberg study. The current 20-pound force requirement allows buckles which require two-hand operation by many adults, and two-hand operation is often awkward and may adversely affect safety in emergency situations. The agency notes that the Canadian standard also specifies a maximum post impact force of 16 pounds. No commenters objected to this proposed change, and it is adopted herein for the reasons explained above.

The preamble to the NPRM did not discuss any other changes to the post-impact testing procedure, because the agency did not intend to propose any changes other than reducing the maximum release force for the buckles. However, section S6.2.2 of Standard No. 213 as published in the NPRM indicated that the self-adjusting sling which is attached to the dummy to simulate a rollover crash situation should be attached only to the dummy's ankles. The Standard currently requires the sling to be attached to the dummy's *wrists* and ankles, and this requirement was inadvertently omitted from the NPRM language. This final rule corrects this omission, so no change is specified for the post-impact testing except the reduction in buckle release force.

Buckle Latching. The NPRM proposed adding the latching performance requirements of sections S4.3(g) and S5.2(g) of Standard No. 209 to Standard No. 213. These procedures test the latching performance of seat belt buckles to ensure that the buckle materials and structure will operate properly after numerous cycles of latchings and unlatchings. As explained in the NPRM, this step should reduce or eliminate the false latching problems experienced by child restraint users. False latching occurs when buckles are apparently latched, but then subsequently pop open. NHTSA believes that most of the false latchings result from poorly designed or cycle degraded latching mechanisms, and that the Standard No. 209 requirements will eliminate latching mechanisms which are poorly designed or subject to cycle degradation.

Most of the commenters who addressed this proposal supported its adoption, although several commenters stated that additional requirements may be needed to ensure that false latching does not continue to be a significant problem. The National Transportation Safety Board stated that it had evidence that brand-new child restraint buckles, not yet subject to material wear, are prone to false latching, and that additional requirements along the lines of the European requirement that latchplates be ejected by a spring located in the buckle when the buckle is not properly latched, may be necessary to prevent false latching. Other suggestions from the commenters included requiring the use of color-coded push buttons to show when the buckle was properly latched and requiring specific warnings in the manufacturer's instruction manuals urging parents to check for false latchings every time they fasten the buckles.

NHTSA has adopted the requirements proposed in the NPRM to reduce the false latching problems. The agency believes that the Standard No. 209 seat belt buckle tests will identify buckles which are subject to false latchings because of materials wear or poor design, because false latching complaints by consumers have been eli-

minated for motor vehicle seat belts and the agency expects that these tests will substantially reduce this problem for child restraint buckles as well. The agency will continue to monitor problems of false latchings, and will consider additional requirements to address that problem if necessary.

Buckle Size. The NPRM proposed to specify a minimum area for the buckle release mechanism, because some of the difficulties reported in opening child restraint buckles were believed to arise from the small size of the buckle release mechanism. As noted earlier, the smaller the area of the push button, the more difficulty there is in applying the forces which must be exerted to open the buckle. Those commenters who addressed this issue supported the proposed requirement that push buttons used on child restraints have a minimum release area of 0.6 square inch, and it is adopted in this final rule.

Belt Length/Shell Width. The NPRM solicited comments on steps which could be taken to address the issues of belt length and shell width. These issues arose after a research report noted that children clad in winter clothes need up to six additional inches of belt webbing, and that many current child restraints do not have this extra belt length. In addition, the report noted that nearly all child restraints are too narrow for the size children they claim to accommodate. The NPRM noted that a long-range solution was for the agency to use additional test dummies to simulate larger children. A possible short-term answer was to conduct the crash tests with the dummies clad in a typical snowsuit.

Several commenters stated that regulatory action was not needed in this area. Child restraint manufacturers generally believe that the industry will adjust belt length and shell width in response to consumer demand, and believe that any regulations at this time would only add costs and research burden without substantially benefiting child safety. The Physicians for Automotive Safety stated that the agency should approach those manufacturers with problems in these areas and request voluntary remedial action, instead of pursuing rulemaking. That group also stated that it knew of only one model of child restraint with problems along these lines. The National Transportation Safety Board stated that the agency should develop regulations in these areas.

Some of the commenters opposed the use of snowsuits on the test dummies because those snowsuits would absorb some of the crash energy. According to these commenters, the agency would, in effect, reduce the severity of the crash tests by so dressing the test dummies.

In view of the above comments rulemaking will be deferred in this area. The agency will continue to monitor the issues of seat shell size and harness webbing length associated with infant and toddler restraints (40 pounds and below) to determine if rulemaking in this area will be necessary in the future.

Editorial Correction. Several commenters noticed that there was a typographical error in section S5.4.3.5(a) of the NPRM. That section referred to testing in accordance with section S6.2.2, while the correct reference was to section S6.2.1. This error is corrected in this final rule.

§571.213 [Amended]

In consideration of the foregoing, Title 49 of the Code of Federal Regulations is amended by revising §571.213 to read as follows:

1. The authority citation for 571 continues to read as follows:

Authority: 15 U.S.C. 1392, 1401, 1403, and 1407; delegation of authority at 49 CFR 1.50.

2. Section S5.4.3.5 is revised to read as follows:

*　　*　　*　　*　　*

S5.4.3.5 *Buckle Release.* Any buckle in a child restraint system belt assembly designed to restrain a child using the system shall:

(a) When tested in accordance with S6.2.1 prior to the dynamic test of S6.1, not release when a force of less than 9 pounds is applied and shall release when a force of not more than 14 pounds is applied;

(b) After the dynamic test of S6.1, when tested in accordance with S6.2.3, release when a force of not more than 16 pounds is applied;

(c) Meet the requirements of S4.3(d)(2) of FMVSS No. 209 (§571.209), except that the minimum surface area for child restraint buckles designed for push button application shall be 0.6 square inch;

(d) Meet the requirements of S4.3(g) of FMVSS No. 209 (§571.209) when tested in accordance with S5.2(g) of FMVSS No. 209; and

(e) Not release during the testing specified in S6.1.

*　　*　　*　　*　　*

3. Section S6.2 is revised to read as follows:

*　　*　　*　　*　　*

S6.2 *Buckle Release Test Procedure.* The belt assembly buckles used in any child restraint system shall be tested in accordance with S6.2.1 through S6.2.4 inclusive.

*　　*　　*　　*　　*

4. Section S6.2.1 is revised to read as follows:

*　　*　　*　　*　　*

S6.2.1. Before conducting the testing specified in S6.1, place the locked buckle on a hard, flat, horizontal surface. Each belt end of the buckle shall be pre-loaded in the following manner. The anchor end of the buckle shall be loaded with a 2-pound force in the direction away from the buckle. In the case of buckles designed to secure a single latch plate, the belt latch plate end of the buckle shall be loaded with a 2-pound force in the direction away from the buckle. In the case of buckles designed to secure two or more latch plates, the belt latch plate ends of the buckle shall be loaded equally so that the total load is 2 pounds, in the direction away from the buckle. For push-button release buckles the release force shall be applied by a conical surface (cone angle not exceeding 90 degrees). For push-button release mechanisms with a fixed edge (referred to in Figure 6 as "hinged button"), the release force shall be applied at the centerline of the button, 0.125 inches away from the movable edge directly opposite the fixed edge, and in the direction that produces maximum releasing effect. For push-button release mechanisms with no fixed edge (referred to in Figure 6 as "floating button"), the release force shall be applied at the center of the release mechanism in the direction that produces the maximum releasing effect. For all other buckle release mechanisms, the force shall be applied on the centerline of the buckle lever or finger tab in the direction that produces the maximum releasing effect. Measure the force required to release the buckle. Figure 6 illustrates the loading for the different buckles and the point where the release force should be applied, and Figure 7 illustrates the conical surface used to apply the release force to push-button release buckles.

*　　*　　*　　*　　*

5. Section S6.2.2 is revised to read as follows:

*　　*　　*　　*　　*

S6.2.2. After completion of the testing specified in S6.1, and before the buckle is unlatched, tie a self-adjusting sling to each wrist and ankle of the test dummy in the manner illustrated in Figure 4.

*　　*　　*　　*　　*

6. Section S6.2.4 is revised to read as follows:

*   *   *   *   *

S6.2.4. While applying the force specified in S6.2.3, and using the device shown in Figure 7 for push-button release buckles, apply the release force in the manner and location specified in S6.2.1 for that type of buckle. Measure the force required to release the buckle.

*   *   *   *   *

7. Section S6.2.5 is deleted.

8. Two new drawings (Figures 6 and 7) are added at the end of §571.213, appearing as follows:

Issued on August 15, 1985.

Diane K. Steed
Administrator

**50 F.R. 33722
August 21, 1985**

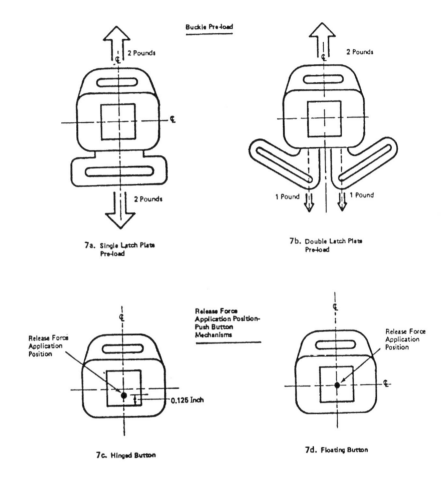

Figure 7. Pre-Impact Buckle Release Force Test Set-up

PART 571; S213—PRE 49

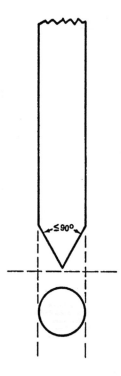

Figure 8. Release Force Application Device — Push Button Release Buckles

PART 571; S213—PRE 50

PREAMBLE TO AN AMENDMENT TO MOTOR VEHICLE SAFETY STANDARD NO. 213

Child Restraint Systems

(Docket No. 74-09; Notice 18)

ACTION: Final rule.

SUMMARY: This rule amends Standard No. 213, *Child Restraint Systems,* by requiring all child restraints equipped with tether straps (other than child harnesses, booster seats, and restraints designed for use by physically handicapped children) to pass the 30 miles per hour (mph) test with the tether strap unattached. This change is being made because survey results consistently show that, in the vast majority of instances, child restraints with tether straps are used by the public without attaching the tether strap to the vehicle. This amendment will ensure that children riding in child restraints with unattached tethers will be afforded crash protection equivalent to that afforded to children riding in child restraints designed without a tether.

This rule also eliminates the requirement that those child restraints pass a 20 mph test with the tether unattached. Since those restraints will now be required to pass the 30 mph test under the same test conditions, it is unnecessary for those restraints to also be tested at a low speed.

Finally, this rule clarifies two items of information required to be included in the instructions accompanying child restraints. These clarifications do not alter the amount of information that must be included in the instructions: they simply explain what the agency intended to require.

EFFECTIVE DATE: August 12, 1986.

SUPPLEMENTARY INFORMATION: Standard No. 213, *Child Restraint Systems* (49 CFR S571.213) currently provides two different test configurations applicable to child restraint systems. First, a 30 mph frontal crash test is conducted for all child restraints. In that test, the restraints are installed according to the child restraint manufacturer's in-structions. This test is referred to as Test Configuration I in section S6.1.2.1.1 of Standard No. 213.

Second, a 20 mph test is conducted for two types of child restraint systems. One type is a child restraint equipped with an anchorage belt. Anchorage belts, more commonly referred to as tether straps, are supplemental belts under to attach the child restraint to the vehicle. The other type of restraint subject to the 20 mph crash test is a child restraint with a fixed or movable surface which helps to restrain the child's forward movement in the event of a crash. This type of child restraint provides protection by the use of its own belt system and a surface which can be used independently of the belt system. Both these types of child restraints are tested with only the vehicle lap belt holding the child restraint to the standard test seat and, in the case of restraints with a fixed or movable surface forward of the child, without attaching the restraint's belt system to hold the test dummy in place. This test, referred to as Test Configuration II in section S6.1.2.1.2 in Standard No. 213, is intended to take account of the possibility that the tether strap or the restraint's belt system will either be misused or not used at all by parents. If this happens, Test Configuration II should ensure that these types of restraints will offer minimal protection even when they are not properly used.

This rulemaking action addresses only the question of restraints with tether straps, and does not affect restraints with fixed or movable surfaces forward of the child. Tether straps have presented a difficult question for the agency since at least 1979. When a tether strap is properly attached, a child restraint equipped with a tether strap will generally offer the best protection for child occupants, particularly those riding in the front seat or involved in side impact crashes.

However, the results of surveys have continually shown that tether straps are not attached by the vast majority of the public. The most recent study available to the agency on this topic (Cynecki and Goryl, "The Incidence and Factors Associated with Child Safety Seat Misuse"; December 1984, DOT HS–806 676) found that nearly 85 percent of child restraints with tether straps were used without properly attaching the tether straps. The Cynecki and Goryl study recommended that the best solution for this problem would be to redesign the restraints to eliminate the need for tether straps.

This same suggestion had been made previously by several commenters in connection with the final rule substantially upgrading the performance requirements fo Standard No. 213; 44 FR 72131, December 13, 1979. At the time of that rulemaking action, however, restraints. The agency decided that it would be inappropriate to issue a rule which would have the effect of requiring a major redesign of most child restraint systems then on the market, especially when the public was just beginning to appreciate the importance of using child restraints. Further, NHTSA expected that proper usage of restraints with tethers would grow as public awareness and knowledge of child restraints grew.

When NHTSA reexamined this decision in light of the Cynecki and Goryl report, the reasoning no longer seemed valid. First, at this time, approximately one-fifth of all new child restraints, including booster seats, are equipped with a tether strap necessary for the protection of the child occupant. Thus, a rule which would have the effect of requiring a redesign of these restraints would have a substantially smaller impact on the child restraint market now than it would have had in 1979.

Second, and most significant, the expectation of increased proper use of tether straps has not been realized. Perhaps the most troubling fact in the Cynecki and Goryl report cited above was that 78 percent of the persons not using the tether strap to attach the child restraint to the vehicle *knew that its use was necessary*. This indicates that, while public awarness and knowledge of child restraints has grown significantly since 1979, that awareness and knowledge has not resulted in increased proper use of tether straps.

Because of its concern for the safety of children riding in motor vehicles, NHTSA tentatively de-

cided that it was no longer reasonable to allow restraints with tethers to be tested in only a 20 mph crash in the way they will be used by the public, that is, without attaching the tether strap. The agency believed that those restraints, like restraints without tethers, should be tested in a 30 mph crash in the way they will be used by the public. This would ensure that all child restraints afforded equivalent protection to children riding therein.

Accordingly, NHTSA published a notice of proposed rulemaking (NPRM) on July 5, 1985; 50 FR 27633, proposing that all child restraints other than child harnesses be tested in the 30 mph crash test when attached to the test seat only by means of the lap belt. This proposal was intended to ensure that restraints with tethers afford the same level of protection to child restraint occupants as do restraints without tethers when tested in the manner both will be used by the public.

That NPRM also proposed some less significant changes to Standard No. 213. These were as follows:

(1) The standard currently specifies that the child restraint be installed in the center seating position during the testing. However, many new vehicles are produced without a front or rear center seating position. This trend raised the concern that the tests were growing less representative of the conditions which would be encountered by the child restraint when it was in use. Accordingly, the NPRM proposed to amend Standard No. 213 to require that child restraints be tested in one of the two outboard seating positions. An anticipated added benefit of this change would be that it would reduce testing costs for the child restraint manufacturers, because two child restraints could be evaluated in the same test.

(2) Standard No. 213 requires that all child restraints equipped with a tether strap be permanently lableed with a notice that the tether strap must be properly secured as specified in the manufacturer's instructions. The NPRM proposed that the phrase "For extra protection in frontal and side impacts" be added in front of that notice. This change would convey the fact that the tether strap was a supplementary safety device, as proposed in the NPRM, while also affirming that additional safety protection is afforded when the tether strap is properly attached.

(3) Two changes were proposed to clarify what was meant in the requirements concerning the in-

stallation instructions to be provided along with the child restraint by the restraint's manufacturer. These were:

(a) The installation instructions are currently required to state that, in most vehicles, the rear center seating position is the safest seating position for installing a child restraint. This statement in the instructions has resulted in numerous inquiries to the agency by consumers wanting to know the safest seating position for vehicles with only two rear outboard seating positions. To eliminate this confusion on the part of the public, the NPRM proposed that the installation instructions be modified to state that, for maximum safety protection, the child restraint should be installed in a rear seating position in vehicles with two rear seating positions and in the center rear seating position in vehicles with three rear seating positions.

(b) The installation instructions in Standard No. 213 also require that child restraint manufacturer to "specify in general terms the types of vehicles, seating positions, and vehicle lap belts with which the system can or cannot be used." This requirement has frequently been erroneously interpreted to mean that child restraint manufacturers are required to state the specific vehicles, specific seating positions, and the specific vehicle lap belts with which a child restraint can or cannot be used. The NPRM proposed an amendment to make clear the agency's intent that the instructions specify the *types* of vehicles (e.g., passenger cars, pickup trucks, vans, buses, etc.), the *types* of seating positions (e.g., front, rear, bench, bucket, side facing, rear facing, folding, etc.) and the *types* of vehicle safety belts (e.g., diagonal, lap-shoulder, emergency locking, etc.) with which the restraint system can or cannot be used.

A total of 15 comments were received on the NPRM. The commenters included vehicle manufacturers, child restraint manufacturers, the National Transportation Safety Board, researchers from two state universities, child safety advocates, and individual consumers. Each of these comments was considered and the most significant ones are addressed below.

Attaching Tether Straps During the 30 MPH Test and the Need for the 20 MPH Test

Before discussing the comments received on this issue, the most significant one raised in the NPRM,

NHTSA believes it would be useful to explain the differences between the different types of child restraints.

1. *Child seats.* A child seat is a child restraint that uses a plastic shell as a frame around the child, and has a shield, belts, or the like attached to the shell to restrain the child in the event of a crash. All but one of the currently produced models of child seats do not need to have an attached tether strap to pass the 30 mph test. However, two of the models which do not need a tether strap to pass the 30 mph test offer a tether strap as an option for extra protection of the child restraint's occupant.

2. *Booster seats.* A booster seat is a platform used to elevate a child in a vehicle. It does not have a frame or any other structural protection behind the child's back or head. Booster seats are designed to be used by older children who have outgrown child seats. By elevating these children, the booster seat allows the child to see out of the vehicle and to use the belt system in the vehicle. About half the current production of booster seats uses a special harness system attached to the vehicle by a tether strap to provide upper torso restraint for the booster seat occupant. The other half of current production of booster seats uses a small shield in front of the child to provide upper torso restraint.

3. *Child harnesses.* A child harness consists of a web of belts which are placed around the child, and is then anchored to the vehicle by a tether strap. Only one model of child harness is currently in production. Child harnesses are tested only in the 30 mph test with the tether attached according to the manufacturer's instructions, and are not subject to the 20 mph test. The reason for this differing treatment for child harnesses are compared to other child restraints is the agency's opinion that child harness tethers are in fact properly used by the public, due to the nature of the device—i.e., if the tether strap is not attached, it would be obvious that the child would be completely unrestrained in the event of a crash.

4. *Restraints for use by physically handicapped children.* These restraints are essentially wheelchairs, some of which fold so that the wheelchair can be positioned in the rear seat of passenger cars. Other restraints are simply devices to tie down a wheelchair while the child is travelling in a van, bus, or similar vehicle. All currently produced child restraints for use by physically handicapped children use their own belt system and tether straps to provide the necessary upper torso restraint. The NPRM did not propose any exemp-

tion for these restraints from the proposed requirement that they pass the 30 mph test without attaching any tether straps. Thus, if the NPRM were adopted as proposed, all of these restraints would have to be redesigned.

This final rule establishes the following requirements for the different types of child restraints. Child seats will not be allowed to have any tether straps attached during the 30 mph test required by Standard No. 213. They will also no longer be required to be tested in the 20 mph test. However, child harnesses, booster seats, and restraints for use by physically handicapped children will be allowed to continue to have tether straps attached during the 30 mph test. The reasoning supporting these decisions is set forth below.

CHILD SEATS

Almost all of the commenters addressing the agency's proposal to require child seats equipped with tether straps to pass the 30 mph test without attaching the tether supported the requirement. The only commenter which opposed this requirement was a child restraint manufacturer, arguing that a change at this time would "cause confusion of dealers and consumers with units that required tethers". The manufacturer further argued that if this change were made, "the Federal government must given child restraint manufacturers some sort of security blanket to protect them from lawsuits and recall of existing units."

NHTSA does not believe it is very likely that either dealers or consumers will be confused by the requirement that child seats with tethers pass the 30 mph test with the tether strap unattached. The new requirement would apply only to child seats manufactured after the effective date of this rule. Child seats manufactured before the effective date of this rule may be sold even if their tether strap must be attached to pass the 30 mph test. Hence, the agency does not see any reason for child seat dealers to be confused by this rule. Moreover, the public will receive the manufacturer's instructions with the child seat explaining how it is to be used. Thus, there does not appear to be any reason for the public to be confused by this rule.

NHTSA does not have any authority to given restraint manufacturers a "security blanket" to protect them from lawsuits or recalls of child seats with tethers. Even if NHTSA believed it was appropriate to protect a manufactuer from

lawsuits in a particular instance, only Congress has authority to do so. A recall of child seats must be based on a determination that the seats either do not comply with the requirements of Standard No. 213 in effect on the date of manufacture of the seat or that the seat contains a safety-related defect, as specified in sections 151 and 152 of the National Traffic and Motor Vehicle Safety Act (15 U.S.C. 1411 and 1412). If either determinatioon were made, the manufacturer is required by Section 154 of the Safety act (15 U.S.C. S1415) to remedy the noncompliance or defect.

For the reasons set forth at length in the NPRM and briefly reiterated at the beginning of this preamble, and because only one child seat model is being produced that requires the tether strap to be attached, NHTSA is adopting the proposed requirement that all child seats pass the 30 mph test without any tether straps attached. This requirement applies to all child seats manufactured after the effective date of this rule.

As an adjunct to this rulemaking, child seats equipped with a tether strap will not longer be subject to the requirement that they also pass a 20 mph test with the tether unattached. Since these child seats will now be subject to the 30 mph test with tether unattached, no purpose would be served by requiring the seats to be tested in a less severe manner under the same conditions.

BOOSTER SEATS

The commenters split on the issue of whether booster seats should be required to pass the 30 mph crash test with the tether unattached. The Insurance Institute of Highway Safety, Chrysler Corporation, the National Transportation Safety Board, and two individuals supported the proposed requirements for the reasons explained in the NPRM. However, the National Child Passenger Safety Association, Physicians for Automotive Safety, the University of Michigan, and researchers associated with the University of North Carolina opposed the proposed requirement. The gist of these opposing comments was as follows: the only means currently available for providing the needed upper torso restraint to booster seat occupants is with either a tether strap and harness or with a short shield in front of the child. A requirement to pass the 30 mph test without an attached tether strap would force manufacturers to equip all booster seats with a short shield. These commenters were concerned about the adequacy of the safety protection afforded to booster seat occupants by these short shields.

The University of Michigan commented that it is currently engaged in a research program to develop an abdominal penetration sensor for the 3-year old dummy currently used in Standard No. 213 testing. They stated that they have undertaken this research because of their concern about the abdominal loading to which the short shield exposes the child during the 30 mph crash test. The University of Michigan concluded its comment by stating that its preliminary tests with a prototype of its abdominal penetration sensor suggests that children are in fact exposed to high abdominal loading by the short shields used on booster seats without tethers. The researchers associated with the University of North Carolina concurred with the University of Michigan on the need to examine the abominal loading associated with booster seats without tethers before mandating that all booster seats be capable of passing the 30 mph test without an attached tether.

The agency is also aware of other concerns which have been expressed by child safety researchers in connection with the short shields used in booster seats without tethers. For example, there is concern that older children could be seriously injured by having their head and neck wrap around the shield, since the shield is not large enough to restrain those parts of the body in a crash situation. This concern was raised in the comments submitted by the National Child Passenger Safety Association. Another concern is that the short shield booster seats do not provide any crotch restraint. It is possible that smaller children could submarine under the short shields on booster seats, leaving these children completely unrestrained in the event of a crash.

NHTSA wishes to emphasize that booster seats without tethers comply with all current requirements of Standard No. 213 using the 3-year old dummy. Nevertheless, the issues raised by the commenters regarding the effectiveness of short shields on booster seats are matters of concern to the agency. Since the short shields used on booster seats without tethers represent the only current alternative to the use of tether straps on booster seats, NHTSA has concluded that it would be an unwise policy to essentially require the use of short shields on booster seats (by adopting the proposed requirements) before the agency has investigated the validity of the above-mentioned safety concerns. If testing showed that short shields did not provide adequate safety protection to children after the agency had essentially required the use of

such shields on all booster seats, this rulemaking would not achieve the agency's goal of improving the protection offered to child restraint occupants. Therefore, it is premature to adopt the proposed requirements as they apply to booster seats.

The agency will investigate the allegations that have been made about the short shields on booster seats. The agency investigation, together with the University of Michigan testing on the abdominal loading imposed by these short shields, should help resolve the stated concerns.

There is also an important distinction between child seats with tethers and booster seats with tethers, which suggests that it is not as imperative to require that booster seats not be permitted to have an attached tether strap during the 30 mph test. Booster seats equipped with tethers are designed to be used *either* with the tether strap attached to the vehicle *or* with a lap-shoulder belt. When a lap-shoulder belt in a vehicle so that it will provide the necessary upper torso support. When upper torso support is provided by a vehicle shoulder belt, it is not necessary to attach the tether strap to provide the necessary upper torso support.

This feature resulted in observed correct usage of booster seats equipped with tethers in 38.0 percent of the total cases in the Cynecki and Goryl report cited above. The tether strap was properly attached in 8.5 percent of the cases, and the lap-shoulder belt was correctly used with the booster seat in 29.5 percent of the observed cases. This 38.0 percent correct usage of booster seats with tethers compares favorably with the 41.2 percent correct usage of child seats not equipped with tethers, and both stand in sharp contrast to the 7.0 percent usage of child seats equipped with tethers.

The reason explained in the NPRM for proposing that tether straps not be attached during the 30 mph test was because of the overwhelming incorrect usage of child restraints with tethers by the public. However, the data available to the agency suggest that booster seats equipped with tethers are used correctly almost as often as child seats without tethers.

CHILD HARNESSES

The NPRM did not propose to change the current treatment for child harnesses in the Standard No. 213 testing. The surveys and data available to the agency have not examined the extent to which child harness tethers are misused by the public.

Moreover, NHTSA believes it would be obvious to users of child harnesses that the failure to attach the tether strap would leave the child completely unrestrained in a crash. The absence of data indicating misuse of child harness tether straps, together with the obvious need to attach these tether straps, resulted in the agency's position that the NPRM should not propose any changes to Standard No. 213 in this regard: that is, child harness would be permitted to have their tether straps attached during the 30 mph test and not be subject to the 20 mph test. No commenters addressed this area of the proposal, and the final rule does not make any changes to the current requirements for child harnesses for the reasons explained above.

CHILD RESTRAINTS FOR PHYSICALLY HANDICAPPED CHILDRES

A number of commenters urged the agency to exempt child restraints designed for handicapped children from the proposal that all child restraints, except child harnesses, pass the 30 mph test in Standard No. 213 without any tether strap attached. A manufacturer of child restraints for physically handicapped children commented: "Now that safe transportation for the handicapped child has become a reality, through the use of restraint harnesses, tether systems, and wheelchairs engineered to meet Standard No. 213, it seems counterproductive for the handicapped population and manufacturers to start over again."

NHTSA did not intend to require any changes to thes restraints, and a statement proposing the continuation of current testing requirements for restraints for physically handicapped children was inadvertently omitted from the NPRM. The agency during the 30 mph test and will not require these restraints to be subjected to the 20 mph test without test with the tether attached. NHTSA has no data showing that these restraints are frequently misused by the public. Additionally, there is no alternative at present to the use of tether straps to provide the necessary upper torso support for physically handicapped children. Hence, any requirement to eliminate the use of tether straps on restraints for physically handicapped children would lessen the protection available for those children. This was not the agency's intent in the NPRM.

OTHER ISSUES

The NPRM proposed that child restraints be installed at one of the two outboard seating positions on the standard seat during the testing. As ex-

plained above, this was proposed to ensure that the testing would be representative of the way in which child restraints would be used by the public. It was also proposed to enable child restraint manufacturers to reduce testing costs by evaluating two child restraint systems in a single test.

The commenters that addressed this proposed change generally opposed it. The University of Michigan commented that there was no basis for the concern expressed in the NPRM that testing in the center seating position might not be representative of the way in which child restraints are used by the public. The University stated: "We know from field experience that those restraints that meet the 30 mph test in the center seating position also effectively protect children in most crashes." Stated differently, child restraints that pass the 30 mph crash test in the center seating position have performed well when installed in the outboard seating positions of vehicles in use. The available data on the performance of child restraint systems indicate that the Standard No. 213 test procedures are representative of the conditions encountered by restraint systems when in use.

Further, one child restraint manufacturer and the University of Michigan stated that the agency's proposed change might increase testing costs, instead of achieving the agency's stated intent of reducing those costs. This could happen because child restraints would be subjected to slightly differing forces produced by asymmetrical lap belt anchorages at the outboard seating positions. Further, it was stated that all child restraints are not symmetrical, and their test performance might be affected by a twist in one direction, but not the other. These facts would mean that all existing models of child restraints would have to be retested to ensure that the restraints would pass the Standard No. 213 requirements when installed at the outboard seating positions. In addition, the child restraints would have to be tested at both the left and right outboard seating positions, because of the differenct forces presented at these different seating locations.

The proposed change to the required seating position for testing child restraints is not adopted in this final rule, because of the reasons set forth in the comments.

The NPRM also proposed that manufacturers be required to insert the phrase "For extra protection in frontal and side impacts" before the notice on the label that tether straps must be attached in accordance with the manufacturer's instructions.

This change was proposed in connection with the proposal to require all child restraints equipped with tethers to pass the 30 mph test without attaching the tethers. The change in the label language was intended to inform the public that the tether strap would offer supplementary safety protection when attached, but that it was not necessary to attach the tether for adequate protection.

BMW commented that the proposed change would have the unintended effect of implying that it was not necessary to use tether straps, and this implication would decrease the already low use of tether straps. The agency believes that the BMW comment has merit. The possibility of decreasing tether usage, combined with the fact that child harnesses, booster seats, and restraints for physically handicapped children may include tether straps, the attachment of which is necessary for adequate protection of the child, have led the agency to conclude that the proposed change to the label language should not be adopted in this final rule.

The other proposed changes were clarifications to the instructions which must accompany each child restraint. No commenters addressed these clarifications, and they are adopted for the reasons explained in the NPRM.

PART 571—[AMENDED]

In consideration of the foregoing, 49 CFR S571.213 is amended as follows:

1. The authority citation for Part 571 continues to read as follows:

AUTHORITY: 15 U.S.C. 1392, 1401, 1403, 1407; delegation of authority at 49 CFR 1.50.

2. S4 is amended by adding the following definition immediately before the definition of "car bed":

S4. Definitions.

"Booster seat" means a child restraint which consist of only a seating platform that does not extend up to provide a cushion for the child's back or head.

3. S5.6.1 is revised to read as follows:

S5.6.1 The instructions shall state that, for maximum safety protection, child restraint systems should be installed in a rear seating position in vehicles with two rear seating positions and in the center rear seating position in vehicles with such a seating position.

4. S5.6.2 is revised to read as follows:

S5.6.2 The instructions shall specify in general terms the types of vehicles, the types of seating positions, and the types of vehicle safety belts with which the system can or cannot be used.

5. S6.1.2.1 is revised to read as follows:

S6.1.2.1 Test configuration.

S6.1.2.1.1 Test configuration I. In the case of each child restraint system other than a child harness, a booster seat with a top anchorage strap, or a restraint designed for use by physically handicapped children, install a new child restraint system at the center seating position of the standard seat assembly in accordance with the manufacturer's instructions provided with the system pursuant to S5.6, except that the restraint shall be secured to the standard vehicle seat using only the standard vehicle lap belt. A child harness, booster seat with a top anchorage strap, or a restraint designed for use by physically handicapped children shall be installed at the center seating position of the standard seat assembly in accordance with the manufacturer's instructions provided with the system pursuant to S5.6.

S6.1.2.1.2 Test configuration II. In the case of each child restraint system which is equipped with a fixed or movable surface described in S5.2.2.2, or a booster seat with a top anchorage strap, install a new child restraint system at the center seating position of the standard seat assembly using only the standard seat lap belt to secure the system to the standard seat.

Issued on February 10, 1986.

Diane K. Steed
Administrator
51 F.R. 5335
February 13, 1986

FEDERAL MOTOR VEHICLE SAFETY STANDARD NO. 213

Child Restraint Systems, Seat Belt Assemblies, and Anchorages
(Docket No. 74-9; Notice 6)

S1. Scope. This standard specifies requirements for child restraint systems used in motor vehicles and aircraft.

S2. Purpose. The purpose of this standard is to reduce the number of children killed or injured in motor vehicle crashes and in aircraft.

S3. Application. This standard applies to child restraint systems for use in motor vehicles and aircraft.

S4. Definitions. ["Booster seat" means a child restraint which consists of only a seating platform that does not extend up to provide a cushion for the child's back or head. (51 F.R. 5335—February 13, 1986. Effective: August 12, 1986)]

"Car bed" means a child restraint system designed to restrain or position a child in the supine or prone position on a continuous flat surface.

"Child restraint system" means any device, except Type I or Type II seat belts, designed for use in a motor vehicle or aircraft to restrain, seat, or position children who weigh 50 pounds or less.

"Contactable surface" means any child restraint system surface (other than that of a belt, belt buckle, or belt adjustment hardware) that may contact any part of the head or torso of the appropriate test dummy, specified in S7, when a child restraint system is tested in accordance with S6.1.

"Representative aircraft passenger seat" means either a Federal Aviation Administration approved production aircraft passenger seat or a simulated aircraft passenger seat conforming to Figure 6.

"Seat orientation reference line" or "SORL" means the horizontal line through Point Z as illustrated in Figure 1A.

"Torso" means the portion of the body of a seated anthropomorphic test dummy, excluding the thighs, that lies between the top of the child restraint system seating surface and the top of the shoulders of the test dummy.

S5. Requirements for child restraint systems certified for use in motor vehicles. Each child restraint certified for use in motor vehicles shall meet the requirements in this section when, as specified, tested in accordance with S6.1.

S5.1 Dynamic performance.

S5.1.1 Child restraint system integrity. When tested in accordance with S6.1, each child restraint system shall:

(a) Exhibit no complete separation of any load bearing structural element and no partial separation exposing either surfaces with a radius of less than ¼ inch or surfaces with protrusions greater than ⅜ inch above the immediate adjacent surrounding contactable surface of any structural element of the system;

(b) If adjustable to different positions, remain in the same adjustment position during the testing as it was immediately before the testing; and

(c) If a front facing child restraint system, not allow the angle between the system's back support surfaces for the child and the system's seating surface to be less than 45 degrees at the completion of the test.

Ref. NHTSA Drawing No. SAS-1000

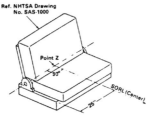

Point Z
90°
SORL (Center)
25°

SORL – SEAT ORIENTATION REFERENCE LINE (HORIZONTAL)

SORL Location on the Standard Seat
FIGURE 1A

S5.1.2 Injury criteria. When tested in accordance with S6.1, each child restraint system that, in accordance with S5.5.2(f), is recommended for use by children weighing more than 20 pounds, shall—

(a) Limit the resultant acceleration at the location of the accelerometer mounted in the test dummy head as specified in Part 572 such that the expression:

$$\left[\frac{1}{t_2 - t_1} \int_{t_1}^{t_2} a\, dt \right]^{2.5} (t_2 - t_1)$$

shall not exceed 1,000, where a is the resultant acceleration expressed as a multiple of g (the acceleration of gravity), and t_1 and t_2, are any two moments during the impacts.

(b) Limit the resultant acceleration at the location of the accelerometer mounted in the test dummy upper thorax as specified in Part 572 to not more than 60 g's, except for intervals whose cumulative duration is not more than 3 milliseconds.

S5.1.3 Occupant excursion. When tested in accordance with S6.1 and adjusted in any position which the manufacturer has not, in accordance with S5.5.2(i), specifically warned against using in motor vehicles, each child restraint system shall meet the applicable excursion limit requirements specified in S5.1.3.1–S5.1.3.3.

[S5.1.3.1 Child restraint systems other than rear-facing ones and car beds. In the case of each child restraint system other than a rear-facing child restraint system or a car bed, the test dummy's torso shall be retained within the system and no portion of the test dummy's head shall pass through the vertical transverse plane that is 32 inches forward of point Z on the standard seat assembly, measured along the center SORL (as illustrated in Figure 1B), and neither knee pivot point shall pass through the vertical transverse plane that is 36 inches forward of point Z on the standard seat assembly, measured along the center SORL. (45 F.R. 67095—October 9, 1980. Effective: 10/7/80)]

S5.1.3.2 Rear-facing child restraint systems. In the case of each rear-facing child restraint system, all portions of the test dummy's torso shall be

retained within the system and no portion of the target point on either side of the dummy's head shall pass through the transverse orthogonal planes whose intersection contains the forward-most and top-most points on the child restraint system surfaces (illustrated in Figure 1C).

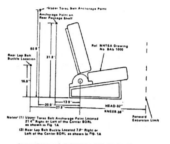

Locations of Additional Belt Anchorage Points and Forward Excursion Limit
FIGURE 1B

S5.1.3.3 Car beds. In the case of car beds, all portions of the test dummy's head and torso shall be retained within the confines of the car bed.

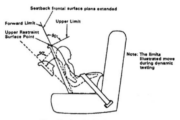

Rear Facing Child Restraint
Forward and Upper Head Excursion Limits
FIGURE 1C

S5.1.4 Back support angle. When a rear-facing child restraint system is tested in accordance with S6.1, the angle between the system's back support surface for the child and the vertical shall not exceed 70 degrees.

S5.2 Force distribution.

S5.2.1 Minimum head support surface—child restraints other than car beds.

S5.2.1.1 Except as provided in S5.2.1.2, each child restraint system other than a car bed shall provide restraint against rearward movement of the head of the child (rearward in relation to the child) by means of a continuous seat back which is an integral part of the system and which—

(a) Has a height, measured along the system seat back surface for the child in the vertical longitudinal plane passing through the longitudinal centerline of the child restraint systems from the lowest point on the system seating surface that is contacted by the buttocks of the seated dummy, as follows:

Weight [1] (in pounds)	Height [2] (in inches)
Less than 20 lb.................................	18
20 lb or more, but not more than 40 lb...............	20
More than 40 lb................................	22

[1] When a child restraint system is recommended under S5.5 (f) for use by children of the above weights.
[2] The height of the portion of the system seat back providing head restraint shall not be less than the above.

(b) Has a width of not less than 8 inches, measured in the horizontal plane at the height specified in paragraph (a) of this section. Except that a child restraint system with side supports extending at least 4 inches forward from the padded surface of the portion of the restraint system provided for support of the child's head may have a width of not less than 6 inches, measured in the horizontal plane of the height specified in paragraph (a) of this section.

(c) Limits the rearward rotation of the test dummy head so that the angle between the head and torso of the dummy specified in S7 when tested in accordance with S6.1 is not more than 45 degrees greater than the angle between the head and torso after the dummy has been placed in the system in accordance with S6.1.2.3 and before the system is tested in accordance with S6.1.

S5.2.1.2 A front facing child restraint system is not required to comply with S5.2.1.1 if the target

point on either side of the dummy's head is below a horizontal plane tangent to the top of the standard seat assembly when the dummy is positioned in the system and the system is installed on the assembly in accordance with S6.1.2.

S5.2.2 Torso impact protection. Each child restraint system other than a car bed shall comply with the applicable requirements of S5.2.2.1 and S5.2.2.2.

S5.2.2.1 (a) The system surface provided for the support of the child's back shall be flat or concave and have a continuous surface area of not less than 85 square inches.

(b) Each system surface provided for support of the side of the child's torso shall be flat or concave and have a continuous surface of not less than 24 square inches for systems recommended for children weighing 20 pounds or more, or 48 square inches for systems recommended for children weighing less than 20 pounds.

(c) Each horizontal cross section of each system surface designed to restrain forward movement of the child's torso shall be flat or concave and each vertical longitudinal cross section shall be flat or convex with a radius of curvature of the underlying structure of not less than 2 inches.

S5.2.2.2 Each forward facing child restraint system shall have no fixed or movable surface directly forward of the dummy and intersected by a horizontal line parallel to the SORL and passing through any portion of the dummy, except for surfaces which restrain the dummy when the system is tested in accordance with S6.1.2.1.2 so that the child restraint system shall conform to the requirements of S5.1.2 and S5.1.3.1.

S5.2.3 Head impact protection.

S5.2.3.1 Each child restraint system, other than a child harness, which is recommended under S5.5.2 (f) for children weighing less than 20 pounds shall comply with S5.2.3.2.

[S5.2.3.2 Each system surface, except for protrusions that comply with S5.2.4, which is contactable by the dummy head when the system is tested in accordance with S6.1 shall be covered with slow recovery, energy absorbing material with the following characteristics:

(a) A 25 percent compression-deflection resistance of not less than 0.5 and not more than 10 pounds per square inch when tested in accordance with S6.3. (45 F.R. 29045. Effective: 5/1/80)]

[(b) A thickness of not less than ½ inch for material having a 25 percent compression-deflection resistance of not less than 1.8 and not more than 10 pounds per square inch when tested in accordance with S6.3. Materials having a 25 percent compression-deflection resistance of less than 1.8 pounds per square inch shall have a thickness of not less than ¾ inch. (45 F.R. 82264—December 15, 1980. Effective: 12/15/80)]

S5.2.4 Protrusion limitation. Any portion of a rigid structural component within or underlying a contactable surface, or any portion of a child restraint system surface that is subject to the requirements of S5.2.3 shall, with any padding or other flexible overlay material removed, have a height above any immediately adjacent restraint system surface of not more than ⅜ inch and no exposed edge with a radius of less than ¼ inch.

S5.3 Installation.

S5.3.1 Each child restraint system shall have no means designed for attaching the system to vehicle seat cushion or vehicle seat back and no component (except belts) that is designed to be inserted between the vehicle seat cushion and vehicle seat back.

S5.3.2 When installed on a vehicle seat, each child restraint system, other than child harnesses, shall be capable of being restrained against forward movement solely by means of a Type I seat belt assembly (defined in S571.209) that meets Standard No. 208 (S571.208), or by means of a Type I seat belt assembly plus one additional anchorage strap that is supplied with the system and conforms to S5.4.

S5.3.3 Car beds. Each car bed shall be designed to be installed on a vehicle seat so that the car bed's longitudinal axis is perpendicular to a vertical longitudinal plane through the longitudinal axis of the vehicle.

S5.4 Belts, belt buckles, and belt webbing.

S5.4.1 Performance requirements. The webbing of belts provided with a child restraint system and used to attach the system to the vehicle or to restrain the child within the system shall—

[(a) After being subjected to abrasion as specified in § 5.1(d) or 5.3(c) of FMVSS No. 209 (§ 571.209), have a breaking strength of not less than 75 percent of the strength of the unabraided webbing when tested in accordance with S5.1(b) of FMVSS No. 209. (45 F.R. 29045—May 1, 1980. Effective: 5/1/80)]

(b) Meet the requirements of S4.3 (e) through (h) of FMVSS No. 209 (S571.209); and

(c) If contactable by the test dummy torso when the system is tested in accordance with S6.1, have a width of not less than 1½ inches when measured in accordance with S5.4.1.1.

S5.4.1.1 Width test procedure. Condition the webbing for 24 hours in an atmosphere of any relative humidity between 48 and 67 percent, and any ambient temperature between 70° and 77° F. Measure belt webbing width under a tension of 5 pounds applied lengthwise.

S5.4.2 Belt buckles and belt adjustment hardware. Each belt buckle and item of belt adjustment hardware used in a child restraint system shall conform to the requirements of S4.3 (a) and S4.3 (b) of FMVSS No. 209 (S571.209).

S5.4.3 Belt Restraint.

S5.4.3.1 General. Each belt that is part of a child restraint system and that is designed to restrain a child using the system shall be adjustable to snugly fit any child whose height and weight are within the ranges recommended in accordance with S5.5.2 (f) and who is positioned in the system in accordance with the instructions required by S5.6.

S5.4.3.2 Direct restraint. Each belt that is part of a child restraint system and that is designed to restrain a child using the system and to attach the system to the vehicle shall, when tested in accordance with S6.1, impose no loads on the child that result from the mass of the system or the mass of the seat back of the standard seat assembly specified in S7.3.

S5.4.3.3 Seating systems. Except for child restraint systems subject to S5.4.3.4, each child restraint system that is designed for use by a child in a seated position and that has belts designed to restrain the child shall, with the test dummy specified in S7 positioned in the system in accordance with S6.1.2.3, provide:

(a) upper torso restraint in the form of:

(i) belts passing over each shoulder of the child; or

(ii) a fixed or movable surface that complies with S5.2.2.1(c), and

(b) lower torso restraint in the form of:

(i) a lap belt assembly making an angle between 45° and 90° with the child restraint seating surface at the lap belt attachment points, or

(ii) a fixed or movable surface that complies with S5.2.2.1(c), and

(c) in the case of each seating system recommended for children over 20 pounds, crotch restraint in the form of:

(i) a crotch belt connectable to the lap belt or other device used to restrain the lower torso, or

(ii) a fixed or movable surface that complies with S5.2.2.1(c).

S5.4.3.4 Harnesses. Each child harness shall:

(a) Provide upper torso restraint, including belts passing over each shoulder of the child;

(b) Provide lower torso restraint by means of lap and crotch belt; and

(c) Prevent a child of any height for which the restraint is recommended for use pursuant to S5.5.2 (f) from standing upright on the vehicle seat when the child is placed in the device in accordance with the instructions required by S5.6.

S5.4.3.5 Buckle Release. [Any buckle in a child restraint system belt assembly designed to restrain a child using the system shall:

(a) When tested in accordance with S6.2.1 prior to the dynamic test of S6.1, not release when a force of less than 9 pounds is applied and shall release when a force of not more than 14 pounds is applied:

(b) After the dynamic test of S6.1, when tested in accordance with S6.2.3, release when a force of not more than 16 pounds is applied;

(c) Meet the requirements of S4,3(d)(2) of FMVSS No. 209 (§ 571.209), except that the minimum surface area for child restraint buckles designed for push-button application shall be 0.6 square inch.

(d) Meet the requirements of S4.3(g) of FMVSS No. 209 (§ 571.209) when tested in accordance with S5.2(g) of FMVSS No. 209; and

(e) Not release during the testing specified in S6.1. (50 F.R. 33722—August 21, 1985. Effective: February 18, 1986)]

S5.5 Labeling.

S5.5.1 Each child restraint system shall be permanently labeled with the information specified in S5.5.2 (a) through (l).

S5.5.2 The information specified in paragraphs (a)-(l) of this section shall be stated in the English language and lettered in letters and numbers that are not smaller than 10 point type and are on a contrasting background.

(a) The model name or number of the system.

(b) The manufacturer's name. A distributor's name may be used instead if the distributor assumes responsibility for all duties and liabilities imposed on the manufacturer with respect to the system by the National Traffic and Motor Vehicle Safety Act, as amended

(c) The statement: "Manufactured in ——," inserting the month and year of manufacture.

(d) The place of manufacture (city and State, or foreign country). However, if the manufacturer uses the name of the distributor, then it shall state the location (city and State, or foreign country) of the principal offices of the distributor.

(e) The statement: "This child restraint system conforms to all applicable Federal motor vehicle safety standards."

(f) One of the following statements, inserting the manufacturer's recommendations for the maximum weight and height of children who can safely occupy the system:

(i) This infant restraint is designed for use by children who weigh _____ pounds or less and whose height is _____ inches or less; or

(ii) This child restraint is designed for use only by children who weigh between _____ and _____ pounds and whose height is _____ inches or less and who are capable of sitting upright alone; or

(iii) This child restraint is designed for use only by children who weigh between _____ and _____ pounds and are between _____ and _____ inches in height.

(g) The following statement, inserting the location of the manufacturer's installation instruction booklet or sheet on the restraint:

WARNING! FAILURE TO FOLLOW EACH OF THE FOLLOWING INSTRUCTIONS CAN RESULT IN YOUR CHILD STRIKING THE VEHICLE'S INTERIOR DURING A SUDDEN STOP OR CRASH.

SECURE THIS CHILD RESTRAINT WITH A VEHICLE BELT AS SPECIFIED IN THE MANUFACTURER'S INSTRUCTIONS LOCATED _____.

(h) In the case of each child restraint system that has belts designed to restrain children using them: SNUGLY ADJUST THE BELTS PROVIDED WITH THIS CHILD RESTRAINT AROUND YOUR CHILD.

(i) In the case of each child restraint system which is not intended for use in motor vehicles at certain adjustment positions, the following statement, inserting the manufacturer's adjustment restrictions.

DO NOT USE THE _____ ADJUSTMENT POSITION(S) OF THIS CHILD RESTRAINT IN A MOTOR VEHICLE.

(j) In the case of each child restraint system equipped with an anchorage strap, the statement: SECURE THE TOP ANCHORAGE STRAP PROVIDED WITH THIS CHILD RESTRAINT AS SPECIFIED IN THE MANUFACTURER'S INSTRUCTIONS.

(k) In the case of each child restraint system which can be used in a rear-facing position, one of the following statements:

(i) PLACE THIS CHILD RESTRAINT IN A REAR-FACING POSITION WHEN USING IT WITH AN INFANT; or

(ii) PLACE THIS INFANT RESTRAINT IN A REAR-FACING POSITION WHEN USING IT IN THE VEHICLE.

(l) An installation diagram showing the child restraint system installed in the right front outboard seating position equipped with a continuous-loop lap/shoulder belt and in the center rear seating position as specified in the manufacturer's instructions.

(m) Child restraints that are certified as complying with the provisions of section 58 shall be labeled with the statement "THIS RESTRAINT IS CERTIFIED FOR USE IN MOTOR VEHICLES AND AIRCRAFT". This statement shall be in red lettering, and shall be placed after the certification statement required by paragraph (e) of this section.

S5.5.3 The information specified in S5.5.2 (g)-(k) shall be located on the child restraint system so that it is visible when the system is installed as specified in S5.6.

S5.6 Installation Instructions. Each child restraint system shall be accompanied by printed instructions in the English language that provide a step-by-step procedure, including diagrams, for installing the system in motor vehicles, securing the system in the vehicles, positioning a child in the system, and adjusting the system to fit the child.

S5.6.1 [The instructions shall state that, for maximum safety protection, child restraint systems should be installed in a rear seating position in vehicles with two rear seating positions and in the center rear seating position in vehicles with such a seating position. (51 F.R. 5335—February 13, 1986. Effective: August 12, 1986.)]

S5.6.2 [The instructions shall specify in general terms the types of vehicles, the types of seating positions, and the types of vehicle safety belts with which the system can or cannot be used. (51 F.R. 5335—February 13, 1986. Effective: August 12, 1986.)]

S5.6.3 The instructions shall explain the primary consequences of noting following the warnings required to be labeled on the child restraint system in accordance with S5.5.2 (g)-(k).

S5.6.4 The instructions for each car bed shall explain that the car bed should position in such a way that the child's head is near the center of the vehicle.

S5.6.5 The instructions shall state that child restraint systems should be securely belted to a vehicle, even when they are not occupied, since in a crash an unsecured child restraint system may injure other occupants.

S5.6.6 Each child restraint system shall have a location on the restraint for storing the manufacturer's instructions.

S5.7 Flammability. Each material used in a child restraint system shall conform to the requirements of S4 of FMVSS No. 302 (S571.302).

S6. Test Conditions and Procedures.

S6.1 Dynamic Systems Test.

S6.1.1 Test Conditions.

S6.1.1.1 The test device is the standard seat assembly specified in S7.3. It is mounted on a dynamic test platform so that the center SORL of the seat is parallel to the direction of the test platform travel and so that movement between the base of the assembly and the platform is prevented. The platform is instrumented with an accelerometer and data processing system having a frequency response of 60Hz channel class as specified in Society of Automotive Engineers Recommended Practice J211a, "Instrumentation for Impact Tests." The accelerometer sensitive axis is parallel to the direction of the test platform travel.

ACCELERATION FUNCTION FOR ΔV = 30MPH.

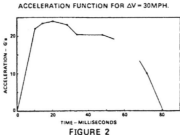

FIGURE 2

S6.1.1.2 The tests are frontal barrier impact simulations and for—

(a) Test configuration I specified in S6.1.2.1.1, are at a velocity change of 30 mph with the acceleration of the test platform entirely within the curve shown in Figure 2.

(b) Test configuration II specified in S6.1.2.1.2, are at a velocity change of 20 mph with the acceleration of the test platform entirely within the curve shown in Figure 3.

ACCELERATION FUNCTION FOR ΔV = 20MPH.

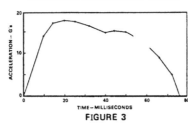

FIGURE 3

S6.1.1.3 Type I seat belt assemblies meeting the requirements of Standard No. 209 (S571.209) and having webbing with a width of not more than 2 inches are attached, without the use of retractors or reels of any kind, to the seat belt anchorage points (illustrated in Figure 1B) provided on the standard seat assembly.

S6.1.1.4 Performance tests under S6.1 are conducted at any ambient temperature from 66° to 78° F and at any relative humidity from 10 percent to 70 percent.

S6.1.2 Dynamic Test Procedure.

S6.1.2.1 Test Configuration.

S6.1.2.1.1 Test Configuration I. [In the case of each child restraint system other than a child harness, a booster seat with a top anchorage strap, or a restraint designed for use by physically handicapped children, install a new child restraint system at the center seating position of the standard seat assembly in accordance with the manufacturer's instructions provided with the system pursuant to S5.6, except that the restraint shall be secured to the standard vehicle seat using only the standard vehicle lap belt. A child harness, booster seat with a top anchorage strap, or a restraint designed for use by physically handicapped children shall be installed at the center seating position of the standard seat assembly in accordance with the manufacturer's instructions provided with the system pursuant to S5.6. (51 F.R. 5335—February 13, 1986. Effective: August 12, 1986.)]

S6.1.2.1.2 Test Configuration II. [In the case of each child restraint system which is equipped with a fixed or movable surface described in S5.2.2.2, or a booster seat with a top anchorage strap, install a new child restraint system at the center seat position of the standard seat assembly using only the standard seat lap belt to secure the system to the standard seat. (51 F.R. 5335—February 13, 1986. Effective: August 12, 1986.)]

S6.1.2.2 Tighten all belts used to attach the child restraint system to the standard seat assembly to a tension of not less than 12 pounds and not more than 15 pounds, as measured by a load cell used on the webbing portion of the belt.

S6.1.2.3 Place in the child restraint any dummy specified in S7 for testing systems for use by children of the heights and weights for which the system is recommended in accordance with S5.6.

S6.1.2.3.1 When placing the 3-year-old test dummy in child restraint systems other than car beds, position the test dummy according to the instructions for child positioning provided by the manufacturer with the system in accordance with S5.6 while conforming to the following:

(a) Place the test dummy in the seated position within the system with the midsagittal plane of the test dummy head coincident with the center SORL of the standard seating assembly, holding the torso upright until it contacts the system's design seating surface.

(b) Extend the arms of the test dummy as far as possible in the upward vertical direction. Extend the legs of the dummy as far as possible in the forward horizontal direction, with the dummy feet perpendicular to the centerline of the lower legs.

(c) Using a flat square surface with an area of 4 square inches, apply a force of 40 pounds, perpendicular to the plane of the back of the standard seat assembly, first against the dummy crotch and then at the dummy thorax in the midsagittal plane of the dummy. For a child restraint system with a fixed or movable surface described in S5.2.2.2 which is being tested under the conditions of test configuration II, do not attach any of the child restraint belts unless they are an integral part of the fixed or movable surface. For all other child restraint systems and for a child restraint system with a fixed or movable surface which is being tested under the conditions of test configuration I, attach all appropriate child restraint belts and tighten them as specified in S6.1.2.4. Attach all appropriate vehicle belts and tighten them as specified in S6.1.2.2. Position each movable surface in accordance with the manufacturer's instructions provided in accordance with S5.6.

(d) After the steps specified in paragraph (c) of this section, rotate each dummy limb downwards in the plane parallel to its midsagittal plane until the limb contacts a surface of the child restraint system or the standard seat. Position the limbs, if necessary, so that limb placement does not inhibit torso or head movement in tests conducted under S6.

S6.1.2.3.2 When placing the 6-month-old dummy in child restraint systems other than car beds, position the test dummy according to the instructions for child positioning provided with the system by the manufacturer in accordance with S5.6 while conforming to the following:

(a) With the dummy in the supine position on a horizontal surface, and while preventing movement of the dummy torso by placing a hand on the center of the torso, rotate the dummy legs upward by lifting the feet until the legs contact the upper torso and the feet touch the head, and then slowly release the legs but do not return them to the flat surface.

(b) Place the dummy in the child restraint system so that the back of the dummy torso contacts the back support surface of the system. For a child restraint system with a fixed or movable surface described in S5.2.2.2 which is being tested under the conditions of test configuration II, do not attach any of the child restraint belts unless they are an integral part of the fixed or movable surface. For all other child restraint systems and for a child restraint system with a fixed or movable surface which is being tested under the conditions of test configuration I, attach all appropriate child restraint belts and tighten them as specified in S6.1.2.4. Attach all appropriate vehicle belts and

tighten them as specified in S6.1.2.2. Position each movable surface in accordance with the manufacturer's instructions provided in accordance with S5.6. If the dummy's head does not remain in the proper position, it shall be taped against the front of the seat back surface of the system by means of a single thickness of 1/4-inch-wide paper masking tape placed across the center of the dummy face.

(c) Position the dummy arms vertically upwards and then rotate each arm downward toward the dummy's lower body until it contacts a surface of the child restraint system or the standard seat assembly, ensuring that no arm is restrained from movement in other than the downward direction, by any part of the system or the belts used to anchor the system to the standard seat assembly.

S6.1.2.3.3 When placing the 6-month-old dummy or 3-year-old dummy in a car bed, place the dummy in the car bed in the supine position with its midsagittal plane perpendicular to the center SORL of the standard seat assembly and position the dummy within the car bed in accordance with instructions for child positioning provided with the car bed by its manufacturer in accordance with S5.6.

S6.1.2.4 If provided, shoulder and pelvic belts that directly restrain the dummy shall be adjusted as follows:

Tighten the belts until a 2-pound force applied (as illustrated in Figure 5) to the webbing at the top of each dummy shoulder and to the pelvic webbing two inches on either side of the torso midsagittal plane pulls the webbing 1/4 inch from the dummy.

S6.1.2.5 Accelerate the test platform to simulate frontal impact in accordance with S6.1.1.2 (a) or S6.1.1.2. (b), as appropriate.

S6.1.2.6 Measure dummy excursion and determine conformance to the requirements specified in S5.1 as appropriate.

S6.2 Buckle release test procedure. [The belt assembly buckles used in the child restraint system shall be tested in accordance with S6.2.1 through S6.2.4 inclusive (50 F.R. 33722—August 21, 1985. Effective: February 18, 1986.)]

S6.2.1 [Before conducting the testing specified in S6.1, place the loaded buckle on a hard, flat, horizontal surface. Each belt end of the buckle shall be pre-loaded in the following manner. The anchor end of the buckle shall be loaded with a

2-pound force in the direction away from the buckle. In the case of buckles designed to secure a single latch plate, the belt latch plate end of the buckle shall be pre-loaded with a 2-pound force in the direction away from the buckle. In the case of buckles designed to secure two or more latch plates, the belt latch plate ends of the buckle shall be loaded equally so that the total load is 2 pounds, in the direction away from the buckle. For push-button release buckles the release force shall be applied by a conical surface (cone angle not exceeding 90 degrees). For push-button release mechanisms with a fixed edge (referred to in Figure 7 as "hinged button"), the release force shall be applied at the centerline of the button, 0.125 inches away from the movable edge directly opposite the fixed edge, and in the direction that produces maximum releasing effect. For push-button release mechanisms with no fixed edge (referred to Figure 7 as "floating button"), the release force shall be applied at the center of the release mechanism in the direction that produces the maximum releasing effect. For all othe buckle release mechanisms, the force shall be applied on the centerline of the buckle lever or finger tab in the direction that produces the maximum releasing effect. Measure the force required to release the buckle. Figure 7 illustrates the loading for the different buckles and the point where the release force should be applied, and Figure 8 illustrates the conical surface used to apply the release force to push-button release buckles. (50 F.R. 33722—August 21, 1985. Effective: February 18, 1986.)]

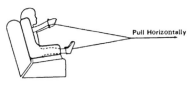

Pull Horizontally

Buckle Release Test

FIGURE 4

S6.2.2 [After completion of the testing specified in S6.1, and before the buckle is un-latched, tie a self-adjusting sling to each wrist and ankle of the test dummy in the manner illustrated

in Figure 4. (50 F.R. 33722—August 21, 1985. Effective: February 18, 1986)]

S6.2.3 Pull the sling horizontally in the manner illustrated in Figure 4 and parallel to the center SORL of the seat assembly and apply a force of 20 pounds in the case of a system tested with a 6 month-old dummy and 45 pounds in the case of a system tested with a 3 year-old dummy.

S6.2.4 [While applying the force specified in S6.2.3, and using the device shown in Figure 8 for push-button release buckles, apply the release force in the manner and location specified in S6.2.1 for that type of buckle. Measure the force required to release the buckle. (50 F.R. 33722—August 21, 1985. Effective: February 18, 1986.)]

S6.3 Head impact protection—energy absorbing material test procedure.

S6.3.1 Prepare and test specimens of the energy absorbing material used to comply with S5.2.3 in accordance with the applicable 25 percent compression-deflection test described in the American Society for Testing and Materials (ASTM) Standard D1056-73, "Standard Specification for Flexible Cellular Materials—Sponge or Expanded Rubber", or D1564-71. "Standard Method of Testing Flexible Cellular Materials— Slab Urethane Foam" or D1565-76 "Standard Specification for Flexible Cellular Materials—Vinyl Chloride Polymer and Copolymer open-cell foams.

S7 Test dummies.

S7.1 Six-month-old dummy. An unclothed "Six-month-old Size Manikin" conforming to Sub-part D of Part 572 of this chapter is used for testing a child restraint system that is recommended by its manufacturer in accordance with S5.6 for use by children in a weight range that includes children weighing not more than 20 pounds.

S7.2 Three-year-old dummy. A three-year-old dummy conforming to Subpart C of Part 572 of this chapter is used for testing a child restraint that is recommended by its manufacturer in accordance with S5.6 for use by children in a weight range that includes children weighing more than 20 pounds.

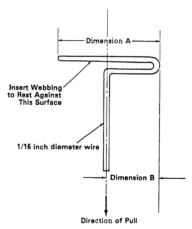

Dimension A

Insert Webbing
to Rest Against
This Surface

1/16 inch diameter wire

Dimension B

Direction of Pull

Dimension A - Width of Webbing Plus 1/8 inch
Dimension B - 1/2 of Dimension A

Webbing Tension Pull Device
FIGURE 5

S7.2.1 Before being used in testing under this standard, the dummy is conditioned at any ambient temperature from 66° F to 78° F and at any relative humidity from 10 percent to 70 percent for at least 4 hours.

S7.2.2 When used in testing under this standard, the dummy is clothed in thermal knit waffle-weave polyester and cotton underwear, a size 4 long-sleeved shirt weighing 0.2 pounds, a size 4 pair of long pants weighing 0.2 pounds and cut off just far enough above the knee to allow the knee target to be visible, and size 7M sneakers with rubber toe caps, uppers of dacron and cotton or nylon and a total weight of 1 pound. Clothing other than the shoes is machine-washed in 160° F to 180° F water and machine dried at 120° F to 140° F for 30 minutes.

S7.3 Standard seat assembly. The standard seat assembly used in testing under this standard are:

(a) For testing for motor vehicle use, a simulated vehicle bench seat, with three seating positions, which is described in Drawing Package SAS-100-1000 (consisting of drawings and a bill of materials); and

(b) For testing for aircraft use, a representative aircraft passenger seat.

S8. Requirements, test conditions, and procedures for child restraint systems manufactured for use in an aircraft. [Each child restraint system manufactured for use in both motor vehicles and aircraft must comply with all of the applicable requirements specified in section S5 and with the additional requirement specified in S8.1 and S8.2.

S8.1 Installation instructions. Each child restraint system manufactured for use in aircraft shall be accompanied by printed instructions in the English language that provide a step-by-step procedure, including diagrams, for installing the system in aircraft passenger seats, securing the system to the seat, positioning a child in the system when it is installed in aircraft, and adjusting the system to fit the child. In the case of each child restraint which is not intended for use in aircraft at certain adjustment positions, the following statement, with the manufacturer's restrictions inserted, shall be included in the instructions.

DO NOT USE THE—ADJUSTMENT POSITION(S) OF THIS CHILD RESTRAINT IN AIRCRAFT.

S8.2 Inversion test. When tested in accordance with S8.2.1 through S8.2.5 and adjusted in any position which the manufacturer has not, in accordance with S8.1, specifically warned against using in aircraft, each child restraint system manufactured for use in aircraft shall meet the requirements of S.8.2.1 through S8.2.6. The manufacturer may, at its option, use any seat which is a representative aircraft passenger seat within the meaning of S4.

S8.2.1 A representative aircraft passenger seat shall be positioned and adjusted so that its horizontal and vertical orientation and its seat back angle are the same as shown in Figure 6.

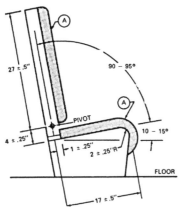

27 ± .5"

90 – 95°

PIVOT

10 – 15°

4 ± .25"

1 ± .25"

2 ± .25"R

A

A

FLOOR

17 ± .5"

FIGURE 6: Simulated Aircraft Passenger Seat

"A" represents a 2- to 3-inch thick polyurethane foam pad, 1.5-2.0 pounds per cubic foot density, over 0.020-inch-thick aluminum pan, and covered by 12- to 14-ounce marine canvas. The sheet aluminum pan is 20 inches wide and supported on each side by a rigid structure. The seat back is a rectangular frame covered with the aluminum sheet and weighing between 14 and 15 pounds, with a center of mass 13 to 16 inches above the seat pivot axis. The mass moment of inertia of the seat back about the seat pivot axis is between 195 and 220 ounce-inch-second². The seat back is free to fold forward about the pivot, but a stop prevents rearward motion. The passenger safety belt anchor points are spaced 21 to 22 inches apart and are located in line with the seat pivot axis. (50 F.R. 15155—April 17, 1985. Effective: April 17, 1985)

S8.2.2 The child restraint system shall be attached to the representative aircraft passenger seat using, at the manufacturer's options, any Federal Aviation Administration approved aircraft safety belt, according to the restraint manufac-

turer's instructions for attaching the restraint to an aircraft seat. No supplementary anchorage belts or tether straps may be attached; however, Federal Aviation Administration approved safety belt extensions may be used.

S8.2.3 In accordance with S6.1.2.3.1 through S6.1.2.3.3, place in the child restraint any dummy specified in S7 for testing systems for use by children of the heights and weights for which the system is recommended in accordance with S5.5 and S8.1.

S8.2.4 If provided, shoulder and pelvic belts that directly restrain the dummy shall be adjusted in accordance with S6.1.2.4.

S8.2.5 The combination of representative aircraft passenger seat, child restraint, and test dummy shall be rotated forward around a horizontal axis which is contained in the median transverse vertical plane of the seating surface portion of the aircraft seat and is located one inch below the bottom of the seat frame, at a speed of 35 to 45 degrees per second, to an angle of 180 degrees. The rotation shall be stopped when it reaches that angle and the seat shall be held in this positon for three seconds. The child restraint shall not fall out of the aircaft safety belt, nor shall the test dummy fall out of the child restraint at any time during the rotation or the three second period. The specified rate of rotation shall be attained in not less than one-half second, and not more than one second, and the rotating combination shall be brought to a stop in not less than one half second and not more than one second.

S8.2.6 Repeat the procedures set forth in S8.2.1 through S8.2.4. The combination of the representative aircraft passenger seat, child restraint, and test dummy shall be rotated sideways around a horizontal axis which is contained in the median longitudinal vertical plane of the seating surface portion of the aircraft seat and is located one inch below the bottom of the seat frame, at a speed of 35 to 45 degrees per second, to an angle of 180 degrees. The rotation shall be stopped when it reaches that angle and the seat

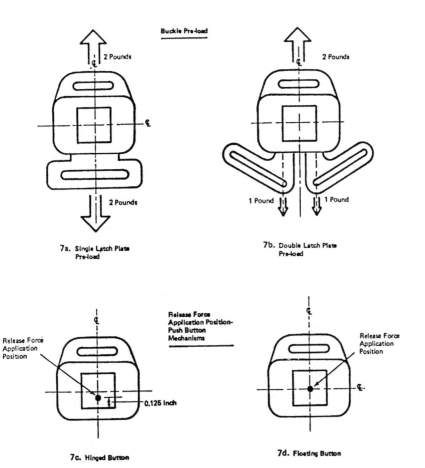

Figure 7. Pre-Impact Buckle Release Force Test Set-up

PART 571; S 213-12

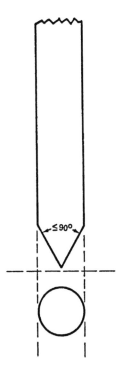

Figure 8. Release Force Application Device — Push Button Release Buckles

PART 571; S 213–13

shall be held in this position for three seconds. The child restraint shall not fall out of the aircraft safety belt, nor shall the test dummy fall out of the child restraint at any time during the rotation or the three second period. The specified rate of rotation shall be attained in not less than one half second and not more than one second, and the rotating combination shall be brought to a stop in not less than one half second and not more than one second. (50 F.R. 15155—April 17, 1985. Effective: April 17, 1985)]

44 F.R. 72131
December 13, 1979

PREAMBLE TO MOTOR VEHICLE SAFETY STANDARD NO. 214

Side Door Strength—Passenger Cars

(Docket No. 2–6; Notice No. 3)

The purpose of this amendment to §571.21 of Title 49, Code of Federal Regulations, is to add a new motor vehicle safety standard that sets minimum strength requirements for side doors of passenger cars. The standard differs in only a few details from the notice of proposed rulemaking published on April 23, 1970 (35 F.R. 6512).

As noted in the proposal of April 23, the percentage of dangerous and fatal injuries in side collisions increases sharply as a maximum depth of penetration increases. With this in mind, the notice of proposed rulemaking stressed the need for a door that offers substantal resistance to intrusion as soon as an object strikes it. The proposal required a door to provide an average crush resistance of 2,500 pounds during the first 6 inches of crush. One comment stated that equivalent protection can be provided by structures further to the interior of the door and that the proper measure of protection is the force needed to deflect the inner door panel rather than that needed to deflect the outer panel. Although inboard mounted structures may be effective in preventing intrusion if the door has a large cross section, with a correspondingly large distance between the protective structure and the inner panel, the standard as issued reflects the determination that doors afford the greatest protection if the crush resisting elements are as close to the outer panel as possible. It follows from this determination that the surface whose crush is to be measured must be the outer panel rather than the inner one. The value specified for the initial crush resistance has, however, been reduced from 2,500 pounds to 2,250 pounds, a value that has been determined to be more appropriate, particularly for lighter vehicles.

Two comments suggested that the crush distance should be the distance traveled by the loading device after an initial outer panel distortion caused by a "pre-load." This suggestion is without merit, in that it would permit use of needlessly light outer panel materials and thereby diminish the distance between the protective elements of the door and the occupants.

The comments revealed a considerable difference of opinion concerning the value and validity of the concept of "equivalent crush resistance." The equivalent crush resistance was to be derived by adding $\frac{1}{4}$ (3000–W) to the average force required to crush the door 12 inches. It had been thought that the resulting bias against heavier vehicles was necessary in that their greater mass would cause them to move sideways less in a collision than lighter vehicles, with more of the impacting force being absorbed by the door. Recent studies, however, show that occupants of heavier vehicles involved in side collisions generally suffer a lower proportion of serious injuries and fatalities than persons in lighter vehicles. In light of these studies and other information, the standard retains the basic crush resistance requirement, but deletes the weight correction factor. Since it is no longer appropriate to use the term "equivalent crush resistance," in its place the standard employs the phrase "intermediate crush resistance." The slightly lower figure of 3,500 pounds has been substituted for the 3,750 pound force proposed in the notice. The effect of the change is to increase slightly the crush resistance required for vehicles having curb weight less than 1,800 pounds, and to decrease it slightly for vehicles weighing more than 1,800 pounds.

Similar reasoning lies behind a change in the requirement for peak crush resistance. The available information does not support a peak crush requirement that increases indefinitely with increasing vehicle curb weight. The standard therefore sets a ceiling of 7,000 pounds to the requirement that the door have a peak crush resistance of twice the vehicle's curb weight. In effect, the requirement is unchanged from the proposal for vehicles weighing less than 3,500 pounds, and is diminished for vehicles exceeding that weight.

Several comments suggested that the vehicle should be tested with all seats in place, since the seats may provide protection against intrusion in side impacts. It is recognized that proper seat design can contribute to occupant safety. The retention of the seat would, however, introduce a variable into the test procedue whose bearing on safety is not objectively measurable at this time. For this reason, the standard adopts the proposed requirement that the vehicle be tested with its seats removed.

It was suggested that the location of force application should be changed. The location has been designated to approximate the weakest section of that part of the door structure likely to be struck by another vehicle. The area designated has been found the most approriate for the bulk of the automobile population.

Effective date: January 1, 1973.

The majority of comments stated that an effective date of September 1, 1971, as initially proposed, would not be feasible. After evaluation of the comments and other information, it has been determined that the structural changes required by the standard will be such that many manufacturers woud be unable to meet the standard if the September 1, 1971, effective date were retained. It has been decided that there is good cause for establishing an effective date more than 1 year after issuance of the rule.

In consideration of the above, Standard No. 214 is adopted as set forth below.

Issued on October 22, 1970.

Douglas W. Toms,
Director.

35 F.R. 16801
October 30, 1970

PREAMBLE TO AN AMENDMENT TO MOTOR VEHICLE SAFETY STANDARD NO. 214

Side Door Strength

(Docket No. 2-6; Notice No. 6)

ACTION: Final Rule.

SUMMARY: The purpose of this notice is to amend Safety Standard No. 214, *Side Door Strength*, to allow manufacturers the option of leaving the seats in a vehicle while its ability to resist external forces pressing inward on its door is tested. This amendment was proposed by the NHTSA in response to a petition for rulemaking from Volvo of America Corporation (44 FR 33444, June 11, 1979). The change is intended to give manufacturers broader design capabilities for improving the safety of vehicle occupants involved in side impact collisions. The performance levels for the alternative requirements are lower than those specified in the notice of proposed rulemaking, due to the agency's consideration of public comments on that notice.

EFFECTIVE DATE: The amendment made by this notice becomes effective upon publication in the FEDERAL REGISTER.

ADDRESSES: Any petitions for reconsideration of this rule should refer to the docket number and notice number and be submitted to the National Highway Traffic Safety Administration, 400 Seventh Street, S.W., Washington, D.C. 20590.

FOR FURTHER INFORMATION CONTACT:
Mr. William Brubaker, Office of Vehicle Safety Standards, National Highway Traffic Safety Administration. (202-426-2242).

SUPPLEMENTARY INFORMATION:
Safety Standard No. 214, *Side Door Strength* (49 CFR 571.214), specifies performance requirements for the side doors of passenger cars to minimize the life-threatening forces caused by intrusion of objects such as other vehicles, poles and tree trunks into the occupant compartment in side-impact accidents. The standard currently specifies three static crush tests (initial, intermediate and peak) to measure the crush resistance of the side doors. The basis for these tests is that early studies concerning side impact protection demonstrated that, in fatal side collisions, most occupants die because of the door structures collapsing inward on them. The static crush tests are intended to ensure that there are strong door structures to limit this intrusion. Under the peak crush test of the standard, the vehicle door may not be deformed more than 18 inches inward when the door is subjected to a force of 7,000 pounds, or two times the curb weight of the vehicle, whichever, is less.

The existing test procedures of the standard specify that the vehicle seats are to be removed during the crush tests. Although it was recognized when the standard was originally promulgated that proper seat design can also reduce the amount of intrusion of side door structures into the occupant compartment, it was determined that this standard should measure the integrity of door structures alone.

Manufacturers have generally incorporated various types of beams in the outer door panels to provide crush resistance in compliance with the standard. Last year, however, Volvo of America Corporation petitioned the agency to allow vehicle seats to remain in the automobile during the crush resistance tests. Volvo stated that it has developed an advanced side impact protection system that incorporates the vehicle seats as an essential component and dispenses with door beams. Test data indicate that the Volvo design provides side impact protection that is equal to or greater than that provided by current production designs.

PART 571; S 214—PRE 3

In response to Volvo's petition, the agency issued a notice of proposed rulemaking to allow manufacturers to adopt this option (44 FR 33444, June 11, 1979). The notice stated that manufacturers should be encouraged to develop innovative designs for improving side impact protection, particularly designs that will improve vehicle fuel economy because of reduced weight. Although not included in Volvo's petition, the proposal specified higher crush resistance levels for vehicles tested with their seats intact (a 16,000-pound peak force).

The criteria were set at levels intended to assure an equivalent or greater level of protection compared to the existing requirements. Agency data show that the seats of some current models contribute 4 to 5 thousand pounds of crush resistance in addition to the crush resistance provided by the doors themselves. Therefore, the higher performance levels were proposed to ensure that the current level of crush resistance that is being obtained by strong door beams will not be degraded.

Nearly all of the twelve comments received in response to the notice supported the proposal to give manufacturers the option of testing with seats installed in the vehicle. A majority of the commenters objected to the higher crush resistance levels for the alternative procedure, however. Only Volkswagen Corporation stated that the standard should not be amended to allow the option. Following is a discussion of these comments.

The Insurance Institute for Highway Safety stated that the proposed amendment would give auto manufacturers a broader range of design alternatives than they currently have to reduce the likelihood of injuries to occupants of vehicles struck in the side. Most commenters made similar statements. Mercedes-Benz of North America noted that manufacturers would be afforded greater latitude in selecting designs to comply with the standard, without sacrificing occupancy protection, and at the same time could reduce vehicle weight.

While agreeing with the concept of the proposed alternative requirement, a large number of commenters felt the proposed performance criteria were too stringent. Peugeot, as well as the Motor Vehicle Manufacturers Association, stated that the current performance levels should apply whether the seats are left in the vehicle during testing or not. American Motors Corporation argued that the proposed crush resistance levels for the alternative procedure are significantly more stringent than existing 214 requirements, and that the NHTSA has not identified any safety need to justify this higher level of performance.

The agency does not agree that the performance levels of the standard should be the same whether the seats are left in the vehicle or are removed. As noted in the proposal, current vehicle seat designs often provide four to five thousand pounds of additional crush resistance above that required by the standard. Further, the standard was originally only intended to test the crush resistance of the doors alone. Therefore, if the performance criteria were the same with and without the seats in the vehicle during the test, manufacturers could reduce the current protection provided by their doors without upgrading their vehicles in other areas. Given the large number of fatalities in side impact accidents, the agency is very concerned that such a degradation of vehicle performance not occur under the alternative test procedure. Therefore, it is the agency's position that there is a substantial safety need to assure that the level of protection provided under the alternative procedure is equivalent to or greater than that provided under the existing test procedure.

Several commenters argued that the data and test results relied upon by the agency to establish the crush resistance levels for the alternative procedure are too limited, and that research should be expanded to include tests of other models prior to establishing the criteria. General Motors stated, for example, that the two vehicles used in NHTSA tests may not be representative of other vehicle designs which could exhibit differing door-to-seat interaction.

The agency disagrees with these contentions. Volvo and Ford Motor Company provided the NHTSA with data from tests they conducted with seats and without seats installed in some of their production vehicles. The agency conducted comparable tests on a Plymouth Volare, and the tests included both bench seats and bucket seats. This and other information substantiate that vehicle seats can and do provide much additional resistance to side door intrusion. These data demonstrate that crush resistance levels should be higher if vehicle seats are left installed during the testing in order to maintain the level of protection currently being provided.

Ford Motor Company argued that the proposed higher performance levels were based on limited tests of current production models, and that the higher performance results achieved in those tests represent built-in reserves by manufacturers above the minimum performance requirements of the standard. Ford stated that the crush resistance criteria of the proposed alternative should not be set at this upper level of performance. Other commenters, including Volvo, also argued that the proposed criteria were too high to allow for production variances. General Motors stated that the proposal does not really remove inhibitions to design innovation due to the increased performance requirements of the proposed alternative procedure. Finally, Rolls-Royce Motors urged that the performance criteria be set low enough that the potential weight savings offered by the proposal can be realized in practice.

After considering these comments, the agency has determined that the crush resistance levels for vehicles tested with their seats intact should be somewhat lower than those specified in the proposal. This will allow for production variances and enable manufacturers to build in a margin of protection above the minimum performance requirements specified in the standard.

In its comments, Volvo Corporation suggested that the intermediate crush resistance level should be set at 4,375 pounds (the proposal specified 7,000 pounds) and that peak crush resistance should be set at 12,000 pounds (the proposal specified 16,000 pounds). Volvo stated that tests of its current production cars that have door beams indicate a spread in intermediate crush resistance of approximately 2,000 pounds. The company noted that an intermediate crush resistance level that is twenty-five percent above the existing requirement would compensate for the addition of seats during testing and at the same time allow manufacturers a sufficient margin to comply with the standard. Volvo also stated that since the seats of some current cars add approximately 4,000 to 5,000 pounds of peak crush resistance, this should be the amount of increase above the existing requirements, i.e., from 7,000 pounds to 12,000 pounds. Although Volvo's preliminary testing of its advanced side impact protection system indicates that the 16,000-pound requirement could be met, the company feels that the margin is not sufficient to allow for production variances.

The agency agrees with Volvo's suggested crush resistance levels, since they should ensure that the level of protection provided under the alternative requirement is at least equivalent to that provided currently. Therefore, these criteria are adopted in this amendment. While it is encouraging that Volvo's advanced system can meet the 16,000-pound peak force specified in the proposal, this may be too high for other manufacturers at the present time, and the agency's primary concern in allowing the alternative test procedure is to avoid any degradation of the protection being provided under the current requirement. The high performance of Volvo's advanced system will be considered very seriously, however, during the planned rulemaking to upgrade side impact protection (an advance notice of proposed rulemaking concerning improving side impact protection was recently issued: 44 FR 70204, December 6, 1979).

As noted above, data indicate that current seat designs contribute approximately 5,000 pounds to the crush resistance capacity of vehicle side structures. Therefore, the 12,000-pound peak force level specified in this amendment will assure that side impact protection is not degraded, but will also allow manufacturers to develop new designs to meet the requirements. As demonstrated by Volvo, manufacturers will be able to develop new side structures and seat designs that will provide over 12,000 pounds of crush resistance without the use of heavy door beams.

Mercedes-Benz of North America commented that the "initial" crush resistance requirement of the proposed alternative should be deleted (paragraph S3.2.1 of the proposal). Mercedes argued that the three-stage static crush tests assign too much significance to the first stage (initial crush resistance), since door reinforcement is necessary primarily to ensure compliance with this initial test. According to Mercedes, the initial resistance is achieved within the first six inches of crush depth (measured at the outer surface of the door), but that this is not more than one-ninth of the total energy absorption when testing without the vehicle seats. When testing with the seats, according to Mercedes, the percentage of energy absorption at the outer surface of the door panel is meaningless with respect to the total energy management and occupant protection.

The agency does not agree with this rationale. The initial crush resistance stage is necessary to ensure that vehicle doors have at least a minimum of structural integrity. This is particularly important because of the risk of occupant ejection if door hinges and latches separate during an accident, allowing the door to fly open. Although seat design can ameliorate intrusion into the occupant compartment to a certain extent, it is important to coordinate door structure and seat design to achieve the optimum occupant protection. Because of the initial crush resistance requirements, manufacturers may not be able to delete door beams altogether in some models. However, manufacturers will be able to use much lighter beams than are currently being used, without a reduction in overall performance.

Several commenters addressed the seat location specified in the proposed alternative requirement. The proposal provided that vehicles must be able to meet the specified crush resistance levels with the vehicle seats located in any position and at any seat back angle in which they are designed to be adjusted. Volvo's petition had requested that the mid, horizontal seat adjustment position be specified. Volkswagen of America stated that the new proposed test procedure, with the seat in any position of its adjustment range, potentially increases the test effort. Volkswagen argued that manufacturers would have the obligation to determine, by a test series, the most adverse test positions of the seat, and that this would be much more costly than the existing requirement.

While it may be true that requiring a vehicle to comply with the seat in any position to which it can be adjusted will require more effort by manufacturers, the agency has determined that this is a necessary aspect of the new procedure. If the vehicle seats are to be used as an integral part of the side impact protection system, it is important that the protection is provided regardless of where the seat is located along its adjustment range.

General Motors stated in its comments that it is reasonable to require demonstrated performance to assure that the occupant seat will assist in limiting side crush in any normal driving position. However, General Motors stated that the same rationale should not apply to seat back angle, and that the normal riding or driving angle established by the manufacturer should be used for compliance purposes. Volvo's comments agreed with General Motors regarding seat back angle.

The agency does not see a distinction between horizontal seat adjustment and seat back angle adjustment. If a particular seat is designed to be adjusted through a range of seat back angles, the vehicle should be able to comply with the requirement of the standard with the seat back at any of its adjustment angles, for the same reasons as noted above for horizontal adjustment. Further, the agency does not believe that the cost of testing will be substantially different if manufacturers are responsible for compliance with the seat in any adjustment position. Manufacturers, in some cases, may be able to determine the "worst case" position for seat location by engineering judgment and analysis prior to testing the vehicle. If a manufacturer has designed the vehicle seat to be an integral part of the side impact protection system, the manufacturer will likely know which position provides the most support and resistance to intrusion (and which provides the least support).

Of the commenters on the proposal, only Volkswagen Corporation was opposed to the proposed alternative test procedure. Volkswagen stated that the proposed requirement is not in keeping with the original purpose of the standard—to prevent intrusion. The company argued that there is a potential for reduced occupant protection in the case of oblique angle or "side-swipe" crashes since a vehicle with a door structure of inferior strength, as compared to current designs, runs the possible risk of door destruction or separation. Volkswagen noted that this could expose vehicle occupants to the risk of ejection.

While the agency shares Volkswagen's concern that the occupant protection being afforded by current vehicle doors not be lessened, it does not believe that the optional test procedure will result in reduced performance. The higher crush resistance requirements for vehicles tested with their seats installed should ensure that the overall protection currently provided is maintained. Moreover, since the initial crush resistance stage is included in the alternative procedure, in spite of comments that it should be deleted, door structures will have to maintain a certain amount of structural integrity. The 2,250-pound initial crush resistance level will ensure that door hinges and latches are of sufficient strength to preclude separation in most cases. Therefore, the agency does not believe that the alternative procedure will lead to increased ejections. The agency does believe, however, that both the current requirement and the alternative requirement should be upgraded. As noted earlier, the agency is presently involved in rulemaking regarding such an upgrade of the standard. The agency does not agree with Volkswagen's contention that the proposed test procedure is not aligned with the original purpose of the standard, since it has been demonstrated that effective seat design can substantially reduce intrusion into the occupant compartment.

The notice proposing this amendment specifically requested comments concerning the effect modifications to side door structures (i.e., lighter door beams or deletion of door beams, altogether) might have on vehicle integrity in frontal and front-angular crashes. In response to this request, Rolls-Royce Motors commented that the door beams used in its vehicles have had a negligible effect on vehicle integrity in frontal crashes. The company added that the requirements of Safety Standard No. 208, *Occupant Crash Protection,* will ensure that manufacturers maintain sufficient structural integrity for front-end crashes even with sophisticated vehicle designs achieving the maximum savings in weight.

American Motors Corporation also stated that the various safety standards requiring frontal impact tests will maintain frontal integrity regardless of modifications to side door structures. Volvo provided data from off-set crash tests involving vehicles both with and without door beams. Both vehicles showed deformation characteristics (damage to vehicle structure) that are within the variances found for current production cars. In light of this information and the fact that there are other safety standards to ensure vehicle integrity in frontal impacts, the agency has concluded that the alternative test procedure set forth in this amendment will have no adverse effect on frontal occupant crash protection.

The agency has reviewed this amendment in accordance with the specifications of Executive Order 12044, "Improving Government Regulations," and the Departmental guidelines implementing that order and determined it has no significant environmental impact and that its economic impact is so minimal as not to require a regulatory evaluation. The amendment will merely provide manufacturers an alternative test procedure for determining compliance with an existing standard. For this reason, also, the agency has determined that an immediate effective date for this amendment is in order.

The engineer and lawyer primarily responsible for the development of this rule are William Brubaker and Hugh Oates, respectively.

In consideration of the foregoing, Safety Standard No. 214 (49 CFR 571.241) is amended as set forth below.

Section S3 (S3 through S3.3) is amended to read as follows and the first sentence of subparagraph S4(a) is deleted.

§ 571.214 Standard No. 214; Side door strength.

* * * * *

S3 *Requirements.* Each vehicle shall be able to meet the requirements of either, at the manufacturer's option, S3.1 or S3.2 when any of its side doors that can be used for occupant egress are tested according to S4.

S3.1 With any seats that may affect load upon or deflection of the side of the vehicle removed from the vehicle, each vehicle must be able to meet the requirements of S3.1.1 through S3.1.3.

S3.1.1 *Initial Crush Resistance.* The initial crush resistance shall not be less than 2,250 pounds.

S3.1.2 *Intermediate Crush Resistance.* The intermediate crush resistance shall not be less than 3,500 pounds.

S3.1.3 *Peak Crush Resistance.* The peak crush resistance shall not be less than two times the curb weight of the vehicle or 7,000 pounds, whichever is less.

S3.2 With seats installed in the vehicle, and located in any horizontal or vertical position to which they can be adjusted and at any seat back angle to which they can be adjusted, each vehicle must be able to meet the requirements of S3.2.1 through S3.2.2.

S3.2.1 *Initial Crush Resistance.* The initial crush resistance shall not be less than 2,250 pounds.

S3.2.2 *Intermediate Crush Resistance.* The intermediate crush resistance shall not be less than 4,375 pounds.

S3.2.3 *Peak Crush Resistance.* The peak crush resistance shall not be less than three and one half times the curb weight of the vehicle or 12,000 pounds, whichever is less.

Issued on March 11, 1980.

Joan Claybrook
Administrator

45 F.R. 17015
March 17, 1980

———————————

MOTOR VEHICLE SAFETY STANDARD NO. 214

Side Door Strength—Passenger Cars

(Docket No. 2-6; Notice No. 3)

S1. Purpose and scope. This standard specifies strength requirements for side doors of a motor vehicle to minimize the safety hazard caused by intrusion into the passenger compartment in a side impact accident.

S2. Application. This standard applies to passenger cars.

S3. Requirements. Each vehicle shall be able to meet the requirements of either, at the manufacturer's option, S3.1 or S3.2 when any of its side doors that can be used for occupant egress are tested according to S4.

S3.1 With any seats that may affect load upon or deflection of the side of the vehicle removed from the vehicle, each vehicle must be able to meet the requirements of S3.1.1 through S3.1.3.

S3.1.1 Initial Crush Resistance. The initial crush resistance shall be not less than 2,250 pounds.

S3.1.2 Intermediate Crush Resistance. The intermediate crush resistance shall not be less than 3,500 pounds.

S3.1.3 Peak crush resistance. The peak crush resistance shall not be less than two times the curb weight of the vehicle or 7,000 pounds, whichever is less.

S3.2 With seats installed in the vehicle, and located in any horizontal or vertical position to which they can be adjusted and at any seat back angle to which they can be adjusted, each vehicle must be able to meet the requirements of S3.2.1 through S3.2.2.

S3.2.1 Initial Crush Resistance. The initial crush resistance shall not be less than 2,250 pounds.

S3.2.2 Intermediate Crush Resistance. The intermediate crush resistance shall not be less than 4,375 pounds.

S3.2.3 Peak Crush Resistance. The peak crush resistance shall not be less than three and one half times the curb weight of the vehicle or 12,000 pounds, whichever is less.

S4. Test procedures. The following procedures apply to determining compliance with section S3:

(a) Place side windows in their uppermost position and all doors in locked position. Place the sill of the side of the vehicle opposite to the side being tested against a rigid unyielding vertical surface. Fix the vehicle rigidly in position by means of tiedown attachments located at or forward of the front wheel centerline and at or rearward of the rear wheel centerline.

(b) Prepare a loading device consisting of a rigid steel cylinder or semi-cylinder 12 inches in diameter with an edge radius of one-half inch. The length of the loading device shall be such that the top surface

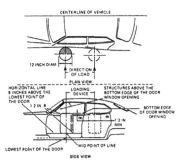

LOADING DEVICE LOCATION AND APPLICATION TO THE DOOR

FIGURE I

of the loading device is at least one-half inch above the bottom edge of the door window opening but not of a length that will cause contact with any structure above the bottom edge of the door window opening during the test.

(c) Locate the loading device as shown in Figure I (side view) of this section so that:

(1) Its longitudinal axis is vertical;

(2) Its longitudinal axis is laterally opposite the midpoint of a horizontal line drawn across the outer surface of the door 5 inches above the lowest point of the door;

(3) Its bottom surface is in the same horizontal plane as the horizontal line described in subdivision (2) of this subparagraph; and

(4) The cylindrical face of the device is in contact with the outer surface of the door.

(d) Using the loading device, apply a load to the outer surface of the door in an inboard direction normal to a vertical plane along the vehicle's longitudinal centerline. Apply the load continuously such that the loading device travel rate does not exceed one-half inch per second until the loading device travels 18 inches. Guide the loading device to prevent it from being rotated or displaced from its direction of travel. The test must be completed within 120 seconds.

(e) Record applied load versus displacement of the loading device, either continuously or in increments of not more than 1 inch or 200 pounds for the entire crush distance of 18 inches.

(f) Determine the initial crush resistance, intermediate crush resistance, and peak crush resistance as follows:

(1) From the results recorded in subparagraph (e) of this paragraph, plot a curve of load versus displacement and obtain the integral of the applied load with respect to the crush distances specified in subdivisions (2) and (3) of this paragraph. These quantities, expressed in inch-pounds and divided by the specified crush distances, represent the average forces in pounds required to deflect the door those distances.

(2) The initial crush resistance is the average force required to deform the door over the initial 6 inches of crush.

(3) The intermediate crush resistance is the average force required to deform the door over the initial 12 inches of crush.

(4) The peak crush resistance is the largest force recorded over the entire 18-inch crush distance.

October 30, 1970
35 F.R. 16801

PREAMBLE TO MOTOR VEHICLE SAFETY STANDARD NO. 216

Roof Crush Resistance—Passenger Cars

(Docket No. 2–6; Notice 5)

The purpose of this amendment to Part 571 of Title 49, Code of Federal Regulations, is to add a new Motor Vehicle Safety Standard 216, (49 CFR § 571.216) that sets minimum strength requirements for a passenger car roof to reduce the likelihood of roof collapse in a rollover accident. The standard provides an alternative to conformity with the rollover test of Standard 208.

A notice of proposed rulemaking on this subject was issued on January 6, 1971 (36 F.R. 166). As noted in that proposal, the strength of a vehicle roof affects the integrity of the passenger compartment and the safety of the occupants. A few comments suggested that there is no significant causal relationship between roof deformation and occupant injuries in rollover accidents. However, available data have shown that for non-ejected front seat occupants in rollover accidents, serious injuries are more frequent when the roof collapses.

The roof crush standard will provide protection in rollover accidents by improving the integrity of the door, side window, and windshield retention areas. Preserving the overall structure of the vehicle in a crash decreases the likelihood of occupant ejection, reduces the hazard of occupant interior impacts, and enhances occupant egress after the accident. It has been determined, therefore, that improved roof strength will increase occupant protection in rollover accidents.

Standard 208 (49 CFR § 571.208), *Occupant Crash Protection*, also contains a rollover test requirement for vehicles that conform to the "first option" of providing complete passive protection. The new Standard 216 issued herewith is intended as an alternative to the Standard 208 rollover test, such that manufacturers may conform to either requirement as they choose. Standard 208 is accordingly amended by this notice; the effect of the amendment, together with the new Standard 216, is as follows:

(1) From January 1, 1972, to August 14, 1973, a manufacturer may substitute Standard 216 for the rollover test requirement in the first option of Standard 208; Standard 216 has no mandatory application.

(2) From August 15, 1973, to August 14, 1977, Standard 216 is in effect as to all passenger cars except those conforming by passive means to the rollover test of Standard 208, but it may continue to be substituted for that rollover test.

(3) After August 15, 1977, Standard 216 will no longer be a substitute for the Standard 208 rollover test. It is expected that as of that date Standard 216 will be revoked, at least with respect to its application to passenger cars.

A few comments stated that on some models the strength required in the A pillar could be produced only by designs that impair forward visibility. After review of strengthening options available to manufacturers, the Administration has concluded that a satisfactory increase in strength can be obtained without reducing visibility.

Some comments suggested that the crush limitation be based on the interior deflection of the test vehicle rather than the proposed external criterion. After comparison of the two methods, it has been concluded that a test based on interior deflection would produce results that are significantly less uniform and more difficult to measure, and therefore the requirement based on

PART 571; S 216—PRE 1

external movement of the test block has been retained.

Several changes in detail have been made, however, in the test procedure. A number of comments stated that the surface area of the proposed test device was too small, that the 10-degree pitch angle was too severe, and that the 5 inches of padded test device displacement was not enough to measure the overall roof strength. Later data available after the issuance of the NPRM (Notice 4) substantiated these comments. Accordingly, the dimensions of the test block have been changed from 12 inches square to 30 inches by 72 inches, the face padding on the block has been eliminated, and the pitch angle has been changed from 10 degrees to 5 degrees.

Several manufacturers asked that convertibles be exempted from the standard, stating that it was impracticable for those vehicles to be brought into compliance. The Administration has determined that compliance with the standard would pose extreme difficulties for many convertible models. Accordingly, manufacturers of convertibles need not comply with the standard; however, until August 15, 1977, they may comply with the standard as an alternative to conformity with the rollover test of Standard 208.

A few comments objected to the optional 5,000-pound ceiling to the requirement that the roof have a peak resistance of 1½ times the unloaded vehicle weight. Such objections have some merit, if the energy to be dissipated during a rollover accident must be absorbed entirely by the crash vehicle. In the typical rollover accident, however, in which the vehicle rolls onto the road shoulder, significant amounts of energy are absorbed by the ground. This is particularly true in heavier vehicles. Some of the heavier vehicles, moreover, would require extensive redesign, at a considerably greater cost penalty than in the case of lighter vehicles, to meet a strength requirement of 1½ times their weight. At the same time, heavier vehicles generally have a lower rollover tendency than do lighter vehicles. On the basis of these factors, it has been determined that an upper limit of 5,000 pounds on

the strength requirement is justified, and it has been retained.

It was requested that the requirement of mounting the chassis horizontally be deleted. It has been determined that the horizontal mounting position contributes to the repeatability of the test procedure and the requirement is therefore retained.

The required loading rate has been clarified in light of the comments. The requirement has been changed from a rate not to exceed 200 pounds per second to a loading device travel rate not exceeding one-half inch per second, with completion of the test within 120 seconds.

A number of manufacturers requested that repetition of the test on the opposite front corner of the roof be deleted. It has been determined that, as long as it is clear that both the left and right front portions of the vehicle's roof structure must be capable of meeting the requirements, it is not necessary that a given vehicle be capable of sustaining successive force applications at the two different locations. The second test is accordingly deleted.

Effective date: August 15, 1973. After evaluation of the comments and other information, it has been determined that the structural changes required by the standard will be such that many manufacturers would be unable to meet the requirements if the January 1, 1973 effective date were retained. It has therefore been found, for good cause shown, that an effective date more than one year after issuance is in the public interest. On or after January 1, 1972, however, a manufacturer may substitute compliance with this standard for compliance with the rollover test requirement of Standard 208.

In consideration of the above, the following changes are made in Part 571 of Title 49, Code of Federal Regulations:

1. Standard No. 208, 49 CFR § 571.208, is amended by adding the following sentence at the end of S5.3, *Rollover:* "However, vehicles manufactured before August 15, 1977, that conform to the requirements of Standard No. 216 (§ 571.216) need not conform to this rollover test requirement."

Effective: August 15, 1973

2. A new § 571.216, Standard No. 216 *Roof Crush Resistance*, is added. . . .

This rule is issued under the authority of sections 103 and 119 of the National Traffic and Motor Vehicle Safety Act, 15 U.S.C. 1392, 1407, and the delegation of authority at 49 CFR 1.51.

Issued on December 3, 1971.

Charles H. Hartman
Acting Administrator

36 F.R. 23299
December 8, 1971

PREAMBLE TO AMENDMENT TO MOTOR VEHICLE SAFETY STANDARD NO. 216

Roof Crush Resistance

(Docket No. 69–7; Notice 29)

The purpose of this notice is to postpone the effective date of the requirements of Standards No. 208, Occupant Crash Protection, and 216, Roof Crush Resistance, applicable to the upcoming model year, from August 15, 1973, to September 1, 1973.

The amendment of the effective date was proposed in a notice published July 17, 1973 (38 F.R. 19049), in response to a petition filed by Chrysler Corporation. Chrysler had stated that the build-out of their 1973 models was in danger of running beyond the August 15 date, due to a variety of factors beyond the company's control. In proposing the postponement of the date, the NHTSA noted that the August 15 date had been chosen to coincide with the normal changeover date and that a delay would not appear to have any effect beyond allowing a slightly prolonged build-out.

The two comments submitted in response to the proposal were both favorable. The agency has not discovered any adverse consequences of a delay which would make it inadvisable, and has therefore decided to postpone the effective date as proposed.

In light of the foregoing, 49 CFR 571.208, Standard No. 208, Occupant Crash Protection, is amended by changing the date of August 14, 1973, appearing in S4.1.1 to August 31, 1973, and by changing the date of August 15, 1973, appearing in S4.1.2 to September 1, 1973. The effective date of 49 CFR 571.216, Standard No. 216, Roof Crush Resistance, is changed from August 15, 1973, to September 1, 1973.

Because this amendment relieves a restriction and imposes no additional burden, an effective date of less than 30 days from the date of issuance is found to be in the public interest.

(Sec. 103, 119, Pub. L. 89–563, 80 Stat. 718, 15 U.S.C. 1392, 1407; delegation of authority at 49 CFR 1.51.)

Issued on August 10, 1973.

James B. Gregory
Administrator

38 F.R. 21930
August 14, 1973

MOTOR VEHICLE SAFETY STANDARD NO. 216

Roof Crush Resistance—Passenger Cars

S1. Scope. This standard establishes strength requirements for the passenger compartment roof.

S2. Purpose. The purpose of this standard is to reduce deaths and injuries due to the crushing of the roof into the passenger compartment in rollover accidents.

S3. Application. This standard applies to passenger cars. However, it does not apply to vehicles that conform to the rollover test requirements (S5.3) of Standard 208 (§ 571.208) by means that require no action by vehicle occupants. It also does not apply to convertibles, except for optional compliance with the standard as an alternative to the rollover test requirements in S5.3 of Standard 208.

S4. Requirements. A test device as described in S5 shall not move more than 5 inches, measured in accordance with S6.4, when it is used to apply a force of 1½ times the unloaded vehicle weight of the vehicle of 5,000 pounds, whichever is less, to either side or the forward edge of a vehicle's roof in accordance with the procedures of S6. Both the left and right front portions of the vehicle's roof structure shall be capable of meeting the requirements, but a particular vehicle need not meet further requirements after being tested at one location.

S5. Test Device. The test device is a rigid unyielding block with its lower surface formed as a flat rectangle 30 inches × 72 inches.

S6. Test Procedure. Each vehicle shall be capable of meeting the requirements of S4 when tested in accordance with the following procedure.

S6.1. Place the sills or the chassis frame of the vehicle on a rigid horizontal surface, fix vehicle rigidly in position, close all windows, close and lock all doors, and secure any convertible top or removable roof structure in place over the passenger compartment.

S6.2 Orient the test device as shown in Figure 1, so that—

(a) Its longitudinal axis is at a forward angle (side view) of 5° below the horizontal, and is parallel to the vertical plane through the vehicle's longitudinal centerline;

(b) Its lateral axis is at a lateral outboard angle, in the front view projection, 25° below the horizodntal;

(c) Its lower surface is tangent to the surface of the vehicle; and

(d) The initial contact point, or center of the initial contact area, is on the longitudinal centerline of the lower surface of the test device and 10 inches from the forwardmost point of that centerline.

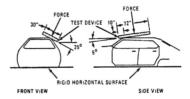

TEST DEVICE LOCATION AND APPLICATION TO THE ROOF

Figure 1

S6.3. Apply force in a downward direction perpendicular to the lower surface of the test device at a rate of not more than one-half inch per second until reaching a force of 1½ times the unloaded vehicle weight of the tested vehicle or 5,000 pounds, whichever is less. Complete the test within 120 seconds. Guide the test device so that throughout the test it moves, without rotation, in a straight line with its lower surface oriented as specified in S6.2(a) through S6.2(d).

S6.4 Measure the distance that the test device moves, *i.e.*, the distance between the original location of the lower surface of the test device and its location as the force level specified in S6.3 is reached.

36 F.R. 23299
December 8, 1971

MOTOR VEHICLE SAFETY STANDARD NO. 217

Bus Window Retention and Release

(Docket No. 2–10; Notice 3)

The purpose of this amendment to § 571.21 of Title 49, Code of Federal Regulations, is to add a new motor vehicle safety standard that establishes minimum requirements for bus window retention and release to reduce the likelihood of passenger ejection in accidents and enhance passenger exit in emergencies.

A notice of proposed rulemaking on this subject was published on August 15, 1970 (35 F.R. 13025). The comments received in response to the notice have been considered in this issuance of a final rule.

For reasons of clarification, the requirements paragraph has been reorganized and the demonstration procedures paragraph has been replaced by a test conditions paragraph. Some of the specifications of the demonstration procedures paragraph are incorporated under the requirements paragraph, and the remainder are retained under the test conditions paragraph. With the exception of the changes discussed below, the reorganization does not affect the substance of the standard.

In altering the window retention requirements, the final rule lowers the force application limit, provides more precise glazing breakage and glazing yield limits, and exempts small windows. With respect to the emergency exit requirements, the standard permits devices other than push-out windows to be used for emergency exits, permits buses with a GVWR of 10,000 pounds or less to utilize devices other than emergency exits for emergency egress, and permits an alternate roof exit when the bus configuration precludes provision of a rear emergency exit. It also raises the force limits for release and extension of emergency exits, deletes the inertial load requirement for the release mechanism, and requires that emergency exit location markings be located within each occupant space adjacent to an exit.

A few changes have been made in the diagram accompanying the standard. Figure 1, "Adjacent Designated Seating Position, Occupant Spaces, and Push-Out Window Relationship," has been deleted from the final rule because the relationship is sufficiently described in the text of the standard. Accordingly, Figures 2 and 3 have been renumbered as Figures 1 and 2, respectively. A new Figure 3, indicating access regions for emergency exits which do not have adjacent seats, has been added. For reasons of clarification, Figures 2a and 2b and Figures 3a and 3b in the proposed rule have been placed beside each other to form Figures 1 and 2 respectively.

The torque in Figures 2a and 2b of the proposed rule has been transferred to the text and has been explained to indicate that the force used to obtain the torque shall be not more than 20 pounds. In addition, the clearance specifications in Figures 1 and 2 have been clarified in the text to require that the lower edge of the force envelope shall be located 5 inches above the seat, or 2 inches above the armrest, if any, whichever is higher. In several instances, minor changes have been made in the labeling without altering the substance of the diagrams.

A number of comments sought changes in the window retention requirements. Two comments requested an exemption for intra-city buses because the probability of rollover accidents would be minimal in slow-speed operation. Urban transit buses are subjected to risks of rollover accidents within the city when they travel at moderate to high speed on intra-urban expressways, and should therefore be covered by the standard. Accordingly, the request for this exemption is denied.

Several comments requested an exemption for small windows. Since there is little likelihood of passenger ejection or protrusion from window openings whose minimum surface dimension measured through the center of the area is less than eight inches, an exemption for windows of this size has been granted.

Two comments asked that the 2,000 pound force application limit in the window retention requirement be lowered. The data indicates that a 1,200-pound limit would be more compatible with the glazing strength. Accordingly, the 2,000-pound force application limit has been lowered to 1,200 pounds.

Several manufacturers stated that they encountered difficulties in ascertaining when the proposed head form penetration limit of the window retention requirement had been reached. After observation of window retention testing, the NHTSA has concluded that the penetration limit as specified in the notice of proposed rulemaking is difficult to determine. For this reason the head form penetration limit has been rephrased in terms of the development of cracks in the glazing and the amount of depression of the glazing surface in relation to its original position.

A number of comments objected to the requirement that at least 75% of the glazing be retained in the window mounting during window retention testing. The NHTSA has determined that the intent of this requirement is already accomplished by the requirement that each window be retained during testing by its surrounding structure in a manner which would prevent passage of a 4-inch sphere, and the requirement is accordingly deleted from the final rule.

With respect to the emergency exit requirements, the standard permits devices other than push-out windows to be used for emergency exits. Upon review of the requirements, it has been determined that devices such as panels and doors which meet the emergency exit requirements would be as effective as push-out windows for emergency egress. Because the Administration has concluded that passenger egress is enhanced when several emergency exits are pro-

vided, the standard requires that in computing whether a bus meets the unobstructed openings area requirements, no emergency exit, regardless of its area, shall be credited with more than 520 square inches of the total area requirement.

A number of motor vehicle manufacturers sought exemption from the emergency exit requirements for smaller vehicles weighing 10,000 pounds or less GVWR, such as limousines and station wagons, which are designed to carry more than 10 persons and are therefore considered to be buses under NHTSA regulations (49 CFR 571.3). Such vehicles are usually provided with numerous doors and windows which provide sufficient unobstructed openings for emergency exit. Therefore the Administration has concluded that the configuration of these vehicles satisfies the intent of the standard with respect to provision of emergency exits, and they are exempted from the emergency exit openings requirements.

The emergency exit requirements have been changed to permit installation of an alternate roof exit when the bus configuration precludes provision of a rear exit, provided that the roof exit meets the release, extension, and identification requirements. The NHTSA has established this alternative in order to allow design flexibility while providing for emergency egress in rollover situations.

A number of comments expressed concern that the proposed maximum force level for release and extension of emergency exits in Figures 2a and b and 3a and b were too low to inhibit inadvertent operation by passengers and suggested that the required maximum force level be raised. After consideration of the goals of facilitating emergency egress and preserving the integrity of the passenger compartment under normal operation, it has been determined that the maximum force levels should be raised from 10 and 30 pounds to 20 and 60 pounds respectively.

One comment submitted the results of testing which indicated that the 30g inertial load requirement for the release mechanism was unnecessarily high. The testing also revealed that the engineering concepts upon which the inertial load requirement is based are not generally applied in the industry and that the requirement would be impracticable. Moreover, an increase in maximum force levels for emergency exit operation in the rule should improve latch integrity. For these reasons, the requirement has been deleted.

The standard requires emergency exit location markings to be placed in certain occupant spaces because of a possible contradiction under the proposed standard between the requirement that the identification markings be located within 6 inches of the point of operation and the requirement that the markings be visible to a seated occupant. The NHTSA has concluded that emergency egress could be hindered if the passenger has difficulty in finding the marking, and that location of the marking outside of an occupant space containing an adjacent seat, which would be permitted under the proposed standard, could create this problem. At the same time it is desirable for the identification and instructions to be located near the point of release. Therefore the final rule requires that when a release mechanism is not located within an occupant space containing an adjacent seat, a label indicating the location of the nearest release mechanism shall be placed within that occupant space.

The temperature condition has been reworded to make it clear, in light of the explanation of usage in § 571.4, that the vehicle must be capable of meeting the performance requirements at any temperature from 70° F. to 85° F.

Effective date: September 1, 1973. After evaluation of the comments and other information, it has been determined that the structural changes required by the standard will be such that many manufacturers will require an effective date of at least fifteen months after issuance. It is therefore found, for good cause shown, that an effective date more than one year from the date of issuance is in the public interest.

In consideration of the above, Standard No. 217, Bus Window Retention and Release, is added to § 571.21 of Title 49, Code of Federal Regulations, as set forth below.

This rule is issued under the authority of sections 103, 112, and 119 of the National Traffic and Motor Vehicle Safety Act, 15 U.S.C. 1392, 1401, 1407, and the delegation of authority at 49 CFR 1.51.

Issued on May 3, 1972.

Douglas W. Toms
Administrator

37 F.R. 9394
May 10, 1972

PREAMBLE TO AMENDMENT TO MOTOR VEHICLE SAFETY STANDARD NO. 217

Bus Window Retention and Release

(Docket 2–10; Notice 4)

The purpose of this notice is to respond to petitions for reconsideration of Motor Vehicle Safety Standard No. 217, Bus Window Retention and Release, in § 571.217 of Title 49, Code of Federal Regulations. The standard was issued on May 10, 1972 (37 F.R. 9394).

International Harvester stated that it manufactures an 18-passenger airport limousine, the "Stageway Coach Conversion", weighing 10,700 pounds GVWR and requested that it be exempted from the requirements of S5.2.1, "Buses with GVWR of more than 10,000 pounds." They emphasized that the 18-passenger model is equipped with 10 side doors, two more than is provided by a 15-passenger, 10,000-pound, version of a similar airport limousine vehicle which they manufacture. The NHTSA has concluded that vehicles which provide at least one door for each three passenger seating positions afford sufficient means of emergency egress regardless of their weight. S5.2.1 has accordingly been amended to provide that buses with a GVWR of more than 10,000 pounds may alternatively meet the unobstructed openings requirement of S5.2 by providing at least one door for each three passenger spaces in the vehicle. The "Stageway Coach Conversion" falls into the category of vehicles covered by this amendment and thus International Harvester's request is granted.

International Harvester, General Motors, and Chrysler all requested a clarification of the S5.1 window retention requirements because they felt it was possible to interpret the paragraph as prohibiting the use of tempered glass for window glazing. Ford also submitted a request for exemption from the window retention requirements for buses under 10,000 pounds GVWR based on its interpretation of S5.1 as precluding the use

of tempered glass. The petitioners stated that tempered glass would shatter under the application of pressure required, and were not certain whether S5.1(b), describing the development of cracks in the glazing, would cover this occurrence. The NHTSA did not intend to prohibit the use of tempered glass, and in order to correct this possible ambiguity, S5.1(b) has been amended to include shattering of the window glazing.

General Motors also requested an interpretation of the method of measuring whether 80 percent of the glazing thickness has developed cracks as described in S5.1(b). The paragraph refers to a measurement through the thickness of glass and not a measurement of the glazing surface area, as GM suggests it could mean. GM also doubted that the percentage of glazing thickness which develops cracks could be measured. The NHTSA has determined that the intent of the language is clear and that performance of this measurement is within the state of the art, so that no change in the language is necessary. The request is therefore denied.

General Motors requested a clarification of the term "minimum surface dimension" in paragraph S5.1(c). The NHTSA agrees that a clarification is necessary to prevent interpretations which may not meet the intent of this standard, and the paragraph has been accordingly amended to specify that the dimension is to be measured through the center of the area of the sheet of glazing.

General Motors stated that it interpreted the head form travel rate specified in S5.1.1 of two inches per minute as a "nominal value" requirement, since no tolerances are given in the standard. The test conditions in a safety standard

PART 571; S 217—PRE 5

represent the performance levels that the product must be *capable* of meeting. They are not instructions either to the manufacturers' or the government's test laboratories, or a requirement that the product should be tested at "exactly" those levels. The manufacturers' tests in this case should be designed to demonstrate that the vehicle would meet the stated requirements *if* tested at two inches per minute. If that is what General Motors means by a "nominal value", its interpretation is correct.

In consideration of the foregoing, Motor Vehicle Safety Standard No. 217, Bus Window

Retention and Release, 49 CFR 571.217, is amended....

Effective date: September 1, 1973.

This notice is issued under the authority of sections 103, 112, and 119 of the National Traffic and Motor Vehicle Safety Act, 15 U.S.C. 1392, 1401, 1407, and the delegation of authority at 49 CFR 1.51.

Issued on August 30, 1972.

Douglas W. Toms
Administrator

37 F.R. 18034
September 6, 1972

PART 571; S 217—PRE 6

PREAMBLE TO AMENDMENT TO MOTOR VEHICLE SAFETY STANDARD NO. 217

Bus Window Retention and Release

(Docket No. 2–10; Notice 5)

The purpose of this notice is to amend Motor Vehicle Safety Standard No. 217, Bus Window Retention and Release, 49 CFR § 571.217, in response to petitions received. Several minor amendments for purposes of clarification have also been made. The standard was published initially on May 10, 1972, (37 F.R. 9394), and amended September 6, 1972 (37 F.R. 18034).

Wayne Corporation has petitioned that the torque limit of 20 inch-pounds for the actuation of rotary emergency exit releases in S5.3.2(a)(3) of the standard is impractical. The Blue Bird Body Company also objected to the requirement, requesting that the limit be raised to 225 inch-pounds in order to avoid inadvertent openings. The NHTSA has decided, based on these petitions, that a maximum torque requirement is redudant, since the force magnitude generally is limited in S5.3.2 to not more than twenty pounds. Accordingly the torque requirement is deleted from the rule.

Blue Bird also requested that Figure 3A, which depicts access region for roof and side emergency exits without adjacent seats in both an upright and overturned bus, be made more explicit.

In response to this request, Figure 3A is being replaced by two figures, one of which depicts a side emergency exit (Figure 3A), and the other a roof emergency exit (Figure 3B). Existing Figure 3B, depicting access regions for a rear exit with a rear shelf or other obstruction behind the rearmost seat, becomes Figure 3C. A new Figure 3D is added to depict rear seat access regions in buses not having a rear shelf or other obstruction behind the rearmost seat, a configuration common to school buses. Paragraph S5.2.1, regarding provision of emergency exits, is amended to make it clear that a required rear exit must meet the requirements of S5.3 through S5.5 when the bust is overturned on either side, with the occupant standing facing the exit, as well as when the bus is upright.

In consideration of the above, Standard No. 217, Bus Window Retention and Release, 49 CFR 571.217, is amended . . . .

Effective date: September 1, 1973.

(Sec. 103, 112, 119, P.L. 89–563, 80 Stat. 718, 15 U.S.C. 1392, 1401, 1407) and the delegation of authority at 49 CFR 1.51.

Issued on February 28, 1973.

Douglas W. Toms
Administrator

38 F.R. 6070
March 6, 1973

Bus Window Retention and Release

This notice amends Federal Motor Vehicle Safety Standard No. 217, "Bus Window Retention and Release" (49 CFR § 571.217), to exempt from the standard buses manufactured for the purpose of transporting persons under physical restraint. The amendment is based on a notice of proposed rulemaking published October 1, 1973 (38 F.R. 27227), following petitions received from the Bureau of Prisons, United States Department of Justice.

The comments received in response to the proposal agreed that buses manufactured for the specified purpose should not be provided with the emergency exits required by Standard No. 217. The standard specifies that buses contain emergency exits operable by bus occupants, requirements which the NHTSA considers obviously incompatible with the need to transport prison inmates. The National Transportation Safety Board (NTSB) commented, however, that compensatory measures should be taken to minimize the likelihood of fire in prison buses, since the probability of safely evacuating a prison bus is less than that of any other type of bus. The NTSB urged that the exemption be limited to diesel-fueled buses, since diesel fuel is less likely to ignite than gasoline.

The NHTSA recognizes the desirability of minimizing the likelihood of fire in buses. However, at the present time it is not practical to expect that all newly manufactured prison buses be equipped with diesel engines, given the apparent immediate need for the exemption. Appropriate rulemaking action can be taken in the future if it appears necessary to mitigate from a safety standpoint the loss of emergency exits in prison buses.

In light of the above, paragraph S3 of section 571.217, Title 49, Code of Federal Regulations (Motor Vehicle Safety Standard No. 217), is amended. . . .

Effective date: June 3, 1974. This amendment imposes no additional burdens on any person and relieves restrictions found to be unwarranted. Accordingly, good cause exists and is hereby found for an effective date less than 180 days from the day of issuance.

(Secs. 103, 112, and 119, Pub. L. 89–563; 80 Stat. 718; 15 U.S.C. 1392, 1491, 1407; delegations of authority at 49 CFR 1.51.)

Issued on April 26, 1974.

James B. Gregory
Administrator

39 F.R. 15274
May 2, 1974

PREAMBLE TO AMENDMENT TO MOTOR VEHICLE SAFETY STANDARD NO. 217

Bus Window Retention and Release

(Docket No. 75–6; Notice 2)

This notice amends Federal Motor Vehicle Safety Standard No. 217, *Bus Window Retention and Release*, 49 CFR 571.217, to clarify the marking requirements for emergency exits on buses. The amendment requires certain markings on all bus emergency exits except manually-operated windows of sufficient size and doors in buses with a GVWR of 10,000 pounds or less.

The amendment was proposed in a notice published April 18, 1975 (40 FR 17266). Comments were received from Chrysler Corporation and General Motors. Chrysler concurred with the proposal. GM, while also concurring, suggested that the wording of the amendment be modified somewhat. The amendment has been reworded to reflect more clearly the intent of this amendment, distinguishing between emergency exits that require markings and those that do not. The NHTSA has determined that special emergency exit markings are unnecessary for doors and manually-operated windows in buses with a GVWR of 10,000 pounds or less. This amendment does not exempt buses with a GVWR of 10,000 pounds or less from complying with the unobstructed openings requirements of S5.2.

It only provides that the openings do not have to be marked as emergency exits. However, specially-installed emergency exits in such buses, such as push-out windows, are not exempted from the marking requirements.

The amendment also allows bus manufacturers the option of designating an emergency door as "Emergency Door" or "Emergency Exit." This will bring Standard No. 217 into conformity with current NHTSA interpretations of the emergency exit marking requirements. However, any emergency exit other than a door must have the designation "Emergency Exit."

Accordingly, S5.5.1 of 49 CFR 571.217, *Bus Window Retention and Release*, is amended . . . .

Effective date: October 16, 1975.

(Secs. 103, 112, 119, Pub. L. 89–563, 80 Stat. 718 (15 U.S.C. 1392, 1401, 1407); delegations of authority at 49 CFR 1.51).

Issued on October 8, 1975.

Gene G. Mannella
Acting Administrator

40 F.R. 48512
October 16, 1975

Effective: October 26, 1976

PREAMBLE TO AMENDMENT TO MOTOR VEHICLE SAFETY STANDARD NO. 217

Bus Window Retention and Release

(Docket NO. 75–3; Notice 2)

This notice amends Federal Motor Vehicle Safety Standard No. 217, *Bus Window Retention and Release*, 49 CFR 571.217, to specify requirements for emergency doors for school buses pursuant to the provisions of section 202 of the Motor Vehicle and Schoolbus Safety Amendments of 1974 (Public Law 93–492, 88 Stat. 1484, 15 U.S.C. 1392). It responds to the congressional mandate to establish standards concerning school bus emergency exits (15 U.S.C. § 1392(i) (1) (A) (i)).

Section 202 requires that certain school bus safety standards be published within 15 months of the passage of the 1974 amendments on October 27, 1974. In addition, these statutory provisions remove the otherwise discretionary authority of the NHTSA to establish lead times for compliance under the general rulemaking provisions of the National Traffic and Motor Vehicle Safety Act by specifying an effective date for the amendment of 9 months from the date of publication of this notice (15 U.S.C. § 1392(i) (1) (B)). The proposed amendments upon which this notice is based were published on February 28, 1975 (40 F.R. 8569).

Many comments were received in response to the proposal to require either one rear emergency door or two side emergency doors in the rear half of the bus passenger compartment. Many objected that the proposal provided for too few emergency doors, and requested requirements for additional side doors and roof exits. Some commenters suggested that push-out windows and the "California" rear exit be required. The agency does not discourage the inclusion of additional emergency exits in school buses so long as they comply with the requirements applicable to non-school bus emergency exits. The NHTSA believes that "California" rear window emergency exits may be preferable in certain circumstances and proposes in this issue of the Federal Register to amend this rule to permit the use of the "California" rear window along with a side door emergency exit in place of the rear door emergency exit. In the alternative, it is proposed to allow this option only on rear-engine-powered school buses. Under either proposal the requirements of the standard would not be met by providing two sidedoor emergency exits. In addition, the subject of roof exits is being considered and could be the subject of future rulemaking. However, roof exit requirements cannot be included in this rulemaking action because of the statutorily imposed deadline on promulgation of these amendments.

A number of comments were received opposing the proposed interlock requirement on the ground that it would prevent restarting the engine after the school bus stalls in a dangerous intersection or a railroad crossing and panicky passengers jam the release mechanism. The intent of this requirement is to prevent the initial starting of the bus engine until the doors have been unlocked, by a key, combination, or the operation of a remote switch at the beginning of the day. The deletion of the phrase "or otherwise inoperable" excludes inadvertent jamming of the door release mechanism from the requirement. The word "locked" has been defined for this purpose as not releasable at the door except by a key or combination. It would include doors openable by a remote switch.

Six comments supported the proposal to require an audible alarm when the ignition is on and the release mechanism of any emergency door is not closed. Five of these, however, objected that an alarm at each door in addition to one in the driver's compartment would be unnecessary and unduly costly. The NHTSA does not agree. The purpose of audible alarms at each door is to indicate which release mechanism is not closed. This is especially critical while the vehicle is in motion, as it will serve to warn the passengers in the area of the possibility that an emergency door could open. In addition, it will serve as a deterrent to tampering by children with the emergency door release mechanisms. Therefore, the requirement that an audible alarm be positioned at each emergency door and at the driver's position has been retained.

Objectives were received to the requirement that the magnitude of force required to activate the emergency door release mechanism be not more than 40 pounds. The NHTSA does not consider that the 40 pound force limit is too high in light of the location and access requirements of this standard. If the maximum force level were substantially lowered, there would be a significant likelihood that emergency door release mechanisms would be inadvertently activated by a passenger.

In addition, the NHTSA has noted the possibility of ambiguity with respect to the wording of paragraph S5.4 of the old standard and S5.4.2 of the proposal. The intent of these paragraphs is to specify conditions applicable to the opening of the exit *after* the release mechanism has been activated. Accordingly, the wording of the two paragraphs has been modified to clearly reflect this intent.

Many school districts and manufacturers objected to the parallelepiped clearance requirement for the emergency doors because of the number of seats that would be eliminated and the costs of redesigning van-type school buses to meet the clearance requirements. In addition, many commenters pointed out that the 12-inch aisle in most school buses precludes effective use of a large exit meeting the proposed requirements.

The NHTSA has determined that these arguments have merit. As a result, the proposed parallelepiped requirements have been modified by reducing the height from 48 inches to 45 inches, reducing the depth from 24 to 12 inches for rear exits in buses over 10,000 lbs GVWR, and to 6 inches for rear exits in buses under 10,000 lbs GVWR. For side exits the depth has been eliminated altogether. Additionally, the forward edge of the side door now coincides with a vertical transverse plane tangent to the rearmost point of the adjacent seat, thus permitting simultaneous exiting of two occupants, between the seat backs and over the seat cushion.

In light of the above, 49 CFR § 571.217, *Bus Window Retention and Release*, is amended . . . .

Effective date: October 26, 1976.

(Secs. 103, 112, 119, Pub. L. 89–563, 80 Stat. 718; Sec. 202, Pub. L. 93-492, 88 Stat. 1484 (15 U.S.C. 1392, 1401, 1407); delegation of authority at 49 CFR 1.50.)

Issued on January 22, 1976.

Howard J. Dugoff
Acting Administrator
41 F.R. 3871
January 27, 1976

Effective: October 26, 1976

PREAMBLE TO AMENDMENT TO MOTOR VEHICLE SAFETY STANDARD NO. 217

Bus Window Retention and Release

(Docket No. 75–3; Notice 4)

This notice amends Standard No. 217, *Bus Window Retention and Release*, to modify the emergency exit requirements of the standard in response to a petition for reconsideration of recent amendments and after consideration of comments on the agency's proposal to specify new performance options and labeling for emergency exits.

PETITION FOR RECONSIDERATION OF NOTICE 2

The National Highway Traffic Safety Administration (NHTSA) recently amended Standard No. 217 (49 CFR 571.217) to provide emergency exit requirements for school buses (41 FR 3871, January 27, 1976 (Notice 2)). Section S5.2.3.1 of the standard (as it becomes effective for school buses on October 26, 1976) specifies that a rear emergency door shall be hinged on the right side. Chrysler Corporation has petitioned for reconsideration of this provision, asking that a manufacturer option be provided so that the rear emergency door or doors on van-type school buses may be hinged on the right or left.

The purpose of specifying that the rear emergency door hinge to the right is based on the NHTSA finding that school buses often operate on rural highways that are bordered by drainage ditches, and that a school bus that leaves the highway and rolls over is likely to come to rest in the right-hand ditch on its right side. When a bus comes to a rest on its side, the emergency door on the rear of the bus is easier to operate, particularly by small children, if it is hinged so that its operation is assisted by gravity.

Chrysler pointed out that the rear emergency door on van-type school buses is often used routinely for loading and unloading passengers. For this reason, Chrysler offers a single rear door that hinges at the left side, so that the door swings out of the way to safely accommodate curb-side loading. In the case of larger buses, routine loading and unloading does not occur through the rear emergency door.

The NHTSA agrees with Chrysler that the common practice of curb-side loading through the rear door of van-type school buses justifies a manufacturer option in selecting the side of the door which should be hinged. On balance, the agency considers that the increase in safety for routine curb-side loading through a left-hinged door would outweight any potential loss of safety benefit for emergency evacuation from a van-type bus that comes to rest on its right side. Accordingly, S5.2.3.1 of the standard is appropriately amended. The agency also takes the opportunity to correct an inadvertent reference to emergency "exit" in S5.2.3.2 when the requirements are actually intended to apply only to an emergency "door."

In a matter unrelated to the Chrysler petition, some uncertainty has arisen over the form of S5.4 as it was revised in Notice 2 to become effective October 26, 1976. Also, the division between buses with a GVWR of 10,000 pounds or less and those with a greater GVWR was imperfectly stated in amending S5.4. For this reason, the amendment of S5.4 is republished in the correct form in this notice. No substantive changes are made in this republication of S5.4.

EMERGENCY EXIT AND LABELING PROPOSAL—NOTICE 3

At the time the amendments just discussed were published, the NHTSA published a proposal to clarify certain emergency exit labeling for all buses, and to replace the established option for school bus emergency exits with a new

PART 571; S 217—PRE 15

option (41 FR 3878, January 27, 1976; Notice 3). Comments were received from the Lanai Road Elementary School Parent-Teachers Association, Gillig Brothers (Gillig), Chrysler Corporation, Mr. Allen Braslow, Crown Coach Corporation (Crown), and International Harvester (IH). No comment was received from manufacturers of transit or intercity buses, or from the manufacturers of body-on-chassis school buses. The National Motor Vehicle Safety Council did not comment on this proposal.

With regard to emergency exit labeling, Mr. Braslow suggested two labeling changes intended to assist bus occupants, as well as a requirement for regular testing of emergency exits in buses in highway service. While the latter suggestion lies beyond the authority of the agency under the National Traffic and Motor Vehicle Safety Act (15 U.S.C. § 1391, et seq.), the agency will consider for future action the suggestion to label all bus exits in the same manner as school bus exits, as well as the suggestion to develop a universal emergency exit insignia with diagramatic instructions. For the moment, the agency is limited by the extent of its proposal, and accordingly, makes final the changes as proposed.

Standard No. 217 requires (effective October 26, 1976) school buses to provide either a rear emergency door or two side emergency doors in satisfaction of the emergency exit requirements. In Notice 3, the agency proposed to modify this option to require either provision of a rear emergency door or, at the option of the manufacturer, provision of a left-side emergency door and a "California rear window" exit at the rear of the bus. This type of rear window exit provides a large (16 by 48 inch) opening which is more easily utilized than a side emergency door if a bus has rolled onto its side. In the alternative, the agency proposed that the option to use a rear window exit only be allowed in rear-engine buses.

The two manufacturers of transit-type school buses supported the new option, but objected to the alternative proposal that would limit use of the option to rear-engine buses. Both Gillig and Crown build mid-engine school buses with essentially the same configuration as rear-engine buses and consider the rear window exit equally useful in these buses. The agency has considered the mid-engine design and agrees with the argument made by Crown and Gillig. Accordingly, the agency amends the standard as proposed to apply the option to all school buses. Crown Coach pointed out that the NHTSA proposal to limit rear-window-exit release mechanisms to a single release would necessitate a change in existing hardware. The NHTSA has investigated the available hardware (consisting in all cases of two release mechanisms that are located within 36 inches of each other) and concludes that the only significant safety hazard in some of the designs is that some require simultaneous operation for release. For this reason, the agency will allow not more than two release mechanisms, provided that the two mechanisms do not have to operate simultaneously to effect release. If new designs present a problem of any nature, further rulemaking will be undertaken.

In accordance with recently enunciated Department of Transportation policy encouraging adequate analysis of the consequences of regulatory action (41 FR 16201, April 16, 1976), the agency herewith summarizes its evaluation of the economic and other consequences of this proposal on the public and private sectors, including possible loss of safety benefits. The option to hinge some rear emergency doors on the right or left, and the option to use a "California rear window" do not involve additional expenditures. The agency estimates that these additional exit arrangements will not significantly reduce the level of safety provided in the affected bus categories. The new requirements for more specific operating instructions for school bus emergency exits are calculated to involve annual costs of about $67,000. Although the agency is unable to quantify the benefit of clearer exit labeling, it is estimated that better instructions will serve to reduce the possibility of death and injury involved in an attempt to use the emergency exits. Therefore, the agency concludes that the amendments should issue as set forth in this notice.

For the benefit of interested persons, it is noted that Docket 75-6 concerning labeling of bus emergency exits is related to this rulemaking.

In consideration of the foregoing, Standard No. 217 (49 CFR 571.217) as it is amended to become effective for school buses on October 26, 1976, is revised. . . .

Effective date: October 26, 1976. The effective date of the amendments numbered 1, 2, 3 and 5 is established as 9 months after the date of issuance of the amendments on which they are based, as required by the Motor Vehicle and Schoolbus Safety Amendments of 1974, Pub. L. 93–492, section 202 (15 U.S.C. 1397(i)(1)(A)). The effective date of the amendment numbered 4 is also established as October 26, 1976, although a manufacturer can meet the requirements at an earlier date if the manufacturer so chooses.

(Sec. 103, 119, Pub. L. 89–563, 80 Stat. 718 (15 U.S.C. 1392, 1407); Sec. 202, Pub. L. 93–492, 88 Stat. 1470 (15 U.S.C. 1392); delegation of authority at 49 CFR 1.50.)

Issued on May 25, 1976.

James B. Gregory
Administrator

41 F.R. 22356
June 3, 1976

PREAMBLE TO AN AMENDMENT TO
FEDERAL MOTOR VEHICLE SAFETY STANDARD NO. 217

Bus Window Retention and Release
(Docket No. 75-03; Notice 7)

ACTION: Final rule.

SUMMARY: This notice makes permanent an interim final rule that modified the agency's school bus emergency exit standard. The interim final rule, which was issued in February 1979, was implemented immediately to increase the availability of passenger vans for use as small school buses at reasonable costs. The interim rule slightly altered several emergency exit requirements in a manner that made it easier to mass produce small buses without significantly affecting the level of safety achieved by those vehicles. Concurrent with the issuance of the interim final rule, the agency solicited comments on the amendments to the standard. This notice responds to the comments and makes the interim rule permanent.

EFFECTIVE DATE: Since this notice makes permanent an existing interim final rule, it is effective immediately.

SUPPLEMENTARY INFORMATION: On February 8, 1979, the agency published an interim final rule and a proposal (44 F.R. 7961) to modify the school bus emergency exit safety standard, Standard No. 217, *Bus Window Retention and Release*. In that notice, the agency made effective immediately some modifications to the school bus emergency exit standard to increase the supply of reasonably priced vehicles suitable for school bus conversion. Among the changes implemented by the interim final rule were a slight decrease in the size of rear emergency exits for vehicles (typically passenger vans) with gross vehicle weight ratings (GVWR) less than 10,000 pounds, and increased flexibility in the location requirements for release mechanisms on the emergency exits of small school buses. The agency concluded at the time the interim rule was issued that the level of safety achieved by small buses would not be diminished by these changes and that the changes would allow more small buses to be mass produced, thereby lowering their prices. The agency also asked in the interim final rule for comments on the advisability of these changes.

In response to the agency's request, Ford, Chrysler, the Center for Auto Safety, and the California Highway Patrol (CHP) submitted comments. The two manufacturers, Ford and Chrysler, both supported the agency's action. The Center and the CHP both opposed the action.

The Center and the CHP both argued that the rear emergency exit in small school buses (passenger vans which have GVWR's less than 10,000 pounds and are used as school buses) should not be reduced in size. The Center stated that the exit should be broad enough for two students to exit simultaneously in case of an emergency. The CHP stressed that reducing the size of the exit would make it too small to permit the exiting of children in wheelchairs.

With respect to the argument that the size of the rear exit should allow room to exit students two abreast, the agency stated in the proposal that this argument, while valid for larger school buses, is not meritorious for school vehicles with GVWR's less than 10,000 pounds. Larger school buses frequently transport 60 or more school children. Accordingly, rapid evacuation of those vehicles in an emergency requires that the students be able to exit two abreast. In order to accomplish this, the agency has required that some space be provided behind the rearmost seat in these buses so that students exiting through the narrow center aisles will have room at the exits to get out two abreast.

In small school buses where the number of students carried frequently is 16 or less, the need for exiting two abreast to achieve rapid evacuation is significantly reduced. In recognition of this factor, the agency has never required bus manufacturers to provide space behind the rear seat of small buses that would allow students to exit two abreast. As a result, the rear seats of small buses are frequently quite near or are against the rear bus wall. Students exiting down a bus aisle, which is normally around 12 inches in width, reach an exit where no space is provided to exit two abreast. Accordingly, any requirement that an exit in small buses be large enough to facilitate exiting two abreast would not accomplish that goal. Small bus manufacturers would need to redesign their bus seat plans in some fashion to provide space behind the rear seat in order to allow exiting two abreast. Such a redesign would significantly decrease the available seating in small buses. Given the fact that evacuating small buses has not been a safety problem, the agency concludes that the cost resulting from the reduced vehicle seating that would be required to accomplish the Center's objectives would far outweigh the benefits. Accordingly, the agency concludes that a broader rear exit is not needed in small school buses.

The CHP objected to the same requirement stating that the new exit door would be too narrow for wheelchairs. The CHP further stated that California has always required wider exits so that wheelchairs can be used in the vehicles.

The agency's new exit requirement is a minimum size requirement for standard school buses. In special instances in which larger exits are desired, such as in buses for carrying the handicapped, the States may require that their buses have such exits. The agency deems that approach to be preferable to its requiring larger exits in all vehicles. The situation with respect to rear door size is analogous to that involving seat back height. The agency requires a minimum seat back height. New York mandates a seat back height greater than the Federal specification. The NHTSA has no objection to the New York requirement and will not object to requirements by other States for wider rear emergency exits. The agency also notes that buses designed for the handicapped constitute a small portion of all buses and usually are equipped with special doors and larger aisles.

The Center also objected to the agency's interpretation that the parallelipiped device used for measuring rear door size could be lifted up to 1-inch to overcome small protrusions near the floor. The agency issued an interpretation permitting this at the time of the implementation of the standard. This interpretation simply reflects real-world conditions. Many doors in vehicles have small door sills or other minor protrusions that sometimes serve necessary functions in the proper operation of the door. These minor protrusions play no significant role in the ability of students to exit from a vehicle in an emergency. Therefore, the agency will not reconsider its interpretation.

The Center objected to the agency's removal of exit release mechanism location and force application requirements for small school buses. The Center agreed that the existing requirements are more appropriate for larger buses, but it insisted that the agency should develop another set of location requirements for smaller buses instead of abandoning the requirements entirely.

The agency is sympathetic to the Center's concerns about this issue. The location of the release mechanism for small school buses in an easily accessible location is important for the rapid evacuation of these vehicles in an emergency. However, the mere setting of location requirements would not ensure that the release mechanisms would be accessible. Due to the limited space in the rear of small buses and the variability of design in those areas, the agency could not readily specify a location which would provide the necessary accessibility. The agency believes that allowing manufacturers the option of locating the release mechanism in any easily accessible location on or near the exit will be more beneficial to achieving the intended safety results than any rigid inflexible location requirement. NHTSA anticipates that product liability concerns and the agency's authority to declare inaccessible release mechanisms to be safety-related defects will suffice to induce the manufacturers to select accessible locations. The agency will closely monitor the location and accessibility of the release mechanisms and, if necessary, use both its defects and rulemaking authority to take corrective action.

Finally, the Center objected to the fact that the agency permitted pull-type release mechanisms.

The Center stated that release mechanism standardization is helpful in assuring the safe evacuation of vehicles.

While the agency agrees that standardization has value in this instance, there are competing ways for achieving standardization in the case of small school buses. One way is to require that small school buses have releases that operate with an upward motion as in larger school buses. Another way is to permit small school buses (which, as noted before, are passenger vans) to have the same pull-type releases that are found in other vans and some cars. The agency doesn't believe that either basis for standardization is clearly superior from a safety standpoint to the other. Further, permitting the use of the pull-type releases will enable the manufacturers to achieve cost savings. Accordingly, the agency declines to adopt the Center's recommendation.

Since this notice makes permanent an existing amendment, it is effective immediately. The agency has reviewed the amendment in accordance with E.O. 12291 and concludes that the rule is not significant under the Department of Transportation's regulatory procedures. In fact, by permitting these changes, more buses can be mass produced, which may result in a small decrease in the cost of complying with the standard. Since the economic impact of this rule is minimal, a regulatory evaluation is not required for this amendment.

The agency has also considered the effect of this rule in relation to the Regulatory Flexibility Act and certifies that it would not have a significant economic impact on a substantial number of small entities. The only economic impact might be a reduction in bus prices. There would similarly be no significant impact on a substantial number of small government jurisdictions and small organizations.

Finally the agency has analyzed this rule for purposes of the National Environmental Policy Act and has determined that it would have no significant impact on the human environment.

Issued on February 10, 1982.

Diane K. Steed
Acting Administrator

47 F.R. 7255
February 18, 1982

MOTOR VEHICLE SAFETY STANDARD NO. 217

Bus Window Retention and Release

S1. Scope. This standard establishes requirements for the retention of windows other than windshields in buses, and establishes operating forces, opening dimensions, and markings for push-out bus windows and other emergency exits.

S2. Purpose. The purpose of this standard is to minimize the likelihood of occupants being thrown from the bus and to provide a means of readily accessible emergency egress.

S3. Application. This standard applies to buses, except buses manufactured for the purpose of transporting persons under physical restraint.

S4. Definitions.

"Push-out window" means a vehicle window designed to open outward to provide for emergency egress.

"Adjacent seat" means a designated seating position located so that some portion of its occupant space is not more than 10 inches from an emergency exit, for a distance of at least 15 inches measured horizontally and parallel to the exit.

"Occupant space" means the space directly above the seat and footwell, bounded vertically by the ceiling and horizontally by the normally positioned seat back and the nearest obstruction of occupant motion in the direction the seat faces.

S5. Requirements.

S5.1 Window Retention. Except as provided in S5.1.2, each piece of window glazing and each surrounding window frame, when tested in accordance with the procedure in S5.1.1 under the conditions of S6.1 through S6.3, shall be retained by its surrounding structure in a manner that prevents the formation of any opening large enough to admit the passage of a 4-inch diameter sphere under a force, including the weight of the sphere, of 5 pounds until any one of the following events occurs:

(a) A force of 1200 pounds is reached.

(b) At least 80% of the glazing thickness has developed cracks running from the load contact region to the periphery at two or more points, or shattering of the glazing occurs.

(c) The inner surface of the glazing at the center of force application has moved relative to the window frame, along a line perpendicular to the undisturbed inner surface, a distance equal to one-half of the square root of the minimum surface dimension measured through the center of the area of the entire sheet of window glazing.

S5.1.1 An increasing force shall be applied to the window glazing through the head form specified in Figure 4, outward and perpendicular to the undisturbed inside surface at the center of the area of each sheet of window glazing, with a head form travel of 2 inches per minute.

S5.1.2 The requirements of this standard do not apply to a window whose minimum surface dimension measured through the center of its area is less than 8 inches.

S5.2 Provision of Emergency Exits. Buses other than school buses shall provide unobstructed openings for emergency exit which collectively amount, in total square inches, to at least 67 times the number of designated seating positions on the bus. At least 40 percent of the total required area of unobstructed openings, computed in the above manner, shall be provided on each side of a bus. However, in determining the total unobstructed openings provided by a bus, no emergency exit, regardless of its area, shall be credited with more than 536 square inches of the total area requirement. School buses shall provide openings for emergency exits that conform to S5.2.3.

S5.2.1 Buses with GVWR of more than 10,000 pounds. Except as provided in S5.2.1.1, buses with a GVWR of more than 10,000 pounds shall meet the unobstructed openings requirements by providing side exits and at least one rear exit that conforms to S5.3 through S5.5. The rear exit shall meet the requirements when the bus is upright and when the bus is overturned on either side, with the occupant standing facing the exit. When the bus configuration precludes installation of an accessible rear exit, a roof exit that meets the requirements of S5.3 through S5.5 when the bus is overturned on either side, with the occupant standing facing the exit, shall be provided in the rear half of the bus.

S5.2.1.1 A bus with GVWR of more than 10,000 pounds may satisfy the unobstructed openings requirement by providing at least one side door for each three passenger seating positions in the vehicle.

S5.2.2 Buses with a GVWR of 10,000 pounds or less. Buses with a GVWR of 10,000 pounds or less may meet the unobstructed openings requirement by providing:

(a) Devices that meet the requirements of S5.3 through S5.5 without using remote controls or central power systems;

(b) Windows that can be opened manually to a position that provides an opening large enough to admit unobstructed passage, keeping a major axis horizontal at all times, of an ellipsoid generated by rotating about its minor axis an ellipse having a major axis of 20 inches and a minor axis of 13 inches; or

(c) Doors.

S5.2.3 School buses.

S5.2.3.1 Each school bus shall comply with either one of the following minimum emergency exit provisions, chosen at the option of the manufacturer:

(a) One rear emergency door that opens outward and is hinged on the right side (either side in the case of a bus with a GVWR of 10,000 pounds or less); or

(b) One emergency door on the vehicle's left side that is in the rear half of the bus passenger compartment and is hinged on its forward side, and a push-out rear window that provides a minimum opening clearance 16 inches high and 48 inches wide. This window shall be releasable by operation of not more than two mechanisms which are located in the high force access region as shown in Figure 3C, and which do not have to be operated simultaneously. Release and opening of the window shall require force applications, not to exceed 40 pounds, in the directions specified in S5.3.2.

S5.2.3.2 The engine starting system of a school bus shall not operate if any emergency door is locked from either inside or outside the bus. For purposes of this requirement, "locked" means that the release mechanism cannot be activated by a person at the door without a special device such as a key or special information such as a combination.

S5.3 Emergency exit release.

S5.3.1 Each push-out window or other emergency exit not required by S5.2.3 shall be releasable by operating one or two mechanisms located within the regions specified in Figure 1, Figure 2, or Figure 3. The lower edge of the region in Figure 1, and Region B in Figure 2, shall be located 5 inches above the adjacent seat, or 2 inches above the armrest, if any, whichever is higher.

S5.3.2 When tested under the conditions of S6, both before and after the window retention test required by S5.1, each emergency exit not required by S5.2.3 shall allow manual release of the exit by a single occupant using force applications each of which conforms, at the option of the manufacturer, either to (a) or (b). The release mechanism or mechanisms shall require for release one or two force applications, at least one of which differs by 90 to 180° from the direction of the initial push-out motion of the emergency exit (outward and perpendicular to the exit surface).

(a) Low-force application.

Location: As shown in Figure 1 or Figure 3.

Type of Motion: Rotary or straight.

Magnitude: Nor more than 20 pounds.

(b) High force application.

Location: As shown in Figure 2 or Figure 3.

Type of Motion: Straight, perpendicular to the undisturbed exit surface.

Magnitude: Not more than 60 pounds.

S5.3.3 When tested under the conditions of S6., both before and after the window retention test required by S5.1, each school bus emergency door shall allow manual release of the door by a single person, from both inside and outside the bus passenger compartment, using a force application that conforms to paragraphs (a) through (c) [except a school bus with a GVWR of 10,000 pounds or less does not have to conform to paragraph (a). (47 F.R. 7255—February 18, 1982. Effective: February 18, 1982.)] Each release mechanism shall operate without the use of remote controls or tools, and notwithstanding any failure of the vehicle's power system. When the release mechanism is not in the closed position and the vehicle ignition is in the "on" position, a continuous warning sound shall be audible at the driver's seating position and in the vicinity of the emergency door having the unclosed mechanism.

(a) Location: Within the high force access region shown in Figure 3A for a side emergency door, and in Figure 3D for a rear emergency door.

(b) Type of motion: Upward from inside the bus; at the discretion of the manufacturer from outside the bus. [Buses with a GVWR of 10,000 pounds or less shall provide interior release mechanisms that operate by either an upward or pull-type motion. The pull-type motion shall be used only when the release mechanism is recessed in such a manner that the handle, lever, or other activating device does not protrude beyond the rim of the recessed receptacle. (47 F.R. 7255—February 18, 1982. Effective: February 18, 1982)]

(c) Magnitude of force: Not more than 40 pounds.

The present S5.4 is renumbered S5.4.1, and the phrase "Each push-out window or other emergency exit shall, after the release mechanism has been operated," is replaced by the phrase "After the release mechanism has been operated, each push-out window or other emergency exit not required by S5.2.3," at the beginning of the paragraph.

S5.4 Emergency exit extension.

S5.4.1 After the release mechanism has been operated, each push-out window or other emer-

gency exit not required by S5.2.3 shall, under the conditions of S6, before and after the window retention test required by S5.1, using the reach distances and corresponding force levels specified in S5.3.2, be manually extendable by a single occupant to a position that provides an opening large enough to admit unobstructed passage, keeping a major axis horizontal at all times, of an ellipsoid generated by rotating about its minor axis an ellipse having a major axis of 20 inches and a minor axis of 13 inches.

S5.4.2 School bus emergency exit extension.

S5.4.2.1 School bus with a GVWR of more than 10,000 pounds. After the release mechanism has been operated, the emergency door of a school bus with a GVWR of more than 10,000 pounds shall, under the conditions of S6, before and after the window retention test required by S5.1, using the force levels specified in S5.3.3, be manually extendable by a single person to a position that permits—

(a) In the case of rear emergency door, an opening large enough to permit unobstructed passage of a rectangular parallelepiped 45 inches high, 24 inches wide, and 12 inches deep, keeping the 45-inch dimension vertical, the 24-inch dimension parallel to the opening, and the lower surface in contact with the floor of the bus at all times; and

(b) In the case of a side emergency door, an opening at least 45 inches high and 24 inches wide. A vertical transverse plane tangent to the rear-most point of a seat back shall pass through the forward edge of a side emergency door.

S5.4.2.1 School Buses Less Than 10,000 Pounds or Less. A school bus with a GVWR of 10,000 pounds or less shall conform to all the provisions of S5.4.2 except that the parallelepiped dimension for the opening of the rear emergency door or doors shall be 45 inches high, 22 inches wide, and 6 inches deep.

S5.5 Emergency exit identification.

S5.5.1 In buses other than school buses, except for windows serving as emergency exits in accordance with S5.2.2(b) and doors in buses with a GVWR of 10,000 pounds or less, each emergency door shall have the designation "Emergency Door" or "Emergency Exit" and each push-out window or other emergency exit shall have the designation "Emergency Exit" followed by concise operating instructions describing each motion necessary to unlatch and open the exit, located within 6 inches of the release mechanism.

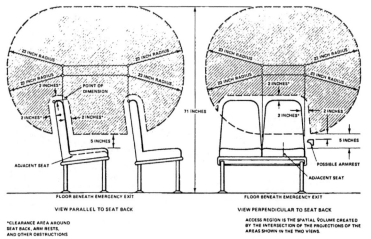

VIEW PARALLEL TO SEAT BACK

VIEW PERPENDICULAR TO SEAT BACK

*CLEARANCE AREA AROUND
SEAT BACK, ARM RESTS,
AND OTHER OBSTRUCTIONS

ACCESS REGION IS THE SPATIAL VOLUME CREATED
BY THE INTERSECTION OF THE PROJECTIONS OF THE
AREAS SHOWN IN THE TWO VIEWS.

FIGURE 1 LOW-FORCE ACCESS REGION FOR EMERGENCY EXITS HAVING ADJACENT SEATS

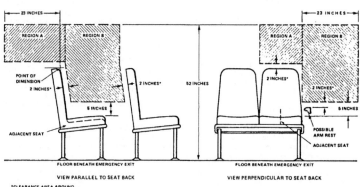

VIEW PARALLEL TO SEAT BACK

VIEW PERPENDICULAR TO SEAT BACK

*CLEARANCE AREA AROUND
SEAT BACK, ARM RESTS,
AND OTHER OBSTRUCTIONS

FIGURE 2 HIGH-FORCE ACCESS REGIONS FOR EMERGENCY EXITS HAVING ADJACENT SEATS

PART 571; S 217-4

LOW AND HIGH-FORCE ACCESS REGIONS FOR EMERGENCY EXITS WITHOUT ADJACENT SEATS

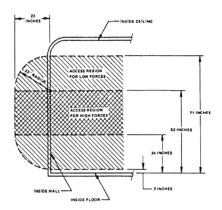

3A. SIDE EMERGENCY EXIT

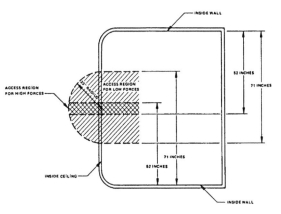

3B. ROOF EMERGENCY EXIT

PART 571; S 217-5

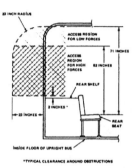

23 INCH RADIUS

ACCESS REGION
FOR LOW FORCES

71 INCHES

ACCESS
REGION
FOR HIGH
FORCES

52 INCHES

REAR SHELF

2 INCHES *

23 INCHES

REAR
SEAT

INSIDE FLOOR OF UPRIGHT BUS

*TYPICAL CLEARANCE AROUND OBSTRUCTIONS

3C. REAR EMERGENCY EXIT WITH REAR OBSTRUCTION

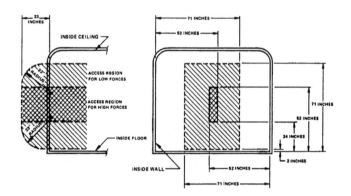

23 INCHES

INSIDE CEILING

23" RADIUS

ACCESS REGION
FOR LOW FORCES

ACCESS REGION
FOR HIGH FORCES

23" RADIUS

INSIDE FLOOR

INSIDE WALL

71 INCHES

52 INCHES

71 INCHES

52 INCHES

24 INCHES

2 INCHES

52 INCHES

71 INCHES

3D. REAR EMERGENCY EXIT WITHOUT REAR OBSTRUCTION

PART 571; S 217-6

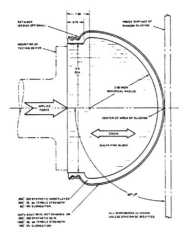

FIGURE 4 HEAD FORM

Examples: (1) Lift to Unlatch
              Push to Open

         (2) Lift Handle and
              Push out to Open

When a release mechanism is not located within an occupant space of an adjacent seat, a label meeting the requirements of S5.5.2 that indicates the location of the nearest release mechanism shall be placed within the occupant space.

Example: Emergency exit instructions located next to seat ahead.

S5.5.2 In buses other than school buses, except as provided in S5.5.2.1, each marking shall be legible, when the only source of light is the normal night-time illumination of the bus interior, to occupants having corrected visual acuity of 20/40 (Snellen ratio) seated in the adjacent seat, seated in the seat directly adjoining the adjacent seat, and standing in the aisle location that is closest to that adjacent seat. The marking shall be legible from each of these locations when the other two corresponding locations are occupied.

S5.5.2.1 If the exit has no adjacent seat, the marking must meet the legibility requirements of S5.5.2 for occupants standing in the aisle location nearest to the emergency exit, except for a roof exit, which must meet the legibility requirements for occupants positioned with their backs against the floor opposite the roof exit.

S5.5.3 School Bus. Each school bus emergency exit provided in accordance with S5.2.3.1 shall have the designation "Emergency Door" or "Emergency Exit," as appropriate, in letters at least 2 inches high, of a color that contrasts with its background, located at the top of or directly above the emergency exit on both the inside and outside surfaces of the bus. Concise operating instructions describing the motions necessary to unlatch and open the emergency exit, in letters at least three-eights of an inch high, of a color that contrasts with its background, shall be located within 6 inches of the release mechanism on the inside surface of the bus.

Example: (1) Lift to Unlatch
              Push to Open

         (2) Lift Handle
              Push Out to Open.

· **S6. Test** conditions.

S6.1 The vehicle is on a flat, horizontal surface.

S6.2 The inside of the vehicle and the outside environment are kept at any temperature from 70° to 85° Fahrenheit for 4 hours immediately preceding the tests, and during the tests.

S6.3 For the window retention test, windows are installed, closed, and latched (where latches are provided) in the condition intended for normal bus operation.

S6.4 For the emergency exit release and extension tests, windows are installed as in S6.3, seats, armrests, and interior objects near the windows are installed as for normal use, and seats are in the upright position.

37 F.R. 9394
May 10, 1972

PART 571; S 217-7-8

PREAMBLE TO MOTOR VEHICLE SAFETY STANDARD NO. 218

Motorcycle Helmets

(Docket No. 72–6; Notice 2)

The purpose of this amendment to Part 571 of Title 49, Code of Federal Regulations, is to add a new Motor Vehicle Safety Standard No. 218, Motorcycle Helmets, 49 CFR § 571.218, that establishes minimum performance requirements for motorcycle helmets manufactured for use by motorcyclists and other motor vehicle users.

A notice of proposed rulemaking on this subject was published on May 19, 1972 (37 F.R. 10097). The comments received in response to the notice have been carefully considered in this issuance of a final rule.

In the previous notice, the NHTSA proposed that, effective September 1, 1974, the performance levels for the impact attenuation requirements be upgraded to that of the Head Injury Criterion (HIC) required by Motor Vehicle Safety Standard No. 208. A number of comments on this subject sought to defer a final determination until further research and additional tests could be conducted. The agency has carefully reviewed the issues raised by these comments and has determined that technical data presently being generated on this matter by several investigations should be considered in upgrading the impact attenuation requirements. Accordingly, a decision on the upgrading will be deferred until after this research has been completed and the results evaluated, and after any appropriate data have been reviewed.

Comments to the docket on the initial impact attenuation requirement ranged from abolishing the time duration criteria of 2.0 milliseconds and 4.0 milliseconds at the 200g and 150g levels, respectively, to increasing these criteria to 2.8 milliseconds at the 200g level and 5.6 milliseconds at the 150g level. One approach taken in regard to this requirement contends that the available test data are insufficient for quantifying time

limits for the relatively short duration accelerations which are involved in helmet testing. Several comments questioned the validity of the proposed time duration limits, since these limits were based on the optional swing-away (as opposed to fixed anvil) test of the American National Standards Institute (ANSI) Standard Z90.1–1966, which was omitted from the most recent issues of the Z90.1 Standard (1971 and 1973) and was not contained in the proposed motorcycle helmet standard. An additional comment points out that helmets designed to meet higher energy impacts than the initial impact attenuation requirement occasionally have difficulty meeting a 2.0 millisecond requirement at the 200g level.

A review of available biomechanical data indicates that the head impact exposure allowed by the 2.0 and 4.0 millisecond limits at the 200g and 150g levels, respectively, is greater than that allowed by other measures of head injury potential. It is the agency's view, moreover, that the best evidence indicates that an increase in the time duration criteria would permit a substantial reduction in the protection provided to the helmet wearer. Since the comments to the docket did not provide any new data or sufficiently compelling arguments which would justify relaxing the proposed limits for tolerable head impact exposure, the 2.0 and 4.0 millisecond criteria are retained as part of the initial impact attenuation criteria.

In response to comments recommending that the allowable weight of the supporting assembly for the impact attenuation drop test be changed to 20% instead of the proposed 10% of the weight of the drop assembly, the NHTSA has determined that such a change would enable more durable testing equipment to be used without any significant effect on test results. Accordingly, this weight limitation has been raised to 20%.

Several comments expressed concern that the proposed 0.04-inch indentation limit included under the penetration test would create problems of measurement. The agency has determined that the intent of this 0.04-inch indentation limit is sufficiently accomplished by the requirement that the striker not contact the surface of the test headform, and the 0.04-inch indentation limit is therefore deleted from the final rule. Further, in consideration of the need to readily detect any contact by the striker, the agency has determined that the contactable surfaces of the penetration test headforms should be constructed of a metal or metallic alloy which will insure detection. Several minor changes in the test conditions for the penetration test have also been made, without altering the substance of those conditions.

A number of comments recommended that where the retention system consists of components which can be independently fastened without securing the complete assembly, such components should not have to individually meet the retention test requirements. Since helmets have a tendency to be thrown off by a crash and motorcyclists sometimes only partially fasten the retention system where such an option exists, the agency has concluded that retention components as well as the entire assembly should meet the test requirements in every fastening mode as specified in the notice of proposed rulemaking.

A number of comments requested that the 105° minimum peripheral vision clearance to each side of the midsagittal plane be increased to 120°. The 105° minimum requirement was proposed because it satisfies a demand by the public for the availability of some helmets which provide added protection to the temporal areas in exchange for a minimal reduction in peripheral vision capability without compromising the safe limits of peripheral vision clearance. A review of available field-of-vision studies and the lack of any evidence to the contrary indicate that 105° minimum clearance to each side of the midsagittal plane provides ample peripheral vision capability. Since the requests for increasing the

minimum clearance to 120° were not accompanied by any supporting data or arguments, the agency has concluded that the standard should allow the additional protection which the 105° minimum clearance would permit and, accordingly, this requirement is retained.

With respect to providing important safety information in the form of labeling, one comment recommended that, due to possible label deterioration, both the manufacturer's identification and the helmet model designation should be permanently marked by etching, branding, stamping, embossing, or molding on the exterior of the helmet shell or on a permanently attached component so as to be visible when the helmet is in use. The NHTSA has determined that the practical effect of this recommendation is accomplished by requiring each helmet to be permanently and legibly labeled. The method to be used to permanently and legibly affix a label for each helmet is therefore left to the discretion of the manufacturer. However, in order that there may be some external, visual evidence of conformity to the standard, the labeling requirement has been further modified to require manufacturer certification in the form of the DOT symbol to appear in permanent form on the exterior of the helmet shell.

One comment recommended that the preliminary test procedures include the application of a 10-pound static test load to the apex of a helmet after it is placed on the reference headform and before the "test line" is drawn to insure that the reference marking will be relatively uniform, thus reducing variances in test results of identical helmets. The agency concurs in this recommendation and it has been included in the standard.

A number of comments objected to the location of the test line. With respect to the proposed requirement that the test line on the anterior portion of a helmet coincide with the reference plane of its corresponding reference headform, it was pointed out that the helmet's brow area would have to be excessively thick in order to meet the impact attenuation criteria at any point less than approximately 1 inch from the brow opening. The data indicate that this objection is valid, and the location of the anterior test line has been modified by placing it 1 inch above and parallel to the reference plane.

A number of comments objected to the proposed requirement that the test line on the posterior portion of a helmet coincide with the basic plane of its corresponding reference headform. The principal objection expressed concern that, by extending the posterior test line to the basic plane, the resulting increase in the posterior surface of a helmet could cause the helmet to impact the wearer's neck where rearward rotation of the head occurs, thereby increasing the potential for injury in certain cases. After further consideration of this aspect of helmet safety, the agency has determined that the location of the test line on the posterior portion of a helmet should be modified by placing it 1 inch below and parallel to the reference plane.

Several comments questioned the sufficiency of the anatomical dimensions and diagrams provided for the reference headforms in the Appendix of the notice of proposed rulemaking. Of these comments, two proposed adopting the dimensional specifications of the existing ANSI Z90.1 headform, while a third recommended the inclusion of an additional reference headform to accommodate their smallest child helmet. The agency has concluded that, in order to promote greater uniformity in testing and more repeatable results, one of the reference headforms should have the dimensional specifications of the readily available Z90.1 headform, the others being scaled proportionally, and that a reference headform for smaller child helmets should be added. Accordingly, the Appendix has been revised to reflect these changes.

Effective date: March 1, 1974.

In consideration of the foregoing, a new Motor Vehicle Safety Standard No. 218, Motorcycle Helmets, is added as § 571.218 of Title 49, Code of Federal Regulations, as set forth below.

(Secs. 103, 112, 119, Public Law 89–563, 80 Stat. 718, 15 U.S.C. 1392, 1401, 1407; delegation of authority at 49 CFR 1.51.)

Issued on August 9, 1973.

James B. Gregory
Administrator

38 F.R. 22390
August 20, 1973

PREAMBLE TO AMENDMENT TO MOTOR VEHICLE SAFETY STANDARD NO. 218

Motorcycle Helmets

(Docket No. 72-6; Notice 3)

The purpose of this notice is to respond to petitions for reconsideration and petitions for rulemaking to amend Motor Vehicle Safety Standard No. 218, *Motorcycle Helmets* (49 CFR 571.218).

Standard No. 218, published on August 20, 1973, (38 F.R. 22390), established minimum performance requirements for helmets manufactured for use by motorcyclists and other motor vehicle users. Pursuant to 49 CFR 553.35, petitions for reconsideration were filed by the Safety Helmet Council of America (SHCA) and Lear-Siegler, Inc., Bon-Aire Division. Additionally, pursuant to 49 CFR 553.31, petitions to amend the standard were filed by the Z-90 Committee of the American National Standards Institute, Midwest Plastics Corp., Approved Engineering Test Laboratories, Bell-Toptex, Inc., Premier Seat and Accessory Co., Safetech Co., Sterling Products Co., Inc., Lanco Division of Roper Corp., American Safety Equipment Corp., and Electofilm, Inc.

In response to information contained in both the petitions for reconsideration and the petitions for rulemaking, the standard is being amended in some minor respects, and its effectiveness is temporarily suspended for helmets that must be tested on headform sizes A, B, and D. Requested changes in other requirements of the standard are denied.

1. Effective date. The NHTSA received comments from Royal Industries/Grant Division, Jefferson Helmets, Inc., and Rebcor, Inc., urging that the March 1, 1974, effective date be reaffirmed and stating that they either have already produced or could produce helmets by that date which meet the standard's requirements. The NHTSA commends these manufacturers for

their outstanding efforts and their positive attitude toward producing safer products.

The parties who submitted petitions, however, all requested some postponement of the standard's effective date. The postponement requests ranged from an indefinite extension to a delay until the manufacturers are able to test helmets to the required headforms, and were sought on the following three grounds: (1) additional time in order to obtain headforms required for reference marking and testing; (2) alleged inadequacy of the headform diagrams provided in the final rule; and (3) inability to find a supplier or forge for the K-1A magnesium alloy required for the impact attenuation test headforms.

As explained in the preamble to the standard, the headforms provided in the Appendix of the notice of proposed rulemaking (May 19, 1972, 37 F.R. 10097), were changed by the agency in order to utilize the readily available Z90.1 headform and to promote greater uniformity in testing and more repeatable results. In view of the fact that the size C headform of the final rule is identical to the Z90.1 headform, is readily available in test laboratories, is used for several ongoing certification programs, and that the other headforms are scaled proportionally, the NHTSA anticipated that competition would motivate both the manufacturers and the test laboratories to take the initiative either to obtain or to produce the other required headforms. It now appears that the problem of finding a supplier or forge for the K-1A magnesium alloy required for the A, B, and D impact attentuation test headforms is substantial enough to justify the requests for a postponement of the standard's effective date for helmets that must be tested on headform sizes A, B, and D.

Because the NHTSA determined that the size C headform would be identical to the Z90.1 headform, the low resonance magnesium alloy (K–1A) specified for making the Z90.1 headform also was specified for headforms required by the standard. Statements that it might be difficult to find suppliers or forges for the material were first made in the petitions on the standard. The NHTSA has determined that other low-resonance magnesium alloys can be substituted for the K–1A type without causing significant variances in the results of any of the helmet tests, so that manufacturers can determine compliance without undue cost penalties even where the K–1A alloy is in short supply. Accordingly, the K–1A alloy is retained as the basic headform material for the standard.

In view of the foregoing considerations with particular emphasis on the fact that testing services through commercial testing laboratories have been readily available for several years for the ANSI Z90.1 Standard headform, which is the size C headform of the standard, the requests for postponing the standard's effective date are denied with respect to helmets that fit headform C.

The petitions for a postponement of the effective date are granted, however, with respect to helmets that must be tested on headforms A, B, and D. A sentence is being added to the Application section of the standard, excepting from its coverage helmets that must be tested on these headform sizes. The second sentence in S6.1.1 of the standard relating to the selection of a reference headform to be used for reference marking should be disregarded until the standard is made effective for helmets that must be tested on headform sizes A, B, and D. To facilitate both the production and availability of headforms, the NHTSA has contracted with the Snell Memorial Foundation to monitor the preparation of detail drawings and model headforms consistent with the requirements of the standard. The drawings and headforms will be included in the docket for public examination upon their completion. A review of the leadtime information provided by the comments to the docket indicates that approximately 8 months of manufacturer leadtime will be needed after the detail dimensional drawings of the A, B, and D head-

forms become available. When the drawings are available, notice to that effect will be published in the Federal Register. The planned effective date for the A, B, and D-size helmets is 8 months from the date of the publication of that notice.

2. Time duration criteria for impact attenuation test. Petitions on the impact attenuation test time duration criteria of paragraphs S5.1(b) ranged from eliminating the time duration criteria of 2.0 milliseconds and 4.0 milliseconds at the 200g and 150g levels, respectively, to increasing these criteria to 3.0 milliseconds at the 200g level and 6.0 milliseconds at the 150g level. None of these petitions raised any issues or submitted any data different from those already considered by the NHTSA. The available biomechanical data indicate that the head impact protection provided to the helmet user by the standard's time duration criteria is greater than that which would result from the proposed changes, and the 2.0 and 4.0 millisecond criteria are retained.

3. Conditioning period. One petitioner requested that the 24-hour conditioning requirement for each of the four impact tests in paragraph S6.3 be modified to "4 to 24 hours," consistent with the requirements of ANSI Z90.1, arguing that 4 hours is sufficient to condition a helmet to the various environmental conditions required for the respective tests without compromising the intent of the standard. Upon further study of this matter, the NHTSA has concluded that, although 4 hours would not be sufficient as a general condition, changing the conditioning period to 12 hours would facilitate product testing without compromising the intent of the standard. Accordingly, paragraph S6.3, "Conditioning," is revised by changing the "24-hour" conditioning requirement to "12 hours" in each place the 24-hour requirement appears.

4. Low temperature conditioning requirement. Three petitioners objected to the −20° F. low temperature conditioning requirement in paragraph S6.3(b) on the basis that the requirement is overly severe. On review of available information, this agency has determined that precise data on the best low temperature requirements for testing are not available. Pending receipt of more specific information, therefore, the cold temperature requirement of 14° F. that has been used up to now by the American National Standards Institute appears to be the most appropriate. Accordingly, paragraph S6.3(b), "Low temperature," is revised by changing the "−20° F." conditioning requirement to "14° F.".

5. *Projections.* One petitioner requested that paragraph S5.5, "Projections," be changed to permit a maximum rigid projection inside the helmet shell of 0.080 in. with a minimum diameter of 0.150 in. The basis for this request is to allow for the use of eyelets and rivets for attachment of snaps for face shields and retention systems. The NHTSA is concerned that due care be exercised with regard to minimizing the injury producing potential of such fasteners. Eyelets and rivets for the attachment of snaps should be designed to form a portion of the continuous surface of the inside of the helmet shell. Where they are so designed, such attachments would not be "rigid projections." Accordingly, no revision to this requirement is necessary.

6. *Labeling.* One petitioner recommended that the labeling requirements in paragraph S5.6 be clarified with the help of manufacturers and other interested parties. Since the petitioner did not specify the points requiring clarification and because no other comments were received on this subject, the NHTSA has determined that no sufficient reasons have been given to change the labeling requirements.

In consideration of the foregoing, 49 CFR 571.218, Motor Vehicle Safety Standard No. 218, *Motorcycle Helmets*, is amended. . . .

Effective date: March 1, 1974.

(Secs. 103, 112, 119, Public Law 89–563, 80 Stat. 718, 15 U.S.C. 1392, 1401, 1407; delegation of authority at 49 CFR 1.51.)

Issued on January 23, 1974.

James B. Gregory
Administrator
39 F.R. 3554
January 28, 1974

PREAMBLE TO AN AMENDMENT TO MOTOR VEHICLE SAFETY STANDARD NO. 218

Motorcycle Helmets

(Docket No. 72-6; Notice 06)

ACTION: Final Rule.

SUMMARY: The purpose of this notice is to amend Safety Standard No. 218, *Motorcycle Helmets*, to extend application of the current requirements to all helmets that can be placed on the size "C" headform. The amendment is an interim rule requiring the certification of all large-size and many small-size helmets, and will be in effect until test headform sizes "A" and "D" have been developed and incorporated in the standard. This extended application of the standard will establish a minimum level of performance for a large number of helmets that are currently not being tested and certified by manufacturers, but which are suitable for testing on the size "C" headform.

EFFECTIVE DATE: May 1, 1980.

ADDRESSES: Any petitions for reconsideration should refer to the docket number and notice number and be submitted to: National Highway Traffic Safety Administration, Nassif Building, 400 Seventh Street, S.W., Washington, D.C. 20590.

FOR FURTHER INFORMATION CONTACT:

Mr. William J. J. Liu, Office of Vehicle Safety Standards, National Highway Traffic Safety Administration, Washington, D.C. 20590 (202-426-2264)

SUPPLEMENTARY INFORMATION: For reasons discussed below, on September 27, 1979, the NHTSA published a notice of proposed rulemaking to require, as an interim measure, the testing and certification of all motorcycle helmets that can be placed on the size "C" headform as described in

Safety Standard No. 218 (44 FR 55612). Only one comment was received in response to that notice, supporting the proposal.

Safety Standard No. 218, *Motorcycle Helmets* (49 CFR 571.218), specifies minimum performance requirements for helmets designed for use by motorcyclists and other motor vehicle users. Currently, the standard is only applicable to a portion of the annual helmet production. Paragraph S3 of the standard provides:

> * * * The requirements of this standard apply to helmets that fit headform size C, manufactured on or after March 1, 1974. Helmets that do not fit headform size C will not be covered by this standard until it is extended to those sizes by further amendments.

"Fitting" is intended to mean something that is neither too small nor too large. It excludes not only helmets that are too small to be placed on the size "C" headform, but also helmets so large that they could be placed on the size "D" headform were it available. As explained below, that headform size is not currently available.

The standard references and describes in its appendix four test headform sizes ("A", "B", "C", and "D"). Currently only test headform size "C" has been developed, and it is identical to the American National Standard specifications for Protective Headgear for Vehicular Users, ANSI Z90.1-1971. The other test headforms are to be scaled proportionately from the ANSI Z90 (size "C") headform. The performance requirements of the standard for helmets fitting other than size C headforms were held in abeyance until these additional headform sizes could be developed (39 FR 3554, January 28, 1974). Because of problems with prototype headforms supplied to NHTSA under contract (the headforms did not meet dimensional tolerances considered acceptable), development of these additional headforms has been delayed over the past years. However, the agency now anticipates that the standard will include requirements for headform sizes "A" and "D" effective April 1, 1982 (size "B" will be deleted from the standard).

Last year, the Safety Helmet Council of America (SHCA) recommended that the agency require certification of all adult-size helmets on the size "C" headform. The SHCA stated that the delay in development of the additional headform sizes has led to confusion and unfair practices since many helmets are reportedly being improperly certified and many other helmets are not being certified that are required to comply with the standard. The agency has stated in the past that only helmets that are subject to compliance with Standard No. 218 should be certified and labeled with the "DOT" symbol. Apparently, some manufacturers have used the "DOT" label on untested helmets for competitive purposes. The SHCA stated that these practices have placed considerable burdens on the integrity of manufacturers of high quality helmets. The organization pointed out that under the ANSI standard only one headform (size "C") was used to test all helmets except child-size helmets, and that approximately 95 percent of current helmet production could and should be tested on the size "C" headform and certified for compliance with Standard No. 218.

The NHTSA Office of Vehicle Safety Standards has investigated the current labeling and certification practices of helmet manufacturers. It was found that most manufacturers currently test only "medium" size helmets on the size "C" headform, yet there is considerable variation among manufacturers as to which helmets are considered medium. Further, the agency found that the percentage of helmets subject to certification under the current applicability of the standard is substantially greater than the 40 percent that manufacturers are now testing on the size "C" headform. (Data from the investigation have been placed in the NHTSA docket under the docket number of this notice.)

As stated earlier, under the existing applicability requirements of the standard, only helmets that "fit" headform size "C" must be certified. Apparently, interpretation of the term "fit" by manufacturers has led to some mislabelings and failures to certify. Under the existing requirements, "helmets that fit headform size C" should be all helmets other than those that must be tested on the other headform sizes. To determine which helmets must be tested on a particular headform size, one follows the procedures of paragraph S6.1.1 of the standard. That paragraph provides in part:

* * * Place the complete helmet to be tested on the reference headform of the largest size specified in the Appendix whose circumference is not greater than the internal circumference of the headband when adjusted to its largest setting, or if no headband is provided to the corresponding interior surface of the helmet.

Using the procedure of paragraph S6.1.1, manufacturers currently need only concern themselves with headform sizes "C" and "D", since small, child-size helmets that could not physically be placed on the size "C" headform would not have to be tested. As to the other helmet sizes, helmets that "fit headform size C" means any helmet that can be placed on the size "C" headform, except those helmets which the manufacturer can demonstrate could be placed on a size "D" headform. To make that demonstration, the manufacturers would have to show that the internal circumference of the helmet headband or the corresponding interior surface of the helmet is larger than the circumference of the size "D" headform. Even though the size "D" headform is not currently available, the dimensions of the headform are specified in the appendix of the standard, from which the manufacturer can make its determination. Regarding small, child-size helmets, the determination whether or not a particular helmet can be placed on the size "C" headform should be based on normal fitting procedures. This means, for example, that undue force should not be applied to forcibly push the headform into the helmet. However, efforts necessary for the ordinary wearing of the helmet should be employed, such as expanding the lower portions of a flexible-shell, full-face helmet. Apparently, many manufacturers have failed to use these procedures for determining which of their helmets "fit" headform size "C" and must be certified.

In light of the improper certification and the noncertification, the unavailability of the additional headform sizes at the present time, the need to ensure the safe performance of the large helmets and the apparent sufficiency of the size "C" headform for testing large helmets, the agency has concluded that the recommendations of the Safety Helmet Council of America have merit. Therefore, this notice amends Safety Standard No. 218 to require all motorcycle helmets that can be placed on the size "C" headform to be certified in accordance with the requirements of the standard. "Placed" is a broader term than "fit" primarily in that the former term does not imply any upper limit on helmet size.

Under these interim requirements, more than 90 percent of current helmet production will be tested on the size "C" headform. Only small, child-size helmets (size "A") will be excluded since they cannot physically be placed on the size "C" headform. As noted in the procedures discussed above, normal fitting procedures are used to determine if a particular helmet can be placed on the size "C" headform, without the use of undue force.

During its investigation, the NHTSA contacted manufacturers whose collective market share exceeds 80 percent of current annual helmet production. All of these manufacturers indicated that 90 percent or more of their helmet production could be placed and tested on the size "C" headform. Many of the manufacturers indicated that they are already testing the majority of their helmets on the size "C" headform for quality-control purposes, even though not required by the standard. Also, it was found that helmet shells and performance characteristics of a particular manufacturer's helmets do not generally vary significantly over the various size ranges of helmets produced.

This amendment is only an interim measure to establish a minimum level of performance for the large number of helmets that are currently not being certified for compliance with Standard No. 218. Testing extra-large helmets on the size "D" headform would require a higher level of performance for those helmets, since the weight of the size "D" headform is greater than that of the size "C" headform. Therefore, development of the size "A" and size "D" headforms has continued, and incorporation of requirements in the standard for these headforms will occur after development is completed. However, until this is accomplished,

the agency believes that the performance level that will be required by testing on the size "C" headform is preferable to an absence of any requirements whatsoever. As stated earlier, the ANSI standard for helmets specifies only one headform size ("C") for testing all helmets. The additional headform sizes were originally specified in Standard No. 218 in response to suggestions from some manufacturers that requirements be more "fine-tuned" for the various helmet sizes.

The agency has concluded that the new requirements will preclude the great majority of unsafe helmets currently on the road. Further, with all adult helmets certified, retailers and consumers will no longer be confused or misled concerning the DOT certification labels found in their helmets, and NHTSA's enforcement activities will become more effective and uniform.

Under these new requirements, extra-large helmets should be tested on the size "C" headform without the use of "shims" or other devices to obtain a secure fit of the helmet on the headform. Agency tests involving extra-large helmets on the size "C" headform show results that correlate well with tests of medium-size helmets on the size "C" headform. (Data from these tests have been placed in the NHTSA docket). Therefore, the agency has concluded that repeatable results can be obtained under the existing procedures with the size "C" headform.

The effective date for extending the applicability of Standard No. 218 to all helmets that can be placed on the size "C" headform is May 1, 1980. The agency's past position has been that it would be "false and misleading," within the meaning of the statute (15 U.S.C. 1397(C)), for a "DOT" symbol to appear without qualification on helmets manufactured before the effective date of the standard. However, since the standard is currently effective for helmets that fit size "C" headforms, and since there is such a widespread variation among manufacturers as to which helmets they consider to fit the size "C" headform, the agency will allow voluntary certification and labeling of helmets prior to May 1, 1980. This, of course, would only apply to helmets that can be placed on the size "C" headform. Small helmets that could not be placed on the headform could not be certified with the "DOT" symbol until after the standard has been amended to include specifications for the size "A" headform. Also, helmets certified and labeled with the "DOT" symbol prior to the May 1, 1980, effective date will be subject to the general enforcement provisions of the National Traffic and Motor Vehicle Safety Act. Therefore, manufacturers will have to exercise "due care" to assure that any helmet they certify in fact complies with the performance requirements of Standard No. 218.

The agency has determined that this amendment does not qualify as a significant regulation under Executive Order 12044, "Improving Government Regulations." A final regulatory evaluation of this amendment has been placed in the docket for the benefit of all interested persons.

The engineer and lawyer primarily responsible for the development of this notice are William J. J. Liu and Hugh Oates, respectively.

In consideration of the above, paragraph S3 of Safety Standard No. 218, *Motorcycle Helmets* (49 CFR 571.218), is amended to read as follows:

§ 571.218 *Standard No. 218; motorcycle helmets.*

*   *   *   *   *

S3. *Application.* This standard applies to helmets designed for use by motorcyclists and other motor vehicle users. The requirements of this standard apply to all helmets that can be placed on the size C headform using normal fitting procedures. Helmets that cannot be placed on the size C headform will not be covered by this standard until it is extended to those sizes by further amendment.

*   *   *   *   *

(The second sentence in S6.1.1 of the standard relating to the selection of a reference headform should be disregarded until the standard is made effective for helmets that must be tested on headform sizes A and D.)

Issued on February 29, 1980.

Joan Claybrook
Administrator

45 F.R. 15179
March 10, 1980

MOTOR VEHICLE SAFETY STANDARD NUMBER 218

Motorcycle Helmets

(Docket No. 72-6; Notice 2)

S1. Scope. This standard establishes minimum performance requirements for helmets designed for use by motorcyclists and other motor vehicle users.

S2. Purpose. The purpose of this standard is to reduce deaths and injuries to motorcyclists and other motor vehicle users resulting from head impacts.

S3. Application. This standard applies to helmets designed for use by motorcyclists and other motor vehicle users. The requirements of this standard apply to all helmets that can be placed on the size C headform using normal fitting procedures. Helmets that cannot be placed on the size

C headform will not be covered by this standard until it is extended to those sizes by further amendment.

S4. Definitions.

"Basic plane" means a plane through the centers of the right and left external ear openings and the lower edge of the eye sockets (Figure 1) of a reference headform (Figure 2) or test headform.

"Midsagittal plane" means a longitudinal plane through the apex of a reference headform or test headform that is perpendicular to the basic plane (Figure 3).

"Reference plane" means a plane above and parallel to the basic plane on a reference head-

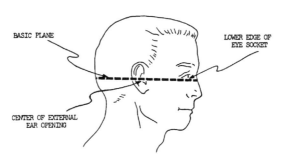

Figure 1

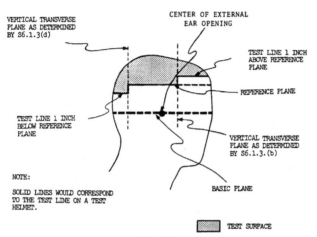

VERTICAL TRANSVERSE
PLANE AS DETERMINED
BY S6.1.3(d)

CENTER OF EXTERNAL
EAR OPENING

TEST LINE 1 INCH
ABOVE REFERENCE
PLANE

REFERENCE PLANE

TEST LINE 1 INCH
BELOW REFERENCE
PLANE

VERTICAL TRANSVERSE
PLANE AS DETERMINED
BY S6.1.3.(b)

NOTE:

SOLID LINES WOULD CORRESPOND
TO THE TEST LINE ON A TEST
HELMET.

BASIC PLANE

▓▓▓ TEST SURFACE

Figure 2

form or test headform (Figure 2) at the distance indicated in the Appendix.

"Reference headform" means a measuring device contoured to the dimensions of one of the four headforms described in the Appendix, with surface markings indicating the locations of the basic, midsagittal, and reference planes, and the centers of the external ear openings.

"Test headform" means a test device contoured to the dimensions of one of the four reference headforms described in the Appendix for all surface areas that contact the helmet, with surface markings indicating the locations of the basic, midsagittal, and reference planes.

"Retention system" means the complete assembly by which the helmet is retained in position on the head during use.

"Helmet positioning index" means the distance in inches, as specified by the manufacturer, from the lowest point of the brow opening at the lateral midpoint of the helmet to the basic plane of a reference headform, when the helmet is

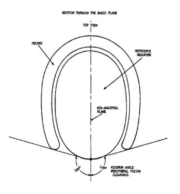

Figure 3

firmly and properly positioned on the reference headform.

S5. Requirements. Each helmet shall meet the requirements of S5.1 through S5.3 when subjected to any conditioning procedure specified in S6.3, and tested in accordance with S7.

S5.1 Impact attenuation. When an impact attenuation test is conducted in accordance with S7.1, all of the following requirements shall be met:

(a) Peak accelerations shall not exceed 400g;

(b) Accelerations in excess of 200g shall not exceed a cumulative duration of 2.0 milliseconds; and

(c) Accelerations in excess of 150g shall not exceed a cumulative duration of 4.0 milliseconds.

S5.2 Penetration. When a penetration test is conducted in accordance with S7.2, the striker shall not contact the surface of the test headform.

S5.3 Retention system.

S5.3.1 When tested in accordance with S7.3:

(a) The retention system or its components shall attain the loads specified without separation; and

(b) The adjustable portion of the retention system test device shall not move more than 1 inch measured between preliminary and test load positions.

S5.3.2 Where the retention system consists of components which can be independently fastened without securing the complete assembly, each such component shall independently meet the requirements of S5.3.1.

S5.4 Configuration. Each helmet shall have a protective surface of continuous contour at all points in or above the test line described in S6.1.3. The helmet shall provide peripheral vision clearance of at least 105° to each side of the midsagittal plane, when the helmet is adjusted as specified in S6.2. The vertex of these angles, shown in Figure 3, shall be at the point on the anterior surface of the reference headform at the intersection of the midsagittal and basic planes. The brow opening of the helmet shall be at least 1 inch above all points in the basic plane that are within the angles of peripheral vision (see Figure 3).

S5.5 Projections. A helmet shall not have any rigid projections inside its shell. Rigid projections outside any helmet's shell shall be limited to those required for operation of essential accessories, and shall not protrude more than 0.19 inch.

S5.6 Labeling.

S5.6.1 Each helmet shall be permanently and legibly labeled, in a manner such that the label(s) can be easily read without removing padding or any other permanent part, with the following:

(1) Manufacturer's name or identification.

(2) Precise model designation.

(3) Size.

(4) Month and year of manufacture. This may be spelled out (e.g., June 1974), or expressed in numerals (e.g., 6/74).

(5) The symbol DOT, constituting the manufacturer's certification that the helmet conforms to the applicable Federal Motor Vehicle Safety Standards. This symbol shall appear on the outer surface, in a color that contrasts with the background, in letters at least ⅜ inch high, centered laterally approximately 1¼ inches from the bottom edge of the posterior portion of the helmet.

(6) Instruction to the purchaser as follows:

"Shell and liner constructed of (identify type(s) of materials).

"Helmet can be seriously damaged by some common substances without damage being visible to the user. Apply only the following: (Recommended cleaning agent, paints, adhesives, etc., as appropriate).

"Make no modifications. Fasten helmet securely. If helmet experiences a severe blow, return it to the manufacturer for inspection, or destroy and replace it." (On an attached tag, brochure, or other suitable means, any additional, relevant safety information should be supplied at the time of purchase).

S5.7 Helmet positioning index. Each manufacturer of helmets shall establish a positioning index for each helmet he manufactures. This index shall be furnished immediately to any person who requests the information, with respect to a helmet identified by manufacturer, model designation, and size.

S6. Preliminary test procedures. Before subjecting a helmet to the testing sequence specified in S7., prepare it according to the following procedures.

S6.1 Reference marking.

S6.1.1 Use a reference headform that is firmly seated with the basic and reference planes horizontal. Place the complete helmet to be tested on the reference headform of the largest size specified in the Appendix whose circumference is not greater than the internal circumference of the headband when adjusted to its largest setting, or if no headband is provided to the corresponding interior surface of the helmet.

S6.1.2 Apply a 10-pound static load normal to the helmet's apex. Center the helmet laterally and seat it firmly on the reference headform according to its helmet positioning index.

S6.1.3 Maintaining the load and position described in S6.1.2, draw a line (hereinafter referred to as "test line") on the outer surface of the helmet coinciding with portions of the intersection of that surface with the following planes, as shown in Figure 2:

(a) A plane 1 inch above and parallel to the reference plane in the anterior portion of the reference headform;

(b) A vertical transverse plane 2.5 inches behind the point on the anterior surface of the reference headform at the intersection of the midsagittal and reference planes;

(c) The reference plane of the reference headform;

(d) A vertical transverse plane 2.5 inches behind the center of the external ear opening in a side view; and

(e) A plane 1 inch below and parallel to the reference plane in the posterior portion of the reference headform.

S6.2 Helmet positioning. Prior to each test, fix the helmet on a test headform in the position that conforms to its helmet positioning index. Secure the helmet so that it does not shift position prior to impact or the application of force during testing.

S6.2.1 In testing as specified in S7.1 and S7.2, place the retention system in a position such that

it does not interfere with free fall, impact, or penetration.

S6.3 Conditioning. Immediately prior to conducting the testing sequence specified in S7., condition each test helmet in accordance with any one of the following procedures:

(a) *Ambient conditions.* Expose to a temperature of 70° F. and a relative humidity of 50% for 12 hours.

(b) *Low temperature.* Expose to a temperature of 14° F. for 12 hours.

(c) *High temperature.* Expose to a temperature of 122° F. for 12 hours.

(d) *Water immersion.* Immerse in water at a temperature of 77° F. for 12 hours.

If during testing, the time out of the conditioning environment for a test helmet exceeds 5 minutes, return the helmet to the conditioning environment for a minimum of 3 minutes for each minute out of the conditioning environment or 12 hours, whichever is less, prior to resumption of testing.

S7. Test conditions.

S7.1 Impact attenuation test.

S7.1.1 Impact attenuation is measured by determining acceleration imparted to an instrumented test headform on which a complete helmet is mounted as specified in S6.2, when it is dropped in guided free fall upon fixed hemispherical and flat steel anvils.

S7.1.2 Each helmet is impacted at four sites with two successive, identical impacts at each site. Two of these sites are impacted upon a flat steel anvil and two upon a hemispherical steel anvil as specified in S7.1.7 and S7.1.8. The impact sites are at any point on the area above the test line described in S6.1.3, and separated by a distance not less than one-sixth of the maximum circumference of the helmet.

S7.1.3 The guided free fall drop heights for the helmet and test headform combination onto the hemispherical anvil and flat anvil are 54.5 inches and 72 inches, respectively.

S7.1.4 Test headforms for impact attenuation testing are constructed of magnesium alloy (K-1A), and exhibit no reasonant frequencies below 3,000 Hz.

S7.1.5 Weight of the drop assembly, as specified in Table I, is the combined weight of the instrumented test headform and supporting assembly for the drop test. The weight of the supporting assembly does not exceed 20% of the weight of the drop assembly. The center of gravity of the combined test headform and supporting assembly lies within a cone with its axis vertical and forming a 10° included angle with the vertex at the point of impact.

TABLE I
WEIGHTS FOR
IMPACT ATTENUATION TEST
DROP ASSEMBLY

Reference Headform Size	Weight (Lbs)*
A	7.8
B	8.9
C	11.0
D	13.4

*Combined weight of instrumented test headform and supporting assembly for drop test.

S7.1.6 The acceleration transducer is mounted at the center of gravity of the combined test headform and supporting assembly with the sensitive axis aligned to within 5% of vertical when the test headform is in the impact position. The acceleration data channel complies with SAE Recommended Practice J211 requirements for channel class 1,000.

S7.1.7 The flat anvil is constructed of steel with a 5-inch minimum diameter impact face, and the hemispherical anvil is constructed of steel with a 1.9-inch radius impact face.

S7.1.8 The rigid mount for both of the anvils consists of a solid mass of at least 300 pounds, the outer surface of which consists of a steel plate with minimum thickness of 1 inch and minimum surface area of 1 ft.²

S7.2 Penetration test.

S7.2.1. The penetration test is conducted by dropping the penetration test striker in guided free fall, with its axis aligned vertically, onto the outer surface of the complete helmet, when mounted as specified in S6.2, at any point above

the test line, described in S6.1.3, except on a fastener or other rigid projection.

S7.2.2 Two penetration blows are applied at least 3 inches apart, and at least 3 inches from the centers of any impacts applied during the impact attenuation test.

S7.2.3 The height of the guided free fall is 118.1 inches, as measured from the striker point to the impact point on the outer surface of the test helmet.

S7.2.4 The contactable surfaces of the penetration test headforms are constructed of a metal or metallic alloy having a Brinell hardness number no greater than 55, which will readily permit detection should contact by the striker occur. The surface is refinished if necessary prior to each penetration test blow to permit detection of contact by the striker.

S7.2.5 The weight of the penetration striker is 6 pounds, 10 ounces.

S7.2.6 The point of the striker has an included angle of 60°, a cone height of 1.5 inches, a tip radius of 0.019 inch (standard 0.5 millimeter radius) and a minimum hardness of 60 Rockwell, C-scale.

S7.2.7 The rigid mount for the penetration test headform is as described in S7.1.8.

S7.3 Retention system test.

S7.3.1 The retention system test is conducted by applying a static tensile load to the retention assembly of a complete helmet, which is mounted, as described in S6.2, on a stationary test headform as shown in Figure 4, and by measuring the movement of the adjustable portion of the retention system test device under tension.

S7.3.2 The retention system test device consists of both an adjustable loading mechanism by which a static tensile load is applied to the helmet retention assembly and a means for holding the test headform and helmet stationary. The retention assembly is fastened around two freely moving rollers, both of which have 0.5 inch diameter and a 3-inch center-to-center separation, and which are mounted on the adjustable

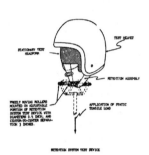

TEST HELMET

STATIONARY TEST HEADFORM

RETENTION ASSEMBLY

FREELY MOVING ROLLERS MOUNTED ON ADJUSTABLE PORTION OF RETENTION SYSTEM TEST DEVICE WITH DIAMETERS 0.5 INCH, AND CENTER-TO-CENTER SEPARATION 3 INCHES.

APPLICATION OF STATIC TENSILE LOAD

RETENTION SYSTEM TEST DEVICE

Figure 4

portion of the tensile loading device (Figure 4). The helmet is fixed on the test headform as necessary to ensure that it does not move during the application of the test loads to the retention assembly.

S7.3.3 A 50-pound preliminary test load is applied to the retention assembly, normal to the basic plane of the test headform and symmetrical with respect to the center of the retention assembly for 30 seconds, and the maximum distance from the extremity of the adjustable portion of the retention system test device to the apex of the helmet is measured.

S7.3.4 An additional 250-pound test load is applied to the retention assembly, in the same manner and at the same location as described in S7.3.3, for 120 seconds, and the maximum distance from the extremity of the adjustable portion of the retention system test device to the apex of the helmet is measured.

APPENDIX

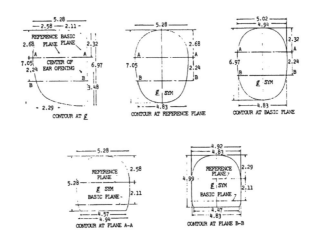

CONTOUR AT ℓ

CONTOUR AT REFERENCE PLANE

CONTOUR AT BASIC PLANE

CONTOUR AT PLANE A-A

CONTOUR AT PLANE B-B

HEADFORM A

ALL DIMENSIONS IN INCHES

PART 571; S 218-7

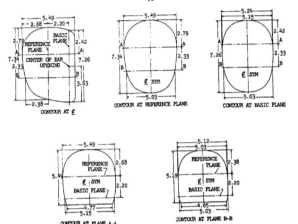

CONTOUR AT ₵

CONTOUR AT REFERENCE PLANE

CONTOUR AT BASIC PLANE

CONTOUR AT PLANE A-A

CONTOUR AT PLANE B-B

HEADFORM B

ALL DIMENSIONS IN INCHES

PART 571; S 218-8

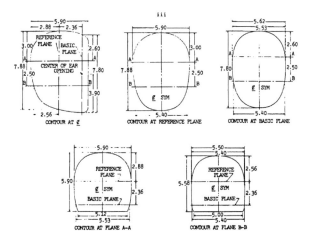

HEADFORM C

ALL DIMENSIONS IN INCHES.

PART 571; S 218-9

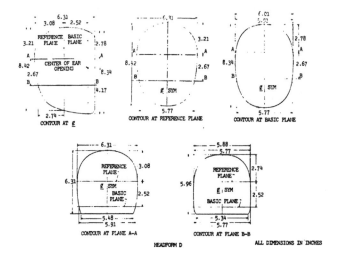

CONTOUR AT ℓ

CONTOUR AT REFERENCE PLANE

CONTOUR AT BASIC PLANE

CONTOUR AT PLANE A–A

CONTOUR AT PLANE B–B

HEADFORM D

ALL DIMENSIONS IN INCHES

38 F.R. 22390
August 20, 1973

PART 571; S 218–10

PREAMBLE TO MOTOR VEHICLE SAFETY STANDARD NO. 219

Windshield Zone Intrusion

(Docket No. 74–21; Notice 2)

This notice establishes a new Motor Vehicle Safety Standard No. 219, 49 CFR 571.219, that regulates the intrusion of vehicle parts from outside the occupant compartment into a defined zone in front of the windshield during a frontal barrier crash test.

The notice of proposed rulemaking on which this issuance is based was issued on May 20, 1974 (39 F.R. 17768). An earlier notice had been issued on August 31, 1972 (37 F.R. 17763), proposing a standard that would prohibit penetration of the protected zone by any part of a vehicle outside of the occupant compartment during a 30-mph frontal impact into a fixed barrier. After further study and an analysis of comments submitted in response to that notice, the NHTSA determined that the initial rule was unnecessarily stringent since its near-total ban on intrusion had the effect of prohibiting entrance into the protected zone or contact with the windshield by small particles such as paint chips and glass which do not represent a danger to the vehicle occupants if they enter the zone and impact the windshield opening with a limited amount of force.

Consequently, in the notice published on May 20, 1974, the proposed standard on windshield zone intrusion was amended to permit penetration by particles, to a depth of no more than one-quarter inch into a styrofoam template in the shape of the protected zone and affixed to the windshield, during a 30-mph frontal barrier crash.

In addition, the amended proposal published May 20, 1974, provided that contact by vehicle parts with the windshield opening in the area below the protected zone, during a 30-mph barrier crash test, would not be prohibited provided

that the inner surface of that portion of the windshield is not penetrated. The procedure for determining the lower edge of the protected zone was also revised.

Standard No. 219, *Windshield Zone Intrusion*, reflects some minor changes incorporated for clarification following publication of the proposed rule on May 20, 1974. First, open-body-type vehicles with fold-down or removable windshields have been added to forward control vehicles as vehicle types to which the standard does not apply. A structurally unsupported windshield, essential to the utility of this vehicle type, typically does not remain in place during a 30-mph frontal barrier crash test, hence the test is impracticable for this type of vehicle.

In addition, the standard provides that its prohibitions against penetration by particles to a depth of more than one-quarter inch into the styrofoam template and penetration of the inner surface of the portion of the windshield below the protected zone do not apply to windshield molding and other components designed to be normally in contact with the windshield. This provision was contained in the proposed standard published August 31, 1972 but omitted from the proposal published May 20, 1974.

The standard as adopted also specifies that the 6.5-inch-diameter rigid sphere employed to determine the lower edge of the protected zone shall weigh 15 pounds, the approximate weight of the head and neck of an average driver or passenger.

Comments submitted by Wayne Corporation and Sheller-Globe Corporation, manufacturers of funeral coaches and ambulances, urged that the standard for windshield zone intrusion contain an exception for such vehicles in view of the low incidence of accidents involving funeral coaches and ambulances, the low volume of production of such vehicles, and the high cost of barrier crash testing. The NHTSA has determined that these arguments are without merit. The manufacturers have presented no evidence to support the contention that funeral coaches and ambulances are involved in fewer accidents in proportion to their numbers than other vehicles. Furthermore, several comments criticizing the allegedly prohibitive costs of compliance with the standard appear to have erroneously assumed that every manufacturer must conduct barrier crash tests. The performance requirement for windshield zone intrusion is set out in S5. of the standard. A manufacturer of funeral coaches and ambulances may, for example, assure itself that the requirement is met by barrier crashing the conventional chassis which is a component of the special vehicle, modified to simulate the dynamic characteristics of the funeral coach or ambulance. Or, the manufacturer may use the design characteristic of the vehicle taking into account the modifications it makes, or information supplied by the chassis manufacturer.

Low volume of production is not an appropriate basis for an exemption. As the NHTSA has maintained in past proceedings where the same argument was advanced, the appropriate means to avoid application of a standard on hardship grounds is a temporary exemption under 49 CFR Part 555.

Finally, the NHTSA is continuing to promote compatibility and economy in barrier crash testing by adopting vehicle loading and dummy restraint requirements in Standard No. 219 identical to those set out in proposed amendments to Standard No. 301, *Fuel System Integrity*, 49 CFR 571.301 (40 F.R. 17036, April 16, 1975). It has therefore required that 50th-percentile test dummies be placed in the seating positions whose restraint system is required to be tested by a dummy under Standard No. 208, *Occupant Crash Protection*, 49 CFR 571.208, and that they may be restrained only by the means that are installed in the vehicle at the respective seating positions.

In consideration of the foregoing, 49 CFR Part 571 is amended by the addition of a new Standard No. 219, 49 CFR 571.219, *Windshield Zone Intrusion. . . .*

Effective date: September 1, 1976.

(Secs. 103, 119, Pub. L. 89–563, 80 Stat. 718 (15 U.S.C. 1392, 1407); delegation of authority at 49 C.F.R. 1.51.)

Issued on June 9, 1975.

James B. Gregory
Administrator

**40 F.R. 25462
June 16, 1975**

PREAMBLE TO AMENDMENT TO MOTOR VEHICLE SAFETY STANDARD NO. 219

Windshield Zone Intrusion

(Docket No. 74–21; Notice 3)

This notice responds to four petitions for reconsideration of the notice published June 16, 1975 (40 FR 25462), which established a new Motor Vehicle Safety Standard No. 219, *Windshield Zone Intrusion*, 49 CFR 571.219, regulating the intrusion of vehicle parts from outside the occupant compartment into a defined zone in front of the windshield during a frontal barrier crash test. The National Highway Traffic Safety Administration (NHTSA) hereby amends Standard No. 219 on the basis of the information and arguments presented by some of the petitioners.

Petitions for reconsideration were received from the Motor Vehicle Manufacturers Association (MVMA), General Motors, Ford, and Jeep. MVMA, General Motors, and Ford requested substitution of the term "daylight opening" for "windshield opening," and General Motors and Jeep requested a change in the effective date of Standard No. 219 from September 1, 1976 to September 1, 1977. In addition, Jeep requested that Standard No. 219 not become applicable until final issuance of Standard No. 212, *Windshield Mounting*, 49 CFR 571.212.

The NHTSA has determined that the petitions of MVMA, General Motors, and Ford requesting substitution of the term "daylight opening" for "windshield opening" have merit, and they are therefore granted. These petitioners requested that the term "windshield opening" be replaced by the term "daylight opening", which is defined in paragraph 2.3.12 of section E, Ground Vehicle Practice, SAE Aerospace-Automotive Drawing Standards, September, 1963. The part of the windshield below the daylight opening is protected by the cowling and instrument panel. There is little likelihood that

in a frontal crash any vehicle component will penetrate the cowling and instrument panel with sufficient force to pose a threat to the vehicle occupants. Therefore, the zone intrusion requirements of Standard No. 219 should only apply to the area of the windshield susceptible to actual penetration by vehicle components in a crash. Accordingly, the term "windshield opening" as it is used in Standard No. 219, is replaced by "daylight opening." The SAE definition of "daylight opening" has been slightly modified to reflect the particular characteristics of Standard No. 219.

The NHTSA has concluded that the petitions of General Motors and Jeep requesting a change in the effective date of Standard No. 219 should be granted in part and denied in part. The economic considerations involved in coordinating the effective date of Standard No. 219 with that of Standard No. 212, *Windshield Mounting*, justify postponement of the effective date to September 1, 1977, for application of Standard No. 219 to all vehicles except passenger cars. However, the effective date of September 1, 1976, will be retained for passenger cars because of their greater susceptibility to the intrusion of vehicle parts against which this standard is designed to protect. This postponement of effective dates also grants in part Jeep's petition requesting that the applicability of Standard No. 219 be postponed until final issuance of Standard No. 212.

In consideration of the foregoing, § 571.219 is amended by revising S4., S5., and S6.1(d) of Standard No. 219, *Windshield Zone Intrusion*, to read as follows:

Effective date: September 1, 1976, for passenger cars; September 1, 1977, for multipurpose

PART 571; S 219—PRE 3

passenger vehicles, trucks, and buses with a GVWR of 10,000 pounds or less.

(Sec. 103, 119, Pub. L. 89–563, 80 Stat. 718 (15 U.S.C. 1392, 1407) ; delegation of authority at 49 CFR 1.51.)

Issued on November 10, 1975.

James B. Gregory
Administrator

40 F.R. 53033
November 14, 1975

PREAMBLE TO AMENDMENT TO MOTOR VEHICLE SAFETY STANDARD NO. 219

Windshield Zone Intrusion

(Docket No. 74–21; Notice 5)

This notice amends Standard No. 219, *Windshield Zone Intrusion*, to exclude walk-in van-type vehicles from the requirements of the standard.

The National Highway Traffic Safety Administration (NHTSA) proposed to exclude walk-in van-type vehicles from the applicability of Standard No. 219 (49 CFR 571.219) in a notice published March 11, 1976 (41 FR 10451). No opposition was registered in response to the proposed rulemaking. The National Motor Vehicle Safety Advisory Council did not take a position on the proposal.

The NHTSA, therefore, amends Standard No. 219 in accordance with the proposal. For the information of all interested persons, the NHTSA considers a "walk-in van-type" vehicle to be only the "step van" city delivery type of vehicle that permits a person to enter the vehicle without stooping.

It has been determined that this amendment will have a negligible economic and environmental impact, since it creates an exemption from existing requirements that is expected to affect relatively few vehicles.

In consideration of the foregoing, paragraph S3 of Standard No. 219 (49 CFR 571.219) is amended . . . .

Effective date: December 16, 1976. Because this amendment relieves a restriction and does not create additional obligations for any person and because it permits the resumption of manufacture of a vehicle type not intended to be covered by the standard, it is found that an immediate effective date is in the public interest.

(Sec. 103, 119, Pub. L. 89–563, 80 Stat. 718 (15 U.S.C. 1392, 1407); delegation of authority at 49 CFR 1.50.)

Issued on December 10, 1976.

Charles E. Duke
Acting Administrator

41 FR 54945
December 16, 1976

PREAMBLE TO AN AMENDMENT TO FEDERAL MOTOR VEHICLE SAFETY STANDARD NO. 219

Windshield Zone Intrusion

(Docket No. 79-14; Notice 2)

ACTION: Final Rule.

SUMMARY: This notice amends two safety standards, Standard No. 212, *Windshield Mounting,* and Standard No. 219, *Windshield Zone Intrusion,* to limit the maximum unloaded vehicle weight at which vehicles must be tested for compliance with these standards. This action is being taken in response to petitions from the Truck Body and Equipment Association and the National Truck Equipment Association asking the agency to amend the standards to provide relief from some of the test requirements for final-stage manufacturers. Many of these small manufacturers do not have the sophisticated test devices of major vehicle manufacturers. The agency concludes that the weights at which vehicles are tested can be lessened while providing an adequate level of safety for vehicles such as light trucks and while ensuring that compliance with these standards does not increase their aggressivity with respect to smaller vehicles.

EFFECTIVE DATE: Since this amendment relieves a restriction by easing the existing test procedure and will not impose any additional burdens upon any manufacturer, it is effective (upon publication).

FOR FURTHER INFORMATION CONTACT:

Mr. William Smith, Crashworthiness Division, National Highway Traffic Safety Administration, 400 Seventh Street, S.W., Washington, D.C. 20590 (202-426-2242)

SUPPLEMENTARY INFORMATION:

On August 2, 1979, the National Highway Traffic Safety Administration published a notice of proposed rulemaking (44 FR 45426) relating to two safety standards: Standard Nos. 212, *Windshield Mounting,* and 219 *Windshield Zone Intrusion.* That notice proposed two options for amending the test procedures of the standards that were designed to ease the compliance burdens of small final-stage manufacturers.

The agency issued the proposal after learning that final-stage manufacturers were frequently unable to certify certain vehicles in compliance with these two safety standards. The problem arises because of weight and center of gravity restrictions imposed upon the final-stage manufacturer by the incomplete vehicle manufacturer. (The final-stage manufacturer typically purchases an incomplete vehicle from an incomplete vehicle manufacturer, usually Ford, General Motors or Chrysler.) The incomplete vehicle usually includes the windshield and mounting but does not include any body or work-performing equipment. Since the incomplete vehicle manufacturer installs the windshield, it represents to the final-stage manufacturer that the windshield will comply with the two subject safety standards. In making this representation, however, the incomplete vehicle manufacturer states that the representation is contingent on the final-stage manufacturer's adherence to certain restrictions. Any final-stage manufacturer that does not adhere to the restrictions imposed by the incomplete vehicle manufacturer must recertify the vehicle based upon its own information, analysis, or tests. The major restrictions imposed by the incomplete vehicle manufacturers on the final-stage manufacturer involve weight and center of gravity limitation. In many instances, these limitations have made it impossible for final-stage manufacturers either to rely on the incomplete vehicle manufacturer's certification or to complete vehicles on the same chassis that they were accustomed to using (prior to the extension of the two safety standards to these vehicle types). As a result, the final-stage manufacturer is faced either with buying the same chassis as before and recertifying them or with buying more expensive chassis with higher GVWR's and less stringent weight and center of gravity limitations.

The agency has tried several different ways to alleviate this problem for the final-stage manufacturer. The NHTSA has met with representatives of the major incomplete vehicle manufacturers to encourage them to respond voluntarily by strengthening their windshield structures and reducing the restrictions that they currently impose upon final-stage manufacturers. The agency also discussed the possibility of its mandating these actions by upgrading Standards Nos. 212 and 219. Ford and General Motors indicated that the making of any major changes in these standards could lead to their deciding to discontinue offering chassis for use in the manufacturing of multi-stage vehicles. They said that such chassis were a very small percentage of their light truck sales and that, therefore, they would not consider it worth the cost to them to make any extensive modifications in their vehicles. NHTSA also asked the incomplete vehicle manufacturers to be sure that they have properly certified their existing vehicles and that they are not imposing unnecessarily restrictive limitations upon final-stage manufacturers. To this agency's knowledge, these vehicle manufacturers have neither undertaken any strengthening of their vehicles' windshield structures nor lessened any of their restrictions.

At the same time that the agency was made aware of the final-stage manufacturers' problems of certifying to these standards, the agency was becoming concerned about the possibility that compliance of some light trucks and vans with these standards might have made the vehicles more aggressive with respect to smaller passenger cars that they might impact. According to agency information, if these standards require a substantial strengthening of vehicle frames, the aggressivity of the vehicles is increased. Therefore, as a result of the agency's concern about aggressivity and its desire to address the certification problems of final-stage manufacturers in a manner that would not lead to a cessation of a chassis sales to those manufacturers, the agency issued the August 1979 proposal. The agency hoped that the proposal would allow and encourage incomplete vehicle manufacturers to reduce their

weight and center of gravity restrictions, thereby easing or eliminating the compliance test burdens of final-stage manufacturers. The agency believed that this could occur using either option, because either would result in vehicles being tested at lower weights. Currently vehicles are tested under both standards at their unloaded vehicle weights plus 300 pounds.

The first option would have required some vehicles whose unloaded vehicle weights exceeded 4,000 pounds to be tested by being impacted with a 4,000 pound moving barrier. The second option proposed by the agency would have required vehicles to be tested at their unloaded vehicle weight up to a maximum unloaded vehicle weight of 5,500 pounds. This option was suggested to the agency by several manufacturers and manufacturer representatives.

Comments on Notice

In response to the agency's notice, nine manufacturers and manufacturer representatives submitted comments. All of the commenters supported some action in response to the problems of final-stage manufacturers. Most of the commenters also suggested that the agency's second alternative solution was more likely to achieve reductions in the restrictions being imposed by incomplete vehicle manufacturers. The first option would have created a new, unproven test procedure, and manufacturers would have been cautious in easing center of gravity or weight restrictions based upon this test procedure. Accordingly, most commenters were not sure that the first option would achieve the desired results. The consensus was, therefore, that the second option should be adopted.

Some manufacturers recommended that both options be permitted allowing the manufacturer to decide how to test its vehicles. The agency does not agree with this recommendation. Not only would it be more difficult and expensive to enforce a standard that has alternative test procedures, but most manufacturers prefer the 5,500 pound weight limit option. The NHTSA concludes that as a result of the comments supporting the 5,500 pound maximum test weight, that this is an acceptable procedure for testing compliance with these two standards. Therefore, the standards are amended to incorporate this procedure.

The major incomplete vehicle manufacturers commenting on the notice suggested that testing vehicles at a maximum weight of 5,500 pounds might provide some immediate relief. None of the major incomplete vehicle manufacturers provided any information concerning how substantial that relief might be. Ford indicated that any relief might be limited.

The agency believes that the incomplete vehicle manufacturers must accept the responsibility for establishing reasonable restrictions upon their incomplete vehicles. The NHTSA has not been provided with sufficient evidence substantiating the statements of the incomplete vehicle manufacturers that their existing restrictions are reasonable. In fact, some evidence indicates that unnecessarily stringent restrictions are being imposed because incomplete vehicle manufacturers do not want to conduct the necessary testing to establish the appropriate weight and center of gravity restrictions. Since this amendment should reduce the severity of the test procedures, the agency concludes that incomplete vehicle manufacturers should immediately review their certification test procedures and reduce the restrictions being passed on to final-stage manufacturers.

Due to changes in the light truck market, there is reason to believe that the incomplete vehicle manufacturers will be more cooperative than when the agency spoke to them before beginning this rulemaking. At that time, light truck sales were still running well. Now that these sales are down, these manufacturers may be more solicitous of the needs of the final-stage manufacturers. If relief is not provided by the incomplete vehicle manufacturers, then the agency will consider taking additional steps, including the upgrading of Standards Nos. 212 and 219 as they apply to all light trucks.

General Motors (GM) questioned one of the agency's rationales for issuing the notice of proposed rulemaking. GM stated that the agency concludes that this action will provide a more appropriate level of safety for the affected vehicles while the initial extension of these standards to the affected vehicles provides, in GM's view, only a slight increase in the level of safety of the vehicles. GM indicates that since the application of these standards to the affected vehicles provides only slight benefits and since this amendment will

reduce those benefits, the standards should not apply to light trucks and vans. The agency disagrees with this suggestion.

The agency is currently reviewing the applicability of many of its safety standards to determine whether they ought to be extended to light trucks and other vehicles. Accident data clearly indicate the benefits that have resulted from the implementation of safety standards to cars. The fatality rate for passenger cars has decreased substantially since the implementation of a broad range of safety standards to those vehicles. On the other hand, light trucks and vans have not had a corresponding reduction in fatality rates over the years. The agency attributes much of this to the fact that many safety standards have not been applied to those vehicles. Since those vehicles are becoming increasingly popular as passenger vehicles, the agency concludes that safety standards must apply to them.

In response to GM's comment that this reduction in the test requirements for Standard Nos. 212 and 219 will remove all benefits derived by having the standards apply to those vehicles, the agency concludes that GM has misinterpreted the effects of this amendment. This amendment will reduce somewhat the compliance test requirements for those light trucks and vans with unloaded vehicle weights in excess of 5,500 pounds. It will not affect light trucks with unloaded vehicle weights below 5,500 pounds. According to agency information, approximately 25 percent of the light trucks have unloaded vehicle weights in excess of 5,500 while the remainder fall below that weight. As a result of weight reduction to improve fuel economy, it is likely that even more light trucks will fall below the 5,500 pound maximum test weight in the future. Therefore, this amendment will have no impact upon most light trucks and vans. In light of the small proportion of light trucks and vans affected by this amendment and considering the potential benefits of applying these standards to all light trucks and vans, the agency declines to adopt GM's suggestion that the standards be made inapplicable to these vehicles.

With respect to GM's question about the appropriate level of safety for light trucks, the agency's statement in the notice of proposed rulemaking was intended to show that the safety of light trucks and vans cannot be viewed without considering the relative safety of lighter vehicles that they may impact. Accordingly, the level of safety that the agency seeks to achieve by this and other safety standards is determined by balancing the interests of the occupants of passenger cars and heavier vehicles.

GM also questioned the agency's statement that vehicle aggressivity may be increased by imposing too severe requirements on these vehicles. GM suggested that no evidence exists that vehicle aggressivity is increased as a result of complying with these standards.

The agency stated in the proposal that it was concerned that compliance with the standards as they now exist might have increased the aggressivity of the vehicles, thereby harming the occupants of passenger cars that are impacted by these larger, more rigid vehicles. The agency is now beginning to examine the full range of vehicle aggressivity problems. The docket for this notice contains a paper recently presented by a member of our staff to the Society of Automotive Engineers on this subject. The agency tentatively concludes, based upon the initial results of our research and analysis, that vehicle aggressivity could be a safety problem and that the agency considers that possibility in issuing its safety standards. The NHTSA notes that Volkswagen applauds the agency's recognition of the vehicle aggressivity factor in safety.

As to GM's argument that compliance with the standards may not have increased vehicle aggressivity, our information on this point came from the manufacturers. The manufacturers indicated that compliance with Standards 212 and 219 requires strengthening the vehicle frame. This makes a vehicle more rigid. Our analysis indicates that making a vehicle more rigid may also make it more aggressive. Therefore, the agency concludes partially on the basis of the manufacturer's information, that compliance with the safety standards as they are written may have increased the aggressivity of the vehicles.

Ford Motor Company suggested that, rather than change these two particular standards, the agency should amend the certification regulation (Part 568) to state that any vehicle that is barrier tested would be required only to comply to an unloaded vehicle weight of 5,500 pounds or less. Ford suggested that this would standardize all of the tests and provide uniformity.

The agency is unable to accept Ford's recommendation for several reasons. First, the certification regulation is an inappropriate place to put a test requirement applicable to several standards. The tests' requirements of the standards should be found in each standard. Second, the Ford recommendation would result in a reduction of the level of safety currently imposed by Standard No. 301, *Fuel System Integrity*.

As we stated earlier and in several other notices, the agency is legislatively forbidden to modify Standard No. 301 in a way that would reduce the level of safety now required by that standard. Even without this legislative mandate, the agency would not be likely to relieve the burdens imposed by Standard No. 301. That standard is extremely important for the prevention of fires during crashes. Compliance of a vehicle with this standard not only protects the occupants of the vehicle that is in compliance but also protects the occupants of vehicles that it impacts. The agency concludes that the standard now provides a satisfactory level of safety in vehicles, and NHTSA would not be likely to amend it to reduce these safety benefits even if such an amendment were possible.

With respect to fuel system integrity, several manufacturers suggested that the agency had underestimated the impact of that standard upon weight and center of gravity restrictions. These commenters indicated that compliance with that standard requires more than merely adding shielding to the fuel systems of the vehicles. The agency is aware that compliance with that standard in certain instances has imposed restrictions upon manufacturers. Nonetheless, the agency continues to believe that as a result of this amendment, the chassis manufacturers will be able to reduce their weight and center of gravity restrictions while still maintaining the compliance of their vehicles with Standard No. 301.

Chrysler commented that the agency should consider including the new test procedure in Standard No. 204 and all other standards that require barrier testing. The agency has issued a notice on Standard No. 204 (44 FR 68470) stating that it was considering a similar test provision for that standard. The agency also is aware that any barrier test requirement imposed upon vehicles subject to substantial modifications by final-stage manufacturers will create problems for the final-stage manufacturers. Accordingly, the agency will consider the special problems of these manufacturers prior to the the issuance of standards that might affect them and will attempt to make the test requirements of the various standards consistent wherever possible.

The agency has reviewed this amendment in accordance with Executive Order 12044 and concludes that it will have no significant economic or other impact. Since the regulation relieves some testing requirements, it may slightly reduce costs associated with some vehicles. Accordingly, the agency concludes that this is not a significant amendment and a regulatory analysis is not required.

In accordance with the foregoing, Volume 49 of the Code of Federal Regulations Part 571 is amended by adding the following sentence to the end of paragraph S6.1(b) of Standard No. 212 (49 CFR 571.212) and paragraph S7.7(b) of Standard No. 219 (49 CFR 571.219).

Vehicles are tested to a maximum unloaded vehicle weight of 5,500 pounds.

The authors of this notice are William Smith of the Crashworthiness Division and Roger Tilton of the Office of Chief Counsel.

Issued on March 28, 1980.

Joan Claybrook
Administrator

45 F.R. 22044
April 3, 1980

MOTOR VEHICLE SAFETY STANDARD NO. 219

Windshield Zone Intrusion

S1. Scope. This standard specifies limits for the displacement into the windshield area of motor vehicle components during a crash.

S2. Purpose. The purpose of this standard is to reduce crash injuries and fatalities that result from occupants contacting vehicle components displaced near or through the windshield.

S3. Application. This standard applies to passenger cars and to multipurpose passenger vehicles, trucks and buses of 10,000 pounds or less gross vehicle weight rating. However, it does not apply to forward control vehicles, walk-in van-type vehicles, or to open body-type vehicles with fold-down or removable windshields.

S4. Definitions.

"Daylight Opening" (DLO) means the maximum unobstructed opening through the glazing surface, including reveal or garnish moldings adjoining the surface, as measured parallel to the outer surface of the glazing material.

"Windshield opening" means the outer surface of the windshield glazing material.

S5. Requirement. When the vehicle traveling longitudinally forward at any speed up to and including 30 mph impacts a fixed collision barrier that is perpendicular to the line of travel of the vehicle, under the conditions of S7, no part of the vehicle outside the occupant compartment, except windshield molding and other components designed to be normally in contact with the windshield, shall penetrate the protected zone template, affixed according to S6, to a depth of more than one-quarter inch, and no such part of a vehicle shall penetrate the inner surface of

that portion of the windshield, within the DLO, below the protected zone defined in S6.

S6. Protected zone template.

S6.1 The lower edge of the protected zone is determined by the following procedure (see Figure 1).

(a) Place a 6.5-inch diameter rigid sphere, weighing 15 pounds, in a position such that it simultaneously contacts the inner surface of the

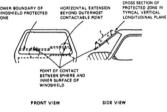

WINDSHIELD PROTECTED ZONE
Figure 1

windshield glazing and the surface of the instrument panel, including padding. If any accessories or equipment such as the steering control system obstruct positioning of the sphere, remove them for the purposes of this procedure.

(b) Draw the locus of points on the inner surface of the windshield contactable by the sphere across the width of the instrument panel. From the outermost contactable points, extend the locus line horizontally to the edges of the glazing material.

(c) Draw a line on the inner surface of the windshield below and one-half inch distant from the locus line.

(d) The lower edge of the protected zone is the longitudinal projection onto the outer surface of the windshield of the line determined in S6.1(c).

S6.2 The protected zone is the space enclosed by the following surfaces, as shown in Figure 1:

(a) The outer surface of the windshield in its precrash configuration.

(b) The locus of points 3 inches outward along perpendiculars drawn to each point on the outer surface of the windshield.

(c) The locus of lines forming a 45° angle with the outer surface of the windshield at each point along the top and side edges of the outer surface of the windshield and the lower edge of the protected zone determined in S6.1, in the plane perpendicular to the edge at that point.

S6.3 A template is cut or formed from Styrofoam, type DB, cut cell, to the dimensions of the zone as determined in S6.2. The template is affixed to the windshield so that it delineates the protected zone and remains affixed throughout the crash test.

S7. Test conditions. The requirement of S5 shall be met under the following conditions:

S7.1 The protected zone template is affixed to the windshield in the manner described in S6.

S7.2 The hood, hood latches, and any other hood retention components are engaged prior to the barrier crash.

S7.3 Adjustable cowl tops or other adjustable panels in front of the windshield are in the position used under normal operating conditions when windshield wiping systems are not in use.

S7.4 The parking brake is disengaged and the transmission is in neutral.

S7.5 Tires are inflated to the vehicle manufacturer's specifications.

S7.6 The fuel tank is filled to any level from 90 to 95 percent of capacity.

S7.7 The vehicle, including test devices and instrumentation, is loaded as follows:

(a) Except as specified in S7.6, a passenger car is loaded to its unloaded vehicle weight plus its rated cargo and luggage capacity weight, secured in the luggage area, plus a 50th-percentile test dummy as specified in Part 572 of this chapter at each front outboard designated seating position and at any other position whose protection system is required to be tested by a dummy under the provisions of Standard No. 208. Each dummy is restrained only by means that are installed for protection at its seating position.

(b) Except as specified in S7.6, a multipurpose passenger vehicle, truck or bus is loaded to its unloaded vehicle weight, plus 300 pounds or its rated cargo and luggage capacity, whichever is less, secured to the vehicle, plus a 50th-percentile test dummy as specified in Part 572 of this chapter at each front outboard designated seating position and at any other position whose protection system is required to be tested by a dummy under the provisions of Standard No. 208. Each dummy is restrained only by means that are installed for protection at its seating position. The load is distributed so that the weight on each axle as measured at the tire-ground interface is in proportion to its GAWR. If the weight on any axle when the vehicle is loaded to its unloaded vehicle weight plus dummy weight exceeds the axle's proportional share of the test weight, the remaining weight is placed so that the weight on that axle remains the same. For the purposes of this section, unloaded vehicle weight does not include the weight of work-performing accessories.

40 F.R. 25462
June 16, 1975

PREAMBLE TO MOTOR VEHICLE SAFETY STANDARD NO. 220

School Bus Rollover Protection

(Docket No. 75-2; Notice 2)

This notice establishes a new motor vehicle safety Standard No. 220, *School Bus Rollover Protection*, 49 CFR 571.220, specifying performance requirements for the structural integrity of the passenger compartment of school buses when subjected to forces that can be encountered in rollovers.

The Motor Vehicle and Schoolbus Safety Amendments of 1974 (the Act) mandate the issuance of Federal motor vehicle safety standards for several aspects of school bus performance, including crashworthiness of the vehicle body and frame. Pub. L. 93-942, section 202 (15 U.S.C. 1392(i)(1)(A)). Based on this mandate and on bus body crashworthiness research (DOT-HS-046-3-694), the NHTSA proposed rollover protection requirements for school buses (40 F.R. 8570, February 28, 1975). Citing statistics on the safety record of school bus operation, several manufacturers questioned whether any standard for school bus rollover protection could be justified.

The Act reflects a need, evidenced in correspondence to the NHTSA from the public, to protect the children who ride in school buses. They and their parents have little direct control over the types of vehicles in which they ride to school, and are not in a position to determine the safety of the vehicles. It is for this reason that the school bus standards must be effective and meaningful.

At the same time, the safety history of school buses does not demonstrate that radical modification of school bus structure would substantially decrease occupant death and injury. As noted in the "School Bus Safety Improvement Program" contract conducted by Ultrasystems, Inc., (DOT-HS-046-3-694) for the NHTSA:

"School buses are a relatively safe mode of human transportation. School bus accident rates and injury/fatality rates on a per-vehicle, per-vehicle-mile, per-passenger-mile, or per-passenger basis are significantly less than for other passenger vehicles. Accidents to school children while enroute to and from school occur primarily in modes other than as school bus passengers. However, school bus safety can and should be improved."

As a practical matter, the amount of structural modification called for in this standard is also limited as a result of the 9-month lead time available to implement the provisions of each school bus standard after its promulgation. The various new requirements imposed in response to the mandate of the Act will require considerable effort by school bus manufacturers to bring their products into conformity in the 9-month period.

The Physicians for Automotive Safety, The National Transportation Safety Board, the Home Insurance Company and other commenters suggested that the NHTSA had ignored the recommendations of the report submitted by Ultrasystems on school bus improvement. The report concluded that the improved school bus design tested by Ultrasystems could withstand a significantly greater load for the same amount of roof crush than existing school bus designs.

In fact, the NHTSA evaluated the test results and Ultrasystem's recommendations carefully. While the percentage of reduction of roof crush would be substantial as a result of the recommended design change, no relationship of this decrease in deflection to improved safety for occupants was established. Ultrasystems reported that increases of $500 in cost and 530 pounds were incurred to achieve several improvements, including those of the vertical roof crush test.

The recommendations also implied increased structural rigidity but did not evaluate its effect on the amount of energy absorbed by vehicle occupants in a crash. Also, Ultrasystems, did not consider the problems of lead time and retooling costs in making its recommendations. The NHTSA continues to consider that its proposal of 5⅛ inches of maximum roof crush under a load equal to 1½ times the vehicle's unloaded weight provides a satisfactory level of occupant crash protection. Available data do not support the conclusion that a 2- or 3-inch reduction of this crush would significantly improve the level of passenger safety in school buses. It is the intention of the NHTSA to continually review accident statistics relating to school bus safety. Accordingly, future upgrading of the standard will be considered should such action be warranted based upon availability of appropriate data.

In response to inquiries from the Motor Vehicle Manufacturers Association and General Motors as to the origin of the 5⅛-inch requirement, the limit is drawn from the existing School Bus Manufacturers Institute requirement for school bus structural integrity (Static Load Test Code for School Bus Body Structure, issued by the School Bus Manufacturers Institute).

In adopting the 5⅛-inch limit found in the present industry standard, the NHTSA is not merely preserving the status quo. While a manufacturer may have designed its products to meet the industry standard in the past, certain of its products presumably performed either better or worse than the nominal design. Conformity to NHTSA standards, in contrast, requires that every vehicle be capable of meeting the 5⅛-inch limit. This means that the manufacturer must design its vehicles to meet a higher level of performance, to provide a compliance margin for those of its products which fall below the nominal design level. Of course, the manufacturer can reduce the compliance-margin problem without redesign by improving the consistency of its manufacturing processes.

The standard requires that, upon the application of vertical downward force to the bus roof equal to 1½ times the vehicle's unloaded weight,

the vehicle roof shall not crush more than 5⅛ inches, and the emergency exits shall be capable of being opened, with the weight applied, and after its release. The National Transportation Safety Board, the Vehicle Equipment Safety Commission (VESC), Mercedes-Benz, and the Action for Child Transportation Safety organization suggested other methods for evaluation of crashworthiness. The NHTSA has considered these, but concludes that the static test specified in this standard provides a reasonable means to determine crashworthiness without unnecessary testing expense.

Based on submitted comments, the standard varies in some respects from the proposal. The sizes of the force application plates used to apply force and the method of application have been revised to simplify the test procedures and equipment, and to spread the force over larger areas of the vehicle roofs of large and small vehicles. The proposal specified a rigid, rectangular force application plate 36 inches wide and 20 inches shorter than the vehicle roof, preventing reliance on the roof end structures for rollover protection in typical body-on-chassis construction. Commenters pointed out that the end structures of the roof are almost certain to bear the weight of a rollover and should be included in a test of a vehicle's crashworthiness. Several manufacturers and other commenters recommended an increase in the size of the force application plate, in order to permit the foremost and rearmost roof "bows" of their buses to absorb a portion of the test load. Ford Motor Company stated it had performed the test as proposed and asserted that the roof of its van-type vehicle, as presently designed could not meet the requirement without an increase in the size of the force application plate to distribute the load over the entire vehicle roof. Chrysler Corporation stated it would find it necessary to discontinue production of small school buses because of redesign costs if the requirements were adopted as proposed.

With a view to the safety record of school buses and the 9-month lead time, the NHTSA concludes that the force application plate can be modified so that an additional "bow" or "bows" bear part of the applied force. It is the NHTSA's view that a change to permit both roof end structures to fully contribute to support of the applied force in the case of buses of more than 10,000 pounds would be a relaxation of current industry practices. Accordingly, the extent of change recommended by the industry is not adopted. The NHTSA concludes that an 8-inch increase in the length of the force application plate is sufficient to allow some portion of the applied force to be absorbed by the end bows of the roof while maintaining adequate crash protection. Therefore, for these buses the width of the plate remains as proposed while the length of the plate is increased 8 inches.

In the case of lighter buses, which are generally of the van type, the NHTSA has increased both the width and length of the plate to encompass the entire roof.

The procedure for applying force through the plate has also been modified in some respects. Many comments objected that the procedure required an expensive, complex hydraulic mechanism that would increase the costs of compliance without justification. The proposal specified an "evenly-distributed vertical force in a downward direction through the force application plate", starting with the plate horizontal. Commenters interpreted these specifications to mean that the vehicle would be required to absorb the energy in evenly-distributed fashion and that the horizontal attitude of the plate must be maintained.

Actually these specifications were included in the proposed method to advise manufacturers of the precise procedures to be employed in compliance testing of their products. Understanding that some manufacturers may choose to achieve the required force application by applying weights evenly over the surface of the plate, the standard specified an "evenly-distributed force" to eliminate other methods (such as a concentrated force at one end of the plate) that could unfairly test the vehicle structure. The horizontal attitude of the plate was also intended to establish a beginning point for testing on which a manufacturer can rely. While these specifications establish the exact circumstances under which vehicles can be tested, a manufacturer can depart from them as long as it can be shown that the vehicle would comply if tested exactly as specified. In place of the perfectly rigid plate called for in the standard, for example, a manu-

facturer could employ a plate of sufficient stiffness to ensure that the test results are not affected by the lack of rigidity.

Some modification of the test procedures has been made for simplification and clarity. To permit placement of the plate on the roof to begin testing without a suspension mechanism, the specification for horizontal attitude is modified to permit the plate to depart from the horizontal in the fore and aft direction only. Some manufacturers considered the initial application of force as an unnecessary complication. However, the initial force application of 500 pounds has been retained in order to permit elimination of inconsequential deformation of the roof structure prior to measurement of the permissible 5⅛ inches of deflection. In instances where the force application plate weighs more than 500 pounds, some type of suspension mechanism could be used temporarily to constrain the load level to the initial value, if the manufacturer decides to conduct his testing exactly as specified in the standard's procedures.

The requirement that force be applied "through the plate" has been changed to "to the plate" in order to avoid a misunderstanding that the vehicle must absorb energy evenly over the surface of its roof.

As proposed by several commenters, the rate of application in pounds per minute has been changed to inches per second, specifically "at any rate not more than ½ inch per second." Manufacturers should understand that "any" in this context is defined by the NHTSA (49 CFR § 571.4) to mean that the vehicle roof must satisfy the requirement at every rate of application within the stated range. General Motors reports that as a practical matter, the effect of speed in rate of application for tests of this nature is not significant in the range of 0.12 inches per second to 1 inch per second.

The requirement that movement "at any point" on the plate not exceed 5⅛ inches has not been modified despite some objections. The NHTSA considers it reasonable that excessive crush not be permitted at the extremities of the plate. Measurement of movement only at the center of the plate, for example, would permit total collapse of the structure in any direction as long as one point on the bus maintained its integrity.

PART 571; S 220—PRE 3

The preparation of the vehicle for the application of force has been modified to specify replacement of non-rigid body mounts with equivalent rigid mounts. The compression of deformable body mounts is unrelated to crashworthiness of the structure and can therefore be eliminated to permit testing of the structure itself.

Accessories or components which extend upward from the vehicle's roof (such as school bus lights) are removed for test purposes. It is also noted that the vehicle's transverse frame members or body sills are supported for test purposes. In response to a question from Blue Bird Body Company, a frame simulator may be used along with any other variations as long as the manufacturer assures himself that the vehicle would conform if tested precisely as specified in the standard.

The vehicle's emergency exits must also be capable of opening when the required force is applied, and following release of the force. As noted in comments, this requirement simulates the use of the exits after a rollover, whether or not the vehicle comes to rest on its roof. The proposed requirement of ability to close these exits is eliminated because such a capability is unnecessary in an emergency evacuation of the bus. For this reason, the requirement has been modified so that a particular test specimen (i.e., a particular bus) will not be required to meet requirements for emergency exits which open following release of force, if the exits have already been tested while the application force is maintained.

With regard to the requirements as a whole, Crown Coach and other manufacturers argued that the application of 1½ times the vehicle's unloaded weight unfairly discriminates against buses with a higher vehicle weight-to-passenger ratio. The NHTSA disagrees, and notes that the relevant consideration in rollover is the weight of the vehicle itself in determining the energy to be absorbed by the structure. In a related area, one manufacturer suggested that the increased weight of the NHTSA's contemplated new standards for school buses would increase unloaded vehicle weight to the point where redesign would be required to meet the rollover standard. The NHTSA has considered this issue and estimates that the only significant new weight would be for improved seating. This weight increase would not substantially increase the severity of the rollover standard.

The State of California suggested consolidation of the rollover standard with the joint strength. While such a consolidation would appear logical for school buses alone, the NHTSA prefers the flexibility of separate standards with a view to their use independently in the future for other vehicle types. For example, the application of vertical force to the vehicle structure may be appropriate in a vehicle for which the joint strength requirement would not be appropriate.

The State of Georgia requested that transit systems transporting school children be exempted from Standard No. 220. This commenter apparently misunderstood the applicability of the standard. It only applies to newly-manufactured vehicles and does not require modification of existing fleets, whether or not operated by a transit authority.

Interested persons should note that the NHTSA has issued a proposal to modify the definition of "school bus" (40 F.R. 40854, September 1, 1975) and that if that definition is adopted the requirements of this standard will apply to all vehicles that fall within the definition, whether or not they fall within the present definition.

In consideration of the foregoing, a new motor vehicle safety standard No. 220, *School Bus Rollover Protection*, is added as § 571.220 of Part 571 of Title 49, Code of Federal Regulations. . . .

Effective date: October 26, 1976.

The effective date of this standard is established as 9 months after the date of its issuance, as required by the Motor Vehicle and Schoolbus Safety Amendments of 1974, Pub. L. 93-492, section 202 (15 U.S.C. 1397(i)(1)(A)).

(Sec. 103, 119, Pub. L. 89-563, 80 Stat. 718 (15 U.S.C. 1392, 1407); § 202, Pub. L. 93-492, 88 Stat. 1470 (15 U.S.C. 1392); delegation of authority at 49 CFR 1.51)

Issued on January 22, 1976.

Howard J. Dugoff
Acting Administrator
41 F.R. 3874
January 27, 1976

PREAMBLE TO AMENDMENT TO MOTOR VEHICLE SAFETY STANDARD NO. 220

School Bus Rollover Protection

(Docket No. 73–3; Notice 7)

(Docket No. 73–20; Notice 10)

(Docket No. 73–34; Notice 4)

(Docket No. 75–2; Notice 3)

(Docket No. 75–3; Notice 5)

(Docket No. 75–7; Notice 3)

(Docket No. 75–24; Notice 3)

This notice announces that the effective dates of the redefinition of "school bus" and of six Federal motor vehicle safety standards as they apply to school buses are changed to April 1, 1977, from the previously established effective dates. This notice also makes a minor amendment to Standard No. 220, *School Bus Rollover Protection*, and adds a figure to Standard No. 221, *School Bus Body Joint Strength*.

The Motor Vehicle and Schoolbus Safety Amendments of 1974 (the Act) mandated the issuance of Federal motor vehicle safety standards for several aspects of school bus performance, Pub. L. 93–492, § 202 (15 U.S.C. § 1392 (i)(1)(A)). These amendments included a definition of school bus that necessitated a revision of the existing definition used by the NHTSA in establishing safety requirements. The Act also specified that the new requirements "apply to each school bus and item of school bus equipment which is manufactured . . . on or after the expiration of the 9-month period which begins on the date of promulgation of such safety standards." (15 U.S.C. § 1392(i)(1)(B)).

Pursuant to the Act, amendments were made to the following standards: Standard No. 301–75, *Fuel System Integrity* (49 CFR 571.301–75), effective July 15, 1976, for school buses not already covered by the standard (40 FR 483521, October 15, 1975); Standard No. 105–75, *Hydraulic Brake Systems* (49 CFR 571.105–75), effective October 12, 1976 (41 FR 2391, January

16, 1976); and Standard No. 217, *Bus Window Retention and Release* (49 CFR 571.217), effective for school buses on October 26, 1976 (41 FR 3871, January 27, 1976).

In addition, the following new standards were added to Part 571 of Title 49 of the Code of Federal Regulations, effective October 26, 1976: Standard No. 220, *School Bus Rollover Protection* (41 F.R. 3874, January 27, 1976); Standard No. 221, *School Bus Body Joint Strength* (41 F.R. 3872, January 26, 1976); and Standard No. 222, *School Bus Passenger Seating and Crash Protection* (41 F.R. 4016, January 28, 1976). Also, the existing definition of "school bus" was amended, effective October 27, 1976, in line with the date set by the Act for issuance of the standards.

The Act was recently amended by Public Law 94–346 (July 8, 1976) to change the effective dates of the school bus standards to April 1, 1977 (15 U.S.C. § 1392(i)(1)(B)). This notice is intended to advise interested persons of these changes of effective dates. In the case of Standard No. 301–75, the change of effective date is reflected in a conforming amendment to S5.4 of that standard. A similar amendment is made in S3 of Standard No. 105–75.

The agency concludes that the October 27, 1976, effective date for the redefinition of "school bus" should be postponed to April 1, 1977, to conform to the new effective dates for the upcoming requirements. If this were not done, the new classes of school buses would be required to meet existing standards that apply to school buses (e.g., Standard No. 108 (49 CFR 571.108)) before being required to meet the new standards. This would result in two stages of compliance, and would complicate the redesign efforts that Congress sought to relieve.

This notice also amends Standard No. 220 in response to an interpretation request by Blue Bird Body Company, and Sheller-Globe Corporation's petition for reconsideration of the standard. Both companies request confirmation that the standard's requirement to operate emergency exits during the application of force to the vehicle roof (S4(b)) does not apply to roof exits which are covered by the force application plate. The agency did not intend to require the operation of roof exits while the force application plate is in place on the vehicle. Accordingly, an appropriate amendment has been made to S4(b) of the standard.

With regard to Standard No. 220, Sheller-Globe also requested confirmation that, in testing its school buses that have a gross vehicle weight rating (GVWR) of 10,000 pounds or less, it may test with a force application plate with dimensions other than those specified in the standard. The standard does not prohibit a manufacturer from using a different dimension from that specified, in view of the NHTSA's expressed position on the legal effect of its regulations. To certify compliance, a manufacturer is free to choose any means, in the exercise of due care, to show that a vehicle (or item of motor vehicle equipment) would comply if tested by the NHTSA as specified in the standard. Thus the force application plate used by the NHTSA need not be duplicated by each manufacturer or compliance test facility. Sheller-Globe, or example, is free to use a force application plate of any width as long as it can certify its vehicle would comply if tested by the NHTSA according to the standard.

In a separate area, the agency corrects the inadvertent omission of an illustration from Standard No. 221 as it was issued January 26, 1976 (41 F.R. 3872). The figure does not differ from that proposed and, in that form, it received no adverse comment.

In accordance with recently enunciated Department of Transportation policy encouraging adequate analysis of the consequences of regulatory action (41 F.R. 16200, April 16, 1976), the agency herewith summarizes its evaluation of the economic and other consequences of this action on the public and private sectors, including possible loss of safety benefits. The changes in effective dates for the school bus standards are not evaluated because they were accomplished by law and not by regulatory action.

The change of effective date for the redefinition of "school bus" will result in savings to manufacturers who will not be required to meet existing school bus standards between October 27, 1976, and April 1, 1977. The agency calculates that the only standard that would not be met would be the requirement in Standard No. 108 for school bus marker lamps. In view of the agency's existing provision for the marking of light school buses in Pupil Transportation Standard No. 17 (23 CFR 1204), it is concluded that the absence of this equipment until April 1, 1977, will not have a significant adverse impact on safety.

The interpretative amendment of Standard No. 220 and the addition of a figure to Standard No. 221 are not expected to affect the manufacture or operation of school buses.

In consideration of the foregoing, Part 571 of Title 49 of the Code of Federal Regulations is amended. . . .

Effective dates:

1. Because the listed amendments do not impose additional requirements of any person, the National Highway Traffic Safety Administration finds that an immediate effective date of August 26, 1976 is in the public interest.

2. The effective date of the redefinition of "school bus" in 49 CFR Part 571.3 that was published in the issue of December 31, 1976 (40 F.R. 60033) is changed to April 1, 1977.

3. The effective dates of Standard Nos. 105–75, 217, 301–75, 220, 221, and 222 (as they apply to school buses) are April 1, 1977, in accordance with Public Law 94–346.

Effective: August 26, 1976

(Sec. 103, 119, Pub. L. 89–563, 80 Stat. 718 (15 U.S.C. 1392, 1407) ; Pub. L. 94–346, Stat. (15 U.S.C. § 1392(i)(1)(B)) ; delegation of authority at 49 CFR 1.50.)

Issued on August 17, 1976.

John W. Snow
Administrator

**41 F.R. 36027
August 26, 1976**

MOTOR VEHICLE SAFETY STANDARD NO. 220

School Bus Rollover Protection

S1. Scope. This standard establishes performance requirements for school bus rollover protection.

S2. Purpose. The purpose of this standard is to reduce the number of deaths and the severity of injuries that result from failure of the school bus body structure to withstand forces encountered in rollover crashes.

S3. Applicability. This standard applies to school buses.

S4. Requirements. When a force equal to 1½ times the unloaded vehicle weight is applied to the roof of the vehicle's body structure through a force application plate as specified in S5., Test procedures—

(a) The downward vertical movement at any point on the application plate shall not exceed 5⅛ inches; and

(b) Each emergency exit of the vehicle provided in accordance with Standard No. 217 (§ 571.217) shall be capable of opening as specified in that standard during the full application of the force and after release of the force, except that an emergency exit located in the roof of the vehicle is not required to be capable of being opened during the application of the force. A particular vehicle (*i.e.*, test specimen) need not meet the emergency opening requirement after release of force if it is subjected to the emergency exit opening requirements during the full application of the force.

S5. Test procedures. Each vehicle shall be capable of meeting the requirements of S4. when tested in accordance with the procedures set forth below.

S5.1 With any non-rigid chassis-to-body mounts replaced with equivalent rigid mounts, place the vehicle on a rigid horizontal surface so that the vehicle is entirely supported by means of the vehicle frame. If the vehicle is constructed without a frame, place the vehicle on its body sills. Remove any components which extend upward from the vehicle roof.

S5.2 Use a flat, rigid, rectangular force application plate that is measured with respect to the vehicle roof longitudinal and lateral center-lines;

(a) In the case of a vehicle with a GVWR of more than 10,000 pounds, 12 inches shorter than the vehicle roof and 36 inches wide; and

(b) In the case of a vehicle with a GVWR of 10,000 pounds or less, 5 inches longer and 5 inches wider than the vehicle roof. For purposes of these measurements, the vehicle roof is that structure, seen in the top projected view, that coincides with the passenger and driver compartment of the vehicle.

S5.3 Position the force application plate on the vehicle roof so that its rigid surface is perpendicular to a vertical longitudinal plane and it contacts the roof at not less than two points, and so that, in the top projected view, its longitudinal centerline coincides with the longitudinal centerline of the vehicle, and its front and rear edges are an equal distance inside the front and rear edges of the vehicle roof at the center-line.

S5.4 Apply an evenly-distributed vertical force in the downward direction to the force application plate at any rate not more than 0.5 inch per second, until a force of 500 pounds has been applied.

S5.5 Apply additional vertical force in the downward direction to the force application plate at a rate of not more than 0.5 inch per second until the force specified in S4 has been applied, and maintain this application of force.

S5.6 Measure the downward movement of any point on the force application plate which occurred during the application of force in accordance with S5.5.

S5.7 To test the capability of the vehicle's emergency exits to open in accordance with S4(b)—

(a) In the case of testing under the full application of force, open the emergency exits as specified in S4(b) while maintaining the force applied in accordance with S5.4 and S5.5; and

(b) In the case of testing after the release of all force, release all downward force applied to the force application plate and open the emergency exits as specified in S4(b).

S6. Test conditions. The following conditions apply to the requirements specified in S4.

S6.1 Temperature. The ambient temperature is any level between 32° F. and 90° F.

S6.2 Windows and doors. Vehicle windows, doors, and emergency exits are in fully-closed position, and latched but not locked.

41 F.R. 3874
January 27, 1976

PREAMBLE TO MOTOR VEHICLE SAFETY STANDARD NO. 221

School Bus Body Joint Strength

(Docket No. 73–34; Notice 3)

This notice establishes a new motor vehicle safety standard, No. 221; *School Bus Body Joint Strength*, 49 CFR 571.221, specifying a minimum performance level for school bus body panel joints.

The Motor Vehicle and Schoolbus Safety Amendments of 1974 (Pub. L. 93–492, 88 Stat. 1470, herein, the Act) require the issuance of minimum requirements for school bus body and frame crashworthiness. This rulemaking is pursuant to authority vested in the Secretary of Transportation by the Act and delegated to the Administrator of the NHTSA, and is preceded by notices of proposed rulemaking issued January 29, 1974 (39 F.R. 2490) and March 13, 1975 (40 F.R. 11738).

One of the significant injury-producing characteristics of school bus accidents, exposure to sharp metal edges, occurs when body panels become separated from the structural components to which they have been fastened. In an accident severe lacerations may result if the occupants of the bus are tossed against these edges. Moreover, if panel separation is great the component may be ejected from the vehicle, greatly increasing the possibility of serious injury.

This standard is intended to lessen the likelihood of these modes of injury by requiring that body joints on school buses have a tensile strength equal to 60 percent of the tensile strength of the weakest joined body panel, as suggested by the Vehicle Equipment Safety Commission (VESC). The NHTSA has determined that this is an appropriate level of performance for body joints and that its application to school buses is both reasonable and practicable. Furthermore, the NHTSA believes that adoption of this standard will provide an effective and meaningful solution to the body panel problem.

It is anticipated that this rule will burden manufacturers only to the extent of requiring the installation of more rivets than are currently used. The NHTSA has reviewed the economic and environmental impact of this proposal and determined that neither will be significant.

In their response to the two NHTSA proposals on this subject, several of the commenters suggested that the standard could be met by reducing the strength of the panel rather than increasing the strength of the joint, and that a minimum joint strength should be required. For several reasons the NHTSA does not believe that a minimum absolute joint strength is desirable at this time. While this standard will tend to increase the overall strength of buses, it is not designed to set minimum body panel strength requirements. Its purpose is to prevent panels from separating at the joint in the event of an accident. In order to deal with the problem of laceration, this regulation must be applicable to both exterior and interior joints. An absolute minimum joint strength requirement would be constrained by the level of performance appropriate for the relatively thin interior panels. Thus, the overall level of performance could not be defined in a meaningful fashion without severely and unnecessarily limiting the manufacturer's flexibility in designing his product. The NHTSA School Bus Rollover Protection Standard (49 CFR 571.220), which specifies requirements for the structural integrity of school bus bodies, should result in a practical lower limit on panel strength and thereby set a practical absolute minimum joint strength.

The NHTSA has no evidence that the mode of failure found in the larger traditional school buses also occurs in smaller, van-type school buses currently manufactured by automobile manufacturers for use as 11- to 17-passenger school buses. Ford Motor Company commented that the mode of injury sought to be prevented by this standard does not occur in accidents involving school buses converted from multipurpose passenger vehicles (vans). Chrysler Corporation suggested that the proposed requirement is inappropriate when applied to vans with "coach" joint construction. Based on these comments, the NHTSA has determined that until information to the contrary appears or is developed these vehicles should not be covered by the requirement. Accordingly, the application of the standard has been limited to school buses with a gross vehicle weight rating over 10,000 pounds.

Several commenters suggested that certain types of joints might not be susceptible of testing in the manner specified in this regulation. Up to this time the NHTSA has not found sufficient evidence in support of that position to justify amending the standard. If information is received indicating that different test methods are required for certain applications, appropriate action will be initiated.

In consideration of the foregoing, a new motor vehicle safety standard, No. 221, *School Bus Body Joint Strength*, is added as § 571.221 of Part 571 of Title 49, Code of Federal Regulations, as set forth below.

Effective date: October 26, 1976.

The effective date of this standard is 9 months after the date of issuance, as required by the Motor Vehicle and Schoolbus Safety Amendments of 1974, Pub. L. 93–492, section 202 (15 U.S.C. 1397(i)(1)(A)).

(Secs. 103, 119, Pub. L. 89–563, 80 Stat. 718 (15 U.S.C. 1392, 1407); § 202, Pub. L. 93–492, 88 Stat. 1470 (15 U.S.C. 1392); delegation of authority at 49 CFR 1.50.)

Issued on January 22, 1976.

Howard J. Dugoff
Acting Administrator

41 F.R. 3872
January 27, 1976

PREAMBLE TO AMENDMENT TO MOTOR VEHICLE SAFETY STANDARD NO. 221

School Bus Body Joint Strength

(Docket No. 73–3; Notice 7)

(Docket No. 73–20; Notice 10)

(Docket No. 73–34; Notice 4)

(Docket No. 75–2; Notice 3)

(Docket No. 75–3; Notice 5)

(Docket No. 75–7; Notice 3)

(Docket No. 75–24; Notice 3)

This notice announces that the effective dates of the redefinition of "school bus" and of six Federal motor vehicle safety standards as they apply to school buses are changed to April 1, 1977, from the previously established effective dates. This notice also makes a minor amendment to Standard No. 220, *School Bus Rollover Protection*, and adds a figure to Standard No. 221, *School Bus Body Joint Strength*.

The Motor Vehicle and Schoolbus Safety Amendments of 1974 (the Act) mandated the issuance of Federal motor vehicle safety standards for several aspects of school bus performance, Pub. L. 93–492, § 202 (15 U.S.C. § 1392(i)(1)(A)). These amendments included a definition of school bus that necessitated a revision of the existing definition used by the NHTSA in establishing safety requirements. The Act also specified that the new requirements "apply to each schoolbus and item of schoolbus equipment which is manufactured . . . on or after the expiration of the 9-month period which begins on the date of promulgation of such safety standards." (15 U.S.C. § 1392(i)(1)(B)).

Pursuant to the Act, amendments were made to the following standards: Standard No. 301–75, *Fuel System Integrity* (49 CFR 571.301–75), effective July 15, 1976, for school buses not already covered by the standard, (40 F.R. 483521, October 15, 1975); Standard No. 105–75, *Hydraulic Brake Systems* (49 CFR 571.105–75), effective October 12, 1976 (41 F.R. 2391, Jan-

uary 16, 1976); and Standard No. 217, *Bus Window Retention and Release* (49 CFR 571.217), effective for school buses on October 26, 1976 (41 F.R. 3871, January 27, 1976).

In addition, the following new standards were added to Part 571 of Title 49 of the Code of Federal Regulations, effective October 26, 1976: Standard No. 220, *School Bus Rollover Protection* (41 F.R. 3874, January 27, 1976); Standard No. 221, *School Bus Body Joint Strength* (41 F.R. 3872, January 26, 1976); and Standard No. 222, *School Bus Passenger Seating and Crash Protection* (41 F.R. 4016, January 28, 1976). Also, the existing definition of "school bus" was amended, effective October 27, 1976, in line with the date set by the Act for issuance of the standards.

The Act was recently amended by Public Law 94–346 (July 8, 1976) to change the effective dates of the school bus standards to April 1, 1977 (15 U.S.C. § 1392(i)(1)(B)). This notice is intended to advise interested persons of these changes of effective dates. In the case of Standard No. 301–75, the change of effective date is reflected in a conforming amendment to S5.4 of that standard. A similar amendment is made in S3 of Standard No. 105–75.

The agency concludes that the October 27, 1976, effective date for the redefinition of "school bus" should be postponed to April 1, 1977, to conform to the new effective dates for the upcoming requirements. If this were not done, the new classes of school buses would be required to meet existing standards that apply to school buses (e.g., Standard No. 108 (49 CFR 571.108)) before being required to meet the new standards. This would result in two stages of compilance, and would complicate the redesign efforts that Congress sought to relieve.

This notice also amends Standard No. 220 in response to an interpretation request by Blue Bird Body Company, and Sheller-Globe Corporation's petition for reconsideration of the standard. Both companies request confirmation that the standard's requirement to operate emergency exits during the application of force to the vehicle roof (S4(b)) does not apply to roof exits which are covered by the force application plate. The agency did not intend to require the operation of roof exits while the force application plate is in place on the vehicle. Accordingly, an appropriate amendment has been made to S4(b) of the standard.

With regard to Standard No. 220, Sheller-Globe also requested confirmation that, in testing its school buses that have a gross vehicle weight rating (GVWR) of 10,000 pounds or less, it may test with a force application plate with dimensions other than those specified in the standard. The standard does not prohibit a manufacturer from using a different dimension from that specified, in view of the NHTSA's expressed position on the legal effect of its regulations. To certify compliance, a manufacturer is free to choose any means, in the exercise of due care, to show that a vehicle (or item of motor vehicle equipment) would comply if tested by the NHTSA as specified in the standard. Thus the force application plate used by the NHTSA need not be duplicated by each manufacturer or compliance test facility. Sheller-Globe, for example, is free to use a force application plate of any width as long as it can certify its vehicle would comply if tested by the NHTSA according to the standard.

In a separate area, the agency corrects the inadvertent omission of an illustration from Standard No. 221 as it was issued January 26, 1976 (41 F.R. 3872). The figure does not differ from that proposed and, in that form, it received no adverse comment.

In accordance with recently enunciated Department of Transportation policy encouraging adequate analysis of the consequences of regulatory action (41 F.R. 16200, April 16, 1976), the agency herewith summarizes its evaluation of the economic and other consequences of this action on the public and private sectors, including possible loss of safety benefits. The changes in effective dates for the school bus standards are not evaluated because they were accomplished by law and not by regulatory action.

The change of effective date for the redefinition of "school bus" will result in savings to manufacturers who will not be required to meet existing school bus standards between October 27, 1976, and April 1, 1977. The agency calculates that the only standard that would not be met would be the requirement in Standard No. 108 for school bus marker lamps. In view of the agency's existing provision for the marking of light school buses in Pupil Transportation Standard No. 17 (23 CFR 1204), it is concluded that the absence of this equipment until April 1, 1977, will not have a significant adverse impact on safety.

The interpretative amendment of Standard No. 220 and the addition of a figure to Standard No. 221 are not expected to affect the manufacture or operation of school buses.

In consideration of the foregoing, Part 571 of Title 49 of the Code of Federal Regulations is amended. . . .

Effective dates:

1. Because the listed amendments do not impose additional requirements of any person, the National Highway Traffic Safety Administration finds that an immediate effective date of August 26, 1976 is in the public interest.

2. The effective date of the redefinition of "school bus" in 49 CFR Part 571.3 that was published in the issue of December 31, 1976 (40 F.R. 60033) is changed to April 1, 1977.

3. The effective dates of Standard Nos. 105–75, 217, 301–75, 220, 221, and 222 (as they apply to school buses) are April 1, 1977, in accordance with Public Law 94–346.

Effective: August 26, 1976

(Sec. 103, 119, Pub. L. 89–563, 80 Stat. 718 (15 U.S.C. 1392, 1407) ; Pub. L. 94–346, Stat. (15 U.S.C. § 1392(i)(1)(B)) ; delegation of authority at 49 CFR 1.50).

Issued on August 17, 1976.

John W. Snow
Administrator

41 F.R. 36027
August 26, 1976

MOTOR VEHICLE SAFETY STANDARD NO. 221

School Bus Body Joint Strength

S1. Scope. This standard establishes requirements for the strength of body panel joints in school bus bodies.

S2. Purpose. The purpose of this standard is to reduce deaths and injuries resulting from the structural collapse of school bus bodies during crashes.

S3. Application. This standard applies to school buses with gross vehicle weight ratings of more than 10,000 pounds.

S4. Definitions.

"Body component" means a part of a bus body made from a single piece of homogeneous material or from a single piece of composite material such as plywood.

"Body panel" means a body component used on the exterior or interior surface to enclose the bus' occupant space.

"Body panel joint" means the area of contact or close proximity between the edges of a body panel and another body component, excluding spaces designed for ventilation or another functional purpose, and excluding doors, windows, and maintenance access panels.

"Bus body" means the portion of a bus that encloses the bus' occupant space, exclusive of the bumpers, the chassis frame, and any structure forward of the forwardmost point of the windshield mounting.

S5. Requirement. When tested in accordance with the procedure of S6, each body panel joint shall be capable of holding the body panel to the member to which it is joined when subjected to a force of 60% of the tensile strength of the weakest joined body panel determined pursuant to S6.2.

S6. Procedure.

S6.1 Preparation of the test specimen.

S6.1.1 If a body panel joint is 8 inches long or longer, cut a test specimen that consists of any randomly selected 8-inch segment of the joint, together with a portion of the bus body whose dimensions, to the extent permitted by the size of the joined parts, are those specified in Figure 1, so that the specimen's centerline is perpendicular to the joint at the midpoint of the joint segment. Where the body panel is not fastened continuously, select the segment so that it does not bisect a spot weld or a discrete fastener.

S6.1.2 If a joint is less than 8 inches long, cut a test specimen with enough of the adjacent material to permit it to be held in the tension testing machine specified in S6.3.

S6.1.3 Prepare the test specimen in accordance with the preparation procedures specified in the 1973 edition of the Annual Book of ASTM Standards, published by the American Society for Testing and Materials, 1916 Race Street, Philadelphia, Pennsylvania 19103.

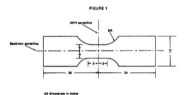

FIGURE 1

All dimensions in inches

S6.2 Determination of minimum allowable strength. For purposes of determining the minimum allowable joint strength, determine the tensile strengths of the joined body components as follows:

(a) If the mechanical properties of a material are specified by the American Society for Testing and Materials, the relative tensile strength for such a material is the minimum tensile strength specified for that material in the 1973 edition of the Annual Book of ASTM Standards.

(b) If the mechanical properties of a material are not specified by the American Society for Testing and Materials, determine its tensile strength by cutting a specimen from the bus body outside the area of the joint and by testing it in accordance with S6.3.

S6.3 Strength test.

S6.3.1 Grip the joint specimen on opposite sites of the joint in a tension testing machine calibrated in accordance with Method E4, Verification of Testing Machines, of the American Society for Testing and Materials (1973 Annual Book of ASTM Standards).

S6.3.2 Adjust the testing machine grips so that the joint, under load, will be in stress approximately perpendicular to the joint.

S6.3.3 Apply a tensile force to the specimen by separating the heads of the testing machine at any uniform rate not less than ⅛ inch and not more than ⅝ inch per minute until the specimen separates.

41 F.R. 3872
January 27, 1976

PREAMBLE TO MOTOR VEHICLE SAFETY STANDARD NO. 222

School Bus Seating and Crash Protection

(Docket No. 73–3; Notice 5)

This notice establishes a new motor vehicle safety Standard No. 222, *School Bus Seating and Crash Protection*, that specifies seating. restraining barrier, and impact zone requirements for school buses.

The Motor Vehicle and Schoolbus Safety Amendments of 1974, Pub. L. 93-492, directed the issuance of a school bus seating systems performance standard (and other standards in seven areas of vehicle performance). The NHTSA had already issued two proposals for school bus seating systems prior to enactment of the 1974 Safety Amendments (the Act) (38 F.R. 4776, February 22. 1973) (39 F.R. 27585, July 30, 1974) and subsequently published two additional proposals (40 F.R. 17855, April 23, 1975) (40 F.R. 47141, October 8, 1975). Each aspect of the requirements was fully considered in the course of this rulemaking activity. Comments received in response to the most recent proposal were limited to a few aspects of the Standard.

The largest number of comments were received on the requirement that school bus passenger seats be equipped with seat belt anchorages at each seating position. The standard relies on compartmentalization between well-padded and well-constructed seats to provide occupant protection on school buses (other than van-type buses). At the same time, seat belt anchorages were proposed so that a greater measure of protection could be gained if a particular user chose to use the anchorages by installation of seat belts together with a system to assure that seat belts would be worn, properly adjusted, and not misused.

Bus operators strongly expressed the view that the presence of seat belt anchorages would encourage the installation of seat belts by school districts without providing the necessary supervision of their use. This association of school bus operators (National School Transportation Association) also questioned the benefits that would be derived from anchorage installation as long as their utilization is not required. In view of these factors, and the indications that in any event only a small fraction of school buses would have belts installed and properly used, the NHTSA concludes that the proposed seat belt anchorage requirement should not be included in this initial school bus seating standard. Further study of the extent to which belts would be installed and properly used should permit more certainty as the basis for any future action.

NHTSA calculations demonstrate that the strength characteristics of the seat specified by the standard to provide the correct amount of compartmentalization also provide the strength necessary to absorb seat belt loads. This means that an operator or school district may safely attach seat belts to the seat frame. even where anchorages are not installed as original equipment. The seat is strong enough to take the force of occupants against the seat back if no belts are utilized. or the force of occupants against seat belts if occupants are restrained by belts attached to the seat frame through the anchorages provided.

The Physicians for Automotive Safety (PAS) requested that lap belts be required in addition to the compartmentalization offered by the seating systems. The agency concluded earlier in this rulemaking procedure that compartmentalization provides satisfactory protection and that a requirement for belts without the assurance of proper supervision of their use would not be an effective means of providing occupant protection.

PAS has not provided data or arguments that would modify this conclusion, and its request is therefore denied.

PAS, relying on testing undertaken at the University of California at Los Angeles in 1967 and 1969, argued that a vertical seat back height of 24 inches above the seating reference point (SRP) is necessary to afford adequate protection against occupant injury. The NHTSA, as noted in its fourth notice of school bus crash protection, based its 20-inch requirement on newer data generated in dynamic and static testing by AMF Corporation of prototype seats designed to meet the proposed requirements of the standard ("Development of a Unitized School Bus", DOT-HS-400969). While the NHTSA does not dispute that a properly constructed, higher seat back provides more protection than a lower seat back, the data support the agency's determination that the 20-inch seat back provides a reasonable level of protection. School bus accident data do not provide substantial evidence of a whiplash injury experience that could justify a 4-inch increase in seat back height. For this reason, the seat back height is made final as proposed.

Several commenters objected to applicability of the standard to school buses with a gross vehicle weight rating (GVWR) of 10,000 pounds or less (light school buses), asserting that the special requirements of the standard for those buses were inappropriate, or unachievable within the 9-month leadtime for compliance mandated by the Act.

Chrysler Corporation requested exclusion of light school buses from this standard for an indefinite period, and Ford Motor Company requested that essentially the same package of standards as already are provided in its van-type multi-purpose passenger vehicles and school bus models be required in the future, with no additional protection. Both companies believe that the relatively small numbers of their vehicles sold as school buses would have to be withdrawn from the market because of the expense of tooling new seating that offers more crash protection than present seating. Wayne Corporation manufactures a light school bus that is not based on a van-type vehicle, and requested that seats used

in its larger models be permitted in smaller models, along with seat belts that comply with Standard No. 209.

The Congressional direction to issue standards for school bus seating systems (15 U.S.C. § 1392(i)(1)(A)(iv)) implies that existing seating and occupant crash protection standards are insufficient for vehicles that carry school children. The NHTSA has proposed a combination of requirements for light school buses that differ from those for heavier buses, because the crash pulse experienced by smaller vehicles is more severe than that of larger vehicles in similar collisions. The standard also specifies adequate numbers of seat belts for the children that the vehicle would carry, because such restraints are necessary to provide adequate crash protection in small vehicles. The requirements applicable to light school buses are considered reasonable, and are therefore included in the final rule as proposed.

In Wayne's case, it is not clear why the seat it has developed for heavier school buses will not serve in its smaller school buses. Seat belts may need to be attached to the floor to support the force specified by Standard No. 210 for anchorages. Also, some interior padding may be necessary to meet the vehicle impact zone requirements of S5.3.1.1(a).

Sheller-Globe Corporation (Sheller) and Wayne considered unreasonable the standard's limitation on maximum distance between a seat's SRP and the rear surface of the seat or restraining barrier forward of the SRP (S5.2). The limitation exists to minimize the distance an occupant travels before forward motion is arrested by the padded structure that compartmentalizes the occupant. The two bus manufacturers contend that they must also comply with State requirements for a minimum distance between seats that results in only 1 inch of tolerance in seating placement.

Section 103(d) of the National Traffic and Motor Vehicle Safety Act provides in part:

(d) Whenever a Federal motor vehicle safety standard . . . is in effect, no State or political subdivision of a State shall have any authority either to establish or continue in effect, with respect to any motor vehicle or item of motor vehicle equipment any safety standard applicable to the same aspect of performance of such vehicle or item of equipment which is not identical to the Federal standard.

It is the opinion of the NHTSA that any State requirement relating to seat spacing, other than one identical to the Federal requirement for maximum spacing of 20 inches from the SRP, is preempted under § 103(d), 15 U.S.C. § 1392(d).

Sheller advocated wider seat spacing for activity buses, because seats are occupied for longer periods of time on road trips. The NHTSA, noting that activity buses are often used on the open highway at high speeds for long periods of time, requests comments on the advisability of specifying a seat belt requirement in place of the seat spacing requirement in the case of these buses.

Much of Sheller and Wayne's concern over tolerances may stem from a misunderstanding of the meaning of "seating reference point" (SRP). As defined by the NHTSA (49 CFR 571.3), the SRP is essentially the manufacturer's design reference point which simulates the pivot center of the human torso and thigh, located in accordance with the SAE Standard J826. Thus the manufacturer calculates, on its seat design seen in side projected view, the pivot center of the human torso and thigh of the potential seat occupant, and then establishes a design reference point that simulates the location of the actual pivot center. The NHTSA has interpreted that this design reference point may be fixed by the manufacturer with reference to the seating structure to simplify calculation of its location in a bus for purposes of measurement and compliance.

Sheller also requested that the "seat performance forward" testing be simplified by eliminating the 8-inch range of locations at which the lower loading bar can be applied against the seat back. As noted in the preamble to Notice 4 of this docket in response to a similar request from Blue Bird Body Company, the NHTSA declines to make this restriction, to discourage the addition of a narrow 2-inch wide structural member at this point simply to meet the requirement. This reasoning remains valid and Sheller's request is denied.

Sheller also asked that the requirement for forward-facing seats be eliminated from the standard, in view of the practice of installing side-facing seats in some buses for handicapped students. The NHTSA designed the seating system in this standard for protection from fore and aft crash forces, and considers it necessary that the seats be forward-facing to achieve the objective of occupant protection. Comments are solicited on whether the provision of this protection in special vehicles is impractical.

The Vehicle Equipment and Safety Commission (VESC) asked for a minimum seat width of 13 inches for each designated seating position, noting that the standard's formula permits seating of 12.67 inches in width. The agency does not believe its standard will encourage seats narrower than those presently provided in school buses, but will watch for any indication that that is occurring. Action can be taken in the future if it appears that seating is being designed to be narrower than at present.

In consideration of the foregoing, a new motor vehicle safety Standard No. 222, *School Bus Seating and Crash Protection*, is added as § 571.222, of Part 571 of Title 49, Code of Federal Regulations. . . .

Effective date: October 26, 1976. The effective date of this standard is established as 9 months after the date of its issuance, as required by the Motor Vehicle and Schoolbus Safety Amendments of 1974, Pub. L. 93–492, section 202 (15 U.S.C. 1397(i)(1)(A)).

(Sec. 103, 119, Pub. L. 89–563, 80 Stat. 718 (15 U.S.C. 1392, 1407); § 202, Pub. L. 93–492, 88 Stat. 1470 (15 U.S.C. 1392); delegation of authority at 49 CFR 1.50).

Issued on January 22, 1976.

Howard J. Dugoff
Acting Administrator

41 F.R. 4016
January 28, 1976

PREAMBLE TO AMENDMENT TO MOTOR VEHICLE SAFETY STANDARD NO. 222

School Bus Seating and Crash Protection

(Docket No. 73–3; Notice 6)

This notice responds to two petitions for reconsideration of Standard No. 222, *School Bus Passenger Seating and Crash Protection*, as it was issued January 22, 1976.

Standard No. 222 (49 CFR 571.222 was issued January 22, 1976 (41 F.R. 4016, January 28, 1976), in accordance with § 202 of the Motor Vehicle and Schoolbus Safety Amendments of 1974, Pub. L. 93–492 (15 U.S.C. § 1392(i)(1)) and goes into effect on October 26, 1976. The standard provides for compartmentalization of bus passengers between well-padded and well-constructed seats in the event of collision. Petitions for reconsideration of the standard were received from Sheller-Globe Corporation and from the Physicians for Automotive Safety (PAS), which also represented the views of Action for Child Transportation Safety, several adult individuals, and several school bus riders.

PAS expressed dissatisfaction with several aspects of the standard. The organization objected most strongly to the agency's decision that seat belts should not be mandated in school buses. PAS disagreed with the agency conclusion (39 F.R. 27585, July 30, 1974) that, whatever the potential benefits of safety belts in motor vehicle collisions, the possibility of their non-use or misuse in the hands of children makes them impractical in school buses without adequate supervision. In support of safety belt installation, PAS cited statistics indicating that 23 percent of reported school bus accidents involve a side impact or rollover of the bus.

While safety belts presumably would be beneficial in these situations, PAS failed to provide evidence that the belts, if provided, would be properly utilized by school-age children. The agency will continue to evaluate the wisdom of

its decision not to mandate belts, based on any evidence showing that significant numbers of school districts intend to provide the supervision that should accompany belt use. In view of the absence of evidence to date, however, the agency maintains its position that requiring the installation of safety belts on school bus passenger seats is not appropriate and denies the PAS petition for reconsideration. The agency continues to consider the reduced hostility of improved seating to be the best reasonable form of protection against injury.

PAS asked that a separate standard for seat belt assembly anchorages be issued. They disagree with the agency's conclusion (41 F.R. 4016) that seat belt anchorages should not be required because of indications that only a small fraction of school buses would have belts installed and properly used. However, PAS failed to produce evidence that a substantial number of school buses would be equipped with safety belts, or that steps would be taken to assure the proper use of such belts. In the absence of such information, the agency maintains its position that a seat belt anchorage requirement should not be included in the standard at this time, and denies the PAS petition for reconsideration.

The NHTSA does find merit in the PAS concern that in the absence of additional guidance, improper safety belt installation may occur. The Administration is considering rulemaking to establish performance requirements for safety belt anchorages and assemblies when such systems are installed on school bus passenger seats.

PAS also requested that the seat back height be raised from the 20-inch level specified by the standard to a 24-inch level. In support of this position, the organization set forth a "common sense" argument that whiplash must be occurring to school bus passengers in rear impact. However, the agency has not been able to locate any quantified evidence that there is a significant whiplash problem in school buses. The crash forces imparted to a school bus occupant in rear impact are typically far lower than those imparted in a car-to-car impact because of the greater weight of the school bus. The new and higher seating required by the standard specifies energy absorption characteristics for the seat back under rear-impact conditions, and the agency considers that these improvements over earlier seating designs wil reduce the number of injuries that occur in rear impact. For lack of evidence of a significant whiplash problem, the PAS petition for a 24-inch seat back is denied.

PAS believed that the States and localities that specify a 24-inch seat back height would be precluded from doing so in the future by the preemptive effect of Standard No. 222 under § 103(f) of the National Traffic and Motor Vehicle Safety Act (15 U.S.C. § 1392(f)):

§ 103 * * * * *

(d) Whenever a Federal motor vehicle safety standard under this subchapter is in effect, no State or political subdivision of a State shall have any authority either to establish, or to continue in effect, with respect to any motor vehicle or item of motor vehicle equipment any safety standard applicable to the same aspect of performance of such vehicle or item of equipment which is not identical to the Federal standard. Nothing in this section shall be construed to prevent the Federal Government or the government of any State or political subdivision thereof from establishing a safety requirement applicable to motor vehicle equipment procured for its own use if such requirement imposes a higher standard of performance than that required to comply with the otherwise applicable Federal standard.

Standard No. 222 specifies a minimum seat back height (S5.1.2) which manufactures may exceed as long as their product conforms to all other requirements of the standards applicable to school buses. It is the NHTSA's opinion that any State standard of general applicability concerning seat back height of school bus seating

would also have to specify a minimum height identical to the Federal requirement. Manufacturers would not be required to exceed this minimum. Thus, the PAS petition to state seat back height as a minimum is unnecessary and has already been satisfied, although it does not have the effect desired by the PAS.

With regard to the PAS concern that the States' seat height requirements would be preempted, the second sentence of § 103(d) clarifies that the limitation on safety regulations of general applicability does not prevent governmental entities from specifying additional safety features in vehicles purchased for their own use. Thus, a State or its political subdivisions could specify a seat back height higher than 20 inches in the case of public school buses. The second sentence does not permit these governmental entities to specify safety features that prevent the vehicle or equipment from complying with applicable safety standards.

With regard to which school buses qualify as 'public school buses' that may be fitted with additional features, it is noted that the agency includes in this category those buses that are owned and operated by a private contractor under contract with a State to provide transportation for students to and from public schools.

Sheller-Globe Corporation (Sheller) petitioned for exclusion from the seating requirements for seating that is designed for handicapped or convalescent students who are unable to utilize conventional forward-facing seats. Typically, side-facing seats are installed to improve entry and egress since knee room is limited in forward-facing seats, or spaces on the bus are specifically designed to accommodate wheelchairs. The standard presently requires that bus passenger seating be forward-facing (S5.1) and conform to requirements appropriate for forward-facing seats. Blue Bird Body Company noted in a March 29, 1976, letter that it also considered the standard's requirements inappropriate for special seating.

The agency has considered the limited circumstances in which this seating would be offered in school buses and concludes that the seat-spacing requirement (S5.2) and the fore-and-aft seat performance requirements (S5.1.3, S5.1.4) are not

PART 571; S 222—PRE 6

appropriate for side-facing seats designed solely for handicapped or convalescent students. Occupant crash protection is, of course, as important for these students as others, and the agency intends to establish requirements suited to these specialized seating arrangements. At this time, however, insufficient time remains before the effective date of this standard to establish different requirements for the seating involved. Therefore, the NHTSA has decided to modify its rule by the exclusion of side-facing seating installed to accommodate handicapped or convalescent passengers.

School bus manufacturers should note that the limited exclusion does not relieve them from providing a restraining barrier in front of any forward-facing seat that has a side-facing seat or wheelchair position in front of it .

Sheller also petitioned for a modification of the head protection zone (S5.3.1.1) that describes the space in front of a seating position where an occupant's head would impact in a crash. The outer edge of this zone is described as a vertical longitudinal plane 3.25 inches inboard of the outboard edge of the seat.

Sheller pointed out that van-type school buses utilize "tumble home" in the side of the vehicle that brings the bus body side panels and glazing into the head protection zone. As Sheller noted, the agency has never intended to include body side panels and glazing in the protection zone. The roof structure and overhead projections from the interior are included in this area of the zone. To clarify this distinction and account for the "tumble home," the description of the head impact zone in S5.3.1.1 is appropriately modified.

In accordance with recently enunciated Department of Transportation policy encouraging adequate analysis of the consequences of regulatory action (41 F.R. 16201; April 16, 1976), the agency herewith summarizes its evaluation of the economic and other consequences of this action on the public and private sectors, including possible loss of safety benefits. The decision to withdraw requirements for side-facing seats used by handicapped or convalescent students will result in cost savings to manufacturers and pur-

chasers. The action may encourage production of specialized buses that would otherwise not be built if the seating were subject to the standard. Because the requirements are not appropriate to the orientation of this seating, it is estimated that no significant loss of safety benefits will occur as a result of the amendment. The exclusion of sidewall, window or door structure from the head protection zone is simply a clarification of the agency's longstanding intent that these components not be subject to the requirements. Therefore no new consequences are anticipated as a result of this amendment.

In an area unrelated to the petitions for reconsideration, the Automobile Club of Southern California petitioned for specification of a vandalism resistance specification for the upholstery that is installed in school buses in compliance with Standard No. 222. Data were submitted on experience with crash pads installed in school buses operated in California. Vandalism damage was experienced, and its cost quantified in the submitted data.

The Automobile Club made no argument that the damage to the upholstery presents a significant safety problem. While it is conceivable that removal of all padding from a seat back could occur and expose the rigid seat frame, the agency estimates that this would occur rarely and presumably would result in replacement of the seat. Because the agency's authority under the National Traffic and Motor Vehicle Safety Act is limited to the issuance of standards that meet the need for motor vehicle safety (15 U.S.C. § 1392(a)), the agency concludes that a vandalism resistance requirement is not appropriate for inclusion in Standard No. 222.

In light of the foregoing, Standard No. 222 (49 CFR 571.222) is amended. . . .

Effective date: October 26, 1976. Because the standard becomes effective on October 26, 1976, it is found to be in the public interest that an effective date sooner than 180 days is in the public interest. Changes in the text of the Code of Federal Regulations should be made immediately.

Effective: October 26, 1976

(Sec. 103, 119, Pub. L. 89–563, 80 Stat. 718 (15 U.S.C. 1392, 1407); delegation of authority at 49 CFR 1.50.)

Issued on July 7, 1976.

James B. Gregory
Administrator

41 F.R. 28506
July 12, 1976

PREAMBLE TO AMENDMENT TO MOTOR VEHICLE SAFETY STANDARD NO. 222

School Bus Seating and Crash Protection

(Docket No. 73–3; Notice 8)

This notice amends Standard No. 222, *School Bus Passenger Seating and Crash Protection*, to delay the effective date for maximum rearward deflection of seats from April 1, 1977, to April 1, 1978.

Standard No. 222 (49 CFR 571.222), as published January 28, 1976 (41 F.R. 4016), established October 27, 1976, as the effective date of the standard, as mandated by the Motor Vehicle and Schoolbus Safety Amendments of 1974 (the Act) (Pub. L. 93–492). Congress subsequently amended the Act by Public Law 94–346 (July 8, 1976) to extend the effective date for the implementation of school bus standards to April 1, 1977.

The NHTSA has promulgated regulations on several aspects of performance mandated by Congress in the Act. These regulations become effective on April 1, 1977. The agency concludes, however, that compliance with one provision of Standard No. 222 by the April 1, 1977, effective date would be impracticable, would result in substantial economic waste, and would not be in the public interest.

Since publication of Standard No. 222, a misunderstanding has arisen within the industry concerning the definition of the term "absorbed" when used in connection with the requirements in sections S5.1.3.4 and S5.1.4.2. The NHTSA explained the term "absorbed" in an interpretation to Thomas Built Buses (July 30, 1976) to mean "receive without recoil." This interpretation requires that returned energy be subtracted from total energy applied to the seat back to calculate energy "absorbed" by the seat back.

School bus manufacturers tested their seats in accordance with the NHTSA definition of "absorbed" and found that the seats continued to comply with the requirements of Standard No. 222 when tested for forward performance (S5.1.3), but these same seats were marginally below the NHTSA requirements for rearward seat deflection. Based upon these test data, petitions have been received from Thomas Built Buses, Blue Bird Body Company, Carpenter Body Works, Wayne Corporation, and Ward School Bus Manufacturing, all requesting a change in rearward performance requirements.

The NHTSA has examined the data submitted by the manufacturers and concludes that the seats upon which the tests were made demonstrate a high probability of meeting most of the requirements of Standard No. 222. Further, the agency concludes that to mandate full compliance with the rearward performance requirements of Standard No. 222 would require extensive retooling and redesign. This could result in substantial economic waste of seats now in production and severe economic hardship for manufacturers.

The NHTSA is particularly concerned that to require full compliance with the rearward performance requirements at this late date might mean that manufacturers would be unable to redesign their seats in time to commence manufacture of completed buses on April 1, 1977. Since single-stage buses produced after April 1, 1977, must meet NHTSA safety requirements in all other respects, they will be substantially safer than buses currently in use. Therefore, the agency finds that it is in the interest of safety to ensure that these safer buses will be available on April 1, 1977, to replace older less safe models. To ensure that safer buses can be marketed without delay, the NHTSA extends the effective date of requirements for maximum rearward deflection of seats to April 1, 1978. It is emphasized that the numerous other requirements for school bus seating, including all other rearward performance requirements, remain in effect, which ensures adequate interior protection as of April 1, 1977, as mandated by Congress. A proposal for minor modification of S5.1.4 (to be published shortly) will permit reinstitution of rearward deflection requirements following the 1-year delay.

Because of the imminent effective date of the school bus safety standards and the lead time required to modify seat design, the NHTSA for good cause finds that notice and public procedure on this amendment are impracticable and contrary to the public interest.

In consideration of the foregoing, S5.1.4(b) of Standard No. 222 (49 CFR 571.222) is amended by the addition, at the beginning of the first sentence, of the following phrase: "In the case of a school bus manufactured on or after April 1, 1978,".

Effective date: December 16, 1976. Because this amendment relieves a restriction and does not impose requirements on any person, it is found, for good cause shown, that an immediate effective date is in the public interest.

(Secs. 103, 119, Pub. L. 89-563, 80 Stat. 718 (15 U.S.C. 1392, 1407); Sec. 202, Pub. L. 93-492, 88 Stat. 1470 (15 U.S.C. 1392); delegation of authority at 49 CFR 1.50.)

Issued on December 10, 1976.

Acting Administrator
Charles E. Duke

41 F.R. 54945
December 16, 1976

PREAMBLE TO AMENDMENT TO MOTOR VEHICLE SAFETY STANDARD NO. 222

(Docket No. 73–3; Notice 12)

This notice amends Standard No. 222, *School Bus Passenger Seating and Crash Protection*, increasing the allowable rearward deflection of seats from 8 to 10 inches. The action is taken in response to petitions that indicated the current rearward deflection requirement is unnecessarily restrictive in that it would require costly retooling of school bus seats with no measurable safety advantage over a somewhat greater deflection distance that would not entail significant retooling. Additionally, a minor modification of the standard is made clarifying the meaning of "absorbed energy" consistent with an agency interpretation of that term.

Effective Date: April 1, 1978.

For further information contact:

Mr. Timothy Hoyt, Crashworthiness Division, National Highway Traffic Safety Administration, 400 Seventh Street, S.W., Washington, D.C. 20590 (202-426-2264).

Supplementary Information: On November 10, 1977, the NHTSA published a notice proposing to amend the rearward deflection requirement of Standard No. 222, *School Bus Passenger Seating and Crash Protection*. The impetus for that proposal came from several petitions from school bus manufacturers claiming that the rearward deflection requirement was unnecessarily restrictive since it would require significant retooling of school bus seats which would not be measurably superior, in terms of safety, to seats designed to meet a slightly greater deflection distance. They stated that seats produced in compliance with a somewhat greater rearward deflection requirement, as opposed to the currently specified 8-inch requirement, would not require retooling. The NHTSA agreed with the petitioners and, accordingly, proposed to increase the allowable rearward deflection of seats from 8 to 10 inches. By the same notice, the NHTSA proposed a minor modification of the standard clarifying the agency's meaning of absorbed energy.

Only one comment was received in response to that notice of proposed rulemaking. The Vehicle Equipment Safety Commission did not submit comments.

The only commenter, Blue Bird Body Company, took issue with the agency's proposed method for limiting rearward seat deflection. It asserted that the requirement expressed in S5.1.4 (c) of the standard should be the only limitation on rearward seat deflection. That section provides that a seat shall not, when tested, come within 4 inches of any portion of another passenger seat.

Blue Bird's comment is not persuasive. The requirement of S5.1.4(c) addresses an entirely separate safety concern than the requirement of S5.1.4(b). Section S5.1.4(b) limits the rearward deflection of a seat, by this notice, to a maximum of 10 inches. That requirement functions as part of the compartmentalization scheme of Standard 222. Limiting the degree of seat back deflection helps to contain a child within the seat structures in the event of an accident. This requirement should be distinguished from that contained in S5.1.4(c), which is intended to ensure that a minimum amount of space remains between seats following an accident so that a child does not become trapped. Since both requirements are necessary to maintain the safety level considered necessary for school buses, Blue Bird's request is denied.

Blue Bird stated in its comments a preference for specifying maximum rearward seat deflection in terms of inches rather than angle. This comment suggests that Blue Bird misinterpreted the statements in the notice of proposed rulemaking as indicating that the NHTSA was contemplating an amendment that would limit the angle of seat deflection. The reference in the notice to a 40° seat angle was made only to justify the proposed 10-inch maximum seat deflection. A 40° seat angle roughly translates to 10 inches of rearward seat deflection. There was no intention to suggest that an angle limitation was under consideration. In fact, the preamble stated that the NHTSA had abandoned, in earlier rulemaking, attempts to adopt an angular measurement owing to the difficulty of making such a measurement.

The agency concludes that the extension of the allowable rearward deflection of seats from 8 to 10 inches assures passenger safety while minimizing the cost impact of compliance with the school bus regulations. Since this amendment relieves a restriction, it should result in no increase in costs.

In consideration of the foregoing, Part 571, of Title 49, CFR, is amended. . . .

The principal authors of this proposal are Timothy Hoyt of the Crashworthiness Division and Roger Tilton of the Office of Chief Counsel.

(Secs. 103, 119, Pub. L. 89–563, 80 Stat. 718 (15 U.S.C. 1392, 1407); Sec. 203, Pub. L. 93–492, 88 Stat. 1470 (15 U.S.C. 1392); delegation of authority at 49 CFR 1.50.)

Issued on March 1, 1978.

Joan Claybrook
Administrator

43 F.R. 9149
March 6, 1976

School Bus Seating and Crash Protection

(Docket No. 73–3; Notice 13)

Action: Final rule.

Summary: This notice makes final an existing interim amendment to Standard No. 222, *School Bus Seating and Crash Protection,* increasing the maximum allowable seat spacing in school buses from 20 to 21 inches. In issuing the original standard, the agency intended that the seats be spaced approximately 20 inches apart (S5.2). However, because of manufacturing tolerances, some school bus manufacturers were spacing their seats at distances less than 20 inches to ensure that the spacing does not exceed the prescribed maximum. A seat spacing specification of 21 inches permits 20-inch spacing of seats by taking manufacturing tolerances into fuller account. This spacing will accommodate large high school students while still ensuring a safe level of school bus seat performance.

Effective date: Since this amendment merely makes final an existing interim rule, it is effective March 29, 1979.

For further information contact:

Mr. Robert Williams, Crashworthiness Division, National Highway Traffic Safety Administration, 400 Seventh Street, S.W., Washington, D.C. 20590 (202) 426–2264.

Supplementary information: On December 22, 1977, the National Highway Traffic Safety Administration issued a proposal to increase the allowable seat spacing in school buses from 20 to 21 inches (42 FR 64136). Concurrently with that proposal, the NHTSA issued an interim final rule permitting buses to be constructed immediately with the increased seat spacing (42 FR 64119). This action was taken to provide the amount of seat spacing in school buses originally intended by the agency and to relieve immediately problems created by the unnecessarily limited seat spacing in buses then being built. The action resulted from numerous complaints by school bus users relating to seat spacing. The proposal and interim final rule responded to petitions from the Wisconsin School Bus Association and the National School Transportation Association asking for increased seat spacing.

The agency received many comments in response to its December 1977 proposal. Most comments favored some extension in the seat spacing allowance in school buses. Commenters differed as to the amount of seat spacing needed to accommodate fully the larger school children. Some commenters suggested that the agency provide still more seat spacing than proposed in the December 22 notice. Other commenters supported the agency's suggested modification.

The agency has reviewed all of the comments and the petitions concerning this issue and has concluded that the proposal and interim rule provide sufficient seat spacing in school buses for all school children. To provide greater seat spacing, as suggested by some commenters, might necessitate changing the seat structures to absorb more energy. See the December proposal for further discussion of this point. The NHTSA does not believe that such a costly change is warranted at this time. The agency notes that as a result of the interim rule seat spacing in buses has become adequate to meet the needs for pupil transportation to and from school. The agency continues, however, to research the proper seating for activity buses and will address that issue in a separate notice as soon as all of the research and analysis is completed.

In accordance with the foregoing, Volume 49 of the Code of Federal Regulations, Part 571, Standard No. 222, *School Bus Seating and Crash Protection*, is amended . . . .

The principal authors of this notice are Robert Williams of the Crashworthiness Division and Roger Tilton of the Office of Chief Counsel.

(Secs. 103, 119, Pub. L. 89–563, 80 Stat. 718 (15 U.S.C. 1392, 1407) ; Sec. 203, Pub. L. 93–492, 88 Stat. 1470 (15 U.S.C. 1392) ; delegation of authority at 49 CFR 1.50.)

Issued on March 21, 1979.

Joan Claybrook
Administrator

44 F.R. 18674–18675
March 29, 1979

PREAMBLE TO AN AMENDMENT TO
FEDERAL MOTOR VEHICLE SAFETY STANDARD NO. 222

Federal Motor Vehicle Safety Standards;
School Bus Passenger Seating and Crash Protection

[Docket No. 73-3; Notice 15]

ACTION: Final rule.

SUMMARY: This notice amends the agency's school bus seating standard to increase seat spacing from 21 to 24 inches. This amendment is being issued to resolve problems experienced by users, i.e., school districts and contract carriers, to the effect that mandatory seat spacing at the prior level inhibited some necessary uses. The agency finds that an additional space seating option will not inhibit safety.

DATE: This amendment is effective March 24, 1983.

SUPPLEMENTARY INFORMATION: Standard No. 222, *School Bus Passenger Seating and Crash Protection*, was one of several standards implemented pursuant to the Motor Vehicle and School Bus Safety Amendments of 1974 (Pub. L. 93-492). The standard regulates the performance aspects of school bus seats. One portion of the standard limits the longitudinal spacing between seats in buses with gross vehicle weight ratings (GVWR) of more than 10,000 pounds. No seat may be positioned more than 21 inches from the seat immediately to the front, measured from the seating reference point to the seat back or restraining barrier located in front of the seat.

The initial version of Standard 222 which became effective on April 1, 1977, limited school bus seat spacing to 20 inches. Soon after school buses began to be produced in compliance with this requirement, users began to experience problems of inadequate spacing. Because of quality control and other production problems affecting seat spacing, manufacturers were spacing seats significantly less than the 20 inches permitted by the standard to ensure compliance. As manufacturers improved their production techniques, seat spacing was extended.

The agency upon examination of its then existing data concluded later that same year that it could extend seat spacing to 21 inches without adversely affecting the compartmentalization concept that was the key to protecting children in the buses. Compartmentalization attempts to protect children between well padded high-backed seats. The agency amended the rule accordingly (42 F.R. 64119, December 22, 1977) and undertook to study further the appropriateness of the required seat spacing.

Both the amendment and improved manufacturer production methods reduced the number of spacing problems significantly. Some problems continue to exist, however, especially concerning buses used to transport children long distances to and from school, or to and from school related events which may be located far from the school itself. The agency has conducted tests to see whether it could improve seat spacing to respond to these continuing problems, without compromise of safety. The tests, which are available in the Technical Reference Section of the agency under H73-3 "School Bus Passenger Seat and Lap Belt Sled Tests," DOT-HS-804985, December 1978, show that seat spacing could be increased up to 24 inches without impairing the concept of compartmentalization. An increase in seat spacing beyond 24 inches might impair the ability of the seats to absorb energy in the manner required by the standard. Accordingly, on February 25, 1982, the agency proposed a further increase in seat spacing to 24 inches (47 F.R. 8231).

The agency received numerous comments in response to the notice of proposed rulemaking. Virtually all of those comments supported the agency's action. In accordance with the comments and the existing agency information, the agency, by this notice, makes final the increased seat spacing to 24 inches.

Three school districts out of the more than 140 commenters on the February notice objected to the increased seat spacing. It appears that these commenters were afraid that the increased seat spacing was mandatory and that this would in turn reduce the seating capacity in their vehicles resulting in the need to purchase additional buses or realign school routes. This understanding is not accurate. The increased seat spacing is merely optional. If a school chooses to have additional spacing in some or all of its buses, up to 24 inches, this would be permitted. Otherwise, schools may continue to purchase buses with seats spaced as they are today. Seat spacing less than 24 inches is completely within the discretion of the school that is purchasing the vehicles.

Commenters to the February notice raised another issue that is somewhat related to seat spacing. They requested more comfortable seats and additional leg room for long distance school buses. These are the vehicles that frequently have been involved in transporting children to and from activities or, in some instances, carry children over long distances to schools in some of the Western States. The commenters in general would prefer to have recliner seats or some other seating system that would be more comfortable for these uses.

The agency has explored the possibility of establishing another optional seating mode in school vehicles that would accommodate the concerns of these commenters. The agency concludes that recliner seats could not provide the same level of safety as provided by existing seat requirements in school buses. Accordingly, the agency declines to adopt this suggestion. NHTSA believes that the seat spacing extension being made today should address adequately the problem of comfort in buses used for school activities.

This amendment is being made effective immediately. It relieves a restriction, and is completely optional, and does not require any manufacturer or purchaser to alter present practices. Further, the agency has learned that many companies and purchasers are waiting for this amendment before purchasing new vehicles. Therefore, an immediate effective date is in the public interest.

Issued on March 17, 1983.

Raymond A Peck, Jr.
Administrator
48 F.R. 12384
March 24, 1983

MOTOR VEHICLE SAFETY STANDARD NO. 222

School Bus Seating and Crash Protection

S1. Scope. This standard establishes occupant protection requirements for school bus passenger seating and restraining barriers.

S2. Purpose. The purpose of this standard is to reduce the number of deaths and the severity of injuries that result from the impact of school bus occupants against structures within the vehicle during crashes and sudden driving maneuvers.

S3. Application. This standard applies to school buses.

S4. Definitions. "Contactable surface" means any surface within the zone specified in S5.3.1.1 that is contactable from any direction by the test device described in S6.6, except any surface on the front of a seat back or restraining barrier 3 inches or more below the top of the seat back or restraining barrier.

"School bus passenger seat" means a seat in a school bus, other than the driver's seat or a seat installed to accommodate handicapped or convalescent passengers as evidenced by orientation of the seat in a direction that is more than 45 degrees to the left or right of the longitudinal centerline of the vehicle.

S4.1 The number of seating positions considered to be in a bench seat is expressed by the symbol W, and calculated as the bench width in inches divided by 15 and rounded to the nearest whole number.

S5. Requirements. (a) Each vehicle with a gross vehicle weight rating of more than 10,000 pounds shall be capable of meeting any of the requirements set forth under this heading when tested under the conditions of S6. However, a particular school bus passenger seat (i.e., test specimen) in that weight class need not meet further requirements after having met S5.1.2 and S5.1.5, or having been subjected to either S5.1.3, S5.1.4, or S5.3.

(b) Each vehicle with a gross vehicle weight rating of 10,000 pounds or less shall be capable of meeting the following requirements at all seating positions other than the driver's seat: (1) The requirements of §§ 571.208, 571.209, and 571.210 (Standard Nos. 208, 209, and 210) as they apply to multipurpose passenger vehicles; and (2) the requirements of S5.1.2, S5.1.3, S5.1.4, S5.1.5, and S5.3 of this standard. However, the requirements of Standard Nos. 208 and 210 shall be met at W seating positions in a bench seat using a body block as specified in Figure 2 of this standard, and a particular school bus passenger seat (i.e., a test specimen) in that weight class need not meet further requirements after having met S5.1.2 and S5.1.5, or having been subjected to either S5.1.3, S5.1.4, S5.3, or § 571.210 (Standard No. 210).

S5.1 Seating requirements. School bus passenger seats shall be forward facing.

S5.1.1 [Reserved]

S5.1.2 Seat back height and surface area. Each school bus passenger seat shall be equipped with a seat back that, in the front projected view, has a front surface area above the horizontal plane that passes through the seating reference point, and below the horizontal plane 20 inches above the seating reference point, of not less than 90 percent of the seat bench width in inches multiplied by 20.

S5.1.3 Seat performance forward. When a school bus passenger seat that has another seat behind it is subjected to the application of force as specified in S5.1.3.1 and S5.1.3.2, and subsequently, the application of additional force to the seat back as specified in S5.1.3.3 and S5.1.3.4:

(a) The seat-back force/deflection curve shall fall within the zone specified in Figure 1;

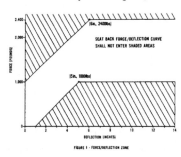

FIGURE I - FORCE/DEFLECTION ZONE

(b) Seat back deflection shall not exceed 14 inches; (for determination of (a) and (b) the force/deflection curve describes only the force applied through the upper loading bar, and only the forward travel of the pivot attachment point of the upper loading bar, measured from the point at which the initial application of 10 pounds of force is attained.)

(c) The seat shall not deflect by an amount such that any part of the seat moves to within 4 inches of any part of another school bus passenger seat or restraining barrier in its originally installed psition;

(d) The seat shall not separate from the vehicle at any attachment point; and

(d) Seat components shall not separate at any attachment point.

S5.1.3.1 Position the loading bar specified in S6.5 so that it is laterally centered behind the seat back with the bar's longitudinal axis in a transverse plane of the vehicle and in any horizontal plane between 4 inches above and 4 inches below the seating reference point of the school bus passenger seat behind the test specimen.

S5.1.3.2 Apply a force of 700W pounds horizontally in the forward direction through the loading bar at the pivot attachment point. Reach the specified load in not less than 5 nor more than 30 seconds.

S5.1.3.3 No sooner than 1.0 second after attaining the required force, reduce that force to 350W pounds and, while maintaining the pivot point position of the first loading bar at the position where the 350W pounds is attained, position a second loading bar described in S6.5 so that it is laterally centered behind the seat back with the bar's longitudinal axis in a transverse plane of the vehicle and in the horizontal plane 16 inches above the seating reference point of the school bus passenger seat behind the test specimen, and move the bar forward against the seat back until a force of 10 pounds has been applied.

S5.1.3.4 Apply additional force horizontally in the forward direction through the upper bar until 4,000W inch-pounds of energy have been absorbed in deflecting the seat back (or restraining barrier). Apply the additional load in not less than 5 seconds nor more than 30 seconds. Maintain the pivot attachment point in the maximum forward travel position for not less than 5 seconds nor more than 10 seconds and release the load in not less than 5 nor more than 30 seconds. (For the determination of S5.1.3.4 the force/deflection curve describes only the force applied through the upper loading bar, and the forward and rearward travel distance of the upper loading bar pivot attachment point measured from the position at which the initial application of 10 pounds of force is attained.)

S5.1.4 Seat performance rearward. When a school bus passenger seat that has another seat behind it is subjected to the application of force as specified in S5.1.4.1 and S5.1.4.2:

(a) Seat back force shall not exceed 2,200 pounds;

(b) In the case of a school bus manufactured on or after April 1, 1978, seat back deflection shall not exceed 10 inches; (For determination of (a) and (b) the force/deflection curve describes only the force applied through the loading bar, and only the rearward travel of the pivot attachment point of the loading bar, measured from the point at which the initial application of 50 pounds of force is attained.)

(c) The seat shall not deflect by an amount such that any part of the seat moves to within 4 inches of any part of another passenger seat in its originally installed position;

(d) The seat shall not separate from the vehicle at any attachment point; and

(e) Seat components shall not separate at any attachment point.

S5.1.4.1 Position the loading bar described in S6.5 so that it is laterally centered forward of the seat back with the bar's longitudinal axis in a transverse plane of the vehicle and in the horizontal plane 13.5 inches above the seating reference point of the test specimen, and move the loading bar rearward against the seat back until a force of 50 pounds has been applied.

S5.1.4.2 Apply additional force horizontally rearward through the loading bar until 2,800W inch-pounds of energy have been absorbed in deflecting the seat back. Apply the additional load in not less than 5 seconds nor more than 30 seconds. Maintain the pivot attachment point in the maximum rearward travel position for not less than 5 seconds nor more than 10 seconds and release the load in not less than 5 seconds nor more than 30 seconds. (For determination of S5.1.4.2 the force/deflection curve describes the force applied through the loading bar and the rearward and forward travel distance of the loading bar pivot attachment point measured from the position at which the initial application of 50 pounds of force is attained.)

S5.1.5 Seat cushion retention. In the case of school bus passenger seats equipped with seat cushions, with all manual attachment devices between the seat and the seat cushion in the manufacturer's designed position for attachment, the seat cushion shall not separate from the seat at any attachment point when subjected to an upward force of five times the seat cushion weight, applied in any period of not less than 1 nor more than 5 seconds, and maintained for 5 seconds.

S5.2 Restraining barrier requirements. Each vehicle shall be equipped with a restraining barrier forward of any designated seating position that does not have the rear surface of another school bus passenger seat within 24 inches of its seating reference point, measured along a horizontal longitudinal line through the seating reference point in the forward direction.

S5.2.1 Barrier-seat separation. The horizontal distance between the restraining barrier's rear surface and the seating reference point of the seat in front of which it is required shall be not more than 24 inches, measured along a horizontal longitudinal line through the seating reference point in the forward direction.

S5.2.2 Barrier position and rear surface area. The position and rear surface area of the restraining barrier shall be such that, in a front projected view of the bus, each point of the barrier's perimeter coincides with or lies outside of the perimeter of the seat back of the seat for which it is required.

S5.2.3 Barrier performance forward. When force is applied to the restraining barrier in the same manner as specified in S5.1.3.1 through S5.1.3.4 for seating performance tests:

(a) The restraining barrier force/deflection curve shall fall within the zone specified in Figure 1;

(b) Restraining barrier deflection shall not exceed 14 inches; (For computation of (a) and (b) the force/deflection curve describes the force applied through the upper loading bar, and only the forward travel of the pivot attachment point of the loading bar, measured from the point at which the initial application of 10 pounds of force is attained.)

(c) Restraining barrier deflection shall not interfere with normal door operation;

(d) The restraining barrier shall not separate from the vehicle at any attachment point; and

(e) Restraining barrier components shall not separate at any attachment point.

S5.3 Impact zone requirements.

S5.3.1 Head protection zone. Any contactable surface of the vehicle within any zone specified in S5.3.1.1 shall meet the requirements of S5.3.1.2 and S5.3.1.3. However, a surface area that has been contacted pursuant to an impact test need not meet further requirements contained in S5.3.

S5.3.1.1 The head protection zones in each vehicle are the spaces in front of each school bus passenger seat which are not occupied by bus sidewall, window, or door structure and which, in relation to that seat and its seating reference point, are enclosed by the following planes;

(a) Horizontal planes 12 inches and 40 inches above the seating reference point;

(b) A vertical longitudinal plane tangent to the inboard (aisle side) edge of the seat;

(c) A vertical longitudinal plane 3.25 inches inboard of the outboard edge of the seat, and

(d) Vertical transverse planes through and 30 inches forward of the reference point.

S5.3.1.2 Head form impact requirement. When any contactable surface of the vehicle within

the zones specified in S5.3.1.1 is impacted from any direction at 22 feet per second by the head form described in S6.6, the axial acceleration at the center of gravity of the head form shall be such that the expression

shall not exceed 1,000 where a is the axial acceleration expressed as a multiple of g (the acceleration due to gravity), and t_1 and t_2 are any two points in time during the impact.

S5.3.1.3 Head form force distribution. When any contactable surface of the vehicle within the zones specified in S5.3.1.1 is impacted from any direction at 22 feet per second by the head form

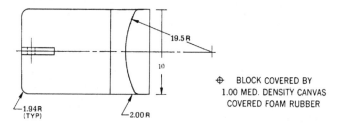

⊕ BLOCK COVERED BY
1.00 MED. DENSITY CANVAS
COVERED FOAM RUBBER

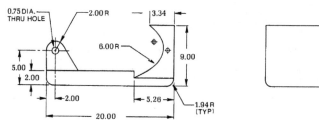

FIGURE 2 – BODY BLOCK FOR LAP BELT

PART 571; S 222-4

described in S6.6, the energy necessary to deflect the impacted material shall be not less than 40 inch-pounds before the force level on the head form exceeds 150 pounds. When any contactable surface within such zones is impacted by the head form from any direction at 5 feet per second, the contact area on the head form surface shall be not less than 3 square inches.

S5.3.2 Leg protection zone. Any part of the seat backs or restraining barriers in the vehicle within any zone specified in S5.3.2.1 shall meet the requirements of S5.3.2.2.

S5.3.2.1. The leg protection zones of each vehicle are those parts of the school bus passenger seat backs and restraining barriers bounded by horizontal planes 12 inches above and 4 inches below the seating reference point of the school bus passenger seat immediately behind the seat back or restraining barrier.

S5.3.2.2. When any point on the rear surface of that part of a seat back or restraining barrier within any zone specified in S5.3.2.1 is impacted from any direction at 16 feet per second by the knee form specified in S6.7, the resisting force of the impacted material shall not exceed 600 pounds and the contact area on the knee form surface shall not be less than 3 square inches.

S6. Test conditions. The following conditions apply to the requirements specified in S5.

S6.1 Test surface. The bus is at rest on a level surface.

S6.2 Tires. Tires are inflated to the pressure specified by the manufacturer for the gross vehicle weight rating.

6.3 Temperature. The ambient temperature is any level between 32 degrees F. and 90 degrees F.

S6.4 Seat back position. If adjustable, a seat back is adjusted to its most upright position.

S6.5 Loading bar. The loading bar is a rigid cylinder with an outside diameter of 6 inches that has hemispherical ends with radii of 3 inches and with a surface roughness that does not exceed 63 micro-inches, root mean square. Then length of the loading bar is 4 inches less than the

width of the seat back in each test. The stroking mechanism applies force through a pivot attachment at the centerpoint of the loading bar which allows the loading bar to rotate in a horizontal plane 30 degrees in either direction from the transverse position.

S6.5.1 A vertical or lateral force of 4,000 pounds applied externally through the pivot attachment point of the loading bar at any position reached during a test specified in this standard shall not deflect that point more than 1 inch.

S6.6 Head form. The head form for the measurement of acceleration is a rigid surface comprised of two hemispherical shapes, with total equivalent weight of 11.5 pounds. The first of the two hemispherical shapes has a diameter of 6.5 inches. The second of the two hemispherical shapes has a 2 inch diameter and is centered as shown in Figure 3 to protrude from the outer surface of the first hemispherical shape. The surface roughness of the hemispherical shapes does not exceed 63 micro-inches, root mean square.

S6.6.1 The direction of travel of the head form is coincidental with the straight line connecting the centerpoints of the two spherical outer surfaces which constitute the head form shape.

S6.6.2 The head form is instrumented with an acceleration sensing device whose output is recorded in a data channel that conforms to the requirements for a 1,000 Hz channel class as specified in SAE Recommended Practice J211a, December 1971. The head form exhibits no resonant frequency below three times the frequency of the channel class. The axis of the acceleration sensing device coincides with the straight line connecting the centerpoints of the two hemispherical outer surfaces which constitute the head form shape.

S6.6.3 The head form is guided by a stroking device so that the direction of travel of the head form is not affected by impact with the surface being tested at the levels called for in the standard.

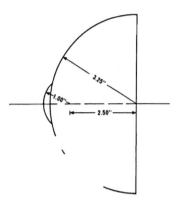

BIHEMISPHERICAL HEAD FORM RADII

3.25"

1.00"

2.50"

FIGURE 3

S6.7 Knee form. The knee form for measurement of force is a rigid 3-inch-diameter cylinder, with an equivalent weight of 10 pounds, that has one rigid hemispherical end with a 1½ inch radius forming the contact surface of the knee form. The hemispherical surface roughness does not exceed 63 micro-inches, root mean square.

S6.7.1 The direction of travel of the knee form is coincidental with the centerline of the rigid cylinder.

S6.7.2 The knee form is instrumented with an acceleration sensing device whose output is recorded in a data channel that conforms to the requirements of a 600 Hz channel class as specified in the SAE Recommended Practice J211a, December 1971. The knee form exhibits no resonant frequency below three times the frequency of the channel class. The axis of the acceleration sensing device is aligned to measure acceleration along the centerline of the cylindrical knee form.

S6.7.3 The knee form is guided by a stroking device so that the direction of travel of the knee form is not affected by impact with the surface being tested at the levels called for in the standard.

S6.8 The head form, knee form, and contactable surfaces are clean and dry during impact testing.

41 F.R. 4016
January 28, 1976

PREAMBLE TO AMENDMENT TO MOTOR VEHICLE SAFETY STANDARD NO. 301

Fuel System Integrity

(Docket No. 70–20; Notice 2)

This notice amends Motor Vehicle Safety Standard No. 301 on fuel system integrity to specify static rollover requirements applicable to passenger cars on September 1, 1975, and to extend applicability of the standard to multipurpose passenger vehicles, trucks, and buses with a GVWR of 10,000 pounds or less on September 1, 1976.

The NHTSA proposed amending 49 CFR 571.301, *Fuel Tanks, Fuel Tank Filler Pipes, and Fuel Tank Connections,* on August 29, 1970, (35 F.R. 13799). Under the proposal the standard would be extended to all vehicles with a GVWR of 10,000 pounds or less. No fuel spillage would be permitted during the standard's tests. As proposed, these would include a spike stop from 60 mph, and a 30 mph frontal barrier crash. Additional tests for vehicles with a GVWR of 6,000 pounds or less would include a rear-end collision with a fixed barrier at 30 mph, and a static rollover test following the frontal barrier crash. With respect to the proposal: the frontal impact and static rollover tests are adopted but with an allowance of fuel spillage of 1 ounce per minute; the spike stop test is not adopted; and the rear-end fixed barrier collision test is being reproposed in a separate rule making action published today to substitute a moving barrier.

The proposal that there be zero fuel spillage was almost universally opposed for cost/benefit reasons. The NHTSA has concluded that the requirement adopted, limiting fuel spillage to 1 ounce per minute, will have much the same effect as a zero-loss requirement. The standard will effectively require motor vehicles to be designed for complete fuel containment, since any spillage allowed by design in the aftermath of

testing could well exceed the limit of the standard. At the same time, the 1-ounce allowance would eliminate concern over a few drops of spillage that in a functioning system may be unavoidable.

Fuel loss will be measured for a 15-minute period for both impact and rollover tests.

The NHTSA proposed a panic-braking stop from 60 mph to demonstrate fuel system integrity. Many commented that this appeared superfluous, increasing testing costs with no performance improvements, since the proposed front and rear impact tests represented considerably higher deceleration loadings than could be achieved in braking. The NHTSA concurs, and has not adopted the panic stop test. The frontal barrier crash at 30 mph has been retained for passenger cars, and extended to multipurpose passenger vehicles, trucks, and buses with a GVWR of 10,000 pounds or less as of September 1, 1976.

The static rollover test was adopted as proposed. It applies to passenger cars as of September 1, 1975, and to multipurpose passenger vehicles, trucks, and buses with a GVWR of 6,000 pounds or less, as of September 1, 1976. The rollover test follows the front barrier crash, and consists of a vehicle being rotated on its longitudinal axis at successive increments of 90°. A condition of the test is that rotation between increments occurs in not less than 1 minute and not more than 3 minutes. After reaching a 90° increment, the vehicle is held in that position for 5 minutes.

The proposed rear-end crash test incorporated a fixed collision barrier. Manufacturers generally favored a moving barrier impact as a closer simulation of real world conditions. The NHTSA concurs and is not adopting a rear end fixed barrier test. Instead, it is proposing a rear-end moving barrier collision test as part of the notice of proposed rulemaking published today.

Under the proposal the vehicle would be loaded to its GVWR with the fuel tank filled to any level between 90 and 100 percent of capacity. Many commenters objected on the grounds that full loading of a vehicle represents an unrealistic condition in terms of actual crash experience. The NHTSA does not agree. Although full loading of a vehicle is not the condition most frequently encountered, it certainly occurs frequently enough that the vehicle should be designed to give basic protection in that condition. The vehicle test weight condition has been adopted as proposed. It should be noted that, in the parallel notice of proposed rulemaking issued today, vehicles would be tested under the weight conditions specified in Standard No. 208, effective September 1, 1975.

In consideration of the foregoing, 49 CFR Part 571.301, Motor Vehicle Safety Standard No. 301, is amended . . . .

Effective date: September 1, 1975. Because of the necessity to allow manufacturers sufficient production leadtime it is found for good cause shown that an effective date later than 1 year after issuance of this rule is in the public interest.

(Sec. 103, 119, Pub. L. 89–563, 80 Stat. 718, 15 U.S.C. 1392, 1407; delegation of authority at 49 CFR 1.51.)

Issued on August 15, 1973.

James B. Gregory
Administrator
38 F.R. 22397
August 20, 1973

PREAMBLE TO AMENDMENT TO MOTOR VEHICLE SAFETY STANDARD NO. 301

Fuel System Integrity

(Docket No. 73–20; Notice 2)

The purpose of this notice is to amend Federal Motor Vehicle Safety Standard No. 301, *Fuel System Integrity*, to upgrade substantially the requirements of the standard by specifying a rear moving barrier crash, a lateral moving barrier crash, and a frontal barrier crash including impacts at any angle up to 30° in either direction from the perpendicular.

A notice of proposed rulemaking published August 20, 1973 (38 F.R. 22417) proposed the imposition of additional testing requirements designed to ameliorate the dangers associated with fuel spillage following motor vehicle accidents. In an amendment to Standard No. 301, published on the same day as the proposal, a frontal barrier crash and a static rollover test were specified. In order to ensure the safety of fuel systems in any possible collision situation, the NHTSA finds it essential to incorporate additional proposed test requirements into the present standard and to make these requirements applicable to all vehicle types with a GVWR of 10,000 pounds or less.

Comments in response to the proposal were received from 29 commenters. Any suggestions for changes of the proposal not specifically mentioned herein are denied, on the basis of all the information presently available to this agency. A number of the issues raised in the comments have been dealt with by the agency in its response to the petitions for reconsideration of the final rule issued on August 20, 1973. In its notice responding to the petitions, the NHTSA considered objections to the use of actual fuel during testing, the specified fuel fill level, the application of the standard to vehicles using diesel fuel, the fuel spillage measuring requirement, and the allegedly more stringent loading requirements

applicable to passenger cars. The type of fuel subject to the standard was also clarified.

Objections were registered by 13 commenters to the proposed inclusion of a dynamic rollover test in the fuel system integrity standard. As proposed, the requirement calls for a measurement of the fuel loss while the vehicle is in motion. Commenters pointed out the exceptional difficulty in measuring or even ascertaining a leakage when the vehicle is rolling over at 30 mph. The NHTSA has decided that the objections have merit, and has deleted the dynamic rollover test. The results of the dynamic rollover do not provide sufficiently unique data with regard to the fuel system's integrity to justify the cost of developing techniques for accurately measuring spillage during such a test, and of conducting the test itself. The NHTSA has concluded that the severity of the other required tests, when conducted in the specified sequence, is sufficient to assure the level of fuel system integrity intended by the agency.

Triumph Motors objected to the use of a 4,000-pound barrier during the moving barrier impacts, asserting that such large barriers discriminate against small vehicles. Triumph requested that the weight of the barrier be the curb weight of the vehicle being tested in order to alleviate the burden on small vehicles. The NHTSA has concluded that no justification exists for this change. The moving barrier is intended to represent another vehicle with which the test vehicle must collide. The use of a 4,000-pound moving barrier is entirely reasonable since vehicles in use are often over 4,000 pounds in weight and a small vehicle is as likely to collide with a vehicle of that size as one smaller. The NHTSA considers it important that vehicle fuel systems be designed in such a way as to withstand impacts from vehicles they are exposed to on the road, regardless of the differences in their sizes.

Jeep and American Motors objected to the effective dates of the proposed requirements and asked that they be extended. Jeep favors an effective date not earlier than September 1, 1979, and American Motors favors a September 1, 1978, effective date. The NHTSA denies these requests. It has found that the time period provided for development of conforming fuel systems is reasonable and should be strictly adhered to considering the urgent need for strong and resilient fuel systems.

Several commenters expressed concern over the impact of the prescribed testing procedures on manufacturers of low-volume specialty vehicles. The NHTSA appreciates the expense of conducting crash tests on low-production vehicles, realizing that the burden on the manufacturer is related to the number of vehicles he manufactures. However, there are means by which the small-volume manufacturer can minimize the costs of testing. He can concentrate test efforts on the vehicle(s) in his line that he finds most difficult to produce in conformity with the standard. These manufacturers should also be aware that an exemption from application of the standard is available where fewer than 10,000 vehicles per year are produced and compliance would subject him to substantial financial hardship.

In responding to the petitions for reconsideration of the amendment to Standard No. 301, published August 20, 1973, the NHTSA revised the fuel system loading requirement to specify Stoddard solvent as the fuel to be used during testing. In accordance with that amendment, the proposed requirement that the engine be idling during the testing sequence is deleted. However, electrically driven fuel pumps that normally run when the electrical system in the vehicle is activated shall be operating during the barrier crash tests.

In order to fulfill the intention expressed in the preamble to the proposal, that simultaneous testing under Standards Nos. 208 and 301 be possible, language has been added to subparagraph S7.1.5 of Standard No. 301 specifying the same method of restraint as that required in Standard No. 208. In its response to petitions for reconsideration of Standard No. 301 (39 F.R. 10586) the NHTSA amended the standard by requiring that each dummy be restrained during testing only by means that are installed in the vehicle for protection at its seating position and that require no action by the vehicle occupant.

Suggestions by several commenters that the application of certain crash tests should be limited to passenger cars in order to maintain complete conformance to the requirements of Standard No. 208 are found to be without merit. Enabling simultaneous testing under several standards, although desirable, is not the most important objective of the safety standards. The NHTSA is aware of the burden of testing costs, and therefore has sought to ease that burden where possible by structuring certain of its standards to allow concurrent testing for compliance. It must be emphasized, however, that the testing requirements specified in a standard are geared toward a particular safety need. Application of the tests proposed for Standard No. 301 to all vehicle types with a GVWR of 10,000 pounds or less is vital to the accomplishment of the degree of fuel system integrity necessary to protect the occupants of vehicles involved in accidents.

No major objections were raised concerning the proposed angular frontal barrier crash, lateral barrier crash, or rear moving barrier crash. On the basis of all information available to this agency, it has been determined that these proposed crash tests should be adopted as proposed.

In consideration of the foregoing, 49 CFR 571.301, Motor Vehicle Safety Standard No. 301, is amended to read as set forth below.

Effective date: September 1, 1975, with additional requirements effective September 1, 1976, and September 1, 1977, as indicated.

(Secs. 103, 119, Pub. L. 89-563, 80 Stat. 718, 15 U.S.C. 1392, 1407; delegation of authority at 49 CFR 1.51.)

Issued on March 18, 1974.

James B. Gregory
Administrator

39 F.R. 10588
March 21, 1974

PREAMBLE TO AMENDMENT TO MOTOR VEHICLE SAFETY STANDARD NO. 301-75

Fuel System Integrity

(Docket No. 73-20; Notice 3)

This notice responds to petitions for reconsideration of the two recent Federal Register notices amending and upgrading Standard No. 301 (39 F.R. 10586; 39 F.R. 10588) and amends the standard in several respects.

On March 21, 1974 two notices were published pertaining to Standard No. 301, *Fuel System Integrity*. One notice (39 F.R. 10586) responded to petitions for reconsideration of an earlier amendment to the standard (38 F.R. 22397), while the other (39 F.R. 10588) substantially upgraded the standard's performance requirements. It was the intention of the NHTSA that the notice upgrading the standard be considered as the final rule and supersede the notice responding to petitions. Hereafter, the notice responding to petitions will be referred to as Notice 1, while the notice upgrading the standard will be referred to as Notice 2.

On October 27, 1974, the Motor Vehicle and Schoolbus Safety Amendments of 1974 (P.L. 93-492) were signed into law. These amendments to the National Traffic and Motor Vehicle Safety Act incorporate Standard No. 301 as it was published in Notice 2 on March 21, 1974. According to the amendment the technical errors which appeared in Notice 2 may be corrected, while future amendments are prohibited from diminishing the level of motor vehicle safety which was established in the notice. The changes contained in this notice conform to these statutory requirements.

Due to an oversight, Notice 2 failed to include two provisions which appeared in Notice 1. The limitation of the standard's application to vehicles which use fuel with a boiling point above 32°F was inadvertently omitted in Notice 2 and is hereby reinstated. Notice 2 also failed to include a provision specifying that vehicles not be

altered during the testing sequences. It was the intent of the NHTSA that damage or other alteration of the vehicle incurred during the barrier crashes not be corrected prior to the static rollover tests. The test requirements are therefore amended to prohibit the alteration of vehicles following each of the specified test impacts.

In order to clarify the manner in which the load is to be distributed during testing of multipurpose passenger vehicles, trucks, and buses, S7.1.5(b) is amended to require that when the weight on one of the axles exceeds its proportional share of the loaded vehicle weight, when the vehicle is loaded only with dummies, the remainder of the required test weight shall be placed on the other axle, so that the weight on the first axle remains the same. The loading specification did not specifically address this contingency.

The requirement that the load be located in the load carrying area of multipurpose passenger vehicles, trucks, and buses during testing is deleted since the agency has determined that such a limitation is consistent with the provision specifying distribution of weight in proportion with the vehicle's gross axle weight ratings.

Petitions for reconsideration were received from eleven petitioners. Although only those comments raising issues found to be significant have been discussed, due consideration has been given to all requests. Any requests not specifically discussed herein are denied.

A substantial number of petitioners objected to the requirement that dummies used during testing be restrained only by passive means installed at the seating positions. Petitioners pointed out that mandatory passive restraint systems proposed in Standard No. 208 have a proposed effective date of September 1, 1976; one year after the September 1, 1975 effective date set for implementation of Standard 301. This would leave a period of time when most dummies would be involved in testing while totally unrestrained. Renault, Jeep, American Motors, Mercedes-Benz, General Motors, and Ford requested that the dummies be restrained during testing by whatever means, active or passive, are installed at the particular seating positions. To provide otherwise, they argued, would unnecessarily expose the dummies to costly damage when subjected to impacts in an unrestrained condition.

The NHTSA finds petitioners' objections meritorious. Although this agency has determined that reliable test results can be best obtained when occupant weight is included in the vehicle during crash testing, the manner in which that weight is installed is subject to additional considerations. The NHTSA has made clear its desire to enable simultaneous testing under more than one standard where the test requirements are compatible. Standards 301 and 208 both require frontal and lateral barrier crash tests which can be conducted concurrently if the vehieles are loaded uniformly. Since Standard 208 provides for crash testing with dummies in vehicles with passive restraint systems, Standard 301 testing of these same vehicles should be conducted with dummies installed in the seating positions provided under Standard 208. The presence of the passive restraints will protect the dummies from unnecessary damage and the required testing for compliance with both standards can be accomplished simultaneously. Where a vehicle is not equipped with passive restraints, and Standard 208 testing is not mandated, weight equal to that of a 50th percentile test dummy should be secured to the floor pan at the front outboard designated seating positions in the vehieles being tested.

Further concern over the damage to which test dummies might be exposed was manifested by Jeep and American Motors. They petitioned for the removal of the dummies prior to the static rollover tests, arguing that their presence serves no safety-related purpose. The NHTSA has granted the request, on the basis of its determination that the dummies would have little or no effect on the fuel system's integrity during the rollover segment of the test procedure.

Jeep and American Motors further suggested that the standard specify that hardware and instrumentation be removed prior to the static rollover test in order to prevent its damage. This request is denied as unnecessary. Standard No. 301 contains no specification for the inclusion of instrumentation during testing. Any instrumentation present in the vehicle is there by decision of the manufacturer to assist him in monitoring the behavior of the fuel system during testing, and must be installed and utilized in such a manner as not to affect the test results. Therefore, as long as the loading requirements of the standard are met, manufacturers may deal with their instrumentation in any fashion they wish, as long as the test results are unaffected.

Volkswagen urged that unrestrained dummies not be required during the rear moving impact test, citing the absence of such a test in Standard 208 and alleging that the integrity of vehicle fuel systems would not be greatly affected by the presence of dummies. This request is denied. The rear moving barrier crash specified in proposed Standard 207, *Seating Systems*, provides for the installation of dummies in the same seating positions as required for Standard 301, thus permitting simultaneous conduct of the rear barrier crashes required by both standards. In order to obtain realistic and reliable test results, occupant weight must be in vehicles during Standard 301 crash testing. The NHTSA has determined that unrestrained dummies would have, at most, slight vulnerability to damage during rear barrier crash tests, since the impact is such that the seats themselves serve as protective restraint mechanisms. It has therefore been concluded that the best method for including occupant weight during rear barrier crash testing is with test dummies.

Notice 2 specified that the parking brake be engaged during the rear moving barrier crash test. Ford requested in its petition for reconsideration that this requirement be changed in order to enable simultaneous rear barrier crash testing with Standard 207 which provides for disengagement of the parking brake in its recent proposal. The NHTSA has decided to grant Ford's request. The condition of the parking brake during this test sequence would not so significantly affect the test results as to warrant retention of a requirement that would prevent simultaneous testing.

The Recreational Vehicle Institute objected to the standard, arguing that it was not cost-effective as applied to motor homes. RVI requested that different test procedures be developed for motor home manufacturers. Specifically it objected to what it suggested was a requirement for unnecessary double testing in situations where the incomplete vehicle has already been tested before the motor home manufacturer receives it. RVI expressed the view that the motor home manufacturer should not have to concern himself with compliance to the extent that he must test the entire vehicle in accordance with the standard's test procedures.

The NHTSA has found the requirements of Standard 301 to be reasonable in that they enforce a level of safety that has been determined necessary and provide adequate lead time for manufacturers to develop methods and means of compliance. The National Traffic and Motor Vehicle Safety Act does not require a manufacturer to test vehicles by any particular method. It does require that he exercise due care in assuring himself that his vehicles are capable of satisfying the performance requirements of applicable standards when tested in the manner prescribed. This may be accomplished, however, by whatever means the manufacturer reasonably determines to be reliable. If the final stage manufacturer of a motor home concludes that additional testing by him of the entire vehicle for compliance is unnecessary, and he has exercised due care in completing the vehicle in a manner that continues its conformity to applicable standards, he is under no obligation to repeat the procedures of the standards.

RVI further pressed its contention that the standard is not cost-beneficial by arguing that the agency has not provided specific data indicating a frequency of fuel system fires in motor homes that would justify the costs imposed by the standard.

Sufficient record evidence has been found to support the conclusion that fuel spillage in the types of crashes with which the standard deals is a major safety hazard. The only basis upon which motor home manufacturers could justify

the exception of their vehicles from Standard 301's requirements would be an inherent immunity from gasoline spillage. The standard establishes a reasonable test of a vehicle's ability to withstand impacts without experiencing fuel loss. If a motor home is designed in such a way as to preclude the spillage of fuel during the prescribed test impacts, compliance with the standard should present no significant hardship.

Volkswagen challenged the cost-benefit rationale of the more extensive performance requirements contained in Notice 2, and proposed that only the rear barrier crash be retained, if sufficient data exists to support its inclusion. The agency has carefully considered the issues raised in the Volkswagen petition. As discussed earlier, Standard 301 has been designed to allow testing for its requirements with some of the same barrier crash tests that are required by other standards: 208, 204, 212, and 207. This should reduce substantially the costs of testing to Standard 301, especially when viewed on a cost-per-vehicle basis. The NHTSA has concluded that the changes necessary for vehicles to comply with the standard are practicable and that the need for such increased fuel system integrity is sufficient to justify the costs.

The Recreational Vehicle Institute also urged that the effective date for motor homes be delayed 1 year beyond the date set for application of the standard to other vehicles. RVI contends that a uniform effective date for all manufacturers will create serious problems for the motor home manufacturer who will not have complying incomplete vehicles available to him until the effective date of the standard.

The NHTSA finds RVI's argument lacking in merit. Adequate lead time has been provided in Standard 301 to allow final stage manufacturers of multistage vehicles to become familiar with the requirements and to assure themselves that chassis and other vehicle components are available sufficiently in advance of the effective date to enable timely compliance. The availability of complying incomplete vehicles is a situation that should properly be resolved in the commercial dealings between motor home manufacturers and their suppliers. If the motor home manufacturer is unable to obtain complying incomplete vehicles far enough in advance of the standard's effective date, he might, for example, work out an arrangement with his supplier whereby the supplier will provide information relating to the manner in which the incomplete vehicle must be completed in order to remain in compliance with all applicable safety standards. The lead time provided in the standards is planned to take into account the needs of persons at each stage of the manufacturing process, including final stage manufacturers.

Jeep, American Motors, and Toyota urged delays in the implementation of various aspects of the standard. Jeep suggested a new schedule for application of the standard's requirements to multipurpose passenger vehicles, trucks, and buses, stating that the current lead time is insufficient to enable completion of necessary design changes and compliance testing. American Motors requested a 1-year delay in the effective date for the static rollover test in order to allow satisfactory completion of the required Environmental Protection Agency 50,000 mile durability test. Once vehicles have completed required EPA testing and certification, their fuel system components cannot be altered. AMC says that it cannot make the design changes necessary for Standard 301 compliance in time to utilize them in this year's EPA tests. AMC also desires a 2-year delay in the frontal angular, rear, and lateral impact tests, alleging that that constitutes the minimum time necessary to produce designs that comply. Toyota asked for a delay in the frontal angular crash test for all passenger vehicles until 1978, in order to allow them sufficient time to develop a satisfactory means of compliance with the specified performance level.

All of these requests are denied. The lead time that has been provided for compliance with Standard 301 is found adequate and reasonable. The rollover requirements have been in rule form for over a year, and the more extensive requirements were proposed more than 3 years in advance of their effective dates. Considering the urgent need for stronger and more durable fuel systems, further delay of the effective dates is not justified. On the basis of all information available, the NHTSA has determined that development of complying fuel systems can be attained in the time allowed. In addition, Congress has expressed in the recently enacted amendments to the National Traffic and Motor Vehicle Safety Act its decision that the effective dates specified in Notice 2 should be strictly adhered to.

Toyota requested that the requirements of the rear moving barrier crash not be imposed on vehicles with station wagon or hatch-back bodies, alleging difficulty in relocation of the fuel tank to an invulnerable position. The request is denied as the NHTSA has determined that satisfaction of the rear barrier crash requirements by station wagons and hatch-backs is practicable and necessary.

Volkswagen raised several objections in its petition to the static rollover test, including assertions that the test does not reflect real world accidents, and that the test procedure is unclear since the direction of rotation is unspecified.

The NHTSA does not consider these arguments to be germane. It is true that the static rollover test, like any "static" test, is not designed as a simulation of the actual behavior of a vehicle in a dynamic crash situation. It is intended rather as a laboratory method of quantitatively measuring the vehicle properties that contribute to safety in a range of crash situations. The NHTSA has found that a vehicle's performance in the static rollover test is directly related to the fuel system integrity that is the goal of the standard, and is an appropriate means of measuring that aspect of performance.

With regard to the direction of rotation, the NHTSA has stipulated that only a certain amount of fuel may escape during a 360° rotation of a vehicle on its longitudinal axis. The vehicle must be capable of meeting this performance level regardless of the direction of its rotation.

British Leyland (in a petition for rulemaking) and Volkswagen requested revision of the aspect of the barrier crash requirement limiting the amount of fuel spillage taking place from impact until motion of the vehicle has ceased. They stated that the current 1-ounce limitation is too difficult to measure in the period while the vehicle is moving and suggested that fuel spillage be averaged over the period from impact until 5 minutes following the cessation of motion.

The NHTSA must deny this request. The purpose of the current limitation on the spillage of fuel during the impact and post-impact motion is to prohibit the sudden loss of several ounces of fuel which might occur, as an example, by the displacement of the filler cap. Simultaneous loss of several ounces of fuel during the impact and subsequent vehicle motion could have a fire-causing potential, because of sparks that are likely to be given off during a skid or metal contact between vehicles.

Chrysler petitioned to have the requirement specifying that the moving barrier be guided during the entire impact sequence deleted in favor of a requirement that would allow the termination of guidance of the barrier immediately prior to impact. They argued that their suggested procedure is more representative of real world impacts.

The request is denied. The condition that there be no transverse or rotational movement of the barrier, which has been in effect since January 1, 1972, eliminates random variations between different tests and therefore makes the standard more repeatable and objective as required by the statute.

Jeep requested clarification that a given vehicle is only required to be subjected to one of the specified barrier impacts followed by a static rollover. This request is granted as it follows the agency's intent and the standard is not specific on that point. Section S6. is amended to require that a single vehicle need only be capable of meeting a single crash test followed by a static rollover.

American Motors submitted a request that the agency finds repetitious of previous petitions, urging that vehicle fluids be stabilized at ambient temperatures prior to testing. In responding to earlier petitions for reconsideration from MVMA and GM in Notice 1, the NHTSA denied a request for temperature specification, stating that it intended that the full spectrum of temperatures encountered on the road be reflected in the test procedure. That continues to be this agency's position.

In light of the foregoing S3., S6., S6.1, S6.3, S7.1.4, and S7.1.5 of Standard No. 301, *Fuel System Integrity*, (49 CFR 571.301) are amended . . .

Effective date: September 1, 1975, with additional requirements effective September 1, 1976 and September 1, 1977, as indicated.

(Secs. 103, 119, Pub. L. 89–563, 80 Stat. 718, 15 U.S.C. 1392, 1407; delegation of authority at 49 CFR 1.51.)

Issued on November 15, 1974.

James B. Gregory
Administrator

39 F.R. 40857
November 21, 1974

PREAMBLE TO AMENDMENT TO MOTOR VEHICLE SAFETY STANDARD NO. 301-75

Fuel System Integrity

(Docket No. 73-20; Notice 6)

This notice amends Standard No. 301, *Fuel System Integrity* (49 CFR 571.301), to specify new loading conditions and to establish a 30-minute fuel spillage measurement period following barrier crash tests.

On April 16, 1975, the NHTSA published a notice (40 F.R. 17036) proposing a revision of the loading conditions and fuel spillage measurement period requirement in Standard 301. The NHTSA also proposed in that notice an extension of the applicability of Standard 301 to school buses with a GVWR in excess of 10,000 pounds. At the request of several Members of Congress, the due date for comments on the school bus proposal was extended to June 26, 1975, and final rulemaking action on it will appear in a later Federal Register notice.

It was proposed that the current 15-minute fuel spillage measurement period be extended to 30 minutes in order to allow more time for leaks to be located and rates of flow to be established. Measurement of fuel loss during only a 15 minute time period is difficult because fuel may be escaping from various parts of the vehicle where it is not readily detectable. Chrysler, American Motors, and General Motors objected to the proposed change and asked that it either not be adopted or that adoption be delayed for one year until September 1, 1976.

The commenters argued that the revision was unnecessary and would involve a change in their testing methods. The NHTSA has fully considered these arguments and does not consider the amendment to prescribe a higher level of performance. It concludes that the 30-minute measurement period is necessary to achieve accurate measurement of fuel loss and assessment of vehicle compliance and accordingly amends

Standard 301 to prescribe the longer period for measurement.

The April 16, 1975, notice also proposed a change in the Standard 301 loading conditions to specify that 50th percentile test dummies be placed in specified seating positions during the frontal and lateral barrier crash tests, and that they be restrained by means installed in the vehicle for protection at the particular seating position. Currently the standard requires (during the frontal and lateral barrier crash tests) ballast weight secured at the specified designated seating positions in vehicles not equipped with passive restraint systems. In vehicles equipped with passive restraints, 50th percentile test dummies are to be placed in the specified seating positions during testing.

In petitions for reconsideration of this amendment to Standard No. 301 (39 F.R. 40857) various motor vehicle manufacturers stated that attachment of such ballast weight to the vehicle floor pans during the barrier crashes would exert unrealistic stresses on the vehicle structure which would not exist in an actual crash. The NHTSA found merit in petitioners' arguments, and its proposed revision of the loading conditions is intended to make the crash tests more representative of real-life situations.

Only Mazda objected to the proposal. It argued that curb weight be prescribed as the loading condition so that it could conduct Standard 301 compliance testing concurrently with testing for Standards No. 212 and 204. The NHTSA does not find merit in Mazda's request as the Standard 301 loading condition is considered necessary to assure an adequate level of fuel system integrity. Since the proposed loading conditions are more stringent than a curb weight condition, manufacturers could conduct compliance testing for Standards 301, 212, and 204 simultaneously. If the vehicle complied with the requirements of Standards 212 and 204 when loaded according to 301 specifications, the manufacturer presumably could certify the capability of the vehicles to comply with the performance requirements of 212 and 204 when loaded to curb weight. It should be noted that the NHTSA is considering amending Standards 212 and 204 to specify the same loading conditions as proposed for Standard 301.

All other commenters supported immediate adoption of the proposed loading conditions. Therefore, the NHTSA adopts the loading conditions as they were proposed in the April 16, 1975, notice.

In consideration of the foregoing, S5.5 and S7.1.6 of Motor Vehicle Safety Standard No.

301, *Fuel System Integrity* (49 CFR 571.301), are amended to read as follows:

Effective date: Because this amendment revises certain requirements that are part of 49 CFR 571.301-75, Motor Vehicle Safety Standard 301-75, effective September 1, 1975, and creates no additional burden upon any person, it is found for good cause shown that an effective date of less than 180 days after publication is in the public interest.

(Secs. 103, 119, Pub. L. 89-563, 80 Stat. 718 (15 U.S.C. 1392, 1407); delegation of authority at 49 CFR 1.51.)

Issued August 1, 1975.

Robert L. Carter
Acting Administrator

40 F.R. 33036
August 6, 1975

PREAMBLE TO AMENDMENT TO MOTOR VEHICLE SAFETY STANDARD NO. 301–75

Fuel System Integrity

(Docket No. 73–20; Notice 7)

This notice responds to a petition for reconsideration of the notice published August 6, 1975 (40 FR 33036), which amended Standard No. 301, *Fuel System Integrity* (49 CFR 571.301), to specify new loading conditions and establish a 30-minute fuel spillage measurement period following a barrier crash test.

American Motors Corporation (AMC) has petitioned for reconsideration of the amendment to S5.5 of Standard No. 301 insofar as it establishes an effective date of September 1, 1975, for the 30-minute fuel spillage requirement. AMC requests that the effective date for the 30-minute fuel spillage measurement time be delayed for 180 days from the date of publication of the rule.

The NHTSA has determined that AMC's petition has merit. AMC argues that the imposition of an effective date 25 days after the publication of the rule is burdensome because the 30-minute spillage requirement is a more stringent requirement than the previous 15-minute requirement and therefore requires additional testing to determine compliance. The NHTSA agrees that 25 days is not enough time to complete the additional testing. However, the effective date will be postponed 12 months instead of the 6 months requested by AMC so that manufacturers will not have to conduct compliance testing for 1976 model vehicles already certified under the old 15-minute spillage requirement. For these reasons the petition of American Motors Corporation is granted.

In S5.5 of Standard No. 301, *Fuel System Integrity*, (49 CFR 571.301), the amendment of August 6, 1975 (40 FR 33036), changing the term "10-minute period" to "25-minute period" effective September 1, 1975, is hereby made effective September 1, 1976.

(Sec. 103, 119, Pub. L. 89–563, 80 Stat. 718 (15 U.S.C. 1392, 1407); delegation of authority at 49 CFR 1.51).

Issued on October 3, 1975.

Gene G. Mannella
Acting Administrator

40 F.R. 47790
October 10, 1975

Effective: October 15, 1975
July 15, 1976

PREAMBLE TO AMENDMENT TO MOTOR VEHICLE SAFETY STANDARD NO. 301–75

Fuel System Integrity

(Docket No. 73–20; Notice 8)

The purpose of this notice is to amend Motor Vehicle Safety Standard No. 301, *Fuel System Integrity* (49 CFR 571.301) to extend the applicability of the standard to school buses with a GVWR in excess of 10,000 pounds. The amendment specifies conditions for a moving contoured barrier crash for school buses in order to determine the amount of fuel spillage following impact.

On October 27, 1974, the Motor Vehicle and Schoolbus Safety Amendments of 1974, amending the National Traffic and Motor Vehicle Safety Act, were signed into law (Pub. L. 93–492, 88 Stat. 1470). Section 103(i)(1)(A) of the Act, as amended, orders the promulgation of a safety standard establishing minimum requirements for the fuel system integrity of school buses. Standard No. 301 currently contains requirements for school buses with a GVWR of 10,000 pounds or less which will become effective beginning September 1, 1976. Larger school buses, which comprise approximately 90 percent of the school bus population, will be included in Standard No. 301 by this amendment.

A proposal to amend Standard No. 301 with respect to school buses, loading conditions, and spillage measurement time was published on April 16, 1975 (40 FR 17036). An amendment to the Standard specifying certain loading conditions and establishing a 30-minute fuel spillage measurement period was published on August 6, 1975 (40 FR 33036). At the request of several members of Congress, the period for comments on the school bus proposals was extended. This notice responds to the comments received with respect to the inclusion of school buses within the requirements of the standard.

Seven manufacturers opposed the requirement of a single impact test by a moving contoured barrier at any point on the school bus body, arguing that such a requirement would necessitate a proliferation of expensive tests in order to ensure compliance at every conceivable point of impact. The NHTSA does not agree. Although not specifying a particular impact point, the test condition allows for testing at the few most vulnerable points of each kind of school bus fuel system configuration. Therefore, only impacts at those points are necessary to determine compliance. On the basis of its knowledge of the bus design, a manufacturer should be able to make at least an approximate determination of the most vulnerable points on the bus body.

Two school bus body manufacturers requested a requirement that the manufacturer who installs the fuel system be responsible for compliance testing, while one chassis manufacturer argued that responsibility for compliance should rest with the final manufacturer. In most cases, if the basic fuel system components are included in the chassis as delivered by its manufacturer, the multistage vehicle regulations of 49 CFR Part 568 require the chassis manufacturer at least to describe the conditions under which the completed vehicle will conform, since it could not truthfully state that the design of the chassis has no substantial determining effect on conformity. Beyond that, however, the NHTSA position is that the decision as to who should perform the tests and who should take the responsibility is best not regulated by the government. The effect of Part 568 is to allow the final-stage manufacturer to avoid primary responsibility for conformity to a standard if it completes the vehicle in accordance with the conditions or instructions furnished with the incomplete vehicle by its manufacturer. Whether it does so is a decision it must make in light of all the circumstances.

This notice extends the proposed exclusion for vehicles that use fuel with a boiling point below 32° F. to school buses having a GVWR greater than 10,000 pounds. Fuel systems using gaseous fuels are not subject to the spillage problems against which this standard is directed.

The Vehicle Equipment Safety Commission requested that school buses be required to undergo static rollover tests and that the engine be running during the tests. Upon consideration, the NHTSA finds that a static rollover test for school buses is impractical in light of the expensive test facility that would be required. A requirement that the engine be running during the impact test would make little difference in the resulting fuel spillage. Since the standard requires that the fuel tank be filled with Stoddard solvent during the impact test, the test vehicle would have to be equipped with an auxiliary fuel system for the engine. The expense of modifying the test vehicle to allow the engine to run during the test would not justify the minimal benefits resulting from a requirement that the engine be running. However, the fuel system integrity of school buses will be continually monitored and analyzed by the NHSTA. Therefore, suggestions such as these may be the subject of future rulemaking.

One school bus body manufacturer cited the infrequency of school bus fires resulting from collisions as a reason for ameliorating or eliminating altogether fuel system integrity requirements for school buses. In promulgating these amendments to Standard No. 301, the NHTSA is acting under the statutory mandate to develop regulations concerning school bus fuel systems. This statute reflects the need, evidently strongly felt by the public, to protect the children who ride in the school buses. They and their parents have little direct control over the types of vehicles in which they ride to school, and are therefore not in a position to determine the safety of the vehicles. Considering the high regard expressed by the public for the safety of its children, the NHTSA finds it important that the school bus standards be effective and meaningful.

The California Highway Patrol expressed the concern that these amendments would preempt State regulations to the extent that the State would be precluded from specifying the location of fuel tanks, fillers, vents, and drain openings in school buses. The standard will unavoidably have that effect, by the operation of section 103(d) of the National Traffic and Motor Vehicle Safety Act. However, although a State may not have regulations of general applicability that bear on these aspects of performance, the second sentence of the same section makes it clear that a State or political subdivision may specify higher standards of performance for vehicles purchased for its own use, although of course the Federal standards must be met in any case.

In addition to provisions directly relating to school buses, this notice clarifies the loading condition amendments in the notice of August 6, 1975, by amending S6.1 to provide for testing with 50th percentile dummies. The wording of S6.1 is identical to that of the proposal.

In light of the foregoing, 49 CFR 571.301, Motor Vehicle Safety Standard No. 301, is amended....

Effective date: July 15, 1976, in conformity with the schedule mandated by the 1974 Amendments to the Traffic Safety Act. However, the effective date of the amendment of S6.1 is October 15, 1975. Because the amendment to that paragraph clarifies the revision of certain requirements which became effective September 1, 1975, it is found for good cause shown that an effective date for the amendment of S6.1 less than 180 days after issuance is in the public interest.

(Sec. 103, 119, Pub. L. 89–563, 80 Stat. 718 (15 U.S.C. 1392, 1407); Sec. 202, Pub. L. 93–492, 88 Stat. 1470 (15 U.S.C. 1392); delegations of authority at 49 CFR 1.51 and 501.8).

Issued on October 8, 1975.

Gene G. Mannella
Acting Administrator

40 F.R. 48352
October 15, 1975

PREAMBLE TO AMENDMENT TO MOTOR VEHICLE SAFETY STANDARD NO. 301–75

Fuel System Integrity

(Docket No. 73–20; Notice 9)

This notice clarifies the effective date of the change in Standard No. 301–75 (49 CFR 571.301–75) from a 15-minute to a 30-minute fuel spillage measurement period following cessation of motion in barrier crash tests.

Until August 1975, S5.4 of Standard No. 301–75 specified a 15-minute fuel spillage measurement period for the barrier crash test requirements that would become effective September 1, 1975. To allow more time for leaks to be located and rates of flow to be established, that period was extended to 30 minutes in Notice 6 (40 FR 33036, August 6, 1975; correction of section numbers at 40 FR 37042, August 25, 1975). Notice 6 set the effective date of the change as September 1, 1975.

In response to a petition for reconsideration filed by American Motors Corporation, the NHTSA in Notice 7 (40 FR 47790; October 10, 1975) delayed for 1 year the effective date of that change, thereby establishing the following scheme: a 15-minute period would be used in applying the standard to vehicles manufactured before September 1, 1976, while a 30-minute measurement period would be used for vehicles manufactured after that date.

In Notice 8, which was published on October 15, 1975 (40 FR 48352), the loading conditions of S6.1 were revised, effective immediately, and the standard was extended to apply to school buses with a GVWR in excess of 10,000 pounds, effective July 15, 1976. Because these amendments were made by republishing the entire text of the standard, it appeared that the effective date of the change from a 15-minute measurement period to a 30-minute measurement period had been advanced from September 1, 1976, to July 15, 1976, for all vehicles. The NHTSA did not intend such an advancement, and this notice amends the standard to reestablish the September 1, 1976, effective date for vehicles other than school buses with a GVWR greater than 10,000 pounds.

The following corrections of Notice 8 are also made: the standard is designated as "Standard No. 301–75" and typographical errors in S6.4 and S7.5.2 are corrected.

In consideration of the foregoing, § 571.301 of 49 CFR Part 571 (Standard No. 301, *Fuel System Integrity*), as published in the issue of October 15, 1975 (40 FR 48352), is redesignated as § 571.301–75 and amended. . . .

Effective dates: As set forth in the standard. Changes indicated in the text of the Code of Federal Regulations should be made immediately.

(Sec. 103, 119, Pub. L. 89–563, 80 Stat. 718 (15 U.S.C. 1392, 1407); Sec. 108, Pub. L. 93–492, 88 Stat. 1470 (15 U.S.C. 1392 note); delegation of authority at 49 CFR 1.50.)

Issued on February 25, 1976.

James B. Gregory
Administrator

41 F.R. 9350
March 4, 1976

PREAMBLE TO AMENDMENT TO MOTOR VEHICLE SAFETY STANDARD NO. 301-75

Fuel System Integrity

(Docket No. 73–03; Notice 07); Docket No. 73–20; Notice 010);
(Docket No. 73–34; Notice 04); (Docket No. 75–02; Notice 03);
(Docket No. 75–03; Notice 05); (Docket No. 75–07; Notice 03);
(Docket No. 75–24; Notice 03)

This notice announces that the effective dates of the redefinition of "school bus" and of six Federal motor vehicle safety standards as they apply to school buses are changed to April 1, 1977, from the previously established effective dates. This notice also makes a minor amendment to Standard No. 220, *School Bus Rollover Protection*, and adds a figure to Standard No. 221, *School Bus Body Joint Strength*.

The Motor Vehicle and Schoolbus Safety Amendments of 1974 (the Act) mandated the issuance of Federal motor vehicle safety standards for several aspects of school bus performance, Pub. L. 93–492, § 202 (15 U.S.C. § 1392(i)(1)(A)). These amendments included a definition of school bus that necessitated a revision of the existing definition used by the NHTSA in establishing safety requirements. The Act also specified that the new requirements "apply to each schoolbus and item of schoolbus equipment which is manufactured . . . on or after the expiration of the 9-month period which begins on the date of promulgation of such safety standards." (15 U.S.C. § 1392(i)(1)(B)).

Pursuant to the Act, amendments were made to the following standards: Standard No. 301–75, *Fuel System Integrity* (49 CFR 571.301–75), effective July 15, 1976, for school buses not already covered by the standard (40 FR 483521, October 15, 1975); Standard No. 105–75, *Hydraulic Brake Systems* (49 CFR 571.105–75), effective October 12, 1976 (41 FR 2391, January 16, 1976); and Standard No. 217, *Bus Window Retention and Release* (49 CFR 571.217), effective for school buses on October 26, 1976 (41 FR 3871, January 27, 1976).

In addition, the following new standards were added to Part 571 of Title 49 of the Code of Federal Regulations, effective October 26, 1976: Standard No. 220, *School Bus Rollover Protection* (41 FR 3874, January 27, 1976); Standard No. 221, *School Bus Body Joint Strength* (41 FR 3872, January 26, 1976); and Standard No. 222, *School Bus Passenger Seating and Crash Protection* (41 FR 4016, January 28, 1976). Also, the existing definition of "school bus" was amended, effective October 27, 1976, in line with the date set by the Act for issuance of the standards.

The Act was recently amended by Public Law 94–346 (July 8, 1976) to change the effective dates of the school bus standards to April 1, 1977 (15 U.S.C. § 1392(i)(1)(B)). This notice is intended to advise interested persons of these changes of effective dates. In the case of Standard No. 301–75, the change of effective date is reflected in a conforming amendment to S5.4 of that standard. A similar amendment is made in S3 of Standard No. 105–75.

The agency concludes that the October 27, 1976, effective date for the redefinition of "school bus" should be postponed to April 1, 1977, to conform to the new effective dates for the upcoming requirements. If this were not done, the new classes of school buses would be required to meet existing standards that apply to school buses (e.g., Standard No. 108 (49 CFR 571.108)) before being required to meet the new standards. This would result in two stages of compliance, and would complicate the redesign efforts that Congress sought to relieve.

This notice also amends Standard No. 220 in response to an interpretation request by Blue Bird Body Company, and Sheller-Globe Corporation's petition for reconsideration of the standard. Both companies request confirmation that the standard's requirement to operate emergency exits during the application of force to the vehicle roof (S4(b)) does not apply to roof exits which are covered by the force application plate. The agency did not intend to require the operation of roof exits while the force application plate is in place on the vehicle. Accordingly, an appropriate amendment has been made to S4(b) of the standard.

With regard to Standard No. 220, Sheller-Globe also requested information that, in testing its school buses that have a gross vehicle weight rating (GVWR) of 10,000 pounds or less, it may test with a force application plate with dimensions other than those specified in the standard. The standard does not prohibit a manufacturer from using a different dimension from that specified, in view of the NHTSA's expressed position on the legal effect of its regulations. To certify compliance, a manufacturer is free to choose any means, in the exercise of due care, to show that a vehicle (or item of motor vehicle equipment) would comply if tested by the NHTSA as specified in the standard. Thus the force application plate used by the NHTSA need not be duplicated by each manufacturer or compliance test facility. Sheller-Globe, for example, is free to use a force application plate of any width as long as it can certify its vehicle would comply if tested by the NHTSA according to the standard.

In a separate area, the agency corrects the inadvertent omission of an illustration from Standard No. 221 as it was issued January 26, 1976 (41 FR 3872). The figure does not differ from that proposed and, in that form, it received no adverse comment.

In accordance with recently enunciated Department of Transportation policy encouraging adequate analysis of the consequences of regulatory action (41 FR 16200, April 16, 1976), the agency herewith summarizes its evaluation of the economic and other consequences of this action on the public and private sectors, including possible loss of safety benefits. The changes in effective dates for the school bus standards are not evaluated because they were accomplished by law and not by regulatory action.

The change of effective date for the redefinition of "school bus" will result in savings to manufacturers who will not be required to meet existing school bus standards between October 27, 1976, and April 1, 1977. The agency calculates that the only standard that would not be met would be the requirement in Standard No. 108 for school bus marker lamps. In view of the agency's existing provision for the marking of night school buses in Pupil Transportation Standard No. 17 (23 CFR 1204), it is concluded that the absence of this equipment until April 1, 1977, will not have a significant adverse impact on safety.

The interpretative amendment of Standard No. 220 and the addition of a figure to Standard No. 221 are not expected to affect the manufacture or operation of school buses.

In consideration of the foregoing, Part 571 of Title 49 of the Code of Federal Regulations is amended. . . .

Effective dates:

1. Because the listed amendments do not impose additional requirements of any person, the National Highway Traffic Safety Administration finds that an immediate effective date of August 26, 1976 is in the public interest.

2. The effective date of the redefinition of 'school bus" in 49 CFR Part 571.3 that was published in the issue of December 31, 1976 (40 FR 60033) is changed to April 1, 1977.

3. The effective dates of Standard Nos. 105–75, 217, 301–75, 220, 221, and 222(as they apply to school buses) are April 1, 1977, in accordance with Public Law 94–346.

(Sec. 103, 119, Pub. L. 89–563, 80 Stat. 718 (15 U.S.C. 1392, 1407); Pub. L. 94–346, Stat. (15 U.S.C. § 1392(i)(1)(B)); delegation of authority at 49 CFR 1.50.)

Issued on August 17, 1976.

John W. Snow
Administrator

41 F.R. 36026
August 26, 1976

MOTOR VEHICLE SAFETY STANDARD NO. 301

Fuel System Integrity

S1. Scope. This standard specifies requirements for the integrity of motor vehicle fuel systems.

S2. Purpose. The purpose of this standard is to reduce deaths and injuries occurring from fires that result from fuel spillage during and after motor vehicle crashes.

S3. Application. This standard applies to passenger cars, and to multipurpose passenger vehicles, trucks, and buses that have a GVWR of 10,000 pounds or less and use fuel with a boiling point above 32° F., and to school buses that have a GVWR greater than 10,000 pounds and use fuel with a boiling point above 32° F.

S4. Definition. "Fuel spillage" means the fall, flow, or run of fuel from the vehicle but does not include wetness resulting from capillary action.

S5. General requirements.

S5.1 Passenger cars. Each passenger car manufactured from September 1, 1975, to August 31, 1976, shall meet the requirements of S6.1 in a perpendicular impact only, and S6.4. Each passenger car manufactured on or after September 1, 1976, shall meet all the requirements of S6, except S6.5.

S5.2 Vehicles with GVWR of 6,000 pounds or less. Each multipurpose passenger vehicle, truck, and bus with a GVWR of 6,000 pounds or less manufactured from September 1, 1976, to August 31, 1977, shall meet all the requirements of S6.1 in a perpendicular impact only, S6.2, and S6.4. Each of these types of vehicles manufactured on or after September 1, 1977, shall meet the requirements of S6, except S6.5.

S5.3 Vehicles with GVWR of more than 6,000 pounds but not more than 10,000 pounds. Each multipurpose passenger vehicle, truck, and bus with a GVWR of more than 6,000 pounds but not more than 10,000 pounds manufactured from September 1, 1976, to August 31, 1977, shall meet the requirements of S6.1 in a perpendicular impact only. Each vehicle manufactured on or after September 1, 1977, shall meet all the requirements of S6, except S6.5.

S5.4 School buses with a GVWR greater than 10,000 pounds. Each school bus with a GVWR greater than 10,000 pounds manufactured on or after April 1, 1977, shall meet the requirements of S6.5.

S5.5 Fuel spillage: Barrier crash. Fuel spillage in any fixed or moving barrier crash test shall not exceed 1 ounce by weight from impact until motion of the vehicle has ceased, and shall not exceed a total of 5 ounces by weight in the 5-minute period following cessation of motion. For the subsequent 25-minute period (for vehicles manufactured before September 1, 1976, other than school buses with a GVWR greater than 10,000 pounds: the subsequent 10-minute period), fuel spillage during any 1-minute interval shall not exceed 1 ounce by weight.

S5.6 Fuel spillage: Rollover. Fuel spillage in any rollover test, from the onset of rotational motion, shall not exceed a total of 5 ounces by weight for the first 5 minutes of testing at each successive 90° increment. For the remaining testing period, at each increment of 90° fuel spillage during any 1-minute interval shall not exceed 1 ounce by weight.

S6. Test requirements. Each vehicle with a GVWR of 10,000 pounds or less shall be capable of meeting the requirements of any applicable barrier crash test followed by a static rollover, without alteration of the vehicle during the test sequence. A particular vehicle need not meet further requirements after having been subjected to a single barrier crash test and a static rollover test.

S6.1 Frontal barrier crash. When the vehicle traveling longitudinally forward at any speed up to and including 30 mph impacts a fixed collision barrier that is perpendicular to the line of travel of the vehicle, or at any angle up to 30° in either direction from the perpendicular to the line of travel of the vehicle, with 50th-percentile test dummies as specified in Part 572 of this chapter at each front outboard designated seating position and at any other position whose protection system is required to be tested by a dummy under the provisions of Standard No. 208, under the applicable conditions of S7, fuel spillage shall not exceed the limits of S5.5. (Effective: October 15, 1975)

S6.2 Rear moving barrier crash. When the vehicle is impacted from the rear by a barrier moving at 30 mph, with test dummies as specified in Part 572 of this chapter at each front outboard designated seating position, under the applicable conditions of S7, fuel spillage shall not exceed the limits of S5.5.

S6.3 Lateral moving barrier crash. When the vehicle is impacted laterally on either side by a barrier moving at 20 mph with 50th-percentile test dummies as specified in Part 572 of this chapter at positions required for testing to Standard No. 208, under the applicable conditions of S7, fuel spillage shall not exceed the limits of S5.5.

S6.4 Static rollover. When the vehicle is rotated on its longitudinal axis to each successive increment of 90°, following an impact crash of S6.1, S6.2, or S6.3, fuel spillage shall not exceed the limits of S5.6.

S6.5 Moving contoured barrier crash. When the moving contoured barrier assembly traveling longitudinally forward at any speed up to and including 30 mph impacts the test vehicle (school bus with a GVWR exceeding 10,000 pounds) at any

point and angle, under the applicable conditions of S7.1 and S7.5, fuel spillage shall not exceed the limits of S5.5.

S7. Test conditions. The requirements of S5 and S6 shall be met under the following conditions. Where a range of conditions is specified, the vehicle must be capable of meeting the requirements at all points within the range.

S7.1 General test conditions. The following conditions apply to all tests:

S7.1.1 The fuel tank is filled to any level from 90 to 95 percent of capacity with Stoddard solvent, having the physical and chemical properties of type 1 solvent, Table I ASTM Standard D484–71, "Standard Specifications for Hydrocarbon Dry Cleaning Solvents."

S7.1.2 The fuel system other than the fuel tank is filled with Stoddard solvent to its normal operating level.

S7.1.3 In meeting the requirements of S6.1 through S6.3, if the vehicle has an electrically driven fuel pump that normally runs when the vehicle's electrical system is activated, it is operating at the time of the barrier crash.

S7.1.4 The parking brake is disengaged and the transmission is in neutral, except that in meeting the requirements of S6.5 the parking brake is set.

S7.1.5 Tires are inflated to manufacturer's specifications.

S7.1.6 The vehicle, including test devices and instrumentation, is loaded as follows:

(a) Except as specified in S7.1.1, a passenger car is loaded to its unloaded vehicle weight plus its rated cargo and luggage capacity weight, secured in the luggage area, plus the necessary test dummies as specified in S6, restrained only by means that are installed in the vehicle for protection at its seating position.

(b) Except as specified in S7.1.1, a multipurpose passenger vehicle, truck, or bus with a GVWR of 10,000 pounds or less is loaded to its unloaded vehicle weight, plus the necessary test dummies, as specified in S6, plus 300 pounds of its rated cargo and luggage capacity weight, whichever is less, secured to the vehicle and dis-

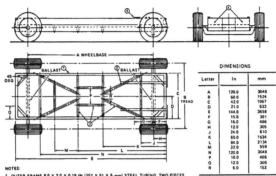

DIMENSIONS

Letter	In	mm
A	120.0	3048
B	60.0	1524
C	42.0	1067
D	21.0	533
E	144.0	3658
F	15.0	381
G	16.0	406
H	12.0	305
J	24.0	610
K	60.0	1524
L	84.0	2134
M	22.0	559
N	120.0	3048
P	16.0	406
Q	12.0	305
R	6.0	152

NOTES:

1. OUTER FRAME 6.0 X 2.0 X 0.19 IN (152 X 51 X 5 mm) STEEL TUBING, TWO PIECES WELDED TOGETHER FOR A 12.0 IN (305 mm) HEIGHT.

2. BALLAST TIE DOWNS.

3. ALL INNER REINFORCEMENTS AND FRAME GUSSETS OF 4.0 X 2.0 X 0.19 IN (102 X 51 X 5 mm) STEEL TUBING.

4. REINFORCE AREAS FOR BOLTING ON FACE PLATES.

FIG. 1—COMMON CARRIAGE FOR MOVING BARRIERS

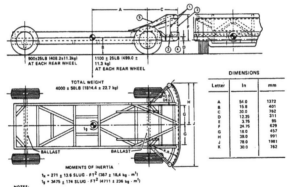

900±25LB (408.2±11.3kg) AT EACH REAR WHEEL

1100 ± 25LB (499.0 ± 11.3 kg) AT EACH REAR WHEEL

TOTAL WEIGHT
4000 ± 50LB (1814.4 ± 22.7 kg)

DIMENSIONS

Letter	In	mm
A	54.0	1372
B	15.8	401
C	30.0	762
D	12.25	311
E	3.75	95
F	24.75	629
G	18.0	457
H	39.0	991
J	78.0	1981
K	30.0	762

MOMENTS OF INERTIA

$I_R = 271 \pm 13.6$ SLUG $\cdot$ FT^2 (367 $\pm$ 18.4 kg $\cdot$ m^2)

$I_g = 3475 \pm 174$ SLUG $\cdot$ FT^2 (4711 $\pm$ 236 kg $\cdot$ m^2)

NOTES:

1. UPPER FRAME 4.0 IN DIA X 0.25 IN WALL (102 mm DIA X 6 mm WALL)STEEL TUBING (THREE SIDES).

2. LOWER FRAME 6.0 IN DIA X 0.50 IN WALL (152 mm DIA X 13 mm WALL) STEEL TUBING.

3. FACE PLATE 0.75 IN (19 mm) THICK COLD ROLLED STEEL.

4. LEADING EDGE 1.0 X 4.0 IN (25 X 102 mm) STEEL BAND, SHARP EDGES BROKEN.

5. ALL INNER REINFORCEMENTS 4.0 X 2.0 X 0.19 IN (102 X 51 X 5 mm) STEEL TUBING.

FIG. 2—COMMON CARRIAGE WITH CONTOURED IMPACT SURFACE ATTACHED

PART 571; S 301-3

tributed so that the weight on each axle as measured at the tire-ground interface is in proportion to its GAWR. If the weight on any axle, when the vehicle is loaded to unloaded vehicle weight plus dummy weight, exceeds the axle's proportional share of the test weight, the remaining weight shall be placed so that the weight on that axle remains the same. Each dummy shall be restrained only by means that are installed in the vehicle for protection at its seating position.

(c) Except as specified in S7.1.1, a school bus with a GVWR greater than 10,000 pounds is loaded to its unloaded vehicle weight plus 120 pounds of unsecured weight at each designated seating position.

S7.2 Lateral moving barrier crash test conditions. The lateral moving barrier crash test conditions are those specified in S8.2 of Standard No. 208, 49 CFR 571.208.

S7.3 Rear moving barrier test conditions. The rear moving barrier test conditions are those specified in S8.2 of Standard No. 208, 49 CFR 571.208, except for the positioning of the barrier and the vehicle. The barrier and test vehicle are positioned so that at impact—

(a) The vehicle is at rest in its normal attitude;

(b) The barrier is traveling at 30 mph with its face perpendicular to the longitudinal centerline of the vehicle; and

(c) A vertical plane through the geometric center of the barrier impact surface and perpendicular to that surface coincides with the longitudinal centerline of the vehicle.

S7.4 Static rollover test conditions. The vehicle is rotated about its longitudinal axis, with the axis kept horizontal, to each successive increment of 90°, 180°, and 270° at a uniform rate, with 90° of rotation taking place in any time interval from 1 to 3 minutes. After reaching each 90° increment the vehicle is held in that position for 5 minutes.

S7.5 Moving contoured barrier test conditions. The following conditions apply to the moving contoured barrier crash test:

S7.5.1. The moving barrier, which is mounted on a carriage as specified in Figure 1, is of rigid construction, symmetrical about a vertical longitudinal plane. The contoured impact surface, which is 24.75 inches high and 78 inches wide, conforms to the dimensions shown in Figure 2, and is attached to the carriage as shown in that figure. The ground clearance to the lower edge of the impact surface is 5.25 ± 0.5 inches. The wheelbase is 120 ± 2 inches.

S7.5.2 The moving contoured barrier, including the impact surface, supporting structure, and carriage, weighs $4,000 \pm 50$ pounds with the weight distributed so that 900 ± 25 pounds is at each rear wheel and 1100 ± 25 pounds is at each front wheel. The center of gravity is located 54.0 ± 1.5 inches rearward of the front wheel axis, in the vertical longitudinal plane of symmetry, 15.8 inches above the ground. The moment of inertia about the center of gravity is:

$$I_x = 271 \pm 13.6 \text{ slug ft}^2$$
$$I_z = 3475 \pm 174 \text{ slug ft}^3$$

S7.5.3 The moving contoured barrier has a solid nonsteerable front axle and fixed rear axle attached directly to the frame rails with no spring or other type of suspension system on any wheel. (The moving barrier assembly is equipped with a braking device capable of stopping its motion.)

S7.5.4 The moving barrier assembly is equipped with G78–15 pneumatic tires with a tread width of 6.0 ± 1 inch, inflated to 24 psi.

S7.5.5 The concrete surface upon which the vehicle is tested is level, rigid, and of uniform construction, with a skid number of 75 when measured in accordance with American Society of Testing and Materials Method E–274–65T at 40 mph, omitting water delivery as specified in paragraph 7.1 of that method.

S7.5.6 The barrier assembly is released from the guidance mechanism immediately prior to impact with the vehicle.

38 F.R. 22397
August 20, 1973

40 F.R. 48352
October 15, 1975

MOTOR VEHICLE SAFETY STANDARD NO. 302

Flammability of Interior Materials—Passenger Cars, Multipurpose Passenger Vehicles, Trucks, and Buses

(Docket No. 3–3; Notice 4)

This notice amends § 575.21 of Title 49 of the Code of Federal Regulations by adding a new motor vehicle safety standard, No. 302, Flammability of Interior Materials. Notices of proposed rulemaking on the subject were published on December 31, 1969 (34 F.R. 20434) and June 26, 1970 (35 F.R. 10460).

As stated in the notice of December 31, 1969, the occurrence of thousands of fires per year that begin in vehicle interiors provide ample justification for a safety standard on flammability of interior materials. Although the qualities of interior materials cannot by themselves make occupants safe from the hazards of fuel-fed fires, it is important, when fires occur in the interior of the vehicle from such sources as matches, cigarettes, or short circuits in interior wiring, that there be sufficient time for the driver to stop the vehicle, and if necessary for occupants to leave it, before injury occurs.

The question on which the public responses to the above notices differed most widely was the burn rate limit to be required. The rate proposed was 4 inches per minute, measured by a horizontal test. Some manufacturers suggested maximum burn rates as high as 15 inches per minute. The Center for Auto Safety, the Textile Fibers and By-Products Association, and the National Cotton Batting Institute, on the other hand, suggested essentially a zero burn rate, or self-extinguishment, requirement, with a vertical rather than a horizontal test. A careful study was made of the available information on this subject, including the burn rates of materials currently in use or available for use, recommendations or regulations of other agencies, and the economic and technical consequences of various possible rate levels and types of tests. A con-

siderable amount of Bureau-sponsored research has been conducted and is continuing on the subject. On consideration of this data, the Bureau has decided to retain the 4-inch-per-minute burn limit, with the horizontal test, in this standard. It has been determined that suitable materials are not available in sufficient quantities, at reasonable costs, to meet a significantly more stringent burn rate by the effective date that is hereby established. The 4-inch rate will require a major upgrading of materials used in many areas, and a corresponding improvement in this aspect of motor vehicle safety. It is important that this standard not hinder manufacturers' efforts to comply with the crash protection requirements that are currently being imposed, and that in the Bureau's judgment are of the greatest importance. Further study will be made, however, of the feasibility of, and justification for, imposing more stringent requirements with a later effective date.

As pointed out in several comments, the problem of toxic combustion by-products is closely related to that of burn rate. Release of toxic gases is one of the injury-producing aspects of motor vehicle fires, and many of the common ways of treating materials to reduce their burn rates involve chemicals that produce highly poisonous gases such as hydrogen chloride and hydrogen cyanide. The problem of setting standards with regard to combustion by-products is difficult and complex, and the subject of continuing research under Bureau auspices. Until enough is known in this area to form the basis for a standard, and to establish the proper interaction between burn rate and toxicity, this uncertainty constitutes an additional reason for not requiring self-extinguishing materials.

The proposal specified a particular commercial gas for the test burn and several comments suggested problems in obtaining the gas for manufacture testing. As is the case with all the motor vehicle safety standards, the test procedures describe the tests that the regulated vehicles or equipment must be capable of passing, when tested by the Bureau, and not the method by which a manufacturer must ascertain that capability. Any gas with at least as high a flame temperature as the gas described in the standard would therefore be suitable for manufacturer testing. To make this point clearer, and to use a more readily available reference point, the standards been reworded to specify a gas that "has a flame temperature equivalent to that of natural gas."

The dimensions of the enclosure within which the test is conducted have been changed from those proposed, in order to provide more draft-free conditions, and consequently more repeatable results. Smaller cabinets, furthermore, evidently are more generally available than larger ones. Again it should be noted that there is no necessity that manufacturers duplicate the dimensions of the test cabinet, as long as they can establish a reasonable basis for concluding that their materials will meet the requirements when tested in such a cabinet.

Several comments questioned the need for specifying the temperature and relative humidity under which the material is conditioned and the test is conducted. The foregoing discussions of the relation of the standard to manufacturer testing apply here also. The specification of temperature and relative humidity for conditioning and testing is made to preclude any arguments, in the face of a compliance test failure, that variations in test results are due to permitted variations in test conditions. The relative humidity specification has been changed from 65 percent, as proposed, to 50 percent. This humidity level represents more closely the conditions encountered in use during fairly dry weather. While it is a slightly more stringent condition, it is one in wide use for materials testing, according to the comments, and is not, in the judgment of the Bureau, a large enough change in the substance of the proposal to warrant further notice and opportunity for comment.

Several comments suggested that the standard should specify the number of specimens to be tested, with averaging of results, as is commonly found in specification-type standards. The legal nature of the motor vehicle safety standards is such, however, that sampling and averaging provisions would be inappropriate. As defined by the National Traffic and Motor Vehicle Safety Act, the standards are minimum performance levels that must be met by every motor vehicle or item of motor vehicle equipment to which they apply. Enforcement is based on independent Bureau testing, not review of manufacturer testing, and manufacturers are required to take legal responsibility for every item they produce. The result, and the intent of the Bureau in setting the standards, is that manufacturers must establish a sufficient margin of performance between their test results and the standard's requirements to allow for whatever variances may occur between items tested and items produced.

The description of portions to be tested has been changed slightly, such that the surface and the underlying materials are tested either separately or as a composite, depending on whether they are attached to each other as used in the vehicle. In the proposal, surface and underlying materials were to be tested separately regardless of how used, an element of complexity found unnecessary for safety purposes.

In response to comments with respect to materials that burn at a decreasing rate, to which the application of the test is not clear, an additional criterion has been added. If material stops burning before it has burned for 60 seconds, and does not burn more than 2 inches, it is considered to meet the requirement.

In consideration of the foregoing, § 571.21 of Title 49, Code of Federal Regulations, is amended by the addition of Standard No. 302, Flammability of Interior Materials.

Effective date: September 1, 1972. Because of the extensive design changes that will be necessitated by this new standard, and the lead-time consequently required by manufacturers to prepare for production, it is found, for good

Effective: September 1, 1972

cause shown, that an effective date later than one year from the issuance of this notice is in the public interest.

Issued on December 29, 1970.

Douglas W. Toms
Director

36 F.R. 289
January 8, 1971

PREAMBLE TO AMENDMENT TO MOTOR VEHICLE SAFETY STANDARD NO. 302

Flammability of Interior Materials

(Docket No. 3–3; Notice 7b)

This notice modifies the test procedures and specimen preparation requirements of Motor Vehicle Safety Standard No. 302, *Flammability of interior materials* (49 CFR 571.302). A notice of proposed rulemaking was issued on May 17, 1973 (38 FR 12934).

Several comments on the notice of proposed rulemaking suggested exempting small components on the basis of size because of the confusion caused by paragraph S4.1. This agency has not found, however, that the exemption of a component on the basis of size is consistent with safety. Rather, NHTSA finds that if a component is too small to produce an acceptable test sample, a test sample consisting of the material from which the component is fabricated should be substituted. Consequently, a new section S4.1.1 has been added to require surrogate testing of such components as switches, knobs, gaskets, and grommets which are considered too small to be effectively tested under the current procedures.

A previous notice of proposed rulemaking (36 FR 9565) suggested a scheme for testing single and composite materials that would allow the testing of certain configurations of vehicle interior materials not taken into account under the present scheme. Examples of such configurations are multi-layered composites and single layers of underlying materials that are neither padding nor cushioning materials. Comments to that notice argued that some aspects of the proposed scheme would require some duplicative testing without providing a measurable safety benefit.

In response to these arguments, it was proposed (38 F.R. 12934) that S4.2 be amended to take into account some omissions in the present scheme and to reduce the complexity of testing single and composite materials. After reviewing the comments, the proposed scheme is adopted. Thus, the standard is amended to require single materials or composites (materials that adhere at every point of contact), any part of which is within ½ inch of the surface of the component, to meet the burn-rate requirements. Materials that are not part of adhering composites are subject to the requirements when tested separately. Those materials that do adhere to adjacent materials at every point of contact are subject to the requirements as composites when tested with the adjacent materials. The concept of "adherence" would replace language presently contained in the standard describing materials as "bonded, sewed, or mechanically attached." An illustrative example is included in the text of the section.

Several comments in response to the notice of proposed rulemaking requested changes in the test cabinet, as did comments in response to previous notices concerning this standard. The NHTSA has evaluated various recommendations and suggestions concerning the cabinet. No changes are proposed in this notice, however, as sufficient justification has not been found for a design change at this time.

Paragraph S5.2.1 of the standard presently provides that materials exceeding ½ inch in thickness are to be cut down to ½ inch in thickness before testing. As described in the notice of proposed rulemaking, cutting certain materials to the prescribed thickness produces a tufted surface upon which a flame front may be propagated at a faster rate than it would be upon the surface of the material before cutting, thereby creating an artificial test condition. In order to avoid this, the requirements for the transmission rate of a flame front are amended in S4.3(a) to exclude surfaces created by cutting.

The notice of proposed rulemaking points out that a related problem has arisen concerning which surfaces of a test specimen should face the flame in the test cabinet. To answer this question and avoid unnecessary test duplication, the test procedures are amended to provide that the surface of the specimen closest to the occupant compartment air space face downward on the test frame. The test specimen is produced by cutting the material in the direction that provides the most adverse test results.

In light of the above, Motor Vehicle Safety Standard No. 302, 49 CFR § 571.302, is amended. . . .

Effective date: Oct. 1, 1975.

(Secs. 103, 119, Pub. L. 89–563, 80 Stat. 718 (15 U.S.C. 1392, 1407); delegation of authority at 49 CFR 1.51.)

Issued on March 17, 1975.

James B. Gregory
Administrator

40 F.R. 14318
March 31, 1975

PREAMBLE TO AMENDMENT TO MOTOR VEHICLE SAFETY STANDARD NO. 302

Flammability of Interior Materials

(Docket No. 3–3; Notice 9)

On March 31, 1975, the National Highway Traffic Safety Administration (NHTSA) issued a notice modifying the test procedures and specimen preparation requirements of Motor Vehicle Safety Standard No. 302, 49 CFR 571.302, *Flammability of interior materials* (40 FR 14318). Petitions for reconsideration of the rule were received from American Motors Corporation, General Motors Corporation, White Motor Corporation, Chrysler Corporation, Volkswagen of America, Inc., Toyota Motor Sales, U.S.A., Inc., Ford Motor Company, and the Motor Vehicle Manufacturers Association of the United States, Inc.

The NHTSA notice established a process of surrogate testing for components which were too small to test without difficulty using the procedures previously prescribed by Standard No. 302. The objections raised to this new process by the petitioners were that (a) the surrogate testing procedure is an entirely new departure, and the public should have been afforded an opportunity for comment, (b) the results of surrogate testing will in certain cases differ from the results of testing the actual component, (c) the creation of a surrogate testing sample of certain materials, such as elastic cord, is impossible, and (d) the dimensions of the surrogate sample are inappropriate.

It should be fully understood that small components which would otherwise be included within the purview of Standard No. 302 are not excluded by virtue of their size. Further, the NHTSA intends to utilize a surrogate testing procedure, among other testing procedures, in the case of small components as the first step in determining whether a safety defect exists pursuant to section 152 of the National Traffic and

Motor Vehicle Safety Act. Since the testing of small components is a more difficult process, the NHTSA concluded in amending Standard No. 302 to include the surrogate testing process that the new requirement was less stringent than that currently required by the standard. Further, by amending the standard the industry could also be fully apprised of one of the methods the NHTSA intended to use to determine whether a section 152 defect existed.

Nonetheless, it appears from the petitions for reconsideration which were received that a number of manufacturers feel that they should be allowed an opportunity for comment. The NHTSA concludes their request is reasonable and the rule, as it relates to surrogate testing, is hereby revoked and is reissued as a notice of proposed rulemaking in this issue of the FEDERAL REGISTER.

A number of the petitioners questioned the need for including any small components within the ambit of Standard No. 302, citing the notice of proposed rulemaking (38 FR 12934, May 17, 1973) which stated that certain small components designed to absorb energy are not fire hazards. Therefore, the petitioners believe the NHTSA has reversed its previous position.

This understanding is correct. As the NHTSA said in the preamble to the proposed amendment to Standard No. 302, issued concurrently with the amendment to the Standard (March 31, 1975, 40 FR 14340):

On May 11, 1973, the NHTSA issued a notice (38 FR 12934) which proposed, inter alia, amending paragraph S4.1 of Standard No. 302 to enumerate the interior components of vehicle occupant compartment which fell within the ambit of the standard.

* * * * * *

PART 571; S 302—PRE 7

Comments to the notice, however, have made clear that the enumeration of components, even with the proposed amendment, will continue to confuse manufacturers required to meet the standard.

* * * * * *

While some materials exposed to the occupant compartment air space are not fire hazards, the burden of ascertaining that fact should properly lie with the manufacturer.

Several petitions also questioned what safety benefits would come from applying the standard to small components. As petitioner American Motors pointed out, the purpose of Standard No. 302 is to provide sufficient time for the occupants of a vehicle to exit in case of an interior fire. Thus, even small components which are highly flammable would hasten the spreading of fires in motor vehicles, resulting in a serious hazard.

Testing procedures. Petitioners pointed out that while the preamble provides that the surface of the specimen closest to the occupant compartment air space face downward on test frame, this is not made entirely clear in the body of the standard itself. The standard is amended to clarify this matter. Likewise, a definition of the term "occupant compartment air space" is added, although this term was used in the notice of proposed rulemaking without raising a problem for those commenting.

Extension of effective date of amendment. Several petitioners asked for an extension of the effective date. As the surrogate testing procedures have been revoked and reissued as a proposed rule, the NHTSA concludes that an extension of the effective date is not necessary.

Redesignation of Docket 3-3; Notice 7. Through a clerical error, two notices were issued with the heading, "Docket 3-3; Notice 7" (July 11, 1973, 38 FR 18564; March 31, 1975, 40 FR 14318). The notice appearing at 38 FR 18564 is hereby redesignated "Notice 7a" and that appearing at 49 FR 14318 is redesignated "Notice 7b."

In consideration of the foregoing, Motor Vehicle Safety Standard No. 302, 49 CFR 571.302, is amended. . . .

Effective date: September 16, 1975.

Because this amendment relieves a restriction, it is found for good cause shown that an immediate effective date is in the public interest.

(Secs. 103, 119, Pub. L. 89-563, 80 Stat. 718 (15 U.S.C. 1392, 1407); delegation of authority at 49 CFR 1.51.)

Issued on September 10, 1975.

James B. Gregory
Administrator

September 16, 1975
40 F.R. 42746

PREAMBLE TO AMENDMENT TO MOTOR VEHICLE SAFETY STANDARD NO. 302

Flammability of Interior Materials

(Docket No. 3–3; Notice 11)

This notice establishes a new section, S3A. *Definitions,* in Motor Vehicle Safety Standard No. 302, 49 CFR 571.302.

On September 16, 1975, the NHTSA published in the Federal Register its response to a petition for reconsideration of Motor Vehicle Safety Standard No. 302, *Flammability of interior materials* (40 FR 42746). The rule established a definition of the term "occupant compartment air space" that was supposed to be added to "S3A. *Definitions.*" The wording of the amendment was faulty, however, since the Definitions section had not yet been established in Standard No. 302. This notice corrects the error by adding that section to the standard.

Petitions have been received from General Motors Corporation, Motor Vehicle Manufacturers Association, American Motors Corporation, and Ford Motor Company requesting that the definition of "occupant compartment air space" in Notice 9 be revoked. These petitions will be addressed in a separate notice. The purpose of this notice is only to promulgate the section heading which was omitted in error from Notice 9.

In light of the above, in place of the amendment numbered 1. in Docket 3–3, Notice 9 (40 FR 42746, September 16, 1975), Motor Vehicle Safety Standard No. 302 is amended by adding a new S3A. *Definitions.* . . .

Effective date: December 4, 1975. Because this amendment is of an interpretative nature and makes no substantive change in the rule, it is found for good cause shown that an immediate effective date is in the public interest.

(Sec. 103, 119 Pub. L. 89–563, 80 Stat. 718 (15 U.S.C. 1392, 1407); delegation of authority at CFR 1.51)

Issued on November 28, 1975.

James B. Gregory
Administrator

40 F.R. 56667
December 4, 1975

MOTOR VEHICLE SAFETY STANDARD NO. 302

Flammability of Interior Materials—Passenger Cars, Multipurpose Passenger Vehicles, Trucks, and Buses

(Docket N. 3-3; Notice 4)

S1. Scope. This standard specifies burn resistance requirements for materials used in the occupant compartments of motor vehicles.

S2. Purpose. The purpose of this standard is to reduce the deaths and injuries to motor vehicle occupants caused by vehicle fires, especially those originating in the interior of the vehicle from sources such as matches or cigarettes.

S3. Application. This standard applies to passenger cars, multipurpose passenger vehicles, trucks, and buses.

S3A. Definitions.

"Occupant compartment air space" means the space within the occupant compartment that normally contains refreshable air. (40 F.R. 42746—September 16, 1975. Effective 9/16/75. 40 F.R. 56667—December 4, 1975. Effective: 12/4/75)

S4. Requirements.

S4.1 The portions described in S4.2 of the following components of vehicle occupant compartments shall meet the requirements of S4.3: Seat cushions, seat backs, seat belts, headlining, convertible tops, arm rests, all trim panels including door, front, rear, and side panels, compartment shelves, head restraints, floor coverings, sun visors, curtains, shades, wheel housing covers, engine compartment covers, mattress covers, and any other interior materials, including padding and crash-deployed elements, that are designed to absorb energy on contact by occupants in the event of a crash.

S4.1.1 Deleted and Reserved.

S4.2 Any portion of a single or composite material which is within ½ inch of the occupant compartment air space shall meet the requirements of S4.3.

S4.2.1 Any material that does not adhere to other material(s) at every point of contact shall meet the requirements of S4.3 when tested separately.

S4.2.2 Any material that adheres to other material(s) at every point of contact shall meet the requirements of S4.3 when tested as a composite with the other material(s). Material A has a non-adhering interface with material B and is tested separately. Part of material B is within ½ inch of the occupant compartment air space, and materials B and C adhere at every point of contact; therefore B and C are tested as a composite. The cut is in material C as shown, to make a specimen ½ inch thick.

Illustrative Example

S4.3(a) When tested in accordance with S5, material described in S4.1 and S4.2 shall not burn, nor transmit a flame front across its surface, at a rate of more than 4 inches per minute.

However, the requirement concerning transmission of a flame front shall not apply to a surface created by the cutting of a test specimen for purposes of testing pursuant to S5.

(b) If a material stops burning before it has burned for 60 seconds from the start of timing, and has not burned more than 2 inches from the point where timing was started, it shall be considered to meet the burn-rate requirement of S4.3(a).

S5. Test procedure.

S5.1 Conditions.

S5.1.1 The test is conducted in a metal cabinet for protecting the test specimens from drafts. The interior of the cabinet is 15 inches long, 8 inches deep, and 14 inches high. It has a glass observation window in the front, a closable opening to permit insertion of the specimen holder, and a hole to accommodate tubing for a gas burner. For ventilation, it has a ½-inch clearance space around the top of the cabinet, ten ¾-inch-diameter holes in the base of the cabinet, and legs to elevate the bottom of the cabinet by three-eighths of an inch, all located as shown in Figure 1.

S5.1.2 Prior to testing, each specimen is conditioned for 24 hours at a temperature of 70° F. and a relative humidity of 50 percent, and the test is conducted under those ambient conditions.

S5.1.3 The test specimen is inserted between two matching U-shaped frames of metal stock 1 inch wide and three-eighths of an inch high. The interior dimensions of the U-shaped frames are 2 inches wide by 13 inches long. A specimen that softens and bends at the flaming end so as to cause erratic burning is kept horizontal by supports consisting of thin, heat resistant wires, spanning the width of the U-shaped frame under the specimen at 1-inch intervals. A device that may be used for supporting this type of material is an additional U-shaped frame, wider than the U-shaped frame containing the specimen, spanned by 10-mil wires of heat-resistant composition at 1-inch intervals, inserted over the bottom U-shaped frame.

S5.1.4 A bunsen burner with a tube of ⅜-inch inside diameter is used. The gas adjusting valve is set to provide a flame, with the tube vertical, of 1½ inches in height. The air inlet to the burner is closed.

S5.1.5 The gas supplied to the burner has a flame temperature equivalent to that of natural gas.

S5.2 Preparation of specimens.

S5.2.1 Each specimen of material to be tested shall be a rectangle 4 inches wide by 14 inches long, wherever possible. The thickness of the specimen is that of the single or composite material used in the vehicle, except that if the material's thickness exceeds ½ inch, the specimen is cut down to that thickness measured from the surface of the specimen closest to the occupant compartment air space. Where it is not possible to obtain a flat specimen because of surface curvature, the specimen is cut to not more than ½ inch in thickness at any point. The maximum available length or width of a specimen is used where either dimension is less than 14 inches or 4 inches, respectively, unless surrogate testing is required under S4.1.1.

S5.2.2 The specimen is produced by cutting the material in the direction that provides the most adverse test results. The specimen is oriented so that the surface closest to the occupant compartment air space faces downward on the test frame.

S5.2.3 Material with a napped or tufted surface is placed on a flat surface and combed twice against the nap with a comb having seven to eight smooth, rounded teeth per inch.

S5.3 Procedure.

(a) Mount the specimen so that both sides and one end are held by the U-shaped frame, and one end is even with the open end of the frame. Where the maximum available width of a specimen is not more than 2 inches, so that the sides of the specimen cannot be held in the U-shaped frame, place the specimen in position on wire supports as described in S5.1.3, with one end held by the closed end of the U-shaped frame.

(b) Place the mounted specimen in a horizontal position, in the center of the cabinet.

(c) With the flame adjusted according to S5.1.4, position the bunsen burner and specimen so that the center of the burner tip is three-fourths of an inch below the center of the bottom edge of the open end of the specimen.

(d) Expose the specimen to the flame for 15 seconds.

(e) Begin timing (without reference to the period of application of the burner flame) when the flame from the burning specimen reaches a point 1½ inches from the open end of the specimen.

(f) Measure the time that it takes the flame to progress to a point 1½ inches from the clamped end of the specimen. If the flame does not reach the specified end point, time its progress to the point where flaming stops.

(g) Calculate the burn rate from the formula

$$B = \frac{60 \times D}{T}$$

Where B = burn rate in inches per minute,

D = length the flame travels in inches, and

T = time in seconds for the flame to travel D inches.

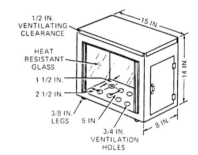

36 F.R. 289
January 8, 1971

Section Three

Part 571—Ruling on Chassis Cabs and Campers Slide-in and Chassis Mount

Part 572—Anthropomorphic Test Dummies

Part 573—Defect and Noncompliance Reports

Part 574—Tire Identification and Recordkeeping

Part 575—Consumer Information Regulations

Part 576—Record Retention

Part 577—Defect and Noncompliance Notification

Part 579—Defect and Noncompliance Responsibility

Part 580—Odometer Disclosure Requirements

Part 581—Bumper Standard

Part 582—Insurance Cost Information Regulations

Part 590—Motor Vehicle Emission Inspections

Department of the Treasury Regulation Relating to Importation of Motor
Vehicles and Items of Motor Vehicle Equipment

MOTOR VEHICLE SAFETY STANDARDS
Notice of Ruling Regarding Chassis-Cabs

Inquiry has been received from persons engaged in the sale of trucks, buses, and multi-purpose vehicles regarding their legal responsibility under the National Traffic and Motor Vehicle Safety Act of 1966 for assuring that vehicles sold by them are in conformity with all applicable motor vehicle safety standards. Such persons commonly purchase chassis-cabs from manufacturers and bodies or work-performing and load-carrying structures from other manufacturers and then combine the chassis-cab with the body or other structure. A regulation is being issued this date by the Federal Highway Administration defining the chassis-cab as a vehicle within the meaning of the Act, requiring that it meet all motor vehicle safety standards applicable on the date of manufacture of the chassis-cab.[1] Under this regulation the manufacturer of a chassis-cab manufactured subsequent to January 1, 1968, will have responsibility for compliance with all applicable motor vehicle safety standards as set forth therein and for certification of such compliance to distributors and dealers.

Section 101(5) of the National Traffic and Motor Vehicle Safety Act defines a "manufacturer" as any person engaged in the "assembling" of motor vehicles. Persons who combine chassis-cabs with bodies or similar structures are, therefore, manufacturers within the meaning of the Act. Inasmuch as the chassis-cab's manufacturer is responsible for compliance with standards under the regulation issued today, persons who add bodies or other structures to such chassis-cab are not considered manufacturers of the chassis-cab and, therefore, will not be responsible for the conformance of the chassis-cab to the standards certified by the manufacturer of the

chassis-cab. In numerous instances the chassis-cab will not be capable of complying with motor vehicle safety standard 108 because it will not be equipped with all items of lighting equipment referred to in such standard. Where vendors combine a chassis-cab which has not been certified to be in conformance with standard 108, with a body or other like structure, such vendor will be responsible for compliance with the lighting standard, and where such vendor sells the combined assemblage to another vendor, certification of compliance with the lighting standard must accompany the vehicle.

We are advised that a substantial inventory of chassis-cabs manufactured prior to the effective date of the initial motor vehicle safety standards and hence not required to comply with the same will be held by manufacturers, distributors, and dealers on January 1, 1968. These vehicles may contain various items of lighting equipment manufactured prior to the effective date of the lighting standard or be designed to accept such equipment. Under these circumstances, it does not appear appropriate to require compliance with the lighting standard when such chassis-cabs, i.e., those manufactured prior to January 1, 1968, are combined with bodies or similar strucutres. Section 108(a)(1) of the Act also prohibits any person from manufacturing for sale or selling any motor vehicle manufactured "after the date any applicable Federal motor vehicle safety standard takes effect under this title unless it is in conformity with such standard ***." Under this provision persons who combine the chassis-cab with a body or other structure will be responsible for (1) compliance of the combined assemblage with any motor vehicle safety standard applicable to the end use of the combined assemblage in effect on the date of manufacture of the chassis-cab, compliance with which has not already been certified

---

[1] See F.R. Doc. 67-15174. Title 23, in Rules and Regulations Section, supra.

PART 571; (RULING)-1

by the chassis-cab manufacturer, and (2) compliance with all applicable standards in effect on the date of manufacturer of the chassis-cab to the extent that the addition of a body or other structure to the chassis-cab affects the chassis-cab's previous conformance with applicable standards.

To insure that the person combining the chassis-cab with the body or other structure has adequate information to enable him to meet the conditions specified above, the regulation being issued concurrently with this ruling requires the chassis-cab manufacturer to affix a label to the chassis-cab which identifies the Federal motor vehicle safety standards with which the chassis-cab fully complies for the principal end uses of such chassis-cab.

Issued in Washington, D.C., on December 29, 1967.

Lowell K. Bridwell,
Federal Highway Administrator

33 F.R. 29
January 3, 1968

FEDERAL MOTOR VEHICLE SAFETY STANDARDS
(FHWA Ruling 68-1)
Notice of Ruling Regarding Campers Slide-in and Chassis-Mount

This ruling is in response to inquiries for a clarification of the applicability of Federal Motor vehicle safety standards to certain items of motor vehicle equipment commonly known as "campers" which are used mostly for recreational purposes.

A "camper" can be described generally as a portable structure designed to be loaded onto, or affixed to, a motor vehicle to provide temporary living quarters for recreation, travel, or other use. The ruling is concerned with two general categories of campers. The first, a "slide-in camper", is placed on, or slides onto a completed vehicle, usually a pickup truck. The second, a "chassis-mount camper", is mounted on a chassis-cab.

In past months the Bureau received a number of written inquiries regarding the applicability of the glazing material standard (No. 205) to slide-in campers. These persons received responses from the Bureau indicating that slide-in campers would have to comply with standard 205 under certain specified conditions. These responses of the Bureau apparently received widespread dissemination in the industry. Subsequently, additional inquiries were received from affected persons asking for clarification of the Bureau's earlier responses with respect to the question of whether standard 205 was applicable to glazing materials contained in slide-in campers sold by the manufacturer of such campers to members of the public and to dealers when not an integral part of the vehicle.

The Bureau has reconsidered this question and determined that the glazing standard is applicable to slide-in campers.

Standard 205 is applicable to "glazing materials for use in passenger cars, multipurpose passenger vehicles, motorcycles, trucks and buses."

The slide-in camper is an item of motor vehicle equipment for use in motor vehicles. As such, glazing materials contained in slide-in campers must comply with standard 205 when such campers are sold as a separate unit as well as when attached to a completed pickup truck. Additionally, manufacturers of slide-in campers must also comply with the certification requirements set forth in section 114 of the National Traffic and Motor Vehicle Safety Act of 1966 (15 U.S.C. 1403).

Review of the Bureau's prior communications with affected persons indicates that such persons, and others who received notice of such communications, could justifiably have concluded that standard 205 was subject to an interpretation which excluded its application to slide-in campers sold directly to consumers or to dealers when not an integral part of the pickup truck. In these circumstances the Bureau does not regard it as appropriate that the interpretation of the applicability of standard 205, which this ruling announces, should be given retroactive effect. Further, in view of such reliance a reasonable time should be afforded affected parties to allow for possible necessary production adjustments. Accordingly, it is determined that with respect to slide-in campers, the interpretation announced by this ruling shall not become effective until July 1, 1968.

With regard to the chassis-mount camper, it is an integral part of the vehicle when attached to a chassis-cab as defined in § 371.3(b), Part 371, Federal Motor Vehicle Safety Standards (33 F.R. 19).

Persons who mount the chassis-mount camper to the chassis-cab are manufacturers of vehicles within the meaning of section 102(3) of the National Traffic and Motor Vehicle Safety Act of 1966 (15 U.S.C. 1392). As such, they are to be guided by the regulation and ruling on chassis-cabs issued December 29, 1967 (33 F.R. 19 and 33 F.R. 29). Under this regulation and ruling persons combining a chassis-cab manufactured on or after January 1, 1968, with a body or like structure (in this case the chassis-mount camper) are responsible for assuring that the completed assemblage complies with all applicable standards in effect on the date of manufacture of the chassis-cab which had not previously been met by the manufacturer of the chassis-cab, and for assuring that previously met standards have not been adversely affected by the addition of the chassis-mount camper.

Issued in Washington, D.C., on March 20, 1968.

Lowell K. Bridwell,
Federal Highway Administrator

33 F.R. 5020
March 26, 1968

PREAMBLE TO PART 572—ANTHROPOMORPHIC TEST DUMMY

(Docket No. 73–8; Notice 2)

The purposes of this notice are (1) to adopt a regulation that specifies a test dummy to measure the performance of vehicles in crashes, and (2) to incorporate the dummy into Motor Vehicle Safety Standard No. 208 (49 CFR § 571.208), for the limited purpose of evaluating vehicles with passive restraint systems manufactured under the first and second restraint options between August 15, 1973, and August 15, 1975. The question of the restraint system requirements to be in effect after August 15, 1975, is not addressed by this notice and will be the subject of future rulemaking action.

The test dummy regulation (49 CFR Part 572) and the accompanying amendment to Standard No. 208 were proposed in a notice published April 2, 1973 (38 F.R. 8455). The dummy described in the regulation is to be used to evaluate vehicles manufactured under sections S4.1.2.1 and S4.1.2.2, (the first and second options in the period from August 15, 1973, to August 15, 1975), and the section incorporating the dummy is accordingly limited to those sections. The dummy has not been specified for use with any protection systems after August 15, 1975, nor with active belt systems under the third restraint option (S4.1.2.3). The recent decision in *Ford* v. *NHTSA*, 473 F. 2d 1241 (6th Cir. 1973), removed the injury criteria from such systems. To make the dummy applicable to belts under the third option, the agency would have to provide additional notice and opportunity for comment.

By invalidating the former test dummy specification, the decision in *Chrysler* v. *DOT*, 472 F. 2d 659 (6th Cir. 1972), affected the restraint options in effect before August 15, 1975, as well as the mandatory passive restraint requirements that were to be effective after that date. A manufacturer who built cars with passive restraints under one of the options would therefore be unable to certify the cars as complying with the standard, as illustrated by the necessity for General Motors to obtain a limited exemption from the standard in order to complete the remainder of a run of 1,000 air-bag equipped cars.

The immediate purpose of this rulemaking is to reconstitute those portions of the standard that will enable manufacturers to build passive restraint vehicles during the period when they are optional. The test dummy selected by the agency is the "GM Hybrid II", a composite developed by General Motors largely from commercially available components. GM had requested NHTSA to adopt the Hybrid II on the grounds that it had been successfully used in vehicle tests with passive restraint systems, and was as good as, or better than, any other immediately available dummy system. On consideration of all available evidence, the NHTSA concurs in this judgment. One fact weighing in favor of the decision is that General Motors has used this dummy to measure the conformity of its vehicles to the passive protection requirements of Standard 208, in preparation for the announced introduction of up to 100,000 air-bag-equipped vehicles during the 1974 model year.

No other vehicle manufacturer has announced plans for the production of passive restraint systems during the optional phase, nor has any other vehicle manufacturer come forward with suggestions for alternatives to Hybrid II. The NHTSA would have considered other dummies had some other manufacturer indicated that it was planning to produce passive restraint vehicles during the option period and that some other dummy had to be selected in order to allow them to proceed with their plans. If there had been any such plans, NHTSA would have made every effort to insure that a test device satisfactory to said manufacturer would have been selected.

This agency recognizes that since various types of dummy systems have been in use under the previous specification, any selection of one dummy, as is required by the *Chrysler* decision, will necessitate readjustments by some manufacturers. However, considering the quantity of GM's production, the scope and advanced state of its passive restraint development program, and the fact that the Hybrid II does not differ radically from other dummies currently in use, in the NHTSA's judgment that dummy represents the best and least costly choice. That conclusion has not been contradicted by the comments to the docket.

The agency will not make any final decision regarding reinstatement of mandatory passive restraint requirements without further notice and opportunity for comment. Should the agency propose mandatory passive restraint requirements, the question of the conformity of the dummy that is chosen with the instructions of the court in *Chrysler* will again be open for comment. The NHTSA strongly encourages the continuance of the dummy test programs mentioned in the comments, in the hope that any problems that may arise can be identified and resolved before the dummy specifications for later periods are issued.

The Hybrid II dummy has been found by NHTSA to be a satisfactory and objective test instrument. In sled and barrier tests conducted by GM with the GM restraint systems and in sled tests conducted by Calspan Corp. on behalf of NHTSA, the Hybrid II has produced results that are consistent and repeatable. This is not to say that each test at the same nominal speed and deceleration has produced identical values.

In testing with impact sleds, and to an even greater extent with crash-tested vehicles, the test environment itself is complex and necessarily subject to variations that affect the results. The test data show, however, that the variance from dummy to dummy in these tests is sufficiently small that a manufacturer would have no difficulty in deciding whether his vehicle would be likely to fail if tested by NHTSA.

The provisions of the dummy regulation have been modified somewhat from those proposed in the notice of proposed rulemaking, largely as a result of comments from GM. Minor corrections have been made in the drawings and materials specifications as a result of comments by GM and the principal dummy suppliers. The dummy specification, as finally adopted, reproduces the Hybrid II in each detail of its design and provides, as a calibration check, a series of performance criteria based on the observed performance of normally functioning Hybrid II components. The performance criteria are wholly derivative and are intended to filter out dummy aberrations that escape detection in the manufacturing process or that occur as a result of impact damage. The revisions in the performance criteria, as discussed hereafter, are intended to eliminate potential variances in the test procedures and to hold the performance of the Hybrid II within the narrowest possible range.

General Motors suggested the abandonment of the definition of "upright position" in section 572.4(c), and the substitution of a setp-up procedure in section 572.11 to serve both as a positioning method for the performance tests and as a measurement method for the dummy's dimensions as shown in the drawings. The NHTSA does not object to the use of an expanded set-up procedure, but has decided to retain the term "upright position" with appropriate reference to the new section 572.11(i).

The structural properties test of section 572.5(c), which had proposed that the dummy keep its properties after being subjected to tests producing readings 25 percent above the injury criteria of Standard No. 208, has been revised to provide instead that the properties must be retained after vehicle tests in accordance with Standard No. 208.

The head performance criteria are adopted as proposed. The procedures have been amended to insure that the forehead will be oriented below the nose prior to the drop, to avoid interference from the nose. In response to comments by the Road Research Laboratory, American Motors, and GM, an interval of at least 2 hours between tests is specified to allow full restoration of compressed areas of the head skin.

The neck performance criteria are revised in several respects, in keeping with GM's recommendations. The pendulum impact surface, shown in Figure 4, has been modified in accordance with GM's design. The zero time point has been specified as the instant the pendulum contacts the honeycomb, the instructions for determining chordal displacement have been modified, and the pulse shape of the pendulum deceleration curve has been differently specified. The maximum allowable deceleration for the head has been increased slightly to 26g. In response to suggestions by the Road Research Laboratory and the Japan Automobile Manufacturers Association (JAMA), as well as GM, a tolerance has been specified for the pendulum's impact velocity to allow for minor variances in the honeycomb material.

With respect to the thorax test, each of the minor procedural changes requested by GM has been adopted. As with the head, a minimum recovery time is specified for the thorax. The seating surface is specified in greater detail, and the test probe orientation has been revised to refer to its height above the seating surface. The test probe itself is expressly stated to have a rigid face, by amendment to section 572.11, thereby reflecting the probes actually used by NHTSA and GM. A rigid face for the probe was also requested by Mercedes Benz.

The test procedures for the spine and abdomen tests are specified in much greater detail than before, on the basis of suggestions by GM and others that the former procedures left too much room for variance. The test fixtures for the spinal test orientation proposed by GM, and its proposed method of load application have been adopted. The parts of the dummy to be assembled for these tests are specifically recited, and an initial 50° flexion of the dummy is also specified. The rates of load application and removal, and the method of taking force readings are each specified. The direction of force application is clarified in response to a comment by Volvo.

The abdomen test is amended with respect to the initial point of force measurement, to resolve a particular source of disagreement between GM's data and NHTSA's. The boundaries of

the abdominal force-deflection curve are modified to accord with the measurements taken by GM subsequent to the issuance of the notice. The rate of force application is specified as not more than 0.1 inch per second. in response to comments by Mercedes Benz, JAMA, and GM.

The test procedures for the knee tests are revised to specify the type of seating surface used and to control the angle of the lower legs in accordance with suggestions by JAMA, the Road Research Laboratory, and GM. The instrumentation specifications of section 572.11 are amended to clarify the method of attachment and orientation of the thorax accelerometers and to specify the channel classes for the chest potentiometer, the pendulum accelerometer, and the test probe accelerometer, as requested by several comments.

The design and assembly drawings for the test dummy are too cumbersome to publish in the *Federal Register*. During the comment period on the April 2 notice, the agency maintained master copies of the drawings in the docket and placed the reproducible mylar masters from which the copies were made with a commercial blueprint facility from whom interested parties could obtain copies. The NHTSA has decided to continue this practice and is accordingly placing a master set of drawings in the docket and the reproducible masters for these drawings with a blueprint facility.

The drawings as adopted by this notice differ only in minor detail from those that accompanied the April 2 notice. The majority of the changes, incorporated into corrected drawings, have already been given to those persons who ordered copies. The letter of June 13, 1973, that accompanied the corrected drawings has been placed in the docket. The June corrections are incorporated into the final drawing package. Additional adjustments are made hereby to reflect better the weight distribution of separated segments of the dummy, to allow other materials to be used for head ballast, and to specify the instrument for measuring skin thickness. The details of these changes are recited in a memorandum incorporated into the drawing package.

Each of the final drawings is designated by the legend "NHTSA Release 8/1/73". Each drawing so designated is hereby incorporated as part of the test dummy specifications of 49 CFR Part 572. Subsequent changes in the drawings will not be made without notice and opportunity for comment.

The incorporation of the Part 572 test dummy into Standard No. 208 makes obsolete several test conditions of the standard that had been adopted to supplement the former test dummy specifications. The location, orientation, and sensitivity of test instrumentation formerly specified by sections S8.1.15 through S8.1.18 are now controlled by Part 572 are are no longer necessary within Standard No. 208. Similarly, the use of rubber components for the head, neck and torso joints as specified in Part 572, supplant the joint setting specifications for those joints in section S8.1.10 of the standard. The NHTSA has determined that the deletion of the above portions of the Standard No. 208 will have no effect on the substantive requirements of the standard and that notice and public procedure thereon are unnecessary.

In consideration of the foregoing, Title 49, Code of Federal Regulations, is amended by the addition of Part 572, Anthropomorphic Test Dummy. . . .

In view of the pressing need for a test dummy to permit the continued development of passive restraint systems, and the fact that it presently only relates to a new option for compliance, the NHTSA finds that there is good cause to adopt an immediate effective date. Accordingly, Part 572 is effective August 1, 1973, and the amendment to Standard 208 is effective August 15, 1973.

Issued under the authority of sections 103 and 119 of the National Traffic and Motor Vehicle Safety Act, P.L. 89-563, 15 U.S.C. 1392, 1407, and the delegation of authority at 38 F.R. 12147.

Issued on July 26, 1973.

James E. Wilson
Associate Administrator
Traffic Safety Programs

38 F.R. 20449
August 1, 1973

PREAMBLE TO AMENDMENT TO PART 572—ANTHROPOMORPHIC TEST DUMMIES

(Docket No. 73–8; Notice 4)

This notice amends Part 572, *Anthropomorphic Test Dummy*, to specify several elements of the dummy calibration test procedures and make minor changes in the dummy design specifications. Part 572 is also reorganized to provide for accommodation of dummies other than the 50th-percentile male dummy in the future.

Part 572 (49 CFR Part 572) establishes, by means of approximately 250 drawings and five calibration tests, the exact specifications of a test device that simulates an adult occupant of a motor vehicle, for use in evaluating certain types of crash protection systems provided in accordance with Standard No. 208, *Occupant Crash Protection* (49 CFR § 571.208). Interested persons are advised that NHTSA Docket Nos. 69–7 and 74–14 concerning Standard No. 208 are related to this rulemaking.

Proposed occupant protection requirements in Standard No. 208 were reviewed by the Sixth Circuit in 1972 ("*Chrysler v. Department of Transportation*," 472 F. 2d 659 (6th Cir. 1972)), and the dummy previously specified for use in testing was invalidated as insufficiently objective. The NHTSA subsequently established new dummy specifications under Part 572 for the limited purpose of qualifying passive restraint systems which manufacturers choose to offer on an optional basis (38 FR 20499, August 1, 1973). After examining test experience with the Part 572 dummy, the NHTSA specified its use in a proposal to mandate passive restraint systems (39 FR 10271, March 19, 1974).

Recently, the agency proposed minor changes in calibration procedures and dummy drawings (40 FR 33462, August 8, 1975) in response to the comments of manufacturers and others on the March 1974 notice. The August 1975 proposal only addressed the issue of dummy objectivity raised by the Sixth Circuit, while issues of dummy similarity to humans, sensitivity to test environment, and dummy positioning in a vehicle have been treated elsewhere (41 FR 29715, July 19, 1976).

It is noted that the most recent Department of Transportation proposals on Standard No. 208 (41 FR 24070, June 14, 1976) reflected a modification of performance requirements that reduce the number and types of tests in which the Part 572 dummy would be used in Standard No. 208 dynamic tests. Specifically, rollover and lateral testing would no longer be required if a lap belt were installed in the front seating positions. The NHTSA's July 1976 proposal noted above would conform existing tests in Standard No. 208 to the modified approach. It would also increase the permissible femur force loads that could be registered on the dummy during impact, and restrict femur force requirements to compressive forces. Interested persons should be aware of these significant potential changes in the use of the dummy in Standard No. 208.

As for the dummy objectivity treated by the proposal that underlies this notice, manufacturers' comments stressed the complexity of the test environment in which the device is used and their uncertainty as to how much the dummy characteristics contribute to the variability that is encountered. In somewhat contradictory fashion, several of the manufacturers repeated requests for a "whole systems" calibration of the dummy that would be conducted under conditions approximating the barrier crash whose complex variables had just been emphasized.

As is the case with any measuring instrument, variations in readings can result from imperfection in the instrument or variations in the phenomenon being measured (in this case, the complex events that occur as a passenger car impacts a barrier at 30 mph, or is impacted laterally by a 4,000-pound moving barrier, or is rolled over). While the "*Chrysler*" court delayed Standard No. 208 so that variation in the dummy's behavior could be corrected, it found the standard (and the dynamic test procedures) practicable and "designed to meet the need for motor vehicle safety" (472 F2d at 674, 675). To meet the need for motor vehicle safety, the dynamic tests are realistic simulations of the actual crash environment. Variations in the precise circumstances to which the dummy is exposed from test to test are expected.

Simulation of such crashes to provide a "whole systems" calibration of the dummy would not be reasonable, however, because of the variations that are inherent in the 30-mph (and the other) impacts. Unless the inputs to the dummy during calibration are precisely controlled, as is the case with the five sub-assembly tests, the "whole systems" calibration would be meaningless. To conduct precisely controlled 30-mph barrier crash tests as part of the dummy calibration procedure would be very expensive, since dummy calibration is normally performed before and after each compliance test. The good results obtained in sub-assembly calibration, and supported by the controlled "whole dummy" test results referred to in the preamble to the proposal, make such a "whole systems" test redundant. The agency concludes that introduction into Part 572 of an extremely expensive and unfamiliar additional calibration is unjustified.

General Motors (GM), Chrysler Corporation, Ford Motor Company, and the Motor Vehicle Manufacturers Association (MVMA) stated that the dummy construction is unsuited to measurements of laterally-imposed force, thereby rendering the dummy unobjective in the "lateral impact environment." While the agency does not agree with these objections, the modified performance levels put forward by the Department of Transportation and the agency would allow manufacturers to install lap belts if they do not wish to undertake lateral or rollover testing. Any manufacturer that is concerned with the objectivity of the dummy in such impacts would provide lap belts at the front seating positions in lieu of conducting the lateral or rollover tests.

Ford and Chrysler argued that the test dummy is insufficiently specified despite the approximately 250 detailed drawings that set forth dummy construction. Their concern seems to be limited to minor contour dimensions that they consider critical to dummy objectivity. To eliminate any such concern the agency will place a specimen of the dummy in the data and drawings package and incorporate it by reference into Part 572.

The MVMA stated that its reading of the docket comments indicated that the dummy cannot be assembled as it is designed. The agency is aware that dimensional tolerances could, at their extremes, "stack up" to cause the need in rare instances for selective fitting of components. Manufacturers can avoid any such problem by reducing the dispersion of tolerances or by select fitting of components to avoid tolerance "stack-up." Of the three dummy manufacturers' comments on this proposal, only Humanoid Systems (Humanoid) listed discrepancies. The agency has reviewed the asserted discrepancies and concludes that the specifications themselves, the manufacturing practices just noted, or the calibration procedures are adequate to resolve the cited problems. To simplify the dummy, certain studs located at the side of the dummy femurs (used for mounting photographic targets and unnecessary to NHTSA test procedures) are deleted because of their potential for reducing repeatability under some circumstances. These studs are designated F/02, G/02, F/25, and G/25.

Bayerische Motorenverken recited test experience that demonstrated different performance characteristics among the products of different dummy manufacturers, although they are all warranted to meet the specifications of the regulations. NHTSA Report DOT-HS-801-861 demonstrates that some manufacturer-warranted dummies did not meet all calibration requirements of Part 572. The agency, however, is not in a position to assume responsibility for the contractual terms established between private parties.

Humanoid noted that experience with the vinyl flesh specification of the dummy led to resolution of aging problems on which it had earlier commented. The company did recommend latitude in vinyl formulation to permit market competition. General Motors also expressed concern that specification of the Part 572 dummy not stifle innovation. Alderson Research Laboratories (ARL) once again asked that the agency specify a one-piece casting in place of the welded head presently specified. The agency sympathizes with this interest in improvement of the dummy manufacturing techniques. However, the dummy is a test instrument crucial to the validity of an important motor vehicle safety standard and as such, it cannot be loosely described for the benefit of innovation.

Volkswagen requested improvement in aging and in storage techniques for the dummy. The agency considers that it has met its responsibilities by specifying calibration tests that will signal improper storage or age-related changes. Further development in this area is within the province of the manufacturers and users. Significant improvements in aging or storage factors will, of course, not be ignored by the agency.

Although Ford and American Motors Corporation (AMC) made no comment on the specifics of the NHTSA proposal, Chrysler Corporation and several other vehicle manufacturers, as well as the dummy manufacturers, supported the proposed changes. The National Motor Vehicle Safety Advisory Council took no position on the proposal. The Vehicle Equipment Safety Commission did not comment on the proposal. Having carefully reviewed all of the comments submitted and additional data compiled by the agency, the changes are adopted, essentially as proposed. The agency proposed modification of the five calibration procedures for dummy subassemblies, along with minor changes in the drawings that describe all components of the dummy.

HEAD

The head calibration involves dropping the head 10 inches so that its forehead strikes a rigid surface and registers acceleration levels that must fall within a certain range. No comments were received on the small relocation of measurement points or the specification of "instant release" of the head, and these modifications are made as proposed.

The proposal included a specification of 250 microinches (rms) for the finish of the steel plate on which the head is dropped. The agency had considered other factors (particularly friction at the skull-skin interface of the dummy forehead) that might affect the accelerometer readings. It was found that, in most instances, the dummy as received from the manufacturer conformed to the specifications. When deviations were encountered, treatment of the head in accordance with manufacturer recommendations eliminated the effect of these factors on results. Comparison of data on 100 head drop tests conducted since issuance of the proposal confirms that conclusion. Ninety-seven percent of these head drops registered readings within the specified limits, with a mean response value of 232g and a standard deviation of 14g, indicating a coefficient of variance of 6 percent. Of the three failures, the response values were 203g, 204g and 263g. All of the drop tests fell within the specified 0.9- to 1.5-ms time range at the 100g level. The surface finish of the drop plate was 63 microinches (rms). In view of this data, it does not appear necessary to adjust either the response range as advocated by Humanoid or the time range as recommended by Ford. The test results, however, support the request by a number of comments to change the proposed 250-microinch finish to a value below 100 microinches (rms). On the basis of the comments and NHTSA test data, the impact plate surface finish is specified as any value in the range from 8 to 80 microinches (rms).

General Motors asked whether coating of the steel plate is permitted. Coating is permitted so long as the 8- to 80-microinch range for the surface is maintained.

Humanoid recommended that any lubrication or surface smoothness introduced by the dummy manufacturers be made uniform in the interests of component interchange. Volkswagen also recommended a skull-to-skin interface finish specification. The NHTSA, however, does not believe that differing procedures for preparation of the skull-skin interface prevent interchange of the heads, and the requests are therefore not granted.

In view of the agency decision to incorporate by reference a specimen of the Part 572 dummy in the drawings and data package, it is also considered unnecessary to specify, as requested by Humanoid, thickness and performance specification for the headform at 45 and 90 degrees from the midsaggital plane. With regards to Humanoid's view that head drop tests are irrelevant to performance of the dummy as a measuring instrument, the agency considers them closely tied to the characteristics of the dummy that affect its repeatability as a measuring device.

Renault and Peugeot recommended consideration of a revision in the test criteria of Standard No. 208, in the case of safety belts, to replace the limitation on head acceleration with a limitation on submarining. The agency considers the present limit on head acceleration a valuable means to limit head loading and neck hyperflexion in belt systems as well as other systems. It is a requirement that is already being met on a production basis by Volkswagen.

Toyota stated that the 10g limit on lateral acceleration during the head drop would be impossible to satisfy. The NHTSA's own test experience did not exhibit any evidence of the noted problem. None of the manufacturers of dummies objected to the proposal, and Alderson Research Laboratories (ARL) supported the 10g limit. It is therefore made final as proposed.

ARL once more requested consideration of the one-piece headform in place of the welded headform presently specified. If, as ARL states, its customers accept and utilize the one-piece casting, the agency does not understand the necessity to modify the specification. ARL's request for consideration of a one-piece neck bracket is subject to the same response. As earlier noted, the justification to "freeze" the dummy specification is clear from its use as a measurement instrument that is the basis of manufacturer compliance with, and agency verification testing to, a major motor vehicle safety standard.

NECK

Comments generally agreed with the proposed changes in the dummy neck calibration (attachment of the head form to the neck, and attachment of the neck to the end of a pendulum which impacts an energy-absorbing element, inducing head rotation which must fall within specified limits). General Motors clarified that its engineers' reason for recommending a non-articulated neck instead of an articulated neck concerned the cost, maintenance, and complexity of the latter's construction. Volkswagen agreed with Sierra Engineering Company (Sierra) that a smaller tolerance for the pendulum's speed at impact should be considered. Humanoid agreed with the agency's view that the articulated neck does not provide the desired level of repeatability at this time. Having considered these comments the agency makes final the proposed location change for the accelerometers, deletion of § 572.7 (c)(5), and clarification of the "t4" point and the 26g level.

Manufacturers made several additional recommendations. Humanoid expressed support of AMC's view that the neck calibration should be conducted at barrier impact velocity. The agency has reviewed these comments and finds that the specified energy levels are adequate for the intended purpose of establishing dynamic response characteristics and the measurement of repeatability of dummy necks under dynamic test conditions. Testing at higher levels would bring other dummy components besides the neck into direct impact interaction, thereby obscuring or completely masking the measured phenomena.

Volkswagen cautioned against an entirely free selection of damping materials because of variation in rebound characteristics produced with different materials that can achieve conforming deceleration time histories. The agency agrees that a limit on rebound should be established to compliment the choice of damping materials and has added such a specification to the end of the text of § 572.7(b).

Humanoid noted interference in the attachment of the neck bracket to the backplate of the sternothoracic structure, due to the presence of a welding bead. The agency has found no interference in the dummies manufactured by two companies and concludes that the interference must be associated with Humanoid's manufacturing technique.

THORAX

The NHTSA proposed several additional specifications for test probe orientation, dummy seating, and limb positioning for the calibration test. The calibration consists of striking the torso of the seated dummy at two speeds with a specified striker to measure thorax resistance, deflection, and hysteresis characteristics. Comments did not object to the changes and they are incorporated as proposed.

The agency also proposed several changes in the drawings for the thorax sub-assembly of the dummy and, without objection, they are made final in virtually the same form. ARL indicated that four heat seals should be used on the zipper. ARL clarified that the longer socket head cap screw is intended to permit sufficient thread engagement, not more latitude in the ballast configuration as stated in the proposal. Humanoid's request to know the clavicle contours that constitute the Part 572 specification is met by placing the dummy specimen in the drawings and data package as earlier noted. Humanoid and Toyo Kogyo suggested an increase in clavicle strength. The agency's experience with the clavicle since the last consideration of this suggestion has been that all dummies are not significantly susceptible to clavicle breakage. Accordingly, the agency does not consider the modification necessary.

The major suggestion by vehicle and dummy manufacturers was a slight revision of the thorax resistance and deflection values, which must not be exceeded during impact of the chest. The present values (1400 pounds and 1.0 inch at 14 fps, 2100 pounds and 1.6 inches at 22 fps) were questioned by GM, which recommends an increase in both resistance and deflection values to better reflect accurate calibration of a correctly designed dummy. Comparable increases were recommended by Humanoid and Sierra. ARL noted that the present values are extremely stringent.

The agency's experience with calibration of the thorax since issuance of the proposal confirms that a slight increase in values is appropriate, although not the amount of increase recommended by the manufacturers. The values have accordingly been modified to 1450 pounds and 1.1 inches at 14 fps, and 2250 pounds and 1.7 inches at 22 fps. The agency does not set a minimum limit on the value as recommended by General Motors, because the interaction of the deflection and resistance force values make lower limits unnecessary. The changes in values should ease ARL's concern about the seating surface, although the agency's own experience does not indicate that a significant problem exists with the present specifications of the surface.

In conjunction with these changes, the agency has reduced the maximum permissible hysteresis of the chest during impact to 70 percent as recommended by GM.

GM requested a clarification of the dummy limb positioning procedures for purposes of thorax impact testing, citing the possibility of limb misadjustment between steps (1) and (4) of § 572.8(d). The agency has added wording to subparagraph (4) to make clear that the limbs remain horizontally outstretched. The agency does not consider GM's suggested wording to be adequate for calibration. For example, the attitude of the test probe at impact is not specified. For this reason, the requested modification is not undertaken.

Humanoid requested clarification of paragraph (7) of § 572.8(d) that specifies measurement of horizontal deflection "in line with the longitudinal centerline of the probe." Humanoid expressed concern that, as the thorax rotated backwards, the horizontal measurement could not be made. A clarification has been added to the cited language.

Humanoid also requested a less temperature-sensitive rib damping material than is presently employed. The NHTSA concludes that its strict limitation on permissible temperature and humidity conditions for calibration testing adequately controls the effects of temperature on this damping material.

LUMBAR SPINE, ABDOMEN

The NHTSA proposed minor modifications of the lumbar spine construction, and several changes in the procedures for lumbar spine calibration, which consists of spine flexion from the upright position, followed by release of the force which was required to attain this deflection, and measurement of the return angle. Manufacturers supported the majority of the changes, and they are made final in this notice. The agency proposed that measurements be taken when "flexing has stopped," and Toyota, noting the difficulty of establishing this point under some circumstances, suggested that the measurement be made 3 minutes after release. This modification is reasonable and is included in the final action.

Testing at NHTSA's Safety Research Laboratory demonstrates the need to clarify proposed § 572.9(c)(3) to specify return of the lumbar spine sufficiently so that it remains in "its initial position in accordance with Figure 11" unassisted. An appropriate further specification has been made.

Humanoid requested that the four-bolt attachment of the push plate be revised to two-bolt attachment in view of Humanoid's practice of providing a two-bolt plate. The agency has undertaken its data collection using four-bolt attachment, and to preserve the uncontested validity of these data, declines to modify the proposed specification.

ARL requested reconsideration of NHTSA's decision to leave unchanged the lumbar cable ball and socket attachment design. The agency has continued to examine test results and cannot conclude that the present attachment design has caused a calibration or compliance problem. Accordingly, ARL's request is denied. An ARL request to limit the reference to the strength requirements of the military specification in the case of lumbar cable swaging is granted. If such a limitation were not specified, the other elements of the military specification might arguably be included in the NHTSA's specification.

Calibration of the abdomen of the dummy is accomplished by application of a specified force to the abdomen while the dummy torso is placed on its back, with a required "force/deflection" curve resulting. The proposal added a range of force application rates to make the procedure more uniform, as well as a 10-pound preload and further specification of the horizontal surface. Manufacturers did not oppose these changes.

Manufacturers did oppose the proposed specification changes that would require the dummy abdominal sac to be sealed. Various reasons unrelated to abdomen performance were listed (e.g., transportation of sealed sac in unpressur-

ized aircraft compartment) and available data show successful calibration in both configurations. In view of the expressed preference for the unsealed design, the leak test has been removed from the drawings, and the vent is retained.

Humanoid requested that the shape of the abdominal insert be modified to conform more closely to the dummy's abdominal cavity. The shape of the insert affects the dummy performance, however, and the agency does not consider a change with unknown consequences advisable at this time. The agency also concludes that Humanoid's request to drop all specification of wall thickness for the abdominal sac is also unadvisable for this reason.

Ford, the MVMA, and Humanoid noted an asymmetry of the dummy pelvic castings and requested a justification for it. The asymmetry is apparently an artifact of the adoption of Society of Automotive Engineers specifications, whose origin is unknown. In the agency's judgment, based on experience with numerous Part 572 dummies and evaluation of test results, no degradation in performance is attributable to the asymmetry. While the agency intends to further review the asymmetry noted, no action will be taken without evidence that the specification affects testing.

LIMBS

Little comment was received on the changes proposed for limb calibration, which consists of impacting the knees of a seated dummy with a test probe of a specified weight at a specified speed and measuring the impact force on the dummy femurs. In response to Toyota's request for clarification, the positioning in accordance with § 572.11 is followed by the leg adjustments specified in § 572.10(c), which have the effect of changing leg position from that achieved under § 572.11.

The proposed specification of vinyl skin thickness over the knee face was supported in comments, although two manufacturers requested that the thickness tolerance be moved upward to thicken the skin somewhat. Humanoid did suggest elimination of the femur calibration as useless, but the agency considers such a control important to repeatable performance of the dummy.

Ford interpreted information contained in contract work undertaken for the NHTSA (DOT-HS-4-00873) to show that femur force loads registered too high in 50 percent of cases conducted under the calibration conditions of the standard. In NHTSA tests of 100 dummy knees on Part 572 dummies (DOT-HS-801 861), the 2,500-pound limit was exceeded only twice. The same data indicated a tendency for the femur to register lower than previously estimated, and a minor reduction of the lower limit is established in this action. The agency considers the small reduction to fall within the ambit of the proposal to improve conditions for calibration.

Ford's and Humanoid's observations with regard to off-center impacts that result in bending or torque have been dealt with in the recent agency proposal to limit femur force requirements of Standard No. 208 to compressive force. As for Humanoid's concern that unacceptable variation is possible in the femur load cell, it is noted that General Motors and Volkswagen have both certified thousands of vehicles based on impact readings taken from this dummy with these femur cells installed.

GENERAL TEST CONDITIONS

The agency proposed minor changes in the general test conditions of § 572.11 that apply to dummy test, such as a minimum period of dummy exposure to the temperature and humidity at which calibration tests are conducted. With correction of accelerometer locations, a clarification of dummy positioning, and an increase of zipper heat seals from three to four, the contemplated changes are made as proposed.

Sierra requested a broader range of humidity conditions for the calibration tests, stating that a range of 10- to 90-percent humidity would not affect results of "performance tests." The company cited freezing and desert heat conditions as reasons for a 6-hour conditioning rather than the 4-hour conditioning proposed by the agency. Humanoid and Toyota also addressed this aspect of the general test conditions. It appears that Sierra misunderstood the temperature and humidity specifications as applicable to vehicle performance tests. This rulemaking action addresses only calibration tests which presumably would be conducted indoors in a temperature-

controlled setting. Because the dummies are not expected to be stored in areas of great temperature extremes prior to calibration testing, the proposed ranges of humidity and temperature conditions are considered to be effective to stabilize the affected dummy properties. While instrumentation would be affected by the 90-percent humidity condition suggested by Sierra, the agency has reduced the lower humidity condition to a 10-percent level in agreement that the change does not affect the ability to calibrate the dummy.

Sierra objected that a dummy manufacturer's warranty of conformity of its products to Part 572 would be complicated by a time specification for temperature and humidity conditioning. The company believed that its customers would require that 4 hours of conditioning occur whether or not the dummy had already stabilized at the correct temperature. The agency sees no reason why a purchaser would insist on a senseless condition but, in any case, has no control over the contractual dealings between the dummy manufacturer and the purchaser. The NHTSA cannot delete necessary stabilizing conditions from its regulations simply because a purchaser wishes to make an unreasonable contractual specification based on it. The same rationale is responsive to Sierra's request for shorter recovery intervals between repeated tests.

Toyota supplied data to demonstrate that more consistent thorax and knee impact tests could be achieved by using cotton pants on the dummy. The agency's data do not agree with Toyota's and no other manufacturer took issue with the agency's proposal to delete all clothing requirements. This deletion is made final as proposed.

ARL asked why the agency's proposed prohibition against painting dummy components is qualified to state "except as specified in this part or in drawings subtended by this part." This qualification simply preserves the agency's opportunity to specify painted components in the future.

No conclusive evidence of preferable storage methods was submitted by commenters. The agency therefore does not specify that the dummy calibrations be preceded by positioning in a specific posture. To avoid the possibility of introducing a variable, however, the eye bolt in the dummy head has been relabeled on the drawings as "not for use in suspending dummy in storage."

Interested persons are advised that the first stage of choosing a replacement foaming agent for the specified Nitrosan are complete. Details are available in document HS–802–030 in the public docket.

In accordance with recently enunciated Department of Transportation policy encouraging adequate analysis of the consequences of regulatory action (41 FR 16200, April 16, 1976), the agency herewith summarizes its evaluation of the economic and other consequences of this action on the public and private sectors, including possible loss of safety benefits. The changes made are all to existing specifications and calibration procedures and are intended as clarifications of specifications already established. Therefore, the cost of the changes are calculated as minimal, consisting at most of relatively small modifications of test equipment and minor dummy components. The number and complexity of calibration tests are not affected by the changes. At the same time, the clarification will improve a manufacturer's ability to conduct compliance tests of safety systems and will thereby contribute to an increase in motor vehicle safety.

Note—

The economic and inflationary impacts of this rulemaking have been carefully evaluated in accordance with Office of Management and Budget Circular A–107, and an Inflation Impact Statement is not required.

In anticipation of the use of dummies other than the 50th-percentile male dummy in compliance testing, the agency takes this opportunity to reorganize Part 572 so that the 50th-percentile dummy occupies only one Subpart.

In consideration of the foregoing, 49 CFR Part 572, *Anthropomorphic Test Dummy*, and the dummy design drawings incorporated by reference in Part 572, are amended . . . .

Effective date: August 8, 1977.

(Sec. 103, 119, Pub. L. 89–563, 80 Stat. 718 (15 U.S.C. 1392, 1407); delegation of authority at 49 CFR 1.50.)

Issued on January 31, 1977.

John W. Snow
Administrator
42 F.R. 7148
February 7, 1977

PREAMBLE TO AMENDMENT TO PART 572—ANTHROPOMORPHIC TEST DUMMIES

(Docket No. 74–14; Notice 11; Docket No. 73–8; Notice 07)

This notice amends occupant crash protection Standard No. 208 and its accompanying test dummy specification to further specify test procedures and injury criteria. The changes are minor in most respects and reflect comments by manufacturers of test dummies and vehicles and the NHTSA's own test experience with the standard and the test dummy.

Date: Effective date—July 5, 1978.

Addresses: Petitions for reconsideration should refer to the docket number and be submitted to: Docket Section, Room 5108, Nassif Building, 400 Seventh Street, S.W., Washington, D.C. 20590.

For further information contact:

> Mr. Guy Hunter,
> Motor Vehicle Programs,
> National Highway Traffic Safety
> Administration,
> Washington, D.C. 20590
> (202 426–2265)

Supplementary information: Standard No. 208, *Occupant Crash Protection* (49 CFR 571.208), is a Department of Transportation safety standard that requires manufacturers to provide a means of restraint in new motor vehicles to keep occupants from impacting the vehicle interior in the event a crash occurs. The standard has, since January 1968, required the provision of seat belt assemblies at each seating position in passenger cars. In January 1972 the requirements for seat belts were upgraded and options were added to permit the provision of restraint that is "active" (requiring some action be taken by the vehicle occupant, as in the case of seat belts) or "passive" (providing protection without action being taken by the occupant).

In a separate notice issued today (42 FR 34289; FR Reg. 77-19137), the Secretary of Transportation has reached a decision regarding the future occupant crash protection that must be installed in passenger cars. The implementation of that decision will involve the testing of passive restraint systems in accordance with the test procedures of Standard No. 208, and this notice is intended to make final several modifications of that procedure which have been proposed for change by the NHTSA. This notice also responds to two petitions for reconsideration of rulemaking involving the test dummy that is used to evaluate the compliance of passive restraints systems.

DOCKET 74–14; NOTICE 05

Notice 5 was issued July 15, 1976 (41 FR 29715; July 19, 1976) and proposed that Standard No. 208's existing specification for passive protection in frontal, lateral, and rollover modes (S4.1.2.1) be modified to specify passive protection in the frontal mode only, with an option to provide passive protection or belt protection in the lateral and rollover crash modes. Volkswagen had raised the question of the feasibility of small cars meeting the standard's lateral impact requirements: A 20-mph impact by a 4,000-pound, 60-inch-high flat surface. The agency noted the particular vulnerability of small cars to side impact and the need to provide protection for them based on the weight of other vehicles on the highway, but agreed that it would be difficult to provide passive lateral protection in the near future. Design problems also underlay the proposal to provide a belt option in place of the existing passive rollover requirement.

Ford Motor Company argued that a lateral option would be inappropriate in Standard No. 208 as long as the present dummy is used for

PART 572—PRE 13

measurement of passive system performance. This question of dummy use as a measuring device is treated later in this notice. General Motors Corporation (GM) supported the option without qualification, noting that the installation of a lap belt with a passive system "would provide comparable protection to lap-shoulder belts in side and rollover impacts." Chrysler did not object to the option, but noted that the lap belt option made the title of S4.1.2.1 ("complete passive protection") misleading. Volkswagen noted that its testing of belt systems without the lap belt portion showed little loss in efficacy in rollover crashes. No other comments on this proposal were received. The existing option S4.1.2.1 is therefore adopted as proposed so that manufacturers will be able to immediately undertake experimental work on passive restraints on an optional basis in conformity with the Secretary's decision.

There were no objections to the agency's proposal to permit either a Type 1 or Type 2 seat belt assembly to meet the requirements, and thus it is made final as proposed.

The NHTSA proposed two changes in the injury criteria of S6 that are used as measures of a restraint system's qualification to Standard No. 208. One change proposed an increase in permissible femur force limits from 1,700 pounds to 2,250 pounds. As clarification that tension loads are not included in measurement of these forces, the agency also proposed that the word "compressive" be added to the text of S6.4. Most commenters were cautionary about the changes, pointing out that susceptibility to fracture is time dependent, that acetabular injury could be exacerbated by increased forces, and that angular applications of force were as likely in the real world as axial forces and would more likely fracture the femur.

The agency is aware of and took into account these considerations in proposing the somewhat higher femur force limit. The agency started with the actual field experience of occupants of GM and Volkswagen vehicles that have been shown to produce femur force readings of about 1,700 pounds. Occupants of these vehicles involved in crashes have not shown a significant

incidence of femur fracture. The implication from this experience that the 1,700-pound figure can safely be raised somewhat is supported in work by Patrick on compressive femur forces of relatively long duration. The Patrick data (taken with aged embalmed cadavers) indicate that the average fracture load of the patella-femur-pelvis complex is 1,910 pounds. This average is considered conservative, in that cadaver bone structure is generally weaker than living human tissues. While these data did not address angular force applications, the experience of the GM and Volkswagen vehicle occupants does suggest that angular force application can go higher than 1,700 pounds.

The agency does not agree that the establishment of the somewhat higher outer limit for permissible femur force loads of 2,250 pounds is arbitrary. What is often ignored by the medical community and others in commenting on the injury criteria found in motor vehicle safety standards is that manufacturers must design their restraint systems to provide greater protection than the criteria specified, to be certain that each of their products will pass compliance tests conducted by the NHTSA. It is a fact of industrial production that the actual performance of some units will fall below nominal design standards (for quality control and other reasons). Volkswagen made precisely this point in its comments. Because the National Traffic and Motor Vehicle Safety Act states that each vehicle must comply (15 U.S.C. § 1392(a)(1)(a)), manufacturers routinely design in a "compliance margin" of superior performance. Thus, it is extremely unlikely that a restraint system designed to meet the femur force load criterion of 2,250 pounds will in fact be designed to provide only that level of performance. With these considerations in mind, the agency makes final the changes as proposed.

While not proposed for change, vehicle manufacturers commented on a second injury criterion of the standard: A limitation of the acceleration experienced by the dummy thorax during the barrier crash to 60g, except for intervals whose cumulative duration is not more than 3 milliseconds (ms). Until August 31, 1977, the agency has specified the Society of Automotive Engineers' (SAE) "severity index" as a substitute for the 60g-3ms limit, because of greater familiarity of the industry with that criterion.

General Motors recommended that the severity index be continued as the chest injury criterion until a basis for using chest deflection is developed in place of chest acceleration. GM cited data which indicate that chest injury from certain types of blunt frontal impact is a statistically significant function of chest deflection in humans, while not a function of impact force or spinal acceleration. GM suggested that a shift from the temporary severity index measure to the 60g-3ms measurement would be wasteful, because there is no "strong indication" that the 60g-3ms measurement is more meaningful than the severity index, and some restraint systems have to be redesigned to comply with the new requirement.

Unlike GM, Chrysler argued against the use of acceleration criteria of either type for the chest, and rather advocated that the standard be delayed until a dummy chest with better deflection characteristics is developed.

The Severity Index Criterion allows higher loadings and therefore increases the possibility of adverse effects on the chest. It only indirectly limits the accelerations and hence the forces which can be applied to the thorax. Acceleration in a specific impact environment is considered to be a better predictor of injury than the Severity Index.

NHTSA only allowed belt systems to meet the Severity Index Criterion of 1,000 instead of the 60g-3ms criterion out of consideration for leadtime problems, not because the Severity Index Criterion was considered superior. It is recognized that restraint systems such as lap-shoulder belts apply more concentrated forces to the thorax than air cushion restraint, and that injury can result at lower forces and acceleration levels. It is noted that the Agency is considering rulemaking to restrict forces that may be applied to the thorax by the shoulder belt of any seat belt assembly (41 FR 54961; December 16, 1976).

With regard to the test procedures and conditions that underlie the requirements of the standard, the agency proposed a temperature range for testing that would be compatible with the temperature sensitivity of the test dummy. The test dummy specification (Part 572, "Anthropomorphic Test Dummy," 49 CFR Part 572) contains calibration tests that are conducted at any temperature between 66° and 78° F. This is because properties of lubricants and nonmetallic parts used in the dummy will change with large temperature changes and will affect the dummy's objectivity as a test instrument. It was proposed that the Standard No. 208 crash tests be conducted within this temperature range to eliminate the potential for variability.

The only manufacturers that objected to the temperature specification were Porsche, Bayerische Motoren Werke (BMW), and American Motors Corporation (AMC). In each case, the manufacturers noted that dynamic testing is conducted outside and that it is unreasonable to limit testing to the few days in the year when the ambient temperature would fall within the specified 12-degree range.

The commenters may misunderstand their certification responsibilities under the National Traffic and Motor Vehicle Safety Act. Section 108(b)(2) limits a manufacturer's responsibility to the exercise of "due care" to assure compliance. The NHTSA has long interpreted this statutory "due care" to mean that the manufacturer is free to test its products in any fashion it chooses, as long as the testing demonstrates that due care was taken to assure that, if tested by NHTSA as set forth in the standard, the product would comply with the standard's requirements. Thus, a manufacturer could conduct testing on a day with temperatures other than those specified, as long as it could demonstrate through engineering calculations or otherwise, that the difference in test temperatures did not invalidate the test results. Alternatively, a manufacturer might choose to perform its preparation of the vehicle in a temporarily erected structure (such as a tent) that maintains a temperature within the specified range, so that only a short exposure during acceleration to the barrier would occur in a higher or lower temperature. To assist any such arrangements, the test temperature condition has been limited to require a stabilized temperature of the test dummy only, just prior to the vehicle's travel toward the barrier.

In response to an earlier suggestion from GM, the agency proposed further specificity in the clothing worn by the dummy during the crash test. The only comment was filed by GM, which argued that any shoe specification other than weight would be unrelated to dummy performance and therefore should not be included in the specification. The agency disagrees, and notes that the size and shape of the heel on the shoe can affect the placement of the dummy limb within the vehicle. For this reason, the clothing specifications are made final as proposed, except that the requirement for a conforming "configuration" has been deleted.

Renault and Peugeot asked for confirmation that pyrotechnic pretensioners for belt retractors are not prohibited by the standard. The standard's requirements do not specify the design by which to provide the specified protection, and the agency is not aware of any aspect of the standard that would prohibit the use of pretensioning devices, as long as the three performance elements are met.

With regard to the test dummy used in the standard, the agency proposed two modifications of Standard No. 208: a more detailed positioning procedure for placement of the dummy in the vehicle prior to the test, and a new requirement that the dummy remain in calibration without adjustment following the barrier crash. Comments were received on both aspects of the proposal.

The dummy positioning was proposed to eliminate variation in the conduct of repeatable tests, particularly among vehicles of different sizes. The most important proposed modification was the use of only two dummies in any test of front seat restraints, whether or not the system is designed for three designated seating positions. The proposal was intended to eliminate the problem associated with placement of three 50th-percentile male dummies side-by-side in a smaller vehicle. In bench seating with three positions, the system would have to comply with a dummy at the driver's position and at either of the other two designated seating positions.

GM supported this change, but noted that twice as many tests of 3-position bench-seat vehicles would be required as before. The company suggested using a simulated vehicle crash as a means to test the passive restraint at the center seat position. The agency considers this approach unrepresentative of the actual crash pulse and vehicle kinematic response (e.g., pitching, yawing) that occur during an impact. To the degree that GM can adopt such an approach in the exercise of "due care" to demonstrate that the center seating position actually complies, the statute does not prohibit such a certification approach.

Ford objected that the dummy at the center seat position would be placed about 4 inches to the right of the center of the designated seating position in order to avoid interference with the dummy at the driver's position. While the NHTSA agrees that a small amount of displacement is inevitable in smaller vehicles, it may well occur in the real world also. Further, the physical dimensions of the dummy preclude any other positioning. With a dummy at the driver's position, a dummy at the center position cannot physically be placed in the middle of the seat in all cases. In view of these realities, the agency makes final this aspect of the dummy positioning as proposed.

GM suggested the modification of other standards to adopt "2-dummy" positioning. The compatibility among dynamic tests is regularly reviewed by the NHTSA and will be again following this rulemaking action. For the moment, however, only those actions which were proposed will be acted on.

As a general matter with regard to dummy positioning, General Motors found the new specifications acceptable with a few changes. GM cautioned that the procedure might not be sufficiently reproducible between laboratories, and Chrysler found greater variation in positioning with the new procedures than with Chrysler's own procedures. The agency's use of the procedure in 15 different vehicle models has shown consistently repeatable results, as long as a reasonable amount of care is taken to avoid the effect of random inputs (see "Repeatability of Set Up and Stability of Anthropometric Landmarks and Their Influence on Impact Response of Automotive Crash Test Dummies." Society of Automotive Engineers, Technical Paper No. 770260, 1977). The agency concludes that, with the minor improvements cited below, the positioning procedure should be made final as proposed.

The dummy is placed at a seating position so that its midsagittal plane is vertical and longitudinal. Volkswagen argued against use of the midsagittal plane as a reference for dummy placement, considering it difficult to define as a practical matter during placement. The agency has used plane markers and plane lines to define the midsagittal plane and has experienced no significant difficulty in placement of the dummy with these techniques. For this reason, and because Volkswagen suggested no simpler orientation technique, the agency adopts use of the midsagittal plane as proposed.

Correct spacing of the dummy's legs at the driver position created the largest source of objections by commenters. Ford expressed concern that an inward-pointing left knee could result in unrealistically high femur loads because of femur-to-steering column impacts. GM asked that an additional 0.6 inch of space be specified between the dummy legs to allow for installation of a device to measure steering column displacement. Volkswagen considered specification of the left knee bolt location to be redundant in light of the positioning specification for the right knee and the overall distance specification between the knees of 14.5 inches.

The commenters may not have understood that the 14.5- and 5.9-inch dimensions are only initial positions, as specified in S8.1.11.1.1. The later specification to raise the femur and tibia centerlines "as close as possible to vertical" without contacting the vehicle shifts the knees from their initial spacing to a point just to the left and right of the steering column.

As for GM's concern about instrumentation, the agency does not intend to modify this positioning procedure to accommodate instrumentation preferences not required for the standard's purposes. GM may, of course, make test modifications so long as it assures, in the exercise of due care, that its vehicles will comply when tested in accordance with the specification by the agency.

In the case of a vehicle which is equipped with a front bench seat, the driver dummy is placed on the bench so that its midsagittal plane intersects the center point of the plane described by the steering wheel rim. BMW pointed out that the center plane of the driver's seating position may not coincide with the steering wheel center and that dummy placement would therefore be unrealistic. Ford believed that the specification of the steering wheel reference point could be more precisely specified.

The agency believes that BMW may be describing offset of the driver's seat from the steering wheel in bucket-seat vehicles. In the case of bench-seat vehicles, there appears to be no reason not to place the dummy directly behind the steering wheel. As for the Ford suggestion, the agency concludes that Ford is describing the same point as the proposal did, assuming, as the agency does, that the axis of the steering column passes through the center point described. The Ford description does have the effect of moving the point a slight distance laterally, because the steering wheel rim upper surface is somewhat higher than the plane of the rim itself. This small distance is not relevant to the positioning being specified and therefore is not adopted.

In the case of center-position dummy placement in a vehicle with a drive line tunnel, Ford requested further specification of left and right foot placement. The agency has added further specification to make explicit what was implicit in the specifications proposed.

Volkswagen suggested that the NHTSA had failed to specify knee spacing for the passenger side dummy placement. In actuality, the specification in S8.1.11.1.2 that the femur and tibia centerlines fall in a vertical longitudinal plane has the effect of dictating the distance between the passenger dummy knees.

The second major source of comments concerned the dummy settling procedure that assures uniformity of placement on the seat cushion and against the seat back. Manufacturers pointed out that lifting the dummy within the vehicle, particularly in small vehicles and those with no rear seat space, cannot be accomplished easily. While the NHTSA recognizes that the procedure is not simple, it is desirable to improve the uniformity of dummy response and it has been accomplished by the NHTSA in several small cars (e.g., Volkswagen Rabbit, Honda Civic, Fiat

Spider, DOT HS–801–754). Therefore, the requests of GM and Volkswagen to retain the method that does not involve lifting has been denied. In response to Renault's question, the dummy can be lifted manually by a strap routed beneath the buttocks. Also, Volkswagen's request for more variability in the application of rearward force is denied because, while difficult to achieve, it is desirable to maintain uniformity in dummy placement. In response to the requests of several manufacturers, the location of the 9-square-inch push plate has been raised 1.5 inches, to facilitate its application to all vehicles.

Volkswagen asked with regard to S10.2.2 for a clarification of what constitutes the "lumbar spine" for purposes of dummy flexing. This refers to the point on the dummy rear surface at the level of the top of the dummy's rubber spine element.

BMW asked the agency to reconsider the placement of the driver dummy's thumbs over the steering wheel rim because of the possibility of damage to them. The company asked for an option in placing the hands. The purpose of the specification in dummy positioning, however, is to remove discretion from the test personnel, so that all tests are run in the same fashion. An option under these circumstances is therefore not appropriate.

Ultrasystems, Inc., pointed out two minor errors in S10.3 that are hereby corrected. The upper arm and lower arm centerlines are oriented as nearly as possible in a vertical plane (rather than straight up in the vertical), and the little finger of the passenger is placed "barely in contact" with the seat rather than "tangent" to it.

Two corrections are made to the dummy positioning procedure to correct obvious and unintended conflicts between placement of the dummy thighs on the seat cushion and placement of the right leg and foot on the acceleration pedal.

In addition to the positioning proposed, General Motors suggested that positioning of the dummy's head in the fore-and-aft axis would be beneficial. The agency agrees and has added such a specification at the end of the dummy settling procedure.

In a matter separate from the positioning procedure, General Motors, Ford, and Renault requested deletion of the proposed requirement that the dummy maintain proper calibration following a crash test without adjustment. Such a procedure is routine in test protocols and the agency considered it to be a beneficial addition to the standard to further demonstrate the credibility of the dummy test results. GM, however, has pointed out that the limb joint adjustments for the crash test and for the calibration of the lumber bending test are different, and that it would be unfair to expect continued calibration without adjustment of these joints. The NHTSA accepts this objection and, until a means for surmounting this difficulty is perfected, the proposed change to S8.1.8 is withdrawn.

In another matter unrelated to dummy positioning, Volkswagen argued that active belt systems should be subject to the same requirements as passive belt systems, to reduce the cost differential between the compliance tests of the two systems. As earlier noted the NHTSA has issued an advance Notice of Proposed Rulemaking (41 FR 54961, December 16, 1976) on this subject and will consider Volkswagen's suggestion in the context of that rulemaking.

Finally, the agency proposed the same belt warning requirements for belts provided with passive restraints as are presently required for active belts. No objections to the requirement were received and the requirement is made final as proposed. The agency also takes the opportunity to delete from the standard the out-of-date belt warning requirements contained in S7.3 of the standard.

RECONSIDERATION OF DOCKET 73–8; NOTICE 04

The NHTSA has received two petitions for reconsideration of recent amendments in its test dummy calibration test procedures and design specifications (Part 572, "*Anthropomorphic Test Dummy*," 49 CFR Part 572). Part 572 establishes, by means of approximately 250 drawings and five calibration tests, the exact specifications of the test device referred to earlier in this notice that simulates the occupant of a motor vehicle for crash testing purposes.

Apart from requests for a technical change of the lumbar flexion force specifications, the petitions from General Motors and Ford contained a repetition of objections made earlier in the rulemaking about the adequacy of the dummy as an objective measuring device. Three issues were raised: lateral response characteristics of the dummy, failure of the dummy to meet the five subassembly calibration limits, and the need for a "whole systems" calibration of the assembled dummy. Following receipt of these comments, the agency published notification in the *Federal Register* that it would entertain any other comments on the issue of objectivity (42 FR 28200; June 2, 1977). General comments were received from Chrysler Corporation and American Motors, repeating their positions from earlier comments that the dummy does not qualify as objective.

The objectivity of the dummy is at issue because it is the measuring device that registers the acceleration and force readings specified by Standard No. 208 during a 30-mph impact of the tested vehicle into a fixed barrier. The resulting readings for each vehicle tested must remain below a certain level to constitute compliance. Certification of compliance by the vehicle manufacturer is accomplished by crash testing representative vehicles with the dummy installed. Verification of compliance by the NHTSA is accomplished by crash testing one or more of the same model vehicle, also with a test dummy installed. It is important that readings taken by different dummies, or by the same dummy repeatedly, accurately reflect the forces and accelerations that are being experienced by the vehicle during the barrier crash. This does not imply that the readings produced in tests of two vehicles of the same design must be identical. In the real world, in fact, literally identical vehicles, crash circumstances, and test dummies are not physically attainable.

It is apparent from this discussion that an accurate reflection of the forces and accelerations experienced in nominally identical vehicles does not depend on the specification of the test dummy alone. For example, identically specified and responsive dummies would not provide identical readings unless reasonable care is exercised in the preparation and placement of the dummy. Such care is analogous to that exercised in positioning a ruler to assure that it is at the exact point where a measurement is to commence. No one would blame a ruler for a bad measurement if it were carelessly placed in the wrong position.

It is equally apparent that the forces and accelerations experienced in nominally identical vehicles will only be identical by the greatest of coincidence. The small differences in body structure, even of mass-produced vehicles, will affect the crash pulse. The particular deployment speed and shape of the cushion portion of an inflatable restraint system will also affect results.

All of these factors would affect the accelerations and forces experienced by a human occenpant of a vehicle certified to comply with the occupant restraint standard. Thus, achievement of identical conditions is not only impossible (due to the inherent differences between tested vehicles and underlying conditions) but would be unwise. Literally identical tests would encourage the design of safety devices that would not adequately serve the variety of circumstances encountered in actual crash exposure.

At the same time, the safety standards must be "stated in objective terms" so that the manufacturer knows how its product will be tested and under what circumstances it will have to comply. A complete lack of dummy positioning procedures would allow placement of the dummy in any posture and would make certification of compliance virtually impossible. A balancing is provided in the test procedures between the need for realism and the need for objectivity.

The test dummy also represents a balancing between realism (biofidelity) and objectivity (repeatability). One-piece cast metal dummies could be placed in the seating positions and instrumented to register crash forces. One could argue that these dummies did not act at all like a human and did not measure what would happen to a human, but a lack of repeatability could not be ascribed to them. At the other end of the spectrum, an extremely complex and realistic surrogate could be substituted for the existing Part 572 dummy, which would act realistically but differently each time, as one might expect different humans to do.

PART 572—PRE 19

The existing Part 572 dummy represents 5 years of effort to provide a measuring instrument that is sufficiently realistic and repeatable to serve the purposes of the crash standard. Like any measuring instrument, it has to be used with care. As in the case of any complex instrumentation, particular care must be exercised in its proper use, and there is little expectation of literally identical readings.

The dummy is articulated, and built of materials that permit it to react dynamically, similarly to a human. It is the dynamic reactions of the dummy that introduce the complexity that makes a check on repeatability desirable and necessary. The agency therefore devised five calibration procedures as standards for the evaluation of the important dynamic dummy response characteristics.

Since the specifications and calibration procedures were established in August 1973, a substantial amount of manufacturing and test experience has been gained in the Part 572 dummy. The quality of the dummy as manufactured by the three available domestic commercial sources has improved to the point where it is the agency's judgment that the device is as repeatable and reproducible as instrumentation of such complexity can be. As noted, GM and Ford disagree and raised three issues with regard to dummy objectivity in their petitions for reconsideration.

Lateral response characteristics. Recent sled tests of the Part 572 dummy in lateral impacts show a high level of repeatability from test to test and reproducibility from one dummy to another ("Evaluation of Part 572 Dummies in Side Impacts"—DOT HS 020 858). Further modification of the lateral and rollover passive restraint requirements into an option that can be met by installation of a lap belt makes the lateral response characteristics of the dummy largely academic. As noted in Notice 4 of Docket 73–8 (42 FR 7148; February 7, 1977), "Any manufacturer that is concerned with the objectivity of the dummy in such [lateral] impacts would provide lap belts at the front seating positions in lieu of conducting the lateral or rollover tests."

While the frontal crash test can be conducted at any angle up to 30 degrees from perpendicular to the barrier face, it is the agency's finding that the lateral forces acting on the test instrument are secondary to forces in the midsagittal plane and do not operate as a constraint on vehicle and restraint design. Compliance tests conducted by NHTSA to date in the 30-degree oblique impact condition have consistently generated similar dummy readings. In addition, they are considerably lower than in perpendicular barrier impact tests, which renders them less critical for compliance certification purposes.

Repeatability of dummy calibration. Ford questioned the dummy's repeatability, based on its analysis of "round-robin" testing conducted in 1973 for Ford at three different test laboratories (Ford Report No. ESRO S–76–3 (1976)) and on analysis of NHTSA calibration testing of seven test dummies in 1974 (DOT–HS–801–861).

In its petition for reconsideration, Ford equated dummy objectivity with repeatability of the calibration test results and concluded "it is impracticable to attempt to meet the Part 572 component calibration requirements with test dummies constructed according to the Part 572 drawing specifications."

The Ford analysis of NHTSA's seven dummies showed only 56 of 100 instances in which all of the dummy calibrations satisfied the criteria. The NHTSA's attempts to reproduce the Ford calculations to reach this conclusion were unsuccessful, even after including the HO3 dummy with its obviously defective neck. This neck failed badly 11 times in a row, and yet Ford apparently used these tests in its estimate of 56 percent compliance. This is the equivalent of concluding that the specification for a stop watch is inadequate because of repeated failure in a stop watch with an obviously defective part. In this case, the calibration procedure was doing precisely its job in identifying the defective part by demonstrating that it did not in fact meet the specification.

The significance of the "learning curve" for quality control in dummy manufacture is best understood by comparison of three sets of dummy calibration results in chronological order. Ford in earlier comments relied on its own "round-robin" testing, involving nine test dummies. Ford stated that none of the nine dummies could pass all of the component calibration requirements. What the NHTSA learned through follow-up questions to Ford was that three of the nine dummies were not built originally as Part 572 dummies, and that the other six were not fully certified by their manufacturers as qualifying as Part 572 dummies. In addition, Ford instructed its contractors to use the dummies as provided whether or not they met the Part 572 specifications.

In contrast, recent NHTSA testing conducted by Calspan (DOT–HS–6–01514, May and June 1977 progress reports) and the results of tests conducted by GM (USG 1502, Docket 73–8, GR 64) demonstrate good repeatability and reproducibility of dummies. In the Calspan testing a total of 152 calibration tests were completed on four dummies from two manufacturers. The results for all five calibration tests were observed to be within the specified performance criteria of Part 572. The agency concludes that the learning curve in the manufacturing process has reached the point where repeatability and reproducibility of the dummy has been fully demonstrated.

Interestingly, Ford's own analysis of its round-robin testing concludes that variations among the nine dummies were not significant to the test results. At the same time, the overall acceleration and force readings did vary substantially. Ford argued that this showed unacceptable variability of the test as a whole, because they had used "identical" vehicles for crash testing. Ford attributed the variations in results to "chance factors," listing as factors placement of the dummy, postural changes during the ride to the barrier, speed variations, uncertainty as to just what part of the instrument panel or other structure would be impact loaded, instrumentation, and any variations in the dynamics of air bag deployment from one vehicle to another.

The agency does not consider these to be uncontrolled factors since they can be greatly reduced by carefully controlling test procedures. In addition, they are not considered to be unacceptable "chance factors" that should be eliminated from the test. The most important advantage of the barrier impact test is that it simulates with some realism what can be experienced by a human occupant, while at the same time limiting variation to achieve repeatability.

As discussed, nominally identical vehicles are not in fact identical, the dynamics of deployment will vary from vehicle to vehicle, and humans will adopt a large number of different seated positions in the real world. The 30-mph barrier impact requires the manufacturer to take these variables into account by providing adequate protection for more than an overly structured test situation. At the same time, dummy positioning is specified in adequate detail so that the manufacturer knows how the NHTSA will set up a vehicle prior to conducting compliance test checks.

"Whole systems" calibration. Ford and GM both suggested a "whole systems" calibration of of the dummy as a necessary additional check on dummy repeatability. The agency has denied these requests previously, because the demonstrated repeatability and reproducibility of Part 572 dummies based on current specification is adequate. The use of whole systems calibration tests as suggested would be extremely expensive and would unnecessarily complicate compliance testing.

It is instructive that neither General Motors nor Ford has been specific about the calibration tests they have in mind. Because of the variables inherent in a high energy barrier crash test at 30 mph, the agency judges that any calibration readings taken on the dummy would be overwhelmed by the other inputs acting on the dummy in this test environment. The Ford conclusion from its round-robin testing agrees that dummy variability is a relatively insignificant factor in the total variability experienced in this type of test.

GM was most specific about its concern for repeatability testing of the whole dummy in its comments in response to Docket 74–14; Notice 01:

Dummy whole body response requirements are considered necessary to assure that a dummy, assembled from certified components, has acceptable response as a completed structure. Interactions between coupled components and subsystems must not be assumed acceptable simply because the components themselves have been certified. Variations in coupling may lead to significant variation in dummy response.

PART 572—PRE 21

There is a far simpler, more controlled means to assure oneself of correct coupling of components than by means of a "whole systems" calibration. If, for example, a laboratory wishes to assure itself that the coupling of the dummy neck structure is properly accomplished, a simple statically applied input may be made to the neck prior to coupling to obtain a sample reading, and then the same simple statically applied input may be repeated after the coupling has been completed. This is a commonly accepted means to assure that "bolting together" the pieces is properly accomplished.

Lumbar spine flexion. The flexibility of the dummy spine is specified by means of a calibration procedure that involves bending the spine through a forward arc, with specified resistance to the bending being registered at specified angles of the bending arc. The dummy's ability to flex is partially controlled by the characteristics of the abdominal insert. In Notice 04, the agency increased the level of resistance that must be registered, in conjunction with a decision not to specify a sealed abdominal sac as had been proposed. Either of these dummy characteristics could affect the lumbar spine flexion performance.

Because of the agency's incomplete explanation for its actions, Ford and General Motors petitioned for reconsideration of the decision to take one action without the other. Both companies suggested that the specification of resistance levels be returned to that which had existed previously. The agency was not clear that it intended to go forward with the stiffer spine flexion performance, quite apart from the decision to not specify an abdomen sealing specification. The purpose for the "stiffer" spine is to attain more consistent torso return angle and to assure better dummy stability during vehicle acceleration to impact speed.

To assure itself of the wisdom of this course of action, the agency has performed dummy calibration tests demonstrating that the amended spine flexion and abdominal force deflection characteristics can be consistently achieved with both vented and unvented abdominal inserts (DOT HS-020875 (1977)).

Based on the considered analysis and review set forth above, the NHTSA denies the petitions of General Motors and Ford Motor Company for further modification of the test dummy specification and calibration procedures for reasons of test dummy objectivity.

In consideration of the foregoing, Standard No. 208 (49 CFR 571.208) is amended as proposed with changes set forth below, and Part 572 (49 CFR Part 572) is amended by the addition of a new sentence at the end of § 572.5, *General Description*, that states: "A specimen of the dummy is available for surface measurements, and access can be arranged through: Office of Crashworthiness, National Highway Traffic Safety Administration, 400 Seventh Street, S.W., Washington, D.C. 20590."

In accordance with Department of Transportation policy encouraging adequate analysis of the consequences of regulatory action (41 FR 16200; April 16, 1976), the Department has evaluated the economic and other consequences of this amendment on the public and private sectors. The modifications of an existing option, the simplification and clarification of test procedures, and the increase in femur force loads are all judged to be actions that simplify testing and make it less expensive. It is anticipated that the "two dummy" positioning procedure may occasion additional testing expense in some larger vehicles, but not the level of expense that would have general economic effects.

The effective date for the changes has been established as one year from the date of publication to permit Volkswagen, the only manufacturer presently certifying compliance of vehicles using these test procedures, sufficient time to evaluate the effect of the changes on the compliance of its products.

The program official and lawyer principally responsible for the development of this amendment are Guy Hunter and Tad Herlihy, respectively.

(Sec. 103, 119, Pub. L. 89–563, 80 Stat. 718 (15 U.S.C. 1392, 1407); delegation of authority at 49 CFR 1.50.)

Issued on June 30, 1977.

Joan Claybrook
Administrator
42 F.R. 34299
July 5, 1977

PREAMBLE TO AMENDMENT TO PART 572—ANTHROPOMORPHIC TEST DUMMIES REPRESENTING SIX-MONTH-OLD AND THREE-YEAR-OLD CHILDREN

(Docket No. 78-09; Notice 4)

ACTION: Final rule.

SUMMARY: This notice is issued in conjunction with new Standard No. 213, *Child Restraint Systems*, which requires child restraint systems to be dynamically tested using anthropomorphic test dummies representing 6-month-old and 3-year-old children. This notice establishes the specifications for the dummies to be used in the child restraint testing. In addition, it sets performance criteria as calibration checks to assure the repeatability of the dummy's performance.

DATES: The amendment is effective upon publication in the Federal Register. December 27, 1979.

ADDRESSES: Petitions for reconsideration should refer to the docket number and be submitted to: Docket Section, Room 5108, National Highway Traffic Safety Administration, 400 Seventh Street, S.W., Washington, D.C. 20590.

FOR FURTHER INFORMATION CONTACT:

Mr. Vladislav Radovich, Office of Vehicle Safety Standards, National Highway Traffic Safety Administration, 400 Seventh Street, S.W., Washington, D.C. 20590 (202-426-2264)

SUPPLEMENTARY INFORMATION:

This notice amends Part 572, *Anthropomorphic Test Dummies*, to establish specifications and performance requirements for two test dummies, one representing a 6-month-old child and the other representing a 3-year-old child. This final rule is issued to supplement new Standard No. 213, *Child Restraint Systems*, published in the *Federal Register* for December 13, 1979 (44 FR 72131). Standard No. 213 evaluates the performance of child restraints in dynamic sled tests using the anthropomorphic test dummies whose specifica-

tions are established in this final rule. Restraints recommended for children weighing 20 pounds or less will be tested with an anthropomorphic dummy representing a 6-month-old child and restraints recommended for children weighing more than 20 pounds, but not more than 50 pounds will be tested with an anthropomorphic dummy representing a 3-year-old child.

On May 18, 1978, NHTSA published a notice of proposed rulemaking for the anthropomorphic test dummy amendment (43 FR 21490) and the child restraint standard (43 FR 21470). The comment closing date for both notices was December 1, 1978. The May 18, 1978, proposal on the anthropomorphic dummies noted that the calibration requirements proposed for the 3-year-old child test dummy were tentative. The agency said it would continue further testing on the calibrations and the results of that work would be placed in the public docket as soon as possible after the testing was completed. Based on the testing, NHTSA tentatively decided to make several minor modifications to the test dummy specifications and calibration requirements to improve the accuracy of the test dummy as a tool for measuring the performance of child restraints. A copy of the modifications was placed in the public docket on September 27, 1978, and the dummy manufacturers and child restraint testing facilities were advised of the modifications. The tentative modifications were published in the *Federal Register* on November 16, 1978 (43 FR 53478).

At the request of the Juvenile Products Manufacturers Association, the agency extended the comment closing date until January 5, 1979, for the portions of the child restraint and test dummy proposals dealing with testing with the anthropomorphic dummies. NHTSA granted the extension because manufacturers were reportedly having problems obtaining the proposed test dummies to conduct their own evaluations. Based on information gathered by the agency about the availability of testing facilities and dummies, the agency concluded that manufacturers could conduct the necessary testings before the extended comment closing date.

On December 21, 1978, NHTSA made available one of the agency's test dummies to General Motors Corp. (GM) for the purpose of resolving certain calibration problems GM reported it had experienced with its own test dummy. All other interested parties also were advised of the availability of the NHTSA test dummy and informed that NHTSA did not plan to issue a final rule on the test dummy proposal until at least mid-summer. The agency said it would review additional testing material submitted to the docket before issuance of the final rule. The final rule issuance date was subsequently rescheduled for October 1979 in the Department's March 1, 1979, Semi-Annual Regulations Agenda (44 FR Part II, 38) and for November 1979 in the August 27, 1979 Agenda (44 FR 50195).

Following issuance of the May 1978 notice of proposed rulemaking, NHTSA conducted additional testing of the test dummies. This testing, completed in July 1979, further confirmed the results of the agency's prior testing which showed the anthropomorphic dummies to be objective test devices. The results of this testing were periodically placed in the public docket so that all interested parties could comment on them.

This final rule is based on the data obtained in the agency's testing, data submitted in the comments, and data obtained from other pertinent documents and test reports. Significant comments submitted to the docket are addressed below.

Infant Test Dummy

The infant test dummy is based on a simple design representing the dimensions and mass distribution characteristics of a 6-month-old child. The test dummy is used to assess the ability of infant restraints to retain their occupants and maintain their structural integrity during dynamic testing. Because of its construction, the dummy cannot be instrumented to measure the forces that would be exerted upon an infant in a crash. NHTSA's tests have shown the infant dummy will reliably and consistently represent the dynamics of an infant during simulated impact tests.

GM, the only party to comment on the specification for the infant test dummy, reported that it had "no significant problem in building or verifying the compliance of the dummy to the proposed specification." To improve the durability of the test dummy, GM recommended adding a wooden form to the head to maintain its geometry and using steel instead of lead for ballast in the test dummy. Since these recommendations should not affect the dummy's performance and should increase its durability, NHTSA has adopted a modified version of the proposed changes. The changes add a plastic form to the dummy's head, since a plastic form is easier to manufacture and duplicate than a wooden form. In addition, a portion of the ballast materials are now required to be steel and aluminum.

The revised design drawings and a construction manual for the infant dummy are available for examination in the NHTSA docket section, which is open from 7:45 a.m. to 4:15 p.m., Monday through Friday. Copies of these documents can be obtained from: Keuffel and Esser Co., 1512 North Danville Street, Arlington, Virginia 22201.

3-Year-Old Child Test Dummy

The test dummy representing a 3-year-old child is based on the Alderson Model VIP-3C test dummy. It was chosen over the other available test dummies representing a 3-year-old child, such as the Sierra 492-03 test dummy, because it has more complete design details, can adequately withstand the test load imposed during impact testing, has more accurate anthropometry and mass distribution, can be easily instrumented for testing, more closely simulates the responses of a child during impact testing and has more consistent head and chest acceleration measurements during impact testing.

As with the infant test dummy, the final rule establishes a complete set of design specifications for the 3-year-old test dummy. For the 3-year-old test dummy, NHTSA has provided: a drawing package containing all of the technical details of the dummy parts and the stages of dummy manufacture; a set of master patterns for all molded and cast parts of the dummy; and a maintenance manual containing instructions for the assembly, disassembly, use, adjustment and maintenance of the dummy. These materials will ensure that manufacturers can accurately and consistently produce the test dummy.

The drawings and the maintenance manual for the 3-year-old test dummy are available for examination at the agency's docket section. Copies of these drawings and the maintenance manual can be obtained from the Keuffel and Esser Co., 1512 North Danville Street, Arlington, Va. 22201. In addition, patterns for all the cast and molded parts are available on a loan basis from the agency's Office of Vehicle Safety Standards, at the address given at the beginning of this notice.

Calibration Requirements

Unlike the infant test dummy, the 3-year-old child test dummy can be instrumented with accelerometers to measure the forces imposed on the dummy during an impact. Thus, in Standard No. 213, *Child Restraint Systems,* the 3-year-old test dummy is used to measure the amount of head and knee excursion and the magnitude of head and chest acceleration allowed by the child restraint.

Since a test dummy is a complex instrument required to measure important parameters, it is essential that the test dummy be properly calibrated to ensure accurate and repeatable results. NHTSA has developed detailed test dummy specifications and instrumentation requirements to ensure that the test dummies are as much as possible identically constructed and identically instrumented. The agency also developed calibration performance requirements that the test dummy must meet in dynamic and static tests. The calibration tests will determine whether the test dummies are uniformly constructed and properly instrumented.

In its comments, GM reported that it was unable to calibrate its 3-year-old test dummies. As mentioned previously, NHTSA loaned GM one of the agency's test dummies for the purpose of resolving the reported calibration problem. Using the NHTSA test dummy equipped with NHTSA's accelerometers, GM was able to meet the peak resultant acceleration requirements set for the dummy's head in specified pendulum impact tests, but was not able to meet the lateral acceleration requirement. When the same dummy was tested with GM's accelerometers, the dummy did not meet any of the head acceleration performance requirements. In the case of the chest calibration performance requirements, the accelerations measured by GM test dummies and the NHTSA test dummy, using both GM's and NHTSA's accelerometers, were within the range set for peak resultant and lateral acceleration.

GM also said that because the agency did not define the term "unimodal" it was not certain that the acceleration measurements that it made complied with the requirement that the acceleration-time curves for the head and chest impacts be unimodal. To clarify the requirement, NHTSA has defined unimodal in the final rule to mean an acceleration curve that only has one prominent peak and has specified that the measured acceleration-time curve during the head and chest impact testing need only be unimodal during a short time period when the accelerations are above a specified level.

GM attributed the calibration problem to resonances in the head and chest of the test dummies. (A resonance is a vibrational state that can magnify the accelerations imposed on the test dummy and thus prevent the accurate measurement of those accelerations.) GM said that because of the possible inaccurate measurements caused by the resonances, the test dummy cannot be used as an objective tool for assessing the performance of child restraint systems.

The calibration testing done for the agency indicates that the acceleration responses for the head and chest pendulum impacts include a limited amount of vibration. Such responses exist to some extent in any acceleration measuring device and are also found in similar pendulum impact tests of the Part 572 adult test dummy. However, dynamic sled tests of child test dummies in child restraint systems have demonstrated that the test dummies produce very repeatable results and do not show the vibrations found in the more severe pendulum impact tests. The agency's calibration tests also show that the test dummies produce very repeatable results. Even in GM tests of its three test dummies equipped with GM's instrumentation, the test dummies produced repeatable results. Such repeatability could not be obtained with resonating systems. Based on a review of GM's and the agency's test data, NHTSA concludes that the GM calibration failures are not attributable to resonances, but are very likely due to the differences, discussed below, in the mounting of the accelerometers in the GM test dummies.

NHTSA recognizes that because of different instrumentation and test procedures, different test facilities may obtain different results in what are essentially the same tests. To reduce such differences, NHTSA proposed requirements to standardize the test and instrumentation procedures. In calibration tests conducted at Calspan

Corporation the measurements of the peak resultant head accelerations and the lateral head acceleration were found to be close to the upper limits of the tentative head calibration requirements (112 g peak resultant acceleration and 5 g lateral acceleration) proposed by the agency. To further accommodate expected differences between different testing facilities, NHTSA has decided to broaden the head acceleration calibration requirements for peak resultant head acceleration to 115 g's and for lateral acceleration to 7 g's.

Instrumentation

Based on a review of GM's and the agency's test data, NHTSA concludes that one of the significant differences between NHTSA's and GM's test dummy is the manner in which the accelerometer mounting plate is attached to the head of the test dummy. Finding what it thought was an incompatibility between the angle of the accelerometer mounting plate bolt and the angle of the surface of the plate that attaches to the dummy's head, GM changed the angle of the surface in its test dummies. However, NHTSA specified the difference in the two angles for an important reason. Having a difference in the angles allows for a firmer attachment of the accelerometer mounting plate to the dummy. The difference in the firmness of the attachment of the accelerometer mounting plate may account for the additional acceleration that occurred in the head calibration tests of the GM test dummies.

GM also asked the agency to set a torque specification for the accelerometer mounting plate bolt. In response to GM's request, the agency has added a torque specification of 10 ft. lbs. to the specifications set out in the maintenance manual for the test dummy.

GM said that another possible source of the difference between the measurements it obtained with its own test dummies and the measurements it made with the NHTSA test dummies could be due to differences in the type and location of the accelerometers in the test dummies. GM noted that the specifications proposed in the rule allow the use of different types of accelerometers by allowing a number of different accelerometer placements within the test dummy.

As explained below, testing done for the agency has shown that the use of different types of accelerometers within the permissible locations does not prevent the test dummy from producing accurate and repeatable results. However, to further reduce the possibility of test differences due to accelerometer placement, the agency has more specifically defined several of the permissible accelerometer mounting locations.

Testing done for the agency at two different facilities to develop the calibration requirements used two types of accelerometers and different accelerometer locations. That testing produced no appreciable differences in test results and showed that different facilities could obtain repeatable results, when the accelerometers are properly mounted.

The agency's test experience with the adult test dummy also shows that minor differences in accelerometer mounting locations do not affect the ability of the test dummy to produce similar and repeatable results. The number of permissible accelerometer locations allowed for the adult test dummy is in some cases larger than the number permitted in the child test dummy. Yet no significant differences in test results for the adult test dummy have been encountered due to accelerometer location.

GM's own test data also indicate that use of different types of properly mounted accelerometers and different mounting locations produces only minor variations in the measurements. GM tested NHTSA's test dummy using two types of accelerometers mounted at different locations within the prescribed tolerances. The average measured acceleration in the chest impact tests varied by only 4 percent between the two types of accelerometers. It was only when GM used the improperly installed accelerometer mounting block in the head impact tests, discussed above, that GM obtained a 14 percent difference in measured accelerations within the NHTSA dummy using two types of accelerometers.

Calibration Procedures

GM also raised questions about the procedures for conducting the chest and head calibration tests. GM said that the sequence of procedures for positioning the dummy for the chest pendulum impact test was ambiguous since it called for the test dummy to be adjusted so that the area on the chest of the dummy immediately adjacent to the impact point is vertical. However, that surface of the dummy is curved and has variable radii. GM also pointed out that when the dummy is moved to the more vertical position, the area that a pendulum strikes the dummy also moves so that the portion of the test dummy's chest which is too rigid might be impacted. NHTSA has changed the dummy's positioning procedures so that a plane tangent to the surface of the chest immediately adjacent to the designated impact area is vertical. The positioning of the pendulum is also changed to ensure that the pendulum consistently strikes the chest at the designated point on the chest.

GM also raised questions about the positioning of the pendulum for the head calibration impact tests. The proposed requirement specified that the impact point for the pendulum was to be measured relative to the top of the dummy's head. GM said that because of differences in the thickness and shape of the dummy's skin, the location of the impact point can vary. GM recommended determining the impact point relative to the head center of gravity reference pins which protrude through the test dummy's skin.

NHTSA has evaluated GM's proposed head impact positioning procedure and decided to adopt a modified version of it. A measurement made from the head center of gravity pins will be used to determine the head impact point to ensure that all test dummies will be struck in the same location during the head impact tests.

GM said that the lumbar spine calibration test was ambiguous because it did not specify either the direction in which the force was to be applied to the lumbar spine or the location on the spine which is to be used to define the direction of force application. GM also pointed out that the procedures erroneously set requirements for femur friction plungers which are not included in the 3-year-old test dummy. NHTSA has corrected the test procedures to specify the direction of force application and deleted the reference to friction plungers.

GM also criticized ambiguities in the specification for the amount of chest deflection. NHTSA has reevaluated the need for a chest deflection specification and has decided to eliminate the requirement, since the chest acceleration test should serve as an adequate calibration test of the dummy's chest.

Repeatability

Ford, GM and the Motor Vehicle Manufacturers Association (MVMA) raised questions about the ability of the 3-year-old test dummy to give repeatable results in crash testing. MVMA proposed that the agency conduct another series of tests to determine the amounts of variances in test results between the same dummy in several tests and between different dummies in the same tests.

MVMA and Ford also recommended that the additional testing also include testing of the proposed Economic Commission for Europe (ECE) test dummy to determine if it would be an objective test device. The agency has not conducted an evaluation of the ECE test dummy since there are no calibration requirements for that test dummy. Without calibration requirements, there is no means to ensure the accuracy of the measurements obtained by the test dummy and therefore it cannot be used as an objective test device.

The agency has already conducted three separate research programs to evaluate the 3-year-old test dummy as an objective test device. As explained below, those programs have shown that the test dummy is an objective device that produces repeatable test results.

During 1977-78, the agency had simultaneous research programs conducted at the University of Michigan's Highway Safety Research Institute and NHTSA's Vehicle Research and Test Center in East Liberty, Ohio to develop and evaluate the calibration performance requirements and test procedures for the 3-year-old test dummy. Four of the 3-year-old test dummies were used in the testing program. Two of the dummies were tested by one laboratory and the other two were tested by the other laboratory. Then the two sets of test dummies were exchanged by the laboratories and subjected to the same calibration tests. By setting up the research program in this manner, the agency was able to determine if the test procedures and calibration performance requirements were repeatable from test dummy to test dummy and from test laboratory to test laboratory. The test results from both research programs showed that the calibration test procedures and performance requirements produced repeatable results.

The repeatability of the test dummy was reaffirmed in further testing conducted between June 1978 and July 1979 at Calspan Corporation. In that research program, four of the 3-year-old test dummies were used with two different types of child restraints—one shield type (Chrysler Mopar) and one plastic shell with integral harness type (GM Love Seat). Each of the four test dummies was subjected to six sled tests at 30 mph in both types of child restraints. The harness type restraint was also subjected to 3 sled tests at 20 mph with the top tether strap unattached.

To determine the repeatability of the test dummies, the head and chest accelerations and the amounts of head and knee excursion experienced by the test dummies were analyzed. That analysis showed that the amount of deviation measured by the same dummy in the different tests was small and similar in nature to the results obtained with Part 572 test dummies representing adults, which have been established as objective test devices.

In addition to examining the results obtained for the same dummy in different tests, the research program also examined the results for each of the four 3-year-old dummies in the same test. Based on previous testing of test dummies representing adults, it was determined that if the absolute deviation of the observed test results for each performance criteria, such as head acceleration, was less than six percent from the mean results, then the dummies had sufficient repeatability. In all but one of the test results, the deviation from the mean was less than six percent. The single exception involved the amount of chest acceleration measurered in the test dummies in the 20 mph tests of an untethered harness-type restraint. In that instance the deviation was only 7.7 percent. The reason for the variation in that test is probably due to the increased movement of the seat because the tether strap was unattached, rather than due to any variability in the test dummy.

Costs

The agency has considered the economic and other impacts of this final rule and determined that this rule is not significant within the meaning of Executive Order 12044 and the Department of Transportation's policies and procedures for implementing that order. The agency's assessment of the benefits and economic consequences of this final rule are contained in a regulatory evaluation which has been placed in the docket. Copies of that regulatory evaluation can be obtained by writing to NHTSA's docket section at the address given in the beginning of this notice.

The cost of the infant test dummy is estimated to be approximately $1,000. The 3-year-old test dummy should cost approximately $4,000. The materials used in the dummies are commercially obtainable. The availability of the test dummy drawing and other specifications means that any manufacturer can produce its own test dummy and does not have to purchase the test dummy from an independent test dummy manufacturer.

Strollee, a child restraint manufacturer, and the Juvenile Products Manufacturers Association asked the agency to reconsider the calibration requirements set for the 3-year-old dummy. They argued that the cost of calibrating the test dummy is approximately $800 to $1,100. Combined with the cost of the sled testing, each test of a car seat could cost approximately $2,000–$3,500. Such costs "would certainly discourage a manufacturer from testing frequently," Strollee said.

The calibration requirements set by this final rule are essential to ensure that the test dummy is an objective test device that will produce repeatable results in dynamic sled tests. So that the requirements would be practicable, the agency established the minimum number of calibration tests possible which would still ensure that the test dummy is properly constructed and properly instrumentated. Each manufacturer, in the exercise of due care, must determine how frequently it will calibrate its test dummy and how frequently it will run tests to determine its child restraint's compliance with Standard No. 213.

In its own testing, the agency has used some test dummies in as many as 15 tests over a 2-3 week period without recalibrating them and has not found any difference in their performance. With other test dummies, the agency has found it necessary to recalibrate them after several tests. However, in its compliance testing the agency will use properly calibrated dummies.

The principal authors of this notice are Vladislav Radovich, Office of Vehicle Safety Standards, and Stephen Oesch, Office of Chief Counsel.

In consideration of the foregoing, Part 572, *Anthropomorphic Test Dummies*, of Title 49 of the Code of Federal Regulations is amended as follows:

1. A new subsection (c) is added . . . Subpart A—General, Section 572.4 Terminology (49 CFR 572.4) to read as follows:

(c) The term "unimodal", when used in Subpart C, refers to an acceleration-time curve which has only one prominent peak.

2. A new Subpart C—Three Year Old Child, is added . . . .

Issued on December 20, 1979.

Joan Claybrook
Administrator

44 F.R. 76527
December 27, 1979

PREAMBLE TO AN AMENDMENT TO PART 572

Anthropomorphic Test Dummies
(Docket No. 78-9, Notice 5; Docket No. 73-8, Notice 9)

ACTION: Final rule.

SUMMARY: This notice amends Part 572, Anthropomorphic Test Dummies, to allow the use of an alternative chemical foaming agent for molding the dummy's flesh parts. In response to a Ford petition, the notice also makes a minor technical amendment to modify one specification in the calibration procedures for the neck of the test dummy representing a 50th percentile male. The effect of the latter amendment is to simplify the calibration test.

DATES: The amendment is effective on June 16, 1980.

ADDRESSES: Petitions for reconsideration should refer to the docket numbers and be submitted to: Docket Section, Room 5108, National Highway Traffic Safety Adminstration, 400 Seventh Street, S.W., Washington, D.C. 20590. (Docket hours: 8:00 a.m. to 4:00 p.m.)

FOR FURTHER INFORMATION CONTACT:
    Mr. Vladislav Radovich, Office of Vehicle
    Standards, National Highway
    Traffic Safety Administration,
    400 Seventh Street, S.W.,
    Washington, D.C. 20590 (202-426-2264)

SUPPLEMENTARY INFORMATION: This notice amends Part 572, Anthropomorphic Test Dummies, to modify the design specification for molding the test dummy's flesh parts to allow the use of an alternative chemical foaming agent, "OBSH/TBPP," to the currently specified "Nitrosan." In response to a petition from the Ford Motor Company, the agency is also making a minor technical amendment to simplify the calibration test for the neck used in the 50th percentile male test dummy. The amendment deletes the current specification and substitutes the specification used in the calibration testing of the recently issued three-year-old child test dummy (44 FR 76527,

December 27, 1979).

The agency published the proposed changes to the flesh molding and neck calibration specifications in the *Federal Register* of December 18, 1978 (43 FR 58843). Only one party, Ford Motor Co., commented on the proposed changes and Ford supported the adoption of both proposed changes.

Molding Specifications

The agency proposed the changes in the molding specification because the sole manufacturer of "Nitrosan," the currently specified chemical foaming agent, has discontinued its production due to the hazardous propensities of the compound during its manufacturing process. Based on an extensive research program to develop and test new chemical foaming agents (which was fully described in the notice of proposed rulemaking), the agency found that test dummy flesh parts made from "OBSH/TBPP" have comparable material properties to those produced with "Nitrosan" and are superior in some respects. Based on an evaluation of the research results, the agency concludes that flesh parts produced from "OBSH/TBPP" can be used for all purposes for which test dummies are required by the applicable safety standards and the dummy performance will be equivalent to the performance of dummies produced with "Nitrosan." Therefore, the agency is amending the regulation to allow the use of "OBSH/TBPP."

Drawings and specifications outlining the formulations for molding dummy flesh parts with the "OBSH/TBPP" compound are available for examination in NHTSA Docket 73-8 and Docket 78-9, Room 5108, 400 Seventh Street, S.W., Washington, D.C. 20590. Copies of these drawings may also be obtained from the Keuffel and Esser Company, 1513 North Danville Street, Arlington, Virginia 22201.

Neck Calibration Requirements

In response to a request from Ford, the agency

proposed an amendment to the pendulum impact test specification established in section 572.7(b) for the calibration of the 50th percentile male test dummy. The amendment would have replaced the current specification with the specification for calibration testing established for the 3-year old child test dummy.

The pendulum neck test found in Subpart B of the standard for the 50th percentile male dummy is intended to measure the bending properties of the dummy's neck. The current test specifies that, during the neck bending procedure, the pendulum shall not reverse direction until "T = 123 ms." This means that from the $t_i m_e$ the pendulum contacts the arresting material which it must strike, the pendulum cannot reverse direction for 123 milliseconds. The original intent of this requirement was to negate the effects of arresting material having rebound characteristics that could force the pendulum to reverse its motion before the bending properties of the neck could be measured. Ford requested a change in this specification because in certain instances the use of a special apparatus may be required to hold the pendulum arm for at least 123 milliseconds after the pendulum has impacted the arresting material.

Research by NHTSA and the industry has shown that when appropriate crushable materials are used in pendulum impact tests, the pendulum does not reverse its motion until the neck has straightened out and the head's center of gravity has returned to its original zero-time position relative to the pendulum. At that time, all measurements of the neck bending characteristics are completed and the pendulum's motion thereafter is inconsequential. In light of this research, the recent addition of Subpart C to Part 572, specifying requirements for the 3-year-old child dummy, modified the language concerning reversal of the pendulum arm during the neck impact test. Section 572.17 of that subpart specifies that "the pendulum shall not reverse direction until the head's center of gravity returns to the original zero time position relative to the pendulum arm." Under this requirement, a dummy user could only use an arresting material for the impact test whose rebound characteristics would not overcome the pendulum's inertia before the head and neck returned to the zero time position.

Since the specification in Subpart C of Part 572 represents a simplification of the pendulum

impact test specified in the current Subpart B, without any degradation of performance characteristics, the agency is amending section 572.7(b) of Subpart B to read as section 572.17(b) of Subpart C.

Costs

The agency has considered the economic and other impacts of this final rule and determined that this rule is not significant within the meaning of Executive Order 12044 and the Department of Transportation's policies and procedures for implementing that order. Based on that assessment, the agency has concluded also that the economic and other consequences of this proposal are so minimal that a regulatory evaluation is not necessary. The impact is minimal since there is no estimated increase in the cost of the test dummies due to the change in the foaming agent and neck calibration specification. In addition, the amendments would have no adverse environmental effects.

The engineer and lawyer primarily responsible for this notice are Vladislav Radovich and Stephen Oesch, respectively.

In consideration of the foregoing, Part 572, Anthropomorphic Test Dummies, of Title 49 of the Code of Federal Regulations is amended as follows:

1. Technical drawing ATD-6070 incorporated by reference in Section 572.15 of Subpart C – 3-Year-Old-Child is amended to add the formulation for "OBSH/TBPP" foaming compound.

2. Technical drawing ATD-7151 incorporated by reference in Section 572.5 of Subpart B – 50th Percentile Male is amended to add the formulation for "OBSH/TBPP" foaming compound.

3. The last sentence of Section 572.7(b) of Subpart B – 50th Percentile Male is amended to read:
   "The pendulum shall not reverse direction until the head's center of gravity returns to the original zero time position relative to the pendulum arm."

Issued on June 9, 1980.

Joan Claybrook
Administrator

45 FR 40595
June 16, 1980

PREAMBLE TO AN AMENDMENT TO PART 572

Anthropomorphic Test Dummies Representing 6-month-old and 3-year-old Children

(Docket No. 78-09; Notice 6)

ACTION: Response to petition for reconsideration.

SUMMARY: This notice grants in part and denies in part a General Motors (GM) petition for reconsideration of the 3-year-old test dummy requirements set in Part 572, Anthropomorphic Test Dummies. GM said it could not calibrate its test dummies because of resonances in the dummies, which prevent accurate acceleration measurements. NHTSA found that GM's calibration problems are due to its failure to comply with all of the design specifications set for the dummy and its use of single axis rather than triaxial accelerometers. In another notice in today's *Federal Register* the agency is proposing to require the use of triaxial accelerometers. This notice also corrects typographical errors in the final rule.

DATES: The amendments are effective on June 26, 1980.

FOR FURTHER INFORMATION CONTACT:

Mr. Vladislav Radovich, Office of Vehicle Safety Standards, National Highway Traffic Safety Administration, 400 Seventh Street, S.W., Washington, D.C. 20590 (202-426-2264)

SUPPLEMENTARY INFORMATION: On December 27, 1979, NHTSA published in the *Federal Register* a final rule amending Part 572, Anthropomorphic Test Dummies, to establish specifications and performance requirements for two test dummies, one representing a 6-month-old child and the other representing a 3-year-old child (44 FR 76527). The dummy is used in testing child restraint systems in accordance with Federal Motor Vehicle Safety Standard No. 213, Child Restraint Systems. General Motors (GM) timely filed a petition for reconsideration concerning the specifications and performance requirements set

for the test dummy representing a 3-year-old child. No other petitions were filed and GM raised no issues concerning the specifications set for the test dummy representing a 6-month-old child.

In its petition, GM again argued that the 3-year-old test dummy is not an objective test device for acceleration measurement because of resonances in the test dummy. GM requested the agency not to use the dummy as an acceleration measurement device until the resonances are eliminated.

GM also asked the agency to revise its accelerometer specifications to require the axes of triaxial accelerometers to intersect at a single point. GM said the change would reduce possible variability between different types of accelerometers. In addition, GM requested a further change in the lumbar spine test procedures to permit the use of either a pull or a push force during the spine calibration tests.

GM also raised questions about the possible use of different signal filtering techniques at different test laboratories. GM said that the use of different filters might account for differences between its testing and testing done for the agency.

NHTSA has evaluated GM's comments and the agency's responses to GM's petition are discussed below. All requests that are not specifically granted below are denied.

Signal Filtering

GM argued that one of the possible reasons for the differences between the test dummy head calibration test results at GM and other laboratories was the use of incorrect filters (devices used in the electronic processing of the acceleration measurements) by some laboratories. Part 572 requires the acceleration measurements to be filtered according to the Society of Automotive Engineers Recommended Practice J211a. Both Calspan Corporation and the agency's Vehicle Research and Test Center (VRTC), which did

PART 572—PRE 31

testing for NHTSA, used the required filter and instrumented their test dummies with triaxial accelerometers. The test results at VRTC were all within the limits set by the agency.

The Calspan test results originally reported to the agency were also within the limits. In rechecking its data, however, Calspan determined that it had made an error in calculating the peak resultant accelerations in the head calibration test. The corrected data showed that in one of the four head calibration tests the peak resultant acceleration was 116 g's, which exceeds the 115 g limit set in Part 572. To evaluate possible variability in the processing of the data by different laboratories, the agency also had HSRI and VRTC process the Calspan data. For the tests which exceeded the calibration limit, there was little variability between the different laboratories, with HSRI measuring 118 g's and VRTC measuring 117.4 g's.

The dummies Calspan used in the calibration testing were subsequently used in sled tests of child restraint systems. In the sled tests, the dummies provided consistent and repeatable acceleration measurements. Since dummies that experience 118 g's in the head calibration test can provide consistent and repeatable acceleration measurements, the agency, in a separate notice appearing in today's *Federal Register*, is proposing to increase the head resultant acceleration calibration limit from 115 to 118 g's.

NHTSA has found that the University of Michigan's Highway Safety Research Institute (HSRI), which instrumented its dummies with single axis accelerometers, did not use the filter required by Part 572, but instead used a filter that deviates from the required filter. To determine whether the use of the HSRI filter made a difference in the calibration tests conducted by that laboratory, the agency had HSRI process the accelerations recorded during its head calibration tests with the correct filter. Using the correct filter, HSRI found that in five of the eighteen head calibration tests the peak resultant acceleration exceeded the limits set in Part 572. In those five tests, the peak resultant acceleration ranged from 115.9 to 119.1 g's.

The peak resultant accelerations and the shape of the acceleration pulses in the HSRI tests that exceeded the calibration limit were smaller than and not the same shape as the measurements made by GM in its tests, which also used test dummies instrumented with single axis accelerometers. In the two sets of data submitted by GM to the docket, the peak resultant accelerations ranged from 119 to 130 g's. In addition, the shape of the GM head acceleration pulse was different than the pulses measured in all the testing done for the agency. In the GM acceleration pulse, there is a brief secondary peak after initial peak is reached. Based on the agency's testing of adult test dummies, such secondary peaks are usually indications of accelerometer vibration resulting from improper installation.

The differences between the GM testing and the testing done for the agency is not attributable to the use of different filters. When all the test data is filtered as specified in the standard, the peak resultant accelerations measured by GM are still greater than those obtained at the other three laboratories. As explained below, use of triaxial accelerometers, rather than the single axis accelerometers used by GM and HSRI, will provide repeatable, complying results in the head calibration test.

Instrumentation

Part 572 allows the use of two different types of accelerometers (single axis and triaxial) in the test dummy and sets different axis intersection requirements for each type of accelerometer. GM asked the agency to apply the axis intersection requirements set for single axis accelerometers to triaxial accelerometers. It said such a requirement would reduce the variability in test measurements resulting from use of different types of accelerometers.

The agency's testing has demonstrated that variability can be sufficiently controlled by use of the existing specification with a triaxial accelerometer. Testing done by GM has also shown that the test dummy can be properly calibrated with triaxial accelerometers. When GM tested one of the agency's test dummies with GM's accelerometer mounting place and single axis accelerometers, the peak lateral accelerations measured in the test dummy's head exceeded the limits currently set in the regulation. Yet when GM tested the same test dummy equipped with triaxial accelerometers placed on the mounting plate required by the design specifications, the test dummy easily met the calibration requirements. Therefore, rather than adopt GM's proposal, the agency is proposing, elsewhere in today's *Federal Register*, to require the use of only triaxial accelerometers.

Resonances

GM said that "the consistent lack of correlation between dummy tests at General Motors and at other laboratories" was attributable to resonances in the test dummy. It said the dummy could not be used as an objective test device until the resonances were eliminated. As explained previously, the variability between different test laboratories can be controlled by the use of triaxial accelerometers.

One reason for the "resonances" in the GM test results may be GM's failure to use dummies that fully comply with the agency's design specifications. The agency's review of some of the blueprints used in the construction of the GM test dummies revealed that GM did not use the accelerometer mounting plate required by the NHTSA design specifications. The mounting plate used by GM was smaller and presumably lighter than the plate specified by the agency. Use of a smaller and lighter plate may have also contributed to the higher acceleration readings obtained by GM.

Thus, the agency denies GM's request not to use the dummy for acceleration measurement and concludes that the 3-year-old test dummy instrumented with triaxial accelerometers is an objective test device for measuring accelerations in child restraints.

Spine Calibration

The calibration requirements for the lumbar spine of the test dummy specify the amount of flexion the spine must experience when force is applied to it. The calibration procedures specify that the applied force is to be applied as a pull force. GM requested the agency to permit the use of a "push" force saying that it "is more convenient to apply in some test set-ups."

When the agency developed the spine calibration tests, both pull and push forces were used to apply force to the spine. However, the testing done by the Highway Safety Research Institute (HSRI) found that use of a push force "proved to be awkward and inconsistent." HSRI also found that use of a pull force was simpler procedure and provided consistent data. Based on the HSRI

testing, the agency has decided to deny GM's request since the use of a pull force provides a simple, repeatable method to measure compliance.

Corrections

In the final rule issued on December 12, 1979, NHTSA amended the instrumentation requirements for the chest to more specifically define several of the accelerometers mounting locations. The revised specifications inadvertently reversed two of the axis mounting locations in the chest. The specifications have been amended in this notice to correct that error.

The test procedure for conducting the head impact test set forth in the final rule contained a typographical error. The tolerance for positioning the test probe was listed as ± 1.1 inches. The regulation has been amended in this notice to specify the correct tolerance of ± 0.1 inches.

The performance requirement for the neck calibration test was incorrectly listed as 84 degrees ± 18 degrees rather than the correct figure of 84 degrees ± 8 degrees. The necessary corrections have been made in this notice to the regulation.

The principal authors of this notice are Vladislav Radovich, Office of Vehicle Safety Standards, and Stephen Oesch, Office of Chief Counsel.

In consideration of the foregoing, Subpart C — 3-Year-Old Child of Part 572, Anthropomorphic Test Dummies, of Title 49 of the Code of Federal Regulations, is amended as follows:

1. Section §572.1(c)(2) is amended to read as follows:

(2) Adjust the test probe so that its longitudinal centerline is at the forehead at the point of orthogonal intersection of the head midsagittal plane and the transverse plane which is perpendicular to the "Z" axis of the head (longitudinal centerline of the skull anchor) and is located 0.6 ± 0.1 inches above the centers of the head center of gravity reference pins and coincides within 2 degrees with the line made by the intersection of horizontal and midsagittal planes passing through this point.

2. The first sentence of section §572.17(b) is amended to read as follows:

(b) When the head-neck assembly is tested in accordance with paragraph (c) of this section, the head shall rotate in reference to the pendulum's longitudinal centerline a total of 84 degrees ± 8 degrees about its center of gravity, rotating to the extent specified in the following table at each indicated point in time, measured from impact, with the chordal displacement measured at its center of gravity.

3. Section §572.21(c) is amended to read as follows:

(c) Accelerometers are mounted in the thorax on the mounting plate attached to the vertical transverse bulkhead shown in the drawing subreferenced under assembly No. SA 103C 030 in drawing SA 103C 001 so that their sensitive axes are orthogonal and their seismic masses are positioned relative to the axial intersection point located in the midsagittal plane 3 inches above the top surface of the lumbar spine and 0.3 inches dorsal to the accelerometer mounting plate surface. Except in the case of triaxial accelerometers, the sensitive axes shall intersect at the axial intersection point. One accelerometer is aligned with its sensitive axis parallel to the vertical bulkhead and midsagittal planes, and with its seismic mass center at any distance up to 0.2 inches to the left, 0.1 inches inferior and 0.2 inches ventral of the axial intersection point. Another accelerometer is aligned with its sensitive axis in the transverse horizontal plane and perpendicular to the midsagittal plane and with its seismic mass center at any distance up to 0.2 inches to the right, 0.1 inches inferior and 0.2 inches ventral to the axial intersection point. A third accelerometer is aligned with its sensitive axis parallel to the midsagittal and transverse horizontal planes and with its seismic mass center at any distance up to 0.2 inches superior, 0.5 inches to the right and 0.1 inches ventral to the axial intersection point. In the case of a triaxial accelerometer, its axes are aligned in the same way that the axes of three separate accelerometers are aligned.

Issued on June 17, 1980.

Joan Claybrook
Administrator

45 FR 43352
June 17, 1980

PREAMBLE TO AN AMENDMENT TO PART 572

Anthropomorphic Test Dummies
(Docket No. 78-09; Notice 8)

ACTION: Response to petitions for reconsideration, final rule and correction.

SUMMARY: This notice amends Subpart C of Part 572, Anthropomorphic Test Dummies, to specify the use of a triaxial accelerometer in the test dummy representing a 3-year-old child. The use of a triaxial accelerometer will eliminate calibration problems associated with single axis accelerometers. The notice also denies petitions filed by Ford Motor Company and General Motors Corporation seeking reconsideration of the agency's June 26, 1980 notice responding to a prior General Motors Corporation petition for reconsideration. Finally, the notice corrects a typographical error in the agency's June 26, 1980 final rule.

DATES: The amendments are effective on December 15, 1980.

ADDRESSES: Petitions for reconsideration should refer to the docket number and be submitted to: Docket Section, Room 5108, National Highway Traffic Safety Administration, 400 Seventh Street, S.W., Washington, D.C. 20590.

FOR FURTHER INFORMATION CONTACT:

Mr. Vladislav Radovich, Office of Vehicle Safety Standards, National Highway Traffic Safety Administration, 400 Seventh Street, S.W., Washington, D.C. 20590 (202-426-2264)

SUPPLEMENTARY INFORMATION: This notice amends Subpart C of Part 572, Anthropomorphic Test Dummies, to change several of the requirements for the test dummy representing a 3-year-old child. The test dummy is used in testing child restraint systems in accordance with Federal Motor Vehicle Safety Standard No. 213, Child Restraint Systems.

The notice amends Subpart C of Part 572 to specify the use of triaxial accelerometers, instead of single axis accelerometers, in the head and chest of the test dummy. In addition the notice increases the upper limit for permissible resultant acceleration in the head calibration test from 115 g's to 118 g's. The agency published a notice proposing these changes in the *Federal Register* for June 26, 1980 (45 FR 43355). Only two parties, Ford Motor Company (Ford) and General Motors Corporation (GM), submitted comments on the proposal. The final rule is based on the data submitted in those comments, data obtained in the agency's testing and data obtained from other pertinent documents. Significant comments submitted to the docket are addressed below.

This notice also denies petitions filed by Ford and GM seeking reconsideration of the agency's June 26, 1980 notice (45 FR 43352) that granted in part and denied in part a prior GM petition for reconsideration.

Finally, this notice corrects a typographical error in an amendment made in the agency's June 26, 1980 notice (45 FR 43352) responding to a prior GM petition for reconsideration.

Resonances

Ford and GM both agree with the agency that the test dummy representing a 3-year-old child is an objective test device for measuring the amount of head and knee excursion that occurs in child restraint system testing using the test dummy. The fundamental disagreement stated in the Ford and GM comments and petitions for reconsideration is whether the test dummy is an objective test device for measuring accelerations in the dummy's head and chest during child restraint testing. GM argues that the test dummy is not an objective device because of the presence of resonances in the head and chest of the test dummy. Ford says that the test dummy "may be a suitable measuring device, when there is no head impact (such as in a shoulder harness type of child restraint)" during child restraint testing. It, however, argues that if there is a head impact in the child restraint testing, then the test dummy's head will resonate.

Ford and GM both argue that the resonances can reinforce or attenuate the measurement of impact forces on the test dummy. Thus, if the test dummy does resonate, the acceleration measured in the test dummy may not represent the actual forces experienced by the test dummy.

Ford argues that the source of the resonance is an oscillation of the urethane skull of the test dummy. Ford included with its petition and comments on the June 26, 1980 proposal the results of several tests in which it struck the head of the test dummy with a rubber mallet. Ford said that regardless of the direction of the impact, the head resonated with a frequency of approximately 200 Hertz (Hz) when it was struck.

The agency has reviewed the Ford and other test data and concluded that the test dummy is an objective test device that can be used for measuring accelerations. As explained below, the agency's conclusion is based on an analysis of the structure of the test dummy's head and chest and the relationship between that structure and the impact response of the test dummy.

Many physical structures, such as the test dummy's head, have a natural or resonating frequency at which they will vibrate when they are driven by a force of the same frequency. When resonance occurs, small variations in the applied force can produce large variations in the measured acceleration, thus preventing accurate measurement of the acceleration. The resonance, however, will not occur if the driving force is of a frequency that is below the natural or resonating frequency of the object being struck.

Analysis of the test dummy shows that the natural or resonating frequency of the head is approximately 128 Hz, while the natural frequency of the accelerometer attachment in the test dummy's head is approximately 255 Hz. The natural resonating frequencies of the test dummy's chest and chest accelerometer attachment are approximately 85 Hz and 185 Hz.

Impacts with hard and unyielding objects, such as the unpadded portion of a car's instrument panel, can create high frequencies, generally up to 1,000 Hz. Impacts with soft and yielding surfaces, such as a padded child restraint, create low frequencies, generally less than 50 Hz.

The test used in Standard No. 213 to evaluate child restraints does not include impacts with hard and unyielding surfaces. In Standard No. 213 testing, the child restraint is placed on a vehicle seat and attached by a lap belt. There is no portion of a vehicle's interior, such as an instrument panel, placed in front of or to the side of the vehicle seat. Thus, during the testing, the dummy will contact the belts or padded surfaces of the child restraint. Since the belts and padded surfaces are yielding and energy-absorbing, contact with them will involve impacts where the frequencies are well below the natural or resonating frequency of the test dummy's head and chest.

Ford raised the issue of whether contact between the head and arms of the dummy during the testing might produce frequencies that will cause the test dummy's head to resonate. Ford said that it had experienced dummy head and arm contact in some of its tests and resonance occurred.

The agency has conducted more than 150 tests of child restraint systems. There have only been 2 tests in which the head of the test dummy struck the toes and resonances occurred. The head-limb contact occurred in those tests because of massive structural failures in the child restraint system.

Although resonances did occur when the head struck the toes, the validity of the acceleration measurement in those tests is irrelevant for determining if the child restraint complied with Standard No. 213, Child Restraint Systems. The structural failure is, by itself, a violation of the standard. The agency had not found head and limb contact affecting acceleration measurements in any child restraint that maintained its structural integrity during the testing.

In the past several years, the agency has conducted 10 tests of the Ford TOT GUARD. In one of those tests, the arm briefly touched the head, but there was no effect on the acceleration measurement. The dummy in those tests was positioned in accordance with the test procedure set out in Standard No. 213. Since the test procedure permits the limbs to be positioned so that they will not inhibit the movement of the head or torso the agency looked at the effect of positioning the dummy's arm in different locations on the shield or the side of the TOT GUARD. None of the different arm positions resulted in head to arm contact affecting acceleration measurement.

Triaxial Accelerometers

Part 572 currently allows the use of either triaxial accelerometers or single axis accelerometers to measure accelerations in the head and chest of the 3-year-old child test dummy. The June 26, 1980 notice (45 FR 43355) proposed specifying the use of only triaxial accelerometers in the test dummy to eliminate calibration problems caused by single axis accelerometers. The agency proposed only using triaxial accelerometers after GM was unable to calibrate its test dummies with single axis accelerometers. In GM's head calibration tests, the peak resultant acceleration exceeded the upper limit set by the regulation.

GM agreed that use of a triaxial accelerometer "may reduce the possibility of exceeding the peak acceleration in the dummy calibration test." It, however, argued that the use of triaxial accelerometers will not solve the problem of resonance. As previously explained, the types of impacts experienced in child restraint testing will not produce resonances. The purpose of requiring the use of triaxial accelerometers is to enable manufacturers to calibrate consistently their test dummies within the acceleration limits set in the regulation.

Ford argued that single axis accelerometers are easier to work with, more reliable and more easily repaired than triaxial accelerometers. The agency is not aware of any data, and Ford supplied none, indicating that triaxial accelerometers are less reliable than single axis accelerometers. Contrary to Ford's assertion, a triaxial accelerometer should be easier to use. The axes and seismic mass center of the triaxial acceleromter (Endevco model 7267C-750) currently used in dummy testing are permanently fixed in a mounting block. With single axis accelerometers, three separate accelerometers must be positioned by each user on a mounting block in order to instrument the dummy. Thus the possibility of variation in mounting location between different users is increased by the use of single axis accelerometers.

Single axis accelerometers are more readily repairable than triaxial accelerometers. The agency, however, has used triaxial accelerometers in

numerous dummy tests for several years and has found that their repair experience is comparable to single axis accelerometers.

Based on all these considerations, the agency has decided to adopt the triaxial accelerometer requirement as proposed.

Calibration Limit

To accommodate minor variation in test measurements between different test laboratories, the agency's June 26, 1980 notice (45 Fr 43355) proposed to slightly increase the permissible resultant acceleration limit for the head calibration test from 115 g's to 118 g's. Neither Ford nor GM opposed this change, so the agency is adopting it as proposed. Although the agency is expanding the upper limit of the calibration range, experience with the Part 572 adult test dummy has shown that manufacturers will develop production techniques to produce test dummies that have acceleration responses that fall within the middle of the specified calibration range.

Correction

The final rule established by the agency's June 26, 1980 notice (45 FR 43352) amended the head calibration head test procedures. The notice inadvertently made the amendment to section 572.1(c)(2) of Part 572 instead of to section 572.16(c)(2). This notice corrects that typographical error and makes the amendment to section 572.16(c)(2).

Costs

The agency has considered the economic and other impacts of this final rule and determined that this rule is not significant within the meaning of Executive Order 12221 and the Department of Transportation's policies and procedures implementing that order. Based on that assessment, the agency has concluded that the economic and other consequences of this rule are so minimal that a regulatory evaluation is not necessary. The impact is minimal since the primary effect of this rule is to bind the agency to using one of the two types of accelerometers formerly permitted by the regulation. The economic impact on manufacturers choosing to purchase triaxial accelerometers needed to instrument the dummy is approximately $2,500.

The agency finds, for good cause shown, that it is in the public interest that the amendments made by this notice have an immediate effective date. The immediate effective date is needed since the test dummy will be used in conducting compliance tests for Standard No. 213, Child Restraint Systems, which goes into effect on January 1, 1981.

The engineer and lawyer primarily responsible for this notice are Vladislav Radovich and Stephen Oesch, respectively.

In consideration of the foregoing, Subpart C of Part 572, Anthropomorphic Test Dummies, of Title 49 of the Code of Federal Regulations is revised to read as follows:

1. The first sentence of section 572.16(b) is revised to read as follows:

(b) When the head is impacted in accordance with paragraph (c) of this section by a test probe conforming to §572.21(a) at 7 fps., the peak resultant acceleration measured at the location of the accelerometer mounted in the headform in accordance with §572.21(b) shall be not less than 95g and not more than 118g.

2. Section 572.21(b) is revised to read as follows:

(b) A triaxial accelerometer is mounted in the head on the mounting block (A/310) located on the horizontal transverse bulkhead as shown in the drawings subreferenced under assembly SA 103C 010 so that its seismic mass centers are positioned as specified in this paragraph relative to the head accelerometer reference point located at the intersection of a line connecting the longitudinal centerlines of the transfer pins in the sides of the dummy head with the midsagittal plane of the dummy head. The triaxial accelerometer is aligned with one sensitive axis parallel to the vertical bulkhead and midsagittal plane and its seismic mass center is located 0.2 inches dorsal to and 0.1 inches inferior to the head accelerometer reference point. Another sensitive axis of the triaxial accelerometer is aligned with the horizontal plane and is perpendicular to the midsagittal plane and its seismic mass center is located 0.1 inch inferior to, 0.4 inches to the right of and 0.9 inch dorsal to the head accelerometer reference point. The third sensitive axis of the triaxial accelerometer is aligned so that it is parallel to the midsagittal and horizontal planes and its seismic mass center is located 0.1 inches inferior to, 0.6 inches dorsal to and 0.4 inches to the right of the head accelerometer reference point. All seismic mass centers shall be positioned within ± 0.05 inches of the specified locations.

3. Section 572.21(c) is revised to read as follows:

(c) A triaxial accelerometer is mounted in the thorax on the mounting plate attached to the vertical transverse bulkhead shown in the drawing subreferenced under assembly No. SA 103C 030 in drawing SA 103C 001 so that its seismic mass centers are positioned as specified in this paragraph relative to the thorax accelerometer reference point located in the midsagittal plane 3 inches above the top surface of the lumbar spine and 0.3 inches dorsal to the accelerometer mounting plate surface. The triaxial accelerometer is aligned so that one sensitive axis is parallel to the vertical bulkhead and midsagittal planes and its seismic mass center is located 0.2 inches to the left of, 0.1 inches inferior to and 0.2 inches ventral to the thorax accelerometer reference point. Another sensitive axis of the triaxial accelerometer is aligned so that it is in the horizontal transverse plane and perpendicular to the midsagittal plane and its seismic mass center is located 0.2 inches to the right of, 0.1 inches inferior to and 0.2 inches ventral to the thorax accelerometer reference point. The third sensitive axis of the triaxial accelerometer is aligned so that it is parallel to the midsagittal and horizontal planes and its seismic mass center is located 0.2 inches superior to, 0.5 inches to the right of and 0.1 inches ventral to the thorax accelerometer reference point. All seismic mass centers shall be positioned within ± 0.05 inches of the specified locations.

4. The document amending Subpart C—Three-Year-Old Child of Part 572, Anthropomorphic Test Dummies, of Title 49 of the Code of Federal Regulations published in the *Federal Register* of June 26, 1980 as 45 FR 43352 is corrected by changing the reference to "Section 571.1(c)(2)" made in the first amendment to the regulation set out on page 43353 to read "572.16(c)(2)."

Issued on December 8, 1980.

Joan Claybrook
Administrator

45 FR 82265
December 15, 1980

PREAMBLE TO AN AMENDMENT TO PART 572

Anthropomorphic Test Dummies
[Docket No. 85-05; Notice 1]

ACTION: Final rule.

SUMMARY: This document amends regulations concerning the National Highway Traffic Safety Administration's specifications for anthropomorphic test dummies by revising sections that state where copies of drawings may be obtained.

EFFECTIVE DATE: June 19, 1985.

SUPPLEMENTARY INFORMATION: The purpose of this notice is to amend Part 572 of Chapter V of Title 49, Code of Federal Regulations by revising §§ 572.5(a), 572.15(a)(1), and 572.25(a), which state where copies of drawings and a construction manual describing the materials and the procedures involved in the manufacturing of anthropomorphic dummies may ·be obtained. The amendment changes the supply source for the drawings and manual from Keuffel and Esser Company to Rowley-Scher Reprographics, Incorporated. This revision is required because of the sale of the Keuffel and Esser Company reproduction facilities to Rowley-Scher Reprographics, Incorporated.

The amendment to Part 572 as set forth below is technical in nature and does not alter existing obligations. This notice simply provides the correct address for obtaining copies of drawings and the construction manuals. The National Highway Traffic Safety Administration therefore finds for good cause that this amendment may be made effective without notice and opportunity for comment, may be made effective within 30 days after publication in the *Federal Register,* and is not subject to the requirements of Executive Order 12291.

In consideration of the foregoing, 49 CFR Part 572 is amended as follows:

1. In § 572.5, paragraph (a) is revised to read as follows: § 572.5 General description.

(a) The dummy consists of the component assemblies specified in Figure 1, which are described in their entirety by means of approximately 250 drawings and specifications that are grouped by component assemblies under the following nine headings:

SA 150 M070—Right arm assembly
SA 150 M071—Left arm assembly
SA 150 M050—Lumbar spine assembly
SA 150 M060—Pelvis and abdomen assembly
SA 150 M080—Right leg assembly
SA 150 M081—Left leg assembly
SA 150 M010—Head assembly
SA 150 M020—Neck assembly
SA 150 M030—Shoulder-thorax assembly

The drawings and specifications are incorporated in this Part by reference to the nine headings, and are available for examination in Docket 73-8, Room 5109, 400 Seventh Street, S.W., Washington, D.C., 20590. Copies may be obtained from Rowley-Scher Reprographics, Inc., 1216 K Street, N.W., Washington, D.C., 20005, attention Mr. Allan Goldberg and Mr. Mark Krysinski ((202) 628-6667). The drawings and specifications are subject to changes, but any change will be accomplished by appropriate administrative procedures, will be announced by publication in the *Federal Register,* and will be available for examination and copying as indicated in this paragraph. The drawings and specifications are also on file in the reference library of the *Federal Register,* National Archives and Records Services, General Services Administration, Washington, D.C.

* * * *

2. In § 572.15, paragraph (a) is revised to read as follows: § 572.15 General description.

(a) (1) The dummy consists of the component assemblies specified in drawing SA 103C 001, which are described in their entirety by means of approximately 122 drawings and specifications that are grouped by component assemblies under the following thirteen headings:

SA 103C 010 Head Assembly
SA 103C 020 Neck Assembly
SA 103C 030 Torso Assembly
SA 103C 041 Upper Arm Assembly Left
SA 103C 042 Upper Arm Assembly Right
SA 103C 051 Forearm Hand Assembly Left
SA 103C 052 Forearm Hand Assembly Right
SA 103C 061 Upper Leg Assembly Left
SA 103C 062 Upper Leg Assembly Right
SA 103C 071 Lower Leg Assembly Left
SA 103C 072 Lower Leg Assembly Right
SA 103C 081 Foot Assembly Left
SA 103C 082 Foot Assembly Right

The drawings and specifications are incorporated in this Part by reference to the thirteen headings and are available for examination in Docket 78-09, Rm 5109, 400 Seventh Street, S.W., Washington, D.C., 20590. Copies may be obtained from Rowley-Scher Reprographics, Inc., 1216 K Street, N.W., Washington, D.C., 20005, attention Mr. Allan Goldberg and Mr. Mark Krysinski ((202) 628-6667).
* * * * *

(3) An Operation and Maintenance Manual (dated May 28, 1976, Contract No. DOT-HS-6-01294) with instructions for the use and maintenance of the test dummies is incorporated in this Part by reference. Copies of the manual can be obtained from Rowley-Scher Reprographics, Inc. All provisions of this manual are valid unless modified by this regulation. This document is available for examination in Docket 78-09.
* * * * *

3. In § 572.25, paragraph (a) revised to read as follows: § 572.25 General description.

(a) The infant dummy is specified in its entirety by means of 5 drawings (No. SA 100I 001) and a construction manual which describe in detail the materials and the procedures involved in the manufacturing of this dummy. The drawings and the manual are incorporated in this Part by reference and are available for examination in Docket 78-09, Room 5109, 400 Seventh Street, S.W., Washington, D.C., 20590. Copies may be obtained from Rowley-Scher Reprographics, Inc., 1216 K Street, N.W., Washington, D.C., 20005, attention Mr. Allan Goldberg and Mr. Mark Krysinski ((202) 628-6667). The drawings and the manual are subject to changes, but any change will be accomplished by appropriate administrative procedures, will be announced by publication in the *Federal Register,* and will be available for examination and copying as indicated in this paragraph. The drawings and manual are also on file in the reference library of the *Federal Register,* National Archives and Records Services, General Services Administration, Washington, D.C.

Issued on April 17, 1985

Diane K. Steed
Administrator

50 F.R. 25422
June 19, 1985

PREAMBLE TO AN AMENDMENT TO PART 572

Anthropomorphic Test Dummies

(Docket No. 74-14; Notice 45)

ACTION: Final Rule.

SUMMARY: This notice adopts the Hybrid III test dummy as an alternative to the Part 572 test dummy in testing done in accordance with Standard No. 208, Occupant Crash Protection. The notice sets forth the specifications, instrumentation, calibration test procedures, and calibration performance criteria for the Hybrid III test dummy. The notice also amends Standard No. 208 so that effective October 23, 1986, manufacturers have the option of using either the existing Part 572 test dummy or the Hybrid III test dummy until August 31, 1991. As of September 1, 1991, the Hybrid III will replace the Part 572 test dummy and be used as the exclusive means of determining a vehicle's conformance with the performance requirements of Standard No. 208.

The notice also establishes a new performance criterion for the chest of the Hybrid III test dummy which will limit chest deflection. The new chest deflection limit applies only to the Hybrid III since only that test dummy has the capability to measure chest deflection.

These amendments enhance vehicle safety by permitting the use of a more advanced test dummy which is more human-like in response than the current test dummy. In addition, the Hybrid III test dummy is capable of making many additional sophisticated measurements of the potential for human injury in a frontal crash.

DATES: The notice adds a new Subpart E to Part 572 effective on October 23, 1986.

This notice also amends Standard No. 208 so that effective October 23, 1986, manufacturers have the option of using either the existing Part 572 test dummy or the Hybrid III test dummy until August 31, 1991. As of September 1, 1991, the Hybrid III will replace the Part 572 test dummy and be used as the exclusive means of determining a vehicle's conformance with the performance requirements of Standard No. 208. The incorporation by reference of certain publications listed in the regulation is approved by the Director of the Federal Register as of October 23, 1986.

SUPPLEMENTARY INFORMATION: In December 1983, General Motors (GM) petitioned the agency to amend Part 572, *Anthropomorphic Test Dummies*, to adopt specifications for the Hybrid III test dummy. GM also petitioned for an amendment of Standard No. 208, *Occupant Crash Protection*, to allow the use of the Hybrid III as an alternative test device for compliance testing. The agency granted GM's petition on July 20, 1984. The agency subsequently received a petition from the Center for Auto Safety to propose making Standard No. 208's existing injury criteria more stringent for the Hybrid III and to establish new injury criteria so as to take advantage of the Hybrid III's superior measurement capability. The agency granted the Center's petition on September 17, 1984. On April 12, 1985 (50 FR 14602), NHTSA proposed amendments to Part 572 and Standard No. 208 that were responsive to the petitioners and which, in the agency's judgment, would enhance motor vehicle safety. Twenty-eight individuals and companies submitted comments on the proposed requirements. This notice presents the agency's analysis of the issues raised by the commenters. The agency has decided to adopt the use of the Hybrid III test dummy and some of the proposed injury criteria. The agency has also decided to issue another notice on the remaining injury criteria to gain additional information about the potential effects of adopting those criteria.

This notice first discusses the technical specifications for the Hybrid III, its calibration requirements, its equivalence with the existing Part 572 test dummy, and the applicable injury criteria. Finally, it discusses the test procedure used to position the dummy for Standard No. 208 compliance testing and the economic and other effects of this rule.

Test Dummy Drawings and Specifications

Test dummies are used as human surrogates for evaluation of the severity of injuries in vehicle crashes. To serve as an adequate surrogate, a test dummy must be capable of simulating human impact responses. To serve as an objective test device, the test dummy must be adequately defined through technical drawings and performance specifications to ensure uniformity in construction, impact response, and measurement of injury in identical crash conditions.

Virtually all of the commenters, with the exception of GM, said that they have not had sufficient experience with the Hybrid III to offer comments on the validity of the technical specifications for the test dummy. Since the issuance of the notice, GM has provided additional technical drawings and a Society of Automotive Engineers-developed user's manual to further define the Hybrid III. These new drawings do not alter the basic nature of the test dummy, but instead provide additional information which will enable users to make sure that they have a correctly designed and correctly assembled test dummy. The user's manual provides information on the inspection, assembly, disassembly, and use of the test dummy. Having the user's manual available will assist builders and users of the Hybrid III in producing and using the test dummy. GM also provided information to correct the misnumbering of several technical drawings referenced in the notice.

In addition, the agency has reviewed the proposed drawings and specifications. While NHTSA believes the proposed drawings are adequate for producing the test dummy, the agency has identified and obtained additional information which should make production and use of the test dummy even more accurate. For example, the agency has obtained information on the range of motions for each moving body part of the test dummy. Finally, to promote the ease of assembly, NHTSA has made arrangements with GM to ensure that the molds and patterns for the test dummy are available to all interested parties. Access to the molds will assist other potential builders and users of the Hybrid III since it is difficult to specify all of the details of the various body contours solely by technical drawings.

The agency has adopted the new drawings and user manual in this rule and has made the necessary corrections to the old drawings. The agency believes that the available drawings and technical specifications are more than sufficient for producing, assembling, and using the Hybrid III test dummy.

Commercial Availability of the Hybrid III

A number of commenters raised questions about the commercial availability of the Hybrid III test dummy, noting problems they have experienced in obtaining calibrated test dummies and the instrumentation for the neck and lower leg of the Hybrid III. For example, Chrysler said that it had acquired two Hybrid III test dummies, but has been unable to obtain the lower leg and neck instrumentation for five months. Likewise, Ford said that it has been unable to obtain the knee displacement and chest deflection measurement devices for the Hybrid III. It also said that of the test dummies it had received, none had sufficient spine stiffness to meet the Hybrid III specifications. Ford claimed to have problems in retaining a stable dummy posture which would make it difficult to carry out some of the specified calibration tests. Subsequent investigation showed that the instability was caused by out-of-specification rubber hardness of the lumbar spine, and was eliminated when spines of correct hardness were used. In addition, Ford said that the necks and ribs of the test dummy would not pass the proposed calibration procedures. Finally, Ford said that the equipment needed for calibrating the dummy is not commercially available.

Although the commenters indicated they had experienced difficulty in obtaining the instrumentation for the Hybrid III's neck and lower legs, they did not indicate that there is any problem in obtaining the instrumentation needed to measure the three injury criteria presently required by Standard No. 208, the head injury criterion, chest acceleration, and femur loading and which are being adopted by this rule for the Hybrid III. For example, Volkswagen said it had obtained Hybrid III test dummies with sufficient instrumentation to measure the same injury criteria as with the Part 572. VW did say it had ordered the additional test devices and instrumentation for the Hybrid III but was told the instrumentation would not be available for six months.

The agency notes that there are now two commercial suppliers of the Hybrid III test dummy, Alderson Research Labs (ARL) and Humanoid Systems. Humanoid has built nearly 100 test dummies and ALR has produced five prototype test dummies as of the end of December 1985. Both manufacturers have indicated that they are now capable of producing sufficient Hybrid IIIs to meet the demand for those dummies. For example, Humanoid Systems said that while the rate of production is dependent on the number of orders, generally three test dummies per week are produced. Thus, in the case of the basic test dummy, there appears to be sufficient commercial capacity to provide sufficient test dummies for all vehicle manufacturers.

As to test dummy instrumentation, the agency is aware that there have been delays in obtaining the new neck, thorax, and lower leg instrumentation for the Hybrid III. However, as Humanoid commented, while there have been delays, the supplies of the needed parts are expected to increase. Even if the supply of the lower leg instrumentation is slow to develop, this will not pose a problem, since the agency is not adopting, at this time, the proposed lower leg injury criteria. In the case of the neck instrumentation, the supply problem should be minimized because each test facility will only need one neck transducer to calibrate all of its test dummies. The neck instrumentation will not be needed for a manufacturer's crash testing since at this time, the agency is not adopting any neck injury criteria. In the case of the instrumentation for measuring thoracic deflection, the supplier has indicated that it can deliver the necessary devices within 3 months of the time an order is placed. As to Ford's comment about calibration test equipment, the agency notes that current equipment used for calibrating the existing Part 572 test dummy can be used, with minor modification, to calibrate the Hybrid III test dummy.

Calibration Requirements

In addition to having complete technical drawings and specifications, a test dummy must have adequate calibration test procedures. The calibration tests involve a series of static and dynamic tests of the test dummy components to determine whether the responses of the test dummy fall within specified performance requirements for each test. The testing involves instrumenting the head, thorax and femurs to measure the test dummy's responses. In addition, there are tests of the neck, whose structural properties may have considerable influence on the kinematics and impact responses of the instrumented head. Those procedures help ensure that the test dummy has been properly assembled and that, as assembled, it will provide repeatable and reproducible results in crash testing. (Repeatability refers to the ability of the same test dummy to produce the same results when subjected to several identical tests. Reproducibility refers to the ability of one test dummy to provide the same results as another test dummy built to the same specifications.)

Lumbar Spine Calibration Test

The technical specifications for the Hybrid III set out performance requirements for the hardness of the rubber used in the lumbar spine to ensure that the spine will have appropriate rigidity. NHTSA's test data show that there is a direct relationship between rubber hardness and stiffness of the spine and

that the technical specification on hardness is sufficient to ensure appropriate spine stiffness. Accordingly, the agency believes that a separate calibration test for the lumbar spine is not necessary. Humanoid supported the validity of relying on the spine hardness specification to assure adequate stability of the dummy's posture, even though it found little effect on the dummy's impact response. Humanoid's support for this approach was based on tests of Hybrid III dummies which were equipped with a variety of lumbar spines having different rubber hardnesses.

Subsequent to issuance of the notice, the agency has continued its testing of the Hybrid III test dummy. Through that testing, the agency found that commercially available necks either cannot meet or cannot consistently meet all of the calibration tests originally proposed for the neck. To further evaluate this problem, NHTSA and GM conducted a series of round robin tests in which a set of test dummies were put through the calibration tests at both GM's and NHTSA's test laboratories.

The test results, which were placed in the docket after the tests were completed, showed that none of the necks could pass all of the originally specified calibration tests.

In examining the test data, the agency determined that while some of the responses of the necks fell slightly outside of the performance corridors proposed in the calibration tests, the responses of the necks showed a relatively good match to existing biomechanical data on human neck responses. Thus, while the necks did not meet all of the calibration tests, they did respond as human necks are expected to respond.

In discussions with GM, the agency learned that the calibration performance requirements were originally established in 1977 based on the responses of three prototype Hybrid III necks. GM first examined the existing biomechanical data and established several performance criteria that reflected human neck responses. GM then built necks which would meet the biomechanically based performance criteria. GM established the calibration tests that it believed were necessary to ensure that the necks of the prototype test dummies would produce the required biomechanical responses. Although extensive performance specifications may have been needed for the development of specially built prototype necks, not all of the specifications appear to be essential once the final design was established for the mass-produced commercial version. Based on the ability of the commercially available test dummies to meet the biomechanical response criteria, NHTSA believes that the GM-derived calibration requirements should be adjusted to reflect the response characteristics of commercially available test dummies and simplified as much as possible to reduce the complexity of the testing.

Based on the results of the NHTSA-GM calibration test series, the agency is making the following changes to the neck calibration tests. In the flexion (forward bending) calibration test, the agency is:

1. increasing the time allowed for the neck to return to its preimpact position after the pendulum impact test from a range of 109–119 milliseconds to a range of 113–128 milliseconds.

2. changing the limits for maximum head rotation from a range of 67°–79° to a range of 64°–78°.

3. expanding the time limits during which maximum moment must occur from a range of 46–56 milliseconds to 47–58 milliseconds.

4. modifying the limits for maximum moment from a range of 72–90 ft-lbs to a range of 65–80 ft-lbs.

5. increasing the time for the maximum moment to decay from a range of 95–105 milliseconds to a range of 97–107 milliseconds.

In the extension (backward bending) calibration test, the agency is:

1. expanding the time allowed for the neck to return to its preimpact position after the pendulum impact test from a range of 157–167 milliseconds to a range of 147–174 milliseconds.

2. changing the limits for maximum head rotation from a range of 94°–106° to a range of 81°–106°.

3. expanding the time limit during which the minimum moment must occur from a range of 69–77 milliseconds to 65–79 milliseconds.

4. modifying the limits for minimum moment from a range of −52 to −63 ft-lbs to a range of −39 to −59 ft-lbs.

5. increasing the time for the minimum moment to decay from the range of 120–144 milliseconds, contained in GM's technical specifications for the Hybrid III, to a range of 120–148 milliseconds.

In reviewing the NHTSA-GM test data, the agency also identified several ways of simplifying the neck's performance requirements. In each case, the following calibration specifications appear to be redundant and their deletion should not affect the performance of the neck. The agency has thus deleted the requirement for minimum moment in flexion and the time requirement for that moment. For extension, the agency has eliminated the limit on the maximum moment permitted and the time requirement for that moment. The agency has

deleted those requirements since the specification on maximum rotation of the neck in flexion and minimum rotation of the neck in extension appear to adequately measure the same properties of the neck. Similarly, the agency has simplified the test by eliminating the pendulum braking requirement for the neck test, since GM's testing shows that the requirement is not necessary to ensure test consistency. Finally, the agency is clarifying the test procedure by deleting the specification in the GM technical drawings for the Hybrid III calling for two pre-calibration impact tests of the neck. GM has informed the agency that the two pre-calibration tests are not necessary.

Based on the NHTSA-GM calibration test data, the agency is making two additional changes to the neck calibration test procedure. Both NHTSA and GM routinely control the calibration pendulum impact speed to within plus or minus one percent. Currently available dummy necks are able to meet the calibration response requirements consistently when the pendulum impact speed is controlled to that level Thus, NHTSA believes that the proposed range of allowable velocities (± 8.5 percent) for the pendulum impact is excessive. Reducing the allowable range is clearly feasible and will help maintain a high level of consistency in dummy neck responses. The agency has therefore narrowed the range of permissible impact velocities to the neck to ± 2 percent. This range is readily obtainable with commercially available test equipment. In reviewing the neck calibration test data, GM and NHTSA noted a slight sensitivity in the neck response to temperature variation. In its docket submission of January 27, 1986, GM recommended controlling the temperature during the neck calibration test to 71° ± 1°. NHTSA agrees that controlling the temperature for the neck calibration test will reduce variability, but the agency believes that a slightly wider temperature range of 69° to 72°, which is the same range used in the chest calibration test, is sufficient.

Neck Durability

Nissan commented that, in sled tests of the two test dummies, the neck bracket of one of the Hybrid III test dummies experienced damage after 10 tests, while the Part 572 test dummy had no damage. The agency believes that Nissan's experience may be the result of an early neck design which has been subsequently modified by GM. (See GM letter of September 16, 1985, Docket 74-14, Notice 39, Entry 28.) The agency has conducted numerous 30 mile per hour vehicle impact tests using the Hybrid III test dummy and has not had any neck bracket failures.

Thorax Calibration Test

As a part of the NHTSA-GM calibration test series, both organizations also performed the proposed calibration test for the thorax on the same test dummies. That testing showed relatively small differences in the test results measured between the two test facilities The test results from both test facilities show that the chest responses of the Hybrid III test dummies were generally within the established biomechanical performance corridors for the chest. In addition, the data showed that the Hybrid III chest responses fit those corridors substantially better than the chest responses of the existing Part 572 test dummy. The data also showed that the chest responses in the high speed (22 ft/sec) pendulum impact test more closely fit the corridors than did the chest responses in the low speed (14 ft/sec) test. In addition, the data showed that if a test dummy performed satisfactorily in the low speed pendulum impact test, it also performed satisfactorily in the more severe high speed test.

Based on those results, GM recommended in a letter of January 27, 1986, (Docket No. 74-14, Notice 39, Entry 41) that only the low speed pendulum impact be used in calibration testing of the Hybrid III chest. GM noted that deleting the more severe pendulum impact test "can lead to increasing the useful life of the chest structure."

Based on the test data, the agency agrees with the GM recommendation that only one pendulum impact test is necessary. NHTSA recognizes that using only the low speed pendulum impact will increase the useful life of the chest. However, the agency has decided to retain the high speed rather than the low speed test. While NHTSA recognizes that the high speed test is more severe, the agency believes the high speed test is more appropriate for a number of reasons. First, the data showed that the high speed chest impact responses compared more closely with the biomechanical corridors than the low speed responses. Thus, use of the high speed test will make it easier to identify chests that do not have the correct biofidelity. In addition, since the higher speed test is more severe it will subject the ribcage to higher stresses, which will help identify chest structural degradation. Finally, the high speed impact test is more representative of the range of impacts a test dummy can receive in a vehicle crash test.

Although the NHTSA-GM test data showed that the production version of the Hybrid III chest had sufficient biofidelity, the data indicated that proposed calibration performance requirements should be lightly changed to account for the wider range in calibration test responses measured in commercially available test dummies. Accordingly, the agency is adjusting the chest deflection requirement to increase the allowable range of deflections from 2.51-2.75 inches to 2.5-2.85 inches. In addition, the agency is adjusting the resistive force requirement from a range of 1186-1298 pounds to a range of 1080-1245 pounds. Also, the hysteresis requirement is being adjusted from a 75-80 percent range to a 69-85 percent range. Finally, the agency is clarifying the chest calibration test procedure by deleting the specification in GM's technical drawing for the Hybrid III that calls for two pre-calibration impact tests of the chest. GM has informed the agency that these tests are not necessary. These slight changes will not affect the performance of the Hybrid III chest, since the NHTSA-GM test data showed that commercially available test dummies meeting these calibration specifications had good biofidelity.

Chest Durability

Testing done by the agency's Vehicle Research and Test Center has indicated that the durability of the Hybrid III's ribs in calibration testing is less than that of the Part 572 test dummy. ("State-of-the-Art Dummy Selection, Volume I" DOT Publication No. HS 806 722) The durability of the Hybrid III was also raised by several commenters. For example, Toyota raised questions about the durability of the Hybrid III's ribs and suggested the agency act to improve their durability.

The chest of the Hybrid III is designed to be more flexible, and thus more human-like, than the chest of the Part 572 test dummy. One of the calibration tests used for the chest involves a 15 mph impact into the chest by a 51.5 pound pendulum; an impact condition which is substantially more severe than a safety belt or airbag restrained occupant would experience in most crashes. The chest of the Hybrid III apparently degrades after such multiple impacts at a faster rate than the chest of the Part 572 test dummy. As the chest gradually deteriorates, the amount of acceleration and deflection measured in the chest are also affected. Eventually the chest will fall out of specification and will require either repair or replacement.

In its supplemental comments to the April 1985 notice, GM provided additional information about the durability of the Hybrid III ribs. GM said that it uses the Hybrid III in unbelted testing, which is the most severe test for the dummy. GM said that the Hybrid III can be used for about 17 crash tests before the ribs must be replaced. GM explained that it does not have comparable data for the Part 572 test dummy since it does not use that test dummy in unbelted tests. GM said, however, that it believes that the durability of the Part 572 test dummy ribs in vehicle crash testing would be comparable to that of the Hybrid III.

Having reviewed all the available information, the agency concludes that both the Hybrid III and existing Part 572 test dummy ribs will degrade under severe impact conditions. Although the Hybrid III's more flexible ribs may need replacement more frequently, particularly after being used in unrestrained testing, the Hybrid III's ribs appear to have reasonable durability. According to GM's data, which is in line with NHTSA's crash test experience, the Hybrid III's ribs can withstand approximately 17 severe impacts, such as found in unrestrained testing, before they must be replaced. Ford, in a presentation at the MVMA Hybrid III workshop held on February 5, 1986, noted that one of its belt-restrained Hybrid III test dummies was subjected to 35 vehicle and sled crashes without any failures. The potential lower durability of the ribs in unrestrained testing should be of little consequence if the Hybrid III test dummy is used in air bag or belt testing.

Chest Temperature Sensitivity

The April 1985 notice said NHTSA tests have indicated that the measurements of chest deflection and chest acceleration by the Hybrid III are temperature sensitive. For this reason, GM's specifications for the Hybrid III recognize this problem and call for using the test dummy in a narrower temperature range (69° to 72° F) to ensure the consistency of the measurements. GM has also suggested the use of an adjustment factor for calculating chest deflection when the Hybrid III is used in a test environment that is outside of the temperature range specified for the chest. While this approach may be reasonable to account for the adjustment of the deflection measurement, there is no known method to adjust the acceleration measurement for variations in temperature. For this reason, the agency is not adopting GM's proposed adjustment factor, but is instead retaining the proposed 69° to 72° F temperature range.

A number of commenters addressed the feasibility and practicability of maintaining that temperature range. BMW said that although it has an enclosed crash test facility, it had reservations about its ability to control the test temperature within the proposed range. Daihatsu said that it was not sure it could assure the test dummy's temperature will remain within the proposed range. Honda said that while it had no data on the temperature sensitivity of the Hybrid III, it questioned whether the proposed temperature range was practical. Mercedes-Benz said it is not practicable to maintain the proposed temperature range because the flood lights necessary for high speed filming of crash tests can cause the test dummy to heat up. Nissan said it was not easy to maintain the current 12 degree range specified for the existing Part 572 test dummy and thus it would be hard to maintain the three degree range proposed for the Hybrid III. Ford also said that maintaining the three degree range could be impracticable in its current test facilities.

Other manufacturers tentatively indicated that the proposed temperature range may not be a problem. VW said the temperature range should not be an insurmountable problem, but more experience with the Hybrid III is necessary before any definite conclusions can be reached. Volvo said it could maintain the temperature range in its indoor test facilities, but it questioned whether outdoor test facilities could meet the proposed specification. Humanoid indicated in its comments, that it has developed an air conditioning system individualized for each test dummy which will maintain a stable temperature in the test dummy up to the time of the crash test.

The agency believes that there are a number of effective ways to address the temperature sensitivity of the Hybrid III chest. The test procedure calls for placing the test dummy in an area, such as a closed room, whose temperature is maintained within the required range for at least four hours before either the calibration tests or the use of the test dummy in a crash test. The purpose of the requirement is to ensure that the primary components of the test dummy have reached the correct temperature before the test dummy is used in a test. As discussed below, analytical techniques can be used to determine the temperature within the test dummy, to calculate how quickly the test dummy must be used in a crash test before its temperature will fall outside the required temperature range.

Testing done by the agency with the current Part 572 test dummy, whose construction and materials are similar to the Hybrid III, has determined how long it takes for various test dummy components to reach the required temperature range once the test dummy is placed in a room within that range. ("Thermal Responses of the Part 572 Dummy to Step Changes in Ambient Temperature" DOT Publication No. HS-801 960, June 1976) The testing was done by placing thermocouples, devices to measure temperature, at seven locations within the dummy and conducting a series of heating and cooling experiments. The tests showed that the thermal time constants (the thermal time constant is the time necessary for the temperature differential between initial and final temperatures to decrease from its original value to 37% of the original differential) varied from 1.2 hours for the forehead to 6.2 hours for the lumbar spine. Using this information it is possible to estimate the time it takes a test dummy originally within the required temperature range to fall out of the allowable range once it has been exposed to another temperature. The rib's thermal time constant is 2.9 hours. This means, for example, that if a test dummy's temperature has been stabilized at 70.5° F and then transferred to a test environment at 65° F, it would take approximately 0.8 hours for the rib temperature to drop to 69° F, the bottom end of the temperature range specified in Part 572.

Thus, the NHTSA test results cited above show that the chest can be kept within the range proposed by the agency if the test dummy is placed in a temperature-controlled environment for a sufficient time to stabilize the chest temperature. Once the chest of the test dummy is at the desired temperature, the test data indicate that it can tolerate some temperature variation at either an indoor or outdoor crash test site and still be within the required temperature range as long as the crash test is performed within a reasonable amount of time and the temperature at the crash site, or within the vehicle, or within the test dummy is controlled close to the 69 to 72 degrees F range. Obviously, testing conducted at extremely high or low temperatures can move the test dummy's temperature out of the required range relatively quickly, if no means are used to maintain the temperature of the test dummy within the required range. However, auxiliary temperature control devices can be used in the vehicle or the test environment to maintain a stabilized temperature prior to the crash test. Therefore, the agency has decided to retain the proposed 69 to 72 degrees F temperature range.

Chest Response to Changes in Velocity

The April notice raised the issue of the sensitivity of the Hybrid III's chest to changes in impact velocities. The notice pointed out that one GM study on energy-absorbing steering columns ("Factors Influencing Laboratory Evaluation of Energy-Absorbing Steering Systems," Docket No. 74-14, Notice 32, Entry 1666B) indicated that the Hybrid III's chest may be insensitive to changes in impact velocities and asked commenters to provide further information on this issue.

Both GM and Ford provided comments on the Hybrid III's chest response. GM said that since the Hybrid III chest is designed to have a more human-like thoracic deflection than the Part 572 test dummy, the Hybrid III's response could be different. GM referenced a study ("System Versus Laboratory Impact Tests for Estimating Injury Hazard" SAE paper 680053) which involved cadaver impacts into energy-absorbing steering columns. The study concluded that the force on the test subject by the steering assembly was relatively constant despite changes in test speeds. GM said that this study indicated that "rather than the Hybrid III chest being insensitive to changes in velocity in steering system tests, it is the Part 572 which is too sensitive to changes in impact velocity to provide meaningful information for evaluating steering systems."

GM also presented new data on chest impact tests conducted on the Hybrid III and Part 572 test dummies. The tests involved chest impacts by three pendulum impact devices with different masses and three impact speeds. GM said that the test results show that "the Hybrid III chest deflection is sensitive to both changes in impact velocity and impactor mass." Ford also noted that the Hybrid III appears sensitive in the range of speed and deflections that are relevant to Standard No. 208 testing with belt-restrained dummies.

Ford noted that the GM testing referenced in the April notice was conducted at higher impact speeds than used in the calibration testing of the Hybrid III. Ford said it agreed with GM that the indicated insensitivity of chest acceleration to speed and load is a reflection of the constant-force nature of the steering column's energy absorption features. After reviewing the information provided by Ford and GM, NHTSA agrees that in an impact with a typical steering column, once the energy-absorbing mechanism begins to function, the test dummy's chest will receive primarily constant force. The lower stiffness of the Hybrid III chests would make it respond in a more human-like manner to these forces than the existing Part 572 test dummy.

Chest Accelerometer Placement

Volvo pointed out that the chest accelerometer of the Hybrid III is located approximately at the center of gravity of the chest, while the accelerometer is higher and closer to the back in the Part 572 test dummy. Volvo said that since the biomechanical tolerance limits for the chest were established using a location similar to that in the Part 572, it questioned whether the acceleration limits should apply to the Hybrid III. Volvo recommended changing the location of the accelerometer in the Hybrid III or using different chest acceleration criteria for the Hybrid III.

The agency recognizes that Hybrid III accelerometer placement should more correctly reflect the overall response of the chest because it is placed at the center of gravity of the chest. However, the dimensional differences between the accelerometer placements in the two test dummies are so small that in restrained crash tests the differences in acceleration response, if any, should be minimal.

Repeatability and Reproducibility

As discussed previously, test dummy repeatability refers to the ability of one test dummy to measure consistently the same responses when subjected to the same test. Reproducibility refers to the ability of two or more test dummies built to the same specifications to measure consistently the same responses when they are subjected to the same test.

Ford said that it is particularly concerned about the repeatability of the chest acceleration and deflection measurements of the Hybrid III and about the reproducibility of the Hybrid III in testing by different laboratories. Ford said that once a test dummy positioning procedure has been established, the agency should conduct a series of 16 car crash tests to verify the repeatability and reproducibility of the Hybrid III.

In its comments, GM provided data showing that the repeatability of the Hybrid III is the same as the existing Part 572 test dummy. Volvo, the only other commenter that addressed repeatability, also said that its preliminary tests show that the Hybrid III has a repeatability comparable to the Part 572. The agency's Vehicle Research and Test Center has also evaluated the repeatability of the Hybrid III and the Part 572 in a series of sled tests. The data from those tests show that the repeatability of the two test dummies is comparable. ("State-of-the-Art Dummy Selection, Volume I" DOT Publication No. HS 806 722.)

GM also provided data showing that the reproducibility of the Hybrid III is significantly better than the Part 572. In its supplemental comments filed on September 16, 1985, GM also said that Ford's proposed 16 car test program was not needed. GM said that "in such test the effects of vehicle build variability and test procedure variability would totally mask any effect of Hybrid III repeatability and reproducibility."

The agency agrees with GM that additional testing is unnecessary. The information Provided by GM and Volvo shows that the repeatability of the Hybrid III is at least as good as the repeatability of the existing Part 572 test dummy. Likewise, the GM data show that the reproducibility of the Hybrid III is better than that of the existing Part 572 test dummy. Likewise, the recent NHTSA-GM calibration test series provides further confirmation that tests by different laboratories show the repeatability and reproducibility of the Hybrid III.

Equivalence of Hybrid III and Part 572

As noted in the April 1985 notice, the Hybrid III and the Part 572 test dummies do not generate identical impact responses. Based on the available data, the agency concluded that when both test dummies are tested in lap/shoulder belts or with air cushions, the differences between the two test dummies are minimal. The agency also said that it knew of no method for directly relating the response of the Hybrid III to the Part 572 test dummy.

The purpose of comparing the response of the two test dummies is to ensure that the Hybrid III will meet the need for safety by adequately identifying vehicle designs which could cause or increase occupant injury. The agency wants to ensure that permitting a choice of test dummy will not lead to a degradation in safety performance.

As mentioned previously, one major improvement in the Hybrid III is that it is more human-like in its responses than the current Part 572 test dummy. The primary changes to the Hybrid III that make it more human-like are to the neck, chest and knee. Comparisons of the responses of the Part 572 and Hybrid III test dummies show that responses of the Hybrid III are closer than the Part 572 to the best available data on human responses. (See Chapter II of the Final Regulatory Evaluation on the Hybrid III.)

In addition to being more human-like, the Hybrid III has increased measurement capabilities for the neck (tension, compression, and shear forces and bending moments), chest (deflection), knee (knee shear), and lower leg (knee and tibia forces and moments). The availability of the extra injury measuring capability of the Hybrid III gives vehicle manufacturers the potential for gathering far more information about the performance of their vehicle designs than they can obtain with the Part 572.

To evaluate differences in the injury measurements made by the Hybrid III and the existing Part 572 test dummy, the agency has reviewed all of the available data comparing the two test dummies. The data come from a variety of sled barrier crash tests conducted by GM, Mercedes-Benz, NHTSA, Nissan, and Volvo. The data include tests where the dummies were unrestrained and tests where the dummies were restrained by manual lap/shoulder belts, automatic belts, and air bags. For example, subsequent to issuance of the April 1985 notice, NHTSA did additional vehicle testing to compare the Part 572 and Hybrid III test dummies. The agency conducted a series of crash tests using five different types of vehicles to measure differences in the responses of the test dummies. Some of the tests were frontal 30 mile per hour barrier impacts, such as are used in Standard No. 208 compliance testing, while others were car-to-car tests. All of the tests were done with unrestrained test dummies to measure their impact responses under severe conditions. The agency's analysis of the data for all of the testing done by NHTSA and others is fully described in the Final Regulatory Evaluation for this rulemaking. This notice will briefly review that analysis.

One of the reasons for conducting the analysis was to address the concern raised by the Center for Auto Safety (CAS) in its original petition and the Insurance Institute for Highway Safety (IIHS) in its comments that the Hybrid III produces lower HIC responses than the existing Part 572 test dummy. As discussed in detail below, the test data do not show a trend for one type of test dummy to consistently measure higher or lower HIC's or femur readings than the other. Based on these test data, the agency concludes that the concern expressed by CAS and IIHS that the use of the Hybrid III test dummy will give a manufacturer an advantage in meeting the HIC performance requirement of Standard No. 208 is not valid.

In the case of chest acceleration measurements, the data again do not show consistently higher or lower measurements for either test dummy, except in the case of unrestrained tests. In unrestrained tests, the data show that the Hybrid III generally measures lower chest g's than the existing Part 572 test dummy. This difference in chest g's measurement is one reason why the agency is adopting the additional chest deflection measurement for the Hybrid III, as discussed further below.

HIC Measurements

The April 1985 notice specifically invited comments on the equivalence of the Head Injury Criterion (HIC) measurements of the two test dummies. Limited laboratory testing done in a University of California at San Diego study conducted by Dr. Dennis Schneider and others had indicated that the Hybrid III test dummy generates lower acceleration responses than either the Part 572 test dummy or cadaver heads in impacts with padded surfaces. The notice explained that the reasons for those differences had not yet been resolved.

In its comments, GM explained that it had conducted a series of studies to address the Schneider results. GM said that those studies showed that the Schneider test results are "complicated by the changing characteristics of the padding material used on his impact surface. As a result, his tests do not substantiate impactor response difference between the Hybrid III head, the Part 572 head and cadaver heads. After examining our reports, Dr. Schneider agreed with the finding that padding degradation resulting from multiple impact exposures rendered an input-response comparison invalid between the cadaver and the dummies." (The GM and Schneider letters are filed in Docket 74-14, General Reference, Entry 556.)

The agency's Vehicle Research and Test Center has also conducted head drop tests of the current Part 572 and Hybrid III heads. The tests were conducted by dropping the heads onto a two inch thick steel plate, a surface which is considerably more rigid than any surface that the test dummy's head would hit in a vehicle crash test. One purpose of the tests was to assess the performance of the heads in an impact which can produce skull fractures in cadavers. The tests found that the response of the Hybrid III head was more human-like at the fracture and subfracture acceleration levels than the Part 572 head. The testing did show that in these severe impacts into thick steel plates, the HIC scores for the Hybrid III were lower than for the Part 572. However, as discussed below, when the Hybrid III is tested in vehicle crash and sled tests, which are representative of occupant impacts into actual vehicle structures, the HIC scores for the Hybrid III are not consistently lower than those of the Part 572 test dummy.

The agency examined crash and sled tests, done by GM, Mercedes-Benz, NHTSA and Volvo, in which both a Hybrid III and the existing Part 572 test dummy were restrained by manual lap/shoulder belts. (The complete results from those and all the other tests reviewed by the agency are discussed in Chapter III of the Final Regulatory Evaluation on the Hybrid III.) The HIC responses in those tests show that the Hybrid III generally had higher HIC responses than the Part 572 test dummy. Although the data show that the Hybrid III's HIC responses are generally higher, in some cases 50 percent higher than the Part 572, there are some tests in which the Hybrid III's responses were 50 percent lower than the responses of the Part 572.

For two-point automatic belts, the agency has limited barrier crash test data and the direct comparability of the data is questionable. The tests using the existing Part 572 test dummy were done in 1976 on 1976 VW Rabbits for compliance purposes. The Hybrid III tests were done in 1985 by the agency's Vehicle Research and Test Center as part of the SRL-98 test series on a 1982 and a 1984 VW Rabbit. Differences in the seats, safety belts, and a number of other vehicle parameters between these model years and between the test set-ups could affect the results. In the two-point automatic belt tests, the data show that the Hybrid III measured somewhat higher head accelerations than the existing Part 572 test dummy. In two-point automatic belts, the differences appear to be minimal for the driver and substantially larger for the passenger. In air bag sled tests, the Hybrid III's HIC responses were generally lower; in almost all the air bag tests, the HIC responses of both the Hybrid III and the Part 572 test dummies were substantially below the HIC limit of 1,000 set in Standard No. 208. Because of the severe nature of the unrestrained sled and barrier tests, in which the uncontrolled movement of the test dummy can result in impacts with different vehicle structures, there was no consistent trend for either test dummy to measure higher or lower HIC responses than the other.

Chest Measurements

For manual lap/shoulder belts, NHTSA compared the results from GM, Mercedes-Benz, NHTSA, and Volvo sled tests, and GM frontal barrier tests. The NHTSA sled test results at 30 and the Volvo sled test results at 31 mph are very consistent, with the mean Hybrid III chest acceleration response being only 2–3 g's higher than the response of the existing Part 572 test dummy. In the 35 mph Volvo sled tests, the Hybrid III chest acceleration response was up to 44 percent higher than the existing Part 572 response. The GM 30 mph sled and barrier test data were fairly evenly divided. In general, the Hybrid III chest acceleration response is slightly higher than that of the existing Part 572 test dummy. The agency concludes from these data that at Standard No. 208's compliance test speed (30 mph) with manual lap/shoulder belts there are no large differences in chest acceleration responses between the two dummies. In some vehicles, the Hybrid III may produce slightly higher responses and in other vehicles it may produce slightly lower responses.

As discussed earlier, the agency has limited test data on automatic belt tests and their comparability is questionable. The Hybrid III chest acceleration responses are up to 1.5 times higher than those for the existing Part 572 test dummy. Only very limited sled test data are available on air bags alone, air bag plus lap belt, and air bag plus lap/shoulder belt. In all cases, the Hybrid III chest acceleration responses were lower than those for the existing Part 572 test dummy.

For unrestrained occupants, the Hybrid III produces predominantly lower chest acceleration responses than the existing Part 572 test dummy in sled and barrier tests, and in some cases the difference is significant. In some tests, the Hybrid III chest acceleration response can be 40 to 45 percent lower than the Part 572 response, although in other tests the acceleration measured by the Hybrid III can exceed that measured by the Part 572 test dummy by 10 to 15 percent.

In summary, the test data indicate the chest acceleration responses between the Hybrid III and the existing Part 572 test dummy are about the same for restrained occupants, but differ for some cases of unrestrained occupants. This is to be expected since a restraint system would tend to make the two dummies react similarly even though they have different seating postures. The different seating postures, however, would allow unrestrained dummies to impact different vehicle surfaces which would in most instances produce different responses. Since the Hybrid III dummy is more human-like, it should experience loading conditions that are more human-like than would the existing Part 572 test dummy. One reason that the agency is adding a chest deflection criterion for the Hybrid III is that the unrestrained dummy's chest may experience more severe impacts with vehicle structures than would be experienced in an automatic belt or air bag collision. Chest deflection provides an additional measurement of potential injury that may not be detected by the chest acceleration measurement.

Femur Measurements

The test data on the femur responses of the two types of test dummies also do not show a trend for one test dummy to measure consistently higher or lower responses than the other. In lap/shoulder belt tests, GM's sled and barrier tests from 1977 show a trend toward lower measurements for the Hybrid III, but GM's more recent tests in 1982–83 show the reverse situation. These tests, however, are of little significance unless there is femur loading due to knee contact. These seldom occur to lap/shoulder belt restrained test dummies. Also, in none of the tests described above do the measurements approach Standard No. 208's limit of 2250 pounds for femur loads. The air bag test data are limited; however, they show little difference between the femur responses of the two test dummies. As would be expected, the unrestrained tests showed no systematic differences, because of the variability in the impact locations of an unrestrained test dummy.

Injury Criteria

Many manufacturers raised objections to the additional injury criteria proposed in the April 1985 notice. AMC, Ford, and MVMA argued that adopting the numerous injury criteria proposed in the April 1985 notice would compound a manufacturer's compliance test problems. For example, Ford said it "would be impracticable to require vehicles to meet such a multitude of criteria in a test with such a high level of demonstrated variability. Notice 39 appears to propose 21 added pass-fail measurements per dummy, for a total of 25 pass-fail measurements per dummy, or 50 pass-fail measurements per test. Assuming these measurements were all independent of one another, and a car design had a 95% chance of obtaining a passing score on each measurement, the chance of obtaining a passing score on all measurements in any single test for a single dummy would be less than 28% and for both dummies would be less than 8%." Ford, Nissan, VW and Volvo also said that with the need for additional measurements, there will be an increase in the number of tests with incomplete data. BMW, while supporting the use of the Hybrid III as a potential improvement to safety, said that the number of measurements needed for the additional injury criteria is beyond the capability of its present data processing equipment.

VW said there is a need to do additional vehicle testing before adopting any new criteria. It said that if current production vehicles already meet the additional criteria then the criteria only increase testing variability without increasing safety. If current vehicles cannot comply, then additional information is needed about the countermeasures needed to meet the criteria. Honda said there are insufficient data to determine the relationship between actual injury levels and the proposed injury criterion.

As discussed in detail below, the agency has decided to adopt only one additional injury criterion, chest deflection, at this time. The agency plans to issue another notice on the remaining criteria proposed in the April 1985 notice to gather additional information on the issues raised by the commenters.

Alternative HIC Calculations

The April 1985 notice set forth two proposed alternative methods of using the head injury criterion

(HIC) in situations when there is no contact between the test dummy's head and the vehicle's interior during a crash. The first proposed alternative was to retain the current HIC formula, but limit its calculation to periods of head contact only. However, in non-contact situations, the agency proposed that an HIC would not be calculated, but instead new neck injury criteria would be calculated. The agency explained that a crucial element necessary for deciding whether to use the HIC calculation or the neck criteria was an objective technique for determining the occurrence and duration of head contact in the crash test. As discussed in detail in the April 1985 notice, there are several methods available for establishing the duration of head contact, but there are questions about their levels of consistency and accuracy.

The second alternative proposed by the agency would have calculated an HIC in both contact and non-contact situations, but it would limit the calculation to a time interval of 36 milliseconds. Along with the requirement that an HIC not exceed 1,000, this would limit average head acceleration to 60 g's or less for any durations exceeding 36 milliseconds.

Almost all of the commenters opposed the use of the first proposed alternative. The commenters uniformly noted that there is no current technique that can accurately identify whether head contact has or has not occured during a crash test in all situations. However, the Center for Auto Safety urged the agency to adopt the proposed neck criteria, regardless of whether the HIC calculation is modified.

There was a sharp division among the commenters regarding the use of the second alternative; although many manufacturers argued that the HIC calculation should be limited to a time interval of approximately 15 to 17 milliseconds (ms), which would limit average long duration (i.e., greater than 15–17 milliseconds) head accelerations to 80–85 g's. Mercedes-Benz, which supported the second alternative, urged the agency to measure HIC only during the time interval that the acceleration level in the head exceeds 60 g's. It said that this method would more effectively differentiate results received in contacts with hard surfaces and results obtained from systems, such as airbags, which provide good distribution of the loads experienced during a crash. The Center for Auto Safety, the Insurance Institute for Highway Safety and State Farm argued that the current HIC calculation should be retained; they said that the proposed alternative would lower HIC calculations without ensuring that motorists were still receiving adequate head protection.

NHTSA is in the process of reexamining the potential effects of the two alternatives proposed by the agency and of the two additional alternatives suggested by the commenters. Once that review has been completed, the agency will issue a separate notice announcing its decision.

Thorax

At present, Standard No. 208 uses an acceleration-based criterion to measure potential injuries to the chest. The agency believes that the use of a chest deflection criterion is an important supplement to the existing chest injury criterion. Excessive chest deflection can produce rib fractures, which can impair breathing and inflict damage to the internal organs in the chest. The proposed deflection limit would only apply to the Hybrid III test dummy, since unlike the existing Part 572 test dummy, it has a chest which is designed to deflect like a human chest and has the capability to measure deflection of the sternum relative to the spine, as well as acceleration, during an impact.

The agency proposed a three-inch chest deflection limit for systems, such as air bags, which symmetrically load the chest during a crash and a two-inch limit for all other systems. The reason for the different proposed limits is that a restraint system that symmetrically and uniformly applies loads to the chest increases the ability to withstand chest deflection as measured by the deflection sensor, which is centrally located in the dummy.

The commenters generally supported adoption of a chest deflection injury criterion. For example, Ford said it supported the use of a chest deflection criterion since it may provide a better means of assessing the risk of rib fractures. Likewise, the Insurance Institute for Highway Safety said the chest deflection criteria "will aid in evaluating injury potential especially in situations where there is chest contact with the steering wheel or other interior components." IIHS also supported adoption of a three-inch deflection limit for inflatable systems and a two-inch limit for all other systems. However, most of the other commenters addressing the proposed chest deflection criteria questioned the use of different criteria for different restraint systems.

GM supported limiting chest deflections to three-inches in all systems. GM said that it uses a two-inch limit as a guideline for its safety belt system testing, but it had no data to indicate that the two-inch limit is appropriate as a compliance limit.

Renault/Peugeot also questioned the three-inch deflection limit for systems that load the dummy symmetrically and two inches for systems that do not. It said that the difference between those systems should be addressed by relocation of the deflection sensors. It also asked the agency to define what constitutes a symmetrical system. VW also questioned the appropriateness of setting separate limits for chest compression for different types of restraint systems. It recommended adoption of a three-inch limit for all types of restraint systems.

Volvo also raised questions about the appropriateness of the proposed deflection criteria. Volvo said that the GM-developed criteria proposed in the April 1985 notice were based on a comparison of accident data gathered by Volvo and evaluated by GM in sled test simulations using the Hybrid III test dummy. Volvo said that the report did not analyze "whether the chest injuries were related to the chest acceleration or the chest deflection, or a combination of both."

The agency recognizes that there are several different types of potential chest injury mechanisms and that it may not be possible to precisely isolate and measure what is the relevant contribution of each type of mechanism to the final resulting injury. However, there is a substantial amount of data indicating that chest deflection is an important contributing factor to chest injury. In addition, the data clearly demonstrate that deflection of greater than three inches can lead to serious injury. For example, research done by Neathery and others has examined the effects of frontal impacts to cadaver chests with an impactor that represents the approximate dimensions of a steering wheel hub. Neathery correlated the measured injuries with the amount of chest deflection and recommended that for a 50th percentile male, chest deflection not exceed three inches. (Neathery, R. F., "Analysis of Chest Impact Response Data and Scaled Performance Recommendations," SAE Paper No. 741188)

Work by Walfisch and others looked at crash tests of lap/shoulder belt restrained cadavers. They found that substantial injury began to occur when the sternum deflection exceeded 30 percent of the available chest depth ("Tolerance Limits and Mechanical Characteristic of the Human Thorax in Frontal and Side Impact and Transposition of these Characteristics into Protective Criteria," 1982 IRCOBI Conference Proceedings). With the chest of the average man being approximately 9.3 inches deep, the 30 percent limit would translate into a deflection limit of approximately 2.8 inches. Since the chest of the Hybrid III test dummy deflects somewhat less than a human chest under similar loading conditions, the chest deflection limit for systems which do not symmetrically and uniformly load the chest, such as lap/shoulder belts, must be set at a level below 2.8 inches to assure an adequate level of protection.

To determine the appropriate level for non-symmetrical systems, the agency first reviewed a number of test series in which cadaver injury levels were measured under different impact conditions. (All of the test results are fully discussed in Chapter III of the Final Regulatory Evaluation on the Hybrid III.) The impact conditions included 30 mph sled tests done for the agency by Wayne State University in which a pre-inflated, non-vented air bag system symmetrically and uniformly spread the impact load on the chest of the test subject. NHTSA also reviewed 30 mph sled tests done for the agency by the University of Heidelberg which used a lap/shoulder belt system, which does not symmetrically and uniformly spread chest loads. In addition, the agency reviewed 10 and 15 mph pendulum impact tests done for GM to evaluate the effects of concentrated loadings, such as might occur in passive interior impacts. The agency then compared the chest deflection results for Hybrid III test dummies subjected to the same impact conditions. By comparing the cadaver and Hybrid III responses under identical impact conditions, the agency was able to relate the deflection measurements made by the Hybrid III to a level of injury received by a cadaver.

The test results show that when using a relatively stiff air bag, which was pre-inflated and non-vented, the average injury level measured on the cadavers corresponded to an Abbreviated Injury Scale (AIS) of 1.5. (The AIS scale is used by researchers to classify injuries an AIS of one is a minor injury, while an AIS of three represents a serious injury.) In tests with the Hybrid III under the same impact conditions, the measured deflection was 2.7 inches. These results demonstrate that a system that symmetrically and uniformly distributes impact loads over the chest can produce approximately three inches of deflection and still adequately protect an occupant from serious injury.

The testing in which the impact loads were not uniformly or symmetrically spread on the chest or were highly concentrated over a relatively small area indicated that chest deflection measured on the Hybrid III must be limited to 2-inches to assure those systems provide a level of protection comparable to that provided by systems that symmetrically spread the load. In the lap/shoulder belt tests, the average AIS was 2.6. The measured deflection for the Hybrid III chest in the same type of impact test was 1.6 inches. Likewise, the results from the

pendulum impact tests showed that as the chest deflection measured on the Hybrid III increased, the severity of the injuries increased. In the 10 mph pendulum impacts, the average AIS was 1.3 and the average deflection was 1.3 inches. In the 15 mph pendulum impacts the average AIS rose to 2.8. Under the same impact conditions, the chest deflection measured on the Hybrid III was 2.63 inches.

Based on these test results NHTSA has decided to retain the two-inch limit on chest deflection for systems that do not symmetrically and uniformly distribute impact loads over a wide area of the chest. Such systems include automatic safety belts, passive interiors and air bag systems which use a lap and shoulder belt. For systems, such as air bag only systems or air bag combined with a lap belt, which symmetrically and uniformly distribute chest forces over a large area of the chest, the agency is adopting the proposed three-inch deflection limit. This should assure that both symmetrical and non-symmetrical systems provide the same level of protection in an equivalent frontal crash.

In addition to the biomechanical basis for the chest deflection limits adopted in this notice, there is another reason for adopting a two-inch deflection limit for systems that can provide concentrated loadings over a limited area of the test dummy. The Hybrid III measures chest deflection by a deflection sensor located near the third rib of the test dummy. Tests conducted on the Hybrid III by NHTSA's Vehicle Research and Test Center have shown that the deflection sensor underestimates chest displacement when a load is applied to a small area away from the deflection sensor. (The test report is filed in Docket No. 74-14, General Reference, Entry 606.)

In a crash, when an occupant is not restrained by a system which provides centralized, uniform loading to a large area, such as an air bag system, the thorax deflection sensor can underestimate the actual chest compression. Thus, in a belt-restrained test dummy, the deflection sensor may read two-inches of deflection, but the actual deflection caused by the off-center loading of a belt near the bottom of the ribcage may be greater than two inches of deflection. Likewise, test dummies in passive interior cars may receive substantial off-center and concentrated loadings. For example, the agency has conducted sled tests simulating 30 mile per hour frontal barrier impacts in which unrestrained test dummies struck the steering column, as they would do in a passive interior equipped car. Measurements of the pre- and post-impact dimensions of the steering wheel rim showed that there was substantial non-symmetrical steering wheel deformation, even though these were frontal impacts. (See, e.g.,

"Frontal Occupant Sled Simulation Correlation, 1983 Chevrolet Celebrity Sled Buck," Publication No. DOT HS 806 728, February 1985.) The expected off-center chest loadings in belt and passive interior systems provide a further basis for applying a two-inch deflection limit for those systems to assure they provide protection comparable to that provided by symmetrical systems.

Use of Acceleration Limits for Air Bag Systems

Two commenters raised questions about the use of an acceleration-based criterion for vehicles which use a combined air bag and lap/shoulder belt system. Mercedes-Benz said that acceleration-based criteria are not appropriate for systems that reduce the deflection of the ribs but increase chest acceleration values. Ford also questioned the use of acceleration-based criteria. Ford said that its tests and testing done by Mercedes-Benz have shown that using an air bag in combination with a lap/shoulder belt can result in increased chest acceleration readings. Ford said it knew of no data to indicate that combined air bag-lap/shoulder belt system loads are more injurious than shoulder belt loads alone. Ford recommended that manufacturers have the option of using either the chest acceleration or chest deflection criterion until use of the Hybrid III is mandatory.

As discussed previously, acceleration and deflection represent two separate types of injury mechanisms. Therefore, the agency believes that it is important to test for both criteria. Although the tests by Mercedes-Benz and Ford show higher chest accelerations, the tests also show that it is possible to develop air bag and lap/shoulder belt systems and meet both criteria. Therefore, the agency is retaining the use of the acceleration-based criterion.

Use of Additional Sensors

Mercedes-Benz said the deflection measuring instrumentation of the Hybrid III cannot adequately measure the interaction between the chest and a variety of vehicle components. Mercedes-Benz said that it is necessary to use either additional deflection sensors or strain gauges. Renault/Peugeot recommended that the agency account for the difference between symmetrical systems and asymmetrical systems by relocating the deflection sensor.

The agency recognizes that the use of additional sensors could be beneficial in the Hybrid III to measure chest deflection. However, such technology would require considerable further development before it could be used for compliance purposes. NHTSA believes that, given the current level of technology, use of a single sensor is sufficient for

the assessment of deflection-caused injuries in frontal impacts.

Femurs

The April 1985 notice proposed to apply the femur injury reduction criterion used with the Part 572 test dummy to the Hybrid III. That criterion limits the femur loads to 2250 pounds to reduce the possibility of femur fractures. No commenter objected to the proposed femur limit and it is accordingly adopted.

Ford and Toyota questioned the need to conduct three pendulum impacts for the knee. They said that using one pendulum impact with the largest mass impactor (11 pounds) was sufficient. GM has informed the agency that the lower mass pendulum impactors were used primarily for the development of an appropriate knee design. Now that the knee design is settled and controlled by the technical drawings, the tests with the low mass impactors are not needed. Accordingly, the agency is adopting the suggestion from Ford and Toyota to reduce the number of knee calibration tests and will require only the use of the 11-pound pendulum impactor.

Hybrid III Positioning Procedure

The April notice proposed new positioning procedures for the Hybrid III, primarily because the curved lumbar spine of that test dummy requires a different positioning technique than those for the Part 572. Based on its testing experience, NHTSA proposed adopting a slightly different version of the positioning procedure used by GM. The difference was the proposed use of the Hybrid III, rather than the SAE J826 H-point machine, with slightly modified leg segments, to determine the H-point of the seat.

GM urged the agency to adopt its dummy positioning procedure. GM said that users can more consistently position the test dummy's H-point using the SAE H-point machine rather than using the Hybrid III. Ford, while explaining that it had insufficient experience with the Hybrid III to develop data on positioning procedures, also urged the agency to adopt GM's positioning procedure. Ford said that since GM has developed its repeatability data on the Hybrid III using its positioning procedure, the agency should use it as well. Ford also said that the use of GM's method to position the test dummy relative to the H-point should reduce variability.

Based on a new series of dummy positioning tests done by the agency's Vehicle Research and Test Center (VRTC), NHTSA agrees that use of the SAE H-point machine is the most consistent method to position the dummy's H-point on the vehicle seat.

Accordingly, the agency is adopting the use of the H-point machine.

In the new test series, VRTC also evaluated a revised method for positioning the Hybrid III test dummy. The testing was done after the results of a joint NHTSA-SAE test series conducted in November 1985 showed that the positioning procedure used for the current Part 572 test dummy and the one proposed in the April 1985 notice for the Hybrid III does not satisfactorily work in all cars. (See Docket 74-14, Notice 39, Entry 39.) The positioning problems are principally due to the curved lumbar spine of the Hybrid III test dummy. In its tests, VRTC positioned the Hybrid III by using the SAE H-point machine and a specification detailing the final position of the Hybrid III body segments prior to the crash test. The test results showed that the H-point of the test dummy could be consistently positioned but that the vertical location of the Hybrid III H-point is ¼ inch below the SAE H-point machine on average. Based on these results, the agency is adopting the new positioning specification for the Hybrid III which requires the H-point of the dummy to be within a specified zone centered ¼ inch below the H-point location of the SAE H-point machine.

GM also urged the agency to make another slight change in the test procedures. GM said that when it settles the test dummy in the seat it uses a thin sheet of plastic behind the dummy to reduce the friction between the fabric of the seat back and the dummy. The plastic is removed after the dummy has been positioned. GM said this technique allows the dummy to be more repeatably positioned. The agency agrees that use of the plastic sheet can reduce friction between the test dummy and the seat. However, the use of the plastic can also create problems, such as dislocating the test dummy during removal of the plastic. Since the agency has successfully conducted its positioning tests without using a sheet of plastic, the agency does not believe there is a need to require its use.

Ford noted that the test procedure calls for testing vertically adjustable seats in their lowest position. It said such a requirement was reasonable for vertically adjustable seats that could not be adjusted higher than seats that are not vertically adjustable. However, Ford said that new power seats can be adjusted to positions above and below the manually adjustable seat position. It said that testing power seats at a different position would increase testing variability. Ford recommended adjusting vertically adjustable seats so that the dummy's hip point is as close as possible to the manufacturer's design

H-point with the seat at the design mid-point of its travel.

The agency recognizes that the seat adjustment issue raised by Ford may lead to test variability. However, the agency does not have any data on the effect of Ford's suggested solution on the design of other manufacturer's power seats. The agency will solicit comments on Ford's proposal in the NPRM addressing additional Hybrid III injury criteria.

Volvo said that the lumbar supports of its seats influence the positioning of the Hybrid III. It requested that the test procedure specify that adjustable lumbar supports should be positioned in their rearmost position. Ford made a similar request. GM, however, indicated that it has not had any problems positioning the Hybrid III in seats with lumbar supports. To reduce positioning problems resulting from the lumbar supports in some vehicles, the agency is adopting Ford's and Volvo's suggestion.

Test Data Analysis

The Chairman of the Society of Automotive Engineers Safety Test Instrumentation Committee noted that the agency proposed to reference an earlier version of the SAE Recommended Practice on Instrumentation (SAE J211a, 1971). He suggested that the agency reference the most recent version (SAE J211, 1980), saying that better data correlation between different testing organizations would result. The agency agrees with SAE and is adopting the SAE J211, 1980 version of the instrumentation Recommended Practice.

Ford and GM recommended that the figures 25 and 26, which proposed a standardized coordinate system for major body segments of the test dummy, be revised to reflect the latest industry practice on coordinate signs. Since those revisions will help ensure uniformity in data analysis by different test facilities, the agency is making the changes for the test measurements adopted in this rulemaking.

Both GM and Ford also recommended changes in the filter used to process electronically measured crash data. GM suggested that a class 180 filter be used for the neck force transducer rather than the proposed class 60 filter. Ford recommended the use of a class 1,000 filter, which is the filter used for the head accelerometer.

NHTSA has conducted all of the testing used to develop the calibration test requirement for the neck using a class 60 filter. The agency does not have any data showing the effects of using either the class 180 filter proposed by GM or the class 1,000 filter proposed by Ford. Therefore the agency has adopted the use of a class 60 filter for the neck transducer during the calibration test. The agency also used a class 60 filter for the accelerometer mounted on the neck pendulum and is therefore adopting the use of that filter to ensure uniformity in measuring pendulum acceleration.

Optional and Mandatory Use of Hybrid III

AMC, Chrysler, Ford, Jaguar and Subaru all urged the agency to defer a decision on permitting the optional use of the Hybrid III test dummy until manufacturers have had more experience with using that test dummy. AMC said it has essentially no experience with the Hybrid III and urged the agency to postpone a decision on allowing the optional use of that test dummy. AMC said this would give small manufacturers time to gain experience with the Hybrid III.

Chrysler also said that it has no experience with the Hybrid III test dummy and would need to conduct two years of testing to be able to develop sufficient information to address the issues raised in the notice. Chrysler said that it was currently developing its 1991 and 1992 models and has no data from Hybrid III test dummies on which to base its design decisions. It said that allowing the optional use of the Hybrid III before that time would give a competitive advantage to manufacturers with more experience with the test device and suggested indefinitely postponing the mandatory effective date.

Ford said that the effective date proposed for optional use of the Hybrid III should be deferred to allow time to resolve the problems Ford raised in its comments and to allow manufacturers time to acquire Hybrid III test dummies. It suggested deferring the proposed optional use until at least September 1, 1989. Ford also recommended that the mandatory use be deferred. Jaguar also said it has not had experience with the Hybrid III and asked that manufacturers have until September 1, 1987, to accumulate information on the performance of the test dummy. Subaru said that it has exclusively used the Part 572 test dummy and does not have any experience with the Hybrid III. It asked the agency to provide time for all manufacturers to gain experience with the Hybrid III, which in its case would be two years, before allowing the Hybrid III as an alternative.

A number of manufacturers, such as GM, Honda, Mercedes-Benz, Volkswagen, and Volvo, that supported optional use of the Hybrid III, urged the agency not to mandate its use at this time. GM asked the agency to permit the immediate optional use of the Hybrid III, but urged NHTSA to provide more

time for all interested parties to become familiar with the test dummy before mandating its use. Honda said that while it supported optional use, it was just beginning to assess the performance of the Hybrid III and needed more time before the use of the Hybrid III is mandated. Mercedes-Benz also supported the use of the Hybrid III as an alternative test device because of its capacity to measure more types of injuries and because of its improved biofidelity for the neck and thorax. However, Mercedes recommended against mandatory use until issues concerning the Hybrid III's use in side impact, the biofidelity of its leg, durability and chest deflection measurements are resolved. Nissan opposed the mandatory us of the Hybrid III saying there is a need to further investigate the differences between the Hybrid III and the Part 572. Toyota said that it was premature to set a mandatory effective date until the test procedure and injury criteria questions are resolved. Volkswagen supported the adoption of the Hybrid III as an alternative test device, but it opposed mandating its use. Volvo supported the optional use of the Hybrid III. It noted that since NHTSA is developing an advanced test dummy, there might not be a need to require the use of the Hybrid III in the interim.

The agency recognizes that manufacturers are concerned about obtaining the Hybrid III test dummy and gaining experience with its use prior to the proposed September 1, 1991, date for mandatory use of that test dummy. However, information provided by the manufacturers of the Hybrid III shows that it will take no longer than approximately one year to supply all manufacturers with sufficient quantities of Hybrid III's. This means that manufacturers will have, at a minimum, more than four years to gain experience in using the Hybrid III. In addition, to assist manufacturers in becoming familiar with the Hybrid III, NHTSA has been placing in the rulemaking docket complete information on the agency's research programs using the Hybrid III test dummy in crash and calibration tests. Since manufacturers will have sufficient time to obtain and gain experience with the Hybrid III by September 1, 1991, the agency has decided to mandate use of the Hybrid III as of that date.

As discussed earlier in this notice, the evidence shows that the Hybrid III is more human-like in its responses to impacts than the existing Part 572 test dummy. In addition, the Hybrid III has the capability to measure far more potential injuries than the current test dummy. The agency is taking advantage of that capability by adopting a limitation on chest deflection which will enable NHTSA to measure a significant source of injury that cannot be measured on the current test dummy. The combination of the better biofidelity and increased injury-measuring capability available with the Hybrid III will enhance vehicle safety.

Adoption of the Hybrid III will not give a competitive advantage to GM, as claimed by some of the commenters, such as Chrysler and Ford. As the developer of the Hybrid III, GM obviously has had more experience with that test dummy than other manufacturers. However, as discussed above, the agency has provided sufficient leadtime to allow all manufacturers to develop sufficient experience with the Hybrid III test dummy. In addition, as discussed in the equivalency section of this notice, there are no data to suggest that it will be easier for GM or other manufacturers to meet the performance requirements of Standard No. 208 with the Hybrid III. Thus GM and other manufacturers using Hybrid III during the phase-in period will not have a competitive advantage over manufacturers using the existing Part 572 test dummy.

Finally, in its comments GM suggested that the agency consider providing manufacturers with an incentive to use the Hybrid III test dummy. GM said that the agency should consider providing manufacturers with extra vehicle credits during the automatic restraint phase-in period for using the Hybrid III. The agency does not believe it is necessary to provide any additional incentive to use the Hybrid III. The mandatory effective date for use of the Hybrid III provides sufficient incentive, since manufacturers will want to begin using the Hybrid III as soon as possible to gain experience with the test dummy before that date.

Optional use of the Hybrid III may begin October 23, 1986. The agency is setting an effective date of less than 180 days to facilitate the efforts of those manufacturers wishing to use the Hybrid III in certifying compliance with the automatic restraint requirements.

Use of Non-instrumented Test Dummies

Ford raised a question about whether the Hybrid III may or must be used for the non-crash performance requirements of Standard No. 208, such as the comfort and convenience requirements of S7.4.3, 7.4.4, and 7.4.5 of the standard. Ford said that manufacturers should be given the option of using either the Part 572 or Hybrid III test dummy to meet the comfort and convenience requirements. The agency agrees that until September 1, 1991, manufacturers should have the option of using either the Part 572 or Hybrid III test dummy. However, since it is important the crash performance requirements and comfort and convenience

requirements be linked together through the use of a single test dummy to measure a vehicle's ability to meet both sets of requirements. Therefore, beginning on September 1, 1991, use of the Hybrid III will be mandatory in determining a vehicle's compliance with any of the requirements of Standard No. 208.

In addition, Ford asked the agency to clarify whether manufacturers can continue to use Part 572 test dummies in the crash tests for Standard Nos. 212, 219, and 301, which only use non-instrumented test dummies to simulate the weight of an occupant. Ford said that the small weight difference and the small difference in seated posture between the two test dummies should have no effect on the results of the testing for Standard Nos. 212, 219, and 301. The agency agrees that use of either test dummy should not affect the test results for those standards. Thus, even after the September 1, 1991, effective date for use of the Hybrid III in the crash and non-crash testing required by Standard No. 208, manufacturers can continue to use, at their option, either the Part 572 or the Hybrid III test dummy in tests conducted in accordance with Standard Nos. 212, 219, and 301.

Economic and Other Impacts

NHTSA has examined the impact of this rulemaking action and determined that it is not major within the meaning of Executive Order 12291 or significant within the meaning of the Department of Transportation's regulatory policies and procedures. The agency has also determined that the economic and other impacts of this rulemaking action are not significant. A final regulatory evaluation describing those effects has been placed in the docket.

In preparing the regulatory evaluation, the agency has considered the comments from several manufacturers that the agency had underestimated the costs associated with using the Hybrid III. Ford said that the cost estimates contained in the April 1985 notice did not take into account the need to conduct sled tests during development work. Ford said that for 1985, it estimated it will conduct 500 sled tests requiring 1000 test dummy applications. Ford also said that NHTSA's estimate of the test dummy inventory needed by a manufacturer is low. It said that it currently has an inventory of 31 Part 572 test dummies and would expect to need a similar inventory of Hybrid III's. In addition, Ford said that NHTSA's incremental cost estimate of $3,000 per test dummy was low. It said that the cost for monitoring the extra data generated by the Hybrid III is $2,700. Ford said that it also would have to incur costs due to upgrading its data acquisition and data processing equipment.

GM said that NHTSA's estimate of a 30-test useful life for the test dummy substantially underestimates its actual useful life, assuming the test dummy is repaired periodically. It said that some of its dummies have been used in more than 150 tests. GM also said that the agency's assumption that a large manufacturer conducts testing requiring approximately 600 dummy applications each year underestimates the actual number of tests conducted. In 1984, GM said it conducted sled and barrier tests requiring 1179 dummy applications. GM said that the two underestimates, in effect, cancel each other out, since the dummies are usable for at least five times as many tests, but they are used four times as often.

Mitsubishi said that its incremental cost per vehicle is $7 rather than 40 cent as estimated by the agency. Mitsubishi explained the reason for this difference is that the price of an imported Hybrid III is approximately two times the agency estimate and its annual production is about one-tenth of the amount used in the agency estimate. Volvo also said the agency had underestimated the incremental cost per vehicle. Volvo said it conducts approximately 500-600 test dummy applications per year in sled and crash testing, making the incremental cost in the range of $15-18 per vehicle based on its export volume to the United States.

NHTSA has re-examined the costs associated with the Hybrid III test dummy. The basic Hybrid III dummy with the instrumentation required by this final rule costs $35,000 or approximately $16,000 more than the existing 572 test dummy. Assuming a useful life for the test dummy of 150 tests, the total estimated incremental capital cost is approximately $107 per dummy test.

To determine the incremental capital cost per test, the agency had to estimate the useful life of the Hybrid III. Based on NHTSA's test experience, the durability of the existing Part 572 test dummy and the Hybrid III test dummy is essentially identical with the exception of the Hybrid III ribs. Because the Hybrid III dummy chest was developed to simulate human chest deflection, the ribs had to be designed with much more precision to reflect human impact response. This redesign uses less metal and consequently they are more susceptible to damage during testing than the Part 572 dummy.

As discussed previously, GM estimates that the Hybrid III ribs can be used in severe unrestrained testing approximately 17 times before the ribs or the damping material needs replacement. In addition, GM's experience shows that the Hybrid III can withstand as many as 150 test applications as long as occasional repairs are made. Ford reported at the previously cited MVMA meeting that one of its belt-restrained Hybrid III test dummies underwent 35 crash tests without any degradation. Clearly, the estimated useful life of the test dummy is highly dependent on the type of testing, restrained or unrestrained, it is used for. Based on its own test experience and the experience of Ford and GM cited above, the agency has decided to use 30 applications as a conservative estimate of the useful life of the ribs. Assuming a life of 30 tests before a set of ribs must be replaced at a cost of approximately $2,000, the incremental per test cost is approximately $70.

The calibration tests for the Hybrid III test dummy have been simplified from the original specification proposed in the April 1985 notice. The Transportation Research Center of Ohio, which does calibration testing of the Hybrid III for the agency, vehicle manufacturers and others estimates the cost of the revised calibration tests is $1528. This is $167 less than the calibration cost for the existing Part 572 test dummy.

Numerous unknown variables will contribute to the manufacturers' operating expense, such as the cost of new or modified test facilities or equipment to maintain the more stringent temperature range of 69° F to 72° F for test dummies, and capital expenditures for lab calibration equipment, signal conditioning equipment, data processing techniques and capabilities, and additional personnel. Obviously, any incremental cost for a particular manufacturer to certify compliance with the automatic restraint requirements of Standard No. 208 will also depend on the extent and nature of its current test facilities and the size of its developmental and new vehicle test programs.

In addition to the costs discussed above, Peugeot raised the issue of a manufacturer's costs increasing because the proposed number of injury measurements made on the Hybrid III will increase the number of tests that must be repeated because of lost data. Since the agency is only adding one additional measurement, chest deflection, for the Hybrid III the number of tests that will have to be repeated due to lost data should not be substantially greater for the Hybrid III than for the Part 572.

NHTSA has determined that it is in the public interest to make the optional use of the Hybrid III test dummy effective in 90 days. This will allow manufacturers time to order the new test dummy to use in their new vehicle development work. Mandatory use of the Hybrid III does not begin until September 1, 1991.

In consideration of the foregoing, Part 572, *Anthropomorphic Test Dummies*, and Part 571.208, *Occupant Crash Protection*, of Title 49 of the Code of Federal Regulations is amended as follows:

Part 572-[AMENDED]

1. The authority citation for Part 572 is amended to read as follows:
Authority: 15 U.S.C. 1392, 1401, 1403, and 1407; delegation of authority at 49 CFR 1.50.

2. A new Subpart E is added to Part 572 to read as follows:
Subpart E—Hybrid III Test Dummy
§ 572.30 *Incorporated materials*
§ 572.31 *General description*
§ 572.32 *Head*
§ 572.33 *Neck*
§ 572.34 *Thorax*
§ 572.35 *Limbs*
§ 572.36 *Test conditions and instrumentation*

§ 572.30 *Incorporated Materials*

(a) The drawings and specifications referred to in this regulation that are not set forth in full are hereby incorporated in this part by reference. The Director of the Federal Register has approved the materials incorporated by reference. For materials subject to change, only the specific version approved by the Director of the Federal Register and specified in the regulation are incorporated. A notice of any change will be published in the *Federal Register*. As a convenience to the reader, the materials incorporated by reference are listed in the Finding Aid Table found at the end of this volume of the Code of Federal Regulations.

(b) The materials incorporated by reference are available for examination in the general reference section of Docket 74-14, Docket Section, National Highway Traffic Safety Administration, Room 5109, 400 Seventh Street, S.W., Washington, DC 20590. Copies may be obtained from Rowley-Scher Reprographics, Inc., 1216 K Street, N.W., Washington, DC 20005 ((202) 628-6667). The drawings and specifications are also on file in the reference library of the Office of the Federal Register, National Archives and Records Administration, Washington, D.C.

§ 572.31 *General description*

(a) The Hybrid III 50th percentile size dummy consists of components and assemblies specified in the Anthropomorphic Test Dummy drawing and specifications package which consists of the following six items:

(1) The Anthropomorphic Test Dummy Parts List, dated July 15, 1986, and containing 13 pages, and a Parts List Index, dated April 26, 1986, containing 6 pages,

(2) A listing of Optional Hybrid III Dummy Transducers, dated April 22, 1986, containing 4 pages,

(3) A General Motors Drawing Package identified by GM drawing No. 78051-218, revision P and subordinate drawings,

(4) Disassembly, Inspection, Assembly and Limbs Adjustment Procedures for the Hybrid III dummy, dated July 15, 1986,

(5) Sign Convention for the signal outputs of Hybrid II dummy transducers, dated July 15, 1986,

(6) Exterior Dimensions of the Hybrid III dummy, dated July 15, 1986.

(b) The dummy is made up of the following component assemblies:

Drawing Number	Revision
78051-61 Head Assembly—Complete—	(T)
78051-90 Neck Assembly—Complete—	(A)
78051-89 Upper Torso Assembly—Complete—	(I)
78051-70 Lower Torso Assembly—Without Pelvic Instrumentation Assembly,	
Drawing No. 78051-59	(C)
86-5001-001 Leg Assembly—Complete (LH)—	
86-5001-002 Leg Assembly—Complete (RH)—	
78051-123 Arm Assembly—Complete (LH)—	(D)
78051-124 Arm Assembly—Complete (RH)—	(D)

(c) Any specifications and requirements set forth in this part supercede those contained in General Motors Drawing No. 78051-218, revision P.

(d) Adjacent segments are joined in a manner such that throughout the range of motion and also under crash-impact conditions, there is no contact between metallic elements except for contacts that exist under static conditions.

(e) The weights, inertial properties and centers of gravity location of component assemblies shall conform to those listed in drawing 78051-338, revision S.

(f) The structural properties of the dummy are such that the dummy conforms to this part in every respect both before and after being used in vehicle test specified in Standard No. 208 of this Chapter (§ 571.208).

§ 572.32 Head

(a) The head consists of the assembly shown in the drawing 78051-61, revision T, and shall conform to each of the drawings subtended therein.

(b) When the head (drawing 78051-61, revision T) with neck transducer structural replacement (drawing 78051-383, revision F) is dropped from a height of 14.8 inches in accordance with paragraph (c) of this section, the peak resultant accelerations at the location of the accelerometers mounted in the head in accordance with 572.36(c) shall not be less than 225g, and not more than 275g. The acceleration/time curve for the test shall be unimodal to the extent that oscillations occurring after the main acceleration pulse are less than ten percent (zero to peak) of the main pulse. The lateral acceleration vector shall not exceed 15g (zero to peak).

(c) *Test Procedure.* (1) Soak the head assembly in a test environment at any temperature between 66° F to 78° F and at a relative humidity from 10% to 70% for a period of at least four hours prior to its application in a test.

(2) Clean the head's skin surface and the surface of the impact plate with 1,1,1 Trichlorethane or equivalent.

(3) Suspend the head, as shown in Figure 19, so that the lowest point on the forehead is 0.5 inches below the lowest point on the dummy's nose when the midsagittal plane is vertical.

(4) Drop the head from the specified height by means that ensure instant release onto a rigidly supported flat horizontal steel plate, which is 2 inches thick and 2 feet square. The plate shall have a clean, dry surface and any microfinish of not less than 8 microinches (rms) and not more than 80 microinches (rms).

(5) Allow at least 2 hours between successive tests on the same head.

§ 572.33 Neck

(a) The neck consists of the assembly shown in drawing 78051-90, revision A and conforms to each of the drawings subtended therein.

(b) When the neck and head assembly (consisting of the parts 78051-61, revision T; -84; -90, revision A; -96; -98; -303, revision E; -305; -306; -307, revision X, which has a neck transducer (drawing 83-5001-008) installed in conformance with 572.36(d), is tested in accordance with paragraph (c) of this section, it shall have the following characteristics:

(1) *Flexion* (i) Plane D, referenced in Figure 20, shall rotate, between 64 degrees and 78 degrees, which shall occur between 57 milliseconds (ms) and 64 ms from time zero. In first rebound, the rotation of plane D shall cross 0 degree between 113 ms and 128 ms.

(ii) The moment measured by the neck transducer (drawing 83-5001-008) about the occipital condyles, referenced in Figure 20, shall be calculated by the following formula: Moment (lbs-ft) = M_y + 0.02875 $_x F_x$ where M_y is the moment measured in lbs-ft by the moment sensor of the neck transducer and F_x is the force measure measured in lbs by the x axis force sensor of the neck transducer. The moment shall have a maximum value between 65 lbs-ft and 80 lbs-ft occurring between 47 ms and 58 ms, and the positive moment shall decay for the first time to 0 lb-ft between 97 ms and 107 ms.

(2) *Extension* (i) Plane D, referenced in Figure 21, shall rotate between 81 degrees and 106 degrees, which shall occur between 72 and 82 ms from time zero. In first rebound, the rotation of plane D shall cross 0 degree between 147 and 174 ms.

(ii) The moment measured by the neck transducer (drawing 83-5001-008) about the occipital condyles, referenced in Figure 21, shall be calculated by the following formula: Moment (lbs-ft) = M_y + 0.02875 $_x F_x$ where M_y is the moment measured in lbs-ft by the moment sensor of the neck transducer and F_x is the force measure measured in lbs by the x axis force sensor of the neck transducer. The moment shall have a minimum value between −39 lbs-ft and −59 lbs-ft, which shall occur between 65 ms and 79 ms., and the negative moment shall decay for the first time to 0 lb-ft between 120 ms and 148 ms.

(3) Time zero is defined as the time of contact between the pendulum striker plate and the aluminum honeycomb material.

(c) *Test Procedure.* (1) Soak the test material in a test environment at any temperature between 69 degrees F to 72 degrees F and at a relative humidity from 10% to 70% for a period of at least four hours prior to its application in a test.

(2) Torque the jamnut (78051-64) on the neck cable (78051-301, revision E) to 1.0 lbs-ft ± .2 lbs-ft.

(3) Mount the head-neck assembly, defined in paragraph (b) of this section, on a rigid pendulum as shown in Figure 22 so that the head's midsagittal plane is vertical and coincides with the plane of motion of the pendulum's longitudinal axis.

(4) Release the pendulum and allow it to fall freely from a height such that the tangential velocity at the pendulum accelerometer centerline at the instance of contact with the honeycomb is 23.0 ft/sec ± 0.4 ft/sec. for flexion testing and 19.9 ft/sec ± 0.4 ft/sec. for extension testing. The pendulum deceleration vs. time pulse for flexion testing shall

conform to the characteristics shown in Table A and the decaying deceleration-time curve shall first cross 5g between 34 ms and 42 ms. The pendulum deceleration vs. time pulse for extension testing shall conform to the characteristics shown in Table B and the decaying deceleration-time curve shall cross 5g between 38 ms and 46 ms.

Table A
Flexion Pendulum Deceleration vs. Time Pulse

Time (ms)	Flexion deceleration level (g)
10...............................	22.50-27.50
20...............................	17.60-22.60
30...............................	12.50-18.50
Any other time above 30 ms.......	29 maximum

Table B
Extension Pendulum Deceleration vs. Time Pulse

Time (ms)	Extension deceleration level (g)
10...............................	17.20-21.20
20...............................	14.00-19.00
30...............................	11.00-16.00
Any other time above 30 ms.......	22 maximum

(5) Allow the neck to flex without impact of the head or neck with any object during the test.

§ 572.34 Thorax

(a) The thorax consists of the upper torso assembly in drawing 78051-89, revision I and shall conform to each of the drawings subtended therein.

(b) When impacted by a test probe conforming to S572.36(a) at 22 fps ± .40 fps in accordance with paragraph (c) of this section, the thorax of a complete dummy assembly (78051-218, revision P) with left and right shoes (78051-294 and -295) removed, shall resist with the force measured by the test probe from time zero of 1162.5 pounds ± 82.5 pounds and shall have a sternum displacement measured relative to spine of 2.68 inches ± .18 inches. The internal hysteresis in each impact shall be more than 69% but less than 85%. The force measured is the product of pendulum mass and deceleration. Time zero is defined as the time of first contact between the upper thorax and pendulum face.

(c) Test procedure. (1) Soak the test dummy in an environment with a relative humidity from 10% to 70% until the temperature of the ribs of the test dummy have stabilized at a temperature between 69° F and 72° F.

(2) Seat the dummy without back and arm supports on a surface as shown in Figure 23.

(3) Place the longitudinal centerline of the test probe so that it is .5 ± .04 in. below the horizontal centerline of the No. 3 Rib (reference drawing number 79051-64, revision A-M) as shown in Figure 23.

(4) Align the test probe specified in S572.36(a) so that at impact its longitudinal centerline coincides within .5 degree of a horizontal line in the dummy's midsagittal plane.

(5) Impact the thorax with the test probe so that the longitudinal centerline of the test probe falls within 2 degrees of a horizontal line in the dummy midsagittal plane at the moment of impact.

(6) Guide the probe during impact so that it moves with no significant lateral, vertical, or rotational movement.

(7) Measure the horizontal deflection of the sternum relative to the thoracic spine along the line established by the longitudinal centerline of the probe at the moment of impact, using a potentiometer (ref. drawing 78051-317, revision A) mounted inside the sternum as shown in drawing 78051-89, revision I.

(8) Measure hysteresis by determining the ratio of the area between the loading and unloading portions of the force deflection curve to the area under the loading portion of the curve.

§572.35 Limbs

(a) The limbs consist of the following assemblies: leg assemblies 86-5001-001 and -002 and arm assemblies 78051-123, revision D, and -124, revision D, and shall conform to the drawings subtended therein.

(b) When each knee of the leg assemblies is impacted by the pendulum defined in S572.36(b) in accordance with paragraph (c) of this section at 6.9 ft/sec ± .10 ft/sec., the peak knee impact force, which is a product of pendulum mass and acceleration, shall have a minimum value of not less than 996 pounds and a maximum value of not greater than 1566 pounds.

(c) Test Procedure. (1) The test material consists of leg assemblies (86-5001-001) left and (-002) right with upper leg assemblies (78051-46) left and

(78051-47) right removed. The load cell simulator (78051-319, revision A) is used to secure the knee cap assemblies (79051-16, revision B) as shown in Figure 24.

(2) Soak the test material in a test environment at any temperature between 66° F to 78° F and at a relative humidity from 10% to 70% for a period of at least four hours prior to its application in a test.

(3) Mount the test material with the leg assembly secured through the load cell simulator to a rigid surface as shown in Figure 24. No contact is permitted between the foot and any other exterior surfaces.

(4) Place the longitudinal centerline of the test probe so that at contact with the knee it is colinear within 2 degrees with the longitudinal centerline of the femur load cell simulator.

(5) Guide the pendulum so that there is no significant lateral, vertical or rotational movement at time zero.

(6) Impact the knee with the test probe so that the longitudinal centerline of the test probe at the instant of impact falls within .5 degrees of a horizontal line parallel to the femur load cell simulator at time zero.

(7) Time zero is defined as the time of contact between the test probe and the knee.

§ 572.36 *Test conditions and instrumentation*

(a) The test probe used for thoracic impact tests is a 6 inch diameter cylinder that weighs 51.5 pounds including instrumentation. Its impacting end has a flat right angle face that is rigid and has an edge radius of 0.5 inches. The test probe has an accelerometer mounted on the end opposite from impact with its sensitive axis colinear to the longitudinal centerline of the cylinder.

(b) The test probe used for the knee impact tests is a 3 inch diameter cylinder that weighs 11 pounds including instrumentation. Its impacting end has a flat right angle face that is rigid and has an edge radius of 0.2 inches. The test probe has an accelerometer mounted on the end opposite from impact with its sensitive axis colinear to the longitudinal centerline of the cylinder.

(c) Head accelerometers shall have dimensions, response characteristics and sensitive mass locations specified in drawing 78051-136, revision A or its equivalent and be mounted in the head as shown in drawing 78051-61, revision T, and in the assembly shown in drawing 78051-218, revision D.

(d) The neck transducer shall have the dimensions, response characteristics, and sensitive axis

locations specified in drawing 83-5001-008 or its equivalent and be mounted for testing as shown in drawing 79051-63, revision W, and in the assembly shown in drawing 78051-218, revision P.

(e) The chest accelerometers shall have the dimensions, response characteristics, and sensitive mass locations specified in drawing 78051-136, revision A or its equivalent and be mounted as shown with adaptor assembly 78051-116, revision D, for assembly into 78051-218, revision L.

(f) The chest deflection transducer shall have the dimensions and response characteristics specified in drawing 78051-342, revision A or equivalent, and be mounted in the chest deflection transducer assembly 87051-317, revision A, for assembly into 78051-218, revision L.

(g) The thorax and knee impactor accelerometers shall have the dimensions and characteristics of Endevco Model 7231c or equivalent. Each accelerometer shall be mounted with its sensitive axis colinear with the pendulum's longitudinal centerline.

(h) The femur load cell shall have the dimensions, response characteristics, and sensitive axis locations specified in drawing 78051-265 or its equivalent and be mounted in assemblies 78051-46 and -47 for assembly into 78051-218, revision L.

(i) The outputs of acceleration and force-sensing devices installed in the dummy and in the test apparatus specified by this part are recorded in individual data channels that conform to the requirements of SAE Recommended Practice J211, JUNE 1980, "Instrumentation for Impact Tests," with channel classes as follows:

(1) Head acceleration—Class 1000
(2) Neck force—Class 60
(3) Neck pendulum acceleration—Class 60
(4) Thorax and thorax pendulum acceleration—Class 180
(5) Thorax deflection—Class 180
(6) Knee pendulum acceleration—Class 600
(7) Femur force—Class 600

(j) Coordinate signs for instrumentation polarity conform to the sign convention shown in the document incorporated by § 572.31(a)(5).

(k) The mountings for sensing devices shall have no resonance frequency within range of 3 times the frequency range of the applicable channel class.

(l) Limb joints are set at lg, barely restraining the weight of the limb when it is extended horizontally. The force required to move a limb segment shall not exceed 2g throughout the range of limb motion.

(m) Performance tests of the same component, segment, assembly, or fully assembled dummy are separated in time by a period of not less than 30 minutes unless otherwise noted.

(n) Surfaces of dummy components are not painted except as specified in this part or in drawings subtended by this part. PART 571 [Amended]

2. The authority citation for Part 571 continues to read as follows:

Authority: 15 U.S.C. 1392, 1401, 1403, 1407; delegation of authority at 49 CFR 1.50.

3. Section S5 of Standard No. 208 (49 CFR 571.208) is amended by revising S5.1 to read as follows:

§ 571.208 [Amended]

S5. *Occupant crash protection requirements.*

S5.1 Vehicles subject to S5.1 and manufactured before September 1, 1991, shall comply with either, at the manufacturer's option, 5.1(a) or (b). Vehicles subject to S5.1 and manufactured on or after September 1, 1991, shall comply with 5.1(b).

(a) Impact a vehicle traveling longitudinally forward at any speed, up to and including 30 mph, into a fixed collision barrier that is perpendicular to the line of travel of the vehicle, or at any angle up to 30 degrees in either direction from the perpendicular to the line of travel of the vehicle under the applicable conditions of S8. The test dummy specified in S8.1.8.1 placed at each front outboard designated seating position shall meet the injury criteria of S6.1.1, 6.1.2, 6.1.3, and 6.1.4.

(b) Impact a vehicle traveling longitudinally forward at any speed, up to and including 30 mph, into a fixed collision barrier that is perpendicular to the line of travel of the vehicle, or at any angle up to 30 degrees in either direction from the perpendicular to the line of travel of the vehicle, under the applicable conditions of S8. The test dummy specified in S8.1.8.2 placed at each front outboard designated seating position shall meet the injury criteria of S6.2.1, 6.2.2, 6.2.3, 6.2.4, and 6.2.5.

3. Section S5.2 of Standard No. 208 is revised to read as follows:

S5.2 Lateral moving barrier crash.

S5.2.1 Vehicles subject to S5.2 and manufactured before September 1, 1991, shall comply with either, at the manufacturer's option, 5.2.1(a) or (b). Vehicles subject to S5.2 and manufactured on or after September 1, 1991, shall comply with 5.2.1(b).

(a) Impact a vehicle laterally on either side by a barrier moving at 20 mph under the applicable conditions of S8. The test dummy specified in S8.1.8.1 placed at the front outboard designated seating position adjacent to the impacted side shall meet the injury criteria of S6.1.2 and S6.1.3.

(b) When the vehicle is impacted laterally under the applicable conditions of S8, on either side by a barrier moving at 20 mph, with a test device specified in S8.1.8.2, which is seated at the front outboard designated seating position adjacent to the impacted side, it shall meet the injury criteria of S6.2.2, and S6.2.3.

4. Section S5.3 of Standard No. 208 is revised to read as follows:

S5.3 *Rollover* Subject a vehicle to a rollover test under the applicable condition of S8 in either lateral direction at 30 mph with either, at the manufacturer's option, a test dummy specified in S8.1.8.1 or S8.1.8.2, placed in the front outboard designated seating position on the vehicle's lower side as mounted on the test platform. The test dummy shall meet the injury criteria of either S6.1.1 or S6.2.1.

5. Section S6 of Standard No. 208 is revised to read as follows:

S6. *Injury Criteria*

S6.1 Injury criteria for the Part 572, Subpart B, 50th percentile Male Dummy.

S6.1.1 All portions of the test dummy shall be contained within the outer surfaces of the vehicle passenger compartment throughout the test.

S6.1.2 The resultant acceleration at the center of gravity of the head shall be such that the expression:

$$\left[\frac{1}{t_2 - t_1} \int_{t_1}^{t_2} a\,dt \right]^{2.5} t_2 - t_1$$

shall not exceed 1,000, where a is the resultant acceleration expressed as a multiple of g (the acceleration of gravity), and t_1 and t_2 are any two points during the crash.

S6.1.3 The resultant acceleration at the center of gravity of the upper thorax shall not exceed 60 g's, except for intervals whose cumulative duration is not more than 3 milliseconds.

S6.1.4 The compressive force transmitted axially through each upper leg shall not exceed 2250 pounds.

S6.2 *Injury criteria for the Part 572, Subpart E, Hybrid III Dummy*

S6.2.1 All portions of the test dummy shall be contained within the outer surfaces of the vehicle passenger compartment throughout the test.

S6.2.2 The resultant acceleration at the center of gravity of the head shall be such that the expression:

$$\left[\frac{1}{t_2 - t_1} \int_{t_1}^{t_2} a\,dt \right]^{2.5} t_2 - t_1$$

shall not exceed 1,000, where a is the resultant acceleration expressed as a multiple of g (the acceleration of gravity), and t_1 and t_2 are any two point during the crash.

S6.2.3 The resultant acceleration calculated from the thoracic instrumentation shown in drawing 78051-218, revision L, incorporated by reference in Part 572, Subpart E of this Chapter, shall not exceed 60g's, except for intervals whose cumulative duration is not more than 3 milliseconds.

S6.2.4 Compression deflection of the sternum relative to spine, as determined by instrumentation shown in drawing 78051-317, revision A, incorporated by reference in Part 572, Subpart E of this Chapter, shall not exceed 2 inches for loadings applied through any impact surfaces except for those systems which are gas inflated and provide distributed loading to the torso during a crash. For gas-inflated systems which provide distributive loading to the torso, the thoracic deflection shall not exceed 3 inches.

S6.2.5 The force transmitted axially through each upper leg shall not exceed 2250 pounds.

6. Section S8.1.8 of Standard No. 208 is revised to read as follows:

S8.1.8 *Anthropomorphic test dummies*

S8.1.8.1 The anthropomorphic test dummies used for evaluation of occupant protection systems manufactured pursuant to applicable portions of paragraphs S4.1.2, 4.1.3, and S4.1.4 shall conform to the requirements of Subpart B of Part 572 of this Chapter.

S8.1.8.2 Anthropomorphic test devices used for the evaluation of occupant protection systems manufactured pursuant to applicable portions of paragraphs S4.1.2, S4.1.3, and S4.1.4 shall conform to the requirements of Subpart E of Part 572 of this Chapter.

7. Section S8.1.9 of Standard No. 208 is revised to read as follows:

S8.1.9.1 Each Part 572, Subpart B, test dummy specified in S8.1.8.1 is clothed in formfitting cotton stretch garments with short sleeves and midcalf length pants. Each foot of the test dummy is equipped with a size 11EE shoe which meets the config-

uration size, sole, and heel thickness specifications of MIL-S-131192 and weighs 1.25 ± 0.2 pounds.

S8.1.9.2 Each Part 572, Subpart E, test dummy specified in S8.1.8.2 is clothed in formfitting cotton stretch garments with short sleeves and midcalf length pants specified in drawings 78051-292 and -293 incorporated by reference in Part 572, Subpart E, of this Chapter, respectively or their equivalents. A size 11EE shoe specified in drawings 78051-294 (left) and 78051-295 (right) or their equivalents is placed on each foot of the test dummy.

8. Section S8.1.13 of Standard No. 208 is revised to read as follows:

S8.1.13 *Temperature of the test dummy*

S8.1.13.1 The stabilized temperature of the test dummy specified by S8.1.8.1 is at any level between 66 degrees F and 78 degrees F.

S8.1.13.2 The stabilized temperature of the test dummy specified by S8.1.8.2 is at any level between 69 degrees F and 72 degrees F.

9. A new fourth sentence is added to section S8.1.3 to read as follows:

Adjustable lumbar supports are positioned so that the lumbar support is in its lowest adjustment position.

10. A new section S11 is added to read as follows:

S11. *Positioning Procedure for the Part 572 Subpart E Test Dummy*

Position a test dummy, conforming to Subpart E of Part 572 of this Chapter, in each front outboard seating position of a vehicle as specified in S11.1 through S11.6. Each test dummy is restrained in accordance with the applicable requirements of S4.1.2.1, 4.1.2.2 or S4.6.

S11.1 *Head.* The transverse instrumentation platform of the head shall be horizontal within ½ degree.

S11.2 *Arms*

S11.2.1 The driver's upper arms shall be adjacent to the torso with the centerlines as close to a vertical plane as possible.

S11.2.2 The passenger's upper arms shall be in contact with the seat back and the sides of torso.

S11.3 *Hands*

S11.3.1 The palms of the driver test dummy shall be in contact with the outer part of the steering wheel rim at the rim's horizontal centerline. The thumbs shall be over the steering wheel rim and attached with adhesive tape to provide a breakaway force of between 2 to 5 pounds.

S11.3.2 The palms of the passenger test dummy shall be in contact with outside of thigh. The little finger shall be in contact with the seat cushion.

S11.4 *Torso*

S11.4.1 In vehicles equipped with bench seats, the upper torso of the driver and passenger test dummies shall rest against the seat back. The midsagittal plane of the driver dummy shall be vertical and parallel to the vehicle's longitudinal centerline, and pass through the center of the steering wheel rim. The midsagittal plane of the passenger dummy shall be vertical and parallel to the vehicle's longitudinal centerline and the same distance from the vehicle's longitudinal centerline as the midsagittal plane of the driver dummy.

S11.4.2 In vehicles equipped with bucket seats, the upper torso of the driver and passenger test dummies shall rest against the seat back. The midsagittal plane of the driver and the passenger dummy shall be vertical and shall coincide with the longitudinal centerline of the bucket seat.

S11.4.3 *Lower torso*

S11.4.3.1 *H-point.* The H-point of the driver and passenger test dummies shall coincide within ½ inch in the vertical dimension and ½ inch in the horizontal dimension of a point ¼ inch below the position of the H-point determined by using the equipment and procedures specified in SAE J826 (Apr 80) except that the length of the lower leg and thigh segments of the H-point machine shall be adjusted to 16.3 and 15.8 inches, respectively, instead of the 50th percentile values specified in Table 1 of SAE J826.

S11.4.3.2 *Pelvic angle.* As determined using the pelvic angle gage (GM drawing 78051-532 incorporated by reference in Part 572, Subpart E, of this chapter) which is inserted into the H-point gaging hole of the dummy, the angle measured from the horizontal on the 3 inch flat surface of the gage shall be 22½ degrees plus or minus 2½ degrees.

S11.5 *Legs.* The upper legs of the driver and passenger test dummies shall rest against the seat cushion to the extent permitted by placement of the feet. The initial distance between the outboard knee clevis flange surfaces shall be 10.6 inches. To the extent practicable, the left leg of the driver dummy and both legs of the passenger dummy shall be in vertical longitudinal planes. Final adjustment to accommodate placement of feet in accordance with S11.6 for various passenger compartment configurations is permitted.

S11.6 *Feet*

S11.6.1 The right foot of the driver test dummy shall rest on the undepressed accelerator with the rearmost point of the heel on the floor surface in the plane of the pedal. If the foot cannot be placed on the accelerator pedal, it shall be positioned perpendicular to the tibia and placed as far forward as possible in the direction of the centerline of the pedal with the rearmost point of the heel resting on the floor surface. The heel of the left foot shall be placed as far forward as possible and shall rest on the floor surface. The left foot shall be positioned as flat as possible on the floor surface. The longitudinal centerline of the left foot shall be placed as parallel as possible to the longitudinal centerline of the vehicle.

S11.6.2 The heels of both feet of the passenger test dummy shall be placed as far forward as possible and shall rest on the floor surface. Both feet shall be positioned as flat as possible on the floor surface. The longitudinal centerline of the feet shall be placed as parallel as possible to the longitudinal centerline of the vehicle.

S11.7 *Test dummy positioning for latchplate access.* The reach envelopes specified in S7.4.4 are obtained by positioning a test dummy in the driver's seat or passenger's seat in its forwardmost adjustment position. Attach the lines for the inboard and outboard arms to the test dummy as described in Figure 3 of this standard. Extend each line backward and outboard to generate the compliance arcs of the outboard reach envelope of the test dummy's arms.

S11.8 *Test dummy positioning for belt contact force.* To determine compliance with S7.4.3 of this standard, position the test dummy in the vehicle in accordance with the requirements specified in S11.1 through S11.6 and under the conditions of S8.1.2 and S8.1.3. Pull the belt webbing three inches from the test dummy's chest and release until the webbing is within 1 inch of the test dummy's chest and measure the belt contact force.

S11.9 *Manual belt adjustment for dynamic testing.* With the test dummy at its designated seating position as specified by the appropriate requirements of S8.1.2, S8.1.3 and S11.1 through S11.6, place the Type 2 manual belt around the test dummy and fasten the latch. Remove all slack from the lap belt. Pull the upper torso webbing out of the retractor and allow it to retract; repeat this operation four times. Apply a 2 to 4 pound tension load to the lap belt. If the belt system is equipped with a tension-relieving device introduce the maximum amount of slack into the upper torso belt that is recommended by the manufacturer for normal use in the owner's manual for the vehicle. If the belt system is not equipped with a tension-relieving device, allow the excess webbing in the shoulder belt to be retracted by the retractive force of the retractor.

Issued on July 21, 1986

Diane K. Steed
Administrator

51 F.R. 26688
July 25, 1986

PART 572—ANTHROPOMORPHIC TEST DUMMIES

Subpart A—General

§ 572.1 Scope. This part describes the anthropomorphic test dummies that are to be used for compliance testing of motor vehicles and motor vehicle equipment with motor vehicle safety standards.

§ 572.2 Purpose. The design and performance criteria specified in this part are intended to describe measuring tools with sufficient precision to give repetitive and correlative results under similar test conditions and to reflect adequately the protective performance of a vehicle, or item or motor vehicle equipment, with respect to human occupants.

§ 572.3 Application. This part does not in itself impose duties or liabilities on any person. It is a description of tools that measure the performance of occupant protection systems required by the safety standards that incorporate it. It is designed to be referenced by, and become a part of, the test procedures specified in motor vehicle safety standards such as Standard No. 208, Occupant Crash Protection.

§ 572.4 Terminology.

(a) The term "dummy," when used in this Subpart A, refers to any test device described by this part. The term "dummy," when used in any other subpart of this part, refers to the particular dummy described in that part.

(b) Terms describing parts of the dummy, such as "head," are the same as names for corresponding parts of the human body.

(c) The term "upright position" means the position of the dummy when it is seated in accordance with the procedures of 572.11(i).

Subpart B—50th Percentile Male

§ 572.5 General description.

(a) The dummy consists of the component assemblies specified in Figure 1, which are described in their entirety by means of approximately 250 drawings and specifications that are grouped by component assemblies under the following nine headings:

SA 150 M070	Right arm assembly
SA 150 M071	Left arm assembly
SA 150 M050	Lumbar spine assembly
SA 150 M060	Pelvis and abdomen assembly
SA 150 M080	Right leg assembly
SA 150 M081	Left leg assembly
SA 150 M010	Head assembly
SA 150 M020	Neck assembly
SA 150 M030	Shoulder-thorax assembly

The drawings and specifications are incorporated in this Part by reference to the nine headings, and are available for examination in Docket 73–8, Room 5109, 400 Seventh Street, S.W., Washington, D.C. 20590. [Copies may be obtained from Rowley-Scher Reprographics, Inc. 1216 K Street, N.W., Washington, D.C. 20005, attention Mr. Allan Goldberg and Mr. Mark Krysinski ((202) 628–6667). The drawings and specifications are subject to changes, but any change will be accomplished by appropriate administrative procedures, will be announced by publication in the *Federal Register*, and will be available for examination and copying as indicated in the paragraph. The drawings and specifications are also on file in the reference library of the *Federal Reigister*, National Archives and Records Services, General Services Administration, Washington, D.C. (50 F.R. 25422–June 19, 1985. Effective: June 19, 1985)]

The drawings and specifications are on file in the reference library of the *Federal Register*, National Archives and Records Service, General Services Administration, Washington, D.C.

(b) Adjacent segments are joined in a manner such that throughout the range of motion and also under crash-impact conditions there is no contact between metallic elements except for contacts that exist under static conditions.

(c) The structural properties of the dummy are such that the dummy conforms to this part in every respect both before and after being used in vehicle tests specified in Standard No. 208 (§ 571.208).

A specimen of the dummy is available for surface measurements, and access can be arranged through: Office of Vehicle Safety Standards, National Highway Traffic Safety Administration, 400 Seventh Street, S.W., Washington, D.C. 20590.

§ 572.6 Head.

(a) The head consists of the assembly shown as number SA 150 M010 in Figure 1 and conforms to each of the drawings subtended by number SA 150 M010.

(b) When the head is dropped from a height of 10 inches in accordance with paragraph (c) of this section, the peak resultant accelerations at the location of the accelerometers mounted in the head form in accordance with § 572.11(b) shall be not less than 210g, and not more than 260g. The acceleration/time curve for the test shall be unimodal and shall lie at or above the 100g level for an interval not less than 0.9 milliseconds and not more than 1.5 milliseconds. The lateral acceleration vector shall not exceed 10g.

(c) Test procedure:

(1) Suspend the head as shown in Figure 2, so that the lowest point on the forehead is 0.5 inches below the lowest point on the dummy's nose when the midsagittal plane is vertical.

(2) Drop the head from the specified height by a means that ensures instant release onto a rigidly supported flat horizontal steel plate, 2 inches thick and 2 feet square, which has a clean, dry surface and any microfinish of not less than 8 microinches (rms) and not more than 80 microinches (rms).

(3) Allow a time period of at least 2 hours between successive tests on the same head.

§ 572.7 Neck.

(a) The neck consists of the assembly shown as number SA 150 M020 in Figure 1 and conforms to each of the drawings subtended by number SA 150 M020.

(b) When the neck is tested with the head in accordance with paragraph (c) of this section, the head shall rotate in reference to the pendulum's longitudinal centerline a total of 68° ± 5° about its center of gravity, rotating to the extent specified in the following table at each indicated point in time, measured from impact, with a chordal displacement measured at its center of gravity that is within the limits specified. The chordal displacement at time T is defined as the straight line distance between (1) the position relative to the pendulum arm of the head center of gravity at time zero, and (2) the position relative to the pendulum arm of the head center of gravity at time T as illustrated by Figure 3. The peak resultant acceleration recorded at the location of the accelerometers mounted in the head form in accordance with § 572.11(b) shall not exceed 26g. The pendulum shall not reverse direction until the head's center of gravity returns to the original zero time position relative to the pendulum arm.

Rotation (degrees)	Time (ms) ± (2 + .08T)	Chordal Displacement (inches ± 0.5)
0	0	0.0
30	30	2.6
60	46	4.8
Maximum	60	5.5
60	75	4.8
30	95	2.6
0	112	0.0

(c) Test procedure:

(1) Mount the head and neck on a rigid pendulum as specified in Figure 4, so that the head's midsagittal plane is vertical and coincides with the plane of motion of the pendulum's longitudinal centerline. Mount the neck directly to the pendulum as shown in Figure 4.

(2) Release the pendulum and allow it to fall freely from a height such that the velocity at impact is 23.5 ±2.0 feet per second (fps), measured at the center of the accelerometer specified in Figure 4.

(3) Decelerate the pendulum to a stop with an acceleration-time pulse described as follows:

(i) Establish 5g and 20g levels on the a−t curve.

(ii) Establish t_1 at the point where the rising a−t curve first crosses the 5g level, t_2 at the point where the rising a−t curve first crosses the 20g level, t_3 at the point where the decaying a – t curve last crosses the 20g level, and t_4 at the point where the decaying a – t curve first crosses the 5g level.

(iii) $t_2 - t_1$ shall be not more than 3 milliseconds.

(iv) $t_3 - t_2$ shall be not less than 25 milliseconds and not more than 30 milliseconds.

(v) $t_4 - t_3$ shall be not more than 10 milliseconds.

(vi) The average deceleration between t_2 and t_3 shall be not less than 20g and not more than 24g.

(vii) Allow the neck to flex without impact of the head or neck with any object other than the pendulum arm.

§ 572.8 Thorax.

(a) The thorax consists of the assembly shown as number SA 150 M030 in Figure 1, and conforms to each of the drawings subtended by number SA 150 M030.

(b) The thorax contains enough unobstructed interior space behind the rib cage to permit the midpoint of the sternum to be depressed 2 inches without contact between the rib cage and other parts of the dummy or its instrumentation, except for instruments specified in subparagraph (d) (7) of this section.

(c) When impacted by a test probe conforming to § 572.11(a) at 14 fps and at 22 fps in accordance with paragraph (d) of this section, the thorax shall resist with forces measured by the test probe of not more than 1450 pounds and 2250 pounds, respectively, and shall deflect by amounts not greater than 1.1 inches and 1.7 inches, respectively. The internal hysteresis in each impact shall be not less than 50 percent and not more than 70 percent.

(d) Test Procedure:

(1) With the dummy seated without back support on a surface as specified in § 572.11(i) and in the orientation specified in § 572.11(i), adjust the dummy arms and legs until they are extended horizontally forward parallel to the midsagittal plane.

(2) Place the longitudinal center line of the test probe so that it is 17.7 ± 0.1 inches above the seating surface at impact.

(3) Align the test probe specified in § 572.11 (a) so that at impact its longitudinal centerline

coincides within 2 degrees of a horizontal line in the dummy's midsagittal plane.

(4) Adjust the dummy so that the surface area on the thorax immediately adjacent to the projected longitudinal center line of the test probe is vertical. Limb support, as needed to achieve and maintain this orientation, may be provided by placement of a steel rod of any diameter not less than one-quarter of an inch and not more than three-eighths of an inch, with hemispherical ends, vertically under the limb at its projected geometric center.

(5) Impact the thorax with the test probe so that its longitudinal centerline falls within 2 degrees of a horizontal line in the dummy's midsagittal plane at the moment of impact.

(6) Guide the probe during impact so that it moves with no significant lateral, vertical, or rotational movement.

(7) Measure the horizontal deflection of the sternum relative to the thoracic spine along the line established by the longitudinal centerline of the probe at the moment of impact, using a potentiometer mounted inside the sternum.

(8) Measure hysteresis by determining the ratio of the area between the loading and unloading portions of the force deflection curve to the area under the loading portion of the curve.

§ 572.9 Lumbar spine, abdomen, and pelvis.

(a) The lumbar spine, abdomen, and pelvis consist of the assemblies designated as numbers SA 150 M050 and SA 150 M060 in Figure 1 and conform to the drawings subtended by these numbers.

(b) When subjected to continuously applied force in accordance with paragraph (c) of this section, the lumbar spine assembly shall flex by an amount that permits the rigid thoracic spine to rotate from its initial position in accordance with Figure 11 by the number of degrees shown below at each specified force level, and straighten upon removal of the force to within 12 degrees of its initial position in accordance with Figure 11.

Flexion (degrees)	Force (± 6 pounds)
0	0
20	28
30	40
40	52

(c) Test procedure:

(1) Assemble the thorax, lumbar spine, pelvic, and upper leg assemblies (above the femur force transducers), ensuring that all component surfaces are clean, dry, and untreated unless otherwise specified, and attach them to the horizontal fixture shown in Figure 5 at the two link rod pins and with the mounting brackets for the lumbar test fixtures illustrated in Figure 6 to 9.

(2) Attach the rear mounting of the pelvis to the pelvic instrument cavity rear face at the four ¼″ cap screw holes and attach the front mounting at the femur axial rotation joint. Tighten the mountings so that the pelvic-lumbar adapter is horizontal and adjust the femur friction plungers at each hip socket joint to 240 inch-pounds torque.

(3) Flex the thorax forward 50° and then rearward as necessary to return it to its initial position in accordance with Figure 11 unsupported by external means.

(4) Apply a forward force perpendicular to the thorax instrument cavity rear face in the midsagittal plane 15 inches above the top surface of the pelvic-lumbar adapter. Apply the force at any torso deflection rate between .5 and 1.5 degrees per second up to 40° of flexion but no further, continue to apply for 10 seconds that force necessary to maintain 40° of flexion, and record the force with an instrument mounted to the thorax as shown in Figure 5. Release all force as rapidly as possible and measure the return angle 3 minutes after the release.

(d) When the abdomen is subjected to continuously applied force in accordance with paragraph (e) of this section, the abdominal force-deflection curve shall be within the two curves plotted in Figure 10.

(e) Test procedure:

(1) Place the assembled thorax, lumbar spine, and pelvic assemblies in a supine position on a flat, rigid, smooth, dry, clean horizontal surface, ensuring that all component surfaces are clean, dry, and untreated unless otherwise specified.

(2) Place a rigid cylinder 6 inches in diameter and 18 inches long transversely across the abdomen, so that the cylinder is symmetrical about the midsagittal plane, with its longi-tudinal centerline horizontal and perpendicular to the midsagittal plane at a point 9.2 inches above the bottom line of the buttocks, measured with the dummy positioned in accordance with Figure 11.

(3) Establish the zero deflection point as the point at which a force of 10 pounds has been reached.

(4) Apply a vertical downward force through the cylinder at any rate between 0.25 and 0.35 inches per second.

(5) Guide the cylinder so that it moves without significant lateral or rotational movement.

§ 572.10   Limbs.

(a) The limbs consist of the assemblies shown as numbers SA 150 M070, SA 150 M071, SA 150 M080, and SA 150 M081 in Figure 1 and conform to the drawings subtended by these numbers.

(b) When each knee is impacted at 6.9 ft/sec. in accordance with paragraph (c) of this section, the maximum force on the femur shall be not more than 2500 pounds and not less than 1850 pounds, with a duration above 1000 pounds of not less than 1.7 milliseconds.

(c) Test procedure:

(1) Seat the dummy without back support on a surface as specified in § 572.11(i) that is 17.3±0.2 inches above a horizontal surface, oriented as specified in § 572.11(i), and with the hip joint adjustment at any setting between 1g and 2g. Place the dummy legs in planes parallel to its midsagittal plane (knee pivot centerline perpendicular to the midsagittal plane) and with the feet flat on the horizontal surface. Adjust the feet and lower legs until the lines between the midpoints of the knee pivots and the ankle pivots are at any angle not less than 2 degrees and not more than 4 degrees rear of the vertical, measured at the centerline of the knee pivots.

(2) Reposition the dummy if necessary so that the rearmost point of the lower legs at the level one inch below the seating surface remains at any distance not less than 5 inches and not more than 6 inches forward of the forward edge of the seat.

(3) Align the test probe specified in § 572.11(a) so that at impact its longitudinal centerline coincides within ±2° with the longitudinal centerline of the femur.

(4) Impact the knee with the test probe moving horizontally and parallel to the midsagittal plane at the specified velocity.

(5) Guide the probe during impact so that it moves with no significant lateral, vertical, or rotational movement.

§ 572.11 Test conditions and Instrumentation.

(a) The test probe used for thoracic and knee impact tests is a cylinder 6 inches in diameter that weighs 51.5 pounds including instrumentation. Its impacting end has a flat right face that is rigid and that has an edge radius of 0.5 inches.

(b) Accelerometers are mounted in the head on the horizontal transverse bulkhead shown in the drawings subreferenced under assembly No. SA 150 M010 in Figure 1, so that their sensitive axes intersect at a point in the midsagittal plane 0.5 inches above the horizontal bulkhead and 1.9 inches ventral of the vertical mating surface of the skull with the skull cover. One accelerometer is aligned with its sensitive axis perpendicular to the horizonal bulkhead in the midsagittal plane and with its seismic mass center at any distance up to 0.3 inches superior to the axial intersection point. Another accelerometer is aligned with its sensitive axis parallel to the horizontal bulkhead and perpendicular to the midsagittal plane, and with its seismic mass center at any distance up to 1.3 inches to the left of the axial intersection point (left side of dummy is the same as that of man). A third accelerometer is aligned with its sensitive axis parallel to the horizontal bulkhead in the midsagittal plane, and with its seismic mass center at any distance up to 1.3 inches dorsal to the axial intersection point.

(c) Accelerometers are mounted in the thorax by means of a bracket attached to the rear vertical surface (hereafter "attachment surface") of the thoracic spine so that their sensitive axes intersect at a point in the midsagittal plane 0.8 inches below the upper surface of the plate to which the neck mounting bracket is attached and 3.2 inches perpendicularly forward of the surface to which

the accelerometer bracket is attached. One accelerometer has its sensitive axis oriented parallel to the attachment surface in the midsagittal plane, with its seismic mass center at any distance up to 1.3 inches inferior to the intersection of the sensitive axes specified above. Another accelerometer has its sensitive axis oriented parallel to the attachment surface and perpendicular to the midsagittal plane, with its seismic mass center at any distance up to 0.2 inches to the right of the intersection of the sensitive axes specified above. A third accelerometer has its sensitive axis oriented perpendicular to the attachment surface in the midsagittal plane, with its seismic mass center at any distance up to 1.3 inches dorsal to the intersection of the sensitive axes specified above. Accelerometers are oriented with the dummy in the position specified in § 572.11(i).

(d) A force-sensing device is mounted axially in each femur shaft so that the transverse centerline of the sensing element is 4.25 inches from the knee's center of rotation.

(e) The outputs of acceleration and forcesensing devices installed in the dummy and in the test apparatus specified by this Part are recorded in individual data channels that conform to the requirements of SAE Recommended Practice J211a, December 1971, with channel classes as follows:

(1) Head acceleration—Class 1000.
(2) Pendulum acceleration—Class 60.
(3) Thorax acceleration—Class 180.
(4) Thorax compression—Class 180.
(5) Femur force—Class 600.

(f) The mountings for sensing devices have no resonance frequency within a range of 3 times the frequency range of the applicable channel class.

(g) Limb joints are set at 1g, barely restraining the weight of the limb when it is extended horizontally. The force required to move a limb segment does not exceed 2g throughout the range of limb motion.

(h) Performance tests are conducted at any temperature from 66° F to 78° F and at any relative humidity from 10 percent to 70 percent after exposure of the dummy to these conditions for a period of not less than 4 hours.

(i) For the performances tests specified in §§ 572.8, 572.9, and 572.10, the dummy is positioned in accordance with Figure 11 as follows:

(1) The dummy is placed on a flat, rigid, smooth, clean, dry, horizontal, steel test surface whose length and width dimensions are not less than 16 inches, so that the dummy's midsagittal plane is vertical and centered on the test surface and the rearmost points on its lower legs at the level of the test surface are at any distance not less than 5 inches and not more than 6 inches forward of the forward edge of the test surface.

(2) The pelvis is adjusted so that the upper surface of the lumbar-pelvic adapter is horizontal.

(3) The shoulder yokes are adjusted so that they are at the midpoint of their anterior posterior travel with their upper surfaces horizontal.

(4) The dummy is adjusted so that the rear surfaces of the shoulders and buttocks are tangent to a transverse vertical plane.

(5) The upper legs are positioned symmetrically about the midsagittal plane so that the distance between the knee pivot bolt heads is 11.6 inches.

(6) The lower legs are positioned in planes parallel to the midsagittal plane so that the lines between the midpoint of the knee pivots and the ankle pivots are vertical.

(j) The dummy's dimensions, as specified in drawing number SA 150 M002, are determined as follows:

(1) With the dummy seated as specified in paragraph (i), the head is adjusted and secured so that its occiput is 1.7 inches forward of the transverse vertical plane with the vertical mating surface of the skull with its cover parallel to the transverse vertical plane.

(2) The thorax is adjusted and secured so that the rear surface of the chest accelerometer mounting cavity is inclined 3° forward of vertical.

(3) Chest and waist circumference and chest depth measurements are taken with the dummy positioned in accordance with paragraph (i), (1) and (2) of this section.

(4) The chest skin and abdominal sac are removed and all following measurements are made without them.

(5) Seated height is measured from the seating surface to the uppermost point on the head-skin surface.

(6) Shoulder pivot height is measured from the seating surface to the center of the arm elevation pivot.

(7) H-point locations are measured from the seating surface to the center of the holes in the pelvis flesh covering in line with the hip motion ball.

(8) Knee pivot distance from the backline is measured to the center of the knee pivot bolt head.

(9) Knee pivot distance from floor is measured from the center of the knee pivot bolt head to the bottom of the heel when the foot is horizontal and pointing forward.

(10) Shoulder width measurement is taken at arm elevation pivot center height with the centerlines between the elbow pivots and the shoulder pivots vertical.

(11) Hip width measurement is taken at widest point of pelvic section.

(k) Performance tests of the same component, segment, assembly, or fully assembled dummy are separated in time by a period of not less than 30 minutes unless otherwise noted.

(1) Surfaces of dummy components are not painted except as specified in this part or in drawings subtended by this part.

Subpart C—Three Year Old Child

Sec.
572.15 General description.
572.16 Head.
572.17 Neck.
572.18 Thorax.
572.19 Lumbar, spine, abdomen and plevis.
572.20 Limbs.
572.21 Test conditions and instrumentation.

Subpart C—Three Year Old Child

§ 572.15 General description.

(a)(1) The dummy consists of the component assemblies specified in drawing SA 103C 001, which are described in their entirety by means of approximately 122 drawings and specifications grouped by component assemblies under the following headings:

SA 103C 010 Head Assembly
SA 103C 020 Neck Assembly
SA 103C 030 Torso Assembly
SA 103C 041 Upper Arm Assembly Left
SA 103C 042 Upper Arm Assembly Right
SA 103C 051 Forearm Hand Assembly Left
SA 103C 052 Forearm Hand Assembly Right
SA 103C 061 Upper Leg Assembly Left
SA 103C 062 Upper Leg Assembly Right
SA 103C 071 Lower Leg Assembly Left
SA 103C 072 Lower Leg Assembly Right
SA 103C 081 Foot Assembly Left
SA 103C 082 Foot Assembly Right

The drawings and specifications are incorporated in this part by reference to the thirteen headings and are available for examination in Docket 78–09, Room 5109, 400 Seventh Street S.W., Washington, D.C. 20590. [Copies may be obtained from Rowley-Scher Reprographics, Inc., 1216 K Street, N.W., Washington, D.C. 20005, attention Mr. Allan Goldberg and Mr. Mark Krysinski ((202) 628–6667). (50 F.R. 25422—June 19,1985. Effective: June 19, 1985)]

(2) The patterns of all cast and molded parts for reproduction of the molds needed in manufacturing of the dummies are incorporated in this part by reference. A set of the patterns can be obtained on a loan basis by manufacturers of the test dummies, or others if need is shown, from the Office of Vehicle Safety Standards, NHTSA, 400 Seventh Street S.W., Washington, D.C. 20590.

(3) [An Operation and Maintenance Manual (dated May 28, 1976, Contract No. DOT-HS-6-01294) with instructions for the use and maintenance of the test dummies is incorporated in this Part by reference. Copies of the manual can be obtained from Rowley-Scher Reprographics, Inc. All provisions of this manual are valid unless modified by this regulation. This document is available for examination in Docket 78-09. (50 F.R. 25422—June 19, 1985. Effective: June 19, 1985)]

(4) The drawings, specifications and the manual are subject to changes, but any change will be accomplished by appropriate administrative procedures and announced by publication in the Federal Register and be available for examination and copying as indicated in this paragraph.

(5) The drawings, specifications, patterns, and manual are on file in the reference library of the Federal Register, National Archives and Records Service, General Services Administration, Washington, D.C.

(b) Adjacent segments are joined in a manner such that throughout the range of motion and also under simulated crash-impact conditions, there is no contact between metallic elements except for contacts that exist under static conditions.

(c) The structural properties of the dummy are such that the dummy conforms to this part in every respect both before and after being used in tests specified by Standard No. 213, Child Restraint Systems (§ 571.213).

§ 572.16 Head.

(a) The head consists of the assembly shown in drawing SA 103C 001 by number SA 103C 010, and conforms to each of the drawings listed under this number on drawing SA 103C 002, sheet 8.

(b) When the head is impacted in accordance with paragraph (c) of this section by a test probe conforming to § 572.21(a) at 7 fps., the peak resultant accelerations measured at the location of the accelerometers mounted in the headform in accordance with § 572.21(b) shall be not less than 95g, and not more than 115g. The recorded acceleration-time curve for this test shall be unimodal at, or above the 50g level and shall lie at, or above that level for an interval not less than 2.0 and not more than 3.0 milliseconds. The lateral acceleration vector shall not exceed 7g.

(c) Test Procedure:

(1) Seat the dummy on a seating surface having a back support as specified in § 572.21(h) and orient the dummy in accordance with § 572.21(h) and adjust the joints of the limbs at any setting between 1g and 2g, which just supports the limbs' weight when the limbs are extended horizontally forward.

(2) Adjust the test probe so that its longitudinal centerline is at the forehead at the point of orthogonal intersection of the head midsagittal plane and the transverse plane which is perpendicular to the "Z" axis of the head (longitudinal centerline of the skull anchor) and is located 0.6 ± .1 inches above the centers of the head center of gravity reference pins and coincides within 2 degrees with the line made by the intersection of horizontal and midsagittal planes passing through this point.

(3) Adjust the dummy so that the surface area on the forehead immediately adjacent to the projected longitudinal centerline of the test probe is vertical.

(4) Impact the head with the test probe so that at the moment of impact the probe's longitudinal centerline falls within 2 degrees of a horizontal line in the dummy's midsagittal plane.

(5) Guide the probe during impact so that it moves with no significant lateral, vertical, or rotational movement.

(6) Allow a time period of at least 20 minutes between successive tests of the head.

§ 572.17   Neck.

(a) The neck consists of the assembly shown in drawing SA 103C 001 as number SA 103C 020, and conforms to each of the drawings listed under this number on drawing SA 103C 002, sheet 9.

(b) When the head-neck assembly is tested in accordance with paragraph (c) of this section, the head shall rotate in reference to the pendulum's longitudinal centerline a total of 84 degrees ± 8 degrees about its center of gravity, rotating to the extent specified in the following table at each indicated point in time, measured from impact, with the chordal displacement measured at its center of gravity. The chordal displacement at time T is defined as the straight line distance between (1) the position relative to the pendulum arm of the head center of gravity at time zero, and (2) the position relative to the pendulum arm of the head center of gravity at time T as illustrated by Figure 3. The peak resultant acceleration recorded at the location of the accelerometers mounted in the headform in accordance with § 572.21(b) shall not exceed 30g. The pendulum shall not reverse direction until the head's center of gravity returns to the original zero time position relative to the pendulum arm.

Rotation (degrees)	Time (ms) ± (2+.08T)	Chordal Displacement (inches ±0.8)
0.............	0	0
30.............	21	2.2
60.............	36	4.3
Maximum	62	5.8
60.............	91	4.3
30.............	108	2.2
0.............	123	0

(c) *Test Procedure:*

(1) Mount the head and neck on a rigid pendulum as specified in Figure 4, so that the head's midsagittal plane is vertical and coincides with the plane of motion of the pendulum's longitudinal centerline. Mount the neck directly to the pendulum as shown in Figure 15.

(2) Release the pendulum and allow it to fall freely from a height such that the velocity at impact is 17.00 ± 1.0 feet per second (fps), measured at the center of the accelerometer specified in Figure 4.

(3) Decelerate the pendulum to a stop with an acceleration-time pulse described as follows:

(i) Establish 5g and 20g levels on the a-t curve.

(ii) Establish t_1 at the point where the a-t curve first crosses the 5g level, t_2 at the point where the rising a-t curve first crosses the 20g level, t_3 at the point where the decaying a-t curve last crosses the 20g level, and t_4 at the point where the decaying a-t curve first crosses the 5g level.

(iii) t_2-t_1, shall be not more than 4 milliseconds.

(iv) t_3-t_2, shall be not less than 18 and not more than 21 milliseconds.

(v) t_4-t_3, shall be not more than 5 milliseconds.

(vi) The average deceleration between t_2 and t_3 shall be not less than 20g and not more than 34g.

(4) Allow the neck to flex without contact of the head or neck with any object other than the pendulum arm.

(5) Allow a time period of at least 1 hour between successive tests of the head and neck.

§ 572.18   Thorax.

(a) The thorax consists of the part of the torso shown in assembly drawing SA 103C 001 by number SA 103C 030 and conforms to each of the applicable drawings listed under this number on drawings SA 103C 002, sheets 10 and 11.

(b) When impacted by a test probe conforming to § 572.21(a) at 13 fps. in accordance with paragraph (c) of this section, the peak resultant accelerations at the location of the accelerometers mounted in the chest cavity in accordance with § 572.21(c) shall be not less than 50g and not more than 70g. The acceleration-time curve for the test shall be unimodal at or above the 30g level and shall lie at or above the 30g level for an interval not less than 2.5 milliseconds and not more than 4.0 milliseconds. The lateral acceleration shall not exceed 5g.

(c) *Test Procedure:*

(1) With the dummy seated without back support on a surface as specified in § 572.21(h) and

oriented as specified in § 572.21(h), adjust the dummy arms and legs until they are extended horizontally forward parallel to the midsagittal plane, the joints of the limbs are adjusted at any setting between 1g and 2g, which just supports the limbs' weight when the limbs are extended horizontally forward.

(2) Establish the impact point at the chest midsagittal plane so that it is 1.5 inches below the longitudinal centerline of the bolt that attaches the top of the ribcage sternum to the thoracic spine box.

(3) Adjust the dummy so that the tangent plane at the surface on the thorax immediately adjacent to the designated impact point is vertical and parallel to the face of the test probe.

(4) Place the longitudinal centerline of the test probe to coincide with the designated impact point and align the test probe so that at impact its longitudinal centerline coincides within 2 degrees with the line formed by intersection of the horizontal and midsagittal planes passing through the designated impact point.

(5) Impact the thorax with the test probe so that at the moment of impact the probe's longitudinal centerline falls within 2 degrees of a horizontal line in the dummy midsagittal plane.

(6) Guide the probe during impact so that it moves with no significant lateral, vertical or rotational movement.

(7) Allow a time period of at least 20 minutes between successive tests of the chest.

§ 572.19 Lumbar spine, abdomen and pelvis.

(a) The lumbar spine, abdomen, and pelvis consist of the part of the torso assembly shown by number SA 103C 030 on drawing SA 103C 001 and conform to each of the applicable drawings listed under this number on drawing SA 103C 002, sheets 10 and 11.

(b) When subjected to continuously applied force in accordance with paragraph (c) of this section, the lumbar spine assembly shall flex by an amount that permits the rigid thoracic spine to rotate from its initial position in accordance with Figure 18 of this subpart by 40 degrees at a force level of not less than 34 pounds and not more than 47 pounds, and straighten upon removal of the force to within 5 degrees of its initial position.

(c) *Test Procedure:* (1) The dummy with lower legs removed is positioned in an upright seated position on a seat as indicated in Figure 18, ensuring that all dummy component surfaces are clean, dry and untreated unless otherwise specified.

(2) Attach the pelvis to the seating surface by a bolt C/328, modified as shown in Figure 18, and the upper legs at the knee axial rotation joints by the attachments shown in Figure 18. Tighten the mountings so that the pelvis-lumbar joining surface is horizontal and adjust the femur ball-flange screws at each hip socket joint to 50 inch pounds torque. Remove the head and the neck and install a cylindrical aluminum adapter 2.0 inches in diameter and 2.80 inches long in place of the neck.

(3) Flex the thorax forward 50 degrees and then rearward as necessary to return to its initial position in accordance with Figure 18 unsupported by external means.

(4) Apply a forward pull force in the midsagittal plane at the top of the neck adapter, so that at 40 degrees of the lumbar spine flexion the applied force is perpendicular to the thoracic spine box. Apply the force at any torso deflection rate between 0.5 and 1.5 degrees per second up to 40 degrees of flexion but no further; continue to apply for 10 seconds the force necessary to maintain 40 degrees of flexion, and record the highest applied force at that time. Release all force as rapidly as possible and measure the return angle 3 minutes after the release.

§ 572.20 Limbs.

The limbs consist of the assemblies shown on drawing SA 103C 001 as Nos. SA 103C 041, SA 103C 042, SA 103C 051, SA 103C 052, SA 103C 061, SA 103C 062, SA 103C 071, SA 103C 072, SA 103C 081, SA 103C 082, and conform to each of the applicable drawings listed under their respective numbers of the drawing SA 103C 002, sheets 12 through 21.

§ 572.21 Test conditions and instrumentation.

(a) The test probe used for head and thoracic impact tests is a cylinder 3 inches in diameter, 13.8 inches long and weighs 10 lbs., 6 ozs. Its impacting end has a flat right face that is rigid and that has an edge radius of 0.5 inches.

(b) Accelerometers are mounted in the head on the mounting block (A/310) located on the horizontal transverse bulkhead shown in the drawings subreferenced under assembly SA 103C 010 so that their sensitive axes are orthogonal and their seismic masses are positioned relative to the axial intersection point. Except in the case of tri-axial accelerometers, the sensitive axes shall intersect at the axial intersection point located at the intersection of a line connecting the longitudinal centerlines of the transfer pins in the sides of the dummy head with the midsagittal plane of the dummy head. One accelerometer is aligned with its sensitive axis parallel to the vertical bulkhead and midsagittal plane, and with its seismic mass center at the midsagittal plane at any distance up to 0.3 inches dorsal and 0.1 inches inferior to the axial intersection point. Another accelerometer is aligned with its sensitive axis in the horizontal plane and perpendicular to the midsagittal plane, and with its seismic mass center at any distance up to 0.2 inches inferior to, 0.4 inches to the right of, and 1 inch dorsal to the axial intersection point (right side of dummy is the same as that of child). A third accelerometer is aligned with its sensitive axis parallel to the midsagittal and horizontal planes, and with its seismic mass center at any distance up to 0.2 inches inferior to, 0.6 inches dorsal to, and 0.4 inches to the right of the axial intersection point. In the case of a tri-axial accelerometer, its axes are aligned in the same way that the axes of three separate accelerometers are aligned.

(c) Accelerometers are mounted in the thorax on the mounting plate attached to the vertical transverse bulkhead shown in the drawings subreferenced under assembly No. SA 103C 030 in drawing SA 103C 001 so that their sensitive axes are orthogonal and their seismic masses are positioned relative to the axial intersection point located in the midsagittal plane 3 inches above the top surface of the lumbar spine and 0.3 inches dorsal to the accelerometer mounting plate surface. Except in the case of tri-axial accelerometers, the sensitive axes shall intersect at the axial intersection point. One accelerometer is aligned with its sensitive axis parallel to the vertical bulkhead and midsagittal planes, and with its seismic mass center at any distance up to 0.2 inches to the right, 0.2 inches inferior and 0.1 inches ventral of the axial intersection point. Another accelerometer is aligned with its sensitive axis in the horizontal transverse plane and perpendicular to the midsagittal plane and with its seismic mass center at any distance up to 0.3 inches to the left, 0.2 inches

inferior and 0.2 inches ventral to the axial intersection point. A third accelerometer is aligned with its sensitive axis parallel to the midsagittal and horizontal planes and with its seismic mass center at any distance up to 0.3 inches superior, 0.6 inches to the right and 0.1 inches ventral to the axial intersection point. In the case of a tri-axial accelerometer, its axes are aligned in the same way that the axes of three separate accelerometers are aligned.

(d) The outputs of accelerometers installed in the dummy, and of test apparatus specified by this part, are recorded in individual data channels that conform to the requirements of SAE Recommended Practice J211a, December 1971, with channel classes as follows:

(1) Head acceleration—Class 1,000.

(2) Pendulum acceleration—Class 60.

(3) Thorax acceleration—Class 180.

(e) The mountings for accelerometers have no resonance frequency less than 3 times the cut-off frequency of the applicable channel class.

(f) Limb joints are set at the force between 1–2g, which just supports the limbs' weight when the limbs are extended horizontally forward. The force required to move a limb segment does not exceeed 2g throughout the range of limb motion.

(g) Performance tests are conducted at any temperature from 66° F to 78° F and at any relative humidity from 10 percent to 70 percent after exposure of the dummy to these conditions for a period of not less than 4 hours.

(h) For the performance tests specified §§ 572.16, 572.18, and 572.19, the dummy is positioned in accordance with Figures 16, 17, and 18 as follows:

(1) The dummy is placed on a flat, rigid, clean, dry, horizontal surface of teflon sheeting with a smoothness of 40 microinches and whose length and width dimensions are not less than 16 inches, so that the dummy's midsagittal plane is vertical and centered on the test surface. For head tests, the seat has a vertical back support whose top is 12.4 ± 0.2 inches above the seating surface. The rear surfaces of the dummy's shoulders and buttocks are touching the back support as shown in Figure 16. For thorax and lumbar spine tests, the seating surface is without the back support as shown in Figures 17 and 18 respectively.

(2) The shoulder yokes are adjusted so that they are at the midpoint of their anterior-posterior travel with their upper surfaces horizontal.

(3) The dummy is adjusted for head impact and lumbar flexion tests so that the rear surfaces of the shoulders and buttocks are tangent to a transverse vertical plane.

(4) The arms and legs are positioned so that their centerlines are in planes parallel to the midsagittal plane.

(i) The dummy's dimensions are specified in drawings No. SA 103C 002, sheets 22 through 26.

(j) Performance tests of the same component, segment, assembly or fully assembled dummy are separated in time by a period of not less than 20 minutes unless otherwise specified.

(k) Surfaces of the dummy components are not painted except as specified in this part or in drawings subtended by this part.

Subpart D—Six Month Old Infant

§ 572.25 General Description.

(a) The infant dummy is specified in its entirety by means of 5 drawings (No. SA 100I 001) and a construction manual which describes in detail the materials and the procedures involved in the manufacturing of this dummy. The drawings and the manual are incorporated in this part by reference and are available for examination in Docket 78-09, Room 5109, 400 Seventh Street S.W., Washington, D.C. 20590. Copies may be obtained from Rowley-Scher Reprographics, Inc. 1216 K Street, N.W. Washington, D.C., 20005, attention Mr. Allan Goldberg and Mr. Mark Krysinski ((202) 628-6667). The drawings and the manual are subject to changes, but any change will be accomplished by appropriate administrative procedures and announced by publication in the *Federal Register* and be available for examination and copying as indicated in this paragraph. The drawings and manual are on file in the reference library of the *Federal Register*, National Archives and Records Services, General Services Administration, Washington, D.C. (50 F.R. 25422—June 19, 1985. Effective: June 19, 1985)

(b) The structural properties and dimensions of the dummy are such that the dummy conforms to this part in every respect, both before and after being used in tests specified by Standard No. 213 (571.213).

[§ 572.30 Incorporated Materials.

(a) The drawings and specifications referred to in this regulation that are not set forth in full are hereby incorporated in this part by reference. The Director of the Federal Register has approved the materials incorporated by reference. For materials subject to change, only the specific version approved by the Director of the Federal Register and specified in the regulation are incorporated. A notice of any change will be published in the *Federal Register*. As a convenience to the reader, the materials incorporated by reference are listed in the Finding Aid Table found at the end of this volume of the Code of Federal Regulations.

(b) The materials incorporated by reference are available for examination in the general reference section of Docket 74-14, Docket Section, National Highway Traffic Safety Administration, Room 5109, 400 Seventh Street, S.W., Washington, D.C. 20590. Copies may be obtained from Rowley-Scher Reprographics, Inc., 1216 K Street, N.W., Washington, D.C.20005 ((202) 628-6667). The drawings and specifications are also on file in the reference library of the Office of the Federal Register, National Archives and Records Administration, Washington, D.C.

§ 572.31 General Description.

(a) The Hybird III 50th percentile size dummy consists of components and assemblies specified in the Anthropomorphic Test Dummy drawing and specification package which consists of the following six items:

(1) The Anthropomorphic Test Dummy Parts List, dated July 15, 1986, and containing 13 pages, and Parts list Index, dated April 26, 1986, containing 6 pages,

(2) A listing of Optional Hybrid III Dummy Transducers, dated April 22, 1986, contained 4 pages

(3) A General Motors Drawing package identified by GM drawing No. 78051-218 revision P and subordinate drawings.

(4) Disassembly, Inspection, Assembly and Limbs Adjustment Procedures for the Hybrid III Dummy, dated July 15, 1986,

(5) Sign Convention for the signal outputs of Hybrid III Dummy Transducers, dated July 15, 1986,

(6) Exterior Dimensions of the Hybrid III Dummy, dated July 15, 1986.

(b) The dummy is made up of the following component assemblies:

Drawing Number		Revision
78051–61	Head Assembly–Complete–	(T)
78051–90	Neck Assembly–Complete–	(A)
78051–89	Upper Torso Assembly–Complete–	(I)
78051–70	Lower Torso Assembly–Without Pelvic Instrumentation Assembly, Drawing Number 78051-59	(C)
86–5001–001	Leg Assembly–Complete (LH)–	
86–5001–002	Leg Assembly–Complete (RH)–	
78051–123	Arm Assembly–Complete (LH)–	(D)
78051–124	Arm Assembly–Complete (RH)–	(D)

(c) Any specifications and requirements set forth in this part supercede those contained in General Motors Drawing No. 78051–218, revision P.

(d) Adjacent segments are joined in a manner such that throughout the range of motion and also under crash-impact conditions, there is no contact between metallic elements except for contacts that exist under static conditions.

(e) The weights, inertial properties and centers of gravity location of component assemblies shall conform to those listed in drawing 78051–338, revision S.

(f) The structural properties of the dummy are such that the dummy conforms to this part in every respect both before and after being used in vehicle test specified in Standard No. 208 of this Chapter (Æ571.208).

§ 572.32 Head.

(a) The head consists of the assembly shown in the drawing 78051–61, revision T, and shall conform to each of the drawings subtended therein.

(b) When the head (drawing 78051–61, revision T) with neck transducer structural replacement (drawing 78051–383, revision F) is dropped from a height of 14.8 inches in accordance with paragraph (c) of this section, the peak resultant accelerations at the location ·of the accelerometers mounted in the head in accordance with 572.36(c) shall not be less than 225g, and not more than 275g. The acceleration/time curve for the test shall be unimodal to the extent that oscillations occurring after the main acceleration pulse are less than ten percent (zero to peak) of the main pulse. The lateral acceleration vector shall not exceed 15g (zero to peak).

(c) Test Procedure. (1) Soak the head assembly in a test environment at any temperature between

66 degrees F to 78 degrees F and at a relative humidity from 10% to 70% for a period of at least four hours prior to its application in a test.

(2) Clean the head's skin surface and the surface of the impact plate with 1,1,1 Trichlorethane or equivalent.

(3) Suspend the head, as shown in Figure 19, so that the lowest point on the forehead is 0.5 inches below the lowest point on the dummy's nose when the midsagittal plane is vertical.

(4) Drop the head from the specified height by means that ensure instant release onto a rigidly supported flat horizontal steel plate, which is 2 inches thick and 2 feet square. The plate shall have a clean, dry surface and any microfinish of not less than 8 microinches (rms) and not more than 80 microinches (rms).

(5) Allow at least 2 hours between successive tests on the same head.

§ 572.33 Neck.

(a) The neck consists of the assembly shown in drawing 78051–90, revision A and conforms to each of the drawings subtended therein.

(b) When the neck and head assembly (consisting of the parts 78051–61, revision T; –84; –90, revision A; –96; –98; –303, revision E; –305; –306; –307, revision X, which has a neck transducer (drawing 83–5001–008) installed in conformance with 572.36(d), is tested in accordance with paragraph (c) of this section, it shall have the following characteristics:

(1) Flexion. (i) Plane D, referenced in Figure 20, shall rotate, between 64 degrees and 78 degrees, which shall occur between 57 milliseconds (ms) and 64 ms from time zero. In first rebound, the rotation of plane D shall cross O degree between 113 ms and 128 ms.

(ii) The moment measured by the neck transducer (drawing 83–5001–008) about the occipital condyles, referenced in Figure 20, shall be calculated by the following formula: Moment (lbs-ft) = $M_y + 0.02875 \times F_x$, where M_y is the moment measured in lbs-ft by the moment sensor of the neck transducer and F_x is the force measure measured in lbs by the x axis force sensor of the neck transducer. The moment shall have a maximum value between 65 lbs-ft occurring between 47 ms and 58 ms, and the positive moment shall decay for the first time to 0 lb-ft between 97 ms and 107 ms.

(2) *Extension.* (i) Plane D, referenced in Figure 21, shall rotate between 81 degrees and 106 degrees, which shall occur between 72 and 82 ms from time zero. In first rebound, the rotation of plane D shall cross 0 degree between 147 and 174 ms.

(ii) The moment measured by the neck transducer (drawing 83-5001-008) about the occipital condyles, referenced in Figure 21, shall be calculated by the following formula: Moment (lbs-ft) = M_y + 0.02875 × F_x, where M_y is the moment measured in lbs-ft by the moment sensor of the neck transducer and F_x is the force measure measured in lbs by the x axis force sensor of the neck transducer. The moment shall have a minimum value between −39 lbs-ft and −59 lbs-ft, which shall occur between 65 ms and 79 ms, and the negative moment shall decay for the first time to 0 lb-ft between 120 ms and 148 ms.

(3) Time zero is defined as the time of contact between the pendulum striker plate and the aluminum honeycomb material.

(c) *Test Procedure.* (1) Soak the test material in a test environment at any temperature between 69 degrees F to 72 degrees F and at a relative humidity from 10% to 70% for a period of at least four hours prior to its application in a test.

(2) Torque the jamnut (78051-64) on the neck cable (78051-301, revision E) to 1.0 lbs-ft ±.2 lbs-ft.

(3) Mount the head-neck assembly, defined in paragraph (b) of this section, on a rigid pendulum as shown in Figure 22 so that the head's midsagittal plane is vertical and coincides with the plane of motion of the pendulum's longitudinal axis.

(4) Release the pendulum and allow it to fall freely from a height such that the tangential velocity at the pendulum accelerometer centerline at the instance of contact with the honeycomb is 23.0 ft/sec ± 0.4 ft/sec. for flexion testing and 19.9 ft/sec ± 0.4 ft/sec. for extension testing. The pendulum deceleration vs. time pulse for flexion testing shall conform to the characteristics shown in Table A and the decaying deceleration-time curve shall first cross 5g between 34 ms and 42 ms. The pendulum deceleration vs. time pulse for extension testing shall conform to the characteristics shown in Table B and the decaying deceleration-time curve shall cross 5g between 38 ms and 46 ms.

Table A
Flexion Pendulum Deceleration vs. Time Pulse

Time (ms)	Flexion deceleration level (g)
10	22.50—27.50
20	17.60—22.60
30	12.50—18.50
Any other time above 30 ms......	29 maximum

Table B
Extension Pendulum Deceleration vs. Time Pulse

Time (ms)	Extension deceleration level (g)
10	17.20—21.00
20	14.00—19.00
30	11.00—16.00
Any other time above 30 ms	22 maximum

(5) Allow the neck to flex without impact of the head or neck with any object during the test.

§ 572.34 Thorax.

(a) The thorax consists of the upper torso assembly in drawing 78051-89, revision I and shall conform to each of the drawings subtended therein.

(b) When impacted by a test probe conforming to Я572.36(a) at 22 fps ± .40 fps in accordance with paragraph (c) of this section, the thorax of a complete dummy assembly (78051-218, revision P) with left and right shoes (78051-294 and -295) removed, shall resist with the force measured by the test probe from time zero of 1162.5 pounds ± 82.5 pounds and shall have a sternum displacement measured relative to spine of 2.68 inches ± .18 inches. The internal hysteresis in each impact shall be more than 69% but less than 85%. The force measured is the product of pendulum mass and deceleration. Time zero is defined as the time of first contact between the upper thorax and pendulum face.

(c) *Test procedure.* (1) Soak the test dummy in an environment with a relative humidity from 10% to 70% until the temperature of the ribs of the test dummy have stabilized at a temperature between 69 degrees F and 72 degrees F.

(2) Seat the dummy without back and arm supports on a surface as shown in Figure 23.

(3) Place the longitudinal centerline of the test probe so that it is .5 in ± .04 in. below the horizontal centerline of the No. 3 Rib (reference drawing number 79051-64, revision A-M) as shown in Figure 23.

(4) Align the test probe specified in S572.36(a) so that at impact it longitudinal centerline coincides within .5 degree of a horizontal line in the dummy's midsagittal plane.

(5) Impact the thorax with the test probe so that the longitudinal centerline of the test probe falls within 2 degrees of a horizontal line in the dummy's midsagittal plane at the moment of impact.

(6) Guide the probe during impact so that it moves with no significant lateral, vertical, or rotational movement.

(7) Measure the horizontal deflection of the sternum relative to the thoracic spine along the line established by the longitudinal centerline of the probe at the moment of impact, using a potentiometer (ref. drawing 78051-317, revision A) mounted inside the sternum as shown in drawing 78051-89, revision I.

(8) Measure hysteresis by determining the ratio of the area between the loading and unloading portions of the force deflection curve to the area under the loading portion of the curve.

§ 572.35   Limbs.

(a) The limbs consist of the following assemblies: leg assemblies 86-5001-001 and -002 and arm assemblies 78051-123, revision D, and -124, revision D, and shall conform to the drawings subtended therein.

(b) When each knee of the leg assemblies is impacted by the pendulum defined in S572.36(b) in accordance with paragraph (c) of this section at 6.9 ft/sec ± .10 ft/sec., the peak knee impact force, which is a product of pendulum mass and acceleration, shall have a minimum value of not less than 996 pounds and a maximum value of not greater than 1566 pounds.

(c) *Test Procedure.* (c) The test material consists of leg assemblies (86-5001-001) left and (-002) right with upper leg assemblies (78051-46) left and (78051-47) right removed. The load cell simulator (78051-319, revision A) is used to secure the knee cap assemblies (79051-16, revision B) as shown in Figure 24.

(2) Soak the test material in a test environment at any temperature between 66 degrees F to 78 degrees F and at a relative humidity from 10% to 70% for a period of at least four hours prior to its application in a test.

(3) Mount the test material with the leg assembly secured through the load cell simulator to a rigid surface as shown in Figure 24. No contact is permitted between the foot and any other exterior surfaces.

(4) Place the longitudinal centerline of the test probe so that at contact with the knee it is colinear within 2 degrees with the longitudinal centerline of the femur load cell simulator.

(5) Guide the pendulum so that there is no significant lateral, vertical or rotational movement at time zero.

(6) Impact the knee with the test probe so that the longitudinal centerline of the test probe at the instant of impact falls within .5 degrees of a horizontal line parallel to the femur load cell simulator at time zero.

(7) Time zero is defined as the time of contact between the test probe and the knee.

§ 572.36   Test Conditions and Instrumentation.

(a) The test probe used for thoracic impact tests is a 6 inch diameter cylinder that weighs 51.5 pounds including instrumentation. Its impacting end has a flat right angle face that is rigid and has an edge radius of 0.5 inches. The test probe has an accelerometer mounted on the end opposite from impact with its sensitive axis colinear to the longitudinal centerline of the cylinder.

(b) The test probe used for the knee impact tests is a 3 inch diameter cylinder that weighs 11 pounds including instrumentation. Its impacting end has a flat right angle face that is rigid and has an edge radius of 0.2 inches. The test probe has an accelerometer mounted on the end opposite from impact with its sensitive axis colinear to the longitudinal centerline of the cylinder.

(c) Head accelerometers shall have dimensions, response characteristics and sensitive mass locations specified in drawing 78051-136, revision A or its equivalent and be mounted in the head as shown in drawing 78051-61, revision T, and in the assembly shown in drawing 78051-218, revision D.

(d) The neck transducer shall have the dimensions, response characteristics, and sensitive axis locations specified in drawing 83-5001-008 or its equivalent and be mounted for testing as shown in drawing 79051-63, revision W, and in the assembly shown in drawing 78051-218, revision P.

(e) The chest accelerometers shall have the dimensions, response characteristics, and sensitive mass locations specified in drawing 78051-136, revision A or its equivalent and be mounted as shown with adaptor assembly 78051-116, revision D for assembly into 78051-218, revision L.

(f) The chest deflection transducer shall have the dimensions and response characteristics specified in drawing 78051-342, revision A or equivalent and be mounted in the chest deflection transducer assembly 87051-317, revision A for assembly into 78051-218, revision L.

(g) The thorax and knee impactor accelerometers shall have the dimensions and characteristics of Endevco Model 7231c or equivalent. Each accelerometer shall be mounted with its sensitive axis colinear with the pendulum's longitudinal centerline.

(h) The femur load cell shall have the dimensions, response characteristics, and sensitive axis locations specified in drawing 78051-265 or its equivalent and be mounted in assemblies 78051-46 and -47 for assembly into 78051-218, revision L.

(i) The outputs of acceleration and force-sensing devices installed in the dummy and in the test apparatus specified by this part are recorded in individual data channels that conform to the requirements of SAE Récommended Practice J211, JUN 1980, "Instrumentation for Impact Tests," with channel classes as follows:

(1) Head acceleration—Class 1000
(2) Neck force—Class 60
(3) Neck pendulum acceleration—Class 60
(4) Thorax and thorax pendulum acceleration—Class 180
(5) Thorax deflection—Class 180
(6) Knee pendulum acceleration—Class 600
(7) Femur force—Class 600

(j) Coordinate signs for instrumentation polarity conform to the sign convention shown in the document incorporated by §572.31(a)(5).

(k) The mountings for sensing devices shall have no resonance frequency within range of 3 times the frequency range of the applicable channel class.

(l) Limb joints are set at 1g, barely restraining the weight of the limb when it is extended horizontally. The force required to move a limb segment shall not exceed 2g throughout the range of limb motion.

(m) Performance tests of the same component, segment, assembly, or fully assembled dummy are separated in time by a period of not less than 30 minutes unless otherwise noted.

(n) Surfaces of dummy components are not painted except as specified in this part or in drawings subtended by this part. (51 F.R. 26688—July 25, 1986. Effective: October 23, 1986)

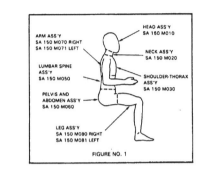

FIGURE NO. 1

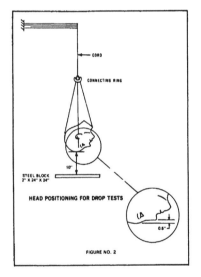

HEAD POSITIONING FOR DROP TESTS

FIGURE NO. 2

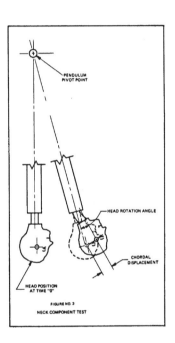

FIGURE NO. 3
NECK COMPONENT TEST

PART 572—ART PAGE 1

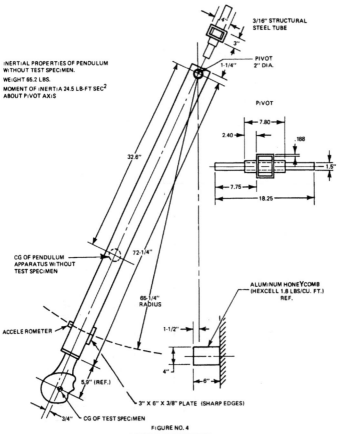

3/16" STRUCTURAL
STEEL TUBE

INERTIAL PROPERTIES OF PENDULUM
WITHOUT TEST SPECIMEN.

WEIGHT 65.2 LBS.

MOMENT OF INERTIA 24.5 LB-FT SEC^2
ABOUT PIVOT AXIS

PIVOT
2" DIA.

1-1/4"

3"

PIVOT

7.80

2.40

.188

1.5"

7.75

18.25

32.6"

CG OF PENDULUM
APPARATUS WITHOUT
TEST SPECIMEN

72-1/4"

65-1/4"
RADIUS

ALUMINUM HONEYCOMB
(HEXCELL 1.8 LBS/CU. FT.)
REF.

1-1/2"

ACCELEROMETER

4"

6"

5.9" (REF.)

3/4"

CG OF TEST SPECIMEN

3" X 6" X 3/8" PLATE (SHARP EDGES)

FIGURE NO. 4
NECK COMPONENT TEST

PART 572—ART PAGE 2

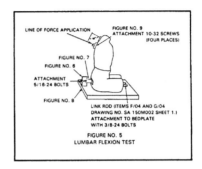

LINE OF FORCE APPLICATION

FIGURE NO. 9
ATTACHMENT 10-32 SCREWS
(FOUR PLACES)

FIGURE NO. 7

FIGURE NO. 6

ATTACHMENT
5/16-24 BOLTS

FIGURE NO. 8

LINK ROD (ITEMS F/04 AND G/04)
DRAWING NO. SA 150M002 SHEET 1.)
ATTACHMENT TO BEDPLATE
WITH 3/8-24 BOLTS

FIGURE NO. 5
LUMBAR FLEXION TEST

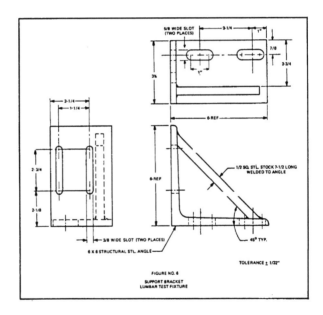

5/8 WIDE SLOT
(TWO PLACES)

3-1/4

1"

7/8

2-3/4

3½

1"

6-REF

2-1/4

1-1/4

2-3/4

6-REF

1/2 SQ. STL. STOCK 7-1/2 LONG
WELDED TO ANGLE

2-1/8

3/8 WIDE SLOT (TWO PLACES)

6 X 6 STRUCTURAL STL. ANGLE

45° TYP.

TOLERANCE ± 1/32"

FIGURE NO. 6
SUPPORT BRACKET
LUMBAR TEST FIXTURE

PART 572—ART PAGE 3

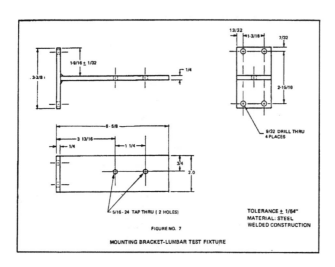

TOLERANCE ± 1/64"
MATERIAL: STEEL
WELDED CONSTRUCTION

FIGURE NO. 7

MOUNTING BRACKET-LUMBAR TEST FIXTURE

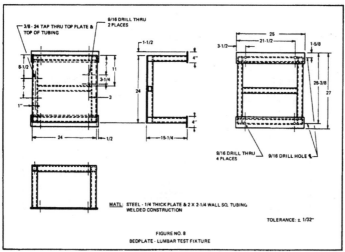

MATL: STEEL - 1/4 THICK PLATE & 2 X 2-1/4 WALL SQ. TUBING
WELDED CONSTRUCTION

TOLERANCE: ± 1/32"

FIGURE NO. 8

BEDPLATE - LUMBAR TEST FIXTURE

PART 572—ART PAGE 4

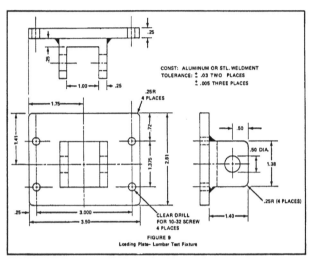

CONST: ALUMINUM OR STL. WELDMENT
TOLERANCE: ± .03 TWO PLACES
± .005 THREE PLACES

FIGURE 9
Loading Plate- Lumbar Test Fixture

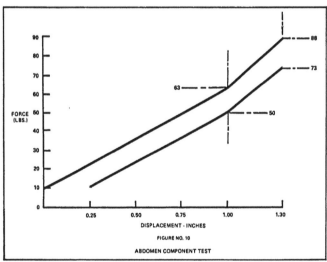

FIGURE NO. 10

ABDOMEN COMPONENT TEST

PART 572—ART PAGE 5

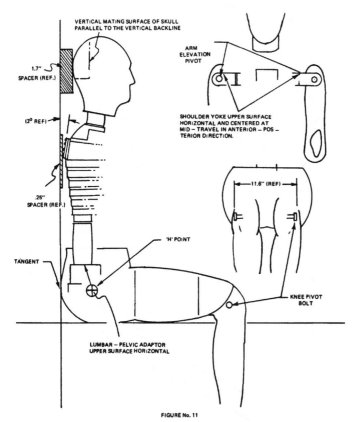

VERTICAL MATING SURFACE OF SKULL
PARALLEL TO THE VERTICAL BACKLINE

1.7"
SPACER (REF.)

(3° REF.)

.25"
SPACER (REF.)

TANGENT

ARM
ELEVATION
PIVOT

SHOULDER YOKE UPPER SURFACE
HORIZONTAL AND CENTERED AT
MID – TRAVEL IN ANTERIOR – POS –
TERIOR DIRECTION.

11.6" (REF)

'H' POINT

KNEE PIVOT
BOLT

LUMBAR – PELVIC ADAPTOR
UPPER SURFACE HORIZONTAL

FIGURE No. 11

UPRIGHT SEATED POSITION FOR LINEAR MEASUREMENTS

PART 572–ART PAGE 6

**Space for figures 12 thru 14
reserved for future use.**

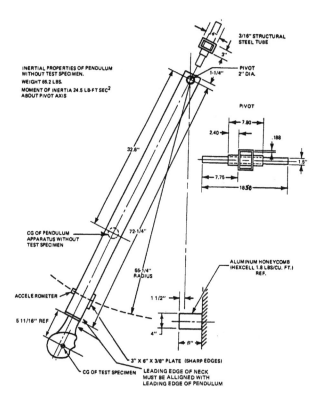

INERTIAL PROPERTIES OF PENDULUM
WITHOUT TEST SPECIMEN.

WEIGHT 65.2 LBS.

MOMENT OF INERTIA 24.5 LB-FT SEC^2
ABOUT PIVOT AXIS

3/16" STRUCTURAL
STEEL TUBE

PIVOT
2" DIA.

1-1/4"

3"

PIVOT

7.80

2.40

.188

7.75

1.5"

18.50

32.6"

CG OF PENDULUM
APPARATUS WITHOUT
TEST SPECIMEN

72-1/4"

65-1/4"
RADIUS

ALUMINUM HONEYCOMB
(HEXCELL 1.8 LBS/CU. FT.)
REF.

1 1/2"

ACCELEROMETER

5 11/16" REF

4"

6"

3" X 6" X 3/8" PLATE (SHARP EDGES)

CG OF TEST SPECIMEN

LEADING EDGE OF NECK
MUST BE ALLIGNED WITH
LEADING EDGE OF PENDULUM

FIGURE NO. 15
NECK COMPONENT TEST

PART 572—ART PAGE 8

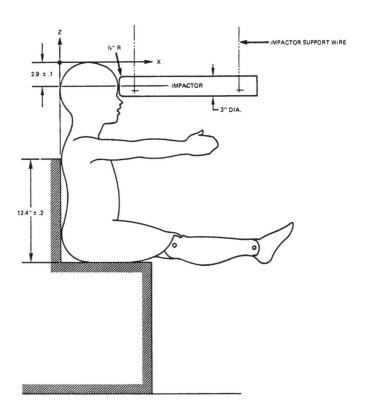

FIGURE NO. 16
HEAD IMPACT TEST

PART 572—ART PAGE 9

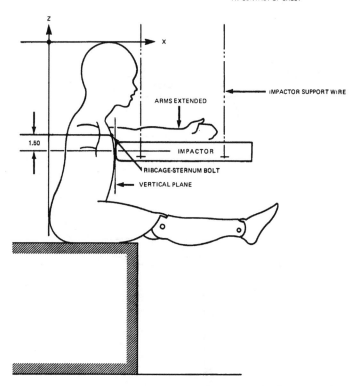

IMPACTOR FACE TO BE VERTICAL ± 2°
AT CONTACT OF CHEST

Z

X

IMPACTOR SUPPORT WIRE

ARMS EXTENDED

1.50

IMPACTOR

RIBCAGE-STERNUM BOLT

VERTICAL PLANE

FIGURE NO. 17
CHEST IMPACT TEST

PART 572—ART PAGE 10

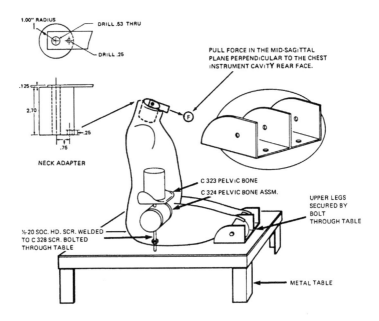

1.00" RADIUS

DRILL .53 THRU

DRILL .25

.125

2.70

.25

.75

NECK ADAPTER

PULL FORCE IN THE MID-SAGITTAL PLANE PERPENDICULAR TO THE CHEST INSTRUMENT CAVITY REAR FACE.

C 323 PELVIC BONE

C 324 PELVIC BONE ASSM.

UPPER LEGS SECURED BY BOLT THROUGH TABLE

½-20 SOC. HD. SCR. WELDED TO C 328 SCR. BOLTED THROUGH TABLE

METAL TABLE

FIGURE NO. 18
LUMBAR-SPINE FLEXION TEST

PART 572—ART PAGE 11-12

PREAMBLE TO PART 573—DEFECT REPORTS

(Docket No. 69–31; Notice No. 2)

On December 24, 1969, a notice of proposed rulemaking entitled, "Defect Reports", was published in the *Federal Register* (34 F.R. 20212). The notice proposed requirements for reports and information regarding defects in motor vehicles, to be submitted to the National Highway Traffic Safety Administration by manufacturers of motor vehicles pursuant to sections 112, 113, and 119 of the National Traffic and Motor Vehicle Safety Act (15 U.S.C. 1401, 1402, and 1407).

The notice requested comments on the proposed requirements. All comments received have been considered and some are discussed below.

Several comments asked whether both the fabricating manufacturer and the importer of imported vehicles were required to comply with all the proposed requirements. A similar question was asked in regard to manufacturers of incomplete vehicles and subsequent manufacturers of the same vehicles. In response to the comments, § 573.3 provides that in the case of imported vehicles, compliance by either the fabricating manufacturer or the importer of the imported vehicle with §§ 573.4 and 573.5 of this part, with respect to a particular defect, shall be considered compliance by both. In the case of vehicles manufactured in two or more stages, compliance by either the manufacturer of the incomplete vehicle or one of the subsequent manufacturers of the vehicle with §§ 573.4 and 573.5 of this part, with respect to a particular defect, shall be considered compliance by both the incomplete vehicle manufacturer and the subsequent manufacturers.

Many comments requested that the time for the initial filing of the direct information report be increased to allow opportunity for the extensive and complex testing often necessary to determine whether a defect is safety-related. As proposed, the time for initially filing the report was within 5 days after the discovery of a defect that the manufacturer subsequently determined to be safety-related. In response to these comments, § 573.4(b) provides that the report shall be submitted by the manufacturer not more than 5 days after he or the Administrator has determined that a defect in the manufacturer's vehicles relates to motor vehicle safety.

Several comments requested the deletion of one or more items of information proposed for inclusion in the defect information report. Objections to providing an evaluation of the risk of accident due to the defect, a list of all incidents related to the defect, and an analysis of the cause of the defect were based on the ground that the information would be inherently speculative. The proposed requirements for these three items of information have been deleted. In place of the list of incidents, § 573.4(c)(6) requires a chronology of all principal events that were the basis for the determination of the existence of a safety-related defect. In accordance with the deletion of the list of incidents, the provision in the proposal requiring quarterly reports to contain information concerning previously unreported incidents has also been deleted.

Several comments stated that the requirement in the proposal for the submission of a copy of all communications sent to dealers and purchasers concerning a safety-related defect would create an unreasonable burden on the manufacturers. The comments reported that the manufacturers would be required to submit to the Administration a large volume of useless correspondence between the manufacturers and individual dealers or purchasers. To mitigate this problem, § 573.4(c)(8) provides that the manufacturers shall submit to the Administration only those communications that are sent to more than one dealer or purchaser. For the same reason, the requirement in § 573.7 that a manufacturer submit a copy of all communications, other than those required under § 573.4(c)(8), regarding any defect, whether or not safety-related, in his vehicles, is also limited to communications sent to more than one person.

Many comments requested that a regular schedule for submitting quarterly reports be established. They suggested that this be accomplished by requiring that the first quarter for submitting a quarterly report with respect to a particular defect be the calendar quarter in which the defect information report for the defect is initially submitted. As proposed, the first quarter began on the date on which the defect information report was initially submitted. Several of these comments also objected to the proposed requirements for submitting both quarterly reports and annual defect summaries on the ground that the latter would be partially redundant. In response to these comments, the proposed requirement for filing a separate series of quarterly reports for each defect notification campaign has been deleted. Instead, § 573.5(a) requires that each manufacturer submit a quarterly report not more than 25 working days after the close of each calendar quarter. The information specified in § 573.5(c) is required to be provided with respect to each notification campaign, beginning with the quarter in which the campaign was initiated. Unless otherwise directed by the Administration, the information for each campaign is to be included in the quarterly reports for six consecutive quarters or until corrective action has been completed on all defective vehicles involved in the campaign, whichever occurs sooner.

The proposed requirement for filing annual summaries has been deleted. Instead, § 573.5 (d) requires that the figures provided in the quarterly reports under paragraph (c) (5), (6), (7), and (8) of § 573.5 be cumulative. In addition, § 573.5(b) requires that each quarterly report contain the total number of vehicles produced during the quarter for which the report is submitted.

Several changes have been made for the purpose of clarification, § 573.4(c)(8) requires that manufacturers submit three copies of the communications specified in that section. In response to questions concerning the use of computers for maintaining owner lists, a reference to computer information storage devices and card files has been added to § 573.6 to indicate that they are suitable. A reference to first purchasers and subsequent purchasers to whom a warranty has been transferred, and any other owners known to the manufacturer, has been added to the same section to make clear that the owner list is required to include both types of purchasers as well as other known owners.

Effective date: October 1, 1971.

Issued on February 10, 1971.

Douglas W. Toms,
Acting Administrator, National Highway Traffic Safety Administration.

36 F.R. 3064
February 17, 1971

PREAMBLE TO AMENDMENT TO PART 573—DEFECT REPORTS

(Docket No. 69–31; Notice 5)

This notice amends the Defect Reports regulation (49 CFR Part 573) to require manufacturers to submit vehicle identification numbers as part of the information furnished by them to the NHTSA. A notice of proposed rulemaking regarding this subject was published November 7, 1972 (37 F.R. 23650).

The purpose of including VIN's in defect reports would be to improve the notification of owners of vehicles involved in safety defect notification campaigns. The State Farm Insurance Company had suggested, for example, that insurance companies could use VIN's to identify vehicles which they insure, and to themselves notify owners of record. The Center for Auto Safety also requested the inclusion of VIN's in defect reports, so it could more readily inform persons who inquire whether particular vehicles were subject to campaigns. Other possible uses, it was noted, would be that State and local inspection facilities could determine, as part of inspection programs, whether particular vehicles had been subjected to campaigns, and if so, whether they had been repaired.

The proposal would have required the submission in the "defect information report" (§ 573.4), within five days of the defect determination, of the vehicle identification number for each vehicle potentially affected by the defect. It also proposed to substitute "line" for "model" as one of the identifying classifications describing potentially affected vehicles.

The comments demonstrated that the vehicle identification number is a useful tool for locating second and later owners of vehicles. In a study conducted by the Ford Motor Company and the State Farm Insurance Company, a fairly significant percentage of owners who either had not received or responded to the initial notification mailed by the manufacturer did respond to subsequent letters sent on the basis of the VIN.

As a result of comments received, however, the NHTSA has decided that vehicle identification numbers should only be required to be supplied in the second "quarterly report", approximately six months after a campaign is initiated, rather than in the defect information report as proposed. Only the VIN's for vehicles not repaired by that date are required to be provided. The NHTSA believes this approach will provide the safety benefits to be derived from having publicly available lists of defective vehicle VIN's and will also reduce duplication and facilitate the agency's efforts to compile and report the information.

The NHTSA requests that vehicle identification numbers be submitted in a form suitable for automatic data processing (magnetic tape, discs, punched cards, etc.) when more than 500 numbers are reported for any single campaign. While not required by this notice, the use of automatic data processing for large campaigns will facilitate the dissemination of the information for the agency. The agency may include specific requirements in this regard at a later time.

The comments argued that the benefits of having VIN's available during the initial stages of a campaign are limited, and that the compilation of identification numbers for every vehicle in a campaign would create significant problems for manufacturers related to conducting campaigns. The NHTSA believes these comments to have merit. It is clear that the chief use of VIN's will be to notify other than first purchasers, i.e., owners of older vehicles, as the names of these owners will not be available to manufacturers. By delaying the furnishing of VIN's until the filing of the second quarterly report, the VIN's reported will represent to a greater degree the names and addresses of second and later owners. The later reporting will also reduce the possibility that first purchasers will receive duplicate notices.

Many comments challenged generally the utility of the VIN in notification campaigns. Other comments complained that insurance companies might abuse the information; for example, by cancelling policies on defective vehicles. Still others believed VIN's to be privileged proprietary information, both taken separately and when combined with other information submitted pursuant to Part 573.

While it is true that the effectiveness of the requirement will depend to an extent upon the voluntary activities of third parties, the NHTSA does not view this as a reason not to issue the requirement. The offers of insurance companies and other groups to participate in notification campaigns appear to be reasonable and properly motivated. There has been no evidence brought to the NHTSA's attention to support the allegations of possible misuse of the information by insurance companies.

The agency also cannot agree that information identifying defective vehicles is or relates to proprietary information. The comments on this point seem to equate what may be embarrassing information with notions of confidentiality.

There is no basis under existing statutory definitions of confidentiality for including within them VIN's or other information identifying defective vehicles.

The proposed substitution of "line" for "model" in the descriptive information for vehicles was opposed in one comment because the term "line" is apparently more suited for passenger cars than other vehicle types. The comment indicated that "model" is a more appropriate term for trucks. In light of this comment, the terms are specified as alternatives in the regulation.

In light of the above, Part 573 of Title 49, Code of Federal Regulations, "Defect Reports," is amended. . . .

Effective date: May 6, 1974.

(Sections 103, 112, 113, and 119, Pub. L. 89–563, 80 Stat. 718; 15 U.S.C. 1392, 1401, 1402, 1407, and the delegation of authority at 49 CFR 1.51 Office of Management and Budget Approved 04–R5628.)

Issued on January 30, 1974.

James B. Gregory
Administrator

39 F.R. 4578
February 5, 1974

PREAMBLE TO AMENDMENT TO PART 573—DEFECT REPORTS

(Docket No. 69–31; Notice 6)

This notice responds to petitions for reconsideration of the amendment of 49 CFR Part 573, "Defect Reports," requiring the submission to NHTSA of the vehicle identification numbers (VIN) of motor vehicles found to contain safety related defects. The amendment was published February 5, 1974 (39 F.R. 4578). Except insofar as granted by this notice, the requests of the petitioners are denied.

Two petitions for reconsideration. one from General Motors Corporation and the other from Chrysler Corporation, were received. Both petitions objected to the requirement that VIN's be reported in the second quarterly report filed subsequent to the initiation of the defect notification campaign. Both pointed out that the NHTSA had stated in the amendment published February 5, 1974, that it was desirable to defer reporting VIN's until six months had passed from the time a notification campaign had begun. Both petitioners argued that the time for filing the second quarterly report is frequently less than six months, and suggested that the third quarterly report rather than the second was the more appropriate quarterly report to contain vehicle identification numbers. General Motors indicated that the average elapsed time from the initiation of a notification campaign to the filing of the second quarterly report is four and one-half months, while the elapsed time until the filing of the third quarterly report is, on the average, seven and one-half months. The NHTSA still believes it reasonable to allow a six-month period from the initiation of the campaign to elapse before VIN's are submitted. Accordingly, the NHTSA has granted the petitions insofar as they request that VIN's be reported in the third quarterly report submitted to NHTSA by the manufacturer.

Chrysler objected to the VIN reporting requirement generally, on the basis that it is unnecessary and will not produce the desired results. It is requested that an evaluation of the usefulness of the requirement be conducted after it is in effect, and that appropriate modifications be made if the requirement fails to achieve the desired results. General Motors requested that NHTSA maintain a public record of requests for VIN's so that future consideration can be given to the extent that the data is useful, and to whom it is useful. The NHTSA believes that public availability of VIN's will facilitate locating and repairing defective vehicles no longer in the hands of first purchasers. At the same time it agrees to conduct an evaluation of the efficacy of the requirement once it is in effect. The extent of usage is a relevant aspect of an evaluation of this type, and the NHTSA sees no prohibition against maintaining a public record of requests for the information.

The amended regulation will be effective August 6, 1974, and as such will require all third quarterly reports submitted to NHTSA on or after that date to contain appropriate vehicle identification numbers. The effective date has been changed from May 6, 1974, as a result of the change requiring the third rather than the second quarterly report to contain VIN's. As a practical matter, VIN's will be required to be reported in the third quarterly report for all defect notification campaigns initiated on or after January 1, 1974 (NHTSA campaign numbers 74–0001 and subsequent campaigns).

In light of the above, 49 CFR Part 573, Defect Reports, is amended by revising § 573.5(e) . . . .

Effective date: August 6, 1974.

(Secs. 103, 112, 113, and 119, Pub. L. 89-563, 80 Stat. 718; 15 U.S.C. 1392, 1401, 1402, 1407, and the delegation of authority at 49 CFR 1.51; Office of Management and Budget approved 04-R5628.)

Issued on May 6, 1974.

Gene G. Mannella
Acting Administrator

39 F.R. 16469
May 9, 1974

PREAMBLE TO AMENDMENT TO PART 573—DEFECT REPORTS

(Docket No. 74-7; Notice 2)

This notice amends Part 573—"Defect Reports" by revoking the requirement that manufacturers of motor vehicles report quarterly to the National Highway Traffic Safety Administration production figures for vehicles manufactured or imported during the calendar quarter. A notice of proposed rulemaking in which this amendment was proposed was published January 15, 1974 (39 FR 1863).

The NHTSA is revoking the requirement for the reporting of quarterly production figures because it has found that the value of the information has not justified the burden on manufacturers of providing it. This amendment will eliminate the need for manufacturers to file quarterly reports unless they are conducting notification campaigns during the calendar quarter.

The notice of proposed rulemaking of January 15, 1974, proposed to extend the applicability of the Defect Reports regulations to include manufacturers of motor vehicle equipment, and to modify the information required to be reported. Since the issuance of this proposal, Congress has amended sections of the National Traffic and Motor Vehicle Safety Act which deal with manufacturers' responsibilities for safety related defects in motor vehicles and motor vehicle equipment. (Pub. L. 93-492, Oct. 27, 1974) These amendments to the Safety Act in part enlarge the responsibilities of manufacturers of motor vehicle equipment for safety related defects. Ultimately the Defect Reports regulations will reflect completely the expanded scope of the statutory amendments. While the language of

the proposed rule of January 15, 1974, is in most cases sufficiently broad to reflect these statutory changes, the scope of the proposal under the previous language of the Safety Act is materially different. Consequently, the NHTSA has decided to issue a further notice, with opportunity for public comment, that specifically reflects the expanded scope of the statutory amendments. This notice will be issued at some time following the effective date (December 26, 1974) of the statutory amendments.

The NHTSA has determined, however, that relief from the production-figures reporting requirements should not be further deferred, and by this notice deletes those requirements from the Defect Reports regulation.

In light of the above, 49 CFR Part 573, Defect Reports, is amended by revoking and reserving paragraph (b) of section 573.5 ("Quarterly reports").

Effective date: December 10, 1974. This amendment relieves a restriction and imposes no additional burden on any person. Consequently good cause exists and is hereby found for an effective date less than 30 days from publication.

(Secs. 108, 112, 113, 119, Pub. L. 89-563, 80 Stat. 718, 15 U.S.C. 1397, 1401, 1402, 1408; delegation of authority at 49 CFR 1.51)

Issued on December 4, 1974.

James B. Gregory
Administrator
39 F.R. 43075
December 10, 1974

PREAMBLE TO AMENDMENT TO PART 573—DEFECT AND NONCOMPLIANCE REPORTS

(Docket No. 74-7; Notice 4)

This notice amends Part 573, *Defect and Noncompliance Reports*, by adding reporting requirements for equipment manufacturers and altering somewhat the requirements for vehicle manufacturers as authorized by the 1974 Motor Vehicle and Schoolbus Safety Amendments. The amended regulation requires the submission of reports to the agency concerning defects and noncompliance with safety standards and specifies the information to be included in those reports.

Effective date: January 25, 1979.

Addresses: Petitions for reconsideration should refer to the docket number and be submitted to: Room 5108, Nassif Building, National Highway Traffic Safety Administration, 400 Seventh Street, S.W., Washington, D.C. 20590.

For further information contact:

Mr. James Murray, Office of Defects Investigation, National Highway Traffic Safety Administration, 400 Seventh Street, S.W., Washington, D.C. 20590 (202-426-2840)

Supplementary information:

This notice amends Part 573, *Defect and Noncompliance Reports*. A notice of proposed rulemaking was published on September 19, 1975 (40 F.R. 43227), proposing new requirements for vehicle and equipment manufacturers regarding submittal to the NHTSA of defect and noncompliance reports as authorized by the Motor Vehicle and Schoolbus Safety Amendments of 1974 (the Amendments) (Pub. L. 93-492).

Sections 151 to 160, or Part B of the Amendments alter the defect notification requirements of the National Traffic and Motor Vehicle Safety Act of 1966 ("the Act") (15 U.S.C. 1381 *et seq.*). These Amendments require manufacturers of motor vehicle replacement equipment to notify purchasers and to remedy any defects or noncompliances following the manufacturer's or the Administrator's determination that the equipment contains either a defect which relates to motor vehicle safety or a noncompliance with an applicable Federal motor vehicle safety standard. Prior to the enactment of these provisions, manufacturers of motor vehicle equipment were responsible under the Act for notification of defects or noncompliances only following a determination by the National Highway Traffic Safety Administrator that the item of equipment contained a safety-related defect or failed to comply (Sec. 113(e), Pub. L. 89-563, 15 U.S.C. 1402).

Comments on the proposal were received from manufacturers, safety organizations, and manufacturer representatives. The Vehicle Equipment Safety Commission did not submit comments. All comments were considered and the most significant ones are discussed below.

I. Scope.

Several manufacturers objected to the scope of the regulation indicating that it exceeded the agency's authority to regulate vehicle and equipment manufacturers. For example, manufacturers alleged that the agency only has authority over safety-related defects and accordingly should restrict the defects mentioned in this section to safety-related defects. Further, many equipment manufacturers apparently thought that they would be required to retain purchaser and owner lists of all vehicles containing items of their equipment.

The intent of this regulation is not to impose upon equipment manufacturers recordkeeping requirements for all equipment that they manufacture. This regulation merely imposes limited recordkeeping requirements for that equipment which is determined to be defective or in noncompliance. In other words, an equipment manufacturer, after discovery of a defect or noncompliance, would ascertain from a vehicle manufacturer the identity of the vehicles and vehicle owners possessing the affected equipment. Notification would then be sent to those owners. The NHTSA would require that the equipment manufacturer retain the records of those sent notice of the defect.

Several manufacturers requested that the agency limit the applicability of this regulation to safety-related defects. They argued that the NHTSA has no authority to require information pertaining to non-safety-related defects. Section 158 of the Act specifically authorizes the agency to require information on any defect, whether or not safety-related, in order to enable it to undertake defect investigations which permit a determination regarding the safety-related nature of the defect. Much of this regulation pertains only to safety-related defects and each section indicates whether it applies to all defects or only those that are safety related.

II. Application.

Many manufacturers complained about the use of the term "direct control" in Section 573.3(a). Some manufacturers contended that the use of the term was unnecessary. Importers contended that they should not be required to submit reports where a defect is identified before the vehicles leave their direct control since the Act considers them to be manufacturers and they would be in direct control of vehicles being imported. The Center for Auto Safety would have the agency drop the term and replace it with "beyond their place of final manufacture."

In the notice of proposed rulemaking, the NHTSA indicated the reasoning for excluding vehicles and equipment within the "direct control" of the manufacturer from the reporting requirements. Vehicles and equipment within the direct control of manufacturers are virtually assured of remedy of any defect or noncompliance, because they are still within the physical possession of the manufacturer. In the NPRM it was noted that direct control does not include in the possession of a dealer or distributor. For vehicles and equipment possessed by those entities, reports concerning defects or noncompliance would be required to be submitted to the agency. The agency declines to adopt the suggestion of the

Center for Auto Safety for reasons explained in the NPRM. The phrase "beyond the place of final manufacture" is not broad enough to handle all instances where vehicles are still within the direct control of the manufacturer. For example, vehicles might be stored on a manufacturer's lot far removed from the place of manufacture. Nonetheless, these vehicles are still within the direct control of the manufacturer. Therefore, the agency concludes that the term "direct control" best accomplishes the objective of providing a limited exclusion from the reporting requirements. The agency agrees with importers that since they are considered manufacturers under the Act, vehicles that manifest defects while they are within their direct control are excluded from the reporting requirements.

Some manufacturers apparently misunderstood the requirements of Section 573.3(d). Manufacturers indicated that reports should be required to be filed either by the brand name owner or the manufacturer, not by both. The section as written permits this. Compliance with the reporting requirements by the brand name owner shall be considered compliance by the manufacturer. Either one is permitted to submit the required reports. The Act treats tire brand name owners as manufacturers. Therefore, the wording of this section has been changed to reflect the responsibility of tire brand name owners.

Several commenters requested that the name of fabricating manufacturers not be submitted since this might cause competitive disadvantage to the brand name or trademark owner. The NHTSA finds it a legitimate need to know the actual manufacturer of a product. That manufacturer could, for example, be manufacturing the same or similar components for other brand name or trademark owners. The agency would need this information to ensure that all potentially defective or noncomplying equipment is remedied.

Many manufacturers complained of the requirements in Section 573.3(f) that reports be filed both by the equipment manufacturer and the vehicle manufacturer where an equipment manufacturer's equipment has been used by more than one vehicle manufacturer. Manufacturers stated that this requirement is duplicative and costly, providing identical information from both sources. The NHTSA stated in the NPRM that this issue had been thoroughly considered prior to the issuance of the NPRM. It has again been explored by the agency in response to these comments and the agency concludes that the dual reporting requirement for the 573.5 report is necessary. Reports submitted by equipment and vehicle manufacturers will have different information in them. In both cases, the information is of importance to the agency in pursuing its defects and noncompliance obligations. Therefore, this requirement has been retained. It should be reaffirmed for clarity that where an equipment manufacturer's equipment is used in vehicles of only one vehicle manufacturer, reports need only be submitted by that vehicle manufacturer.

On a related matter, the NHTSA agrees that reports required under Section 573.6 need not be filed by both vehicle and equipment manufacturers. These reports need only be filed by the manufacturer undertaking the recall. Section 573.3(f) has been amended to reflect this change.

Other commenters on this section indicated their disapproval of the shared responsibility for remedying defects and noncompliance between vehicle and equipment manufacturers. Section 573.3 places certain reporting responsibilities upon both equipment and vehicle manufacturers, depending upon the nature of the defect. For the most part, vehicle manufacturers are responsible for reports relating to defects or noncompliance in their vehicles while equipment manufacturers are responsible for reports on their defective or noncomplying equipment. In those instances where a defect or noncompliance is discovered in equipment installed in the vehicles of more than one vehicle manufacturer, both the equipment and vehicle manufacturers must report. Equipment manufacturers suggested that vehicle manufacturers should be responsible for defects and noncompliance reports while vehicle manufacturers want to place the burdens upon equipment manufacturers. The NHTSA adopted the present scheme of shared responsibility between vehicle and equipment manufacturers for compliance with agency regulations in response to the 1974 Amendments. Congress indicated in those amendments that equipment and vehicle manufacturers should share the burden of rem-

edying defects in their equipment and vehicles. The NHTSA concludes that the reporting requirements outlined in this regulation implement the basic intent of those Amendments.

III. Definitions.

Many commenters objected to the definitions of original and replacement equipment. Further, some of these commenters indicated that the NHTSA had little, if any, authority to place responsibility on an original equipment manufacturer, since Section 159 of the Act makes the vehicle manufacturers responsible for original equipment. The NHTSA has deleted the definitions of original and replacement equipment from Part 573 since both terms are defined in Part 579. The NHTSA notes that with respect to the authority to place responsibility for defects or noncompliance upon original equipment manufacturers rather than the vehicle manufacturer, Section 159 states that the Act's defect and noncompliance scheme of responsibility shall be controlling unless otherwise provided by regulation. Therefore, the NHTSA does have the authority to shift the responsibility from the vehicle manufacturer to the equipment manufacturer if it determines that such alteration will advance the efficiency of enforcement actions. Part 579, *Defect and Noncompliance Responsibility*, outlines the responsibilities of the various manufacturers and defines "replacement" and "original" equipment.

Commenters also requested that the agency define the term "safety-related defect" so as to clarify the agency's intent in this area. The NHTSA has in the past rejected requests to establish a specific definition of safety-related defect. Whether or not a defect is safety-related depends upon a variety of factors and must be ascertained based upon the circumstances of each separate case. Thus, a specific definition cannot feasibly be created.

Ford Motor Company argued that the agency's preambular discussion tended to indicate that the definition of "first purchaser for purposes other than resale" would include the dealer or distributor. This was not the intent of the regulation. "First purchaser" is based on a similar statutory term and has been used by the agency for years with a specific meaning. The first purchase occurs where the purchaser does not buy the vehicle with the purpose of reselling it. Obviously, sale of a vehicle to a dealer presupposes that the dealer intends to resell the vehicle to the ultimate consumer or purchaser. Therefore, sale to a dealer would not constitute the sale to the first purchaser for purposes other than resale. The use of the term first-purchaser list in the preamble of the proposal in reference to the lists required to be retained by equipment manufacturers was a colloquial use of the term rather than its more precise meaning under the Act.

IV. Defect and noncompliance information reports.

Prestolite Company interpreted the requirements of Section 573.5(a) to mean that they would be required to file a report with the NHTSA every time a defective piece of equipment was brought to their attention, since there is no specific definition of safety-related defect. This they suggested would be a burdensome requirement. Such a requirement is not the intent of this regulation. A manufacturer submits a report to the NHTSA when either it or the agency makes a determination under Section 151 or 152 of the Act that a defect related to motor vehicle safety in fact exists. A failure of a single piece of equipment may not occasion the finding of a safety-related defect. Further, some equipment failures might have no adverse safety effects. Therefore, every failure of equipment will not necessarily require a report to the NHTSA. It is incumbent upon the agency and each manufacturer to make a good faith determination concerning the safety relatedness of any defect before a report under this paragraph is filed.

International Harvester (IH) suggested that a manufacturer should not have to file a report if it intends to file a petition for inconsequentiality. The NHTSA does not agree with this position. The agency needs to know of potential safety-related defects or noncompliances at the earliest possible time. If a manufacturer intends to file a petition for inconsequentiality, it should indicate such in the report as part of the information supplied in accordance with subparagraph (c) (8).

Many manufacturers objected to the 5-day requirement in Section 573.5(b) under which information must be submitted within 5 working days

after a safety-related defect or noncompliance has been discovered. Manufacturers suggested increasing the number of working days and changing the word "submitted" to "mailed." Ford requested that the 5-day period not begin until written notification is received from the NHTSA for agency-initiated determinations.

The agency does not find persuasive arguments for altering the existing 5-working day requirement. The NHTSA needs this information as rapidly as possible to aid expeditious notification and recall. Not all information need be supplied within the 5 working days if some of it is unavailable. The regulation clearly states that any unavailable information would be submitted later as it becomes available. The NHTSA also considers it unnecessary to change the word "submitted" to "mailed." The term 'submitted" is broader than "mailed." Information may be submitted by mailing it or delivering it to the agency in person. If mailed, it must be mailed within 5 working days.

With respect to the alleged insufficient time to prepare information in 5 working days, the NHTSA notes that this requirement has existed in Part 573 for several years. Since the requirement has operated smoothly for that period of time, the agency declines to adopt recommendations that would change it.

The NHTSA declines to adopt Ford's recommendation concerning agency-initiated determinations. Agency initiated defect or noncompliance determinations are made after thorough investigations conducted by the NHTSA. A manufacturer is aware of these ongoing investigations, and therefore, it should not be unnecessarily burdened or surprised when the NHTSA makes a determination. Since the need for expeditious action exists after an agency determination and the manufacturer is aware of a pending agency decision, the NHTSA considers it adequate that a manufacturer submit the report in 5 working days after receipt of either written or oral agency notification.

Several equipment manufacturers contended that the requirements of paragraph (c) (2) would impose additional burdens upon them to mark the equipment that they manufacture. Paragraph (c) (2) requires defect and noncompliance reports to contain certain information that identifies the defective or noncomplying equipment. For example, they argued that the requirements for the date of manufacture of the affected equipment would be burdensome since much of their equipment is not dated according to time of manufacture. Therefore, they suggested that the NHTSA only require date of manufacture information when it is known.

It is important to remember that Part 573 is for the most part a reporting regulation. It is not a recordkeeping or labeling regulation. A manufacturer, under the regulation, only supplies to the NHTSA that information which is available to it. In the case of date of manufacture of equipment, the equipment manufacturer in most instances need not label its equipment in such a manner as to identify its date of manufacture. The regulation merely directs a manufacturer to supply such information to the NHTSA in its reports. Obviously, if a manufacturer does not know the dates of manufacture, it would be unable to supply them to the agency. However, a manufacturer must supply the approximate dates of manufacture if that information is available.

Manufacturers should note that the manufacturing date requirement is included in the regulation for the benefit of the equipment manufacturer. If that manufacturer knows the approximate dates when a defective piece of equipment was produced, then its recall can be limited to equipment manufactured during those dates. On the other hand, a manufacturer without such information might be required to undertake a more extensive recall of its equipment to ensure that all defective products are recalled.

The Center for Auto Safety requested that the NHTSA require motor vehicle manufacturers to submit the vehicle identification numbers (VIN) of vehicles involved in any recall activity. The NHTSA does not require this information in the Part 573.5 reports because the agency normally has no need at the time of the reports issuance for such information. The agency does require the VIN's to be submitted in the Part 573.6 reports for those vehicles that are uncorrected in a manufacturer's recall. In these instances, the agency uses the information to supplement a manufacturer's recall efforts. Until such time as a manufacturer determines that some vehicles are uncorrected however, the agency usually has little use for VIN information on all recalled vehicles. In those limited instances when VIN information is necessary at the time of submission of the Part 573.5 report, the agency has the ability to request it from a manufacturer.

In regard to paragraph (c)(3), several manufacturers objected to the requirement that the precise number of vehicles or equipment in each category be reported. These manufacturers stated that often this information is not known. The NHTSA agrees and therefore modifies the section to require the submittal of this information when it is known. Chrysler suggested that the agency require the numbers of affected vehicles to be submitted by GVWR breakdown rather than by model. The agency disagrees with this recommendation since it usually undertakes recalls based upon model classification, not upon GVWR categories. Therefore, the submission of information based upon a GVWR classification would not be as useful as a classification based upon vehicle model.

Atlas Supply Company suggested that the agency not require the information specified in paragraph (c)(4) since, for tire manufacturers, tires are destroyed, making the required calculations difficult. Paragraph (c)(4) requires the provision of information that estimates the percentage of defective or noncomplying equipment on vehicles. The NHTSA considers estimates of the amount of affected vehicles or equipment to be necessary to obtain an idea of the scope of the defect or noncompliance problem. Since the section merely requires an estimate, the agency does not consider this to place a difficult burden upon manufacturers.

Many manufacturers complained about the requirements of paragraph (c)(6) which requires the submission of information upon which the determination was made that a safety-related defect exists. These manufacturers indicated that it would impose unreasonable burdens upon manufacturers by requiring them to retrieve a large amount of information in a short period of time and to retain vast amounts of data. The intent of this section is to provide a summary to the NHTSA of the information upon which a manufacturer based his defect determination. This information, since it has been used by a manufacturer for its determination of a defect, should be readily available to it. The NHTSA notes that the submission of summary information is intended to reduce a manufacturer's burdens. However, the specificity and clarity of information must be maintained, and the agency might require further information if the summary information is inadequate. The NHTSA has reworded the paragraph somewhat to indicate that it is only necessary to submit a summary of the information upon which the determination was based.

Several manufacturers suggested that the requirement for submission of noncompliance test data in paragraph (c)(7) would require them to conduct tests and submit details of test procedures to the agency. This paragraph requires only that manufacturers supply the results and data of tests, if any are conducted, upon which a noncompliance determination was based. Test procedures need not be submitted. If a noncompliance determination is made on information other than tests, then that information would be submitted.

Manufacturers claimed that they would be unable to submit a plan for remedy as required by paragraph (c)(8) in the required 5 working days. The NHTSA needs to have an indication of a manufacturer's plan for remedy as soon as possible. Like all of the information required by this section, the plan need not be extensively detailed in the initial 5-working day period and is subject to modification if subsequent circumstances warrant a change. In other words, a manufacturer is not binding itself to only those items established in the plan submitted during the first 5 days after a defect or noncompliance has been determined to exist. The NHTSA has amended the wording of this paragraph somewhat to indicate that a copy of a manufacturer's plan for remedying a defect or noncompliance will be made public in the NHTSA docket.

The Center for Auto Safety argued that paragraph (c)(9) should require actual copies of the defect or noncompliance notice bulletins or communications, not representative copies. The reason the NHTSA used the terminology contained in the notice is that in some instances a manufacturer has a multiple mailing of one communication. To require actual copies of multiple mailings would require copies of each of these identical communications. Therefore, the agency allows a representative copy (e.g., one actual copy) of such information. The NHTSA concludes that this requirement fulfills the agency's need for accurate copies.

V. Quarterly defect reports.

Many manufacturers disagreed with the agency's scheme for quarterly defect reports outlined in Section 576.6. Equipment manufacturers suggested that vehicle manufacturers should be responsible for these reports, while vehicle manufacturers asserted that the equipment manufacturers are better able to accomplish the reporting requirements. The NHTSA requires any manufacturer, either vehicle or equipment, undertaking a recall to comply with the quarterly reporting requirement. This report tells the agency the status of recalls, and therefore, is best accomplished by the party conducting the recall. The NHTSA declines to adopt suggestions that would change this scheme.

Subparagraph (b)(6) requires the submission of information on the number of vehicles or equipment that is determined to be unreachable. Several manufacturers argued for deletion of this information suggesting that it was impossible to ascertain why certain vehicles or equipment are unreachable. The manufacturer need only give the reasons why vehicles are unreachable when such information is available to him. This information aids the agency in understanding the effectiveness of a recall. The agency can determine from this data the number of vehicles still in use that were not corrected by a manufacturer and why.

VI. Purchaser and owner lists.

The intent of this section was misunderstood by a number of commenters. Many manufacturers, both equipment and vehicle, indicated that this requirement burdened them with new record-keeping requirements far beyond those currently in existence. This is not the case. For example, Part 573.7(a) requires vehicle manufacturers to maintain lists of owners of vehicles involved in a notification campaign, not all vehicles produced. General recordkeeping requirements for vehicle and equipment manufacturers are found in the Act and in the agency's regulations in Part 576. These general recordkeeping requirements are not affected by this regulation.

Equipment manufacturers strenuously objected to paragraph (c) as placing huge recordkeeping burdens upon them while achieving little in the way of benefits. The agency does not find these arguments persuasive. The recordkeeping requirement in this paragraph is limited. The agency has reworded this section to clarify an equipment manufacturer's recordkeeping requirements. This requirement does not mandate an equipment manufacturer to make and retain a list of all purchasers of its equipment as the equipment is sold. Equipment manufacturers will be required to retain a list of individuals, dealers, distributors and manufacturers determined by the manufacturer or the agency to be in possession of potentially defective or noncomplying equipment. This limited requirement is within the authority granted by Section 112(b) of the Act. The list would be compiled during the course of a defect or noncompliance campaign. If an equipment manufacturer is unable to find those in possession of its equipment, no list is required to be retained. The burden imposed by this requirement is minimal since it merely requires that manufacturers retain some information that will, by necessity, be generated should they be required to conduct either a defect or noncompliance campaign.

With respect to paragraph (b), tire manufacturers indicated that each tire does not have a different identification number and therefore the paragraph should be amended somewhat to reflect this. The agency agrees and has modified the language accordingly.

VII. Notices, bulletins, and other communications.

Many manufacturers objected to the requirements in Section 573.8 as being too broad and beyond the scope of the NHTSA's authority. This section requires the submission of information concerning defects in equipment and vehicles. Further, the manufacturers recommended that the parentheticals be deleted from the section and

that the term "defect' be changed to "safety-related defect." The agency does not agree with these comments.

First, the agency needs information concerning any defect in a manufacturer's product, not just those defects that a manufacturer deems to be safety-related. The Act contemplates a two-pronged approach to defects determinations. Either a manufacturer or the agency can make such a determination. For the agency to carry out its half of that responsibility, it needs information pertaining to all defects so that it can then judge for itself whether a defect is in fact safety related. To require only information pertaining to manufacturer-determined safety-related defects, would in effect mean that manufacturers would not be required to submit defect information to the agency until such time as that manufacturer had made a safety-related defect determination. This would stymie the agency's ability to make independent judgments concerning defects that is necessary for proper enforcement of the Act. In the past year, the NHTSA has made several safety-related defect determinations on the basis of information routinely submitted by manufacturers concerning defects that they had not considered safety-related. For example, some Airstream Trailers and White Trucks were recalled when the agency discovered safety-related problems that were mentioned in those companies' technical bulletins. Therefore, the agency needs all types of defect information, not just information that manufacturers determine to be safety-related.

Second, the parentheticals were added to this section to help clarify the type of information intended to be covered by its requirements. These lists are not all-inclusive. The NHTSA concludes, however, that they do clarify the type of information the agency seeks to obtain from a manufacturer, and therefore, they will be retained in the regulation.

The agency has deleted from Section 573.8 all references to noncompliances. All noncompliances must be reported to the agency under Part 573.5 (c)(9). Therefore, it is unnecessary to include references to noncompliances in this paragraph.

In response to the allegations that the agency has no authority to require submittal of defect information, whether or not safety related, Section 158 of the Act specifically grants the agency that authority.

VIII. Address for submitting required reports and other information.

The address listed in Part 573.9 has been altered to reflect the new agency organization and authority for enforcement actions.

In accordance with agency policy, the NHTSA has considered the costs and benefits of this requirement. The agency concludes that the regulation will help enforcement of defect and noncompliance cases by ensuring that adequate information is submitted to the NHTSA. The costs to both industry and government of the regulation will be less than $5 million annually.

The principal authors of this notice are James Murray of the Office of Defects Investigation and Roger Tilton of the Office of Chief Counsel.

In consideration of the foregoing, Part 573, *Defect and Noncompliance Reports*, of Volume 49 of the Code of Federal Regulations is amended. . . .

(Secs. 108, 112, 119, Pub. L. 89–563, 80 Stat. 718; Secs. 102, 103, 104, Pub. L. 93–492; 88 Stat. 1470; 15 U.S.C. 1397, 1401, 1408, 1411–1420; delegation of authority at 49 CFR 1.50.)

Issued on December 18, 1978.

Joan Claybrook
Administrator

**43 F.R. 60165–60169
December 26, 1978**

PREAMBLE TO AN AMENDMENT TO PART 573

Defect and Noncompliance Reports
(Docket No. 74-7; Notice 7)

ACTION: Final Rule.

SUMMARY: The purpose of this final rule is to amend 49 CFR Part 573—*Defect and Noncompliance Reports,* to delete certain reporting requirements for motor vehicle or motor vehicle equipment manufacturers conducting a defect or noncompliance notification campaign. Under this rule, motor vehicle manufacturers no longer have to submit, in the third quarterly report to the agency, the vehicle identification number (VIN) for each vehicle for which corrective measures have not been completed. Other quarterly report information requirements are also deleted or clarified, based on the agency's experience since 1974 with this portion of the defect and noncompliance reports.

EFFECTIVE DATE: January 6, 1986

SUPPLEMENTARY INFORMATION: Part 573—*Defect and Noncompliance Reports,* includes requirements for manufacturers to report to NHTSA safety-related defects and nonconformities with Federal motor vehicle safety standards, to maintain lists of purchasers and owners notified of defective and noncomplying motor vehicles and items of equipment, and to provide the agency with quarterly reports on the progress of defect and noncompliance notification campaigns. The quarterly reports must contain specified information and be submitted for six consecutive quarters after initiation of a campaign, unless corrective action is completed earlier.

This rule amends only section 573.6 of Part 573 which sets forth the information required to be submitted to the agency in these quarterly reports. The notice of proposed rulemaking, which was issued on March 27, 1985 (50 FR 12056), proposed to delete or clarify certain information requirements in the third quarterly report. This amendment was proposed in response to a petition by the Motor Vehicle Manufacturers Association (MVMA). The agency received comments on the proposal from nine motor vehicle manufacturers and the MVMA. All comments supported the proposal as lessening an administrative and cost burden. The agency is adopting the changes as proposed.

First, the rule deletes the requirement in section 573.6(b)(7) that manufacturers submit, in the third quarterly report to the agency, the VIN for each vehicle for which corrective measures have not been completed. All commenters supported this change, stating that the deletion of these VIN's from the third quarterly report would lessen the administrative and cost burdens of producing the information and would not adversely affect the progress of safety campaigns. In addition, all commenters agreed that these VIN's would be supplied to the agency, if requested, within a reasonable time.

As stated in the proposal, this rule will not change the agency's practice of assisting any individual vehicle owner who requests recall information about a particular vehicle or item of equipment. The agency will continue to provide information to enable the owner to contact the appropriate office of the manufacturer.

Second, this rule also deletes the requirement in section 573.6(b)(4) that each quarterly report include the number of vehicles or items of equipment estimated to contain the defect. This total number is initially supplied to NHTSA under the requirements of section 573.5 which states that the manufacturer's first report must include information specifically identifying the vehicles or items of equipment potentially containing the defect or noncompliance, and the percentage of those vehicles or equipment items estimated to actually contain the defect or noncompliance.

The agency's purpose in having this number updated in the quarterly reports has been to determine the potential size of notification campaigns. Ford Motor Company stated that updated information could be sent, if needed, within 10 working days. Ford added that information requiring supplier analysis on returned components would take longer. The agency concludes that updated estimates in the quarterly reports are no longer necessary. NHTSA will continue to receive quarterly report information on the number of vehicles or items of equipment involved in the notification campaign under section 573.6(b)(3). The requirement in section 573.6(b)(4) is therefore deleted in the rule.

Third, commenters also agreed with the proposed amendment to the language in section 573.6(b)(5) which clarifies the agency's intent that the number of vehicles and equipment items inspected and repaired and the number inspected and determined not to need repair should be separately reported. The rule adopts this clarification.

Fourth, the rule deletes the requirement in section 573.6(c) concerning the correction of errors in quarterly reports. Under this section, manufacturers must submit revised information in quarterly reports when they determine that an original report contained incorrect data concerning the number of vehicles or items of equipment (1) involved in a notification campaign, (2) estimated to contain the defect, or (3) determined to be unreachable for inspection for any reason. The agency does not believe submittal of this information on a regular basis is necessary and commenters agreed, adding the data could be supplied if necessary, upon request from NHTSA.

In their comments, Ford requested that the final sentence of section 573.6(b)(6) be deleted. This section requires that the number of vehicles or items of equipment, which are determined to be unreachable for inspection due to export, theft, scrapping, failure to receive notification, or other reasons, be reported to NHTSA. The last sentence of the section requires that the number of vehicles or items of equipment in each of these categories be specified. The agency did not propose in the March notice that this sentence be deleted, because this information is utilized by the agency. For example, NHTSA keeps track of the number of owners who were unreachable to assist the agency in determining whether renotification to new owners is necessary or whether additional types of notification should be adopted. Moreover, the manufacturers currently receive notice of whether a vehicle or equipment item has been exported, stolen, or scrapped by return postcard, from the person notified of the campaign. Therefore, this requirement is not changed.

In consideration of the foregoing, 49 CFR Part 573 is amended as follows:

1. The authority citation for Part 573 is revised to read as follows:

AUTHORITY: 15 U.S.C. 1397, 1401, 1408, 1411-20; delegation of authority at 49 CFR 1.50.

2. Section 573.6 is revised to read as follows:

Section 573.6 *Quarterly Reports*

(a) Each manufacturer who is conducting a defect or noncompliance notification campaign to manufacturers, distributors, dealers, or purchasers, shall submit to NHTSA a report in accordance with paragraphs (b) and (c) of this section, not more than 25 working days after the close of each calendar quarter. Unless otherwise directed by the NHTSA, the information specified in paragraphs (b)(1) through (5) of this section shall be included in the quarterly report, with respect to each notification campaign, for each of six consecutive quarters beginning with the quarter in which the campaign was initiated (i.e., the date of initial mailing of the defect or noncompliance notification to owners) or corrective action has been completed on all defective or noncomplying vehicles or items of replacement equipment involved in the campaign, whichever occurs first.

(b) Each report shall include the following information identified by and in the order of the subparagraph headings of this paragraph.

(1) The notification campaign number assigned by NHTSA.

(2) The date notification began and the date completed.

(3) The number of vehicles or items of equipment involved in the notification campaign.

(4) The number of vehicles and equipment items which have been inspected and repaired and the number of vehicles and equipment items inspected and determined not to need repair.

(5) The number of vehicles or items of equipment determined to be unreachable for inspection due to export, theft, scrapping, failure to receive notification, or other reasons (specify). The number of vehicles or items of equipment in each category shall be specified.

(c) Information supplied in response to the paragraphs (b)(4) and (5) of this section shall be cumulative totals.

Issued on: December 31, 1985.

Diane K. Steed
Administrator
51 F.R. 397
January 6, 1986

PART 573—DEFECT AND NONCOMPLIANCE REPORTS

(Docket No. 74-7; Notice 4)

Sec.

573.1   Scope.

573.2   Purpose.

573.3   Application.

573.4   Definitions.

573.5   Defect and noncompliance information report.

573.6   Quarterly report.

573.7   Owner lists.

573.8   Notices, bulletins, and other communications.

573.9   Address for submitting required reports and other information.

[AUTHORITY: 15 U.S.C. 1397, 1401, 1408, 1411–20; delegation of authority at 49 CFR 1.50. (51 F.R. 397—January 6, 1986. Effective: January 6, 1986)]

§ 573.1 Scope.

This part specifies requirements for manufacturers to maintain lists of purchasers and owners of defective and noncomplying motor vehicles and motor vehicle original and replacement equipment, and for reporting to the National Highway Traffic Safety Administration defects in motor vehicles and motor vehicle equipment, for reporting noncomformities to motor vehicle safety standards, for providing quarterly reports on defect and noncompliance notification campaigns, and for providing copies to NHTSA of communications with distributors, dealers, and purchasers regarding defects and noncompliances.

§ 573.2 Purpose.

The purpose of this part is to inform NHTSA of defective and noncomplying motor vehicles and items of motor vehicle equipment, and to obtain information for NHTSA on the adequacy of manufacturers' defect and noncompliance notification campaigns, on corrective action, on owner response, and to compare the defect incidence rate among different groups of vehicles.

§ 573.3 Application.

(a) This part applies to manufacturers of complete motor vehicles, incomplete motor vehicles, and motor vehicle original and replacement equipment, with respect to all vehicles and equipment that have been transported beyond the direct control of the manufacturer.

(b) In the case of a defect or noncompliance determined to exist in a motor vehicle or equipment item imported into the United States, compliance with §§ 573.5 and 573.6 by either the fabricating manufacturer or the importer of the vehicle or equipment item shall be considered compliance by both.

(c) In the case of a defect or noncompliance determined to exist in a vehicle manufactured in two or more stages, compliance with §§ 573.5 and 573.6 by either the manufacturer of the incomplete vehicle or any subsequent manufacturer of the vehicle shall be considered compliance by all manufacturers.

(d) In the case of a defect or noncompliance determined to exist in an item of replacement equipment (except tires) compliance with §§ 573.5 and 573.6 by the brand name or trademark owner shall be considered compliance by the manufacturer. Tire brand name owners are considered manufacturers (15 U.S.C. 1419(1)) and have the same reporting requirements as manufacturers.

(e) In the case of a defect or noncompliance determined to exist in an item of original equipment used in the vehicles of only one vehicle manufacturer, compliance with §§ 573.5 and 573.6 by either the vehicle or equipment manufacturer shall be considered compliance by both.

(f) In the case of a defect or noncompliance determined to exist in original equipment installed in the vehicles of more than one vehicle manufacturer, compliance with § 573.5 is required of the equipment manufacturer as to the equipment item, and of each vehicle manufacturer as to the vehicles in which the equipment has been installed. Compliance with § 573.6 is required of the manufacturer who is conducting a recall campaign.

§ 573.4 Definitions.

For purposes of this part:

"Act" means the National Traffic and Motor Vehicle Safety Act of 1966, as amended (15 U.S.C. 1391 *et seq.*).

"Administrator" means the Administrator of the National Highway Traffic Safety Administration or his delegate.

"First purchaser" means first purchaser for purposes other than resale.

§ 573.5 Defect and noncompliance information report.

(a) Each manufacturer shall furnish a report to the NHTSA for each defect in his vehicles or in his items of original or replacement equipment that he or the Administrator determines to be related to motor vehicle safety, and for each noncompliance with a motor vehicle safety standard in such vehicles or items of equipment which either he or the Administrator determines to exist.

(b) Each report shall be submitted not more than 5 working days after a defect in a vehicle or item of equipment has been determined to be safety-related, or a noncompliance with a motor vehicle safety standard has been determined to exist. Information required by paragraph (c) of this section that is not available within that period shall be submitted as it becomes available. Each manufacturer submitting new information relative to a previously submitted report shall refer to the notification campaign number when a number has been assigned by the NHTSA.

(c) Each manufacturer shall include in each report the information specified below.

(1) The manufacturer's name: The full corporate or individual name of the fabricating manufacturer and any brand name or trademark owner of the vehicle or item of equipment shall be spelled out, except that such abbreviations as "Co." or "Inc.," and their foreign equivalents, and the first and middle initials of individuals may be used. In the case of a defect or noncompliance determined to exist in an imported vehicle or item of equipment, the agent designated by the fabricating manufacturer pursuant to section 110(e) of the National Traffic and Motor Vehicle Safety Act (15 U.S.C. 1399(e)) shall be also stated. If the fabricating manufacturer is a corporation that is controlled by another corporation that assumes responsibility for compliance with all requirements of this part the name of the controlling corporation may be used.

(2) Identification of the vehicles or items of motor vehicle equipment potentially containing the defect or noncompliance.

(i) In the case of passenger cars, the identification shall be by the make, line, model year, the inclusive dates (month and year) of manufacture, and any other information necessary to describe the vehicles.

(ii) In the case of vehicles other than passenger cars, the identification shall be by body style or type, inclusive dates (month and year) of manufacture, and any other information necessary to describe the vehicles, such as GVWR or class for trucks displacement (cc) for motorcycles, and number of passengers for buses.

(iii) In the case of items of motor vehicle equipment, the identification shall be by generic name of the component (tires, child seating systems, axles, etc.), part number, size and function if applicable, the inclusive dates (month and year) of manufacture, and any other information necessary to describe the items.

(3) The total number of vehicles or items of equipment potentially containing the defect or noncompliance, and where available the number of vehicles or items of equipment in each group identified pursuant to paragraph (c)(2) of this section.

(4) The percentage of vehicles or items of equipment specified pursuant to paragraph (c) (2) of this section estimated to actually contain the defect or noncompliance.

(5) A description of the defect or noncompliance, including both a brief summary and a detailed description with graphic aids as necessary, of the nature and physical location (if applicable) of the defect or noncompliance.

(6) In the case of a defect, a chronology of all prinicipal events that were the basis for the determination that the defect related to motor vehicle safety, including a summary of all warranty claims, field or service reports, and other information, with their dates of receipt.

(7) In the case of a noncompliance, the test results or other data on the basis of which the manufacturer determined the existence of the noncompliance.

(8) A description of the manufacturer's program for remedying the defect or noncompliance. The manufacturer's program will be available for inspection in the public docket, Room 5109, Nassif Building, 400 Seventh St., SW., Washington, D.C. 20950.

(9) A representative copy of all notices, bulletins, and other communications that relate directly to the defect or noncompliance and are sent to more than one manufacturer, distributor, dealer, or purchaser. These copies shall be submitted to the NHTSA not later than 5 days after they are initially sent to manufacturers, distributors, dealers, or purchasers. In the case of any notification sent by the manufacturer pursuant to Part 577 of this chapter, the copy of the notification shall be submitted by certified mail.

§ 573.6 Quarterly reports.

[(a) Each manufacturer who is conducting a defect or noncompliance notification campaign to manufacturers, distributors, dealers, or purchasers, shall submit to NHTSA a report in accordance with paragraphs (b) and (c) of this section, not more than 25 working days after the close of each calendar quarter. Unless otherwise directed by the NHTSA, the information specified in paragraphs (b) (1) through (b) (5) of this section shall be included in the quarterly report, with respect to each notification campaign, for each of six consecutive quarters beginning with the quarter in which the campaign was initiated (i.e., the date of initial mailing of the defect or noncompliance notification to owners) or corrective action has been completed on all defective or noncomplying vehicles or items of replacement equipment involved in the campaign, whichever occurs first.

(b) Each report shall include the following information identified by and in the order of the subparagraph headings of this paragraph.

(1) The notification campaign number assigned by NHTSA.

(2) The date notification began and the date completed.

(3) The number of vehicles or items of equipment involved in the notification campaign.

(4) The number of vehicles and equipment items which have been inspected and repaired and the numbe of vehicles and equipment items inspected and determined not to need repair.

(5) The number of vehicles or items of equipment determined to be unreachable for inspection due to export, theft, scrapping, failure to receive notification, or other reasons (specify). The number of vehicles or items of equipment in each category shall be specified.

(c) Information supplied in response to the paragraphs (b) (4) and (b) (5) of this section shall be cumulative totals. (51 F.R. 397—January 6, 1986. Effective: January 6, 1986)]

§ 573.7 Purchaser and owner lists.

(a) Each manufacturer of motor vehicles shall maintain, in a form suitable for inspection such as computer information storage devices or card files, a list of the names and addresses of the registered owners, as determined through State motor vehicle registration records or other sources, or the most recent purchasers where the registered owners are unknown, for all vehicles involved in a defect or noncompliance notification campaign initiated after the effective date of this part. The list shall include the vehicle identification number for each vehicle and the status of remedy with respect to each vehicle, updated as of the end of each quarterly reporting period specified in § 573.6. Each list shall be retained, beginning with the date on which the defect or noncompliance information report required by § 573.5 is initially submitted to the NHTSA, for 5 years.

(b) Each manufacturer (including brand name owners) of tires shall maintain, in a form suitable for inspection such as computer information storage devices or card files, a list of the names and addresses of the first purchasers of his tires for all tires involved in a defect or noncompliance notification campaign initiated after the effective date of this part. The list shall include the tire identification number of all tires and shall show the status of remedy with respect to each owner involved in each notification campaign, updated as of the end of each quarterly reporting period specified in § 573.6. Each list shall be retained, beginning with the date on which the defect information report is initially submitted to the NHTSA, for 3 years.

(c) For each item of equipment involved in a defect or noncompliance notification campaign initiated after the effective date of this part, each manufacturer of motor vehicle equipment other than tires shall maintain, in a form suitable for inspection, such as computer information storage devices or card files, a list of the names and addresses of each distributor and dealer of such manufacturer, each motor vehicle or motor vehicle equipment manufacturer and most recent purchaser known to the manufacturer to whom a potentially defective or noncomplying item of equipment has been sold, the number of such items sold to each, and the date of shipment. The list shall show as far as is practicable the number of items remedied or returned to the manufacturer and the dates of such remedy or return. Each list shall be retained, beginning with the date on which the defect report required by § 573.5 is initially submitted to the NHTSA for 5 years.

§ 573.8   Notices, bulletins, and other communications.

Each manufacturer shall furnish to the NHTSA a copy of all notices, bulletins, and other communications (including warranty and policy extension communiques and product improvement bulletins), other than those required to be submitted pursuant to § 573.5(c) (9), sent to more than one manufacturer, distributor, dealer, or purchaser, regarding any defect in his vehicles or items of equipment (including any failure or malfunction beyond normal deterioration in use, or any failure of performance, or any flaw or unintended deviation from design specifications), whether or not such defect is safety-related. Copies shall be submitted monthly, not more than 5 working days after the end of each month.

§ 573.9   Address for submitting required reports and other information.

All required reports and other information, except as otherwise required by this part, shall be submitted to the Associate Administrator for Enforcement, National Highway Traffic Safety Administration, Washington, D.C. 20590.

43 F.R. 60169
December 26, 1978

PREAMBLE TO PART 574—TIRE IDENTIFICATION AND RECORDKEEPING

(Docket No. 70–12; Notice No. 5)

On November 10, 1970, the National Highway Safety Bureau (now the National Highway Traffic Safety Administration, or NHTSA) published the Tire Identification and Recordkeeping Regulations (35 F.R. 18116). Thereafter, pursuant to § 553.35 of the rulemaking procedures (49 CFR Part 553, 35 F.R. 5119), petitions for reconsideration or petitions for rulemaking were filed by the American Retreaders' Association, Inc., the Armstrong Rubber Co., Bandag Inc., the National Tire Dealers & Retreaders Association, Inc., the Goodyear Tire & Rubber Co., the Lee Tire and Rubber Co., Chrysler Corp., the Rubber Manufacturers Association, Ford Motor Co., the Kelly-Springfield Tire Co., Pirelli Tire Corp., the B. F. Goodrich Co., Uniroyal Tire Co., Cooper Tire & Rubber Co., Michelin Tire Corp., the Firestone Tire & Rubber Co., White Motor Corp., Bert Schwarz-S&H Inc., and the Truck Trailer Manufacturers Association. Several petitioners requested the opportunity to demonstrate difficulties they were having meeting the regulation as issued, and as a result a public meeting was held December 21, 1970. Notice of the meeting was published in the *Federal Register* (35 F.R. 19036) and the transcript of the meeting is in the public docket. The substance of the petitions and comments made at the meeting have been considered. Certain parts of the Tire Identification and Recordkeeping Regulation are hereby amended.

The definition of "Tire brand name owner" in § 574.3(c) is changed to make it clear that a person manufacturing a brand name tire that he markets himself is not a brand name owner for the purposes of this regulation.

The regulation is amended to except from its requirements tires manufactured for pre-1948 vehicles. This exception is consistent with the Federal Motor Vehicle Safety Standard for passenger car tires (Standard No. 109).

After consideration of the comments in the petitions concerning the tire identification number requirements, several changes have been made.

1. Section 574.5 is amended to specify the numbers and letters to be used in the identification number.

2. Figures 1 and 2 are modified to allow three-quarters of an inch, instead of one-half inch, between the DOT symbol and the identification number and between the second and third grouping. Tires with cross section width of 6 inches or less may use $\frac{5}{32}$-inch letters. The DOT symbol may be located to the right of the identification number as well as above, below, or to the left of the identification number. Retreaders, as well as new tire manufacturers, may locate the DOT symbol above, below, to the left, or to the right of the identification number. The minimum depth of the identification number has been changed from 0.025 inch to 0.020 inch, measured from the surface immediately surrounding the characters.

3. The second grouping, identifying the tire size, has been changed with respect to retreaded tires to provide that if a matrix is used for processing the retreaded tire the code must identify the matrix used. The change requiring retreaded tire identification numbers to contain a matrix code rather than a size code was made because, in the event of a defect notification, the matrix would be a more meaningful method of identifying the suspect tires and it was considered impracticable to require retreaders to include the tire size in the tire-identification number.

4. The third grouping, for identifying the significant characteristics of the tire, has been changed to provide that if a tire is manufactured

PART 574—PRE 1

for a brand name owner the code shall include symbols identifying the brand name owner, which shall be assigned by the manufacturer rather than by the NHTSA. Manufacturers are required to provide the NHTSA with the symbols assigned to brand name owners upon the NHTSA's request. This change should result in a shorter identification number and allow manufacturers greater flexibility in the use of the third grouping.

Standard No. 109 presently requires that passenger car tires contain a DOT symbol, or a statement that the tire complies with the standard, on both sidewalls of the tire between the section width and the bead. The requirement in Standard No. 109 is being changed by notice published in this issue (36 F.R. 1195 to provide that the DOT symbol may be on either sidewall, in the location specified by this regulation. The requested change that the DOT symbol be allowed on tires for which there is no applicable standard in effect is denied, since such use would tend to give consumers the impression those tires were covered by a Federal standard.

Several petitioners requested that other DOT symbols (located as required by the present Standard No. 109) be permitted to remain on the tire along with the three-digit manufacturer's code number assigned pursuant to that standard. The Tire Identification and Record-keeping regulation does not prohibit the continued use of the symbol and code number provided the numbers are not close enough to the identification number to be confused with it. In no event should the three-digit number, formerly required by Standard No. 109, immediately follow the tire identification number.

As a result of petitions by vehicle manufacturers the requirement in § 574.10 that vehicle manufacturers maintain the record of tires on each vehicle shipped has been changed to eliminate the requirement that this information be maintained by identification number. It would evidently be extremely difficult and expensive for the vehicle manufacturer to record each tire identification number. Vehicle manufacturers have stated that their present system provides records that enable them to notify the purchaser of a vehicle that may contain suspect tires.

Several petitioners requested that the effective date of the regulation be extended beyond May 1, 1971. The 1970 amendment to the National Traffic and Motor Vehicle Safety Act requires that the provisions relating to maintaining records of tire purchasers shall be effective not later than 1 year after the date of enactment of these amendments (May 22, 1971). It has been determined that in view of the complexities involved in establishing the recordkeeping system required and the effect of the same on existing processes, good cause exists for making the regulations effective on the latest date manufacturers are required by statute to maintain records. It is further determined that a May 22, 1971, effective date is in the public interest.

Effective date: May 22, 1971.

Issued on January 19, 1971.

Douglas W. Toms,
Acting Administrator, National Highway Traffic Safety Administration.

36 F.R. 1196
January 26, 1971

PREAMBLE TO AMENDMENT TO PART 574—TIRE IDENTIFICATION AND RECORDKEEPING

(Docket No. 70–12; Notice No. 9)

Amendment to Figure 2 Concerning the Location of the Tire Identification Number for Retreaded Tires

The purpose of this amendment is to provide retreaders with an alternative location for the placement of the tire identification number.

On January 26, 1971, the National Highway Traffic Safety Administration published Docket No. 70-12, Notice No. 5, a revised version of the Tire Identification and Record Keeping Regulation, 49 CFR Part 574 (36 F.R. 1196). Section 574.5 requires retreaders to permanently mold or brand into or onto one sidewall a tire identification number in the manner specified in Figure 2 of the regulation. Figure 2 requires that the tire identification number be located in the area of the shoulder between the tread edge and the maximum section width of the tire. The regulation specified this location because, generally, it is the area upon which retreaders apply new retread material.

Bandag, Inc., has petitioned for rulemaking to allow the tire identification to be below the section width of the tire. The petition requests this relief because the Bandag process only affects the tread surface, a comparatively smooth surface is needed for application of the identification number, and many casings have no smooth area between the tread edge and the maximum section width.

Therefore, in view of the above, Figure 2 of Part 574 (36 F.R. 1200) is hereby amended as set forth below to require that the tire identification number be on one sidewall of the tire, either on the upper segment between the maximum section width and the tread edge, or on the lower segment between the maximum section width and bead in a location such that the number will not be covered by the rim flange when the tire is inflated. In no event should the number be on the surface of the scuff rib or ribs.

Effective date: May 22, 1971.

Because this amendment relieves a restriction and does not impose any additional burden on any person it is found that notice and public procedure thereon are unnecessary and impracticable, and that, for good cause shown, an effective date less than 30 days after the date of issuance is in the public interest.

Issued on May 21, 1971.

Douglas W. Toms
Acting Administrator

PREAMBLE TO AMENDMENT TO PART 574—TIRE IDENTIFICATION AND RECORD KEEPING

(Docket No. 70–14; Notice 15)

The purpose of this amendment to Part 574 of Title 49, Code of Federal Regulations, is to provide that the second group of symbols within the tire identification number shall, in the case of new tires, be assigned at the option of the manufacturer rather than conforming to the tire size code presently found in Table I of the regulation.

Under the present system, even if the presently unassigned symbols "O" and "R" are used, a maximum of 900 tire size codes can be assigned. Due to the many new tire sizes being introduced, it is necessary to change the system to allow more flexibility. Therefore, Table I is herewith deleted, new tire manufacturers are allowed to assign their own two-digit code for the tire size, and retreaders are allowed to use either a self-assigned matrix code or a self-assigned tire size code. Each new tire manufacturer will still be required to use a two-symbol size code and to maintain a record of the coding system used, which shall be provided to the National Highway Traffic Safety Administration upon written request. It is recommended but not required that manufacturers use the code sizes previously assigned by this agency for active sizes, and re-use the codes for obsolete sizes when additional size codes are needed.

A notice of proposed rulemaking on this subject was published on June 16, 1972 (37 F.R. 11979). The comments received in response to the notice have been considered in the issuance of this final rule. The rule is issued as it appeared in the proposal including the letter "T" inadvertently omitted from the proposal.

Three of the tire manufacturers who commented favored the proposed change, and the National Tire Dealers and Retreaders Association, the Japan Automobile Manufacturers Association and The European Tyre and Rim Technical Organisation commented without objection to the proposed change.

Bandag, Inc., a retreader of tires, objected to the proposed change on the grounds that allowing tire manufacturers to assign their own tire size code would remove one of the methods a retreader has to determine the tire size of a casing to be retreaded.

Mercedes-Benz of North America and Volkswagen of America did not favor the change because of the possibility of confusion for the vehicle manufacturer that equips its vehicle with several manufacturers' tires.

The principal objection raised by Bandag should be considerably alleviated by an amendment to Standard No. 109 (36 F.R. 24824) under consideration, which would require tire manufacturers to place the actual tire size, as well as other pertinent information, between the section width and the bead of the tire so that the information will be less susceptible to obliteration during use or removal during the retreading process.

With respect to the comment by Mercedes-Benz of North America and Volkswagen of America, it was concluded that because the existing system does not provide enough symbols to meet the anticipated introduction of new tire sizes, the proposed change is necessary. Mercedes' recommendation that "G", "Q", "S", and "Z" be added or that a three-digit size code be used was rejected, because the additional symbols suggested are difficult to apply to the tire, and the addition of a third symbol would, according to the tire manufacturers, be impractical and inefficient.

A list of the tire size codes assigned up to this time is published in the general notice section of this issue of the *Federal Register* (37 F.R. 23742). The NHTSA urges tire manufacturers to use

PART 574—PRE 5

these existing codes for tire sizes presently being produced and to work within their tire and rim associations to make code assignments for new tire sizes on an industry-wide basis and reuse obsolete size codes wherever possible. In this way the usefulness of the tire size code to the vehicle manufacturer will be maintained.

In consideration of the foregoing, in Part 574 of Title 49, Code of Federal Regulations, Table I is deleted and § 574.5 is amended . . . .

Effective date: November 8, 1972.

Because this amendment relieves a restriction, and because of the immediate need for the introduction of new tire size codes, it is found for good cause shown that an effective date less than 30 days from the date of issuance is in the public interest.

Issued under the authority of sections 103, 112, 113, 119 and 201 of the National Traffic and Motor Vehicle Safety Act, 15 U.S.C. 1392, 1401, 1402, 1407 and 1421, and the delegation of authority at 49 CFR 1.51.

Issued on October 31, 1972.

Charles H. Hartman
Acting Administrator

37 F.R. 23727
November 8, 1972

PREAMBLE TO AMENDMENT TO PART 574—TIRE IDENTIFICATION AND RECORD KEEPING

(Docket No. 71-18; Notice 7)

This notice amends Standard No. 119, *New pneumatic tires for vehicles other than passenger cars*, 49 CFR 571.119, to specify lettering sizes and modified treadwear indicator requirements for tires. In addition, it amends Part 574, *Tire Identification*, 49 CFR 574, to permit the labeling of certain tires with the symbol DOT prior to the effective date of the standard. This notice also responds to petitions for reconsideration of Standard 119's effective date by maintaining the present date of March 1, 1975.

To avoid a costly production shutdown on the effective date to engrave tire molds with the DOT compliance symbol required by the standard, the National Highway Traffic Safety Administration (NHTSA) proposed a modification of the Part 574 prohibition on the symbol's use prior to the effective date (39 F.R. 3967, January 31, 1974). The Rubber Manufacturers Association and five tire manufacturers agreed that the DOT should be engraved on tire molds prior to the effective date, but objected to the expense of covering the DOT with a label stating that "no Federal motor vehicle safety standard applies to this tire," when the DOT appears on tires which (presumably) satisfy Standard 119 requirements. Firestone pointed out that the large label size could obscure other label information. Goodrich noted that, as proposed, the DOT could be molded on tires which met no standard and could mislead a user if the label fell off.

The NHTSA will not permit the appearance of the DOT compliance symbol on any item of motor vehicle equipment to which no standard is applicable. The terms "applicability" and "applies" have only one meaning for Federal motor vehicle safety standards: that the vehicle or equipment concerned is subject to a safety standard. To permit use of the DOT symbol on vehicles or items of motor vehicle equipment to which no standard applies would confuse the meaning of the symbol and the concept of compliance.

In response to Firestone and Goodrich, the NHTSA has modified the lettering size on the label and limited use of the DOT symbol to tires for which a standard has been issued. With the small lettering size, the rubber labels used on retread tires can be applied over the DOT symbol in fulfillment of the requirement. Another method which manufacturers did not mention but which would be permissible is the removal of the DOT at the same time imperfections are buffed off the tire.

All comments on the proposal objected to the specific location requirements for treadwear indicators based on the concept of even tread wear across the tread width. Goodyear demonstrated in a meeting with the NHTSA Tire Division on February 13, 1974, and detailed in its submission to the Docket, the difficulty in equating ideal tire wear with actual road experience. They recommended the simpler concept that a tire has worn out when any major tread groove has only $\frac{2}{32}$ in tread remaining. The NHTSA has concluded that treadwear indicators must be placed at the discretion of the manufacturer to give a person inspecting the tire visual indication of whether the tire has worn to a certain tread depth. Accordingly, the lateral location requirements for treadwear indicators have been deleted from the standard.

There was no discussion of the lettering size and depth proposal, and these proposals are adopted as proposed.

The comments requested reconsideration of the standard's March 1, 1975, effective date (published February 1, 1974, 39 F.R. 4087), asserting the need for 18 months of lead time following

PART 574—PRE 7

publication of this notice to engrave tire molds as required by the standard. The NHTSA has found that 11 months is sufficient leadtime to accomplish these changes, and accordingly these petitions are denied.

To correct an inadvertent omission in the amendment of Standard No. 119 in response to petitions for reconsideration (39 F.R. 5190, February 11, 1974), superscripts are added to Table III entries for "All other, A, B, C, D range tires".

In consideration of the foregoing, Parts 571 and 574 of Title 49, Code of Federal Regulations, are amended. . . .

Effective date: Standard No. 119 amendments: March 1, 1975. Part 574 amendment: April 3, 1974. Because the Part 574 amendment creates no additional burden, and because modification of tire molds must begin immediately, it is found for good cause shown that an effective date less than 180 days after issuance is in the public interest.

(Secs. 103, 112, 119, 201, Pub. L. 89-563, 80 Stat. 718; 15 U.S.C. 1392, 1401, 1407, 1421; delegation of authority at 49 CFR 1.51.)

Issued on March 28, 1974.

James B. Gregory
Administrator

39 F.R. 12104
April 3, 1974

PREAMBLE TO AMENDMENT TO PART 574—
TIRE IDENTIFICATION AND RECORDKEEPING

(Docket No. 70–12; Notice 19)

This notice amends the Tire Identification and Recordkeeping regulation, 49 CFR Part 574, to establish an optional universal registration format for tire registration forms. It also requires manufacturers of new tires to redirect registration forms of other manufacturers of new tires which have been forwarded to them in error.

On March 9, 1973, the NHTSA issued a notice of proposed rulemaking (38 F.R. 6398) proposing a universal registration form for tire identification and record keeping. The notice was issued in response to requests from multi-brand tire dealers who were faced with a multiplicity of different forms and procedures for tire registration. Currently, the regulation merely requires manufacturers and retreaders to supply a "means" of registration. The proposed rule also envisioned that a copy of the form would be provided to the first purchaser and that manufacturers and retreaders would be required to redirect registration forms which had been forwarded to them in error.

All comments received in response to the notice were sympathetic to the problems faced by the multi-brand dealers, and the majority were willing to provide a "universal form" if requested by a dealer.

Most manufacturers, however, pointed out that their exclusive dealerships had received training in the use of the current form, as had their own personnel, and that a total change-over would work a hardship without a concomitant benefit for single-brand dealers. In view of these comments, NHTSA has decided to promulgate the universal registration format, which appears as Fig. 3, as an optional format to be followed if requested by a dealer and as a guide if a dealer prefers to supply his own forms.

The proposal to require tire manufacturers and retreaders to forward all misdirected registration forms within 30 days was universally opposed by new-tire manufacturers, who stated that they are currently participating in a voluntary but limited program for forwarding these misdirected forms. Furthermore, new-tire manufacturers believe they should not be responsible for misdirected retreaded tire registration forms, as there are over 5,000 tire retreaders in the country and such a task would be formidable. One new-tire manufacturer indicated that he had received over 15,000 misdirected retreaded tire registration forms during January 1973. The docket contained only one submission from the retreading industry, and it did not deal with the problem of misdirected forms.

It also appears from the comments received and other information available to NHTSA that new-tire manufacturers maintain a computer-based registration process, while only approximately 25% of the retreading industry utilizes computers for this purpose. Thus, the requirement for forwarding all misdirected forms would fall heavily on both segments of the industry, new-tire manufacturers in that most misdirected forms appear to be sent to them and retreaders in that a majority are ill-equipped to carry out the forwarding functions.

Therefore, rather than issue an all-inclusive forwarding requirement at this time, NHTSA has decided to require only that new-tire manufacturers redirect new tire registraiton forms erroneously forwarded to them. Further, the NHTSA has determined that a 90-day forwarding period will be sufficient, rather than the 30 days originally proposed. It is expected that the use of the manufacturer's logo on the universal registration format and increased vigilance on the part of the industry will substantially curtail the number of misdirected forms. If it later appears that tire registrations are not being properly received, the NHTSA intends to take further action in this area.

The notice proposed that tire manufacturers furnish their dealers with duplicate copies of the registration form so that a copy could be given to consumers at the time of purchase. This provision was objected to by all new-tire manufacturers and the retreaders' association. In their view, the increased expense served no viable function as Part 574 currently requires all purchasers to be notified by certified mail of safety defects. They argued that the possession of a duplicate registration form would not aid the purchaser in the case of recall. The manufacturers also said that the completion of registration forms is often reserved until the end of the day or other slack time, and further that the consumer automatically receives a copy of his tire identification number on the guarantee if one is given.

The NHTSA finds these arguments to have merit, and the requirement to give the purchaser a copy of the registration form is deleted from the final rule.

In consideration of the foregoing, 49 CFR 574.7 is amended....

Effective date: September 3, 1974.

(Secs. 103, 112, 113, 119, 201, Pub. L. 89–563, 80 Stat. 718, 15 U.S.C. 1392, 1401, 1402, 1407, 1421; delegation of authority at 49 CFR 1.51.)

Issued on May 28, 1974.

James B. Gregory
Administrator

39 F.R. 19482
June 3, 1974

PREAMBLE TO AMENDMENT TO PART 574—TIRE IDENTIFICATION AND RECORDKEEPING

(Docket No. 70-12; Notice 21)

This notice amends 49 CFR Part 574 to provide that the Universal Registration Forms supplied by dealers must conform in size and be similar in format to Figure 3 of the regulation.

On June 2, 1974, 49 CFR Part 574 was amended to require a Universal Registration Format when tire registration forms are supplied by manufacturers to dealers (39 F.R. 19482). Three petitions for reconsideration were received in response to this notice. All three, Michelin Tire Corporation, Rubber Manufacturers Association, and the Firestone Tire and Rubber Company, requested that the regulation be amended to require that dealer-supplied registration forms also conform in size and be similar in format to Figure 3 of the regulation. The petitioners pointed out that registration handling methodology has been standardized throughout the industry, and that the use of different sizes and formats would be costly and inefficient. The NHTSA concurs in this assessment, and therefore amends 49 CFR 574.7(a) to require that the dealer-supplied forms must conform in size and be similar in format to Figure 3.

In addition, Firestone petitioned to revise Figure 3 slightly and to extend the effective date of the amendment to 120 days after the response to the petitions for reconsideration. Since 49 CFR 574.7 currently requires only that the forms be "similar" to Figure 3, Firestone's proposed modification is authorized by the regulation and no amendment to the standard is needed. Firestone's request to extend the effective date of the standard is denied, as NHTSA has determined sufficient lead time was available from the date the amendment was issued to prepare forms.

In consideration of the foregoing, the last sentence of 49 CFR 574.7(a) is amended. . . .

Effective date: November 1, 1974.

(Secs. 103, 112, 113, 119, 201, Pub. L. 89-563, 80 Stat. 718, 15 U.S.C. 1392, 1401, 1402, 1407, 1421; delegation of authority at 49 CFR 1.51.)

Issued on October 29, 1974.

James B. Gregory
Administrator
39 F.R. 38658
November 1, 1974

This notice corrects the authority citations to Part 574, *Tire Identification and Recordkeeping*, and makes other small corrections of citations in the text of the regulation to reflect statutory amendments. This correction is being made to conform the statutory authority citations to the existing statute.

Effective dates: Since these technical corrections do not affect the responsibilities under the regulation, they are made effective December 26, 1978.

For further information contact:

Roger Tilton, Office of Chief Counsel, National Highway Traffic Safety Administration, 400 Seventh Street, S.W., Washington, D.C. 20590 (202-426-2992).

Supplementary information: Since issuance of the Tire Identification and Recordkeeping regulation, several changes have been made to the agency's authorizing statute that require NHTSA to correct the authority citations of the regulation. While authority citatoins found in NHTSA's regulations and standards are not parts of the rules, they are useful to those who wish to review the legislative background of the rulemaking action. Therefore, NHTSA corrects the authority citations for clarity and to provide information to those who are interested.

The agency also corrects Part 574.2 and 574.8 by altering the existing reference to section 113. Section 113 was the safety defect and noncompliance notification section of the National Traffic and Motor Vehicle Safety Act of 1966 (Pub. L. 89-563). Section 102 of the 1974 Motor Vehicle and Schoolbus Safety Amendments (Pub. L. 93-492) transferred the notification provisions from section 113 to section 151 and 152 of the Safety Act, as amended (15 U.S.C. 1411 and 1412). Since the regulation currently refers to the old Act rather than the Act as amended, the agency is correcting the affected provisions of the regulation to bring them up to date.

Since this notice simply corrects references in the regulation and its authority citations without altering any of its substantive provisions, the Administrator finds that notice is unnecessary and that an immediate effective date is in the public interest.

In consideration of the foregoing, Volume 49 of the Code of Federal Regulations, Part 574, *Tire Identification and Recordkeeping*, is amended. . . .

(Secs. 103, 108, 112, 119, 201, Pub. L. 89-563, 80 Stat. 718 (15 U.S.C. 1392, 1397, 1401, 1407, 1421); Secs. 102, 103, 104, Pub. L. 93-492, 88 Stat. 1470 (15 U.S.C. 1397, 1401, 1411-1420); delegation of authority at 49 CFR 1.50).

Issued on December 18, 1978.

Joan Claybrook
Administrator

43 F.R. 60171
December 26, 1978

Action: Amendment of rule.

Summary: Congress has recently amended the National Traffic and Motor Vehicle Safety Act of 1966 (the Safety Act) to exempt manufacturers of retreaded tires from the registration requirements of the Act. This notice makes conforming amendments to the regulations implementing the tire registration requirements of the Act. The amendment is being published as a final rule without notice and opportunity for comment and is effective immediately, rather than 180 days after issuance, since the agency lacks discretion on the manner implementing this Congressional mandate.

Effective date: February 8, 1979.

For further information contact:

Arturo Casanova, Office of Vehicle Safety Standards, National Highway Traffic Safety Administration, 400 Seventh Street, S.W., Washington, D.C. 20590 (202) 426–1715.

Supplementary information: Congress has recently enacted the Surface Transportation Assistance Act of 1978, P.L. 95–599. Section 317 of that Act amends the Safety Act by exempting manufacturers of retreaded tires from the registration requirements of section 158(b) of the Safety Act.

This amendment modifies the requirements of Part 574 to specify that manufacturers of retreaded tires are not subject to the mandatory registration requirements set forth in that Part. Manufacturers of retreaded tires are free to continue voluntarily registering the tires, and the agency encourages these manufacturers to provide some means for notifying purchasers in the event of a recall of tires that do not comply with federal safety standards or contain a safety-related defect. However, this choice will be left to the individual retreaders.

The remaining obligations of retreaders under Part 574 are set forth in §§ 574.5 and 574.6, which provisions are not affected by this amendment. Those sections require that the retreader label contain certain information on its tires. These provisions allow a retreader who determines that some of its tires do not comply with a Federal safety standard or contain a safety-related defect to warn the public of that fact, and indicate the label numbers of the affected tires.

Since Congress has amended the Safety Act to exempt the manufacturers of retreaded tires from the registration requirements, this amendment of Part 574 is published without notice and opportunity for comment. The Administrator finds good cause for foregoing these procedures in this instance, because Congress has specifically mandated this action, and the agency has no authority to disregard a legislative mandate. For the same reason, this amendment is effective immediately, rather than 180 days after issuance.

The agency has reviewed the impacts of this amendment and determined that they will reduce costs to the manufacturers. Further, the agency has determined that the amendment is not a significant regulation within the meaning of Executive Order 12044.

The program official and attorney principally responsible for the development of this amendment are Arturo Casanova and Stephen Kratzke, respectively.

In consideration of the foregoing, 49 CFR Part 574, Tire Identification and Recordkeeping, is amended . . . .

AUTHORITY: Sections 103, 108, 112, 119, 201, Pub. L. 89–563, 80 Stat. 718 (15 U.S.C. 1392, 1397, 1401, 1407, 1421); secs. 102, 103, 104, Pub. L. 93–492, 88 Stat. 1470 (15 U.S.C. 1411–1420); Stat. 2689 (15 U.S.C. 1418); delegation of authority at 49 CFR 1.51.

Issued on January 31, 1979.

Joan Claybrook
Administrator

44 F.R. 7963
February 8, 1979

PREAMBLE TO AN AMENDMENT TO PART 574

Tire Identification and Recordkeeping;
Interim Final Rule and Request for Comments

(Docket No. 70-12; Notice 24)

ACTION: Interim final rule and request for comments.

SUMMARY: In October 1982, Congress adopted an amendment to the National Traffic and Motor Vehicle Safety Act of 1966 (the Safety Act) regarding tire registration requirements of 49 CFR Part 574, *Tire identification and recordkeeping.* Those requirements are intended to provide tire manufacturers and brand name owners with the names of tire purchasers so that the purchasers can be notified in the event that their tires are determined to contain a safety defect or to fail to comply with a safety standard.

The amendment prohibits this agency from requiring independent tire dealers and distributors (i.e., those whose business is not owned or controlled by a tire manufacturer or brand name owner) to comply with the existing tire registration requirements in Part 574. All other tire dealers and distributors must continue to comply with those requirements.

The prohibition regarding independent dealers and distributors is self-executing and became effective on the date of enactment, October 15, 1982. In place of the existing requirements, the amendment directed the Secretary of Transportation to require each of those dealers and distributors to furnish a registration form to each tire purchaser after the dealer or distributor has first filled in the tire identification number(s) of the tire(s) sold on the form. Purchasers wishing to register their tires may then do so by filling in their name on the form and mailing the completed form to the tire manufacturer or brand name owner. Because the new

statutory requirements regarding registration of tires sold by independent dealers and distributors are not self-executing, they do not affect those dealers and distributors until this agency has issued and put into effect a rule adopting those requirements. This rule accomplishes that result.

The Safety Act amendment also requires that the agency specify the format and content of the forms to be used in complying with the new requirements. This rule sets forth those specifications.

DATES: This rule is effective beginning June 20, 1983.

SUPPLEMENTARY INFORMATION: Prior to the enactment of the Motor Vehicle Safety and Cost Savings Authorization Act of 1982 (hereinafter referred to as the Authorization Act) (Pub. L. 97-311), all tire dealers and distributors were required by 49 CFR Part 574, *Tire identification and recordkeeping,* to register all sales of new tires. Under that regulation, NHTSA required dealers and distributors to write specified information (i.e., the purchaser's name and address, the dealer's name and address, and the identification numbers of the tires) on a registration form and send the completed form to the tire manufacturer, brand name owner (hereinafter referred to as "tire manufacturer") or its designee.

Tire registration provisions of the Authorization Act. Compliance with the requirement for mandatory registration was uneven. While virtually all tires on new vehicles were registered, slightly less than half of all replacement tires were registered. In its report on the Authorization Act, the House Committee on Energy and Commerce found that dealers and distributors whose business was owned or controlled [1]by a tire manufacturer registered between 80 and 90 percent of the replacement tires they sold. However, dealers and distributors whose businesses were not owned or controlled by a tire manufacturer (hereinafter collectively referred to as "independent dealers") registered only 20 percent of the replacement tires that they sold (*Id.* at 8).

In an effort to improve the registration rate for the tires sold by independent dealers, Congress included a tire registration provision in the Authorization Act. That provision amended section 158(b) of the National Traffic and Motor Vehicle Safety Act of 1966 (hereinafter referred to as "Safety Act") (15 U.S.C. 1381 *et seq.*) to prohibit the Secretary of Transportation from requiring independent dealers to comply with the Part 574 requirements for mandatory registration. (The Secretary's authority under the Safety Act has been delegated to the NHTSA Administrator, 49 CFR 1.50.) Dealers and distributors other than independent dealers (hereinafter collectively referred to as "non-independent dealers") remain subject to these requirements.

The prohibition concerning independent dealers was self-executing (i.e., its effectiveness was not conditioned on prior action by this agency) and became effective on the date of enactment of the Authorization Act, October 15, 1982. Thus, even without any amendment by the agency to Part 574, its requirements for mandatory registration ceased on October 15 to have any effect insofar as they apply on their face to independent dealers.

In place of the mandatory registration process, Congress directed that a voluntary process be established for independent dealers. Section 158(b) (2) (B) provides

The Secretary shall require each dealer and distributor whose business is not owned or controlled by a manufacturer of tires to furnish the first purchaser of a tire with a registration form (containing the tire identification number of the tire) which the purchaser may complete and return directly to the manufacturer of the tire. The contents and format of such forms shall be established by the Secretary and shall be standardized for all tires. Sufficient copies of such forms shall be furnished to such dealers and distributors by manufacturers of tires.

Under the voluntary process, the primary responsibility for registering tires sold by independent dealers is shifted from the dealer to the purchaser. NHTSA is mandated by section 158(b) (2) (B) to require the independent dealer to (1) fill in the identification number(s) of the tire(s) sold to a purchaser on a registration form and then (2) hand the form to the purchaser. If the purchaser wishes to register the tires, he or she may do so by filling in his or her name and address, adding postage and sending the completed form to the tire manufacturer or its designee.

In addition, NHTSA is required by section 158(b) (3) to evaluate the effect of the switch to voluntary tire registration on the registration rate for tires sold by independent dealers. That evaluation must be conducted at the end of the two year period following the effective date of the Authorization Act, i.e., October 15, 1984. In the evaluation, the agency is required to assess the efforts of the independent dealers to encourage consumers to register their tires and the extent of the dealers' compliance with the voluntary registration procedures established by this notice. NHTSA is required also to determine whether to impose any additional requirements on dealers for the purpose of promoting higher registration levels.

The agency has received several telephone inquiries from independent dealers as to whether, notwithstanding the amendments to section 158(b), they could elect to continue following the requirements for mandatory registration. It does not appear that the independent dealers have this option. Section 158(b) (2) (B) specifies that the agency "shall require *each* . . . (independent dealer) to furnish the first purchaser of a tire with a registration form (containing the tire identification number of the tire) which the purchaser may complete and return directly to the manufacturer of the tire." However, nothing in the section appears to preclude the purchaser from voluntarily giving the form back to the dealer for transmission to the manufacturer or his designee. Comments are requested on the issues raised by these inde-

---

[1] As explained in the House Report on the Authorization Act, " 'company owned and controlled' means a significant component of direct equity ownership of the dealer or distributor which gives that party, as a factual matter, effective control of the business. Thus, it would not encompass buy-sell agreements, mortgages, notes, franchise agreements or similar financial arrangements which a tire company may have with a dealer or distributor." H.R. Rep No. 576, 97th Cong. 2d Sess. 8–9 (1982).

pendent dealers as well as on the reasons why some independent dealers desire the opportunity to continue mandatory registration.

Congress made no provision for immediate replacement of mandatory registration by voluntary registration. Unlike the amendment prohibiting the agency from requiring independent dealers to follow the mandatory registration process, the amendment concerning voluntary registration is not self-executing. Before voluntary registration can be initiated, the agency must first issue a rule requiring participation by the independent dealers in the voluntary registration process and put that rule into effect.

New standardized registration forms. In addition to setting forth such a requirement, this rule also specifies the content, format and size of the registration forms to be used by the independent dealers. This aspect of the rule responds to the directive in section 158(b) (2) (B) for the standardization of such forms. NHTSA wishes to emphasize that this rule does not require standardization of the forms used by nonindependent dealers. Tire manufacturers need not make any change in the forms which they have been providing those dealers.

In selecting interim requirements standardizing the content, format and size of registration forms to be provided to or used by independent dealers, NHTSA has made the minimum changes to Part 574 necessary to comply with section 158(b) (2). This approach will minimize both the burdens of this rulemaking and the period during which independent dealers are not subject to any registration requirements.

The new standardized forms would be very similar to the forms which the manufacturers have been providing dealers over the last eight years. Since 1974, Part 574 has specified the type of information for which blanks and titles are to appear on registration forms. (§ 574.7(a) (1)–(3)). This information includes the name and address of the tire purchaser, the tire identification number, and the name and address of the dealer or other means by which the manufacturer could identify the dealer. This rule would require the new registration forms for independent dealers to have blanks and titles for the same information.

This rule also adopts as mandatory the format specifications which have appeared as a suggested guide in Part 574. Those specifications have been generally followed since 1974 without any complaints from either manufacturers or dealers.

In recognition of the shift of primary responsibility for registering tires from the independent dealer to the purchaser, this rule substitutes a new reminder on the form. The old reminder warned the dealer that registration of tires was required by Federal law. The new reminder informs the purchaser that completing and mailing the form will enable the tire manufacturer to contact him or her directly in the event that the tire is recalled for safety reasons, i.e., if the tire is determined to contain a safety defect or to fail to comply with an applicable safety standard.

Both a mailing address and a statement about appropriate postage must be printed on each form. The House report states that the form is to be presented to the purchaser in a manner suitable for mailing. (H.R. Rep. No. 576, 97th Cong. 2d Sess. 8 (1982)). Thus, the form itself must be mailable without the necessity of the purchasers providing an envelope. Forms provided by the manufacturers must be preaddressed to either the manufacturer or its designee. As to postage, the form must bear the statement that first class postage is required. This notation will ensure that the purchaser realizes that post card postage is not sufficient. If insufficient postage were placed on the form, it would not be delivered and the tire would not be registered. The need for first class postage is explained below.

This rule standardizes the size of the form so that all forms will be mailable using a single stamp of the same class of postage. The suggested guide in Part 574 specifies dimensions of 3¼ inches in width and 7⅜ inches in length. This rule does not adopt those dimensions because, under existing postal regulations, a form 3¼ inches by 7⅜ inches is too small to be mailed unless enclosed in an envelope. Since NHTSA does not wish to require manufacturers to provide self-addressed envelopes, the agency has adopted the dimensions in the postal regulations for cards mailable without envelopes under first class postage as the dimensions for the registration forms. Thus, the forms must be rectangular; not less than .007 inches thick; more than 3½ inches, but not more than 6¼ inches wide; more than 5 inches, but not more than 11½ inches long. If any of those maxima were exceeded, a single, first class stamp would not be sufficient postage. The agency has not adopted a post card-sized form due to uncertainty whether such a form would be large enough to permit the easy, legible recording of all of the necessary information.

Finally, the mandatory format requirements include a requirement that the form must show the manufacturer's name to prevent confusion of dealers and purchasers. This will enable the independent dealer to determine the brand of tire for which a particular form is to be used for registration purposes. This requirement is necessary since independent dealers often sell several different brands of tires. Since the dealer will have as many different types of registration forms as it has different brands of tires for sale, the dealer must have some way of identifying the appropriate form. The name may appear either in the mailing address or anywhere else on the form.

Continued use of old registration forms. During the limited period that this interim rule is in effect, the agency will provide the option of using existing forms instead of the new standardized ones. Election of that option is conditioned upon the tire purchaser's being provided not only with a form bearing the tire identification numbers and the dealer's name and address, but also with an envelope that is suitable for mailing the form, bears the same reminder to consumers required on the new forms, and is addressed to the tire manufacturer or its designee.

Source of registration forms. Under the requirements for mandatory registration requirements which previously applied to independent dealers, those dealers were permitted to use either the registration forms provided by the tire manufacturers or use forms obtained from other sources. The latter type of form was typically one purchased from a clearinghouse. The clearinghouse forms were not manufacturer specific (i.e., did not bear any mark or information identifying a particular tire manufacturer or brand name) and thus could be used to register any manufacturer's tires. When the forms of a clearinghouse were completed, they were returned to the clearinghouse. The clearinghouse would then forward them to appropriate manufacturers.

Except under the circumstances described above in the discussion of the temporary continued use of existing forms, the amendments to section 158(b)

and their legislative history compel an end to the practice of using forms which are not addressed to the manufacturer or its designee. Forms may continue to be addressed to an intermediary such as a clearinghouse if that intermediary has been designated by a tire manufacturer to serve as an initial recipient or as an ultimate repository for registration forms. Further, the amendments require standardization of the forms to be used by independent dealers. Hence, while independent dealers are still permitted to obtain registration forms from a source other than the tire manufacturers, those forms must comply with all of the requirements applicable to forms provided by manufacturers.

Responsibility for filling out and mailing registration form. The responsibility for completing the registration forms would be divided between independent tire dealers and purchasers. The tire dealer would be required to fill in the identification number of each tire sold and his name and address or some other unique identifier like a code number. The necessity for having the dealer's name and address arises from the statutorily-required evaluation of the voluntary registration requirements. In order to conduct that evaluation, the agency will need information on the registration rates for tires sold by individual independent dealers. This information will aid NHTSA in identifying different levels of registration among dealers and evaluate the reasons underlying those differences. The simplest and most effective way of ensuring the recording of the dealer's names and addresses is to require the recording of the information by the party who can most accurately provide it. A dealer's proper name and address are obviously better known to that dealer than to his customers. Further, through the use of an inexpensive rubber stamp, the dealer can record that information on a form much more easily and quickly than a tire purchaser can.

After the dealer has filled in this information and handed the card (and envelope under the option for using existing forms) to the tire purchaser, it is the purchaser's responsibility to complete the registration process. If a purchaser wishes to register his new tire, he must fill in his name and address, place the appropriate postage on the form (or envelope) and mail it.

Other issues. Any questions concerning the classification of a particular dealer as independent or otherwise should be addressed in writing to the Chief Counsel, NHTSA, at the street address given above. The legislative history cited early in this notice provides some guidance on this point. NHTSA notes that it is possible for motor vehicle dealers to be considered tire dealers in certain situations, as specified in 49 CFR 574.9. Whether a new motor vehicle dealer is required to follow the procedures for mandatory or voluntary registration depends on whether the dealer is owned or controlled by a tire manufacturer. The agency believes that most motor vehicle dealers would be considered independent dealers for the purposes of Part 574. These motor vehicle dealers are reminded that they should provide the motor vehicle purchaser with a voluntary tire registration form at the time they deliver the new vehicle to the purchaser, and with the identification number(s) of all of the vehicle's tires and the dealer's name and address entered on the form.

Enforcement of the new provisions of Part 574 would be carried out under sections 108–110 of the Safety Act. Failure to comply with the new provisions would be a violation of section 108(a) (2) (D) which prohibits failure to comply with any order or other requirement applicable to any manufacturer, distributor or dealer pursuant to Part B of the Safety Act. Section 109(a) provides that a civil penalty of $1,000 may be assessed for each violation of section 108. Under section 110(a), the agency could seek an injunction against a violator of section 108 to prevent further violations.

The information collection requirements contained in this rule have been submitted to the Office of Management and Budget (OMB) for its approval, pursuant to the requirements of the Paperwork Reduction Act of 1980 (44 U.S.C. 3501 et seq.). A notice will be published in the Federal Register when OMB approves this information collection.

As noted above, this rule is being issued as an interim final rule, without prior notice and opportunity for comment. NHTSA believes that there is good cause for finding that notice and comment rulemaking is impracticable and contrary to the public interest in this instance. The absence of any tire registration requirements for independent dealers has created an emergency necessitating immediate action.

The agency is concerned that, until a rule regarding voluntary registration can be implemented, registration of tires sold by independent dealers may fall well below the 20 percent rate which existed prior to the enactment of the Authorization Act on October 15. As long as this situation lasts, substantial numbers of tire purchasers may be unable to register their tires. Although some efforts are being made by independent dealers to continue to follow the mandatory registration process, the agency does not have any indication how widespread or successful those efforts are. Purchasers whose tires are unregistered will not receive direct notification from the manufacturer of those tires in the event that the tires are found to contain a safety defect or to fail to comply with an applicable standard. Ignorant of the safety problem, the purchasers will continue to drive on tires presenting a threat to their safety and that of other motorists.

Providing opportunity for comment is also unnecessary to a substantial extent. Many of the new provisions of Part 574 were expressly mandated by Congress.

Nevertheless, this agency is providing an opportunity to comment on this notice during the 45 days following its publication in the Federal Register. Those comments will be carefully considered since the agency does not intend to maintain this rule as the permanent final rule on voluntary registration. A permanent final rule will be issued not later than October 14, 1983.

NHTSA seeks comments from all interested parties on what requirements should be included in the permanent final rule. Pursuant to a contract with the agency, American Institutes for Research in the Behavioral Sciences has explored ways of more effectively structuring and wording the voluntary registration forms to induce as many purchasers as possible to complete their forms and send them to the manufacturers. Copies of the results of the Institute's work have been placed in the docket. Comments are requested on that work. Comments are also requested on the feasibility of using post card sized forms. The agency is uncertain whether those forms would provide sufficient space to permit the easy, legible recording of the requisite information. If so, then this alternative appears attractive since the lower postal rate for such cards could induce a higher rate of registration by purchasers.

The results of the contract study on registration forms and all comments submitted in response to this notice will be considered by the agency in selecting the provisions to include in the permanent final rule. If, after examining the study, the agency determines that the registration forms for independent dealers should be significantly altered, a notice of proposed rulemaking will be issued to ensure full comment on those changes.

The requirements of this rule become effective 30 days after the date on which it is published in the *Federal Register*. The 30-day period provides adequate time for tire manufacturers to print and distribute the new voluntary registration forms (or envelopes, under the option for using existing forms) to the independent dealers. Since this rule requires no change to the forms provided to or used by nonindependent dealers, manufacturers and nonindependent dealers may continue to use their current forms.

NHTSA has analyzed the impacts of this action and determined that it is neither "major" within the meaning of Executive Order 12291 nor "significant" within the meaning of the Department of Transportation regulatory policies and procedures. The requirements concerning the registration forms for independent dealers will impose minimally higher costs on tire manufacturers. Compared to the costs and administrative burdens to independent dealers of complying with the Part 574 requirements for mandatory registration, independent dealers should achieve slight savings under this rule. Requirements for nonindependent dealers are not changed by this rule. Consumers purchasing tires from independent dealers will now have to pay 20 cents for postage if they wish to register those tires. The bearing of this cost by consumers has been mandated by Congress. For these reasons, a full regulatory evaluation has not been prepared.

The agency has also considered the impacts of this action on small entities, and determined that this rule will not have a significant economic impact on a substantial number of those small entities. The agency believes that few if any of the tire manufacturers are small entities. Although many dealers are considered to be small entities, this rule will not have a significant impact on them. The requirements for tire manufacturers are unchanged except that the size, content and cost of the registration forms they supply to independent dealers would be slightly different. No change at all is made in the requirements for nonindependent dealers. Independent dealers will realize minimal savings from this rule. Small organizations and governmental units which purchase tires from independent dealers will have to pay postage to register those tires. However, those costs will not be significant.

All interested persons are invited to comment on this interim final rule. It is requested but not required that 10 copies be submitted.

All comments must be limited not to exceed 15 pages in length. Necessary attachments may be appended to these submissions without regard to the 15 page limit. This limitation is intended to encourage commenters to detail their primary arguments in a concise fashion.

If a commenter wishes to submit certain information under a claim of confidentiality, three copies of the complete submission, including purportedly confidential information, should be submitted to the Chief Counsel, NHTSA, at the street address given above, and seven copies from which the purportedly confidential information has been deleted should be submitted to the Docket Section. A request for confidentiality should be accompanied by a cover letter setting forth the information specified in the agency's confidential business information regulation (49 CFR Part 512).

All comments received before the close of business on the comment closing date indicated above will be considered, and will be available for examination in the docket at the above address both before and after that date. To the extent possible, comments filed after the closing date will also be considered. However, the rulemaking action may proceed at any time after that date, and comments received after the closing date and too late for consideration in regard to the action will be treated as suggestions for future rulemaking. The NHTSA will continue to file relevant material as it becomes available in the docket after the closing date, and it is recommended that interested persons continue to examine the docket for new material.

Those persons desiring to be notified upon receipt of their comments in the rules docket should enclose, in the envelope with their comments, a self-addressed stamped post card. Upon

PART 574—PRE 22

receiving the comments, the docket supervisor will return the post card by mail.

List of Subjects in 49 CFR 574

Consumers protection, Motor vehicle safety, Motor vehicles, Rubber and rubber products, Tires.

PART 574—(Amended)

In consideration of the foregoing, the following amendments are made to Part 574, Tire Identification and Recordkeeping, of Title 49 of the Code of Federal Regulations:

1. Section 574.1 is revised to read as follows:

§574.1 Scope.

This part sets forth the method by which new tire manufacturers and new tire brand name owners shall identify tires for use on motor vehicles and maintain records of tire purchasers, and the method by which retreaders and retreaded tire brand name owners shall identify tires for use on motor vehicles. This part also sets forth the methods by which independent tire dealers and distributors shall record, on registration forms, their names and addresses and the identification number of the tires sold to tire purchasers and provide the forms to the purchasers, so that the purchasers may report their names to the new tire manufacturers and new tire brand name owners, and by which other tire dealers and distributors shall record and report the names of tire purchasers to the new tire manufacturers and new tire brand name owners.

2. Section 574.3 is amended by adding a new paragraph (c) (1) immediately after "*Definitions used in this part.*" and redesignating existing paragraphs (c) (1) through (c) (4) as paragraphs (c) (2) through (c) (5):

§ 574.3 Definitions.

* * * * *

(c) * * *

(1) "Independent" means, with respect to a tire distributor or dealer, one whose business is not owned or controlled by a tire manufacturer or brand name owner.

* * * * *

3. Section 574.7 is revised to read as follows:

§ 574.7 Information requirements—new tire manufacturers, new tire brand name owners.

(a) (1) Each new tire manufacturer and each new tire brand name owner (hereinafter referred to in this section and § 574.8 as "tire manufacturer") or its designee, shall provide tire registration forms to every distributor and dealer of its tires which offers new tires for sale or lease to tire purchasers.

(2) Each tire registration form provided to independent distributors and dealers pursuant to paragraph (a) (1) of this section shall comply with either paragraph (a) (2) (A) or (B) of this section.

(A) Each form shall contain space for recording the information specified in paragraphs (a) (5) (A) through (a) (5) (C) of this section and shall conform in content and format to Figures 3a and 3b. Each form shall be:

(i) Rectangular;

(ii) Not less than .007 inches thick;

(iii) Greater than 3½ inches, but not greater than 6⅛ inches wide; and

(iv) Greater than 5 inches, but not greater than 11½ inches long.

(B) Each form shall comply with the same requirements specified in paragraph (a) (4) of this section for forms provided to distributors and dealers other than independent distributors and dealers.

(3) Each tire manufacturer or designee which does not give an independent distributor or dealer forms complying with paragraph (a) (2) (A) of this section shall give that distributor or dealer envelopes for mailing forms complying with paragraph (a) (2) (B) of this section. Each envelope shall bear the name and address of the tire manufacturer or its designee and the reminder set forth in Figure 3a.

(4) Each tire registration form provided to distributors and dealers, other than independent distributors and dealers, pursuant to paragraph (a) (1) of this section shall be similar in format and size to Figure 4 and shall contain space for recording the information specified in paragraph (a) (5) (A) through (a) (5) (C) of this section.

(5) (A) Name and address of the tire purchaser.

(B) Tire identification number.

(C) Name and address of the tire seller or other means by which the tire manufacturer can identify the tire seller.

(b) Each tire manufacturer shall record and maintain, or have recorded and maintained for it by a designee, the information from registration forms which are submitted to it or its designee. No tire manufacturer shall use the information on the registration forms for any commercial purpose detrimental to tire distributors and dealers. Any tire manufacturer to which registration forms are mistakenly sent shall forward those registration forms to the proper tire manufacturer within 90 days of the receipt of the forms.

(c) Each tire manufacturer shall maintain, or have maintained for it by a designee, a record of each tire distributor and dealer that purchases tires directly from the manufacturer and sells them to tire purchasers, the number of tires purchased by each such distributor or dealer, the number of tires for which reports have been received from each such distributor or dealer other than an independent distributor or dealer, the number of tires for which reports have been received from each such independent distributor or dealer, the total number of tires for which registration forms have been submitted to the manufacturer or its designee, and the total number of tires sold by the manufacturer.

(d) The information that is specified in paragraph (a)(5) of this section and recorded on registration forms submitted to a tire manufacturer or its designee shall be maintained for a period of not less than three years from the date on which the information is recorded by the manufacturer or its designee.

4. Section 574.8 is revised to read as follows:

§ 574.8 Information requirements—tire distributors and dealers.

(a) *Independent distributors and dealers.* (1) Each independent distributor and each independent dealer selling or leasing new tires to tire purchasers or lessors (hereinafter referred to in this section as "tire purchasers") shall provide each tire purchaser at the time of sale or lease of the tire(s) with a tire registration form.

(2) The distributor or dealer may use either the registration forms provided by the tire manufacturers pursuant to § 574.7(a) or registration forms obtained from another source. Forms obtained from other sources shall comply with the requirements specified in § 574.7(a) for forms provided by tire manufacturers to independent distributors and dealers.

(3) Before giving the registration form to the tire purchaser, the distributor or dealer shall record in the appropriate spaces provided on that form:

(A) The entire tire identification number of the tire(s) sold or leased to the tire purchaser; and

(B) The distributor's or dealer's name and address or other means of identification known to the tire manufacturer.

(4) Multiple tire purchases or leases by the same tire purchaser may be recorded on a single registration form.

(b) *Other distributors and dealers.* (1) Each distributor and each dealer, other than an independent distributor or dealer, selling new tires to tire purchasers shall submit the information specified in § 574.7(a)(5) to the manufacturer of the tires sold, or to its designee.

(2) Each tire distributor and each dealer, other than an independent distributor or dealer, shall submit registration forms containing the information specified in § 574.7(a)(5) to the tire manufacturer, or person maintaining the information, not less often than every 30 days. However, a distributor or dealer which sells less than 40 tires, of all makes, types and sizes during a 30-day period may wait until he or she sells a total of 40 new tires, but in no event longer than six months, before forwarding the tire information to the respective tire manufacturers or their designees.

(c) Each distributor and each dealer selling new tires to other tire distributors or dealers shall supply to the distributor or dealer a means to record the information specified in § 574.7(a)(5), unless such a means has been provided to that distributor or dealer by another person or by a manufacturer.

(d) Each distributor and each dealer shall immediately stop selling any group of tires when so directed by a notification issued pursuant to sections 151 and 152 of the Act (15 U.S.C. 1411 and 1412).

Issued on April 21, 1983.

Raymond A. Peck, Jr.,
Administrator
48 F.R. 22572
May 19, 1983

PREAMBLE TO AN AMENDMENT TO PART 574

Tire Code Marks Assigned to New Tire Manufacturers

ACTION: Publication of tire code marks assigned to new tire manufacturers.

SUMMARY: The NHTSA last published a complete listing of the tire code marks assigned to new tire manufacturers in 1972. Since that time, there have been several additions and changes in names and addresses for the assigned code marks. This publication will inform the public of those additions and changes.

SUPPLEMENTARY INFORMATION: Section 574.5 of the Title 49, Code of Federal Regulations, requires tire manufacturers to mold a tire identification number onto or into the sidewall of each tire they manufacture. In the case of new tires, the first two digits of the tire identification number are the code mark assigned to the manufacturer. This code mark identifies the manufacturer and the plant where the tire was manufactured.

The NHTSA published a complete listing of the tire codes at 37 FR 342, January 11, 1972. This list enables interested members of the public to identify the manufacturer and place of manufacture of any new tire.

Since 1972, there have been several changes in the names of the manufacturers and the plant addresses for the assigned code marks. Further, there have been some 150 additional code marks assigned for new tires since the 1972 publication. Accordingly, this updated listing of the assigned code marks for new tires is being published to bring the public up-to-date with the revisions and new code numbers which have been assigned since the publication of the 1972 list.

Issued on June 8, 1983.

Kennerly H. Digges,
*Acting Associate Administrator
for Rulemaking*
48 F.R. 27635
June 16, 1983

PREAMBLE TO AN AMENDMENT TO PART 574

Tire Identification and Recordkeeping

[Docket No. 70-12; Notice 25]

ACTION: Final rule.

SUMMARY: This final rule sets forth the requirements relating to the registration of new tires sold by independent dealers and distributors. Recording the names and addresses of the first purchasers and transmitting this information to the manufacturers will make it possible for those purchasers to be contacted in the event that the tires are recalled by the manufacturers for safety reasons. These requirements supersede those contained in the interim final rule on this subject published in the May 19, 1983, edition of the Federal Register.

This rule primarily clarifies some aspects of the provisions of the interim final rule concerning the tire registration form to be provided by the tire manufacturers to the independent dealers. These changes, which were made to maximize the registration of tires sold through independent dealers, are as follows:

(1) The size of the registration form to be given to the consumer by independent dealers has been reduced, so that only a 13-cent postcard stamp need be affixed to the registration form. The interim final rule had specified that a first-class-mail-sized card be used for the registration form. This change was made to minimize the costs for consumers to register their tires.

(2) The statement in the upper left corner of that registration form, informing the tire purchaser of the importance of completing and returning the form, has been modified so as to be more comprehensible and more effective at motivating the purchaser to register his or her tires.

(3) Instructions to the tire purchaser have been added, so that the purchaser will print instead of write his or her name on the registration form.

(4) That portion of the registration form which is to be filled in by the independent dealer (i.e., the portion for filling in suitable identification of the dealer and the tire identification number(s) of the tire(s) sold) must be shaded with a 10-percent screen tint. This change was made to emphasize to the tire purchaser the limited amount of information which the purchaser must fill in to register his or her tires.

EFFECTIVE DATE: The changes made by this notice become effective March 25, 1984. As of that date, the tire manufacturers will be required to provide registration forms in compliance with this rule, and they must cease their distribution of the forms specified by the interim final rule. Independent dealers may continue to use the forms specified by that rule until their existing supplies of that form are exhausted or until April 1, 1984, whichever comes first.

SUPPLEMENTARY INFORMATION

Background
Motor Vehicle Safety and Cost Savings Authorization Act of 1982

The Motor Vehicle Safety and Cost Savings Authorization Act of 1982 (hereinafter referred to as "the Authorization Act") amended the National Traffic and Motor Vehicle Safety Act of 1966 (hereinafter referred to as "the Safety Act") by requiring this agency to change its tire registration requirements insofar as they applied to independent tire dealers and distributors. (This class of dealers and distributors is defined below.) These requirements are set forth in 49 CFR Part 574, *Tire Identification and Recordkeeping.* Before the Authorization Act became effective, Part 574 required all tire dealers and distributors to comply with the mandatory registration system. Under the system, dealers and distributors were required to record certain information (i.e., the tire purchaser's name and address, seller's name and address, and the identification number(s) of the tire(s) sold) on a registration form and send the completed form to the tire manufacturer or the brand-name owner (hereinafter collectively referred to as "tire manufacturers") or a designee of the tire manufacturer.

The tire registration requirements were adopted pursuant to requirements in the Safety Act intended to insure that tire purchasers could be notified if their tires are recalled for safety reasons, either because they contain a safety-related defect or because they do not comply with an applicable safety standard. The purchasers of unregistered tires would not be directly notified in those instances and would instead unknowingly continue to drive on unsafe tires.

On examining the rate of tire registration, Congress found a substantial difference between the rates for tires sold by independent dealers (dealers and distributors whose business is not owned or controlled by a tire manufacturer) and those sold by nonindependent dealers (dealers and distributors whose business is owned or controlled by a tire manufacturer). Independent dealers, who handle slightly less than half of the replacement tires sold annually, registered about 20 percent of the tires they sold. Nonindependent dealers, whose sales account for the balance of annual replacement tire sales, registered between 80 and 90 percent of their tires.

Given the importance of tire registration to safety, Congress determined that an alternative method of registration should be instituted for tires sold by independent dealers. Accordingly, it included provisions in the Authorization Act prohibiting the Secretary of Transportation from requiring independent dealers to comply with the mandatory registration requirements. (In view of the high rate of registration of tires sold by nonindependent dealers, Congress did not mandate any change in the application of the mandatory registration requirements to those dealers.) The prohibition regarding independent dealers was self-executing (i.e., its effectiveness was not conditioned on any prior rulemaking or other implementing action by this agency) and became effective on the date that the Authorization Act became law, October 15, 1982.

In lieu of requiring independent dealers to comply with the mandatory registration process, Congress directed that they comply with a voluntary registration process to be established by the Secretary. Under the voluntary process, the primary responsibility for registering tires sold by independent dealers is borne by the purchaser instead of the dealer. NHSTA is mandated by the Safety Act, as amended by the Authorization Act, to require that independent dealers (1) fill in the tire identification number(s) of the tire(s) sold to a purchaser on a registration form and then (2) give the form to the purchaser. If the purchaser wishes to register the tires, he or she may do so by filling in his or her name and address, adding postage, and sending the form to the tire manufacturer or its designee.

To ascertain whether the changes mandated by the Authorization Act have the desired effect of increasing the registration rate of tires sold by independent dealers, Congress directed NHTSA to conduct an evaluation covering the 2-year period ending October 14, 1984. Upon completion of the evaluation, NHTSA must determine the extent to which independent dealers have encouraged purchasers to register their tires and the extent to which those dealers have complied with the voluntary tire registration procedures. Further, the agency is required to determine whether to impose any additional requirements on the independent dealers or the manufacturers for the purpose of promoting higher levels of tire registration.

The provision in the Authorization Act mandating a voluntary registration system for independent dealers was not self-executing. Thus, the voluntary system could not become effective until NHTSA issued a rule establishing that system. An interim final rule doing so was published at 48 Fed. Reg. 22572, May 19, 1983, and became effective June 20, 1983.

Interim Final Rule

The interim final rule imposed the following requirements on the various parties:

Tire manufacturers. Except as noted, new registration forms had to be provided for independent dealers. All of those forms were required to be identical in format and content and within the size range specified in the interim final rule. Alternatively, the manufacturer could provide independent dealers with preaddressed envelopes in which tire purchasers could mail the mandatory registration forms. In either case, the manufacturer would have to maintain a record of all returned registration forms for at least 3 years after receipt.

No change was made in the requirements regarding forms provided to nonindependent dealers.

Tire dealers and distributors which sell tires to other dealers and distributors. These parties are required to give the purchasing dealer or distributor the registration forms provided by the tire manufacturers so that that dealer or distributor can comply with the applicable tire registration requirements. The new forms must be provided to independent dealers.

Nonindependent dealers. No changes were made to the tire registration requirements applicable to these parties. They are still required to follow the mandatory tire registration system formerly applicable to all tire dealers. Thus, the nonindependent dealers must record the purchaser's name and address, the tire identification number(s) of the tire(s) sold, and a suitable identification of themselves as the selling dealer on a tire registration form, and return the completed forms to the tire manufacturers or their designees.

Independent dealers. These dealers were required by the interim final rule to record the tire identification number(s) of the tire(s) sold, along with their name and address, on a registration form and give the form to the tire purchaser.

The interim final rule sought comments on the issues raised by the requirements specified therein, and specifically asked commenters to address the issue of adopting the registration form devised by the American Institute for Research in the Behavioral Sciences pursuant to a contract with the agency.

Final Rule

After considering the comments on the interim final rule, NHTSA has decided to retain most of the requirements in that rule. Several changes have been made to the requirements regarding the forms to be provided to independent dealers. These changes are relatively minimal and do not disturb the essential continuity of the voluntary registration requirements. Accordingly, both the tire manufacturers and the independent dealers should be able to implement the voluntary registration system as amended by this rule with minimal disruption to the practices they have been following since the interim final rule became effective.

Voluntary Tire Registration Procedures

Several commenters stated that independent dealers that wish to continue following the mandatory tire registration requirements should be permitted to do so. The premise underlying these comments is that mandatory registration, when properly implemented, is the most effective means of insuring that virtually all replacement tires are registered.

While NHTSA does not disagree with the premise of these commenters, the agency is not free to adopt their suggestion. Section 158(b)(2)(B) of the Safety Act specifies that this agency

> ...shall require *each*...(independent dealer) to furnish the first purchaser with a registration form (containing the tire indentification number of the tire) which the purchaser may complete and return directly to the manufacturer of the tire. (Emphasis added.)

This mandate to the agency is completely inclusive, directing the agency to make the voluntary registration procedures applicable not simply to independent dealers in general, but to "each" independent dealer. Further, this mandate is not offset by any express authority to make exceptions.

As a practical as well as a legal matter, independent dealers may nevertheless register the tires they sell if they first comply with the voluntary registration procedures. Independent dealers are not prohibited from filling in the information required by the voluntary procedures on the forms specified by those procedures, furnished the forms to tire purchasers, and then offering to fill in the balance of the information and mail the form to the manufacturer.

Based on the comments, it appears that some commenters are confused about the status of motor vehicle dealers under the mandatory and voluntary registration procedures. The preamble to the interim final rule mentioned motor vehicle dealers only very briefly because they are minimally affected by the voluntary registration procedures. The preamble stated that there are two situations in which motor vehicle dealers are considered to be tire dealers and are required to register the tires on the vehicles as specified in section 574.9. In these situations, the preamble noted that whether the motor vehicle dealer would be required to follow the mandatory or voluntary registration procedures would depend on whether the motor vehicle dealer's business was owned or controlled by a tire manufacturer. Since such ownership or control seems highly improbable, the preamble stated that the motor vehicle dealer would in all likelihood have to follow the voluntary registration procedures.

The discussion in that notice left some commenters uncertain whether the original equipment tires on new vehicles were subject to mandatory or voluntary registration procedures. This uncertainty apparently arose because the interim final rule made no mention of the mandatory tire registration requirements that have been applicable to original-equipment tires since 1971. No mention of these requirements was made, since the notice did not propose to amend section 574.10, which specifies the actions to be taken by motor vehicle manufacturers to register their original-equipment tires.

The two situations to which the interim final rule's preamble referred are those situations in which the motor vehicle dealer, as opposed to the motor vehicle manufacturer, is responsible for registering tires. These situations, which are relatively infrequent, are set forth in section 574.9. First, if a motor vehicle dealer sells a used vehicle or leases a vehicle for more than 60 days, and the vehicle is equipped with new tires, the dealer must register the tires on the vehicle. Second, if a motor vehicle dealer sells a new vehicle and the vehicle is equipped with tires other than those shipped with the vehicle by the motor vehicle manufacturer, the motor vehicle dealer must register the tires on the vehicle. The interim final rule was intended to make clear that motor vehicle dealers whose business is not owned or controlled by a tire manufacturer should follow the voluntary registration procedures in those two rare types of situations, when the vehicle dealer is responsible for registering the tires on the vehicle.

One commenter urged that NHTSA delete the requirement that independent dealers record their name and address on the registration form before giving that form to the tire purchaser. This commenter noted that Congress stated the Authorization Act's voluntary registration provisions had been adopted partially for the purpose

of reducing the burdens which mandatory registration procedures paced on independent dealers. Further, the commenter asserted that the Authorization Act requires only that the independent dealers record the tire identification number on the registration form, and that the absence of any mention of further specific information to be filled in by independent dealers is evidence that Congress did not intend those dealers to have to fill in any information other than the identification number. Finally, this commenter noted that NHTSA had indicated in the preamble to the interim final rule that the dealer's name and address was needed on the registration form to aid the agency in evaluating the voluntary registration process. This commenter stated that it would be sufficient for evaluation purposes for the registration forms used by independent dealers to show simply that they came from that class of dealers, instead of identifying a specific independent dealer. It was further suggested that this information would be all that was needed for the agency to determine the extent to which voluntary registration had been successful at increasing the rate of tire registration for tires sold by independent dealers.

Similarly, two tire manufacturers commented that a manufacturer should not be required any longer to maintain records which show, for each of its tires sold by an independent dealer, the identity of that particular dealer. They argued that manufacturers should only be required to maintain registration for independent dealers as a group. These commenters also asserted that this information was all that the agency needed to determine whether or not voluntary registration had successfully increased the registration rate for tires sold by independent dealers.

The preamble to the interim final rule may not have adequately explained the full breadth of the evaluative task which Congress instructed the agency to perform. In order to conduct a proper evaluation which not only reports the aggregate results of the voluntary registration program but also attempts to explain those results, the agency will need to be able to determine registration rates for individual dealers. With that ability, the agency can differentiate dealers with high rates from dealers with low ones and then proceed to attempt to assess the reasons for those differences. Having performed that analysis, the agency would be in a position to provide Congress with insight about the impact of the voluntary registration program. It would also enable the agency to determine what additional requirements, if any, should be adopted to improve the registration program. NHTSA may find that those improvements can be more effectively obtained by enforcing the requirements established by this notice than by imposing additional requirements on all independent dealers.

NHTSA believes that it has authority under the Authorization Act to require independent dealers to record not only the tire identification numbers but also their names and addresses on registration forms. There is no express prohibition against the agency's requiring dealers to fill in more than the tire identification numbers. While the Authorization Act makes no mention of requiring dealers to fill in their names and addresses, the agency does not regard that fact as dispositive. The Authorization Act does not, in fact, specify that the dealer's name and address is to be filled in by either the dealer or the purchaser. Since there isn't any clear indication that it was Congress' intent that this information no longer be required, the agency will not infer such intent from Congress' decision not to assign that task expressly to any particular party. It appears that Congress has left the question of that assignment to NHTSA's discretion. Since the names and addresses of dealers have long been recorded on registration forms and since that information is needed to enable the agency to conduct an effective evaluation, this agency believes that it should continue to be recorded. In view of the fact that dealers are more likely than purchasers to provide this information accurately, and since dealers can easily resort to the expandiency of a stamp bearing their name and address, NHTSA reaffirms its decision to assign the task of filling in that information to the dealers.

As to the tire manufacturers, the burden on them regarding the identity of specific independent dealers is simply to continue doing what they have been doing since 1971, i.e., maintaining registration records for each dealer. The agency believes that continued maintenance of these records is warranted by the value of dealer-specific information to the evaluation and to tire recall campaigns. In fact, the agency recently issued a special order to nine tire manufacturers to obtain information on the registration rates for individual independent dealers. The agency will continue to monitor those rates.

Several commenters suggested that the agency, when conducting its evaluation of the effect of the voluntary registration program on the registration rate, determine its own baseline for registration of tires sold by independent dealers before that program began. The commenters urged that the agency not adopt the 20-percent rate mentioned in the legislative history of the Authorization Act. In lieu of that figure, the commenters offered several lower ones, including a figure of 7 percent. The agency intends to determine its own baseline. The special order mentioned above will provide the information necessary for that determination.

Registration Forms

In selecting the registration form to be used by independent dealers under the interim final rule, the agency consciously sought to find a form that would satisfy all of the statutory requirements for the voluntary registration system, while making as few changes as possible to existing forms being used under the mandatory registration system. This conservative approach was necessary because the amendments to the Vehicle Safety Act did not provide adequate time to follow normal rulemaking procedures and seek comments on more far-reaching changes.

To determine outside the strictures of a rigid time schedule what type of form would be most effective in inducing tire purchasers to register their tires, NHTSA contracted with American Institute for Research in the Behavioral Sciences (AIRBS) to conduct a study. AIRBS designed a postcard-size registration form separated into two parts by a line of perforation. The top part, which would be detached and retained by the purchaser, would contain a message explaining the importance of tire registration to the purchaser and motivating the purchaser to register the tires by sending the form to the manufacturer. On the reverse of the top side, there would be a space where the purchaser could record the registration information and save it for his or her personal records.

The bottom part of the AIRBS registration form would be the part that would be sent to the tire manufacturer. On one side would be the manufacturer's preprinted address. On the other would be space for filling in the tire registration information.

The agency placed the AIRBS study and form in the public docket and requested in the interim final rule that interested persons comment on the contractor's recommendations. Several commenters addressed the desirability of adopting the AIRBS form as the registration form to be used by independent dealers. Many commenters stated that a postcard-sized form was too small to allow the necessary information to be legibly recorded. One commenter argued that the AIRBS form would not be any more effective at encouraging consumers to register their tires than the simple one-part card mandated in the interim final rule, and that the AIRBS form might actually be more confusing. Another commenter objected to be used the AIRBS form because the perforated edge of the portion of the form to be returned to the manufacturer could not be automatically fed through a microfilming machine. The same commenter also argued that the printing costs for the AIRBS form would be about 12 percent higher than those for the form mandated in the interim final rule.

After considering these comments, NHTSA has decided not to adopt the AIRBS form. That form poses a number of potential problems which neither AIRBS nor the agency foresaw. Further, NHTSA does not believe that use of a two-part form is necessary. AIRBS stated in its study that the reason for its recommending a two-part form was its belief that the space available on a single-part form was insufficient to allow the printing of the motivational message to the consumer, the instructions, and the necessary registration information with type and spacing large enough to permit easy reading. In the agency's own judgment, the single-part form mandated by this final rule will not be overly crowded, will avoid the potential problems which commenters attributed to the two-part form, and will be almost as successful in motivating consumers to register their tires as would the two-part form.

However, the agency has adopted the AIRBS recommendation that the registration forms provided to consumers be postcard size. It will be less expensive for tire purchasers to use 13-cent postcard stamps to mail registration forms of that size, and this low cost might motivate some purchasers who would not otherwise do so to register their tires. The maximum dimensions permitted by the U.S. Postal Service for a postcard are 4¼ by 6 inches. This area is, in NHTSA's judgment, sufficient to permit the motivational message and the space for recording the required information to appear on the same size of the card, without being overly crowded or difficult to read. Given the importance of encouraging consumers to return the completed tire registration forms, and the likely effectiveness of lower postage costs at encouraging consumers to return the forms, this rule specifies that the registration forms be of the dimensions permitted for using postcard stamps.

Some other minor changes are made in this notice to the registration form required by the interim final rule. First, the motivational message has been changed so that it is now identical to that recommended by AIRBS. The AIRBS message provided stronger encouragement to send the form to the manufacturer and will be readily understood by consumers.

Second, the agency has decided to require the form to include instructions to the tire purchaser to print his or her name and address on the form. Those instructions were inadvertently omitted from the interim final rule. They have now been added at the urging of several of the commenters.

One commenter requested that tire manufacturers be allowed to divide the spaces for recording the purchaser's name and address into little boxes so that each letter or number would be printed in a separate box. According to this commenter, this approach would help insure accurate transcription by the manufacturer of the information on the registration forms. Based on its assessment of the AIRBS study, the agency has decided not to adopt this change. AIRBS indicated to this agency that the use of boxes discourages people from filling in information on forms and that the return rate for the registration forms would therefore be higher if boxes were not used.

Third, NHTSA is adopting a requirement that contrasting shading be used for the area of the form containing the blanks to be completed by the independent dealer and that a white background be used for the areas to be completed by the tire purchasers. AIRBS recommended this requirement in its study as a means of emphasizing to the tire purchaser the minimal quantity of information which he or she must record in order to register his or her tires. AIRBS indicated that the shading could be achieved by using a 10-percent screen tint. The tinted forms would be inexpensive to produce and still easily readable by data processors.

One manufacturer commented that independent dealers should be required to enter both their name and address and their dealer identification number assigned by the manufacturer on the registration form. The dealer identification number is a unique identifier assigned by a tire manufacturer to each dealer selling that manufacturer's tires. This commenter asserted that requiring the dealer identification number to be placed on the registration forms would greatly simplify the data-processing task for the manufacturer as it recorded the information from the registration forms sent in by tire purchasers.

NHTSA agrees that such a requirement would simplify the manufacturers' task, but only at the cost of significantly complicating the registration responsibilities of the independent dealers. The dealer identification numbers assigned to a particular dealer are not coordinated among the various tire manufacturers. Thus, an independent dealer which sells tires produced by seven different manufacturers would have seven different dealer identification numbers assigned to it. The interim final rule required independent dealers to record their name and address on the registration form. This could be done simply by purchasing and using a rubber stamp with the dealer's name and address on it. If the final rule were amended to require the dealer to also record its dealer identification number, and the independent dealer sold seven different manufacturers' tires (as in the example above), the dealer would either have to fill in its name, address, and identification number by hand on each registration form or buy seven different rubber stamps. If it chose to purchase seven different rubber stamps, the dealer would also have to be certain that it used the appropriate stamp for each manufacturer's registration form. If the dealer used the wrong dealer identification number on a manufacturer's registration form, it would complicate the manufacturer's data-processing task. After considering these facts, NHTSA has decided not to adopt this comment, and the independent dealers remain subject to the requirement that they record their name and address on the registration form before giving the form to the tire purchaser.

Other Issues

Several commenters objected to the language in the interim final rule stating that enforcement of this regulation would be under the authority of sections 108-110 of the Safety Act (15 U.S.C. 1397-99) and that each violation could subject the violator to a penalty of $1,000. These commenters noted that the Committee report on the Authorization Act stated an expectation that independent dealers which failed to comply with the voluntary registration requirements would not have to pay the maximum penalty unless there was a clear, continuous pattern of violations.

The statutory provisions recited in the interim final rule are consistent with the committee report. Section 109 of the Safety Act provides that the amount of any penalty imposed by the agency should reflect consideration of the size of the business which committed the violation and of the gravity of the violation. As a matter of practice, the agency makes a distinction in its enforcement activities between isolated violations and continuous patterns of violations. The agency will continue to make this distinction and thus will be following the guidance in the committee report.

Some commenters urged that the agency permit continued use of registration forms addressed to clearinghouses. These forms, which were permitted under mandatory registration, were generic instead of manufacturer-specific (i.e., they did not bear any mark or information identifying them for use in registering a particular manufacturer's tires and thus could be used to register any manufacturer's tires. The tire dealer would fill in the manufacturer or brand-name owner identified on the tire to be registered, and send the forms to a clearinghouse. The clearinghouse would then forward the information to the appropriate manufacturer or brand-name owner.

As explained in the preamble to the interim final rule, the amendments to section 158(b) of the Safety Act and their legislative history compel an end to the practice of using forms which are not addressed to a specific manufacturer or its designee. Section 158(b) requires that the purchaser be able to send the form directly to the manufacturer of the tire, and that the forms used by independent dealers be standardized for all tires. Hence, the agency cannot permit continued use of forms which are not manufacturer-specific and which are not addressed to a particular manufacturer or its designee.

One commenter asked that dealers be allowed to continue to use the forms mandated by the interim final rule until the supply was exhausted. The interim final rule permitted the continued use of the forms used under mandatory registration as long as the manufacturers provided pre-addressed envelopes in which to enclose those forms. To minimize the expenses and disruption associated with the transition from the interim final rule to this final rule, independent dealers will be permitted to continue using the forms specified by the interim final rule until their existing supplies are exhausted, or until April 1, 1984, whichever comes first. As of the effective date of this rule, the manufacturers will be required to provide registration forms in compliance with this rule, and distribution of the forms specified under the interim final rule must be ended.

A related issue was raised in a petition which Cooper Tire & Rubber Company ("Cooper") submitted for reconsideration of the interim final rule. Cooper currently has a no-charge warranty program for two tire lines. As part of that program, Cooper has printed a booklet and registration form. The form, which was developed and printed before the interim final rule was issued, contains a different motivational statement than was mandated by the interim final rule. Further, it does not contain a notation to affix first-class postage on the reverse side. Cooper reported that it had achieved a 66-percent registration rate for the two tire lines, using its own registration forms.

After considering these minor variations, the agency has decided that this Cooper registration form can be considered as complying with the requirements of the interim final rule. It is significant that Cooper prepared and began distributing these forms in December 1982, before the interim final rule had been published. From the interval of January 1, 1983, to June 20, 1983, Cooper achieved a 66-percent registration rate for tires sold by independent dealers, when there were no registration requirements applicable to independent dealers. This suggests that the Cooper form has been effective at motivating consumers to return that form, and achieving higher tire registration rates is the goal of the change in tire registration procedures.

NHTSA wishes to emphasize that Cooper was in a unique postion, and that permitting the variations in the Cooper form from that mandated by the interim final rule does not mean that the agency will countenance variations from the form prescribed by this final rule. This form has been developed after considering the AIRBS study, and it is important that it be used in connection with tire registration, to insure that the NHTSA evaluation of the voluntary tire registration system is conducted with an effective standardized registration form.

One commenter suggested that there would be a stronger incentive for consumers to register their tires if the agency were to require the manufacturers to prepay the postage for the registration forms. Adopting such a requirement was one of the actions which the House committee report indicated could be adopted after the 2-year evaluation period if the agency determined that further steps were necessary to achieve adequate registration rates. The implication of this discussion in the report is that the requirement may not be adopted at an earlier time. Accordingly, the agency is not adopting a requirement for prepaid postage.

Several commenters stated that the 30-day period between the publication of the interim final rule and its effective date was inadequate to allow the necessary registration forms to be printed and distributed to all of the manufacturer's independent dealers. Accordingly, they asked that a longer leadtime period be established for this final rule. The agency understands that it is asking the manufacturers to move very expeditiously to print and distribute the voluntary registration forms. NHTSA believes that short leadtime periods are necessary due to the importance of registration and to the requirement to conduct an evaluation of voluntary registration 2 years after passage of the Authorization Act. At the same time, the agency wishes to make some accommodation of the request for additional leadtime. Accordingly, the agency is specifying an effective date of 45 days after publication of this notice. This date will still require expeditious action by the manufacturers, but does provide 2 more weeks than were allowed for the interim final rule.

The information-collection requirements contained in this rule have been submitted to and approved by the Office of Management and Budget (OMB), pursuant to the requirements of the Paperwork Reduction Act of 1980 (44 U.S.C. 3501 et seq.). Those requirements have been approved through May 31, 1985 (OMB #2127-0050). All printed registration forms must display this OMB clearance number and expiration date in the upper right-hand corner of the form.

NHTSA has analyzed the impacts of this rule and determined that it is neither "major" within the meaning of Executive Order 12291 nor "significant" within the meaning of the Department of Transportation regulatory policies and procedures. The changes in the requirements for the registration forms to be provided by tire manufacturers to independent dealers will impose minimally higher costs on those manufacturers. Compared to the costs and administrative burdens imposed on independent dealers under mandatory registration, those dealers should achieve a slight savings under this rule. Consumers purchasing tires from independent dealers will now have to pay for postage if they wish to register their new tires. The assumption of that cost by consumers was mandated by Congress. For this reason, a full regulatory evaluation has not been prepared.

The agency has also considered the impacts of this rule on small entities, as required by the Regulatory Flexibility Act. NHTSA believes that few, if any, of the tire manufacturers are small businesses. Although many of the dealers could be considered small businesses, this rule will not have a significant impact on them. As noted above, they may experience a slight savings as compared to the mandatory registration requirements. The requirements for tire manufacturers are unchanged, except for some minor changes which they must make to the registration forms to be provided to independent dealers. Small organizations and governmental units will have to bear the minor expense of paying postage for any new tires they register. Based on the foregoing, I certify that this rule will not have a significant economic impact on a substantial number of small entities.

In consideration of the foregoing, the following amendments are made to Part 574, Tire Identification and Recordkeeping, of Title 49 of the Code of Federal Regulations.

1. Section 574.3 is amended by adding a new paragraph (c)(1) immediately after *"Definitions used in this part."* and redesignating existing paragraphs (c)(1) through (c)(4) as paragraphs (c)(2) through (c)(5):

§ 574.3 **Definitions.**

* * * * *

(c) * * *

(1) "Independent" means, with respect to a tire distributor or dealer, one whose business is not owned or controlled by a tire manufacturer or brand name owner.

* * * * *

3. Section 574.7 is revised to read as follows:

§ 574.7 **Information requirements—new tire manufacturers, new tire brand name owners.**

(a)(1) Each new tire manufacturer and each new tire brand name owner (hereinafter referred to in this section and § 574.8 as "tire manufacturer") or its designee, shall provide tire registration forms to every distributor and dealer of its tires which offers new tires for sale or lease to tire purchasers.

(2) Each tire registration form provided to independent distributors and dealers pursuant to paragraph (a)(1) of this section shall contain space for recording the information specified in paragraphs (a)(4)(A) through (a)(4)(C) of this section and shall conform in content and format to Figures 3a and 3b. Each form shall be:

(A) Rectangular;

(B) Not less than .007 inches thick;

(C) Greater than 3 1/2 inches, but not greater than 6 1/4 inches wide; and

(D) Greater than 5 inches, but not greater than 6 inches long.

(3) Each tire registration form provided to distributors and dealers, other than independent distributors and dealers, pursuant to paragraph (a)(1) of this section shall be similar in format and size to Figure 4 and shall contain space for recording the information specified in paragraphs (a)(4)(A) through (a)(4)(C) of this section.

(4)(A) Name and address of the tire purchaser.

(d) The information that is specified in paragraph (a)(4) of this section and recorded on registration forms submitted to a tire manufacturer or its designee shall be maintained for a period of not less than three years from the date on which the information is recorded by the manufacturer or its designee.

4. Section 574.8 is revised to read as follows:

§ 574.8 **Information requirements—tire distributors and dealers.**

(b) *Other distributors and dealers.* (1) Each distributor and each dealer, other than an independent distributor or dealer, selling new tires to tire purchasers shall submit the information specified in § 574.7(a)(4) to the manufacturer of the tires sold, or to its designee.

(2) Each tire distributor and each dealer, other than an independent distributor or dealer, shall submit registration forms containing the information specified in § 574.7(a)(4) to the tire manufacturer, or person maintaining the information, not less often than every 30 days. However, a distributor or dealer which sells less than 40 tires, of all makes, types and sizes during a 30-day period may wait until he or she sells a total of 40 new tires, but in no event longer than six months, before forwarding the tire information to the respective tire manufacturers or their designees.

(c) Each distributor and each dealer selling new tires to other tire distributors or dealers shall supply to the distributor or dealer a means to record the information specified in § 574.7(a)(4), unless such a means has been provided to that distributor or dealer by another person or by a manufacturer.

Issued on February 3, 1984.

Diane K. Steed
Administrator

49 FR 4755
February 8, 1984

PREAMBLE TO AN AMENDMENT TO PART 574

Tire Identification and Recordkeeping

[Docket No. 84-07; Notice 2]

ACTION: Final rule.

SUMMARY: This rule amends Part 574 to give retreaders of tires for motor vehicles other than passenger cars an option during the retreading process of either removing the original manufacturer's DOT symbol from the sidewall of the finished retread or leaving that symbol on the tire. This action is taken because NHTSA has determined that no significant safety interest is served by requiring that retreaders remove the original manufacturer's DOT symbol as part of the retreading process. That requirement, which did not expressly appear in Part 574, resulted from unforeseen events and from unexpected effects of the language in Part 574. This rule avoids imposing unnecessary costs on these retreaders without degrading the safety of the tires or the safety value of the information available to consumers.

EFFECTIVE DATE: February 15, 1985.

SUPPLEMENTARY INFORMATION: The Federal Motor Vehicle Safety Standards require that a DOT symbol appear on the sidewall of most new and retreaded tires as a means of certifying compliance with the performance requirements of the applicable safety standard. Thus, the DOT symbol must appear on new tires for use on passenger cars which are subject to Standard No. 109, new tires for use on vehicles other than passenger cars which are subject to Standard No. 119, and retreaded passenger-car tires which are subject to Standard No. 117. (For the sake of easy reference, tires for use on motor vehicles other than passenger cars will be referred to as "non-car tires"

throughout the rest of this preamble.)

Regulations issued under the National Traffic and Motor Vehicle Safety Act expressly prohibit the presence of the DOT symbol on tires not subject to a Federal safety standard. 49 CFR Part 574, *Tire Identification and Recordkeeping*, provides, in pertinent part: "The DOT symbol shall not appear on tires to which no Federal Motor Vehicle Safety Standard is applicable..." (574.5). Since retreaded non-car tires are the only new or retreaded tires not subject to a Federal safety standard, they are the only tires subject to that prohibition.

NHTSA adopted the language in § 574.5 because of its concern that the appearance of the DOT symbol on tires to which no safety standard was applicable would confuse consumers. That is, NHTSA believed that consumers could mistakenly conclude that the tires in question met some applicable Federal requirements, when, in fact, there were no such requirements.

However, although the agency's concern in adopting the prohibition in § 574.5 was with the addition of a DOT symbol to a tire that was not subject to any Federal safety standard, the language of the prohibition was broader. It did not simply state that manufacturers cannot add the DOT symbol to tires to which no Federal safety standard is applicable. It stated that the DOT symbol "shall not appear" on such tires. The breadth of that language gave rise to a duty not only to refrain from adding a DOT symbol to tires to which no safety standard was applicable, but also to remove an original manufacturer's symbol when, as in the case of retreaded non-car tires, the tires were subject to a safety standard when new but are not subject to any standard when retreaded.

In no other circumstances under the Safety Act, such as in the remanufacturing of a vehicle, is a person required to remove a previous manufacturer's certification. Additionally, the agency learned that most non-car tire retreaders had not been removing the original manufacturer's DOT symbol.

NHTSA tentatively concluded that there was no safety or informational value associated with the requirement that non-car tire retreaders remove the original manufacturer's DOT symbol. Accordingly, the agency published a notice of proposed rulemaking on this subject at 49 FR 20880, May 17, 1984. That notice explained in detail the origins of the prohibition in § 574.5, and the bases for the agency's tentative conclusions that no safety or informational purposes were served by the requirement that retreaders of non-car tires remove the original manufacturer's DOT symbol from the sidewall of the tire. Further, the notice noted that although NHTSA had received over 10,000 consumer complaints regarding non-car tires since 1976, not one of those complaints related to the presence or absence of the DOT symbol on a retreaded non-car tire. The hypothetical consumer confusion which NHTSA thought might occur has in fact *not* occurred with respect to retreaded non-car tires. Accordingly, NHTSA proposed that the prohibition in § 574.5 be replaced by language which would give non-car tire retreaders the option of removing the original manufacturer's DOT symbol or leaving it on the finished retread, while emphasizing the those retreaders were still prohibited from adding a new DOT symbol to the sidewall of retreaded non-car tires.

Three commenters responded to the notice of proposed rulemaking. All three supported the agency's proposal to eliminate the requirement that non-car tire retreaders remove the original manufacturer's DOT symbol. One of the commenters suggested that the agency move beyond its proposed option for these retreaders to remove or not remove the original manufacturer's DOT symbol, and instead require that any non-car tires with a DOT symbol on the sidewall retain that DOT symbol after the retreading is completed.

The agency has not been persuaded by this comment, for the reasons expressed in the proposal. To repeat, the value of the DOT symbol on a worn tire carcass in assessing the probable performance capabilities of a retreaded tire is not very significant. Intervening factors, such as latent problems with the carcass, inadvertent damage to the carcass during the retreading process, the amount of old tread not buffed off during the retreading, and the application and design of the new tread are of far greater significance in determining the performance of the retread than is the condition of the carcass when the tire was new. Those retreaders which choose to retain the original manufacturer's DOT symbol on the sidewall are free to do so, and those retreaders which choose to remove the original manufacturer's DOT symbol are also free to do so, since NHTSA has concluded that the symbol has so little significance for purchasers of retreaded non-car tires. Hence, the proposed change to the language in § 574.5 is hereby adopted, for the reasons set forth in the proposal.

NHTSA has analyzed this rule and determined that it is neither "major" within the meaning of Executive Order 12291 nor "significant" within the meaning of the Department of Transportation regulatory policies and procedures. The impact of this rule is simply to authorize a practice which has been followed by most non-car tire retreaders for the last 7 years (i.e., not removing the original manufacturer's DOT symbol). No additional paperwork or costs will be imposed as a result of this rule. No cost savings are expected, either, since this rule merely authorizes existing practices. Since the impacts associated with the rule are so minimal, a full regulatory evaluation has not been prepared.

NHTSA has also analyzed this rule in accordance with the Regulatory Flexibility Act. Based on that analysis, I certify that this amendment will not have a significant economic impact on a substantial number of small entities. This rule does not impose any additional burden on tire retreaders, because it merely authorizes a practice most of them have followed, i.e., leaving the original manufacturer's DOT symbol on the sidewall of the finished retread. Those retreaders which have not followed that practice will be able to reduce their costs slightly by leaving that symbol on the sidewall, if they choose. Small organizations and small governmental jurisdictions which purchase retreaded non-car tires will not be affected by this rule. To the extent that this rule might produce some cost savings for the retreaders by allowing them not to buff off the original manufacturer's DOT symbol, those savings are already reflected in the prices charged for most retreaded non-car tires. Hence, no significant savings are expected for small entities as a result of this rule. A full Regulatory Flexibility Analysis has not been prepared for this rule.

Finally, the agency has considered the environmental implications of this rule in accordance with the National Environmental Policy Act and determined that this rule will have no effect on the human environment.

LIST OF SUBJECTS IN 49 CFR PART 574: Labeling, motor-vehicle safety, motor vehicles, reporting and recordkeeping requirements, rubber and rubber products, tires.

In consideration of the foregoing, 49 CFR § 574.5 is amended by revising the introductory text to read as follows:

574.5 *Tire identification requirements.*

Each tire manufacturer shall conspicuously label on one sidewall of each tire it manufactures, except tires manufactured exclusively for mileage-contract purchasers, by permanently molding into or onto the sidewall, in the manner and location specified in Figure 1, a tire identification number containing the information set forth in paragraphs (a) through (d) of this section. Each tire retreader, except tire retreaders who retread tires solely for their own use, shall conspicuously label one sidewall of each tire it retreads by permanently molding or branding into or onto the sidewall, in the manner and location specified in Figure 2, a tire identification number containing the informa-

tion set forth in paragraphs (a) through (d) of this section. In addition, the DOT symbol required by Federal Motor Vehicle Safety Standards shall be located as shown in Figures 1 and 2. The DOT symbol shall not appear on tires to which no Federal Motor Vehicle Safety Standard is applicable, except that the DOT symbol on tires for use on motor vehicles other than passenger cars may, prior to retreading, be removed from the sidewall or allowed to remain on the sidewall, at the retreader's option. The symbols to be used in the tire identification number for tire manufacturers and retreaders are; "A, B, C, D, E, F, H, J, K, L, M, N, P, R, T, U, V, W, X, Y, 1, 2, 3, 4, 5, 6, 7, 8, 9, 0." Tires manufactured or retreaded exclusively for mileage-contract purchasers are not required to contain a tire identification number if the tire contains the phrase "for mileage contract use only" permanently molded into or onto the tire sidewall in lettering at least ¼ inch high.

*　　*　　*　　*　　*

Issued on January 10, 1985.

Diane K. Steed
Administrator
50 FR 2287
January 16, 1985

PREAMBLE TO AN AMENDMENT TO PART 574

Tire Code Marks Assigned to New Tire Manufacturers

ACTION: Publication of tire code marks assigned to new tire manufacturers.

SUMMARY: The agency first published a complete listing of the tire code marks assigned to new tire manufacturers in 1972. The second publication of this listing in June 1983 added an additional 150 code marks. Since that last publication, there have been several additions and changes in names and addresses for the assigned code marks. This publication will inform the public of those additions and changes as reported to the agency.

SUPPLEMENTARY INFORMATION: Section 574.5 of Title 49, Code of Federal Regulations, requires tire manufacturers to mold a tire identification number into or onto the sidewall of each tire they manufacture. In the case of new tires, the first two digits of the tire identification number are the code mark assigned to the manufacturer. This code mark identifies the tire manufacturer and the plant where the tire was manufactured.

The NHTSA first published a complete listing of the tire codes at 37 FR 342, January 11, 1972. This list enables interested members of the public to identify the manufacturer and place of manufacture of any new tire. The NHTSA published an updating of the tire codes at 48 FR 27635, June 16, 1983, adding some 150 additional code marks assigned to new tire manufacturers since the 1972 publication.

This update listing of the assigned code marks for new tire manufacturers is being published to bring the public up to date with the revisions and new code numbers which have been assigned since the publication of the 1983 list.

Issued on March 11, 1985.

Barry Felrice
Associate Administrator
for Rulemaking

50 FR 10880
March 18, 1985

M8   Premier Tyres Limited, Kalamassery, Kerala State, India

Y8   Bombay Tyres International Limited, Hay Bunder Road, Bombay, Maharashtra, India 400 033

C9   Seven Star Rubber Company, Ltd., 2-1 Chang-Swei Road, Pin-Tou Hsiang, Chang-Hua, Taiwan, R.O.C.

F9   Dunlop New Zealand, Limited, P.O. Box 40343, Upper Hutt, New Zealand

H9   Reifen-Berg, 5000 Koln 80 (Mulheim), Clevischer Ring 134, West Germany

J9   P.T. Intirub, 454 Cililitan, P.O. Box 2626, Besar, Jakarta, Indonesia

K9   Natier Tire & Rubber Co., Ltd., 557, Shan Chiao Road, Sec. 1, Shetou, Changhua, Taiwan, R.O.C. 511

M9   Uniroyal Tire Corporation, Uniroyal Research Center, Middlebury, CT 06749

N9   Cia Pneus Tropical, Km105/BR, 324, Centro Industrial Desubae 44100, Feira de Santana, Bahia, Brazil

P9   MRF, Ltd., P.B. No. 1 Ponda, Goa 403 401, India

T9   MRF, Ltd., Thiruthani Road, Ichiputhur 631 060, Arkonam, India

U9   Cooper Tire & Rubber Company, 1689 South Green Street, Tupelo, MS 38801

V9   M & R Tire Co., 309 Main Street, Watertown, MA 02172

Reported Name Change
New Tire Manufacturers

Code	Old Name	New Name
AA	General Tire & Rubber Co. One General Street Akron, Ohio 44329	GenCorp Inc. One General Street Akron, OH 44329
BB	B.F. Goodrich Tire Company 5400 E. Olympic Blvd. Los Angeles, CA 90022	B.F. Goodrich Tire Company Department 6517 P.O. Box 31 Miami, OK 74354
LK	Uniroyal Croyden, S.A. Carrera 7A, No. 22-1 Cali, Colombia	Productora Nacional de Llantas, S.A. Carrera 7A, No. 22-1 Cali, Colombia
WT	Madras Rubber Factory, Ltd. 175/1 Mount Road Madras, India	Madras Rubber Factory, Ltd. Tiruvottiyur High Road Madras 600 019 India
H2	Sam Yang Tire Mfg. Co., Ltd. Song Jung Eup Junnam, Korea	Kumho & Co., Inc. 555 Sochon-Ri Songjung-Eup Kwangsan-Kun Chonnam, Korea

MISCELLANEOUS NEW TIRE MANUFACTURERS TRANSACTIONS
As Reported to NHTSA

Manufacturer	Code	Remark
Armstrong Rubber Company	CE	Plant closed 4/3/81
Bridgestone Tire Company	LH	Purchased from UNIROYAL as of 6/13/82
Ceat, S.p.A.	HU	Sold to Pirelli Tire Corp. in May 1984
Cooper Tire & Rubber Company	U9	Purchased from Pennsylvania Tire & Rubber on 1/25/84
Dayton Tire & Rubber Company	DC	Purchased from Dunlop on 11/1/75
Dunlop Olympic Tyres	DT,DU,WM,W4	Merger of Dunlop and Olympic on 4/29/81
Dunlop Tire & Rubber Corp.	DF, DH, DJ, DP, WN	Plants closed
ditto	DT, DU, WM W4	Plants sold to Dunlop Olympic on 4/29/81
ditto	DC	Plant sold to Firestone T&R on 11/1/75
Firestone Tire & Rubber	DC	Purchased from Dunlop T&R on 11/1/75
ditto	VV	Plant sold to Viskafors Gummifabrik in April 1980
General Tire & Rubber Company	LV	Purchased from Mansfield-Denman on 11/30/78
B.F. Goodrich Company	BJ	Plant sold 12/79
ditto	BK	Plant sold 1/80
ditto	BM	Plant sold to Olympic in 7/75
ditto	BN	Plant sold 8/81
ditto	BP	Plant sold 5/78
Nitto Tire Company, Ltd.	N3	Plant sold to Ryoto Tire Co., Ltd., on 1/23/80
Olympic Tire & Rubber Co., Pty., Ltd.	WM, W4	Sold to Dunlop Olympic on 4/29/81
ditto	WN	Plant closed in 1978

MISCELLANEOUS NEW TIRE MANUFACTURERS TRANSACTIONS
As Reported to NHTSA
(Continued)

Manufacturer	Code	Remark
Pennsylvania Tire & Rubber of Mississippi	WK	Plant sold to Cooper T&R on 1/24/84
Pirelli Tire Corporation	HU	Plant purchased from Ceat, S.p.A. in May 1984
Ryoto Tire Company	N3	Plant purchased from Nitto Tire Company on 1/23/80
SAMYAND Tire, Inc.	XU	Plant closed in 1976
UNIROYAL, Inc.	LH	Plant sold to Bridgestone Tire Company on 6/13/82
Viskafors Gummifabrik AB	VV	Plant purchased from Firestone T&R in April 1980

PART 574—TIRE IDENTIFICATION AND RECORDKEEPING

(Docket No. 70-12; Notice No. 5)

Sec.
574.1 Scope.
574.2 Purpose.
574.3 Definitions.
574.4 Applicability.
574.5 Tire identification requirements.
574.6 Identification mark.
574.7 Information requirements—tire manufacturers, brand name owners, retreaders.
574.8 Information requirements—tire distributors and dealers.
574.9 Requirements for motor vehicle dealers.
574.10 Requirements for motor vehicle manufacturers.

§ 574.1 Scope.

This part sets forth the method by which new tire manufacturers and new tire brand name owners shall identify tires for use on motor vehicles and maintain records of tire purchasers, and the method by which retreaders and retreaded tire brand name owners shall identify tires for use on motor vehicles. This part also sets forth the methods by which independent tire dealers and distributors shall record, on registration forms, their names and addresses and the identification number of the tires sold to tire purchasers and provide the forms to the purchasers, so that the purchasers may report their names to the new tire manufacturers and new tire brand name owners, and by which other tire dealers and distributors shall record and report the names of tire purchasers to the new tire manufacturers and new tire brand name owners.

§ 574.2 Purpose.

The purpose of this part is to facilitate notification to purchasers of defective or nonconforming tires, pursuant to sections 151 and 152 of the National Traffic and Motor Vehicle Safety Act of 1966, as amended (15 U.S.C. 1411 and 1412)

(hereafter the Act), so that they may take appropriate action in the interest of motor vehicle safety.

§ 574.3 Definitions.

(a) *Statutory definitions.* All terms in this part that are defined in section 102 of the Act are used as defined therein.

(b) *Motor vehicle safety standard definitions.* Unless otherwise indicated, all terms used in this part that are defined in the Motor Vehicle Safety Standards, part 571 of this subchapter (hereinafter the Standards), are used as defined therein.

(c) *Definitions used in this part.* (1) "Mileage contract purchaser" means a person who purchases or leases tire use on a mileage basis.

[(2)] "Independent" means, with respect to a tire distributor or dealer, one whose business is not owned or controlled by a tire manufacturer or brand name owner.

[(3)] "New tire brand name owner" means a person, other than a new tire manufacturer, who owns or has the right to control the brand name of a new tire or a person who licenses another to purchase new tires from a new tire manufacturer bearing the licensor's brand name.

[(4)] "Retreaded tire brand name owner" means a person, other than a retreader, who owns or has the right to control the brand name of a retreaded tire or a person who licenses another to purchase retreaded tires from a retreader bearing the licensor's brand name.

[(5)] "Tire purchaser" means a person who buys or leases a new tire, or who buys or leases for 60 days or more a motor vehicle containing a new tire for purposes other than resale.

§ 574.4 Applicability.

This part applies to manufacturers, brand name owners, retreaders, distributors, and dealers of new and retreaded tires for use on motor vehicles manufactured after 1948 and to manufacturers and dealers of motor vehicles manufactured after 1948. However, it does not apply to persons who retread tires solely for their own use.

§ 574.5 Tire Identification requirements.

[Each tire manufacturer shall conspicuously label on one sidewall of each tire it manufactures, except tires manufactured exclusively for mileage-contract purchasers, by permanently molding into or onto the sidewall, in the manner and location specified in Figure 1, a tire identification number containing the information set forth in paragraphs (a) through (d) of this section. Each tire retreader, except tire retreaders who retread tires solely for their own use, shall conspicuously label one sidewall of each tire it retreads by permanently molding or branding into or onto the sidewall, in the manner and location specified in Figure 2, a tire identification number containing the information set forth in paragraphs (a) through (d) of this section. In addition, the DOT symbol required by Federal Motor Vehicle Safety Standards shall be located as shown in Figures 1 and 2. The DOT symbol shall not appear on tires to which no Federal Motor Vehicle Safety Standard is applicable, except that the DOT symbol on tires for use on motor vehicles other than passenger cars may, prior to retreading, be removed from the sidewall or allowed to remain on the sidewall, at the retreader's option. The symbols to be used in the tire identification number for tire manufacturers and retreaders are: "A, B, C, D, E, F, H, J, K, L, M, N, P, R, T, U, V, W, X, Y, 1, 2, 3, 4, 5, 6, 7, 8, 9, 0." Tires manufactured or retreaded exclusively for mileage-contract purchasers are not required to contain a tire identification number if the tire contains the phrase "for mileage contract use only" permanently molded into or onto the tire sidewall in lettering at least one-quarter inch high. (50 F.R. 2288—January 16, 1985. Effective: February 15, 1985)]

(a) First grouping. The first group, of two or three symbols, depending on whether the tire is new or retreaded, shall represent the manufacturer's assigned identification mark (see § 574.6).

(b) Second grouping. For new tires, the second group, of no more than two symbols, shall be used to identify the tire size. For retreaded tires, the second group, of no more than two symbols, shall identify the retread matrix in which the tire was processed or a tire size code if a matrix was not used to process the retreaded tire. Each new tire manufacturer and retreader shall maintain a record of each symbol used, with the corresponding matrix or tire size and shall provide such record to NHTSA upon written request.

(c) Third grouping. The third group, consisting of no more than four symbols, may be used at the option of the manufacturer or retreader as a descriptive code for the purpose of identifying significant characteristics of the tire. However, if the tire is manufactured for a brand name owner, one of the functions of the third grouping shall be to identify the brand name owner. Each manufacturer or retreader who uses the third grouping shall maintain a detailed record of any descriptive or brand name owner code used, which shall be provided to the Bureau upon written request.

(d) Fourth grouping. The fourth group, of three symbols, shall identify the week and year of manufacture. The first two symbols shall identify the week of the year using "01" for the first full calendar week in each year. The final week of each year may include not more than 6 days of the following year. The third symbol shall identify the year. (Example: 311 means the 31st week of 1971, or Aug. 1 through 7, 1971; 012 means the first week of 1972, or Jan. 2 through 8, 1972.) The symbols signifying the date of manufacture shall immediately follow the optional descriptive code (paragraph (c) of this section). If no optional descriptive code is used the symbols signifying the date of manufacture shall be placed in the area shown in Figures 1 and 2 for the optional descriptive code.

§ 574.6 Identification mark.

To obtain the identification mark required by § 574.5(a), each manufacturer of new or retreaded motor vehicle tires shall apply after November 30, 1970, in writing to Tire Identification and Record-keeping, National Highway Traffic Safety Administration, 400 Seventh Street SW., Washington, D.C. 20590, identify himself as a manufacturer of new tires or retreaded tires, and furnish the following information:

(a) The name, or other designation identifying the applicant, and his main office address.

(b) The name, or other identifying designation, of each individual plant operated by the manufacturer and the address of each plant, if applicable.

(c) The type of tires manufactured at each plant, e.g., passenger car tires, bus tires, truck tires, motorcycle tires, or retreaded tires.

3. Section 574.7 is revised to read as follows:

§ 574.7 Information requirements—new tire manufacturers, new tire brand name owners.

(a) (1) Each new tire manufacturer and each new tire brand name owner (hereinafter referred to in this section and § 574.8 as "tire manufacturer") or its designee, shall provide tire registration forms to every distributor and dealer of its tires which offers new tires for sale or lease to tire purchasers.

(2) Each tire registration form provided to independent distributors and dealers pursuant to paragraph (a) (1) of this section shall contain space for recording the information specified in paragraphs (a) [(4)] (A) through (a) [(4)] (C) of this section and shall conform in content and format to Figures 3a and 3b. Each form shall be:

[(a) Rectangular;

(b) Not less than .007 inches thick;

(c) Greater than 3½ inches, but not greater than 6⅛ inches wide; and

(d)] Greater than 5 inches, but not greater than [6] inches long.

[(3)] Each tire registration form provided to distributors and dealers, other than independent distributors and dealers, pursuant to paragraph (a) (1) of this section shall be similar in format and size to Figure 4 and shall contain space for

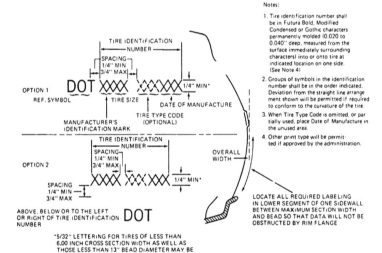

FIGURE 1—IDENTIFICATION NUMBER FOR NEW TIRES

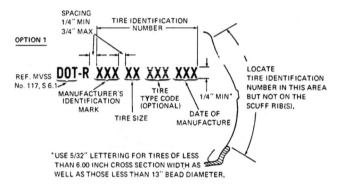

OPTION 1

REF. MVSS
No. 117, S 6.1

SPACING
1/4" MIN
3/4" MAX

TIRE IDENTIFICATION
NUMBER

DOT-R XXX XX XXX XXX

MANUFACTURER'S
IDENTIFICATION
MARK

TIRE
TYPE CODE
(OPTIONAL)

TIRE SIZE

DATE OF
MANUFACTURE

1/4" MIN*

LOCATE
TIRE IDENTIFICATION
NUMBER IN THIS AREA
BUT NOT ON THE
SCUFF RIB(S).

*USE 5/32" LETTERING FOR TIRES OF LESS
THAN 6.00 INCH CROSS SECTION WIDTH AS
WELL AS THOSE LESS THAN 13" BEAD DIAMETER.

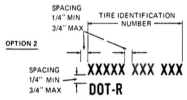

OPTION 2

SPACING
1/4" MIN
3/4" MAX

TIRE IDENTIFICATION
NUMBER

SPACING
1/4" MIN
3/4" MAX

XXXXX XXX XXX
DOT-R

ABOVE, BELOW OR TO THE LEFT
OR RIGHT OF TIRE IDENTIFICATION
NUMBER.

NOTES:

1. Tire identification number shall be in "Futura
Bold, Modified, Condensed or Gothic" characters permanently molded (0.020 to 0.040"
deep, measured from the surface immediately
surrounding characters) into or onto tire at
indicated location on one side.
(See Note 4)

2. Groups of symbols in the identification number shall be in the order indicated. Deviation
from the straight line arrangement shown will
be permitted if required to conform to the
curvature of the tire.

3. When Tire Type Code is omitted, or partially
used, place Date of Manufacture in the unused
area.

4. Other print type will be permitted if approved
by the Administration.

FIGURE 2—IDENTIFICATION NUMBER FOR RETREADED TIRES

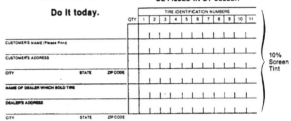

IMPORTANT A

In case of a recall, we can reach you only if we have your name and address. You MUST send in this card to be on our recall list.

SHADED AREAS MUST
BE FILLED IN BY SELLER

Do It today.

QTY	TIRE IDENTIFICATION NUMBERS										
	1	2	3	4	5	6	7	8	9	10	11

10%
Screen
Tint

CUSTOMER'S NAME (Please Print)

CUSTOMER'S ADDRESS

CITY   STATE   ZIP CODE

NAME OF DEALER WHICH SOLD TIRE

DEALER'S ADDRESS

CITY   STATE   ZIP CODE

A Preprinted tire manufacturer's name—unless the manufacturer's name appears on reverse side of the form.

Fig. 3a—Registration form for independent distributors and dealers— tire identification number side

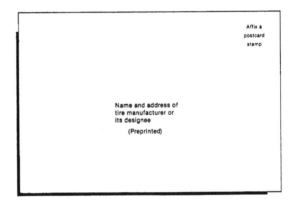

Affix a
postcard
stamp

Name and address of
tire manufacturer or
its designee

(Preprinted)

Fig. 3b—Registration form for independent distributors and dealers— address side

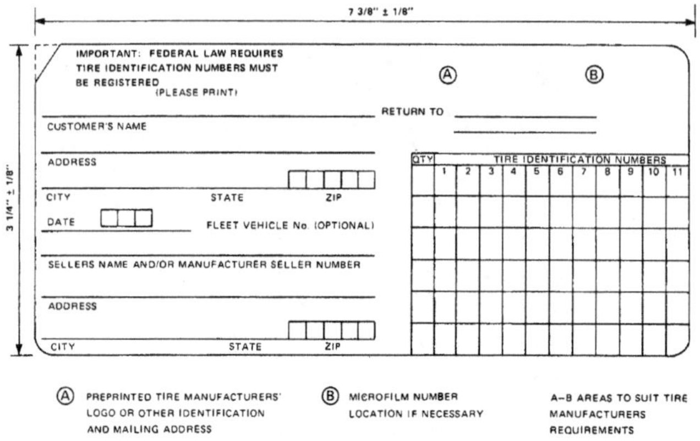

Fig. 4—UNIVERSAL FORMAT

recording the information specified in paragraphs (a) [(4)] (A) through (a) [(4)] (C) of this section.

[(4)] (A) Name and address of the tire purchaser.

(B) Tire identification number.

(C) Name and address of the tire seller or other means by which the tire manufacturer can identify the tire seller.

(b) Each tire manufacturer shall record and maintain, or have recorded and maintained for it by a designee, the information from registration forms which are submitted to it or its designee. No tire manufacturer shall use the information on the registration forms for any commercial purpose detrimental to tire distributors and dealers. Any tire manufacturer to which registration forms are mistakenly sent shall forward those registration forms to the proper tire manufacturer within 90 days of the receipt of the forms.

(c) Each tire manufacturer shall maintain, or have maintained for it by a designee, a record of each tire distributor and dealer that purchases tires directly from the manufacturer and sells them

to tire purchasers, the number of tires purchased by each such distributor or dealer, the number of tires for which reports have been received from each such distributor or dealer other than an independent distributor or dealer, the number of tires for which reports have been received from each such independent distributor or dealer, the total number of tires for which registration forms have been submitted to the manufacturer or its designee, and the total number of tires sold by the manufacturer.

(d) The information that is specified in paragraph (a) [(4)] of this section and recorded on registration forms submitted to a tire manufacturer or its designee shall be maintained for a period of not less than three years from the date on which the information is recorded by the manufacturer or its designee.

§ 574.8 Information requirements—tire distributors and dealers.

(a) *Independent distributors and dealers.* (1) Each independent distributor and each independent dealer selling or leasing new tires to tire purchasers or lessors (hereinafter referred to in this section as "tire purchasers") shall provide each tire purchaser at the time of sale or lease of the tire(s) with a tire registration form.

(2) The distributor or dealer may use either the registration forms provided by the tire manufacturers pursuant to § 574.7(a) or registration forms obtained from another source. Forms obtained from other sources shall comply with the requirements specified in § 574.7(a) for forms provided by tire manufacturers to independent distributors and dealers.

(3) Before giving the registration form to the tire purchaser, the distributor or dealer shall record in the appropriate spaces provided on that form:

(A) The entire tire identification number of the tire(s) sold or leased to the tire purchaser; and

(B) The distributor's or dealer's name and address or other means of identification known to the tire manufacturer.

(4) Multiple tire purchases or leases by the same tire purchaser may be recorded on a single registration form.

(b) *Other distributors and dealers.* (1) Each distributor and each dealer, other than an independent distributor or dealer, selling new tires to tire purchasers shall submit the information specified in § 574.7(a) [(4)] to the manufacturer of the tires sold, or to its designee.

(2) Each tire distributor and each dealer, other than an independent distributor or dealer, shall submit registration forms containing the information specified in § 574.7(a) [(4)] to the tire manufacturer, or person maintaining the information, not less often than every 30 days. However, a distributor or dealer which sells less than 40 tires, of all makes, types and sizes during a 30-day period may wait until he or she sells a total of 40 new tires, but in no event longer than six months, before forwarding the tire information to the respective tire manufacturers or their designees.

(c) Each distributor and each dealer selling new tires to other tire distributors or dealers shall supply to the distributor or dealer a means to record the information specified in § 574.7(a) [(4)], unless such a means has been provided to that distributor or dealer by another person or by a manufacturer.

(d) Each distributor and each dealer shall immediately stop selling any group of tires when so

directed by a notification issued pursuant to sections 151 and 152 of the Act (15 U.S.C. 1411 and 1412).

§ 574.9   Requirements for motor vehicle dealers.

(a) Each motor vehicle dealer who sells a used motor vehicle for purposes other than resale, or who leases a motor vehicle for more than 60 days, that is equipped with new tires or newly retreaded tires is considered, for purposes of this part, to be a tire dealer and shall meet the requirements specified in § 574.8.

(b) Each person selling a new motor vehicle to first purchasers for purposes other than resale, that is equipped with tires that were not on the motor vehicle when shipped by the vehicle manufacturer is considered a tire dealer for purposes of this part and shall meet the requirements specified in § 574.8.

§ 574.10   Requirements for motor vehicle manufacturers.

Each motor vehicle manufacturer, or his designee, shall maintain a record of tires on or in each vehicle shipped by him to a motor vehicle distributor or dealer, and shall maintain a record of the name and address of the first purchaser for purposes other than resale of each vehicle equipped with such tires. These records shall be maintained for a period of not less than three years from the date of sale of the vehicle to the first purchaser for purposes other than resale.

Interpretation

Under section 113(f) of the National Traffic and Motor Vehicle Safety Act (15 U.S.C. 1402(f)) and Part 574, it is the tire manufacturer who has the ultimate responsibility for maintaining the records of first purchasers. Therefore, it is the tire manufacturer or his designee who must maintain these records. The term "designee," as used in the regulation, was not intended to preclude multiple designees; if the tire manufacturer desires, he may designate more than one person to maintain the required information. Furthermore, neither the Act nor the regulation prohibits the distributor or dealer from being the manufacturer's designee, nor do they prohibit a distributor or dealer from selecting someone to be the manufacturer's designee provided the manufacturer approves of the selection.

With respect to the possibility of manufacturers using the maintained information to the detriment of a distributor or dealer, NHTSA will of course investigate claims by distributors or dealers of alleged misconduct and, if the maintained information is being misused, take appropriate action.

36 F.R. 4783
March 12, 1971

36 F.R. 13757
July 24, 1971

36 F.R. 16510
August 21, 1971

PREAMBLE TO TIRE CODE MARKS ASSIGNED TO NEW TIRE MANUFACTURERS

The purpose of this notice is to publish the code numbers assigned to new-tire manufacturers under the Tire Identification and Recordkeeping Regulation, 49 CFR Part 574 (36 F.R. 1196).

The Tire Identification and Recordkeeping Regulation (hereafter Part 574) requires that new .tires manufactured after May 22, 1971, be marked with a. two-symbol manufacturer's code, and that retreaded tires be marked with a three-symbol manufacturer's code. The manufacturer's code is the first grouping within the tire identification number (after the symbol "DOT" or "R" where required).

Under Part 574 a separate code number is assigned to each manufacturer's plant. Table 1 of the notice lists the code numbers assigned and the manufacturer that received each code number. Table 2 lists the same information by

manufacturer. Codes assigned to retreaders will be available for inspection in the Docket Section, Room 5217, 400 Seventh Street SW., Washington, D.C. 20590.

The codes assigned to new-tire manufacturers replace the three-digit code numbers required on new brand-name passenger car tires manufactured prior to May 22, 1971, under Standard No. 109. (The list of numbers assigned under Standard No. 109 was published in the *Federal Register* of July 2, 1968, 34 F.R. 11158.)

Issued on April 14, 1971.

Rodolfo A. Diaz,
Acting Associate Administrator,
Motor Vehicle Programs.

36 F.R. 7539
April 21, 1971

PREAMBLE TO TIRE SIZE CODES

The purpose of this notice is to publish an updated list of tire size codes assigned by the National Highway Traffic Safety Administration in accordance with the Tire Identification and Record Keeping regulation, 49 CFR Part 574 (36 F.R. 1196).

The Tire Identification and Record Keeping regulation requires that a tire identification number be placed on new and retreaded tires, and that the second grouping of the number be a code that identifies the tire size or, in the case of a retreaded tire, the tire matrix. New tire manufacturers have up to now been required to use a specific tire size code assigned to the tire size by the NHTSA. Because of the number of new tire sizes being introduced into the market, the possible combinations of letters and numbers have been virtually exhausted.

In order to accommodate new tire sizes the regulation is being amended by notice published elsewhere in this issue (37 F.R. 23727), to allow each tire manufacturer to assign a two-symbol size code of his own choice, rather than having the number assigned by the agency. However, it is urged that manufacturers maintain the assigned tire size code for existing tire sizes, and that they reuse obsolete tire size codes for new sizes wherever possible.

For convenience of reference, an updated list of the tire size codes assigned by the NHTSA is published below for the information and guidance of tire manufacturers.

This notice is issued under the authority of sections 103, 113, 119, 201 and 1402, 1407, 1421 and 1426; and the delegations of authority at 49 CFR 1.51 and 49 CFR 501.8.

Issued on October 26, 1972.

Robert L. Carter
Associate Administrator
Motor Vehicle Programs

38 F.R. 23742
November 8, 1972

Table 1. List of Alpha-Numeric Code Assignments to New Tire Manufacturers
(Based on the following Alpha-numeric code with letters: ABCDEFHJKLMNPTUVWXY
and Nos. 123456789)

Code No.	New Tire Manufacturers	Code No.	New Tire Manufacturers
AA	The General Tire Co.	B7	Michelin Tire Corp., 2306 Industrial Road, Dothan, Alabama 36301.
AB	The General Tire Co.	B8	Cia Brasiliera de Pneumaticos Michelin Ind., Estrada Da Cachamorra 5000, 23000 Campo Grande, Rio De Janeiro, Brazil.
AC	The General Tire Co.		
AD	The General Tire Co.		
AE	The General Tire Co. (Spain).	B9	Michelin Tire Corp., 2520 Two Notch Road, P.O. Box 579, Lexington, S. Carolina 29072.
AF	The General Tire Co. (Portugal).		
AH	The General Tire Co. (Mexico).	CA	The Mohawk Rubber Co.
AJ	Uniroyal, Inc.	CB	The Mohawk Rubber Co.
AK	Uniroyal, Inc.	CC	The Mohawk Rubber Co.
AL	Uniroyal, Inc.	CD	Alliance Tire & Rubber Co., Ltd. (Israel).
AM	Uniroyal, Inc.	CE	The Armstrong Rubber Co.
AN	Uniroyal, Inc.	CF	The Armstrong Rubber Co.
AP	Uniroyal, Inc.	CH	The Armstrong Rubber Co.
AT	Avon Rubber Co. (England).	CJ	Inoue Rubber Co., Ltd. (Japan).
AU	Uniroyal, Ltd. (Canada).	CK	Not assigned.
AV	The Sieberling Tire & Rubber Co.	CL	Not assigned.
AW	Samson Tire & Rubber Co., Ltd. (Israel).	CM	Continental Gummiwerke A.G. (Germany).
AX	Phoenix Gummiwerke A.G. (Germany).	CN	Continental Gummiwerke A.G. (France).
AY	Phoenix Gummiwerke A.G. (Germany).	CP	Continental Gummiwerke A.G. (Germany).
A1	Manufacture Francaise Pneumatiques Michelin, Poitiers, France.	CT	Continental Gummiwerke A.G. (Germany).
		CU	Continental Gummiwerke A.G. (Germany).
A2	Lee Tire & Rubber Co., Anhanguera Highway, Kilometer 128, Sao Paulo, Brasil.	CV	The Armstrong Rubber Co.
		CW	The Toyo Rubber Industry Co., Ltd. (Japan).
A3	General Tire & Rubber Co., Mount Vernon, Illinois 62864.	CX	The Toyo Rubber Industry Co., Ltd. (Japan).
		CY	McCreary Tire & Rubber Co.
A4	Hung-A Industrial Co., Ltd., 42 JyonPo-Dong Pusanjin-Ku, Pusan, Korea.	C1	Michelin (Nigeria) Ltd., Port Harcourt, Nigeria.
A5	Debickie Zaklady Opon Samochodowych "Stomil," A1.1 Maja 1, 39-200 Debica, Poland.	C2	Kelly Springfield Companhia Goodyear Do Brasil, Km-128 Americana, Sao Paulo, Brazil.
A6	Apollo Tires Ltd., Jos. Anne M.C.Road, Cochin 682016, Kerala, India.	C3	McCreary Tire & Rubber Co., 3901 Clipper Road, Baltimore, Maryland 21211.
A7	Thai Bridgestone Tire Co. Ltd., Tambol Klong-1, Amphur Klong Luang. Changwad Patoom, Thani, Thailand.	C4	Armstrong Rubber Co., Eagle Bend Industrial Park, Clinton, Tennessee.
		C5	Poznanskie Zaklady Opon Samochodowych "STOMIL," ul. Starolecka 18, Poznan, Poland.
A8	P.T. Bridgestone Tire Co. Ltd., Desa Harapan Jaya-Bekasi, Km27-Jawa Barat, Indonesia.	C6	Mitas NP Praha 10-Zahradni Mesto, Komarovova 1900, Praque, Czechoslovakia.
A9	General Tire & Rubber Co., 927 S. Union, St., Bryan, Ohio 44350.	C7	Ironsides Tire & Rubber Co., 2500 Grassland Drive, Louisville, Ky 40299.
BA	The B. F. Goodrich Co.	C8	Bridgestone Hsin Chu Plant, Chung Yi Rubber Industrial Co. Ltd., No. 1 Chuang Ching Road, Taiwan.
BB	The B. F. Goodrich Co.		
BC	The B. F. Goodrich Co.		
BD	The B. F. Goodrich Co.	[C9	Seven Star Rubber Company, Ltd, 2-1 Chang-Swei Road, Pin-Tou Hsiang, Chang-Hua, Taiwan, R.O.C.]
BE	The B. F. Goodrich Co.		
BF	The B. F. Goodrich Co.		
BH	The B. F. Goodrich Co. (Canada).	DA	The Dunlop Tire & Rubber Corp.
BJ	The B. F. Goodrich Co. (Germany).	DB	The Dunlop Tire & Rubber Corp.
BK	The B. F. Goodrich Co. (Brazil).	DC	The Dunlop Tire & Rubber Corp. (Canada).
BL	The B. F. Goodrich Co. (Colombia).	DD	The Dunlop Tire & Rubber Corp. (England).
BM	The B. F. Goodrich Co. (Australia).	DE	The Dunlop Tire & Rubber Corp. (England).
BN	The B. F. Goodrich Co. (Philipines).	DF	The Dunlop Tire & Rubber Corp. (England).
BP	The B. F. Goodrich Co. (Iran).	DH	The Dunlop Tire & Rubber Corp. (Scotland).
BT	Semperit Gummiwerke A.G. (Austria).	DJ	The Dunlop Tire & Rubber Corp. (Ireland).
BU	Semperit Gummiwerke A.G. (Ireland).	DK	The Dunlop Tire & Rubber Corp. (France).
BV	IRI International Rubber Co.	DL	The Dunlop Tire & Rubber Corp. (France).
BW	The Gates Rubber Co.	DM	The Dunlop Tire & Rubber Corp. (Germany).
BX	The Gates Rubber Co.	DN	The Dunlop Tire & Rubber Corp. (Germany).
BY	The Gates Rubber Co.	DP	The Dunlop Tire & Rubber Corp. (England).
B1	Manufacture Francaise Pneumatiques Michelin, LaRoche Sur Yon, France.	DT	The Dunlop Tire & Rubber Corp. (Australia).
		DU	The Dunlop Tire & Rubber Corp. (Australia).
B2	Dunlop Malaysian Industries Berhad, Selangor, Malaysia.	DV	Vredestein (The Netherlands).
		DW	Vredestein (The Netherlands).
B3	Michelin Tire Mfg. Co. of Canada Ltd., Bridgewater, Nova Scotia.	DX	Vredestein Radium (The Netherlands).
		DY	Denman Rubber Manufacturing Co.
B4	Taurus Hungarian Rubber Works, 1965 Budapest, Kerepesi UT17, Hungary.	D1	Viking-Askim-1800 Askim, Norway.
B5	Olsztynskie Zaklady Opon Samochodowych "STOMIL," A1.Zwyciestwa 71, Olsztyn, Poland.	D2	Dayton Tire & Rubber Co., P.O. Box 1000, La Vergne, Tennessee 37086.
B6	Michelin Tire Corp., P.O. Box 5049, Spartanburg, S. Carolina 29304.	D3	United Tire & Rubber Co., Northam Ind. Park Cobourg, Ontario, Canada K9A 4K2.

(Rev. 1/16/85)

PART 574; (TIRE CODE)-1

Code No.	New Tire Manufacturers
D4	Dunlop India Ltd., P.O. Box Sahaganj, Dist. Hooghly, West Bengal, India.
D5	Dunlop India Ltd., Ambattur, Madras-600053, India.
D6	Borovo, Ygoslavenski Kombinat Gume i Obose, Borovo, Yugoslavia.
D7	Dunlop South Africa Ltd., Ladysmith plant 151, Helpmekaar Road, Danskraal Ind. sites, Rep. of S. Africa.
D8	Dunlop South Africa Ltd., Durban Plant 265, Sydney Road, 4001 Durban, Rep. of S. Africa.
D9	United Tire & Rubber Co., Ltd., 275 Belfield Road, Rexdale, Ontario, Canada, M9 W 5C6.
EA	Metzeler A.G. (Germany).
EB	Metzeler A.G. (Germany).
EC	Metzeler A.G. (Germany).
ED	Okamoto Riken Gomu Co., Ltd. (Japan).
EE	Nitto Tire Co., Ltd. (Japan).
EF	Hung Ah Tire Co., Ltd. (Korea).
EH	Bridgestone Tire Co., Ltd. (Japan).
EJ	Bridgestone Tire Co., Ltd. (Japan).
EK	Bridgestone Tire Co., Ltd. (Japan).
EL	Bridgestone Tire Co., Ltd. (Japan).
EM	Bridgestone Tire Co., Ltd. (Japan).
EN	Bridgestone Tire Co., Ltd. (Japan).
EP	Bridgestone Tire Co., Ltd. (Japan).
ET	Sumitomo Rubber Industries, Ltd. (Japan).
EU	Sumitomo Rubber Industries, Ltd. (Japan).
EV	Kleber-Colombes Co. (France).
EW	Kleber-Colombes Co. (France).
EX	Kleber-Colombes Co. (France).
EY	Kleber-Colombes Co. (France).
E1	Chung Hsin Industrial Co. Ltd., Taichong Hsin, Taiwan.
E2	Industria de Pneumatico Firestone SA, Sao Paulo, Brazil.
E3	Seiberling Tire & Rubber Co., P.O. Box 1000, La Vergne, Tennessee 37086.
E4	Firestone of New Zealand, Papanuvi, Christ Church 5, New Zealand.
E5	Firestone South Africa (Pty) Ltd., P.O. Box 992, Port Elizabeth 6000, S. Africa.
E6	Firestone Tunisie SA, Boite Postale 55, Menzel-Bourguiba, Tunisia.
E7	Firestone East Africa Ltd., P.O. Box 30429, Nairobi, Kenya.
E8	Firestone Ghana Ltd., P.O. Box 5758, Accra, Ghana.
E9	Firestone South Africa (Pty), P.O. Box 496, Brits 0250, South Africa.
FA	The Yokohama Rubber Co., Ltd. (Japan).
FB	The Yokohama Rubber Co., Ltd. (Japan).
FC	The Yokohama Rubber Co., Ltd. (Japan).
FD	The Yokohama Rubber Co., Ltd. (Japan).
FE	The Yokohama Rubber Co., Ltd. (Japan).
FF	Michelin Tire Corp. (France).
FH	Michelin Tire Corp. (France).
FJ	Michelin Tire Corp. (France).
FK	Michelin Tire Corp. (France).
FL	Michelin Tire Corp. (France).
FM	Michelin Tire Corp. (France).
FN	Michelin Tire Corp. (France).
FP	Michelin Tire Corp. (Algeria).
FT	Michelin Tire Corp. (Germany).
FU	Michelin Tire Corp. (Germany).
FV	Michelin Tire Corp. (Germany).
FW	Michelin Tire Corp. (Germany).
FX	Michelin Tire Corp. (Belgium).
FY	Michelin Tire Corp. (The Netherlands).
F1	Michelin Tyre Co. Ltd., Baldovie Dundee, Scotland.
F2	CA Firestone Venezolana, Valencia, Venezuela.
F3	Manufacture Francaise Des Pneumatic Michelin, Roanne, France.
F4	Fabrica De Pneus Fapobol, Sari Rua Azevedo Coutinho 39-1.0, Oporto, Portugal.

Code No.	New Tire Manufacturers
F5	Fate S.A.I.C.I., Avda Alte Blanco Encalada 3003, Buenos Aires, Argentina.
F6	General Fabrica Espanola (Firestone Owned) Torrelavega Plant, Spain.
F7	General Fabrica Espanola (Firestone Owned) Puente San Miguel Plant, Spain.
F8	Vikrant Tyres Ltd., K.R.S. Road, Mysore (Karnataka State) India.
[F9	Dunlop New Zealand, Limited, P.O. Box 40343, Upper Hutt, New Zealand]
HA	Michelin Tire Corp. (Spain).
HB	Michelin Tire Corp. (Spain).
HC	Michelin Tire Corp. (Spain).
HD	Michelin Tire Corp. (Italy).
HE	Michelin Tire Corp. (Italy).
HF	Michelin Tire Corp. (Italy).
HH	Michelin Tire Corp. (Italy).
HJ	Michelin Tire Corp. (United Kingdom).
HK	Michelin Tire Corp. (United Kingdom).
HL	Michelin Tire Corp. (United Kingdom).
HM	Michelin Tire Corp. (United Kingdom).
HN	Michelin Tire Corp. (Canada).
HP	Michelin Tire Corp. (South Vietnam).
HT	CEAT (Italy).
HU	CEAT (Italy).
HV	CEAT (Italy).
HW	Withdrawn.
HX	The Dayton Tire & Rubber Co.
HY	The Dayton Tire & Rubber Co.
H1	De La SAFE Neumaticos Michelin, Valladolid, Spain.
H2	SamYang Tire Mfg. Co. Ltd., Song Jung Plt., Junnam, Korea.
H3	Sava Industrija Gumijevih, 64,000 Kranj, Yugoslavia.
H4	Bridgestone-Houfu, Yamaguchi-ken, Japan.
H5	Hutchinson-Mapa, 45120 Chalette Sur Loing, France.
H6	Shin Hung Rubber Co. Ltd., 156 Sang Pyong-Dong Junju, Kyung Nam, Korea.
H7	Li Hsin Rubber Industrial Co. Ltd., 42 Yuan Lu Road, Sec. 1, Taiwan, China.
H8	Firestone, 2600 South Council Road, Oklahoma City, OK. 73124.
[H9	Reifen-Berg, 5000 Koln 80 (Mulheim), Clevischer Ring 134, West Germany]
JA	The Lee Tire & Rubber Co.
JB	The Lee Tire & Rubber Co.
JC	The Lee Tire & Rubber Co.
JD	The Lee Tire & Rubber Co.
JE	The Lee Tire & Rubber Co.
JF	The Lee Tire & Rubber Co.
JH	The Lee Tire & Rubber Co.
JJ	The Lee Tire & Rubber Co.
JK	The Lee Tire & Rubber Co.
JL	The Lee Tire & Rubber Co.
JM	The Lee Tire & Rubber Co.
JN	The Lee Tire & Rubber Co.
JP	The Lee Tire & Rubber Co.
JT	The Lee Tire & Rubber Co.
JU	The Lee Tire & Rubber Co. (Canada).
JV	The Lee Tire & Rubber Co. (Canada).
JW	The Lee Tire & Rubber Co. (Canada).
JX	Lee Tire & Rubber Co. (Canada).
JY	Lee Tire & Rubber Co. (Argentina).
J1	Phillips Petroleum Co., Bartlesville, OK 74004.
J2	Bridgestone Singapore Co. Ltd., 2 Jurong Port Road, Jurong Town, Singapore 22, Singapore.
J3	Gumarne Maja, Puchov, Czechoslovakia.
J4	Rubena N.P., Nachod, Czechoslovakia.
J5	Lee Tire & Rubber Co., State Rt. 33, Box 799, Logan, Ohio 43138.
J6	Jaroslavl Tire Co., Jaroslavl, USSR.
J7	R&J Mfg. Corp., 1420 Stanley Dr., Plymouth, Indiana 46563.

PART 574; (TIRE CODE)-2

Code No.	New Tire Manufacturers
J8	DaChung Hua Rubber Ind. Co., Shanghai Tire Plant, 839 Hanyshan Rd., Shanghai, China.
[J9	P.T. Intirub, 454 Cililitan, P.O. Box 2626, Besar, Jakarta, Indonesia]
KA	Lee Tire & Rubber Co. (Australia).
KB	Lee Tire & Rubber Co. (Australia).
KC	Lee Tire & Rubber Co. (Brazil).
KD	Lee Tire & Rubber Co. (Colombia).
KE	Lee Tire & Rubber Co. (Republic of Congo).
KF	Lee Tire & Rubber Co. (France).
KH	Lee Tire & Rubber Co. (Germany).
KJ	Lee Tire & Rubber Co. (Germany).
KK	Lee Tire & Rubber Co. (Greece).
KL	Lee Tire & Rubber Co. (Guatemala).
KM	Lee Tire & Rubber Co. (Luxembourg).
KN	Lee Tire & Rubber Co. (India).
KP	Lee Tire & Rubber Co. (Indonesia).
KT	Lee Tire & Rubber Co. (Italy).
KU	Lee Tire & Rubber Co. (Jamaica).
KV	Lee Tire & Rubber Co. (Mexico).
KW	Lee Tire & Rubber Co. (Peru).
KX	Lee Tire & Rubber Co. (Philippines).
KY	Lee Tire & Rubber Co. (Scotland).
K1	Phillips Petroleum Co., 1501 Commerce Drive, Stow, Ohio 44224.
K2	Lee Tire & Rubber Co., Madisonville, KY 42431.
K3	Kenda Rubber Industrial Co. Ltd., Yuanlin, Taiwan.
K4	Uniroyal S.A., Queretaro, Qte. Mexico.
K5	VEB Reifenkombinat Furstenwalde, GDR-124 Furstenwalde-Sud, Trankeweg Germany.
K6	Lee Tire & Rubber Co., One Goodyear Blvd., Lawton, Oklahoma.
K7	Lee Tire & Rubber Co., Camino Melipilla KM16, Maipu Box 3607, Santiago, Chile.
K8	Kelly Springfield Tire Co., Peti Surat 49, Shah, Alam, Sehngor, Malaysia.
[K9	Natier Tire & Rubber Co., Ltd., 557 Shan Chiao Road, Sec. 1, Shetou, Changhua, Taiwan, R.O.C. 511]
LA	Lee Tire & Rubber Co. (South Africa).
LB	Lee Tire & Rubber Co. (Sweden).
LC	Lee Tire & Rubber Co. (Thailand).
LD	Lee Tire & Rubber Co. (Turkey.)
LE	Lee Tire & Rubber Co. (Venezuela.)
LF	Lee Tire & Rubber Co. (England).
LH	Uniroyal, Inc. (Australia).
LJ	Uniroyal, Inc. (Belgium).
LK	Uniroyal, Inc. (Columbia).
LL	Uniroyal, Inc. (France).
LM	Uniroyal, Inc. (Germany).
LN	Uniroyal, Inc. (Mexico).
LP	Uniroyal, Inc. (Scotland).
LT	Uniroyal, Inc. (Turkey).
LU	Uniroyal, Inc. (Venezuela).
LV	Mansfield-Denman-General Co., Ltd. (Canada).
LW	Trelleborg Rubber Co., Inc. (Sweden).
LX	Mitsuboshi Belting, Ltd. (Japan).
LY	Mitsuboshi Belting, Ltd. (Japan).
L1	Goodyear Taiwan Ltd., Taipei, Taiwan, Rep. of China.
L2	Wuon Poong Industrial Co., Ltd., 112-5 Sokong-Dong, Chung-Ku, Seoul, Korea.
L3	Tong Shin Chemical Products Co., Ltd., Seoul, Korea.
L4	Cipcmp Intreprinderea De Anvelope, Danubiana, Romania.
L5	Lassa Lastik Sanayi VeTicaret, A.S. Fabrikas, Kosekoy, P.K. 250 Izmit, Turkey.
L6	Modi Rubber Limited, Modipurnam Plant, Meerut UP250110, India.
L7	Cipcmp Intreprinderea De Anvelope, Zalau, Romania.
L8	Dunlop Zimbabwe Ltd., Donnington, Bulawayo, Zimbabwe.

Code No.	New Tire Manufacturers
MA	The Goodyear Tire & Rubber Co.
MB	The Goodyear Tire & Rubber Co.
MC	The Goodyear Tire & Rubber Co.
MD	The Goodyear Tire & Rubber Co.
ME	The Goodyear Tire & Rubber Co.
MF	The Goodyear Tire & Rubber Co.
MH	The Goodyear Tire & Rubber Co.
MJ	The Goodyear Tire & Rubber Co.
MK	The Goodyear Tire & Rubber Co.
ML	The Goodyear Tire & Rubber Co.
MM	The Goodyear Tire & Rubber Co.
MN	The Goodyear Tire & Rubber Co.
MP	The Goodyear Tire & Rubber Co.
MT	The Goodyear Tire & Rubber Co.
MU	The Goodyear Tire & Rubber Co. (Argentina)
MV	The Goodyear Tire & Rubber Co., (Australia)
MW	The Goodyear Tire & Rubber Co. (Australia).
MX	The Goodyear Tire & Rubber Co. (Brazil).
MY	The Goodyear Tire & Rubber Co. (Colombia).
M1	Goodyear Maroc S.A. Casablanca, Morocco.
M2	Goodyear Tire & Rubber Co., Madisonville, KY 42431.
M3	Michelin Tire Corp., 730 S. Pleasantburg Drive, Greenville, S. Carolina 29602.
M4	Goodyear Tyre & Rubber Co., Logan, Ohio 43138.
M5	Michelin Tire Mfg. Co. of Canada Ltd., P.O. Box 5000, Kentville, Nova Scotia B4NV36.
M6	Goodyear Tire & Rubber Co., One Goodyear Blvd., Lawton, OK 73504.
M7	Goodyear DeChile S.A.I.C., Camino Melipilla K.M.16 Maipu, P.O. Box 3607, Santiago, Chile.
[M8	Premier Tyres Limited, Kalamassery, Kerala State, India]
[M9	Uniroyal Tire Corporation, Uniroyal Research Center, Middlebury, CT 06749]
NA	The Goodyear Tire & Rubber Co. (Republic of Congo).
NB	The Goodyear Tire & Rubber Co. (England).
NC	The Goodyear Tire & Rubber Co. (France).
ND	The Goodyear Tire & Rubber Co. (Germany).
NE	The Goodyear Tire & Rubber Co. (Germany).
NF	The Goodyear Tire & Rubber Co. (Greece).
NH	The Goodyear Tire & Rubber Co.
NJ	The Goodyear Tire & Rubber Co. (Luxembourg).
NK	The Goodyear Tire & Rubber Co. (India.)
NL	The Goodyear Tire & Rubber Co. (Indonesia).
NM	The Goodyear Tire & Rubber Co. (Italy).
NN	The Goodyear Tire & Rubber Co. (Jamaica).
NP	The Goodyear Tire & Rubber Co. (Mexico).
NT	The Goodyear Tire & Rubber Co. (Peru).
NU	The Goodyear Tire & Rubber Co (Philippines).
NV	The Goodyear Tire & Rubber Co. (Scotland).
NW	The Goodyear Tire & Rubber Co. (South Africa).
NX	The Goodyear Tire & Rubber Co. (Sweden).
NY	The Goodyear Tire & Rubber Co. (Thailand).
N1	Maloja AG Pneu Und Gummiwerke, Ormalinger-strasse Gelterkinden, Switzerland, CH 4460.
N2	Hurtubise Nutread, 525 Vickers Street, Tona-wanda, N.Y. 14150.
N3	Ryoto Tire Co., Ltd., Kuwana Plant, 2400 Arano Nakagami, Tohin-Cho Inabe-Gun, Mie-ken. Japan.
N4	Cipcmp Intreprinderea De Anvelope, Victoria, Romania.
N5	Pneumant, VEB Reifenwerk Riesa, Paul-Greifzu-Strasse 20, 84 Riesa, Germany.
N6	Pneumant VEB Reifenwerk Heidenau Haudtstrass. 44 GDR, 8312 Heidenau, Germany.
N7	Cipcmp Intreprinderea De Anvelope, Caracal, Romania.
N8	Lee Tire & Rubber Co. (Goodyear, Malaysia Ber-had), Peti Surat 49, Shah Alam, Selengor, Malaysia.

Code No.	New Tire Manufacturers
[N9 ____	Cia Pneus Tropical, Km105/BR, 324, Centro Industrial Desubae 44100, Feira de Santana, Bahia, Brazil]
PA ____	The Goodyear Tire & Rubber Co. (Turkey).
PB ____	The Goodyear Tire & Rubber Co. (Venezuela).
PC ____	The Goodyear Tire & Rubber Co. (Canada).
PD ____	The Goodyear Tire & Rubber Co. (Canada).
PE ____	The Goodyear Tire & Rubber Co. (Canada).
PF ____	The Goodyear Tire & Rubber Co. (Canada).
PH ____	The Kelly-Springfield Tire Co.
PJ ____	The Kelly-Springfield Tire Co.
PK ____	The Kelly-Springfield Tire Co.
PL ____	The Kelly-Springfield Tire Co.
PM ____	The Kelly-Springfield Tire Co.
PN ____	The Kelly-Springfield Tire Co.
PP ____	The Kelly-Springfield Tire Co.
PT ____	The Kelly-Springfield Tire Co.
PU ____	The Kelly-Springfield Tire Co.
PV ____	The Kelly-Springfield Tire Co.
PW ____	The Kelly-Springfield Tire Co.
PX ____	The Kelly-Springfield Tire Co.
PY ____	The Kelly-Springfield Tire Co.
P1 ____	Gislaved Gummi Fabriken, 33200 Gislaved, Sweden.
P2 ____	Kelly Springfield, Madisonville, Ky. 42431.
P3 ____	Skepplanda Gummi AB, 440-40 Alvangen, Sweden.
P4 ____	Kelly Springfield, Route 33, Logan, Ohio 43138.
P5 ____	General Popo S.A., Central Camionera, Zona Industrial, San Luis Potosi S.L.P., Mexico.
P6 ____	Kelly Springfield Tire Co,. One Goodyear Blvd., Lawton, OK 73504.
P7 ____	Kelly Springfield, Camino Melipilla K.M.16, Maipu, P.O. Box 3607, Santiago, Chile.
P8 ____	China National Chemicals Import & Export Corp., Shandong Branch, Quingdao 97 Cangtai Rd., China.
[P9 ____	MRF, Ltd., P.B. No. 1 Ponda, Goa 403401, India]
TA ____	The Kelly-Springfield Tire Co.
TB ____	The Kelly-Springfield Tire Co. (Argentina).
TC ____	The Kelly-Springfield Tire Co. (Australia).
TD ____	The Kelly-Springfield Tire Co. (Australia).
TE ____	The Kelly-Springfield Tire Co. (Brazil).
TF ____	The Kelly-Springfield Tire Co. (Colombia).
TH ____	The Kelly-Springfield Tire Co. (Republic of Congo).
TJ ____	The Kelly-Springfield Tire Co. (England).
TK ____	The Kelly-Springfield Tire Co. (France).
TL ____	The Kelly-Springfield Tire Co. (Germany).
TM ____	The Kelly-Springfield Tire Co. (Germany).
TN ____	The Kelly-Springfield Tire Co. (Greece).
TP ____	The Kelly-Springfield Tire Co. (Guatemala).
TT ____	The Kelly-Springfield Tire Co. (Luxembourg).
TU ____	The Kelly-Springfield Tire Co. (India).
TV ____	The Kelly-Springfield Tire Co. (Indonesia).
TW ____	The Kelly-Springfield Tire Co. (Italy).
TX ____	The Kelly-Springfield Tire Co. (Jamaica).
TY ____	The Kelly-Springfield Tire Co. (Mexico).
T1 ____	Hankook Tire Mfg. Co., Ltd., Seoul, Korea.
T2 ____	Ozos (Uniroyal) A.G., Olsztyn, Poland.
T3 ____	Debickie Zattidy Opon Samochodowych, Stomil, Debica, Poland (Uniroyal).
T4 ____	S.A. Carideng (Rubber Factory), Jan Rosierlaan 114, B 3760 Lanaken, Belgium.
T5 ____	Tigar Pirot, 18300 Pirot, Yugoslavia.
T6 ____	Hulera Tornel S.A., Sta. Lucia 198 Fracc. Ind. San Antonio, Mexico, 16, D.F.
T7 ____	Hankook Tire Mfg. Co. Inc., Daejun Plant, 658-1 Sukbong-RI, Daeduk-kun, Choongchung Namdo, Korea.
T8 ____	Goodyear Tire & Rubber Co., Goodyear Malaysia Berhad, Peti Surat 49, Shah Alam, Selangor, Malaysia.
[T9 ____	MRF, Ltd., Thiruthani Road, Ichiputhur 631 060, Arkonam, India]
UA ____	The Kelly-Springfield Tire Co. (Peru).
UB ____	The Kelly-Springfield Tire Co. (Philippines).

Code No.	New Tire Manufacturers
UC ____	The Kelly-Springfield Tire Co. (Scotland).
UD ____	The Kelly-Springfield Tire Co. (South Africa).
UE ____	The Kelly-Springfield Tire Co. (Sweden).
UF ____	The Kelly-Springfield Tire Co. (Thailand).
UH ____	The Kelly-Springfield Tire Co. (Turkey).
UJ ____	The Kelly-Springfield Tire Co. (Venezuela).
UK ____	The Kelly-Springfield Tire Co. (Canada).
UL ____	The Kelly-Springfield Tire Co. (Canada).
UM ____	The Kelly-Springfield Tire Co. (Canada).
UN ____	The Kelly-Springfield Tire Co. (Canada).
UP ____	Copper Tire & Rubber Co.
UT ____	Copper Tire & Rubber Co.
UU ____	Carlisle Tire & Rubber Division of Carlisle Corp.
UV ____	Kyowa Rubber Industry Co., Ltd. (Japan).
UW ____	Not assigned.
UX ____	Not assigned.
UY ____	Not assigned.
U1 ____	Lien Shin Tire Co. Ltd., 20 Chung Shan Road, Taipei, Taiwan.
U2 ____	Sumitomo Rubber Industries Ltd., Shirakawa City, Fukoshima Pref. Japan (Dunlop).
U3 ____	Miloje Zakic, 3700 Krusevac, Yugoslavia.
U4 ____	Geo. Byers Sons, Inc., 46 East Town Street, Columbus, Ohio 43215.
U5 ____	Farbentabriken Bayer GMBH, D 5090 Leverkusen, West Germany.
U6 ____	Pneumant-VEB Reifenwerk Dresden, GDR-8040 Dresden, Mannheimer Strasse Germany.
U7 ____	Pneumant-VEB Reifenwerk Neubrandenburg GDR-20 Neubrandenberg, Germany.
U8 ____	Hsin Fung Factory of Nankang Rubber Corp. Ltd., 399 Hsin Shing Road, Yuan San, Taiwan.
[U9 ____	Cooper Tire & Rubber Company, 1689 South Green Street, Tupelo, MS 38801]
VA ____	The Firestone Tire & Rubber Co.
VB ____	The Firestone Tire & Rubber Co.
VC ____	The Firestone Tire & Rubber Co.
VD ____	The Firestone Tire & Rubber Co.
VE ____	The Firestone Tire & Rubber Co.
VF ____	The Firestone Tire & Rubber Co.
VH ____	The Firestone Tire & Rubber Co.
VJ ____	The Firestone Tire & Rubber Co.
VK ____	The Firestone Tire & Rubber Co.
VL ____	The Firestone Tire & Rubber Co. (Canada).
VM ____	The Firestone Tire & Rubber Co. (Canada).
VN ____	The Firestone Tire & Rubber Co. (Canada).
VP ____	The Firestone Tire & Rubber Co. (Italy).
VT ____	The Firestone Tire & Rubber Co. (Spain).
VU ____	Withdrawn.
VV ____	The Firestone Tire & Rubber Co. (Sweden).
VW ____	The Firestone Tire & Rubber Co. (Japan).
VX ____	The Firestone Tire & Rubber Co. (England).
VY ____	The Firestone Tire & Rubber Co. (Wales).
V1 ____	Livingston Tire Shop, North Main Street, Hubbard, Ohio 44425.
V2 ____	Volzhsky Tire Plant, Volzhsk 404103, USSR.
V3 ____	Tahsin Rubber Tire Co. Ltd., Tuchen Village Taipei, Hsieng, Taiwan.
V4 ____	Ohtsu Tire & Rubber Co., Miyakonojo City, Miyazaki Pref., Japan (Firestone).
V5 ____	Firestone Tire & Rubber Co., Mexico City, Mexico.
V6 ____	Firestone Tire & Rubber Co., Cuernavaca, Mexico.
V7 ____	Voronezhsky Tire Plant, Voronezh 494034 USSR.
V8 ____	Boras Gummi Fabrik A.B. Dockvagenl, S502 38 Boras, Sweden (Mac Ripper Tire and Rubber Company).
[V9 ____	M & R Tire Co., 309 Main Street, Watertown, MA 02172]
WA ____	The Firestone Tire & Rubber Co. (France).
WB ____	The Firestone Tire & Rubber Co. (Costa Rica).
WC ____	The Firestone Tire & Rubber Co. (Australia).
WD ____	The Firestone Tire & Rubber Co. (Switzerland).

Code No.	New Tire Manufacturers
WE	Withdrawn.
WF	The Firestone Tire & Rubber Co. (Spain).
WH	The Firestone Tire & Rubber Co. (Sweden).
WJ	The Firestone Tire & Rubber Co. (Australia).
WK	Pennsylvania Tire & Rubber Company of Mississippi.
WL	The Mansfield Tire & Rubber Co.
WM	Olympic Tire & Rubber Co. Pty., Ltd. (Australia).
WN	Olympic Tire & Rubber Co Pty., Ltd. (Australia).
WP	Schenuit Industries, Inc.
WT	Madras Rubber Factory, Ltd. (India).
WU	Not Assigned.
WV	Not Assigned.
WW	Not Assigned.
WX	Not Assigned.
WY	Not Assigned.
W1	Firestone Tire & Rubber Co., P.O. Box 1000, La Vergne, Tennessee 37086.
W2	Firestone Tire & Rubber Co., Wilson, N. Carolina 27893.
W3	Vredestein Doetinchem B.V., Doetinchem, The Netherlands (B.F. Goodrich).
W4	Dunlop Tyres, Somerton, Victoria, Australia.
W5	Firestone Argentina SAIC, Antartida, Argentina, 2715 Llavollol, Buenos Aires, Argentina.
W6	Firestone Tire & Rubber Co., P.O. Box 1355, Commerce Center, Makati, Risal, Philippines.
W7	Firestone Portuguesa S.A.R.L., Apartado 3, Alcochete, Portugal.
W8	Firestone Tire & Rubber Co. Ltd., P.O. Box Prakanong 11/118, Bangkok, Thailand.
W9	Industrie De Pneumaticos Firestone S.A., Caixa Postal 2505, Rio De Janeiro, Brazil.
XA	Pirelli Tire Corp. (Italy).
XB	Pirelli Tire Corp. (Italy).
XC	Pirelli Tire Corp. (Italy).
XD	Pirelli Tire Corp. (Italy).
XE	Pirelli Tire Corp. (Italy).

Code No.	New Tire Manufacturers
XF	Pirelli Tire Corp. (Spain).
XH	Pirelli Tire Corp. (Greece).
XJ	Pirelli Tire Corp. (Turkey).
XK	Pirelli Tire Corp. (Brazil).
XL	Pirelli Tire Corp. (Brazil).
XM	Pirelli Tire Corp. (Argentina).
XN	Pirelli Tire Corp. (England).
XP	Pirelli Tire Corp. (England).
XT	Veith-Pirelli A.G. (Germany).
X1	Tong Shin Chemical Products, Co. Inc., Seoul, Korea.
X2	Hwa Fong Rubber Ind. Co. Ltd., 45 Futsen Road, Yuanlin, Taiwan.
X3	Belotserkovsky Tire Plant, Belaya Tserkov, 256414, U.S.S.R.
X4	Pars Tyre Co., (Pirelli), Saveh, Iran.
X5	JK Industries Ltd., Kankroli, Udaipur District, Rajasthan, India.
X6	Bobruysky Tire Plant, Bobruysk 213824 U.S.S.R.
X7	Chimkentsky Tire Plant, Chimkent 486025 U.S.S.R.
X8	Dnepropetrovsky Tire Plant, Dnepropetrovsk 320033 U.S.S.R.
X9	Moscovsky Tire Plant, Moscow 109088 U.S.S.R.
X0	Nizhnekamsky Tire Plant, Nishnekamsk 423510 U.S.S.R.
Y1	Companhia Goodyear DoBrasil, KM-128 Americana, Sao Paulo, Brasil.
Y2	Dayton Tire Co., Wilson, N. Carolina 27893.
Y3	Seiberling Tire & Rubber Co., Wilson, N. Carolina 27893.
Y4	Dayton Tire & Rubber Co., 345-15th St. S.W., Barberton, Ohio (Firestone).
Y5	Tsentai Rubber Factory, 27 Chung Shan Rd., E.I. Shanghai, China.
Y6	I.T. International Sdn. Bhd., P.O. Box 100 Alor Setar Kedah, Malaysia.
Y7	Bridgestone Tire Co., (U.S.A.) Ltd., I-24 Waldron Dr., La Vergne, Tenn.
Y8	Bombay Tyres International Limited, Hay Bunder Road, Bombay, Maharashtra, India 400 033

(Rev. 1/16/85)

Miscellaneous New Tire Manufacturers Transactions
As Reported to NHTSA

Manufacturer	Code	Remark
Armstrong Rubber Company	CE	Plant closed 4/3/81
Bridgestone Tire Company	LH	Purchased from UNIROYAL as of 6/13/82
Ceat, S.p.a.	HU	Sold to Pirelli Tire Corp. in May 1984
Cooper Tire & Rubber Company	U9	Purchased from Pennsylvania Tire & Rubber on 1/24/84
Dayton Tire & Rubber Company	DC	Purchased from Dunlop on 11/1/75
Dunlop Olympic Tyres	DT, DU, WM, W4	Merger of Dunlop and Olympic on 4/29/81
Dunlop Tire & Rubber Corp.	DF, DH, DJ, DP, WN	Plants closed
ditto	DT, DU, WM, W4	Plants sold to Dunlop Olympic on 4/29/81
ditto	DC	Plant sold to Firestone T&R on 11/1/75
Firestone Tire & Rubber	DC	Purchased from Dunlop T&R on 11/1/75
ditto	VV	Plant sold to Viskafors Gummifabrik in April 1980
General Tire & Rubber Company	LV	Purchased from Mansfield-Denman on 11/30/78
B.F. Goodrich Company	BJ	Plant sold 12/79
ditto	BK	Plant sold 1/80
ditto	BM	Plant sold to Olympic in 7/75
ditto	BN	Plant sold 8/81
ditto	BP	Plant sold 5/78
Nitto Tire Company, Ltd.	N3	Plant sold to Ryoto Tire Co., Ltd. on 1/23/80
Olympic Tire & Rubber Co., Pty., Ltd.	WM, W4	Sold to Dunlop Olympic on 4/29/81
ditto	WN	Plant closed in 1978
Pennsylvania Tire & Rubber of Mississippi	WK	Plant sold to Cooper T&R on 1/24/84
Pirelli Tire Corporation	HU	Plant purchased from Ceat, S.p.a. in May 1984
Ryoto Tire Company	N3	Plant purchased from Nitto Tire Company on 1/23/80
SAMYAND Tire, Inc.	XU	Plant closed in 1976
UNIROYAL, Inc.	LH	Plant sold to Bridgestone Tire Company on 6/13/82
Viskafors Gummifabrik AB	VV	Plant purchased from Firestone T&R in April 1980

TABLE 3. TIRE SIZE CODES

Tire Size Code	Tire Size Designation [1]	Tire Size Code	Tire Size Designation [1]	Tire Size Code	Tire Size Designation [1]
AA	4.00-4	B7	5.00 R 12	D4	6.00 R 13
AB	3.50-4	B8	5.20-12	D5	6.2-13
AC	3.00-5	B9	5.20-12 LT	D6	6.20-13
AD	4.00-5	CA	5.20 R 12	D7	6.40-13
AE	3.50-5	CB	5.30-12	D8	6.40-13 LT
AF	6.90-6	CC	5.50-12	D9	6.40 R 13
AH	3.00-8	CD	5.50-12 LT	EA	6.50-13
AJ	3.50-6	CE	5.50 R 12	EB	6.50-13 LT
AK	4.10-6	CF	5.60-12	EC	6.50-13 ST
AL	4.50-6	CH	5.60-12 LT	ED	6.50 R 13
AM	5.30-6	CJ	5.60 R 12	EE	6.70-13
AN	6.00-6	CK	5.9-12	EF	6.70-13 LT
AP	3.25-8	CL	5.90-12	EH	6.70 R 13
AT	3.50-8	CM	6.00-12	EJ	6.9-13
AU	3.00-7	CN	6.00-12 LT	EK	6.90-13
AV	4.00-7	CP	6.2-12	EL	7.00-13
AW	4.80-7	CT	6.20-12	EM	7.00-13 LT
AX	5.30-7	CU	6.90-12	EN	7.00 R 13
AY	5.00-8	CV	23.5 X 8.5-12	EP	7.25-13
A1	H60-14	CW	125-12	ET	7.25 R 13
A2	4.00-8	CX	125 R 12	EU	7.50-13
A3	4.80-8	CY	125-12/5.35-12	EV	135-13
A4	5.70-8	C1	135-12	EW	135 R 13
A5	16.5 X 6.5-8	C2	135 R 12	EX	135-13/5.65-13
A6	18.5 X 8.5-8	C3	135-12/5.65-12	EY	145-13
A7	CR70-14	C4	145-12	E1	145 R 13
A8	2.75-9	C5	145 R 12	E2	145-13/5.95-13
A9	4.80-9	C6	145-12/5.95-12	E3	150 R 13
BA	6.00-9	C7	155-12	E4	155-13
BB	6.90-9	C8	155 R 12	E5	155 R 13
BC	3.50-9	C9	155-12/6.15-12	E6	155-13/6.15-13
BD	4.00-10	DA	4.80-10	E7	160 R 13
BE	3.00-10	DB	3.25-12	E8	165-13
BF	3.50-10	DC	3.50-12	E9	165 R 13
BH	5.20-10	DD	4.50-12 LT	FA	165-13/6.45-13
BJ	5.20 R 10	DE	5.00-12 LT	FB	165/70 R 13
BK	5.9-10	DF	7.00-12	FC	170 R 13
BL	5.90-10	DH	5.00-13	FD	175-13
BM	6.50-10	DJ	5.00-13 LT	FE	175 R 13
BN	7.00-10	DK	5.00 R 13	FF	175-13/6.95-13
BP	7.50-10	DL	5.20-13	FH	175/70 R 13
BT	9.00-10	DM	5.20 R 13	FJ	185-13
BU	20.5 X 8.0-10	DN	5.50-13	FK	185 R 13
BV	145-10	DP	5.50-13 LT	FL	185-13/7.35-13
BW	145 R 10	DT	5.50 R 13	FM	185/70 R 13
BX	145-10/5.95-10	DU	5.60-13	FN	195-13
BY	4.50-10 LT [2]	DV	5.60-13 LT	FP	195 R 13
B1	5.00-10 LT	DW	5.60 R 13	FT	195/70 R 13
B2	3.00-12	DX	5.90-13	FU	D70-13
B3	4.00-12	DY	5.90-13 LT	FV	B78-13
B4	4.50-12	D1	5.90 R 13	FW	BR78-13
B5	4.80-12	D2	6.00-13	FX	C78-13
B6	5.00-12	D3	6.00-13 LT	FY	7.50-12

[1] The letters "H", "S", and "V" may be included in the tire size designation adjacent to or in place of a dash without affecting the size code for the designation.

[2] As used in this table the letters at the end of the tire size indicate the following: LT—Light Truck, ML—Mining & Logging, MH—Mobile Home, ST—Special Trailer.

TABLE 2. TIRE SIZE CODES—Continued

Tire Size Code	Tire Size Designation [1]	Tire Size Code	Tire Size Designation [1]	Tire Size Code	Tire Size Designation [1]
F1	140 R 12	J3	175 R 14	L5	E78–14
F2	6.5–13	J4	185–14	L6	ER78–14
F3	185/60 R 13	J5	185 R 14	L7	F78–14
F4	A70–13	J6	185/70 R 14	L8	FR78–14
F5	A78–13	J7	195–14	L9	G78–14
F6	CR78–13	J8	195 R 14	MA	GR78–14
F7	2.25–14	J9	195/70 R 14	MB	H78–14
F8	2.75–14	KA	205–14	MC	HR78–14
F9	3.00–14	KB	205 R 14	MD	J78–14
HA	6.70–14 LT	KC	215–14	ME	JR78–14
HB	165–14 LT	KD	215 R 14	MF	205–14 LT
HC	2.50–14	KE	225–14	MH	G80–24.5
HD	5.00–14 LT	KF	225 R 14	MJ	H80–24.5
HE	5.20–14	KH	620 R 14	MK	7–14.5
HF	5.20 R 14	KJ	690 R 14	ML	8–14.5
HH	5.50–14 LT	KK	AR78–13	MM	9–14.5
HJ	5.60–14	KL	195–14 LT	MN	6.60 R 15
HK	5.90–14	KM	185–14 LT	MP	2.00–15
HL	5.90–14 LT	KN	A80–22.5	MT	2.25–15
HM	5.90 R 14	KP	B80–22.5	MU	2.50–15
HN	6.00–14	KT	C80–22.5	MV	3.00–15
HP	6.00–14 LT	KU	D80–22.5	MW	3.25–15
HT	6.40–14	KV	E80–22.5	MX	5.0–15
HU	6.40–14 LT	KW	F60–14	MY	5.20–15
HV	6.45–14	KX	C60–14	M1	5.5–15
HW	6.50–14	KY	J60–14	M2	5.50–15 L
HX	6.50–14 LT	K1	L60–14	M3	5.50–15 LT
HY	6.70–14	K2	F80–22.5	M4	5.60–15
H1	6.95–14	K3	G80–22.5	M5	5.60 R 15
H2	7.00–14	K4	H80–22.5	M6	5.90–15
H3	7.00–14 LT	K5	J80–22.5	M7	5.90–15 LT
H4	7.00 R 14	K6	A80–24.5	M8	6.00–15
H5	7.35–14	K7	B80–24.5	M9	6.00–15 L
H6	7.50–14	K8	BR78–14	NA	6.00–15 LT
H7	7.50–14 LT	K9	D70–14	NB	6.2–15
H8	7.50 R 14	LA	DR70–14	NC	6.40–15
H9	7.75–14	LB	E70–14	ND	6.40–15 LT
JA	7.75–14 ST	LC	ER70–14	NE	6.40 R 15
JB	8.00–14	LD	F70–14	NF	6.50–15
JC	8.25–14	LE	FR70–14	NH	6.50–15 L
JD	8.50–14	LF	G70–14	NJ	6.50–15 LT
JE	8.55–14	LH	GR70–14	NK	6.70–15
JF	8.85–14	LJ	H70–14	NL	6.70–15 LT
JH	9.00–14	LK	HR70–14	NM	6.70 R 15
JJ	9.50–14	LL	J70–14	NN	6.85–15
JK	135–14	LM	JR70–14	NP	6.9–15
JL	135 R 14	LN	L70–14	NT	7.00–15
JM	135–14/5.65–14	LP	LR70–14	NU	7.00–15 L
JN	145–14	LT	C80–24.5	NV	7.00–15 LT
JP	145 R 14	LU	D80–24.5	NW	7.10–15
JT	145–14/5.95–14	LV	E80–24.5	NX	7.10–15 LT
JU	155–14	LW	F80–24.5	NY	7.35–15
JV	155 R 14	LX	G77–14	N1	7.50–15
JW	155–14/6.15–14	LY	B78–14	N2	7.60–15
JX	155/70 R 14	L1	C78–14	N3	7.60 R 15
JY	165–14	L2	CR78–14	N4	7.75–15
J1	165 R 14	L3	D78–14	N5	7.75–15 ST
J2	175–14	L4	DR78–14	N6	8.00–15

TABLE 3. TIRE SIZE CODES—Continued

Tire Size Code	Tire Size Designation [1]	Tire Size Code	Tire Size Designation [1]	Tire Size Code	Tire Size Designation [1]
N7	8.15-15	T9	205/70 R 14	WB	11.00-15
N8	8.20-15	UA	215/70 R 14	WC	2.25-16
N9	8.25-15	UB	H60-15	WD	2.50-16
PA	8.25-15 LT	UC	E60-15	WE	3.00-16
PB	8.45-15	UD	F60-15	WF	3.25-16
PC	8.55-15	UE	FR60-15	WH	3.50-16
PD	8.85-15	UF	G60-15	WJ	5.00-16
PE	8.90-15	UH	GR60-15	WK	5.10-16
PF	9.00-15	UJ	J60-15	WL	5.50-16 LT
PH	9.00-15 LT	UK	L60-15	WM	6.00-16
PJ	9.15-15	UL	4.60-15	WN	6.00-16 LT
PK	10-15	UM	2.75-15	WP	6.50-16
PL	10.00-15	UN	2.50-9	WT	6.50-16 LT
PM	7.50-15 LT	UP	2.50-10	WU	6.70-16
PN	7.00-15 TR	UT	5.00-9	WV	7.00-16
PP	8.25-15 TR	UU	6.7-10	WW	7.00-16 LT
PT	9.00-15 TR	UV	C70-15	WX	7.50-16
PU	7.50-15 TR	UW	D70-15	WY	7.50-16 LT
PV	125-15	UX	DR70-15	W1	8.25-16
PW	125 R 15	UY	E70-15	W2	9.00-16
PX	125-15/5.35-15	U1	ER70-15	W3	10-16
PY	135-15	U2	F70-15	W4	8.25-16 LT
P1	135 R 15	U3	FR70-15	W5	9.00-16 LT
P2	135-15/5.65-15	U4	G70-15	W6	11.00-16
P3	145-15	U5	GR70-15	W7	19-400 C
P4	145 R 15	U6	H70-15	W8	165-400
P5	145-15/5.95-15	U7	HR70-15	W9	235-16
P6	155-15	U8	J70-15	XA	185-16
P7	155 R 15	U9	JR70-15	XB	19-400 LT
P8	155-15/6.35-15	VA	K70-15	XC	G45C-16
P9	165-15	VB	KR70-15	XD	E50C-16
TA	165-15 LT	VC	L70-15	XE	F50C-16
TB	165 R 15	VD	LR70-15	XF	7.00-16 TR
TC	175-15	VE	17-400 TR	XH	7.50-16 TR
TD	175 R 15	VF	185-300 TR	XJ	8.00-16.5
TE	175-15/7.15-15	VH	185-300 LT	XK	8.75-16.5
TF	175/70 R 15	VJ	AR78-15	XL	9.50-16.5
TH	180-15	VK	BR78-15	XM	10-16.5
TJ	185-15	VL	C78-15	XN	12-16.5
TK	185 R 15	VM	D78-15	XP	185 R 16
TL	185/70 R 15	VN	E78-15	XT	4.50-17
TM	195-15	VP	ER78-15	XU	2.00-17
TN	195 R 15	VT	F78-15	XV	2.25-17
TP	205-15	VU	FR78-15	XW	2.50-17
TT	205 R 15	VV	G78-15	XX	2.75-17
TU	215-15	VW	GR78-15	XY	3.00-17
TV	215 R 15	VX	H78-15	X1	3.25-17
TW	225-15	VY	HR78-15	X2	3.50-17
TX	225 R 15	V1	J78-15	X3	6.50-17
TY	235-15	V2	JR78-15	X4	6.50-17 LT
T1	235 R 15	V3	L78-15	X5	7.00-17
T2	J80-24.5	V4	LR78-15	X6	7.50-17
T3	ER60-15	V5	N78-15	X7	8.25-17
T4	D78-13	V6	17-15 (17-380 LT)	X8	7.50-17 LT
T5	A78-15	V7	17-400 LT	X9	225/70 R 14
T6	DR70-13	V8	11-15	YA	G50C-17
T7	HR60-15	V9	11-16	YB	H50C-17
T8	E60-14	WA	L84-15	YC	195/70 R 15

TABLE 3. TIRE SIZE CODES—Continued

Tire Size Code	Tire Size Designation [1]	Tire Size Code	Tire Size Designation [1]	Tire Size Code	Tire Size Designation [1]
YD	4.20-18	2F	9.00-20	4J	13.5-24.5
YE	8-17.5 LT	2H	9.4-20	4K	7.00-20 ML
YF	11-17.5	2J	10.00-20	4L	7.50-20 ML
YH	7-17.5	2K	10.3-20	4M	8.25-20 ML
YJ	8-17.5	2L	11.00-20	4N	9.00-20 ML
YK	8.5-17.5	2M	11.1-20	4P	10.00-20 ML
YL	9.5-17.5	2N	11.50-20	4T	10.00-22 ML
YM	10-17.5	2P	11.9-20	4U	10.00-24 ML
YN	14-17.5	2T	12.00-20	4V	11.00-20 ML
YP	9-17.5	2U	12.5-20	4W	11.00-22 ML
YT	205/70 R 15	2V	13.00-20	4X	11.00-24 ML
YU	2.25-18	2W	14.00-20	4Y	11.00-25 ML
YV	2.50-18	2X	6.50-20 LT	41	12.00-20 ML
YW	2.75-18	2Y	7.00-20 LT	42	12.00-21 ML
YX	3.00-18	21	13/80-20	43	12.00-24 ML
YY	3.25-18	22	14/80-20	44	12.00-25 ML
Y1	3.50-18	23	2.75-21	45	13.00-20 ML
Y2	4.00-18	24	3.00-21	46	13.00-24 ML
Y3	4.50-18	25	2.50-21	47	13.00-25 ML
Y4	6.00-18	26	2.75-20	48	14.00-20 ML
Y5	7.00-18	27	10.00-22	49	14.00-21 ML
Y6	7.50-18	28	11.00-22	5A	14.00-24 ML
Y7	8.25-18	29	11.1-22	5B	14.00-25 ML
Y8	9.00-18	3A	11.9-22	5C	10.3-20 ML
Y9	10.00-18	3B	12.00-22	5D	11.1-20 ML
1A	11.00-18	3C	14.00-22	5E	12.5-20 ML
1B	6.00-18 LT	3D	11.50-22	5F	9-22.5 ML
1C	6.00-20 LT	3E	4.10-18	5H	9.4-22.5 ML
1D	L50C-18	3F	4.10-19	5J	10-22.5 ML
1E	7.00-18 LT	3H	7-22.5	5K	10.3-22.5 ML
1F	12-19.5	3J	8-22.5	5L	11-22.5 ML
1H	2.00-19	3K	8.5-22.5	5M	11-24.5 ML
1J	2.25-19	3L	9-22.5	5N	14-17.5 ML
1K	2.50-19	3M	9.4-22.5	5P	15-19.5 ML
1L	2.75-19	3N	10-22.5	5T	15-22.5 ML
1M	3.00-19	3P	10.3-22.5	5U	16.5-19.5 ML
1N	3.25-19	3T	11-22.5	5V	16.5-22.5 ML
1P	3.50-19	3U	11.1-22.5	5W	18-19.5 ML
1T	4.00-19	3V	11.5-22.5	5X	18-22.5 ML
1U	11.00-19	3W	11.9-22.5	5Y	19.5-19.5 ML
1V	9.5-19.5	3X	12-22.5	51	23-23.5 ML
1W	10-19.5	3Y	12.5-22.5	52	18-21 ML
1X	11-19.5	31	15-22.5	53	19.5-21 ML
1Y	7-19.5	32	16.5-22.5	54	23-21 ML
11	7.5-19.5	33	18-22.5	55	6.00-13 ST
12	8-19.5	34	215/70 R 15	56	7.35-14 ST
13	9-19.5	35	225/70 R 15	57	8.25-14 ST
14	14-19.5	36	185/60 R 13	58	7.35-15 ST
15	15-19.5	38	9.00-24	59	8.25-15 ST
16	16.5-19.5	38	10.00-24	6A	12.00-22 ML
17	18-19.5	39	11.00-24	6B	4.30-18
18	19.5-19.5	4A	12.00-24	6C	3.60-19
19	6.00-20	4B	14.00-24	6D	3.00-20
2A	6.50-20	4C	3.50-7	6E	4.25-18
2B	7.00-20	4D	3.00-4	6F	MP90-18
2C	7.50-20	4E	12.5-24.5	6H	3.75-19
2D	8.25-20	4F	11-24.5	6J	MM90-19
2E	8.5-20	4H	12-24.5	6K	3.25-7

TABLE 3. TIRE SIZE CODES—Continued

Tire Size Code	Tire Size Designation [1]	Tire Size Code	Tire Size Designation [1]	Tire Size Code	Tire Size Designation [1]
6L	2.75-16	8N	2-22½	0T	Not Assigned
6M	4.00-16	8P	2¼-15	0U	BR60-13
6N	7.9	8T	2¼-16	0V	15.00-20
6P	25X 7.50-15	8U	2¼-17	0W	16.00-20
6T	27X 8.50-15	8V	2¼-18	0X	12/80-20
6U	27X 9.50-15	8W	2¼-19	0Y	14/80-24
6V	29X 12.00-15	8X	2¼-19 R	01	15.5/80-20
6W	31X 13.50-15	8Y	2¼-20	02	13-22.5
6X	31X 15.50-15	81	2½-8	03	21-22.5
6Y	C70-14	82	2½-9	04	9/70-22.5
61	Not Assigned	83	2½-16	05	10/70-22.5
62	Not Assigned	84	2½-17	06	11/70-22.5
63	Not Assigned	85	2½-18	07	12/70-22.5
64	Not Assigned	86	2½-19	08	13/70-22.5
65	Not Assigned	87	2½-19 R	09	7.25/75-17.5
66	3.40-5	88	2¾-9	10	8.00/75-17.5
67	4.10-4	89	2¾-16	20	8.75/75-17.5
68	4.10-5	9A	2¾-17	30	9.50/75-17.5
69	175-14 LT	9B	2¾-17 R	40	7.25/75-16.5
7A	11-14	9C	3-10	50	8.00/75-16.5
7B	E78-14 LT	9D	3-12	60	8.75/75-16.5
7C	G78-15 LT	9E	21 x 4	70	9.50/75-16.5
7D	H78-15 LT	9F	22 x 4½	80	6.70-14 C
7E	180 R 15	9H	15.50-20	90	7-17.5 C
7F	185-16 LT	9J	18.50-20	RA	125-12 C
7H	205-16 LT	9K	19.50-20	RB	125-13 C
7J	215-16 LT	9L	2¼-14	RC	125-14 C
7K	F78-16 LT	9M	2½-20	RD	125-15 C
7L	H78-16 LT	9N	2¾-16 R	RE	135-12 C
7M	L78-16 LT	9P	2¾-18	RF	135-13 C
7N	135 R 10	9T	10-20	RH	135-14 C
7P	6.95-14 LT	9U	11-24	RJ	135-15 C
7T	7-14.5 MH	9V	11.25-24	RK	145-10 C
7U	8-14.5 MH	9W	15 x 4½-8	RL	145-12 C
7V	9-14.5 MH	9X	14.75/80-20	RM	145-13 C
7W	4.25/85-18	9Y	23 x 5	RN	145-14 C
7X	A78-14	91	25 x 6	RP	145-15 C
7Y	7.50-18 MPT	92	15 x 4½-8	RT	155-12 C
71	10.5-18 MPT	93	18 x 7-8	RU	155-13 C
72	12.5-18 MPT	94	21 x 8-9	RV	155-14 C
73	12.5-20 MPT	95	23 x 9-10	RW	155-15 C
74	14.5-20 MPT	96	27 x 10-12	RX	A60-13
75	10.5-20 MPT	97	2.00-15 TR	RY	C60-15
76	10.5-20	98	2.50-15 TR	R1	155-16 C
77	8.25-10	99	3.00-15 TR	R2	165-13 C
78	150 R 12	0A	GR60-14	R3	165-16 C
79	150 R 14	0B	560 x 165-11	R4	175-13 C
8A	1¾-19	0C	680 x 180-15	R5	175-15 C
8B	1¾-19¾	0D	8.55-15 ST	R6	175-16 C
8C	2-12	0E	3.50-14	R7	185-13 C
8D	2-16	0F	3.25-14	R8	185-15 C
8E	2-17	0H	3.50-15	R9	195-15 C
8F	2-17 R	0J	AR70-13	A0	195-16 C
8H	2-18	0K	B60-13	B0	205-15 C
8J	2-19	0L	245/60 R 14	C0	215-14 C
8K	2-19 R	0M	255/60 R 15	D0	215-15 C
8L	2-19¾	0N	2¾-15	E0	225-14 C
8M	2-22	0P	2.50-20	F0	225-15 C

TABLE 3. TIRE SIZE CODES—Continued

Tire Size Code	Tire Size Designation [1]	Tire Size Code	Tire Size Designation [1]	Tire Size Code	Tire Size Designation [1]
H0	225–16 C	BR	LR60–15	VR	13/80–24
J0	235–14 C	CR	ER60–15	WR	175–16 C
K0	235–15 C	DR	D60–13	XR	195–16 C
L0	235–16 C	ER	C60–13	YR	BR70–13
M0	21–400 C	FR	D60–14	1R	185–15 LT
N0	3.50–20	HR	175/70 R 14	2R	13–22.5 ML
P0	3.75–15	JR	MN90–18	3R	MR70–15
T0	3.60–18	KR	MR90–18	4R	E60–26.5
U0	3.00–10 C	LR	4.25–19	5R	6.7–12
V0	4.00–10 C	MR	230–15	6R	5.4–14
W0	4.00–8 C	NR	5.4–10	7R	7.4–14
X0	4.50–8 C	PR	ER60–13	8R	5.4–16
Y0	265/60 R 14	TR	FR60–14	9R	4.60–18
AR	215/60 R 15	UR	C60C–15		

**36 F.R. 7539
April 21, 1971**

PREAMBLE TO PART 575—CONSUMER INFORMATION

Action on Petitions for Reconsideration—Amendment

Regulations requiring manufacturers of passenger cars and motorcycles to provide information on vehicle stopping distance (§ 375.101), tire reserve load (§ 375.102), and acceleration and passing ability (§ 375.106) were issued by the Federal Highway Administrator and published in the *Federal Register* on January 25, 1969 (34 F.R. 1246). Several petitions for reconsideration of these regulations were received. In response to these petitions, and in order to clarify and simplify the requirements and the information to be provided to purchasers, these regulations are hereby amended and reissued in the form set forth below.

§ 375.101 *Vehicle stopping distance.* This section required that manufacturers state the tire size, type and size of brakes, method of brake actuation and auxiliary brake equipment, and maximum loaded and lightly loaded vehicle weights. The effect of stating these requirements was to greatly restrict the grouping of vehicles and options that was permitted for the purposes of furnishing information. It has been determined that in order to reduce the required number of different information documents, manufacturers should be permitted to group vehicles at their discretion, as long as each vehicle in the group can meet or exceed the performance levels indicated, and the vehicles in each group are identified in the terms by which they are normally described to the public. The requirement for specific descriptive information is therefore deleted.

Since the information must be valid for all vehicles in the group to which it applies, the requirement that it refer to the smallest tire size offered has been found unnecessary, and deleted. It has also been determined that variations in stopping distances between different vehicles at 30 mph are not as meaningful for comparison purposes as those at 60 mph, and therefore information is required only for the latter speed.

It should be noted that the regulations establish the conditions under which the performance level represented by the information provided can be met or exceeded by every vehicle to which the information applies. They do not establish the procedures by which manufacturers should generate the information, although those procedures are to be inferred from the regulations. For example, both sections contain the condition that wind velocity is zero. This does not mean that manufacturers' tests must be conducted under still air conditions; it means that the performance level established must be attainable by all vehicles in the group under those conditions. One obvious method of satisfying the condition from the manufacturer's standpoint is to conduct verification tests under adverse wind conditions (tailwind for braking, headwind for acceleration). As another example, the condition that ambient temperature be between 32°F and 100°F means that the information presented must be attainable by all vehicles in the group at all temperatures within that range (when other conditions are as stated).

The amended section requires that stopping distances be those attainable without lock-up on any wheel. This condition is the most meaningful from a safety standpoint, since steering control tends to be lost when wheels are locked. Several petitioners submitted data showing minimal differences in maximum and lightly loaded vehicle weight stopping distances to support their request for substitution of a single test weight. Their results, however, were apparently derived from tests conducted with locked wheels, under which conditions stopping distance becomes a function largely of vehicle velocity and the friction coefficient between the tire and the road, and has no relationship to vehicle weight. It is believed that the condition of no wheel lock-up will result in data showing meaningful differences in stopping distances test weights. Accordingly, the requirement of information covering these two vehicle weight conditions is retained, and petitions on this point are denied.

The section as issued required performance information for a partially failed service brake subsystem ("emergency brake system") only at maximum loaded vehicle weight. It has been determined that in some cases the most adverse condition may occur at lighter loads. The amended rule therefore requires information for "the most adverse combination of maximum or lightly loaded vehicle weight and complete loss of braking in one or the other of the vehicle brake subsystems."

Several petitioners suggested that information be limited to one test weight, instead of requiring it for both lightly loaded and maximum loaded vehicle weight. It has been determined, however, that information on both conditions may reveal vehicles having superior brake balance, and the advantage of anti-skid or load proportioning devices, and also aid purchasers who travel mainly in one or the other of the loading conditions. The petitions to that effect are therefore denied.

§ 375.102 *Tire reverse load.* The section required that manufacturers state the number of passengers and the cargo and luggage weight for two different loading conditions, and the actual vehicle weight within a range of no more than 100 pounds under those conditions. These requirements restricted the grouping of vehicles and options that was permitted for the purposes of furnishing information. It has been determined that in order to reduce the required number of different information documents, manufacturers should be permitted to group vehicles by recommended tire size designations regardless of weight. as long as the reserve load figure is met or exceeded by every vehicle in the group. The requirements for providing weight and loading information are therefore deleted.

Section 375.102 as issued required that reverse load figures be provided for the vehicle at normal vehicle weight (2 or 3 persons and no luggage)as well as maximum loaded vehicle weight. It also required the furnishing of a "tire over-

load percentage", the percentage difference between the load rating of a tire at recommended inflation pressures for normal vehicle weight and the load on the tire at maximum loaded vehicle weight. Several petitions suggested that the providing of these various percentage figures would tend to confuse persons to whom the information is furnished, and therefore decrease its usefulness to the consumer. Representatives of consumer groups have also suggested, in earlier proceedings concerning the consumer information regulations, that for maximum usability the information should be as simple and clear as possible. In light of these considerations, it has been determined that the tire reserve load figure provided should be limited to a single percentage for each recommended tire size designation, at maximum loaded vehicle weight and the manufacturer's recommended inflation pressure. The requirements for tire reserve load at normal vehicle weight and for tire overload percentage accordingly are deleted.

Two further changes in the calculation methods have been made for simplicity and clarity. Instead of using the actual load on each wheel as the basis for calculation, the wheel load figure is changed to one-half of each axle's share of the maximum loaded vehicle weight. This reflects the method used in Standard No. 110 for determining the vehicle maximum load on the tire. Also. the denominator of the fraction representing the tire reserve load percentage is changed from the load on the wheel to the load rating of the tire. A tire with a load rating of 1500 pounds, for example, used with a wheel load of 900 pounds, would have a reserve load percentage of 40% (600/1500 × 100) rather than 66⅔% (600/900 × 100). The former figure has been determined to be somewhat more meaningful in cases of large reserve loads.

§ 375.106 *Acceleration and passing ability.* The section as issued required that times be provided for acceleration from 20 to 35 mph and from 50 to 80 mph, and times and distances for prescribed passing maneuvers involving two lane changes. On the basis of petitions submitted, and further consideration of the need for simplicity and clarity in the information presented, it has been determined that the most useful information would be in the form of passing distances and times for a simple straight-line passing maneuver at low and high speeds. In order to eliminate the difficulties of conducting a uniform passing maneuver involving a long pace vehicle and a limiting of the passing speed precisely to a specified level, the information required is to be derived on the basis of a time-distance plot of vehicle performance at maximum acceleration from 20 to 35 and 50 to 80 miles per hour.

For reasons discussed above in regard to section 375.101, the requirement of providing the weight of the vehicle is deleted from this section.

Because the amended section does not require information relating to an actual passing maneuver, but only that based on two straight-line acceleration maneuvers with a simple graphic computation, the exception of manufacturers of 500 or fewer vehicles annually from certain of the requirements is removed from this section.

Several petitioners contended that the requirement that information be provided under the condition of full-power operation of a vehicle air conditioner would lead to variable, non-repeatable results. This may be true of the results achieved in manufacturers' tests. The information presented is not, however, to be simply the results of manufacturers' tests, but rather a minimum level of performance that can be met or exceeded by every vehicle to which the information applies. Manufacturers are free, therefore, to adjust the data to account for any variation in results that might be encountered. The degradation of acceleration ability by the use of an air conditioner may be significant in some cases, and therefore it is important from the standpoint of safety that it be reflected in the information provided. The petitions to the contrary are accordingly denied.

Some petitioners objected to the required use of a correction factor to ambient conditions in accordance with SAE Standard J816a, pointing out that the factor was designed to be applicable exclusively to engine dynamometer testing and not to road testing of vehicles. The contention has been found to have merit. In the section as amended, ranges of ambient conditions of temperature, dry barometric pressure, and relative humidity are provided, and the information is required to be valid at all points within those ranges.

In addition to the above, a new paragraph (c), containing specific definitions, is added to section 375.2, Definitions.

In order to allow adequate time for manufacturers to prepare the information, the three sections are effective for vehicles manufactured on or after January 1, 1970.

In consideration of the above, 49 CFR §§ 375.101, 375.102, and 375.106 are amended, and a new paragraph (c) is added to § 375.2, to read as set forth below. This notice of action on petitions for reconsideration is issued under the authority of sections 112 and 119 of the National Traffic and Motor Vehicle Safety Act (15 U.S.C. 1401, 1407) and the delegation of authority by the Secretary of Transportation to the Federal Highway Administrator, 49 CFR 1.4(c).

Issued: May 19, 1969.

F. C. Turner
Federal Highway Administrator

SUBPART A—GENERAL

Sec.

375.1 Scope.
375.2 Definitions.
375.3 Matter Incorporated by reference.
375.4 Applicability.
375.5 Separability.
375.6 Requirements.

SUBPART B—CONSUMER INFORMATION ITEMS

375.101 Vehicle Stopping Distance.
375.102 Tire reserve load.
375.103 Reserved.
375.104 Reserved.
375.105 Reserved.
365.106 Acceleration and passing ability.

May 23, 1969
34 F.R. 8112

PREAMBLE TO AMENDMENT TO PART 575—CONSUMER INFORMATION

Amended regulations concerning the furnishing of consumer information for motor vehicles, 49 CFR §§ 375.101, 102, 106, were published in the *Federal Register* of May 23, 1969 (34 F.R. 8112). Sections 375.101, *Vehicle Stopping Distance*, and 375.106, *Acceleration and Passing Ability*, in subsections (d)(7) and (d)(1)(vii) respectively, specified that the information provided shall be valid for road surfaces with a skid number of 70, as measured in accordance with American Society for Testing and Materials Method E-274 at 40 miles per hour, omitting water delivery as specified in paragraph 7.1 of that Method.

Several petitions for reconsideration have been received, requesting that the skid number condition be set at higher level because there are only a limited number of test tracks presently with surfaces of that low a skid number. It is recognized that the level of 70 may be somewhat lower than many existing test track and road surfaces. It has been determined, in light of the petitions received, that the skid number condition can be set at a somewhat higher level without detracting from the value of the information provided or the enforceability of the regulations. Accordingly, the figure "70" in sections 375.101(d)(7) and 375.106(d)(1)(vii) is hereby changed to "75".

One petitioner requested a delay in the effective date of the regulation because of difficulties in obtaining equipment for the measurement of skid number. In light of the relaxation of the skid number requirement embodied in this notice, and the possibility of temporarily leasing either measuring equipment or test facilities, evidenced by fact that only one such request was received, the request for a delay in effective date is denied.

Since this amendment relaxes a requirement and imposes no additional burden on any person, notice and opportunity for comment thereon are unnecessary and the amendment is incorporated into the above-referenced regulations without change in the effective date. This notice of amendment in response to petitioners for reconsideration is issued under the authority of sections 112 and 119 of the National Traffic and Motor Vehicle Safety Act (15 U.S.C. 1402, 1407) and the delegation of authority by the Secretary of Transportation to the Federal Highway Administrator, 49 CFR § 1.4(c).

Issued on July 14, 1969.

F. C. Turner
Federal Highway Administrator

34 F.R. 11974
July 16, 1969

PREAMBLE TO AMENDMENT TO PART 575—CONSUMER INFORMATION

Regulations requiring manufacturers of motor vehicles to provide information to consumers concerning performance characteristics of their vehicles were published on January 25, 1969 (34 F.R. 1246), and amended on May 23, 1969 (34 F.R. 8112). By notice of July 11, 1969 (34 F.R. 11501) it was proposed that the regulations be amended to require manufacturers to provide the information to prospective purchasers, as well as those who have already bought a vehicle, and also to provide the information to the Administrator 30 days before the information is required to be provided to purchasers.

No general objections to the proposed amendment were received. One manufacturer objected to the requirement of providing copies to the Administrator 30 days in advance, on the basis that this did not allow sufficient lead time from the date of the proposal. In light of the fact that the information required to be provided consists only of performance figures that the manufacturer is certain can be exceeded by its vehicles, that the information must be provided in large quantities to dealers by January 1, 1970, and that no other manufacturers evidenced difficulty in meeting the December 1 date, the objection is found not to be meritorious.

The Automobile Manufacturers Association made two suggestions for changes to the regulation, both of which have been accepted and incorporated into the regulation. One change adds language to make it clear that the locations at which the information is to be provided are outlets with which the manufacturer has some legal connection. The other is that the date on which information relating to newly introduced vehicles is required is the "announcement date", on which dealers are authorized to display and sell the vehicles.

The proposal stated that three copies should be submitted to the Administrator by December 1, 1969. It has been determined that in light of the need for immediate processing and the large amount of information that will be received at that time, a somewhat larger number of copies will be needed. The number of copies has been changed, accordingly, from three to ten. Since the additional burden on automotive manufacturers of providing these copies appears to be insubstantial, a further notice of proposed rulemaking is found to be unnecessary. Other minor changes in wording are made for clarity.

Effective Dates: Subsections (a) and (b) of § 375.6, Requirements, are effective January 1, 1970. Subsection (c) of that section is effective December 1, 1969.

In light of the foregoing, Subpart A—General, of 49 CFR Part 375 is amended to read as set forth below. This amendment is issued under the authority of sections 112 and 119 of the National Traffic and Motor Vehicle Safety Act (15 U.S.C. 1401, 1407), and the delegation of authority from the Secretary of Transportation to the Federal Highway Administration, 49 CFR § 1.4(c).

Issued on October 16, 1969.

E. H. Holmes, Acting
Federal Highway Administrator

34 F.R. 17108
October 22, 1969

PREAMBLE TO AMENDMENT TO PART 575—CONSUMER INFORMATION

Motorcycle Brake Burnishing Requirement

On May 23, 1969, the Federal Highway Administration published 49 CFR § 375.101, Vehicle Stopping Distance, of the Consumer Information Regulations (34 F.R. 8112). Paragraph (e)-(1)(ii) of that section, describing the burnishing procedures for motorcycles, is as follows: "Same as for passenger cars, except substitute 30 m.p.h. for 40 m.p.h. and 150° F. for 250° F., and maintain hand lever force to foot lever force ratio of approximately 1 to 2."

A manufacturer has stated that such a burnishing procedure, which was drawn from a draft SAE Recommended Practice, would be inappropriate for its vehicles, and suggests that the required burnishing procedures should be that recommended by the manufacturer. Since it appears that a uniform burnishing procedure suitable for all motorcycles has not yet been developed, the suggestion is found to have merit, to the extent that manufacturers have recommended such procedures. A general burnishing procedure must still be specified, however, for the purpose of determining compliance of those vehicles for which the manufacturers have not made a procedure publicly available. Accordingly, subparagraph (e)(1)(ii) of section 375.101 is hereby amended to read as follows:

"*Motorcycles.* Adjust and burnish brakes in accordance with manufacturer's recommendations. Where no burnishing procedures have been recommended by the manufacturer, follow the procedure specified above for passenger cars, except substitute 30 m.p.h. for 40 m.p.h. and 150° F. and 250° F., and maintain hand lever force to foot lever force ratio of approximately 1 to 2."

The Consumer Information regulations require manufacturers to submit information to the FHWA by December 2, 1969, and it is important, therefore, that this amendment to the regulations be made effective without delay. The regulations require only that the manufacturers submit information to purchasers (and to the FHWA) as to performance levels that can be met or exceeded by their vehicles, and it is not necessary that vehicles be retested as long as they perform as well under the manufacturers' own burnishing procedures as under the previously specified ones. Manufacturers are, of course, free to provide new performance figures at any time, under the procedures specified in Part 375. If in a particular case a manufacturer determines that its vehicles may not be able to meet the performance figures provided when its own recommended burnishing procedures are utilized, and is not able to provide new and appropriate figures within the time specified, it should include a notation to that effect at the time that the figures are first provided to the FHWA. The vehicles in question will not be considered to be in violation of the regulations if they meet the performance figures provided under the previously specified burnishing procedures, and if new and corrected figures are provided under section 375.101, as amended, not later than September 1, 1970.

Because of the importance of providing to consumers by January 1, 1970, the probability that few if any manufacturers will be adversely affected by the amendment, and the provisions for relief included herein, notice and public procedure thereon are found to be impracticable, unnecessary, and contrary to the public interest, and the amendment described above is made effective on publication in the *Federal Register.*

This amendment is issued under the authority of sections 112 and 119 of the National Traffic and Motor Vehicle Safety Act of 1966 (15 U.S.C. 1401, 1407), and the delegation of authority from the Secretary of Transportation to the Federal Highway Administrator, 49 CFR § 1.4(c).

Issued on November 24, 1969.

F. C. Turner
Federal Highway Administrator

34 F.R. 18865
November 26, 1969

PREAMBLE TO AMENDMENT TO PART 575—CONSUMER INFORMATION

(Availability Requirements)

The purpose of this notice is to amend section 575.6 of the Consumer Information Regulations (49 CFR Part 575) to require that the information supplied pursuant to Subpart B of the Regulations be provided in sufficient quantity to permit retention by prospective customers or mailing to them upon request. A notice of proposed rulemaking was published on January 14, 1971 (36 F.R. 557), proposing to carry out the legislative mandate of P.L. 91-625 (84 Stat. 262). That legislation was designed to remedy difficulties resulting from the current practice of making consumer information available only in the showroom, by permitting the Secretary to require that the information be provided in a printed format which could be retained by customers who visit the showroom or mailed to others upon their request.

A limited number of comments were received in response to the Notice, some of which merely expressed support for the additional requirement. The Chrysler Cororation requested that the amendment be clarified to provide that temporary unavailability would not constitute a failure to comply with the regulations. As is noted in the Notice of proposed rulemaking, the uncertainty of demand makes it difficult to establish precise standards as to what is "sufficient." It has been determined, therefore, that any further specification of this provision would be inappropriate at this time. It is intended that manufacturers and dealers will cooperate to take all reasonable steps to ensure that a continuous supply of the information is available.

The Chrysler Corporation further requested that the regulation clearly indicate that a reasonable charge can be made for the materials. The legislative history of P.L. 91-625 indicates that a major purpose of the amendment was to make consumer information more easily available to consumers in making their purchase. A charge for consumer information on several makes and models of vehicles could present the car shopper with as great an obstacle to availability of information as is the case with the present system. In view of this purpose and the general aim of the consumer information regulations to provide for as wide a dissemination of information as possible, it has been determined that the retention copies should be provided without charge.

In consideration of the above, 49 CFR 576.6(b) is amended. . . .

Effective date: January 1, 1972.

Issued on September 28, 1971.

Douglas W. Toms
Administrator

36 F.R. 19310
October 2, 1971

PART 575—PRE 11-12

PREAMBLE TO AMENDMENT TO PART 575—CONSUMER INFORMATION

(Truck-Camper Loading)

(Docket No. 71-7; Notice 5)

This notice reissues the portion of 49 CFR § 571.126, Motor Vehicle Safety Standard No. 126, *Trucker-Camper Loading*, that was previously applicable to truck manufacturers as a consumer information regulation, 49 CFR § 575.103, *Truck-Camper Loading*. It also responds to petitions for reconsideration of Standard No. 126 on issues that are not addressed in Notice 4, which is published in this issue (37 F.R. 26605).

Petitions for reconsideration of Standard No. 126 (37 F.R. 16497) were filed by Chrysler Corporation (Chrysler), Ford Motor Company (Ford), General Motors Corporation (GM), Jeep Corporation (Jeep), Motor Vehicle Manufacturers Association (MVMA) Recreational Vehicle Institute, Inc. (RVI) and Toyota Motor Sales USA, Inc. (Toyota).

In response to information contained in some of the petitions, the portions of the standard previously applicable to truck manufacturers are being reissued under this notice as a consumer information regulation for the reasons stated in Notice 4. Minor amendments are also made to the regulation on the basis of some of the petitions while the Administrator has declined to grant requested relief from other requirements of the regulation.

1. *Effective date.* GM has petitioned for a delayed effective date. As a truck manufacturer, GM feels that additional lead time is required "to develop, process, and print the necessary information on an orderly basis." The Administration has found for good cause shown that an effective date earlier than 180 days after issuance of Standard No. 126 was in the public interest; however, to allow truck manufacturers sufficient time for testing to determine cargo

center of gravity locations the effective date of the requirements applicable to truck manufacturers is being extended 2 months, until March 1, 1973.

2. *Definitions and information.* As discussed in Notice 4 Ford objected to the definition of "cargo weight rating" and the term "total load". Standard No. 126 has been amended to meet Ford's objections, and similar changes are made in the terminology of the new truck consumer information regulation.

Ford also suggests that the phrase "any additional weight carried in or on the camper" should be substituted for "the weight of camper cargo, and the weight of passengers in the camper" in paragraph S5.2.1(d) of Standard No. 126, now § 575.103(e)(3). It believes the suggested language would be more meaningful to the average user and that the present language could be construed as endorsing the carrying of passengers in campers. Ford's request is denied. The NHTSA considers that the specificity of references to cargo and passengers is more meaningful to consumers than the general reference to "any additional weight". Further, given the prevalence of carrying passengers in campers, the NHTSA does not believe that the present language can realistically be considered to have a significant effect on this practice.

Both Ford and GM objected to the paragraph requiring the manufacturer to furnish trailer towing recommendations, on the grounds of vagueness and lack of prior notice and opportunity to comment. The NHTSA concurs, and is deleting this requirement.

Ford suggests that paragraph S5.2.1(a) of Standard No. 126 (now § 575.103(e)(1) should be revised to make clear that the slide-in camper

PART 575—PRE 13

also has a center of gravity designation determined in accordance with the regulation, which falls within the boundaries specified by the vehicle manufacturer. Since campers manufactured before the effective date of the regulation may be mounted on trucks manufactured after March 1, 1973, Ford's suggestion has not been adopted.

GM has petitioned that a warning be required to accompany the regulation's information, stating that the longitudinal center of gravity is only one of the many factors affecting the overall performance of a vehicle and that other factors concerning vehicle handling should be considered by the operator. The NHTSA denies GM's petition on this point. Proper loading and load distribution in truck-camper combinations is a highly significant handling factor, and such a warning might cause a truck operator to feel the loading information presented is of little significance. The regulation does not, however, prohibit GM or other manufacturers from furnishing such additional warnings if they see fit.

GM has also asked for a confirmation of its assumption that "the pictorial representation of the recommended longitudinal center of gravity zone for the cargo weight rating need not be to scale but can be generalized so long as the longitudinal boundaries of the zone are clearly set forth." The NHSTA agrees with this interpretation.

Effective Date: March 1, 1973.

In consideration of the foregoing, 49 CFR Part 575 is amended by adding a new § 575.103, *Truck-camper Loading. . . .*

This notice is issued pursuant to the authority of sections 112 and 119 of the National Traffic and Motor Vehicle Safety Act of 1966 (15 USC 1401, 1407) and the delegation of authority at 49 CFR 1.51.

Issued on December 6, 1972.

Douglas W. Toms
Administrator

37 F.R. 26607
December 14, 1972

PREAMBLE TO AMENDMENT TO PART 575—CONSUMER INFORMATION

Truck-Camper Loading

(Docket No. 71–7; Notice 6)

This notice responds to petitions for reconsideration of 49 CFR § 575.103, *Truck-camper loading*, with amendments extending the effective date to April 1, 1973, and allowing optional wording of certain statements until October 1, 1973.

On December 14, 1972, Part 575 of Title 49, Code of Federal Regulations, was amended by adding § 575.103 *Truck-camper loading* (37 F.R. 26607). The amendment was in essence that portion of Federal Motor Vehicle Safety Standard No. 126, *Truck-camper loading* that applied to manufacturers of trucks accommodating slide-in campers, as originally published on August 15, 1972 (37 F.R. 16497). Pursuant to 49 CFR § 553.35, petitions for reconsideration of § 575.103 have been filed by General Motors Corporation and International Harvester Company. Ford Motor Company has asked for a clarification.

In response to information contained in these petitions the regulation is being amended in certain respects, and a new effective date of April 1, 1973 adopted. Requested changes in other requirements of the regulation are denied.

1. Effective date: Both petitioners request delay of the effective date of the regulation for at least 60 days, until May 1, 1973 at the earliest. One reason for the request is that petitioners had printed their manuals on the basis of the notice of August 15, 1972, and that the additional time is needed to print new materials conforming to modified texts published on December 14, 1972. General Motors also states that the additional time is needed to prepare and disseminate data in a manner meeting the requirement that it be available to prospective purchasers. While data has been prepared for each truck, it has not yet been consolidated into a single sheet or pamphlet

suitable for showroom display and availability. The requests of both petitioners reflect the probability that the material will not be submitted to the Administrator at least 30 days before it is available to prospective purchasers, as required by § 575.6(c), and the possibility that the data will not be ready by March 1, 1973.

The NHTSA has determined that good cause has been shown for postponement of the effective date until April 1, 1973. This agency recognizes, however, that the minor textual changes made in the December notice create problems of conformity for those manufacturers who in good faith relied on the August notice in ordering materials. Accordingly, the regulation is being amended to allow the earlier wording on an optional basis until October 1, 1973. These amendments permit use of the phrase "total load" instead of "total cargo load" in paragraph (e)(3) where it twice appears, and the legend "Aft End of Cargo Area" for "Rear End of Truck Bed" in Figure 1, Truck Loading Information. The word "rating" appearing on the last line of paragraph (e)(5) is properly "ratings" as printed in the August notice, and a correction is made. Further, the NHTSA considers it important that a manufacturer fulfill the requirements of § 575.6(b) by making information available to prospective purchasers when trucks manufactured on or after April 1, 1973 are placed on sale. Considering the short lead time between December 14, 1972 and February 1, 1973 and the intervening holidays, the NHTSA will not take enforcement action with respect to the furnishing of information under §§ 575.103 and 575.6(c) prior to April 1, 1973, if manufacturers provide information to this agency as required by those sections not later than the date by which the information must be provided to prospective purchasers.

2. *Administrative Procedure Act.* Harvester believes that the Administrative Procedure Act was violated in that interested persons were not provided an opportunity to comment upon providing information under Part 575 prior to enactment of § 575.103. The NHTSA views Harvester's comment as a narrow construction of the requirements of the Act, and disagrees with petitioner's conclusion. The content of § 575.103 was proposed on April 9, 1971 (36 F.R. 6837) and adopted as a safety standard on August 15, 1972 (37 F.R. 16497). Pursuant to petitions for reconsideration from Chrysler Corporation, Ford Motor Company, General Motors, Jeep Corporation, and Motor Vehicle Manufacturers' Association that Standard No. 126 would be more appropriate as a consumer information regulation, the NHTSA adopted § 575.103 on December 14, 1972 with content virtually identical to that issued in the previous August. Thus the agency considers it has met 5 USC § 553 by providing notice of the terms and substance of the rule, and an opportunity to comment. It is true that notice was not provided on the specific issue that distinguishes the consumer information regulation from a motor vehicle safety standard (*i.e.,* availability of information to a prospective purchaser and the agency at specified time periods), but the NHTSA considers this issue a minor one in relation to the regulation as a whole for which adequate notice was given. In view of the weight of comment that the standard should properly be a consumer information regulation, no further notice was deemed necessary. The NHTSA has already in this notice indicated its willingness to liberally interpret § 575.6(c) because of the time factor involved.

3. *Clarification.* Ford Motor Company has asked for a clarification of the term "weight of occupants" used to compute "cargo weight rating", as defined by the regulation. Specifically, Ford inquires whether the weight is that of a 95th percentile male—that of an "occupant" as defined by § 571.3(b)—or that of a person weighing 150 pounds, the figure applicable to other consumer information regulations and used in the safety standards.

The NHTSA intended "weight of occupants" to be the "normal occupant weight" figure of 150 pounds specified in Motor Vehicle Safety Standard No. 110 rather than that of a 95th percentile male, which is greater. To clarify this, the phrase, "computed as 150 pounds times the number of designated seating positions," is added to the regulation.

In consideration of the foregoing, 49 CFR § 575.103, *Truck-camper loading,* is amended . . . .

Effective date: April 1, 1973.

(Sec. 112 and 119, Pub. L. 89–563; 80 Stat. 718, 15 USC. 1401, and 1407; delegation of authority at 49 CFR 1.51.)

Issued on February 12, 1973.

Douglas W. Toms
Administrator

38 F.R. 4400
February 14, 1973

PREAMBLE TO AMENDMENT TO PART 575—CONSUMER INFORMATION

Subpart A—General

(Docket No. 73–5; Notice 1)

This notice amends the definition section of the regulation on Federal motor vehicle consumer information reflecting previous amendments to definitions in the Federal motor vehicle safety standards.

The definitions of "brake power unit" and "lightly loaded vehicle weight" in 49 CFR § 575.2(c) have been obsoleted by recent amendments to these terms in Motor Vehicle Safety Standard No. 105a, *Hydraulic Brake Systems* (37 F.R. 17970). "Brake power unit" has been redefined to more accurately describe the characteristics of the component concerned. The term "curb weight" used in defining "lightly loaded vehicle weight" has been replaced by "unloaded vehicle weight" (as defined in § 571.3) as a more precise description of vehicle condition. Finally, "Maximum sustained vehicle speed" should be grammatically a speed "attainable" rather than "obtainable".

Effective date: February 28, 1973. Since these amendments are primarily a matter of form and have no significant effect on substantive requirements, it is found for good cause that notice and public procedure thereon is unnecessary, and an immediate effective date is in the public interest.

(Sec. 112, 119 Pub. L. 89–563, 80 Stat. 718, 15 U.S.C. 1401, 1407; delegation of authority at 49 CFR 1.51.)

Issued on February 21, 1973.

Douglas W. Toms
Administrator

38 F.R. 5338
February 28, 1973

PREAMBLE TO AMENDMENT TO PART 575—CONSUMER INFORMATION

Subpart A—General

(Docket 72–24; Notice 2)

This notice amends 49 CFR 575, Consumer Information, to require manufacturers to identify specially-configured vehicles not available for purchase by the general public as "special vehicles" in the information submitted to the NHTSA under § 575.6(c).

A notice of proposed rulemaking to this effect was published on November 8, 1972 (37 F.R. 23732). As noted in that proposal, inclusion of these vehicles in compilations or rankings published by this agency as consumer information serves no beneficial purpose, and could confuse the consumer.

No comments opposed the proposal. General Motors Corporation commented that the amendment should more clearly indicate that the special vehicle identification requirements only apply to the information supplied to NHTSA under § 575.6(c). The new section reflects this suggestion.

Ford Motor Company agreed with GM that the special vehicle identification is useful in information supplied to NHTSA. Ford also suggested, however, that consumer information on special vehicles need not be included at all in the information supplied "on location" to prospective purchasers in accordance with § 575.6(b). The NHTSA does not have information at present to support or repudiate this suggestion, which is beyond the scope of the proposal. If Ford or any other person wishes to petition for rulemaking on this subject, the agency will consider it for possible future rulemaking.

In response to an implied question by Truck Body and Equipment Association, Inc., the amendment does not change the applicability of the Consumer Information regulations, as set forth in Subpart B of Part 575.

In consideration of the foregoing, 49 CFR Part 575, Consumer Information, is amended....

Effective date: June 11, 1973.

(Secs. 112, 119, Pub. L. 89–563, 80 Stat. 718, 15 U.S.C. 1401, 1407; delegation of authority at 49 CFR 1.51.)

Issued on May 1, 1973.

James E. Wilson
Acting Administrator

38 F.R. 11347
May 7, 1973

PREAMBLE TO AMENDMENT TO PART 575—CONSUMER INFORMATION

(Docket No. 25, Notice 8)

This notice establishes a Consumer Information regulation on Uniform Tire Quality Grading. The notice is based on proposals published March 7, 1973 (38 F.R. 6194), and August 14, 1973 (38 F.R. 21939). An earlier proposal, published September 21, 1971 (36 F.R. 18751) was later withdrawn (April 21, 1972; 37 F.R. 7903). Comments submitted in response to these proposals have been considered in the preparation of this notice.

The regulation will require tire manufacturers and brand name owners to provide relative grading information for 13-, 14- and 15-inch tire size designations for tire traction, treadwear, and high speed performance. The respective grades will be molded into or onto the tire sidewall, contained in a label affixed to each tire, and provided for examination by prospective purchasers in a form retainable by them at each location where tires are sold. The requirements are effective with respect to passenger cars when they are equipped with new tires bearing quality grades.

Treadwear: The regulation requires each tire to be graded for treadwear performance using numbers which indicate the percentage of treadwear the tire will produce when compared to the treadwear obtained from a "control tire" specified in the regulation. Each tire will be graded with either the number "60", representing treadwear performance less than 80 percent of the control tire's, or the number "80", "120", "160" or "200", representing at least that percentage of control tire wear. The grades are fewer in number and represent broader performance ranges than those proposed, as a result of comments that the proposed grades were too numerous and would not take into account inherent differences in tire performance.

The method for obtaining treadwear grades is essentially that proposed in the notice of March 7, 1973. Treadwear grades will be determined by using a convoy of up to four identical passenger cars with one vehicle equipped with four identical control tires, and each of the remaining vehicles equipped with four identical manufacturer's tires (candidate tires) having the same nominal rim diameter as the control tire. The NHTSA intends that the convoy vehicles be driven as similarly as possible with respect to such factors as steering and braking. The vehicles are run for 16,000 miles over a surface that will produce control tire wear equal to between 65 and 85 percent of original tread depth. The proposal had suggested that the tires be worn to 90 percent of tread depth. This percentage has been reduced to prevent the tires from being worn below their treadwear indicators. The proposal had further suggested that candidate tires be loaded to 100 percent of the load specified for their inflation pressure in the 1972 Tire and Rim Association Yearbook. In response to comments that vehicles are rarely loaded to that extent in practice, the load has been changed to 90 percent of the load specified for the inflation pressure in the 1972 Tire and Rim Association Yearbook. The NHTSA believes the road test method for measuring treadwear to be the most satisfactory that is presently available. Moreover, the method has been used for many years by tire manufacturers to evaluate the treadwear potential of newly developed tire designs and compounds.

Many comments agreed that a 16,000-mile road test was appropriate for grading the treadwear of radial tires. Some comments urged, however, that only a 12,000-mile test be specified for bias and bias/belted tires. The NHTSA has

PART 575—PRE 21

not accepted this recommendation as it believes the comparative data for candidate tires of different construction types will necessarily be more accurate if the comparisons are based on the same degree of control tire wear.

Certain comments referred to the existing national energy shortage, requesting that the agency take into account the problems presented by the shortage in the final requirements. The NHTSA recognizes the degree of energy that will be necessary to perform the appropriate grading tests, particularly with respect to the test for treadwear grading. Research has been undertaken and will continue with a view to reducing the energy needs to establish treadwear performance without adversely affecting the validity of test results. The NHTSA invites suggestions or proposals in this regard, including supportive data, directed to the establishment of alternative methods or tests for grading tire treadwear.

Traction: Each tire will bear a traction grade of "90", "105", or "120", representing at least that percentage of control tire performance. The test for obtaining traction grades is similar to that proposed on March 7, 1973. It utilizes a two-wheeled test trailer built essentially to specifications in American Society of Testing and Materials E-274-70, *Skid Resistance of Paved Surfaces Using a Full-Scale Tire.* The test consists of towing the trailer over specified wet test surfaces, equipped first with identical control tires, and then with identical candidate tires of the same rim diameter as the control tire. The average coefficient of friction is computed when one trailer wheel is locked on each of the two surfaces at 20, 40, and 60 miles per hour. The grade, similarly to the treadwear grade, is the comparative difference between candidate and control tire performance. The final rule differs from the notice in that the proposed traction grade representing less than 90 percent of control tire performance has not been included. This results from the notice proposing to amend Motor Vehicle Safety Standard No. 109 (49 CFR 571.109) (38 F.R. 31841; November 19, 1973) to require all passenger car tires to achieve at least this level of control tire performance. The NHTSA expects that this requirement will become effective on the effective date of this

regulation, thereby necessitating the deletion of the grade. The other grades specified differ from those proposed to the extent that the range between grades has been increased to better allow for inherent gradations in actual tire performance.

Many comments urged that grading for tire traction not be established at this time. The comments argued that the current state of the art has not advanced to the point where reliable and reproducible results can be obtained using the proposed two-wheel trailer method.

The NHTSA believes the traction test issued by this notice, utilizing the two-wheeled trailer, is an objective procedure, capable of producing repeatable results, and is therefore satisfactory for the purpose of measuring and grading straight-line, wet-surface braking traction. In this regard, on the basis of information received from General Motors, that company is presently using the identical methodology in the specifications for tire traction for its "TPC" specification tire. This tire is presently manufactured by numerous domestic tire companies. Moreover, grading tire traction is a necessary adjunct, in the view of NHTSA, to grading tire treadwear, for it is commonly known that treadwear and traction performance result from diverse tire properties. The two tests, therefore, serve as a check that manufacturers will not design tires that perform well in one area at the expense of performance in the other. The minimum traction performance requirement recommended by the comments as a substitute for traction grading is insufficient, in the view of NHTSA, to serve this function alone.

Many comments stated that traction test surfaces should be defined by test surface composition and skid number, rather than by skid number alone as proposed. It was argued that without a surface specification, reversals in tire performance may occur. The NHTSA agrees that the inclusion of precise surface specifications may improve the reliability of traction test results. It has not adopted such specifications in this notice as they have not been previously proposed. However, recent developments have been made in the establishment of test surfaces by the Federal Highway Administration of the Department of Transportation. Test surfaces developed

PART 575—PRE 22

by that agency are proposed in a notice issued concurrently with this notice (1061) for later inclusion in the regulation.

Some comments argued that the description of this grading parameter as "traction" was misleading, as the proposed test dealt only with wet braking traction and not dry pavement or cornering traction. They suggested therefore that the grading parameter be referred to as braking or stopping traction, or as "wet-surface traction." The NHTSA does not dispute that these other traction properties are important aspects of tire traction, and expects to add these performance aspects to the traction grading scheme when appropriate test procedures are developed. The NHTSA does not believe, however, that the description of the existing test as "traction" is misleading. The terminology suggested by the comments, in the view of NHTSA, would be over technical and unnecessary.

High speed performance: High speed performance grades of "A", "B", or "C" are required to be affixed to each tire based on its performance on the high speed laboratory test wheel which is presently used in testing for conformity to Motor Vehicle Safety Standard No. 109. The test utilized is as proposed—an extension of the Standard No. 109 high speed performance test. A tire will be graded "C" if it only passes the Standard No. 109 test. In order to achieve a grade of "B", the tire must run without failure an additional $\frac{1}{2}$ hour at 425 rpm and two additional hours, one at 450 rpm and the other at 475 rpm. To achieve a grade of "A" the tire must be run without failure an additional hour at 500 rpm and another hour at 525 rpm. The NHTSA has recently revised the criteria for tire failure in Standard No. 109 (38 F.R. 27050; September 28, 1973) and the revised criteria are the criteria included in this rule.

The principal comment regarding the proposed high speed grading format was that it should consist of only two grades—one recommended for general use and the other for use by emergency vehicles. The comments argued that further grading of high speed performance was unnecessary and would promote high speed driving. The NHTSA views the suggested 2-grade scheme as rendering any high speed grade meaningless for most consumers. Essentially, it provides no information other than conformity to Standard No. 109. The NHTSA believes driving habits with respect to speed do differ among the driving population and that the grading scheme should be based on that consideration.

Control Tires: Both treadwear and traction grades are based on comparative results using a control tire specified in the rule. The control tires are 2-ply, rayon tires of bias construction, in sizes 6.50 x 13, 7.75 x 14, and 8.55 x 15. The control tire in each specified rim diameter will be used in testing all candidate tires having that rim diameter. The precise specifications for the tires are identical to those proposed.

Control tires will be manufactured pursuant to NHTSA contract and will be used in NHTSA compliance testing. They will be made available to the industry for testing purposes, and the NHTSA will accept, for purposes of compliance tests, results based upon their performance. The agency may consider manufacturers who use different test devices to have failed to exercise the due care contemplated by the National Traffic and Motor Vehicle Safety Act should their tires fail to perform to the specified grades when subject to agency tests.

The final rule modifies certain aspects of the proposed rule apart from the grading tests. In response to several comments, labels are not required to be affixed to the tread surface of tires which are furnished as original equipment on new vehicles. These vehicles are generally driven before sale, and labels on the tire tread surface are therefore of questionable value. Information on these tires will still be required to be otherwise furnished with the vehicle, and available for retention by prospective purchasers. The NHTSA did not, however, agree with comments recommending that the affixed label requirement be deleted entirely. Tires are frequently on display in sales outlets, and the affixed label will provide consumers with the clearest understanding of the grades applicable to a particular tire.

The grades molded onto the tire sidewall are required to be placed between the shoulder and the maximum section width, rather than between the maximum section width and the bead as proposed. The NHTSA believes the grades should apply only to the original tire, and the placement of grades above the maximum section width

PART 575—PRE 23

increases the likelihood that grades will be removed if the tire is retreaded.

Certain comments expressed the view that providing information for tires placed on new vehicles and furnishing that information to the NHTSA 30 days before the vehicles are available to the public is difficult to accomplish because of the variety of tire and vehicle combinations involved. The NHTSA does not believe sufficient justification has been shown for deleting these requirements. While some modification may be necessary to existing manufacturer practices, the NHTSA cannot agree that the regulation presents unmanageable problems for manufacturers.

Effective date: September 1, 1974. The NHTSA has issued this notice pursuant to an order of the United States District Court for the District of Columbia. That order specifies that the regulation take effect on September 1, 1974.

In light of the above, sections 575.4 and 575.6 are revised, and a new section 575.104 "Uniform Tire Quality Grading", is added in Chapter V, Title 49, Code of Federal Regulations. . . .

(Secs. 103, 112, 119, 201, 203; Pub. L. 89-563, 80 Stat. 718, 15 U.S.C. 1392, 1401, 1407, 1421, 1423; delegation of authority at 49 CFR 1.51.)

Issued on December 28, 1973.

James B. Gregory
Administrator

39 F.R. 1037
January 4, 1974

PREAMBLE TO AMENDMENT TO PART 575—CONSUMER INFORMATION REQUIREMENTS

(Docket No. 25; Notice 11)

This notice revokes the Uniform Tire Quality Grading regulation published January 4, 1974 (39 F.R. 1037), and responds to petitions for reconsideration received with respect to the regulation.

The Uniform Tire Quality Grading regulation specified the use of "control tires" in the establishment of grades for treadwear and traction. The NHTSA expected that control tires would be manufactured by an industry source pursuant to NHTSA contract, and would be available for both industry and government use. A solicitation for a proposal to manufacture control tires was advertised to the domestic tire industry. Two proposals were received. Each, however, has been determined to be nonresponsive to the solicitation, which has accordingly been cancelled.

Due to the failure of NHTSA to procure a control tire, the agency must revoke the Uniform Tire Quality Grading regulation in its present form. The revocation of the regulation renders moot the petitions for reconsideration received.

On May 2, 1974, an order was entered by the United States District Court for the District of Columbia in the case of *Nash* v. *Brinegar* (Civil Action No. 177–73) requiring the NHTSA to issue, by June 15, 1974, a notice of proposed rulemaking for a revised Uniform Tire Quality Grading regulation having a proposed effective date of May 1, 1975.

In light of the above, § 575.104 "Uniform Tire Quality Grading" of Chapter V, Title 49, Code of Federal Regulations, is revoked, effective

(Secs. 103, 112, 119, 201, 203; Pub. L. 89–563, 80 Stat. 718, 15 U.S.C. 1392, 1401, 1407, 1421, 1423; delegation of authority at 49 CFR 1.51.)

Issued on May 6, 1974.

Gene G. Mannella
Acting Administrator

39 F.R. 16469
May 9, 1974

PREAMBLE TO AMENDMENT TO PART 575—CONSUMER INFORMATION

(Docket No. 74-18; Notice 2)

This notice amends Part 575, Consumer Information, so that the requirement that manufacturers have consumer information available in showrooms does not apply to special vehicles not available to the general public.

On April 26, 1974, the National Highway Traffic Safety Administration proposed to amend Part 575 to provide consumers with information for only those vehicles which they were eligible to purchase (39 F.R. 14728). The proposal, which was in response to a petition from Ford Motor Company, stated that information concerning special vehicles would continue to be made available to eligible purchasers. Comments concerning the proposal were received from American Motors Corporation, General Motors Corporation and Chrysler Corporation. All comments favored the proposal.

In consideration of the foregoing, 49 CFR 575.7 is amended. . . .

Effective date: March 13, 1975. Because the amendment relieves a restriction, it is found for good cause shown that an effective date immediately upon publication is in the public interest.

(Secs. 103, 112, 114, 203, Pub. L. 89-563, 80 Stat. 718, 15 U.S.C. 1392, 1401, 1407, 1423; delegation of authority at 49 CFR 1.51.)

Issued on March 7, 1975.

Noel C. Bufe
Acting Administrator

40 F.R. 11727
March 13, 1975

PREAMBLE TO AMENDMENT TO PART 575—CONSUMER INFORMATION

(Docket No. 25; Notice 17)

This notice establishes Uniform Tire Quality Grading Standards. The notice is based on proposals published June 14, 1974 (39 F.R. 20808, Notice 12), August 9, 1974 (39 F.R. 28644, Notice 14), and January 7, 1975 (40 F.R. 1273, Notice 15). Comments submitted in response to these proposals have been considered in the preparation of this notice.

A rule on this subject was issued on January 4, 1974 (39 F.R. 1037). It was revoked on May 9, 1974 (39 F.R. 16469), due to the inability of the NHTSA to obtain from the tire industry "control tires" which were to have been used as the basis for determining the comparative performance grades for treadwear and traction.

The rule issued today requires manufacturers to provide grading information for new passenger car tires in each of the following performance areas: treadwear, traction, and temperature resistance. The respective grades are to be molded into or onto the tire sidewall, contained in a label affixed to each tire (except for OEM tires), and provided for examination by prospective purchasers in a form retainable by them at each location where tires are sold.

TREADWEAR

Treadwear grades are based on a tire's projected mileage (the distance which it is expected to travel before wearing down to its treadwear indicators) as tested on a single, predetermined test run of approximately 6400 miles. A tire's treadwear grade is expressed as the percentage which its projected mileage represents of a nominal 30,000 miles, rounded off to the nearest lower 10% increment. For example, a tire with a projected mileage of 24,000 would be graded "80", while one with a projected mileage of 40,000 would be graded "130".

The test course has been established by the NHTSA in the vicinity of San Angelo, Texas, as described in Appendix A. It is the same as that discussed at the public briefings on this subject which took place July 23 and July 29, 1974, except that the direction of travel has been reversed on the northwest loop to increase safety by reducing the number left turns. The course is approximately 400 miles long, and each treadwear test will require 16 circuits. It is anticipated that both the industry, at each manufacturer's option, and the agency will perform treadwear tests on this course; the former for establishing grades, and the latter for purposes of compliance testing, i.e., testing the validity of the grades assigned. To arrange for allocations of test time at the site, industry members should contact the NHTSA facility manager, P.O. Box 6591, Goodfellow Air Force Base, San Angelo, Texas 76901; telephone (915) 655–0546. While manufacturers are not required to test on the site, it would be to their advantage to do so, since the legal standard against which compliance with the rule will be measured is a tire's performance in government tests on that course.

The method of determining projected mileages is essentially that proposed in Notice 12 as modified by Notices 14 and 15 in this docket. The treadwear performance of a candidate tire is measured along with that of course monitoring tires (CMTs) if the same general construction type (bias, bias-belted, or radial) used to monitor changes in course severity. The CMTs are tires procured by the NHTSA—one group each of the three general types—which are made available by the agency for purchase and use by regulated persons at the test site. To obtain course monitoring tires, regulated persons should contact the NHTSA facility manager at the above address.

Each test convoy consists of one car equipped with four CMTs and three or fewer other cars equipped with candidate tires of the same construction type. (Candidate tires on the same axle are identical, but front tires on a test vehicle may differ from rear tires as long as all four are of the same size designation.) After a two-circuit break-in period, the initial tread depth of each tire is determined by averaging the depth measured at six equally spaced locations in each groove. At the end of every two circuits (800 miles), each tire's tread depth is measured again in the same way, the tires are rotated, vehicle positions in the convoy are rotated, and wheel alignments are readjusted if necessary. At the end of the 16-circuit test, each tire's overall wear rate is calculated from the nine measured tread depths and their corresponding mileages-after-break-in as follows: The regression line which "best fits" these data points is determined by applying the method of least squares as described in Appendix C; the wear rate is defined as the absolute value of the slope of the regression line, in mils of tread depth per 1000 miles. This wear rate is adjusted for changes in course severity by a multiplier consisting of the base wear rate for that type of course monitoring tire divided by the measured average of the wear rates for the four CMTs in that convoy. A candidate tire's tread depth after break-in (minus 62 mils to account for wearout when the treadwear indicators are reached) divided by its adjusted wear rate and multiplied by 1000, plus 800 miles, yields its projected mileage. The projected mileage is divided by 30,000 and multiplied by 100 to determine the percentage which, when rounded off, represents the candidate tire's treadwear grade.

A discussion of the NHTSA response to the comments on treadwear grading follows.

Duration of break-in period and test. The 400 mile break-in period originally proposed in Notice 12 was extended in Notice 15 to 800 miles, to permit the rotation of each tire between axles after 400 miles. The Rubber Manufacturers Association (RMA) suggested that a 1600-mile break-in, by permitting each tire to be rotated

once through each position on the test car, would provide more reliable results. An analysis of variance in a study conducted by the NHTSA showed no significant variations in wear from one side of a car to the other. Further, a review of data from extensive testing on the San Angelo course showed no anomalies or consistent variations in wear rate occurring after the first 800 miles. The NHTSA is convinced that the 800-mile break-in period is sufficient to allow a tire to establish its equilibrium inflated shape and stabilize its wear rate. Therefore, the RMA suggestion has not been adopted.

Many of the comments to Notice 12 suggested that testing distances greater than 6400 miles are necessary for accurate tread life projections. Testing to 40%, 50%, and even 90% of wearout was urged. Unfortunately, only the submission of North American Dunlop was accompanied by substantive data. These data, showing non-linear wear rates, were of questionable validity because the tires were not broken in prior to testing and because the data were collected by different test fleets in different parts of the country. Nonetheless, as a result of the large number of adverse comments, the NHTSA requested further information from all knowledgeable and concerned parties to document and substantiate the position that a longer treadwear test is necessary. The additional data were requested in a written inquiry to the RMA and in Notice 15. Because of the need to limit test time, test cost, and fuel consumption, the objective was to determine the minimum test distance which can reliably predict ultimate tire treadwear life.

The responses to these requests have been reviewed and analyzed. Again, the NHTSA finds the industry data and conclusions that greater testing distances are necessary lacking in rigor and completeness. In most cases, the conditions of the industry tests were not disclosed or did not coincide with the prescribed control procedures. Serious doubt is cast upon the conclusions because of inadequate information on one or more of the following test conditions: changes in weather and season, course severity, conformity with prescribed break-in period, mileage between

Effective: January 1, 1976
July 1, 1976
January 1, 1977
July 1, 1977

readings, method of projected mileage, size of convoy, number of tires tested, and uniformity and frequency of tread depth measurement.

A controlled test program recently completed by the NHTSA was designed to test the hypothesis that the rate of wear of tires is constant after an 800-mile break-in. The design and conclusions of the test are discussed in detail in a paper by Brenner, Scheiner, and Kondo ("Uniform Tire Quality Grading; Effect of Status of Wear on Tire Wear Rate," *NHTSA Techncial Note T-1014*, March, 1975—General Reference entry no. 42 in this docket.) The general conclusions of the test are: (1) that the inherent rate of wear of tires, after an 800 mile break-in period, is constant and (2) that the projected tread life for a tire estimated from a 6,400-mile test after 800-mile break-in is accurate for all three tire types. Accordingly, the 6,400 mile test period has been retained.

Grading based on minimum performance. The RMA expressed strong disagreement with any system in which treadwear grades are based on a tire line's *minimum* projected mileage on the San Angelo test course, urging instead that the average performance of a line is a more appropriate grade. The RMA suggested further that the proposed grading system "ignores the bell-shaped distribution curve which describes any performance characteristics and would require the downgrading of an entire line of tires until no portion of the distribution curve fell below any selected treadwear grade, notwithstanding that the large bulk of a given group of tires was well above the grade."

The NHTSA rejects the arguments and the position taken by the industry on this issue. It is precisely the fact that, in industrial processes involving production of large numbers of items, the products group themselves into the so-called bell-shaped or normal distribution which allows for measurement of central tendency and variation and forms the basis of scientific quality control.

Tests performed by the NHTSA and described in the paper cited above have shown conclusively that different production tires exhibit considerable differences in their variability about their respective average values. Thus, two different tire brands might have identical average values for treadwear, but differ markedly in their variance or standard deviation. These differences would probably be attributable to differences in process and quality control.

Recognition of differences in inherent variability among tire manufacturers and tire lines is of the utmost importance to the consumer. The average or mean measure of a group of tires does not provide sufficient information to enable the consumer to make an informed choice. If one tire on a user's car wears out in 10,000 miles, the fact that the "average" tire of that type wears to 25,000 miles in the same driving environment does not alter his need to purchase a new tire. Ideally, the consumer might be provided with more information if he were given a measure of the mean (central tendency) and standard deviation (variability) for each tire type, but the complexity and possible confusion generated by such a system would negate its advantages. In the NHTSA's judgment, the most valuable single grade for the consumer is one corresponding to a level of performance wihch he can be reasonably certain is exceeded by the universe population for that tire brand and line.

As with the other consumer information regulations issued by this agency, a grade represents a minimum performance figure to which every tire is expected to conform if tested by the government under the procedures set forth in the rule. Thus, any manufacturer in doubt about the performance capabilities of a line of his tires is free to assign a lower grade than what might actually be achieved, and he is expected to ensure that substantially all the tires marked with a particular grade are capable of achieving it.

Homogeneity of course monitoring tires. Another aspect of the Notice 12 proposal which generated much controversy is the adoption by the NHTSA of production tires for use as course monitoring tires. The commenters suggested that changes in course severity be monitored instead by tires manufactured under rigidly specified conditions to ensure homogeneity. Because variations in the performance of course monitoring tires are reflected in treadwear projections for all candidate tires, it follows that the more homogeneous the universe of the monitoring tires, the more precisely the performance of the candidate tires can be graded. The NHTSA is in complete accord with the industry's desire to minimize the variability of tires chosen for course monitoring. The development of specifications for special "control tires", in which materials, processing, and other conditions are rigidly controlled to a degree beyond that possible for mass production, will continue. The NHTSA hopes to work with the tire industry to reduce the variability of course monitoring tires to the maximum extent possible. However, it should be noted that an earlier version of this regulation had to be revoked due to the difficulty in obtaining such "control tires." Recent tests (summarized in the paper cited above) demonstrate that implementation of a viable treadwear grading system need not be delayed further, pending development of special tires. In these tests, the current radial CMTs—Goodyear Custom Steelgards chosen from a single, short production run—show a coefficient of variation (standard deviation of wear rate divided by mean) of 4.9%. This degree of uniformity is commensurate with universally accepted criteria for test control purposes. Hence, grading of radial tires may be started immediately. The tentatively adopted bias and bias-belted CMTs showed coefficients of variation of 7.3% and 12.4%, respectively. Existing test data indicate that the NHTSA will be able to identify and procure other tires of these two construction types, exhibiting homogeneity comparable to the current radial CMTs, in time for testing in accordance with the implementation schedule set out below. In any event, the variability of course monitoring tires will be taken into account by the NHTSA in connection with its compliance testing. At worst, the degree of grading imprecision associated with CMT variability will be no greater than one-half the levels measured for the current bias and bias-belted tire lots, because the standard deviation for the average of a set of four tires is equal to one-half that of the universe

standard deviation. It is the NHTSA's judgment that treadwear grades of this level of precision will provide substantially more meaningful information to the prospective tire buyer than is currently available.

To make efficient use of the available CMTs, the NHTSA expects to conduct treadwear tests with used CMTs, as well as with new ones. This will not affect any mileage projections, because the inherent wear rate of tires is constant after break-in. Test results will be discarded if the treadwear indicators are showing on any of the CMTs at the end of a test.

The need for three separate course monitoring tires. Many commenters suggested that a single CMT of the bias-ply type be used, arguing that the use of a different CMT for each general construction type would create three separate treadwear rating systems. These suggestions appear to result from a misunderstanding of the role of the course monitoring tires. They are not used as yardsticks against which candidate tires are graded. Instead, they are used to monitor changes in the severity of the test course. Experiments performed by the NHTSA (Brenner, F.C. and Kondo, A., "Elements in the Road Evaluation of Tire Wear", *Tire Science and Technology*, Vol. I, No. 1, Feb. 1973, p. 17—General Reference entry no. 17 in this docket) show that changes in test course severity will affect tires of differing construction types to differing degrees. For example, the improvement in projected tread life from the severest to the mildest test courses in the experiments was 12% for bias tires, yet it was 91% for bias-belted tires and 140% for radial tires. In fact, a variety of factors influence course severity, each having different relative effects on the various tire types. Therefore, the use of a single course monitoring tire on courses of varying severity, or even on a given course whose severity is subject to variation due to weather and road wear, would not permit the correct adjustment of measured wear rates for environmental influences. Only with a CMT for each construction type can a single, uniform treadwear grading system be established.

Expression of treadwear grades. The system of treadwear grading proposed in Notice 12 specified six grades, as follows:

Grade X (projected mileage less than 15,000)
Grade 15 (projected mileage at least 15,000)
Grade 25 (” ” ” ” 25,000)
Grade 35 (” ” ” ” 35,000)
Grade 45 (” ” ” ” 45,000)
Grade 60 (” ” ” ” 60,000)

Among the objections to this proposal was that small differences in actual treadwear in the vicinity of grade boundaries would be misrepresented as large differences because of the breadth of the predetermined categories. The NHTSA was also concerned that the broad categories could in some cases reduce the desirable competitive impact of the treadwear grading system if tires of substantially differing treadwear performance were grouped in the same grade. For these reasons, a relatively continuous grading system was proposed in Notice 15, in which tires would be graded with two digit numbers representing their minimum projected mileages in thousands of miles as determined on the San Angelo test course. The major objection to both of these proposals was that grades expressing projected mileages would lead consumers to expect every tire to yield its indicated mileage. The manufacturers were especially concerned that this would subject them to implied warranty obligations, despite the disclaimer on the label. The NHTSA remains convinced that treadwear grades which are directly related to projected mileages are the most appropriate way of expressing treadwear performance. To overcome any possible misinterpretation by consumers, the grading system established today is changed from that of Notice 15 to indicate relative performance on a percentage basis, as described above. This decision is based in part upon the fact that testing performed to date on the San Angelo course has given projected mileages that are generally higher than those the average user will obtain; i.e., it appears to be a relatively mild course.

Wheel alignment procedure. Test vehicle wheel alignment procedures received considerable comment. Notice 12 proposed alignment to vehicle manufacturer's specifications after vehicle loading. Notice 15 proposed that this be done before loading, and that the measurements taken after loading be used as a basis for setting alignment for the duration of the test. The majority of the commenters strongly favored a return to the original procedure. The NHTSA takes particular cognizance of the fact that those commenters who have actually tried both procedures in testing at San Angelo find the procedure of Notice 12 to be satisfactory and practicable, and that of Notice 15 to be unusable. NHTSA representatives at San Angelo have reported satisfactory operation on a variety of vehicles using the originally proposed procedure, and have not observed any uneven tire wear that would indicate alignment problems. For these reasons, the final rule prescribes alignment procedures which are identical with those proposed in Notice 12.

Tire rotation procedure. Several commenters objected to using the proposed "X" rotation procedure for testing radial tires. The NHTSA is aware that this procedure differs from that recommended by many groups for consumers' use. While some vehicle and tire manufacturers recommend that radial tires be rotated only fore-aft, others recommend no rotation at all and yet others are silent on the subject. The primary reason for these other methods appears to be to improve passenger comfort by reducing vibration. No data have been submitted, however, to suggest that the proposed method has any adverse or uneven effect on radial tire wear. Further, this method has the advantage, for treadwear testing, of balancing out any side-to-side or axle wear differences attributable to the vehicle or to the course. Accordingly, the proposed tire rotation method has been adopted without change.

Choice of grooves to be measured. Some commenters suggested that treadwear projections be calculated from measurements of the most worn grooves on candidate tires, rather than from the averages of measurements made in all grooves.

It was argued that, because many States require replacement of passenger car tires when treadwear indicators appear in any two adjacent grooves, the proposed method of calculation would yield misleadingly high projections. Analysis of projections based on both methods (Brenner, F.C. and Kondo, A., "Patterns of Tread Wear and Estimated Tread Life," *Tire Science and Technology*, Vol. 2, No. 1, 1973—General Reference entry no. 27 in this docket) shows a high correlation between the resulting tire rankings. Because the treadwear grading system established today is based on relative performance, there is no disadvantage in adopting the proposed method. On a related issue, the E.T.R.T.O. pointed out that some grooves near the tire shoulder which are designed only for esthetic reasons exhibit practically no wear, and suggested that measurements be made only in those grooves which contain treadwear indicators. This suggestion has been adopted.

Calculation of projected mileage. Several methods for calculating the tire wear rates to be used in determining projected mileages were considered. Notice 12 proposed calculating the geometric mean of the wear rates measured for each 800-mile increment. This approach was rejected because the geometric mean is extremely sensitive to inaccurate readings in any single measurement. Use of the arithmetic mean of the incremental wear rates appears to be the general industry practice. Unfortunately, however, the intermediate readings have no effect on such a calculation, because the result is a function only of the initial tread depth (after break-in) and that measured 6,400 miles later. Therefore, a wear rate calculated by the industry method is extremely sensitive to errors in these two measurements. In Notice 15, the NHTSA proposed that wear rate be calculated by the least-squares regression method, as described above. This approach has the advantage of weighting all measurements and minimizing the effect of inaccurate readings, so it has been adopted.

Differing tires on a single test vehicle. Uniroyal and the E.T.R.T.O. argued that each test convoy vehicle should be equipped with four identical tires; the reason given was that otherwise, the performance of a candidate tire would be a function of the tires chosen by the NHTSA for use on the other axle of the test vehicle during compliance testing. The NHTSA is unaware of any data that support this position. The rule adopted today requires that all vehicles in a single convoy be equipped with tires of the same general construction type, and that all tires on a single vehicle be of the same size designation. In extensive testing at San Angelo with this procedure, none of the suggested undesirable variations has been observed.

Differing test vehicles in a single convoy. Several commenters suggested that the rule specify that all vehicles in a given convoy be identical, to reduce variations in projected treadlife. The NHTSA is in complete agreement with the premise that those variables which can be identified and which can affect treadwear results should be controlled as closely as is feasible. Variations in vehicle type, however, do not appear to produce significant variations in treadwear projections. Nevertheless, to minimize such variations, tires will be tested for compliance only on vehicles for which they are available as original equipment or recommended replacement options. Where practical, all vehicles in a given convoy will be of the same make. However, to test tires designed for the range of wheel sizes available, the suggested method would require a proliferation of course monitoring tires, one for each combination of wheel size and construction type. Therefore, the suggestion has not been adopted.

Accuracy of tread depth measurements. The RMA suggested that the interval between measurements be increased to 1,600 miles to reduce the effects of measurement error. However, if this interval were used instead of 800 miles, only five readings would be obtained in the 6,400 mile treadwear test, so errors in any one reading would result in a greater overall error. A recently completed study (Kondo, A. and Brenner,

F.C., "Report on Round-Robin Groove Depth Measuring Experiment," *NHTSA Technical Note T-1012*, March 1975—General Reference entry no. 44 in this docket) shows that variations among measurements of the same tread depth by different operators do not present a serious problem. The study found that the only significant variations in measurement results occur as a result of differences in measuring techniques between different laboratories. Since these techniques are consistent within a given laboratory, the different laboratories arrive at the same results of the slope of the tread depth regression line that is the basis of the treadwear grade.

TRACTION

Traction grades are based on a tire's traction coefficient as measured on two wet skid pads, one of asphalt and one of concrete. Because a method for producing identical skid test surfaces at different sites has not yet been developed, the NHTSA has established two skid pads, described in Appendix B, near the treadwear test course in San Angelo. These pads represent typical highway surfaces. The asphalt surface has a traction coefficient, when tested wet using the American Society for Testing and Materials (ASTM) E 501 tire, of 0.50 ±0.10. The concrete surface was described in Notice 12 as having a traction coefficient, when similarly tested, of 0.47 ± 0.05. Due to surface polishing, this coefficient has declined and stabilized at 0.35 ± 0.10. As with the treadwear course, these pads are available for use by manufacturers as well as the agency. For allocations of test time, industry members should contact the NHTSA facility manager at the above address.

Before each candidate tire test, the traction coefficient of each surface is measured with two ASTM tires to monitor variations in the surface, using a two-wheeled test trailer built in accordance with ASTM Method E-274-70. The candidate tire's traction coefficient is similarly measured on each surface, and then adjusted by adding a fixed coefficient (0.50 for asphalt, 0.35

for concrete) and subtracting the average coefficient obtained from measurements with the two ASTM tires.

The tire industry's major objection to the proposed rule was that, with four possible grades for traction, two tires might be graded differently without a meaningful difference in their performance. The RMA suggested a scheme with two grade categories above a minimum requirement. The rule issued today, by setting two threshold levels of performance, establishes three grades: "0", for performance below the first threshold; "*", for performance above the first threshold; and "**", for performance above the second threshold. The NHTSA is convinced that the grades thus defined reflect significant differences in traction performance.

Firestone suggested that further testing may demonstrate that only one pad is necessary to give the best and most consistently repeatable results. However, the ranking of a group of tires based on their performance on one surface can differ from their ranking on another surface. In fact, one tire manufacturer suggested that an additional surface of low coefficient be included in the testing scheme for this reason. The NHTSA agrees that an additional surface may increase the utility of the traction grading system, and anticipates a proposal to implement this suggestion in the future.

The suggestion of Pirelli, that measurements be made during the period between 0.5 and 1.5 seconds after wheel lockup instead of the period between 0.2 and 1.2 seconds, has been adopted. To permit more efficient use of the skid pads, the rule specifies a test sequence which differs slightly from that originally proposed: instead of being tested repeatedly on the asphalt pad and then repeatedly on the concrete pad, each tire is run alternately over the two pads. A change in paragraph (f)(2)(i)(A) permits tires to be conditioned on the test trailer as an alternative to conditioning on a passenger car. Another change facilitates the use of trailers with instrumentation on only one side, which had been inadvertently precluded by the wording of the proposed rule.

TEMPERATURE RESISTANCE

The major objection to the proposed high speed performance grading scheme was that it was neither necessary nor beneficial to the consumer. Several commenters pointed out that Standard No. 109 specifies testing a tire against a laboratory wheel at a speed corresponding to 85 mph, and argued that certification of a tire to this minimum requirement provides the consumer with adequate information about its performance at all expected driving speeds. They suggested that only one higher grade be established, for tires designed to be used on emergency vehicles. Some commenters indicated that, as proposed, the rule seemed to condone or even encourage the unsafe operation of motor vehicles above legal speed limits. To preclude this misinterpretation, the third tire characteristic to be graded has been renamed "temperature resistance". The grade is indicative of the running temperature of the tire. Sustained high temperature can cause the material of the tire to degenerate and reduce tire life, and excessive temperature can lead to sudden tire failure. Therefore, the distinctions provided by three grades of temperature resistance are meaningful to the consumer. Except for the name change, this aspect of quality grading has been adopted as proposed. A grade of "C" corresponds to the minimum requirements of Standard No. 109. "B" indicates completion of the 500 rpm test stage specified in paragraph (g)(9), while "A" indicates completion of the 575 rpm test range.

PROVISION OF GRADING INFORMATION

Several commenters objected to the proposed tread label requirement, suggesting that point-of-sale material such as posters and leaflets could provide the consumer with adequate information about tire grades. For the reasons discussed in Notice 12, the NHTSA is convinced that labels affixed to the tread of the tire are the only satisfactory method of providing complete information to replacement tire purchasers. Therefore, the scheme for transmitting quality grading information to consumers, combining sidewall mold-

ing, tread labels, and point-of-sale materials, has been adopted substantially as proposed. A change in paragraph (d)(1)(ii) clarifies the respective duties of vehicle manufacturers and tire manufacturers to provide information for prospective purchasers.

Several vehicle manufacturers requested that new vehicles not be required to be equipped with graded tires until six months after the date that tires must be graded. These commenters appear to have misunderstood the scope of the quality grading standard. The NHTSA expects that tires which comply with the standard will appear on new vehicles as inventories of ungraded tires are depleted. Part 575.6 requires of the vehicle manufacturer only that he provide the specified information to purchasers and prospective purchasers when he equips a vehicle with one or more tires manufactured after the applicable effective date of this rule.

The NHTSA has determined that an Inflationary Impact Statement is not required pursuant to Executive Order 11821. Industry cost estimates and an inflation impact review are filed in public Docket No. 25. This review includes an evaluation of the expected cost of the rule.

In consideration of the foregoing, a new § 575.104, "Uniform Tire Quality Grading Standards" is added to 49 CFR Part 575. . . .

Effective dates. For all requirements other than the molding requirement of paragraph (d)(1)(i)(A): January 1, 1976, for radial ply tires; July 1, 1976, for bias-belted tires; January 1, 1977, for bias ply tires. For paragraph (d)(1)(i)(A): July 1,1976, for radial ply tires; January 1, 1977, for bias-belted tires; July 1, 1977, for bias-ply tires.

(Secs. 103, 112, 119, 201, 203; Pub. L. 89–563, 80 Stat. 718 (15 U.S.C. 1392, 1401, 1407, 1421, 1423); delegation of authority at 49 CFR 1.51.)

Issued on May 20, 1975.

James B. Gregory
Administrator

40 F.R. 23073
May 28, 1975

PREAMBLE TO AMENDMENT TO PART 575—CONSUMER INFORMATION

(Docket No. 25; Notice 18)

This notice republishes, with minor changes, paragraphs (e)(1)(v) and (f)(2)(i)(B), Figure 2, and the appendices of § 575.104, *Uniform Tire Quality Grading Standards*, which was published May 28, 1975 (40 F.R. 23073; Notice 17).

In describing the rims on which candidate tires are to be mounted, Notice 17 inadvertently referred to the Appendix to Standard No. 110. On February 6, 1975, the definition of "test rim" in Standard No. 109 was amended and the Appendix to Standard No. 110 was deleted (Docket No. 74-25; Notice 2; effective August 5, 1975). Under the new definition, a "test rim" may be any of several widths, only one of which is equal to that listed under the words "test rim width" in Table I of the Appendix to Standard No. 109. Paragraphs (e)(1)(v) and (f)(2)(i)(B) are corrected to specify the rim mounting scheme in terms of the new definition.

As Figure 2 was published in the Federal Register, the words "DOT Quality Grades" appeared as the Figure's title. In fact, the words are a part of the text which must appear on each tread label required by paragraph (d)(1)(B), and accordingly the figure is republished with the correct title.

The treadwear test course described in Appendix A is changed so that the loops are traveled in the following order: south, east, and northwest. This change is designed to increase safety by reducing the number of left turns. The table of key points and mileages is revised to reflect the change. Corresponding changes are made in the numbers used to designate these points in the text and in Figure 3.

To prevent the bunching of test vehicles at STOP signs and thereby increase safety, the speed to which vehicles must decelerate when abreast of the direction sign is changed in Appendix A to read "20 mph".

The reference to Figure 2 in the second paragraph of Appendix B is corrected to indicate that the asphalt skid pad is depicted in Figure 4. The shading of the skid pads is corrected to correspond to the description in the text.

The first two paragraphs of Appendix C, *Method of Least Squares*, were omitted. Those paragraphs are now inserted and the graph is designated as Figure 5.

In consideration of the foregoing, paragraphs (e)(1)(v) and (f)(2)(i)(B), Figure 2, and the appendices to § 575.104 of Title 49, Code of Federal Regulations, are republished. . . .

(Secs. 103, 112, 119, 201, 203; Pub. L. 89-563, 80 Stat. 718 (15 U.S.C. 1392, 1401, 1407, 1421, 1423); delegation of authority at 49 CFR 1.51.)

Issued on June 25, 1975.

James B. Gregory
Administrator

40 F.R. 28071
July 3, 1975

PREAMBLE TO AMENDMENT TO PART 575—CONSUMER INFORMATION

(Docket No. 75–27; Notice 2)

This notice amends Standard No. 105–75, *Hydraulic Brake Systems*, 49 CFR 571.105–75, to revise the parking brake test procedure (S7.7). In addition, this notice amends Subpart B of Part 575, *Consumer Information*, 49 CFR § 575.101, by replacing the present test procedures in that section for passenger car testing with equivalent procedures from Standard No. 105–75.

The NHTSA proposed a modification of the parking brake test procedures in Standard No. 105–75 to permit a reapplication of the parking brake if the first application of the brake failed to hold the vehicle stationary on the test incline. Toyo Kogyo requested the modification as representative of normal driver action (in cases where the application appears to be insufficient to hold the vehicle), justifying the change as necessary to permit new vehicle components to stretch or "set" during the initial application as occurs in any vehicle delivered to a purchaser. The NHTSA agreed that reapplication would be a reasonable test procedure and proposed a revision of S7.7.

Comments were received from Toyo Kogyo, General Motors, American Motors Corporation, and Chrysler Corporation in support of the change. No comments were received that objected to the proposal. The standard is amended accordingly.

The NHTSA also proposed that the consumer information item requiring publication of the stopping ability of passenger cars and motorcycles (49 CFR § 575.101) be modified for passenger cars so that test data developed under Standard No. 105–75 could be the basis for the required consumer information. The existing test procedures of the consumer information item would be replaced by Standard No. 105–75 test procedures, and a transition period until January 1, 1977, would be provided to allow manufacturers latitude in adopting the new procedures.

The Motor Vehicle Manufacturers Association (MVMA), Chrysler Corporation, American Motors Corporation, Ford Motor Company, and General Motors Corporation supported the modifications. The MVMA and Ford pointed out an inadvertent omission in the proposal of a required change in the present loading specification (maximum loaded vehicle weight) to the Standard No. 105–75 loading specification (gross vehicle weight rating (GVWR)). No comments opposed the modification, and the consumer information item is therefore amended as proposed, with the additional modification noted by the MVMA and Ford. The transition period for use of either loading specification conforms to the transition period for use of either test procedure (until January 1, 1977). The MVMA asked for a June 1, 1977, date for transition to the new loading specification but did not explain the need for more time. The NHTSA will consider any data on this subject submitted by the MVMA.

With regard to test loading, Chrysler Corporation repeated a request for revision of the loading conditions of Standard No. 105–75. The request was earlier submitted improperly as a petition for reconsideration of an NHTSA action which did not deal with test loading (40 F.R. 24525, June 9, 1975). Section 553.35 of NHTSA regulations (49 CFR 553.35) allows petitions for reconsideration of rules issued by the NHTSA, but in this case no rule was issued on test loading that could form the basis for reconsideration. The NHTSA discussed Chrysler's request at a meeting with Chrysler officials on August 21, 1975. Based on the limited information presented by Chrysler at that meeting, the

PART 575—PRE 89

NHTSA has concluded that a reduction in test weight would not be justified. At the meeting it was agreed that Chrysler would submit any additional data it had in support of the request. To date no data have been received, and the NHTSA cannot meaningfully reconsider Chrysler's request without further data.

The NHTSA also proposed modification of the means for establishing the skid number of the surface on which stopping distance tests are conducted in Standard No. 105–75, Standard No. 121, *Air Brake Systems*, Standard No. 122, *Motorcycle Brake Systems*, and the Consumer Information Item on brake performance. Comments received were not in agreement on how to accomplish the transition from the former ASTM method to the new one. The skid number proposal will therefore be treated separately at a later date so that its resolution will not delay this amendment of the parking brake and consumer information item test procedures.

In consideration of the foregoing, amendments are made in Chapter V of Title 49, Code of Federal Regulations. . . .

Effective date: January 6, 1976. Because these amendments, to the extent that they impose new substantive requirements, are made optional for an interim period, and because manufacturers must plan future testing based on the test procedures as they exist in the present standard, it is found for good cause shown that an immediate effective date is in the public interest.

(Sec. 103, 119 Pub. L. 89–563, 80 Stat. 718 (15 U.S.C. 1392, 1407); delegation of authority at 49 CFR 1.51).

Issued on December 31, 1975.

James B. Gregory
Administrator

**41 F.R. 1066
January 6, 1976**

Effective: April 1, 1976

PREAMBLE TO AMENDMENT TO PART 575—CONSUMER INFORMATION
(Docket No. 76-1; Notice 2)

This notice amends 49 CFR 567 and 575 to allow manufacturers an alternative method of referring purchasers to appropriate consumer information tables.

On January 22, 1976, the National Highway Traffic Safety Administration issued in the Federal Register (40 FR 3315) a notice which proposed amending 49 CFR 575, Consumer Information, and 49 CFR 567, Certification, to allow the consumer information document provided to the purchaser of a vehicle to refer the reader to the vehicle's certification label to determine which information applied to that vehicle. This information, which relates to the performance characteristics of the vehicle, is required to be made available to purchasers by 49 CFR 575.6(a). Currently, if the document containing this information also contains information relating to other vehicles, the document itself must clearly indicate which information is applicable to the vehicle purchased. The NHTSA proposal was made in response to a petition from the General Motors Corporation which suggested that the proposed alternative procedure would for some companies be a more efficient and less costly method of accomplishing the purposes of the regulation.

Comments in support of the proposal were received from General Motors Corporation, American Motors Corporation, Chrysler Corporation and Ford Motor Company. No comments in opposition were received.

Based on the petition of General Motors and the comments concerning the notice of proposed rulemaking, the NHTSA concludes that allowing an alternative method of designating the appropriate consumer information tables would reduce the possibility of error and lessen the cost to the manufacturer.

In consideration of the foregoing, Parts 567 and 575 of Title 49, Code of Federal Regulations, are amended....

Effective date: April 1, 1976. Because the procedures established herein are optional and impose no increased burden on any party, it is found for good cause shown that an immediate effective date is in the public interest.

(Sec. 103, 112, 114, 119, Pub. L. 80-563, 80 Stat. 718 (15 U.S.C. 1392, 1401, 1403, 1407); delegation of authority at 49 CFR 1.50.)

Issued on: March 26, 1976.

James B. Gregory
Administrator

41 F.R. 13923
April 1, 1976

PREAMBLE TO AMENDMENT TO PART 575—CONSUMER INFORMATION

(Docket No. 75–27; Notice 4)

This notice amends Standard No. 105–75, *Hydraulic Brake Systems*, and Standard No. 122, *Motorcycle Brake Systems*, to modify the means for establishing the frictional resistance of the surface on which stopping distance tests are conducted. A similar amendment is made to Part 575, *Consumer Information*, of Title 49 of the Code of Federal Regulations.

The National Highway Traffic Safety Administration (NHTSA) proposed the change in Standard No. 105–75 (49 CFR 571.105–75), Standard No. 121, *Air Brake Systems* (49 CFR 571.121), Standard No. 122 (49 CFR 571.122), and the Consumer Information Regulations (49 CFR 575.101) in response to a petition from British-Leyland Motors Limited (40 FR 45200, October 1, 1975). The existing test procedure in these regulations has specified use of the American Society for Testing and Materials (ASTM) E–274–65T procedure, using an ASTM E249 tire that is no longer manufactured.

Responses were received on the proposed ASTM change from White Motor Corporation (White), Mack Trucks, Inc. (Mack), Freightliner Corporation (Freightliner), Ford Motor Company (Ford), General Motors Corporation (GM), Chrysler Corporation (Chrysler), American Motors Corporation (AMC), and International Harvester (IH). The National Motor Vehicle Safety Advisory Council made no comment on the proposal.

Most commenters supported use of the new test procedure and tire, although they differed in recommendations for correlating the reading produced under the new procedure with that produced under the old procedure. Manufacturers are presently certifying compliance to brake standards on test surfaces with a satisfactory reading under the old procedure, and they should be able to continue testing and certifying compliance on the same surface without any increase in the severity of the tests. To accomplish this transition, the correlation in readings between the procedures has been determined, and the difference is reflected in a change of the dry surface value from "skid number" 75 to "skid number" 81.

Freightliner urged postponement of any action until it could be supported by "adequate and statistically reliable test data." AMC also recommended that the NHTSA do nothing "until the industry has had sufficient time to evaluate and verify the performance of the ASTM E501 test tire on all types of surfaces."

The change in procedure is prompted by the ASTM decision to utilize a new tire in ascertaining the frictional coefficient of test surfaces. As a result the old tire is no longer manufactured and only the new tire is available for skid number measurement. Manufacturers have conducted comparative tests with the new tire to determine the correlation between the readings given by the two tires. Neither Freightliner nor AMC submitted data showing that the agency's proposal to adjust the dry surface skid number upwards is unjustified. Only Mack submitted data and it supported the NHTSA and Federal Highway Administration test data that have been placed in the docket. General Motors considered the agency's proposed upward adjustment to be the maximum desirable based on its data. International Harvester, Chrysler, and Ford supported the change in dry surface skid number without qualification, and White suggested that a skid number of 85 be utilized. The agency finds that the AMC and Freightliner requests for further delay are unjustified.

Ford and Freightliner asked that the skid number for the lower coefficient (wet) surface also be adjusted. The agency's purpose in proposing the adjustment is limited to changes necessary to avoid a modification of the test surfaces or an increase in the severity of performance levels specified under the safety standards. The NHTSA earlier concluded that change of the wet surface specification was unnecessary, and no evidence has been supplied that would modify the earlier determination.

General Motors noted that an editorial change to the newer ASTM procedure does not appear in early publications of that procedure. To put all interested persons on notice of the editorial change, the NHTSA has included the change in its references to the ASTM E274-70 procedure.

Freightliner asserted that the newer procedure included modification of a formula that justified a larger upwards adjustment than that proposed by the agency. Actually, the modifications only corrected an error in the earlier formula which had no effect on the determination of frictional coefficient. Manufacturers either utilized a test trailer that obviated the need for calculations using the formula, or were aware of the error and corrected for it in their calculations. Thus the adjustment requested by Freightliner is not warranted.

In accordance with recently-enunciated Department of Transportation policy encouraging adequate analysis of the consequences of regulatory action (41 FR 16201, April 16, 1976), the agency herewith summarizes its evaluation of the economic and other consequences of this amendment on the public and private sectors, including possible loss of safety benefit. Because the new references to procedures and a test tire are expected to accord with existing practices, the amendment is judged not to have any significant impact on costs or benefits of the standards and consumer information item that are modified by the change.

Standard No. 121, *Air Brake Systems*, is presently subject to judicial review under Section 105(a) of the National Traffic and Motor Vehicle Safety Act (15 U.S.C. Section 1394(a)). The U.S. Court of Appeals hearing the petition for review has indicated that it prefers to review the standard as it presently exists, without unnecessary amendment. To the degree possible, the agency is complying with that request and therefore, in the case of Standard No. 121, will delay the update of ASTM procedure until review is completed.

It is noted that this change in procedure for ascertaining the frictional resistance of the test surface does not invalidate data collected using the older procedure, and manufacturers can presumably certify on the basis of stopping distance tests conducted on surfaces measured by the old tire.

In consideration of the foregoing, amendments are made in Chapter V of Title 49, Code of Federal Regulations. . . .

Effective date: June 14, 1976. Because the older test tire is no longer manufactured, and because the amendment of procedure and test tire is intended only to duplicate the existing procedure and tire, this amendment creates no additional requirements for any person, and an immediate effective date is found to be in the public interest.

(Sec. 103, 119, Pub. L. 89-563, 80 Stat. 718 (15 U.S.C. 1392, 1407); delegation of authority at 49 CFR 1.50.)

Issued on June 8, 1976.

James B. Gregory
Administrator

41 F.R. 24592
June 17, 1976

Uniform Tire Quality Grading

(Docket No. 25; Notice 24)

Action: Final rule.

Summary: This notice announces the effective dates for implementation of a uniform tire quality grading regulation with respect to bias and bias-belted tires, as authorized by Section 203 of the National Traffic and Motor Vehicle Safety Act of 1966. This notice also responds to comments on, and makes final, proposals concerning course monitoring tires and labeling as well as to petitions for reconsideration of the rule.

Effective date: For all requirements, other than the molding requirement of paragraph (d)(1)(i)(A), the effective dates are: March 1, 1979 for bias ply tires, and September 1, 1979 for bias-belted tires.

For paragraph (d)(1)(i)(A), the molding requirement, the effective dates are: September 1, 1979 for bias ply tires, and March 1, 1980 for bias-belted tires. No effective date is established at this time for radial tires.

Addresses: Petitions for reconsideration of the tire labeling amendments should refer to the docket number and be submitted to: Room 5108, Nassif Building, 400 Seventh Street S.W., Washington, D.C. 20590.

For further information contact:

Dr. F. Cecil Brenner, Office of Automotive Ratings, National Highway Traffic Safety Administration, 400 Seventh Street, S.W., Washington, D.C. 20590 (202) 426–1742.

Supplementary information: On May 28, 1975 (40 FR 23073), the NHTSA published as a final rule a regulation pertaining to Uniform Tire Quality Grading (UTQG) as authorized by the National Traffic and Motor Vehicle Safety Act of 1966 (the Act) (15 U.S.C. 1381 *et seq.*). The purpose of this regulation is to alleviate confusion in the purchase of passenger car tires and to provide simple comparative data upon which an informed tire selection can be made by consumers. Under the regulation, tires will be graded in three areas of performance: treadwear, traction, and temperature resistance.

Implementation of the regulation was delayed pending litigation of the validity of its grading procedures. In *B.F. Goodrich et al v. Department of Transportation*, 541 F.2d 1178 (6th Cir., 1976), the court upheld for the most part the agency's approach to tire quality grading. The court remanded for further agency consideration, however, two aspects of the regulation. First, the court suggested that the NHTSA reexamine the labeling requirements of the regulation to ensure that sufficient warnings would be provided to consumers to avoid the misapplication of the label information. Second, the court remanded to the agency the matter of the selection of course monitoring tires, for the agency to complete its testing and selection of the three course monitoring tires or, if this had already been accomplished, for reopening of the record to permit a brief period of industry comment on the selections. The court upheld the rule in all other respects.

Pursuant to the remand in the *B. F. Goodrich* decision, the agency issued two proposals; one to modify labeling requirements and the other announcing the selection of the course monitoring tires. Comments were received from several manufacturers and manufacturer representatives. This notice responds to those comments.

In response to the publication of the UTQG regulation (May 28, 1975) (40 FR 23073), the agency received several petitions for reconsideration. The agency announced that these petitions would not be immediately answered owing to the ongoing litigation involving the regulation (40 FR 57806). Since the challenge to the regulation has now been disposed of by the court, this notice responds fully to those petitions for reconsideration.

I. Labeling (Notice 21).

On December 13, 1976, the NHTSA published a notice of proposed rulemaking to revise the traction and temperature resistance labeling requirements of UTQG (49 CFR 575.104). That notice was in response to the decision in the *B. F. Goodrich* case.

The petitioners in the *B. F. Goodrich* case argued that the then existing labeling requirements would be misleading in several respects pertaining to traction testing and temperature resistance. The court remanded those issues to the agency for further consideration, suggesting the addition to the labels of clarifying warnings. The agency's December 13, 1976 notice proposed warnings in accordance with the court's decision that would ensure that UTQG label information would not be misconstrued.

The NHTSA received seven comments in response to the notice of proposed rulemaking. Most of these comments favored the warnings proposed by the agency with several comments proposing minor editorial changes for clarity. The agency has altered somewhat the final version of these warnings in consideration of the comments. The Vehicle Equipment Safety Commission did not submit comments.

Treadwear Labeling

The Rubber Manufacturers Association (RMA) recommended in its comments that the agency modify the treadwear example in Figure 2 which explains that tires rated at 200 will achieve twice the mileage as tires rated at 100. RMA indicated that few if any commercially available tires could achieve such a rating. Accordingly, they suggested that the example show that a tire rated 150 would wear 1½ times as well as a tire graded 100.

The agency considers RMA's suggestion to have merit. Initially, the 200 figure was selected for the example because it facilitates understand-

ing of the treadwear grading concept since it speaks in terms of round numbers (e.g., a tire grade 200 wears twice as well as a tire grade 100). However, since few tires can achieve such a rating, the example would have little practical application. Therefore, the agency modifies the example to reflect that 150 represents a treadlife 1½ times as good as that represented by the grade of 100.

Traction Labeling

Goodyear Tire and Rubber Company, Firestone Tire and Rubber Company, and the RMA suggested in their comments that the NHTSA amend the traction information in Figure 2 of the label to indicate that the tires were tested under controlled conditions on specified government test surfaces. The agency believes that this information is useful to prevent misleading the consumer and amends Figure 2 accordingly.

General Motors Corporation (GM) recommended that the agency add further warnings to the traction information that would indicate that actual traction results would differ depending upon tread depth, road surface, and speed. GM contended that the proposed warning did not sufficiently detail the extent of the limitations upon the use of these traction data.

The NHTSA is concerned that the warnings printed in the tire information be kept to the absolute minimum in length while ensuring adequate consumer information. If warnings and tire information become so lengthy as to become burdensome upon the consumer to read, it is possible that the information would go unused. The agency has determined that the statement in the warning that a tire was "measured under controlled conditions on specified government test surfaces" indicates that the test results were achieved under highly specified conditions. Clearly, changes in any of the test conditions could affect the traction results. This meaning is obvious from the present wording of the warning and further elaboration would needlessly lengthen the tire information. Therefore, the agency declines to adopt GM's suggested modification.

The agency has reached the position that the clarity of the traction grading information might be enhanced by the use of the letters A, B, and C in place of the symbols **, *, and O presently employed to denote traction grades. A proposal to modify the traction grading system by substitution of the letters A, B, and C for the present traction symbols is published concurrently with this notice in the proposed rule section of the *Federal Register*.

Temperature Resistance Labeling

Several commenters suggested that the tire temperature warning be clarified to indicate that excessive speed, underinflation, or excessive loading, either alone or in combination, can result in temperature increases and possible tire failure. The commenters suggested this change because heat build-up can occur at normal speeds when there is tire underinflation or overloading. The current proposal, however, implies that heat build-up would only occur at excessive speeds. The NHTSA agrees with this suggestion and modifies the temperature warning accordingly.

The RMA suggested that the label elaborate on the meaning of the temperature grades C, B, and A. The grades C, B, and A represent comparative differences in a tire's ability to withstand the generation of heat without suffering structural degeneration and potential tire failure. Although the grades C, B, and A in themselves do not inform a consumer of the specific amount of difference between tires in the three grades, the grades do convey to the consumer the fact that one tire performs better than the other in this specific test. To specify more exactly the amount of difference in heat dissipation represented by each grade or the technical nature of the test involved would merely confuse many people not versed in the technical nature of the test. Therefore, the agency has determined that the temperature grading method should be retained as it is. The NHTSA notes further that the court in the *B. F. Goodrich* case examined this aspect of temperature grading and found it to be adequate.

Miscellaneous Labeling

Several commenters requested that the agency implement a labeling system similar to that employed by the Federal Trade Commission (FTC) under the Magnuson-Moss Warranty Act (Pub. L. 93–637). The FTC in its regulations (16 CFR Part 702) permits the display of warranty information in any of four locations. The commenters to Notice 21 suggested that the agency should adopt the FTC's approach since Congress could not have intended that our regulations be more burdensome than those imposed under the Magnuson-Moss Warranty Act (Warranty Act).

The purpose of the Warranty Act is to ensure the open display of warranty data in order to provide consumers an opportunity to make buying choices based upon available warranties. The purpose of UTQG is similar but not identical to the Warranty Act. UTQG, like the Warranty Act, is intended to provide information to the consumer permitting him or her to make a rational choice in the selection of a product—specifically tires. Beyond the warranty data, however, the UTQG will dispel some of the inaccuracies and otherwise misleading information currently extant in the tire marketing business.

Congress considered tire retailing procedures to be a substantial problem. Accordingly, the Congress enacted a special provision in the National Traffic and Motor Vehicle Safety Act of 1966 to provide information to consumers on these products. The agency considers this specific mandate to justify the requirement that grading information be provided in several locations. At present, grading information must be contained on the tire sidewall (49 CFR 575.104 (d)(1)(i)(A)), on a label affixed to the tread surface (49 CFR 575.104(d)(1)(i)(B)), and in the information furnished under CFR 575.6(a) and (c) to motor vehicle purchasers and to prospective purchasers of vehicles or tires (49 CFR 575.104(d)(1)(ii) and (iii)). The provision of UTQG information in several locations will ensure the broadest possible dissemination of this information to consumers.

Further, unlike many other consumer goods that can be adequately handled by the Warranty Act, tires deserve additional consumer safeguards owing to their varied methods of marketing and their importance to traffic safety. Many consumer goods are purchased only as a single final unit from a retail outlet (e.g., small appliances). Tires, on the other hand, can be purchased individually or can come, as in the case of original equipment, as a component of another retail product (a motor vehicle). Accordingly, the need for maximum dissemination of information through several labeling locations is increased by the varied methods of tire retailing. The crucial role of tires in motor vehicle safety makes it imperative that information on tire quality be brought to the attention of consumers regardless of the marketing method employed.

The agency has previously carefully assessed its requirements for labeling in compliance with UTQG. In that assessment the agency determined that the Congressional mandate coupled with the unique nature of tire marketing warranted the labeling requirements established by the NHTSA. Further, the court in the *B. F. Goodrich* case upheld this labeling approach. Therefore, the agency declines to adopt the modification suggested by the commenters concerning the establishment of alternative labeling rather than mandatory labeling in several locations.

With regard to the wisdom of the UTQG labeling system in comparison with Warranty Act provisions, it is instructive that the FTC Chairman concluded in a September 16, 1977 letter to Goodyear that "it is apparent that the Uniform Tire Quality Grading System will produce useful, reliable information for the buying public." The letter contained no suggestions for improvement of the UTQG regulation, or that the UTQG regulation is in conflict with the Warranty Act.

On a matter of general application to the information label issue, Goodyear recommended that the agency ensure that the tire grading information will be presented to the tire purchaser. To achieve this goal, Goodyear suggested that the tire retailer be required to display the information. Without such a requirement they argued, tire grading information would not be useful.

The agency agrees that the provision of information in an easily identifiable and readily accessible location is necessary to the success of the tire grading concept. This is one of the reasons that the agency has been insistent about requiring the display of this information in a uniform fashion. The NHTSA encourages the open display of this information but remains convinced that the requirement that tires contain a label on the tire tread explaining the grading system is necessary for purposes of informing the public of tire grading. This label cannot be removed from the tire prior to sale. It is noted that a proposal to modify the requirements for this label is published concurrently with this notice in the proposed rule section of the *Federal Register*.

II. Course Monitoring Tires

On February 14, 1977, the agency issued a notice of proposed rulemaking that tentatively selected the course monitoring tires (CMT's) to be used for treadwear testing (42 FR 10320; February 22, 1977). The CMT's are run on the treadwear test course simultaneously with candidate tires in order to provide an index of course variability that allows the adjustment of treadwear results for such variability. The agency had previously selected the CMT's for radial tires. The court in *B. F. Goodrich* suggested that the NHTSA select all three of the CMT's concurrently including bias ply and bias-belted CMT's which the agency had previously not selected. The court further suggested that the agency permit a short comment period to receive responses on the agency CMT selections.

Most of the comments to this proposal did not question the selection of tires chosen by the NHTSA. Rather, the comments focused upon alleged inadequacies in the NHTSA rulemaking procedures and the statistical analysis employed by the agency to determine the coefficients of variation (COV) for the tires selected. Several commenters criticized aspects of the UTQG procedures previously determined to be valid by the court in the *B. F. Goodrich* case.

Adequacy of NHTSA Data

B. F. Goodrich and several other commenters argued that the agency did not provide ample time for meaningful comment to the notice announcing the selection of CMT's. These commenters alleged that the agency did not submit data to the docket in a timely fashion nor in complete form. For example, they argued that over 2,000 pages of data were docketed on February 14, 1977, which could have been placed in the docket as it was generated through the months of testing.

The agency placed in the public docket on February 14, 1977, more than 2000 pages of data accumulated through tests of the course monitoring tires. The notice announcing the CMT selections was issued simultaneously, and both the data and the notice were promptly brought to the industry's attention, even though the notice was not published by the *Federal Register* until February 22. Thus, the industry was given somewhat more than the 30-day comment period to analyze and evaluate the data. Commenters should note that the court in the *B. F. Goodrich* case considered that a 30-day comment period would be sufficient to permit adequate comment on the agency announcement of the CMT selections.

The agency did not submit the data pertaining to the CMT selections to the docket in a piecemeal fashion as the commenters suggested should be done for several reasons. First, until all the data were generated and reviewed by the agency no decision could be made concerning the adequacy, in light of the court's mandate, of the CMT's initially selected by the agency. Only after accumulating a mass of data from many tests could the agency be sure of its selections and accordingly go forward with a notice making public its selections. To have released this information prior to the actual determination of the adequacy of the chosen tires would have been premature.

A second reason for waiting to release the information was the ongoing litigation on the subject of UTQG. The court's remand did not formally reach the agency until the mandate issued on December 3, 1976. Since further agency rulemaking action depended upon the outcome of the *B. F. Goodrich* case, the NHTSA considered it necessary to receive the final mandate of the court prior to continuing with its rulemaking effort with respect to UTQG. Upon receipt of the mandate of the court, the agency began rulemaking in compliance with the remand. Rulemaking proceeded expeditiously even though petitioners in the *B. F. Goodrich* case had filed a petition for certiorari.

A further criticism by the commenters concerned an alleged continued withholding by the agency of data necessary for informed comments on the CMT selections. Several commenters stated that the data in the docket contain omis-

sions. For example, the numbered data do not progress in a serial manner.

The agency has not withheld relevant information from the docket as the commenters suggest. The extent that the numbered data (test numbers) do not proceed in a serial manner results from the inclusion of the docket only of those tests involved with the computation of the coefficients of variation (COV). The COV's were computed from the first 6,400-mile cycle (after an 800-mile break-in) of the CMT, as prescribed in the UTQG regulation. Subsequent cycles run on the same CMT were not run for purposes of computing the COV. Therefore, subsequent test cycles of the same tires were deleted from the docketed data so as not to be confused with the computation of the COV's. All of the data upon which the agency based its determinations pertaining to the COV's were placed in the docket.

A further argument of the commenters was that the agency failed to include an analysis of the data indicating how our conclusions concerning COV's were achieved. The agency has used an established method for the determination of the coefficients of variation. The method chosen is an accepted statistical technique. The NHTSA does not consider it necessary to reproduce underlying, routine computations when each set of data is put into the docket.

In connection with the alleged lack of information in the docket, several commenters suggested that the NHTSA make further submissions to the docket concerning the test procedures used by the agency in testing the CMT's. The existing rule on UTQG contains the test procedures for conducting treadwear tests, and the *B. F. Goodrich* case upheld these test procedures. When the agency tests CMT's, the procedures outlined in the rule are, of course, rigidly followed. No other information relevant to the conduct of these tests exists to be placed in the docket.

Some commenters argued that the NHTSA should make public some of the test variables in existence on the days tests were conducted. For example, they suggested that weather could have an impact upon test results and, therefore, records of such weather conditions should be made available to them. The agency did not maintain such records, for the simple reason that the CMT procedure is specifically intended to account for all such variables. Of course, data such as weather conditions, can be determined from the information contained in the docket. The test data list the date each test was run. If parties care to gather extraneous data for their own purposes, weather information for the days in question can be obtained by contacting a weather service. It should be noted that many major tire manufacturers test in Southwest Texas. Indeed, Goodyear has stated in a brochure which describes its San Angelo proving ground, that "the San Angelo area presents the most ideal conditions for tire testing in the United States." (Docket 25, GR 86.)

The RMA requested as part of their comments that, since further information should in their opinion be placed in the docket, the agency extend the comment period. The agency, as stated above, placed all pertinent information in the docket, obviating the need for an extended comment period. Further, NHTSA procedures for requesting extensions, 49 CFR 553.19, require that such a request be submitted not less than 10 days before expiration of the comment period in accordance with those procedures. Instead, the RMA included a request for extension in the body of their docket comment. It should be noted that, while the procedurally defective request was not granted, the agency has continued to accept and consider the comments of the RMA and others that have been received well after the comment closing date.

Several commenters suggested that the NHTSA publish the base course wear rates for the CMT's chosen by the agency. Publication of these wear rates, the commenters argued, was necessary for their testing of the CMT's and thus for meaningful comments on Notice 22. The agency disagrees that it is necessary to have the base course wear rates for purposes of commenting upon the tires selected by the agency as CMT's. It is the coefficient of variation experienced in the testing that is relevant to their selection as monitors of the course, and the base course wear rate is irrelevant to this consideration.

Since the commenters desired the publication of these figures, albeit irrelevant to the selection of the CMT's, the agency hereby makes them public. The wear rates for the bias ply tire

(Armstrong Surveyor 78) and for the bias-belted tire (General Jumbo 780) are 9.00 mils and 6.00 mils per 1,000 miles, respectively. Since these figures have no impact upon the selection of CMTs announced in Notice 22, no comment period is required as a result of the publication of the base course wear rates.

Firestone submitted two NHTSA technical papers for inclusion in the Docket. These papers have been modified by Firestone's underlining without other comment. These papers are included in the docket even though they are not relevant to the present UTQG regulation.

Possible Radial Wear Rate Problem

In Notice 22, the agency stated that the data appeared to indicate that the wear rate for some radial tires may not be constant. The NHTSA concluded, therefore, that radials would not be included for the time being under the UTQG rule, since computations made under that rule contemplate a constant adjusted wear rate for projection purposes. Industry commenters objected to this treatment of radials and argued that the agency should not proceed with any of the grading requirements unless it proceeds with them all simultaneously.

These commenters cited the *B. F. Goodrich* case which remanded the course monitoring tire issue to the agency, because a selection of all of the CMT's had not been made prior to the establishment of an effective date for the implementation of the rule to all tire types. The commenters interpreted this court mandate to mean that the agency was required to proceed with the promulgation of grading requirements for all three tire types concurrently. The agency does not interpret the court decision in that manner.

The 6th Circuit Court remanded to the agency the issue of the selection of the CMT's. It should be noted that at the time of the court decision the agency had not selected the bias and bias-belted CMT's even though it had established the effective dates for all tire types. Moreover, the court noted that the selection of the radial CMT had been based upon a series of tests (reported in NHTSA Technical Note T–1014) which were flawed by a problem not clearly identified or explained. The court's conclusion, therefore, was that it was inappropriate to schedule the effective date for compliance of tires with UTQG when the NHTSA had not given notice and invited comment on its selection of the CMT's. This mandate of the court does not prohibit the promulgation of the rule in phases, however.

The court's opinion stated that it would be inappropriate to require grading of a tire when all of the procedures (in this case the CMT selection) had not been chosen, and commented upon, for that tire. The court did not, in the opinion of the NHTSA, state that the agency could not proceed with rulemaking on some tire types pending further study of the application of the rule to another tire type. Therefore, the agency does not find merit in the position of the commenters who allege that the agency must proceed with a rule for all tire types at the same time.

The agency has responded to the remand in Notice 22 by announcing the selection of all CMT's. That notice gave the industry adequate time to comment upon the agency's selections. However, until possible problems concerning the testing of radials are resolved, the agency will not set an effective date for the application of the rule to radial tires. As long as an effective date applicable to the grading of radials is not established prior to the establishment of grading procedures for that tire, the NHTSA can implement the rule with respect to the other tire types and is not in violation of the court's remand.

Several commenters argued that regardless of the court mandate, the NHTSA should not go forward with tire grading for two tire types while excluding radials. The commenters asserted that altered test procedures for radials could result in different tests or a different test course for radial tires which would make comparisons between them and the other tire types meaningless.

By this comment, it is apparent that some people may have misunderstood the agency's earlier notice announcing the possible problem with radials. The problem that may attend the grading of radial tires is one of computing the wear rate after the 6400-mile test has been completed, since there is some evidence suggesting that these tires may not wear at a constant rate after only an 800-mile break-in. No comparable

problem has been found for bias and bias-belted tires. Ample data have been generated demonstrating that the wear rates for bias ply and bias-belted tires are constant after an 800-mile break-in. At present there are no plans to alter the test course or the actual test procedures. If changes were considered necessary in either the test course or procedures, careful attention would then be given to their impact upon the comparative nature of the grades given other tire types. The agency would not implement test procedures for radial tires that differ from the procedures used for bias and bias-belted tires without affording adequate time for comment upon such test procedures and without carefully evaluating comments received on such test procedures.

The agency would like to note that with respect to the issue of radials, it was stated in the earlier notice that an *apparent* problem had been discovered with radials. The agency is not yet convinced that this problem does exist. However, until such time as further analysis can be accomplished, the NHTSA considers it prudent to proceed cautiously with the implementation of the UTQG requirements for radial tires.

Several commenters questioned the validity of the test procedures for testing treadwear. Goodyear stated that the driving instructions are unclear and, in particular, the braking procedure is not good. They stated further that the spacing in convoys was dangerously close on corners. Cooper Tire Company stated that the tests could not be repeated within statistically acceptable margins of error and, therefore, would be unenforceable.

The NHTSA does not agree with these comments questioning the validity of the test methodology. The agency has determined that these procedures provide a viable testing technique which can be duplicated for enforcement purposes. Further, the court in *B. F. Goodrich* upheld the test methodology. Accordingly, the agency sees no need to modify the test procedures.

Goodyear also argued that the test course has been changed since the last update of the rule by the agency. For example, they argued that some stop signs are now yield signs. On a test course of this size and nature, minor modifications of road signs are to be expected with certain regularity. The regulation only lists "key points" to assist regulated parties, and has updated the regulation to reflect changes in these key points and will continue to do so. The minor changes in the test track which have occurred since the last publication of the regulation are included in this notice.

The agency notes that with respect to sign changes in the treadwear course, such minor changes have no significant impact on tire grading. The use of CMT's is designed to reduce the effects, if any, of the course variables, including course markings. Therefore, the agency considers that minor changes in the road markings which will occur from time to time should have no impact upon the comparative ratings of tires. Nevertheless, the NHTSA will make every effort to update the regulation periodically to reflect changed course markings.

III. Effective dates

Several commenters asserted that the agency must propose effective dates to give the industry time to comment on the appropriateness of such dates. Notice 22 did not propose effective dates for the implementation of the regulation to bias and bias-belted tires. The agency has established the effective dates for all provisions other than the molding requirement as seven months from the publication of the final rule in the case of bias ply tires and 13 months from publication in the case of bias-belted tires. An additional six months has been provided in each case for the revision of tire molds. The issue of effective dates was litigated in the *B. F. Goodrich* case. The court there held that the implementation lead time as chosen by the agency was sufficient. The determination was based upon an evaluation of the capacity of the treadwear course and traction skid pads in relation to the number of tires to be tested. Therefore, since the agency has not modified the test procedure in any manner, there is no need to raise again the issue of effective dates as long as the agency allows the same lead time as was held valid by the court. Moreover, as noted in the court's opinion, the agency will closely monitor the actual use of the treadwear course and traction skid pads and will exercise its discretion to extend the lead time periods if it should become necessary to do so in the future.

Cooper Tire Company stated that changing the order of implementation of the requirements requires a reassessment of the effective date requirements. For example, radial tires no longer will be the first tire type to be tested. According to Cooper, a manufacturer may be harmed by the change in the order of implementation and further study of the effective dates is thus warranted.

The agency does not agree that a change in the order of implementation of the grading regulation for different tire types requires total reconsideration of the effective dates. As set forth in this notice and in Notice 22, bias ply will be the first tire construction type required to be graded. A count by NHTSA staff of the number of passenger tire lines set forth in a standard reference, "1977 Tread Design Guide" (published by the Tire Information Center, Commack, New York), excluding winter treads (snow tires) and duplicates of the same tread design, indicates that of some 1139 tire lines on the market, approximately 431 are radials, 408 are bigs-ply, and the remaining 300 are bias-belted. Therefore, if ample time was provided in the previous rule for the testing of radials, and the court held that the lead time was sufficient, there certainly should be sufficient lead time to test bias ply tires which are fewer in number. Although this change may create greater test burdens for individual manufacturers, it will not impair the ability of the test facilities to accommodate tire grading.

IV. Statistical Comments

The RMA criticized the NHTSA's statistical analysis of the data upon which the coefficients of variation were derived. The RMA submitted a paper written by Dr. Shelemyahu Zacks purporting to discredit the NHTSA's analysis. Through this paper the RMA suggested that the coefficients of variation (COV) were larger than the agency had indicated.

The analysis done by the NHTSA was conducted according to statistically acceptable procedures, but the NHTSA concluded that it would be prudent to obtain an impartial review of both the Zacks' and the NHTSA's analyses of the COV's. The agency contracted with a noted statistician, Dr. Herbert Solomon, who reviewed the agency's procedures in view of Dr. Zacks' criticisms of those procedures and concluded that the agency was correct in its method of computation of the COV's. The full text of both the Zacks and Solomon papers as well as the agency's analyses of the former are in the public docket.

Subsequent to the Solomon report, the RMA submitted several comments intended to refute the accuracy of the report. In particular, the RMA contended that the use by NHTSA of "n" ("n"=sample size), rather than "n-1", as the divisor in computing the sample standard deviation was incorrect and produced an inaccurately low COV. After careful review of this question, the agency has concluded that the use of "n" in the formula for the sample standard deviation is a proper statistical approach as a step in the process of determining the sample COV. Moreover even if the alternative "n-1" formula were adopted, the resulting COV's of 4.74, 3.08, and 2.70 for bias, belted bias, and radial tires respectively would still fall within the 5% coefficient of variation which was approved by the court in the *B. F. Goodrich* case. The RMA's other contentions were also carefully reviewed and were found to be invalid and to reiterate much of the information contained in earlier RMA comments. Therefore, the agency declines to adopt the statistical approach proffered by the RMA as well as the other recommendations of the RMA that attend their method of statistical analysis.

B. F. Goodrich submitted a statistical study by its engineering staff of models of the wear behavior of tires. (C. Thomas Wright, "The Adequacy of Linear Models in Tread Life Testing"). The agency's analysis of the study revealed that significant errors in the study accounted for Wright's differences with the linear model employed in the regulation. The agency analysis was placed in the docket, and B. F. Goodrich subsequently filed a rebuttal to the analysis. Review by the agency of that rebuttal confirms that Wright's differences with the regulation's linear model involve his failure to observe conventional statistical precepts.

Uniroyal submitted comments suggesting that the NHTSA testing procedure did not adequately consider the effects of actual driving conditions upon tire grades. Uniroyal conducted a random sampling of tires on automobiles in parking lots. The conclusion of that study was that tires wear at varied rates depending upon the type of car, size of tire, load on the tire, and many other variables. Uniroyal suggested that its results indicated that it would have to test unlimited combinations of its tires to ensure correct grading.

The NHTSA has always stated that UTQG does not give an exact measurement of a tire's life under all conditions. The agency realizes that tire life will vary depending upon a number of conditions. The court in *B. F. Goodrich* also recognized this fact when it stated that no test designed to grade millions of tires will be perfect. Few measuring techniques are. However, for this reason the agency cautions individuals concerning misapplication of the grading information.

The Uniroyal survey yields results that are to be expected but that have no impact upon the validity of the UTQG test procedures. The test procedures for UTQG control most of the variables. The course, speed, drivers, stopping conditions, and many other variables are controlled for tire testing purposes. For those environmental variables beyond the control of the agency, the NHTSA uses the CMT to measure their effect. The Uniroyal study did not control these variables. Accordingly, it does not present an accurate picture of comparative data between tire lines. The agency has determined that comparing different tires under similar conditions on the treadwear course and traction skid pads does yield excellent comparative data. Therefore, the agency discounts the value of the Uniroyal study for purposes of questioning the validity of UTQG testing. The Uniroyal study merely indicates that the public must be cautioned against the misuse of grades provided on the tires. The NHTSA concludes that the warnings provided on the grading label information provide sufficient cautionary advice to the consumer.

Cooper Tire Company ran computer tests intended to show that the same tire might receive different grades with any two tire treadwear tests. According to Cooper this indicated that the UTQG requirements are unenforceable.

It has been argued in the past that enforcement testing for many of the agency's regulations and standards depends upon a test of a single piece of equipment or motor vehicle and accordingly the results cannot be projected to all vehicles or equipment. In other words, the commenters suggest that a noncompliance in one vehicle or item of motor vehicle equipment does not mean that all vehicles are defective.

The agency's enforcement actions pertaining to all standards have been conducted, in the past, using a variety of data. A failure of equipment or a vehicle to reach a performance standard during an agency enforcement test indicates a potential noncompliance. The agency then goes to the manufacturer of the affected vehicle or equipment and requests the results of the manufacturer's tests or other data upon which he based his certification of compliance with the standard. A similar method of enforcement is contemplated for UTQG.

V. Petitions for Reconsideration.

On May 28, 1975, the NHTSA published the final UTQG rule. In response to that rule, several petitions for reconsideration were received by the agency. A response to these petitions for reconsideration was delayed pending the outcome of the litigation in the *B. F. Goodrich* case. Several of the issues raised in the petitions have been answered in that litigation or in subsequent notices issued by the agency. The NHTSA will now respond to those issues raised in the petitions and not previously addressed.

Several tire manufacturers commented that the lead time allowed prior to the effective date of the regulation was not adequate. The Japan Automobile Tire Manufacturers' Association, Inc. argued that there were significant time problems in the shipment of tires to the United States for treadwear testing on our test course and transmission of the resultant data back to Japan.

The issue of lead time was litigated in the *B. F. Goodrich* case. The court upheld the agency's proposed lead time. Since the agency does not propose to reduce the amount of lead time from that proposed in 1975, there should be no problem with meeting the effective date of the regulation.

Automobile manufacturers argued that they need more lead time than tire manufacturers since the specificity of the data required in the owner's manual forces them to wait until they receive the newly graded tires before printing the manuals. On a related point, many of the manufacturers suggested that the agency require in the owner's manual only general tire grading information. They argued that this is necessary because frequently manufacturers are unable to obtain the tire with which they normally equip their cars. In such an event, they would have to print a new owner's manual containing the new tire information and would be required by Part 575 of our regulations to submit a copy of this new information to the NHTSA 30 days prior to its issuance.

The agency has determined that the automobile manufacturers should operate under the same lead time constrictions as the tire manufacturers. Therefore, the effective date of the requirements applicable to the tire manufacturers shall also be applicable to the automobile manufacturers. This will ensure complete dissemination of grading information at the earliest possible time.

The agency has concluded that the manufacturer's suggestion to provide only general tire information in the owner's manual has merit. It would be cumbersome for a manufacturer to submit to the agency for 30-day review its owner's manual information every time a change in tires was contemplated or required. The agency considers it sufficient for purposes of informing consumers, for manufacturers to provide general grading information in the owner's manual. This information would explain the grading system, giving the cautionary warnings to the consumer concerning the possible misuse of the UTQG information. The consumer could then be directed to look at the tire sidewall for the particular grading of the tire. The rule has been amended to reflect this modification.

The Motor Vehicle Manufacturers Association (MVMA) and GM argued that the temperature resistance grading system would be misleading to consumers. Both suggested a two grade approach to temperature testing using the "high speed" designation for tires designed to operate under those conditions. The agency does not agree that the temperature information will be misleading. The implementation of the proposed warnings on the misuse of the temperature information should prevent any potential for consumer misunderstanding. The agency notes further that the court upheld the existing temperature resistance test.

Several manufacturers suggested that the NHTSA exempt the space saver tire from the UTQG requirements. They argued that this tire is designed for a limited life and for a special use only and, therefore, should not be required to comply with the regulation.

The NHTSA agrees that the space saver tire and other temporary use spare tires should be exempt from the requirements of the regulation. These tires are of reduced size or are inflatable. They are designed so that as installed in the vehicle, they reduce vehicle weight and create more vehicle interior space. Since the useful life of these tires is frequently limited to 2,000 miles, it would be inappropriate to require them to comply with the treadwear requirements. The agency amends the regulation to indicate that the space saver and temporary use spare tires are exempted from the regulation's requirements.

Volkswagen and the European Tyre and Rim Technical Organisation (ETRTO) argued that the treadwear information would confuse the public and be misused. ETRTO argued further that treadwear grading has nothing to do with safety and should be deleted from the requirements.

The treadwear labeling requirements are proper and were upheld by the court. Accordingly, the agency declines to change or delete those requirements as suggested by the manufacturers. Further, the agency notes that the UTQG regulation is promulgated under a special authorization of the Act (15 U.S.C. 1423). It is a consumer information regulation issued at the behest of the Congress.

On a related matter of labeling, ETRTO also requested that the words "treadwear", "temperature", and "traction" not be required to be molded into the sidewall owing to the expense of that operation. Once again, the 6th Circuit upheld the agency on its proposed labeling requirements while suggesting additional warnings to prevent the misuse of that information. The NHTSA requires the use of the words "traction", "treadwear", and "temperature", because these words will help avoid confusion as to the meaning of the symbols molded onto the tire sidewall.

ETRTO also suggested that NHTSA extend the effective dates for the traction requirements since the standard test trailer can not accommodate small tires. The agency declines to extend the effective date for the implementation of the requirements. However, small tires are being excluded from the requirements until such time as a test trailer is equipped to test them.

Dunlop recommended that the lowest of the three possible tire traction grades be eliminated, on grounds that an open-ended grade would allow production of tires with extremely poor traction in order to obtain higher treadwear or temperature resistance grades. In effect, Dunlop was requesting a minimum traction standard. The agency has an outstanding proposal that would establish such a minimum standard (38 FR 31841); November 19, 1973) and will respond to Dunlop's request by means of the separate rulemaking.

Dunlop suggested that the agency permit the tire information to be molded onto the tire in two tiers using smaller size lettering. Currently the regulation requires that the information be molded into the sidewall in either one or three tiers using $\frac{1}{4}$ inch lettering. Dunlop argued that some of their tires are too small to permit the display of information printed in one tier without conflicting with other information molded on the sidewall. Further, they stated that the depth of their tires was such that three tiers of information would not easily fit on them.

The exclusion of the smallest tires from the UTQG requirements for the time being may alleviate this problem since these are the tires that present the greatest problems concerning available space for sidewall molding. Nonetheless the agency amends the regulation to reduce the print size of the required molding from $\frac{1}{4}$ inch to $\frac{5}{32}$ inch. Finally, the NHTSA can see no reason not to permit the molding of information into the sidewall in two tiers. Accordingly, the agency amends the regulation establishing a format for two tier information.

In a comment by ETRTO, it was suggested that the agency clarify its position with respect to the use of front wheel drive and rear wheel drive vehicles in a convoy for treadwear testing. The regulation states that the vehicles used will be rear wheel drive vehicles, but the preamble (Notice 17) stated that testing would be accomplished by the use of vehicles for which the tires were designed, which might include front wheel drive vehicles. In accordance with the regulation which was issued in 1975 and upheld by the court, the agency has determined that only rear wheel drive vehicles will be used for treadwear testing. This removes the possibility that any vehicle variations between front and rear wheel drive vehicles will affect the tire test results.

In accordance with Department policy encouraging adequate analysis of the consequences of regulatory action, the agency has evaluated the anticipated economic and other consequences of this amendment on the public and private sectors. The agency has determined that the regulation will benefit tire consumers by affording them more detailed information upon which to make informed tire purchases. The regulation will thus reduce some of the existing confusing claims associated with tire marketing.

As the purpose of UTQGs is to help the consumer make an informed choice in the purchase of passenger car tires, the agency will soon initiate action to evaluate whether the rule is meeting this goal. It is planned that surveys will be undertaken to determine how easily understandable and meaningful the grades are to purchasers, how the grades are utilized in purchase decisions and any measurable economic effect that may occur both within the passenger tire industry and to consumers as a result of the rule. The emphasis will be on the utility of the grading system to consumers. Major points of interest of the consumer survey will be the extent to which consumers use the grading system in their purchase decisions, the extent to which it has increased their knowledge and awareness of the characteristice of various tire constructions and tire lines and whether they feel the grading system is valid and worthwhile.

Effective date finding: Under section 203 of the Act, the Congress stated that the regulation should become effective not sooner than 180 days nor later than one year from the date that the rule is issued. Based upon this direction and other agency findings concerning required lead time for grading tires, the agency has determined, and the Court has upheld, that phased implementation of the rule in essentially 6-month intervals is appropriate.

The program official and lawyer principally responsible for the development of this rulemaking document are Dr. F. Cecil Brenner and Richard Hipolit, respectively.

In consideration of the foregoing Part 575.104 of Title 49 of the Code of Federal Regulations, is amended. . . .

(Secs. 103, 112, 119, 201, 203; Pub. L. 89–563, 80 Stat. 718 (15 U.S.C. 1392, 1401, 1407, 1421, 1423); delegation of authority at 49 CFR 1.50.)

Issued on July 12, 1978.

Joan Claybrook
Administrator

43 F.R. 30542
July 17, 1978

Temperature for Tire Testing

Action: Final rule.

Summary: This notice establishes a uniform tire testing temperature for the test requirements of the Uniform Tire Quality Grading regulation and the Federal motor vehicle safety standard for non-passenger-car tires. This amendment simplifies existing requirements by permitting various tire tests to be conducted at the same temperature.

Effective date: July 17, 1978.

For further information contact:

Arturo Casanova III, Crash Avoidance Division, Office of Vehicle Safety Standards, National Highway Traffic Safety Administration, 400 Seventh Street, S.W., Washington, D.C. 20590 (202) 426–1715.

Supplementary information: The National Highway Traffic Safety Administration (NHTSA) proposed on March 3, 1977 (42 FR 12207), to amend the ambient temperature conditions for tire testing contained in Standard No. 119, *New Pneumatic Tires for Vehicles Other Than Passenger Cars* (49 CFR 571.119), and in Part 575, *Uniform Tire Quality Grading* (49 CFR 575.104) (UTQG). The purpose of this proposed amendment was to harmonize existing tire testing temperatures as requested by the Goodyear Tire and Rubber Company. The ambient temperatures were previously specified as follows:

Standard No. 109: "100±5° F."

Standard No. 119: "any temperature . . . up to 100° F."

UTQG: "at 105° F."

In the notice of proposed rulemaking, the agency proposed to amend Standard No. 119 and UTQG to reflect the tire temperature utilized in Standard No. 109 (100±5° F.). As an alternative method of expressing the test temperature, the NHTSA proposed to amend the standards to specify "any temperature up to 95° F.

Five comments were received in response to that proposal. All comments favored the proposed amendment that would have instituted a 100±5° F. temperature. The Vehicle Equipment Safety Commission did not take a position on this proposal.

After consideration of the issues involved in the proposal and review of the comments, the agency has determined that the test temperature should be expressed as "any temperature up to 95° F." Accordingly, Standard No. 119 and UTQG are amended to specify temperature testing at "any temperature up to 95° F." It is the NHTSA's opinion that the 95° F. test temperature is in effect the same test temperature as would be achieved by using the 5-degree tolerance (100±5).

The NHTSA has often stated in interpretations on similar issues that the use of tolerances in safety standards reflects a misunderstanding of the legal nature of the safety standards. Standards are not instructions, but performance levels that vehicles or equipment are required by law to be capable of meeting. Any tolerance in this context would be meaningless and misleading, since it would merely have the effect of stating a performance level that the equipment must meet when tested by the government, but in a confusing manner.

Recognizing that no measurement is perfectly precise, a manufacturer's tests should be designed to show, using tire testing temperature as an example, that his tires will comply with the requirements at exactly 95° F. This may be done in at least two ways: (1) by using a test method that corresponds so closely to the required temperature that no significant differences could occur as a result of differences between the actual temperature and the specified one, or (2) by determining which side of the specified temperature is adverse to the product tested, and being sure that the actual temperature of the test differs from the specified one on the adverse·side.

The amendment of Standard No. 119 and UTQG to reflect the 95° F. temperature creates a different temperature phraseology for those standards than exists in Standard No. 109 which still has the 100±5° F. temperature. As stated earlier, the NHTSA considers the Standard No. 109 temperature tolerance to mean in actuality "any temperature up to 95° F." However, since modification of that standard was not proposed in the earlier notice, the agency does not amend it in this final rule. However, the agency intends to issue an interpretive amendment that will amend Standard No. 109 to adopt the alternative expression for tire temperature testing (any temperature up to 95° F.) unless objections are received.

In accordance with Departmental policy encouraging analysis of the impact of regulatory actions upon the public and private sectors, the agency has determined that this modification will result in no appreciable safety gains or losses. These amendments may result in slightly lower costs for tire temperature testing since all temperatures will be uniform.

Since these amendments relieve restrictions and impose no additional burdens, it is found for good cause shown that an immediate effective date is in the public interest.

In consideration of the foregoing, . . . amendments are made in Parts 571 and 575 of Title 49, Code of Federal Regulations.

The program official and lawyer principally responsible for the development of this rulemaking document are Arturo Casanova and Roger Tilton, respectively.

(Secs. 103, 112, 119, 201, 203, Pub. L. 89–563, 80 Stat. 718 (15 U.S.C. 1392, 1401, 1421, 1423); delegation of authority at 49 CFR 1.50.)

Issued on July 12, 1978.

Joan Claybrook
Administrator

43 F.R. 30541
July 17, 1978

This notice amends the Uniform Tire Quality Grading (UTQG) Standards to revise the grading symbols used to indicate traction grades and responds to a petition for reconsideration of the effective dates for the information requirement regarding first purchasers of motor vehicles. The notice, further, responds to petitions for reconsideration submitted by the Rubber Manufacturers Association and The Goodyear Tire & Rubber Company, regarding an amendment of the tire testing temperature employed in the UTQG regulation and the non-passenger-car tire safety standards, which established a single test temperature for the performance requirements of the two standards. The notice also withdraws a NHTSA proposal to modify the tread label requirements of the Uniform Tire Quality Grading Standard. These actions are intended to aid consumer understanding of the UTQG grading system and facilitate industry tire testing.

Effective date: October 23, 1978.

For further information contact:

Dr. F. Cecil Brenner, Office of Automotive Ratings, National Highway Traffic Safety Administration, 400 Seventh Street, S.W., Washington, D.C. 20590, (202) 426–1740.

Supplementary information: On July 17, 1978, (43 FR 30542), NHTSA republished the UTQG Standards (49 CFR 575.104) to assist the consumer in the informed purchase of passenger car tires. (Docket No. 25, Notice 24). The standard requires that manufacturers and brand name owners provide simple comparative data on tire performance, which can be considered by purchasers in selecting between competing tire lines. Concurrently, with issuance of the final rule, the agency proposed modifications of the standard's provisions relating to traction grading symbols and tread labels (43 FR 30586; July 17, 1978).

Traction Grading Symbols

The notice of proposed rulemaking (43 FR 30586), issued concurrently with the republished final rule, proposed revision of the symbols used to denote tire traction grades. The agency invited comment on the use of an A, B, C hierarchy of traction grades in place of the **, *, 0 system now required by paragraph (d)(2)(ii).

The Automobile Club of New York commented that the proposed traction grading symbols would be "far more meaningful to consumers" than the asterisks and zeros used in the existing regulation. The National Tire Dealers & Retreaders Association viewed the letter grading proposal as an improvement, and, in response to Notice 24, the Metropolitan Dade County, Florida, Office of the Consumer Advocate approved of an A, B, C grading system as falling within the experience of all consumers.

The only negative comment came from Atlas Supply Company which expressed concern that, if consumers are warned, as the rule requires, that tires with a C traction grade may have poor traction performance, they may assume that a C temperature resistance grade likewise denotes poor temperature resistance qualities. Atlas recommended that the lowest traction grade be abolished completely and that only the symbols A and B be used to represent traction grades.

In fact, the agency is currently considering promulgation of a tire traction safety standard which would set a minimum performance level such that tires falling within the lowest UTQG traction performance grade would not comply with the safety standard (43 FR 11100; March 16, 1978, and 38 FR 31841; November 19, 1973). Pending issuance of such a standard, however, consumers should not be misled as to the nature of the C temperature grade, since the explanation of the grading system, to be furnished under the standard, specifically states that the C grade indicates a level of performance which meets the applicable Federal safety standard.

The agency has concluded that the A, B, C grading symbols for traction performance will be an aid to consumer understanding of the UTQG system due to the general familiarity with letter grading systems and the hierarchy inherently associated with these symbols. Consumer comprehension of the grading system will also be improved by eliminating the need to use three different sets of symbols. The symbols A, B, and C are, therefore, adopted to represent traction grades under the UTQG Standard.

Tread Label Requirements

The existing UTQG regulation provides that each passenger car tire, other than one sold as original equipment on a new vehicle, shall have affixed to its tread surface a label indicating the specific treadwear, traction, and temperature grades for that tire, as well as a general explanation of the grading system. In its July 17, 1978 notice of proposed rulemaking (43 FR 30586), the agency proposed to amend section 575.104 (d)(1)(i)(B) of the standard, to require only general grading information on the tread label, while retaining a separate requirement that specific grades be molded on the tire sidewall. The tread label would have been modified to include a statement referring the consumer to the tire sidewall for the actual grades of the particular tire. The notice also proposed that specific tire grades be supplied, at the manufacturer's option, on either tread labels or on the sidewall during the six-month period prior to the effective dates of the molding requirement.

In commenting on the notice, Goodyear argued that provision of specific grading information on the tread label would not be feasible and would add to the cost of implementation of the standard. American Motors Corporation commented that provision of specific grades in two places would be redundant and an unnecessary expense.

However, Michael Peskoe, an individual involved in early development of the standard, argued that the tread labeling requirement is not redundant, since tire sidewall molding was intended primarily to supply a permanent record

of the tire grades, to be considered when replacing the tires, rather than to convey information to the prospective purchaser. He also stated that, with regard to cost and feasibility considerations, tire specific identification labels, bearing information such as tire line and size, are already in widespread use within the industry to aid in the distribution of tires. Therefore, the burden of adding the specific UTQG grades for the particular tire classification should be minimal.

The Automobile Club of New York and Mr. Peskoe commented that provision of specific tire grades only on the sidewall would hinder use of the information in the situation, common in tire dealerships and service stations, where tires are displayed on racks, sidewall to sidewall. Tires would have to be removed from the display rack before the grades molded on the sidewall could be observed. The problem would be compounded where the purchaser wishes to compare the grades on several tires.

While NHTSA is concerned with keeping the cost of the UTQG regulation at a minimum, existing tire labeling and marketing practices lead the agency to the conclusion that tread labels containing specific tire grading information should continue to be required for replacement tires. The agency had earlier determined that identification of specific tire grades on tread labels is feasible and involves a very limited cost to manufacturers and consumers. Tire-specific tread labels have been demonstrated to be an integral and necessary part of the regulation's plan for getting useful information to tire purchasers. The proposal to require only general grading information on tire tread labels is, therefore, withdrawn.

Effective Dates for Point of Sale Information

Notice 24 set March 1, 1979, in the case of bias-ply tires and September 1, 1979, in the case of bias-belted tires, as effective dates for all UTQG requirements except the molding requirements of paragraph (D)(1)(i)(A). The molding requirements applicable to bias and bias-belted tires were made effective September 1, 1979, and March 1, 1980, respectively.

The purpose of this delayed phase-in schedule for tire sidewall molding is to provide manufacturers with extra time to prepare new tire molds containing grading information. However, the delay in effective dates for tire molding had the unintended effect of creating a six-month interval between the time vehicle manufacturers must provide point of sale information on tire quality grading to prospective purchasers, and first purchasers of motor vehicles (49 CFR 575.104(d) (1)(ii) and (iii)) and the date on which grading information actually must appear on the tires sold. In the case of information to be furnished to first purchasers under paragraph (d)(1)(iii), potential for confusion exists since consumers will be referred to the tire sidewall for specific tire grades, when in many cases, molds will not yet have been modified for the tire lines being supplied.

To correct this situation, American Motors Corporation has petitioned NHTSA to reconsider the effective dates for paragraph (d)(1) (iii). American Motors has recommended that the effective dates for paragraph (d)(1)(iii) be amended to correspond to those of paragraph (d)(1)(i)(A), the molding requirement. The agency has already recognized the difficulties involved in providing specific grades for original equipment tires through the use of tread labels (39 FR 1037; January 4, 1974) or point of sale information (43 FR 30547; July 17, 1978). To better coordinate the availability of specific tire grading information on tire molds and the provision of explanatory information through vehicle owner's manuals, American Motors' petition for reconsideration is granted. The effective dates for paragraph (d)(1)(iii) are changed to September 1, 1979, for bias-ply tires and March 1, 1980, for bias-belted tires.

Paragraph (d)(1)(ii) of the regulation requires that vehicle and tire manufacturers furnish to prospective purchasers an explanation of the UTQG grading system. Although this provision also takes effect six months prior to the tire molding requirements, the agency has concluded that no corresponding change in effective dates is necessary. Paragraph (d)(1)(ii) provides for the availability of valuable information to prospective tire purchasers, since specific grading information will be available on replacement tires sold during the six-month phase-in period. Further, the paragraph contains no potentially confusing

reference to the tire sidewall as does paragraph (d)(1)(iii). Prospective vehicle purchasers who obtain the information prior to the sidewall molding effective dates will be given the opportunity to familiarize themselves in advance with the new grading system.

Temperature for Tire Testing

On March 3, 1977 (42 FR 12207), NHTSA proposed to amend Standard No. 119, *New Pneumatic Tires for Vehicles Other Than Passenger Cars* (49 CFR 571.119), and the UTQG Standards to establish the same ambient temperature for tire testing in both standards, to allow more efficient use of tire test facilities. The notice proposed "any temperature up to 95° F" and "100±5° F" as alternative means of phrasing the new, identical test temperature.

After consideration of comments, the agency determined that the ambient test temperature should be expressed as "any temperature up to 95° F" (43 FR 30541; July 17, 1978). NHTSA received petitions for reconsideration from the Rubber Manufacturers Association (RMA) and The Goodyear Tire & Rubber Company, recommending that the test temperatures for Standard No. 119 and the UTQG regulation include tolerances and be specified as "100° F±5° F." As NHTSA has frequently stated in past notices on these and other standards (e.g., 40 FR 47141; October 8, 1975), such a recommendation reflects a misunderstanding of the legal nature of motor vehicle standards. NHTSA standards are not instructions to test engineers, but performance levels that vehicles and equipment must be capable of meeting. The use of a tolerance range in this context is confusing since it creates ambiguity as to the performance level required.

Establishment of a precise performance requirement, expressed without a tolerance, still recognizes that measurement techniques cannot be controlled perfectly. Given a specified performance level, manufacturers can design their tests to assure compliance in at least two ways: (1) by using a test procedure that conforms so closely to the specified measurement that no significant variations could occur, or (2) by determining which side of the specified level is adverse to the product being tested, and targeting test conditions so that any deviation will occur on the adverse side. In this case, a tire manufacturer may use an ambient temperature slightly above 95° F to demonstrate, through adverse conditions, that its tire would comply at the specified temperature.

In its petition for reconsideration, Goodyear commented that all test laboratories should employ the same ambient temperature conditions. However, such uniformity is not advantageous in a regulatory context, since government compliance testing and manufacturers' laboratory evaluations are undertaken for different purposes.

Goodyear also argued that a fixed 95° F test temperature and a "100±5° F" tolerance range do not establish "in effect the same test temperature", as stated in the agency's July 17, 1978 notice (43 FR 30541). A fixed 95° F requirement is, in fact, from the manufacturers' perspective identical to a "100±5° F" provision, since, given a controlled variation in test conditions of 5° F in either direction from the target temperature, manufacturers seeking to assure compliance with a 95° F requirement will set their test target temperature at 100° F. For these reasons, the petitioners' recommendation of a "100±5° F" test temperature is rejected.

The RMA and Goodyear petitions noted that the open-ended nature of the requirement "any temperature up to 95° F" appeared to require that tires be capable of attaining specified performance levels when tested at temperatures ranging from 95° F to sub-zero conditions. The RMA petition stated as its primary concern the possibility, under the UTQG system, that a tire could be conditioned at a higher temperature than that at which it is tested for temperature resistance. Such inconsistency could, the RMA suggested, result in the tire being underinflated during testing.

The agency has concluded that the ambient temperature specification "at 95° F" more accurately describes the fixed temperature which the agency intended to establish than does the open-ended provision "any temperature up to 95° F." Standard No. 119 and the UTQG

Standards are, therefore, amended by substitution of a fixed temperature requirement of 95° F in place of 'any temperature up to 95° F.'

To the extent that the RMA and Goodyear petitions for reconsideration are not granted by this amendment, the petitions are denied.

In accordance with Departmental policy encouraging analysis of the impact of regulatory actions upon the public and private sectors, the agency has determined that these actions will have no appreciable negative impact on safety. Since the modification of effective dates relieves a restriction, and the change in grading symbols will result in no new burdens, no additional costs will be imposed on manufacturers or the consumer. Withdrawal of the tread labeling proposal imposes no new costs not contemplated in issuance of the UTQG Standards. The new temperature phraseology has absolutely no effect on the tire performance requirements, but will eliminate any possible ambiguity in the standards' meaning. For these reasons, the agency hereby finds that this notice does not have significant impact for purposes of the internal review.

Effective date: In view of the need for a fixed temperature requirement to allow tire performance testing to proceed, and the ongoing preparation by the industry for implementation of the UTQG system, the agency finds that an immediate effective date for the amendments to Standard No. 119 and the UTQG regulation is in the public interest.

In consideration of the foregoing, the following amendments are made in Part 575 and 571....

(Sec. 103, 112, 119, 201, 203, Pub. L. 89–563, 80 Stat. 718 (15 U.S.C. 1392, 1401, 1421, 1423); delegation of authority at 49 CFR 1.50.))

Issued on October 23, 1978.

Joan Claybrook
Administrator

43 F.R. 50430–50440
October 30, 1978

PREAMBLE TO AMENDMENT TO PART 575—CONSUMER INFORMATION

Uniform Tire Quality Grading

(Docket No. 25, Notice 31)

Action: Final rule and establishment of effective dates.

Summary: This notice announces the effective dates for application of the Uniform Tire Quality Grading (UTQG) regulation to radial tires and discusses comments on previously announced testing and analysis of radial tire treadwear under the road test conditions of the UTQG regulation. This notice also interprets the effect of the thirty-day stay of the UTQG effective dates, granted by the U.S. Court of Appeals for the Sixth Circuit, and corrects an inadvertant error in the text of the regulation.

Effective date: For all requirements other than the molding requirement of paragraph (d)(1)(i)(A) and the first purchaser requirement of paragraph (d)(1)(iii), the effective date for radial tires is April 1, 1980.

For paragraph (d)(1)(i)(A), the molding requirement, and paragraph (d)(1)(iii), the first purchaser requirement, the effective date for radial tires is October 1, 1980.

For further information contact:

Dr. F. Cecil Brenner, Office of Automotive Ratings, National Highway Traffic Safety Administration, 400 Seventh Street, S.W., Washington, D.C. 205 (202) 426–1740.

Supplementary information: Acting under the authority of the National Traffic and Motor Vehicle Safety Act of 1966 (the Act) (15 U.S.C. 1381, et seq.), the NHTSA republished as a final rule the UTQG Standards, establishing a system for grading passenger car tires in the performance areas of treadwear, traction and temperature resistance (43 FR 30542); July 17, 1978). The regulation will provide consumers with useful, comparative data upon which to base informed decisions in the purchase of tires. Extensive rulemaking preceded the July 17th notice, and a comprehensive discussion of the regulation's purpose and technical justification may be found in a series of earlier Federal Register notices (40 FR 23073; May 28, 1975; 39 FR 20808; June 14, 1974); 39 FR 1037; January 4, 1974; 36 FR 18751; September 21, 1971).

The July 17 notice also established effective dates for application of the regulation to bias and bias-belted tires. Establishment of an effective date for radial tires was deferred pending further analysis of test results relating to the treadwear properties of radials. Questions concerning the two other performance areas of the standard, traction and temperature resistance had previously been resolved, and therefore are not discussed in this notice.

On November 2, 1978, NHTSA issued a notice (43 FR 51735; November 6, 1978) announcing the availability for inspection of the results of the agency's test program for radial tires and NHTSA's analysis of the test results (Docket 25; Notice 28). A thirty-day period, later extended to 45 days (43 FR 57308; December 7, 1978), was provided for public comment on the data and analysis. After examination of all comments received, NHTSA has concluded that an effective date for grading of radial tires under the UTQG system can and should be established at this time.

Need for Grading of Radial Tires

In response to Notice 28, several commenters pointed out the importance of extending the UTQG Standards to radial tires at the earliest possible date. The Federal Trade Commission (FTC), while recognizing the establishment of a credible system for grading bias and bias-belted tires as a substantial accomplishment, commented that extension of the system to radial tires will be of special significance to the public. The FTC, the Center for Auto Safety (CFAS), and Consumer's Union noted the increasing share of the tire market represented by radial tires, which now account for approximately half of the replacement tire market and an even higher percentage of original equipment sales. CFAS noted that NHTSA's test data revealed significant differences in treadwear properties among radial tires of different manufacturers. In fact, it is likely, based on the data, that some radial tires may yield twice the mileage of those of other manufacturers.

CFAS and the City of Cleveland's Office of Consumer Affairs commented on the need, exemplified by the recent recall of 14.5 million radials by one domestic tire manufacturer, to make safety a factor in the purchase of radial tires. The City of Cleveland reported encountering consumer frustration with present tire marketing practices and expressed concern that inability on the part of consumers to ascertain the quality of tires they are buying may lead to careless and ill-advised purchasing decisions and unsafe operating practices. NHTSA agrees and has seen no new arguments that suggest Congress' directive for establishing a uniform system for grading motor vehicle tires should not be fulfilled by the contemplated method.

Extent of NHTSA Radial Tire Testing

General Motors Corporation and the Rubber Manufacturers Association (RMA) contended that NHTSA's tests of radial tire treadwear were inadequate as a basis for extension of the UTQG regulation to radial tires. General Motors argued that radial tire treadwear does not become constant after tires are broken in, but continues to vary upward and downward, as evidenced by comparing adjusted wear rates in the final 6,400 miles of NHTSA's 38,400-mile radial tire treadwear test with the averages of adjusted wear rates from several 6,400-mile test series. The RMA stated its position that radial tire wear rates continue to decline in the later stages of tire life, pointing to NHTSA and RMA test data on the subject. Both General Motors and the RMA contended that, given the nature of radial tire treadwear, NHTSA must test some radial

tires to actual wearout to confirm that treadwear projections based on 6,400-mile tests correlate closely with actual tire treadlife.

NHTSA has not suggested that radial tire treadwear is precisely constant after break-in. Rather the agency's position, as stated in Notice 28, is that radial tire treadwear after break-in can be adequately described by a straight line fitted to a series of data points representing tread depth against miles traveled, thereby providing an adequate basis for treadwear projections. Variations in wear rate of the type noted by General Motors and the RMA cause a sinuous fluctuation in wear pattern which can be closely approximated by a straight line projection of treadwear based on the first 6,400 miles of testing.

NHTSA chose not to run tested tires to actual wearout because such tests are expensive and time consuming, and accurate projections of treadlife are possible with tires which have substantial wear, but are not worn out. For these reasons, projecting radial tire treadlife from tests run short of wearout is common in the industry (e.g., "A Statistical Procedure for the Prediction of Tire Tread Wear Rate and Tread Wear Rate Differences" by Dudley, Bower, and Reilly of the Dunlop Research Centre) and is, the agency has concluded, a reliable means of determining tire treadwear properties of radial, bias, and bias-belted tires.

Accuracy of the Treadwear Grading Procedure for Radial Tires

General Motors, Michelin Tire Corporation, and the RMA commented that the existing UTQG procedures does not project the treadlife of radial tires with a sufficient degree of accuracy, based on the data submitted to the rulemaking docket in connection with Notice 28. General Motors and the RMA noted that treadwear projections calculated only from wear rates observed in the initial 6,400-mile test sequence differed in some cases by one or two UTQG grade levels from projections based on wear rates from later 6,400-mile test cycles or from averages of several test cycles. These commenters noted that the range of such differences was slightly higher when individual tires were compared rather than the averages of four-tire sets. Michelin expressed concern that the regulation would create an impression of equality among tires which in reality vary in quality. General Motors suggested that projections based on later test cycles or averages established over a longer test period would provide a more accurate projection of actual treadlife.

NHTSA established the 6,400-mile test sequence, with an 800-mile break-in, after considering the adequacy of the data which could be obtained over that test distance and the expenditure of money and resources required for additional testing. The grades arrived at by projecting from later test series or combinations of series were generally consistent with the results obtained in the first 6,400 miles of testing, and those variations which did occur were relatively minor.

As noted by the U.S. Court of Appeals for the Sixth Circuit in *B. F. Goodrich Co. v. Department of Transportation*, 541 F.2d 1178 (1976), no system designed to grade millions of tires can be expected to approach perfection. Considering the present absence of tire quality information in the market place, the agency has concluded that the UTQG treadwear grading procedure provides reasonable accuracy when applied to radial tires and will be of significant value to tire consumers in making purchasing decisions.

General Motors commented that tire grades should be assigned based on the lowest mileage projected for any tire among a set of four candidate tires and not on the average projected mileage of a four tire set. The UTQG regulation states that each tire will be capable of providing at least the level of performance represented by the UTQG grades assigned to it. UTQG grades based solely on either average grade levels or on the projected mileage of a particular tested tire would not provide an adequate basis for consumer reliance on the grading information. In determining accurate treadwear grades for tire lines, manufacturers must consider the population variability evidenced in their tire testing.

Validity of the CMT Adjustment Procedure

The UTQG regulation accounts for environmental influences on candidate tire wear rates during testing by means of an adjustment factor derived by comparing the wear rates of concurrently run course monitoring tires (CMT's) with

an established CMT base course wear rate (BCWR) (49 CFR 575.104(d)(2)). In Notice 28, NHTSA explained how the same adjustment procedure could be used to correct for a measurement anomaly that generates the appearance of a higher wear rate for radial tires in the first 4,000 miles of testing following the 800-mile break-in. In response to Notice 28, CFAS reviewed the UTQG adjustment procedure, as it applies to radial tires, and commented that this procedure is the proper method for grading radials. However, Michelin and the RMA, in their comments on that notice, suggested that the CMT adjustment procedure may be invalid for radial tires, both in the context of wear rate changes and as a control on environmental factors.

The RMA argued that NHTSA has not provided supporting data for its theory that the shift in radial tire wear rate during the initial phases of treadlife is caused by changes in tire geometry as the tire attains its equilibrium shape. However, detailing the underlying mechanism of the apparent change in wear rate is incidental to the fact that radial tire wear rates do stabilize in a consistent fashion, permitting use of the CMT adjustment to project treadlife with reasonable accuracy.

The RMA contended that wear patterns of certain radial tires differ markedly from the apparent accelerated pattern observed by NHTSA during the first 4,000 miles of treadlife after the 800-mile break-in, and that NHTSA's test of several tire brands provided an inadequate basis to draw conclusions about radial tires in general. Michelin, although citing no data on the subject, commented that an accelerated wear pattern in the early stages of treadlife may not exist in all radial tires to the same degree.

NHTSA's test of radial tire treadwear, reported in Notice 28, included ten different tire brands, selected to include a wide range of prices and materials, as well as both domestic and foreign manufacture. This sample constitutes a reasonable and adequate basis upon which to draw conclusions concerning tires available on the American market. In spite of the wide variety of radial designs included in NHTSA's test, the agency found the wear rate patterns of the tires studied to be remarkably consistent in the initial

6,400-miles of testing, after the 800-mile break-in. This consistency is exemplified by treadwear projections in the paper "Test of Tread Wear Grading Procedure—the Course Monitoring Tire Adjustment on Radial Tire Wear Rates", by Brenner and Williams (Docket 25, General Reference No. 105), which compared estimates of tread life for nine sets of candidate tires based on data from the first 6,400 miles of testing after break-in, with estimates based on data from 6,400 to 38,400 miles of testing. The projections computed from these data sets did not differ significantly, indicating that the UTQG adjustment procedure accurately accounted for the initial wear rate characteristics of all tires tested.

Based on this test experience, the agency believes that the data from its tests and analysis of that data has demonstrated that the wear patterns exhibited by radial tires early in their treadlives are sufficiently consistent to permit accurate projection of treadwear based on the existing UTQG test procedure. NHTSA plans to closely monitor testing at the San Angelo course to insure that the UTQG test procedure accommodates future developments in tire technology and continues to provide an accurate basis for treadwear grading.

On the question of consistency beyond the initial 4,000 miles of testing, both Michelin and the RMA argued that not all tires tested by NHTSA responded to environmental factors in an identical manner, as demonstrated by comparing graphs of unadjusted candidate tire wear rates by test cycle with graphs of data from concurrently run CMT's. The RMA also noted that graphic representations of radial tire adjusted wear rates per test cycle were not always horizontal, but in some cases sloped somewhat upward or downward.

Close examination of the graphs of unadjusted candidate tire wear rates and CMT wear rates indicates that the wear rates fluctuated in a reasonably parallel fashion in all but an insignificant number of cases. NHTSA has never contended that every tire of every brand must behave in a perfectly consistent manner before a valid grading system can be established. NHTSA finds that the level of consistency exhibited by the tested tires is sufficient to confirm the validity of the CMT approach as a reasonably fair and reasonably reliable means of radial tire grading.

With regard to the slope of the adjusted wear rate curves, NHTSA has applied a test of independence to this data to determine if the adjusted wear rates of the tested tires were dependent on the test cycle. In no case was the slope significantly different from zero at the 95 percent confidence level. In fact, of the curves which slanted to any measurable degree, sixteen had a slightly positive slope and seventeen had a slightly negative slope, as would be expected if the true slope were zero. This analysis suggests that CMT and candidate tires continue to wear in a consistent fashion beyond the initial phase of testing.

The RMA's comments suggest that some confusion may exist as to whether CMT's are to be reused for testing after an initial 6,400-mile test cycle after break-in. Since radial tires, including CMT's, exhibit an apparent change in wear pattern during this initial phase of treadlife, when measured by a tread depth gauge, the CMT adjustment procedure will be accurate only if new candidate tires are run with new CMT's so that the wear rate change occurs in all tires simultaneously.

Radial CMT's were run beyond the initial 6,400-mile cycle in NHTSA's testing announced in Notice 28, in order to provide an extended comparison of CMT's and candidate tires run concurrently. In its UTQG compliance testing, however, NHTSA will use new radial CMT's, broken-in in accordance with 49 CFR 575.104 (d) (2) (v), for each 6,400-mile test.

Also on the issue of the CMT adjustment procedure, the RMA commented that NHTSA's test data indicate a coefficient of variation (COV) for radial CMT's of over 5 percent, the standard upheld in the *B. F. Goodrich* case as the agency's target for the maximum permissible level of variability for these tires. Much of the data cited by the RMA on this point involved test cycles beyond the initial 6,400-mile cycle, after break-in. Data on the variability of CMT's at test distances beyond 6,400 miles, after break-in, are irrelevant to the UTQG system, since, as noted above, radial CMT's will not be reused after an initial 6,400-mile test cycle.

In examining data from the initial test cycle, the RMA combined wear rates from several test vehicles and then developed COV's from that data, thereby interjecting vehicle variability into the computation. Vehicle variability, while unrelated to the properties of the tire, has the effect of inflating coefficients of variation. When this extraneous factor is removed from the computation, the test data indicate a COV well within the acceptable 5 percent level.

Michelin expressed concern that running CMT's of a standard size with candidate tires of differing sizes may lead to inaccuracy in the adjustment of data. National Bureau of Standards Technical Note 486, "Some Problems in Measuring Tread Wear of Tires," by Spinner and Barton (Docket 25, General Reference No. 4), compared projected mileages for three sizes of radial and bias-ply tires of several manufacturers run under different road conditions. Data in the report suggest that tires of different sizes react similarly to differing external conditions. Therefore, the practical burden of providing a different CMT for each size of candidate tire may be avoided.

Finally, General Motors and the RMA asserted that, in order to facilitate comparisons among radial, bias, and bias-belted tires, BCWR's must be established by running the three types of CMT's concurrently to limit the influence of environmental variables on the test results. The RMA also contended that a BCWR cannot be established without running CMT's to actual wearout.

NHTSA established BCWR's through experience with tires of all three construction types in over 5 million tire miles of testing over a two year period. In the course of this extensive testing, each tire type can be expected to have encountered a random mix of environmental conditions resulting in a similar net impact on treadwear.

Other Comments

Michelin commented that the regulation's procedure of rotating tires among different positions on a test vehicle, but not between vehicles, precludes the detection of vehicle mechanical problems which could affect grading. Adequate preventive maintenance of test vehicles is the primary safeguard against distortion of data by vehicle malfunctions. Additionally, an analysis of variance of the data obtained in a convoy or on a vehicle provides another effective method of detecting a malfunction. (See, "Elements in the Road Evaluation of Tire Wear", by Brenner and Kondo, Docket 25; General Reference No. 17). NHTSA does not believe that rotation of tires among vehicles would significantly improve on these existing techniques.

General Motors noted that several tires studied by NHTSA had to be removed from the test due to failure or uneven wear prior to actual wearout and suggested that the agency must account for these anomalies before proceeding with rulemaking.

Early in the course of rulemaking on UTQG, NHTSA concluded that considerations of cost and consumer understanding required some limitation on the number of grading categories in which UTQG information would be presented. Based on examination of numerous comments in the rulemaking docket, the agency concluded that treadwear, traction, and temperature resistance are the tire characteristics of greatest importance to consumers. For this reason, information on subjects such as evenness of tread wear and susceptibility to road hazard damage, while of value to consumers, is not provided under the regulation. NHTSA will consider General Motors comment, however, as a suggestion for possible future rulemaking.

The RMA noted several minor computational and other errors in the previously referred to paper by Brenner and Williams (Docket 25, General Reference No. 105), submitted to the docket in connection with Notice 28. Some of these errors were corrected by a subsequent submission to the docket (Docket 25, General Reference No. 105A). In any case, the errors were of a non-substantive nature and had no impact on the agency's rulemaking process and decisions.

Impact of the Thirty Day Stay
of Effective Dates

On January 19, 1979, the U.S. Court of Appeals for the Sixth Circuit, in the case *B. F. Goodrich Co. v. Department of Transportation* (No. 78-3392), granted a thirty-day stay of the effective dates for application of the UTQG regulation to bias and bias-belted tires. The regulation was scheduled to become effective March 1, 1979 for bias-ply tires and September 1, 1979 for bias-belted tires, with the exception of the sidewall molding requirements of paragraph (d)(1)(i)(A) and the first purchaser requirements of paragraph (d)(1)(iii) which were to become effective September 1, 1979 and March 1, 1980 for bias and bias-belted tires, respectively.

NHTSA interprets the Sixth Circuit's action as postponing the effective dates of the UTQG regulation one month to April 1, 1979 for bias-ply tires and October 1, 1979 for bias-belted tires. However, the effective dates for the molding requirements of paragraph (d)(1)(i)(A) and the first purchaser requirements of paragraph (d)(1)(iii) are postponed to October 1, 1979 for bias-ply tires and April 1, 1980 for bias-belted tires to allow manufacturers time to convert tire molds. This postponement of effective dates has been taken into account in establishing effective dates for application of the regulation to radial tires, to assure adequate lead time for completion of tire testing.

In accordance with Departmental policy encouraging adequate analysis of the consequences of regulatory actions, the agency has evaluated the anticipated economic, environmental and other consequences of extending the UTQG regulation to include radial tires and has determined that the impact of this action is fully consistent with impacts evaluated in July 1978 in establishing effective dates for bias and bias-belted tires. Based on the authority of Section 203 of the Act,

previous agency findings concerning required lead time for grading tires, and the decision of the U.S. Court of Appeals for the Sixth Circuit in *B. F. Goodrich*, the NHTSA hereby establishes radial tire effective dates consistent with the basic six-month phase-in schedule announced on July 17, 1978 (43 FR 30542) for bias and bias-belted tires.

In an unrelated matter, NHTSA's FEDERAL REGISTER notice announcing effective dates for application of the UTQG Standards to bias and bias-belted tires (43 FR 30542); July 17, 1978) contained an inadvertent error in use of the word "of" rather than the intended word "are" in the first sentence of the third section of Figure 2 of the regulation. This error is corrected by substitution of the word "are" in place of "of" in Figure 2.

In consideration of the foregoing, the Uniform Tire Quality Grading Standards (49 CFR 575.104), are amended . . . .

The program official and lawyer principally responsible for the development of this rulemaking document are Dr. F. Cecil Brenner and Richard J. Hipolit, respectively.

(Sec. 103, 112, 119, 201, 203; Pub. L. 89–563, 80 Stat. 718 (15 U.S.C. 1392, 1401, 1407, 1421, 1423); delegation of authority at 49 CFR 1.50.)

Issued on March 9, 1979.

Joan Claybrook
Administrator

44 F.R. 15721–15724
March 15, 1979

PREAMBLE TO AN AMENDMENT TO PART 575—CONSUMER INFORMATION

Uniform Tire Quality Rating

(Docket No. 25; Notice 35)

ACTION: Final rule.

SUMMARY: This notice amends the Uniform Tire Quality Grading (UTQG) Standards through minor modifications in the format of tire tread labels used to convey UTQG information. The modifications are intended to assure that tires are labeled with the correct UTQG grades, to permit flexibility in the design of labels, and to facilitate consumer access to the grading information.

EFFECTIVE DATE: December 1, 1979.

FOR FURTHER INFORMATION CONTACT:

Dr. F. Cecil Brenner, Office of Automotive Ratings, National Highway Traffic Safety Administration, 400 Seventh Street, S.W., Washington, D.C. 20590 (202-426-1740).

SUPPLEMENTARY INFORMATION: On January 8, 1979, NHTSA published a request for public comment (44 F.R. 1814) on a petition for rulemaking submitted by Armstrong Rubber Company asking that the UTQG regulation be amended to permit tire grading information and explanatory material concerning the UTQG system to be furnished to consumers by means of two separate tire tread labels rather than the single label called for in the regulation (49 CFR 575.104(d) (1) (i) (B)). Armstrong, joined by Atlas Supply Company, contended that the chance of mislabeling tires would be reduced, if UTQG grades could be placed on the same label with tire identification information. However, practical limitations exist on the size of tread labels which can be effectively applied and retained on the tire tread surface. Some manufacturers reportedly encountered difficulty in fitting tire identification information, UTQG grades, and required UTQG explanatory information on a single label. For this reason, Armstrong and Atlas suggested that UTQG explanatory information be furnished on a

separate label adjacent to a label containing UTQG grades and tire identification information.

In view of the favorable comments received in response to NHTSA's request for comment on the Armstrong petition, the agency proposed to modify the tread label format requirements to employ a two-part label format (44 F.R. 30139; May 24, 1979). NHTSA proposed that Part I of the label contain a display of the UTQG grades applicable to the particular tire while Part II would contain the general explanation of the grading system. At the manufacturer's option Parts I and II could appear on separate labels. To assure that the labels would be legible to consumers, the notice also proposed requirements for orientation of the label text and minimum type size.

Commenters on the proposal were in general agreement that flexibility in the design of tire tread labels is a desirable goal. While some manufacturers expressed the opinion without explanation that two-part labels would be impractical for their operations, others welcomed the proposal as a means of dealing with label size limitations.

Some commenters favored retention of the original label format pointing out that the proposed label would be slightly longer than its predecessor and arguing that the proposed label would isolate the tire grades from the explanatory material. Some industry sources expressed the opinion that the proposed changes would be of no benefit to consumers.

NHTSA disagrees with these criticisms of the proposal. The new format should increase the length of the label by only a fraction of an inch, if at all, and should not pose a problem to manufacturers wishing to employ a single label. The separation of the grades from the explanatory material should not create confusion since the two parts could be separated by no more than one inch in any case. The agency has reached the conclusion that displaying grades for all three performance categories together on Part I of the label will in fact benefit consumers by facilitating access to the information.

Maximum retainability will be assured with the new format since manufacturers may choose to employ two labels if they are unable to fit all of the necessary information on a single label of a manageable size. Similarly, the possibility of mislabeling will be reduced, because the two-part option makes it possible in all cases to include applicable UTQG grades on tire identification labels. For these reasons, NHTSA has determined to adopt the proposed two-part label format with minor modifications.

Several commenters suggested that orientation of the tread label text should not be specified in the regulation since flexibility in label design would be reduced by such a requirement. However, NHTSA has concluded that since most manufacturer's tire identification labels are arranged with lines of type running perpendicular to the tread circumference, tires are most likely to be displayed so that labels with this orientation will be easily readable by consumers. Therefore, the agency has chosen to retain the proposed requirement regarding label text orientation.

Goodyear Tire & Rubber Company suggested the possibility of printing Part I of the proposed label below Part II, when both parts are contained on a single tread label. NHTSA finds this suggestion unacceptable because the UTQG grades would be difficult to locate if preceded by a body of textual material.

Goodyear also commented on several occasions that specifying a minimum type size for the printing of labels would be of no benefit since many factors other than type size, such as letter style, spacing, and format, contribute to legibility. NHTSA agrees that a minimum type size requirement alone is insufficient to assure the readability of labels. For this reason, NHTSA has chosen to withdraw its proposed minimum type size requirement at this time. The agency will, however, continue to monitor industry compliance with the labeling requirements to ascertain whether a comprehensive set of requirements is necessary to assure that tread labels will be legible to consumers.

The agency has found considerable merit in another Goodyear suggestion, to delete the range of possible grades adjacent to the categories "TRACTION" and "TEMPERATURE" on Part II of the label. These letters were originally included on the label to provide a display on which the grade attributable to a particular tire could be marked. Since grades will now be marked on Part I of the label, the range of possible grades in Part II is superfluous and has been deleted from the required format. If, however, manufacturers wish to display the array of grades on both Part I and Part II of their labels, NHTSA has no objection to this practice.

Goodyear was joined by General Tire & Rubber Company in requesting that NHTSA clarify whether the three category headings, "TREADWEAR," "TRACTION," and "TEMPERATURE," in Part I of the proposed label must be laid out side by side, across the label, or one below the other, down the label. In the interest of flexibility, the regulation makes either of these layouts acceptable, although the relative order of the categories must be maintained to permit easy reference to the explanatory material.

Similarly, several manufacturers recommended that the regulations permit grades to be displayed either to the right of or directly below the grading category to which they apply. Again, to facilitate efficient label design, the regulation permits the use of either of these locations for the display of grades.

Industry commenters asked that NHTSA clarify whether the use of lower case letters in the label text, as set out in Figure 2 of the regulation, precludes manufacturers from printing labels using all capital letters in the label text. The regulation has been modified to permit the optional use of all capital letters in printing the text of Figure 2.

NHTSA wishes to confirm Firestone Tire & Rubber Company's understanding that the words "Part I" and "Part II" appearing in Figure 2 as proposed are for reference purposes only and need not be printed on the tread label. General and the Rubber Manufacturers Association called NHTSA's attention to certain typographical errors in the proposed Figure 2 text, which have been corrected in the amendment as adopted.

Several manufacturers suggested that the original label format be permitted as an option, or that, as a minimum, waste be avoided by allowing labels printed with the original format to be used up regardless of the adoption of a new label format. NHTSA considers the new two-part label format to be superior to the original format in terms of clarity and readability. Therefore, the agency has concluded that universal conversion to the new format is desirable. However, since manufacturers have expended significant resources in efforts to comply with the original labeling requirement, NHTSA will permit the use of labels employing the original format, at the manufacturers option, until October 1, 1980. This period of flexibility should permit any labels already printed to be used up and allow a smooth transition to the new format.

Since this amendment will increase manufacturers' flexibility in complying with the UTQG labeling requirements, and since the transition to the new labeling format will be phased in so as to avoid economic waste, the agency has found that this notice does not have significant impact for purposes of internal review. In view of the fact that some manufacturers may still be in the process of obtaining labels for their bias-belted tire lines, this amendment will become effective December 1, 1979.

Issued on November 20, 1979.

Joan Claybrook,
Administrator
44 F.R. 68475
November 29, 1979

PREAMBLE TO AN AMENDMENT TO PART 575—CONSUMER INFORMATION

Uniform Tire Quality Rating

(Docket No. 25; Notice 37)

ACTION: Final rule; correction.

SUMMARY: This notice corrects an inadvertent error in the text of the National Highway Traffic Safety Administration's (NHTSA) final rule modifying the tread label format used under the Uniform Tire Quality Grading (UTQG) Standards (49 CFR 575.104).

SUPPLEMENTARY INFORMATION: On November 29, 1979, NHTSA published a notice (44 F.R. 68475) making minor modifications in the final format of tire tread labels used to convey UTQG information to consumers. That notice contained an inadvertent error in the text of Figure 2 of the regulation in that the words "one and one-half" were substituted for the words "one and a half" under the heading "Treadwear" in Part II of the tread label text. The notice is therefore revised to reflect the intended wording.

F.R. Doc. 79–36522 appearing at 44 F.R. 68475 is corrected at page 68477 in the third column as follows:

Figure 2, Part II of the Uniform Tire Quality Grading Standards, 49 CFR 575.104, is corrected by substitution of the words "one and a half" in place of the words "one and one-half" under the heading "Treadwear".

Issued on January 22, 1980.

Michael M. Finkelstein,
*Associate Administrator
for Rulemaking*

45 F.R. 6947
January 31, 1980

PREAMBLE TO PART 575—CONSUMER INFORMATION REGULATIONS
UNIFORM TIRE QUALITY GRADING

(Docket No. 25; Notice 38)

ACTION: Interpretation.

SUMMARY: This notice clarifies the procedure to be used under the Uniform Tire Quality Grading (UTQG) Standards in measuring tread depth of tires without circumferential grooves or with a limited number of grooves. The regulation's provision for measurement of tread depth in tire grooves has given rise to questions concerning the proper means of measurement for such tires. This notice is intended to facilitate testing of tires of this type.

EFFECTIVE DATE: This interpretation is effective immediately.

FOR FURTHER INFORMATION CONTACT:

Mr. Richard Hipolit, Office of the Chief Counsel, National Highway Traffic Safety Administration, 400 Seventh Street, S.W., Washington, D.C. 20590 (202–426–1834).

SUPPLEMENTARY INFORMATION:

The UTQG Standards (49 CFR 575.104) require the grading of passenger car tires on three performance characteristics: treadwear, traction and temperature resistance. In setting forth the procedure to be followed in evaluating treadwear performance, the regulation states that, after an 800-mile break-in, tires are to be run for 6,400 miles over a designated course, with tread depth measurements to be taken every 800 miles. The regulation specifies that tread depth is measured at six equally spaced points in each tire groove other than shoulder grooves, avoiding treadwear indicators. Tire grooves are typically arranged symmetrically around the center of the tread.

On May 24, 1979, the National Highway Traffic Safety Administration (NHTSA) published in the *Federal Register* (44 FR 30139) an interpretation that tires designed for year round use do not qualify as "deep tread, winter-type snow tires,"

which are excluded from the coverage of the UTQG regulation by 49 CFR 575.104(c). In response to this interpretation, the Goodyear Tire & Rubber Company commented to NHTSA (Docket 25; Notice 32–011) that a technical problem may exist in the measurement of tread depth of tires for year round use since circumferential grooves are absent in the designs of many such tires.

NHTSA is aware that certain other standard tire designs, as well as year round designs, may incorporate lugs, discontinuous projections molded in the tread rubber, separated by voids, in place of ribs defined by circumferential grooves. In other cases, the limited number of grooves on the tire could lead to inaccurate results if measurements were made in only those grooves.

To assure accurate tread depth measurements on tires lacking circumferential grooves, and tires with fewer than four grooves, measurements are to be made along a minimum of four circumferential lines equally spaced across the tire tread surface. These lines are to be symmetrically arranged around a circumferential line at the center of the tread. The outermost line on each side of the circumferential tread centerline is to be placed within one inch of the shoulder.

Measurements are to be made at six equally spaced points along each line. If the design of the tire is such that, on a particular circumferential line, six equally spaced points do not exist at which groove or void depth exceeds by $\frac{1}{16}$th of an inch the distance from the tread surface to the tire's treadwear indicator, measurements are not to be taken along that line. If measurements cannot be taken on four equally-spaced, symmetrically-arranged lines, the requirement for equal spacing does not apply. Measurements in that case are to be taken along a minimum of four lines, with an equal number of symmetrically arranged measured lines on either side of the tread centerline.

NHTSA recognizes that, due to the implementation schedule of the regulation, certain manufacturers may have already conducted treadwear tests on tires falling within the scope of this interpretation. The Agency does not object to the use in grading of treadwear data generated prior to the publication date of this notice, if such data was acquired using a test method varying only in minor, non-substantive respects from the method described in this interpretation.

The principal author of this notice is Richard J. Hipolit of the Office of Chief Counsel.

Issued on March 24, 1980.

Joan Claybrook
Administrator

45 F.R. 23441
April 7, 1980

PREAMBLE TO AMENDMENTS TO PART 575—CONSUMER INFORMATION REGULATIONS; UNIFORM TIRE QUALITY GRADING

(Docket No. 25; Notice 39)

ACTION: Final Rule.

SUMMARY: This notice amends the Uniform Tire Quality Grading (UTQG) Standards to exclude from the requirements of the regulation tires produced in small numbers, which are not recommended for use on recent vehicle models. The amendment is intended to reduce costs to consumers and reduce regulatory burdens on industry in an area where the purchase of tires based on comparison of performance characteristics is limited.

EFFECTIVE DATE: This amendment is effective immediately.

FOR FURTHER INFORMATION CONTACT:

Dr. F. Cecil Brenner, Office of Automotive Ratings, National Highway Traffic Safety Administration, 400 Seventh Street, S.W., Washington, D.C. 20590 (202-426-1740).

SUPPLEMENTARY INFORMATION:

The UTQG Standards 49 CFR § 575.104 are intended to enable consumers to make an informed choice in the purchase of passenger car tires through the use of comparative performance information relating to tire treadwear, traction and temperature resistance. The standards apply to new pneumatic tires for use on passenger cars manufactured after 1948. Deep tread, winter-type snow tires, space-saver or temporary use spare tires, and tires with nominal rim diameters of 10 to 12 inches have been excluded from the application of the regulation (49 CFR § 575.104(c)).

Several tire manufacturers and dealers have informed the National Highway Traffic Safety Administration (NHTSA) that a small class of tires exists for which marketplace competition based on performance characteristics is extremely limited. These tires, which are purchased for use on vehicles manufactured after 1948 but nonetheless considered by their owners to be classic or antique, are produced in small numbers in a wide variety of designs and sizes. Purchasers of these tires are reportedly concerned primarily with appearance, authenticity, and availability rather than tire performance.

Information supplied by Intermark Tire Company indicates that a similar limited market exists for tires used on older vehicles requiring tire sizes no longer employed as original equipment on new vehicles. Intermark petitioned NHTSA to remove these tires from the coverage of the regulation on the basis that little market competition exists in their sale and that availability is the primary factor in the purchase of this class of tire.

In order to reduce costs to consumers and eliminate the need for industry to grade the multiplicity of small lines of tires in which comparative performance information would have limited value, NHTSA published a notice proposing to remove certain limited production tires from the application of the UTQG regulation (45 FR 807; January 3, 1980). Four criteria, were specified to define limited production tires. First the annual production by the tire's manufacturer of tires of the same design and size could not exceed 15,000 tires. Second, if the tire were marketed by a brand name owner, the annual purchase by the brand name owner could not exceed 15,000 tires. Third, the tire's size could not have been listed as a manufacturer's recommended size designation for a new motor vehicle produced or imported into this country in quantities greater than 10,000 during the preceding calendar year. Fourth, the annual production by the tire's manufacturer, or the total annual-purchase by the tire's brand name owner, if applicable, of different tires otherwise meeting the criteria for limited production tires could not exceed 35,000 tires. The proposal also clarified that differences in design would be determined on the basis of structural characteristics, materials and tread pattern, rather than cosmetic differences.

Commenters on the proposal, including the Rubber Manufacturers Association, the National Tire Dealers and Retreaders Association, Dunlop Limited, Intermark, Kelsey Tire Company and McCreary Tire and Rubber Company agreed that tire quality grading should not be required for limited production tires. Among the reasons stated for support of the proposal were expected cost savings to industry and the consumer and the special consideration affecting the purchase of these tires. After consideration of these comments, the agency has adopted the proposed amendment with minor modification.

Intermark pointed out a possible anomalous situation which could result from the wording of subparagraph (c) (2) (iv) of the proposal. That provision placed a 35,000 tire limit on a manufacturer's total annual production of tires meeting the limited production criteria, *or*, in the case of tires marketed under a brand name, on the total annual purchase of limited production tires by a brand name owner. Thus, under this commenter's reading of (c) (2) (iv), 40,000 tires meeting the criteria of subparagraphs (c) (2) (i), (ii), and (iii) could be produced by a manufacturer, sold in groups of 10,000 to four different brand name owners, and still qualify as limited production tires. At the same time, another manufacturer could produce 40,000 tires meeting the first three criteria for sale in its own company outlets and be required to grade the tires. To make it clear that the 35,000 tire limitation on manufacturer's production applies whether or not the tires are marketed by a brand name owner, subparagraph (c) (2) (iv) has been modified by substituting the word "and" for "or."

Kelsey Tire Company asked how the criteria would apply to tires which are produced abroad in large numbers but are imported in quantities which would fall within the unit limitations of subparagraphs (c) (2) (i), (ii), and (iv) of the proposal. To make clear that the criteria are to be applied to foreign tires only insofar as they are imported in this country, subparagraphs (c) (2) (i) and (iv) have been modified to refer to 'annual domestic production or importation into the United States by the tire's manufacturer." The reference to "importation . . . by the tire's manufacturer" includes in the total all tires entering the United States for sale under the name of the manufacturer, regardless of the shipping or title arrangements made by the manufacturer with distributors. Similarly, subparagraphs (c) (2) (ii) and (iv) have been modified to clarify the status of tires purchased by brand name owners.

McCreary and Intermark argued that the unit restrictions on production of tires meeting the criteria are too restrictive and should be eliminated or eased significantly. McCeary predicted that the total number of classic car tires produced by individual manufacturers will grow, although production runs of individual designs and sizes will remain small. Intermark contended that production limitations unfairly penalize efficient manufacturers and that a new vehicle recommended size designation provision such as proposed subparagraph (c) (2) (iii) would be sufficient to define the intended class of limited production tires.

NHTSA considers the stated limitations broad enough to encompass the "classic" car tire market as it is presently constituted. With regard to the larger production runs of tires in outdated sizes, NHTSA believes that the production of tires in numbers greater than the proposed limitations is suggestive of wider availability and resulting increased competition which would make UTQG information of greater value. Further, relaxing or eliminating unit restrictions could result in the exclusion from the application of the standard of high performance or racing tires which are not recommended as original equipment. The agency believes that comparative tire grading information should be available to purchasers of tires of this type. NHTSA will monitor the limited production tire market to determine whether future market changes require revision of the 35,000 tire limitation.

Pursuant to E.O. 12044, "Improving Government Regulation," and implementing departmental guidelines, the agency has considered the effects of this amendment. It reaffirms its earlier determination that the amendment is not significant and that the effects are so minimal as not to warrant preparation of a regulatory evaluation. NHTSA has determined that these amendments will result in modest cost savings to industry and consumers, while having no appreciable effect on safety or the environment.

Because this amendment relieves a restriction and because the agency desires to minimize any possible interruption in tire production pending the effective date of this amendment, the amendment is effective immediately.

In consideration of the foregoing, 49 CFR § 575.104(c) is amended to read:

§ 575.104 Uniform tire quality grading standards.

*   *   *   *   *

(c) *Application.*

(1) This section applies to new pneumatic tires for use on passenger cars. However, this section does not apply to deep tread, winter-type snow tires, space-saver or temporary use spare tires, tires with nominal rim diameters of 10 to 12 inches, or to limited production tires as defined in paragraph (c)(2) of this section.

(2) "Limited production tire" means a tire meeting all of the following criteria, as applicable:

(i) The annual domestic production or importation into the United States by the tire's manufacturer of tires of the same design and size as the tire does not exceed 15,000 tires;

(ii) In the case of a tire marketed under a brand name, the annual domestic purchase or importation into the United States by a brand name owner of tires of the same design and size as the tire does not exceed 15,000 tires;

(iii) The tire's size was not listed as a vehicle manufacturer's recommended tire size designation for a new motor vehicle produced in or imported into the United States in quantities greater than 10,000 during the calendar year preceding the year of the tire's manufacture; and

(iv) The total annual domestic production or importation into the United States by the tire's manufacturer, and in the case of a tire marketed under a brand name, the total annual domestic purchase or purchase for importation into the United States by the tire's brand name owner, of tires meeting the criteria of subparagraphs (c)(2)(i), (ii), and (iii) of this section, does not exceed 35,000 tires.

Tire design is the combination of general structural characteristics, materials, and tread pattern, but does not include cosmetic, identifying or other minor variations among tires.

The principal authors of this notice are Dr. F. Cecil Brenner of the Office of Automotive Ratings and Richard J. Hipolit of the Office of Chief Counsel.

Issued on March 24, 1980.

Joan Claybrook
Administrator

45 F.R. 23442
April 7, 1980

ACTION: Final rule.

SUMMARY: This notice amends the Consumer Information Regulations by deletion of the requirement that manufacturers supply information on acceleration and passing ability to vehicle first purchasers and prospective purchasers. The notice also revises the timing of manufacturers' submissions of performance data to the National Highway Traffic Safety Administration (NHTSA). These modifications, which were proposed in response to a General Motors Corporation petition for rulemaking, are intended to lessen regulatory burdens on industry, while providing performance data in a manner more useful to consumers.

EFFECTIVE DATES: The amendment of section 575.6(d) is effective June 1, 1981. The deletion of section 575.106 is effective immediately, July 7, 1980.

FOR FURTHER INFORMATION CONTACT:
Ivy Baer, Office of Automotive Ratings, National Highway Traffic Safety Administration, 400 Seventh Street, S.W., Washington, D.C. 20590 (202-426-1740)

SUPPLEMENTARY INFORMATION: The Consumer Information Regulations (49 CFR Part 575) provide first purchasers and prospective purchasers with performance information relating to the safety of motor vehicles and tires. This information is intended to aid consumers in making comparative purchasing decisions and in the safe operation of vehicles. General Motors Corporation petitioned NHTSA to delete requirements for consumer information on passenger car and motorcycle stopping distance (49 CFR 575.101), passenger car tire reserve load (49 CFR 575.102), and passenger car and motorcycle acceleration and passing ability

(49 CFR 575.106), on the basis that this information is of limited value to consumers. In response to this petition, NHTSA proposed (44 FR 15748; March 15, 1979) to delete the requirement for acceleration and passing ability information and to limit the application of the tire reserve load provisions to vehicles with significant cargo capacity, thus dropping the requirement for most passenger cars. NHTSA also proposed that vehicle manufacturers submit performance data to the agency at least 90 days before model introduction, compared to the 30-day advance submission which had been required (49 CFR 575.6).

Timing of Data Submission

The primary purpose of the advance submission to NHTSA is to permit the agency to compile and disseminate performance data in a comparative format for use by prospective vehicle purchasers. A major criticism of the consumer information program in the past has been that comparative information reached the consumer too late in the model year to be of real value in choosing between competing vehicles. A 90-day advance submission would permit the agency to assemble and distribute comparative information early in the model year, when it would be of greatest value to consumers.

Some industry commenters questioned the need for earlier submission of data on the basis that agency delays in publishing the data will result in comparative information being available late in the model year, in spite of the earlier submission. Other manufacturers argued that consumer interest in the information is limited in any case. General Motors suggested that vehicle design changes during the model year rapidly outdate the information, further limiting its value.

However, the Center for Auto Safety (CFAS) commented that it receives numerous requests from consumers for comparative information on motor vehicles. CFAS also pointed out the popularity of comparative motor vehicle information on the rare occasions when such information is made available by independent publishers. NHTSA has concluded that consumer interest in comparative performance information would be substantial if the information were made available in a timely manner. Further, NHTSA has determined that few running design changes during the model year are so major as to significantly affect the performance characteristics covered by the consumer information regulations.

The success of the Environmental Protection Agency in publishing its popular fuel economy guides in a timely manner indicates that publication of vehicle information by NHTSA early in the model year is practical. However, based on past experience, it appears that a 90-day advance submission is the minimum leadtime necessary for NHTSA to publish and distribute the information.

Some manufacturers indicated they may have difficulty providing accurate performance information 90 days in advance of model introduction due to the possibility of last minute design changes. However, American Motors Corporation commented that a 90-day advance submission requirement would pose no problem at new model introduction, although it would inhibit running changes during the model year. In view of the importance of supplying comparative information early in the model year, NHTSA has adopted the proposed 90-day advance submission requirement for model introduction. However, to avoid delaying the introduction of product improvements, the 30-day notice period has been retained for changes occurring during the model year.

Tire Reserve Load

In response to General Motors' petition, NHTSA proposed modifying the tire reserve load information requirement to limit its application to trucks and multipurpose passenger vehicles with a gross vehicle weight rating of 10,000 pounds or less, and to passenger cars with a maximum cargo capacity of 25 cubic feet or more. The regulation had applied to all passenger cars, but not to trucks or multipurpose passenger vehicles.

Comments from many industry and consumer sources recommended deleting the tire reserve load information requirement completely. CFAS

commented that consumer interest in tire reserve load information has been limited. Many comments from car, truck and recreational vehicle manufacturers expressed concern that presenting information on tire reserve load may encourage vehicle overloading by misleading consumers into thinking that vehicles have additional load carrying capacity. Several commenters suggested that Federal Motor Vehicle Safety Standards 110 and 120 provide the appropriate means of ensuring that vehicles are equipped with tires of adequate size and load rating.

A recent study conducted for NHTSA (Docket 79-02, Notice 1-016) indicates that tire reserve load is an important factor in preventing passenger car tire failure. Additional information is being gathered on this subject and the agency is planning to propose amendment of Federal Motor Vehicle Safety Standard 110 to require a minimum tire reserve load on passenger cars. Preliminary analysis suggests that a tire reserve load percentage of 10% or greater is necessary to provide an adequate safety margin.

NHTSA has found that presently available information is not sufficient to justify extension of the tire reserve load requirements to light trucks and multipurpose passenger vehicles at this time. However, in view of the safety implications of tire reserve load for passenger cars and in the absence of a requirement for minimum tire reserve load, NHTSA believes that information on this subject should be available to passenger car purchasers and owners. The agency has concluded that provision of tire reserve load information in its present form does not encourage vehicle overloading, since a warning against loading vehicles beyond their stated capacity must accompany the information.

For these reasons, NHTSA has determined that the existing requirement for tire reserve load information must remain in effect at least until the completion of rulemaking on the possible amendment of Federal Motor Vehicle Safety Standard 110. If the provision of tire reserve load information no longer appears necessary then, the agency will reconsider the status of tire reserve load as a consumer information item. At this time, however, NHTSA withdraws the proposal to modify the tire reserve load consumer information requirements.

Acceleration and Passing Ability

The final aspect of NHTSA's proposal was deletion of acceleration and passing ability (49 CFR 575.106) from the consumer information requirements. The acceleration and passing ability provision required information on the distance and time needed to pass a truck traveling at 20 mph and at 50 mph. The passing vehicle was permitted to attain speeds of up to 35 mph and 80 mph in the respective maneuvers.

In proposing deletion of this requirement, NHTSA felt that the national interest in energy conservation had substantially diminished consumer demand for rapid acceleration capability. Further, the high speed driving permitted by the test procedures appeared to contradict the safety and energy saving policies behind the national 55-mph speed limit. Commenters on the proposal, including American Motors, CFAS, General Motors and Volkswagen of America, unanimously agreed that the acceleration and passing ability provision was no longer of interest to consumers and had become inconsistent with national goals. Section 575.106 has, therefore, been deleted from the consumer information regulations.

NHTSA's regulatory evaluation, conducted pursuant to E.O. 12044, "Improving Government Regulations" and departmental guidelines, indicates that the amendments are not significant. They decrease the regulatory burden on industry, while having no appreciable negative impact on safety. A copy of the regulatory evaluation can be obtained from the Docket Section, Room 5108, National Highway Traffic Safety Administration, 400 Seventh Street, S.W., Washington, D.C. 20590. Also, the amendments will have no measurable effect on the environment.

Because the amendments as they pertain to acceleration and passing ability relieve a restriction, and to avoid any unnecessary costs in complying with this requirement, the deletion of section 575.106 is effective immediately. So that useful performance information can be provided to consumers for model year 1982 vehicles, the amendment to section 575.6 is effective June 1, 1981.

In consideration of the foregoing, 49 CFR Part 575, Consumer Information Regulations, is amended as follows:

1. Section 575.6(d) is amended to read:

§575.6 Requirements

* * * * *

(d) In the case of all sections of Subpart B, other than §575.104, as they apply to information submitted prior to new model introduction, each manufacturer of motor vehicles shall submit to the Administrator 10 copies of the information specified in Subpart B of this part that is applicable to the vehicles offered for sale, at least 90 days before it is first provided for examination by prospective purchasers pursuant to paragraph (c) of this section. In the case of §575.104, and all other sections of Subpart B as they apply to post-introduction changes in information submitted for the current model year, each manufacturer of motor vehicles, each brand name owner of tires, and each manufacturer of tires for which there is no brand name owner shall submit to the Administrator 10 copies of the information specified in Subpart B of this part that is applicable to the vehicles or tires offered for sale, at least 30 days before it is first provided for examination by prospective purchasers pursuant to paragraph (c) of this section.

2. Section 575.106 is deleted.

The principal authors of this proposal are Ivy Baer of the Office of Automotive Ratings and Richard J. Hipolit of the Office of the Chief Counsel.

Issued on July 7, 1980.

Joan Claybrook
Administrator

45 FR 47152
July 14, 1980

PREAMBLE TO AN AMENDMENT TO PART 575

Consumer Information Regulations
Uniform Tire Quality Grading
(Docket No. 25; Notice 4)

ACTION: Final rule.

SUMMARY: This notice amends the Uniform Tire Quality Grading (UTQG) Standards to provide for the testing of metric tires, tires with inflation pressures measured in kilopascals. Since the original UTQG test requirements were written prior to the introduction of metric tires and specified inflation pressures measured in pounds per square inch, modification of the regulation is now necessary to identify inflation pressures applicable to metric tires. The notice also makes technical changes in the UTQG traction test procedure to facilitate efficient use of test facilities.

EFFECTIVE DATE: The amendments are effective immediately.

FOR FURTHER INFORMATION CONTACT:
Dr. F. Cecil Brenner, Office of Automotive Ratings, National Highway Traffic Safety Administration, 400 Seventh Street, S.W., Washington, D.C. 20590, 202-426-1740

SUPPLEMENTARY INFORMATION: The UTQG standards prescribe test procedures for evaluation of the treadwear, traction, and temperature resistance properties of passenger car tires. Grades based on these are used by consumers to evaluate the relative performance of competing tire lines. Test procedures for all three performance categories were established specifying inflation pressures in pounds per square inch.

Following the introduction of metric tires with inflation pressures measured in kilopascals, the National Highway Traffic Safety Administration (NHTSA) recognized the need to add metric inflation pressures to the UTQG test procedures. The agency proposed (44 F.R. 56389; October 1, 1979; Notice 34) that for purposes of traction testing,

metric tires would be inflated and tire loads determined using a prescribed inflation pressure of 180 kPa. Under the proposal, other tires would continue to be tested at an inflation pressure of 24 psi. NHTSA's notice also proposed modification of the temperature resistance test procedure to provide, in the case of metric tires, for use of inflation pressures 60 kPa less than the tires' maximum permissible inflation pressure.

In response to comments, NHTSA modified the original proposal (45 F.R. 35408; May 27, 1980; Notice 40) to include treadwear testing in the proposed modifications and to incorporate a table indicating treadwear, traction, and temperature resistance test inflation pressures for tires with various maximum permissible inflation pressures in kilopascals and pounds per square inch. In the proposed table, different test inflation pressures were specified for tires with differing maximum permissible inflation pressures.

The agency also proposed, in Notice 34, modification of the traction test procedure to permit the adjustment of candidate tire test results with standard tire test results obtained either before or after the candidate tire test sequence, so long as all data to be compared were collected within the same two-hour period. This change was intended to promote efficient use of the traction test facilities by permitting data from more than one candidate tire test sequence to be adjusted by comparison with the same standard tire sequence.

Upon examination of additional data, NHTSA concluded that a three-hour period could be employed without affecting the accuracy of the test results. Use of a three-hour period would permit more than one candidate tire test sequence to be run both before and after the corresponding standard tire test sequence. A three-hour period for comparative testing was proposed in Notice 40. Having received no negative comments on the traction test sequence proposal as stated in that notice, NHTSA has determined that the amendment will be adopted as proposed.

On the proposed changes to provide for testing of metric tires, Goodyear Tire & Rubber Company noted that the table of test inflation pressures proposed in Notice 40 calls for variations in the prescribed test inflation pressure depending on the maximum permissible inflation pressure of the tested tire. The original traction procedure specified a single test inflation pressure for all tires. Goodyear expressed concern that such a change could affect test results and, consequently, tire grades, and require wasteful additional testing to confirm grades already assigned. Goodyear recommended that NHTSA adopt the amendment proposed in Notice 34 that all metric tires be tested using the inflation pressure 180 kPa and all other tires be tested using the original 24 psi inflation pressure.

NHTSA agrees that unnecessary costs associated with the UTQG Standard should be avoided. For this reason, the agency has determined that reference to traction testing will be deleted from the table of test inflation pressures, and the addition of the metric traction test inflation pressure of 180 kPa proposed in Notice 34 will be adopted instead. Those aspects of Notice 40 pertaining to treadwear and temperature resistance testing of metric tires will be adopted as proposed in that notice.

Pursuant to Executive Order 12044, "Improving Government Regulations," and implementing Departmental guidelines, the agency has considered the effects of these amendments. NHTSA reaffirms its earlier determination that the amendments are not significant and that the effects are so minimal as not to warrant preparation of a regulatory evaluation. NHTSA has determined these amendments will result in modest cost savings to industry and consumers, while having no appreciable effect on safety or the environment.

Because these amendments will facilitate the efficient and accurate completion of testing presently underway, the amendments are effective immediately.

In consideration of the foregoing, 49 CFR §575.104 is amended as follows:

1. In section 575.104(e)(2)(ii) by substitution of the words "the applicable pressure specified in Table 1 of this section," in place of the words "an inflation pressure 8 pounds per square inch less than its maximum permissible inflation pressure."

2. In section 575.104 (f) (2) (i) (B) and (D) by addition of the words, "or, in the case of a tire with inflation pressure measured in kilopascals, to 180 kPa" following the words "to 24 psi."

3. In section 575.104(f)(2)(vii) by addition of the following sentence, at the end thereof: "The standard tire traction coefficient so determined may be used in the computation of adjusted traction coefficients for more than one candidate tire."

4. In section 575.104 (f)(2)(viii) by addition of the words, "or, on the case of a tire with inflation pressure measured in kilopascals, the load specified at 180 kPa," following the words "at 24 psi," and by addition of the sentences, "Candidate tire measurements may be taken either before or after the standard tire measurements used to compute the standard tire traction coefficient. Take all standard tire and candidate tire measurements used in computation of a candidate tire's adjusted traction coefficient within a single three hour period" following the first sentence thereof.

5. In section 575.104 (g) (1) by substitution of the words "the applicable pressure specified in Table 1 of this section," in place of the words "2 pounds per square inch less than its maximum permissible inflation pressure."

6. In section 575.104(g)(3) by substitution of the words "the applicable pressure specified in Table 1 of this section," in place of the words "2 pounds per square inch less than the maximum permissible inflation pressure."

7. In section 575.104(g)(6) by substitution of the words "applicable inflation pressure specified in Table 1 of this section," in place of the words "inflation pressure that is 8 pounds per square inch less than the tire's maximum permissible inflation pressure."

8. In section 575.104(g)(8) by substitution of the words "the applicable pressure specified in Table 1 of this section," in place of the words "2 pounds per square inch less than that the tire's maximum permissible inflation pressure."

9. By addition of the following table at the conclusion of the text of that section:

Table 1.—Test Inflation Pressures

Maximum permissible inflation pressure	32 lb/in²	36 lb/in²	40 lb/in²	240 kPa	280 kPa	300 kPa
Pressure to be used in tests for treadwear and in determination of tire load for temperature resistance testing.	24	28	32	180	220	180
Pressure to used for all aspects of temperature resistance testing other than determination of tire load.	30	34	38	220	260	220

The principal authors of this notice are Dr. F. Cecil Brenner of Office of Automotive Ratings and Richard J. Hipolit of the Office of Chief Counsel.

Issued on October 15, 1980.

Joan Claybrook
Administrator

45 FR 70273
October 23, 1980

PREAMBLE TO AN AMENDMENT TO PART 575

Consumer Information Regulations; Uniform Tire Quality Grading

(Docket No. 25; Notice 45)

ACTION: Final rule.

SUMMARY: This notice amends the Uniform Tire Quality Grading Standards to permit tire grades to be molded on the tire sidewall beginning at any time up to six months after introduction of a new tire line. This amendment, which was proposed in response to a petition from Atlas Supply Company, is intended to avoid disruption of production while tire grades are determined. The notice also extends the deadline for conversion to new format tire tread labels in order to permit unused supplies of old-format labels to be used up.

EFFECTIVE DATE: August 15, 1981.

SUPPLEMENTARY INFORMATION:

Background

On January 26, 1981, the National Highway Traffic Safety Administration (NHTSA) published a notice of proposed rulemaking (46 F.R. 8063; Docket 25, Notice 44) proposing amendment of the sidewall molding and tread labeling requirements of the Uniform Tire Quality Grading (UTQG) Standards (49 CFR 575.104). In response to a petition for rulemaking filed by Atlas Supply Company, NHTSA proposed a four month phase-in period for molding of UTQG grades on the sidewalls of tires of newly introduced tire lines. Under the regulation as originally issued, all covered tires were required to have UTQG grades molded on the sidewall (49 CFR 575.104(d)(1)(i)(A)). Atlas, with

support from the Goodyear Tire & Rubber Company and the General Tire & Rubber Company, requested that initial production runs of new tire lines be exempted from the molding requirement pending determination of UTQG grades.

The notice of proposed rulemaking also responded to a petition for rulemaking submitted by Armstrong Rubber Company. Armstrong had requested that the deadline for conversion to the new UTQG tread label format established in Docket 25, Notice 35 (44 F.R. 68475; November 29, 1979) be extended at least nine months to permit supplies of old-format labels to be used up. In response to Armstrong's petition, NHTSA proposed that the deadline for conversion to the new format be extended from October 1, 1980, until April 1, 1982.

As indicated in the Notice of Intent published by NHTSA on April 9, 1981, (46 F.R. 21203), NHTSA is currently reviewing the requirements of the Uniform Tire Quality Grading System regulatory program, to determine the degree to which it accurately and clearly provides meaningful information to consumers in accordance with the requirements of 15 U.S.C. 1423. Proposed rulemaking or further action on this question will be published within thirty days of this notice.

Proposed Rulemaking—Decision

NHTSA received several comments from tire and motor vehicle manufacturers on the proposed amendments. After review of these comments, the agency has concluded that, while amendment of the regulation is warranted, several changes in the specifics of the proposal are desirable.

Proposed Rulemaking—Comments

Support for the concept of a temporary exemption from the UTQG molding requirements for new tire lines was indicated by both tire and motor vehicle industry sources. The Rubber Manufacturers Association (RMA) commented that such an exemption would resolve difficulties associated with grading new tire lines, and save costs to manufacturers, while not significantly affecting the distribution of grading information to the public.

Ford Motor Company expressed its opinion that a temporary exemption would make good economic sense by permitting full utilization of production facilities while UTQG grades are determined. Full utilization of equipment was a primary goal of the Atlas petition, which expressed concern that a substantial investment in tire molds would be unproductive while UTQG testing was conducted using a small initial sample of tires.

Goodyear also expressed general support for the proposal, since it would permit UTQG grades to be based on testing of production tires. Goodyear noted that while UTQG testing of prototype tires is possible, testing of production tires is desirable because of the greater variety of sizes available for testing.

While supporting the proposal for a molding exemption period, tire industry commenters uniformly agreed that the four-month period proposed by NHTSA would be inadequate. Goodyear, Atlas, and the RMA agreed that a six-month period would be preferable. These commenters viewed four months as the period in which grades could be determined and molds stamped under optimal conditions. However, these sources pointed out that unexpected delays in tire selection, testing, data analysis, retesting, or stamping could easily extend beyond the four-month period. Atlas' comments suggested that the potential for delay is even greater where multiple sources of supply are involved. In order to allow for potential uncontrollable delays of this nature, NHTSA has determined that the period for introduction of molded grades on new tire lines will be extended to six months from the date production commences.

NHTSA's notice of proposed rulemaking on this subject contained a proposed requirement that motor vehicle manufacturers affix to the window of each of their vehicles equipped with tires exempted from the molding requirement a sticker containing tire-specific UTQG information. This proposal was intended to assure that prospective vehicle purchasers have access to UTQG information. Tire-specific grades for original equipment tires are not available on tread labels or in vehicle manufacturers' point of sale information. However, the window sticker proposal was uniformly opposed by motor vehicle and tire industry commenters.

General Motors Corporation, Chrysler Corporation, Volkswagen of America, Inc., and Goodyear all argued that significant assembly line problems would result from adoption of a window sticker requirement. Comments received from these manufacturers indicated that several lines of tires are frequently used as original equipment on a single vehicle model and, under the proposal, more than one tire line without molded grades could be available for use in an assembly plant at one time.

Given this diversity of tire use, commenters pointed out, assembly line personnel would have to inspect each vehicle and determine whether ungraded tires were being used. These employees would then have to determine the correct UTQG window sticker to be affixed to the vehicle. Under such a system, labeling errors would be likely in the absence of costly and time-consuming reinspection. Alternatively, expensive special parts identification and storage programs could be undertaken to track ungraded tires through the plant and affix the appropriate labels when the tires are used.

Several commenters argued that such a labeling program would be unreasonably burdensome and expensive in comparison to the benefits which would be expected from such a program. Ford Motor Company estimated that UTQG window stickers would result in an annual cost to that company of $50,000. General Motors (GM) estimated that window stickers could be affixed at a cost of $.50 per car if used on all cars it produced. According to GM, this cost would be much higher in the limited application contemplated by the proposal, due to increased scheduling and inspection costs.

At the same time, General Motors, Chrysler, and Goodyear argued that the major importance of UTQG is in the replacement market and that tire grades seldom influence new car purchases. GM pointed out that it establishes its own performance criteria for original equipment tires beyond the UTQG performance categories, and that in this way vehicle purchasers are assured of getting suitable tires regardless of molded UTQG grades.

While Ford suggested several alternatives to the window sticker proposal, the other commenters addressing the issue recommended that no accommodation at all is necessary for ungraded original equipment tires. In this regard, Goodyear noted that the estimate used in the notice of proposed rulemaking that no more than five percent of original equipment tires would be ungraded was probably high and the actual figure will likely be considerably below that estimate. NHTSA is also aware that in the event a vehicle purchaser is interested in UTQG information on original equipment tires temporarily exempted from the molding requirement, UTQG information would be readily available from local tire dealers and other sources. In view of the above considerations, NHTSA has determined that the proposed UTQG window sticker is unnecessary and unduly burdensome and the proposal for such a sticker is withdrawn.

NHTSA's notice of proposed rulemaking also proposed a sunset provision for the molding requirement change. This provision would have automatically terminated the molding exemption at the end of three years, unless the agency determined that an extension were necessary. Goodyear and the RMA pointed out in their comments that a sunset provision is unnecessary, since the agency already has the authority to review and amend the regulation at any time, if it appears that the exemption is not working as planned. In fact, Atlas recommended that the agency review the effect of the amendment no later than 18 months after its effective date.

Goodyear noted that, if the sunset provision is adopted, unforeseen delays in completion of NHTSA's review could lead to disruptions in the event the three-year sunset period expires before the review process can be completed and the exemption extended. While NHTSA plans to monitor the effect of the molding exemption and will propose any necessary modifications, the agency has concluded that the proposed sunset provision is unnecessary and potentially disruptive. Therefore, the sunset provision is withdrawn.

Finally, only one commenter expressed an opinion on the proposal to extend the deadline for conversion to the new tread label format. As discussed in Armstrong's petition on this subject, the original October 1, 1980, effective date appeared appropriate at the time it was established. However, a sudden market shift toward radial tires resulted in unused supplies of old-format labels for bias-belted tires. In order to permit existing stocks of labels to be used, NHTSA proposed extension of the deadline for conversion to the new label format until April 1, 1982.

Goodyear complained that it had scrapped unused supplies of old-format labels when the format change took effect and argued that extension of the deadline at this time would not be fair and equitable. Goodyear went on, however, to state its preference that the deadline for conversion be eliminated altogether in the interest of efficient use of available materials.

NHTSA regrets that Goodyear found it necessary to dispose of a quantity of old-format labels which could not be used up prior to the October 1 deadline. However, the agency believes that such economic waste would only be compounded by requiring disposal of labels which may have been retained by other manufacturers. At the same time, complete elimination of the conversion deadline could indefinitely delay conversion to the new label format, which the agency considers superior. For these reasons, the deadline for conversion to the new tread label format is extended until April 1, 1982. Of course, manufacturers and brand name owners wishing to use new-format labels prior to that date are free to do so.

Several commenters stressed the need to act quickly on the proposed amendments in order to avoid production disruptions and economic penalties which may be encountered in the planned introduction of new tire lines. Since the changes outlined above relieve restrictions and have these beneficial effects, they are made effective immediately upon publication.

NHTSA has evaluated these amendments and found that their effect would be to provide minor cost savings for tire manufacturers and brand name owners.

Accordingly, the agency has determined that the amendments are not a major rule within the meaning of Executive Order 12291 and are not significant for purposes of Department of Transportation policies and procedures for internal review of proposals. The agency has further determined that the cost savings are not large enough to warrant preparation of a regulatory evaluation under the procedures. The agency has also determined that the amendments, which relieve restrictions and provide minor cost savings, will not significantly affect a substantial number of small entities. Finally, the agency has concluded that the environmental consequences of the amendments will be minimal.

Issued on July 30, 1981.

Raymond A. Peck, Jr.
Administrator
46 F.R. 41514
August 17, 1981

PREAMBLE TO AN AMENDMENT TO PART 575

Consumer Information Regulations
(Docket No. 79-02; Notice 5)

ACTION: Final rule.

SUMMARY: This notice amends the Consumer Information Regulations to permit amendment of previously submitted motor vehicle performance information at any time up to 30 days prior to new model introduction. This amendment is intended to reduce regulatory burdens on industry by allowing greater flexibility in the implementation of pre-introduction product changes.

EFFECTIVE DATE: June 1, 1982.

SUPPLEMENTARY INFORMATION: The Consumer Information Regulations (49 CFR Part 575) require that manufacturers of motor vehicles and tires provide prospective purchasers and first purchasers with information on the performance of their products in the areas of vehicle stopping ability (49 CFR §575.101), vehicle tire reserve load (49 CFR §575.102), truck camper loading (49 CFR §575.103), and uniform tire quality grading (49 CFR §575.104). In addition to the requirements that information be furnished directly to consumers, manufacturers are required to submit information to the National Highway Traffic Safety Administration (NHTSA) prior to the introduction of new vehicle models and tire lines or modification of existing lines. This advance submission requirement is intended to permit the agency to compile the information supplied by various manufacturers in a comparative format for distribution to consumers.

As originally issued, and presently in force, the regulation requires that all information be submitted to NHTSA at least 30 days prior to the date on which the information is made available to prospective purchasers (49 CFR §575.6(d)). The regulation requires that information must be made available to prospective purchasers not later than the day on which the manufacturer first authorizes the subject product to be put on public display and sold to consumers (49 CFR §575.6(c)).

To enable NHTSA to compile the information in a comparative booklet for distribution early enough in the model year to be useful to most consumers, the agency amended the regulations to require that motor vehicle manufacturers submit information at least 90 days in advance of new model introduction (45 F.R. 47152; July 14, 1980). The 30-day period was retained for post-introduction vehicle changes and for tire quality grading information. The amendment was originally scheduled to take effect June 1, 1981, but the effective date was postponed until June 1, 1982 (46 F.R. 29269; June 1, 1981), to allow consideration of a petition from Ford Motor Company requesting greater flexibility in the requirement.

Ford contended that the 90-day advance submission requirement could create hardships for manufacturers when last minute pre-introduction product changes, resulting from component supply difficulties or other factors, affect the performance characteristics covered by Part 575. In such a situation, a manufacturer could be forced to delay introduction of a vehicle model until a new 90-day advance notice period had been completed. To avoid this result, Ford recommended that manufacturers be permitted to amend initial pre-introduction submissions at any time prior to 30 days before model introduction. NHTSA responded with a notice of proposed rulemaking to permit such revisions in the event of unforeseeable pre-introduction modifications in vehicle design or equipment (46 F.R. 4054; August 10, 1981; Docket 79-02; Notice 4). This proposal was among the deregulatory measures discussed in the Administration's notice of intent on measures to aid the auto industry.

NHTSA received comments from seven motor vehicle manufacturers and importers in response to the notice of proposed rulemaking. All commenters agreed that the proposed amendment would be an improvement over the established 90-day requirement, in that greater flexibility would be provided in the introduction of necessary product changes. As noted by Ford, the amendment would facilitate implementation of product development and marketing schedules, while still providing information adequate for NHTSA's purposes. NHTSA agrees and has determined that the proposed amendment should be adopted with one modification.

General Motors and Volkswagen of America, Inc. commented that limiting changes in performance information to those resulting from "unforeseeable" product changes is inappropriate. Volkswagen argued that only the manufacturer can adequately judge whether product changes are unforeseeable, and that agency attempts to enforce such a requirement could lead to undesirable consequences. Moreover, a manufacturer acting in good faith could be faced with a dilemma if the manufacturer is unable to conclude that a needed product change was unforeseeable, although in fact it had not been anticipated in a particular instance. (Docket 79-02, Notice 4, No. 004). General Motors argued that cost factors alone are a sufficient incentive to manufacturers to avoid last minute product changes and therefore no foreseeability standard is necessary to insure that changes are made in good faith. General Motors suggested that if any qualifier is thought necessary, "unforeseen" or "unanticipated" would be preferable. (Docket 79-02, Notice 4, No. 007).

NHTSA continues to believe that some provision is necessary to assure that only good faith product changes form the basis for modifications of pre-introduction submissions. However, NHTSA does not wish to inhibit product changes which the agency may believe could have been foreseen, but honestly were not. To avoid this result, the agency has concluded that "unforeseen" rather than "unforeseeable" is a more appropriate description of the types of product changes which would justify amendments of pre-introduction consumer information submissions.

Volkswagen and General Motors also commented that the 90-day advance submission requirement is unnecessary and that the original 30-day period should be retained. Volkswagen contended that the agency could not use the manufacturers' submissions until 30 days prior to model introduction in any case because the data would be subject to change. Volkswagen also suggested that manufacturers could circumvent the 90-day requirement by making minimal performance claims in their initial submissions and amending the information at a later date. General Motors commented that the further in advance information is submitted, the less accurate it will be, and that the successful publication of the Environmental Protection Agency's fuel economy guide establishes the feasibility of publishing comparative information with a brief advance submission period.

NHTSA's past experience indicates that 30 days is inadequate for this agency to compile, publish and distribute a useful comparative booklet. Moreover, any design or equipment related inaccuracies inherent in a 90-day advance submission can be corrected under the amendment adopted in this notice. While it is true that the agency could not publish and distribute the information until the period for amendment of initial submissions expired, the agency could compile the information and begin the publishing process, incorporating any necessary changes prior to printing. Comments submitted by Yamaha Motor Corporation, U.S.A. (Docket 79-02, Notice 4, No. 001), suggest that the number of required changes will be small. Finally, the type of abuse noted by Volkswagen would be precluded under the amended regulation because the type of revision described would not have been necessitated by unforeseen product changes.

Commenters also suggested rescinding the advance submission requirement completely or rescinding the stopping distance and tire reserve load provisions. Still other commenters recommended that the agency reassess the costs and benefits of the Consumer Information Regulations as a whole. The rationale for these recommendations centered on the alleged lack of consumer interest in the information and the limited amount of information provided under the program.

As noted by commenters, NHTSA has proposed rescission of the requirement that auto manufacturers provide tire reserve load information to the public and the agency (46 F.R. 47100; September 24, 1981). However, in conjunction with the Administration's efforts to ease regulatory burdens on the auto industry, the agency wishes to maintain a functioning consumer information program as a possible substitute for mandatory safety regulations. As part of the agency's ongoing program to identify and eliminate unnecessary regulatory burdens, NHTSA plans to review the benefits of and need for the Consumer Information Regulations as a component of the agency's total regulatory program. If this review indicates that the consumer information program is not useful and cost-beneficial, the future of the regulation will be addressed in a later rulemaking proceeding.

NHTSA has evaluated this relieving of a restriction and found that its effect will be to provide minor cost savings for motor vehicle manufacturers. Accordingly, the agency has determined that the action is not a major rule within the meaning of Executive Order 12291 and is not significant for purposes of Department of Transportation policies and procedures for internal review of regulatory actions. The agency has further determined that the cost savings are so minimal as to not warrant preparation of a regulatory evaluation under the procedures. The agency certifies pursuant to the Regulatory Flexibility Act that the action will not have a significant economic impact on a substantial number of small entities because the cost savings will be modest and few, if any, motor vehicle manufacturers can be considered small entities within the meaning of the statute. Finally, the agency has concluded that the environmental consequences of the proposed change will be of such limited scope that they clearly will not have a significant effect on the quality of the human environment.

Issued on February 11, 1982.

Raymond A. Peck, Jr.
Administrator

47 F.R. 7257
February 18, 1982

PREAMBLE TO AN AMENDMENT TO PART 575

Consumer Information Regulations
(Docket No. 81-09; Notice 2)

ACTION: Final rule.

SUMMARY: This notice amends the Consumer Information Regulations by revocation of the requirement that motor vehicle manufacturers provide information on passenger car tire reserve load. The National Highway Traffic Safety Administration has concluded that this information is without value to consumers, and that deletion of the requirement will avoid unnecessary regulatory burdens on industry.

EFFECTIVE DATE: This amendment is effective immediately.

SUPPLEMENTARY INFORMATION: The Consumer Information Regulations (49 CFR Part 575) require that manufacturers of motor vehicles and tires provide consumers with information on the performance of their products under various performance criteria. In the case of motor vehicle manufacturers, information is required in the areas of passenger car and motorcycle stopping distance (49 CFR §575.101), passenger car tire reserve load (49 CFR §575.102), and truck camper loading (CFR §575.103). National Highway Traffic Safety Administration (NHTSA) regulations require that motor vehicle manufacturers supply the required performance information in writing to first purchasers of their motor vehicles at the time of delivery (49 CFR §575.6(a)) and that the information be made available for examination by prospective purchasers at each location where the vehicles to which it applies are sold (49 CFR §575.6(c)). The information must also be submitted in advance to NHTSA (49 CFR §575.6(d)).

On September 24, 1981, NHTSA published in the *Federal Register* a proposal to delete from the Consumer Information Regulations the requirement for provision of information on passenger car tire reserve load (46 F.R. 47100; Docket No. 81-09, Notice 1). Tire reserve load is the difference between a tire's stated load rating and the load imposed on the tire at maximum loaded vehicle weight. This difference is expressed as a percentage of tire load rating under the regulation.

NHTSA's proposal noted that a NHTSA analysis, "The Relationship Between Tire Reserve Load Percentage and Tire Failure" (Docket No. 81-09, Notice 1, No. 002), had concluded that no relationship exists between tire reserve load percentage and tire failure rate. This analysis was based on the results of a study prepared for NHTSA by Chi Associates, "Statistical Analysis of Tire Failure vs. Tire Reserve Load Percentage" (Docket No. 81-09, Notice 1, No. 001), using tire reserve load data obtained from eight automobile manufacturers under special order from this agency. The proposal also noted the lack of major differences among manufacturers' reported tire reserve load percentages, and the safeguards against overloading contained in Federal Motor Vehicle Safety Standard No. 110 (FMVSS No. 110), Tire Selection and Rims.

In response to its proposal to delete the requirement for tire reserve load information, NHTSA received comments from seven motor vehicle manufacturers and importers. The commenters were unanimous in their support of the agency's proposal. Comments received generally focused on the lack of benefit to consumers resulting from provision of tire reserve load information.

Several commenters noted the lack of any proven safety benefit from the tire reserve load regulation. Two commenters, Ford Motor Company and Volkswagen of America, Inc., cited the above mentioned NHTSA analysis in support of the proposition that tire reserve load is an invalid predictor of tire failure (Docket No. 81-09, Notice 1, Nos. 004 and 006). General Motors Corporation (Docket No. 81-09, Notice 1, No. 007) and American Motors Corporation (Docket No. 81-09, Notice 1, No. 008, referencing its prior comment, Docket No. 79-02, Notice 1, No. 012) argued that FMVSS No. 110 is sufficient to protect against the installation of tires with inadequate load carrying capacity.

American Motors also pointed out that much of the information required under the tire reserve load regulation is redundant of information which must be included on glove compartment placards pursuant to FMVSS No. 110. In this regard, information on recommended tire size designation and recommended inflation pressure for maximum loaded vehicle weight, required under paragraphs (c)(2) and (3) of the tire reserve load regulation (49 CFR §575.102(c)(2) and (3)) is essentially the same as that required under paragraphs s4.3(c) and (d) of FMVSS No. 110 (49 CFR §575.110, s4.3(c) and (d)).

Several commenters argued that not only is tire reserve load information lacking in safety value, but it may actually pose a danger to highway safety. Renault USA, Inc., Volkswagen, General Motors and American Motors all expressed concern that provision of tire reserve load information would mislead consumers into loading their vehicles beyond gross vehicle weight ratings (Docket No. 81-09, Notice 1, Nos. 003, 006, 007, 008). Renault and American Motors also noted that the tire reserve load regulation fails to take into account the effect of inflation pressure, thus further limiting the usefulness of the regulation and creating additional potential hazards resulting from improper tire inflation.

Chrysler Corporation and General Motors emphasized the minimal consumer interest in tire reserve load information (Docket No. 81-09, Notice 1, Nos. 005 and 007). As evidence of this minimal interest, both manufacturers noted the lack of consumer requests for point of sale information currently available.

Some cost savings are likely to result to automobile manufacturers as a result of deletion of this requirement. General Motors pointed out that, even if tire reserve load is dropped from the consumer information regulations, manufacturers will still be required to print and distribute booklets containing information on vehicle stopping distance and thus cost savings will be limited (Docket No. 81-09, Notice 1, No. 007). However, Ford commented that elimination of the tire reserve load provision would result in some savings in manpower and computer time (Docket No. 81-09, Notice 1, No. 004). Similarly, Volkswagen noted that manufacturers' booklet publication costs would be reduced and reporting requirements simplified if the proposed amendment were adopted (Docket No. 81-09, Notice 1, No. 006).

In view of the lack of benefits of the tire reserve load information requirements, the potential for reduction of unnecessary regulatory burdens by deletion of these requirements, and the other considerations discussed above, NHTSA has concluded that the tire reserve load requirements of the Consumer Information Regulations should be revoked. In order to avoid continued imposition of unnecessary regulatory burdens, this amendment relieving a restriction is made effective immediately.

Several commenters also suggested rescinding the vehicle stopping distance information requirement of the regulation, thereby eliminating all requirements for vehicle specific consumer information applicable to passenger cars. While beyond the scope of this rulemaking proceeding, NHTSA is reviewing the benefits of and need for other aspects of the Consumer Information Regulations in connection with a petition for rulemaking submitted by General Motors. If this review indicates that vehicle stopping distance information is not useful, the potential deletion of this requirement will be made the subject of a future rulemaking proceeding.

NHTSA has evaluated this relieving of a restriction and found that its effect would be to provide minor cost savings for motor vehicle manufacturers. Accordingly, the agency has determined that this action is not a major rule within the meaning of Executive Order 12291 and is not significant for purposes of Department of Transportation policies and procedures for internal review of regulatory actions. The agency has further determined that the cost savings are minimal and do not warrant preparation of a regulatory evaluation under the procedures.

The agency certifies, pursuant to the Regulatory Flexibility Act, that this action will not "have a significant economic impact on a substantial number of small entities," and that a Regulatory Flexibility Analysis was therefore not required. Few, if any, motor vehicle manufacturers can be considered small entities within the meaning of the statute. Small organizations and small government jurisdictions will not be significantly affected by this action. These entities could be affected by the action as motor vehicle purchasers. However, the agency has determined that tire reserve load information is not of value to purchasers. Moreover, possible cost savings associated with the action will be minor in the case of individual purchasers.

Issued on May 28, 1982.

Raymond A. Peck, Jr.
Administrator

47 F.R. 24593
June 7, 1982

PREAMBLE TO AN AMENDMENT TO PART 575

Consumer Information Regulations; Uniform Tire Quality Grading
(Docket No. 25; Notice 46)

ACTION: Interim final rule and request for comments.

SUMMARY: This notice makes several technical amendments to the test procedures in the regulation on Uniform Tire Quality Grading (UTQG). The UTQG regulation specifies that the tire rim size and tire loading used in testing individual tires are to be determined by using Table 1, Appendix A of Federal Motor Vehicle Safety Standard No. 109, New pneumatic tires. Since the portion of Table 1, Appendix A relied upon by the UTQG regulation was deleted in a previous agency rulemaking, effective June 15, 1982, reliance upon that Appendix will no longer be appropriate after that date. This notice replaces the references to Appendix A with equivalent methods for determining rim size and tire loading.

DATES: This amendment is effective June 15, 1982.

SUPPLEMENTARY INFORMATION: The Uniform Tire Quality Grading (UTQG) regulation (49 CFR 575.104) requires that manufacturers and brand name owners of passenger car tires provide consumers with information on the treadwear, traction and temperature resistance of their tires. This information is to be generated in accordance with procedures specified in the regulation.

Two parameters specified in the test procedures are the proper test rim width for each tire, and the load under which the tire is to be tested. The UTQG regulation refers to Appendix A of Federal Motor Vehicle Safety Standard No. 109 (FMVSS 109) for the determination of rim size to be used for testing purposes. Table 1 of Appendix A provides a complete listing of tire sizes available

in this country and for each size indicates the proper test rim size and maximum loads at various tire pressures.

The UTQG regulation also refers to Appendix A of FMVSS 109 for the determination of tire load. The tire load for temperature resistance testing is the load specified in Appendix A of FMVSS 109 for the tire pressure listed in Table 1 of the UTQG regulation. Thus, load is currently determined by obtaining the tire pressure from Table 1 of the UTQG regulation and finding the load for that pressure level in Appendix A. The tire load for treadwear and traction testing is determined in the same way, except that the load level found in Appendix A is multiplied by 85 percent.

Beginning on June 15, 1982, reliance upon Appendix A of FMVSS 109 to determine rim size and tire load for UTQG testing will no longer be possible. On that date, the agency's amendment (December 17, 1981; 46 F.R. 61473) deleting Table 1 of Appendix A will become effective. As FMVSS 109 is currently written, the tire manufacturers and brand name owners must submit the rim size information to NHTSA for incorporation in Table 1. Under the amendment, they will be able to satisfy FMVSS 109 by either securing the incorporation of the information in a publication of a standardization organization like the Tire and Rim Association or one of its foreign counterparts or by submitting it to the agency, their dealers, and others who request it, without the need for the information's incorporation in any other document.

As to tire load information, the tire manufacturers and brand name owners currently calculate loads for pressure levels ranging from 16 to 40 pounds per square inch in most cases and submit the information to NHTSA for incorporation in Table 1. After June 14, they need determine the load only for a single pressure level, the maximum one. The responsibilities of the manufacturers and brand name owners under amended FMVSS 109 regarding load information may be satisfied in the same fashion as their responsibilities regarding rim size.

The deletion of Table 1 of Appendix A was intended to reduce an unnecessary regulatory burden placed by FMVSS 109 on the tire industry and the agency. The action was not intended to make any change in the UTQG test procedures. However, the deletion of Table 1 of Appendix A necessitates amending the UTQG regulation so that rim size and tire load can be determined without reference to that appendix.

This notice provides the means for making those determinations. The rim size to be used for UTQG testing is the same size specified by the tire manufacturer or brand name owner in a publication of a standardization association or in a submission directly to the agency. This provision does not in any way change the rim size used for UTQG testing. Instead, it simply changes the source of obtaining the rim size information.

As to tire loading, the UTQG testing will henceforth rely upon mathematical calculation involving a tire's maximum load, as molded on its sidewall, instead of relying upon information submitted by the manufacturer or brand name owner to any organization or agency. Under the new procedure, the maximum load is multiplied by a factor, ranging from .851 to .887 depending on the tire's maximum inflation pressure, and the result is rounded. The rounded result is used for temperature resistance testing. For treadwear and traction testing, the rounded result is multiplied by 85 percent. In most instances, this procedure produces the same load as is currently obtained by reference to Table 1 of Appendix A. In those instances in which the load is different, the degree of difference is so slight that the difference will not have any practical effect on the UTQG test results.

The agency finds good cause for issuing these amendments without prior notice and comment. The agency believes that prior notice and comment are unnecessary. The revisions are technical and editorial in nature. In most instances, the revisions produce no changes in the procedures under which tires are tested for UTQG purposes. In the few instances in which there will be a change, the change is so slight as to be substantively insignificant. Although the agency has concluded that prior notice and comment are unnecessary, it has decided to go beyond the minimum requirements of the Administrative Procedures Act and provide a 60-day comment period on these amendments. For the same reasons set forth above and to permit continued implementation of the UTQG regulation, the agency finds good cause for making the revisions effective immediately.

Since this proceeding is merely intended to allow the continued implementation of the UTQG regulation without any change in the manner of implementation, NHTSA has determined that this proceeding does not involve a major rule within the meaning of Executive Order 12291 or a significant rule within the meaning of the Department of Transportation regulatory procedures. Further, there are virtually no economic impacts of this action so that preparation of a full regulatory evaluation is unnecessary.

The Regulatory Flexibility Act does not require the preparation of flexibility analyses with respect to rulemaking proceedings, such as this one, for which prior notice and comment is not required by the Administrative Procedures Act. If the requirement for preparation of such analyses were applicable, the agency would certify that this action would not have a significant economic impact on a substantial number of small entities. As noted above, this action will make essentially no change in the implementation of the UTQG regulation.

NHTSA has concluded that this action will have essentially no environmental consequences and therefore that there will be no significant effect on the quality of the human environment.

Interested persons are invited to submit comments on the agency's action announced above and on any other topics relevant to this notice. It is requested but not required that 10 copies be submitted.

All comments must be limited not to exceed 15 pages in length. Necessary attachments may be appended to these submissions without regard to the 15-page limit. This limitation is intended to encourage commenters to detail their primary argument in a concise fashion.

If a commenter wishes to submit certain information under a claim of confidentiality, three copies of the complete submission, including purportedly confidential information, should be submitted to the Chief Counsel, NHTSA, at the street address given above, and seven copies from which the purportedly confidential information has been deleted should be submitted to the Docket Section. Any claim of confidentiality must be supported by a statement demonstrating that the information falls within 5 U.S.C. section 552(b)(4), and that disclosure of the information is likely to result in substantial competitive damage; specifying the period during which the information must be withheld to avoid that damage; and showing that earlier disclosure would result in that damage. In addition, the commenter or, in the case of a corporation, a responsible corporate official authorized to speak for the corporation must certify in writing that each item for which confidential treatment is required is in fact confidential within the meaning of section (b)(4) and that a diligent search has been conducted by the commenter or its employees to assure that none of the specified items have previously been disclosed or otherwise become available to the public.

All comments received before the close of business on the comment closing date indicated above will be considered, and will be available for examination in the docket at the above address both before and after that date. To the extent possible, comments filed after the closing date will also be considered. However, the rulemaking may proceed at any time after that date, and comments received after the closing date and too late for consideration in regard to the action will be treated as suggestions for future rulemaking. NHTSA will continued to file relevant material as it becomes available in the docket after the closing date; it is recommended that interested persons continue to examine the docket for new material. Those persons desiring to be notified upon receipt of their comments in the rulemaking docket should enclose, in the envelope with their comments, a self-addressed stamped postcard. Upon receiving the comments, the docket supervisor will return the postcard by mail.

Issued on June 11, 1982.

Raymond A. Peck, Jr.
Administrator

47 F.R. 25930
June 15, 1982

PREAMBLE TO AN AMENDMENT TO PART 575

Consumer Information Regulations; Uniform Tire Quality Grading
(Docket No. 25; Notice 48)

ACTION: Interim final rule and request for comments.

SUMMARY: This notice makes a technical correction to the test procedures used in Uniform Tire Quality Grading (UTQG). A recently issued amendment to those procedures inadvertently omitted certain factors to be used in determining the load under which tires are to be tested for traction. This notice corrects the prior amendment. This notice also provides that, for a two-year period, tires whose test loads would change significantly as a result of the use of the treadwear, temperature resistance and traction load factors shall continue to be tested at the loads used in UTQG testing prior to June 14, 1982. The agency intends this notice to ensure that test loads will not significantly change from previously specified loads.

EFFECTIVE DATE: The UTQG amendment is effective on August 12, 1982.

SUPPLEMENTARY INFORMATION: Under the UTQG system, tires sold in this country are tested and grades are assigned for treadwear, traction, and temperature resistance. Prior to June 15, 1982, the UTQG Standards provided that the tire rim size and test loads used for UTQG testing were to be obtained from the tire tables of Appendix A to Federal Motor Vehicle Safety Standard No. 109, New pneumatic tires. However, those tables were deleted from FMVSS 109 effective June 15, 1982. In order to provide a substitute means for determining rims and test loads for all three performance characteristics, NHTSA published an interim final rule on June 15, 1982 (47 F.R. 25930). The June 15 notice specified alternative methods for determining test rim sizes and test loads, without having to refer to the now-deleted tire tables of Standard 109.

Of relevance here is the new procedure for determining test loads. That procedure requires multiplying the maximum tire load appearing on the tire's sidewall by certain specified factors.

The agency's June 15 correction notice inadvertently omitted factors for traction testing. The factors which were listed in that notice were those appropriate for treadwear and temperature resistance testing only. Therefore, the agency is now correcting the table set forth in the June 15 notice to include the factors to be used in UTQG traction testing. The agency has selected these factors, like those specified in the June 15 notice for treadwear and temperature resistance testing, in an attempt to produce approximately the same test load as was previously specified by reference to the tire tables of Standard 109. The agency believes that for most tire types and sizes, this procedure will produce tire load specifications which differ from loads specified by the old procedure by less than 10 pounds. The agency believes that this difference will not be large enough to produce significant differences in test results, but invites comment on this point.

The agency has identified 14 individual tire sizes which would have differences of more than 10 pounds in test loads under the load factors for treadwear, temperature resistance or traction testing under UTQG. These discrepancies apparently result from differences in the manner in which various tire companies determine maximum tire loads and "design" loads. For these 14 tires, the agency is specifying as an interim measure that the loads previously determined by reference to the tire tables may continue to be used for a period of two years. The two-year period will permit the tire manufacturers to make any design changes they feel necessary in these tires. While the agency believes that those 14 tire sizes represent the only tires now sold in the U.S. with load discrepancies of greater than 10 pounds, there may be others. Commenters are requested to inform the agency of any additional tires for which such a discrepancy exists. These tires will be added to that list when final action is taken on the interim final rule.

The agency finds good cause for issuing this amendment without prior notice and comment. The agency believes that prior notice and comment are unnecessary, since the revisions are technical and editorial in nature. They are intended to allow the continued implementation of the UTQG regulation in the same manner as it was before June 15, 1982. Although the agency has concluded that prior notice and comment are unnecessary, it has decided to go beyond the minimum requirements of the Administrative Procedures Act and provide a comment period on this amendment. For the same reasons set forth above and to permit continued implementation of the UTQG regulation, the agency finds good cause for making the revisions effective immediately.

Since this amendment is not intended to cause any significant change in implementation of the UTQG regulation as it existed on June 14, 1982, NHTSA has determined that this proceeding does not involve a major rule within the meaning of Executive Order 12291 or a significant rule within the meaning of the Department of Transportation regulatory procedures. Further, there are virtually no economic impacts of this action so that preparation of a full regulatory evaluation is unnecessary.

The Regulatory Flexibility Act does not require the preparation of flexibility analyses with respect to rulemaking proceedings, such as this one, since the agency certifies that this action would not have a significant economic impact on a substantial number of small entities. As noted above, this action will make essentially no change in the implementation of the UTQG regulation.

NHTSA has concluded that this action will have essentially no environmental consequences and therefore that there will be no significant effect on the quality of the human environment.

Interested persons are invited to submit comments on the agency's action announced above and on any other topics relevant to this

notice. It is requested but not required that 10 copies be submitted.

All comments must be limited not to exceed 15 pages in length. Necessary attachments may be appended to these submissions without regard to the 15-page limit. This limitation is intended to encourage commenters to detail their primary argument in a concise fashion.

If a commenter wishes to submit certain information under a claim of confidentiality three copies of the complete submission, including purportedly confidential information, should be submitted to the Chief Counsel, NHTSA, at the street address given above, and seven copies from which the purportedly confidential information has been deleted should be submitted to the Docket Section. Any claim of confidentiality must be supported by a statement demonstrating that the information falls within 5 U.S.C. section 552(b)(4), and that disclosure of the information is likely to result in substantial competitive damage; specifying the period during which the information must be withheld to avoid that damage; and showing that earlier disclosure would result in that damage. In addition, the commenter or, in the case of a corporation, a responsible corporate official authorized to speak for the corporation must certify in writing that each item for which confidential treatment is required is in fact confidential within the meaning of section (b)(4) and that a diligent search has been conducted by the commenter or its employees to assure that none of the specified items have previously been disclosed or otherwise become available to the public.

All comments received before the close of business on the comment closing date indicated above will be considered, and will be available for examination in the docket at the above address both before and after that date. To the extent possible, comments filed after the closing date will also be considered. However, the rulemaking may proceed at any time after that date, and comments received after the closing date and too late for consideration in regard to the action will be treated as suggestions for future rulemaking. NHTSA will continue to file relevant material as it becomes available in the docket after the closing date; it is recommended that interested persons continue to examine the docket for new material. Those persons desiring to be notified upon receipt of their comments in the rulemaking docket should enclose, in the envelope with their comments, a self-addressed stamped postcard. Upon receiving the comments, the docket supervisor will return the postcard by mail.

Raymond A. Peck, Jr.
Administrator

47 F.R. 34990
August 12, 1982

Issued on August 5, 1982.

PREAMBLE TO AN AMENDMENT TO PART 575

Consumer Information Regulations
Uniform Tire Quality Grading

[Docket No. 25; Notice 52]

ACTION: Final rule.

SUMMARY: This notice suspends, on an interim basis, the treadwear grading requirements of the Uniform Tire Quality Grading Standards (UTQGS). No change is made in the requirements of grading the traction and temperature resistance performance of new tires except for a minor change in the format for molding those grades on tires.

The UTQGS treadwear grading requirements are intended to aid consumers in assessing the value of new tires in terms of relative treadwear performance. This suspension is being adopted because available information and analysis indicate that the treadwear grades are apparently not only failing to aid many consumers, but also are affirmatively misleading them in their selection of new tires. The unreliability of the treadwear grades arises from two major sources. One is the variability of treadwear test results, which could be caused by either the lack of sufficient measures in the treadwear test procedures to ensure repeatability, or by the inherent complexity of the structure of individual tires themselves, which would preclude reproducibility of test results and, thus, comparative examination between or among tires. The other major source of unreliability is substantial differences among the practices of the tire manufacturers in translating test results into grades.

The agency has identified a wide variety of presently uncontrolled and perhaps uncontrollable sources of variability in the treadwear test procedure, and believes that other sources remain to be discovered. Although some or all of these sources may ultimately be found to be controllable to the extent that the variability in test results is reduced to acceptable levels, considerable research must be completed before the agency can determine whether or how that can be achieved. Much of the necessary research has already been initiated. When the research is completed, the agency will determine whether the suspension of treadwear grading should be lifted.

The agency is also amending Part 575 to change the format for molding grades on the sidewalls of new tires. The new format, which would include traction and temperature resistance grades but not treadwear grades, must be used on new tires produced in molds manufactured after (180 days after publication in the *Federal Register*). The agency expects and directs that manufacturers will cease printing tire labels and consumer information materials which include treadwear grades described or characterized as having been determined by or under the UTQGS procedures of the United States Government.

As a result of the amendments adopted by this notice, consumers will cease to be misled by unreliable treadwear grade information. In addition, the costs of implementing the treadwear grading program will no longer be imposed on the manufacturers and consumers.

DATES: The suspension of the existing requirements relating to treadwear grades, and the new alternative provision specifying the format for the molding of only traction and temperature resistance information on new tires are effective February 7, 1983. The provision requiring use of the new format is effective for tires produced in molds manufactured on or after August 8, 1983.

SUPPLEMENTARY INFORMATION: Section 203 of the National Traffic and Motor Vehicle Safety Act requires the Secretary of Transportation to prescribe a "uniform quality grading system for motor vehicle tires." As explained in that section, this system is intended to "assist the consumer to make an informed choice in the purchase of motor vehicle tires." The uniform tire quality grading standards (UTQGS) became effective April 1, 1979, for bias tires; October 1, 1979, for bias belted tires; and April 1, 1980, for radial tires. UTQGS requires manufacturers and brand name owners of passenger car tires to test and grade their tires according to their expected performance in use with respect to the properties of treadwear, traction, and temperature resistance, and provide consumers with information regarding those grades.

Treadwear Testing and Grading Process

This notice focuses on the treadwear grades. Unlike grades for the properties of traction and temperature resistance, the treadwear grades have never been intended to promote safety. Their essential value has always been to aid consumers in selecting new tires by informing them of the performance expectations of tread life for each tire offered for sale, so that they can compare on a common basis the relative value of one tire versus another. Although these grades are not intended to be used for predicting the actual mileage that a particular tire will achieve, the relevance and effectiveness of the grades depend directly on the accuracy of the projections of tread life derived from tests and assigned by grades.

The grades are based on a tire's projected mileage (the distance which it is expected to travel before wearing down to its treadwear indicators) as tested on a single, predetermined course laid out on public roads near San Angelo, Texas. Each treadwear test consists of 16 circuits of the approximately 400 mile long course. A tire's tread depth is measured periodically during the test. Based upon these measurements, the tire's projected mileage is calculated. A tire's treadwear grade is expressed as the percentage which its projected mileage represents of a nominal 30,000 miles. For example, a tire with a projected mileage of 24,000 would be graded "80,"

(i.e., 24,000 is 80 percent of 30,000 miles), while one with a projected mileage of 39,000 would be graded "130," (i.e., 39,000 is 130 percent of 30,000, rounded).

Because the measured treadwear upon which grades are based occurs under outdoor road conditions, any comparison between candidate tire performances must involve a standardization of results by correction for the particular environmental conditions of each test. To do this, the treadwear performance of a candidate tire is measured in all cases in conjunction with that of a so-called "course monitoring tire" (CMT) of the same construction type. The treadwear of the standardized CMT's is measured to reflect and monitor changes in course severity due to factors such as road surface wear and environmental conditions. The actual measured treadwear of the candidate tire is adjusted on the basis of the actual measured treadwear on the CMT's run in the same convoy, and the resulting adjusted candidate tire treadwear is used as the basis for assigning the treadwear grade.

To promote their uniformity, the CMT's are selected from a single production lot manufactured at a single plant, under more stringent quality control measures (set by contract with NHTSA) than would otherwise apply to production tires.

Each test convoy consists of one car equipped with four CMT's and three or fewer other cars equipped with candidate tires of the same construction type. Candidate tires on the same axle are identical, but front tires on a test vehicle may differ from rear tires as long as all four are of the same size designation. After a two-circuit break-in period, the initial tread depth of each tire is determined by averaging the depth measured in each groove at six equally spaced locations around the circumference of the tire. At the end of every two circuits (800 miles), each tire's tread depth is measured again, the tires are rotated on the car, and wheel alignments may be readjusted as needed to fall within the ranges of the vehicle manufacturer's specifications. At the end of the 16-circuit test, each tire's overall wear rate is calculated from the nine measured tread depths and their corresponding mileages after break-in by using a regression line technique.

Part 575 requires that the treadwear grading information be disseminated in three ways. First, the actual grade must be molded onto the sidewall of each tire. Second, the grade and an explanation of the treadwear grading process must appear on a paper label affixed to the tire tread. Third, the grade and the same explanation must be included in materials made available to prospective purchasers and first purchasers of new motor vehicles and tires.

Agency's Recent Actions

The basis and validity of the UTQGS has been a longstanding source of controversy and uncertainty within the agency and among interested parties. In view of the manifest potential conflict between the clear desirability of a valid, effective program to enable more informed consumer choice in the marketplace and the potential for serious adverse effect on the marketplace of an inadequate or potentially misleading programmatic result, the agency responded to its own enforcement uncertainties, described more fully below, by reviewing the current state of knowledge concerning the UTQGS, and addressing the specific sources of variability already identified.

Variability due to treadwear test procedures. In response to longstanding concerns about the variability and unreliability of the treadwear test results and grades and about the underlying causes of these problems, the agency conducted a review in May 1982 of treadwear test procedures being used by the tire testing companies in San Angelo. That review confirmed the existence of numerous uncontrolled sources of potential variability in treadwater test results. The potential cumulative effect of those sources would produce test result variability approaching the unacceptable magnitude long asserted by many tire manufacturers. The high level of test result variability could result in tires with better actual treadwear performance being graded as inferior to tires with worse actual performance, or vice versa.

The review did not, however, address in detail the relative significance of the various sources of variability. That question and the ultimate question of whether the identified sources of variability can be sufficiently controlled so as to bring the overall amount of variability down to an acceptable level can be answered only after

extensive research and testing.

Among the sources of variability discussed in the review were the weight scales intended to assure the proper loading of the cars used in the testing convoys, errors or inconsistencies introduced by variations in the amount of force applied to the probes used to measure tread depth and tendencies of measuring personnel to "search" for tread depth measurements consistent with expected rates of treadwear, discrepancies in the level of the training of technicians, fairly wide tolerances on critical alignment settings, unquantifiable variations in vehicle weights and weight distribution and suspension modification, and variations in driver techniques and in weather conditions on the course.

Each of the specific identified sources of such variability is discussed in detail below.

Variability due to grade assignment practices. Following the initial implementation of UTQGS, the agency sent a special order to the tire manufacturers to obtain information regarding their practices for translating treadwear test results into grades. The response indicated wide variation within the industry regarding those practices. Some manufacturers evaluated data by applying statistical procedures to estimate the percentage of their production which would equal or exceed a particular grade. Other manufacturers did not use such a procedure, relying instead on business and engineering judgment in assigning grades. The agency tentatively concluded that these differing practices created the substantial likelihood that different manufacturers, although faced with similar test results, would assign different grades to their tires. Accordingly, NHTSA issued a notice of proposed rulemaking requesting comment on a standardized process for translating test results into grades. (46 F. R. 10429, February 2, 1981). Commenters generally criticized the proposed process, particularly for its failure to account properly for undergrading. The agency is continuing its efforts aimed at developing a uniform procedure for translating test results into treadwear grades. However, until this problem is resolved, the unreliability of treadwear grades is compounded by the fact that the relationship between test results and assigned grades is not a constant one from manufacturer to manufacturer.

Variability inherent in the nature of tire structure. A potential for an unquantified degree of variability is inherent in the differences between seemingly identical (i.e., in terms of brand, line, size, and manufacturing lot) tires. The potential arises from the complex combination of a variety of factors, including the materials, designs, and manufacturing procedures, that go into the production of tires. The materials include the rubber composition and various reinforcing materials such as rayon, steel, polyester, etc., which themselves are developed from complicated manufacturing processes. The design of a tire includes such factors as the $_{cross}$ section shape, the orientation and structure of the reinforcing materials, the tread design, and the construction (bias, bias-belted, or radial). The manufacturing procedures include the processes employed during manufacturing and the conditions such as temperatures and times of vulcanization. Separately and together, these variables can have a significant effect on tread life.

In the production of tires, the manufacturers use a variety of techniques in an attempt to control all of these variables and to achieve a consistent level of quality and performance for their different products. The success of these efforts varies from tire line to tire line, lot to lot, and from manufacturer to manufacturer. The complexity of the entire process will inevitably lead to some variation in performance, including treadwear performance between nominally identical tires.

NOTICE OF PROPOSED RULEMAKING

Based on the assertions and submissions of the tire manufacturers and the agency's review of the test procedures and of its own enforcement data, the agency tentatively concluded in July 1982 that treadwear grading under UTQGS should be suspended pending completion of research regarding the extent to which the sources of variability could be isolated and reduced. Accordingly, it issued a notice of proposed rulemaking to obtain both written comments and oral testimony on suspending treadwear grading (47 F.R. 30084, July 12, 1982) and to schedule a public meeting August 12, 1982. The agency

stated that it was issuing the proposal principally to avoid the dissemination of information potentially misleading to consumers and secondarily to minimize the imposition of unwarranted compliance costs on industry and consumers. The agency noted its concern that the treadwear grading was not only failing to achieve its statutory goal of informing consumers, but also affirmatively misleading them.

In defending UTQGS against earlier judicial challenges, NHTSA had taken the position that the treadwear test procedure was adequately specified to ensure that test result variability was limited to acceptable levels. See *B.F. Goodrich* v. *Department of Transportation*, 541 F. 2d 1178 (6th Cir. 1976) (hereinafter referred to as *"Goodrich I"*); and *B.F. Goodrich* v. *Department of Transportation*, 592 F. 2d 322 (6th Cir. 1979). For example, the agency had stated in the *Goodrich I* litigation that variables in the testing procedure are controlled and taken into account, principally through the selection of a single test course and the use of CMT's. With respect to certain potential sources of variability, the agency stated that their effects on treadwear testing and grading would be minimal. The agency indicated in its suspension proposal that it could no longer make the same representations. These statements have been further undermined by information now available to the agency.

The notice summarized the material relied upon by the agency in making its tentative conclusions, including the information and arguments submitted by the tire manufacturers. Firestone Tire and Rubber Company, for example, found that treadwear test results could vary up to 30 percent even for CMT tires, which are specially manufactured for maximum homogeneity. That company also pointed out several possible causes of the variability, including variability in test vehicles and driver techniques as well as deficiencies in the details of the test procedures themselves. General Tire and Rubber Company reported additional sources of variability, including vehicle wheel alignment, weight distribution, and test course environmental factors. B.F. Goodrich Company stated that differences in tire tread composition between candidate tires being tested and the CMT's could be a major source of variability. As a group, the tire manufacturers generally contended that the variability of the test results is too great to permit meaningful treadwear grading or compliance testing. The agency's own preliminary research confirms this conclusion and supports the need for the suspension.

The proposal also discussed the agency's enforcement data and described at length the review conducted by NHTSA of the treadwear testing companies. The agency emphasized that the list of sources of variability mentioned in the review was not exhaustive, but intended merely to be illustrative of the types of possible such sources and of the difficulties which exist in seeking to establish a treadwear test procedure that could produce valid, repeatable results. The agency found that the combination of the examined sources represented a potential for test result variability of serious dimensions. Each potential source of variability was described and the potential effect of them on test results was estimated. For example, effects of ± 34 or 35 points were estimated for two sources of variability and ± 14 points for another.

Summary of Comments on Proposal

Written comments and oral testimony were received from a variety of sources, although the most detailed ones were from tire manufacturers. While there was a division of opinion regarding the merits of the proposal, most commenters favored the suspension. Proponents of the suspension included tire manufacturers, several tire manufacturers' associations, tire dealers, a motor vehicle manufacturer, some consumers, and a public interest group. Proponents agreed with the agency's statement that the treadwear test results and grades were so variable and unreliable as to confuse and mislead consumers. They also listed again the factors that they thought were causing the variability. Some proponents suggested that the problems are so serious that simple suspension was inadequate. They urged that the agency go further and rescind the treadwear provisions altogether.

Opponents of the proposed suspension included one tire manufacturer, a tire dealer, a public interest group, a county consumer protection agency, and a number of consumers. The tire manufacturer argued that the treadwear grade information was sufficiently correlated with actual differences in tire performance to be helpful to those consumers who use that information. It acknowledged that there was variability in the treadwear test results and differences in the grade assignment practices, but contended that these problems could be satisfactorily controlled through further identified changes in UTQGS. The manufacturer argued that even if there were difficulties in enforcing the current treadwear requirements, the overall value of the comparative treadwear information justified retention of the requirements while the enforcement problems were addressed. The public interest group argued that NHTSA was ignoring its statutory mandate, as interpreted by that group, in contemplating a suspension of treadwear grading. That opponent argued further that the agency had artificially narrowed the options under consideration in this rulemaking proceeding.

Two tire testing companies submitted detailed comments regarding their testing practices. They generally argued that the problems discussed in the agency's review of testing companies did not apply to them. One asserted further that the suspension would have a severe economic impact in the San Angelo, Texas area, where treadwear tests are conducted. The San Angelo Chamber of Commerce concurred in that assessment.

Summary of Suspension Decision

NHTSA has decided to suspend the treadwear provisions of UTQGS because available information and analysis indicate that the treadwear grades are apparently not only failing to aid many consumers, but are also affirmatively misleading them in their selection of new tires. The capacity of these grades to mislead consumers arises principally from variability in treadwear test results unrelated to actual differences in measured or projected performance, and secondarily from differences among manufacturers in their translation of test results into grades. In its proposal, the agency identified some of the wide variety of uncontrolled sources of variability in the insufficiently specific treadwear test procedures. The agency has been able to quantify the effect of only some of those sources. Other sources are believed to exist and continue to be discovered. Indeed, the tire manufacturer opposing the suspension reported only last November its discovery of a "major unreported source of variability." (Letter from R. H. Snyder, Uniroyal Tire Company, to Raymond Peck, NHTSA Administrator, November 12, 1982, Docket 25, Notice 47, No. 090).

In their comments to the agency, the opponents of the suspension did not controvert the premise of the agency that there is substantial variability in test results and that there are specific identified sources of much of that variability. The tire manufacturer opposing suspension conceded that test result variability and differences in grading practices can be so large as to result in changes between the order in which tires are ranked based on test results and the order in which they are ranked based on grades. Indeed, comparisons of the agency's own compliance test data and grades assigned by the tire manufacturers indicate that these ranking changes occur with some frequency and can be substantial. Moreover, the opponents did not deny that there were significant problems with enforcing the treadwear requirements of Part 575 as they are now written.

Where the rank order of measured performances or assigned grades changes, it is clear that only one of such differing results can in fact be objectively correct and valid. Any such change in ranking thus represents a clear and present danger that grades can be affirmatively misleading. Resulting purchasing decisions based on such incorrect grades are not merely wrong, but represent instances in which the government-created program of consumer assistance through the dissemination of objective comparative information has in fact affirmatively misled the consumers which are intended to be assisted.

Although the sources of variability may ultimately be controllable to the extent that the variability and unreliability derived from treadwear test results and grades are reduced to lower, more acceptable levels, considerable research must be completed before that is even a possibility. Even if such research were now complete, it is not clear at this point how much of the current test-derived variability and unreliability could be eliminated. Much of the necessary research has already been initiated.

When the research is completed, the agency will address the question of whether the problems can be reduced to the point that it can begin considering whether to reinstate the UTQGS treadwear system.

Rational for Suspension Decision

Magnitude of the Overall Variability
and Reliability Problem

Available data demonstrate that the treadwear test results can vary substantially and that the treadwear grades assigned by the manufacturers are unreliable for the purposes of comparing tires. Data submitted by the tire manufacturers indicate that subjecting tires of a particular type and line to the same tests on separate occasions produces differences in test results of up to 80 points. The agency's own compliance test data include examples of significant test result variability.[1]

Moreover, in addition to test result variability, the process of assigning grades can and demonstrably has introduced other unacceptable levels of uncertainty as far as the consumer is concerned. Treadwear grades are often not a reliable indicator of the relative tread life of tires because the order in which tires are ranked on the basis of test results can differ significantly from the order in which they are ranked on the basis of grades. The magnitude of these crossovers (i.e., changes in rank) can be

[1] The agency believes that the enforcement data are a particularly significant source of information since the data comprise the most complete set of test results available. They reflect consistent application of test procedures under the direction of a single party, the agency, under circumstances involving the greatest incentive of any interested party to minimize variability in data, the exigencies of the certainty required for enforcement purposes. In fact, to attempt to resolve doubts as to variability, the agency has in fact refined its enforcement test procedures to a greater extent than is required by Part 575. For example, all enforcement tests are conducted by a single contractor, eliminating the influence of differences between test facilities. Highly accurate electronic scales are used to determine wheel loads. Very precise wheel alignment equipment is used. That equipment has been operated by the same skilled technicians for all compliance tests since mid-1981. Thus, NHTSA believes that statements regarding test variability which are based on these enforcement data could tend only to understate the variability experienced by others in testing tires and assigning grades.

substantial, as is shown in a graph which B. F. Goodrich constructed by plotting the agency's enforcement data against the grades assigned by the tire manufacturers for the same tires. (This is the same graph shown on page III-2 of the agency's regulatory evaluation for this rulemaking action and is similar to one prepared by Uniroyal.) Goodrich's graph includes information on radial ply tires primarily, although it also covers tires of other construction types. There are numerous examples in the graph of tires whose test results fell within a 10 point range, but whose assigned grades were spread over an 80 to 100 point range. Some tires had average test results which were 10 points below those of other tires, but were assigned grades as much as 60 or 70 points higher. Some tires assigned the same grade had average test results that were scattered over a 100 point range. These phenomena are not restricted to a particular portion of the graph, but exist throughout, from the left side where bias ply and bias belted tires predominate to the right side where radial ply tires predominate.

The magnitude and pervasiveness of the crossovers and grading quirks means that the treadwear grades have the capacity for more than simply confusing consumers about the relative performance of tires exhibiting nearly the same performance. The possibility exists for confusion even between some tires in the lower third percentile and some tires in the upper third percentile of treadwear performance. Thus, whether a prospective purchaser seeking the particular size (i.e., diameter) of tire appropriate for his or her vehicle is looking at the entire spectrum of construction types, or is focusing on a single construction type only, there is a significant possibility that the person may be misled about the relative performance of tires. The possibility is greatest in the latter case, since the smaller the difference in actual performance between tires under consideration, the greater the probability that test variability and crossovers will cause the grades of those tires to be misleading about the relative performance of those tires. The ranges in grades for particular construction types are not very large when compared with the magnitude of the problems created by test variability and crossovers. Treadwear grades typically range from 60-120 (a

60 point range) for bias ply tires of all sizes, 90-150 (a 60 point range) for bias belted tires of all sizes, 120-200 (an 80 point range) for 13 inch diameter radial ply tires, 160-220 (a 60 point range) for 14 inch diameter radial ply tires, and 170-220 (a 50 point range) for 15 inch radial ply tires. The ranges for radials are particularly relevant since radials account for most original equipment tires on new cars and a substantial majority of replacement tires for used cars.

It is considered especially significant that the occurrence of such rank changes is not uncommon. For examples for each of a majority of the tires in Goodrich's graph, other tires could be found in the graph which had a lower assigned grade but which, based on compliance test results, exhibited superior performance.

Although the agency recognizes that the graphs submitted by Goodrich and Uniroyal reflect, in part, manufacturer-to-manufacturer differences in grade assignment procedures and not just variability in test results, the agency considers the analyses made using the graphs to be significant since they point out the extent to which consumers may in fact be misled about treadwear grades. In its analysis, Uniroyal calculated a correlation coefficient of 0.763 for the two variables (test results and grades),[2] and a similar rank order correlation. The coefficient of 0.763 implies that only about 58 percent (the square of the correlation coefficient) of the variation in tire treadwear grades can be explained by actual differences in treadwear

[2]Using a slightly different data base, B. F. Goodrich calculated a correlation coefficient of 0.78 between the agency's enforcement trest results and assigned grades.

[3]While the argument has been made that this aspect of variability should not be taken into account because it is entirely within the control of the grading manufacturer, the agency is not able to conclude from the data before it that any actually assigned grade is without basis in test data. In implementing the statute to determine whether the sanctions imposed by the statute and agency regulation should be applied to given manufacturers, the agency has been forced to conclude that all assigned grades so reviewed have been reasonable, based on agency and manufacturer supporting data. Under such circumstances, the agency finds that the overwhelming policy purpose of the UTQGS to inform consumers of comparative tire data, in a meaningful way (i.e., one that is valid, reasonably accurate, and objectively verifiable for enforcement purposes) in order to affect their tire purchase decisions, requires that this uncertainty also be taken into account.

performance. The agency estimates that as many as 10 of the 40 percentage points of unexplained variability may be due to differences in grade assignment practices.[3]

In reaching its decision that currently documented levels of variability are unreasonable and cannot sustain retention of the UTQGS treadwear grading requirements in their present form, the agency has been guided by two principal conclusions: first, the rank order of test results and the rank order of assigned grades can and do change with repeated testing under currently allowable procedures. This result has also taken place when the agency's own, far more carefully controlled compliance efforts are the basis for the test.

Second, the levels of certainty and predictability which the agency expected would be achieved over time and which the agency so represented to the courts which have upheld UTQGS against charges of unacceptable uncertainty, have not been achieved in fact.

At a minimum, the agency concludes that such a level of potential rank order change, under applicable test procedures, is unacceptable. The agency also concludes that unless the level of certainty previously asserted by the government in litigation can be verified to exist, the continued integrity of the process is undermined to a separate and unsupportable degree.

Agency research is thus primarily directed to the determination of the degree to which these effects can be eliminated.

Specific Sources of Variability

The agency's proposal described a variety of potential sources of variability in the treadwear test results based on a review of testing being done in San Angelo. The tire manufacturers supporting the suspension, and the commenting tire testing companies generally agreed that many of such sources contributed to test result variability. While some commenters, especially two testing companies describing in detail their own testing practices, disputed the magnitude of the variability that could be caused by several of the sources, it remains uncontroverted that the sources identified in the proposal are potential contributors to variability.

One such testing company objected to the inference it drew from the proposal that the agency believed that the testing companies as a group were to blame for the variability in the test results. That company also stated its belief that the proposal unfairly criticized the practices of testing companies as though all such companies followed identical practices. The agency recognizes, and reaffirms its conclusions, that the primary source of test variability lies in the shortcomings of the test procedures themselves. Further, it rejects any implication that the testing companies were improperly following such procedures.

The agency emphasizes that the list of sources in the proposal was not exhaustive. The proposal specifically noted that the list was included for illustrative purposes only. It was recognized that additional research would likely reveal other sources, of the indisputable and undisputed levels of variability. Indeed, the record of comments has provided information regarding several previously unmentioned sources of variability, e.g., tire/wheel rim width combinations and the effect of rubber's high coefficient of thermal expansion on tire groove depth measurement.

The following specific sources of variability have been confirmed by the agency as a result of the current rulemaking proceeding.

Problems of instrumentation—scales. Some testing companies use scales that are designed for weighing objects up to 20,000 pounds. Scales are rarely accurate below 10 percent of their maximum measuring capacity. Since the loads being weighed for UTQGS purposes are less than half that level, the potential for inaccurately loading the tires on the test cars is obvious. This problem is compounded by the inability of many such scales to provide readings more precise than at 5 pound intervals. The combination of these factors could lead to significant potential measurement errors.

Using a ratio of 1:4 between changes in load and changes in treadwear, the agency stated in its proposal that a 20 to 30 pound error in measuring a 700 to 800 pound load could cause test results errors of ± 20 to 34 points in a tire with a treadwear grade of 200. The two tire testing companies submitting detailed comments stated that their own scales are regularly calibrated, and that maximum weighing errors of not more than 10 pounds could be expected under such circumstances. One of the companies also argued that the ratio between load changes and treadwear changes is actually closer to 1:1. The agency cannot now determine with certainty the correct ratio between changes in tire load and changes in treadwear. Even assuming such actual ratio may be lower than 1:4, the agency believes that scale miscalibration is a factor that can potentially contribute significantly to variability in treadwear test results.

—*tread depth probes.* Tire testing companies currently measure tread depth by means of either mechanical gauges with dial indicators or electronic devices which translate probe displacement into a voltage reading in mils or thousandths of an inch. NHTSA's tests of measurement devices produced measurement errors of between 3 and 5 mils for electronic probes and up to 10 mils for mechanical gauges, with the magnitude of error appearing to depend on the amount of the pressure placed on the probe. Variations in pressure can be caused by differences in strength or technique among personnel or even by the gradual effect of fatigue on a given technician. The resulting measurement differences on tires graded from 160 to 200 can cause treadwear grading errors of ± 2 to 3 points. The two tire testing companies argued that measurement errors of 10 mils were in fact difficult to achieve and would not normally be expected to occur. The agency concurs that the typical such error would be expected to be less than 10 mils, but concludes that variation in the pressure placed on the probes remains one of the potential sources which collectively has produced high levels of test variability.

Electronic probes are subject to other sources of measurement error. The lack of temperature compensation in some of the electronic probes can cause drifts in both the zero reading and the gain. One tire testing company did note that its electronic probes are attached directly to a computer, and asserted that they are capable of measuring accurately over a wide range of temperatures. While such drift can be corrected for in such a process, the agency has determined that such corrections are not in fact routinely sought or made by testing companies in general. Further, any change in probe force at the bottom of the groove for tires with varying hardness will generate different tread depth réadings

depending on the spring constant, the amount of deflection used in the design, and the shape of the tip on the electronic probe. The use of uncalibrated springs produces additional measurement differences.

—*wheel alignment equipment and procedures.* The agency has determined that treadwear is very sensitive to wheel alignment, much more so than had previously been understood by interested parties. One of the two tire testing companies agreed with this proposition. B. F. Goodrich supported this proposition by asserting that 4/32nds of an inch increase in toe-in can decrease tread life by 15 to 30 percent. Since Part 575 permits the wheels to be aligned anywhere within the vehicle manufacturers' specified range of acceptable alignments, differences in toe-in are possible. Armstrong Rubber Company cited various vehicle manufacturer specifications which had a minimum-to-maximum range of from 5/32nds to 14/32nds of an inch.

The comments on the proposal reveal that the use of different toe-in settings for a given vehicle can and do occur. Some testing companies align wheels to the minimum toe-in setting within the acceptable range while others align to the mid-point of the range. Indeed, practices of the two commenting tire testing companies vary in precisely this fashion, with one aligning to the minimum point and the other to the mid-point.

Differences in wheel alignment may also occur as a result of differences in the frequency of wheel alignment and in the skill of the technicians who perform the alignments. The two tire testing commenters asserted that they use accurate alignment equipment and well-trained personnel. Assuming this to be true for these particular companies, however, does not remove wheel alignment as a potential source of variability even with respect to their testing. As noted above, the wheel alignment practices of these two companies vary significantly. Further, for these as well as the other tire testing companies, the problem of maintaining the alignment equipment in proper adjustment is a formidable one. Although all testers have suitable alignment equipment, their success in using it to achieve accurate results depends on the skill of the technicians operating it, the calibration of the equipment, and the frequency of alignment during a test.

PART 575; PRE 117

Problems of measurement. The agency believes that several measurement problems contribute to variability as well. Observed but currently unquantifiable measurement errors occur as a result of information feedback during testing, i.e., access by measuring personnel to the previous day's tread depth measurements and resulting conscious or unconscious bias to parallel or duplicate those measurements. The agency also believes error to be caused by the documented practice of some testing companies to establish an absolute level of coefficient of variation, i.e., the degree of variability among the separate measurements of depth in the same groove around the circumference of the tire. Some technicians tend to "hunt" for groove depths as uniform as possible around the circumference of the tire, on the understandable but not factually supportable or recognizable assumption that such variation should be minimized.

One tire testing company indicated in its comments that it took steps to avoid these sources of variability. Even assuming this company is fully successful in that effort, the agency believes that such problems exist for other testing companies, and would compromise the success of the program unless all companies were equally successful.

Problems with vehicle maintenance and use. The agency continues to believe that factors relating to the test cars produce substantial variability. One of these factors is the wide variation found in the approaches of the testing companies to achieving a proper vertical load on a tire. Some testing companies allow the weight to be placed forward of the front wheels, rearward of the rear wheels, or even on the vehicle exterior. In addition, some but not all companies place heavy deer guards on the front of their test cars.[4]

The overloading of some test cars also produces unquantifiable effects on treadwear test results. Some testing companies load their cars to whatever weight is required to achieve the appropriate load level for a test tire. As a result, the gross vehicle weight rating for the specific

---

[4]Some tire testing companies stated that weight is removed from their cars to compensate for the deer guards. However, the agency did not observe any accurate means of weight compensation.

cars themselves may be exceeded, necessitating the use of special springs or shims to reestablish normal ride height. Such heavy loads can cause the cars to bottom out, while the variations in springs create differences in roll stiffness and weight transfer among vehicles of the same type.

Each of these practices introduces changes in the handling characteristics of the cars and in different polar moments of inertia, between and among wheels, vehicles, and the entire test fleet. These factors would produce different rates of tire wear as the cars corner, accelerate, or decelerate.

The two commenting tire testing companies indicated that they attempt to control these sources of variability. However, there is no evidence that those efforts are fully successful, and agency observations indicate that the other companies are not in practice as careful as those two companies.

Problems with drivers and weather conditions. The agency found in its review that drivers of the test cars varied significantly in their skill and driving techniques. These differences are reflected in the frequency and severity of accelerations and decelerations. Further, the agency believes that adverse weather conditions may affect driving techniques and thereby treadwear. One tire testing company indicated that it carefully sought to limit these sources of variability. However, not all testing companies have adopted the same measures. In addition, adverse weather conditions cannot be controlled.

CMT tread composition. Most CMT's·do not currently have tread composition similar to that of most candidate tires. As a result, a substantial question has been raised as to whether the use of the CMT measurements in fact validly compensate for environmental effects upon candidate tire wear. The last two lots of radial CMT's contained about 30 percent natural rubber. Most tires produced in the U.S. do not contain any natural rubber, while some Japanese tires contain substantial quantities of it. the presence of a significant percentage of natural rubber in CMT's is important since natural rubber is more sensitive to temperature changes than the current tread compounds used in tires, and in general wears at a faster rate in hot weather than the current materials do. Thus, where the CMT in use contains a large percentage of natural rubber and the candidate tires do not, candidate tires graded in hot weather would be expected to have higher grades than those graded in cool weather.

The significance of CMT tread composition appears to be borne out by a report from B. F. Goodrich. That company stated that candidate tires made of compounds similar to that of the CMT's received more consistent ratings than those whose compounds were less similar. B. F. Goodrich's analysis indicates also that the latter tires can receive different relative rankings.

Wheel rim width. Armstrong asserted in comments that the tolerance permitted on rim widths to be used with a given size of tire is a significant source of variability. The agency lacks any corrobative information with respect to this previously unrecognized problem, but will address the issue as another potential source of variability as efforts continue to complete research on treadwear testing variability.

Grade assignment practices. There are significant differences among the tire manufacturers in the procedures they use to translate treadwear test results into grades. These differences arise partially from the differing degree of conservatism that the various manufacturers exercise in selecting a grade for a group of tires so as to ensure that the performance of all tires in the group exceed that grade as required by Part 575 (See discussion above).

Uniroyal Petition

On January 21, 1983, Uniroyal petitioned the agency to make three significant changes to the treadwear test procedures. These changes involve a new procedure for running CMT's, the rotation of candidate tires through each wheel position in a four-car convoy, and a doubling of the break-in period.

The agency has completed its preliminary review of this petition and, in view of the pendency of the current proceeding, has also taken it into account as if it were a supplementary filing to the docket.[5]

Under the Uniroyal petition, CMT's would no longer be run in the same convoys as candidate tires, but in a separate convoy using CMT's exclusively. The CMT's would be rotated through each position in the CMT convoy. This procedure is claimed to substantially reduce vehicle and driver related sources of variability, while reducing costs. However, its validity depends upon the accuracy of Uniroyal's conclusion that the course environment factors measured by the CMT process do not produce rapidly changing treadwear effects, i.e., that the course environment effect on treadwear changes slowly, if at all.

Similarly, the rotation of candidate tires through each position in the test convoys is claimed by Uniroyal to greatly reduce driver and vehicle related variability for those tires. All vehicles in a convoy would be nominally identical. No front wheel drive vehicles could be used due, according to Uniroyal, to "load distribution problems." Uniroyal does not state how it would deal with the problem of declining number of rear wheel drive models being produced, and the difficulty in matching all tire lines with the limited number of those models.

Finally, Uniroyal found that the break-in effect for new tires occurred beyond the 800-mile period currently specified in the regulations. It stated that establishing a longer period would provide a more accurate estimate of treadwear rates.

NHTSA regards Uniroyal's petition as further evidence of the necessity for suspending the treadwear provisions of UTQGS while the agency conducts research and testing to determine the feasibility of reducing variability to more acceptable levels. Uniroyal has revealed yet another previously unidentified factor, barometric pressure, apparently capable of contributing significantly to the variability of test results. Although Uniroyal has proposed several changes which it believes would substantially reduce certain sources of variability, it does not suggest how other factors identified in its petition are to be addressed.

Those factors are barometric pressure, temperature, and wet road surfaces. Uniroyal supplied information indicating that the manner

[5]The disposition at this time of the pending notice of rulemaking does not, of course, affect the pendency of this petition before the agency, since only a suspension of the UTQGS is involved. The petition will thus be treated both as a comment to the current proposal and as a petition directed toward the modification of the suspended portion of the UTQGS and a request for their reinstatement as so modified.

in which temperature differences affect treadwear is more complicated than previously supposed. While some compounds wear more rapidly as temperature increases, Uniroyal reported the example of a tire which wore more rapidly as temperature decreased. Further, the degree of temperature affect was substantial. While Uniroyal's testing showed that one family of tires was only slightly affected by an eight-degree average temperature difference, that same difference caused a 20 percent change in wear rate for another family of tires. Further, Uniroyal noted that wet road surfaces could significantly affect the rate of treadwear and admitted that some allowance must be made for this phenomenon, but didn't indicate how that might be accomplished.

Much of the work done by Uniroyal in support of its proposal is similar to the agency's ongoing research, and it may be that the agency's efforts will lead to the development of test procedures similar to those suggested by Uniroyal. However, Uniroyal's work does not obviate the need for NHTSA to complete its own research and testing and make its own judgments about the changes that might be made to the test procedures. The agency cannot now conclude that Uniroyal's proposal would reduce test variability to acceptable levels. Much more research and testing would be necessary before the agency could even consider proposing to adopt those or any other significant changes.

Not only would the agency need to address the significance of the failure of Uniroyal's proposals to address certain sources of variability, but it would also need to examine the implications of Uniroyal's proposals which in some cases go well beyond those suggested by Uniroyal in its petition. For example, Uniroyal's proposal for rotating candidate tires through each of 16 wheel positions on test convoys would necessitate a doubling of the mileage driven by treadwear testing convoys from 6,400 miles to 12,800 miles (16 x 800). The additional expense and time necessary to conduct such extended testing would be substantial.

Further, although Uniroyal urges the making of substantial and fundamental changes to the treadwear test procedures and the theory underlying those procedures, it argues, without providing the basis for that argument, that there would not be any necessity for retesting all tires in accordance with the modified procedures. Uniroyal apparently contemplates a marketplace in which some tires that were tested and graded under the existing, inadequate procedures are offered for sale side-by-side with others that are tested under new, revised procedures. Thus, Uniroyal would allow the continued dissemination of misleading treadwear information.

In the agency's judgment, the need to make these types of substantial and fundamental changes would render wholesale retesting and suspension unavoidable. The inescapable conclusion from the necessity of making these changes is that the grades generated under the existing procedures are unreliable and should not be presented to the public as a basis for choosing between alternative tires. Further, since the grades that would be assigned to a particular tire if tested under the current and new procedures would differ, the grades would be inherently incompatible. As a matter of responsibility to the consumer and of fairness, the agency could not contemplate the simultaneous use of two fundamentally different yardsticks to measure the treadwear performance of tires.

To avoid this situation, all tires would have to be retested and regraded. To provide time for the completion of these activities and to ensure that substantial numbers of tires graded under the existing procedures are not still in the marketplace when the tires graded under the new ones are introduced, a suspension of the treadwear testing requirements would be necessary.

Inadequacy of Alternatives

NHTSA considered several alternative courses of action in reaching its decision. In addition to suspending the treadwear grading provisions of Part 575, the agency considered rescinding them. NHTSA also considered retaining the provisions intact while it conducted its research and attempted to determine whether modifications to the test procedures and grade assignment practices could reduce variability to acceptable levels for UTQGS purposes.

Rescission. Several commenters argued that the problems with the treadwear grading program were so substantial and intractable that rescission of the treadwear provisions was the only appropriate step for the agency to take at this time. While the agency believes that the problems now identified with respect to the UTQGS treadwear ratings are extensive and serious, that some of them can be addressed only after substantial research, and that some or all may not be fully solved even then, it is convinced there is a substantial possibility that its planned research could eventually lead to amendments that would reduce identified treadwear test result variability to acceptable levels. For example, if the agency were able to develop an appropriate procedure for rotating all tires among the cars in a test convoy, the contribution of vehicle and driver effects to test result variability might be greatly reduced. Similarly, the agency's development and adoption of statistical procedures that would bring uniformity to the translation of test results into grades might contribute significantly to reliable treadwear grading.

In such a case, any remaining variability could more confidently be able to be considered attributable to the inherent complexity of tires themselves. At that stage, a failure to attain significant improvements in the repeatability or reproducibility of tests might well force the agency to the conclusion that no grading system based on measured and projected treadwear could be possible.

Precisely because of the levels of uncertainty now understood to exist as a result of test result variability, however, the agency is not now able to assess whether or not this will likely be the case. Absent some further evidence on this point, and taking into account the positive benefits to the consumer and the orderly working of the market place which a properly functioning UTQGS treadwear system would produce, the agency is unwilling to rescind the program of treadwear rating entirely at this time.

Continue treadwear grading and make improvements in treadwear grading process as they are developed. While conceding that there are variability problems, several commenters argued that the treadwear grades are still sufficiently useful to warrant their retention. They argued further that the agency should simply proceed to make available changes to the treadwear testing procedures and adopt other changes as they are developed. One commenter argued that if the treadwear grading information were more accurate than the information which previously existed in the marketplace, the agency was obligated to continue treadwear grading.

NHTSA believes that the critical issue is in this case not merely whether the treadwear grading provisions are currently fulfilling their statutory objective, that of assisting consumers to make informed choices in purchasing new tires, but of equal or greater importance whether such provisions may to the contrary be affirmatively frustrating the achievement of that objective. As interpreted by the 6th Circuit Court of Appeals, the UTQGS provisions in section 203 of the Act do not contemplate "theoretical perfection" in providing such assistance. *Goodrich I,* at 1189. It calls only for "reasonably fair and reasonably reliable grading procedures." *Id.* The agency believes that this is an appropriate statement of the principal underlying test of certainty which the procedures should satisfy. Procedures which fail to meet that test will tend inappropriately to increase the sales of some tires and decrease those of other tires through inaccurately representing the relative performance of either or both.

In the agency's view, it appears that the current procedures fail to meet that reasonableness test on several counts. Such procedures are not reasonably reliable because of the excessive magnitude of the overall variability.

Moreover, the grades produced under the treadwear grading procedures are not merely imperfect, they appear to be affirmatively misleading.

These problems are not minor. They do not affect only those tires which differ moderately in performance. As noted above in the discussion of the overall variability and reliability problem, the rank reversals produced by the procedures can be substantial and are not uncommon. Tires which are significantly superior to others in performance may be graded significantly below those tires, and vice versa. Tires whose test results show performance differences of up to 100 points may be assigned the same grade.

Thus, while some consumers might be aided in choosing between some tires, particularly those with very substantial differences (greater than 100 points) in treadwear performance, there appears to be a significant likelihood that consumers choosing among closer performing tires will be misled. The agency believes that most consumers fall into the latter category. As noted above, the threshold considerations of tire size and tire construction type should lead most persons considering the purchase of a new tire to look at a universe of potential candidate tires for purchase whose treadwear grades differ by significantly less than 100 points. Accordingly, it appears that the treadwear grading procedures are neither reasonably fair to the tire manufacturers nor reasonably reliable in guiding those consumers who will in fact be purchasing tires for a given vehicle.

The agency believes that the unreasonableness of the level of reliability of the current treadwear grading procedures is compounded by the possibility that many of the identified sources of variability, and thus the overall level of variability, might eventually be able to be significantly reduced, after a period of research and testing, at costs that are not prohibitive.

The agency regulatory evaluation discusses a wide range of possible changes that the agency believes could ultimately reduce test-induced variability to more acceptable levels. Among these are requirements for calibration of alignment equipment, tighter specifications for alignment, load distribution, tire-rim width matchings and CMT composition, prohibition against information feedback, standardization of equipment calibration and tread measurement procedures, limitations on driver acceleration rates and cornering techniques, limitations on tire temperature during tread depth measurement, standardization or elimination of deer guards, standardized statistical procedure for grade assignment, and rotation of candidate and CMT's tires among test cars. The actions which appear at this point to hold the greatest potential for improving the reliability of the grades are adoption of the grade assignment procedure, rotation of the tires, more precise specification of wheel of alignment, and specification of the composition of CMT's.

The relative importance of many of these factors is currently unknown. As a result, it is not possible to determine or assess what actual result

in improved repeatability may be achievable, and how or at what level such an improved result might be determined to be acceptable. However, the agency believes that together such factors contribute substantially to the variability of treadwear test results and unreliability of the resulting grades. The agency's research efforts are expected to provide information about the relative importance of individual sources of variability and the degree to which each source can be controlled.

The agency expects that its research and testing will also provide an indication of the cost of implementing controls on these factors. Based on the costs of the current procedures, the agency has no current basis for concluding whether the costs associated with effective controls would be reasonable either separately or collectively. The current cost of treadwear testing is an average of $.09 per tire. Based on indications from Goodyear that the retail markups for manufacturing costs may be 100 percent, that testing cost would have an $.18 retail price effect, against a retail price of $40 to $70 for a new tire. Thus, for example, a doubling of testing expenses would bring the retail price effect of testing costs up to an average of only $.36 per tire, a presumptively reasonable economic impact in and of itself.

As to the suggestion that the agency immediately commence to make changes in the treadwear testing procedures and make other changes as they are developed, the agency emphasizes that its research and testing have not proceeded sufficiently to enable it to determine either precisely how to define and implement the individual changes or which of those changes will make enough to a contribution to reducing overall variability to warrant adoption. The agency does not believe that the few currently acknowledged options would make a significant change in the overall level of variability. Identifying the range of necessary and appropriate changes will require iterative testing, given the interplay of the many sources of variability.

The issue of adopting an appropriate statistical procedure to standardize the assignment of grades bears special mention. Although the agency has already proposed such a procedure (46 F.R. 10429, February 2, 1981), commenters on that proposal pointed out a variety of shortcomings, particularly with respect to its failure to properly account for undergrading. No commenter in the present rulemaking proceeding has suggested that the procedure as proposed in February 1981 be adopted at this time. The agency is continuing its analysis of the extent and nature of the changes which might be made to the proposal.

The agency does not agree with the suggestion by a public interest group that the mere possibility that the current treadwear grading information may be better than pre-UTQGS information on treadwear would justify continuation of treadwear grading during the period of any further review. In NHTSA's judgment, it is not clear whether and to what extent the UTQGS treadwear information would in fact be superior to any or all information previously available for distinguishing between tires on the basis of expected tread life. To the degree that the UTQGS system is arguably superior in format and direct comparability among tire lines or manufacturers, however, such apparent advantage derives entirely from those aspects of the system which the agency has found to be most flawed: the accuracy and validity of the UTQGS value as expressed in the grade. Stated differently, it is precisely that aspect of the UTQGS which distinguishes it from market claims of manufacturers which also introduces the clear probability that false information is being disseminated by or under the auspices of the government itself. The probable objective falsity of at least some of the information now being disseminated through UTQGS converts the clarity and apparent simplicity of the UTQGS reporting format from an asset to its most damaging liability. Fully cognizant of the view expressed by this commenter that some information, or a less than perfect-functioning system, is better than no information or no system at all, the agency cannot agree. The agency concludes that the government has a superior duty not to participate in such an effort to the probable detriment of consumers, who have every reason to demand, and must necessarily be expected to assume, that such participation implies and connotes a higher level of certainty than the agency can now find in this well-intentioned effort. Given the shortcomings of the UTQGS system as now understood, price differentials and information voluntarily supplied by the manufacturers as to probable treadwear performance may be as useful to consumers as the current grades.[6]

After weighing the possible benefits of the current grades against the potentially extensive problems created by those grades in their effects on consumers and tire manufacturers, NHTSA concludes that the appropriate course of action is suspension pending completion of its research and testing program.

The agency believes that continuing to require the tire manufacturers to comply with the treadwear grading requirements in the interim is not appropriate, because of the above discussed impossibility of enforcing those requirements in an objective way. NHTSA noted in its proposal that the wide variability in its compliance test results prevented the agency from concluding with any certainty whether tires were incapapble of achieving the grades assigned to them. Commenters on the proposal did not controvert the agency's statements on this point.

In the agency's opinion, requiring the tire manufacturers and consumers to continue to bear the costs of treadwear testing during the time necessary to complete the research and testing concerning test procedure improvements would be unreasonable and unwarranted since the treadwear grading program is apparently neither reasonably fair to the tire manufacturers nor reasonably reliable as a guide to consumers. Although the cost per tire is not large, those costs total approximately $10 million annually.

Amendments Adopted by This Notice

This notice adopts several amendments relating to the treadwear grading provisions of Part 575. Most important, it adopts a suspension of those provisions effective the date that this notice is published in the *Federal Register.*

---

[6]To compound the agency's dilemma on this point, the number of consumers potentially aided by treadwear grading information, and thus the number of consumers potentially misled by an invalid result, is apparently fairly limited. According to information submitted by Uniroyal at the public meeting, only 30 percent of consumers surveyed by them even knew about the UTQGS information, after their promotional efforts, and only 60 percent of those consumers stated they would plan to use that information in making their next tire purchase. Thus, only 18 percent of consumers are potentially benefited, or potentially misled, by the treadwear information.

On that date, manufacturers will no longer be required to submit treadwear grading information to this agency or to disseminate it to consumers through moldings on the side of new tires, paper labels on the treads on new tires, or consumer information materials. The only information that would be required to be submitted or disseminated on or after that date would be traction and temperature resistance grading information.

The agency believes there is ample justification for an immediate effective date. The suspension relieves a restriction and will aid in ending as quickly as is reasonably practicable the possibility that consumers will be misled by the treadwear grading information.

The agency is not requiring that manufacturers immediately cease disseminating treadwear information already printed or embodied on tires or tire molds, through the means formerly required by Part 575. Such a requirement would be impracticable. The greatest problem is associated with the molding of treadwear information on the tires. Discontinuation of that practice would necessitate making changes to the molds being used to produce new tires. Specifically, the manufacturers would have to fill in the indentations used to print the word "TREADWEAR" and the appropriate grade on the sidewall of each new tire. The total cost to the tire industry of making those changes to all molds would be approximately $11 million. Instead of requiring that all molds be changed simultaneously, the agency is requiring that all tires produced in molds manufactured after (180 days after publication in the *Federal Register*), use a format which provides for the molding of only traction and temperature resistance grades on new tires.

, Although the manufacturers could cease printing labels and consumer information materials containing treadwear information almost immediately, they are confronted with the problem of existing inventories of labels and materials containing that information. The agency has decided to allow the manufacturers to exhaust those inventories. The agency expects that after the effective date of this suspension, the labels and materials printed and used by the manufacturers to comply with the UTQGS provisions of Part 575 will not contain that information. The continued printing of labels and materials that set forth the treadwear grades without revealing the suspension of the treadwear requirements, or the absence of any participation by the government in procedures to use similar tests or measurement systems as a basis for warranties or other forms of representation as to treadwear expectancy, would be doubly misleading, i.e., it could be misleading as to the relative performance of tires, but also would be misleading as to the current existence of a government sanctioned system for grading treadwear.

The agency believes that the publicity given this notice will minimize the likelihood that consumers will be misled as a result of the continued molding of treadwear information on some new tires and the continued dissemination for a relatively short period of treadwear information by means of labels and other materials. Probable media coverage of the agency's conclusions in taking this action should reduce the extent of any consumer reliance on them. Further, consumers would be even less likely to rely on the grades after the existing inventories of those lables and materials are exhausted. After then, only the grade would appear on the tire. There would not be any explanatory information concerning the development or meaning of the grade. As the molds are replaced, even the treadwear grade would disappear from the tire, during the pendency of this suspension.

Status of Research

As NHTSA noted in its proposal, it has begun several research activities aimed at reducing the variability of treadwear test results. The agency is proceeding diligently to complete these activities. One program discussed above would attempt to establish the relationship between treadwear, tire inflation pressure, and load. The program to develop this relationship is partially completed, with final results expected by the end of February. If such a relationship could be established, it could aid future research to determine the effects of rotating tires through all positions in test car convoys. Rotating tires in this fashion would tend to minimize the variability that is caused by differences in vehicles and in driver techniques. A contract to test the validity of the rotation concept is expected to be awarded by late spring of this year.

Another program is aimed at establishing the effect of reducing tolerances on permitted test vehicle loading configurations, wheel alignment, driver techniques, and tread depth measurement techniques. A contract for this program is expected to be awarded soon.

A third program will attempt to quantify the individual sources of treadwear test variability through a statistical analysis of existing enforcement data. This research program has already begun and should be completed by the end of February.

Research planned for the future includes an attempt to achieve greater accuracy in test equipment, to specify test vehicle maintenance procedures, and to account for differences in the testing and tread depth measurement environment. A contract for this work is expected to be awarded by late summer of this year.

Issued on February 1, 1983.

Raymond A Peck, Jr.
Administrator
48 F. R. 5690
February 7, 1983

PREAMBLE TO AN AMENDMENT TO PART 575

Customer Information Regulations;
Uniform Tire-Quality Grading

[Docket No. 80-14; Notice 8]
[Docket No. 25; Notice 54]

ACTION: Final rule.

SUMMARY: This notice amends the Uniform Tire Quality Grading Standards (UTQGS) by revising the procedure used to establish tire loads under which temperature-resistance tests are conducted. This amendment is being issued to make test loads under the temperature-resistance test consistent with test loads specified for the high-speed test in Federal Motor Vehicle Safety Standard (FMVSS) 109. It is anticipated that this amendment will assure that UTQGS temperature-resistance tests and FMVSS 109 high-speed tests may, to the maximum possible extent, be conducted together.

DATE: This amendment is effective July 1, 1984. Certain minor technical amendments in the notice are effective immediately on publication.

SUPPLEMENTARY INFORMATION: On December 17, 1981, NHTSA amended FMVSS 109, which establishes performance requirements for new automobile tires, by deleting the tire tables in Appendix A of that standard. Information in these tables was previously used, among other purposes, to specify tire test loads under the UTQGS. Therefore, with the deletion of the tire tables of FMVSS 109, it was necessary to establish alternative procedures for determining UTQGS test loads. Interim procedures were established by NHTSA on June 15 and August 12, 1982, in 47 FR 25930 and 34990, and public comment was invited on the adopted technical approaches. On August 19, 1982, the agency issued a notice of proposed rulemaking, inviting further public comment on other possible approaches to be used in specifying test loads under the UTQGS. See 47 FR 36260.

This notice establishes these procedures in final form.

The UTQGS establish procedures for testing tires to evaluate their traction, temperature resistance, and tread-wear performance. (On February 7, 1983, NHTSA suspended the tread-wear portion of the UTQGS, pending the completion of research intended to determine the causes of the high levels of test variability found in tread-wear test results, and to reduce that variability. (See 48 FR 5690.)) The test procedures specify loads to be placed on the tire. Those loads differ for each of the three types of tests. Prior to the deletion of the FMVSS 109 tire tables, temperature-resistance tests were conducted at the maximum load specified in those tables for a tire pressure 8 pounds per square inch (psi) below the tire's maximum inflation pressure. Tread-wear tests were conducted at 85 percent of the load for temperature-resistance testing. Traction tests were conducted at 85 percent of the maximum load specified in the tire tables for tire pressures of 24 psi or 180 kilopascals, as appropriate.

With the deletion of the tire tables, the agency developed a range of numerical factors which relate a tire's maximum load rating, as stated on the tire's sidewall, to the appropriate test load. Rather than relying on the tables, manufacturers or others conducting tests under the UTQGS would simply multiply the maximum load by the factor to determine the test load. This procedure resulted in at most a 10-pound change in the load at which tests were conducted, for all but a small number of tires. For these remaining tires, the agency provided that tests would be conducted at the same load as was done prior to June 15 (relying on the tire tables), until July 1, 1984. After that date, test loads would be determined by using the load factors.

Shortly after the load-factor procedure was established the Rubber Manufacturers Association and the Cooper Tire Company raised objections to it. These parties pointed out that prior to the deletion of the tire tables, a single test could be used to demonstrate compliance with high-speed requirements under FMVSS 109 and temperature-resistance testing under the UTQGS. However, after the deletion of the tire tables, slightly different loads would be specified for those two purposes. (When the tire tables were deleted, NHTSA specified a single test-load factor of 88 percent of the tire's maximum load for high-speed testing under FMVSS 109.)

On August 19, 1982, NHTSA issued a notice of proposed rulemaking, inviting comment on methods for restoring equivalent load specifications for purposes of high-speed testing under FMVSS 109 and temperature-resistance testing under the UTQGS. The agency proposed three possible methods for achieving this result, and requested that commenters present any other alternatives they felt appropriate. The three NHTSA alternatives were:

(1) To amend the UTQGS temperature-resistance test by deleting the load factors and specifying a single 88-percent factor, as was done with FMVSS 109.

(2) To amend the FMVSS 109 high-speed test by deleting the 88-percent factor and adopting the series of load factors used in the UTQGS temperature-resistance test.

(3) To amend FMVSS 109 and the UTQGS by relying on load information published by industry standardization organizations such as the Tire and Rim Association and The European Tyre and Rim Technical Organization. This approach would be much the same as the procedure previously followed by the agency in relying on the FMVSS 109 tire tables.

Virtually all comments received on the agency's notice of proposed rulemaking recommended adopting the third alternative, since it is the closest to past practice and would assure that test data derived under the pre-June 15 procedures would still be valid. Also, some tire manufacturers felt this option would minimize the "load range creep" phenomenon, in which tire manufacturers were encouraged by vehicle manufacturers to increase incrementally the load rating of a tire, thus permitting the use of a smaller, less expensive tire for a given automobile. These increases could ultimately result in overloaded tire operation. The tire manufacturers felt that the existence of tabulated load information would discourage the load creep phenomenon. On the other hand, the European Tyre and Rim Technical Organization favored the first alternative (testing at 88 percent of maximum load), due to the simplicity of that approach.

NHTSA has concluded that the first alternative is preferable, and is herein amending the UTQGS accordingly. That alternative has the advantage of being the simplest to use, and has been shown to work well in FMVSS 109. The agency is concerned that adoption of alternative 3 could result in the reinstitution of NHTSA tire tables. Information on tires not listed by one of the standardization organizations would be submitted to NHTSA under that alternative. However, commenters requested that information on such tires be published by NHTSA to make it available to all interested parties, thereby resulting in new tire tables, albeit on a smaller scale. The possibilities also exist of inconsistent data entries for tires appearing in more than one table and omissions of certain tires from all tables. The undesirability of this unwieldy system is clear and the disadvantages of the continued reliance on tire tables was discussed fully in the notices involving the deletion of the FMVSS 109 tire tables.

With regard to the load range creep phenomenon, the agency does not agree that the third alternative would discourage such actions to any greater degree than would the other alternatives. Under the third option, all a manufacturer would have to do to change a tire's load rating would be to submit new information to a standardization organization. Further, the agency has ample authority to deal with this problem and will take appropriate action to prevent such actions where safety would be jeopardized.

In the case of the second option, amending FMVSS 109 to adopt varying load factors would disrupt testing programs under that standard which have worked well for the past year using the 88-percent load criterion. Further, adopting the varying load factors is slightly more complex than using the single 88-percent factor. Therefore, the agency considers option 1 to be the preferable alternative.

Adoptive alternative 1 will produce no changes in tire testing under FMVSS 109. However, the

Rubber Manufacturers Association points out that adoption of this alternative will increase tire test loads for UTQGS purposes by from 1 to 3 percent for certain tires.

For the vast majority of currently produced tires (p-metric sizes with maximum inflation pressure of 240 kilopascals), the increase in test load is approximately 1.6 percent. An increase in load of this small a magnitude is insufficient to affect temperature-resistance grades. Also, the majority of tires are graded "C" for temperature resistance, a grade which merely signifies minimum compliance with the high-speed test of FMVSS 109. Therefore, increasing the test loads for UTQGS temperature-resistance purposes (which should theoretically make that test more stringent) will not affect the grades of those tires. Therefore, the amendments promulgated herein should impact only a very small number of tires. To the extent that the adoption of identical test loads for the FMVSS 109 high-speed test and the UTQGS temperature-resistance test permits the two tests to be run together, this amendment will produce an overall reduction in testing costs.

This amendment is being made effective on July 1, 1984, to coincide with the effective date for test-load factors for traction and tread-wear testing for all tires, as specified in the August 2, 1982, Federal Register notice.

Two minor amendments are also being promulgated in this notice for which, due to their technical nature, the agency finds good cause for making effective immediately. The first of these adds three size designations to table 2A of the UTQGS, as requested by the Japanese Automobile Tire Manufacturers Association. This addition will avoid (until July 1, 1984) having to test these tires at significantly different test loads than those specified through the FMVSS 109 tire tables. The second technical amendment clarifies that the traction-test pavement-wetting procedure is that specified in the 1979 version of American Society for Testing and Materials Method E 274.

Since this rule should not cause any significant change in implementation of the UTQG regulation, NHTSA has determined that this proceeding does not involve a major rule within the meaning of Executive Order 12291 or a significant rule within the meaning of the Department of Transportation regulatory procedures. Further, there are no significant economic impacts of this action, so that preparation of a full regulatory evaluation is unnecessary.

The agency has also considered the impacts of this rule in accordance with the Regulatory Flexibility Act. I certify that this action will not have a significant economic impact on a substantial number of small entities. As noted above, this action will make essentially no change in the implementation of the UTQG regulation.

NHTSA has concluded that this action will have essentially no environmental consequences and therefore that there will be no significant effect on the quality of the human environment.

Part 575—CONSUMER INFORMATION REGULATIONS

In consideration of the foregoing, 49 CFR Part 575 is amended as follows:

1. Section 575.104(g)(6) is revised to read as follows:

*   *   *   *   *

(g)   *   *   *   *   *

(6) Press the tire against the test wheel with a load of 88 percent of the tire's maximum load rating as marked on the tire sidewall.

2. Section 575.104(h)(1) is revised to read as follows:

(h) *Determination of test load.* To determine test loads for purposes of paragraphs (e)(2)(iii) and (f)(2)(viii), follow the procedure set forth in paragraphs (h)(2) through (5) of this section.

3. Table 2 of section 575.104 is amended by deleting the words "and temperature resistance" from the heading of the middle column of the table.

4. Table 2A of section 575.104 is amended by adding the following new entries at the bottom of the table:

Tire size designation	Temp resistance Max. pressure			Traction	Tread-wear Max. pressure		
	32	36	40		32	36	40
5.20-14	695	785	855	591	591	667	727
165-15	915	1015	1105	779	779	863	939
185/60 R 13	845	915	980	719	719	778	833

5. The references to "ASTM Method E 274-70" in sections 575.104(f)(1)(iii) and (f)(1)(iv) are deleted and replaced by "ASTM Method E 274-79."

Issued on March 5, 1984.

Diane K. Steed
Administrator

48 FR 8929
March 9, 1984

PREAMBLE TO AN AMENDMENT TO PART 575

Consumer Information Regulations
Operation of Utility Vehicles on Paved Roadways
(Docket No. 82-20; Notice 2)

ACTION: Final rule.

SUMMARY: This final rule adds a new requirement to the *Consumer Information Regulations*, applicable to "utility vehicles", i.e., multipurpose passenger vehicles which have a short wheelbase and special features for occasional off-road use. Some of these special features cause utility vehicles to handle and maneuver differently from ordinary passenger cars under certain driving conditions. A driver who is unaware of the differences and who makes sharp turns or abrupt maneuvers when operating utility vehicles on paved roads may lose control of the vehicle or rollover. To inform drivers of the handling differences between utility vehicles and passenger cars, this amendment requires manufacturers to place a prescribed sticker on the windshield, dashboard or some other prominent location of the vehicle to alert operators. In addition, the new regulation requires manufacturers to include information in the vehicle *Owner's Manual* concerning the proper method of on- and off-road driving for utility vehicles.

DATES: This amendment is effective September 1, 1984.

SUPPLEMENTARY INFORMATION: This notice amends the *Consumer Information Regulations* (49 CFR 575) to add a new requirement applicable to "utility vehicles"—multipurpose passenger vehicles (49 CFR 571.3) which have a short wheelbase and special features for occasional off-road operation. This new regulation addresses a safety concern resulting from a possible lack of owner awareness about the proper handling and operation of utility vehicles. These vehicles have features which cause them to handle and maneuver differently than ordinary passenger cars under certain on-pavement driving conditions. Those features include: short wheelbase, narrow track,

high ground clearance, high center of gravity, stiff suspension system and, often, four-wheel drive. Examples of utility vehicles in current production include: AMC Jeeps, Chevrolet Blazer, Ford Bronco, Dodge Ram Charger, Toyota Land Cruiser, and the GMC Jimmy.

Because of the drivers' apparent unfamiliarity with the unique characteristics of these vehicles (their higher center of gravity, narrower track and stiffer suspensions), utility vehicles are more likely to go out of control or roll over than passenger cars during sharp turns or abrupt maneuvers on paved roads, especially at high speeds. Certain research studies appear to indicate that utility vehicles are disproportionately represented in rollover accidents than are passenger cars, and that the rates of death and disabling injury per accident could be twice as high for utility vehicles. (These studies are discussed more fully in this notice.)

In response to these factors, the agency issued a notice of proposed rulemaking on December 30, 1982 (47 FR 58323) to require a new consumer information regulation which would require manufacturers to alert utility vehicle drivers of the unique handling characteristics of these vehicles. As noted in that proposal, the agency believes that the differences in safety statistics and apparent performance with regard to utility vehicles are likely influenced by the lack of awareness by utility vehicle drivers concerning the operational characteristics of these vehicles, especially under conditions approaching the limits of vehicle performance. The occurrence of accidents at observed rates makes it clear that operators do not understand or appreciate the need for adjusting their driving habits to coincide with physical differences between utility vehicles and ordinary passenger cars.

The proposed amendment to the *Consumer Information Regulations* specified a prescribed sticker which manufacturers would be required to place in a prominent vehicle location to alert drivers concerning the special handling characteristics of utility vehicles. Additionally, the proposed regulation specified that manufacturers would be required to include information in the vehicle *Owner's Manual* concerning the proper method of handling and maneuvering these vehicles when driven on paved roads.

There were twenty comments to the notice of proposed rulemaking. Nearly all of these supported promulgation of the proposed new regulation, in principle. However, many commenters did not accept the agency's basis for the rulemaking and nearly all of the comments recommended various changes. The following is a discussion of the major comments, along with agency's response and final conclusions.

Basic Premise of the New Regulation

The proposal cited a study conducted by the Highway Safety Research Institute of the University of Michigan which found that utility vehicles rollover at a rate at least five times higher than that experienced by the average passenger car ("On Road Crash Experience of Utility Vehicles", see NHTSA Docket 82-20). In addition, the proposal noted that NHTSA fatal accident report data indicate that on a statistical basis, given a rollover accident, occupants are more likely to be killed in utility vehicles than in passenger cars (probability twice as high). Several manufacturers took strong exception to the Michigan study and challenged its scientific accuracy in certain regards, citing statements by the study's author that it was not a definitive project. Although these manufacturers did not oppose the proposed new regulation, they strongly objected to using the cited research as support for the regulation. Several manufacturers also stated that the proposal focused too narrowly on the physical characteristics of utility vehicles and failed to take into account the driver and environmental factors which affect the safety operation of these vehicles.

The agency did not intend to imply that it is only the unique physical characteristics of utility vehicles which are responsible for the great number of accidents in these vehicles. The basic premise of the new regulation, as evidenced by

statements in the proposal, is that drivers are apparently unaware of the unique handling characteristics of these vehicles as compared to ordinary passenger cars, and that this coupling of unique vehicle attributes and lack of awareness is apparently a large part of the problem.

Regarding the research cited in the proposal, the agency also did not intend to imply that further study would not be advantageous or that the Michigan study is an exhaustive, definitive statement concerning the actual accident experience of utility vehicles. However, the agency does believe that the information from the Michigan study, together with NHTSA's own data and other research cited below, is sufficiently reliable to indicate that utility vehicles are involved in a substantial number of accidents which appear to be related to their unique handling characteristics, of which their operators may not be fully aware.

In addition to the research mentioned in the proposal, the agency also notes the following information which has been submitted to the Docket concerning this proceeding: "A Comparison of the Crash Experience of Utility Vehicles, Pickup Trucks and Passenger Cars," Reinfurt, et al., Highway Safety Research Center, University of North Carolina, September 1981: "Analysis of Fatal Rollover Accidents in Utility Vehicles," S. R. Smith, NHTSA, February 1982; "Insurance Losses Personal Injury Protection Coverage, Passenger Cars, Vans, Pickups, and Utility Vehicles, 1979-1981 Models," HLDI, 1-18-1, September 1982. These studies also indicated significant rollover accident experience with utility vehicles. While it may be true that these studies do not quantify the contributions of the various possible causes of this accident experience (vehicle characteristics, driver characteristics, vehicle use, environmental factors, etc.), the agency believes that this research does indicate a serious problem which should be brought to the attention of vehicle owners and which can be alleviated by the dissemination of information to alert vehicles owners and drivers.

Application

Several commenters requested changes and clarifications in the definition of "utility vehicle" as set forth in the proposal's application section. The proposal specified the following:

"This Section applies to multipurpose passenger vehicles which have special features for occasional off-road operation ('utility vehicles')."

Commenters noted that the utility vehicles at issue typically have a wheelbase of 110 inches or less and recommended that this specification be added to the definition so that other vehicles are not inadvertently included in the regulation's application. Manufacturers were particularly concerned that certain vehicles such as long wheel base utility trucks like the General Motors "Suburban" line, motor homes and multi-use recreational vehicles would be included even though they do not have the same rollover propensities as utility vehicles. The Insurance Institute for Highway Safety argued that the application of the rule should be limited to those vehicles most likely to present rollover concerns. The agency generally agrees with these concerns. As noted in the proposal, the vehicles which are intended to be covered are those with relatively short wheelbases, narrow tracks, high ground clearances, high centers of gravity and stiff suspensions. The proposal also mentioned four-wheel drive as a characteristic of utility vehicles. While four-wheel drive is typically a characteristic of those vehicles, it was mentioned in the proposal only because it is descriptive of the majority of vehicles at issue. Four-wheel drive in and of itself, however, has very little to do with the rollover propensities involved in this rulemaking, and the agency did not intend to include a vehicle simply because it had four-wheel drive if it did not also have the characteristics which necessitate alerting drivers to special handling methods.

After reviewing these comments and information concerning the vehicles at issue, the agency has determined that the definition should include a 110-inch wheel base specification in order to segregate those vehicles which are disproportionately involved in rollover accidents. Thus, as specified in this new regulation, utility vehicles are multipurpose passenger vehicles which have a wheel base of 110 inches or less and special features for occasional off-road operation (which may or may not include four-wheel drive).

One manufacturer recommended that the new regulation also apply to four-wheel drive light pickup trucks (GVWR of 8,500 pounds or less) as well as to utility vehicles. The manufacturer did not supply any information, however, indicating that the same accident experience occurs with respect to light pickup trucks. Moreover, data before the agency do not indicate that this vehicle class has a different rollover experience than ordinary passenger cars. Therefore, the fact that certain pickup trucks have four-wheel drive does not seem to be sufficient reason for including this vehicle type in the standard's application. As noted earlier, there is no indication that four-wheel drive alone leads to the rollover propensities which are the subject of this rulemaking action. The agency will continue to monitor the accident experience of these vehicles, however, to determine if they should be included in the standard at some time in the future.

Sticker Location

The proposal preceding this new regulation specified that manufacturers shall affix a sticker to "the instrument panel, windshield frame or in some other location in each vehicle prominent and visible to the driver", to alert drivers concerning the special handling characteristics of utility vehicles. Several commenters requested that this requirement specifically include the driver's sun visor as an acceptable location for the required sticker. One commenter stated that the warning should be of a more permanent nature than a sticker affixed to the windshield or instrument panel. That commenter stated that, if the sticker is located on the instrument panel, it should be behind the plastic lens so that it cannot be removed, arguing that the sticker should remain permanently affixed so that subsequent vehicle owners are made aware of "the vehicle's sensitivity to certain maneuvers."

The agency considers the driver's sun visor to be a "prominent" location in a vehicle, and is modifying the language of this requirement to specifically mention that vehicle location. The agency agrees that the sticker should be of a permanent nature, but does not believe that it is necessary at this time to require the sticker to be placed, for example, behind the plastic lens of the instrument panel. There is no wish to place design restrictions on manufacturers, but the agency does intend for the sticker to be permanently affixed in a prominent position and readily visible to drivers. Stickers similar to the placard required in FMVSS 110 would be considered adequate.

Sticker and Manual Language

A majority of the commenters recommended clarification and changes in the prescribed language for the warning sticker and information in the vehicle *Owner's Manual*. The proposal specified that the sticker shall have the language prescribed "or similar language", and included the following caveat:

"The language on the sticker required by this paragraph may be modified as is desired by the manufacturer to make it appropriate for a specific vehicle design, to ensure that consumers are adequately informed concerning the unique propensities of a particular vehicle model."

As proposed, this caveat was not applicable to the language required in the vehicle *Owner's Manual*. Numerous commenters requested that this flexibility be allowed for the *Owner's Manual* as well. One commenter stated that there is no way the sticker can *"ensure"* consumers are adequately informed. One commenter requested that manufacturers be allowed to place the required information in any section of their *Owner's Manual* they choose, rather than in the "introduction" and "on-pavement" driving sections as prescribed in the proposal. Several commenters also suggested that the word "rollover" be specifically included in the required warnings, on the basis that "loss of control" does not sufficiently describe the hazard.

The agency agrees that language flexibility may be useful for the *Owner's Manual* as well as for the prescribed sticker, in order to ensure that consumers are adequately informed concerning the unique characteristics of a particular vehicle design. That modification is made in this notice. The agency believes that the objection to use of the word "ensure" in the specified caveat is a matter of semantics since the agency's intent is that manufacturers make every attempt to adequately inform its customers. It was for this reason that the language flexibility is being allowed. The agency also agrees that use of the word "rollover" in the sticker and *Owner's Manual* might more accurately describe the possible consequences of sharp turns or abrupt maneuvers than the phrase "loss of control" used alone. Accordingly, that word is added to the language specified in this notice. Finally, the agency agrees that manufacturers should be allowed to place the required "on-

pavement" driving information in any prominent location of their *Owner's Manual* they desire, rather than only in a section specifically labeled "on pavement driving". However, the agency believes that the specified introductory statement must be included in the Manual's introduction (or preface) so that any person consulting the *Manual* will be aware that driving guidelines are included in the *Manual*.

One commenter requested that the required information be allowed in a supplement to the *Owner's Manual*, i.e., a separate pamphlet. The agency has no objections to additional, or comprehensive supplements which further describe driving methods and operating procedures for utility vehicles (one manufacturer currently provides such a Supplement). However, the agency believes that the two prescribed (or similar) statements should be placed in the general *Owner's Manual* since some operators might be more likely to consult the *Manual*, which includes all information concerning their vehicles, than they would supplements. Further, the required statements are short and should not be onerous to manufacturers.

Effective Date

The proposal specified that the new regulation, if promulgated, would become effective 60 days after publication of a final rule. Several manufacturers stated that their *Owner's Manuals* are typically updated only at the beginning of a new model year and that longer than 60 days is needed to comply with the requirements of the regulation. After considering these comments, the agency has concluded that the new regulation should become effective September 1, 1984, coincidental with the typical introduction of new models. This is longer than the 60-days leadtime specified in the proposal and should allow all manufacturers sufficient time to comply with the requirements.

NHTSA has examined the impacts of this new regulation and determined that this notice does not qualify as a major regulation within the meaning of *Executive Order 12291* or as a significant regulation under the Department of Transportation regulatory policies and procedures. The agency has also determined that the economic and other impacts of this rule are so minimal that a regulatory evaluation is not required. The prescribed sticker and additional information required in the vehicle *Owner's Manual* will result in only minimal costs for vehicle manufacturers and will not likely result in any cost increase for consumers.

The agency also considered the impacts of this rule under the precepts of the *Regulatory Flexibility Act*. I hereby certify that the regulation will not have a significant economic impact on a substantial number of small entities. As just discussed, the cost of the required sticker and information will be extremely small. Accordingly, there will be virtually no economic effect on any small organiza-tions or governmental units which purchase utility vehicles. Moreover, few, if any, vehicle manufac-turers would qualify as small entities under the Act.

Issued on May 7, 1984.

Diane K. Steed
Administrator
49 F.R. 20016
May 11, 1984

PREAMBLE TO AN AMENDMENT TO PART 575

Consumer Information Regulations
Operation of Utility Vehicles on Paved Roadways
[Docket No. 82-20; Notice 3]

ACTION: Final rule, response to petitions for reconsideration.

SUMMARY: This final rule responds to petitions for reconsideration filed by American Motors Corporation and Subaru of America, Inc., with regard to the agency's requirement that manufacturers of utility vehicles inform drivers of those vehicles of the propensity of such vehicles to rollover. American Motors and Subaru pointed out in their petitions that the scope of this requirement includes certain passenger car derivatives such as the AMC Eagle and the Subaru four-wheel drive vehicles which do not have the operating characteristics which were the focus of the rule. Therefore, the agency is herein clarifying the regulations to exempt passenger car derivatives.

EFFECTIVE DATE: This amendment is effective September 1, 1984.

SUPPLEMENTARY INFORMATION: On May 11, 1984, NHTSA amended its Consumer Information Regulations (49 CFR 575) to add a new requirement applicable to "utility vehicles"—multipurpose passenger vehicles (49 CFR 571.3) which have a short wheelbase (110 inches or less) and special features for occasional off-road operation. See 49 FR 20016. This new regulation addresses a safety concern resulting from a possible lack of owner awareness about the proper handling and operation of utility vehicles have features which causes them to handle and maneuver differently than ordinary passenger cars under certain on-pavement driving conditions. Those features include: short wheelbase, narrow track, high ground clearance, high center of gravity, stiff suspension

system and, often, four-wheel drive. Examples of utility vehicles in current production which were cited in the agency's final rule include: AMC Jeeps, Chevrolet Blazer, Ford Bronco, Dodge Ram Charger, Toyota Land Cruiser, and the GMC Jimmy.

On June 11, 1984, the agency received petitions for reconsideration of the utility vehicle labeling rule from American Motors Corporation and Subaru of America, Inc. Both manufacturers pointed out that although the preamble to the agency's final rule indicated that the rule was intended to apply to a class of vehicles with attributes which might tend to increase the likelihood of vehicle rollover (high center of gravity, narrow track, stiff suspension, etc.), the actual language of the rule applied to certain vehicles without these attributes. In particular, these manufacturers were concerned that the labeling requirements would apply to their four-wheel drive vehicles which are derived from passenger cars, i.e., the American Motors Eagle and the Subaru four-wheel drive station wagons, sedans, and Brat. Both manufacturers requested that the agency clarify the scope of the rule to exclude these vehicles.

Since the American Motors and Subaru vehicles in question are certified as multipurpose passenger vehicles under 49 CFR Part 567, have a wheelbase of 110 inches or less and have four wheel drive, they would fall within the "utility vehicle" definition in the Consumer Information Regulations, and would therefore be subject to the rollover warning label requirements. However, the manufacturers are correct in pointing out that the main thrust of the agency's May 11 rule was to regulate the more traditional types of utility vehicles, such as the Jeep CJ series and the Toyota Land Cruiser.

To assess the appropriateness of subjecting the Eagle and Subaru model lines to the labeling requirements, the agency analyzed its accident data to determine the frequency of involvement in fatal rollover accidents for various types of vehicles. Fatality data were obtained from the agency's Fatal Accident Reporting System, while vehicle registration information was obtained from R. L. Polk data. The rollover rate for the Eagle is much lower than that for the more traditional utility vehicles, and is, in fact, lower than that for all passenger cars. This data strongly supports the American Motors argument that the Eagle should not be subject to the labeling rule. The case for the Subaru vehicles is less clear, since their rollover fatality rate is between that of passenger cars and the more traditional utility vehicles. However, the Subaru four-wheel drive vehicles have a rollover fatality rate which is virtually identical to that of their two-wheel drive counterparts, which are not subject to the labeling requirement, and is still only about ¼ that of more traditional utility vehicles. Subaru submitted data with its reconsideration petition indicating that the handling characteristics of the Subaru four-wheel drive vehicles are on a par with those of passenger cars, and superior to those of more traditional utility vehicles. Therefore, the agency is exempting passenger car derivative multipurpose passenger vehicles from the rollover labeling requirements. These vehicles are typically based upon a passenger car chassis, then modified to have certain attributes common to trucks or utility vehicles. The Subaru and Eagle vehicles are the only vehicles currently sold in the Untied States which fall within this exemption.

The amendments promulgated herein are effective September 1, 1984, to coincide with the effective date of the May 11 labeling rule. The agency finds good cause for making this amendment effective less than 180 days after publication. The amendment relieves an inappropriate restriction, avoiding the need to provide warning information in vehicles which do not pose an unusual risk of rollover.

NHTSA has examined the impacts of this new regulation and determined that this notice does not qualify as a major regulation within the meaning of Executive Order 12291 or as a significant regulation under the Department of Transportation regulatory policies and procedures. The agency has also determined that the economic and other impacts of this rule are so minimal that a regulatory evaluation is not required. The rule merely exempts a small number of vehicles from the labeling rule, which imposed minimal costs. The agency also considered the impacts of this rule under the percepts of the Regulatory Flexibility Act. I hereby certify that the regulation will not have a significant economic impact on a substantial number of small entities. The cost of the required sticker and information will be extremely small, and only a small number of vehicles are being exempted. Accordingly, there will be virtually no economic effect on any small organizations or governmental units which purchase utility vehicles. Moreover, few, if any, vehicle manufacturers would qualify as small entities under the Act.

In consideration of the foregoing, paragraph 575.105(b) is amended to read as follows:

§575.105 Utility Vehicles

(b) *Application.* This section applies to multipurpose passenger vehicles (other than those which are passenger car derivatives) which have a wheelbase of 110 inches or less and special features for occasional off-road operation ("Utility vehicles").

Issued on August 6, 1984.

Diane K. Steed
Administrator

49 FR 32069
August 10, 1984

PREAMBLE TO AN AMENDMENT TO PART 575

Uniform Tire Quality Grading Standards
Effective Dates for Reimplementation of Treadwear Grading

[Docket No. 25; Notice 58]

ACTION: Final rule.

SUMMARY: This rule sets forth the effective dates for the reimplementation of the treadwear grading requirements under this agency's Uniform Tire Quality Grading Standards (UTQGS). Those requirements were suspended after the agency found high levels of variability in treadwear test data and grade assignment practices. The United States Court of Appeals for the District of Columbia Circuit vacated the agency's suspension of the treadwear grading requirements on April 24, 1984.

In response to the court, NHTSA published a notice on August 13, 1984, proposing dates on which tires would again be required to comply with the treadwear grading requirements. Subsequently, the agency learned that there were some problems with reimplementing treadwear grading for bias belted tires by the proposed dates. Therefore, the agency published a notice on September 12, 1984, asking for public comment on what effect, if any, this newly discovered information should have on the proposed schedule for reimplementing treadwear grading for bias belted tires.

Despite these agency actions to reinstate treadwear grading, the U.S. Court of Appeals issued an order on September 27, 1984, finding NHTSA in violation of its April 24 order, and directing the agency to either reinstate the treadwear grading requirements in full "forthwith" or to apply to that court for a modification of the mandate and provide a reasonably prompt reimplementation schedule. NHTSA filed an application for a modification of the mandate on October 11, 1984. On October 31, 1984, the U.S. Court of Appeals granted NHTSA's application and ordered NHTSA to reimplement treadwear grading in accordance with the schedule proposed by NHTSA in its October 11 filing. That same schedule is set forth in this rule.

DATES: In the case of bias ply tires, requirements that treadwear information be included on paper labels affixed by tire manufacturers to tire treads and for the submission of consumer information brochures to NHTSA for review are reimplemented effective December 15, 1984. Those brochures are required to be distributed to prospective purchasers by tire dealers effective January 15, 1985. Requirements regarding the molding of treadwear grades on tire sidewalls become effective again on May 15, 1985.

In the case of bias belted tires, requirements that treadwear information be included on paper labels and for the submission of the consumer information brochures to NHTSA for review are reimplemented effective March 1, 1985. The brochures must be distributed to prospective purchasers effective April 1, 1985. The requirements regarding the molding of treadwear grades on tire sidewalls become effective again on August 1, 1985.

In the case of radial tires, requirements that treadwear information be included on paper labels and for the submission of the consumer information brochures to NHTSA for review are reimplemented effective April 1, 1985. The brochures must be distributed to prospective purchasers effective May 1, 1985. The treadwear grades must be molded on the sidewall of all new radial tires manufactured on or after September 1, 1985.

In the case of vehicle manufacturers, the requirements to include treadwear grading information in the vehicle consumer information are reimplemented effective September 1, 1985.

The amendments made to the UTQGS by this rule are effective December 19, 1984. This action is taken to permit those manufacturers which choose to do so to comply with the treadwear grading requirements before the mandatory reimplementation dates listed above.

SUPPLEMENTARY INFORMATION: NHTSA suspended treadwear grading requirements under the UTQGS at 48 FR 5690, February 7, 1983. This action was announced after the agency found high levels of variability in treadwear test results and in the grade assignment practices of the various tire manufacturers. This variability resulted in a substantial likelihood that treadwear information being provided to the public under this program would be misleading, i.e., that the assigned grades could, in many instances, incorrectly rank the actual treadwear performance of different tires.

On April 24, 1984, the United States Court of Appeals for the District of Columbia Circuit vacated the agency's suspension of the treadwear grading requirements in *Public Citizen v. Steed*, 733 F.2d 93. NHTSA interpreted the court's action as requiring the agency to reimplement the treadwear grading requirements at the earliest reasonable time. To comply with this interpretation of the court order, NHTSA published a notice of proposed rulemaking at 49 FR 32238, August 13, 1984. That proposal set forth the following dates for reimplementing treadwear grading requirements:

AUGUST 13 SCHEDULE

	Bias Ply and Bias Belted Tires	Radial Tires
Tire manufacturers complete testing	NoVember 15, 1984	June 15, 1985
Affix paper labels and submit brochures to NHTSA for reView	December 15, 1984	July 15, 1985
Distribute brochures to the public	January 15, 1985	August 15, 1985
Modify all molds to include treadwear	May 15, 1985	December 15, 1985
Include treadwear grading in vehicle manufacturer's consumer information booklet	September 1, 1985	

The reason for proposing different reimplementation dates for bias ply and bias belted tires, on the one hand, and radial tires, on the other, was the need to procure new course monitoring tires (CMT's, for the radial tires. As of that date, NHTSA believed that its existing supply of bias ply and bias belted CMT's would be adequate for testing those tire types. This fact would allow the manufacturers to begin their testing very quickly, which would in turn allow the treadwear grading requirements to be reimplemented more quickly.

However, shortly after publication of that notice, the agency determined that its existing supply of bias belted CMT's showed unacceptably high levels of variability, and concluded that it would be inappropriate to use such tires as CMT's. A notice announcing these determinations was published at 49 FR 35814, September 12, 1984. This notice asked for public comment on what effect, if any, this newly discovered information would have on the dates proposed for the reimplementation of treadwear grading for bias belted tires.

Despite these agency actions to reimplement treadwear grading, the U.S. Court of Appeals issued an order on September 27, 1984, finding the agency in violation of the court's April 24 order. The court gave the agency a choice of either immediately reinstating treadwear grading in full, or, within 14 days of September 27, applying to the court for a modification of its earlier order and providing the court with a reasonably prompt schedule for reimplementing the treadwear grading requirements.

In accordance with this order, NHTSA applied for a modification of the court's April 24 mandate on October 11, 1984. This application was aecompanied by a proposed schedule for reimplementing treadwear grading and an affidavit in support thereof. The schedule which the agency proposed to the court is shown on the next page.

This schedule was the same as that proposed in the August 13 notice for reimplementing treadwear grading for vehicle manufacturers and for bias ply tires. However, it accelerated the reimplementation of treadwear grading by 3 1/2 months from what had been proposed for radial tires in the August 13 notice, and postponed the proposed dates for bias belted tires by 2 1/2 months. In formulating this revised schedule, NHTSA considered all nine comments received on the August 13 notice, and the one comment it

PROPOSED SCHEDULE

	Bias Ply Tires	Bias Belted Tires	Radial Tires
Tire manufacturers complete testing	November 7, 1984	February 1, 1985	March 1, 1985
Affix paper labels and submit brochures to NHTSA for review	December 15, 1984	March 1, 1985	April 1, 1985
Distribute brochures to the public	January 15, 1985	April 1, 1985	May 1, 1985
Modify all molds to include treadwear	May 15, 1985	August 1, 1985	September 1, 1985
Include treadwear grading in vehicle manufacturer's consumer information booklet		September 1, 1985	

received on the September 12 notice. The agency received an additional comment regarding the September 12 notice on October 12, the comment closing date for that notice. That additional comment was not considered by the agency in preparing its October 11 application.

The court issued an order on October 31, 1984, granting NHTSA's application for a modification of the court's earlier mandate, and ordered the agency to reimplement treadwear grading according to the schedule proposed by the agency in its October 11 application. This final rule implements the court's October 31 order.

Comments received on previous notices. As noted above, all but one of the comments received in response to the agency's August 13 and September 12 notices were considered while the agency formulated the revised schedule for reimplementing treadwear grading which was submitted to the court on October 11. What follows is a brief explanation of the agency's response to the more significant comments.

The petitioners in the U.S. Court of Appeals submitted their motion to enforce judgment, which they filed with the court, as a comment to the agency on its August 13 proposed schedule. The essential allegation of that motion was that the August 13 schedule was not reasonably prompt. NHTSA responded to this allegation in considerable detail in the application and affidavit in support thereof filed with the court on October 11. Rather than repeat this lengthy response herein, this rule incorporates by reference the application and affidavit filed October 11 as the agency response to petitioners' comments. Copies of the application and affidavit are available in Docket No. 25, Notice 58, and any interested persons are advised to contact the Docket Section to obtain a copy of those documents.

Several tire manufacturers commented that the August 13 notice was unclear as to whether the agency would permit tire manufacturers to modify their molds to show treadwear grading prior to the dates by which they were required to modify all their molds. These manufacturers stated that they wanted to modify some of their molds before the effective dates when they had to have all of their molds modified. This issue arises because of amendments made to the UTQGS in connection with the agency's suspension of the treadwear grading requirements. Since NHTSA had concluded that there was a substantial likelihood that treadwear information would be misleading, the UTQGS were amended to prohibit the sidewalls of tires from showing any treadwear grades. As long as that prohibition, contained in 49 CFR §575.104(i)(2)(ii), remains in effect, tire manufacturers may not legally begin converting their molds to show the treadwear grades on the sidewalls of their tires.

NHTSA wishes to encourage the manufacturers to reimplement the treadwear grading requirements as expeditiously as possible, to comply with the decision in *Public Citizen v. Steed*, supra. The agency intended to allow manufacturers to implement any of the necessary steps, including not just the molding of the grades on the sidewall, but also paper labels and the submission and distribution of consumer information brochures, as soon as was feasible. If some requirements can be satisfied by a particular manufacturer prior to an effective date specified in this rule, it would serve no interest to prohibit that manufacturer from disseminating treadwear grading information to consumers. Hence, a manufacturer is permitted to comply with any of these reimplemented treadwear grading requirements in advance of the effective dates specified herein. These dates represent the agency's best judgment as to the earliest dates by which it would be reasonable to require *all* tires to again comply with the treadwear grading requirements. However, manufacturers may comply with the requirements of this notice sooner than the mandatory effective dates, if they wish. To make this intent more clear, a statement has been added to the DATES section to the effect that the amendments made by this rule take effect upon publication. This action immediately removes the prohibition on molding treadwear grades on the sidewalls of tires, which was a part of the action taken by NHTSA in connection with the decision to suspend treadwear grading.

Most tire manufacturers also indicated that they could meet the dates proposed in the August 13 notice for reimplementing treadwear grading for radial tires, albeit "with some difficulty". This notice accelerates that schedule by shortening the time available for the agency's completion of its tasks while retaining the proposed amount of time following these tasks for the manufacturers to achieve compliance. This acceleration was made possible as the result of CMT's being made available to the agency more quickly, and the agency accelerating its own testing. The time periods allowed to the manufacturers for completing each step of the reimplementation process (3 months for testing, 1 month to print paper labels and draft the consumer information brochure to be submitted to NHTSA for its review, 1 month to distribute the brochures to all dealers, and 6 months to modify all molds) will require the manufacturers to move expeditiously, but are reasonable for completing each of the needed steps.

One manufacturer asked for additional time in reimplementing treadwear grading for radial tires imported from other countries. The comment stated that there is a logistical problem in shipping the tires for testing into the U.S., clearing them through customs, shipping the tires to Texas for testing, conducting the tests and evaluating the data, printing the labels in the U.S. and shipping them overseas, and finally affixing the paper labels to the tires for sale before shipping them into the United States to be offered for sale. The comment concluded by requesting an additional 2 months period for affixing paper labels to imported radial tires, and for an additional 1 month to modify all molds to include the treadwear grade.

NHTSA considered these logistical problems. However, the agency believes that radial tires to be imported into the United States can be shipped early enough so that the tires will be in Texas for testing very early, since the foreign producers are well aware of the logistical burdens confronting them. The testing and analysis for these tires would then be among the first completed on radial tires. While the agency agrees that it is more difficult for manufacturers of imported tires to reimplement treadwear grading than manufacturers of domestic tires, the agency believes that the time allotted for reimplementing is feasible and reasonable for all manufacturers. Accordingly, the schedule set forth in this final rule establishes the same dates for compliance with radial tire treadwear grading requirements for both foreign- and domestically-produced tires.

The comments on the proposed dates for reimplementing treadwear grading for bias ply tires all indicated that those dates were feasible, and those dates have been adopted as proposed.

Three manufacturers asked in their comments for an additional month for testing bias belted tires. That would be the same period of time allotted for testing radial tires. The August 13 notice proposed to allow only 2 months for testing bias belted tires, since there are only about 350 bias belted tire designs. Radial tires, for which 3 months were proposed for testing, are produced in about 1,400 designs. Hence, the difference in the number of tires to be tested suggested to NHTSA that bias belted tire testing could be completed in less time than would be needed for radial tire testing. The commenters asking for additional testing time for bias belted tires did not provide any evidence that the proposed 2 months for testing bias belted tires was insufficient. Absent such evidence, NHTSA has no basis for concluding that the proposed 2-month period for testing is insufficient. Accordingly, this final rule adopts the proposed 2-month testing period for bias belted tires.

The only comment addressing the proposed date for reimplementing treadwear grading requirements for vehicle manufacturers stated that the proposed September 1, 1985, date was acceptable as long as the agency had a final rule published by March 1, 1985. This rule is published well in advance of that date.

Impact analyses. NHTSA has determined that this final rule is neither "major" within the meaning of Executive Order 12291 nor "significant" within the meaning of the Department of Transportation regulatory policies and procedures. The treadwear grading is being reimplemented in its current form as a result of the court decision in *Public Citizen v. Steed,* supra, and the dates set forth herein for reimplementation were ordered to be established by the same court in its October 31, 1984, order. The agency is required to comply with those court orders. Most of the analysis in the regulatory evaluation which accompanied the agency's suspension of treadwear (Docket No. 25; Notice 52) is still applicable to this rule. In that regulatory evaluation, NHTSA estimated that the costs of treadwear grading were about $10 million annually to tire manufacturers and brand name owners. That is equivalent to less than 6 cents per tire. These costs are well below the level for classifying a rule as a major action. A separate regulatory evaluation has not been prepared for this rule, because the costs and impacts of treadwear grading set forth in the regulatory evaluation accompanying the suspension of treadwear grading are still the agency's estimate of the effects of treadwear grading.

Pursuant to the Regulatory Flexibility Act, the agency has considered the impacts of this rule on small entities. I hereby certify that this rule will not have a significant economic impact on a substantial number of small entities. Therefore, a regulatory flexibility analysis is not required. NHTSA concluded that few, if any, of the manufacturers and brand name owners are small entities. To the extent that any of these parties are small entities, the additional costs imposed by reimplementing treadwear grading for passenger-car tires are slightly less than 6 cents per tire ($10 million total costs/178 million passenger car tires produced annually). This does not constitute a significant economic impact. Small organizations and small governmental units will be minimally affected in their tire purchases as a result of the minimal additional costs imposed by reimplementing treadwear grading. Further, those minimal costs will have minimal impacts on the costs and sales for any tire dealers which might qualify as small entities.

NHTSA has also considered the environmental impacts of this rule. While it is possible that reimplementation of treadwear testing may have some negative effects on the environment around the Texas test course in terms of increased fuel consumption and increased noise and air pollution, NHTSA has concluded that the environmental consequences of this rule are of such limited scope that they will clearly not have a significant effect on the quality of the human environment.

Effective date. As noted above, the amendments made by this rule are effective as of the date this rule is published in the *Federal Register.* NHTSA has taken this step so that the tire manufacturers and brand name owners who wish to reimplement any portion of the treadwear grading requirements in advance of the dates by which they are required to do so may follow that course of action. Prior to the effective date of these amendments, §575.104(i) prohibits manufacturers from molding treadwear grades on the sidewalls of tires. Manufacturers and brand name owners which are unable or unwilling to reimplement treadwear grading in advance of the mandatory compliance dates specified herein will not be affected by an immediate voluntary compliance date for these amendments, because they are not required to reimplement before the mandatory compliance dates. There is also a public interest in complying with the court orders as soon as possible. For these reasons, NHTSA has concluded that there is good cause for specifying an immediate effective date for the amendments made by this rule.

In consideration of the foregoing, 49 CFR §575.104 is amended as follows:

1. By revising paragraph (i) and adding new paragraphs (j), (k), and (l) to read as follows:

* * * * *

(i) *Effective dates for treadwear grading requirements for radial tires.*

(1) Treadwear labeling requirements of §575.104 (d)(1)(i)(B)(2) apply to tires manufactured on or after April 1, 1985.

(2) Requirements for NHTSA review of treadwear information in consumer brochures, as specified in paragraph 575.6(d)(2), are effective April 1, 1985.

(3) Treadwear consumer information brochure requirements of paragraph 575.6(c) are effective May 1, 1985.

(6) Treadwear sidewall molding requirements of §575.104(d)(1)(i)(A) apply to tires manufactured on or after September 1, 1985.

(j) *Effective dates for treadwear grading requirements for bias ply tires.*

(1) Treadwear labeling requirements of §575.104 (d)(1)(i)(B)(2) apply to tires manufactured on or after December 15, 1984.

(2) Requirements for NHTSA review of treadwear information in consumer brochures, as specified in paragraph 575.6(d)(2), are effective December 15, 1984.

(3) Treadwear consumer information brochure requirements of paragraph 575.6(c) are effective January 15, 1985.

(4) Treadwear sidewall molding requirements of §575.104(d)(1)(i)(A) apply to tires manufactured on or after May 15, 1985.

(k) *Effective dates for treadwear grading requirements for bias belted tires.*

(1) Treadwear labeling requirements of §575.104 (d)(1)(i)(B)(2) apply to tires manufactured on or after March 1, 1985.

(2) Requirements for NHTSA review of treadwear information in consumer brochures, as specified in paragraph 575.6(d)(2), are effective March 1, 1985.

(3) Treadwear consumer information brochure requirements of paragraph 575.6(c) are effective April 1, 1985.

(4) Treadwear sidewall molding requirements of §575.104(d)(1)(i)(A) apply to tires manufactured on or after August 1, 1985.

(l) *Effective date for treadwear information requirements for vehicle manufacturers.*

Vehicle manufacturer treadwear information requirements of §§575.6(a) and 575.104(d)(1)(iii) are effective September 1, 1985.

2. By deleting Figure 6.

Issued on December 14, 1984.

Diane K. Steed
Administrator

49 F.R. 49293
December 19, 1984

PART 575—CONSUMER INFORMATION

SUBPART A—GENERAL

§ 575.1 Scope.

This part contains Federal Motor Vehicle Consumer Information Regulations established under section 112(d) of the National Traffic and Motor Vehicle Safety Act of 1966 (15 U.S.C. 1401(d)) (hereinafter "the Act").

§ 575.2 Definitions.

(a) *Statutory definitions.* All terms used in this part that are defined in section 102 of the Act are used as defined in the Act.

(b) *Motor Vehicle Safety Standard definitions.* Unless otherwise indicated, all terms used in this part that are defined in the Motor Vehicle Safety Standards, Part 571 of this subchapter (hereinafter "The Standards") are used as defined in the Standards without regard to the applicability of a standard in which a definition is contained.

(c) *Definitions used in this part.*

"Brake power unit" means a device installed in a brake system that provides the energy required to actuate the brakes, either directly or indirectly through an auxiliary device, with the operator action consisting only of modulating the energy application level.

"Lightly loaded vehicle weight" means—

(1) For a passenger car, unloaded vehicle weight plus 300 pounds (including driver and instrumentation), with the added weight distributed in the front seat area.

(2) For a motorcycle, unloaded vehicle weight plus 200 pounds (including driver and instrumentation), with added weight distributed on the saddle and in saddle bags or other carrier.

"Maximum loaded vehicle weight" is used as defined in Standard No. 110.

"Maximum sustained vehicle speed" means that speed attainable by accelerating at maximum rate from a standing start for 1 mile.

"Skid number" means the frictional resistance measured in accordance with American Society for Testing and Materials Method E-274 at 40 miles per hour, omitting water delivery as specified in paragraph 7.1 of that Method.

§ 575.3 Matter Incorporated by reference.

The incorporation by reference provisions of § 571.5 of this subchapter applies to this part.

§ 575.4 Application.

(a) *General.* Except as provided in paragraphs (b) through (d) of this section, each section set forth in Subpart B of this part applies according to its terms to motor vehicles and tires manufactured after the effective date indicated.

(b) *Military vehicles.* This part does not apply to motor vehicles or tires sold directly to the Armed Forces of the United States in conformity with contractual specifications.

(c) *Export.* This part does not apply to motor vehicles or tires intended solely for export and so labeled or tagged.

(d) *Import.* This part does not apply to motor vehicles or tires imported for purposes other than resale.

§ 575.5 Separability.

If any section established in this part or its application to any person or circumstances is held invalid, the remainder of the part and the application of that section to other persons or circumstances is not affected thereby.

§ 575.6 Requirements.

(a) At the time a motor vehicle is delivered to the first purchaser for purposes other than resale, the manufacturer of that vehicle shall provide to that purchaser, in writing and in the English language, the information specified in Subpart B of this part that is applicable to that vehicle and its tires. The document provided with a vehicle may contain more than one table, but the document must either (1) clearly and unconditionally indicate which of the tables apply to the vehicle with which it is provided, or (2) contain a statement on its cover referring the reader to the vehicle certification label for specific information concerning which of the tables apply to that vehicle. If the manufacturer chooses option (2), the vehicle certification label shall include such specific information.

Example 1: Manufacturer X furnishes a document containing several tables, which apply to various groups of vehicles that it produces. The document contains the following notation on its front page: "The information that applies to this vehicle is contained in Table 5." The notation satisfies the requirement.

Example 2: Manufacturer Y furnishes a document containing several tables as in Example 1, with the following notation on its front page:
Information applies as follows:

Model P. 6-cylinder engine—Table 1.

Model P. 8-cylinder engine—Table 2.

Model Q—Table 3.

This notation does not satisfy the requirement, since it is conditioned on the model or the equipment of the vehicle with which the document is furnished, and therefore additional information is required to select the proper table.

(b) At the time a motor vehicle tire is delivered to the first purchaser for a purpose other than resale, the manufacturer of that tire, or in the case of a tire marketed under a brand name, the brand name owner, shall provide to that purchaser the information specified in Subpart B of this part that is applicable to that tire.

(c) Each manufacturer of motor vehicles, each brand name owner of tires, and each manufacturer of tires for which there is no brand name owner shall provide for examination by prospective purchasers, at each location where its vehicles or tires are offered for sale by a person with whom the manufacturer or brand name owner has a contractual, proprietary, or other legal relationship, or by a person who has such a relationship with a distributor of the manufacturer or brand name owner concerning the vehicle or tire in question, the information specified in Subpart B of this part that is applicable to each of the vehicles or tires offered for sale at that location. The information shall be provided without charge and in sufficient quantity to be available for retention by prospective purchasers or sent by mail to a prospective purchaser upon his request. With respect to newly introduced vehicles or tires, the information shall be provided for examination by prospective purchasers not later than the day on which the manufacturer or brand name owner first authorizes those vehicles or tires to be put on general public display and sold to consumers.

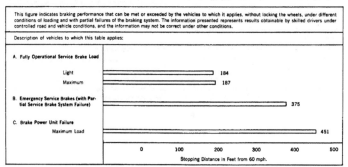

FIGURE 1

(d) (1) (i) Except as provided in paragraph (d) (1) (ii) of this section in the case of all sections of Subpart B, other than § 575.104, as they apply to information submitted prior to new model introduction, each manufacturer of motor vehicles shall submit to the Administrator 10 copies of the information specified in Subpart B of this part that is applicable to the vehicles offered for sale, at least 90 days before it is first provided for examination by prospective purchasers pursuant to paragraph (c) of this section. (2) In the case of § 575.104, and all other sections of Subpart B as they apply to post-introduction changes in information submitted for the current model year, each manufacturer of motor vehicles, each brand name owner of tires, and each manufacturer of tires for which there is no brand name owner shall submit to the Administrator 10 copies of the information specified in Subpart B of this part that is applicable to the vehicles or tires offered for sale, at least 30 days before that information is first provided for examination by prospective purchasers pursuant to paragraph (c) of this section.

(ii) Where an unforeseen pre-introduction modification in vehicle design or equipment results in a change in vehicle performance for a characteristic included in Subpart B of this part, a manufacturer of motor vehicles may revise information previously furnished under (d) (1) (i) of this section by submission to the Administrator of 10 copies of revised information reflecting the performance changes, at least 30 days before information on the subject vehicles is first provided to prospective purchasers pursuant to paragraph (c) of this section.

(2) In the case of § 575.104, and all other sections of Subpart B as they apply to post-introduction changes in information submitted for the current model year, each manufacturer of motor vehicles, each brand name owner of tires, and each manufacturer of tires for which there is no brand name owner shall submit to the Administrator 10 copies of the information specified in Subpart B of this part that is applicable to the vehicles or tires offered for sale, at least 30 days before it is first provided for examination by prospective purchasers pursuant to paragraph (c) of this session.

§ 575.7 Special vehicles.

A manufacturer who produces vehicles having a configuration not available for purchase by the general public need not make available to ineligible purchasers, pursuant to § 575.6(c), the information for those vehicles specified in Subpart B of this part, and shall identify those vehicles when furnishing the information required by § 575.6(d).

§ 575.101 Vehicle stopping distance.

(a) *Purpose and scope.* This section requires manufacturers of passenger cars and motorcycles to provide information on vehicle stopping distances under specified speed, brake, loading and pavement conditions.

(b) *Application.* This section applies to passenger cars and motorcycles manufactured on or after January 1, 1970.

(c) *Required information.* Each manufacturer shall furnish the information in (1) through (5) below, in the form illustrated in Figure 1, except that with respect to (2) and (3) below, a manufacturer whose total motor vehicle production does not exceed 500 annually is only required to furnish performance information for the loaded condition. Each motorcycle in the group to which the information applies shall be capable, under the conditions specified in paragraph (d), and utilizing the procedures specified in paragraph (e), of performing at least as well as the information indicates. Each passenger car in the group to which the information applies shall be capable of performing at least as well as the information indicates, under the test conditions and procedures specified in S6 and S7 of Standard No. 105–75 of this chapter (49 CFR 571.105–75) or, in the case of passenger cars manufactured before January 1, 1977, and at the option of the manufacturer, under the conditions specified in paragraph (d) of this section and the procedures specified in Paragraph (e) of this section.

If a vehicle is unable to reach the speed of 60 miles per hour (mph), the maximum sustained vehicle speed shall be substituted for the 60 mph speed in the requirements specified below, and in the presentation of information as in Figure 1, with an asterisked notation in essentially the following form at the bottom of the figure: "The maximum speed attainable by accelerating at maximum rate from a standing start for 1 mile." The weight requirements indicated in paragraphs (c)(2), (3), and (4) of this section are modified for the motorcycles (and at the option of the manufacturer, in the case of passenger cars manufactured before January 1, 1977) by the fuel tank condition specified in paragraph (d) (4) of this section.

(1) *Vehicle description.* The group of vehicles to which the table applies, identified in the terms by which they are described to the public by the manufacturer.

(2) *Minimum stopping distance with fully operational service brake system.* The minimum stopping distance attainable, expressed in feet,

PART 575–3

from 60 mph, using the fully operational service brake system—

(A) In the case of a motorcycle, at lightly loaded and maximum loaded vehicle weight; and

(B) In the case of a passenger car, at lightly loaded vehicle weight and at gross vehicle weight rating (GVWR), except for a passenger car manufactured before January 1, 1977, and tested, at the option of the manufacturer, under the conditions and procedures of paragraphs (d) and (e) of this section, which passenger car shall be tested at lightly loaded vehicle weight and at maximum loaded vehicle weight.

(3) *Minimum stopping distance with partially failed service brake system.* (Applicable only to passenger cars with more than one service brake subsystem.) The minimum stopping distance attainable using the service brake control, expressed in feet, from 60 mph, for the most adverse combination of GVWR or lightly loaded vehicle weight and partial failure as specified in S5.1.2 of Standard No. 105–75 of this chapter. However, a passenger car manufactured before January 1, 1977, and tested, at the option of the manufacturer, under the conditions and procedures of paragraphs (d) and (e) of this section, shall be tested at maximum loaded vehicle weight instead of GVWR.

(4) *Minimum stopping distance with inoperative brake power assist unit or brake power unit.* (Applicable only to passenger cars equipped with brake power assist unit or brake power unit.) The minimum stopping distance, expressed in feet, from 60 mph, using the service brake system, tested in accordance with the requirements of S5.1.3 of Standard No. 105–75 of this chapter. However, in the case of a passenger car manufactured before Janaury 1, 1977, vehicle loading may, at the option of the manufacturer, be maximum loaded vehicle weight in place of the GVWR loading specified under S5.1.3 of Standard No. 105–75.

(5) *Notice.* The following notice: "This figure indicates braking performance that can be met or exceeded by the vehicles to which it applies, without locking the wheels, under different conditions of loading and with partial failures of the braking system. The information presented represents results obtainable by skilled drivers under controlled road and vehicle conditions, and the information may not be correct under other conditions."

(d) *Conditions.* The data provided in the format of Figure 1 shall represent a level of performance that can be equalled or exceeded by each vehicle in the group to which the table applies, under the following conditions, utilizing the procedures set forth in (e) below:

(1) Stops are made without lock-up of any wheel, except for momentary lock-up caused by an automatic skid control device.

(2) The tire inflation pressure and other relevant component adjustments of the vehicle are made according to the manufacturer's published recommendations.

(3) For passenger cars, brake pedal force does not exceed 150 pounds for any brake application. For motorcycles, hand brake lever force applied 1¼ inches from the outer end of the lever does not exceed 55 pounds, and foot brake pedal force does not exceed 90 pounds.

(4) Fuel tank is filled to any level between 90 and 100 percent of capacity.

(5) Transmission is in neutral, or the clutch disengaged, during the entire deceleration.

(6) The vehicle begins the deceleration in the center of a straight roadway lane that is 12 feet wide, and remains in the lane throughout the deceleration.

(7) The roadway lane has a grade of zero percent, and the road surface has a skid number of 81, as measured in accordance with American Society for Testing and Materials (ASTM) Method E–274–70 (as revised July, 1974) at 40 mph, omitting the water delivery specified in paragraphs 7.1 and 7.2 of that Method.

(8) All vehicle openings (doors, windows, hood, trunk, convertible tops, etc.) are in the closed position except as required for instrumentation purposes.

(9) Ambient temperature is between 32°F and 100°F.

(10) Wind velocity is zero.

(e) *Procedures.*

(1) Burnish.

(i) Passenger cars. Burnish brakes once prior to first stopping distance test by conducting 200 stops from 40 mph (or maximum sustained vehicle speed if the vehicle is incapable of reaching 40 mph) at a deceleration rate of 12 fpsps in normal driving gear, with a cooling interval between stops, accomplished by driving at 40 mph for a sufficient distance to reduce brake temperature to 250°F, or for one mile, whichever occurs first. Readjust brakes according to manufacturer's recommendations after burnishing.

(ii) Motorcycles. Adjust and burnish brakes in accordance with manufacturer's recommendations. Where no burnishing procedures have been recommended by the manufacturer, follow the procedures specified above for passenger cars, except substitute 30 mph for 40 mph and 150° F for 250°F, and maintain hand lever force to foot lever force ratio of approximately 1 to 2.

(2) Ensure that the temperature of the hottest service brake is between 130°F and 150°F prior to the start of all stops (other than burnishing stops), as measured by plug-type thermocouples installed according to SAE Recommended Practice J843a, June 1966.

(3) Measure the stopping distance as specified in (c) (2), (3), and (4), from the point of application of force to the brake control to the point at which the vehicle reaches a full stop.

§ 575.102 [Reserved].

§ 575.103 Truck-camper loading.

(a) *Scope.* This section requires manufacturers of trucks that are capable of accommodating slide-in campers to provide information on the cargo weight rating and the longitudinal limits within which the center of gravity for the cargo weight rating should be located.

(b) *Purpose.* The purpose of this section is to provide information that can be used to reduce overloading and improper load distribution in truck-camper combinations, in order to prevent accidents resulting from the adverse effects of these conditions on vehicle steering and braking.

(c) *Application.* This section applies to trucks that are capable of accommodating slide-in campers.

(d) *Definitions.* "Camper" means a structure designed to be mounted in the cargo area of a truck, or attached to an incomplete vehicle with motive power, for the purpose of providing shelter for persons.

"Cargo weight rating" means the value specified by the manufacturer as the cargo-carrying capacity, in pounds, of a vehicle, exclusive of the weight of occupants, computed as 150 pounds times the number of designated seating positions.

"Slide-in camper" means a camper having a roof, floor and sides, designed to be mounted on and removable from the cargo area of a truck by the user.

(e) *Requirements.* Except as provided in paragraph (f) of this section each manufacturer of a truck that is capable of accommodating a slide-in camper shall furnish the information specified in (1) through (5) below:

(1) A picture showing the manufacturer's recommended longitudinal center of gravity zone for the cargo weight rating in the form illustrated in Figure 1. The boundaries of the zone shall be such that when a slide-in camper equal in weight to the truck's cargo weight rating is installed, no gross axle weight rating of the truck is exceeded. Until October 1, 1973 the phrase "Aft End of Cargo Area" may be used in Figure 1 instead of "Rear End of Truck Bed".

FIGURE 1 TRUCK LOADING INFORMATION

(2) The truck's cargo weight rating.

(3) The statements: "When the truck is used to carry a slide-in camper, the total cargo load of the truck consists of the manufacturer's camper weight figure, the weight of installed additional camper equipment not included in the manufacturer's camper weight figure, the weight of camper cargo, and the weight of passengers in the camper. The total cargo load should not ex-

ceed the truck's cargo weight rating and the camper's center of gravity should fall within the truck's recommended center of gravity zone when installed." Until October 1, 1973 the phrase "total load" may be used instead of "total cargo load".

(4) A picture showing the proper match of a truck and slide-in camper in the form illustrated in Figure 2.

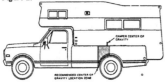

FIGURE 2 - EXAMPLE OF PROPER TRUCK AND CAMPER MATCH

(5) The statements: "Secure loose items to prevent weight shifts that could affect the balance of your vehicle. When the truck camper is loaded, drive to a scale and weigh on the front and on the rear wheels separately to determine axle loads. Individual axle loads should not exceed either of the gross axle weight ratings (GAWR). The total of the axle loads should not exceed the gross vehicle weight rating (GVWR). These ratings are given on the vehicle certification label that is located on the left side of the vehicle, normally the dash, hinge pillar, door latch post, or door edge next to the driver. If weight ratings are exceeded, move or remove items to bring all weights below the ratings."

(f) If a truck would accommodate a slide-in camper but the manufacturer of the truck recommends that the truck not be used for that purpose, the information specified in paragraph (e) shall not be provided but instead the manufacturer shall provide a statement that the truck should not be used to carry a slide-in camper.

§ 575.104 Uniform Tire Quality Grading Standards.

(a) *Scope.* This section requires motor vehicle and tire manufacturers and tire brand name owners to provide information indicating the relative performance of passenger car tires in the areas of treadwear, traction, and temperature resistance.

(b) *Purpose.* The purpose of this section is to aid the consumer in making an informed choice in the purchase of passenger car tires.

(c) *Application.* (1) This section applies to new pneumatic tires for use on passenger cars. However, this section does not apply to deep tread, winter-type snow tires, space-saver or temporary use spare tires, tires with nominal rim diameters of 10 to 12 inches, or to limited production tires as defined in paragraph (c)(2) of this section.

(2) "Limited production tire" means a tire meeting all of the following criteria, as applicable:

(i) The annual domestic production or importation into the United States by the tire's manufacturer of tires of the same design and size as the tire does not exceed 15,000 tires;

(ii) In the case of a tire marketed under a brand name, the annual domestic purchase or importation into the United States by a brand name owner of tires of the same design and size as the tire does not exceed 15,000 tires;

(iii) The tire's size was not listed as a vehicle manufacturer's recommended tire size designation for a new motor vehicle produced in or imported into the United States in quantities greater than 10,000 during the calendar year preceding the year of the tire's manufacturer; and

(iv) The total annual domestic production or importation into the United States by the tire's manufacturer, and in the case of a tire manufacturer, and in case of a tire marketed under a brand name, the total annual domestic purchase or purchase for importation into the United States by the tire's brand name owner, of tires meeting the criteria of paragraphs (c)(2) (i), (ii), and (iii) of this section, does not exceed 35,000 tires.

Tire design is the combination of general structural characteristics, materials, and tread pattern, but does include cosmetic, identifying or other minor variations among tires.

(d) *Requirements.*

(1) *Information.*

(i) Each manufacturer of tires, or in the case of tires marketed under a brand name, each brand name owner, shall provide grading information for each tire of which he is the manufacturer or brand name owner in the manner set forth in paragraphs (d) (1) (i) (A) and (d) (1) (i) (B) of this section. The grades for each tire shall be only those specified in paragraph (d) (2) of this section. Each tire shall be able to achieve the level of performance represented by each grade with which it is labeled. An individual tire need not, however, meet further requirements after having been subjected to the test for any one grade.

(A) Except for a tire line, manufactured within the first six months of production of the tire line, each tire shall be graded with the words, letters, symbols, and figures specified in paragraph (d) (2) of this section, permanently molded into or onto the tire sidewall between the tire's maximum section width and shoulder in accordance with one of the methods in Figure 1.

(B) (1) Each tire manufactured before October 1, 1980, other than a tire sold as original equipment on a new vehicle, shall have affixed to its tread surface in a manner such that it is not easily removable a label containing its grades and other information in the form illustrated in Figure 2, Part II, bearing the heading "DOT QUALITY GRADES." The treadwear grade attributed to the tire shall be either imprinted or indelibly stamped on the label adjacent to the description of the treadwear grade. The label shall also depict all possible grades for traction and temperature resistance. The traction and temperature resistance performance grades attributed to the tire shall be indelibly circled. However, each tire labeled in conformity with the requirements of paragraph (d)(1)(i)(B)(2) of this section need not comply with the provisions of this paragraph.

(2) Each tire manufactured on or after October 1, 1980, other than a tire sold as original equipment on a new vehicle, shall have affixed to its tread surface so as not to be easily removable a label or labels containing its grades and other information in the form illustrated in Figure 2, Parts I and II. The treadwear grade attributed to the tire shall be either imprinted or indelibly stamped on the label containing the material in Part I of Figure 2, directly to the right of or below the word "TREAD-WEAR". The traction and temperature resistance performance grades attributed to the tire shall be indelibly circled in an array of the potential grade letters (ABC) directly to the right of or below the words "TRACTION" and "TEMPERATURE" in Part I of Figure 2. The words "TREAD-WEAR," "TRACTION," and "TEMPER-ATURE," in that order, may be laid out vertically or horizontally. The text part of Part II of Figure 2 may be printed in capital letters. The text of Part I and the text of Part II of Figure 2 need not appear on the same label, but the edges of the two texts must be positioned on the tire tread so as to be separated by a distance of no more than one inch. If the text of Part I and the text of Part II are placed on separate labels, the notation "See EXPLAN-ATION OF DOT QUALITY GRADES" shall be added to the bottom of the Part I text, and the words "EXPLANATION OF DOT QUALITY GRADES" shall appear at the top of the Part II text. The text of Figure 2 shall be oriented on the tire tread surface with lines of type running perpendicular to the tread circumference. If a label bearing a tire size designation is attached to the tire tread surface and the tire size designation is oriented with lines of type running perpendicular to the tread circumference, the text of Figure 2 shall read in the same direction as the tire size designation.

(ii) In the case of information required in accordance with § 575.6(c) to be furnished to prospective purchasers of motor vehicles and tires, each vehicle manufacturer and each tire manufacturer or brand name owner shall as part of that information list all possible grades for traction and temperature resistance, and restate verbatim the explanations for each performance area specified in Figure 2. The information need not be in the same format as in Figure 2. In the case of a tire manufacturer or brand name owner, the information must indicate clearly and unambiguously the grade in each performance area for each tire of that manufacturer or brand name owner offered for sale at the particular location.

(iii) In the case of information required in accordance with § 575.6(a) to be furnished to the first purchaser of a new motor vehicle, other than a motor vehicle equipped with bias-ply tires manufactured prior to October 1, 1979, and April 1, 1980, and a radial-ply tire manufactured prior to October 1, 1980, each manufacturer of motor vehicles shall as part of the information list all possible grades for traction and temperature resistance and restate verbatim the explanation for each performance area specified in Figure 2. The information need not be in the format of Figure 2, but it must contain a statement referring the reader to the tire sidewall for the specific tire grades for the tires with which the vehicle is equipped.

(2) *Performance.*

(i) *Treadwear.* Each tire shall be graded for treadwear performance with the word "TREADWEAR" followed by a number of two of three digits representing the tire's grade for treadwear, expressed as a percentage of the NHTSA nominal treadwear value, when tested in accordance with the conditions and procedures specified in paragraph (e) of this section. Treadwear grades shall be multiples of 10 (e.g., 80, 150).

(ii) *Traction.* Each tire shall be graded for traction performance with the word "TRAC-TION," followed by the symbols C, B, or A (either asterisks or 5-pointed stars) when the tire is tested in accordance with the conditions and procedures specified in paragraph (f) of this section.

(A) The tire shall be graded C when the adjusted traction coefficient is either:

(1) 0.38 or less when tested in accordance with paragraph (f) (2) of this section on the asphalt surface specified in paragraph (f) (1) (i) of this section, or

(2) 0.26 or less when tested in accordance with paragraph (f) (2) of this section on the concrete surface specified in paragraph (f) (1) (i) of this section.

(B) The tire may be graded B only when its adjusted traction coefficient is both:

(1) More than 0.38 when tested in accordance with paragraph (f) (2) of this section on the asphalt surface specified in paragraph (f) (1) (i) of this section, and

(2) More than 0.26 when tested in accordance with paragraph (f) (2) of this section on the concrete surface specified in paragraph (f) (1) (i) of this section.

(C) The tire may be graded A only when its adjusted traction coefficient is both:

(1) More than 0.47 when tested in accordance with paragraph (f) (2) of this section on the asphalt surface specified in paragraph (f) (1) (i) of this section, and

(2) More than 0.35 when tested in accordance with paragraph (f) (2) of this section on the concrete surface specified in paragraph (f) (1) (i) of this section.

(iii) *Temperature resistance.* Each tire shall be graded for temperature resistance performance with the word "TEMPERATURE" followed by the letter A, B, or C, based on its performance when the tire is tested in accordance with the procedures specified in paragraph (g) of this section. A tire shall be considered to have successfully completed a test stage in accordance with this paragraph if, at th end of the test stage, it exhibits no visual evidence of tread, sidewall, ply, cord, innerliner or bead separation, chunking, broken cords, cracking or open splices a defined in § 571.109 of this chapter, and the tire pressure is not less than the pressure specified in paragraph (g) (1) of this section.

(A) The tire shall be graded C if it fails to complete the 500 rpm test stage specified in paragraph (g) (9) of this section.

(B) The tire may be graded B only if it successfully completes the 500 rpm test stage specified in paragraph (g) (9) of this section.

(C) The tire may be graded A only if it successfully completes the 575 rpm test stage specified in paragraph (g) (9) of this section.

(e) *Treadwear grading conditions and procedures.*— (1) *Conditions.* (i) Tire treadwear performance is evaluated on a specific roadway course approximately 400 miles in length, which is established by the NHTSA both for its own compliance testing and for that of regulated persons. The course is designed to produce treadwear rates that are generally representative of those encountered in public use for tires of differing construction types. The course and driving procedures are described in Appendix A to this section.

(ii) Treadwear grades are evaluated by first measuring the performance of a candidate tire on the government test course, and then correcting the projected mileage obtained to account for environmental variations on the basis of the performance of course monitoring tires of the same general construction type (bias, bias-belted, or radial) run in the same convoy. The three types of course monitoring tires are made available by the NHTSA at Goodfellow Air Force Base, San Angelo, Tex., for purchase by any persons conducting tests at the test course.

(iii) In convoy tests each vehicle in the same convoy, except for the lead vehicle, is throughout the test within human eye range of the vehicle immediately ahead of it.

(iv) A test convoy consists of no more than four passenger cars, each having only rear-wheel drive.

[(v) On each convoy vehicle, all tires are mounted on identical rims of design or measuring rim width specified for tires of that size in accordance with 49 CFR 571.109, § 4.4.1(a) or (b), or a rim having a width within − 0 to + 0.50 inches of the width listed. (47 F.R. 25931—June 15, 1982. Effective: June 15, 1982.)]

(2) *Treadwear grading procedure.* (i) Equip a convoy with course monitoring and candidate tires of the same construction type. Place four course monitoring tires on one vehicle. On each other vehicle, place four candidate tires that are identical with respect to size designations. On each axle, manufacturer and line.

(ii) Inflate each candidate and each course monitoring tire the applicable pressure in Table 1 of this section.

[(iii) Load each vehicle so that the load on each course monitoring and candidate tire is 85 percent of the test load specified in § 575.104(h). (47 F.R. 25931—June 15, 1982. Effective: June 15, 1982.)]

(iv) Adjust wheel alignment to that specified by the vehicle manfuacturer.

(v) Subject candidate and course monitoring tires to "break-in" by running the tires in convoy for two circuits of the test roadway (800 miles). At the end of the first circuit, rotate each vehicle's tires by moving each front tire to the same side of the rear axle and each rear tire to the opposite side of the front axle.

(vi) After break-in, allow the tires to cool to the inflation pressure specified in paragraph (e) (2) (ii) of this section or for 2 hours, whichever occurs first. Measure, to the nearest 0.001 inch, the tread depth of each candidate and course monitoring tire, avoiding tread-wear indicators, at six equally spaced points in each groove. For each tire compute the average of the measurements. Do not include those shoulder grooves which are not provided with treadwear indicators.

(vii) Adjust wheel alignment to the manufacturer's specifications.

(viii) Drive the convoy on the test roadway for 6,400 miles. After each 800 miles:

(A) Following the procedure set out in paragraph (e) (2) (vi) of this section, allow the tires to cool and measure the average tread depth of each tire;

(B) Rotate each vehicle's tires by moving each front tire to the same side of the rear axle and each rear tire to the opposite side of the front axle.

(C) Rotate the vehicles in the convoy by moving the last vehicle to the lead position. Do not rotate driver position within the convoy.

(D) Adjust wheel alignment to the vehicle manufacturer's specifications, if necessary.

(ix) Determine the projected mileage for each candidate tire as follows:

(A) For each course monitoring and candidate tire in the convoy, using the average tread depth measurements obtained in accordance with paragraphs (e) (2) (vi) of this section and the corresponding mileages as data points, apply the method of least squares as described in Appendix C of this section to determine the estimated regression line of y on x given by the following formula:

$$y = a + \frac{bx}{1000}$$

where:

y = average tread depth in mils,

x = miles after break-in,

a = y intercept of regression line (reference tread depth) in mils, calculated using the method of least squares; and

b = the slope of the regression line in mils of tread depth per 1,000 miles, calculated using the method of least squares. This slope will be negative in value. The tire's wear rate is defined as the absolute value of the slope of the regression line.

(B) Average the wear rates of the four course monitoring tires as determined in accordance with paragraph (e) (2) (ix) (A) of this section.

(C) Determine the course severity adjustment factor by dividing the base wear rate for the course monitoring tire (see note below) by the average wear rate for the four course monitoring tires determined in accordance with paragraph (e) (2) (ix) (B) of this section.

NOTE.—The base wear rates for the course monitoring tires will be furnished to the purchaser at the time of purchase.

(D) Determine the adjusted wear rate for each candidate tire by multiplying its wear rate determined in accordance with paragraph (e) (2) (ix) (A) of this section by the course severity adjustment factor determined in accordance with paragraph (e) (2) (ix) (C) of this section.

(E) Determine the projected mileage for each candidate tire using the following formula:

$$\text{Projected mileage} = \frac{1000\,(a-62)}{b'} + 800$$

where:

a = y intercept of regression line (reference tread depth) for the candidate tire as determined in accordance with paragraph (e) (2) (ix) (A) of this section.

b' = the adjusted wear rate for the candidate tire as determined in accordance with paragraph (e) (2) (ix) (D) of this section.

(F) Compute the percentage of the NHTSA nominal treadwear value for each candidate tire using the following formula:

$$P = \frac{\text{Projected Mileage}}{30,000} \times 100$$

Round off the percentage to the nearest lower 10% increment.

(f) *Traction grading conditions and procedures.*—(1) *Conditions.* (i) Tire traction performance is evaluated on skid pads that are established, and whose severity is monitored, by the NHTSA both for its compliance testing and for that of regulated persons. The test pavements are asphalt and concrete surfaces constructed in accordance with the specifications for pads "C" and "A" in the "Manual for the Construction and Maintenance of Skid Surfaces," National Technical Information Service No. DOT-HS-800-814. The surfaces have locked wheel traction coefficients when evaluated in accordance with paragraphs (f) (2) (i) through (f) (2) (vii) of this section of 0.50±0.10 for the asphalt and 0.35±0.10 for the concrete. The location of the skid pads is described in Appendix B to this section.

(ii) The standard tire is the American Society for Testing and Materials (ASTM) E 501 "Standard Tire for Pavement Skid Resistance Tests."

(iii) The pavement surface is wetted in accordance with paragraph 3.5, "Pavement Wetting System," of ATSM Method E 274-[79], "Skid Resistance of Paved Surfaces Using a Full-Scale Tire."

(iv) The test apparatus is a test trailer built in conformity with the specifications in paragraph 3, "Apparatus," of ASTM Method E 274-[79], and instrumented in accordance with paragraph 3.3.2 of that Method, except that "wheel load" in paragraph 3.2.2 and tire and rim specifications in paragraph 3.2.3 of that Method are as specified in the procedures in paragraph (f) (2) of this section for standard and candidate tires.

(v) The test apparatus is calibrated in accordance with ASTM Method F 377-74, "Standard Method for Calibration of Braking Force for Testing of Pneumatic Tires" with the trailer's tires inflated to 24 psi and loaded to 1,085 pounds.

(vi) Consecutive tests on the same surface are conducted not less than 30 seconds apart.

(vii) A standard tire is discarded in accordance with ASTM Method E 501.

(2) *Procedure.* (i) Prepare two standard tires as follows:

(A) Condition the tires by running them for 200 miles on a pavement surface.

(B) Mount each tire on a rim of design or measuring rim width specified for tires of its size in accordance with 49 CFR 571.109, § 4.4.1(a) or (b), or a rim having a width within −0 to +0.50 inches of the width listed. Then inflate the tire to 24 psi, or, in the case of a tire with inflation pressure measured in kilopascals, to 180 kPa.

(C) Statically balance each tire-rim combination.

(D) Allow each tire to cool to ambient temperature and readjust its inflation pressure to 24 psi, or, in the case of a tire with inflation pressure measured in kilopascals, to 180 kPa.

(ii) Mount the tires on the test apparatus described in paragraph (f) (1) (iv) of this section and load each tire to 1,085 pounds.

(iii) Tow the trailer on the asphalt test surface specified in paragraph (f) (1) (i) of this section at a speed of 40 mph, lock one trailer wheel, and record the locked-wheel traction coefficient on the tire associated with that wheel between 0.5 and 1.5 seconds after lockup.

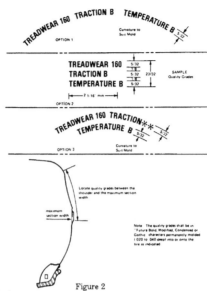

Figure 2

[Part 1] DOT Quality Grades

Treadwear

The treadwear grade is a comparative rating based on the wear rate of the tire when tested under controlled conditions on a specified government test course. For example, a tire graded 150 would wear one and a half (1½) times as well on the government course as a tire graded 100. The relative performance of tires depends upon the actual conditions of their use, however, and may depart significantly from the norm due to variations in driving habits, service practices, and differences in road characteristics and climate.

Traction

The traction grades, from highest to lowest, are A, B, and C, and they represent the tire's ability to stop on wet pavements as measured under controlled conditions on specified government test surfaces of asphalt and concrete. A tire marked C may have poor traction performance. *WARNING:* The traction grade assigned to this tire is based on braking (straightahead) traction tests and does not included cornering (turning) traction.

Temperature

The temperature grades of A (the highest), B, and C, representing the tire's resistance to the generation of heat and its ability to dissipate heat when tested under controlled conditions on a specified indoor laboratory test wheel. Sustained high temperature can cause the material of the tire to degenerate and reduce tire life, and excessive temperature can lead to sudden tire failure. The grade C corresponds to a level of performance which all passenger car tires must meet under the Federal Motor Vehicle Safety Standard No. 109. Grades B and A represent higher levels of performance on the laboratory test wheel than the minimum required by law. *WARNING:* The temperature grade for this tire is established for a tire that is properly inflated and not overloaded. Excessive speed, under-inflation, or excessive loading either separately or in combination, can cause heat buildup and possible tire failure.

[Part II] All Passenger Car Tires Must Conform to Federal Safety Requirements in Addition to These Grades.

(iv) Repeat the test on the concrete surface, locking the same wheel.

(v) Repeat the tests specified in paragraphs (f) (2) (iii) and (f) (2) (iv) of this section for a total of 10 measurements on each test surface.

(vi) Repeat the procedures specified in paragraphs (f) (2) (iii) through (f) (2) (v) of this section, locking the wheel associated with the other tire.

(vii) Average the 20 measurements taken on the asphalt surface to find the standard tire traction coefficient for the asphalt surface. Average the 20 measurements taken on the concrete surface to find the standard tire traction coefficient for the concrete surface. The standard tire traction coefficient so detèrmined may be used in the computation of adjusted traction coefficients for more than one candidate tire.

(viii) Prepare two candidate tires of the same construction type, manufacturer, line, and size designation in accordance with paragraph (f) (2) (i) of this section, mount them on the test apparatus, and test one of them according to the procedures of paragraph (f) (2) (ii) through (v) of this section, except load each tire to 85% of the test load specified in § 575.104(h).

(ix) Compute a candidate tire's adjusted traction coefficient for asphalt (μ_a) by the following formula:

μ_a = Measured candidate tire coefficient for asphalt + 0.50

− Measured standard tire coefficient for asphalt

(x) Compute a candidate tire's adjusted traction coefficient for concrete (μ_c) by the following formula:

μ_c = Measured candidate tire coefficient for concrete + 0.35

− Measured standard tire coefficient for concrete

(g) *Temperature resistance grading.* (1) Mount the tire on a rim of design or measuring rim width specified for tires of its size in accordance with 49 CFR 571.109, § 4.4.1(a) or (b) CFR 571.109, § 4.4.1(a) or (b) and inflate it to the applicable pressure specified in Table 1 of this section.

(2) Condition the tire-rim assembly to any temperature up to 95°F for at least 3 hours.

(3) Adjust the pressure again to the applicable pressure specified in Table 1 of this section.

(4) Mount the tire-rim assembly on an axle, and press the tire tread against the surface of a flat-faced steel test wheel that is 67.23 inches in diameter and at least as wide as the section width of the tire.

(5) During the test, including the pressure measurements specified in paragraphs (g) (1) and (g) (3) of this section, maintain the temperature of the ambient air, as measured 12 inches from the edge of the rim flange at any point on the circumference on either side of the tire at any temperature up to 95°F. Locate the temperature sensor so that its readings are not affected by heat radiation, drafts, variations in the temperature of the surrounding air, or guards or other devices.

(6) [Press the tire against the test wheel with a load of 88 percent of the tire's maximum load rating as marked on the tire sidewall. (48 F.R. 8929—March 9, 1984. Effective: July 1, 1984)]

(7) Rotate the test wheel at 250 rpm for 2 hours.

(8) Remove the load, allow the tire to cool to 95°F or for 2 hours, whichever occurs last, and readjust the inflation pressure to the applicable pressure specified in Table 1 of this section.

Table 1.—Test Inflation Pressures

Maximum permissible inflation pressure	32 lb/in²	36 lb/in²	40 lb/in²	240 kPa	280 kPa	300 kPa
Pressure to be used in tests for treadwear treadwear and in determination of tire load for temperature resistance testing	24	28	32	180	220	180
Pressure to be used for all aspects of aspects of temperature resistance testing other than determination of tire load	30	34	38	220	260	220

(9) Reapply the load and without interruption or readjustment of inflation pressure, rotate the test wheel at 375 rpm for 30 minutes, and then at successively higher rates in 25 rpm increments, each for 30 minutes, until the tire has run at 575 rpm for 30 minutes, or to failure, whichever occurs first.

(h) *Determination of test load.* [(1) To determine test loads for purposes of paragraphs (e) (2) (iii) and (f) (2) (viii), follow the procedure set forth in paragraphs (h) (2) through (5) of this section. (48 F.R. 8929—March 9, 1984. Effective: July 1, 1984)]

(2) Determine the tire's maximum inflation pressure and maximum load rating both as specified on the tire's sidewall.

(3) Determine the appropriate multiplier corresponding to the tire's maximum inflation pressure, as set forth in Table 2.

(4) Multiply the tire's maximum load rating by the multiplier determined in paragraph (3). This is the tire's calculated load.

(5) Round the product determined in paragraph (4) (the calculated load) to the nearest

multiple of ten pounds or, if metric units are used, 5 kilograms. For example, 903 pounds would be rounded to 900 and 533 kilograms would be rounded to 535. This figure is the test load.

TABLE 2*

Maximum inflation pressure	Multiplier to be used for treadwear testing	Multiplier to be used for traction testing
32 psi	.851	.851
36 psi	.870	.797
40 psi	.883	.753
240 psi	.866	.866
280 psi	.887	.804
300 psi	.866	.866

* NOTE: Prior to July 1, 1984, the multipliers in the above table are not to be used in determining loads for the tire size designations listed below in Table 2A. For those designations, the load specifications in that table shall be used in UTQG testing during that period. These loads are the actual loads at which testing shall be conducted and should not be multiplied by the 85 percent factors specified for treadwear and traction testing.

Table 2A

Tire Size Designation	Temp Resistance Max Pressure			Traction	Treadwear Max Pressure		
	32	36	40		32	36	40
145/70 R13	615	650	685	523	523	553	582
155/70 R13	705	740	780	599	599	629	663
165/70 R13	795	835	880	676	676	710	748
175/70 R13	890	935	980	757	757	795	833
185/70 R13	990	1040	1090	842	842	884	926
195/70 R13	1100	1155	1210	935	935	982	1029
155/70 R14	740	780	815	629	629	663	693
175/70 R14	925	975	1025	786	786	829	871
185/70 R14	1045	1100	1155	888	888	935	982
195/70 R14	1155	1220	1280	982	982	1037	1088
155/70 R15	770	810	850	655	655	689	723
175/70 R15	990	1040	1090	842	842	884	927
185/70 R15	1100	1155	1210	935	935	982	1029
5.60–13	725	810	880	616	616	689	748
[5.20–14	695	785	855	591	591	667	727
165–15	915	1015	1105	779	779	863	939
185/60 R13	845	915	980	719	719	778	833

(48 F.R. 8929—March 9, 1984. Effective: July 1, 1984)]

[(i) *Effective dates for treadwear grading requirements for radial tires.*

(1) Treadwear labeling requirements of §575.104 (d)(1)(i)(B)(2) apply to tires manufactured on or after April 1, 1985.

(2) Requirements for NHTSA review of treadwear information in consumer brochures, as specified in paragraph 575.6(d)(2), are effective April 1, 1985.

(3) Treadwear consumer information brochure requirements of paragraph 575.6(c) are effective May 1, 1985.

(6) Treadwear sidewall molding requirements of §575.104(d)(1)(i)(A) apply to tires manufactured on or after September 1, 1985.

(j) *Effective dates for treadwear grading requirements for bias ply tires.*

(1) Treadwear labeling requirements of §575.104 (d)(1)(i)(B)(2) apply to tires manufactured on or after December 15, 1984.

(2) Requirements for NHTSA review of treadwear information in consumer brochures, as specified in paragraph 575.6(d)(2), are effective December 15, 1984.

(3) Treadwear consumer information brochure requirements of paragraph 575.6(c) are effective January 15, 1985.

(4) Treadwear sidewall molding requirements of §575.104(d)(1)(i)(A) apply to tires manufactured on or after May 15, 1985.

(k) *Effective dates for treadwear grading requirements for bias belted tires.*

(1) Treadwear labeling requirements of §575.104 (d)(1)(i)(B)(2) apply to tires manufactured on or after March 1, 1985.

(2) Requirements for NHTSA review of treadwear information in consumer brochures, as specified in paragraph 575.6(d)(2), are effective March 1, 1985.

(3) Treadwear consumer information brochure requirements of paragraph 575.6(c) are effective April 1, 1985.

(4) Treadwear sidewall molding requirements of §575.104(d)(1)(i)(A) apply to tires manufactured on or after August 1, 1985.

(l) *Effective date for treadwear information requirements for vehicle manufacturers.*

Vehicle manufacturer treadwear information requirements of §§575.6(a) and 575.104(d)(1)(iii) are effective September 1, 1985. (49 F.R. 49293—

December 19, 1984. Effective: see Preamble to Docket No. 25; Notice 58)]

§ 575.105   Utility Vehicles

(a) *Purpose and scope.* This section requires manufacturers of utility vehicles to alert drivers that the particular handling and manuvering characteristics of utility vehicles require special driving practices when those vehicles are operated on paved roads.

(b) *Application.* This section applies to multipurpose passenger vehicles (other than those which are passenger car derivatives) which have a wheelbase of 110 inches or less and special features for occasional off-road operation ("Utility vehicles").

(c) *Required information.* Each manufacturer shall prepare and affix a vehicle sticker as specified in paragraph 1 of this subsection and shall provide in the vehicle Owner's Manual the information specified in paragraph 2 of this subsection.

   (1) A sticker shall be permanently affixed to the instrument panel, windshield frame, driver's side sun visor, or in some other location in each vehicle prominent and visible to the driver. The sticker shall be printed in a typeface and color which are clear and conspicuous. The sticker shall have the following or similar language:

This is a multipurpose passenger vehicle which will handle and maneuver differently from an ordinary passenger car, in driving conditions which may occur on streets and highways and off road. As with other vehicles of this type, if you make sharp turns or abrupt maneuvers, the vehicle may rollover or may go out of control and crash. You should read driving guidelines and instructions in the Owner's Manual, and WEAR YOU SEATBELTS AT ALL TIMES.

The language on the sticker required by paragraph (1) and in the Owner's Manual, as required in paragraph (2), may be modified as is desired by the manufacturer to make it appropriate for a specific vehicle design, to ensure that consumers are adequately informed concerning the unique propensities of a particular vehicle model.

(2) (i) The vehicle Owner's Manual shall include the following statement in its introduction.

As with other vehicles of this type, failure to operate this vehicle correctly may result in loss of control or an accident. Be sure to read "on-pavement" and "off-road" driving guidelines which follow.

(i) The vehicle Owner's Manual shall include tbeifollowing or similar statement:

Utility vehicles have higher ground clearance and a narrower track to make them capable of performing in a wide variety of off-road applications. Specific design characteristics give them a higher center of gravity than ordinary cars. An advantage of the higher ground clearance is a better view of the road allowing you to anticipate problems. They are not designed for cornering at the same speeds as conventional 2-wheel drive vehicles any more than low-slung sports cars are designed to perform satisfactorily under off-road conditions. If at all possible, avoid sharp turns or abrupt maneuvers. As with other vehicles of this type, failure to operate this vehicle correctly may result in loss of control or vehicle rollover.

§ 575.106   Deleted

34 F.R. 8112
May 23, 1969

APPENDIX A

Treadwear Test Course and
Driving Procedures

INTRODUCTION

The test course consists of three loops of a total of 400 miles in the geographical vicinity of Goodfellow AFB, San Angelo, Texas.

The first loop runs south 143 miles through the cities of Eldorado, Sonora, and Juno, Texas, to the Camp Hudson Historical Marker, and returns by the same route.

The second loop runs east over Farm and Ranch Roads (FM) and returns to its starting point.

The third loop runs northwest to Water Valley, northeast toward Robert Lee and returns via Texas 208 to the vicinity of Goodfellow AFB.

ROUTE

The route is shown in Figure 3. The table identifies key points by number. These numbers are encircled in Figure 3 and in parentheses in the descriptive material that follows.

Southern Loop

The course begins at the intersection (1) of Ft. McKavitt Road and Paint Rock Road (FM 388) at the northwest corner of Goodfellow AFB.

Drive east via FM 388 to junction with Loop Road 306 (2). Turn right onto Loop Road 306 and proceed south to junction with US 277 (3). Turn onto US 277 and proceed south through Eldorado and Sonora (4), continuing on US 277 to junction with FM 189 (5). Turn right onto FM 189 and proceed to junction with Texas 163 (6). Turn left onto Texas 163, proceed south to Camp Hudson Historical Marker (7) and onto the paved shoulder. Reverse route to junction of Loop Road 306 and FM 388 (2).

Eastern Loop

From junction of Loop Road 306 and FM 388 (2) make right turn onto FM 388 and drive east to junction with FM 2334 (13). Turn right onto FM 2334 and proceed south across FM 765 (14) to junction of FM 2334 and US 87 (15). Make U-turn and return to junction of FM 388 and Loop Road 306 (2) by the same route.

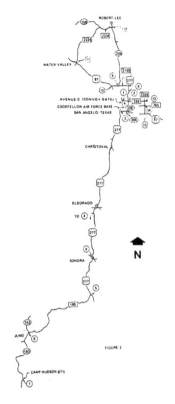

FIGURE 3

Northwestern Loop

From junction of Loop Road 306 and FM 388 (2), make right turn onto Loop Road 306. Proceed onto US 277, to junction with FM 2105(8). Turn left onto FM 2105 and proceed west to junction with US 87 (10). Turn right on US 87 and proceed northwest to the junction with FM 2034 near the town of Water Valley (11). Turn right onto FM 2034 and proceed north to Texas 208 (12). Turn right onto Texas 208 and proceed south to junction with FM 2105 (9). Turn left onto FM 2105 and proceed east to junction with US 277 (8). Turn right onto US 277 and proceed south onto 306 to junction with 388 (2). Turn right onto 388 and proceed to starting point at junction of Ft. McKavitt Road and FM 388 (1).

DRIVING INSTRUCTIONS

The drivers shall run at posted speed limits throughout the course unless an unsafe condition arises. If such condition arises, the speed should be reduced to the maximum safe operating speed.

BRAKING PROCEDURES AT STOP SIGNS

There are a number of intersections at which stops are required. At each of these intersections a series of signs is placed in a fixed order as follows:

Sign Legend

Highway Intersection 1000 (or 2000) Feet

S T O P   A H E A D

Junction X X X

Direction Sign (Mereta →)

S T O P   or   Y I E L D

PROCEDURES

1. Approach each intersection at the posted speed limit.

2. When abreast of the S T O P   A H E A D sign, apply the brakes so that the vehicle decelerates smoothly to 20 mph when abreast of the direction sign.

3. Come to a complete stop at the S T O P sign or behind any vehicle already stopped.

KEY POINTS ALONG TREADWEAR
TEST COURSE, APPROX. MILEAGES,
AND REMARKS

		Mileages	Remarks
1	Ft. McKavitt Road & FM 388	0	
2	FM 388 & Loop 306 ..	3	STOP
3	Loop 306 & US 277 ..	10	
4	Sonora	72	
5	US 277 & FM 189 ...	88	
6	FM 189 & Texas 163 .	124	
7	Historical Marker ... (Camp Hudson)	143	U-TURN
4	Sonora	214	
3	Loop 306 & US 277 ..	276	
2	FM 388 & Loop 306 .	283	
13	FM 388 & FM 2334 ..	290	STOP
14	FM 2334 & FM 765 ..	292	STOP
15	FM 2334 & US 87 ...	295	U-TURN
14	FM 2334 & FM 765 ..	298	STOP
13	FM 388 & FM 2334 ..	300	STOP/YIELD/ BLINKING RED LIGHT
2	FM 388 & Loop 306 .	307	STOP/YIELD/ BLINKING RED LIGHT
8	US 277 & FM 2105 ..	313	
9	FM 2105 & Texas 208	317	STOP
10	FM 2105 & US 87 ...	320	STOP
11	FM 2034 & US 87 ...	338	
12	FM 2034 & Texas 208	362	YIELD
9	FM 2105 & Texas 208	387	
8	FM 2105 & US 277 ..	391	YIELD/STOP
2	FM 388 & Loop 306 .	397	
1	Ft. McKavitt Road & FM 388	400	

APPENDIX B

Traction Skid Pads

Two skid pads have been laid on an unused runway and taxi strip on Goodfellow AFB. Their location is shown in Figure 4.

The asphalt skid pad is 600 ft x 60 ft and is shown in black on the runway in Figure 4. The pad is approached from either end by a 75 ft ramp followed by 100 ft. of level pavement. This arrangement permits the skid trailers to stabilize before reaching the test area. The approaches are shown on the figure by the hash-marked area.

The concrete pad is 600 ft x 48 ft and is on the taxi strip. The approaches to the concrete pad are of the same design as those for the asphalt pads.

A two lane asphalt road has been built to connect the runway and taxi strip. The road is parallel to the northeast-southwest runway at a distance of 100 ft. The curves have super-elevation to permit safe exit from the runway at operating speeds.

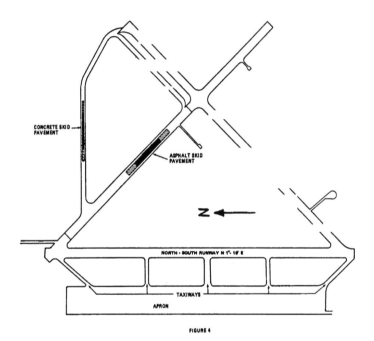

CONCRETE SKID PAVEMENT

ASPHALT SKID PAVEMENT

NORTH - SOUTH RUNWAY N 1°- 18′ E

TAXIWAYS

APRON

FIGURE 4

PART 575-19-20

APPENDIX C

Method Of Least Squares

The method of least squares is a method of calculation by which it is possible to obtain a reliable estimate of a true physical relationship from a set of data which involve random error. The method may be used to establish a regression line that minimizes the sum of the squares of the deviations of the measured data points from the line. The regression line is consequently described as the line of "best fit" to the data points. It is described in terms of its slope and its "y" intercept.

The graph in Figure 5 depicts a regression line calculated using the least squares method from data collected from a hypothetical tread-wear test of 6,400 miles, with tread depth measurements made at every 800 miles.

In this graph, (x_j, y_j) $[j = 0, 1, \ldots 8]$ are the individual data points representing the tread depth measurements (the overall average for the tire with 6 measurements in each tire groove) at the beginning of the test (after break-in and at the end of each 800-mile segment of the test.

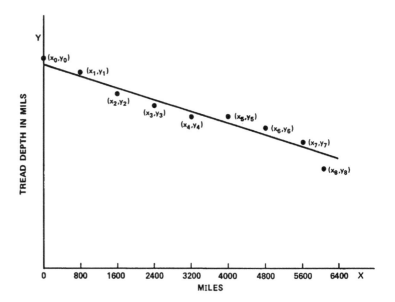

Figure 5

The absolute value of the slope of the regression line is an expression of the mils of tread worn per 1,000 miles, and is calculated by the following formula:

$$b = 1000 \; \frac{\left(\sum_{j=0}^{8} X_j Y_j - \frac{1}{9} \sum_{j=0}^{8} X_j \sum_{j=0}^{8} Y_j \right)}{\sum_{j=0}^{8} X_j^2 - \frac{1}{9} \left(\sum_{j=0}^{8} X_j \right)^2}$$

The "y" intercept of the regression line (a) in mils is calculated by the following formula:

$$a = \frac{1}{9} \sum_{j=0}^{8} Y_j - \frac{b}{9000} \sum_{j=0}^{8} X_j$$

PREAMBLE TO PART 576—RECORD RETENTION
(Docket No. 74–31; Notice 1)

This notice establishes an immediate temporary requirement for retention by motor vehicle manufacturers of records concerning malfunctions that may be related to motor vehicle safety.

By a separate notice published today, 39 FR 30048, the NHTSA proposes to establish permanent requirements for the retention of records by manufacturers. The proposed rule would require motor vehicle manufacturers to retain for 5 years all records in their possession relating to failures, malfunctions, or flaws that could be a causative factor in accidents or injuries. These records are needed in agency investigations of possible defects related to motor vehicle safety, or of nonconformity to the safety standards and regulations. A fuller discussion of the proposal is contained in that notice.

The NHTSA finds it important that existing records and those that may be generated or acquired while this rulemaking is under consideration not be disposed of prior to the permanent effectiveness of the rule. In order to maintain the status quo, therefore, this rule is issued to be effective immediately upon posting for public inspection at the *Federal Register*. For the reasons stated, pursuant to 5 U.S.C. 553(b), notice and public procedure thereon with respect to this interim notice are found to be impracticable and contrary to the public interest. This rule in its present form will be effective only until action is taken upon the proposed permanent rule issued concurrently.

In light of the foregoing, a new Part 576, *Record Retention*, is added to Title 49, Code of Federal Regulations.

Effective date: August 15, 1974.

AUTHORITY: Sec. 108, 112, 113, 119, Pub. L. 89–563, 80 Stat. 718, 15 U.S.C. 1397, 1401, 1402, 1407; delegation of authority at 49 CFR 1.51.

Issued on August 13, 1974.

James B. Gregory
Administrator
39 F.R. 30045
August 20, 1974

PART 576—RECORD RETENTION

(Docket No. 74-13; Notice 1)

Sec.

576.1   Scope.

576.2   Purpose.

576.3   Application.

576.4   Definitions.

576.5   Basic Requirement.

576.6   Records.

576.7   Retention.

576.8   Malfunctions Covered.

§ 576.1 Scope. This part establishes requirements for the retention by motor vehicle manufacturers of complaints, reports, and other records concerning motor vehicle malfunctions that may be related to motor vehicle safety.

§ 576.2 Purpose. The purpose of this part is to preserve records that are needed for the proper investigation, and adjudication or other disposition, of possible defects related to motor vehicle safety and instances of nonconformity to the motor vehicle safety standards and associated regulations.

§ 576.3 Application. This part applies to all manufacturers of motor vehicles, with respect to all records generated or acquired after August 15, 1969.

§ 576.4 Definitions. All terms in this part that are defined in the Act are used as defined therein.

§ 576.5 Basic Requirements. Each manufacturer of motor vehicles shall retain as specified in § 576.7 all records described in § 576.6 for a period of 5 years from the date on which they were generated or acquired by the manufacturer.

§ 576.6 Records. Records to be retained by manufacturers under this part include all documentary materials, films, tapes, and other information-storing media that contain information concerning malfunctions that may be related to motor vehicle safety. Such records include, but are not limited to, communications from vehicle users and memoranda of user complaints; reports and other documents related to work performed under, or claims made under, warranties; service reports or similar documents from dealers or manufacturer's field personnel; and any lists, compilations, analyses, or discussions of such malfunctions contained in internal or external correspondence of the manufacturer.

§ 576.7 Retention. Duplicate copies need not be retained. Information may be reproduced or transferred from one storage medium to another (*e.g.*, from paper files to microfilm) as long as no information is lost in the reproduction or transfer, and when so reproduced or transferred the original form may be treated as a duplicate.

§ 576.8 Malfunctions covered. For purposes of this part, "malfunctions that may be related to motor vehicle safety" shall include, with respect to a motor vehicle or item of motor vehicle equipment, any failure or malfunction beyond normal deterioration in use, or any failure of performance, or any flaw or unintended deviation from design specifications, that could in any reasonably foreseeable manner be a causative factor in, or aggravate, an accident or an injury to a person.

39 F.R. 30045
August 20, 1974

PREAMBLE TO PART 577—DEFECT NOTIFICATION

(Docket No. 72-7; Notice 2)

This notice establishes a new regulation covering notifications of motor vehicle safety defects and nonconformity to safety standards. The notice proposing these regulations was published May 17, 1972 (37 F.R. 9783).

The regulation is intended to improve the response of owners in vehicle notification campaigns. Data which the NHTSA has been receiving on the completion rates of notification campaigns show a wide range of completion rates, with campaigns involving newer vehicles, and more serious safety problems, having higher completion rates than others. In many campaigns, however, the rate is alarmingly low.

An examination of the notifications sent by manufacturers reveals wide disparity in emphasis. Although precise evaluation of the impact of notification letters is difficult, due to its being largely subjective, the NHTSA is of the opinion that many notifications have tended to deemphasize the safety problems involved. Some of these notification letters are questionably within the requirements of the National Traffic and Motor Vehicle Safety Act, and litigation on a case by case basis to improve them is practicable. These regulations are intended to ensure that all notification letters contain sufficient information, as determined by NHTSA, to properly notify purchasers.

The regulation applies to manufacturers of incomplete and complete motor vehicles, and motor vehicle equipment. In the case of vehicles manufactured in two or more stages, compliance by any one of the manufacturers of the vehicle is considered compliance by all. This provision is based on similar language in the Defect Reports regulation (Part 573 of this chapter), and is included in response to comments received.

The regulation requires the notification to contain substantially the information specified in the proposal. It requires each notification to begin with a statement that it is sent pursuant to the requirements of the National Traffic and Motor Vehicle Safety Act. The NHTSA did not concur with comments to the effect that the inclusion of this statement would not promote the purpose of the regulation. The regulation requires the notification to state that the manufacturer, or the National Highway Traffic Safety Administrator, as the case may be, has determined that a defect relating to motor vehicle safety (or a noncompliance with a motor vehicle safety standard) exists in the vehicle type, or item of motor vehicle equipment, with which the notification is concerned. When the manufacturer (or the Administrator) has, as part of his determination, also found that the defect may not exist in each such vehicle or equipment item, he may include a statement to that effect. The NHTSA has decided to allow such statements based on comments that many defects in fact do not exist in each vehicle or equipment item of the group whose owners are notified.

The manufacturer must also describe the defect, evaluate the risk it poses to traffic safety, and specify measures which the recipient should take to have it remedied. In each case, the regulation requires information which the NHTSA has determined will meet these objectives. In describing the defect, the manufacturer must indicate the vehicle system or particular items of equipment affected, describe the malfunction that may occur, including operating conditions that may cause it to occur, and precautions the purchaser should take to reduce the likelihood of its occurrence. In providing that the vehicle system affected be mentioned, the regulation reflects comments to the effect that listing each particular part involved would be too technical to be useful to most consumers.

In evaluating the risk to traffic safety, the manufacturer must indicate if vehicle crash is the potential result, and whatever warning may occur. Where vehicle crash is not the potential result, the manufacturer must indicate the general type of injury which the defect can cause. Although many comments protested that it was impossible to predict a specific type of injury, the NHTSA believes that manufacturers can easily foresee the general type of injury, such as asphyxiation, that can result from those defects which are not expected to result in crashes.

In stating measures to be taken to repair the defect, the requirements differ in the case where the manufacturer's dealers repair the vehicle free of charge to the purchaser, where the manufacturer merely offers to pay for the repair, and where he refuses to pay for the repair. The purpose of this distinction is to provide information sufficient to have adequate repairs made in each case.

Where the manufacturer's dealers repair the vehicle free of charge, the notification must include a general description of the work involved, the manufacturer's estimate of when his dealers will be supplied with parts and instructions, and his estimate of the time reasonably necessary to perform the labor involved in correcting the defect. The agency's position is that consumers are entitled to know approximately when their cars will be repaired and how much labor is needed in order for the repair to be made. The NHTSA realizes that dealers frequently retain vehicles longer than the actual work involved, due to difficulties in scheduling repairs. However, manufacturers are free to impart this information to consumers under the regulation. Some comments objected to requiring manufacturers to provide information on when replacement parts will be available, on the basis that manufacturers cannot know, at the time a notification is issued, precisely when parts deliveries will be made to dealers. To include this information, it is argued, would therefore delay the issuance of the notification. The NHTSA has modified the proposed language to allow manufacturers to "estimate" when corrective parts will be available. The estimate would be based on the manufacturer's knowledge at the time the notification is sent, thereby eliminating any reasons for delay.

When manufacturers do not provide for repairs to be made by dealers, the notification is required to contain, in addition, full lists of parts and complete instructions on making the repairs. The regulation also requires the manufacturer to recommend, generally, where the vehicle should be repaired, and manufacturers are free to make general and specific recommendations. This requirement reflects the intent of the proposal that manufacturers who believe particular repairs may require special expertise should indicate that fact to purchasers.

When the manufacturer does not offer to pay for repairs, he must, in addition, include full cost information on necessary parts. The notice would have required the retail cost of all parts, and information on labor charges of the manufacturer's dealers in the general area of the purchaser. In response to comments, the cost information is limited to the suggested retail price of parts. Manufacturers have indicated they do not set actual prices of parts, but do have suggested list prices. With respect to labor charges, manufacturers have indicated that labor charges vary, and that requiring them to ascertain exact charges would delay issuance of notifications. The NHTSA believes these comments to be well-founded, and has dropped the proposed requirements regarding labor charges. Consumers will still have information on costs of parts, and time necessary for repairs to be performed, from which they can obtain a fair idea of the cost of a repair.

The regulations prohibit the notification from stating or implying that the problem is not a defect, or that it does not relate to motor vehicle safety. Moreover, in those cases where the notification is sent pursuant to the direction of the Administrator, it cannot state or imply that the manufacturer disagrees with the Administrator's finding. Many comments opposed these requirements on the basis that they unconstitutionally limited manufacturers' freedom of speech. The NHTSA emphatically rejects this contention. Notification letters are not intended to serve as forums where manufacturers can argue that problems are not safety-related or dispute the Administration's findings. Their purpose is to unambiguously and adequately induce owners to remedy a potentially hazardous situation. The

NHTSA is of the opinion that there is ample precedent that allows the Federal government to require manufacturers to warn purchasers in a particular manner that certain products they manufacture may be hazardous. If a manufacturer does not believe that his condition is a safety-related defect, he is not required by law to notify owners at all. It is only when he determines that a defect exists that he must notify in accordance with the regulations. Similarly, when the Administrator has made the finding that a certain product is defective, the manufacturer can administratively and judicially challenge this determination as provided in the National Traffic and Motor Vehicle Safety Act before sending a notification.

The NHTSA received other objections to the proposed requirements. Numerous tire manufacturers argued that parts of the regulation dealing with repairs of defects are inappropriate when applied to them, since repairs generally meant replacement. Certain manufacturers of lighting equipment argued that notification requirements should not apply to them at all. The NHTSA disagrees with both of these contentions. In the case of tire manufacturers, the NHTSA believes that the requirements can be followed. If the repair of a defective tire entails its replacement, this can certainly be stated within the regulatory scheme. Similarly, lighting equipment manufacturers are responsible for defects to the same extent as manufacturers of other equipment. The NHTSA rejects completely the argument that no lighting failures can be considered safety-related because of the millions of lights that burn out every year without resulting in accidents. The question in each case is not whether a failure may occur, but whether a defect exists, and whether the defect may cause a hazardous situation to arise.

The notice of proposed rulemaking would have prohibited manufacturers from making statements contemporaneous with the notification that disagreed with its conclusions. This proposal has not been adopted. After careful consideration, the NHTSA has determined that its inclusion is probably unnecessary. The agency's position is that if notification letters clearly and unambiguously describe and evaluate defects in accordance with this regulation, other statements by manufacturers will not normally affect reactions of consumers.

Certain comments requested that manufacturers be allowed to state in the notification that it does not constitute an admission of liability or wrongdoing. The regulation does not preclude the making of such statements, as the agency has concluded that their inclusion will not significantly deter owners from having repairs made.

One comment suggested that the notification be required to contain a postage-free card by which consumers could notify manufacturers when vehicles had been sold or otherwise disposed of. While the NHTSA believes this practice would be advantageous in improving notification campaigns, it has concluded that such a requirement would be outside the scope of the regulation, which is limited to notifications to first purchasers and warranty holders.

Certain comments objected to the regulations on the ground that they prescribed a rigid format in an area where each case must be treated separately, and thus where flexibility was required. The NHTSA has modified to some extent the proposed restrictions on format. Manufacturers are free, within the limits established, to compose notifications to fit each case. As issued, these regulations do not require rigid, inflexible letters (only the first two sentences must contain specific statements in a set order), but require that manufacturers include certain important items of information. It is hoped that manufacturers in meeting these requirements will provide required information in easily understandable form.

In light of the above, a new Part 577, "Defect Notification" is added to Chapter V of Title 49, Code of Federal Regulations, to read as set forth as below.

Effective date: March 26, 1973. Because these requirements are not technical in nature, and do not require lead times for compliance, good cause exists, and is hereby found, for an effective date less than 180 days from the day of issuance.

Issued on January 17, 1973.

Douglas Toms
Administrator

38 F.R. 2215
January 23, 1973

PREAMBLE TO AMENDMENT TO PART 577—DEFECT NOTIFICATION

(Docket No. 72-7; Notice 3)

This notice responds to petitions for reconsideration of the Defect Notification regulations, published January 23, 1973 (38 FR 2215). Petitions were received from the Firestone Tire and Rubber Company, Chrysler Corporation, the Motor and Equipment Manufacturers' Association, and the Recreational Vehicle Institute. A petition was also received from the Wagner Electric Company. Although not received within 30 days of the regulation's publication (49 CFR 553.35), it has been considered in the preparation of this notice. Insofar as this notice does not grant the requests of the petitioners, they are hereby denied.

The Firestone Tire and Rubber Company has petitioned for reconsideration of section 577.6, "Disclaimers", which prohibits manufacturers from starting or implying that the notification does not involve a safety related defect. Firestone requested that the provision, for Federal Constitutional reasons, be dropped from the rule. This request is denied. The NHTSA does not believe, for the reasons set forth in the notice of January 23, 1973 (38 FR at 2216), that the provision is violative of the Constitution.

Chrysler Corporation has requested that the phrase, "his dealers" be modified in section 577.4-(e)(1)(ii), which requires the manufacturer to estimate the date by which his dealers will be supplied with corrective parts and instructions. It argues that the phrase "his dealers" could be interpreted to mean all dealers, regardless of whether all of the manufacturer's dealers are involved in the campaign. This request is denied. Neither section 113 of the Safety Act nor the regulation require a notification campaign to extend to all of the manufacturer's dealers, whether or not they have any involvement in a particular campaign. The NHTSA does not believe that the phrase "his dealers", when read in context, means all of the manufacturer's dealers.

Chrysler also asks that special requirements be specified for the notification of "noncompliance non-operational defects", citing as an example the improper placement of the VIN plate under Motor Vehicle Safety Standard No. 115. Chrysler states that existing provisions of the regulation dealing with malfunctions (specifically 577.4-(c)(2), (c)(3), (c)(4)), and evaluating the risk to traffic safety (sections 577.4(d), (d)(1), (d)(1)(i), (d)(1)(ii), (d)(2)) are not pertinent to these defects. This request is denied. The NHTSA does not believe that separate requirements for notification of the type of defect described by Chrysler are either necessary or desirable. If a particular defect does not involve a malfunction, to be in compliance with the regulation a manufacturer should, in response to the appropriate provisions of the regulation, indicate that to be the case. The NHTSA believes this approach will notify purchasers of the defect as effectively as separate, more specific requirements. The NHTSA does not agree that the relationship to safety of these types of defects should not be evaluated in notification letters, similarly to other defects.

The Motor and Equipment Manufacturers Association (MEMA objects to the requirements of sections 577.4(e)(2) (vi) and 577.4(e)(3)(vi) that the manufacturer recommend whom the purchaser should have perform necessary repair work, and requests that these provisions be deleted. MEMA argues that the requirement is anti-competitive in that it sanctions the steering of consumers to vehicle dealerships for repairs, to the detriment of the independent repair industry, even when the manufacturer does not pay for the repair. MEMA argues that original equipment replacement parts are frequently more expensive than competitively produced parts, resulting in added costs to owners. It argues also that limiting repairs to dealers precludes the use of the full domestic repair industry, which should be utilized fully given the magnitude of recent notification campaigns.

While the NHTSA appreciates the concern of this association in not being precluded from a large market, the NHTSA believes the requirement as issued to be consistent with the National Traffic and Motor Vehicle Safety Act and the need for motor vehicle safety. The NHTSA has, in issuing the requirement, indicated that manufacturers should indicate to purchasers when special expertise may be necessary to correct defects. The repairs in issue do not involve normal maintenance, but constitute defects whose proper repair is essential to the safety of the nation's highways. Frequently these repairs involve a higher degree of expertise and familiarity with a particular vehicle than that required to perform normal maintenance. If such expertise will more likely be found at dealerships, in the view of the vehicle manufacturer, the NHTSA believes that opinion should be imparted to purchasers.

Moreover, even if the NHTSA deleted the requirement the manufacturer could if he desired, consistently with the regulation, recommend a repair facility. The NHTSA would not prohibit the making of such a recommendation, for it is responsive to the statutory requirement that the notification contain a statement of the measures to be taken to repair the defect (15 U.S.C. 1402(c)). Moreover, the argument that the regulation stifles competition does not appear to have merit. In the event the manufacturer does not bear the cost of repair, the regulation (§ 577.4(e)(3)(i)) requires the manufacturer to provide the purchaser with the suggested list price of repair parts. As a consequence, purchasers will be provided with information with which they can "shop", with full knowledge, for the least expensive repair facilities. The petition is accordingly denied.

The Recreational Vehicle Institute (RVI) has petitioned that the requirements of both section 577.4(a), requiring an opening statement that the notification is sent pursuant to the Act, and section 577.6, prohibiting disclaimers, be deleted. RVI argues such requirements may result in delay by manufacturers in determining that defects

exist, forcing the use of administrative and legal procedures before purchasers are notified. The agency cannot accept the position that the notification should be diluted because of possible evasion by manufacturers. The NHTSA believes that the need that notification letters fully inform purchasers outweighs the possible problems caused by manufacturers delaying their notifications to purchasers until forced to notify them. The request is denied.

RVI points out that section 577.4 seems to assume that defects will be evidenced by some form of mechanical failure. It asks, therefore, whether a safety-related defect can exist where proper corrective action to avoid an occurrence or possible occurrence is appropriate maintenance or operational use. RVI also requests, if NHTSA adheres to its present position regarding these issues, that it undertake rulemaking to define "safety related defect". For the following reasons, these requests are denied. There is no intent in the regulation to limit the concept of safety related defects to those involving mechanical failures. As stated above, in reply to the petition from Chrysler, non-mechanical defects can be the basis of defect notification, and purchasers can be fully notified of them under the present regulatory scheme. Moreover, the NHTSA believes any attempt to precisely define safety related defect would be ill-advised. Whether a defect exists depends solely on the facts of each particular situation. The fact that such determinations may encompass a wide variety of factual situations, and may consequently be difficult to make, does not mean that it is necessary, desirable, or even possible to replace the decision with a simple formula. The NHTSA believes, on the contrary, that the relatively broad definition of defect contained in the Safety Act is best suited to the wide variety of defective conditions that may arise.

RVI has also pointed out that references to a manufacturer's dealers in section 577.4(e), specifying measures to be taken to repair the defect, overlook the fact that manufacturers' dealers may not always provide service facilities, or that manufacturers may use service facilities other than dealers. The NHTSA agrees with RVI, and has therefore modified the provisions of that section to include "other service facilities of the manufacturer", as well as his dealers.

RVI requested that the regulation be amended to permit compliance by either a component manufacturer or a vehicle manufacturer, when the defect involves a specific component. RVI also requested that compliance be permitted by either the vehicle alterer or the complete vehicle manufacturer in cases involving altered vehicles. The regulations do not prohibit the sending of notification letters by persons other than the vehicle manufacturer. Accordingly, no modification of the regulation is called for. However, manufacturers who do utilize the services of others in meeting requirements still bear the ultimate responsibility for compliance with the regulation under the National Traffic and Motor Vehicle Safety Act.

The Wagner Electric Company has requested that the provisions of the regulation regarding manufacturers of motor vehicle equipment (excluding tires) be reconsidered in light of the fact that, under present marketing procedures, it is difficult or impossible for such manufacturers to notify jobbers, installers, dealers, or consumers. The notification required by the regulation is directed at the notification sent to retail purchasers and not that sent to distributors or dealers of the manufacturer. The notification of the latter is subject only to the statutory provision of section 113 of the Safety Act (15 U.S.C. 1402). Moreover, manufacturers of equipment (other than tires) who do not have the names of first purchasers are not required to notify them either under the National Traffic and Motor Vehicle Safety Act or the regulation. There is consequently no need for modification of the regulation for the reasons presented by Wagner, and its request is accordingly denied.

In light of the above, Part 577 of Title 49, Code of Federal Regulations, "Defect Notification", is amended . . . .

Effective date: April 17, 1973. These amendments impose no additional burdens on any person, and serve only to clarify the application of existing requirements to specific situations. Accordingly, notice and public procedure thereon are unnecessary, and good cause exists for an effective date less than thirty days from the day of publication.

(Sec. 108, 112, 113, 119, Pub. L. 89–563, 80 Stat. 718 as amended, sec. 2, 4, Pub. L. 91–265, 84 Stat. 262 (15 U.S.C. 1397, 1401, 1402, 1408); delegation of authority at 49 CFR 1.51)

Issued on April 10, 1973.

James E. Wilson
Acting Administrator

38 F.R. 9509
April 17, 1973

PREAMBLE TO AMENDMENT TO PART 577—DEFECT NOTIFICATION

(Docket No. 74–42; Notice 2)

This notice amends 49 CFR Part 577, *Defect Notification*, to require that bilingual notification be sent to owners in certain cases, and to clarify the wording manufacturers are required to use to indicate their determination that a safety-related defect exists.

A notice of proposed rulemaking on this subject was published on November 25, 1974, (39 F.R. 41182) and an opportunity afforded for comment. The Center for Auto Safety had questioned the efficacy of defect notification campaigns in Puerto Rico conducted in the English language since the primary language of that Commonwealth is Spanish. A National Highway Traffic Safety Administration (NHTSA) survey in Puerto Rico confirmed that there was a need for bilingual defect notification. It was proposed that whenever the address of the purchaser is in either the Commonwealth of Puerto Rico or the Canal Zone the notification be sent in both the English and Spanish languages.

The notice also proposed clarifying § 577.4(e)(1) so that the second paragraph of a notification letter could no longer be written to reflect a manufacturer's belief that the cause of a defect is an item other than that which he manufactured.

Only Chrysler Corporation and Firestone Tire and Rubber Company commented on bilingual notification. Both stated that it was not necessary for the Canal Zone. Firestone also felt that the requirement to translate the notification would delay its mailing, and voiced the belief that NHTSA must express the exact wording in Spanish for § 577.4(a) and (b). Chrysler commented that it had been providing bilingual notification to owners of automobiles purchased in Puerto Rico but that extensive and burdensome data-processing reprogramming would be required to identify owners of vehicles originally purchased on the mainland and later taken to Puerto Rico.

The NHTSA believes that the language problem is a significant factor in the below-average response to notification campaigns in Puerto Rico, and that owner response rate to campaigns in the Canal Zone will improve if notifications are provided in Spanish as well as English. Information from the Census Bureau indicates that more than 50% of the residents of each area speak Spanish as their primary language. Translation may delay mailing to these areas a few days, but this is deemed inconsequential compared with the benefits to be derived by an improved response to campaigns. This agency does not consider that it need specify the exact wording in Spanish of § 577.4(a) and (b). If it appears that manufacturers are providing ambiguous statements it will consider the matter further. Finally, since section 153(a)(1) of the National Traffic and Motor Vehicle Safety Act, 15 U.S.C. 1413(a)(1), requires notification to be sent to the person who is registered under State law as the owner of the vehicle to be campaigned, Chrysler's comments on reprogramming of data do not appear to have merit.

This notice also amends § 577.4(b)(1), which presently requires the second sentence of the notification to state that the manufacturer has determined that a defect which relates to motor vehicle safety exists in its motor vehicles or motor vehicle equipment. Certain notification letters have characterized the defect as existing in a vehicle or item of equipment not manufactured by the manufacturer making the determination. The intent of the section is that a manufacturer of motor vehicles would state its determination that the defect exists in the motor vehicle it manufactures, while a manufacturer of motor vehicle equipment would state its determination that the defect exists in the motor vehicle equipment it manufactures. If the manufacturer believes the cause of the defect to be an item other than that which he manufactured, that information can be imparted in the other parts of the notification, but not in the second paragraph where the content is specifically prescribed.

Kelsey-Hayes Company and Skyline Corporation commented on the proposal to clarify § 577.4 (b) (1). Both objected to it, feeling that the present regulation is adequate and that the mandatory statement may be prejudicial. However, in the opinion of this agency, manufacturers with limited experience in composing notification letters have in many cases misinterpreted § 577.4 (b) (1). Clarification of the sentence should eliminate mistakes.

In consideration of the foregoing, Part 577 of Title 49, Code of Federal Regulations, *Defect Notification*, is amended. . . .

Effective date: September 14, 1975.

(Sec. 108, 112, 113, 119, Pub. L. 89–563, 80 Stat. 718; sec. 2, 4, Pub. L. 91–265, 84 Stat. 262 (15 U.S.C. 1397, 1401, 1402, 1407); delegation of authority at 49 CFR 1.51.)

Issued on June 10, 1975.

James B. Gregory
Administrator

**40 F.R. 25463
June 16, 1975**

PREAMBLE TO AMENDMENT TO PART 577—DEFECT NOTIFICATION

(Docket No. 75–10; Notice 2)

This notice amends 49 CFR Part 577, "Defect Notification," to conform to §§ 151 through 160 of the National Traffic and Motor Vehicle Safety Act (the Act) (Pub. L. 93–492, 88 Stat. 1470, October 27, 1974; 15 U.S.C. 1411–1420).

The amendments of Part 577 were published as a notice of proposed rulemaking in the *Federal Register* on May 6, 1975 (40 FR 19651). Approximately 30 comments were received from vehicle and equipment manufacturers, equipment distributors, trade associations representing these groups, and the Center for Auto Safety. The National Motor Vehicle Safety Advisory Council did not take a position on this proposal. Interested persons are advised that NHTSA Dockets 75–30 (Defect and Noncompliance Responsibility), 75–31 (Petitions for Hearing on Notification and Remedy of Defects or Failure to Comply), and 74–7 (Defect and Noncompliance Reporting) are relevant to the subject matter of this rulemaking.

The agency is amending its earlier notification procedures to reflect the major expansion of manufacturer responsibilities under the Motor Vehicle and Schoolbus Safety Amendments of 1974 to notify vehicle and equipment owners or purchasers of noncompliances with safety standards and of defects that relate to motor vehicle safety (hereinafter referred to as defects), chief of which is that remedy shall be without charge in most cases.

The new regulation specifies the content, timing, and form of notification that complies with the requirements set forth in § 153 of the Act. Distinctions among notifications that arise under different circumstances are set forth in detail. Provisions concerning disclaimers in the notification and conformity to the statutory requirements are carried over from the former Part 577.

Comments on the proposal were generally in agreement with the revision of the regulation, in recognition that the revision reflects responsibilities already a matter of law. Several questions were raised with regard to the authority for or wisdom of specific provisions of the proposed regulation, and these are discussed below.

Motor vehicle manufacturers and the Motor Vehicle Manufacturers Association (MVMA) expressed strong support for modification of the statutory definitions of "original equipment" and "replacement equipment" that allocate responsibility for notification and remedy between vehicle and equipment manufacturers. The agency has issued a separate proposal to redistribute responsibility (40 FR 56930, December 5, 1975) which addresses the issues raised. Resolution of that proposal will be responsive to the issues raised by the MVMA and vehicle manufacturers. To simplify any future action in this area, the two terms are no longer set forth in Part 577.

In the definitions section of the regulation, the phrase "in good faith" has been added to the definition of "first purchaser" to conform to its meaning under § 108(b)(1) of the Act.

The Recreational Vehicle Industry Association (RVIA) requested that vehicle alterers be permitted to meet (assume) the obligations of manufacturers for notification and remedy on a voluntary basis. Without notice and opportunity for comment on this idea, the agency does not consider it wise to modify the regulation as suggested by the RVIA.

NOTIFICATION PURSUANT TO A MANUFACTURER'S DETERMINATION

Section 151 of the Act provides that a manufacturer who determines in good faith that a defect or noncompliance exists in its products "shall furnish notification to the Secretary and to owners, purchasers, and dealers in accordance with section 153, and he shall remedy the defect or failure to comply in accordance with section 154."

Section 577.5 of Part 577 provides for manufacturer-initiated notifications in accordance with § 151. The section specifies, among other things, that a statement appear in the notification that the manufacturer has determined that a defect or noncompliance exists in identified vehicles or equipment. An additional statement may be made to indicate that the problem may not exist in each such vehicle or item of equipment. The MVMA and American Motors Corporation (AMC) believed that a better approach would be to state that the defect or noncompliance exists in some, but not all, vehicles or items of equipment (if such is the case), and that an owner should bring his vehicle in for inspection in any case. The agency does not believe that either the MVMA or AMC has an expertise in this area and declines to adopt the suggested modification.

Paragraph (e) of § 577.5 requires a clear description of the defect or noncompliance, including, among other things,

(e) ***

(2) A description of any malfunction that may occur. The description of a noncompliance with an applicable standard shall include the difference between the performance of the noncomplying vehicle or item of replacement equipment and the performance specified by the standard;

The MVMA viewed the phrase "any malfunction" as overbroad and ambiguous, in that a manufacturer would be held to correctly anticipate a malfunction, whether or not related to safety or the noncompliance. The agency agrees that such a description would go beyond the purpose of the notification and therefore has narrowed somewhat the language proposed.

Vehicle manufacturers and the MVMA argued that the second sentence of paragraph (e)(2) should be deleted because an exact description of the difference in performance due to noncompliance would be too technical for comprehension by most owners, require extensive and expensive

testing in some cases that would delay notification, and be the basis for a technical violation of the regulation. The agency believes that the description is valuable to vehicle or equipment owners in understanding the noncompliance, but agrees that a detailed description could delay notification unnecessarily. Accordingly, the phrase "in general terms" is added to modify the required description.

The Center for Auto Safety (the Center) believed that the statement required by (e) to minimize the chances of an accident before remedy failed to mention prior warnings that the vehicle's operating characteristics might offer. While prior warning is adequately covered by the "evaluation of risk" statement made regarding the possibility of vehicle crash (paragraph (f)(1)(ii)), the agency has added a comparable requirement to paragraph (f)(2) (that covers "non-crash" type defects and noncompliances).

The Specialty Equipment Manufacturers Association objected that any evaluation of the risk to motor vehicle safety would be speculative and therefore was unjustified. This requirement, however, is based on the specific requirement of § 153(a) of the Act, and cannot be eliminated.

The Center believed that the evaluation of risk to motor vehicle safety is a discretionary statement that need not be made by a manufacturer. This is not the case. Section 577.5 is a requirement that the information (b) through (g) be listed and, under paragraph (f), the evaluation must either describe the crash hazard or be a description of the "general type of injury to occupants, or [others], that can result."

Paragraph (g) of § 577.5, dealing with measures to be taken by the owner, proved to be the greatest source of comments on the proposal. The paragraph is divided into subparagraphs dealing with notification of remedy without charge and notification of remedy for which the manufacturer will charge. This distinction is based on § 154(a)(4) of the Act which limits the "remedy without charge" to vehicles or equipment first purchased no more than 8 years (3 years in the case of tires) before notification in accordance with §§ 151 or 152.

Paragraph (g)(1) specifies requirements both for notification when the remedy must be undertaken and also notification when the manufacturer voluntarily decides to remedy without charge. The MVMA and General Motors (GM) felt that manufacturers undertaking voluntary remedy should not be subjected to the same notification requirements as those manufacturers required to remedy. The agency distinguishes between the separate duties of notification and remedy, however, and notes that the notification requirements of § 153 of the Act contain no exceptions for older vehicles and equipment. The MVMA's abbreviated list of requirements for a voluntary remedy do not fulfill the requirements of § 153. For example, § 153(a)(2) requires that the notification contain an evaluation of the risk to motor vehicle safety.

It is the agency's philosophy that a manufacturer undertaking a remedy should provide the same information to the owner whether or not the remedy is undertaken voluntarily. In this way, an owner will be apprised of the information necessary to make informed decision. Also, events beyond the manufacturer's control will not be able to negate the remedy without agency or manufacturer's knowledge. For these reasons, the agency does not modify the requirements as suggested.

Aside from the general suitability of paragraph (g)(1)'s requirements for a voluntary remedy, manufacturers raised more specific questions about the separate provisions.

International Harvester Company (IH) asserted with regard to paragraph (g)(1)(i) that no basis existed for the exception of replacement equipment from the right to refund as a means of remedy. In the agency's view, § 154(2)(B) of the Act clearly limits the remedy for items of replacement equipment to either repair or replacement.

IH objected to the requirements that the earliest date for repair set under paragraph (g)(1)(ii) be premised on anticipated receipt by dealers of necessary parts for repair. The company pointed out that some repair parts would not typically be forwarded to a dealer for repair until a specific request has arisen. The agency would like to clarify that the "earliest date" can be established as a certain number of days following inspection of the defective or noncomply-

ing vehicle. Thus a manufacturer need only calculate the time that it would take to get the parts to the dealer following an inspection and then state that the earliest date for repair will follow the date of inspection by that amount.

AMC argued that the requirement for a general description of the work and amount of time involved in a repair without charge by the manufacturer's dealer exceeded the authority of the Act and is unnecessary when the manufacturer undertakes repair. The same argument was made with regard to paragraphs (g)(1)(v) and (vi). The agency disagrees, and notes that the specific authority listed in § 153(a) is "in addition to such other matters as the Secretary may prescribe by regulation." As for the need for a general description, it is concluded that the owner would value knowledge of the time involved and the nature of the repair that is involved, to correctly weigh the gravity of the problem. Correspondingly, the offer of replacement or refund is more helpful to the owner if it includes the detail that has been specified.

In paragraph (g)(1)(iv), the MVMA asked for parallelism with the construction of paragraph (g)(1)(iii). It is accomplished by the addition of "or its dealers" following the word "manufacturer." IH suggested the addition of 'authorized service centers" to the list, but this is unnecessary in view of the NHTSA's interpretation of "dealer" to include an authorized service center.

The Center, Mack Trucks, and Crane Carrier Corporation (CCC) commented on paragraph (g)(1)(iv)'s requirement that the method or basis for a manufacturer's assessment of depreciation be specified. The two manufacturers suggested use of a retailer's price guide as the basis. The Center suggested that a method for determination of depreciation be devised by a panel of industry, government, and consumer representatives. The legislative history indicates that retailer price guides should not be the sole criterion, and thus the Mack and CCC recommendations are not adopted. Until there is some indication that the manufacturers' chosen methods of assessment are unreasonable, the agency does not consider it necessary to exercise its authority in this area, and the Center's suggestion is also not adopted.

The greatest objections were raised regarding the statement advising an owner how to inform the NHTSA if he believes that the notification or remedy is inadequate, or that the remedy was untimely or not made in accordance with the notification. PACCAR, AMC, Chrysler, GM, IH, the RVIA, and the MVMA considered the statement to be, in some respects, beyond the agency's statutory authority and not contemplated by Congress. As earlier noted, § 153 is prefaced by a general grant of authority to the agency to specify the contents of the notification.

The agency has considered the objections, in any case, particularly in view of the decision to require the same notification in the case of voluntary and mandatory remedy notices. It is concluded that modification of the statements to reflect the exact terms of § 154(a)(6) is appropriate.

Manufacturers objected to the language of paragraph (g)(1)(vii)(C) that invites owner complaints if a remedy is not effected within a reasonable period. The agency considers timeliness to be an aspect of whether a manufacturer has failed or is unable to provide a remedy as specified in § 153(a)(6) of the Act. The agency does agree that remedy by replacement or refund should not be limited to the first 60 days, since it might follow a failure to repair within that 60-day period. In conforming to § 154(b)(1), the agency substitutes "tender" for "first attempt." Also reference to extension by the Administrator of the 60-day repair period has been added to paragraph (g)(1)(vii)(C)(1).

GM suggested that an additional statement be made to owners, advising them of recourse available with the manufacturer if the dealer's response is unsatisfactory. The agency considers this desirable but, without the benefit of notice and opportunity for comment, declines to make this addition. Paragraph (g)(1), of course, only sets forth what the manufacturer "shall include" in its notification, and it may make such additional statements as it deems necessary.

There was no comment on the second part of § 577.5 that deals with manufacturer notices in which remedy without charge is not required and is not volunteered. Accordingly, the paragraph is adopted as proposed.

NOTIFICATION PURSUANT TO ADMINISTRATOR'S DETERMINATION

Section 577.6 provides for Administration-ordered notifications in accordance with § 152. Paragraphs (a), (b), and (c) set forth requirements for the three types of notification contemplated by the Act. Manufacturers made no comment on the requirements for notification ordered by the Administrator in the first instance, and paragraph (a) is accordingly made final as proposed.

PACCAR objected to provisional notification as placing an unreasonable burdén on the manufacturer, rendering any court decision in its favor meaningless. Section 155(b) of the Act clearly contemplates such an order, however, and the regulations consequently do provide for it.

Comments were received on the proposed content of the provisional notification. The MVMA pointed out that the requirement in paragraph (b)(2) should be clarified to permit a statement that the defect or non-compliance may not occur in all the described vehicles. The agency agrees and adds a paragraph similar to § 577.5(d).

With regard to the proposed paragraph (b)(4), the MVMA asked that reference to a "United States District Court" be broadened to "the Federal courts" and that the statement make clear that the NHTSA and not the court is ordering provisional notification. The agency concurs in these clarifications and they are made where appropriate in the final rule.

The requirements of paragraphs (b)(5), (6), and (7) provide for a description of the Administrator's determination, his evaluation of the hazard, and the recommended measures to avoid unreasonable hazard resulting from the defect or noncompliance. Fiat requested that the description, evaluation, and recommended measures be provided by the NHTSA. As specified in the requirements, it is the "Administrator's stated basis" that must be described, and the measures "stated in his order" that must be listed. The agency intends to include in each order a description, evaluation, and list of measures that permit quotation or paraphrase by the manufacturer.

Chrysler and the MVMA asked that a manufacturer be permitted more latitude to explain its position than provided for in paragraph (b)(8). The agency has considered this request, and concludes that extensive advocacy of the manufacturer's position would detract from the intent of the provisional notification to put the owner on notice of potential problems. The Chrysler and MVMA suggestion is therefore not adopted.

In the required statement dealing with availability of remedy and reimbursement in the event the court upholds the Administrator's determination (paragraph (b)(9)), Chrysler argued that the suggestion of reimbursement would generate poor customer relations if a repair were sought or undertaken during pendency of a court proceeding in which the manufacturer prevailed. The agency is aware of the possibility for some misunderstanding but is certain that the provisional notification was intended by the Congress to encourage owners to consider repair or other corrective action while the manufacturer contests the determination. For this reason, the notice of possible reimbursement remains in the regulation. The first statement in (b)(9)(i) has been clarified in one minor respect.

The MVMA requested that the phrase "for repair" be substituted for "in repairing" to permit manufacturers to make clear that reimbursement would only cover the repairs that were reasonable and necessary to correct the defect or noncompliance. The NHTSA believes that the term "reasonable and necessary" makes clear what repairs would be reimbursed should the court uphold an Administrator's determination.

The MVMA asked, and the agency agrees, that the reimbursement statement be qualified by the limitations that appear in the statute.

Paragraph (b)(10) requires a statement whether, in the manufacturer's opinion, a repair of the defect or noncompliance is possible. GM asked that "feasible" be substituted for "possible" and the agency makes the change in agreement that it more clearly reflects the judgement made by a manufacturer in choosing its preferred remedy. The MVMA and Chrysler made the more basic objection that (b)(10) assumes that a defect or noncompliance exists prior to the court's ruling, and that it requires unjustified effort to develop repair parts and facilities before a decision is reached on the validity of the Ad-

ministrator's determination. The agency is of the view that the level of detail specified is justified in these cases and necessary to fulfill the purpose of provisional notification contemplated by Congress. The agency has modified the wording to make clear that reimbursement for expenses are limited to those necessary and reasonable for repair.

With regard to proposed paragraph (b)(12), the MVMA asked that only notification and not remedy be mentioned. There will be a discussion of remedy in the notification, however, and the owner should be encouraged to inquire further as to this aspect of the notification.

Firestone and the Automotive Parts and Accessories Association felt that the regulations should apply to the agency and that it should be required to advise the owner, purchaser, and dealer in the event its determination is not upheld by the courts. The statutory scheme being implemented by Part 577 concerns manufacturer obligations under §§ 151 through 160 of the Act to notify and remedy safety problems in vehicles. The agency does not consider an expansion of the regulations beyond this purpose as appropriate. Nothing, of course, prevents the manufacturer from making such a notice to the owner or others.

Paragraph (c) of § 577.6 deals with final notification following a court decision in the Administrator's favor, and it is adopted, with corrections similar to those made in the other sections. Because the MVMA objected to reference to being "upheld in a proceeding in a United States District Court" as the basis for the post-litigation order, the agency has substituted the language of the Act. Also, reference to "a date" on which provisional notification was ordered is corrected to "the date" to reflect that it will in all cases be a specific date.

TIME AND MANNER OF NOTIFICATION

The major problem with regard to the time and manner of notification concerned the statutory requirement (§ 153(c)(1)) that notification be,

§ 153 * * *

(c) * * *

(1) in the case of a motor vehicle, by first class mail to each person who is registered

under State law as the owner of such vehicle and whose name and address is reasonably ascertainable by the manufacturer through State records or other sources available to him;

PACCAR, Volkswagen, and IH expressed their doubts that all State records would be available or that alternative services would provide timely information. The agency has incorporated the statutory requirements in this regulation word-for-word and, on that basis, declines to modify it. As for the suggestion that "reasonably ascertainable" be defined, it is the agency's view that the phrase is only given meaning by the separate factual situations that arise. The agency cannot agree with PACCAR that records are not "reasonably ascertainable" simply by virtue of delay in retrieving them.

Sheller-Globe Corporation asked if certified mail would be considered the equivalent of first class mail for meeting the requirements. As a school bus manufacturer, Sheller-Globe wanted certainty of notification to school districts and other customers. The NHTSA does not consider them equivalent in view of relevant legislative history. Congress considered the U.S. Postal Service regulation that prohibits forwarding of certified mail and they concluded that first class mail would be a superior means of obtaining notification.

With regard to the maximum times permitted for issuance of notification, the Center asked that the period be reduced to 30 days in the case of all Administration-ordered notifications. Some manufacturers asked that the 30-day period for provisional notification be expanded to 60 days. B.F. Goodrich stated that notification letters cannot be printed in advance of actual mailing, because the date for earliest remedy must be included in the letter. The agency has weighed the conflicting views, and concludes that a 60-day period is justified for administration-ordered recalls. The provisional notification requirement is amended accordingly.

IH suggested that public notice of defects or noncompliances in items of replacement equipment would be adequate, and that notice to the most recent purchaser should be optional. The

agency has simply conformed its regulation to the statutory requirements of § 153(c).

OTHER MATTERS

The MVA suggested that the disclaimer section of the regulation could be clarified by an additional paragraph permitting manufacturer statements that a notification does not "constitute an admission by the manufacturer that it has been guilty of negligence or other wrong doing." The agency views this statement as exactly the type of disclaimer that could contribute to a reader's decision not to take action in response to notification and accordingly declines to adopt the MVMA recommendation.

With regard to the MVMA concern that technical violations of the regulations not be pursued as a violation of the Act under § 577.9, the agency expects to continue to enforce the Act and its regulations in a reasonable manner, calculated to avoid arbitrariness or irrationality.

After-market equipment manufacturers and their associations expressed the view that the notification scheme was unworkable for notice to equipment purchasers, that wear of parts in normal use conflicted with the concept of safety-related defects, and that the 8-year period for remedy without charge was too long. Also, the establishment of a cut-off based on the date of retail sale appeared impractical, because records of these transactions are not maintained. As a response, the agency notes that the regulation conforms to the statute's language and clearly expressed Congressional intent. Experience to date with the requirements does not demonstrate that they are in fact unworkable. The issues of improper installation and remanufactured parts were not addressed by the statute, and resolution of these issues wil require some experience with situations as they arise.

The RVIA asked that the agency exercise its authority to require the submission to manufacturers by dealers of the names and addresses of purchasers. The agency takes this recommendation under advisement but, as it is beyond the scope of Part 577, does not act on it in this notice.

In consideration of the foregoing, Part 577, "Defect Notification," of Title 49, Code of Federal Regulations, is renamed "Defect and Non-compliance Notification" and is amended to read as set forth below.

Effective date: June 28, 1977.

(Secs. 108, 112, 119, Pub. L. 89–563, 80 Stat. 718; Sec. 102, 103, 104, Pub. L. 93–492, 88 Stat. 1470 (15 U.S.C. 1397, 1401, 1407, 1411–1420; delegation of authority at 49 CFR 1.50)

Issued on December 22, 1976.

John W. Snow
Administrator

41 F.R. 56813
December 30, 1976

PREAMBLE TO AN AMENDMENT TO PART 577

Defect and Noncompliance Notification
(Docket No. 80-17; Notice 1)

ACTION: Final rule.

SUMMARY: This notice amends the defect and non-compliance notification regulation to require that manufacturers include the agency's toll free Auto Safety Hotline number in their defect and noncompliance notification letters. The amendment is being made to provide a means of easy access to the agency by consumers who may have complaints about the recall and remedy of their vehicles or equipment. Since it is a minor technical amendment, it is being made effective immediately without notice or opportunity for comment.

EFFECTIVE DATE: January 22, 1981.

FOR FURTHER INFORMATION CONTACT:
Mr. James Murray, Office of Defects Investigation, National Highway Traffic Safety Administration, 400 Seventh Street, S.W., Washington, D.C. 20590, 202-426-2840

SUPPLEMENTARY INFORMATION: This notice makes a minor technical amendment to Part 577, Defect and Noncompliance Notification, to require manufacturers conducting recall campaigns to include the agency's toll free Auto Safety Hotline number in the notification letters.

Existing notification letters are required to state that a consumer may contact the agency if he or she feels that remedy of a defect or noncompliance is not being made without charge or in a reasonable time. Manufacturers also frequently include their address and a toll free number that consumers can call to complain to the manufacturer about the status of a remedy. The agency believes that the use of manufacturer toll free numbers is a good idea and has decided that the agency's toll free number should also be included in the letter. This will provide easy access for consumers to the agency for reporting any complaints concerning the recall or remedy of their vehicles. It also will provide timely information to our Enforcement office pertaining to the compliance with our regulations by the manufacturers.

Since this is a minor technical amendment and will result in little impact upon manufacturers, the agency finds for good cause shown that it is in the interest of safety to make the amendment effective immediately without notice and opportunity for comment.

In consideration of the foregoing, Title 49 of the Code of Federal Regulations, Part 577, Defect and Noncompliance Notification, is amended by revising the introductory sentence in paragraph 577.5(g)(1)(vii) to read as follows:

(vii) A statement informing the owner that he or she may submit a complaint to the Administrator, National Highway Traffic Safety Administration, 400 Seventh Street, S.W., Washington, D.C. 20590 or call the toll free Auto Safety Hotline at 800-426-9393 (Washington, D.C. area residents may call 426-0123), if the owner believes that—

*   *   *   *

The principal authors of this notice are Mr. James Murray of the Office of Defects Investigations and Roger Tilton of the Office of Chief Counsel.

Issued on January 14, 1981.

Joan Claybrook
Administrator

46 FR 6971
January 22, 1981

PART 577—DEFECT AND NONCOMPLIANCE NOTIFICATION

(Docket No. 72-7; Notice 2)

Sec.
577.1 Scope.
577.2 Purpose.
577.3 Application.
577.4 Definitions.
577.5 Notification pursuant to a manufacturer's determination.
577.6 Notification pursuant to the Administrator's determination.
577.7 Time and manner of notification.
577.8 Disclaimers.
577.9 Conformity to statutory requirements.

AUTHORITY: Secs. 108, 112, 119, Pub. L. 89-563; 80 Stat. 718; Secs. 102, 103, 104, Pub. L. 93-492; 88 Stat. 1470 (15 U.S.C. 1397, 1401, 1408, 1411-1420; delegations of authority at 49 CFR 1.51 and 49 CFR 501.8)

§ 577.1 Scope.

This part sets forth requirements for notification to owners of motor vehicles and replacement equipment about the possibility of a defect which relates to motor vehicle safety or a noncompliance with a Federal motor vehicle safety standard.

§ 577.2 Purpose.

The purpose of this part is to ensure that notifications of defects or noncompliances adequately inform and effectively motivate owners of potentially defective or noncomplying motor vehicles or items of replacement equipment to have such vehicles or equipment inspected and, when necessary, remedied as quickly as possible.

§ 577.3 Application.

This part applies to manufacturers of completed motor vehicles, incomplete motor vehicles, and replacement equipment. In the case of vehicles manufactured in two or more stages, compliance by either the manufacturer of the incomplete vehicle, any subsequent manufacturer, or the manufacturer of affected replacement equipment shall be considered compliance by each of those manufacturers.

§ 577.4 Definitions.

For purposes of this part:

"Act" means the National Traffic and Motor Vehicle Safety Act of 1966, as amended, 15 U.S.C. 1391 et seq.

"Administrator" means the Administrator of the National Highway Traffic Safety Administration or his delegate.

"First purchaser" means the first purchaser in good faith for a purpose other than resale.

"Owners" include purchaser.

§ 577.5 Notification pursuant to a manufacturer's determination.

(a) When a manufacturer of motor vehicles or replacement equipment determines that any motor vehicle or item of replacement equipment produced by him contains a defect which relates to motor vehicle safety, or fails to conform to an applicable Federal motor vehicle safety standard, he shall provide notification in accordance with paragraph (a) of § 577.7, unless the manufacturer is exempted by the Administrator (pursuant to section 157 of the Act) from giving such notification. The notification shall contain the information specified in this section. The information required by paragraphs (b) and (c) of this section shall be presented in the form and order specified. The information required by paragraphs (d) through (g) of this section may be presented in any order. Notification sent to an owner whose address is in either the Commonwealth of Puerto Rico or the Canal Zone shall be written in both English and Spanish.

(b) An opening statement: "This notice is sent to you in accordance with the requirements of the National Traffic and Motor Vehicle Safety Act."

(c) Whichever of the following statements is appropriate:

(1) "(Manufacturer's name or division) has determined that a defect which relates to motor vehicle safety exists in (identified motor vehicles, in the case of notification sent by a motor vehicle manufacturer; identified replacement equipment, in the case of notification sent by a replacement equipment manufacturer);" or

(2) "(Manufacturer's name or division) has determined that (identified motor vehicles, in the case of notification sent by a motor vehicle manufacturer; identified replacement equipment, in the case of notification sent by a replacement equipment manufacturer) fail to conform to Federal Motor Vehicle Safety Standard No. (number and title of standard)."

(d) When the manufacturer determines that the defect or noncompliance may not exist in each vehicle or item of replacement equipment, he may include an additional statement to that effect.

(e) A clear description of the defect or noncompliance, which shall include—

(1) An identification of the vehicle system or particular item(s) of motor vehicle equipment affected.

(2) A description of the malfunction that may occur as a result of the defect or noncompliance. The description of a noncompliance with an applicable standard shall include, in general terms, the difference between the performance of the noncomplying vehicle or item of replacement equipment and the performance specified by the standard;

(3) A statement of any operating or other conditions that may cause the malfunction to occur; and

(4) A statement of the precautions, if any, that the owner should take to reduce the chance that the malfunction will occur before the defect or noncompliance is remedied.

(f) An evaluation of the risk to motor vehicle safety reasonably related to the defect or noncompliance.

(1) When vehicle crash is a potential occurrence, the evaluation shall include whichever of the following is appropriate:

(i) A statement that the defect or noncompliance can cause vehicle crash without prior warning; or

(ii) A description of whatever prior warning may occur, and a statement that if this warning is not heeded, vehicle crash can occur.

(2) When vehicle crash is not a potential occurrence, the evaluation must include a statement indicating the general type of injury to occupants of the vehicle, or to persons outside the vehicle, that can result from the defect or noncompliance, and a description of whatever prior warning may occur.

(g) A statement of measures to be taken to remedy the defect or noncompliance, in accordance with paragraph (g)(1) or (g)(2) of this section, whichever is appropriate.

(1) When the manufacturer is required by the Act to remedy the defect or noncompliance without charge, or when he will voluntarily so remedy in full conformity with the Act, he shall include—

(i) A statement that he will cause such defect or noncompliance to be remedied without charge, and whether such remedy will be by repair, replacement, or (except in the case of replacement equipment) refund, less depreciation, of the purchase price.

(ii) The earliest date on which the defect or noncompliance will be remedied without charge. In the case of remedy by repair, this date shall be the earliest date on which the manufacturer reasonably expects that dealers or other service facilites will receive necessary parts and instructions. The manufacturer shall specify the last date, if any, on which he will remedy tires without charge.

(iii) In the case of remedy by repair through the manufacturer's dealers or other service facilities:

(A) A general description of the work involved in repairing the defect or noncompliance; and

(B) The manufacturer's estimate of the time reasonably necessary to perform the labor required to correct the defect or noncompliance.

(iv) In the case of remedy by repair through service facilities other than those of the manufacturer or its dealers:

(A) The name and part number of each part that must be added, replaced, or modified;

(B) A description of any modifications that must be made to existing parts which shall also be identified by name and part number;

(C) Information as to where needed parts will be available;

(D) A detailed description (including appropriate illustrations) of each step required to correct the defect or noncompliance;

(E) The manufacturer's estimate of the time reasonably necessary to perform the labor required to correct the defect or noncompliance; and

(F) The manufacturer's recommendations of service facilities where the owner should have the repairs performed.

(v) In the case of remedy by replacement, a description of the motor vehicle or item of replacement equipment that the manufacturer will provide as a replacement for the defective or noncomplying vehicle or equipment.

(vi) In the case of remedy by refund of purchase price, the method or basis for the manufacturer's assessment of depreciation.

(vii) A statement informing the owner that he or she may submit a complaint to the Administrator, National Highway Traffic Safety Administration, 400 Seventh Street, S.W., Washington, D.C. 20590 or call the toll-free Auto Safety Hotline at 800-424-9393 (Washington D.C. area residents may call 426-0123), if the owner believes that—

(A) The manufacturer, distributor, or dealer has failed or is unable to remedy the defect or noncompliance without charge.

(B) The manufacturer has failed or is unable to remedy the defect or noncompliance without charge—

(1) (In the case of motor vehicles or items of replacement equipment, other than tires) within a reasonable time, which is not longer than 60 days in the case of repair after the owner's first tender to obtain repair following the earliest repair date specified in the notification, unless the period is extended by the Administrator.

(2) (In the case of tires) after the date specified in the notification on which replacement tires will be available.

(2) When the manufacturer is not required to remedy the defect or noncompliance without charge and he will not voluntarily so remedy, the statement shall include—

(i) A statement that the manufacturer is not required by the Act to remedy without charge.

(ii) A statement of the extent to which the manufacturer will voluntarily remedy, including the method of remedy and any limitations and conditions imposed by the manufacturer on such remedy.

(iii) The manufacturer's opinion whether the defect or noncompliance can be remedied by repair. If the manufacturer believes that repair is possible, the statement shall include the information specified in paragraph (g) (1) (iv) of this section, except that—

(A) The statement required by paragraph (g) (1) (iv) (A) of this section shall also indicate the suggested list price of each part.

(B) The statement required by paragraph (G) (1) (iv) (C) of this section shall also indicate the manufacturer's estimate of the date on which the parts will be generally available.

§ 577.6 Notification pursuant to Administrator's determination.

(a) *Manufacturer-ordered-notification.* When a manufacturer is ordered pursuant to section 152 of the Act to provide notification of a defect or noncompliance, he shall provide such notification in accordance with §§ 577.5 and 577.7, except that the statement required by paragraph (c) of § 577.5 shall indicate that the determination has been made by the Administrator of the National Highway Traffic Safety Administration.

(b) *Provisional notification.* When a manufacturer does not provide notification as required by paragraph (a) of this section, and an action concerning the Administrator's order to provide such notification has been filed in a United States District Court, the manufacturer shall, upon the Administrator's further order, provide in accordance with paragraph (b) of § 577.7 a provisional notification containing the information specified in this paragraph, in the order and, where specified, the form of paragraphs (b)(1) through (b)(12) of this section.

(1) An opening statement: "This notice is sent to you in accordance with the requirements of the National Traffic and Motor Vehicle Safety Act."

(2) Whichever of the following statements is appropriate:

(i) "The Administrator of the National Highway Traffic Safety Administration has determined that a defect which relates to motor vehicle safety exists in (identified motor vehicles, in the case of notification sent by a motor vehicle manufacturer; identified replacement equipment, in the case of notification sent by a replacement equipment manufacturer);" or

(ii) "The Administrator of the National Highway Traffic Safety Administration has determined that (identified motor vehicles, in the case of notification sent by a motor vehicle manufacturer; identified replacement equipment, in the case of notification sent by a replacement equipment manufacturer) fail to conform to Federal Vehicle Safety Standard No. (number and title of standard)."

(3) When the Administrator determines that the defect or noncompliance may not exist in each such vehicle or item of replacement equipment, the manufacturer may include an additional statement to that effect.

(4) The statement: "(Manufacturer's name or division) is contesting this determination in a proceeding in the Federal courts and has been required to issue this notice pending the outcome of the court proceeding."

(5) A clear description of the Administrator's stated basis for his determination, as provided in this order, including a brief summary of the evidence and reasoning that the Administrator relied upon in making his determination.

(6) A clear description of the Administrator's stated evaluation as provided in his order of the risk to motor vehicle safety reasonably related to the defect or noncompliance.

(7) Any measures that the Administrator has stated in his order should be taken by the owner to avoid an unreasonable hazard resulting from the defect or noncompliance.

(8) A brief summary of the evidence and reasoning upon which the manufacturer relies in contesting the Administrator's determination.

(9) A statement regarding the availability of remedy and reimbursement in accordance with paragraph 9(i) or 9(ii) below, whichever is appropriate.

(i) When the purchase date of the vehicle or item of equipment is such that the manufacturer is required by the Act to remedy without charge or to reimburse the owner for reasonable and necessary repair expenses, he shall include—

(A) A statement that the remedy will be provided without charge to the owner if the Court upholds the Administrator's determination.

(B) A statement of the method of remedy. If the manufacturer has not yet determined the method of remedy, he shall indicate that he will select either repair, replacement with an equivalent vehicle or item of replacement equipment, or (except in the case of replacement equipment) refund, less depreciation, of the purchase price; and

(C) A statement that, if the Court upholds the Administrator's determination, he will reimburse the owner for any reasonable and necessary expenses that the owner incurs (not in excess of any amount specified by the Administrator) in repairing the defect or noncompliance following a date, specified by the manufacturer, which shall not be later than the date of the Administrator's order to issue this notification.

(ii) When the manufacturer is not required either to remedy without charge or to reimburse, he shall include—

(A) A statement that he is not required to remedy or reimburse, or

(B) A statement of the extent to which he will voluntarily remedy or reimburse, including the method of remedy if then known, and any limitations and conditions on such remedy or reimbursement.

(10) A statement indicating whether, in the manufacturers opinion, the defect or noncompliance can be remedied by repair. When the manufacturer believes that such remedy is feasible, the statement shall include:

(i) A general description of the work and the manufacturer's estimate of the costs involved in repairing the defect or noncompliance;

(ii) Information on where needed parts and instructions for repairing the defect or noncompliance will be available, including the manufacturer's estimate of the day on which they will be generally available;

(iii) The manufacturer's estimate of the time reasonably necessary to perform the labor required to correct the defect or noncompliance; and

(iv) The manufacturer's recommendations of service facilities where the owner could have the repairs performed, including (in the case of a manufacturer required to reimburse if the Administrator's determination is upheld in the court proceeding) at least one service facility for whose charges the owner will be fully reimbursed if the Administrator's determination is upheld.

(11) A statement that further notice wil be mailed by the manufacturer to the owner if the Administrator's determination is upheld in the court proceeding; and

(12) An address of the manufacturer where the owner may write to obtain additional information regarding the notification and remedy.

(c) *Post-litigation notification.* When a manufacturer does not provide notification as required in paragraph (a) of this section and the Administrator prevails in an action commenced with respect to such notification, the manufacturer shall, upon the Administrator's further order, provide notification in accordance with paragraph (b) of § 577.7 containing the information specified in paragraph (a) of this section, except that—

(1) The statement required by paragraph (c) of § 577.5 shall indicate that the determination has been made by the Administrator and that his determination has been upheld in a proceeding in the Federal courts; and

(2) When a provisional notification was issued regarding the defect or noncompliance and the manufacturer is required under the Act to reimburse—

(i) The manufacturer shall state that he will reimburse the owner for any reasonable and necessary expenses that the owner incurred (not in excess of any amount specified by the Administrator) for repair of the defect or noncompliance of the vehicle or item of equipment on or after the date on which provisional notification was ordered to be issued and on or before a date not sooner than the date on which this notification is received by the owner. The manufacturer shall determine and specify both dates.

(ii) The statement required by paragraph (g) (1) (vii) of § 577.5 shall also inform the owner that he may submit a complaint to the Administrator if the owner believes that the manufacturer has failed to reimburse adequately.

PART 577-5

(3) If the manufacturer is not required under the Act to reimburse, he shall include—

(i) A statement that he is not required to reimburse, or

(ii) When he will voluntarily reimburse, a statement of the extent to which he will do so, including any limitations and conditions on such reimbursement.

§ 577.7 Time and manner of notification.

(a) The notification required by § 577.5 shall—

(1) Be furnished within a reasonable time after the manufacturer first determines the existence of a defect which relates to motor vehicle safety, or of a noncompliance.

(2) Be accomplished—

(i) In the case of a notification required to be sent by a motor vehicle manufacturer, by first class mail to each person who is registered under State law as the owner of the vehicle and whose name and address are reasonably ascertainable by the manufacturer through State records or other sources available to him. If the owner cannot be reasonably ascertained, the manufacturer shall notify the most recent purchaser known to the manufacturer.

(ii) In the case of a notification required to be sent by a replacement equipment manufacturer—

(A) By first class mail to the most recent purchaser known to the manufacturer, and

(B) (Except in the case of a tire) if determined by the Administrator to be necessary for motor vehcile safety, by public notice in such manner as the Administrator may determine after consultation with the manufacturer.

(iii) In the case of a manufacturer required to provide notification concerning any defective or noncomplying tire, by first class or certified mail.

(b) The notification required by any paragraph of § 577.6 shall be provided:

(1) Within 60 days after the manufacturer's receipt of the Administrator's order to provide the notification, except that the notification shall be furnished within a shorter or longer period if the Administrator incorporates in his order a finding that such period is in the public interest; and

(2) In the manner and to the recipients specified in paragraph (a) of this section.

§ 577.8 Disclaimers.

(a) A notification sent pursuant to § 577.5 or § 577.6 regarding a defect which relates to motor vehicle safety shall not, except as specifically provided in this part, contain any statement or implication that there is no defect, that the defect does not relate to motor vehicle safety, or that the defect is not present in the owner's vehicle or item of replacement equipment.

(b) A notification sent pursuant to § 577.5 or § 577.6 regarding a noncompliance with an applicable Federal motor vehicle safety standard shall not, except as specifically provided in this part, contain any statement or implication that there is not a noncompliance or that the noncompliance is not present in the owner's vehicle or item of replacement equipment.

§ 577.9 Conformity to statutory requirements.

A notification that does not conform to the requirements of this part is a violation of the Act.

38 F.R. 2215
January 23, 1973

PREAMBLE TO PART 579—DEFECT AND NONCOMPLIANCE RESPONSIBILITY

(Docket No. 75–30; Notice 2)

This notice issues a new regulation, Part 579, *Defect and Noncompliance Responsibility.* The purpose of the regulation is to allocate between motor vehicle and equipment manufacturers the responsibilities under the 1974 Motor Vehicle and Schoolbus Safety Amendments for recalling and remedying defective or noncomplying motor vehicles and equipment. The regulation makes tire manufacturers responsible for original equipment tires as well as tires sold as replacement equipment. Otherwise, the regulation adopts the responsibility scheme in the 1974 Amendments. With this notice, the agency defers final action on its proposal concerning the responsibilities of original equipment manufacturers that supply equipment to five or more vehicle manufacturers. Effective date: September 30, 1978.

Addresses: Petitions for reconsideration should refer to the docket number and be submitted to: Room 5108, Nassif Building, 400 Seventh Street, S.W., Washington, D.C. 20590.

For further information contact:

Mr. James Murray, Office of Defects Investigation, National Highway Traffic Safety Administration, 400 Seventh Street, S.W., Washington, D.C. 20590 (202–426–2840).

This notice issues a new regulation, Part 579, *Defect and Noncompliance Responsibility.* A notice of proposed rulemaking was published on December 5, 1975 (40 F.R. 56930) proposing some reallocation between motor vehicle and equipment manufacturers of the responsibilities for safety-related defects and noncompliances with safety standards. These responsibilities include the duty to notify purchasers of any safety-related defects or noncompliances with safety standards and to make remedy without charge to the purchaser. Currently, the allocation of defect and noncompliance responsibility is governed by section 159(2) of the National Traffic

and Motor Vehicle Safety Act of 1966, as amended, (the Act) (15 U.S.C. 1419(2)).

The Act authorizes the agency to allocate equitably responsibility for defects and noncompliances between equipment and vehicle manufacturers. The substance of the agency's 1975 NPRM was to shift the burdens of compliance somewhat from the vehicle to the equipment manufacturer. As the NPRM on this issue stated, the legislative history of the Act indicates that the Congress intended for the agency to ensure that its defect and noncompliance regulations reflect the realities of the relationship between equipment and vehicle manufacturers.

Comments were received from equipment and vehicle manufacturers and from their representatives. All comments were considered. The Vehicle Equipment Safety Commission did not submit comments.

General Motors Corporation suggested that section 579.1 be changed to indicate that the regulation applies only to Part B of the Act, Discovery, notification, and remedy of motor vehicle defects, not to Part A, General provisions. Since this regulation exercises the authority granted by section 159 of the Act and that section specifically states that it applies only to Part B of the Act, the agency has incorporated GM's recommended change.

The Midland Ross Corporation suggested that the agency add several minor definitions to the list of definitions. They suggested, for example, that the agency define phrases such as "an item of motor vehicle equipment," and "an item of defective or noncomplying equipment."

With respect to "motor vehicle equipment," the agency notes that the term is defined in the Act at section 102(4). Since the agency does not intend to alter that definition, the term is not defined in this section.

"Defective and noncomplying equipment" also does not require definition for purposes of this section, since "noncomplying equipment" obviously means equipment that does not comply with an applicable Federal motor vehicle safety standard. "Defective equipment," on the other hand, cannot be defined in a fashion that would be appropriate for all cases. Whether equipment is defective in a manner that requires action under the Act would depend upon the type of the equipment involved as well as the nature and extent of the defect. As such, "defective" is a legal determination made on a case-by-case basis and the term, therefore, cannot be absolutely defined in advance.

Many manufacturers complained about NHTSA's definition of "original equipment." The Eaton and Bendix Corporations, for example, indicated that they thought NHTSA had violated its authority to issue regulations with respect to this term. They suggested that section 159 does not grant sufficient latitude for the agency to alter the Act's definitions to the extent found in the regulation. The agency disagrees. The language in section 159, "Except as otherwise provided in regulations of the Secretary," and the legislative history of that section very clearly permit the agency to modify the definitions of section 159 of the Act if the agency determines that it would be in the interest of an equitable distribution of enforcement responsibilities upon the various manufacturers. In this instance, the agency has determined that the minor definitional changes included in this regulation will better meet the needs of both the agency and the manufacturers for efficient recalls and remedies.

Several commenters questioned the term "express authorization" as it is used in Part 579.4 (a)(2). The agency stated in the NPRM preamble that express authorization was not limited to written authorization and that "any type of express authorization given by the vehicle manufacturer for the installation of equipment should be sufficient to make the manufacturer responsible for that equipment." The preamble went on to state that "what constitutes adequate authorization will depend upon the facts of each case." Since the issuance of the preamble, nothing has occurred that leads to a simplified defini-

tion of the term 'express authorization." Therefore, the agency declines to adopt a definition for this term and restates that it depends upon the circumstances of each case.

Several commenters indicated that proposed paragraph (1) under section 579.4(a) was overbroad in that it required a vehicle manufacturer to be responsible for equipment manufactured by him even when that equipment was not installed by him or at his direction. NHTSA agrees with these commenters and has deleted paragraph (1) from that section and renumbered the section accordingly.

Section 579.4(b) defines 'replacement equipment" to include tires. The commenters on this paragraph, Goodyear and Firestone, agreed with this definition. They stated that they thought it appropriate for tire manufacturers to be responsible for defects and noncompliances in their equipment.

With respect to the application of this regulation to the tire manufacturers, several misunderstandings occurred. Fruehauf Corporation indicated that the fabricating manufacturer of a tire should be the one responsible for the recall of those tires and not the brand name owner. The agency has held the brand name manufacturer responsible in the past for tire identification and recordkeeping (Part 574). The Act in section 159(1) holds brand name owners of tires responsible for defects and noncompliances by specifying that the brand name owner shall be deemed the manufacturer of the tires. The agency sees no reason to alter this established pattern of responsibility. However, a fabricating manufacturer and brand name manufacturer might establish by contract that the fabricating manufacturer would conduct all notification and recall campaigns.

In the preamble to the NPRM, the agency erroneously stated that tire manfacturers were required to retain the names and addresses of the owners of vehicles upon which their tires were mounted as original equipment. Tire manufacturers pointed out that this was inaccurate. Part 574 requires tire manufacturers to retain lists of people to whom their tires were sold, including vehicle manufacturers. The vehicle manufacturer would have the names of the owners of the vehicles upon which potentially defective or non-complying tires were mounted and, if necessary, would supply that list to a tire manufacturer undertaking a recall campaign.

Proposed Part 579.5(a) and (b) received very few comments. Commenters to these provisions suggested only minor modifications in their language. GM and the Motor Vehicle Manufacturers Association suggested that the term "safety-related" be added to both sections before defect to indicate that manufacturers only had responsibilities for such defects. Under the Act, manufacturers need only recall and remedy defects that are in fact determined to be safety-related. Accordingly, the agency agrees with the commenters and amends the language of the section accordingly.

GM stated that the last part of paragraph (a) of proposed section 579.5 is unnecessary. That part of the sentence that read "installed on or in the vehicle at the time of its delivery to the first purchaser" is identical to the sentence in section 579.4(a) that defines original equipment. Therefore, its inclusion at this point is redundant and unnecessary. The agency has modified the section by the deletion of that portion of the sentence.

NHTSA is publishing this regulation without taking final action on proposed section 579.5(c), and is modifying 579.5(a) to delete all reference to paragraph (c). Paragraph (c) would have placed defect and noncompliance responsibilities upon equipment manufacturers that supplied equipment to five or more vehicle manufacturers. This action is being taken without making any substantive determination on the merits of paragraph (c). A subsequent notice will deal with that paragraph and the comments thereon. However, due to the delay in the issuance of this Part and mindful of the fact that the modified definitions are important to the agency's enforcement scheme, NHTSA has determined that it is in the interest of efficiency to adopt the definitions sections of this regulation as proposed with some minor modifications, while retaining a responsibility section that basically retains the same responsibility provisions as the Act.

The agency has reviewed this regulation with respect to its potential costs and other impacts and has determined that any costs or other impacts will be minimal.

Accordingly, Title 49 of the Code of Federal Regulations is amended by the addtion of Part 579 . . . .

(Secs. 103, 108, 112, 113, Pub. L. 89–563, 80 Stat. 718, Sec. 102, Pub. L. 93–492, 88 Stat. 1470 (15 U.S.C. 1392, 1397, 1401, 1411–1420; delegation of authority at 49 CFR 1.50.)

Issued on August 24, 1978.

Joan Claybrook
Administrator

43 F.R. 38833–38834
August 31, 1978

PART 579—DEFECT AND NONCOMPLIANCE AND RESPONSIBILITY

Sec.

579.1 Scope.

579.2 Purpose.

579.3 Application.

579.4 Definitions.

579.5 Defect and noncompliance responsibility.

§ 579.1 Scope.

This part sets forth the responsibilities under Part B of the Act of manufacturers for safety-related defects and noncompliances with Federal motor vehicle safety standards in motor vehicles and items of motor vehicle equipment.

§ 579.2 Purpose.

The purpose of this part is to facilitate the notification of owners of defective and non-complying motor vehicles and items of motor vehicle equipment, and the remedy of defective and noncomplying vehicles and items of equipment, by equitably reapportioning the responsibility for safety-related defects and noncompliances with Federal motor vehicle safety standards among manufacturers of motor vehicles and motor vehicle equipment.

§ 579.3 Application.

This part applies to all manufacturers of motor vehicles and motor vehicle equipment.

§ 579.4 Definitions.

(a) "Original equipment" means an item of motor vehicle equipment (other than a tire) which was installed in or on a motor vehicle at the time of its delivery to the first purchaser if—

(1) The item of equipment was installed on or in the motor vehicle at the time of its delivery to a dealer or distributor for distribution; or

(2) The item of equipment was installed by the dealer or distributor with the express authorization of the motor vehicle manufacturer.

(b) "Replacement equipment" means—

(1) Motor vehicle equipment other than original equipment as defined in paragraph (a) of this section; and

(2) Tires.

(c) "The Act" means the National Traffic and Motor Vehicle Safety Act of 1966, as amended.

§ 579.5 Defect and noncompliance responsibility.

(a) Each manufacturer of a motor vehicle shall be responsible for any safety-related defect or any noncompliance determined to exist in the vehicle or in any item of original equipment.

(b) Each manufacturer of an item of replacement equipment shall be responsible for any safety-related defect or any noncompliance determined to exist in the equipment.

43 F.R. 38835
August 31, 1978

PREAMBLE TO PART 580—ODOMETER DISCLOSURE REQUIREMENTS

(Docket No. 72–31; Notice 2)

The purpose of this notice is to establish a regulation that will require a person who transfers ownership in a motor vehicle to give his buyer a written disclosure of the mileage the vehicle has traveled. The regulation carries out the directive of section 408(a) of the Motor Vehicle Information and Cost Savings Act, Public Law 92–513, 86 Stat. 947, and completes the provisions of the Act under Title IV, Odometer Requirements.

The regulation was first proposed in a notice published in the *Federal Register* on December 2, 1972 (37 F.R. 25727). As a result of numerous comments on the proposal, the regulation as issued today differs in some respects from its initial form.

As stated in the proposal, the agency's goals were to link the disclosure statement as closely as possible to the documents required for transfer of ownership, so that buyers and sellers would know of the need for disclosure, and to do so in a manner that would not introduce an additional document into motor vehicle transactions. The agency therefore proposed the use of the certificate of title as the document for odometer disclosure.

Upon review of the comments, it became evident that in most jurisdictions it would not be feasible to use the title certificate to convey odometer information. The main drawback to its use lies in the prevalence of state laws providing that if a vehicle is subject to a lien, the title is held by the lienholder. As a result, it appears that in a majority of cases private parties selling motor vehicles do not have possession of a certificate of title, and convey their interest by other means.

In those States that permit the owner of a vehicle subject to a lien to retain the title, the lienholder will be unable to make the odometer disclosure on the title if he attempts to sell the vehicle after repossession. In many States, furthermore, the title certificate is not large enough to contain an adequate odometer disclosure, and the existing data processing and filing equipment would not accommodate an enlarged certificate.

There appears to have been some apprehension that the Federal government intended to compel the States to amend their certificates of title. The Act does not, however, confer any authority over the States in this regard. Even if the regulation were to require transferor disclosure on the title, the States could decline to provide a form for disclosure on the title. This voluntary aspect of the States' participation is a further impediment to the use of the title certificate.

After review of the problems created by the use of the certificate of title, the agency has decided that the purposes of the Act are better served by prescribing a separate form as the disclosure document in most cases. Section 580.4 has been amended accordingly. To avoid the need for duplicate State and Federal disclosures in States having odometer disclosure laws or regulations, the section permits the State form to be used in satisfaction of the Federal requirement, so long as it contains equivalent information and refers to the existence of a Federal remedy.

It should be noted that although the certificate of title is no longer required to be used for disclosure, it can still be used as the disclosure document if it contains the required information and if it is held by the transferor and given by him to the transferee. The basic concept is that the disclosure must be made as part of the transfer, and not at some later time.

In addition to the changes from the proposal represented by the change from the certificate of title to a separate form, there are other differences from the proposal in the regulation. For purposes of convenience, the following discussion treats the amended sections in sequence.

In section 580.3, the proposed definition of transferor might in some jurisdictions include a person who creates a security interest in a vehicle. This type of transaction was not intended to be regulated, and the definitions have been amended accordingly.

In section 580.4, in addition to the changes discussed above, other modifications have been made. In response to a comment suggesting that the disclosure would be made after the purchaser had become committed to buying the vehicle, the order of § 580.4(a) has been rearranged to specify that the odometer disclosure is to be made before the other transfer documents are executed.

The items listed under § 580.4(a) have been increased to allow for additional identification of the vehicle and owner that would be necessary on a separate disclosure document. If the disclosure is a part of another document, however, § 580.4(a)(1) provides that items (2) through (4) need not be repeated if found elsewhere in the document. A number of comments noted that the items under (a) might often be redundant.

A new paragraph (b) has been inserted in § 580.4 to require a reference to the sanctions provided by the Act. No specific form is required, but the inclusion of such a statement is considered essential to notify the transferee of the reason why he is being given the odometer information.

The former paragraph (b) of § 580.4 has been renumbered as (c), and the alternative methods for odometer disclosure discussed above are found as paragraphs (d) and (e).

A new section, § 580.5, Exemptions, has been added in response to a number of comments that objected to the application of the requirements to categories of vehicles for which the odometer is not used as a guide to value. Buses and large trucks, for example, are routinely driven hundreds of thousands of miles, and their main-

tenance records have traditionally been relied on by buyers as the principal guide to their condition. The NHTSA is in agreement with the position taken by Freightliner, White, and the National Association of Motor Bus Operators, and has therefore created an exemption for larger vehicles. The exemption applies to vehicles having gross vehicle weight ratings of more than 16,000 pounds.

A second category of exempt vehicles has been created for antique vehicles, whose value is a function of their age, condition, and scarcity, and for which the odometer mileage is irrelevant. A third exempt category consists of vehicles that are not self-propelled, such as trailers, most of which are not equipped with odometers.

Several vehicle manufacturers stated that the proposal would require them to give disclosure statements to their distributors and dealers, and that such a requirement would be both burdensome and pointless. Upon consideration of the nature of manufacturer-dealer transactions, it has been decided to exempt transfers of new vehicles that occur prior to the first sale of the vehicle for purposes other than resale.

The odometer disclosure form set forth in § 580.6 has been reworded to make it clearer. Space for additional information about the vehicle and owner has been included so that the vehicle will be readily identifiable if the disclosure statement becomes separated from the other transfer documents. In accordance with the instructions of the Act, the transferor is directed to state that the mileage is unknown if he knows that the actual mileage differs from the mileage shown on the odometer. Although several comments suggested that the true mileage, if known, should be stated, such a statement is not provided for in the Act and would not afford the buyer with reliable information about the vehicle.

The effective date proposed in the notice was to have been six months after issuance. Two States, perhaps under the impression that they were required to change their forms, requested an additional six months. Other comments, notably that of the National Automobile Dealers Association, urged an immediate effective date in order to make the disclosure requirements coincide with the effectiveness of the other parts of Title IV of the Act. Upon consideration of the important contribution the disclosure requirements make to the effectiveness of the Act's other provisions, it has been decided that an effective date earlier than six months after issuance is advisable.

Accordingly, the regulation is to become effective March 1, 1973. Although it is likely that most private persons will remain unaware of the disclosure requirements for some time after March 1, 1973, a person who does not know of the requirement will not have "intent to defraud" under section 409(a) of the Act and will therefore not be subject to liability solely because he has failed to make the required statement. The persons most immediately affected by the disclosure requirements are commercial enterprises such as dealers and wholesalers, and of these the largest group, represented by NADA, has already indicated its desire for an early effective date. The earlier effective date is therefore considered appropriate.

In consideration of the foregoing, a new Part 580, Odometer Disclosure Requirements, is added to Title 49, Code of Federal Regulations, to read as set forth below.

Issued under the authority of section 408(a) of the Motor Vehicle Information and Cost Savings Act, P.L. 92–513, 86 Stat. 947, and the delegation of authority at 49 C.F.R. 1.51.

Issued on January 23, 1973.

Douglas W. Toms,
Administrator.

38 F.R. 2978
January 31, 1973

PREAMBLE TO PART 580—ODOMETER DISCLOSURE REQUIREMENTS

(Docket No. 77–03; Notice 2)

This notice amends the odometer disclosure statement that must be executed upon each sale of a motor vehcile. The former statement often proved confusing and was sometimes used in a misleading manner. The amended statement is clearer and less likely to be misused.

Effective date: January 1, 1978.

For further information contact:

Kathleen DeMeter, Office of the Chief Counsel, National Highway Traffic Safety Administration, 400 Seventh Street, SW., Washington, D.C. 20590 (202–426–1834).

Supplementary information: The disclosure statement is required by 49 CFR Part 580, Odometer Disclosure Requirements, a regulation issued by the National Highway Traffic Safety Administration (NHTSA) to implement the requirements of the Motor Vehicle Information and Cost Savings Act (Pub. L. 92–513, as amended by Pub. L. 94–364; 15 U.S.C. 1901–1991). The regulation, which has been in effect since March 1, 1973, requires each transferor of a motor vehicle to give the transferee a written statement attesting to the accuracy of the vehicle's odometer.

Experience with the regulation has shown several respects in which it should be improved. In response to a petition for rulemaking submitted by the National Automobile Dealers Association, and in recognition of the need for improvements in the disclosure statement, the NHTSA issued a notice on February 9, 1977 (42 F.R. 9045) which proposed changes in the form and content of the odometer disclosure statement.

Differences between proposed and final rule. The final rule differs from the proposed rule in several respects. The notice had proposed to require the disclosure form to include the last license plate number, State and year. In view of the number of commenters who stated that this information was not needed to identify a vehicle or to trace a vehicle's history, the agency has decided to delete this requirement from the final rule.

The notice proposed a substantial enlargement of the disclosure form, including a certification that the odometer was either not altered, or altered for repair or replacement purposes only. This certification had been proposed in response to the NADA petition, and drew few critical comments. Two commenters raised Fifth Amendment questions concerning these additional boxes. The Department of Health, Education, and Welfare's Office of Consumer Affairs noted that these alternative certifications might give rise to possible violations of the transferor's right against self-incrimination since a willful false certification may amount to an admission of a violation of the Act. The NHTSA, however, believes that no Fifth Amendment problem could arise. In cases dealing with this issue the Supreme Court has held that where the dominant purpose of a record-keeping requirement is to compel criminals to keep incriminating records, the statute is invalid and the 5th Amendment may be invoked. However, where the record-keeping requirements have an independent purpose and do not involve a selective group which is inherently suspect of criminal activities, the statute is valid and the 5th Amendment may not be invoked. All businessmen, as well as all consumers, who sell automobiles would be required to execute odometer disclosure statements. Statements are not required only of those individuals who are most often found to tamper with odometers. The primary purpose of a statement is to inform a potential buyer of the car's mileage so that he may have an index to the condition and value of the vehicle. The fact that individuals who tamper with vehicle odometers would be executing incriminating records is not the dominant purpose of this requirement. Consequently, these provisions will be retained in the final rule with one minor change suggested by a commenter. In view of the fact that these certifications actually involve three separate statements, instead of two as indicated in the NPRM, the NHTSA had decided to divide the second certification into two: first, that the odometer was altered and the mileage is identical to that before repair; and second, that the odometer was altered and reset to zero, with a statement of the mileage on the original odometer or the odometer before repair.

Several commenters suggested that the transferee's name and address should be provided in a disclosure statement, in addition to his signature. This would provide a useful tool in tracing the vehicle's history and consequently, the NHTSA has decided to require that this information be included.

With the gradual conversion to the metric system now going on in the United States, the regulation has also been changed to provide for odometer readings that are expressed in kilometers where the vehicle records the distance traveled in metric units.

The bulk of the comments received were favorable. The primary objection was that the proposed final effective date of April 15, 1977, did not allow adequate time for new forms to be prepared and printed. In addition, it would have increased costs because it would not have allowed sufficient time for stocks of the present form to be depleted. In response to these comments, the agency has adopted an effective date of January 1, 1978.

One of the original goals of NHTSA was to link the disclosure statement as closely as possible to the documents required for transfer of ownership, so that buyers and sellers would know of the need for disclosure. To accomplish this goal in a manner that would not introduce an additional document into motor vehicle transactions, the agency proposed to use the certificate of title as the document for odometer disclosure.

The comments to that initial proposal persuaded the agency that providing the odometer reading on the title would not be feasible as the sole method of disclosure. NHTSA still believes, however, that placing odometer information on the certificate of title will be useful both to consumers and to law enforcement officials. This belief is substantiated by a recent resolution of the National Association of Attorneys General, which endorsed odometer information on State certificates of title as the most effective means to ensure a permanent record of the mileage history of a motor vehicle, and by the development by the American Association of Motor Vehicle Administrators of model procedures for the disclosure of odometer information on vehicle titles. Such a record would be easily accessible to governmental enforcement agencies as well as prospective purchasers of used motor vehicles.

The notice of February 7, 1977, proposed to allow the use of a State document containing odometer disclosure information if the State document contained "all" of the information required on the Federal form. A comment from the Attorney General of Ohio pointed out that it would be difficult for States to include "all" of the odometer information on their titles because of the limited space available. Consequently, NHTSA has decided to revise § 580.4(f) to accommodate those States that provide odometer information on their titles by establishing a procedure under which States can have their titles approved for use as odometer disclosure statements. In view of the utility of titles and their limited space, the procedure would permit shortening the odometer provisions on the title where necessary. Although a shorter disclosure might sacrifice clarity to a degree, the agency regards this as an acceptable price for gaining the benefits of a combined title and odometer disclosure.

States that wish to have their certificates of title satisfy the Federal odometer disclosure requirements must meet the basic provisions of the Federal requirement, with the following exceptions:

(1) The citation to the Federal law may be deleted in favor of a reference to State law. The reference provisions could then state that "Federal and State regulations require you to state the odometer mileage upon transfer of ownership. (Citation to State law instead of Federal law)."

(2) The initial statement of the odometer reading and the following alternate certifications should be included on the title. States may, however, condense that information as long as none of the certifications are lost. An example of such condensation could be "I certify to the best of my knowledge that the odometer reading is _____ and reflects the actual mileage of the vehicle described herein or (check if applicable).

☐ 1. The amount of mileage stated is in excess of 99,999 miles, or

☐ 2. The odometer reading is not the actual mileage."

3. The transferee's signature must still appear on the title but it need not expressly indicate acknowledgement of receipt of the disclosures.

4. The certification that the odometer was either not altered or altered for repair or replacement purposes may be deleted.

All deviations on the certificate of title from the Federal requirements must be approved by the NHTSA prior to the use of State titles as substitutes for the Federal form. The exceptions noted above are to be used by the States only as guides in preparing conforming titles. In order for the citizens of a State to use the certificate of title as their odometer disclosure form, the Administrator of the State Department of Motor Vehicles must first request an exemption from the provision of the disclosure requirement by submitting such request in writing with a copy of the proposed certificate of title. The NHTSA will then notify the Administrator of its decision to accept or refuse the request and the reasons for its decision. Upon receipt of the NHTSA's acceptance of the request for an exemption, the State may proceed with a campaign to notify consumers, dealers and distributors of such acceptance. It shall be the State's responsibility to publicize that its title may be used in place of the odometer disclosure statement.

Additional comments. One commenter asked whether there would be specifications for size. There are none, with the understanding that all print should be legible to the naked eye. Another commenter suggested that section 580.4(c)(3) be changed to add the word "believed" so that the

reading would be "I hereby certify that to the best of my knowledge the odometer reading as stated above is believed NOT to be the actual mileage. . . ." NHTSA considers this addition unnecessary because the certification already states "to the best of my knowledge."

A commenter proposed that the form should be amended to say that the names and addresses of prior owners are available from a State agency. NHTSA has determined that this should not be added. The addresses are not available from some State agencies and such a provision would therefore be of limited utility. Another addition that was suggested was to add a reference to the minimum damages and attorneys fees available under the Federal law. This was proposed to alert consumers to the fact that certain impediments to enforcement, such as the expense of lawyers and proof of actual damages, are removed by the Act. These references, like any other additions desired by the States or transferors, may be added, but will not be required due to space limitations and to a determination that they are not necessary if there is sufficient publicity of the law.

An individual commented that the seller should be allowed to estimate the amount of mileage difference and explain the error. There is certainly no prohibition against a seller doing so, but NHTSA sees no benefit to be gained in requiring this. A buyer can, and certainly should, request such information, but anyone who has violated the Act will, nonetheless, not provide a truthful statement of the mileage difference or the reason for that difference. The result could be that a buyer is unknowingly led into reliance on this false statement, whereas an independent check of his own could have produced the truth.

It was suggested that positive introductory statements be used for the certification sections. The commenter noted that in its experience, when a positive introductory statement is lacking, the seller fails to check any box. Its proposal would modify the statement as follows: "I _____ _____ state that the odometer now reads _____ miles and I hereby certify that to the best of my knowledge the odometer reading as stated above reflects the actual mileage of the vehicle described below, unless one of the following statements is checked.

☐ (1) I hereby certify that the odometer reading reflects the amount of mileage in excess. . . ."

☐ (2) I hereby certify that to the best of my knowledge the odometer reading as stated above is NOT . . ."

The NHTSA has not experienced the failure to check a box when a positive introductory statement is lacking and consequently, will retain the statement in the proposed rule. Should it become evident that a positive introductory statement is needed, further rulemaking will be undertaken. It should be noted that the form suggested by this commenter would significantly shorten the length of this provision, thus it would be an acceptable alternative only where the odometer disclosure is on the certificate of title.

A suggestion was made to provide a notice that an auxiliary odometer had been used in the vehicle. The auxiliary odometer would interrupt the operation of the regular odometer and cause it to register less than the vehicle's actual mileage. The seller would therefore be required by the present language of the regulation to notify the buyer of the odometer error. In view of this, NHTSA considers it unnecessary to refer specifically to an auxiliary odometer.

It was also suggested that the owner of a vehicle be allowed to replace or adjust the odometer to reflect actual mileage. The commenter noted that occasionally odometers jump ahead 10, 20, or 30 thousand miles and if the odometer cannot be altered to read the actual mileage instead of the mileage on the odometer·before repair or replacement, the trade-in value would be drastically decreased to the harm of the owner. NHTSA believes that the few cases in which the odometer malfunctions and rolls forward too fast are too slight to justify this provision. Such a provision would create a loophole for those who wanted to roll back their odometer and then claim that it

was rolling over too fast and they had to fix it by moving it backward. Anyone whose odometer did jump could replace or repair the odometer, set it to zero so that a buyer would not be misled by the odometer reading, and upon sale provide a statement to the buyer that the mileage is NOT actual and that the actual mileage is less than that shown on the odometer or on the repair or replacement sticker. More importantly, it should be noted that the repair and replacement provisions, wherein the owner is required to reset the odometer to the mileage before repair or replacement or to zero, are part of the Motor Vehicle Information and Cost Savings Act (section 407(a)). Consequently, they are not susceptible to change by NHTSA, but only by Congress.

Requests by commenters that odometer readings be required on registration forms, that statements be required to be retained, and that manufacturers be required to furnish 6 digit odometers are not applicable to this rulemaking action. It should be noted that a retention requirement for odometer disclosure statements will be issued soon and that a proposed rule requiring tamper-proof odometers which indicate when they have exceeded 100,000 miles or kilometers was issued on December 7, 1976. The proposed effective date of the latter rule is September 1, 1979.

In consideration of the foregoing, Part 580, Odometer Disclosure Requirements is amended. . .

The lawyer principally responsible for this rule is Kathleen DeMeter.

(Sec. 408, Pub. L. 92–513, 86 Stat. 962, as amended by Pub. L. 94–364, 90 Stat. 983 (15 U.S.C. 1988); delegation of authority at 49 CFR 501.8(i).)

Issued on July 25, 1977.

Joan Claybrook
Administrator

42 F.R. 38906–38908
August 1, 1977

PREAMBLE TO PART 580—ODOMETER DISCLOSURE REQUIREMENTS

(Docket No. 77-06; Notice 2)

The Secretary of Transportation is authorized by the Motor Vehicle Information and Cost Savings Act to specify requirements for retention of odometer statements by dealers and distributors of motor vehicles. This notice prescribes the manner in which this information should be retained. The intended effect of this regulation is to afford the government and aggrieved parties documentation necessary to prove a violation of the Act, and to pinpoint exactly where the violation occurred.

Effective date: March 9, 1978.

For further information contact:

Kathleen DeMeter, Office of Chief Counsel, National Highway Traffic Safety Administration, 400 Seventh Street, SW., Washington, D.C. 20590 (202-426-1834).

Supplementary information: The Motor Vehicle Information and Cost Savings Act (Pub. L. 92-513, 86 Stat. 947-963, 15 U.S.C. 1901-1999) directed the Secretary of Transportation to issue regulations to require each transferor of a motor vehicle to give the transferee a written statement of the mileage shown on the vehicle's odometer and to advise the transferee if the mileage shown on the odometer was known to be different from the vehicle's actual mileage. A regulation was issued pursuant to section 408 of the Act to prescribe the manner of disclosure (49 CFR Part 580), but the Secretary chose not to exercise the authority given him under subsection 408(a) to specify the manner in which such information was to be retained.

The 1976 amendments to the Act (Pub. L. 94-364, 90 Stat. 981) conferred extensive investigative powers upon the Secretary. One effect of these new powers is to enhance the value of a record retention requirement as an investigatory tool. The disclosure statement plays an important role in the investigation of odometer tampering and fraud. In order to prove that an odometer has been rolled back or otherwise tampered with in violation of the Act, it must be possible to ascertain the amount of actual mileage the vehicle has been driven. An effective way of discovering this information is by examining previous odometer mileage statements required to be executed by all owners in the chain of title.

To enhance the ability of the statement to protect all future transferees a notice of proposed rulemaking (NPRM) was issued on November 1, 1977, which would not only require the dealers and distributors to retain for four years the statements issued to them but would also require them to retain for four years a copy of each statement which they issued. Such retentions would afford the government and aggrieved parties the necessary documentation to prove a violation of the Act, and also to pinpoint exactly where that violation occurred. All of the comments submitted in response to the NPRM have been considered and the most significant ones are discussed below.

The final rule is almost identical to the NPRM. The NPRM proposed that odometer mileage statements be retained in chronological order. The final rule permits mileage statements to be retained in an order appropriate to the business requirements of each dealer and distributor. A majority of commenters objected to the chronological order provision. A number of other methods of filing were suggested, such as by vehicle identification number and alphabetical order by the customer's last name. Due to the wide variety of methods of filing presently used, the NHTSA believes that a single mandated method of filing would result in unnecessary cost and duplication. Therefore, the new section permits dealers and distributors to retain odometer mileage statements in a manner consistent with their existing recordkeeping procedures. The section requires that however the recordkeeping system is organized, it must permit a systematic retrieval of odometer statements.

One commenter suggested that a longer lead-time was necessary to accommodate changes in filing procedures. However, since recordkeeping requirements need not be changed, there should be no lead time problems.

Several commenters objected to the scope of the rule. There appeared to be some confusion among the commenters as to whether the rule applied to insurance companies, manufacturers and financial institutions. The final rule applies to all dealers and distributors of motor vehicles. A "dealer" is defined in section 402 of the Act as "any person who has sold 5 or more motor vehicles in the past 12 months to purchasers who in good faith purchase such vehicles for purposes other than resale." A "distributor" is defined in the same section as "any person who has sold 5 or more vehicles in the past 12 months for resale." Given these definitions, a manufacturer would be a "distributor." However, § 580.5 of Title 49, Code of Federal Regulations specifically exempt manufacturers who sell vehicles to dealers from the requirements of executing disclosure statements. Section 583.7 of this final rule has been reworded to make it clear that only those "dealers" and "distributors" who are required to execute disclosure statements must retain them. Financial institutions and insurance companies do not fall within any of the exemptions set forth in § 580.5, so they must execute and retain the statements unless the transfers involve vehicles that are so badly damaged that they cannot be returned to the road. In such transfers, the agency has ruled that the damaged vehicles are no longer "motor vehicles" for purposes of the disclosure regulations.

In light of the foregoing, Part 580, Odometer Disclosure Requirements, of Title 49, Code of Federal Regulations, is amended as set forth below.

The lawyer principally responsible for this rule is Kathleen DeMeter.

The rule does not require any persons to create additional records or to alter their business practices apart from keeping records they might once have discarded. In view of the expected benefits to the Department's enforcement program, it is found for good cause that the rule may be issued with an immediate effective date.

(Secs. 408, 414, Pub. L. 92–513, 86 Stat. 947, as amended Pub. L. 94–364, 90 Stat. 981 (15 U.S.C. 1988, 1990(d)); delegation of authority at 49 CFR 1.50(f).).

Issued on March 7, 1978.

Joan Claybrook
Administrator

43 F.R. 10921–10922
March 16, 1978

PREAMBLE TO AMENDMENT TO PART 580—ODOMETER DISCLOSURE REQUIREMENTS

(Docket No. 77-06; Notice 4)

ACTION: Final rule.

SUMMARY: This notice allows States to use an abbreviated odometer disclosure statement on all motor vehicle ownership documents. The existing regulation permitted the shortened form to be used merely on the certificate of title. The purpose of this expansion is to increase State usage of odometer disclosure statements.

DATE: The effective date is the date of publication in the Federal Register.

FOR FURTHER INFORMATION CONTACT:

Kathleen DeMeter, Office of Chief Counsel, National Highway Traffic Safety Administration, 400 Seventh Street, S.W., Washington, D.C. 20590. (202-426-1834).

SUPPLEMENTARY INFORMATION: Section 408 of the Motor Vehicle Information and Cost Savings act (15 U.S.C. 1988) requires each transferor of a motor vehicle to provide to the transferee a written disclosure of the distance travelled by the vehicle. 49 CFR Part 580 prescribes the information to be included on the disclosure statement. On August 1, 1977, NHTSA amended the odometer disclosure statement (42 FR 38906). The amended statement is clearer than the former statement and less likely to be misused, but it is also longer.

NHTSA has urged the States to include the odometer statement on the title. Six States had included the original statement. In commenting on the longer statement, several States observed that the title, with its size limitations, presented more problems with inclusion of the odometer statement than did other documents relating to the transfer and ownership of motor vehicles. Because of this, the 1977 amendment specifically allowed a shortened form to be used on certificates of title, but not on other ownership documents.

On May 7, 1979, the NHTSA issued a notice of proposed rulemaking in which it granted a petition by the American Association of Motor Vehicle Administrators (AAMVA) to amend the Federal odometer disclosure requirements to allow the abbreviated form to be used on ownership documents other than the certificate of title (44 FR 28032). The AAMVA emphasized that many of the State documents used to evidence ownership of motor vehicles are too small to accommodate the additional information required. They argued that States should not have to rely on separate odometer forms for these transfers but should be allowed to use the shortened form on all documents which evidence ownership, not only on the certificate of title.

Seven States responded to the notice of proposed rulemaking. Comments were received from the motor vehicle departments in Virginia, Washington, Delaware, Wisconsin, New Jersey, Texas, and Oregon. Most comments were favorable. The Virginia Division of Motor Vehicles asked that the short form be acceptable on all applications for title. The more State documents that contain mileage information the more difficult it will be for odometer rollbacks to go undetected. Consequently, the NHTSA encourages the use of the short form on applications for title as well as certificates of title.

Washington and Wisconsin suggested respectively that the introductory paragraph citing the Federal law be deleted or shortened due to document size limitations. The August 1, 1977, amendment to the disclosure form noted that a reference to State law may be substituted for the citation to the Federal law.

Consistent with this interpretation, it is the agency's opinion that the actual law need not be cited if a warning statement appears such as that suggested by Washington, "Warning False Statements Violate Federal Law."

PART 580; PRE-11

The Texas State Department of Highways and Public Transportation offered the only negative comments to the proposal. It argued that a purchaser who finances a motor vehicle could not execute a form on the certificate of title at the time of sale because the certificate is held by a bank or financial institution as security. Although the Texas comment illustrates the difficulties of trying to require the use of titles for odometer disclosure, the amendment is permissive and would not require Texas to change its practices in any way.

In accordance with Executive Order 12044, the regulation has been reviewed for environmental and economic impacts. It has been determined that the cost of implementing this regulation will be minimal. There are no additional requirements.

The regulation permits States to provide certain information on ownership documents but does not require them to do so. There are no environmental or other economic impacts, therefore, this regulation is not significant.

Issued on December 20, 1979.

Joan Claybrook
Administrator, National
Highway Traffic Safety
Administration

45 F.R. 784
January 3, 1980

PREAMBLE TO AN AMENDMENT TO PART 580

Odometer Disclosure Requirements

[Docket No. 81-13; Notice 2]

ACTION: Final rule.

SUMMARY: This rule amends 49 CFR Part 580 to exempt from the Odometer Disclosure Requirements all sales of new motor vehicles by a motor vehicle manufacturer directly to any agency of the United States. The purpose of this exemption, which is being issued pursuant to a petition by, General Motors Corporation, is to relieve manufacturers of the burden of complying with this requirement.

EFFECTIVE DATE: December 20, 1982.

SUPPLEMENTARY INFORMATION: Since March 1, 1973, a regulation (49 CFR Part 580) has been in effect which requires the transferor of a motor vehicle to make written disclosure to the transferee concerning the odometer reading and its accuracy. This regulation lists four exceptions where the transferor need not disclose the vehicle's mileage.

On December 10, 1981, in response to a petition from General Motors Corporation, NHTSA published (46 F.R. 60482) a Notice of Proposed Rulemaking (NPRM) which proposed creating a fifth category of exempt transactions. That category consists of all sales in conformity with contractual specifications of motor vehicles by a manufacturer directly to any agency of the United States. GM noted that most of a vehicle manufacturer's transfers are already exempt from the disclosure requirements and this exemption would merely extend the existing exemption. GM stressed that the disclosure requirements were designed to protect consumers against odometer fraud in retail transactions. The conditions lending themselves to fraud in the retail market are, GM argued, nonexistent in manufacturer-to-government sales.

Two comments were received in response to the NPRM. Chrysler Corporation supported the proposed change without qualification. PACCAR, Inc. supported the concept of the additional exemption and the rationale behind it, but expressed reservations about the unsettled issue of NHTSA's authority to promulgate any exemption to the odometer disclosure regulation. PACCAR noted correctly that two Federal District Courts have invalidated the exemption for trucks over 16,000 GVWR on the basis that the NHTSA is not authorized to make any exemptions to the law.

Section 408 (a) of the Motor Vehicle Information and Cost Savings Act (15 U.S.C. 1988) states that the Secretary of the Department of Transportation shall prescribe rules requiring transferors to give written mileage disclosures to transferees in connection with the transfer of ownership of a motor vehicle. It is the interpretation of NHTSA that this grant of rulemaking authority empowers the agency to also make exceptions to the requirement where it is shown that no mileage statement is necessary. NHTSA recognizes that there is a conflict between its interpretation of the Act and the interpretation of the United States District Courts for the Districts of Nebraska and Idaho. While these decisions are not binding precedent in other Federal courts, they may, however, be used as guidance and followed should the issue arise in the future with respect to the same or one of the other exemptions. Therefore, NHTSA has advised interested persons of the two court opinions and their conflict with the current language of the regulation and forewarned them that the issue has not been resolved. NHTSA is proceeding with this rulemaking action on the basis that its interpretation is correct, but is also advising manufacturers to consult with their legal counsel to determine what course of action will most effectively protect their legal rights.

Issued on October 5, 1982.

Raymond A. Peck, Jr.
Administrator
47 F. R. 51884
November 18, 1982

PART 580—ODOMETER DISCLOSURE REQUIREMENTS

(Docket No. 72-31; Notice 2)

§ 580.1 Scope.

This part prescribes rules requiring the transferor of a motor vehicle to make written disclosure to the transferee concerning the odometer mileage and its accuracy, and requiring the retention of odometer mileage statements by motor vehicle dealers and distributors, as directed by section 408(a) and 414(b) of the Motor Vehicle Information and Cost Savings Act, Pub. L. 92-513, as amended by Pub. L. 94-364.

§ 580.2 Purpose.

The purpose of this part is to provide each purchaser of a motor vehicle with odometer information to assist him in determining the vehicle's condition and value, and to preserve records that are needed for the proper investigation, and adjudication or other disposition, of possible violations of the Motor Vehicle Information and Cost Savings Act.

§ 580.3 Definitions.

All terms defined in Sections 2 and 402 of the Act are used in their statutory meaning. Other terms used in this part are defined as follows:

"Transferor" means any person who transfers his ownership in a motor vehicle by sale, gift, or any means other than by creation of a security interest.

"Transferee" means any person to whom the ownership in a motor vehicle is transferred by purchase, gift, or any means other than by creation of a security interest.

§ 580.4 Disclosure of odometer information.

(a) Before executing any transfer of ownership document, each transferor of a motor vehicle shall furnish to the transferee a written statement signed by the transferor, containing the following information:

(1) The odometer reading at the time of transfer;

(2) The date of transfer;

(3) The transferor's name and current address;

(4) The transferee's name and current address; and

(5) The identity of the vehicle, including its make, model, year, and body type, and its vehicle identification number.

(b) In addition to the information provided under paragraph (a) of this section, the statement shall refer to the Motor Vehicle Information and Cost Savings Act and shall state that incorrect information may result in civil liability and civil or criminal penalties.

(c) In addition to the information provided under paragraphs (a) and (b) of this section,

(1) The transferor shall certify that to the best of his knowledge the odometer reading reflects the actual miles or kilometers the vehicle has been driven; or

(2) If the transferor knows that the odometer reading reflects the amount of mileage in excess of the designed mechanical odometer limit of 99,999 miles/kilometers, he shall include a statement to that effect; or

(3) If the transferor knows that the odometer reading differs from the number of miles/kilometers the vehicle has actually traveled and that the difference is greater than that caused by odometer calibration error, he shall include a statement that the odometer reading is not the actual mileage, and should not be relied upon.

(d) In addition to the information provided under paragraphs (a), (b) and (c) of this section, the transferor shall certify that:

(1) The odometer was not altered, set back, or disconnected while in the transferor's possession, and he has no knowledge of anyone else doing so;

(2) The odometer was altered for repair or replacement purposes while in the transferor's possession, and the mileage registered on the repaired or replacement odometer was identical to that before such service; or

(3) The odometer was altered for repair or replacement purposes, the odometer was incapable of registering the same mileage, it was reset to zero, and the mileage on odometer before repair was _____ miles/kilometers.

(e) The transferee shall acknowledge receipt of the disclosure statement by signing it.

(f) (1) If the laws or regulations of the State in which the transfer occurs require the odometer disclosure to be made on the certificate of title or other State documents which evidences ownership, the transferor may make the disclosure required by this section by executing the State certificate of title or such other ownership document. In order to utilize the above documents as substitutes for the Federal odometer disclosure statement, they must contain essentially the same information required by paragraphs (a), (b), (c) and (e) of this section. If the information contained thereon varies in any way from that required for the Federal form, the State must obtain approval from the National Highway Traffic Safety Administration before its certificate of title or other ownership document can be used as a substitute for the Federal form. Such approval may be obtained by submitting a copy of the proposed document to the Office of the Chief Counsel, National Highway Traffic Safety Administration, 400 Seventh Street, S.W., Washington, D.C. 20590.

(2) The NHTSA shall respond to the State's request within 30 days of receipt of such request.

(3) If a document, other than the certificate of title, provided under the laws or regulations of the State in which the transfer occurs contains all of the statements required by this section, the transferor may make the disclosure required by this section either by executing the State document or by executing the disclosure form specified in § 580.6.

(g) If there is no State document as described in paragraph (f) of this section, the transferor shall make the disclosure required by this section by executing the disclosure form specified in § 580.6.

§ 580.5 Exemptions.

Notwithstanding the requirements of § 580.4—

(a) [A transferor of any of the following motor vehicles need not disclose the vehicle's odometer mileage;

(1) A vehicle having a gross vehicle weight rating, as defined in § 571.3 of this chapter, or of more than 16,000 pounds;

(2) A vehicle that is not self-propelled;

(3) A vehicle that is 25 years old or older; or

(4) A vehicle sold directly by the manufacturer to any agency of the United States in conformity with contractual specifications. (47 F.R. 51884—November 18, 1982. Effective: December 20, 1982)]

(b) A transferor of a new vehicle prior to its first transfer for purposes other than resale need not disclose the vehicle's odometer mileage.

§ 580.6 Disclosure form.

ODOMETER MILEAGE STATEMENT

(Federal regulations require you to state the odometer mileage upon transfer of ownership. An inaccurate or untruthful statement may make you liable for damages to your transferee, for attorney fees, and for civil or criminal penalties, pursuant to sections 409, 412, and 413 of the Motor Vehicle Information and Cost Savings Act of 1972 (Pub. L. 92-513, as amended by Pub. L. 94-364).

I, _____ state that the
       (transferor's name, Print)
odometer of the vehicle described below now reads _____ miles/kilometers.

☐ (1) I hereby certify that to the best of my knowledge the odometer reading as stated above reflects the actual mileage of the vehicle described below.

☐ (2) I hereby certify that to the best of my knowledge the odometer reading as stated above reflects the amount of mileage in excess of designed mechanical odometer limit of 99,999 miles/kilometers of the vehicle described below.

☐ (3) I hereby certify that to the best of my knowledge the odometer reading as stated above is NOT the actual mileage of the vehicle described below, and should not be relied upon.

Make	Model	Body type
Vehicle identification number		Year

Check one box only.

☐ (1) I hereby certify that the odometer of said vehicle was not altered, set back, or disconnected while in my possession, and I have no knowledge of anyone else doing so.

☐ (2) I hereby certify that the odometer was altered for repair or replacement purposes while in my possession, and that the mileage registered on the repaired or replacement odometer was identical to that before such service.

☐ (3) I hereby certify that the repaired or replacement odometer was incapable of registering the same mileage, that it was reset to zero, and that the mileage on the original odometer or the odometer before repair was _____ miles.

Transferor's Address (seller) _____
                                              (Street)

_____
(City)   (State)   (ZIP Code)
Transferor's Signature (seller) _____
Date of Statement _____

Transferee's Name and Address (buyer) _____

_____
                        (Street)

_____
(City)   (State)   (ZIP Code)

Receipt of copy Acknowledged _____

_____
            (Transferee's signature, buyer)

§ 580.7   Odometer mileage statement retention.

Each dealer or distributor of a motor vehicle who is required by this Part to execute an odometer disclosure statement shall retain for four years each odometer mileage statement which he receives. He shall also retain for four years a photostat, carbon, or other facsimile copy of each odometer mileage statement which he issues. He shall retain each odometer mileage statement at his primary place of business in an order that is appropriate to this business requirements and that permits systematic retrieval. The statement may be reproduced (e.g., photocopies or put on microfilm) as long as no information or identifying marks such as signatures are lost in the reproduction.

38 F.R. 2978
January 31, 1973

PREAMBLE TO PART 581—BUMPER STANDARD

(Docket No. 74–11; Notice 12; Docket No. 73–19; Notice 9)

This notice establishes a new bumper standard, limiting damage to vehicle bumpers and other vehicle surfaces in low-speed crashes.

The standard, 49 CFR Part 581, is issued under the authority of Title I of the Motor Vehicle Information and Cost Savings Act, Public Law 92–513, 15 U.S.C. 1901–1991. In addition to specifying limitations on damage to non-safety-related components and vehicle surface areas, it also incorporates the safety requirements currently contained in Federal Motor Vehicle Safety Standard No. 215, *Exterior Protection.*

Since the enactment of the Motor Vehicle Information and Cost Savings Act, the NHTSA has issued four proposals to establish a front and rear end damage ability standard that fulfills the objectives espoused in the law. Title I *(Bumper Standards)* directs the NHTSA to develop standards which "shall seek to obtain the maximum feasible reduction of costs to the public and to the consumer. . . ." Improving the damage resistance of a vehicle in low-speed impact situations will, in the opinion of Congress, save the consumer a significant amount of money.

During the past several years of ongoing rulemaking in the bumper area, the NHTSA has continued to conduct studies and examine input from all interested persons. The most recent proposal was published March 12 of this year (40 FR 11598). After thoroughly reviewing the available data and comments submitted to the docket, the NHTSA has concluded that the provisions contained in the March notice would constitute a large step towards accomplishing the goals described in Title I.

On January 2, 1975, the NHTSA proposed a reduction in the impact speeds specified in Standard 215 and proposed in Part 581 (40 FR 10). The NHTSA's proposal was based primarily on the results of two agency-sponsored studies which indicated that the cost and weight of many current production bumpers, in light of inflation and fuel shortages, made the bumpers no longer cost-beneficial. Information presented at public hearings on the notice and comments submitted to the docket brought to light additional data which the NHTSA carefully examined. After reviewing its previous studies in light of this new evidence, the agency concluded that the 5-mph protection level (and the 3-mph corner impact level associated with it) should not be reduced. In its March 12, 1975, notice (40 FR 11598) the NHTSA fully explained this decision. Comments have been received from Toyo Kogyo, Volkswagen, Nissan, Motor Vehicle Manufacturers Association, Chrysler, General Motors, Toyota, and Gulf & Western urging the NHTSA to reconsider its rejection of the lower impact test speeds proposed in January.

For the reasons discussed in the March *Federal Register* notice the NHTSA has determined that the pendulum and barrier impact speeds should not be reduced and should remain at 5 mph.

General Motors (GM) submitted two documents, dated January 9, 1976, and January 15, 1976, which analyzed the costs and benefits of 1974 bumper systems based on field surveys conducted in Fort Wayne, Indiana and Milford, Michigan. The conclusion reached by GM in these studies was that the 1974 model year bumper systems were not cost-beneficial. They requested, based on the result of this study, that any raising of the current bumper standard requirements be delayed until longer-term benefit-cost analyses are made.

The NHTSA has examined this study and has concluded that the proposed Part 581 damageability standard, which will upgrade the bumper requirements, should be implemented in accordance with the time schedule set forth in this notice. GM in its study has chosen to analyze the cost-effectiveness of bumper systems designed solely for safety component protection. The costs considered by GM have been those occasioned not only by damage to safety-related components, but to non-safety-related vehicle areas, as well. While it may be true that a bumper system that is designed primarily for safety component protection will also provide some degree of protection against non-safety-related damage, it is unreasonable to evaluate the cost-effectiveness of such a system on its capability to perform outside its primary design function. A bumper system designed to comply with Title I would necessarily provide protection to both safety and non-safety-related components and would thereby reduce the degree of damage suffered by most 1974 model vehicles involved in front and rear impacts. The cost-effectiveness of a Title I system, thus, cannot be realistically measured by an examination of 1974 systems which have been designed to provide a lower level of damage protection.

GM gathered data only on its own 1974 model cars and concluded that the impact of Standard 215 on all vehicles has not been cost-beneficial. Conclusions based on such limited data, however, are not sufficient reason for suspending further rulemaking to improve the damage protection capabilities of bumpers. As explained in the March 12, 1975, notice, considerable data have been presented indicating that the bumper systems on some current-model automobiles are heavier and costlier than necessary. This unnecessary weight not only adds to the initial costs, but also increases the life-time operating costs of the vehicle. The use of such bumpers, it has been concluded, has been the result of unnecessary design choices by motor vehicle manufacturers. Studies conducted by the NHTSA and Houdaille Industries, Inc., a bumper manufacturer, indicate that bumper systems utilizing current technology and designed to meet the Part 581 damageability requirements need not weigh any more than pre-standard-215 bumper systems. Basing future rulemaking on the results of a cost-benefit analysis utilizing bumper systems that have not been optimized would be unreasonable.

In the March 12, 1975, notice, the NHTSA proposed alternative effective dates for implementation of the initial Part 581 test requirements. The applicable requirements call for restricted surface damage except to components that actually contact the impact ridge of the pendulum test device or that fasten such components to the vehicle chassis frame. Commenters were asked to address the feasibility of satisfying the proposed damage criteria by September 1, 1976, September 1, 1977, or September 1, 1978. Chrysler said it could meet the prescribed damage level by September 1, 1976, but only if certain modifications in the test requirements were made. Volvo also stated that it could comply by September 1976, but warned of a significant cost penalty. Toyo Kogyo and British Leyland stated they could meet a September 1, 1977 effective date. Toyo Kogyo, however, commented that this would occasion high development costs. British Leyland, on the other hand, said that it could satisfy an earlier effective date, but only at significant cost. American Motors, Ford, and Toyota urged a September 1, 1978, effective date saying that amount of lead time was necessary to obtain compliance.

The Insurance Institute for Highway Safety, the National Association of Independent Insurers, and State Farm urged a 1976 effective date citing the need for regulation of damage to vehicle components and surface areas aside from those directly related to safety. The Insurance Institute supported its request for a 1976 effective date by stating that many existing cars are substantially able to meet the initial Part 581 requirements.

In the NHTSA's view, adoption of a 1976 or 1977 effective date would impose serious lead time problems on a number of manufacturers. Based upon information submitted by the automobile industry, bringing vehicles into compliance by September 1, 1976 or 1977, if possible at all, would entail the expenditure of large sums of money for redesign and retooling. A September 1, 1978 effective date would assure satisfactory compliance with the Part 581 requirements and would avoid the high costs that would occur as a result of an earlier effective date.

The NHTSA has, therefore, concluded that a September 1, 1978, effective date should be adopted for implementation of the initial Part 581 damageability requirements. This amount of lead time appears necessary for all manufacturers to come into conformity with the provisions.

Toyo Kogyo, American Motors, Motor Vehicle Manufacturers Association, Chrysler, and Ford urged a delay in the proposed September 1, 1979 effective date for implementation of the "no damage" bumper requirements. Toyo Kogyo requested a 1983 effective date, while the other manufacturers suggested that no upgraded requirements be scheduled until field data have been gathered indicating the success of the interim requirements. The National Association of Independent Insurers, anxious for early implementation of the full range of bumper performance requirements, supported adoption of the proposed 1979 effective date.

The NHTSA has examined all of these comments and has concluded that the September 1, 1979 effective date should be adopted. This would provide a lead time of approximately 4 years, which appears sufficient to bring the vehicles into compliance. Awaiting the results of field data related to the interim requirements is not practicable. The information currently before the agency indicates that the proposed 1979 surface damage limitation is a substantial step towards achieving the level of bumper efficiency described by Congress in the Cost Savings Act. Waiting for the accumulation and analysis of additional information would unnecessarily and unreasonably delay the implementation of Part 581, a standard the agency is directed by law to promulgate.

The NHTSA has proposed in several past notices the adoption of test requirements that would allow the manufacture of vehicles with soft exterior surfaces. Currently, the Standard No. 215 exterior protection standard prohibits contact with Planes A and B of the pendulum test device since those areas represent parts of the vehicle that house safety components such as headlamps. Most vehicles constructed with soft exterior surfaces would not be able to comply with the Standard No. 215 requirements since by their very nature they would yield to the impact of the pendulum. The quality of soft face bumper systems which is not taken into account

by the Planes A and B prohibition is that such systems can be constructed in a manner that assures return of the system to its original contours following an impact. The NHTSA proposal would permit contact with the planes at limited force and pressure levels. These force and pressure limitations were intended to assure that the bumper system would yield in a collision to a degree that would minimize damage to the other vehicle's components.

Comments to the proposal to allow contact with Planes A and B focused on that provision's test conditions and its specification of pressure limitations. According to commenters, the prescribed instrumentation of Planes A and B is not practicable since it would be costly with allegedly unreliable test results.

British Leyland, Renault, and Peugeot wanted the agency to clarify the rule by specifying that no instrumentation is necessary on the pendulum where there is no contact during testing with Planes A and B. This fact should be clear based on prior interpretations given by the NHTSA. It has been stated many times in the past that a manufacturer need only exercise due care in assuring that his vehicle would comply with the requirement of a standard when tested by the NHTSA in the manner prescribed. The manufacturer need not conduct the tests prescribed in the standard in order to satisfy this duty. Depending upon the circumstances there may be other means by which he can certify his vehicles' compliance. In the case at issue, the instrumented pendulum would only serve to assure that impact with the planes would not exceed the stated maximum levels. If there is no contact with these planes then obviously the instrumentation would serve no purpose.

Volvo suggested that the provision permitting Planes A and B contact not be added to the standard until a measuring device can be better defined. American Motors, however, presented a suggestion that it contended would significantly simplify the test procedure without diminishing the desired level of vehicle protection. It suggested that the 200-psi limitation be deleted and that a force limitation of 2000 pounds on the combined surfaces of Planes A and B above the impact ridge and 2000 pounds total force on Plane A below the impact ridge be adopted.

American Motors stated that the 200-psi specification was unnecessary in light of the damage limitations contained in the standard.

The initial Part 581 damage criteria [proposed to go into effect September 1, 1976, or 1977, or 1978 (made effective by this notice for September 1, 1978)] presented some problems for Volkswagen, American Motors, Chrysler, Volvo, and Ford with respect to the areas in which damage would be permissible. The proposed section (S5.3.8) limits change to surface areas and safety components, but permits damage to the bumper face bar. The manufacturers argued that damage should also be permitted to cosmetic filler panels, bumper guards, nerf strips, license plate brackets, stone shields, and other components which are not specifically part of the vehicle body. The support for this position is that these components appear to be included in the proposal's description of items that would not be subject to damage limitation during the interim period.

The relevant language of S5.3.8 states that vehicles shall have no damage except to the bumper face bar and the components and associated fasteners that directly attach the bumper face bar to the chassis frame. The bumper face bar is defined as any component of the bumper system that contacts the impact ridge of the pendulum test device. Stone shields and cosmetic filler panels would not be excepted from the damage criteria unless they directly attach the bumper face bar to the chassis frame. Based upon the information currently before the agency, it has determined that neither stone shields nor filler panels are intended to serve such a function.

Bumper guards and nerf strips which are located in a position where they are contacted by the impact ridge of the test device would be considered as a bumper face bar with the lateral metal component (commonly known as a bumper) considered as a component that directly attaches the bumper face bar to the vehicle chassis frame. This reasoning would also apply to bumper systems that have a layer of plastic, rubber, or some other material covering the underlying load bearing structure. The covering material would be considered the bumper face bar and the underlying structure would be considered a component that attaches the face bar to the chassis frame.

Toyo Kogyo commented that the damage criteria contained in S5.3.8 would necessitate the addition of 13 pounds to the bumper which would change the emission rank of some cars and thereby increase their fuel consumption from 4 to 8 percent. The cost of counteracting the increased fuel consumption would, according to Toyo Kogyo, range from $100 to $200 per car.

The additional lead time allowed by the September 1, 1978 date for implementation of the initial damage criteria should enable Toyo Kogyo to concentrate its efforts on minimizing any increase in the weight of complying vehicles.

State Farm expressed concern over the application of the S5.3.8 damage criteria to vehicles with soft face systems. They asserted that allowing damage to the bumper face bar and associated components would, in the case of soft face bumper systems, permit damage to the entire front and rear end of the vehicle. This could occur since some soft-face construction utilizes a single large component in the front and rear of the vehicle that takes on the appearance of the vehicle body, but by definition would be the bumper face bar. It was State Farm's suggestion that damage be permitted only to those portions of the bumper face bar that actually come in contact with the impact ridge of the pendulum test device. This would in their opinion avoid the possibility of widespread damage to areas not actually contacted.

The NHTSA finds State Farm's concern unfounded. The 2000-pound total force limitation to the combined surfaces of Planes A and B of the pendulum test device will have the effect of preventing any substantial damage to the areas mentioned by State Farm. For this reason, the NHTSA denies State Farm's request to revise the language of S5.3.8.

Ford Motor Company criticized the provision prohibiting breakage or release of fasteners or joints (S5.3.9) as unreasonable. It asserted that efficient production requires keeping to a minimum the efforts involved in installing moldings and insignia. Of importance, in their opinion, is assuring that the moldings and insignia resist "popping" on rough roads and during minor parking lot impacts. However, they assert that the performance level that would be achieved by

PART 581—PRE 4

S5.3.9 is unreasonably high since, in their view, moldings which pop off can be easily reinstalled with minimal cost and inconvenience to the car owner.

The NHTSA disagrees with Ford's argument. To allow the type of damage described by Ford would be partially to defeat the effectiveness of the standard. Ornaments that fall off and trim strips that pop off must be repaired if the value of the vehicle is to be maintained. The time and money invested by an individual who must obtain such a repair following a relatively minor collision can be avoided if the manufacturer is required to comply with the performance level of S5.3.9. The NHTSA disagrees with Ford's assessment of the time, cost, and effort involved in obtaining such repairs. The agency has therefore determined that to carry out the Congressional intent to reduce the cost of low-speed accidents, it must require ornaments and trim strips to be immune from damage under the test conditions of the standard.

There were numerous comments on the damageability requirements proposed to go into effect on September 1, 1979. Many of the manufacturers suggested a change in the maximum dent limitation (S5.3.11) and requested that a certain amount of bumper set be allowed. In its March 12 notice, the NHTSA proposed to limit damage to the bumper face bar to permanent dents no greater than ⅜ inch from the original contour. The proposed ⅜-inch deviation was based on a Louis Harris & Associates survey of public reactions to bumper damage at various depths. This survey was commissioned by Houdaille Industries, Inc., a manufacturer of bumpers.

International Nickel Co. and Toyota requested that the provision be revised to allow a ¾-inch deviation from the original bumper contour. In light of the results of the Harris survey, which indicated that consumers did not consider damage to be significant until the dents reached a depth of ¼ to ½ inch, the NHTSA denies their request and adopts the proposed ⅜-inch limitation. To allow deviations to a depth of ¾ inch would be to disregard the results of the survey by permitting damage which would be considered significant by many consumers. This would undercut achievement of the purpose of the Part

581 bumper standard to reduce consumer loss of time and money.

Toyo Kogyo, American Motors, International Nickel, and Houdaille urged that the provision (S5.3.11) be amended to permit a certain degree of bumper set. It was pointed out that the impact to a bumper during testing can result in two types of contour change, dent and set. Bumper set is an overall movement or flattening of the bumper face bar which when minor is rarely detectable by the unaided human eye. Under the currently proposed provision the ⅜-inch deviation limitation would apply to both setting and denting, with the total of these two types of deviations limited to ⅜ inch. Thus, the permissible degree of dent deviation would actually be less than ⅜ inch. Compliance with such a requirement would, according to commenters, result in the production of heavier and more costly bumper systems.

Since the NHTSA has based its ⅜-inch deviation limitation on consumer reaction to a dent of that depth, it agrees with commenters that a certain degree of bumper set could be permitted in addition to dent without visibly altering the level of allowable bumper damage. Minor set is generally imperceptible. Thus, allowing it to occur during impact tests would not significantly reduce the level of performance currently assured in the proposed provision. The NHTSA hereby amends Part 581 to permit ¾ inch of bumper set in addition to dents of ⅜ inch.

Consumers Union asserted that the NHTSA should not require near-zero level of damage on all cars since such a regulation would prevent manufacturers from offering as an option cars with cheap, lightweight, expendable bumpers which meet the standard's other requirements. The NHTSA finds no merit in this suggestion and for the following reasons denies the request. First of all, to make compliance with the "no damage" provisions optional would be to disregard the mandate of Congress in the Cost Savings Act, which instructs the agency to promulgate a standard that will reduce consumer costs occasioned by bumper damage. Second, cars produced with lower performance bumpers would be less expensive than those meeting the Part 581 criteria. They might, therefore, seem more appealing to consumers who are unaware

PART 581—PRE 5

of the costly damage that might be incurred during low-speed collisions. The purpose of Title I of the Cost Savings Act is to protect consumers from such an eventuality. Third, mass production is the factor that will keep manufacturing costs at a low level. If only some vehicles are constructed with damage-resistant bumpers, the cost of those vehicles is likely to be higher than necessary because of this factor.

Nationwide Mutual Insurance Co. and the National Association of Independent Insurers expressed concern that the ⅜-inch deviation limitation was too lenient. Nationwide felt that the ⅜-inch deviation constituted a relaxation of the NHTSA's previous position that only a dimple should be allowed to the bumper. The NHTSA has concluded, based on the Harris survey, that a dent ⅜ inch in depth would be inconsequential to most car owners. Prescribing such a deviation as the maximum allowable in a 5-mph barrier or pendulum impact is, therefore, in keeping with the goal of reducing economic loss occasioned by low-speed collisions.

The National Association of Independent Insurers suggested that the ⅜-inch deviation be upgraded to require that the dent extend over a minimum area in a dishing fashion which would be less noticeable. This suggestion is rejected since the ⅜-inch provision has been fully supported as providing a damage level that fulfills the goals of Title I. In addition, prescribing a dishing effect as a necessary element for compliance would not take into account the various types of impacts to which a vehicle is subject.

State Farm urged that the prohibition against separations of surface materials, paint, polymeric coatings, or other materials from the surface to which they are bonded be extended to cover the bumper face bar during barrier impact tests. Under the current proposal these surface damage limitations would apply only to parts of the vehicle other than the bumper face bar. State Farm asserted that the limitation of application of the no-surface-damage requirements to vehicle surfaces other than the bumper face bar was intended to accommodate the pendulum impact. They therefore see no justification for applying the same limitation during barrier impact testing.

The NHTSA denies State Farm's request. While both barrier and pendulum impacts can cause some chipping or flaking of chrome or soft-face material (depending upon the type of system being tested), such damage is insignificant. Application of a no-surface-damage requirement to the bumper face bar would probably result in manufacturers having to upgrade their plating process or use more sophisticated covering materials to assure compliance. This could result in significant cost increases with little, if any, increase in benefits.

Both State Farm and British Leyland requested that S7.1.1 of Part 581 be clarified to indicate that the pendulum impacts from 16 and 20 inches are intended to be inclusive. Since compliance with the pendulum impact requirements at any height between 16 and 20 inches would necessitate meeting the damage criteria at heights infinitesimally close to 16 and 20 inches, the clarification requested by these commenters is insubstantial. The NHTSA, however, amends S7.1.1 to include the 16- and 20-inch heights as subject to the damage criteria, since some persons apparently considered it unclear.

Chrysler requested a modification of the Part 581 longitudinal pendulum impact test to specify that the required pendulum impacts be at least 12 inches apart laterally and 1 inch apart vertically from any prior impact. The request is denied, since such a modification would prohibit more than one hit in the same area of the bumper. Under the current Part 581 proposal, an impact within 12 inches laterally must be separated from any prior impact by 2 inches, vertically. Based upon available accident data, the NHTSA has concluded that a vehicle will be involved in an average of approximately 2 to 3 bumper collisions at speeds of 5 mph or less in its 10-year life. On an individual vehicle basis, the distribution or the area of the bumper affected by these impacts cannot be predicted. In order to assure a performance level that corresponds with real-world conditions, the NHTSA has determined that each bumper must be capable of meeting the prescribed damage criteria when subjected to more than one pendulum impact in the same area of the bumper.

A substantial number of comments were received from individuals concerned that the Part

PART 581—PRE 6

581 bumper standard might in some way limit the recycling of bumpers in the aftermarket. This concern is unfounded, since the requirements contained in Part 581 ensure that a wide variety of materials can continue to be used in bumper systems. The provisions in no way restrict the use of metals in bumper systems.

Chrysler argued that the pendulum test device should be used only as a means of assuring uniform bumper height. In its opinion, the pendulum impact test does not constitute an appropriate means of evaluating bumper damageability since the pendulum is rigid, heavy, and aggressive.

The NHTSA does not find Chrysler's argument meritorious. To delete the pendulum impact test as a means of establishing bumper damageability resistance would be to lower considerably the proposed level of performance currently contained in Part 581. The pendulum impact requirements assure that a vehicle is capable of involvement in various types of low-speed collisions without sustaining significant damage. They impose localized stresses at various points on the bumper face bar while the barrier impacts only establish a vehicle's overall ability to withstand impacts at specified energy levels, assuring the basic strength of the front and rear bumper. In order to satisfy its Congressional mandate by reducing the economic loss occasioned by low-speed collision damage, the NHTSA has concluded that the Part 581 bumper standard must prescribe test requirements that measure a vehicle's damageability characteristics in both barrier and pendulum-type stress situations.

In light of the foregoing, Title 49, Code of Federal Regulations, is amended . . . .

1. Federal Motor Vehicle Safety Standard No. 215, *Exterior Protection* (49 CFR 571.215), is revoked.

2. A new Part 581, *Bumper Standard*, is added to read as set forth below.

Effective date: September 1, 1978.

(Sec. 103, 119, Pub. L. 89–563, 80 Stat. 718 (15 U.S.C. 1392, 1407) ; sec. 102, Pub. L. 92–513, 86 Stat. 947 (15 U.S.C. 1912) delegation of authority at 49 CFR 1.51.)

Issued on February 27, 1976.

James B. Gregory,
Administrator, National Highway
Traffic Safety Administration

41 F.R. 9346
March 4, 1976

PREAMBLE TO AMENDMENT TO PART 581—BUMPER STANDARD

(Docket No. 74–11; Notice 17; Docket No. 73–19; Notice 14)

This notice responds to petitions for reconsideration of the March 4, 1976, Federal Register notice (41 FR 9346) establishing a new bumper standard that limits damage to vehicle bumpers and other vehicle surfaces in low-speed crashes.

Effective Date: September 1, 1978.

Address: Petitions should be submitted to: Administrator, National Highway Traffic Safety Administration, 400 Seventh Street, S.W., Washington, D.C. 20590.

For Further Information Contact:

Tim Hoyt, Office of Crashworthiness,

Motor Vehicle Programs,

National Highway Traffic Safety Administration,

Washington, D.C. 20590 (202–426–2264).

Supplementary Information:

The standard, 49 CFR Part 581, issued under the authority of Title I of the Motor Vehicle Information and Cost Savings Act, Public Law 92–513, 15 U.S.C. 1901–1991, limits damage to non-safety related components and vehicle surfaces and incorporates the safety-related damage criteria of the current Standard No. 215, *Exterior Protection* (49 CFR Part 571.215). Under the new standard, all vehicles manufactured on or after September 1, 1978, must be capable of undergoing prescribed pendulum and barrier crash tests while experiencing damage only to the vehicle bumper and those components that attach it to the vehicle frame. Vehicles manufactured on or after September 1, 1979, must be capable of undergoing the same tests while experiencing no damage to vehicle exterior surfaces except on the bumper, where dents not exceeding ⅜ inch and set not exceeding ¾ inch may occur.

Petitions for reconsideration were received from General Motors (GM), Ford, Chrysler, American Motors Corporation (AMC), Gulf &

Western, Nissan, and Leyland Cars. The issues raised by petitioners focused primarily on Part 581's cost-benefit basis, its leadtime, and its damage criteria.

GM, Ford, Chrysler, AMC, Nissan, and Gulf & Western stated that the National Highway Traffic Safety Administration (NHTSA) failed to present evidence that Part 581 would be cost beneficial. Ford stated that the record supporting Part 581 gives no assurance that the public will realize incremental savings once the standard is implemented. Chrysler, Nissan, and Gulf &Western cited cost and weight increases which they alleged would impose additional burdens on car owners over and above those presently experienced. AMC complained that the provision for escalating the bumper requirements after one year would result in costly and complex bumper designs, since such a schedule would prohibit the optimization of bumper systems.

Petitioners requested that the agency demonstrate that the requirements of Part 581 will provide cost savings greater than those currently provided by Standard No. 215, *Exterior Protection*. It was suggested by GM, AMC, and Ford that the agency undertake field studies to gather data to support the Part 581 standard. Several manufacturers suggested that implementation of Part 581 be postponed until such time as a field study is completed.

Petitioners' arguments have been raised in past comments to Federal Register notices proposing a Part 581 bumper standard. The NHTSA found them unpersuasive then and hereby rejects them once again. The NHTSA and Houdaille Industries conducted cost benefit studies on compliance with the Part 581 bumper requirements. The studies indicate that bumper systems using current technology and designed to meet the standard's requirements will provide a favorable

PART 581—PRE 9

cost-benefit ratio. Petitioners have not presented evidence that effectively disputes the conclusions reached in these studies.

Conducting field studies as a means of gathering evidence to support implementation of the Part 581 standard is unrealistic and would not demonstrate as accurately as the Houdaille and NHTSA studies the positive cost-saving potential of the standard. Many manufacturers are continuing to comply with the current Standard 215 bumper requirements by means of inefficient, unoptimized bumpers. Data gathered on these systems thus would not indicate the full possibilities of bumpers specifically designed to meet the Part 581 requirements in an efficient manner. Once manufacturers start utilizing the technology and materials available to them the full benefits of the Part 581 bumper standard can be realized. Until such time, however, manufacturers have it within their power to cause field study results to be misleading and unrepresentative of the potential of Part 581.

The NHTSA has ample evidence in the record that manufacturers are capable of meeting the requirements of Part 581. It also has evidence that compliance can be achieved in a cost-efficient manner. There has been no evidence presented by any of the petitioners that the standard would have a negative cost-benefit impact if met in the ways outlined by Houdaille and the NHTSA in their studies. The agency therefore rejects the cost-benefit objections raised by petitioners.

AMC requested additional leadtime to meet the requirements of Part 581. It contended that it needs 36 months' leadtime to comply with Part 581. It asked that the initial effective date of the standard be delayed until September 1, 1979.

The NHTSA finds AMC's request without merit. The 30-month leadtime for the initial requirements and the 42-month leadtime for the final requirements is considered adequate for compliance. No other manufacturers have expressed concern over attaining the level of performance prescribed for 1978, and evidence in the record indicates that most vehicles already come close to satisfying the specified damage criteria. The request of AMC is therefore denied.

General Motors objected in its petition to the prescribed escalation of the bumper requirements

for September 1, 1979, only 1 year after the standard's initial effective date. It stated that compliance with two sets of bumper requirements within such short period of time would result in unrecoverable costs relating to research, design, development, and tooling, and would inhibit the feasibility of optimizing its bumper systems.

Ford Motor Company stated that it plans to redesign its passenger cars for 1981 due to the requirements of the Energy Policy and Conservation Act (Pub. L. 94–163) and associated legislation. Ford explained that compliance with Part 581 will entail some redesign. It therefore requested that the bumper standard's effective date be delayed until September 1, 1980, so that these necessary redesigning efforts can be accomplished simultaneously.

The agency has found both General Motors' and Ford's requests persuasive. It has therefore issued a notice proposing to delay for 1 year the implementation of the second phase of bumper requirements from September 1, 1979, until September 1, 1980. This action does not conform exactly to Ford's request. However, the NHTSA does not know of any vehicles that would require major design changes until implementatoin of the more stringent second phase requirements.

Filler panels and stone shields were identified in the March 4, 1976, final rule as exterior vehicle surfaces that must experience no damage as a result of the prescribed test impacts. GM, Chrysler, and AMC objected to this interpretation of the level of damage resistibility filler panels and stone shields must achieve. GM contended that these components are part of the bumper system and provide the transition between the bumper face bar and body panels. It stated that bumper stroke causes unavoidable surface scratches, abrasions, and displacements, which could be eliminated only by using expensive materials and mounting techniques. Chrysler pointed out that filler panels are designed to flex during bumper impacts and may not return to exactly their original contour. According to AMC, however, once a deformed bumper is repaired following an impact, the flexible filler panel will return to its original contour. All three manufacturers requested that filler panels be permitted to sustain some degree of damage during testing.

The agency has reexamined the role of filler panels and stone shields in the bumper system and finds that although they do not actually hold the bumper to the vehicle frame, they are cosmetic components that are part of the entire system that performs the task of attaching the bumper to the frame of the car.

The NHTSA has concluded that permitting damage to filler panels and stone shields will not significantly degrade the level of performance required for vehicles manufactured after September 1, 1978. The flexibility of the filler panel and stone shield material enables it to withstand deforming impacts without permanently losing its shape, but, as long as the bumper and components attaching it to the vehicle frame are permitted to sustain damage as a result of impacts, the filler panel and stone shield may likewise sustain some degree of damage. Since these components are less visible than the bumper itself, the small amount of damage that they will incur will normally not be as significant as that allowed to the bumper. Therefore, filler panels and stone shields on vehicles manufactured from September 1, 1978, to August 31, 1979, will be permitted to sustain damage during the prescribed test impacts. This, in essence, grants the requests of petitioners. The agency will address in an upcoming notice the application of damage criteria to stone shields and filler panels on vehicles manufactured after September 1, 1979.

Ford and Chrysler charged that the Part 581 damage criteria are impracticable and lacking in objectivity. Specifically, they objected to the criteria that allow no separations or deviations, and require certain systems to operate in a normal manner. According to petitioners, these criteria are not objective since the requirements of no separation and no deviations can be interpreted as meaning that even the most microscopic deviations and separations are prohibited, or alternatively that only those deviations that are readily apparent are prohibited. With regard to the requirement that certain systems operate in a normal manner, petitioners stated that the meaning of "normal" is unclear and can be interpreted differently by different people. Ford and Chrysler expressed concern that the agency will

interpret the meaning of these damage criteria in a manner conflicting with their interpretation. To resolve the situation to which it is objecting, Chrysler suggested that the requirements be revised to allow minimal and inconsequential deviations, while Ford suggested that the agency withdraw S5.3.2 and S5.3.5 and parts of S5.3.3, S5.3.8, S5.3.10, and S5.3.11 pending development of objective criteria to enable manufacturers to predict accurately whether their vehicles will comply.

The agency understands the petitioners' concerns, but finds that a simple interpretation of the cited requirements is adequate to satisfy their objections. The damage criteria allowing no deviations and no separations are not intended to apply to microscopic changes in the vehicle following test impacts. The types of deviations and separations addressed by Part 581 are those that are perceptible without the use of sophisticated magnifying or measuring equipment. What is required is that the vehicle not reflect any normally observable changes in the stated areas following the prescribed test procedure. Damage that is only identifiable by use of microscopically-oriented equipment would not be considered as prohibited under Part 581.

With regard to the requirement that a vehicle's hood, trunk, and doors operate in the normal manner, the standard is simply providing that these systems continue to operate following the test impacts in the same manner as they did before the impacts. This requirement has been a part of Standard No. 215, *Exterior Protection*, since its implementation on September 1, 1972. No compliance controversies have ever arisen concerning it.

Leyland Cars and AMC requested that the requirements of S5.3.11, allowing no more than 3/4-inch set and 3/8-inch dent to the bumper face bar, be made applicable to the component that backs up the bumper face bar. Leyland Cars explained that some of its bumpers are covered by a rubber or plastic molding which, under Part 581, would be considered as the bumper face bar. It requested that the component over which the molding is placed be permitted to sustain the same degree of set allowed for the bumper face bar. AMC asked that the component underlying the molding be permitted to experience dents up to ⅜-inch as is the bumper face bar.

The NHTSA finds petitioners' concerns unfounded. The prohibition against set and denting applies to vehicle exterior surfaces. From the description of the component supplied by Ford and Chrysler it appears that it is completely covered by the molding and is not an exterior surface area of the vehicle. Therefore, it may experience damage during test impacts. The molding enveloping the reinforcement would represent the exterior surface that is subject to the requirements of S5.3.11.

Nissan and Gulf & Western objected to the prescribed limitations on set and denting contained in S5.3.11. Nissan requested that the damage criteria be revised to allow ½-inch dent and 1-inch set, instead of the currently required ⅜-inch dent and ¾-inch set. It was Nissan's contention that such a revision would cause only a slight change in the appearance of a damaged vehicle, while enabling a considerable change in a vehicle's cost and weight. Gulf & Western alleged that there was no economic justification for the ⅜-inch dent and ¾-inch set requirements since they are based solely upon a public opinion poll. It requested that the Part 581 requirements not be implemented until an economic justification is presented.

The NHTSA finds both Nissan's and Gulf & Western's requests lacking in merit. A survey conducted by Louis Harris & Associates of public reaction to various degrees of bumper damage showed that a significant number of people consider ½-inch dents to be damage they would repair. Based upon this information and cost and weight data contained in the various studies upon which the agency relied in the formulation of the standard, it has been determined that the amendment requested by Nissan would adversely affect the results to be achieved by implementation of the Part 581 bumper standard. The results of the Harris survey have definite economic significance in that those individuals indicating that a certain degree of damage was significant enough that they would have it repaired were providing the pollster with cost data. Damage that is repaired will have a financial impact on the car owner. By the same token,

damage that is detectable and thereby have an economic impact on the car owner. These cost factors were all considered in deciding on the ⅜- and ¾-inch damage limitations. For these reasons, the requests of Nissan and Gulf & Western are denied.

Chrysler objected to the procedure prescribed for measuring the depth of bumper dents (S5.3.11(b)), charging that it is unreasonable, inaccurate, and lacks objectivity. Chrysler alleged that the end points of the straight line described in the test procedure for connecting the bumper contours adjoining the contact area are locations that are subjective on bumper face bars with compound curvature. It also charged that the specified measurement method lacks objectivity and can be used only for determining the depth of dents in flat surfaces. Chrysler requested that the agency clarify the provision.

Although the objections raised by Chrysler illustrate that some configurations are more difficult to measure than others, it is the agency's judgement that the method described in S5.3.11(b) is valid and still the most feasible means of determining the extent of damage. Location of the end points of the straight line used to measure the depth of bumper dents does not, in the opinion of the NHTSA, pose a problem. In order to establish the exact location of the end points, the manufacturer may either paint or chalk the pendulum test device. In this way, the pendulum will leave a mark on the precise area of contact.

With regard to Chrysler's objections concerning the measurement of dents, it should be noted that the straight line measurement technique is not necessarily a test procedure. Rather, the language specifying that a deviation from original contour not exceed ⅜-inch when measured from a straight line connecting the bumper contour adjoining the contact area should be considered a definition of a dent. Deformations outside the contact area on the bumper surface, such as recessions of a larger area of the bumper, are defined as set.

The agency realizes that the measurement of dent and set on some bumpers with complex curvature may not be a simple procedure. In such cases, the testers must use measurement procedures that will enable them to accurately measure the degree of dent the bumper has incurred. In situations involving a concave face bar, a reference line can be established by placing a straight line across the area of contact prior to impact. After completion of the actual impact the change in bumper contour can be measured from the previously established reference line. In situations involving a convex face bar, or more complex surfaces, it may be necessary for the manufacturer to remove the bumper following impact in order to compare it with an unimpacted bumper, or to make a cast of the preimpact bumper for comparison with the bumper for comparison with the bumper following the prescribed testing.

Chrysler also requested that S5.3.11 be amended to specify that bumper set be measured relative to the vehicle frame in perpendicular, parallel, and vertical directions with respect to the vehicle's longitudinal centerline. It stated that such a revision would reduce the task of measuring permanent set to a reasonable level.

The NHTSA denies this request since Chrysler has presented no information indicating that the currently prescribed measurement procedure is unfeasible. The agency knows of no reason why reference lines relative to the vehicle frame cannot be established from which bumper set can be measured. To adopt Chrysler's suggested method for measurement would unduly complicate the procedure since determination of the vehicle longitudinal centerline is complex.

GM charged that the NHTSA's definition of bumper face bar may include license plate brackets that are attached to the vehicle bumper, since these components may contact the impact ridge of the pendulum test device. If identified as the bumper face bar, these license plate brackets would be required to meet the level of performance prescribed for bumpers. According to GM, such a result would be extremely costly. License plate brackets capable of complying with the bumper damage criteria would be expensive to produce as well as to replace. This, in GM's opinion, would have a negative cost-benefit impact.

While the NHTSA agrees that license plate brackets should not be required to meet the damage criteria of the bumper face, the NHTSA believes that it is good design practice to locate license plates in an area other than the bumper face. However, recognizing the limited space available on the front of some cars for license plate placement, the NHTSA is reluctantly willing to grant GM's petition on this point. The agency will, in the future, review industry practice on the placement of license plates on new automobiles in an effort to determine if future rulemaking on this matter would be desirable.

AMC requested in its petition that the NHTSA amend the requirements limiting the total force on planes A and B to 2,000 pounds (S5.3.7) to permit a force of 2,000 pounds on plane A below the impact ridge and a force of 2,000 pounds on the combined surfaces of planes A and B above the impact ridge. AMC based its request on the premise that the current requirement allows the full 2,000-pound force to be exerted either above or below the impact ridge of the test device. It pointed out that the NHTSA stated in an earlier notice that the 2,000-pound limit would prevent any substantial damage to the vehicle. Based upon this, AMC argued that allowing 2,000 pounds of force both above and below the impact ridge would not expose those surface areas to any greater force than would be allowed under the current requirements.

The NHTSA disagrees with AMC's contention. The force limitation contained in Part 581 is intended to assure that the primary force of the impact is directed at the bumper face bar. Although all 2,000 pounds of allowable force could be directed to the area either above or below the impact ridge, this total amount of force would not be a significant damage factor. However, if the areas covered by planes A and B were allowed to sustain a total force of 4,000 pounds, the focus of primary force on the bumper face bar would not be assured and the type of aggressive bumper system Part 581 is designed to prevent could be utilized. AMC's request is therefore denied.

AMC requested that Part 581 be amended to include a provision appearing in the January 2, 1975, proposal (40 FR 10) that stated a vehicle need not meet further requirements after having been subjected to either the longitudinal pendulum impacts followed by the barrier impacts, or the corner pendulum impacts.

The agency has stated in past notices that a vehicle will be involved in an average of three low-speed collisions in its 10-year life. There is no way to predict which portion of the bumper will be affected in these impacts. Therefore, it was decided that vehicles should be required to meet the prescribed damage criteria when subjected to the entire series of test impacts. To provide otherwise would be to establish a level of performance lower than necessary to protect a vehicle from the full range of potentially damaging impacts it is likely to incur during its on-road life. It was for this reason that the provision appearing in the January 2. 1975, proposal was not adopted. It is for this same reason that the agency denies AMC's request.

The text of the Title I bumper standard has in previous notices and the March 4, 1976, final rule been published in the format of a motor vehicle safety standard. Since the bumper standard is actually an entire part within Chapter V of the Code of Federal Regulations, the format must be changed in order that it may be properly codified. The content of the standard will remain the same. This notice, however, revises the numbering system so that it conforms to the Code of Federal Regulations format.

The principal authors of this notice are Guy Hunter, Office of Crashworthiness, and Karen Dyson, Office of Chief Counsel.

In light of the foregoing, 49 CFR Part 581, is amended and recodified. . . .

Effective date: September 1, 1978.

(Sec. 103, 119, Pub. L. 89–563, 80 Stat. 718 (15 U.S.C. 1392, 1407); sec. 102, Pub. L. 92–513, 86 Stat. 947 (15 U.S.C. 1912); delegation of authority at 49 CFR 1.50.)

Issued on May 4, 1977.

Joan Claybrook
Administrator

42 F.R. 24056
May 12, 1977

PART 581—PRE 14

PREAMBLE TO AMENDMENT TO PART 581—BUMPER STANDARD

(Docket No. 73–19; Notice 19 & Docket No. 74–11; Notice 22)

This notice corrects an inadvertent error in the notice that changed the format of Part 581, *Bumper Standard*, so that its numbering system conformed to the Code of Federal Regulations format (42 FR 24056; May 12, 1977). In that notice, the new numbering was not totally incorporated into the body of the regulation.

For further information contact:

Mr. Tim Hoyt
Office of Crashworthiness
Motor Vehicle Programs
National Highway Traffic Safety
Administration
Washington, D.C. 20590
202–426–2264

Supplemental information: On May 12, 1977, the National Highway Traffic Safety Administration published a Federal Register notice (42 FR 24056; FR Doc. 77–13235) responding to petitions for reconsideration of the March 4, 1976, notice (41 FR 9346) establishing a new bumper standard. The May notice also changed the format of Part 581. The text of the bumper standard was previously published in the format of a motor vehicle safety standard. Since the standard is actually an entire part within Chap-

ter V of the Code of Federal Regulations its numbering system was revised in order that it could be properly codified.

When Part 581 was published with its revised format, only the section headings were properly renumbered. The texts of the various sections were inadvertently left unchanged. This notice revises the section references in the body of the regulation to conform to the new format.

The principal author of this notice is Karen Dyson, Office of Chief Counsel.

In accordance with the foregoing, changes should be made to 49 CFR Part 581, *Bumper Standard.* . . .

(Sec. 103, 119, Pub. L. 89–563, 80 Stat. 718 (15 U.S.C. 1392, 1407); sec. 102, Pub. L. 92–513, 86 Stat. 947 (15 U.S.C. 1912); delegations of authority at 49 CFR 1.50 and 49 CFR 501.8.)

Issued on July 26, 1977.

Robert L. Carter
Associate Administrator
Motor Vehicle Programs

42 F.R. 38909
August 1, 1977

PREAMBLE TO PART 581—BUMPER STANDARD

(Docket No. 73–19; Notice 24)

This notice responds to a request from Ford Motor Company for further interpretation of the bumper damageability requirements of Part 581, *Bumper Standard*, and announces the photographic procedure NHTSA will use as an aid in determining whether damage to filler panels and stone shields (shielding panels) is normally observable for purposes of compliance with the standard. This interpretation assists manufacturers in ascertaining whether contemplated bumper designs will provide a level of performance consistent with the requirements of Part 581. This notice also corrects an inadvertent error in the previously announced effective dates for Phase I of the bumper requirements.

Date: This interpretation and the correction to Part 581 are effective immediately.

For further information contact:

Mr. Richard Hipolit, Office of Chief Counsel, 400 Seventh Street, S.W., Washington, D.C. 20590 (202–426–9512)

Supplementary information: NHTSA has established, through issuance of Part 581, Bumper Standard (49 CFR Part 581), requirements for the impact resistance of vehicles in low speed collisions. The effective dates of Part 581 are September 1, 1978, for components other than the bumper face bar and certain associated fasteners (Phase I), and September 1, 1979 for all vehicle components (Phase II). On May 15, 1978, the agency published a notice (43 FR 20804) summarizing its interpretation of various aspects of the Part 581 damage resistance requirements as they relate to vehicle exterior surfaces. Ford Motor Company has asked for additional clarification of the requirement of paragraphs 581.5(c)(10) and (11) of the standard, in a June 22, 1978, request for interpretation that has been placed in the public docket.

APPLICATION OF THE DAMAGE CRITERIA TO BUMPER FACE BARS AND ATTACHED COMPONENTS

The Phase II requirements prohibit permanent deviations from the original contours of vehicle exterior surfaces following pendulum and barrier impacts. An exception is made for the "bumper face bar," whose surface is permitted ¾-inch deviation from its original contour and position relative to the vehicle frame (set) and a ⅜-inch deviation from its original contour on areas of contact with the barrier face or the impact ridge of the pendulum test device (dent) (§ 581.(c)(11)). Bumper face bar is defined in § 581.4 as "any component of the bumper system that contacts the impact ridge of the pendulum test device." NHTSA has stated that this definition includes components of a multipiece bumper which are connected as part of the same load bearing structure to a bumper system component which is contacted either by the pendulum test device or the test barrier (43 F.R. 20804; May 15, 1978).

Ford has inquired as to the applicability of this definition of bumper face bar to a variety of components such as directional signals and shielding panels, which may be mounted to a load bearing structure while themselves performing no structural function. Components which do not perform a load bearing function are not necessarily components of the bumper system (and potentially bumper face bar) solely as the result of their incidental mounting on or near a load bearing structure of the bumper system. Components must be examined on a case-by-case basis to determine whether they constitute components of the bumper system.

The agency stated in a previous notice that shielding panels are considered a component of the bumper system and thus will qualify as bumper face bar if contacted in testing (43 F.R. 20804; May 15, 1978). The same would be true of other cosmetic components directly associated with the bumper system's function such as manufacturing cut-out patches and tape strips the primary funtcion of which is to hide protrusions, primary function of which is to hide protrusions, fasteners, or other unsightly aspects of the

Illumination devices, e.g., fog lamps and directional signals, are not associated with the bumper system's function and could not qualify as components of the bumper system, even if contacted by the pendulum test device or barrier.

Still other components could be considered components of the bumper system, depending on their application in a particular vehicle design. For example, a grille, which would generally be associated with the vehicle body, could perform a protective function as a component of a bumper system in a soft-face configuration, and could therefore qualify as a component of the bumper system.

The agency recognizes that components mounted to a bumper face bar, but not themselves considered face bar because they are not part of the bumper system or are not impacted in testing, will necessarily move with the set of the bumper face bar, although they do not qualify for the permissible 3/4-inch set allowance of (c) (11) (i). However, the stricter damage limitations of paragraph 581.5(c)(10), applicable to such components, are actually limited to "normally observable changes in the started area following the prescribed test procedures" (42 F.R. 24058; May 12, 1977). "[M]ovement of small patches covering manufacturing process cut-outs on the face bar" and movement of shielding panels with the set of the bumper are not considered normally observable (43 F.R. 20804; May 15, 1978). Similarly, non-bumper (e.g., fog-lamps) and other bumper system components (e.g., tape strips), attached to or built into a bumper face bar but not contactable by the test device, would not be considered to have normally observable damage when they simply move with

the set of the face bar. Such movement would, however, be normally observable if the function of the mounted component were impaired, e.g., by misalignment, in the case of a fog lamp beam, to the extent that it would not be adjustable to its normal aim.

The thin, polymeric tape strips described above typically are adhesively bonded to the surface areas of the bumper face bar. The impact of the pendulum test device or test barrier with the bumper face bar may cause distortions on portions of the face bar not directly impacted during testing and cause localized separation on these tape strips from the face bar surface, in the form of wrinkling or bubbling.

The agency had previously stated that, "while both barrier and pendulum impacts can cause some chipping or flaking of chrome or soft-face material (depending on the type of system being tested), such damage is significant" (41 F.R. 9346; March 4, 1976). This reasoning also governs minor damage to tape strips, such as wrinkling or bubbling, so long as the strips are contactable and thus qualify as bumper face bar. This interpretation would apply equally whether the damage happened to fall at the area of impact or elsewhere on the face bar.

Any component of the bumper system which can be contacted by the impact ridge of the pendulum test device in any permissible pendulum stroke is considered bumper face bar for testing of that bumper system, whether or not it was actually contacted in a particular test sequence. Further, the interpretation concerning non-contactable but load bearing components of multipiece bumpers discussed above, although originally announced in the context of metal bumpers (43 F.R. 20804; May 15, 1978), would also govern a multipiece bumper assembly equipped with plastic or rubber bumper guards or nerf strips. Thus, all load bearing components of the bumper assembly, whether plastic, rubber, or metal would be considered bumper face bar and be entitled to a 3/4-inch set if they are connected as a part of the same load bearing structure.

MEASUREMENT OF DAMAGE TO THE BUMPER FACE BAR

Paragraph 581.5(c)(11) provides:

Thirty minutes after completion of each pendulum and barrier impact test, the bumper face bar shall have—

(i) No permanent deviation greater than ¾ inch from its original contour and position relative to the vehicle frame; and

(ii) No permanent deviation greater than ⅜ inch from its original contour on areas of contact with the barrier face or the impact ridge of the pendulum test device measured from a straight line connecting the bumper contours adjoining any such contact area.

Ford has inquired as to the measurement techniques the agency will use in determining compliance with these damage limitations. NHTSA has previously recognized that "the measurement of dent and set on some bumpers with complex curvature may not be a simple procedure" (42 F.R. 24056; May 12, 1977). In many cases there may be more than one procedure by which damage can be accurately measured. Innovations in measurement techniques may be needed as new bumper designs are developed. Therefore, while the agency can express the basic measurement geometry (which appears to be Ford's basic concern) that establish compliance with the damage limits, it cannot specify a particular method to be used in measuring those distances in all cases.

Ford requested resolution of the inadvertent inconsistency between agency statements in the May 1978 interpretation that "the two types of deviation are additive in an area of contact with the barrier face or impact ridge" but that "the localized deviation permitted by paragraph (ii) is measured taking any contour in the area of impact and measuring its movement from its location prior-to-impact to post-impact." The first statement accurately represents that the deviations are additive in the area of contact with the barrier or pendulum. The second statement failed to make the different and intended point that the contour of the contact area is measured from the contour previous to contact, but only after movement of the surface position and contour relative to the vehicle frame attributable to

set has been subtracted. It should be noted that contour change attributable to set must result from a generalized flattening of the bumper surface outside the area of contact. Otherwise the concept of dent would be indistinguishable from contour set.

The agency rejects Ford's suggestion to merely measure the contour in the contact area in relation to the surrounding contour following impact. The best example of why the original contour must serve as the baseline is the case in which the contact area consisted of a ⅜-inch protrusion from the surrounding area prior to impact and a ⅜-inch depression in relationship to the surrounding contour following impact. The resulting dent would actually be ¾-inch deep.

Ford further recommended that all dent measurements be made in vertical sections of the plane of impact which produced the dent. Recognizing the need for flexibility in the measurement of complex bumper configurations, Ford has withdrawn this portion of its request for interpretation.

Ford has questioned the portion of NHTSA's previous interpretation (43 F.R. 20804; May 15, 1978) which stated that dent may be measured "along any dimension, i.e., width, length, depth," from any line connecting the adjacent bumper contours. The agency has decided that the ⅜-inch dent limitation of § 581.5(c)(11)(ii) should presently be limited to depth measurements only. Development of the Phase II face-bar contour requirements and studies which formed the basis for the ⅜-inch dent requirements during the rulemaking proceeding focused primarily on limitation of the depth of deviations. A ⅜-inch dent limitation measured in any direction might, at this time, impose an unanticipated burden in some cases and perhaps restrict the flexibility of manufacturers in selecting bumper systems for different model sizes which provide a suitable balance among the interrelated considerations of damage resistance, weight reduction, and cost. Should future testing and bumper design developments indicate that further face-bar dent limitations would be beneficial, such a requirement will be the subject of a future rulemaking notice.

Finally, Ford has asked whether there can be more than one contact area for purposes of measuring damage resulting from a particular impact. It is clear that multiple areas of contact between the bumper face bar and the impact ridge or test barrier may exist, thus creating multiple areas in which dent may occur. Given the complexity of some bumper designs, it would be unrealistic and impractical to require that all damage incurred in an impact be combined for measurement purposes. Deviations caused by impact at non-contiguous locations on the bumper system will be treated as separate contact areas, and damage in each of these areas will be measured separately, without reference to any other area of contact.

PHOTOGRAPHIC PROCEDURES TO AID IN EVALUATING DAMAGE TO SHIELDING PANELS

NHTSA's previous interpretation of the Part 581 requirements (43 F.R. 20804; May 15, 1978) addressed the problem of judging damage to vehicle shielding panels for purposes of determining compliance with paragraph 581.5(c)(10). That provision addresses all exterior surfaces other than bumper face bar and prohibits permanent deviation from original contours or separation of materials from the surface to which they are bonded. The interpretation reiterated that the agency does not consider damage to shielding components to be in violation of the standard if that damage is not "normally observable." In the case of shielding panels, damage not visible in good quality, photographic prints of the suspect area would not be considered by the agency to be "normally observable." The notice indicated that the Office of Vehicle Safety Compliance (OVSC), formerly the Office of Standards Enforcement, would establish standard procedures by which NHTSA would take its evaluative photographs.

While NHTSA originally stated that 8 by 10 inch photographic prints would be employed, the agency has concluded that the use of contact prints of that size may present practical difficulties due to the limited availability and unwieldiness of large cameras. Further study of existing photographs indicates that 4 by 5 inch contact prints are adequate for the agency's testing.

Upon completion of impact tests in accordance with the test procedures of paragraph 581.7, OVSC photographs shielding panel areas that may have experienced permanent deviation or separation of materials.

View Camera. OVSC uses a standard 4 by 5 inch View Camera with focal length of 127 mm, a maximum aperture of f/4.7, a coated lens, and available shutter speeds of 1 second to 1/400 second.

Film. OVSC uses type 52 Pola Pan 4 by 5 inch film for Polaroid prints.

Illumination. OVSC takes the photographs indoors using the following illumination procedures: (11) illuminating the area to be photographed with crosslighting using two 1,000-watt photofloods lamp for main light, and one 1,000-watt photoflood lamp for fill-in light; and (2) positioning the photoflood lamps so that the light rays strike the subject area at a 45° angle from a distance of 10 feet from the area being photographed.

Camera position. OVSC positions the camera at a distance of 6 feet from the center of the suspect area and utilizes ground glass focusing to properly focus the camera for that distance. Photographs are taken both at 90° and 45° angles relative to the suspect area.

Exposure. OVSC utilizes a General Electric, DeJur or Weston photoelectric exposure meter to determine the exposure requirements. Light readings are taken by measuring the intensity of reflected light from a Kodak Gray Card placed upon the area to be photographed. The meter is placed near enough to the subject (gray card) to indicate the average reflected light (at least within a distance equal to the width of the subject being photographed). A light reading is obtained and set opposite the film speed which is indicated on the meter so that the f/stop or the aperture settings and shutter speeds coincide. The correct camera setting is read directly from the meter.

PART 581—PRE 20

Photographic print. OVSC produces 4 by 5 inch black and white photographic contact prints from the Polaroid film.

Examination of contact print. OVSC examines the completed contact print with the unaided eye for compliance with 581.5(c)(10).

CORRECTION OF PHASE I EFFECTIVE DATES

On May 12, 1977, NHTSA published a *Federal Register* notice (42 F.R. 24056) responding to petitions for reconsideration and revising the format of Part 581 as originally announced on March 4, 1976 (41 F.R. 9346). Those notices inadvertently indicated that the Phase I exterior surface requirements, now contained in paragraph 581.5 (c)(8), would apply to vehicles manufactured from September 1, 1978 to August 1, 1979. The requirements of paragraph 581.5(c)(8) actually apply to vehicles manufactured until August 31, 1979, and the regulation is therefore corrected to reflect the intended effective dates.

In consideration of the foregoing, the date 'August 1, 1979," contained in 49 CFR § 581.5 (c)(8), is hereby corrected to read "August 31, 1979."

The program official and lawyer principally responsible for this document are Nelson Gordy and Richard Hipolit, respectively.

(Secs. 103, 119, Pub. L. 89–563, 80 Stat. 718 (15 U.S.C. 1392, 1407); sec. 102, Pub. L. 92–513, 86 Stat. 947 (15 U.S.C. 1912); delegation of authority at 49 CFR 1.50).

Joan Claybrook
Administrator

43 F.R. 40229–40232
September 11, 1978

Bumper Standard
(Docket No. 73-19; Notice 29)

ACTION: Final rule.

SUMMARY: This notice amends the Bumper Standard to reduce the test impact speeds required by that standard to 2.5 mph for longitudinal front and rear barrier and pendulum impacts and 1.5 mph for corner pendulum impacts. The notice also amends the damage resistance criteria of the standard to eliminate limitations on the damage which may be incurred by the bumper face bar and associated components and fasteners in bumper testing.

The agency finds that under this action net benefits will accrue to the public and to the nation's consumers. This action is thus required by the mandate of the Motor Vehicle Information and Cost Savings Act that any bumper standard issued under that statute "seek to obtain the maximum feasible reduction in costs to the public and to the consumer," taking into account the costs and benefits of implementation, effects on insurance and legal costs, savings in consumer time and inconvenience and considerations of health and safety.

Any reduction in costs related to bumper systems, including savings from reduced fuel consumption, will exceed any reduction in benefits which may occur because of increases in damage, insurance costs, delay and inconvenience, and other matters. This action will thus increase and seek to maximize the net consumer and public benefits of the standard. The agency also finds that this action will cause no reduction in vehicle safety.

EFFECTIVE DATE: July 4, 1982.

SUPPLEMENTARY INFORMATION: The "Part 581 Bumper Standard" (49 CFR Part 581) specifies levels of damage resistance performance which passenger motor vehicles must provide in low speed collisions. Bumper performance is measured in test impacts with both a fixed collision barrier and a pendulum test device. Bumpers must meet damage criteria which preclude any damage at all to vehicle exterior surfaces, which ensure protection of various safety-related components of the vehicle, and which allow only minimal damage to the bumper itself.

Background

The history of the Part 581 bumper standard has been long, extremely controversial and fraught with uncertainty. The current action is the culmination of years of study, analysis and agency action and reaction.

Federal Motor Vehicle Safety Standard 215

In its initial efforts in the field of bumper regulation, the National Highway Traffic Safety Administration (NHTSA) issued Federal Motor Vehicle Safety Standard (FMVSS) 215, *Exterior Protection,* under the National Traffic and Motor Vehicle Safety Act (the Safety Act). 15 U.S.C. 1381 *et seq.* As initially implemented on September 1, 1972, that standard imposed requirements which prohibited damage to specified safety-related components and systems, e.g., headlights and fuel systems, in a series of perpendicular barrier impacts, at 5.0-mph for front and 2.5-mph for rear bumper systems.

One year later, several new requirements became effective under FMVSS 215. First, rear barrier impact speeds were increased from 2.5-mph to 5.0-mph. Second, the standard specified 5.0-mph perpendicular front and rear pendulum impacts and 3.0-mph corner front and rear pendulum impacts. Third, a bumper height requirement was in fact established by specifying that the longitudinal pendulum impacts must be

PART 581; PRE 23

made between a height of 16-20 inches. (The corner pendulum impacts were limited to a height of 20 inches until September 1, 1975, when the standard specified that they must be made within the same 16-20 inch height range.)

Motor Vehicle Information and Cost Savings Act

On October 20, 1972, Congress enacted the Motor Vehicle Information and Cost Savings Act, ("the Act"). 15 U.S.C. 1901 *et seq.* The stated purpose of Title I of the Act is to "reduce economic losses associated with low speed collisions of motor vehicles." 15 U.S.C. 1901(b). Section 102(a) directed the Secretary of Transportation[1] to promulgate bumper standards in accordance with the criteria of section 102(b) which requires that such standards—

seek to obtain the maximum feasible reduction of costs to the public and to the consumer, taking into account:

(A) the cost of implementing the standard and the benefits attainable as the result of implementation of the standard;

(B) the effect of implementation of the standard on the cost of insurance and prospective legal fees and costs;

(C) savings in terms of consumer time and inconvenience; and

(D) considerations of health and safety, including emission standards.

15 U.S.C. 1912 (b)(1)

The Act also provides that the bumper standards must not conflict with motor vehicle safety standards issued under the Safety Act. 15 U.S.C. 1912(b)(2).

Adoption of the Part 581 Standard

Pursuant to both the new authority of the Act and that of the Safety Act, NHTSA established the Part 581 Bumper Standard in 1976. 41 Fed. Reg. 9,346 (March 4, 1976). As adopted, this

standard combined the safety features of FMVSS 215 with new damage resistance criteria intended to promote consumer cost savings.

The Part 581 standard established compliance test procedures which consist of a series of five test impacts on both the front and the rear bumper. Each test series includes one longitudinal barrier impact, two longitudinal pendulum impacts and two corner pendulum impacts.

The Part 581 standard sets forth substantive requirements in terms of damage resistance criteria which took effect in two stages. The first stage, or "Phase I" of the Part 581 standard, became effective on Setpember 1, 1978, on which date FMVSS 215 was *ipso facto* revoked. Phase I incorporated the former FMVSS 215 safety criteria, and added new damage resistance criteria which prohibited damage to all exterior vehicle surfaces, e.g., sheet metal, *other than* the bumper face bar and related components and fasteners.

More stringent damage resistance criteria, known as the "Phase II" criteria, became effective one year later, on September 1, 1979. The Phase II criteria expanded Part 581 by also imposing limits on the amount of "dent" and "set" damage which could be sustained by the bumper face bar itself in the same series of test impacts. "Dent" refers to permanent deviation from the original contour of the bumper face bar in areas of contact with the barrier face or the impact ridge of the pendulum test device. "Set" refers to permanent deviation of the bumper from its original contour and position relative to the vehicle frame. Phase II limited allowable dent to 3/8 inch, and set to 3/4 inch, each as measured thirty minutes after completion of each test impact.

Early Proposals and Evaluations of the Bumper Standard

1973

NHTSA initially proposed a Part 581 standard in August 1973, while FMVSS 215 was in force, but after the passage of the Act. This 1973 proposal would have required protection against damage in 5.0-mph test impacts. 38 Fed. Reg. 20,899 (August 3, 1973).

1975

NHTSA thereafter issued a second Part 581 proposal, in January 1975. This revised proposal

---

[1]The authority of the Secretary to promulgate safety standards has been delegated to the NHTSA Administrator. 49 CFR 1.51(a).

would not only have reduced (at least temporarily) the impact speeds required by FMVSS 215, but also would have reduced the damage resistance criteria contained in the Part 581 proposal still pending from 1973. 40 Fed. Reg. 10 (January 2, 1975). These proposed reductions were based primarily on the results of two intervening agency-sponsored studies, which indicated that the cost and weight of many of the then-current production bumpers had made such bumpers no longer cost-beneficial. The 1975 proposal would also have reduced the number of longitudinal pendulum impacts from six front and six rear, to three front and three rear.

After considering information and arguments submitted in response to the August 1973 and January 1975 proposals, the agency issued yet another proposal in March 1975. 40 Fed. Reg. 11,598 (March 12, 1975). At that time, the agency withdrew the January 1975 proposal regarding test speeds, and proposed instead only to amend the still pending 1973 proposal to reduce the number of longitudinal pendulum impacts to two front and two rear.

1976

The agency finally promulgated the Part 581 Bumper Standard in March 1976, specifying 5.0-mph test impact speeds and requiring a total of five barrier and pendulum impact tests for the front bumper and five for the rear.

1977

In 1977, however, NHTSA issued two further rulemaking proposals. The first would have delayed the effective date of the Phase II damage criteria one year. 42 Fed. Reg. 10,862 (February 24, 1977). The second, which replaced the first, proposed three alternatives: (1) a one-year delay of Phase II; (2) a one-year delay with a consumer information program on bumper performance in the interim; and (3) an indefinite delay of Phase II and substitution of the information program. 42 Fed. Reg. 30,655 (June 16, 1977). These proposals were withdrawn by the agency in November of that same year. 42 Fed. Reg. 57,979 (November 7, 1977).

Also in 1977, NHTSA decided to undertake a series of long term studies of its existing and proposed rulemaking efforts. As a part of this initiative, it began a multi-year evaluation of the Part 581 Bumper Standard. This evaluation which was released in April 1981, is discussed in detail below.

1978

In 1978, and after the effective date of the 5.0-mph, Phase I standard, the Senate Appropriations Committee included in its report on the fiscal year 1979 Appropriations Act for the Department of Transportation a directive that NHTSA conduct studies and analyses reevaluating to the maximum extent feasible the question of the level of bumper damage resistance which would be most cost-beneficial to the consumer. The Committee further directed the agency to modify the Part 581 standard (i.e., the standard to which this current rulemaking is addressed) in accordance with the results of such analyses. S. Rep. No. 938, 95th Cong., 2d Sess. 25 (1978).

1979

In February 1979, the agency completed a Preliminary Analysis which concluded that 2.5-mph bumpers offered approximately $77 more net benefits than 5.0-mph bumpers. In March 1979, the agency published an advance notice of proposed rulemaking seeking public comment on its February analysis. The notice indicated that the responses would be used to aid NHTSA in preparing a final report to the Senate Appropriations Committee and in determining the possible need for changes in the Part 581 standard.

In June 1979, NHTSA published a "Final Assessment of the Bumper Standard." That document estimated the net benefits of alternative bumper standards specifying test impact speeds of 2.5 mph, 5.0 mph, and 7.5 mph. The agency at that time concluded that a standard specifying 5.0-mph impact speeds should be retained since it was believed to provide slightly more lifetime vehicle net benefits ($39) than one specifying 2.5-mph impact speeds. In December 1979, the agency updated its assessment based on comments received from the automotive and insurance industries. It concluded that the advantage of the 5.0-mph standard over the 2.5-mph standard was less than previously thought, offering only $11-29 more lifetime vehicle net benefits than a standard specifying 2.5-mph speeds.

In late 1980, during the final days of the 96th Congress, a House-Senate conference committee reported out a bill which would have statutorily reduced the test speed in the Part 581 standard to 2.5 mph for a two-year period. H. R. Rep. No. 1371, 96th Cong., 2d Sess. 25 (1980). Sharp differences of opinion regarding the relative merits of the agency's two 1979 bumper analyses were highlighted in the Congressional debates. See, e.g., Senate debate of September 25, 1980, 126 Cong. Rec. S13499-501. However, Congress adjourned without taking final action on the bill.

1981

In April 1981, NHTSA published a notice of intent to review the Part 581 standard and propose again to modify the requirements of the Part 581 Bumper Standard. 46 Fed. Reg. 21,203 (April 9, 1981).

Also in April 1981, NHTSA completed and published its "Evaluation of the Bumper Standard," which it had begun in 1977. Based upon continually developing data and analyses, this report addressed in still further detail the costs and benefits of each phase of the agency's bumper requirements, beginning with the initial FMVSS 215 standard. The April 1981 Evaluation incorporated newly developed data from various agency studies on insurance claims for vehicles manufactured since the Part 581 standard took effect, on the incidence and extent of low speed collision damage, and on bumper costs. Unlike previous studies, the Evaluation separately analyzed front and rear bumpers. It found that regulated front bumpers tended to be cost effective while rear bumpers were not. This study, in accordance with both the Senate's 1978 directive and the provisions of Executive Order 12291, formed the basis for the agency's undertaking the current rulemaking.

Current Rulemaking

October 1981 Proposal and Analysis

On October 1, 1981, NHTSA published a notice of proposed rulemaking (the NPRM) seeking comments on nine different alternatives for amending Part 581. 46 Fed. Reg. 48,262. The proposals ranged from one reducing the test impact speed to 2.5 mph for rear bumpers only to one eliminating all test impact requirements for front and rear bumpers except as necessary to maintain a height requirement. Specifically, the nine alternatives were as follows:

— Alternative IA would have reduced the test impact speeds for rear bumpers only to 2.5 mph for longitudinal impacts and to 1.5 mph for corner impacts. It would have maintained the test impact speed for front bumpers at 5.0 mph and would have maintained the Phase II damage resistance criteria. (5.0 mph front/2.5 mph rear, Phase II)

— Alternative IB would have made the changes included in alternative IA and substituted Phase I damage resistance criteria for Phase II criteria for front and rear bumpers. (5.0-mph/2.5-mph, Phase I)

— Alternative IIA would have eliminated the damage resistance criteria for rear bumpers only, with the exception of the criterion that is intended to ensure uniform bumper height by requiring bumper contact with a pendulum test device within a specified height range. It would have maintained the 5.0-mph test impact speed and Phase II criteria for front bumpers. (5.0 mph/height only, Phase II)

— Alternative IIB would have made the changes included in alternative IIA and substituted Phase I criteria for Phase II criteria for the front bumper. (5.0 mph/height only, Phase I)

— Alternative IIIA would have reduced the test impact speed for front and rear bumpers to 2.5 mph for longitudinal impacts and 1.5 mph for corner impacts. It would have retained the Phase II damage criteria. (2.5 mph/2.5 mph, Phase II)

— Alternative IIIB would have made the changes included in alternative IIIA and substituted Phase I criteria for Phase II criteria for front and rear bumpers. (2.5 mph/2.5 mph, Phase I. This alternative is referred to below as the 2.5-mph/ 2.5-mph alternative.)

— Alternative IVA would have reduced the test impact speed for front bumpers to 2.5 mph for longitudinal impacts and 1.5 mph for corner impacts. It would also have eliminated the damage criteria for rear bumpers with the exception of the bumper height criterion. (2.5 mph/height only, Phase I)

— Alternative IVB would have made the changes included in alternative IVA and substituted Phase I criteria for Phase II criteria for front bumpers. (2.5 mph/height only, Phase I)

—Alternative V would have eliminated the damage resistance criteria for front and rear bumpers, with the exception of the bumper height criterion. (height only/height only)

The alternatives set forth in the NPRM were developed during the preparation of a Preliminary Regulatory Impact Analysis (PRIA) (Docket 73-19, Notice 27, No. 011).[2] The PRIA which was published for public comment simultaneously with the NPRM, built upon all of the agency's earlier evaluations and assessments. To encourage close scrutiny of the PRIA and the NPRM, and in recognition of the limited empirical data on several important issues, the agency specifically requested comment on 25 detailed questions which were set forth in the NPRM.

Using the present Part 581 standard for comparison, the PRIA estimated the changes in costs and benefits that were likely to occur if the standard were modified in each of the ways set forth in the October notice of proposed rulemaking. The PRIA concluded that the differences in probable net benefits among several alternative bumper standards were small. The results of the PRIA suggested that while 5.0-mph bumper requirements had in fact reduced lifetime repair costs for cars, they also had increased both car purchase prices and fuel consumption. The 5.0-mph bumper requirements had in fact reduced lifetime repair costs for cars, they also had increased both car purchase prices and fuel consumption. The 5.0-mph bumper requirements were found to have decreased insurance company claims payments and overhead, but also to have increased the manufacturing costs of car companies.

Public Meetings

The agency conducted two public meetings on the NPRM on October 22 and November 12, 1981, in fulfillmment of the statory requirement that

all interested persons be given an opportunity to present orally data, views and arguments on the October 1981 NPRM. The agency scheduled two separate meetings instead of a single extended one in response to a request by insurance industry representatives. Those representatives requested an opportunity to introduce data relating to suggested new compliance technologies whose use would reportedly allow the existing requirements of the Part 581 standard to be retained with little if any modification, but at greatly reduced economic cost. In the notice announcing the meetings, the agency urged all interested parties to provide technical and economic data that would help focus the issues at the first public meeting, and indicated that the second meeting would be used to allow others to respond to testimony at the first meeting. 46 FR 48958 (October 5, 1981).

The views and arguments advanced by responding parties with substantial economic interests at stake, e.g., the insurance and automotive manufacturing industries, were similar to those previously expressed in response to earlier analyses, proposals, and requests for comments. However, commenters did submit significant new data on several issues, including those relating to the cost and weight of bumpers providing different levels of protection.

Positions of Interested Parties

Time impact speed. Insurance industry representatives, generally joined by consumer representatives, expressed their support for retaining the current Part 581 requirements, based upon assertions of favorable benefit and cost analyses of the current standard, safety considerations, and the legislative history of the Act. Insurance representatives further contended that the legislative history indicates a Congressional intent that bumper standards be established at a level of 5.0 mph. They strongly opposed the option of adopting Regulation No. 42 of the United Nations Economic Commission for Europe (ECE).[3]

---

[2]In preparing the PRIA, the agency also considered the possibility of raising, as well as lowering the required test impact speeds. The 1979 Final Assessment stated that a 7.5-mph bumper would have marginally greater net benefits than a 5.0-mph bumper. However, the Executive Summary for that document indicated that the conclusions regarding the 7.5-mph bumper were based on substantially less data than were the conclusions regarding the 5.0-mph bumper and thus that the conclusions about the 7.5-mph bumper were far less reliable. Subsequently obtained data and analyses have not provided any basis for placing more credence in those three-year-old conclusions about 7.5-mph bumpers.

[3]ECE Regulation No. 42 requires that a car's safety systems continue to operate normally after the car has been impacted by a pendulum or moving barrier on the front and rear longitudinally at 4 kilometers per hour (about 2.5 mph) and on a front and rear corner at 2.5 kilometers per hour (about 1.5 mph) at 455 mm (about 18 inches) above the ground under loaded and unloaded conditions. See discussion under "Harmonization," below.

Some insurance industry commenters contended that the record in this proceeding is insufficient to support any reduction of the damage resistance or safety requirements of the Bumper Standard below current levels. These commenters, joined by an organization presenting arguments on behalf of consumers, argued (1) that in order to amend the standard the agency must be able to establish affirmatively that any selected alternative is one which uniquely meets the statutory criteria of the Act and the Safety Act, in a manner superior to any and all others, and (2) that on the record the agency is not able to make such a finding with respect to any particular alternative.

Auto industry commenters overwhelmingly supported the alternative proposing reduction of test impact speeds to 2.5 mph in longitudinal impacts and 1.5 mph in corner impacts, and substitution of Phase I damage criteria for Phase II criteria. Among the reasons stated in support of this alternative were assertions of cost-benefit analyses for that alternative more favorable to the consumer, the results of the agency's prior analyses, the similarity of this alternative to ECE Regulation No. 42, the greater relevance of the 2.5-mph design speed to the speed of the typical parking lot collision, and the enhanced prospects of gathering field data on the relative merits of 2.5-mph and 5.0-mph bumpers.

Three foreign manufacturers stated that they favored adoption of the requirements of ECE Regulation No. 42, but that the 2.5-mph/2.5-mph alternative was their second choice because of its similarity to the European standard. Several other manufacturers, while not advocating the adoption of the ECE requirements as such, noted the desirability of harmonizing United States and European bumper requirements. Some domestic and foreign automakers expressed reservations about adoption of the ECE standard in its entirety, but advocated adopting certain aspects of that standard, such as eliminating the fixed barrier test or establishing a single permissible bumper height.

A trade association representing materials suppliers registered its support for the 5.0-mph/5.0-mph standard, asserting that the standard provides the added advantage of affording actual protection at speeds above 5.0 mph. One bumper component manufacturer proposed the additional alternative of lowering the pendulum impact speed to 2.5 mph, while retaining the 5.0-mph impact speed for barrier tests. That commenter contended that the pendulum test, which concentrates force on a particular area of the bumper, is a disproportionately severe test which prevents use of optimum 5.0-mph bumper designs.

A number of private individuals also submitted views on the proposed alternatives. The majority of those commenting favored retention of existing Part 581 requirements, although apparently some comments were based on factual representations contained in media reports of the rulemaking proceeding, instead of the data and issues actually under review. See, e.g., Docket 73-19, Notice 27, No. 209. Insurance industry and public interest commenters claimed that public opinion favors the 5.0-mph/5.0-mph standard, and that significant, if not determinative weight should be given to such alleged preferences.

Phase I-Phase II damage resistance requirements. Several commenters specifically addressed the issue of differences between the Phase I and Phase II damage criteria. Automakers addressing the issue uniformly favored return to the Phase I criteria. Two manufacturers advocated elimination of all criteria addressed to damage to non-safety components. The insurance industry generally favored retention of the Phase II criteria, as did a component parts manufacturer, although one insurance industry commenter advocated consideration of permitting nonself-restoring energy absorbing devices.

Other test procedure modifications. Commenters discussed several other alternative approaches to the Phase I-Phase II issue, including merely amending the bumper standard test procedures. One modification discussed by several commenters would allow manual repositioning of bumper or shielding-panel components during testing. Both insurance and auto industry commenters agreed that manual repositioning would be a desirable modification of the bumper system test procedure. However, some auto industry commenters also stated that eliminating the Phase II damage criteria would serve to alleviate much of the need for manual repositioning.

Three vehicle manufacturers and one component supplier recommended limiting the number of pendulum test impacts so that the bumper standard test procedure would more closely approximate real life experience. These commenters advocated reducing the number of pendulum impacts to one longitudinal impact and one corner impact per bumper, or to one longitudinal and two corner impacts per bumper.

For additional details concerning comments on the NPRM, see the appendix to this notice.

Agency Decision

Drawing on the best available data, public comments submitted in response to the October 1981 NPRM, and comments presented at NHTSA's public meetings on October 22 and November 12, 1981, NHTSA has now completed a Final Regulatory Impact Analysis (FRIA) of the bumper standard alternatives. Docket 73-19, Notice 29, No. 001. Careful consideration was given to the data and analyses contained in the FRIA and all comments received in the rulemaking proceeding. Responses to all significant comments are contained either in this notice or the FRIA. Based on its review of all of these materials, the agency has decided to adopt the 2.5-mph/2.5-mph, Phase I alternative. The alternative reduces to 2.5 mph the front and rear longitudinal barrier and pendulum impacts for testing compliance with the safety and damage resistance criteria and substitutes Phase I damage resistance criteria for Phase II criteria.

In the agency's judgment, neither costs savings nor safety considerations warrant the retention of the current standard. Indeed, the agency believes that the changes in the damage resistance criteria and the compliance test speed are necessary in order to comply with the requirements of the Act that the standard seek to provide the maximum feasible reduction in costs to the public and the consumer.

As discussed in more detail below and in the FRIA, the extensive data analyzed by the agency and the reasoned assumptions made by the agency after opportunity for public comment have led the agency to the firm conclusion that the current 5.0-mph/5.0-mph standard does not meet the statutory requirements. Stated simply, the current standard does not provide or seek to provide the maximum feasible reductions in cost. Therefore, the agency has determined that the current standard can no longer be retained in accordance with the Act. Similarly, it is clear that a standard imposing a height-only requirement

for front and rear bumper systems would provide fewer net benefits than other alternatives considered in this rulemaking proceeding.

The agency recognizes that no single remaining alternative is dramatically superior in terms of net benefits over the wide ranges of reasoned assumptions made about the values of certain important variables. However, after careful comparison of the current standard and the specific proposed alternatives under ranges of assumptions, the agency concludes that the 2.5-mph/2.5-mph, Phase I alternative best satisfies the statutory criterion that the bumper standard "seek to obtain the maximum feasible reduction of costs to the public and to the consumer."

The agency has concluded that the alternatives involving differential front and rear impact speed requirements are less desirable because of uncertainties surrounding the effects of impacts between bumpers with different levels of aggressivity. These alternatives received no support among commenters. Alternatives involving height-only requirements for rear bumpers appeared to provide slightly less net benefits than the 5.0-mph/2.5-mph and 2.5-mph/2.5-mph alternatives under most sets of assumptions considered.

Alternatives which have higher impact speed requirements and would produce essentially the same net benefits, differ from the selected alternative principally in that they make an even trade of additional dollars saved in avoided damage for additional dollars spent for damage protection at such higher speeds. Those alternatives would thus fail to meet the test of the statutory criteria with respect to "maximum feasible reduction of costs." The initial direct costs to consumers of the selected alternative are less than those of that alternative which would in the agency's judgment be most likely to provide comparable net benefits, the 5.0-mph/2.5-mph alternative.

The agency has also concluded that reducing the impact speed to 2.5 mph and eliminating the Phase II damage criteria will not have an adverse effect on safety. Such amendments will have no discernible effect on the number of accidents, deaths or injuries that occur annually.

The new standard adopted in this notice will provide greater latitude and incentive for car manufacturers to improve bumpers through the innovative use of new designs and materials, while conforming to the clear Congressional directive that the agency promulgate and enforce a minimum performance standard seeking maximum feasible reductions in cost. Also, the chosen alternative best advances the goal of harmonization with international standards while meeting applicable statutory requirements.

Pursuant to Executive Order 12291, the agency has concluded that there is a strong and reasonable basis in the record of this rulemaking proceeding for the factual conclusions and choices of data and methodologies underlying the selection of the 2.5-mph/2.5-mph alternative.

Agency Rationale

The sharply opposed positions of the commenters on the many complex technical, analytical and policy issues raised in this proceeding provide dramatic evidence of the difficulty which the agency has faced in reaching this decision. The primary issues involved in the agency's decision are as follows.

Resolution of uncertainty. The Act directs not only that a bumper standard be adopted and maintained, but also that such standard be set at the particular level of performance which "seeks to provide the maximum feasible reduction of costs to the public and to the consumer," taking into account specified elements of costs and benefits.

On several of the issues presented in choosing among the various alternatives, the agency was confronted with uncertainties arising either from conflicts among data or from the absence or limited nature of relevent, reliable data.

Because of the prior history of the standard and the sequence of technology used by manufacturers to comply over time, field performance data under real world conditions are sharply limited to empirical data on two types of systems, as discussed elsewhere in this notice. As a result, the combination of the specificity of the statutory language and the limited nature of the data available has left the agency certain of the need to act, but marginally less certain as to which of the available alternatives and which means of analysis of such alternatives will produce the result most in conformity with the intent of Congress.

For several years, the agency has been taking all prudent steps to obtain more data to reduce uncertainty with respect to the appropriate standard and to analyze and account for the possible effects of remaining uncertainties on certain key variables. In a number of areas, more reliable data could not be developed by the agency. In the PRIA, the agency carefully identified and explained the assumptions it made in those areas and invited public scrutiny and comment. To ensure full discussion of all of the issues presented, the agency asked detailed questions regarding those assumptions in the October 1981 NPRM.

The agency's assumptions were the subject of extensive public comment. The agency received over two hundred comments from a full spectrum of interested parties and sought to gather all available data on the subject of this proceeding. New data, estimates and arguments were received which have assisted the agency in adjusting and refining its analysis of the standard and the alternatives.

The agency believes that sufficient information exists to make all determinations required by applicable statutory criteria. The uncertainties confronting the agency now are significantly less than those which existed when the current standard was promulgated. The agency knows far more now about the benefits and costs of bumper standards with various levels of performance requirements than it did then. In the agency's judgment, there is no reasonable prospect of obtaining more definitive data under the continued application of the existing Part 581 standard.

The record is most clear on the issue of the present standard's noncompliance with the criteria in the Act. If the agency were now setting a bumper standard for the first time, it could not justify establishing a 5.0-mph/5.0-mph standard. The existing 5.0-mph standard provides significantly less net benefit to the public and consumers than would several of the proposed alternatives with less stringent performance requirements.

The record and empirical data before the agency are less definitive with respect to some aspects of the agency's assessment of the proposed alternative standards. Some uncertainty continues to exist with respect to several issues, including the proper economic value to be assigned to delay and inconvenience, the number of relevant low-speed impacts which a car may be expected to sustain over its lifetime, the proper economic value to be assigned to damage which car owners themselves elect not to repair, the proper factor to be applied to determine the relationship between increases in bumper weight and resulting increases in the weight of other vehicle systems and structures to accommodate the heavier bumpers (secondary weight), and the extent of weight reductions which would accrue if various alternative standards were adopted.

NHTSA has explored these areas of uncertainty to the limits of available data and appropriate analytical techniques. Ultimately, the agency has relied in these areas upon inferences from available data, informed judgment about engineering, technical, economic and legal matters, and the informed and expert opinion of commenters on the issue of which alternative level of performance requirements will best achieve the policy objectives set forth in both the Cost Savings and Safety Acts.

The agency has subjected its interim findings and conclusions to sensitivity analyses, to identify and isolate the most significant (i.e., outcome determinative) variables and to determine the levels of confidence which may be placed on the values ultimately assigned to such variables. Where NHTSA could not with certainty assign a single value to a variable determined to be significant, the agency in all cases employed a range of values based upon the best available information. Those ranges generally include the values recommended by the commenters. The use of these ranges permitted the agency to examine the sensitivity of the results of its analysis and ensure the integrity of the outcome.

Finally, the agency identified the sets of assumptions it believes are most probable, and subjected each of its comparative analyses to various combinations of such values. These choices and related assumptions are discussed below in this notice and in greater detail in the FRIA itself.

Selection of test speeds, cost savings considerations—threshold factors. In its efforts to ensure the fullest consideration of the current standard and the proposed alternatives, NHTSA analyzed the net benefits of the standard and each alternative both by the use of average values and the use of extreme values for those variables about which there was either a significant measure of uncertainty or sharp and irreconcilable differences of opinion among the commenters. Some of the extreme assumptions were favorable to the current standard, while others were favorable to a reduced standard. The extreme values so analyzed represent in most cases neither a probable nor a reasonable outcome of events. Such analysis illustrates the most extreme of the possible outcomes in order to ensure the fullest consideration of the results of the agency's action.

Under the three sets of those extreme assumptions deemed to be the more reasonable by the agency, the net benefit calculation was found to favor a reduced standard. In these comparisons, all but one alternative proposal proved superior to the 5.0-mph/5.0-mph standard in terms of net benefits. See Table X-9 of the FRIA.

Only under the fourth set of extreme assumptions considered by the agency did the current standard yield more net benefits than did the alternatives. See Table X-9 of the FRIA. However, the agency considers it virtually impossible that the factual elements of that combination of assumptions could occur in reality, in large part because of inherent contradictions in economic or behavioral results that would be associated with such alignment. See Chapter XI of the FRIA.

Therefore, the agency can not, consistent with its statutory mandate, retain the existing standard.

Similarly, alternative V, which would have eliminated all but the height requirement for both front and rear bumpers, also is found to fail to maximize net benefits to the consumer under the range of combinations of assumptions considered. No set of assumptions or average set forth in Tables X-9 and X-10 of the FRIA showed superior net benefits for alternative V. Accordingly, this alternative has been rejected by the agency.

Given the relatively flat nature of the cost and benefit curves over the range between the 5.0-mph/2.5-mph and 2.5-mph/height-only alternatives, the choice among the remaining alternatives is more difficult. Particular sets of assumptions would suggest the superiority of various alternatives which retain some level of front bumper impact requirements but which would eliminate all impact requirements, and retain only a height requirement, for rear bumpers. However, any such apparent superiority in each case occurs only in the unique event of one combination of assumptions. Viewed as a whole, the data and probabilities associated with all combinations of assumptions preclude any reasonable finding that an alternative is superior where the range of necessary factual preconditions is so narrow.

First, under the sets of assumptions considered by the agency to be most likely or representative, the 2.5-mph/unregulated alternative cannot be found to be the alternative which is most likely to maximize net benefits. See Table XI-4 of the FRIA. Under all three sets of assumptions in that table considered by the agency to represent the most likely or average values for disputed elements of fact, the 2.5-mph/unregulated alternative provides fewer net benefits than does the 2.5-mph/2.5-mph alternative. Under two of those sets of facts, the net benefits of the 2.5-mph/unregulated alternative are also inferior to those of the 5.0-mph/2.5-mph alternative.

Second, while the net benefits of the 5.0-mph/unregulated alternative are closer to those of the 5.0-mph/2.5-mph and 2.5-mph/2.5-mph alternatives, they are still inferior. The net benefits of that alternative exceed those of the 2.5-mph/2.5-mph alternative in only one instance in Tables X-9, X-10 and XI-4. In several instances, the 5.0-mph/unregulated alternative yields less net benefits than does either the 5.0-mph/2.5-mph or 2.5-mph/2.5-mph alternative.

Finally, there is another consideration which leads to the rejection of the 5.0-mph/unregulated alternative. Any alternative not providing front and rear impact protection at the same speed raises uncertainty about the aggressivity results or other effects of differential requirements.

Among the alternatives having differential requirements, the 5.0-mph/unregulated alternative has the most extreme differential. Since there are not any hard data on the effects of this differential, those effects could not be factored into the net benefit calculations in the FRIA. However, the agency's engineering judgment leads it to the conclusion that implementing a standard with such a differential would cause front bumpers to be more aggressive than rear bumpers. This aggressivity differential would cause rear ends of cars to receive greater but presently unquantifiable

levels of damage in car-to-car collisions than they would if the impact speed requirements were identical.

The amount of any such additional rear end damage would offset in whole or in part any incremental benefits derived from requiring front bumpers to comply with more stringent requirements. Since these possibilities are not reflected in the net benefit figures for alternatives with differential front and rear impact speeds in Chapters X and XI of the FRIA, such net benefit figures would have to be considered overstated in the event that differential requirements were imposed.

The agency notes that implementing a standard with different front and rear bumper requirements could tend, in a front-to-rear collision between two cars, to have the undesirable effect of subsidizing some of the damage costs of the driver of the striking vehicle, who is most likely to be deemed under law to be at fault in causing the collision.

Finally, although commenters differed on the actual effects of differential impact speed requirements for front and rear bumpers, no commenter advocated adoption of a bumper standard requiring different test impact speeds, and some manufacturers suggested that consumer expectations would make bumpers subject to height-only requirements unacceptable in the marketplace.

Selection of test speeds, cost savings considerations—final decision. The considerations discussed above and the requirement in section 102 that the agency's standard seek to maximize cost reductions thus necessitated the determination by the agency of which of the remaining alternatives, i.e., the 5.0-mph/2.5-mph and 2.5-mph/2.5-mph alternatives, would seek to provide the greatest superiority in net benefits.

Based on the analysis in the FRIA, the agency concludes that the 2.5-mph/2.5-mph alternative more fully satisfies all aspects of the statutory mandate than does the 5.0-mph/2.5-mph alternative. The agency's choice between these two alternatives was reached after comparing the estimated results of implementing these alternatives under all examined sets of extreme assumptions, as well as under those sets of assumptions deemed by the agency most representative or most likely to occur. Under the sets of extreme assumptions in Table X-9 of the FRIA, an equal number of sets support the choice of each of these two alternatives.

However, when the highly unlikely fourth set of assumptions in Table X-9 is discarded, and the net benefits developed using the first three sets of assumptions in lines 1 through 3 of that table are averaged to represent equal probabilities of outcome for each of the sets of facts (See line 1 of Table XI-4), the 2.5-mph/2.5-mph alternative is clearly superior. This alternative yields $42 in net benefits relative to the current standard, compared with $33 in net benefits for the 5.0-mph/2.5-mph alternative.

The agency's direct comparison of these two alternatives in Table XI-4 under other sets (lines 2 and 3 of that table) of assumptions discloses that the 2.5-mph/2.5-mph and the 5.0-mph/2.5-mph alternatives would yield varying net benefits that do not differ greatly.

The agency has noted above the absence of hard data that would be desirable in determining precise values for some of the variables involved in projecting costs and benefits. It is important to note, however, that the variables about which the sharpest disagreements of fact have arisen in the record, e.g., the frequency of low speed accidents, the value of delay and inconvenience, and the appropriate factor to apply to arrive at secondary weight, are in fact also those variables which are the least significant to the outcome of the agency's net benefit calculations. For example, as shown in Table XI-2 of the FRIA, using the value for *each* of these variables which most favors retaining the current standard would reduce the net benefits of the 2.5-mph/2.5-mph alternative by only $4-12 over the life of the car. A shift in the values assigned to these variables would thus be least likely to produce a change in the outcome of the agency's determinative net benefit calculations. Thus, the variables about which the greatest controversy has arisen are in most cases also those which are least important in the decision-making process.

In selecting this alternative, the agency was also guided by its conclusion that where two or more alternatives yield net benefits or ranges of net benefits which are difficult to distinguish, the cost savings goal of the Act is most fully satisfied by selecting the alternative with the requirements which impose the lowest direct, immediate costs.

The 2.5-mph/2.5-mph alternative is the one which imposes the least direct, immediate costs on the consumer, i.e., the least increase in the cost of a new car. To illustrate this point, if the unregulated bumper is considered the baseline, the agency's analysis indicates that the increase in direct immediate cost to the consumer for bumper system components alone would be $21-41 for a car equipped to comply with the 2.5-mph/2.5-mph alternative, but $30-58, or 50 percent higher, for a car equipped to comply with the 5.0-mph/2.5-mph alternative. The choice of the 2.5-mph/2.5-mph alternative over the 5.0-mph/2.5-mph alternative reduces the direct bumper component cost increases by $9-17, and the difference would be even greater if secondary weight costs were considered. See Table VII-8 of the FRIA.

Selection of the alternative with less stringent requirements, and thus lower immediate costs, avoids forcing consumers to spend more in purchasing a new car in order to obtain what would only eventually, if at all, amount to equivalent net savings or benefits.

If the agency did not select the alternative with the lower immediate costs, the consumer would be required to spend additional money in pursuit of benefits whose occurrence and amount are less certain. The agency believes that the consumer is best served by an approach which in close cases favors the more certain over the less certain equivalent net benefit. NHTSA believes that this interpretation of the Act most fully implements the objectives of the Congress and of Executive Order 12291 and represents the soundest public policy.

The agency also must recognize, and if possible implement, the apparent distinction made in the Act between obtaining the "maximum feasible reduction of costs to the *public* and to the *consumer*" (emphasis added). The legislative history of the Act does not suggest a reason for the apparent distinction between the public at large and those who may purchase cars. One possible interpretation of this distinction is that Congress meant to seek the maximum possible benefits for the public in general, including those not purchasing cars. Once the agency has determined that the net benefits of the 5.0-mph/2.5-mph and 2.5-mph/2.5-mph alternatives are close, the agency believes that the only interpretation which would give appropriate weight to the statutory distinction between the "public" and "consumer" would be the alternative which better permits the marketplace to work efficiently and to produce innovative designs, the implementation of which will reduce overall costs to society as well as the purchasers of new cars.

Several automobile manufacturers and component suppliers commented that reduction of the test impact speed to 2.5 mph would facilitate use of new components and technologies, including plastics, ultra-high strength steel, and single-unit bumper systems. NHTSA believes that such design flexibility would be beneficial to the public at this time for several reasons. Innovation could result in more effective bumpers at lower cost to the public than would otherwise be available. Innovation and variety will allow individual consumers to apply their own individual value determinations on such important issues as the cost of delay and inconvenience, by opting to purchase more protection than would be cost-beneficial to the consuming public at large under the Act. Innovation, variety and a range of implemented choices in the marketplace will permit the agency to monitor cost and benefit trends and collect data about different performance levels of bumpers in the future.

The 2.5-mph/2.5-mph alternative will permit more innovation than the 5.0-mph/2.5-mph alternative because the former allows wider design freedom. Moreover, the 2.5-mph/2.5-mph alternative will increase the economic incentive of the manufacturers to retool because the parts for the new designs could be used on both the front and rear bumper systems of a vehicle. Without such innovation and retooling, the designs of bumpers are more likely to remain static, at least in the short run, and the benefits of innovative designs will be unrealized or significantly delayed.

There are other considerations that support the selection of the 2.5-mph/2.5-mph alternative. As noted above, any alternative specifying the same front and rear impact speed is deemed preferable to alternatives involving differential front and rear test impact speeds since an alternative with symmetrical requirements would not raise uncertainty about the effects of differential requirements. Further, a bumper standard requiring differential front and rear

impact speeds would lead to increased production costs and an increase in replacement part inventories as a result of probable losses in commonality of front and rear bumper components. Reduced commonality in a mass production market would be likely to increase the consumer cost of new vehicles and replacement parts.

In view of these differences between the alternatives and the probable consequences of the selection of each, the policies and requirements of the Act favor the choice of the 2.5-mph/2.5-mph alternative. As noted later in this preamble, the goal of section 102 is not to provide maximum protection against damage in low-speed collisions without regard to the cost of such protection. Instead, the goal is to reduce front and rear end damage in low-speed collisions under a statutory criterion and specific considerations that, when read together, indicate the most appropriate result is the one that minimizes the total consumer and public expenditure related to such damage and its prevention. The agency believes that the distinctions it has drawn between and the choices it has made among the alternatives are fully consistent with, and required in furtherance of, the policies of the Act.

Selection of test speeds; safety considerations. As discussed in more detail later, adoption of the 2.5-mph/2.5-mph alternative will not have any measurable effect on the risk that future accidents might be caused by safety components which malfunction due to damage incurred in prior low-speed collisions and which are left unrepaired. Available data indicate that very few accidents occur as a result of malfunctioning of those vehicle components which are subject to the safety criteria of the bumper standard. The agency concludes that far fewer accidents could be attributed, and only by speculation, to a failure to repair such components after they had been damaged in the only type of collision relevant to this discussion, i.e., one which might occur at an impact speed between 2.5 mph and 5.0 mph.

Similarly, the agency concludes that reducing the bumper standard test speeds will not increase the risk that safety components damaged in such low-speed collisions will cause injury in subsequent accidents caused by other factors. The only safety-related system that is covered by the safety criteria of the Part 581 bumper standard and that might contribute to injury in the event of an accident is the fuel system. However, the data relied upon by one commenter addressing this issue predated the effective date of FMVSS 301, Fuel System Integrity. That safety standard provides protection, independent of and substantially superior to that of the bumper standard, against the risk that fuel leaks will create a safety hazard in an accident.

The agency concludes also that reducing the test speeds for the safety criteria will not measurably affect the high-speed crash energy management of cars. The difference in the energy management capability of 5.0-mph bumpers and 2.5-mph bumpers is negligible at crash speeds such as those (30 mph) specified in the safety standards regulating the crashworthiness of new cars.

Finally, NHTSA concludes that reducing the bumper standard test impact speeds will neither create inconsistencies with any of the safety standards nor make compliance with those standards more difficult.

Corner impact speeds. It should be noted that selection of a 2.5-mph test impact speed for longitudinal impacts also necessitates the selection of a 1.5-mph corner impact requirement. The 1.5-mph corner impact speed represents an equivalent proportional reduction in the 3.0-mph corner impact speed in the current standard as compared to the reduction from 5.0 mph to 2.5 mph for longitudinal impacts. The agency has always established corner impact speeds at lower levels due to the greater damage potential of corner pendulum impacts relative to longitudinal pendulum impacts at the same speed. The greater relative severity of the corner impact results from the concentration of crash force on a single location, which is inherent in a corner impact, and the fact that impact absorbing devices are designed to provide maximum protection in the more common longitudinal impacts. If the proportional relationship of the longitudinal and corner impact speeds were not maintained, the effort to maximize net benefits would be frustrated.

Phase I versus Phase II. Making a choice between Phase I and Phase II damage resistance criteria was also difficult because of the limited empirical data available for comparing performance under the two sets of criteria. Phase I of the Part 581 standard remained in effect for only one model year (MY), 1979, and available information indicates that many manufacturers proceeded directly to bumper designs intended to meet the Phase II requirements prior to their effective date. The information that is available from surveys of vehicle owners and from insurance files indicates no discernible difference between the net benefits of MY 1974-78 and MY 1980 bumpers. Even if this information did reveal a difference, there are no data which the agency could use to determine the relative contributions of Phase I and Phase II to those benefits.

No compliance testing of MY 1979 models was conducted by NHTSA. The agency's compliance test results for MY 1980 suggest greater levels of protection for MY 1980 cars than is found in empirical data on real world damage experience for Phase II bumpers. The agency believes that in such cases agency decisions must be more strongly influenced by real world data since they reflect actual experience and are more reliable indicators of future real world experience. The insurance claim and survey data reflect the myriad variations in accident conditions and circumstances encountered in actual driving. In contrast, the compliance tests involve a limited and idealized set of conditions and circumstances. Those tests were necessarily chosen by the agency with the knowledge that they were imperfect surrogates from which to predict on-road experience.

Those commenters addressing the issue generally noted the cost and weight savings available by deleting the Phase II requirements. Commenters also pointed out that the increased use of non-metallic face bars has decreased the visibility of dent and set and thus greatly changed the circumstances under which such damage must be evaluated. Moreover, as suggested in the comments, deletion of Phase II would eliminate present difficulties in evaluating minor damage in compliance testing. The agency has been unable to determine that there are any net benefits associated with the Phase II damage criteria, independent of impact speed requirements.

The agency has also noted and taken into account the factual information and assertions submitted by representatives of the insurance industry concerning the possible use of more economical compliance technology such as nonself-restoring energy absorbers. The use of such technology is prevented by the current Phase II requirements. The availability of such technology on new bumper systems is a desirable result, independent of the impact speed requirement imposed by the bumper standard. Retaining the Phase II requirements would inhibit the further development of such technology.

Finally, the agency took into account the importance of distinguishing in its analyses among favorable net benefit results attributable to impact speed reduction only, those results attributable to action with respect to Phase II only, and those results attributable to both aspects of the decision. Factual data exist in the record only with respect to the first and third of these areas. Thus, any attribution of benefits to the Phase II requirements would be too speculative as a basis for agency decision. The agency believes that the probable effect of its current decision will be the introduction of bumper systems exhibiting at least some characteristics of 5.0-mph, Phase I bumpers. Bumper face bars and reinforcements designed for 5.0-mph impacts, and therefore most probably capable of affording even greater actual protection as a result of over-design to ensure compliance, will undoubtedly continue to be used in at least some new cars in the short term. Effectively, 5.0-mph, Phase I bumpers will thus be produced under the new standard, on an interim basis and for some portion of the new car fleet. The performance of these cars can and will be monitored closely by the agency to estimate the actual effects of the shift to Phase I criteria.

For all of these reasons, the agency has concluded that the Phase II criteria are not justified and that those criteria should be deleted from the standard.

Removal of optional equipment during compliance testing. Several commenters contended that existing Part 581 test procedures restrict the installation of certain optional equipment prior to sale of a vehicle to a first purchaser. Although one domestic manufacturer stated that its optional equipment sales were not restricted, other automobile and equipment manufacturers commented that existing test procedures inhibit installation of fog lamps, running lights, and headlamp washers. Commenters recommended dealing with this problem by removing such equipment prior to

testing, exempting such items from the protective criteria, or limiting testing to standard equipment only.

NHTSA believes that the safety value of optional equipment such as fog lamps has yet to be demonstrated conclusively. To the extent that the equipment does serve a safety function, permitting its removal during testing would encourage its installation and thereby promote safety. Further, distinguishing between optional equipment installed before the purchase of a new car and that installed after such purchase serves little purpose, since equipment installed after purchase would be just as likely to be damaged in a low-speed collision. Moreover, such a distinction unfairly discriminates in favor of aftermarket suppliers at the expense of manufacturers and dealers wishing to attach equipment prior to the sale of new cars. The agency also notes that possible cost savings from factory installation of optional equipment are lost if such installation is discouraged by the test requirements. For these reasons, the agency has amended the standard to permit removal of fog lamps, running lights, other optional equipment attached to the bumper face bar, and headlamp washers prior to testing.

Harmonization. The Trade Agreements Act of 1979 (19 U.S.C. 2532(2)), requires that the agency consider harmonization with international standards in its regulatory actions. In the present context, ECE Regulation No. 42 is relevant.

NHTSA has formally endorsed enhanced efforts at harmonization between and among international standards in presentations to the Group of Experts on the Construction of Vehicles (Working Party 29) which operates under the ECE's Inland Transport Committee. Explicit harmonization of a United States bumper standard with the ECE regulation could have some positive economic effects since domestic manufacturers might experience lower costs due to reduced need for differentiation in design and equipment between cars for sale in this country and cars for export. In addition, European manufacturers subject to the ECE regulation could experience similar reduced costs.

This consideration, however, cannot be deemed to be controlling where United States law creates specific performance or policy criteria for regulatory action. With regard to ECE Regulation No. 42, NHTSA has concluded that the Act imposes specific criteria relating to cost savings which the ECE regulation does not address. Further, it is noted that the Act mandates the bumper standards issued thereunder be drafted so that they regulate performance instead of directly regulating bumper design. Certain provisions of the ECE regulation would impose statutorily impermissible design restrictions on vehicles produced for sale in this country. Finally, NHTSA has concluded that potential bumper mismatch problems could result from substituting the height requirement specified in that regulation for the requirement in the Part 581 Bumper Standard. NHTSA will continue to pursue the question of harmonization in appropriate forums, but at this time merely notes that the 2.5-mph/ 2.5-mph, Phase I alternative selected in this rulemaking is far more compatible with the ECE regulation than the current Part 581 standard or the 5.0-mph/2.5-mph alternative.

Number of pendulum impacts. Some commenters suggested that the number of pendulum test impacts required by the standard be reduced. However, given the likelihood that some cars may incur more than two low-speed bumper impacts in their lifetime, and the possibility that all such impacts may be either longitudinal or corner impacts and may involve the same bumper, the agency has concluded that the current procedure is appropriate to assure that each bumper is able to withstand the impacts to which it may in fact be subjected over its lifetime.

Public opinion survey. Some commenters alleged that public opinion strongly favors the retention of bumper requirements at current levels and should control the agency's decision in this rulemaking. As evidence of public opinion, two commenters cited a survey conducted by the Opinion Research Corporation, Inc., (ORC) for the Insurance Institute for Highway Safety.

NHTSA disagrees with the commenters' suggestion about public opinion. First, the level of bumper standards established by the agency under the Act cannot be determined merely on the basis of what members of the public understand to be the relevant facts and issues, or what they themselves would prefer. The Congress has determined the public policy which must be applied by the agency, and the agency's decision must be reached in accordance with the statutory criteria. Those criteria do not include public

preferences as such, although as noted in the FRIA, adequately demonstrated public preference may be relevant to assessments of future market demand and the response options available to the auto manufacturing and insurance industries.

Second, the agency does not believe that the ORC survey provides reliable evidence on public preferences regarding economic values associated with bumper alternatives before the agency. An analysis of the text of the survey discloses that the structure and specific questions asked did not compensate for the public's general lack of detailed information concerning the costs and benefits of bumpers. Yet the survey asked a variety of questions which could be meaningfully answered only by persons knowledgeable about such matters. Also, many of the specific questions may have inadvertently encouraged respondents to give inflated estimates of the value of the current bumper standard. For these and other reasons discussed in chapter III of the FRIA, the agency regards the ORC survey as an inconclusive indicator of informed public opinion.

Legal issues. Some commenters advocating retention of the current standard have questioned the adequacy of the record in this proceeding to provide a basis for decision and have challenged in advance the legal soundness of any decision to amend the standard.

In this rulemaking proceeding, the agency has compiled voluminous materials over a period of years which have been used in analyzing competing alternatives. Through the notice and comment process and two public meetings, the agency has received over two hundred comments from a full spectrum of interested parties and has gathered all available data on the subject of this proceeding. New data, estimates and arguments have been received which have assisted the agency in refining its analysis of the standard.

As noted above, the agency recognizes that a degree of uncertainty is present in some of its calculations and conclusions by virtue of the absence of conclusive real world data relating to certain categories of benefits and costs. However, this lack of factual certainty no more absolves the agency of its duty under section 102 of the Act to ensure that a bumper standard exists which in fact complies with the requirement to seek maximum feasible reductions in cost than could similar uncertainties have arguably absolved the agency of its duty to issue a standard in the first instance. Under the Act, the agency is directed to adopt and maintain a standard. That standard is further required to meet certain stautory criteria. Implicit in this and any similar statutory mechanism is both a prohibition against rescinding an existing standard altogether and maintaining a standard which, on the basis of a developing evidentiary foundation is found either not to have any net benefits, or to have fewer net benefits than any one or more different standards. As noted above, explicit instructions to the effect were directed to the agency in 1978.

The agency does not accept an expansive view of the limitations imposed on the agency's action in this proceeding by the Act, as inferred by some commenters from the provisions of the Act itself. The agency is cognizant of the relevant statutory criteria imposed by this organic Act and has acted in accordance with them.

The statute does not require, and the legislative history does not support, an inference of Congressional intent that the agency be completely certain regarding the relevant factual issues before it conducts rulemaking under this Act. To the contrary, the Act, its legislative history and Congressional action to date have emphasized the presence of significant uncertainty on all of the relevant issues discussed in this notice. Recognition of the uncertainty may be seen in, for example, the wording of the criterion in section 102 governing the setting of the level of the bumper standard. The agency is not required to establish a standard that *produces* the maximum feasible reduction in costs, but one that "seeks to obtain" such a reduction. The agency has always considered itself bound to proceed with continuing review and rulemaking even in the presence of uncertainty. This conclusion and interpretation of the statute is consistent with the agency's actions since enactment, and is explicity reasserted in this notice.

The statute also does not mandate that the standard be set so as to require the use of the most protective bumpers which can be produced. From the beginning of its action under the provisions of the Act, the agency has always recognized that such bumpers would be so expensive to produce and replace that their use would involve a net economic loss for consumers. 38 Fed. Reg. 20,899 (August 3, 1973). As the agency also noted in that notice, rulemaking under the Act involves the balancing of many factors to determine what level of performance is most beneficial to the public and the consumer.

As the agency interprets the Act and its history, the purpose of the Act's bumper provisions is to secure cost savings for the public and the consumer. The bumper provisions address the issues of the costs of damage in low-speed collisions and the costs of avoiding that damage and authorize and direct the agency to set standards that minimize the combined total of these costs to the public and the consumer. The goal of seeking cost savings is promoted by setting the standards and as appropriate adjusting them toward the level where the marginal benefits equal marginal costs. That is, if raising bumper performance from its unregulated level yields more incremental benefits, reflected in damage reduction, than the incremental costs of increased damage protection, the standard should be raised. The impact speed requirements should be raised to the point where the incremental increase in damage avoided equals the incremental increase in costs. This is the point at which the cost savings or net benefits are maximized.

Raising the requirements above that point of equality would not provide the public and consumer with any additional cost savings. Two possibilities exist regarding the relationship of incremental benefits and costs above the .point. One is that incremental benefits will be less than the incremental costs at all points above the point of equality. In that event, raising the requirements above the point of equality would reduce the cost savings achievable at that point. The other possibility is a variation on the first in that incremental benefits will equal or at least appear to equal incremental costs over some range of requirement levels immediately above the point of equality. The FRIA suggests that there may be a range over which incremental benefits and costs appear to be roughly equal. Setting requirements within such a range would not, however, increase cost savings, and would thus be of questionable validity. It would result in a simple trading of dollars, that is, receiving only as much in reduced damage as one pays for increased damage protection.

In this rulemaking action, NHTSA has determined that the 2.5-mph/2.5-mph alternative is more likely than the current standard and the other alternatives to be the point of equality, that is, where the incremental benefits first equal the incremental costs. Accordingly, the agency has selected that alternative as the new standard. As noted above, setting a higher standard would not increase the savings to the public and consumers. A higher standard would only increase the direct, immediate costs which each new car purchaser must bear.

Some commenters have asserted that a 5.0-mph test impact speed is necessary to satisfy the expectations voiced in Congress during deliberations on the Act. While these expectations are relevant, the determinative fact in all instances must be what the Congress in fact did through legislative action. In the Act, the Congress did not set a particular standard, but instead adopted the maximum feasible cost reduction criterion, and required that bumper standards be set in accordance with it. The criterion is a deliberately flexible one which permits and even requires that bumper standards be adjusted based on available information.

Some commenters suggested that the agency is legally bound to maintain the Part 581 Bumper Standard at its present level because the standard incorporates the safety criteria of former FMVSS 215. One insurer asserted that the criteria in section 103(a) of the Safety Act must form a basis for any decision to amend the Bumper Standard. Those criteria require that safety standard be practicable, be stated in objective terms, and meet the need for motor vehicle safety. 15 U.S.C. 1392(a). Another insurer cited the legislative history of the Act in support of the proposition that Congress intended safety considerations to be controlling in establishing bumper standards.

Given the hybrid nature of the Part 581 Standard, this rulemaking action was initiated under the concurrent authority of the Act and the Safety Act. Without deciding whether the criteria established for safety standards under section 103 necessarily be applied in all cases under the Act where any safety relationship can be asserted, the agency has concluded, based on the discussion in this notice and the FRIA, that its actions in this proceeding are in all respects in accordance with the applicable criteria of the Safety Act itself.

By the same token, this action does not conflict with safety standards promulgated under the Safety Act. To the extent that bumper standards may be considered to be safety standards, the 5.0-mph safety criteria of Part 581 have been determined to be unsupported, even under the Safety Act criteria, and are amended by this notice. Reducing the test speed does not make compliance with any safety standard more difficult. The changes made by this rulemaking action do not necessitate any change in efforts to comply with existing safety standards. To the degree that pedestrian impact protection is a relevant safety consideration, current agency research on the subject suggests the possibility of an adverse safety consequence from bumpers designed for impact speeds of 5.0 mph or higher.

The Final Regulatory Impact Analysis

NHTSA's FRIA estimates the changes in costs and benefits likely to result from amending the Bumper Standard. In assessing the relative merits of the alternative bumper standard amendments described in the notice of proposed rulemaking in this proceeding, NHTSA has considered all available evidence and viewpoints in order to quantify and analyze the various factors relevant to determining bumper system net benefits.

As discussed in the agency's FRIA, the primary measure of benefits of the Part 581 Bumper Standard is the economic cost of the damage avoided by use of a bumper designed to provide protection at a higher impact speed. In the agency's FRIA, this cost was determined for each alternative standard by computing the cost of repaired damage and unrepaired damage. The cost of damage was computed by first using the results of vehicle owner surveys and insurance company claim files to estimate the frequency of damage to bumper systems. This figure was then analyzed in terms of the projected effectiveness of that bumper system in preventing damage, as estimated from insurance records and by use of engineering judgment.

Reduced levels of savings representing the value of damage which the vehicle owner decides not to have repaired were determined by first estimating the repair costs for unrepaired damage described by car owners. NHTSA then reduced the repair cost by a range of values to reflect the fact that the damage was not repaired, the effect of vehicle age on the value of that damage, and the absence of any out-of-pocket expenses incurred by the car owner.

The agency's calculation of benefits also took into account insurance cost savings beyond the value of the damage avoided by the bumper system, i.e., through savings in administrative expense. Savings in consumer time and inconvenience resulting from damage avoidance at various levels of bumper damage resistance were also considered as benefits of bumper regulation. Such savings include the value of time saved at the scene of a low-speed accident, reduced time and expense in obtaining repair estimates, and savings in the avoided cost of obtaining alternative transportation while collision damage is repaired. Finally, although not subject to quantification in the agency's economic analysis, the agency considered the possible beneficial or adverse effects of bumper requirements on vehicle safety.

A very important cost impact of bumper regulation is the increase in new car prices attributable to the use of bumper systems providing greater levels of damage resistance. This cost consists of the cost of the bumper system itself and the cost of upgrading other vehicle components to support the additional weight of more damage resistant systems (i.e., the cost of secondary weight). The FRIA examines the changes in such costs that would result from adopting test speeds below those in the current standard. The costs used in the agency's FRIA represent the marginal change in costs resulting from changing from the current bumper standard to an alternative standard requiring lower levels of bumper performance. Costs are calculated in terms of actual cost to the consumer. Finance charges associated with that portion of the vehicle purchase price attributable to the bumper are considered and taken into account as appropriate.

In addition to the effect on the initial cost of purchasing a car, the added operating cost of driving a car with a heavier bumper system has been considered. The agency has estimated the additional fuel costs incurred in carrying the extra primary and secondary weight associated with bumper systems providing greater levels of damage resistance performance. Costs and benefits to be accrued in the future have been discounted to reflect their value in current dollars. Results of the FRIA have been stated in terms of positive or negative net benefits for the

various alternative standards, as compared to the costs and benefits of the current 5.0-mph/5.0-mph standard. See chapters X and XI of the FRIA.

In the agency's analysis, several factual issues are of particular importance, and the data and opinion evidence relied upon by the agency are summarized in greater detail below.

Frequency of bumper-related collisions. As noted above, benefits derived from the damage avoidance properties of bumpers are computed by estimating first the frequency of bumper-related collisions, and then the ability of the bumper system to protect the car in those collisions. Levels of protection thus computed yield benefits in terms of the costs which would otherwise have been incurred in connection with the avoided damage.

In 1970, the Ford Motor Company conducted a survey of actual observed damage to Ford cars in parking lots. Based on that survey, earlier NHTSA analyses estimated that the average car experienced 3.63 low-speed collisions involving its bumpers during its lifetime.

In the PRIA, the agency estimated the frequency of unreported, low-speed collisions at a lower number, based on the results of a telephone survey of principal operators of cars. That survey was conducted for NHTSA by Westat, Inc.

The agency's October 1981 NPRM specifically requested that commenters address the issue of the best method of estimating such low-speed collision frequency. Responding commenters disagreed on the relative merits of the cited damage frequency estimates. While car manufacturers argued for the use of figures derived from the Westat study, insurers generally favored higher estimates. Commenters addressing this issue generally expressed the view that the actual figure for low speed collision frequency would be somewhere below the figure of 3.63 lifetime impacts estimated from the parking lot surveys by Ford.

The agency agrees with commenters that the Ford survey is inadequate for use in the current context, by virtue of various factors, including its concentration on urban areas. The agency believes that the Westat survey, and the comments to the record by interested parties represent superior, and the best available, data on low-speed accident frequency. They have been considered in the computation of this factor. NHTSA has considered the possible use of crash recorders on cars to assess accident frequency, but finds that this approach would be prohibitively expensive and not technically feasible at this time. For these reasons, the FRIA incorporates a range of values for low-speed accident frequency, using as the bounds of the range the highest estimate provided in the comments and the lower estimate derived from the Westat survey data.

Bumper system effectiveness. On the question of the effectiveness of bumper systems designed to provide protection at differing impact speeds, estimates used in the PRIA were based on comparisons by agency experts between the performance of cars with Part 581 bumpers and with pre-standard cars. The agency was able in the PRIA to make extensive use of field data to determine the effectiveness of bumpers designed to provide protection in 5.0-mph impacts. NHTSA was able also to supplement insurance industry data on reported accidents with Westat survey data on damage incurred in unreported accidents.

However, no similar data on the effectiveness of bumpers designed to provide protection at other impact speeds exists. As a result, the agency was forced to rely in its PRIA on data concerning MY 1973 rear bumpers for its estimates of 2.5-mph bumper effectiveness. These were the only bumpers ever sold in this country which were required to provide 2.5-mph protection. As an alternative and cross-check, the agency also considered in the PRIA estimates which had been developed for use in the June 1979 Final Assessment of the Bumper Standard, and which were based on engineering judgment of the agency's experts regarding the relative effectiveness of various bumper systems. The use of these estimates was supported by the insurance industry in its review of the 1979 Assessment.

Using this methodology, the agency estimated that 2.5-mph bumpers would achieve 67 percent of the effectiveness of 5.0-mph bumpers in low-speed collisions. That is, 2.5-mph bumpers would be two-thirds as effective in preventing damage as 5.0-mph bumpers would be.

Car and insurance industry commenters joined in arguing the unreliability of estimates based on the performance experience of MY 1973 rear bumpers. They stressed the lack of comparability between these early bumpers and the 2.5-mph bumper systems which would be produced today,

citing the absence of any uniform height requirement for MY 1973 bumpers, the actual similarity of MY 1973 bumpers to unregulated bumpers of prior years, the increased uniformity among bumper designs in the present vehicle fleet, and other factors related to the vehicle fleet mix. NHTSA agrees with commenters that data on MY 1973 rear bumpers fail to provide an accurate approximation of current 2.5-mph designs. NHTSA has concluded therefore that the methodology employing MY 1973 rear bumper data should not be used in estimating current levels of bumper effectiveness.

NHTSA has considered relying upon European data relating to the performance experience of bumpers designed in compliance with ECE Regulation No. 42 to assess the effectiveness of 2.5-mph bumpers but has concluded that adequate data of that type are not available. Although alternative data sources were specifically sought in NHTSA's October 1981 NPRM, no field data on the effectiveness of alternative systems in other countries were introduced into the record by commenters. Moreover, European bumpers are required to be designed to meet a safety standard only, and are tested under different procedures than American bumpers. Finally, differences in fleet composition and average vehicle weight, as well as the greater frequency of urban driving in Europe, would limit the relevance of data based on vehicles in use abroad to predicted vehicle experience in American driving conditions.

Insurance industry commenters presented to the record certain laboratory tests undertaken on production vehicles alleged to have been equipped with 2.5-mph bumpers, i.e., pickup trucks and multipurpose passenger vehicles not subject to the Part 581 requirements. NHTSA has concluded, based on the evidence in the record, that the damage levels reported in the insurance industry tests are not sufficiently relevant to predict 2.5-mph bumper performance. The tests reported upon were of limited scope, and no data have been introduced or are known to the agency from which to conclude that the bumper systems tested were designed to, or would in fact, comply with the Part 581 requirements in 2.5-mph barrier and pendulum impacts. Moreover, a commenter from the auto industry pointed out an instance in which the insurance claim frequency for a car equipped with a Part 581 bumper was actually higher than for its counterpart, the four-wheel drive, multipurpose passenger vehicle version which was equipped with an unregulated bumper. The agency has therefore concluded that estimates based on extrapolation from field data better account for factors such as crash angle, impact speed, frequency of occurrence and vehicle fleet mix. Thus, NHTSA makes use in the FRIA of the 67 percent effectiveness figure employed in the 1979 Assessment, but now applies this factor to the superior lifetime damage estimates derived from the 1981 Evaluation.

Primary bumper costs and weight. With respect to the increase in costs associated with bumper systems providing greater levels of damage protection, many motor vehicle manufacturers submitted previously unavailable estimates of the cost and weight penalties associated with providing bumpers meeting current 5.0-mph performance requirements, as compared with the cost of complying with a 2.5-mph, Phase I requirement or with the ECE Regulation No. 42 bumper requirement.

The agency estimates in the FRIA that the primary cost differences between 5.0-mph and 2.5-mph Phase II bumper systems can be best expressed as a range from $18 to $35. The corresponding range of weight differences is estimated to be from 15 to 33 pounds. The $18 to $35 and 15 to 33 pound ranges are based on estimates submitted to NHTSA by the manufacturers and reflect the range of representative cost and weight savings estimates submitted.

In their submissions to the rulemaking docket, the manufacturers generally did not identify all changes in design or components that would take place if the bumper standard were reduced to 2.5 mph/2.5 mph. Certain changes were specifically noted, however. Manufacturers stated that such a reduction would allow the removal of self-restoring, heavy duty energy absorbers and noted that they would probably make that change. Some manufacturers also identified reducing face bar thickness and removing some reinforcements as being among the changes possible if the standard were reduced.

Although the estimates of cost and weight for 2.5-mph bumper systems included in the FRIA generally agree with current estimates of representative manufacturers, and are consistent with those confidential submissions made in response to the 1979 advance notice of proposed

rulemaking, other independent estimates have been generated which indicate that even greater weight reductions are possible if the Part 581 bumper standard were reduced to 2.5 mph/2.5 mph. For example, the 1979 Final Assessment cited a weight reduction estimate of 43 pounds developed by a design engineer under contract with NHTSA. Since the 43 pound figure was developed in reference to cars averaging 3,350 pounds in weight, the appropriate value applicable to the lighter average car produced today would be less. Assuming that weight loss in primary bumper weight would be proportional to total vehicle weight, the appropriate figure for today's cars would be approxiamtely 36-37 pounds. Notwithstanding the higher value thus represented, the upper range set forth in the FRIA is 33 pounds. If the higher figures of 36-37 pounds were used, the weight and cost differential between 5.0-mph and 2.5-mph bumpers, and thus the benefits of the lower impact speed, would be even greater.

In addition, other independent cost studies submitted as evidence in the record indicate that the actual costs for all manufacturers of components such as energy absorbers may in fact be higher than cost estimates by the car manufacturers who submitted data on this point. See, for example, Docket No. 81-07 Notice 1, No. 006. If the cost avoided by removing such energy absorbers from a car were as high as $48, instead of the $20 estimated in confidential submissions responding to the 1979 advance notice of proposed rulemaking (as updated to reflect the weight of current cars), the additional cost savings of reducing the Part 581 standard to 2.5 mph/2.5 mph would be increased by $28, thereby enhancing the cost reduction attributable to that alternative. In this case, although the result may be to underestimate the benefit of the lower standard, the agency has chosen to use in the FRIA the lower cost and weight estimates submitted by the manufacturers who commented in response to the NPRM, since such lower values produce benefit calculations less favorable to the regulatory result urged by the car manufacturers involved.

Secondary weight and cost. On the subject of secondary weight, NHTSA relied in its PRIA on methodologies developed by the Transportation Systems Center (TSC) of Cambridge, Massachusetts, and General Motors. The TSC methodology assumes that, in the case of vehicles with unitized bodies, the vehicle body will not be affected by changes in bumper weight. This methodology results in a secondary weight factor of .5; that is, one half pound of secondary weight will be added to the rest of the vehicle for each pound of added bumper weight. The General Motors methodology, based on actual component weights of MY 1974 General Motors products, assumes that all the weight of a unitized vehicle body is affected by secondary weight. This methodology results in a secondary weight factor of about 1.0.

The agency has concluded, based on all comments received, that the assumptions of the TSC methodology concerning vehicles with unitized bodies are extreme. One manufacturer submitted an estimate of secondary weight based on its analysis of its most efficient new car designs. That analysis indicates a secondary weight factor of 0.7 (i.e., seven-tenths of a pound added for each pound of added bumper system weight). Since all of these were new designs for which secondary weight factors may be lower than for for the fleet as a whole, the agency considers that this estimate most likely represents the lower bound of secondary weight factors in the current vehicle fleet. Older, existing production car designs, which would also be affected by a reduced standard, would be likely to have a secondary weight factor of 1.0 or higher. The agency has concluded that there is no adequate basis to establish a higher value than that based upon actual component weight analysis, and accordingly the agency makes use of both the .7 and 1.0 factors in the FRIA.

Only two commenters addressed the issue of the cost of secondary weight. Both commenters suggested that NHTSA's estimate of $.72 per pound in the PRIA represents the lower bound of possible secondary weight costs, since it was based only on the cost per pound of structural components and did not include cost effects on weight dependent subsystems such as tires and brake linings. However, the agency believes that while changes such as upgrading brake linings or marginally increasing tire size to accommodate increased bumper weight will undoubtedly occur to some extent, they are impossible to quantify in terms of dollar costs on the record before the agency. Thus, the agency continues to use only the cost of major structural materials such as

cold-rolled steel and aluminum to reflect secondary weight cost more conservatively. Because of an error discovered by the agency in its original computation of the markup factor used in the PRIA, the agency has now corrected the cost of secondary weight and uses $.60 per pound in the FRIA.

Use of consumer costs instead of manufacturer variable costs. In calculating for the FRIA the cost savings available from modified bumper requirements, NHTSA considered manufacturers' variable cost savings, but not reductions attributable to savings on dealer markup, which represent some additional potential consumer savings. Several motor vehicle manufacturers endorsed NHTSA's inclusion of variable cost savings in its analysis and projected savings of 10 to 30 percent resulting from reducing the Bumper Standard impact speed level to 2.5 mph. However, the manufacturers also commented that consumer cost (which includes dealer markup), rather than variable cost, is a more realistic determinant of the cost of bumper regulation.

The agency believes that use of consumer costs is more consistent with the requirements of the Act. Using the newly submitted cost savings estimates supplied by the auto manufacturers, and the agency's independent analysis of the reasonableness of these estimates based on the use of teardown studies, NHTSA stated cost savings in terms of consumer costs in its FRIA. The FRIA employs a sensitivity analysis to assess the effect on consumer prices of various possible bumper standard alternatives.

Finance charges. In its PRIA, NHTSA added the cost of new car finance charges to the cost of current bumper systems. While several auto industry sources saw no difficulty with consideration of finance changes from the standpoint of economic theory, certain representatives of the auto and insurance industries noted that the principal of a car loan, in addition to the interest, should have been discounted to estimate true consumer savings. The agency agrees that the approach used in the PRIA overstated consumer savings because of the failure to discount the loan principal also. In estimating new car costs in the FRIA, the agency has discounted both the principal and the interest of new car loans.

Percentage of new car purchases which are financed. One commenter argued that the agency overestimated the percentage of vehicle purchases which are financed, and the duration of the financing obtained. However, the agency's figures on loan duration and percentage of new car sales financed are based on the latest available information from the Federal Reserve Board. The commenter based its alternate suggested percentage figure on data which included used car sales, which are less frequently financed. Moreover, to the extent that a small percentage of new car sales are not financed through consumer credit, e.g., fleet sales, these sales are nonetheless commonly financed through business borrowing at an even higher interest rate. Thus, the agency has not changed its analysis in response to this comment.

Retooling costs. Comments by one domestic manufacturer at NHTSA's public meeting on bumpers indicated that that company would incur a one time retooling cost of one million dollars if the present bumper standard were amended to reduce the test impact speed. Another major domestic manufacturer contended that this cost is irrelevant because, if it were not economically favorable to manufacturers to retool, such expenses would not be incurred. The agency has concluded that in computing overall societal costs of the regulation, this expense is relevant and should be considered. However, retooling costs have already been included in the agency's estimates of new car costs and thus are not addressed as a separate item in the FRIA.

Fuel consumption. In addition to the initial expense of purchasing a bumper system providing increased damage resistance performance, more stringent bumper standards which require heavier systems increase vehicle operating expenses. The added weight of the bumpers causes an increase in fuel consumption. As discussed above, projected weight savings from reduction of the bumper standard test impact speed to 2.5 mph would be significant, even for smaller cars. In its PRIA, NHTSA estimated that each additional pound of weight adds 1.1 gallons to the lifetime fuel consumption of a passenger vehicle. Some commenters accepted this fuel penalty figure as a reasonable approximation. One manufacturer advocated use of a higher figure. However, the source of the 1.1 gallon estimate, a major domestic auto manufacturer, revised its estimate downward to 1.0 gallons per pound, based on

testing and simulation studies on new, lighter weight cars. The agency is using this revised lower figure to be conservative in its estimates of benefits associated with the proposed alternatives to the current 5.0-mph standard.

NHTSA in its PRIA used a projected 1982 fuel cost in 1981 dollars of $1.60 per gallon in calculating the cost of the fuel consumed in carrying additional bumper weight, with small additional real price increases (in terms of 1981 dollars) in subsequent years. The four major domestic automakers concurred in the use of this figure in comments on the notice of proposed rulemaking. However, figures in the latest Department of Energy (DOE) and Data Resources, Inc. (DRI) forecasts suggest that an estimate of $1.28 per gallon more accurately reflects current pricing trends. Accordingly, the agency has used this figure as the 1982 average price in the FRIA.

Discount rate. For purposes of its PRIA, NHTSA used a discount rate of 10 percent in assessing the current value of future costs and benefits. This rate has been established by the Office of Management and Budget for use in Government analyses. Since, however, it is arguable that a statutory mandate to consider actual costs and benefits would require the agency to at least analyze the actual discount rate as well in reaching its conclusions, such an analysis was undertaken. See Table III-6 of the FRIA. Although one commenter suggested a lower figure, NHTSA has concluded that, given the insensitivity of net benefits to changes in the discount rate, the 10 percent rate is appropriate at this time. This figure represents a compromise between competing schools of thought as defined in economic literature, and has been used in past agency regulatory analyses. Its continued use facilitates the comparison of costs and benefits of different regulatory actions. Thus, the 10 percent figure has been retained as the basis for the discount rate used throughout the FRIA, in estimating the current value of both costs and benefits.

Lifetime distribution of accident frequency. NHTSA based its discounting in the PRIA on the assumption that accident frequency is distributed over a vehicle's lifetime, in proportion to the number of miles traveled each year by the vehicle. Car manufacturers differed on the validity of this assumption, with some contending that accident rates are higher for older vehicles. If this were true, then the net benefits of reducing the bumper standard would be even greater than estimated by the agency in the FRIA. However, NHTSA has concluded that the evidence presented on actual distribution of accidents over vehicle lifetime is not sufficiently reliable to attempt more specific yearly estimates, because, among other things, it includes both high- and low-speed accidents and the correlation between these types of accidents has not been established. Thus, the agency continues to use its original assumption on this point.

Effect of non-bumper related design changes on repair costs. A member of the insurance industry contended that not all increases in damage-per-claim figures occurring since implementation of the bumper standard should be attributed to the standard. According to that commenter, new components, such as rectangular headlamps and one-piece plastic front-end panels, which have come into use since implementation of Federal bumper standards, have added to damage-per-claim figures used by NHTSA to assess the effect of the bumper standard. Commenters made no showing regarding the costs of the various front-end components, the extent of their use in given model years, or the frequency and extent of their damage. Further, as several auto industry commenters noted, the increased complexity of the 5.0-mph bumper system makes that system more expensive to repair or replace when damaged in an impact above its design speed of 5.0 mph. Thus, the record provides no objective basis for the agency to modify its analysis.

Value of unrepaired damage. In the PRIA, NHTSA valued the cost of unrepaired damage at the full cost to repair that damage. However, several auto manufacturers commented that such damage should be valued at some lesser figure or should not be counted at all. One manufacturer placed the figure at not more than 50 percent of the cost to repair the damage. The agency's 1979 Final Assessment placed the figure at 75 percent. NHTSA has concluded that unrepaired damage clearly imposes some cost. The value of this cost, however, would necessarily vary with the age of the car, other cumulative damage, whether or not bumper-related, and other factors. NHTSA believes that a range of 50 to 75 percent of the full cost of repair represents a reasonable balancing of competing considerations and has used such a range to approximate the value of unrepaired damage in the FRIA.

One commenter suggested that consumer tolerance for cosmetic vehicle damage increases, and the value of such damage should therefore decrease, with vehicle age. However, the agency has no way of assessing this effect and therefore considers it too speculative to include in the FRIA. Therefore, the agency has not amended its calculations in response to this comment.

Current versus future technology. Throughout the consideration of bumper effectiveness, cost, and weight, the agency has been faced with the alternatives of relying on historical data based on the experience of previous model year vehicles, or on calculations based on present or future technologies. The difficulty of the choice is apparent. The former approach has the advantage of greater and superior empirical data, but may not fully account for the most recent advances in design or materials technology. The latter approach may more fully reflect state current and future conditions, but the absence of any empirical or field data introduces significantly greater elements of uncertainty.

Insurance industry and consumer representatives criticized the agency's analysis for relying on bumper designs used in the late 1970's instead of the best bumper technology available today. These commenters contended that state-of-the-art bumpers in use on the latest vehicle models are lighter, more efficient, and cost less than bumpers on earlier models and are more representative of bumpers which will be used in the future. An insurance industry representative and one component supplier commented that new technologies involving use of plastics could positively affect the net benefits of 5.0-mph bumpers. Motor vehicle manufacturers countered that use of a representative current bumper system as the basis for cost and weight estimates is more realistic, because it is more reflective of immediate cost/benefit impacts and because styling considerations frequently limit the use of the most efficient bumper design available.

The agency believes that analysis of the bumper regulation should be based on real world conditions and that it is unrealistic to assume that the most advanced technology will be used in all cases. While the use of alternative technologies could affect costs and benefits if such technologies were widely adopted, no evidence has been presented that cost, styling, production or other constraints would permit universal acceptance of these new technologies. More important, even if designs more efficient in terms of costs and weight were chosen to represent 5.0-mph bumpers in the FRIA, the effect of this change on the FRIA outcome would be negated in large part by the necessary parallel assumption that bumper systems offering lower levels of protection would also be designed and implemented at the most efficient levels possible. Therefore, NHTSA has concluded that projections of bumper net benefits must continue to be based on data relating to real world bumper systems.

Insurance premium increases. Many comments submitted by insurance industry sources and others noted that insurance premiums would increase if the bumper standard impact speed were lowered from its current levels. Insurers generally concurred that the level of such collision insurance premium increase would be 10 percent if the bumper standard test impact speed were reduced to 2.5 mph. The agency has reviewed in detail the cost of increased collision damage costs and the increased administrative overhead burden that would be incurred. Based on this analysis and on the assumption that only actual cost increases would be approved by state regulatory bodies for pass through and recovery in the form of rate increases the agency can not agree that such estimates are accurate. NHTSA accounts for insurance cost increases through estimates in the FRIA of increased collision damage costs and administrative overhead.

Effect on insurance companies, bumper component suppliers, and new car dealers. The agency's October 1981 notice of proposed rulemaking requested comments on the effect which amendment of the bumper standard would have on the insurance industry and bumper component suppliers. Members of these industries did not respond to this inquiry, except with regard to the insurance premium estimates noted above. Motor vehicle manufacturers addressing this point generally concluded that effects on related industries would not be major. Although one source predicted a reduction in the dollar sales volume of bumper component parts, increased sales of replacement parts would tend to offset to some extent the lower per unit cost of bumper replacement parts.

One industry which did claim a major interest in this proceeding was the automobile retail sales industry, as represented by the National Automobile Dealers Association. That organization pointed out the devastating effects on its membership of the recent depressed automotive retail sales market and provided data indicating the effect on car sales of price increases similar in magnitude to those resulting from the Part 581 Bumper Standard.

Consumer time and inconvenience. Several commenters addressed issues relevant to the consideration by NHTSA, as mandated by the Act, of the value of consumer time and inconvenience related damage incurred in low-speed collisions. NHTSA's PRIA incorporated a figure of $26 per incident as the value of consumer time and inconvenience associated with assessment and repair of low-speed collision damage. Insurance industry and consumer representative commenters presented results of a survey conducted for that industry by Opinion Research Corporation which seemed to suggest that a much higher per accident value should be placed on time and inconvenience. However, NHTSA has concluded that the results of this survey do not require revision of the agency's estimates of the value of delay and inconvenience.

Commenters citing the Opinion Research survey placed values of $150 to $200 per incident on the delay and inconvenience resulting from low-speed accidents, in contrast to NHTSA's PRIA estimate of $26. However, review of the survey results suggests that these estimates may include the value of repair costs to be borne by consumers, i.e., the deductible amount of the consumers' collision insurance, usually $100, a cost accounted for elsewhere in NHTSA's analysis.

Also, the Opinion Research survey focused attention on the delay and inconvenience involved in having collision damage repaired. NHTSA's estimates are based on average time loss for all accidents, including those in which damage was minimal and/or not repaired. The survey included questions which could be accurately answered only by persons with detailed knowledge of the costs and benefits of bumper systems. Moreover, apparent biases in some of the survey questions may have inflated survey respondents' estimates of the value of damage avoidance. When the effect of the above noted factors is accounted for, the insurance industry and consumer representative commenters' estimates and the NHTSA estimate do not differ greatly.

Some automobile industry and consumer representatives commented that the agency's estimate of $10 per incident for the cost of alternate transportation while low-speed collision damage is repaired may be too low. A consumer organization commented that the agency underestimated the time lost at the scene of an accident and in obtaining repair estimates. It suggested that NHTSA had also understated the expense of being without a car while collision damage is repaired. It should be noted that the agency's Analysis counts savings in delay and inconvenience for all accidents, whether or not damage is actually repaired. Since damage is not always repaired, the agency's figures translate into a higher per accident savings for those accidents where repairs are actually made. Nevertheless, after consideration of the comments on these issues, the agency has now used, and has performed a sensitivity analysis using, a range of costs for time and inconvenience of $26 to $50 in the FRIA.

Safety issues. Insurance industry and other commenters expressed concern that reduction of the test impact speed requirements of the standard would pose a risk to vehicle safety due to increased damage to safety-related components. As evidence of the safety impact of bumper regulation, one insurance industry commenter cited a study in which it examined accident claims involving rear impacts to MY 1973 and 1974 vehicles. According to this commenter, the results of this study indicate reductions in trunk lid and taillamp damage on certain models when the bumper standard for rear bumpers was upgraded in MY 1974. This commenter also noted reductions in trunk lid, trunk latch and tailpipe damage on some models in data from NHTSA's driver survey, although the commenter concluded that the survey was of such limited scope as to preclude the drawing of significant conclusions. The commenter asserted that components of the type protected by the Bumper Standard do affect safety in that, even if their malfunction does not actually cause an accident, it increases the risk to occupants once

an accident occurs, e.g., through leaking fuel from a damaged fuel system.

Several auto industry sources commented that current bumper requirements do not provide significant safety benefits. One major domestic manufacturer cited studies conducted by Westat and Indiana University's Institute for Research in Public Safety (Docket No. 73-19, Notice 27, No. 041) in support of its assertion that only one percent of accidents are caused by safety component malfunctions which could have resulted from low-speed collision damage. This commenter contended, moreover, that the nature of these malfunctions (e.g., lamps not working) does not permit the inference that even this low incidence of contribution to accident causation is attributable to collisions, but is instead more commonly experienced as a result of maintenance neglect (e.g., failure to replace burned-out bulbs). As a result, the commenter argues that low-speed collision damage is a minuscule factor in motor vehicle safety. Another major manufacturer also commented that the bumper standard's connection to safety is tenuous, and that there is no evidence that safety would be compromised by amendment of the bumper standard requirements. Other automakers commented that a 2.5-mph bumper standard would be adequate in any event to protect vehicle safety components.

Other commenters asserted that 5.0-mph bumper requirements may in fact have a net adverse effect on vehicle safety. An auto industry trade association commented that the extra weight and rigidity of more damage resistant bumpers could adversely affect crash deformation characteristics and rates of crush and energy absorption so as to reduce potential levels of occupant protection in higher speed collisions. Another auto industry commenter argued that while 5.0-mph bumpers do not contribute significantly to safety through protection of safety components, the added weight of those bumpers necessarily reduces accident avoidance capability by adversely affecting braking and cornering performance.

Finally, the agency's own developing research into pedestrian impact protection indicates a clear possibility of conflict between affording enhanced safety protection in this area and increasing or even maintaining the current bumper standard.

After consideration of the extensive discussion of this issue in the record of this proceeding, including the Indiana University study referenced above, NHTSA has concluded both that no safety based justification exists for the current 5.0-mph bumper requirements, and that relaxation of the impact speed requirements would not compromise any known safety consideration. In the agency's judgment, a safety need for 5.0-mph bumpers has never been demonstrated, either before issuance of the FMVSS 215 and Part 581 standards or by subsequent experience. Moreover, the argument that protection of safety systems in low-speed collisions is important for purposes of vehicle crashworthiness as well as crash avoidance is not convincing in view of the fact that the only Part 581 criterion which contributes significantly to crashworthiness, i.e., the criterion relating to the fuel system, is now protected much more effectively by FMVSS 301.

NHTSA has also considered the energy management consequences of this action with respect to compliance with the applicable FMVSS requirements relating to occupant crash protection and fuel system integrity. Insurance industry commenters noted that the crash energy of a 2.5-mph collision is one quarter that of a 5.0-mph collision. Thus, it was suggested that 2.5-mph bumpers would be less effective in managing crash energy than 5.0-mph bumpers. However, a number of motor vehicle manufacturers commented that in the 30.0-mph barrier impact used to determine compliance with various crashworthiness FMVSS, the vehicle bumper absorbs only a small percentage of the crash energy, generally less than 5 percent. Moreover, some manufacturers commented that reduction of the bumper test impact speed requirements would permit removal of space consuming and aggressive energy absorbers and stiff frame rails which may actually inhibit design of vehicles for efficient high-speed energy management. Also, reduction of bumper test impact requirements could lead to reduced aggressivity of the impacting vehicle in side collisions.

After review of comments received, NHTSA has concluded that reduction of bumper test impact requirements would not have a negative effect on high-speed crash energy management. The amount of energy generated in a 5.0-mph barrier impact is less than three percent of that generated in a 30.0-mph barrier crash. The energy generated in a 2.5-mph barrier impact is one percent of 30.0-mph crash energy. Thus, although 5.0-mph bumpers may absorb more energy than 2.5-mph bumpers, the difference is negligible in a 30.0-mph barrier impact. Moreover, as suggested by commenters, the 5.0-mph bumper requirements may inhibit efficient vehicle energy management design. NHTSA has concluded that 5.0-mph bumpers make no significant contribution to occupant crash protection or to protection of fuel system components which may be damaged in high-speed crashes.

Thus, the agency's action does not conflict with any existing safety standards.

Other Issues

Accounting for vehicle size in testing. NHTSA requested that commenters consider whether the test procedure adequately accounts for vehicle size differences. While some commenters suggested that car size is a factor in damage resistance, those commenters expressing an opinion on the issue commented that the existing test requirements adequately account for these effects. Those requirements adjust test pendulum weight to the mass of the vehicle tested. Commenters also noted that size and weight differences among cars are decreasing as downsizing progresses. Thus, change in the test procedures to account for vehicle size differences does not appear to be warranted.

Manual repositioning of bumper system components during testing. Several commenters suggested the desirability of allowing manual repositioning of bumper or shielding-panel components during testing. These commenters suggested that such a procedure would reduce costs, increase design flexibility, promote the use of new technologies, and reduce the subjectivity now inherent in the evaluation of shielding-panel damage. However, some auto manufacturers also stated that eliminating the Phase II damage resistance requirements would alleviate much of the need for manual repositioning. Since the Phase II criteria are being replaced by Phase I criteria, and manual repositioning might introduce uncertainties into the test procedure, the agency has decided not to permit manual repositioning.

Bumper height. On the issue of bumper height, several auto manufacturers commented that the height requirements of the standard account for a substantial portion of the benefits of the standard. One automaker referred to matching heights as the single most important requirement of the standard. A major insurer, however, contended that a matching requirement associated with an "ineffective" impact speed of 2.5 mph would be meaningless. This commenter also contended that only 49 percent of reported accidents are bumper-to-bumper accidents.

Of course, a significant proportion of reported accidents would be side impacts, rollovers, and single vehicle collisions rather than bumper-to-bumper impacts. Therefore, it does not necessarily follow that damage incurred in non-bumper-to-bumper accidents is attributable to bumper mismatch. Moreover, unreported accidents would be expected to include a higher proportion of bumper-to-bumper accidents than would reported accidents because bumper-to-bumper contact would prevent significant damage in a number of cases. Thus, a number of bumper-to-bumper accidents would not appear in the figures for reported accidents.

Finally, the agency notes that the height of some vehicle structural components may be determined by the height of the bumper. To the degree that uniform side structural members, additional levels of protection may result in side impact collisions from matching of bumpers and frame rails. NHTSA concludes that the height requirement is a useful component of the bumper regulation. Height standardization is maintained under the amendment announced in this notice.

One commenter advocated lowering the prescribed bumper height to less than 16 inches, the current low bound for pendulum testing. This commenter contended that low bumpers would optimize pedestrian protection characteristics, minimize aerodynamic drag, and reduce injuries in side impacts. NHTSA will consider the contribution of bumper height in connection with ongoing research in the areas of pedestrian protection and side impacts. However, until such time as the effects of bumper height in these areas can be fully evaluated, the very high transition cost of converting existing vehicle designs and the desirability of consistency with bumper heights of the existing vehicle fleet makes it preferable that the present height requirements be maintained.

Effective date. Some automobile manufacturers commented on the need for expeditious action to amend the standard. One manufacturer noted that final action by March 1982 would permit bumper system modifications to be made in time for introduction of model year 1983 vehicles. Another commented on the long leadtimes necessary for introduction of product changes. Yet another stated that an effective date for bumper standard amendments in the near future would permit incorporation of bumper system changes in a new vehicle model currently in the design stage. In view of these considerations, and because this action relieves a restriction, NHTSA has determined that good cause exists to make this amendment effective 45 days from the date of publication of this notice in the *Federal Register*.

Requirements for Analyses

NHTSA has determined that this proceeding involves a major rule within the meaning of Section 1, paragraph (b)(1), of Executive Order 12291 in that it is likely to result in an annual effect on the economy of $100 million or more. The agency estimates that current bumper requirements add between $140 to $200 to the cost of a new car compared to the cost of a car with unregulated bumpers. The reduction of test impact speed requirements for each of the roughly 11 million vehicles expected to be sold in this country annually is likely to result in an impact on the economy far exceeding $100 million. For this same reason, this action is considered significant for purposes of Department of Transportation procedures for internal review of regulatory actions. The agency's FRIA for this action has been placed in the public docket. Copies may be obtained by contacting the Docket Section, Room 5108, National Highway Traffic Safety Administration, 400 Seventh Street, S.W., Washington, D.C. 20590.

Pursuant to the Regulatory Flexibility Act, the agency has considered in its FRIA the impact of this rulemaking action on small entities. The agency certifies that this action will not have a significant economic impact on a substantial number of small entities. Therefore, a regulatory flexibility analysis is not required for this action. The agency has concluded that few, if any, manufacturers of motor vehicles and bumper components or vehicle insurers are small entities. New car dealers will not be significantly affected because this action is unlikely to significantly affect new car sales levels for individual dealerships. To the extent that such sales may be affected, the effect would be positive. While increased car collision damage repairs may result from this action, the impact on individual repair shops is not expected to be significant. Again, the effect would be positive.

The economic effects of this action on small organizations and governmental units will generally be the same as those on the general public. As purchasers of new cars, these organizations and units will experience the same increase in net benefits. While this action could result in a minor increase in police time spent at the scene of some low-speed accidents, this effect is not expected to be significant.

In developing this final rule, NHTSA considered the bumper standard promulgated by the International Standards Organization and adopted by the ECE. However, the agency found that standard to be inappropriate for use in this country since it does not adequately deal with consumer cost savings considerations as required by the Act.

NHTSA has prepared an Environmental Assessment of the likely environmental consequences of this proposal. This Assessment has been placed in the public rulemaking docket (Docket 73-19; Notice 27, No. 004). Based on this Assessment, the agency has concluded that this action will not have a significant effect on the human environment and that, for this reason, an Environmental Impact Statement will not be prepared for this action.

Issued on May 14, 1982.

_____
Raymond A. Peck, Jr.
Administrator

**47 F.R. 21820
May 20, 1982**

Appendix

The following is a summary of the more major comments submitted in response to the notice of proposed rulemaking and discussed in more general terms in the preamble of this notice. This summary is organized in broad terms according to the interest groups from which the comments were received. Responses to these comments are set forth in the preamble to the final rule and in the FRIA.

Insurance Industry and Consumer
Representative Comments _

In commenting on the issue of low-speed damage frequency, insurance industry and consumer representatives criticized the Westat survey on a number of grounds. The Insurance Institute for Highway Safety (IIHS) and Consumers Union contended that the survey understates damage frequency due to memory weaknesses on the part of survey respondents. IIHS also noted that nonprincipal drivers were not surveyed directly and cited discrepancies between the original Westat survey and a follow-up survey emphasing operators of later model vehicles. Allstate Insurance Company contended that the Westat survey cannot be used to make judgments about the effects of changing the bumper standard on the frequency of damage to safety components because the sample size is too limited, and that the survey is not representative because it covers only unreported damage. Allstate advocated use of a higher estimate, although not as high as that suggested by the Ford survey results. IIHS also suggested that use of the Westat survey improperly accounts for accidents reported to police. State Farm Mutual Automobile Insurance Company contended that the study understates the number of low-speed impacts due to the probable existence of impacts with parked vehicles, and of accidents not reported to the person interviewed.

On the issue of bumper effectiveness, IIHS and the Highway Loss Data Institute (HLDI) supplied results of laboratory tests on current vehicles not required to meet the Part 581 standard, i.e., pickup trucks and multipurpose passenger vehicles. These commenters reported substantially poorer bumper performance on these vehicles, which, according to these commenters, would comply with a 2.5-mph bumper requirement.

IIHS also argued that vehicle size is a major determinant of the amount and frequency of crash-related property damage. Thus, IIHS contended NHTSA's assessment of bumper effectiveness is biased in favor of older, unregulated vehicles because the more recent vehicle mix includes greater numbers of more damage prone smaller vehicles. Moreover, IIHS argued, imports are more frequently involved in property damage accidents than are domestically produced vehicles, further biasing the analysis against later model years which include a larger percentage of imported vehicles.

The American Insurance Association and State Farm contended that the discount rate of 10 percent applied by the agency to determine the present value of future expenditures is too high. Since bumpers represent an investment which displaces other consumption, these commenters argued that a more accurate discount rate would be 4 percent. Allstate commented that the discounting factor should be applied to inflated costs rather than current costs.

On the subject of delay and inconvenience, the Center for Auto Safety (CFAS) placed the cost of a rental vehicle, which may be required while low-speed collision damage is repaired, at $24 to $30 per day. CFAS estimated that consumers use 1.6 gallons of gasoline in obtaining a single damage repair estimate and that each such estimate now costs $35 on the average. CFAS also contended that the agency underestimated the lost lost at the scene of an accident and in obtaining repair estimates.

An insurance industry representative submitted data from a public opinion poll which, according to the commenter, demonstrates overwhelming public support for the 5.0-mph bumper standard. The commenter also asserted that this poll indicates people are willing to pay for the higher levels of protection provided by the 5.0-mph bumper standard. CFAS also argued that the public supports the 5.0-mph bumper requirements.

The insurance industry argued that ECE Regulation No. 42 is irrelevant and inappropriate to requirements of the Cost Savings Act, primarily because it does not address the issue of protection against economic damage. According to the insurance industry, the ECE requirements amount to merely a weaker version of FMVSS 215. Moreover, this source contended the ECE standard focuses in part on design rather than performance characteristics, and thus is not in accordance with United States statutory requirements for issuance of performance standards.

Librerty Mutual Insurance Company commented that the current Part 581 requirements do not adequately account for vehicle dive, which can contribute to bumper underride in accident situations. Presumably, dive-induced mismatch damage would be increased under ECE requirements.

On the issue of new technologies, IIHS argued that new materials, i.e., polycarbonite plastics, which could significantly reduce the weight of bumpers meeting current 5.0-mph requirements are available at this time. State Farm advocated the possible use of sacrificial components, i.e., components which must be adjusted or replaced after a collision, as a means of reducing bumper cost and weight.

Auto Industry Comments

In addressing the question on the issue of low speed collision frequency, General Motors Corporation and Ford Motor Company commented that studies conducted by Ford overstate damage frequency, principally due to their emphasis on vehicles used in urban areas. These commenters suggested that the Westat survey is a more reliable source of data because it is more current and is based on a more representative sampling system.

Chrysler Corporation, American Motors Corporation, and Volkswagen of America, Inc. commented that neither the Ford nor West data provide an adequate means of assessing low-speed collision frequency. These commenters suggested that use of crash recorders or other controlled tests is necessary to generate data.

In questioning the value of MY-1973 bumpers in assessing 2.5-mph bumper effectiveness, several commenters pointed out that MY-1973 bumpers were not subject to a pendulum impact test and thus were not required to be of a uniform height. Commenters noted that MY-1973 rear bumpers were essentially the same as MY-1972 bumpers, but with stronger mounting brackets. This comment is consistent with State Farm's comment

that its research revealed no difference in performance between MY-1973 and 1972 rear bumpers. Some commenters also concluded that new 2.5-mph bumpers would perform better in the current vehicle mix than did MY-1973 bumpers in previous years, due to the increased uniformity of current bumper designs. General Motors, Ford, and Chrysler joined in attacking the relevance of laboratory tests as a means of assessing the relative performance of bumpers, stating that such tests have never been correlated to real world conditions.

American Motors suggested that NHTSA consider the European experience with 2.5-mph bumpers under ECE Regulation No. 42. However, General Motors commented that its German subsidiary reported an absence of field data on the effectiveness of 2.5-mph bumpers in Europe. Moreover, General Motors contended that the European bumper standard is purely a safety standard and that bumpers designed to meet that standard would not be representative of future American 2.5-mph designs. In General Motors' opinion, the estimates used in NHTSA's 1979 Final Assessment provide the best available information on bumper effectiveness at alternative design speeds.

Several auto industry sources argued that unregulated bumpers produced in the future would provide greater levels of damage resistance performance than pre-standard bumpers. The factor most commonly cited in support of this contention was that consumer expectations would require that bumpers provide higher levels of performance. Insurance cost considerations, international harmonization, and experience in designing improved bumpers were also cited as contributing to the prospects for improved performance from future unregulated bumpers. Certain auto industry sources estimated that unregulated bumpers would exceed 1.5-mph performance and, at least initially, provide performance approximating that available under a 2.5-mph Phase I standard or ECE Regulation No. 42.

In discussions of bumper cost and weight savings from use of 2.5-mph bumpers, estimates of overall weight savings ranged from 8 lbs. for Volkswagen to over 38 lbs. for Volvo of America Corporation. Ford reported weight savings of 34 lbs. for its European Escort model compared to its American counterpart as a result of differing bumper requirements. Associated cost savings of roughly $35 were estimated by several manufacturers.

On the related issue of secondary weight, a recent General Motors analysis of seventeen late model front-wheel drive vehicles produced a secondary weight factor of .72. General Motors stated that this factor was used in the design process of its recent "X" and "J" car models. Toyota Motor Company also estimated a secondary weight factor of .7 for its current models. Renault agreed that the correct secondary weight factor is greater than .5. Comments received from Ford, Chrysler, and American Motors all contended that a secondary weight factor of 1.0 would be appropriate for NHTSA's analysis.

The fuel penalty factor of 1.1 gallons of fuel consumed for each additional pound of bumper weight, used in NHTSA's Preliminary Regulatory Impact Analysis, was based on testimony presented by General Motors before Congress. General Motors, in its comments on the notice of proposed rulemaking on bumper standard amendments, revised its estimate downward to 1.0 gallon of fuel per pound of vehicle weight. However, several other motor vehicle manufacturers commented the 1.1 gallon figure is reasonable. Chrysler noted that a higher figure could be used.

Chrysler estimated the increased cost to repair 5.0-mph bumpers as compared to 2.5-mph bumpers at between $70 and $90. BMW of North America, Inc. cited an analysis prepared by a West German technical institute which found that at impact speeds of 18 kph (approximately 11 mph) and higher, repair costs for American-made bumpers are greater than for European bumpers due to more expensive bumper shock absorbers and body components. BMW also noted a West German insurance study reporting that the great majority of all collisions occur at speeds above 11 kph.

General Motors and Ford commented that NHTSA's figure for the hourly value of lost time is too high, General Motors contending that the figure should be somewhere between the average hourly wage rate and the minimum wage. Ford argued that a figure of $3.50, roughly half the average hourly earnings figure, would be more accurate. This figure is consistent with a Consumer's Research report which concluded that commuters are willing to pay 42 percent of an hour's wage to save one hour of travel time. Regarding the cost of alternate transportation while collision damage is being repaired, Ford concurred in the agency's estimate of $10 per incident. Volkswagen commented that the figure seemed too low, and General Motors suggested that the agency consider the actual cost of rental vehicles.

Chrysler expressed the opinion that insurance premiums would decrease due to a reduction in bumper repair costs if the performance requirements of the standard were lowered. Ford commented that insurance industry premium discounts and surcharges based on vehicle damage claims experience provide a significant marketplace incentive to manufacturers to design vehicles providing better damage resistance performance.

Daimler-Benz AG, Renault, and Peugeot S.A. cited cost and consistency considerations as the basis for their positions in support of the ECE standard. Other commenters suggested that cost savings, e.g., savings in tooling and testing costs, would result from harmonization. Renault estimated weight savings of 14-15 kg. for its vehicles equipped with bumpers designed to meet the ECE standard.

Volkswagen and American Motors discussed at length their position that the fixed-barrier impact test should be dropped from the standard. ECE Regulation No. 42 does not require a fixed-barrier test. According to Volkswagen, elimination of the barrier test would reduce testing costs, promote international harmonization, and make the standard more equitable. Volkswagen criticized the barrier test as unreliable, unsophisticated, and adding nothing to the standard. American Motors contended that the pendulum test alone would be sufficient, since it assures height standardization and proper bumper geometry to minimize override, and the versatile positioning of the pendulum permits testing of the entire bumper system. American Motors suggested that the pendulum test could be run with the vehicle idling to provide a test relevant to dynamic situations. Volvo suggested the alternative of employing the ECE test procedure with damage criteria taken from the Part 581 standard.

Volkswagen and BL Technology Ltd. pointed out that the ECE standard provides for pendulum impact at a single height rather than within a height range as is the case with the Part 581 standard. BL Technology contended that the ECE height requirement should be adopted in this country to promote harmonization and reduce costs. BL Technology also noted that the single height requirement permits reduced vertical bumper width thereby improving engine cooling. However, Volkswagen argued there is little difference between the Eruopean and United States' height requirements in terms of benefits and that the Part 581 requirement should be retained to avoid possible mismatch with vehicles already in use.

On the subject of Phase I versus Phase II damage criteria Ford and General Motors questioned the cost-effectiveness of the Phase II requirements. General Motors argued that NHTSA's analysis overstates the benefits of the Phase II standard because the agency overestimates the effectiveness of Phase II bumpers in impacts at speeds of 5.0 mph or below. General Motors added that NHTSA must consider the 5 lbs. of additional weight and resulting $6 additional fuel cost imposed by the Phase II requirements. Information supplied by Volvo and the Bureau of Labor statistics suggests that initial consumer costs of between $10 and $15 result from the Phase II requirements. Ford contended that no true Phase I bumpers have ever been produced because model year 1979 vehicles represented a transition period between FMVSS 215 and Part 581, Phase II.

Ford contended that the pendulum test is not appropriate for assessing damage resistance properties of the bumper itself due to its concentration of force in particular locations. This test, in combination with the Phase II criteria may, according to Ford, require use of expensive energy absorbers even if the test impact speed were lowered to 2.5 mph. Although Davidson Rubber Division commented that the Phase II criteria posed no problem for soft face systems, that manufacturer at the same time advocated reduction of the pendulum impact speed to 2.5 mph. BL Technology and General Motors commented that return to Phase I criteria would encourage design innovation and the use of new, lighter weight materials. Mitsubishi Motors Corporation favored the Phase I criteria because bumper deformation would improve the crash energy management characteristics of the bumper system.

Ford also noted objectivity problems in evaluating bumper damage under the Phase II criteria. Finally, Ford argued that the increased use of rubber and polymeric bumper materials has changed consumer perceptions and reduced the visibility of and concern about minor dents and similar damage which was inherent in the use of chrome-plated bumpers.

Two auto manufacturers advocated dropping not only the damage criteria applicable to the bumper system itself, but all criteria limiting damage to the exterior surfaces of the vehicle. Saab-Scania of America, Inc. made this suggestion in the context of a possible decision to retain the 5.0-mph test impact speed requirement. Toyota's comment noted vehicle cost and weight could be reduced by eliminating the exterior surface protection requirements.

Commenters addressing the issue differed on the extent of manual repositioning which should be permitted. Ford recommended permitting manual repositioning which could be performed without special equipment or experience. Volkswagen favored manual repositioning without tools, while Chrysler suggested that manual repositioning without "special" tools be permitted.

On the question of new technologies, Ford and Volkswagen commented that relaxation of the bumper standard requirements would permit use of fiberglass bumpers, plastic face bars, rubber mountings, and ultrahigh strength steel components which could result in cost and weight savings, increased styling flexibility and improved aerodynamic characteristics. Davidson Rubber offered compressible plastics, i.e., foam or honeycomb materials, as examples of materials which could be used if the standard requirements were lowered. C&F Stamping Company, Inc. cited plastics and single-unit bumper systems. American Motors commented that return to Phase I would increase usage of SMC Components. Chrysler noted the potential for cost and weight savings from ultrahigh strength steel if Phase II criteria were eliminated. One component supplier, Molnar Industries, Inc. noted the availability of fiber reinforced plastic bumpers which it contended may make lowering the bumper standard requirements unnecessary.

47 F.R. 21820
May 20, 1982

PREAMBLE TO AN AMENDMENT TO PART 581
Bumper Standard
[Docket No. 73-19; Notice 32]

ACTION: Interpretive amendment.

SUMMARY: The Part 581 Bumper Standard specifies that certain equipment be removed from a vehicle before testing. This notice clarifies the wording of a May 20, 1982, amendment to make it clear that (1) no change was intended in the requirement as it related to trailer hitches and license plate brackets, i.e., that all trailer hitches and license plate brackets are removed, whether or not they are optional equipment, and (2) all running lights and fog lamps which are optional equipment should be removed, whether or not they are mounted on the bumper face bar.

EFFECTIVE DATE: September 23, 1983.

SUPPLEMENTARY INFORMATION: Section 581.6(a)(5) of the Bumper Standard specifies that certain equipment be removed from a vehicle before testing. Prior to the most recent amendment, the section specified that trailer hitches and license plate brackets be removed from the vehicle. The standard was amended in a notice published in the Federal Register (46 FR 48262) on May 20, 1982, which, among other things, expanded the specified equipment that is removed to include headlamp washers and certain optional equipment, i.e., running lights, fog lamps, and equipment mounted on the bumper face bar. The section was revised to read:

Trailer hitches, license plate brackets, running lights, fog lamps, other optional equipment mounted on the bumper face bar and headlamp washers are removed from the vehicle.

The amended section might be read to be more restrictive than the former section as it relates to trailer hitches and license plate brackets, i.e., that only trailer hitches and license plate brackets which are optional equipment must be removed. This notice clarifies the wording of that amendment to make it clear that no change was intended in the requirement as to these types of equipment. Thus, this notice makes it clear that all trailer hitches and license brackets must be removed. The agency neither proposed nor intended any change in the requirement as it relates to those types of equipment.

Another possible question of interpretation under the amended section is whether all running lights and fog lamps which are optional equipment should be removed, or only those which are mounted on the bumper face bar. This notice clarifies the wording of the amendment to make it clear that running lights and fog lamps which are optional equipment should be removed, whether or not they are mounted on the bumper face bar.

This amendment is an interpretive amendment which does not change the substantive requirements of the Bumper Standard in any respect. Accordingly, it is found for good cause shown that notice and comment are unnecessary and that an immediate effective date is in the public interest.

In consideration of the foregoing, 49 CFR Part 581 is amended as follows:

§581.6 [Amended]

Section 581.6(a)(5) is revised to read:

(a) * * *

(5) Trailer hitches, license plate brackets, and headlamp washers are removed from the vehicle. Running lights, fog lamps, and equipment mounted on the bumper face bar are removed from the vehicle if they are optional equipment.

Issued on September 19, 1983.

Diane K. Steed
Deputy Administrator

48 FR 43331
September 23, 1983

PART 581—BUMPER STANDARD

(Docket No. 74-11; Notice 12; Docket No. 73-19; Notice 9)

§ 581.1 Scope. This standard establishes requirements for the impact resistance of vehicles in low speed front and rear collisions.

§ 581.2 Purpose. The purpose of this standard is to reduce physical damage to the front and rear ends of a passenger motor vehicle from low speed collisions.

§ 581.3 Application. This standard applies to passenger motor vehicles other than multipurpose passenger vehicles.

§ 581.4 Definitions. All terms defined in the Motor Vehicle Information and Cost Savings Act, P.L. 92-513, 15 U.S.C. 1901-1991, are used as defined therein.

"Bumper face bar" means any component of the bumper system that contacts the impact ridge of the pendulum test device.

§ 581.5 Requirements.

(a) [Each vehicle shall meet the damage criteria of §§ 581.5(c) (1) through 581.5 (c) (9) when impacted by a pendulum-type test device in accordance with the procedures of § 581.7(b), under the conditions of § 581.6, at an impact speed of 1.5 m.p.h., and when impacted by a pendulum-type test device in accordance with the procedures of § 581.7(a) at 2.5 m.p.h., followed by an impact into a fixed collision barrier that is perpendicular to the line of travel of the vehicle, while traveling longitudinally forward, then longitudinally rearward, under the conditions of § 581.6, at 2.5 m.p.h." (47 F.R. 2182—May 20, 1982. Effective: July 4, 1982)]

(b) [Reserved.]

(c) Protective criteria.

(1) Each lamp or reflective device except license plate lamps shall be free of cracks and shall comply with applicable visibility requirements of S4.3.1.1 of Standard No. 108 (§ 571.108 of this part). The aim of each headlamp shall be adjustable to within the beam aim inspection limits specified in Table 2 of SAE Recommended Practice J599b, July 1970, measured with a mechanical aimer conforming to the requirements of SAE Standard J602a, July 1970.

(2) The vehicle's hood, trunk, and doors shall operate in the normal manner.

(3) The vehicle's fuel and cooling systems shall have no leaks or constricted fluid passages and all sealing devices and caps shall operate in the normal manner.

(4) The vehicles' exhaust system shall have no leaks or constrictions.

(5) The vehicle's propulsion, suspension, steering, and braking systems shall remain in adjustment and shall operate in the normal manner.

(6) A pressure vessel used to absorb impact energy in an exterior protection system by the accumulation of gas pressure or hydraulic pressure shall not suffer loss of gas or fluid accompanied by separation of fragments from the vessel.

(7) The vehicle shall not touch the test device, except on the impact ridge shown in Figures 1 and 2, with a force that exceeds 2000 pounds on the combined surfaces of Planes A and B of the test device.

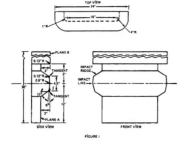

FIGURE 1

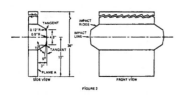

FIGURE 2

(8) The exterior surfaces shall have no separations of surface materials, paint, polymeric coatings, or other covering materials from the surface to which they are bonded, and no permanent deviations from their original contours 30 minutes after completion of each pendulum and barrier impact, except where such damage occurs to the bumper face bar and the components and associated fasteners that directly attach the bumper face bar to the chassis frame.

(9) Except as provided in § 581.5(c) (8), there shall be no breakage or release of fasteners or joints.

(10) Reserved.

(11) Reserved.

§ 581.6 Conditions. The vehicle shall meet the requirements of § 581.5 under the following conditions:

(a) *General.*

(1) The vehicle is at unloaded vehicle weight.

(2) The front wheels are in the straight ahead position.

(3) Tires are inflated to the vehicle manufacturer's recommended pressure for the specified loading condition.

(4) Brakes are disengaged and the transmission is in neutral.

(5) [Trailer hitches, license plate brackets, and headlamp washers are removed from the vehicle. Running lights, fog lamps, and equipment mounted on the bumper face bar are removed from the vehicle if they are optional equipment. (48 F.R. 43331—September 23, 1983. Effective: September 23, 1983)]

(b) *Pendulum test conditions.* The following conditions apply to the pendulum test procedures of § 581.7(a) and § 581.7(b):

(1) The test device consists of a block with one side contoured as specified in Figure 1 and Figure 2 with the impact ridge made of A1S1 4130 steel hardened to 34 Rockwell "C." The impact ridge and the surfaces in Planes A and B of the test device are finished with a surface roughness of 32 as specified by SAE Recommended Practice J449A, June 1963. From the point of release of the device until the onset of rebound, the pendulum suspension system holds Plane A vertical, with the arc described by any point on the impact line lying in a vertical plane

(for § 581.7(a), longitudinal; for § 581.7(b), at an angle of 30° to a vertical longitudinal plane) and having a constant radius of not less than 11 feet.

(2) With Plane A vertical, the impact line shown in Figures 1 and 2 is horizontal at the same height as the test device's center of percussion.

(3) The effective impacting mass of the test device is equal to the mass of the tested vehicle.

(4) When impacted by the test device, the vehicle is at rest on a level rigid concrete surface.

(c) Barrier Test Condition. At the onset of a barrier impact, the vehicle's engine is operating at idling speed in accordance with the manufacturer's specification. Vehicle systems that are not necessary to the movement of the vehicle are not operating during impact.

§ 581.7 Test Procedures.

(a) Longitudinal Impact Test Procedures.

(1) Impact the vehicle's front surface and its rear surface two times each with the impact line at any height from 16 to 20 inches, inclusive, in accordance with the following procedure.

(2) For impacts at aheight of 20 inches, place the test device shown in Figure 1 so that Plane A is vertical and the impact line is horizontal at the specified height.

(3) For impacts at a height between 20 inches and 16 inches, place the test device shown in Figure 2 so that Plane A is vertical and the impact line is horizontal at a height within the range.

(4) For each impact, position the test device so that the impact line is at least 2 inches apart in vertical direction from its position in any prior impact, unless the midpoint of the impact line with respect to the vehicle is to be more than 12 inches apart laterally from its position in any prior impact.

(5) For each impact, align the vehicle so that it touches, but does not move, the test device, with the vehicle's longitudinal centerline perpendicular to the plane that includes Plane A of the test device and with the test device inboard of the vehicle corner test positions specified in § 581.7(b).

(6) Move the test device away from the vehicle, then release it to impact the vehicle.

(7) Perform the impacts at intervals of not less than 30 minutes.

(b) Corner impact test procedure.

(1) Impact a front corner and a rear corner of the vehicle once each with the impact line at a height of 20 inches and impact the other front corner and the other rear corner once each with the impact line at any height from 16 to 20 inches, inclusive, in accordance with the following procedure.

(2) For an impact at a height of 20 inches, place the test device shown in Figure 1 so that Plane A is vertical and the impact line is horizontal at the specified height.

(3) For an impact at a height between 16 inches and 20 inches, place the test device shown in Figure 2 so that Plane A is vertical and the impact line is horizontal at a height within the range.

(4) Align the vehicle so that a vehicle corner touches, but does not move, the lateral center of the test device with Plane A of the test device forming an angle of 60 degrees with a vertical longitudinal plane.

(5) Move the test device away from the vehicle, then release it to impact the vehicle.

(6) Perform the impacts at intervals of not less than 30 minutes.

41 F.R. 9346
March 4, 1976

PREAMBLE TO PART 582—INSURANCE COST INFORMATION REGULATION

(Docket 74-40; Notice 2)

This notice establishes an insurance cost information regulation pursuant to the Motor Vehicle Information and Cost Savings Act (15 U.S.C. 1901 *et seq.*). The regulation is based upon a notice of proposed rulemaking published November 4, 1974 (39 F.R. 38912) and comments submitted in response to the notice.

The regulation will require automobile dealers to distribute to prospective purchasers information which compares differences in insurance costs for different makes and models of passenger motor vehicles based upon differences in their damage susceptibility and crashworthiness. In the absence of insurance cost information that reflects damageability and crashworthiness, this rule does not, at the present time, have an effect on automobile dealers. Damage susceptibility and crashworthiness studies currently being conducted by the NHTSA are expected to influence the insurance rate structure by providing data which will enable the insurance industry to take these factors into account. As this occurs, the NHTSA will prepare comparative indices for the dealers to distribute to prospective purchasers.

Several comments on the proposed rulemaking discussed the merits of the Motor Vehicle Information and Cost Savings Act and are therefore beyond the scope of this rulemaking. Other comments offered methods for performing the damage susceptibility and crashworthiness studies. These comments have been forwarded to the technical staff performing the studies. Two comments suggested minor changes in the text of the regulation for clarity and to make the proposed regulation more consistent with the purposes of the Act. These suggestions have been adopted in the final regulation. Their effect is that the insurance cost information disseminated by the dealers would be in the form of comparative indices, based on differences in damage susceptibility and crashworthiness, rather than simply the insurance premium rate which is determined by many factors.

One comment expressed the view that providing this information to consumers within 30 days after its publication in the *Federal Register* was an excessive burden upon the dealers. The NHTSA does not believe that sufficient justification for this position has been made in light of the need to provide the information to the consumer in time for it to be of use to him in purchasing an automobile.

Therefore, a new Part 582, *Insurance Cost Information*, is added in Chapter V, Title 49, Code of Federal Regulations, to read as set forth below.

Effective date: Although the final rule is effective February 1, 1975, as specified in the Cost Savings Act, the dates when automobile dealers will be required to distribute insurance cost information are dependent upon NHTSA progress in developing such information and will be published at a later date in the *Federal Register*.

(Sec. 201(c), P. L. 92-513, 86 Stat. 947 (15 U.S.C. 1941(e)); delegation of authority at 49 CFR 1.51).

Issued on January 31, 1975.

James B. Gregory
Administrator
40 F.R. 4918
February 3, 1975

PART 582—INSURANCE COST INFORMATION REGULATIONS

§ 582.1 **Scope.** This part requires automobile dealers to make available to prospective purchasers information reflecting differences in insurance costs for different makes and models of passenger motor vehicles based upon differences in damage susceptibility and crashworthiness, pursuant to section 201(e) of the Motor Vehicle Information and Cost Savings Act (15 U.S.C. 1941(e)), herein "the Cost Savings Act."

§ 582.2 **Purpose.** The purpose of this part is to enable prospective purchasers to compare differences in auto insurance costs for the various makes and models of passenger motor vehicles based upon differences in damage susceptibility and crashworthiness, and to realize any savings in collision insurance resulting from differences in damageability, and any savings in medical payment insurance resulting from differences in crashworthiness.

§ 582.3 **Definitions.**

(a) *Statutory definitions.* All terms used in this part which are defined in section 2 of the Cost Savings Act are used as so defined.

(b) *Definitions used in this part.*

(1) "Automobile dealer" means any person who engages in the retail sale of new or used automobiles as a trade or business.

(2) "Collision insurance" means insurance that reimburses the insured party for physical damage to his property resulting from automobile accidents.

(3) "Insurance cost" means the insurance premium rate, as expressed in appropriate indices, for collision and medical payment, including personal injury protection in no-fault states.

(4) "Medical payment insurance" means insurance that reimburses the insured party for medical expenses sustained by himself, his family, and his passengers in automobile accidents.

§ 582.4 **Requirements.**

(a) Each automobile dealer shall provide the insurance cost information specified in § 582.5 for examination by prospective purchasers at each location where he offers vehicles for sale.

(b) The information shall be provided without charge and in sufficient quantity to have it available for retention by prospective purchasers, within 30 days after its publication in the *Federal Register.*

(c) The information shall be in English and, if a significant portion of the prospective purchasers do not speak English, in the non-English language most widely spoken by prospecive purchasers.

§ 582.5 **Insurance cost information form.**

The insurance cost information provided pursuant to section 582.4 shall be presented as follows: [Form to be specified].

40 F.R. 4918
February 3, 1975

PREAMBLE TO PART 585—AUTOMATIC RESTRAINT PHASE-IN REPORTING REQUIREMENTS

(Docket No. 74-14; Notice 43)

ACTION: Final rule.

SUMMARY: On April 12, 1985, NHTSA issued a notice proposing a number of amendments to Standard No. 208, *Occupant Crash Protection*. Based on its analysis of the comments received in response to that notice, the agency has decided to take the following actions: retain the oblique crash test for automatic restraint equipped cars, adopt some New Car Assessment Program test procedures for use in the standard's crash tests, provide in the standard for a due care defense with respect to the automatic restraint requirement, and require the dynamic testing of manual lap/shoulder belts in passenger cars. This notice also creates a new Part 585 that sets reporting requirements regarding compliance with the automatic restraint phase-in requirements of the standard.

EFFECTIVE DATE: The amendments made by this notice will take effect on May 5, 1986, except the requirement for dynamic testing of manual safety belts in passenger cars will go into effect on September 1, 1989, if the automatic restraint requirement is rescinded.

SUPPLEMENTARY INFORMATION

Background

On July 11, 1984 (49 FR 28962), the Secretary of Transportation issued a final rule requiring automatic occupant protection in all passenger cars. The rule is based on a phased-in schedule beginning on September 1, 1986, with full implementation being required by September 1, 1989. However, if before April 1, 1989, two-thirds of the population of the United States are covered by effective state mandatory safety belt use laws (MULs) meeting specified criteria, the automatic restraint requirement will be rescinded.

More specifically, the rule requires:

• Front outboard seating positions in passenger cars manufactured on or after September 1, 1986, for sale in the United States, will have to be equipped with automatic restraints based on the following schedule:

• Ten percent of all cars manufactured on or after September 1, 1986.

• Twenty-five percent of all cars manufactured on or after September 1, 1987.

• Forty percent of all cars manufactured on or after September 1, 1988.

• One hundred percent of all cars manufactured on or after September 1, 1989.

• During the phase-in period, each car that is manufactured with a system that provides automatic protection to the driver without the use of safety belts and automatic protection of any sort to the passenger will be given an extra credit equal to one-half car toward meeting the percentage requirement. In addition, each car which provides non-belt automatic protection solely to the driver will be given a one vehicle credit.

• The requirement for automatic restraints will be rescinded if MULs meeting specified conditions are passed by a sufficent number of states before April 1, 1989, to cover two-thirds of the population of the United States. The MULs must go into effect no later than September 1, 1989.

In the July 1984 notice, the Secretary identified various issues requiring additional rulemaking. On April 12, 1985, the agency issued two notices setting forth proposals on all of those issues. One notice (50 FR 14589), which is the basis for the final rule being issued today, proposed: reporting requirements for the phase-in, deletion of the oblique test, alternative calculations of the head injury criterion (HIC), allowing the installation of manual belts in convertibles, use of the New Car Assessment Program (NCAP) test procedures, and adoption of a due care defense. The notice also proposed the dynamic testing of manual lap/shoulder belts for passenger cars, light trucks and light vans. The second notice (50 FR 14602) set forth the agency's proposals on the use of the Hybrid III test dummy and additional injury criteria. NHTSA has not yet completed its analysis of the comments and issues raised by the Hybrid III proposal or the proposal regarding convertibles and dynamic testing of safety belts in light trucks and light vans. The agency will publish a separate *Federal Register* notice announcing its decision with regard to these issues when it has completed its analysis.

Oblique Crash Tests

Standard No. 208 currently requires cars with automatic restraints to pass the injury protection criteria in 30 mph head-on and oblique impacts into a barrier. The April 1985 notice contained an extensive discussion of the value of the oblique test and requested commenters to provide additional data regarding the safety and other effects of deleting the requirements.

The responses to the April notice reflected the same difference of opinion found in the prior responses on this issue. Those favoring elimination of the test argue that the test is unnecessary since oblique crash tests generally show lower injury levels. They also said the additional test adds to the cost of complying with the standard—although manufacturers differed as to the extent of costs. Four manufacturers suggested that any cost reduction resulting from elimination of the test would be minimal, in part because they will continue to use the oblique tests in their restraint system developmental programs, regardless of what action the agency takes. Another manufacturer, however, said that while it would continue to use oblique testing during its vehicle development programs, the elimination of the oblique test in Standard No. 208 would result in cost and manpower savings. These savings would result because the parts used in vehicles for certification testing must be more representative of actual production parts than the parts used in vehicles crashed during development tests.

Those favoring retention of the test again emphasized that the test is more representative of real-world crashes. In addition, they said that occupants in systems without upper torso belts, such as some air bag or passive interior systems, could experience contact with the A-pillar and other vehicle structures in the oblique test that they would not experience in a head-on test. Although, again, there were conflicting opinions on this issue—one manufacturer said that oblique tests would not affect air bag design, while other manufacturers argued that the oblique test is necessary to ensure the proper design of air bag systems. The same manufacturer that said air bag design would not be affected by the oblique test, emphasized that vehicles with 2-point automatic belts or passive interiors, "may show performance characteristics in oblique tests that do not show up on perpendicular tests." Similarly, one manufacturer said that oblique tests will not result in test dummy contact with the A-pillar or front door—while another manufacturer argued that in the oblique test contact could occur with the A-pillar in vehicles using non-belt technologies.

After examining the issues raised by the commenters, the agency has decided to retain the oblique tests. There are a number of factors underlying the agency's decision. First, although oblique tests generally produce lower injury levels, they do not consistently produce those results. For example, the agency has conducted both oblique and frontal crash tests on 14 different cars as part of its research activities and NCAP testing. The driver and passenger HIC's and chest acceleration results for those tests show that the results in the oblique tests are lower in 31 of the 38 cases for which data were available. However, looking at the results in terms of vehicles, 6 of the 14 cars had higher results, exclusive of femur results, in either passenger or driver HIC's or chest accelerations in the oblique tests. The femur results in approximately one-third of the measurements were also higher in the oblique tests. Accident data also indicate that oblique impacts pose a problem. The 1982 FARS and NASS accident records show that 14 percent of the fatalities and 22 percent of the AIS 2-5 injuries occur in 30 degree impacts.

The agency is also concerned that elimination of the oblique test could lead to potential design problems in some automatic restraint systems. For example, air bags that meet only a perpendicular impact test could be made much smaller. In such a case, in an oblique car crash, the occupant would roll off the smaller bag and strike the A-pillar or instrument panel. Similarly, the upper torso belt of an automatic belt system could slip off an occupant's shoulder in an oblique crash. In belt system with a tension-relieving device, the system will be tested with the maximum amount of slack recommended by the vehicle manufacturer, potentially increasing the possibility of the upper torso belt slipping off the occupant's shoulder. In the case of passive interiors, an occupant may be able to contact hard vehicle structures, such as the A-pillar, in oblique crashes that would not be contacted in a perpendicular test. If the A-pillar and other hard structures are not designed to provide protection in oblique crashes then there would be no assurance, as there presently is, that occupants would be adequately protected. Thus, the oblique test is needed to protect unrestrained occupants in passive interiors, and to ensure that air bags and automatic or manual safety belts are designed to accommodate some degree of oblique impact.

The agency recognizes that retention of the oblique test will result in additional testing costs for manufacturers. The agency believes, however, that there are a number of factors which should minimize those costs. First, even manufacturers opposing retention of the oblique test indicated that they will continue to perform oblique crash tests to meet their own internal requirements as well as to meet the oblique test requirements of the Standard No. 301, *Fuel System Integrity*. Since the oblique tests of Standard No. 208 and Standard No. 301 can be run simultaneously, the costs resulting from retention of the oblique crash test requirements of Standard No. 208 should not be significant.

Dynamic Testing of Manual Belts

The April notice proposed that manual lap/shoulder belts installed at the outboard seating positions of the front seat of four different vehicle types comply with the dynamic testing requirements of Standard No. 208. Those requirements provide for using test dummies in vehicle crashes for measuring the level of protection offered by the restraint system. The four vehicle types subject to this proposal are passenger cars, light trucks, small van-like buses, and light multipurpose passenger vehicles (MPV's). (The agency considers light trucks, small van-like buses, and light MPV's to be vehicles with a Gross Vehicle Weight Rating (GVWR) of 10,000 pounds or less and an unloaded vehicle weight of 5,500 pounds or less. The 5,500 pound unloaded vehicle weight limit is also used in Standard No. 212, *Windshield Retention*, and Standard No. 219, *Windshield Zone Intrusion*. The limit was adopted in those standards on April 3, 1980

'(45 FR 22044) to reduce compliance problems for final-stage manufacturers. Readers are referred to the April 1980 notice for a complete discussion of the 5,500 pound limit.)

Currently, manual belts are not subject to dynamic test requirements. Instead they must be tested in accordance with Standard No. 209, *Seat Belt Assemblies*, for strength and other qualities in laboratory bench tests. Once a safety belt is certified as complying with the requirements of Standard No. 209, it currently may be installed in a vehicle without any further testing or certification as to its performance in that vehicle. The safety belt anchorages in the vehicle are tested for strength in accordance with Standard No. 210, *Seat Belt Assembly Anchorages*.

The April 1985 notice also addressed the issue of tension-relieving devices on manual belts. Tension-relieving devices are used to introduce slack in the shoulder portion of a lap-shoulder belt to reduce the pressure of the belt on an occupant or to effect a more comfortable "fit" of the belt to an occupant. The notice proposed that manufacturers be required to specify in their vehicle owner's manuals the maximum amount of slack they recommend introducing into the belt under normal use condition. Further, the owner's manual would be required to warn that introducing slack beyond the maximum amount specified by the manufacturer could significantly reduce the effectiveness of the belt in a crash. During the agency's dynamic testing of manual belts, the tension-relieving devices would be adjusted so as to introduce the maximum amount of slack specified in the owner's manual.

The agency proposed that the dynamic test requirement for passenger cars take effect on September 1, 1989, and only if the Secretary determines that two-thirds of the population is covered by effective safety belt use laws, thereby rescinding the automatic restraint requirement. Should such a determination be made, it is important that users of manual belts be assured that their vehicles offer the same level of occupant protection as if automatic restraints were in their vehicles. Absent a rescission of the automatic restraint requirement, application of the dynamic testing requirements to manual safety belts in passenger cars would be unnecessary since those belts would not be required in the outboard seating positions of the front seat. In the case of light trucks, light MPV's and small van-like buses, the agency proposed that the dynamic test requirement take effect on September 1, 1989. The proposed effective date for light trucks, light MPV's and van-like buses was not conditional, because those vehicles are not covered by the automatic restraint requirement and will likely continue to have manual safety belts.

Adoption of the requirement

As discussed in detail below, the agency has decided to adopt a dynamic test requirement for safety belts used in passenger cars. The agency is still analyzing the issues raised in the comments about dynamic testing for safety belt systems in other vehicles and will announce its decision about safety belt systems in light trucks, MPV's and buses at a later date.

Most of the commenters favored adopting a dynamic test requirement for manual belts at least with respect to passenger cars, although many of those commenters raised questions about the lead-time needed to comply with the requirement. Those opposing the requirement argued that the field experience has shown that current manual belts provide substantial protection and thus a dynamic test requirement is not necessary. In addition, they argued that dynamic testing would substantially increase a manufacturer's testing costs, and its testing workload. One commenter said that because of the unique nature of the testing, it could not necessarily be combined with other compliance testing done by a manufacturer. The same commenter argued that vehicle downsizing, cited by the agency as one reason for dynamically testing belts, does not create safety problems since the interior space of passenger cars has remained essentially the same as it was prior to downsizing. The commenter also argued there is no field evidence that the use of tension-relieving devices in safety belts, the other reason cited by the agency in support of the need to test dynamically manual safety belts, is compromising the performance of safety belts.

The agency strongly believes that current manual belts provide very substantial protection in a crash. The Secretary's 1984 automatic protection decision concluded that current manual safety belts are at least as effective, and in some cases, more effective than current automatic belt designs. That conclusion was based on current manual safety belts, which are not certified to dynamic tests. However, as discussed in the April 1985 notice, the agency is concerned that as an increasing number of vehicles are reduced in size for fuel economy purposes and as more tension-relieving devices are used on manual belts, the potential for occupant injury increases. The agency agrees that downsizing efforts by manufacturers have attempted to preserve the interior space of passenger cars, while reducing their exterior dimensions. Preserving the interior dimensions of the passenger compartment means that occupants will not be placed closer to instrument panels and other vehicle structures which they could strike in a crash. However, the reduction in exterior dimensions can result in a lessening of the protective crush distance available in a car. Thus the agency believes it is important to ensure that safety belts in downsized vehicles will perform adequately. In the case of tension-relieving devices, agency tests of lap/shoulder belt restrained test dummies have shown that as more slack is introduced into a shoulder belt, the injuries measured on the test dummies increased. Thus, as discussed in detail later in this notice, the agency believes it is important to ensure that safety belts with tension-relievers provide adequate protection when they are used in the manner recommended by vehicle manufacturers. This is of particular concern to the agency since the vast majority of new cars (nearly all domestically-produced cars) now are equipped with such devices. For those reasons, the agency is adopting the dynamic test requirement.

The adoption of this requirement will ensure that each and every passenger car, as compared to the vehicle population in general, offers a consistent, minumum level of protection to front seat occupants. By requiring dynamic testing, the standard will assure that the vehicle's structure, safety belts, steering column, etc., perform as a unit to protect occupants, as it is only in such a test that the synergistic and combination effects of these vehicle component can be measured. As discussed in detail in the Final Regulatory Evaluation (FRE), vehicle safety improvements will result from dynamic testing; and, as discussed later in this notice, such improvements can often be made quickly and at low cost.

The agency recognizes that manufacturers may have to conduct more testing than they currently do. However, the dynamic testing of manual belts in passenger cars, as with testing of automatic restraints, can be combined with other compliance tests to reduce the overall number of tests. The agency notes that in its NCAP tests, it has been able to combine the dynamic testing of belts with measuring the vehicle's compliance with other standards. The agency has followed the same practice in its compliance tests. For example, the agency has done compliance testing for Standard Nos. 208, 212, 219, and 301 in one test. The agency would, of course, recognize a manufacturer's use of combined tests as a valid testing procedure to certify compliance with these standards.

Effective Date

Two commenters argued that the requirement should become effective as soon as practical. As discussed in the April 1985 notice, the agency proposed an effective date of September 1, 1989, because it did not want to divert industry resources away from designing automatic restraints for passenger cars. The agency continues to believe it would be inappropriate to divert those resources for the purposes of requiring improvements on manual belt systems that might not be permitted in passenger cars.

Other commenters asked for a delay in the effective date – one asked for a delay until September 1, 1991, while another asked that the effective date be set 2-3 years after the determination of whether a sufficient number of States have passed effective mandatory safety belt use laws. NHTSA does not agree there is a need to delay the effective date beyond September 1, 1989 for passenger cars. Commenters argued that the time span between any decision on rescission of the automatic restraint requirements (as late as April 1, 1989) and the effective date of the dynamic testing of manual belts (September 1, 1985) is too short to certify manual belts.

The agency believes there is sufficient leadtime for passenger cars. Most of the vehicle components in passenger cars necessary for injury reduction management are the same for automatic restraint vehicles and dynamically tested manual belt vehicles. Additionally, as indicated and discussed in the April notice, approximately 40 percent of the passenger cars tested in the agency's 35 mph (NCAP) program meet the injury criteria specified in Standard No. 208, even though a 35 mph crash involves 36 percent more energy than the 30 mph crash test required by Standard No. 208. In addition, the FRE shows that with relatively minor vehicle and/or restraint system changes some safety belt systems can be dramatically improved. This is further evidence that development of dynamically tested manual belts for passenger cars in 30 mph tests should not be a major engineering program. Thus, a delay in the effective date for passenger cars is not needed.

Webbing tension-relieving devices

With one exception, those manufacturers who commented on the proposal concerning tension-relieving devices supported testing safety belts adjusted so that they have the amount of slack recommended by the manufacturer in the vehicle owner's manual. However, one manufacturer and two other commenters objected to the provision related to dynamic

testing with the tension-relieving device adjusted to the manufacturer's maximum recommended slack position. The manufacturer objected to a dynamic test that would require any slack at all to be introduced into the belt system, on the grounds that uncontrolled variability would be introduced into the dynamic test procedure, which would then lack objectivity. The manufacturer asserted that it might have to eliminate all tension-relieving devices for its safety belts.

The agency's proposed test procedure was intended to accommodate tension-relieving devices since they can increase the comfort of belts. At the same time, the proposal would limit the potential reduction in effectiveness for safety belt systems with excessive slack. The agency does not agree that this test procedure need result in the elimination of tension-relieving devices from the marketplace. As mentioned earlier, other manufacturers supported the proposal and did not indicate they would have to remove tension-relieving devices from their belt systems. The commenter opposing the requirement did not show that injury levels cannot be controlled within the specified injury criteria by testing with the recommended amount of slack, as determined by the manufacturer. The recommended slack could be very small or at any level selected by the manufacturer as appropriate to relieve belt pressure and still ensure that the injury reduction criteria of Standard No. 208 would be met. As a practical matter, most tension-relievers automatically introduce some slack into the belt for all occupants. Testing without such slack would be unrealistic.

The two other commenters objected to the proposal that manual belt systems using tension-relieving devices meet the injury criteria with only the specified amount of slack recommended in the owner's manual. They stated that most owners would not read the instructions in the owner's manual regarding the proper use of the tension-relieving device. They said an occupant could have a false sense of adequate restraint when wearing a belt system adjusted beyond the recommended limit.

The agency's views on allowing the use of tension relievers in safety belts were detailed in the April 1985 notice. The agency specifically noted the effectiveness of a safety belt system could be compromised if excessive slack were introduced into the belt. However, the agency recognizes that a belt system must be used to be effective at all. Allowing manufacturers to install tension-relieving devices makes it possible for an occupant to introduce a small amount of slack to relieve shoulder belt pressure or to divert the belt away from the neck. As a result, safety belt use is promoted. This factor should outweigh any loss in effectiveness due to the introduction of a recommended amount of slack in normal use. This is particularly likely in light of the requirement that the belt system, so adjusted, must meet the injury criteria of Standard No. 208 under 30 mph test conditions. Further, the inadvertent introduction of slack into a belt system, which is beyond that for normal use, is unlikely in most current systems. In addition, even if too much slack is introduced, the occupant should notice that excessive slack is present and a correction is needed, regardless of whether he or she has read the vehicle's owner's manual.

Exemption from Standard Nos. 203 and 204

One commenter suggested that vehicles equipped with dynamically tested manual belts be exempt from Standard Nos. 203, *Impact Protection for the Driver from the Steering Control Systems,* and 204, *Steering Column Rearward Displacement.* The agency does not believe such an exemption would be appropriate because both those standards have been shown to provide substantial protection to belted drivers.

Latching procedure in Standard No. 208

One commenter asked that Standard No. 208 be modified to include a test procedure for latching and adjusting a manual safety belt prior to the belt being dynamically tested. NHTSA agrees that Standard No. 208 should include such a procedure. The final rule incorporates the instructions contained in the NCAP test procedures for adjusting manual belts, as modified to reflect the introduction of the amount of slack recommended by the vehicle manufacturer.

Revisions to Standard No. 209

The notice proposed to exempt dynamically tested belts from the static laboratory strength tests for safety belt assemblies set forth in S4.4 of Standard No. 209. One commenter asked that such belts be exempted from the remaining requirements of Standard No. 209 as well.

NHTSA agrees that an additional exemption from some performance requirements of Standard No. 209 is appropriate. Currently, the webbing of automatic belts is exempt from the elongation and other belt webbing and attachment hardware requirements of Standard No. 209, since those belts have to meet the injury protection criteria of Standard No. 208 during a crash. For dynamically-tested manual belts,

NHTSA believes that an exemption from the webbing width, strength and elongation requirements (sections 4.2(a)-(c)) is also appropriate, since these belts will also have to meet the injury protection requirements of Standard No. 208. The agency has made the necessary changes in the rule to adopt that exemption.

The agency does not believe that manual belts should be exempt from the other requirements in Standard No. 209. For example, the requirements on buckle release force should continue to apply, since manual safety belts, unlike automatic belts, must be buckled every time they are used. As with retractors in automatic belts, retractors in dynamically tested manual belts will still have to meet Standard No. 209's performance requirements.

Revisions to Standard No. 210

The notice proposed that dynamically tested manual belts would not have to meet the location requirements set forth in Standard No. 210, *Seat Belt Assembly Anchorages.* One commenter suggested that dynamically tested belts be completely exempt from Standard No. 210; it also recommended that Standard No. 210 be harmonized with Economic Commission for Europe (ECE) Regulation No. 14. Two other commenters suggested using the "out-of-vehicle" dynamic test procedure for manual belts contained in ECE Regulation No. 16, instead of the proposed barrier crash test in Standard No. 208.

The agency does not believe that the "out-of-vehicle" laboratory bench test of ECE Regulation No. 16 should be allowed as a substitute for a dynamic vehicle crash test. The protection provided by safety belts depends on the performance of the safety belts themselves, in conjunction with the structural characteristics and interior design of the vehicle. The best way to measure the performance of the safety belt/vehicle combination is through a vehicle crash test.

The agency has already announced its intention to propose revisions to Standard No. 210 to harmonize it with ECE Regulation No. 14; therefore the commenters' suggestions concerning harmonization and exclusion of dynamically tested safety belts from the other requirements of Standard No. 210 will be considered during that rulemaking. At the present time, the agency is adopting only the proposed exclusion of anchorages for dynamically tested safety belts from the location requirements, which was not opposed by any commenter.

Belt Labelling

One commenter objected to the proposal that dynamically tested belts have a label indicating that they may be installed only at the front outboard seating positions of certain vehicles. The commenter said that it is unlikely that anyone would attempt to install a Type 2 lap shoulder belt in any vehicle other than the model for which it was designed. The agency does not agree. NHTSA believes that care must be taken to distinguish dynamically tested belt systems from other systems, since misapplication of a belt in a vehicle designed for use with a specific dynamically tested belt could pose a risk of injury. If there is a label on the belt itself, a person making the installation will be aware that the belt should be installed only in certain vehicles.

Use of the Head Injury Criterion

The April 1985 notice set forth two proposed alternative methods of using the head injury criterion (HIC) in situations when there is no contact between the test dummy's head and the vehicle's interior during a crash. The first proposed alternative was to retain the current HIC calculation for contact situations. However, in non-contact situations, the agency proposed that a HIC would not be calculated, but instead new neck injury criteria would be calculated. The agency explained that a crucial element necessary for deciding whether to use the HIC calculation or the neck criteria was an objective technique for determining the occurrence and duration of head contact in the crash test. As discussed in detail in the April 1985 notice, there are several methods available for establishing the duration of head contact, but there are questions about their levels of consistency and accuracy.

The second alternative proposed by the agency would have calculated a HIC in both contact and non-contact situations, but it would limit the calculation to a time interval of 36 milliseconds. Along with the requirement that a HIC not exceed 1000, this would limit average head acceleration to 60g's or less.

Almost all of the commenters opposed the use of the first proposed alternative. The commenters uniformly noted that there is no current technique that can accurately identify whether head contact has or has not occurred during a crash test in all situations. However, one commenter urged the agency to adopt the proposed neck criteria, regardless of whether the HIC calculation is modified. There was a sharp division among the commenters on the second proposed alternative. Manufacturers commenting on

the issue uniformly supported the use of the second alternative; although many manufacturers argued that the HIC calculation should be limited to a time interval of approximately 15 to 17 milliseconds (ms), which would limit average head accelerations to 80-85 g's. Another manufacturer, who supported the second alternative, urged the agency to measure HIC only during the time interval that the acceleration level in the head exceeds 60 g's. It said that this method would more effectively differentiate results received in contacts with hard surfaces and results obtained from systems, such as airbags, which provide good distribution of the loads experienced during a crash. Other commenters argued that the current HIC calculation should be retained; they said that the proposed alternatives would lower HIC calculations without ensuring that motorists were still receiving adequate head protection.

NHTSA is in the process of reexamining the potential effects of the two alternatives proposed by the agency and of the two additional alternatives suggested by the commenters. Once that review has been completed, the agency will issue a separate notice announcing its decision.

NCAP Test Procedures

The April 1985 notice proposed adopting the test procedures on test dummy positioning and vehicle loading used in the agency's NCAP testing. The commenters generally supported the adoption of the test procedures, although several commenters suggested changes in some of the proposals. In addition, several commenters argued that the new procedures may improve test consistency, but the changes do not affect what they claim is variability in crash test results. As discussed in the April 1985 notice, the agency believes that the test used in Standard No. 208 does produce repeatable results. The proposed changes in the test procedures were meant to correct isolated problems that occurred in some NCAP tests. The following discussion addresses the issues raised by the commenters about the specific test procedure changes.

Vehicle test attitude

The NPRM proposed that when a vehicle is tested, its attitude should be between its "as delivered" condition and its "loaded" condition. (The "as delivered" condition is based on the vehicle attitude measured when it is received at the test site, with 100 percent of all its fluid capacities and with all its tires inflated to the manufacturer's specifications. For passenger cars, the "loaded" condition is based on the vehicle's attitude with a test dummy in each front outboard designated seating position, plus carrying the cargo load specified by the manufacturer).

One commenter said that the weight distribution, and therefore the attitude, of the vehicle is governed more by the Gross Axle Weight Rating (defined in 49 CFR Part 571.3) than the loading conditions identified by the agency. The commenter recommended that the proposal not be adopted. Another commenter said that the agency should adopt more specific procedures for the positioning of the dummy and the cargo weight. For example, that commenter recommended that the "cargo weight shall be placed in such manner that its center of gravity will be coincident with the longitudinal center of the trunk, measured on the vehicle's longitudinal centerline." The commenter said that unless a more specific procedure is adopted, a vehicle's attitude in the fully loaded condition would not be constant.

The agency believes that a vehicle attitude specification should be adopted. The purpose of the requirement is to ensure that a vehicle's attitude during a crash test is not significantly different than the fully loaded attitude of the vehicle as designed by the manufacturer. Random placement of any necessary ballast could have an effect on the test attitude of the vehicle. If these variables are not controlled, then the vehicle's test attitude could be affected and potential test variability increased.

NHTSA does not agree that the use of the Gross Axle Weight Rating (GAWR) is sufficient to determine the attitude of a vehicle. The use of GAWR only defines the maximum load-carrying capacity of each axle rather than in effect specifying a minimum and maximum loading as proposed by the agency. In addition, use of the GAWR may, under certain conditions, make it necessary to place additional cargo in the passenger compartment in order to achieve the GAWR loading. This condition is not desirable for crash testing, since the passenger compartment should be used for dummy placement and instrumentation and not ballast cargo. Thus the commenter's recommendation is not accepted.

The other commenter's recommendations regarding more specific test dummy placement procedures for the outboard seating positions were already accommodated in the NPRM by the proposed new S10.1.1, *Driver position placement*, and S10.1.2, *Passenger position placement*. Since those proposals adequately describe dummy placement in these positions, they are adopted.

NHTSA has evaluated the commenter's other sug-

gestion for placing cargo weight with its center of gravity coincident with the longitudinal center of the trunk. The agency does not believe that it is necessary to determine the center of gravity of the cargo mass, which would add unnecessary complexity to the test procedure, but does agree that the cargo load should be placed so that it is over the longitudinal center of the trunk. The test procedures have been amended accordingly.

Open window

One commenter raised a question about the requirement in S8.1.5 of Standard No. 208 that the vehicle's windows are to be closed during the crash test. It said adjustment of the dummy arm and the automatic safety belt can be performed only after an automatic belt is fully in place, which occurs only after the door is closed. Therefore, the window needs to be open to allow proper arm and belt placement after the door is closed.

NHTSA agrees that the need to adjust the slack in automatic and dynamically-tested manual belts prior to the crash test may require that the window remain open. The agency has modified the test procedure to allow manufacturers the option of having the window open during the crash test.

Seat back position

One commenter recommended that proposed S8.1.3, *Adjustable seat back placement*, be modified. The notice proposed that adjustable seat backs should be set in their design riding position as measured by such things as specific latch or seat track detent positions. The commenter suggested two options. The first option would be to allow vehicle manufacturers to specify any means they want to determine the seat back angle and the resulting dummy torso angle. As its second option, the commenter recommended that if the agency decides to adopt the proposal, it should determine the "torso angle with a H-point machine according to SAE J826." The commenter said that depending on how the torso angle is established, different dummy torso angles could result in substantial adjustment deviations that can affect seat back placement.

The purpose of the requirement is to position the seat at the design riding position used by the manufacturer. The agency agrees with the commenter that manufacturers should have the flexibility to use any method they want to specify the seat back angle. Thus, the agency has made the necessary changes to the test procedure.

Dummy placement

One commenter made several general comments about dummy placement. It agreed that positioning is very important and can have an influence on the outcome of crash tests. It argued that both the old and the proposed procedures are complicated and impractical to use. The commenter claims this sitution will become more complicated if the Hybrid III is permitted, since the positioning must be carried out within a narrow temperature range (3°F) for the test dummy to remain in calibration.

The commenter also believes that the positioning of the dummy should relate to vehicle type. It said that the posture and seating position of a vehicle occupant will not be the same in a van as in a sports car. For example, it said it has tried the proposed positioning procedures and found that they can result in an "unnatural" position for the dummy in a sports vehicle. The commenter argued that this "unnatural" position would then lead to a knee bolster design which would perform well in a crash test, but would likely not provide the same protection to a real occupant because of difference in positioning. The commenter recommended that the old positioning procedure be retained and the new procedure be provided as an option for those manufacturers whose vehicles cannot be adequately tested otherwise.

Because consistency in positioning the dummy is required prior to test, NHTSA believes that a single set of procedures should apply. As discussed in the April 1985 notice, the agency proposed the new procedures because of positioning problems identified in the NCAP testing. Allowing the use of the old positioning procedures could lead to sources of variability, thus negating a major objective of the procedures. The commenter's suggestion is therefore not adopted. The agency also notes that during its NCAP testing, which has involved tests of a wide variety of cars (including sports cars), trucks and MPV's, NHTSA has not experienced the "unnatural" seating position problem cited by the commenter.

Knee pivot bolt head clearance

Two commenters said that the proposal did not specify the correct distance between the dummy's knees, as measured by the clearance between the knee pivot bolt heads. The commenters are correct that the distance should be 11¾ inches rather than the proposed value of 14½ inches. The agency has corrected the number in the final rule.

Foot rest

One commenter believes that a driver of cars equipped with foot rests typically will place his or her left foot on the foot rest during most driving and therefore this position should be used to simulate normal usage. The commenter said that using the foot rest will minimize variations in the positioning of the left leg, thus improving the repeatability of the test. In a discussion with the commenter, the agency has learned that the type of foot rest the commenter is referring to is a pedal-like structure where the driver can place his or her foot.

For vehicles without foot rests, the commenter recommended the agency use the same provisions for positioning the left leg of the driver as are used for the right leg of the passenger. It noted that positioning the driver's left leg, as with the passenger's right leg, can be hampered by wheelwell housing that projects into the passenger compartment and thus similar procedures for each of those legs should be used.

NHTSA agrees that in vehicles with foot rests, the test dummy's left food should be positioned on the foot rest as long as placing the foot there will not elevate the test dummy's left leg. As discussed below, the agency is concerned that foot rests, such as pads on the wheelwell, that elevate the test dummy's leg can contribute to test variability. The agency also agrees that the positioning procedures for the driver's left leg and the passenger's right leg should be similar in situations where the wheelwell housing projects into the passenger compartment and has made the necessary changes to the test procedure.

Wheelwell

One commenter believes that the wheelwell should be used to rest the dummy's foot. It said that positioning the test dummy's foot there is particularly appropriate if the wheelwell has a design feature, such as a rubber pad, installed by the manufacturer for this purpose.

NHTSA disagrees that the dummy's foot should be rested on the wheelwell housing. The agency is concerned that elevating the test dummy's leg could lead to test variability by, among other things, making the test dummy unstable during a crash test. Although the wheelwell problem is similar to the foot rest problem, placement of the test dummy's foot on a separate, pedal-like foot rest can be accomplished while retaining the heel of the test dummy in a stable position on the floor. That is not the case with pads located on the wheelwell.

Another commenter also said that the proposed procedure for positioning the test dummy's legs in vehicles where the wheelwell projected into the passenger compartment was unclear as to how the centerlines of the upper and lower legs should be adjusted so that both remain in a vertical longitudinal plane. In particular, it was concerned that in a vehicle with a large wheelhousing, it may not be possible to keep the left foot of the driver test dummy in the vertical longitudinal plane after the right foot has been positioned. It believes that the procedure should specify which foot position should be given priority; it recommended that the position of the right leg be required to remain in the plane, while bringing the left leg as close to the vertical longitudinal plane as possible. The agency agrees that maintaining the inboard leg of the test dummy in the vertical plane is more easily accomplished since it will not be blocked by the wheelwell. The agency has modified the test procedure to specify that when it is not possible to maintain both legs in the vertical longitidinal plane, that the inboard leg must be kept as close as possible to the vertical longitudinal plane and the outboard leg should be placed as close as possible to the vertical plane.

Lower leg angle

One commenter argued that proposed sections on lower leg positioning (S10.1.2.1 (b) and S10.1.2.2 (b)) will not result in a constant positioning of the test dummy's heels on the floor pan, thus causing differences in the lower leg angles. It stated that the lower leg angles will affect the femur load generated at the moment the foot hits the toe board during a collision. The commenter therefore proposed that the test procedure be revised to include placing a 20 pound load on the test dummy's knee during the foot positioning procedure. The commenter did not, however, explain the basis for choosing a force of 20 pounds.

NHTSA believes that use of the additional weight loading and settling procedure proposed by the commenter will add an unnecessary level of complexity to the test procedure without adding any corresponding benefit. The positioning of the test dummy's heel has not been a problem in the agency's NCAP tests. Accordingly, the agency is not adopting the commenter's recommendation.

Shoulder adjustment

One commenter asked the agency to specify that the shoulders of the test dummy be placed at their lowest adjustment position. While the shoulders are slightly adjustable, the agency believes that specifying an adjustment position is unnecessary. The agency's test experience has shown that the up and down movement of the shoulders is physically limited by the test dummy's rubber "skin" around the openings where the arms are connected to the test dummy's upper torso.

Dummy lifting procedure

One commenter was concerned about the dummy lifting proposed in (Section S10.4.1, Dummy Vertical Upward Displacement). It said that if the dummy lifting method is not standardized, test results could be affected by allowing variability in the position of the dummy's H point (the H point essentially represents the hip joint) through use of different lifting methods. It recommended use of a different chest lifting method to avoid variability in the subsequent positioning of the test dummy H-point.

The agency is not aware of any test data indicating that the use of different lifting methods is a significant source of variability. As long as a manufacturer follows the procedures set forth in S10.4.1 in positioning the test dummy, it can use any lifting procedure it wants.

Dummy settling load

One commenter was concerned about the proposed requirements for dummy settling (S10.4.2, *Lower torso force application*, and S10.4.5, *Upper torso force application*). The commenter believes that the proposals are inadequate because they do not prescribe the area over which to apply the load used to settle the test dummy in the seat. The commenter said that if the proposed 50 pound settling force is applied to an extremely small contact area, then the dummy may be deformed. It recommended that the load be applied to a specified area of 9 square inches on the dummy. In addition, it recommended that the agency specify the duration of the 50 lb. force application during the adjustment of the upper torso; it suggested a period of load application ranging from 5 to 10 seconds.

NHTSA and others have successfully used the proposed settling test procedures in their own tests without having any variability problems. Unless abnormally small contact areas are employed, or extremely short durations are used, standard laboratory practices should not result in any such problems. The agency believes that further specifying the area and timing of the force application is not necessary.

Dummy head adjustment

One commenter pointed out that it is impossible to adjust the head according to S10.6, Head Adjustment, because the Part 572 test dummy does not have a head adjustment mechanism. The agency agrees and has deleted the provision.

Additional dummy settling and shoulder belt positioning procedures

One commenter suggested a substantial revised dummy settling procedure and new procedures for positioning of the shoulder belt. NHTSA believes that its proposed procedures sufficiently address the setting and belt position issues. In addition, the commenter did not provide any data to show that variability would be further reduced by its suggested procedures. A substantial amount of testing would be needed to verify if the commenter's suggested test procedures do, in fact, provide any further decrease in variability than that obtained by the agency's test procedures. For those reasons, the agency is not adopting the commenter's suggestions for new procedures.

Due Care

In the April 1985 notice, the agency proposed amending the standard to state that the due care provision of section 108(b)(2) of the National Traffic and Motor Vehicle Safety Act (15 U.S.C. 1397(b)(2)) applies to compliance with the standard. Thus, a vehicle would not be deemed in noncompliance if its manufacturer establishes that it did not have reason to know in the exercise of due care that such vehicle is not in conformity with the standard.

Commenters raised a number of questions about the proposal, with some saying that the agency needed to clarify what constitutes "due care," others recommending that the agency reconsider the use of "design to conform" language instead of due care and another opposing the use of any due care provision.

A number of commenters, while supporting the use of a due care provision, said that the proposal provides no assurance that a manufacturer's good faith effort will be considered due care. They said that the agency should identify the level of testing and analysis necessary to constitute due care. Another commenter emphasized that in defining due care, the agency must ensure that a manufacturer uses recognized statistical procedures in determining that its products comply with the requirements of the standard.

Another group of commenters requested the agency to reconsider its decision not to use "design to conform" language in the standard; they said that the agency's concerns about the subjectivity of a "design to conform" language are not greater and could well be less than that resulting from use of due care language.

One commenter opposed the use of any due care language in the standard. It argued that the National Traffic and Motor Vehicle Safety Act requires the agency to set objective performance requirements in its standards. When a manufacturer determines that it has not met those performance requirements, then the manufacturer is under an obligation to notify owners and remedy the noncomplying vehicles. It argued that the proposed due care provision, in effect, provides manufacturers with an exemption from the Vehicle Safety Act recall provisions.

As discussed in the July 1984 final rule and the April 1985 notice, the agency believes that the test procedure of Standard No. 208 produces repeatable results in vehicle crash tests. The agency does, however, recognize that the Standard No. 208 test is more complicated than NHTSA's other crash test standards since a number of different injury measurements must be made on the two test dummies used in the testing. Because of this complexity, the agency believes that manufacturers need assurance from the agency that, if they have made a good faith effort in designing their vehicles and have instituted adequate quality control measures, they will not face the recall of their vehicles because of an isolated apparent failure to meet one of the injury criteria. The adoption of a due care provision provides that assurance. For the reasons discussed in the July 1984 final rules, the agency still believes use of a due care provision is a better approach to this issue than use of a design to conform provision.

As the agency has emphasized in its prior interpretation letters, a determination of what constitutes due care can only be made on a case-by-case basis. Whether a manufacturer's action will constitute due care will depend, in part, upon the availability of test equipment, the limitations of available technology, and above all, the diligence evidenced by the manufacturer.

Adoption of a due care defense is in line with the agency's long-standing and well-known enforcement policy on test differences. Under this long standing practice if the agency's testing shows noncompliance and a manufacturer's tests, valid on their face, show complying results, the agency will conduct an inquiry into the reason for the differing results. If the agency concludes that the difference in results can be explained to the agency's satisfaction, that the agency's results do not indicate an unreasonable risk to safety, and that the manufacturer's tests were reasonably conducted and were in conformity with standard, then the agency does not use its own tests as a basis for a finding of noncompliance. Although this interpretation has long been a matter of public record, Congress, in subsequent amendments of the Vehicle Safety Act, has not acted to alter that interpretation. The Supreme Court has said that under those circumstances, it can be presumed that the agency's interpretation has correctly followed the intent of the statute. (*See United States* v. *Rutherford*, 442 U.S. 544, 544 n. 10 (1979))

Phase-In

Attribution rules

With respect to cars manufacturered by two or more companies, and cars manufactured by one company and imported by another, the April 1985 notice proposed to clarify who would be considered the manufacturer for purposes of calculating the average annual production of passenger cars for each manufacturer and the amount of passenger cars manufacturered by each manufacturer that must comply with the automatic restraint phase-in requirements. In order to provide maximum flexibility to manufacturers, while assuring that the percentage phase-in goals are met, the notice proposed to permit manufacturers to determine, by contract, which of them will count, as its own, passenger cars manufactured by two or more companies or cars manufactured by one company and imported by another.

The notice also proposed two rules of attribution in the absence of such a contract. First, a passenger car which is imported for purposes of resale would be attributed to the importer. The agency intended that this proposed attribution rule would apply to both direct importers as well as importers authorized by the vehicle's original manufacturer. (In this context, direct importation refers to the importation of cars which are originally manufactured for sale outside the U.S. and which are then imported without the manufacturer's authorization into the U.S. by an importer for purposes of resale. The Vehicle Safety Act requires that such vehicles be brought into conformity with Federal motor vehicle safety standards.) Under the second proposed attribution rule, .a passenger car manufactured in the United States by more than one manufacturer, one of which also

markets the vehicle, would be attributed to the manufacturer which markets the vehicle.

These two proposed rules would generally attribute a vehicle to the manufacturer which is most responsible for the existence of the vehicle in the United States, i.e., by importing the vehicle or by manufacturing the vehicle for its own account as part of a joint venture, and marketing the vehicle. (Importers generally market the vehicles they import.) All commenters on these proposals supported giving manufacturers the flexibility to determine contractually which manufacturer would count the passenger car as its own. The commenters also supported the proposed attribution rules. Therefore, the agency is adopting the provisions as proposed.

Credit for early phase-in

The April 1985 notice proposed that manufacturers that exceeded the minimum percentage phase-in requirements in the first or second years could count those extra vehicles toward meeting the requirements in the second or third years. In addition, manufacturers could also count any automatic restraint vehicles produced during the one year preceding the first year of the phase-in. Since all the commenters addressing these proposals supported them, the agency is adopting them as proposed. The agency believes that providing credit for early introduction will encourage introduction of larger numbers of automatic restraints and provide increased flexibility for manufacturers. In addition, it will assure an orderly build-up of production capability for automatic restraint equipped cars as contemplated by the July 1984 final rule.

One commenter asked the agency to establish a new credit for vehicles equipped with non-belt automatic restraints at the driver's position and a dynamically-tested manual belt at the passenger position. The commenter requested that such a vehicle receive a 1.0 credit. The commenter also asked the agency to allow vehicles equipped with driver-only automatic restraint systems to be manufactured after September 1, 1989, the effective date for automatic restraints for the driver and front right passenger seating positions in all passenger cars. In its August 30, 1985 notice (50 FR 35233) responding to petitions for reconsideration of the July 1984 final rule on Standard No. 208, the agency has already adopted a part of the commenter's suggestion by establishing a 1.0 vehicle credit for vehicles equipped with a non-belt automatic restraint at the driver's position and a manual lap/shoulder belt at the passenger's position. For reasons detailed in the July 1984 final rule, the agency believes that the automatic restraint requirement should apply to both front outboard seating positions beginning on September 1, 1989, and is therefore not adopting the commenter's second suggestion.

Phase-In Reporting Requirements

The April 1985 notice proposed to establish a new Part 585, *Automatic Restraint Phase-in Reporting Requirements*. The agency proposed requiring manufacturers to submit three reports to NHTSA, one for each of the three automatic restraint phase-in periods. Each report, covering production during a 12-month period beginning September 1 and ending August 31, would be required to be submitted within 60 days after the end of such period. Information required by each report would include a statement regarding the extent to which the manufacturer had complied with the applicable percentage phase-in requirement of Standard No. 208 for the period covered by the report; the number of passenger cars manufactured for sale in the United States for each of the three previous 12-month production periods; the actual number of passenger cars manufactured during the reporting production (or during a previous production period and counted toward compliance in the reporting production period) period with automatic safety belts, air bags and other specified forms of automatic restraint technology, respectively; and brief information about any express written contracts which concern passenger cars produced by more than one manufacturer and affect the report.

One commenter questioned the need for a reporting requirement, saying that the requirement was unnecessary since manufacturers must self-certify that their vehicles meet Standard No. 208. The agency believes that a reporting requirement is needed for the limited period of the phase-in of automatic restraints so that the agency can carry out its statutory duty to monitor compliance with the Federal motor vehicle safety standards. During the phase-in, only a certain percentage of vehicles are required to have automatic restraints. It would be virtually impossible for the agency to determine if the applicable percentage of passenger cars has been equipped with automatic restraints unless manufacturers provide certain production information to the agency. NHTSA is therefore adopting the reporting requirement.

The same commenter said that requiring the report to be due 60 days after the end of the production year can be a problem for importers. The commenter said that production records may accompany the vehicle, which may not actually reach the United States until ,30 or 45 days after the production year ends. The commenter asked the agency to provide an appeal process to seek an extension of the period to file the report. The agency believes that the example presented by the commenter represents a worst case situation and complying with the 60 day requirement should not be a problem for manufacturers, including importers. However, to eliminate any problems in worst case situations, the agency is amending the regulation to provide that manufacturers seeking an extension of the deadline to file a report must file a request for an extension at least 15 days before the report is due.

Calculation of average annual production

The agency also proposed an alternative to the requirement that the number of cars that must be equipped with automatic restraints must be based on a percentage of each manufacturer's average annual production for the past three model years. The proposed alternative would permit manufacturers to equip the required percentage of its actual production of passenger cars with automatic restraints during each affected year. Since all commenters addressing this proposal supported it, the agency is adopting it as an alternative means of compliance, at the manufacturer's option. In the case of a new manufacturer, the manufacturer would have to calculate the amount of passenger cars required to have automatic restraints based on its production of passenger cars during each of the affected years. Since the agency has decided to adopt the alternative basis for determining the production quota, it has made the necessary conforming changes in the reporting requirements adopted in this notice.

One commenter also requested the agency to clarify whether a manufacturer does have to include its production volume of convertibles when it is calculating the percentage of vehicles that must meet the phase-in requirement. The automatic restraint requirement applies to all passenger cars. Thus, a manufacturer's production figures for passenger car convertibles must be counted when the manufacturer is calculating its phase-in requirements.

Retention of VINs

In order to keep administrative burdens to a minimum, the agency proposed that the required report need not use the VIN to identify the particular type of automatic restraint installed in each passenger car produced during the phase-in period. Since that information could be necessary for purposes of enforcement, however, the agency proposed to require that manufacturers maintain records until December 31, 1991, of the VIN and type of automatic restraint for each passenger car which is produced during the phase-in period and is reported as having automatic restraints. Although direct import cars are not required to have a US-format VIN number, those cars would still have a European-format VIN number and thus direct importers would be required to retain that VIN information. (The agency is considering a petition from Volkswagen requesting that direct import cars be required to have US-format VINs.)

The reason for retaining the information until 1991 is to ensure that such information would then be available until the completion of any agency enforcement action begun after the final phase-in report is filed in 1990. The agency believes this requirement meets the needs of the agency, with minimal impacts on manufacturers, and therefore is adopting it as proposed. One commenter asked whether a manufacturer is required to keep the VIN information as a separate file or whether keeping the information as a part of its general business records is sufficient. As long as the VIN information is retrievable, it may be stored in any manner that is convenient for a manufacturer.

In consideration of the foregoing, 49 CFR Part 571.208 is amended as follows:

The authority citation for Part 571 would continue to read as follows:

Authority: 15 U.S.C. 1392, 1401, 1403, 1407; delegation of authority at 49 CFR 1.50.

1. Section S4.1.3.1.2 is revised to read as follows:

S4.1.3.1.2 Subject to S4.1.3.4 and S4.1.5, the amount of passenger cars, specified in S4.1.3.1.1 complying with the requirements of S4.1.2.1 shall be not less than 10 percent of:

(a) the average annual production of passenger cars manufactured on or after September 1, 1983, and before September 1, 1986, by each manufacturer, or

(b) the manufacturer's annual production of passenger cars during the period specified in S4.1.3.1.1.

2. Section 4.1.3.2.2 is revised to read as follows:

S4.1.3.2.2 Subject to S4.1.3.4 and S4.1.5, the amount of passenger cars specified in S4.1.3.2.1 complying with the requirements of S4.1.2.1 shall be not less than 25 percent of:

(a) the average annual production of passenger cars manufactured on or after September 1, 1984,

and before September 1, 1987, by each manufacturer, or

(b) the manufacturer's annual production of passenger cars during the period specified in S4.1.3.2.1.

3. Section 4.1.3.3.2 is revised to read as follows:

S4.1.3.3.2 Subject to S4.1.3.4 and S4.1.5, the amount of passenger cars specified in S4.1.3.3.1 complying with the requirements of S4.1.2.1 shall not be less than 40 percent of:

(a) the average annual production of passenger cars manufactured on or after September 1, 1985, and before September 1, 1988, by each manufacturer or

(b) the manufacturer's annual production of passenger cars during the period specified in S4.1.3.3.1.

4. Section S4.1.3.4 is revised to read as follows:

S4.1.3.4 *Calculation of complying passenger cars.*

(a) For the purposes of calculating the numbers of cars manufactured under S4.1.3.1.2, S4.1.3.2.2, or S4.1.3.3.2 to comply with S4.1.2.1:

(1) each car whose driver's seating position complies with the requirements of S4.1.2.1(a) by means not including any type of seat belt and whose front right seating position will comply with the requirements of S4.1.2.1(a) by any means is counted as 1.5 vehicles, and

(2) each car whose driver's seating position complies with the requirements of S4.1.2.1(a) by means not including any type of seat belt and whose right front seat seating position is equipped with a manual Type 2 seat belt is counted as one vehicle.

(b) For the purposes of complying with S4.1.3.1.2, a passenger car may be counted if it:

(1) is manufactured on or after September 1, 1985, but before September 1, 1986, and

(2) complies with S4.1.2.1.

(c) For the purposes of complying with S4.1.3.2.2, a passenger car may be counted if it:

(1) is manufactured on or after September 1, 1985, but before September 1, 1987, and

(2) complies with S4.1.2.1, and

(3) is not counted toward compliance with S4.1.3.1.2

(d) For the purposes of complying with S4.1.3.3.2, a passenger car may be counted if it:

(1) is manufactured on or after September 1, 1985, but before September 1, 1988, and

(2) complies with S4.1.2.1, and

(3) is not counted toward compliance with S4.1.3.1.2 or S4.1.3.2.2.

5. A new section S4.1.3.5 is added to read as follows:

S4.1.3.5 *Passenger cars produced by more than one manufacturer.*

S4.1.3.5.1 For the purposes of calculating average annual production of passenger cars for each manufacturer and the amount of passenger cars manufactured by each manufacturer under S4.1.3.1.2, S4.1.3.2.2 or S4.1.3.3.2, a passenger car produced by more than one manufacturer shall be attributed to a single manufacturer as follows, subject to S4.1.3.5.2:

(a) A passenger car which is imported shall be attributed to the importer.

(b) A passenger car manufactured in the United States by more than one manufacturer, one of which also markets the vehicle, shall be attributed to the manufacturer which markets the vehicle.

S4.1.3.5.2 A passenger car produced by more than one manufacturer shall be attributed to any one of the vehicle's manufacturers specified by an express written contract, reported to the National Highway Traffic Safety Administration under 49 CFR Part 585, between the manufacturer so specified and the manufacturer to which the vehicle would otherwise be attributed under S4.1.3.5.1.

6. A new section S4.6 is added to read as follows:

S4.6 *Dynamic testing of manual belt systems.*

S4.6.1 If the automatic restraint requirement of S4.1.4 is rescinded pursuant to S4.1.5, then each passenger car that is manufactured after September 1, 1989, and is equipped with a Type 2 manual seat belt assembly at each front outboard designated seating position pursuant to S4.1.2.3 shall meet the frontal crash protection requirements of S5.1 at those designated seating positions with a test dummy restrained by a Type 2 seat belt assembly that has been adjusted in accordance with S7.4.2.

S4.6.2 A Type 2 seat belt assembly subject to the requirements of S4.6.1 of this standard does not have to meet the requirements of S4.2(a)-(c) and S4.4 of Standard No. 209 (49 CFR 571.209) of this Part.

7. S7.4.2 is revised to read as follows:

S7.4.2 *Webbing tension relieving device.* Each vehicle with an automatic seat belt assembly or with a Type 2 manual seat belt assembly that must meet S4.6 installed in a front outboard designated seating position that has either manual or automatic devices permitting the introduction of slack in the webbing of the shoulder belt (e.g., "comfort clips" or "window-shade" devices) shall:

(a) comply with the requirements of S5.1 with the shoulder belt webbing adjusted to introduce the maximum amount of slack recommended by the manufacturer pursuant to S7.4.2.(b);

(b) have a section in the vehicle owner's manual that explains how the tension-relieving device works and specifies the maximum amount of slack (in inches) recommended by the vehicle manufacturer to be introduced into the shoulder belt under normal use conditions. The explanation shall also warn that introducing slack beyond the amount specified by the manufacturer can significantly reduce the effectiveness of the shoulder belt in a crash; and

(c) have an automatic means to cancel any shoulder belt slack introduced into the belt system by a tension-relieving device each time the safety belt is unbuckled or the adjacent vehicle door is opened, except that open-body vehicles with no doors can have a manual means to cancel any shoulder belt slack introduced into the belt system by a tension-relieving device.

8. Section 8.1.1(c) is revised to read as follows:

S8.1.1(c) *Fuel system capacity.* With the test vehicle on a level surface, pump the fuel from the vehicle's fuel tank and then operate the engine until it stops. Then, add Stoddard solvent to the test vehicle's fuel tank in an amount which is equal to not less than 92 and not more than 94 percent of the fuel tank's usable capacity stated by the vehicle's manufacturer. In addition, add the amount of Stoddard solvent needed to fill the entire fuel system from the fuel tank through the engine's induction system.

9. A new section 8.1.1(d) is added to read as follows:

S8.1.1(d) *Vehicle test attitude.* Determine the distance between a level surface and a standard reference point on the test vehicle's body, directly above each wheel opening, when the vehicle is in its "as delivered" condition. The "as delivered" condition is the vehicle as received at the test site, with 100 percent of all fluid capacities and all tires inflated to the manufacturer's specifications as listed on the vehicle's tire placard. Determine the distance between the same level surface and the same standard reference points in the vehicle's "fully loaded condition." The "fully loaded condition" is the test vehicle loaded in accordance with S8.1.1(a) or (b), as applicable. The load placed in the cargo area shall be centered over the longitudinal centerline of the vehicle. The pretest vehicle attitude shall be equal to either the as delivered or fully loaded attitude or between the as delivered attitude and the fully loaded attitude.

10. S7.4.3 is revised by removing the reference to "S10.6" and replacing it with a reference to "S10.7."

11. S7.4.4 is revised by removing the reference to "S10.5" and replacing it with a reference to "S10.6."

12. S7.4.5 is revised by removing the reference to "S8.1.11" and replacing it with a reference to "S10."

13. Section 8.1.3 is revised to read as follows:

S8.1.3 *Adjustable seat back placement.* Place adjustable seat backs in the manufacturer's nominal design riding position in the manner specified by the manufacturer. Place each adjustable head restraint in its highest adjustment position.

14. Sections 8.1.11 through 8.1.11.2.3 are removed.

15. Sections 8.1.12 and 8.1.13 are redesignated 8.1.11 and 8.1.12, respectively.

16. Section 10 is revised to read as follows:

S10 *Test dummy positioning procedures.* Position a test dummy, conforming to Subpart B of Part 572 (49 CFR Part 572), in each front outboard seating position of a vehicle as specified in S10.1 through S10.9. Each test dummy is:

(a) not restrained during an impact by any means that require occupant action if the vehicle is equipped with automatic restraints.

(b) restrained by manual Type 2 safety belts, adjusted in accordance with S10.9, if the vehicle is equipped with manual safety belts in the front outboard seating positions.

S10.1 *Vehicle equipped with front bucket seats.* Place the test dummy's torso against the seat back and its upper legs against the seat cushion to the extent permitted by placement of the test dummy's feet in accordance with the appropriate paragraph of S10. Center the test dummy on the seat cushion of the bucket seat and set its midsagittal plane so that it is vertical and parallel to the centerline of the vehicle.

S10.1.1 *Driver position placement.*

(a) Initially set the knees of the test dummy 11¾ inches apart, measured between the outer surfaces of the knee pivot bolt heads, with the left outer surface 5.9 inches from the midsagittal plane of the test dummy.

(b) Rest the right foot of the test dummy on the undepressed accelerator pedal with the rearmost point of the heel on the floor pan in the plane of the pedal. If the foot cannot be placed on the accelerator pedal, set it perpendicular to the lower leg and place it as far forward as possible in the direction of the geometric center of the pedal with the rearmost point of the heel resting on the floor pan. Except as prevented by contact with a vehicle surface, place the right leg so that the upper and lower leg centerlines fall, as close as possible, in a vertical longitudinal plane without inducing torso movement.

(c) Place the left foot on the toeboard with the rearmost point of the heel resting on the floor pan as close as possible to the point of intersection of the planes described by the toeboard and the floor pan. If the foot cannot be positioned on the toeboard, set it

perpendicular to the lower leg and place it as far forward as possible with the heel resting on the floor pan. Except as prevented by contact with a vehicle surface, place the left leg so that the upper and lower leg centerlines fall, as close as possible, in a vertical plane. For vehicles with a foot rest that does not elevate the left foot above the level of the right foot, place the left foot on the foot rest so that the upper and lower leg centerlines fall in a vertical plane.

S10.1.2 *Passenger position placement.*

S10.1.2.1 *Vehicles with a flat floor pan/toeboard.*

(a) Initially set the knees 11¾ inches apart, measured between the outer surfaces of the knee pivot bolt heads.

(b) Place the right and left feet on the vehicle's toeboard with the heels resting on the floor pan as close as possible to the intersection point with the toeboard. If the feet cannot be placed flat on the toeboard, set them perpendicular to the lower leg centerlines and place them as far forward as possible with the heels resting on the floor pan.

(c) Place the right and left legs so that the upper and lower leg centerlines fall in vertical longitudinal planes.

S10.1.2.2 *Vehicles with wheelhouse projections in passenger compartment.*

(a) Initially set the knees 11¾ inches apart, measured between outer surfaces of the knee pivot bolt heads.

(b) Place the right and left feet in the well of the floor pan/toeboard and not on the wheelhouse projection. If the feet cannot be placed flat on the toeboard, set them perpendicular to the lower leg centerlines and as far forward as possible with the heels resting on the floor pan.

(c) If it is not possible to maintain vertical and longitudinal planes through the upper and lower leg centerlines for each leg, then place the left leg so that its upper and lower centerlines fall, as closely as possible, in a vertical longitudinal plane and place the right leg so that its upper and lower leg centerlines fall, as closely as possible, in a vertical plane.

S10.2 *Vehicle equipped with bench seating.* Place a test dummy with its torso against the seat back and its upper legs against the seat cushion, to the extent permitted by placement of the test dummy's feet in accordance with the appropriate paragraph of S10.1.

S10.2.1 *Driver position placement.* Place the test dummy at the left front outboard designated seating position so that its midsagittal plane is vertical and parallel to the centerline of the vehicle and so that the midsagittal plane of the test dummy passes through the center of the steering wheel rim. Place the legs, knees, and feet of the test dummy as specified in S10.1.1.

S10.2.2 *Passenger position placement.* Place the test dummy at the right front outboard designated seating position as specified in S10.1.2, except that the midsagittal plane of the test dummy shall be vertical and longitudinal, and the same distance from the vehicle's longitudinal centerline as the midsagittal plane of the test dummy at the driver's position.

S10.3 *Initial test dummy placement.* With the test dummy at its designated seating position as specified by the appropriate requirements of S10.1 or S10.2, place the upper arms against the seat back and tangent to the side of the upper torso. Place the lower arms and palms against the outside of the upper legs.

S10.4 *Test dummy settling.*

S10.4.1 *Test dummy vertical upward displacement.* Slowly lift the test dummy parallel to the seat back plane until the test dummy's buttocks no longer contact the seat cushion or until there is test dummy head contact with the vehicle's headlining.

S10.4.2 *Lower torso force application.* Using a test dummy positioning fixture, apply a rearward force of 50 pounds through the center of the rigid surface against the test dummy's lower torso in a horizontal direction. The line of force application shall be 6½ inches above the bottom surface of the test dummy's buttocks. The 50 pound force shall be maintained with the rigid fixture applying reaction forces to either the floor pan/toeboard, the 'A' post, or the vehicle's seat frame.

S10.4.3 *Test dummy vertical downward displacement.* While maintaining the contact of the horizontal rearward force positioning fixture with the test dummy's lower torso, remove as much of the 50 pound force as necessary to allow the test dummy to return downward to the seat cushion by its own weight.

S10.4.4 *Test dummy upper torso rocking.* Without totally removing the horizontal rearward force being applied to the test dummy's lower torso, apply a horizontal forward force to the test dummy's shoulders sufficient to flex the upper torso forward until its back no longer contacts the seat back. Rock the test dummy from side to side 3 or 4 times so that the test dummy's spine is at any angle from the vertical in the 14 to 16 degree range at the extremes of each rocking movement.

S10.4.5 *Upper torso force application.* With the test dummy's midsagittal plane vertical, push the upper torso against the seat back with a force of 50 pounds applied in a horizontal rearward direction along a line that is coincident with the test dummy's midsagittal plane and 18 inches above the bottom surface of the test dummy's buttocks.

S10.5 *Placement of test dummy arms and hands.* With the test dummy positioned as specified by S10.3 and without inducing torso movement, place the arms, elbows, and hands of the test dummy, as appropriate for each designated seating position in accordance with S10.3.1 or S10.3.2. Following placement of the arms, elbows and hands, remove the force applied against the lower half of the torso.

S10.5.1 *Driver's position.* Move the upper and the lower arms of the test dummy at the driver's position to their fully outstretched position in the lowest possible orientation. Push each arm rearward, permitting bending at the elbow, until the palm of each hand contacts the outer part of the rim of the steering wheel at its horizontal centerline. Place the test dummy's thumbs over the steering wheel rim and position the upper and lower arm centerlines as close as possible in a vertical plane without inducing torso movement.

S10.5.2 *Passenger position.* Move the upper and the lower arms of the test dummy at the passenger position to fully outstretched position in the lowest possible orientation. Push each arm rearward, permitting bending at the elbow, until the upper arm contacts the seat back and is tangent to the upper part of the side of the torso, the palm contacts the outside of the thigh, and the little finger is barely in contact with the seat cushion.

S10.6 *Test dummy positioning for latchplate access.* The reach envelopes specified in S7.4.4 are obtained by positioning a test dummy in the driver's seat or passenger's seat in its forwardmost adjustment position. Attach the lines for the inboard and outboard arms to the test dummy as described in Figure 3 of this standard. Extend each line backward and outboard to generate the compliance arcs of the outboard reach envelope of the test dummy's arms.

S10.7 *Test dummy positioning for belt contact force.* To determine compliance with S7.4.3 of this standard, position the test dummy in the vehicle in accordance with the appropriate requirements specified in S10.1 or S10.2 and under the conditions of S8.1.2 and S8.1.3. Pull the belt webbing three inches from the test dummy's chest and release until the webbing is within 1 inch of the test dummy's chest and measure the belt contact force.

S10.9 *Manual belt adjustment for dynamic testing.* With the test dummy at its designated seating position as specified by the appropriate requirements of S8.1.2, S8.1.3 and S10.1 through S10.5, place the Type 2 manual belt around the test dummy and fasten the latch. Remove all slack from the lap belt. Pull the upper torso webbing out of the retractor and allow it to retract; repeat this operation four times. Apply a 2 to 4 pound tension load to the lap belt. If the belt system is equipped with a tension-relieving device introduce the maximum amount of slack into the upper torso belt that is recommended by the manufacturer for normal use in the owner's manual for the vehicle. If the belt system is not equipped with a tension relieving device, allow the excess webbing in the shoulder belt to be retracted by the retractive force of the retractor.

17. S11 is removed.

18. S4.1.3.1.1, S4.1.3.2.1, S4.1.3.3.1, S4.1.4 and S4.6.1 are revised by adding a new second sentence to S4.1.3.1.1, S4.1.3.2.1, S4.1.3.3.1 and S4.1.4 and a new second sentence to S4.6.1 to read as follows:

A vehicle shall not be deemed to be in noncompliance with this standard if its manufacturer establishes that it did not have reason to know in the exercise of due care that such vehicle is not in conformity with the requirement of this standard.

19. S8.1.5 is amended to read as follows:

Movable vehicle windows and vents are, at the manufacturer's option, placed in the fully closed position.

20. S7.4 is amended to read as follows:

S7.4. *Seat belt comfort and convenience.*

(a) *Automatic seat belts.* Automatic seat belts installed in any vehicle, other than walk-in van-type vehicles, which has a gross vehicle weight rating of 10,000 pounds or less, and which is manufactured on or after September 1, 1986, shall meet the requirements of S7.4.1, S7.4.2, and S7.4.3.

(b) *Manual seat belts.*

(1) *Vehicles manufactured after September 1, 1986.* Manual seat belts installed in any vehicle, other than manual Type 2 belt systems installed in the front outboard seating positions in passenger cars or manual belts in walk-in van-type vehicles, which have a gross vehicle weight rating of 10,000 pounds or less, shall meet the requirements of S7.4.3, S7.4.4, S7.4.5, and S7.4.6.

(2) *Vehicles manufactured after September 1, 1989.*

(i) If the automatic restraint requirement of S4.1.4 is rescinded pursuant to S4.1.5, then manual seat belts installed in a passenger car shall meet the requirements of S7.1.1.3(a), S7.4.2, S7.4.3, S7.4.4, S7.4.5, and S7.4.6.

(ii) Manual seat belts installed in a bus, multipurpose passenger vehicle and truck with a gross vehicle weight rating of 10,000 pounds or less, except for walk-in van-type vehicles, shall meet the requirements of S7.4.3, S7.4.4, S7.4.5, and S7.4.6.

571.209 *Standard No. 209, Seat belt assemblies.*

1. A new S4.6 is added, to read as follows:

S4.6 *Manual belts subject to crash protection requirements of Standard No. 208.*

(a) A seat belt assembly subject to the requirements of S4.6.1 of Standard No. 208 (49 CFR Part 571.208) does not have to meet the requirements of S4.2 (a)-(c) and S4.4 of this standard.

(b) A seat belt assembly that does not comply with the requirements of S4.4 of this standard shall be permanently and legibly marked or labeled with the following language:

This seat belt assembly may only be installed at a front outboard designated seating position of a vehicle with a gross vehicle weight rating of 10,000 pounds or less.

571.210 *Standard No. 210, Seat Belt Assembly Anchorages.*

1. The second sentence of S4.3 is revised to read as follows:

Anchorages for automatic and for\ dynamically tested seat belt assemblies that meet the frontal crash protection requirement of S5.1 of Standard No. 208 (49 CFR Part 571.208) are exempt from the location requirements of this section.

PART 585 – AUTOMATIC RESTRAINT PHASE-IN REPORTING REQUIREMENTS

1. Chapter V, Title 49, Transportation, the Code of Federal Regulations, is amended to add the following new Part:

PART 585 – AUTOMATIC RESTRAINT PHASE-IN REPORTING REQUIREMENTS

Secs.

585.1 Scope.
585.2 Purpose.
585.3 Applicability.
585.4 Definitions.
585.5 Reporting requirements.
585.6 Records.
585.7 Petition to extend period to file report.

Authority: 15 U.S.C. 1392, 1407; delegation of authority at 49 CFR 1.50.

585.1 *Scope.*

This section establishes requirements for passenger car manufacturers to submit a report, and maintain records related to the report, concerning the number of passenger cars equipped with automatic restraints in compliance with the requirements of S4.1.3 of Standard No. 208, *Occupant Crash Protection* (49 CFR Part 571.208).

585.2 *Purpose.*

The purpose of the reporting requirements is to aid the National Highway Traffic Safety Administration in determining whether a passenger car manufacturer has complied with the requirements of Standard No. 208 of this Chapter (49 CFR 571.208) for the installation of automatic restraints in a percentage of each manufacturer's annual passenger car production.

585.3 *Applicability.*

This part applies to manufacturers of passenger cars.

585.4 *Definitions.*

All terms defined in section 102 of the National Traffic and Motor Vehicle Safety Act (15 U.S.C. 1391) are used in their statutory meaning.

"Passenger car" is used as defined in 49 CFR Part 571.3.

"Production year" means the 12-month period between September 1 of one year and August 31 of the following year, inclusive.

585.5 *Reporting requirements.*

(a) *General reporting requirements.*

Within 60 days after the end of each of the production years ending August 31, 1987, August 31, 1988, and August 31, 1989, each manufacturer shall submit a report to the National Highway Traffic Safety Administration concerning its compliance with the requirements of Standard No. 208 for installation of automatic restraints in its passenger cars produced in that year. Each report shall –

(1) Identify the manufacturer;

(2) State the full name, title and address of the official responsible for preparing the report;

(3) Identify the production year being reported on;

(4) Contain a statement regarding the extent to which the manufacturer has complied with the requirements of S4.1.3 of Standard No. 208;

(5) Provide the information specified in 585.5(b);

(6) Be written in the English language; and

(7) Be submitted to: Administrator, National Highway Traffic Safety Administration, 400 Seventh Street, S.W., Washington, D.C. 20590.

(b) *Report content.*

(1) *Basis for phase-in production goals.*

Each manufacturer shall provide the number of passenger cars manufactured for sale in the United States for each of the three previous production years, or, at the manufacturer's option, for the current production year. A new manufacturer that is, for the first time, manufacturing passenger cars for sale in the United States must report the number of passenger cars manufactured during the current production year.

(2) *Production.*

Each manufacturer shall report for the production year being reported on, and each preceding production year, to the extent that cars produced during the preceding years are treated under Standard No. 208 as having been produced during the production year being reported on, the following information:

(i) the number of passenger cars equipped with automatic seat belts and the seating positions at which they are installed,

(ii) the number of passenger cars equipped with air bags and the seating positions at which they are installed, and

(iii) the number of passenger cars equipped with other forms of automatic restraint technology, which shall be described, and the seating positions at which they are installed.

(3) *Passenger cars produced by more than one manufacturer.*

Each manufacturer whose reporting of information is affected by one or more of the express written contracts permitted by section S4.1.3.5.2 of Standard No. 208 shall:

(i) Report the existence of each contract, including the names of all parties to the contract, and explain how the contract affects the report being submitted,

(ii) Report the actual number of passenger cars covered by each contract.

585.6 *Records.*

Each manufacturer shall maintain records of the Vehicle Identification Number and type of automatic restraint for each passenger car for which information is reported under 585.5(b)(2), until December 31, 1991.

585.7 *Petition to extend period to file report.*

A petition for extension of the time to submit a report must be received not later than 15 days before expiration of the time stated in 585.5(a). The petition must be submitted to: Administrator, National Highway Traffic Safety Administration, 400 Seventh Street, SW, Washington, DC 20590. The filing of a petition does not automatically extend the time for filing a report. A petition will be granted only if the petitioner shows good cause for the extension and if the extension is consistent with the public interest.

Issued on March 18, 1986

Diane K. Steed
Administrator

51 F.R. 9801
March 21, 1986

PART 585—AUTOMATIC RESTRAINT PHASE-IN REPORTING REQUIREMENTS
(Docket No. 74-14; Notice 43)

Authority: 15 U.S.C. 1392, 1407; delegation of authority at 49 CFR 1.50.

585.1 Scope.

This section establishes requirements for passenger car manufacturers to submit a report, and maintain records related to the report, concerning the number of passenger cars equipped with automatic restraints in compliance with the requirements of S4.1.3 of Standard No. 208, *Occupant Crash Protection* (49 CFR Part 571.208).

585.2 Purpose.

The purpose of the reporting requirements is to aid the National Highway Traffic Safety Administration in determining whether a passenger car manufacturer has complied with the requirements of Standard No. 208 of this Chapter (49 CFR 571.208) for the installation of automatic restraints in a percentage of each manufacturer's annual passenger car production.

585.3 Applicability.

This part applies to manufacturers of passenger cars.

585.4 Definitions.

[(a) All terms defined in section 102 of the National Traffic and Motor Vehicle Safety Act (15 U.S.C. 1391) are used in their statutory meaning.

(b) "Passenger car" means a motor vehicle with motive power, except a multipurpose passenger vehicle, motorcycle, or trailer, designed for carrying 10 persons or less.

(c) "Production year" means the 12-month period between September 1 of one year and August 31 of the following year, inclusive. **(51 F.R. 37028—October 17, 1986. Effective: November 17, 1986)]**

585.5 Reporting requirements.

(a) *General reporting requirements.*

Within 60 days after the end of each of the production years ending August 31, 1987, August 31, 1988, and August 31, 1989, each manufacturer shall submit a report to the National Highway Traffic Safety Administration concerning its compliance with the requirements of Standard No. 208 for installation of automatic restraints in its passenger cars produced in that year. Each report shall—

(1) Identify the manufacturer;

(2) State the full name, title and address of the official responsible for preparing the report;

(3) Identify the production year being reported on;

(4) Contain a statement regarding the extent to which the manufacturer has complied with the requirements of S4.1.3. of Standard No. 208;

(5) Provide the information specified in 585.5(b);

(6) Be written in the English language; and

(7) Be submitted to: Administrator, National Highway Traffic Safety Administration, 400 Seventh Street, S.W., Washington, D.C. 20590.

(b) *Report content.*

(1) *Basis for phase-in production goals.* Each manufacturer shall provide the number of passenger cars manufactured for sale in the United States for each of the three previous production years, or, at the manufacturer's option, for the current production year. A new manufacturer that is, for the first time, manufacturing passenger cars for sale in the United States must report the number of passenger cars manufactured during

the current production year. [For the purpose of the reporting requirements of this Part, a manufacturer may exclude its production of convertibles, which do not comply with requirements of S4.1.2.1 of Part 571.208 of this Chapter, from the report of its production volume of passenger cars manufactured for sale in the United States. (51 F.R. 37028—October 17, 1986. Effective: November 17, 1986)]

(2) *Production.* Each manufacturer shall report for the production year being reported on, and each preceding production year, to the extent that cars produced during the preceding years are treated under Standard No. 208 as having been produced during the production year being reported on, the following information:

    (i) the number of passenger cars equipped with automatic seat belts and the seating positions which they are installed,

    (ii) the number of passenger cars equipped with air bags and the seating positions at which they are installed, and

    (iii) the number of passenger cars equipped with other forms of automatic restraint technology, which shall be described, and the seating positions at which they are installed.

(3) *Passenger cars produced by more than one manufacturer.* Each manufacturer whose reporting of information is affected by one or more of the express written contracts permitted by section S4.1.3.5.2 or Standard No. 208 shall:

    (i) Report the existence of each contract, including the names of all parties to the contract, and explain how the contract affects the report being submitted,

    (ii) Report the actual number of passenger cars covered by each contract.

585.6 Records.

Each manufacturer shall maintain records of the Vehicle Identification Number and type of automatic restraint for each passenger car for which information is reported under 585.5(b)(2), until December 31, 1991.

585.7 Petition to extend period to file report.

A petition for extension of the time to submit a report must be received not later than 15 days before expiration of the time stated in 585.5(a). The petition must be submitted to Administrator, National Highway Traffic Safety Administration, 400 Seventh Street, SW, Washington, D.C. 20590. The filing of a petition does not automatically extend the time for filing a report. A petition will be granted only if the petitioner shows good cause for the extension and if the extension is consistent with the public interest.

Issued on March 18, 1986.

Diane K. Steed
Administrator

F.R. 51 9801
March 21, 1986

PREAMBLE TO PART 590—MOTOR VEHICLE EMISSIONS INSPECTION CRITERIA

(Docket No. 72-24; Notice 2)

This notice issues a regulation to establish emissions inspection criteria for a diagnostic inspection demonstration projects funded pursuant to the Motor Vehicle Information and Cost Savings Act (15 U.S.C. 1901, *et seq.*). The regulation is based upon a notice of proposed rulemaking published June 11, 1974 (39 F.R. 20501) and upon comments submitted in response to the notice, and is issued in consultation with the Administrator of the Environmental Protection Agency.

Under Title 15 U.S.C., Section 1962(a), a State may obtain a grant from the Federal government for the purpose of establishing and operating a diagnostic inspection demonstration project. The purpose of the grant program is to explore the feasibility of using diagnostic test devices to conduct diagnostic safety and emission inspection of motor vehicles. The demonstration projects are also designed to help the Federal and State governments determine the best means of structuring safety and emissions inspection programs. Pursuant to the requirements of section 1962(b), this rule establishes emissions inspection criteria to be met by projects funded under this program. The criteria established govern the manner of operation of five Federally-funded State diagnostic inspection demonstration projects to be conducted in Alabama, Arizona, the District of Columbia, Puerto Rico, and Tennessee, and do not, in themselves, impose requirements on any other State or upon any individual.

The subject most commonly discussed in the comments was whether a loaded test mode or a high speed no load test mode would be more effective than the basic idle-only mode inspection procedure in detecting vehicles with very high emission levels and in diagnosing problems. Because this program calls for demonstration projects and is in the nature of a feasibility study, the NHTSA considers that the most appropriate course is to compare the alternative procedures and, in this way, generate data which may ultimately resolve the question. Accordingly, the States will be allowed to choose between loaded-mode and no-load inspection procedures. For similar reasons no-load inspection procedures will include both low and high speed measurements until such time as the data collected indicates that unloaded high-speed measurements are unwarranted.

Since one of the major purposes of the program is to determine whether this type of inspection is both feasible and cost beneficial, the criteria do not specify that the emission levels be the lowest attainable, but represent a fair balance between low rejection rates which would result in limited program effectiveness and high rejection rates which would result in adverse public reaction. In the event that the actual rejection rate varies significantly from our estimate of approximately 30 percent, the emissions criteria will be modified to bring the rate to the desired level. Because the emission criteria are less stringent than those permitted under the Federal Emission Certification Test criteria, it is not anticipated that conflicting requirements on engine design will result from their application in this program.

Two comments were addressed to the point that the mechanical dynamometer suggested for use in the loaded mode inspection may not simulate normal road loading as well as an electric dynamometer. The purpose of the dynamometer is to provide an adequate load to the engine to allow detection of carburetor main and power circuit malfunctions and ignition misfiring under load. Because this function does not require true road load duplication NHTSA does not consider that the more expensive electric dynamometer should be required.

General Motors Corporation suggested that oxides of nitrogen (NO_x) measurement be included in the emission inspection criteria. The Environmental Protection Agency recommended waiting until such time as NO_x controlled vehicles account for a more significant part of the vehicle population in order to make such a program meaningful. NO_x measuring instruments suitable for this type of inspection have not been developed to a point where low cost, reliable instruments are readily available. Furthermore, tuning a car without NO_x controls tends to increase the NO_x emissions slightly while reducing the hydrocarbon and carbon monoxide emissions. Therefore, NHTSA agrees with the EPA that until newer vehicles with NO_x control devices begin to account for a more substantial part of the overall vehicle population, the level of reduction of emissions of oxides of nitrogen that might be obtained is not large enough to warrant the inclusion of NO_x inspection at this time.

While the criteria developed in this rulemaking would be appropriate for emissions inspection of light duty trucks and other light duty vehicles, NHTSA has decided not to include these vehicles in the data pool for the demonstration projects. The rule requires that the idle speed of the vehicle at the time of inspection must not be more than 100 rpm greater than that recommended by the manufacturer. The purpose of this requirement is to ensure that

high idle speeds are not masking excessive idle carbon monoxide levels. At the suggestion of the American Motors Corporation the units of measure for proposed emission levels are more specifically identified than in the notice of proposed rulemaking. The unit of measurement of carbon monoxide concentration is Mole percent, while that for hydrocarbon concentration is ppm as hexane.

Therefore, a new Part 590, Motor Vehicle Emission Inspections, is added in Chapter V, Title 49, Code of Federal Regulations. . . .

Effective date: This part becomes effective July 5, 1975. The notice of proposed rulemaking had proposed an effective date 30 days after issuance of the final rule. Because the five States that have received grants have all developed their emission inspection in accordance with the proposed criteria, they will not be adversely affected by an immediate effective date. Good cause is accordingly found for an immediate effective date.

(Section 302(b)(1), Pub. L. 92–513, 86 Stat 947, 15 U.S.C. 1901; delegation of authority at 49 CFR 1.51.)

Issued on June 5, 1975.

James B. Gregory
Administrator

40 F.R. 24904
June 11, 1975

PART 590—EMISSION INSPECTIONS

Sec.
590.1 Scope.

590.2 Purpose.

590.3 Applicability.

590.4 Definitions.

590.5 Requirements.

590.6 No-load Inspection.

590.7 Loaded-mode inspection.

590.8 Inspection conditions.

§ 590.1 Scope.

This part specifies standards and procedures for motor vehicle emission inspections by State or State-supervised diagnostic inspection demonstration projects funded under Title III of the Motor Vehicle Information and Cost Savings Act (15 U.S.C. 1901, *et seq.*).

§ 590. Purpose.

The purpose of this part is to support the development of effective regulation of automobile exhaust emissions and thereby improve air quality, by establishing appropriate uniform procedures for diagnostic emission inspection demonstration projects.

§ 590.3 Applicability.

This part does not impose requirements on any person. It is intended to be utilized by State diagnostic inspection demonstration projects operating under Title III of the Cost Savings Act for diagnostic emission inspections of passenger cars powered by spark-ignition engines.

§ 590.4 Definitions.

All terms used in this part that are defined in 49 CFR Part 571, Motor Vehicle Safety Standards, are used as defined in that Part.

§ 590.5 Requirements.

A diagnostic inspection demonstration project shall test vehicles in accordance with either the no-load inspection criteria specified in section 590.6, or the loaded-mode inspection criteria specified in section 590.7.

§ 590.6 No-load inspection.

(a) *Criteria.* The vehicle must meet the following criteria when tested by the no-load inspection method.

(1) The vehicle's idle speed, measured with the transmission in the position recommended by the manufacturer for adjusting the idle speed, shall not be more than 100 rpm higher than the idle speed recommended by the manufacturer.

(2) Concentrations of emission samples taken from each exhaust outlet shall not exceed the following levels:

(i) For model years 1967 and earlier: hydrocarbons (HC) 1200 ppm as hexane, and carbon monoxide (CO) 9.0 mole percent.

(ii) For model years 1968 through 1973: HC 600 ppm as hexans, and CO 7.0 mole percent.

(b) *Method.* No-load inspection is conducted by measuring two emission samples from each exhaust outlet. The first emission sample is collected with the vehicle's transmission in neutral and the engine operating at 2250 rpm. The second sample is collected with the vehicle's transmission in the position recommended by the manufacturer for adjusting the idle speed, and the engine idling.

§ 590.7 Loaded-mode inspection.

(a) *Criteria.* When the loaded-mode inspection is conducted, concentrations of the emission samples taken from each exhaust outlet for each of the three phases of the driving cycle in Table I, conducted in the sequence indicated, shall not exceed the levels given in Table II. For the purpose of determining the weight classification of a motor vehicle for the loaded-mode inspection, 300 pounds are added to the vehicle's unladen curb weight.

TABLE I

Curb weight plus 300	Driving cycle (speed-load combination)		
lbs	1st phase high cruise	2d phase low cruise	3d phase idle
3,801 lbs and up	48 to 50 mi/h at 27 to 30 hp	32 to 35 mi/h at 10 to 12 hp	At idle.
2,801 to 3,800 lbs	44 to 46 mi/h at 21 to 24 hp	29 to 32 mi/h at 8 to 10 hp	Do.
2,000 to 2,800 lbs	36 to 38 mi/h at 13 to 15 hp	22 to 25 mi/h at 4 to 6 hp	Do.

TABLE II

High cruise	Low cruise	Idle
1967 and earlier model years		
HC 900 ppm as hexane	HC 900 ppm as hexane	HC 1,200 ppm as hexane
CO 4.5 mole percent	CO 5.5 mole percent	CO 9.0 mole percent
1968 through 1973		
HC 450 ppm as hexane	HC 450 ppm as hexane	HC 600 ppm as hexane
CO 3.75 mole percent	CO 4.25 mole percent	CO 7.0 mole percent

(b) *Method.* Loaded-mode inspection for the first two phases of the driving cycle described in Table I is conducted by measuring the levels of emission concentrations from each exhaust outlet of a motor vehicle operated on a chassis dynamometer, with the vehicle's transmission in the setting recommended by the vehicle manufacturer for the speed-load combination being tested. For the idle phase, vehicles with automatic transmissions are tested in drive, and vehicles with standard transmissions are tested in neutral.

§ 590.8 Inspection conditions.

(a) The vehicle engine is at its normal operating temperature, as specified by the vehicle manufacturer.

(b) An engine speed indicator with a graduated scale from zero to at least 2500 rpm is used for the unloaded inspection procedure.

(c) The equipment used for analyzing the emission concentration levels—

(1) Has a warm-up period not to exceed 30 minutes;

(2) Is able to withstand sustained periods of continuous use;

(3) Has a direct and continuous meter readout that allows readings for concentration levels of carbon monoxide (CO) from 0–10 mole percent, and of hydrocarbon (HC from 0–2000 ppm as hexane; and if used for the loaded-mode inspection, has at least one additional expanded direct and continuous readout for concentration levels of carbon monoxide and of hydrocarbon, such as from 0–5 mole percent and from 0–1000 ppm as hexane respectively;

(4) Has an accuracy of better than ±5% of the full scale reading for each concentration range;

(5) Permits a reading for each emission concentration level, within 10 seconds after

the emission sample has been taken, that is not less than 90% of the final reading; and

(6) Has a calibration system using a standard gas, or an equivalent mechanical or electrical calibration system which itself is based on a standard gas.

40 F.R. 24904
June 11, 1975

PREAMBLE TO DEPARTMENT OF THE TREASURY REGULATION RELATING TO IMPORTATION OF MOTOR VEHICLES AND ITEMS OF MOTOR VEHICLE EQUIPMENT

On April 10, 1968, Public Law 90-283 was enacted to amend the National Traffic and Motor Vehicle Safety Act of 1966 (15 U.S.C. 1391-1409) by adding a new section 123. This section provides a procedure whereby the Secretary of Transportation is authorized, upon petition by a manufacturer of 500 or less vehicles annually, to temporarily exempt such vehicles from certain Federal motor vehicle safety standards. The procedures for temporary exemption of such vehicles adopted by the Department, as published in the *Federal Register* on September 26, 1968 (33 F.R. 14457), require each exempted vehicle to bear a label or tag permanently affixed containing certain information including a statement listing the safety standards for which an exemption has been obtained. Since vehicles so exempted will no longer bear the "valid certification as required by section 114 of the National Traffic and Motor Vehicle Safety Act of 1966 (15 U.S.C. 1403)" which is required by 19 CFR 12.80(b)(1) if a motor vehicle offered for importation is not to be refused entry, it is deemed desirable to amend 19 CFR 12.80(b) to allow entry of exempted vehicles bearing the exemption labels or tags required under the regulations of the Department of Transportation (23 CFR 217.13).

In addition, the Automobile Manufacturer's Association, Inc., on behalf of itself and its member companies, has made a showing of the necessity of importing and using for purposes of test or experiment for a limited time on the public roads, of a limited number of nonconforming motor vehicles manufactured outside the United States. The Association has requested an amendment of 19 CFR 12.80(b)(2)(vii) which currently, among other things, allows the importation of such vehicles for such purposes only upon a declaration by the importer that these vehicles will not be licensed for use on the public roads.

In consideration of the foregoing, § 12.80(b) is amended as follows:

Subparagraph (b)(1) is amended by changing the period following the words "so labelled or tagged", to a comma and (b)(2)(vii) is amended to read as follows:

§ 12.80 Federal Motor vehicle safety standards.

* * * * *

. (b) * : *

(1) * * * or (iii) (for vehicles only which have been exempted by the Secretary of Transportation from meeting certain safety standards) it bears a label or tag permanently affixed to such vehicle which meets the requirements set forth in the regulations of the Department of Transportation, 23 CFR 217.13.

(2) * * *

(vii) The importer or consignee is importing such vehicle or equipment item solely for the purposes of show, test, experiment, competition, repairs or alterations and that such vehicle or equipment item will not be sold or licensed for use on the public roads: Provided, That vehicles imported solely for purposes of test or experiment may be licensed for use on the public roads for a period not to exceed one year, where such use is an integral part of tests or experiments for which such vehicle is being imported, upon condition that the importer attach to the declaration description of the tests or experiments for which the vehicle is being imported, the period of time during which it is estimated that it will be necessary to test the vehicle on the public roads, and the disposition to be made of the vehicle after completion of the tests or experiments.

* * * * *

(Sec. 108, 80 Stat. 722, 15 U.S.C. 1397)

Since the first amendment is necessitated to conform to regulations of the Department of

Transportation presently in effect and the second will affect a very limited number of persons with a legitimate interest in road testing non-conforming vehicles, notice and public procedure thereon is not considered necessary and good cause is found for dispensing with the delayed effective date provision of 5 U.S.C. 553(d). Therefore, the amendments shall be effective upon publication in the *Federal Register*.

[SEAL]

Lester D. Johnson
Commissioner of Customs

Approved: November 29, 1968.
Joseph M. Bowman,
Assistant Secretary
of the Treasury.
Approved: December 9, 1968.
Lowell K. Bridwell,
Federal Highway Administrator.

33 F.R. 18577
December 14, 1968

PREAMBLE TO AMENDMENT TO DEPARTMENT OF THE TREASURY REGULATION RELATING TO IMPORTATION OF MOTOR VEHICLES AND ITEMS OF MOTOR VEHICLE EQUIPMENT

(T.D. 71–122)

A notice was published in the *Federal Register* on February 18, 1971 (36 F.R. 3121), that it was proposed to amend § 12.80 of the Customs Regulations (19 CFR 12.80) to make the following substantive changes:

1. To provide that motor vehicles and motor vehicle equipment brought into conformity under bond, shall not be sold or offered for sale until the bond is released;

2. To make clear that the term motor vehicle as used in § 12.80 refers to a motor vehicle as defined in the National Traffic and Motor Vehicle Safety Act of 1966;

3. To require a declaration of conformance accompanied by a statement of the vehicle's original manufacturer as evidence of original compliance;

4. To require that declarations filed under paragraph (c) of § 12.80 be signed by the importer or consignee; and

5. To add a bond requirement for the production of a declaration of original compliance and a declaration of conformity after manufacture.

Interested persons were given an opportunity to submit relevant data, views, or arguments. No comments were received. The amendments as proposed, with minor editorial changes, are hereby adopted as set forth below to become effective 30 days after the date of publication in the *Federal Register*.

Robert V. McIntyre,
Acting Commissioner of Customs.

APPROVED: April 22, 1971.

Eugene T. Rossides,
Assistant Secretary of the Treasury.

APPROVED: May 3, 1971.

Douglas W. Toms,
Acting Administrator, National Highway Traffic Safety Administration.

36 F.R. 8667
May 11, 1971

DEPARTMENT OF THE TREASURY REGULATION RELATING TO IMPORTATION OF MOTOR VEHICLES AND ITEMS OF MOTOR VEHICLE EQUIPMENT

Notice of a proposal to add § 12.80 to Part 12 of the Customs Regulations to prescribe regulations providing for the admission or refusal of motor vehicles or items of motor vehicle equipment which are offered for importation into the United States and which are subject to Federal motor vehicle safety standards promulgated by the Department of Transportation in 49 CFR Part 571, pursuant to the provisions of the National Traffic and Motor Vehicle Safety Act of 1966, was published in the *Federal Register* for November 30, 1967 (32 F.R. 16432). Interested persons were given an opportunity to submit relevant data, views, or arguments in writing regarding the proposed regulations. All comments received have been carefully considered.

In response to those comments, in addition to several minor changes, the first paragraph of § 12.80(b) has been amended to provide for the entry, without written declaration, of motor vehicles and items of motor vehicle equipment intended for export and so labeled. A new provision is also added (§ 12.80(b) (2) (iv)) to provide for the entry, upon written declaration, of new vehicles intended for resale which do not fully conform to the safety standards because of the absence of readily attachable equipment items: *Provided,* That the importer or consignee undertakes to attach the missing items before such vehicles are offered to the general public for sale. Finally, the importation of nonconforming vehicles for competition purposes will be permitted under § 12.80(b) (2) (vii) if the vehicle will not be licensed for use on the public roads.

Part 12 is accordingly amended to add a new centerhead and section as follows:

Motor Vehicles and Motor Vehicle Equipment Manufactured on or after January 1, 1968

§ 12.80 Federal motor vehicle safety standards.

(1) *Standards prescribed by the Department of Transportation.* Motor vehicles and motor vehicle equipment manufactured on or after January 1, 1968, offered for sale, or introduction or delivery for introduction in interstate commerce, or importation into the United States are subject to Federal Motor Vehicle Safety Standards (hereafter referred to in this section as "safety standards") prescribed by the Secretary of Transportation under sections 103 and 119 of the National Traffic and Motor Vehicle Safety Act of 1966. (15 U.S.C. 1392, 1407) as set forth in regulations in 49 CFR Part 571. A motor vehicle hereafter referred to in this section as "vehicle" or item of motor vehicle equipment (hereafter referred to in this section as "equipment item"), manufactured on or after January 1, 1968, is not permitted entry into the United States unless (with certain exceptions set forth in paragraph (b) of this section) it is in conformity with applicable safety standards in effect at the time the vehicle or equipment item was manufactured.

(b) *Requirements for entry and release.*

(1) Any vehicle or equipment item offered for importation into the customs territory of the United States shall not be refused entry under this seciton if (i) it bears a certification label affixed by its original manufacturer in accordance with section 114 of the National Traffic and Motor Vehicle Safety Act of 1966 (15 U.S.C. 1403) and regulations issued thereunder by the Secretary of Transportation (49 CFR Part 567) (in the case of a vehicle, in the form of a label or tag permanently affixed to such vehicle or in the case of an equipment item, in the form of a label or tag on such item or on the outside of a container in which such item is delivered), or (ii) it is intended solely for export, such vehicle or equipment

item and the outside of its container, if any, to be so labeled and tagged, or (iii) (for vehicles only which have been exempted by the Secretary of Transportation from meeting certain safety standards) it bears a label or tag permanently affixed to such vehicle which meets the requirements set forth in the regulations of the Department of Transportation, 49 CFR 555.13.

(2) Any such vehicle or equipment item not bearing such certification or export label shall be refused entry unless there is filed with the entry, in duplicate, a declaration signed by the importer or consignee which states that:

(i) Such vehicle or equipment item was manufactured on a date when there were no applicable safety standards in force, a verbal declaration being acceptable at the option of the district director of customs for vehicles entering at the Canadian and Mexican borders; or

(ii) Such vehicle or equipment item was not manufactured in conformity with applicable safety standards but has since been brought into conformity, such declaration to be accompanied by the statement of the manufacturer, contractor, or other person who has brought such vehicle or equipment item into conformity which describes the nature and extent of the work performed; or

(iii) Such vehicle or equipment item does not conform with applicable safety standards, but that the importer or consignee will bring such vehicle or equipment item into conformity with such safety standards, and that such vehicle or equipment item will not be sold or offered for sale until the bond (required by paragraph (c) of this section) shall have been released; or

(iv) Such vehicle is a new vehicle being imported for purposes of resale which does not presently conform to all applicable safety standards because readily attachable equipment items are not attached, but that there is affixed to its windshield a label stating the safety standard with which and the manner in which such vehicle does not conform and that the vehicle will be brought into conformity by attachment of such equipment items before it will be offered for sale to the first purchaser for purposes other than resale; or

(v) The importer or consignee is a non-resident of the United States, importing such vehicle or equipment item primarily for personal use or for the purpose of making repairs or alterations to the vehicle or equipment item, for a period not exceeding 1 year from the date of entry, and that he will not resell it in the United States during that time: PROVIDED, That persons regularly entering the United States by a motor vehicle at the Canadian and Mexican borders may apply to the district director of customs for an appropriate means of identification to be affixed to such vehicle which will serve in place of the declaration required by this paragraph; or

(vi) The importer or consignee is a member of the armed forces of a foreign country on assignment in the United States, or is a member of the Secretariat of a public international organization so designated pursuant to 59 Stat. 669 on assignment in the United States, or is a member of the personnel of a foreign government on assignment in the United States who comes within the class of persons for whom free entry of motor vehicles has been authorized by the Department of State and that he is importing such vehicle or equipment item for purposes other than resale; or

(vii) The importer or consignee is importing such vehicle or equipment item solely for the purpose of show, test, experiment, competition, repairs or alterations and that such vehicle or equipment item will not be sold or licensed for use on the public roads: PROVIDED: That vehicles imported solely for purposes of test or experiment may be licensed for use on the public roads for a period not to exceed one year, where such use is an integral part of tests or experiments for which such vehicle is being imported, upon condition that the importer attach to the declaration a description of the tests or experiments for which the vehicle is being imported, the period of time during which it is estimated that it will be necessary to test the vehicle on the public roads, and the disposition to be made of the vehicle after completion of the tests or experiments.

(viii) Such vehicle which is not manufactured primarily for use on the public roads is not a "motor vehicle" as defined in section 102 of the National Traffic and Motor Vehicle Safety Act of 1966 (15 U.S.C. 1391); or

(ix) Such vehicle was manufactured in conformity with applicable safety standards, such declaration to be accompanied by a statement of the vehicle's original manufacturer as evidence of original compliance.

(3) Any declaration given under this section (except an oral declaration accepted at the option of the district director of customs under subparagraph (2)(i) of this paragraph) shall state the name and United States address of the importer or consignee, the date and the entry number, a description of any equipment item, the make and model, engine serial, and body serial numbers of any vehicle or other identification numbers, and the city and State in which it is to be registered and principally located if known, and shall be signed by the importer or consignee. The district director of customs shall immediately forward the original of such declaration to the National Highway Traffic Safety Administration of the Department of Transportation.

(c) *Release under bond.* If a declaration filed in accordance with paragraph (b) of this section states that the entry is being made under circumstances described in paragraph (b)(2)(iii), or under circumstances described in paragraph (b)(2)(ii) or (ix) of this section where the importer at time of entry does not submit a statement in support of his declaration of conformity the entry shall be accepted only if the importer gives a bond on Customs Forms 7551, 7553, or 7595 for the production of either a statement by the importer or consignee that the vehicle or equipment item described in the declaration filed by the importer has been brought into conformity with applicable safety stand-

ards and identifying the manufacturer, contractor, or other person who has brought such vehicle or equipment item into conformity with such standards and describing the nature and extent of the work performed or a statement of the vehicle manufacturer certifying original conformity. The bond shall be in the amount required under § 25.4(a) of this chapter. Within 90 days after such entry, or such additional period as the district director of customs may allow for good cause shown, the importer or consignee shall deliver to both the district director of customs, and the National Highway Traffic Safety Administration a copy of the statement described in this paragraph. If such statement is not delivered to the district director of customs for the port of entry of such vehicle or equipment item within 90 days of the date of entry or such additional period as may have been allowed by the district director of customs for good cause shown, the importer or consignee shall deliver or cause to be delivered to the district director of customs those vehicles or equipment items, which were released in accordance with this paragraph. In the event that any such vehicle or equipment item is not redelivered within 5 days following the date specified in the preceding sentence, liquidated damages shall be assessed in the full amount of a bond given on Form 7551. When the transaction has been charged against a bond given on Form 7553, or 7595, liquidated damages shall be assessed in the amount that would have been demanded under the preceding sentence if the merchandise had been released under a bond given on Form 7551.

(d) *Merchandise refused entry.* If a vehicle or equipment item is denied entry under the provisions of paragraph (b) of this section, the district director of customs shall refuse to release the merchandise for entry into the United States and shall issue a notice of such refusal to the importer or consignee.

(e) *Disposition of merchandise refused entry into the United States; redelivered merchandise.* Vehicles or equipment items which are denied entry under paragraph (b) of this section or which are redelivered in accordance with paragraph (c) of this section and which are not exported under customs supervision within 90 days from the date of notice of refusal of admission or date of redelivery shall be disposed of under customs laws and regulations; *Provided, however,* That any such disposition shall not result in an introduction into the United States of a vehicle or equipment item in violation of the National Traffic and Motor Vehicle Safety Act of 1966.

(Sec. 623, 46 Stat. 759, as amended, sec. 108, 80 Stat. 722; 19 U.S.C. 1623; 15 U.S.C. 1397)

Since motor vehicles and items of motor vehicle equipment subject to the standards prescribed in 49 CFR Part 571, may shortly be in transit to United States ports of entry, it is important that these regulations be put into effect at the earliest possible date. It is therefore found that the ad-vance publication requirement under 5 U.S.C. 553 is impracticable and good cause is found for adopting these regulations effective upon publication in the *Federal Register*.

(SEAL)

Lester D. Johnson
Commissioner of Customs

APPROVED: January 2, 1968.

Matthew J. Marks,
Acting Assistant Secretary
of the Treasury

APPROVED: January 5, 1968.

Alan S. Boyd
Secretary of Transportation

33 F.R. 360
January 10, 1968

M.V. IMPORT-4

Lightning Source UK Ltd.
Milton Keynes UK
UKHW050155271118
332756UK00030B/43/P